Special Features

MOLECULAR BIOLOGY OF THE CELL

SECOND EDITION

MOLECULAR BIOLOGY OF
THE CELL
SECOND EDITION

Bruce Alberts • Dennis Bray
Julian Lewis • Martin Raff • Keith Roberts
James D. Watson

Garland Publishing, Inc.
New York & London

TEXT EDITOR: Miranda Robertson

GARLAND STAFF

Managing Editor: Ruth Adams
Project Editor: Alison Walker
Production Coordinator: Perry Bessas
Designer: Janet Koenig
Copy Editors: Lynne Lackenbach and Shirley Cobert
Editorial Assistant: Māra Abens
Art Coordinator: Charlotte Staub
Indexer: Maija Hinkle

Bruce Alberts received his Ph.D. from Harvard University and is currently Chairman of the Department of Biophysics and Biochemistry at the University of California Medical School in San Francisco. *Dennis Bray* received his Ph.D. from the Massachusetts Institute of Technology and is currently a Senior Scientist in the Medical Research Council Cell Biophysics Unit at King's College London. *Julian Lewis* received his D.Phil. from Oxford University and is currently a Senior Scientist in the Imperial Cancer Research Fund Developmental Biology Unit, Dept. of Zoology, Oxford University. *Martin Raff* received his M.D. degree from McGill University and is currently a Professor in the Biology Department at University College London. *Keith Roberts* received his Ph.D. from Cambridge University and is currently Head of the Department of Cell Biology at the John Innes Institute, Norwich. *James D. Watson* received his Ph.D. from Indiana University and is currently Director of the Cold Spring Harbor Laboratory. He is the author of *Molecular Biology of the Gene* and, with Francis Crick and Maurice Wilkins, won the Nobel Prize in Medicine and Physiology in 1962.

Library of Congress Cataloging-in-Publication Data

Molecular biology of the cell / Bruce Alberts ... [et al.]—2nd ed.
p. cm.
Includes bibliographies and index.
ISBN 0-8240-3695-6.—ISBN 0-8240-3696-4 (pbk.)
1. Cytology. 2. Molecular biology. I. Alberts, Bruce.
[DNLM: 1. Cells. 2. Molecular Biology. QH 581.2 M718]
QH581.2.M64 1989
574.87—dc19
DNLM/DLC
for Library of Congress 88-38275
CIP

Published by Garland Publishing, Inc.
717 Fifth Avenue, New York, NY 10022

Printed in the United States of America

15 14 13 12 11 10 9 8 7 6

Preface to the Second Edition

More than 50 years ago, E.B. Wilson wrote that "the key to every biological problem must finally be sought in the cell." Yet, until very recently, cell biology was usually taught to biology majors as a specialized upper-level course based largely on electron microscopy. And in most medical schools, many cell-biological topics, such as the mechanisms of endocytosis, chemotaxis, cell movement, and cell adhesion, were hardly taught at all—being regarded as too cellular for biochemistry and too molecular for histology. With the recent dramatic advances in our understanding of cells, however, cell biology is beginning to take its rightful place at the center of biological and medical teaching. In an increasing number of universities, it is a required full-year course for all undergraduate biology and biochemistry majors and is becoming the organizing theme for much of the first-year medical curriculum. The first edition of *Molecular Biology of the Cell* was written in anticipation of these much needed curriculum reforms and in the hope of catalyzing them. We will be gratified if the second edition helps to accelerate and spread these reforms.

In revising the book, we have found few cases where recent discoveries have proved old conclusions wrong. But in the six years since the first edition appeared, a fantastically rich harvest of new information about cells has uncovered new connections and in many places forced a radical change of emphasis. The present revision, therefore, goes deep: every chapter has undergone substantial changes, many have been almost entirely rewritten, and two new chapters—on the control of gene expression and on cancer—have been added.

Some commentators on the first edition, especially teachers, wanted more detailed discussion of the experimental evidence that supports the concepts discussed. We did not wish to disrupt the the conceptual flow or to enlarge an already very large book; but we agree that it is crucial to give students a sense of how advances are made. To this end, John Wilson and Tim Hunt have written a Problems Book to accompany the central portion of the main text (Chapters 5–14). As explained in the Note to the Reader on page xxxv, each section of *The Problems Book* covers a section of *Molecular Biology of the Cell* and takes as its principal focus a set of experiments drawn from the relevant research literature. These are the basis for a series of questions—some easy and some hard—that are meant to involve the reader actively in the reasoning that underlies the process of discovery.

The second edition, like the first, has been a long time in the making. As before, each chapter has been passed back and forth between the author who wrote the first draft and the other authors for criticism and extensive revision, so that every part of the book represents a joint composition; Tim Hunt and John Wilson often helped in this process. In addition, outside experts were invited to make suggestions for revision and, in a few cases, to contribute material for the text, which the authors reworked to fit with the rest of the book. We are especially indebted to James Rothman (Princeton University) for his contribution to Chapter 8 and to Jeremy Hyams (University College London), Tim Mitchison (University of California, San Francisco), and Paul Nurse (University of Oxford) for their contributions to Chapter 13. All sections of the revised text were read by outside experts, whose comments and suggestions were invaluable; a list of acknowledgments is on page xxxiii.

Miranda Robertson again played a major part in the creation of a readable book, by insisting that every page be lucid and coherent and rewriting many pages that were not. We are also indebted to the staff at Garland Publishing, and in particular to Ruth Adams, Alison Walker, and Gavin Borden, for their kindness, good humour, efficiency, and unfailingly generous support during the four years that it has taken to prepare this edition. Special thanks go to Carol Winter, for her painstaking care in typing the entire book and preparing the disks for the printer.

Finally, to our wives, families, colleagues, and students we again offer our gratitude and apologies for several years of impositions and neglect; without their help and tolerance this book could not have been written.

Preface to the First Edition

There is a paradox in the growth of scientific knowledge. As information accumulates in ever more intimidating quantities, disconnected facts and impenetrable mysteries give way to rational explanations, and simplicity emerges from chaos. Gradually the essential principles of a subject come into focus. This is true of cell biology today. New techniques of analysis at the molecular level are revealing an astonishing elegance and economy in the living cell and a gratifying unity in the principles by which cells function. This book is concerned with those principles. It is not an encyclopedia but a guide to understanding. Admittedly, there are still large areas of ignorance in cell biology and many facts that cannot yet be explained. But these unsolved problems provide much of the excitement, and we have tried to point them out in a way that will stimulate readers to join in the enterprise of discovery. Thus, rather than simply present disjointed facts in areas that are poorly understood, we have often ventured hypotheses for the reader to consider and, we hope, to criticize.

Molecular Biology of the Cell is chiefly concerned with eucaryotic cells, as opposed to bacteria, and its title reflects the prime importance of the insights that have come from the molecular approach. Part I and Part II of the book analyze cells from this perspective and cover the traditional material of cell biology courses. But molecular biology by itself is not enough. The eucaryotic cells that form multicellular animals and plants are social organisms to an extreme degree: they live by cooperation and specialization. To understand how they function, one must study the ways of cells in multicellular communities, as well as the internal workings of cells in isolation. These are two very different levels of investigation, but each depends on the other for focus and direction. We have therefore devoted Part III of the book to the behavior of cells in multicellular animals and plants. Thus developmental biology, histology, immunobiology, and neurobiology are discussed at much greater length than in other cell biology textbooks. While this material may be omitted from a basic cell biology course, serving as optional or supplementary reading, it represents an essential part of our knowledge about cells and should be especially useful to those who decide to continue with biological or medical studies. The broad coverage expresses our conviction that cell biology should be at the center of a modern biological education.

This book is principally for students taking a first course in cell biology, be they undergraduates, graduate students, or medical students. Although we assume that most readers have had at least an introductory biology course, we have attempted to write the book so that even a stranger to biology could follow it by starting at the beginning. On the other hand, we hope that it will also be useful to working scientists in search of a guide to help them pick their way through a vast field of knowledge. For this reason, we have provided a much more thorough list of references than the average undergraduate is likely to require, at the same time making an effort to select mainly those that should be available in most libraries.

This is a large book, and it has been a long time in gestation—three times longer than an elephant, five times longer than a whale. Many people have had a hand in it. Each chapter has been passed back and forth between the author who wrote the first draft and the other authors for criticism and revision, so that each chapter represents a joint composition. In addition, a small number of outside experts contributed written material, which the authors reworked to fit with the rest of the book, and all the chapters were read by experts, whose comments and corrections were invaluable. A full list of acknowledgments to these contributors and readers for their help with specific chapters is appended. Paul R. Burton (University of Kansas), Douglas Chandler (Arizona State University), Ursula Goodenough (Washington University), Robert E. Pollack (Columbia University), Robert E. Savage (Swarthmore College), and Charles F. Yocum (University of Michigan) read through all or some of the manuscript and made many helpful suggestions.

The manuscript was also read by undergraduate students, who helped to identify passages that were obscure or difficult.

Most of the advice obtained from students and outside experts was collated and digested by Miranda Robertson. By insisting that every page be lucid and coherent, and by rewriting many of those that were not, she has played a major part in the creation of a textbook that undergraduates will read with ease. Lydia Malim drew many of the figures for Chapters 15 and 16, and a large number of scientists very generously provided us with photographs: their names are given in the figure credits. To our families, colleagues, and students we offer thanks for forbearance and apologies for several years of imposition and neglect. Finally, we owe a special debt of gratitude to our editors and publisher. Tony Adams played a large part in improving the clarity of the exposition, and Ruth Adams, with a degree of good-humored efficiency that put the authors to shame, organized the entire production of the book. Gavin Borden undertook to publish it, and his generosity and hospitality throughout have made the enterprise of writing a pleasure as well as an education for us.

Contents in Brief

List of Topics

Introduction to the Cell

PART I

CHAPTER 1 The Evolution of the Cell

CHAPTER 2 Small Molecules, Energy, and Biosynthesis

CHAPTER 3

Macromolecules: Structure, Shape, and Information

CHAPTER 4

How Cells Are Studied

PART 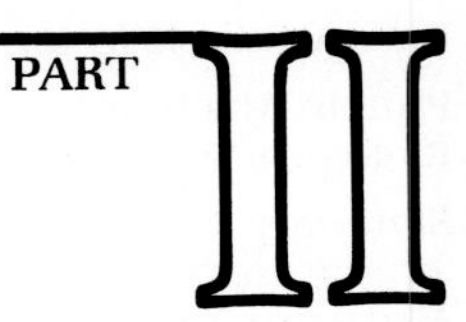

The Molecular Organization of Cells

CHAPTER 5

Basic Genetic Mechanisms

CHAPTER 7

Energy Conversion: Mitochondria and Chloroplasts

CHAPTER 8

Intracellular Sorting and the Maintenance of Cellular Compartments

CHAPTER 10

Control of Gene Expression

CHAPTER 11 The Cytoskeleton

PART 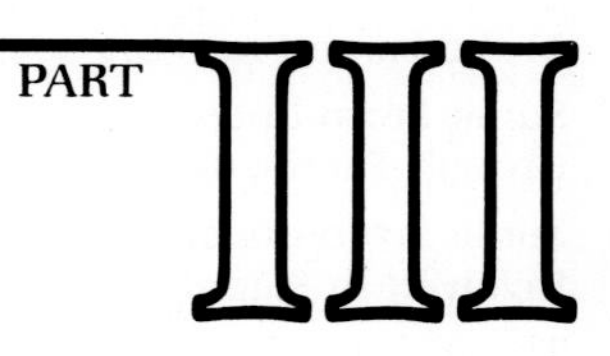

From Cells to Multicellular Organisms

CHAPTER 15

Germ Cells and Fertilization

CHAPTER 18

The Immune System

CHAPTER 19

The Nervous System

CHAPTER 20

Special Features of Plant Cells

Special Features

Acknowledgments

In writing this book we have benefited greatly from the advice of many biologists. In addition to those who advised us by telephone and those who helped with the first edition, we would like to thank the following for their written advice in preparing this edition:

Chapter 1 Hans Bode (University of California, Irvine), Tom Cavalier-Smith (Kings College, London), Elaine Robson (University of Reading, U.K.).

Chapter 2 Efraim Racker (Cornell University), Harry van der Westen (Wageningen, The Netherlands), Richard Wolfenden (University of North Carolina).

Chapter 3 Charles Cantor (Columbia University), Russell Doolittle (University of California, San Diego), Steven Harrison (Harvard University), Richard Wolfenden (University of North Carolina).

Chapter 4 Charles Cantor (Columbia University), Steven Harrison (Harvard University), Richard Henderson (MRC Laboratory of Molecular Biology, Cambridge, U.K.), George Ratcliffe (Oxford University), Peter Shaw (John Innes Institute, Norwich, U.K.).

Chapter 5 Tim Hunt (Cambridge University), Marilyn Kozak (University of Pittsburgh), Thomas Lindahl (Imperial Cancer Research Fund, South Mimms, U.K.), Harold Varmus (University of California, San Francisco).

Chapter 6 Mark Bretscher (MRC Laboratory of Molecular Biology, Cambridge, U.K.), Lewis Cantley (Harvard University), Stuart Cull-Candy (University College London), Anthony Gardner-Medwin (University College London), Walter Gratzer (Kings College London), Ari Helenius (Yale University), Richard Henderson (MRC Laboratory of Molecular Biology, Cambridge, U.K.), Regis Kelly (University of California, San Francisco), Mark Marsh (Institute of Cancer Research, London), Samuel Silverstein (Columbia University), Wilfred Stein (Hebrew University), Peter Walter (University of California, San Francisco), Judy White (University of California, San Francisco).

Chapter 7 Martin Brand (Cambridge University), Leslie Grivell (University of Amsterdam), Richard McCarty (Cornell University), David Nicholls (University of Dundee), Gottfried Schatz (University of Basel), Alison Smith (John Innes Institute, Norwich, U.K.).

Chapter 8 Regis Kelly (University of California, San Francisco), Stuart Kornfeld (Washington University, St. Louis), Paul Lazarow (Rockefeller University), Vishu Lingappa (University of California, San Francisco), George Palade (Yale University), James Rothman (Princeton University), Gottfried Schatz (University of Basel), Kai Simons (EMBO Laboratory, Heidelberg), Alex Varshavsky (Massachusetts Institute of Technology), Peter Walter (University of California, San Francisco), William Wickner (University of California, Los Angeles).

Chapter 9 Pierre Chambon (University of Strasbourg), Sarah Elgin (Washington University, St. Louis), Gary Felsenfeld (National Institutes of Health, Bethesda), Christine Guthrie (University of California, San Francisco), Harold Weintraub (Hutchinson Cancer Center, Seattle), Keith Yamamoto (University of California, San Francisco).

Chapter 10 Pierre Chambon (University of Strasbourg), Enrico Coen (John Innes Institute, Norwich, U.K.), Gary Felsenfeld (National Institutes of Health, Bethesda), Ira Herskowitz (University of California, San Francisco), Tim Hunt (Cambridge University), Robert Roeder (Rockefeller University), Harold Weintraub (Hutchinson Cancer Center, Seattle), John Wilson (Baylor University), Keith Yamamoto (University of California, San Francisco).

Chapter 11 Roger Cooke (University of California, San Francisco), Marc Kirschner (University of California, San Francisco), Michael Klymkowsky (University of Colorado, Boulder), Mark Mooseker (Yale University), Tom Pollard (Johns Hopkins University), Joel Rosenbaum (Yale University), Lewis Tilney (University of Pennsylvania), Klaus Weber (Max Planck Institute for Biophysical Chemistry, Göttingen).

Chapter 12 Henry Bourne (University of California, San Francisco), Philip Cohen (University of Dundee), Graham Hardie (University of Dundee), Daniel Koshland (University of California, Berkeley), Alex Levitzki (Hebrew University), Robert Mishell (University of Birmingham, U.K.), Anne Mudge (University College London), Michael Schramm (Hebrew University), Zvi Sellinger (Hebrew University), Tom Vanaman (University of Kentucky).

Chapter 13 Robert Brooks (King's College London), Leland Hartwell (University of Washington, Seattle), John Heath (Oxford University), Tim Hunt (Cambridge University), Jeremy Hyams (University College London), Marc Kirschner (University of California, San Francisco), Tim Mitchison (University of California, San Francisco), Andrew Murray (University of California, San Francisco), Paul Nurse (Oxford University).

Chapter 14 Michael Bennett (Albert Einstein College of Medicine), Benny Geiger (Weizmann Institute), Daniel Goodenough (Harvard Medical School), Barry Gumbiner (University of California, San Francisco), Richard Hynes (Massachusetts Institute of Technology), Tom Jessell (Columbia University), Louis Reichardt (University of California, San Francisco), Erkki Ruoslahti (La Jolla Cancer Research Foundation), Masatoshi Takeichi (Kyoto University), Robert Trelstad (UMDNJ—Robert Wood Johnson Medical School), Anne Warner (University College London).

Chapter 15 Adelaide Carpenter (University of California, San Diego), James Crow (University of Wisconsin, Madison), David Epel (Stanford University), Tim Hunt (Cambridge University), James Maller (University of Colorado Medical School), John Maynard Smith (University of Sussex), Anne McLaren (University College London), Montrose Moses (Duke University), Lewis Tilney (University of Pennsylvania), Victor Vacquier (University of California, San Diego).

Chapter 16 Marianne Bronner-Fraser (University of California, Irvine), Robert Horvitz (Massachusetts Institute of Technology), James Hudspeth (University of California, San Francisco), Philip Ingham (Imperial Cancer Research Fund, Oxford), Ray Keller (University of California, Berkeley), Cynthia Kenyon (University of California, San Francisco), Judith Kimble (University of Wisconsin, Madison), Tom Kornberg (University of California, San Francisco), Mark Krasnow (Stanford University), Peter Lawrence (MRC Laboratory of Molecular Biology, Cambridge, U.K.), Gail Martin (University of California, San Francisco), Patrick O'Farrell (University of California, San Francisco), Jonathan Slack (Imperial Cancer Research Fund, Oxford).

Chapter 17 Michael Banda (University of California, San Francisco), Alan Boyde (University College London), Michael Dexter (Paterson Institute for Cancer Research, Manchester, U.K.), Charles Emerson (University of Virginia), Howard Green (Harvard University), David Housman (Massachusetts Institute of Technology), Norman Iscove (Ontario Cancer Institute, Toronto), Anne Mudge (University College London), Michael Solursh (University of Iowa), Jim Till (Ontario Cancer Institute, Toronto), Fiona Watt (Imperial Cancer Research Fund, London).

Chapter 18 Fred Alt (Columbia University), Peter Lachmann (MRC Center, Cambridge, U.K.), Avrion Mitchison (University College London), William Paul (National Institutes of Health, Bethesda), Ronald Schwartz (National Institutes of Health, Bethesda), David Standring (University of California, San Francisco).

Chapter 19 Tim Bliss (National Institute for Medical Research, London), Michael Brown (Oxford University), Steven Burden (Massachusetts Institute of Technology), Gerald Fischbach (Washington University, St. Louis), James Hudspeth (University of California, San Francisco), Trevor Lamb (Cambridge University), Dale Purves (Washington University, St. Louis), James Schwartz (Columbia University), Charles Stevens (Yale University), Michael Stryker (University of California, San Francisco).

Chapter 20 Brian Gunning (Australian National University, Canberra), Andy Johnston (John Innes Institute, Norwich, U.K.), Clive Lloyd (John Innes Institute, Norwich, U.K.), Freiderick Meins (Freiderich Miescher Institut, Basel), Scott Stachel (University of California, Berkeley) Andrew Staehelin (University of Colorado, Boulder), Anthony Trewavas (Edinburgh University), Virginia Walbot (Stanford University), Patricia Zambryski (University of California, Berkeley).

Chapter 21 Michael Bishop (University of California, San Francisco), John Cairns (Harvard School of Public Health), Ruth Ellman (Institute of Cancer Research, Sutton, U.K.), Hartmut Land (Imperial Cancer Research Fund, London), Bruce Ponder (Institute of Cancer Research, Sutton, U.K.).

The cover photograph shows a set of human chromosomes at metaphase. Chromatin is shown red, fluorescently stained with propidium iodide, while the centromeres are revealed with an anti-kinetochore antiserum. The photograph was supplied by Bill Brinkley and is taken from Merry, D.E., Pathak, S., Hsu, T.C. and Brinkley, B.R. *Am. J. Hum. Genet.* 37:425–430, 1985.

A Note to the Reader

Although the chapters of this book can be read independently of one another, they are arranged in a logical sequence of three parts. The first three chapters of **Part I** cover elementary principles and basic biochemistry. They can serve either as an introduction for those who have not studied biochemistry or as a refresher course for those who have. Chapter 4, which concludes Part I, deals with the principles of the main experimental methods for investigating cells. It is not necessary to read this chapter in order to understand the later chapters, but a reader will find it a useful reference.

Part II represents the central core of cell biology and is concerned mainly with those properties that are common to most eucaryotic cells, beginning with the fundamental molecular mechanisms of heredity and concluding with cell adhesion and the extracellular matrix.

Part III follows the behavior of cells in the construction of multicellular organisms, starting with the formation of eggs and sperm and ending with the disruption of multicellular organization that occurs in cancer.

Chapter 4 includes several tables giving the dates of crucial developments along with the names of the scientists involved. Elsewhere in the book the policy has been to avoid naming individual scientists. The authors of major discoveries, however, can usually be identified by consulting the **lists of references** at the end of each chapter. These references frequently include the original papers in which important discoveries were first reported. **Superscript numbers** that accompany the text headings refer to the numbered citations in the reference lists, providing a convenient means of following up specific topics.

Throughout the book, **boldface type** has been used to highlight key terms at the point in a chapter where the main discussion of them occurs. This may or may not coincide with the first appearance of the term in the text. *Italics* are used to set off important terms with a lesser degree of emphasis.

The Problems Book is designed as a companion volume that will help the reader appreciate the elegance, the ingenuity, and the surprises of research. It provides problems to accompany the central portion of this book (Chapters 5-14). Each chapter of problems is divided into sections that correspond to the sections of the main textbook, the principal focus of each section being a set of research-oriented problems derived from the scientific literature.

Most of the research problems illustrate points in the main text and are flagged with a symbol in the margin next to the relevant concept heading. Thus 5–4 in the margin of this text refers to Problem 4 in Chapter 5 of *The Problems Book*. In addition, each section of *The Problems Book* begins with a set of short fill-in-the-blank and true-false questions intended to help the reader review the vocabulary and main concepts of the relevant topic. *The Problems Book* should be useful for homework assignments and as a basis for class discussion. It could even provide ideas for exam questions.

MOLECULAR BIOLOGY OF
THE CELL
SECOND EDITION

A sense of scale. These scanning electron micrographs, taken at progressively higher magnifications, show bacterial cells on the point of an ordinary domestic pin. (Courtesy of Tony Brain and the Science Photo Library.) ▶

Introduction to the Cell

I

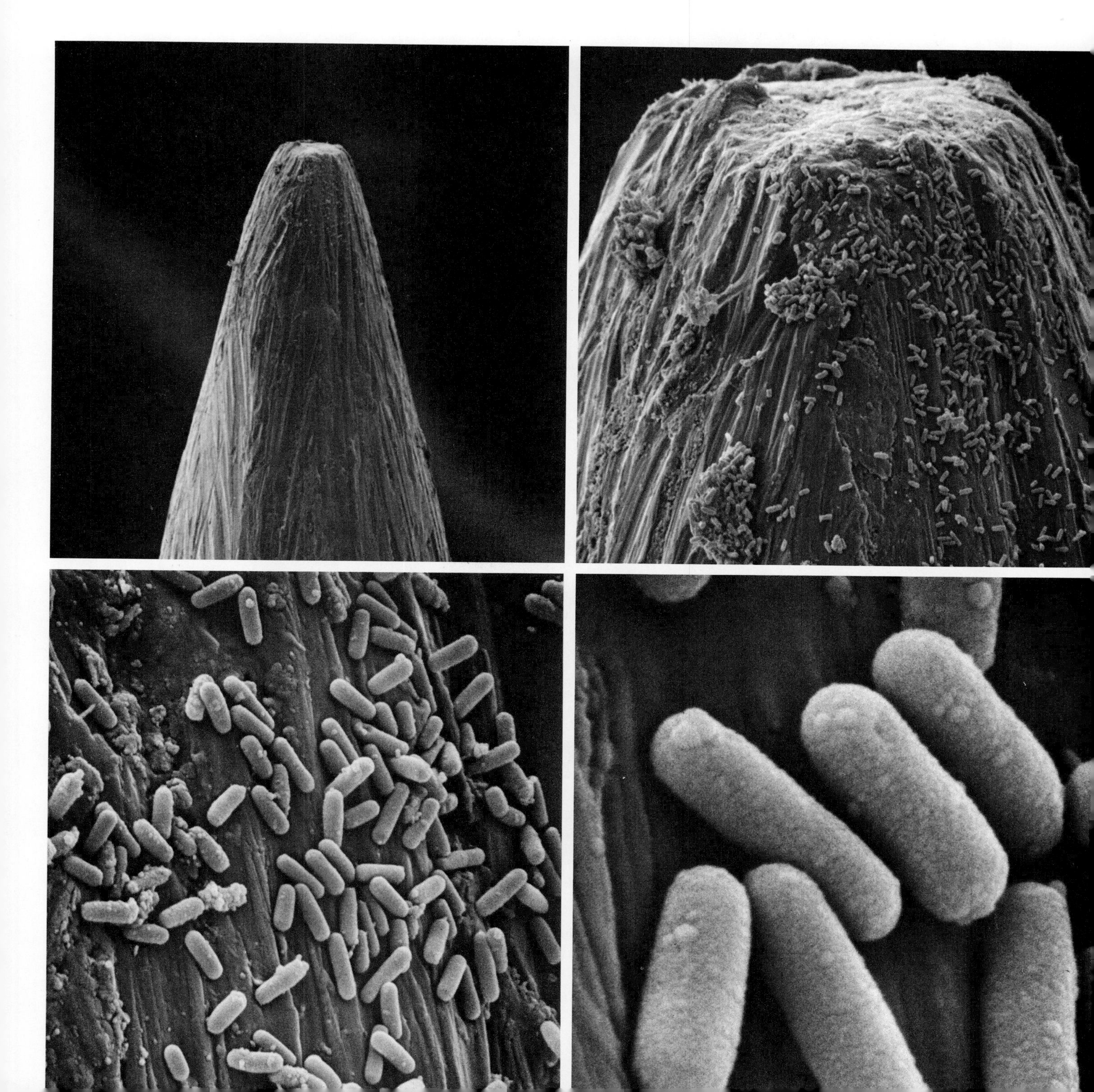

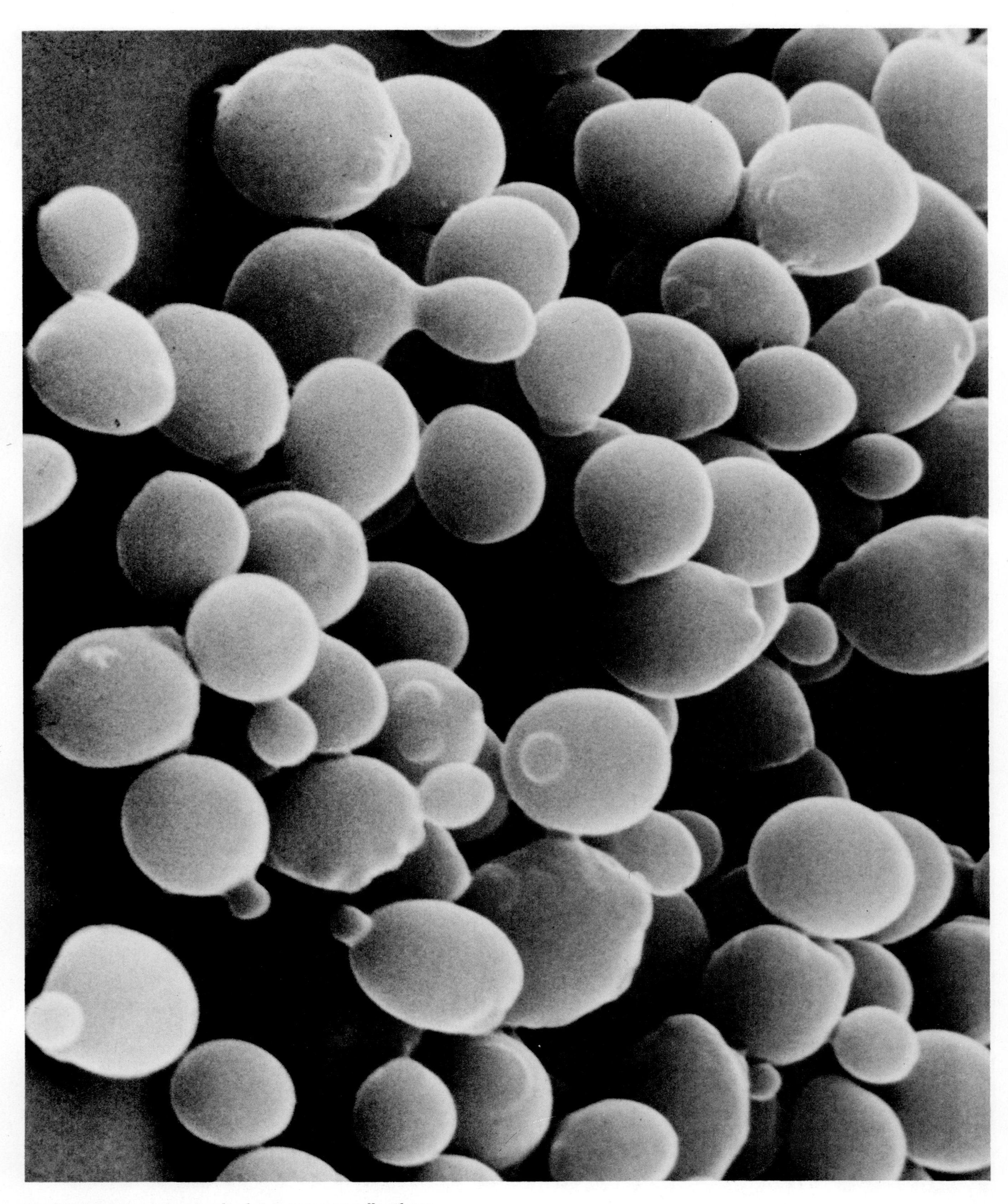

Scanning electron micrograph of growing yeast cells. These
unicellular eucaryotes bud off small daughter cells as they multiply.
(Courtesy of Ira Herskowitz and Eric Schabatach.)

The Evolution of the Cell

1

All living creatures are made of cells—small membrane-bounded compartments filled with a concentrated aqueous solution of chemicals. The simplest forms of life are solitary cells that propagate by dividing in two. Higher organisms, such as ourselves, are like cellular cities in which groups of cells perform specialized functions and are linked by intricate systems of communication. Cells occupy a halfway point in the scale of biological complexity. We study them to learn, on the one hand, how they are made from molecules and, on the other, how they cooperate to make an organism as complex as a human being.

All organisms, and all of the cells that constitute them, are believed to have descended from a common ancestor cell by *evolution.* Evolution involves two essential processes: (1) the occurrence of random *variation* in the genetic information passed from an individual to its descendants and (2) *selection* in favor of genetic information that helps its possessors to survive and propagate. Evolution is the central principle of biology, helping us to make sense of the bewildering variety in the living world.

This chapter, like the book as a whole, is concerned with the progression from molecules to multicellular organisms. It discusses the evolution of the cell, first as a living unit constructed from smaller parts and then as a building block for larger structures. Through evolution, we introduce the cell components and activities that are to be treated in detail, in broadly similar sequence, in the chapters that follow. Beginning with the origins of the first cell on earth, we consider how the properties of certain types of large molecules allow hereditary information to be transmitted and expressed and permit evolution to occur. Enclosed in a membrane, these molecules provide the essentials of a self-replicating cell. Following this, we describe the major transition that occurred in the course of evolution, from small bacteriumlike cells to much larger and more complex cells such as are found in present-day plants and animals. Lastly, we suggest ways in which single free-living cells might have given rise to large multicellular organisms, becoming specialized and cooperating in the formation of such intricate organs as the brain.

Clearly, there are dangers in introducing the cell through its evolution: the large gaps in our knowledge can be filled only by speculations that are liable to be wrong in many details. We cannot go back in time to witness the unique molecular events that took place billions of years ago. But those ancient events have left many traces for us to analyze. Ancestral plants, animals, and even bacteria are preserved as fossils. Even more important, every modern organism provides evidence of the character of living organisms in the past. Present-day biological molecules, in particular, are a rich source of information about the course of evolution, revealing fundamental similarities between the most disparate of living organisms and allowing us to map out the differences between them on an objective universal scale. These molecular similarities and differences present us with a problem like that which confronts the literary scholar who seeks to establish the original text of an ancient author by comparing a mass of variant manuscripts that have been corrupted through repeated copying and editing. The task is hard, and the evidence is incomplete, but it is possible at least to make intelligent guesses about the major stages in the evolution of living cells.

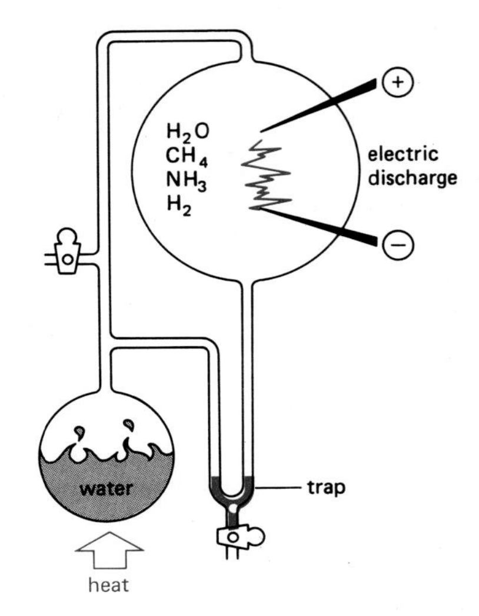

Figure 1–1 A typical experiment simulating conditions on the primitive earth. Water is heated in a closed apparatus containing CH_4, NH_3, and H_2, and an electrical discharge is passed through the vaporized mixture. Organic compounds accumulate in the U-tube trap.

From Molecules to the First Cell[1]

Simple Biological Molecules Can Form Under Prebiotic Conditions

The conditions that existed on the earth in its first billion years are still a matter of dispute. Was the surface initially molten? Did the atmosphere contain ammonia, or methane? Everyone seems to agree, however, that the earth was a violent place with volcanic eruptions, lightning, and torrential rains. There was little if any free oxygen and no layer of ozone to absorb the ultraviolet radiation from the sun.

Simple organic molecules (that is, molecules containing carbon) are likely to have been produced under such conditions. The best evidence for this comes from laboratory experiments. If mixtures of gases such as CO_2, CH_4, NH_3, and H_2 are heated with water and energized by electrical discharge or by ultraviolet radiation, they react to form small organic molecules—usually a rather small selection, each made in large amounts (Figure 1–1). Among these products are a number of compounds, such as hydrogen cyanide (HCN) and formaldehyde (HCHO), that readily undergo further reactions in aqueous solution (Figure 1–2). Most important, the four major classes of small organic molecules found in cells—*amino acids, nucleotides, sugars,* and *fatty acids*—are generated.

Although such experiments cannot reproduce the early conditions on the earth exactly, they make it plain that the formation of organic molecules is surprisingly easy. And the developing earth had immense advantages over any human experimenter; it was very large and could produce a wide spectrum of conditions. But above all, it had much more time—hundreds of millions of years. In such circumstances it seems very likely that, at some time and place, many of the simple organic molecules found in present-day cells accumulated in high concentrations.

HCHO	formaldehyde
HCOOH	formic acid
HCN	hydrogen cyanide
CH_3COOH	acetic acid
NH_2CH_2COOH	glycine
$CH_3CH(OH)COOH$	lactic acid
$NH_2CH(CH_3)COOH$	alanine
$NH(CH_3)—CH_2COOH$	sarcosine
$NH_2—C(=O)—NH_2$	urea
$NH_2CH(CH_2COOH)COOH$	aspartic acid

Figure 1–2 A few of the compounds that might form in the experiment described in Figure 1–1. Compounds shown in color are important components of present-day living cells.

Polynucleotides Are Capable of Directing Their Own Synthesis

Simple organic molecules such as amino acids and nucleotides can associate to form large *polymers.* One amino acid can join with another by forming a peptide bond, and two nucleotides can join together by a phosphodiester bond. The repetition of these reactions leads to linear polymers known as **polypeptides** and **polynucleotides,** respectively. In present-day living organisms, polypeptides—known as *proteins*—and polynucleotides—in the form of both *ribonucleic acids (RNA)* and *deoxyribonucleic acids (DNA)*—are commonly viewed as the most important constituents. A restricted set of 20 amino acids constitute the universal building blocks of the proteins, while RNA and DNA molecules are each constructed from just four types of nucleotides. One can only speculate as to why these particular sets of monomers were selected for biosynthesis in preference to others that are chemically similar.

The earliest polymers may have formed in any of several ways—for example, by the heating of dry organic compounds or by the catalytic activity of high concentrations of inorganic polyphosphates. Under laboratory conditions the products of similar reactions are polymers of variable length and random sequence in which the particular amino acid or nucleotide added at any point depends mainly on chance (Figure 1–3). Once a polymer has formed, however, it can influence the formation of other polymers. Polynucleotides, in particular, have the ability to

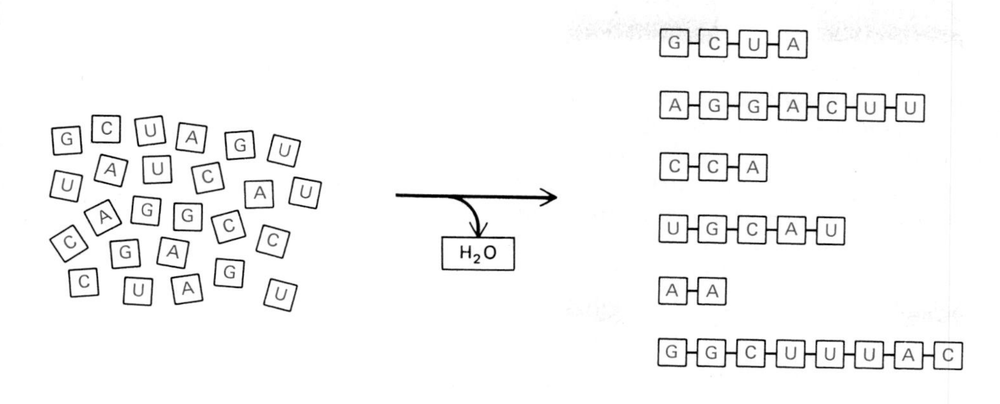

Figure 1–3 Nucleotides of four kinds (here represented by the single letters A, U, G, and C) can undergo spontaneous polymerization with the loss of water. The product is a mixture of polynucleotides that are random in length and sequence.

specify the sequence of nucleotides in new polynucleotides by acting as *templates* for the polymerization reactions. A polymer composed of one nucleotide (for example, polyuridylic acid, or poly U) can serve as a template for the synthesis of a second polymer composed of another type of nucleotide (polyadenylic acid, or poly A). Such templating depends on the fact that one polymer preferentially binds the other. By lining up the subunits required to make poly A along its surface, poly U promotes the formation of poly A (Figure 1–4).

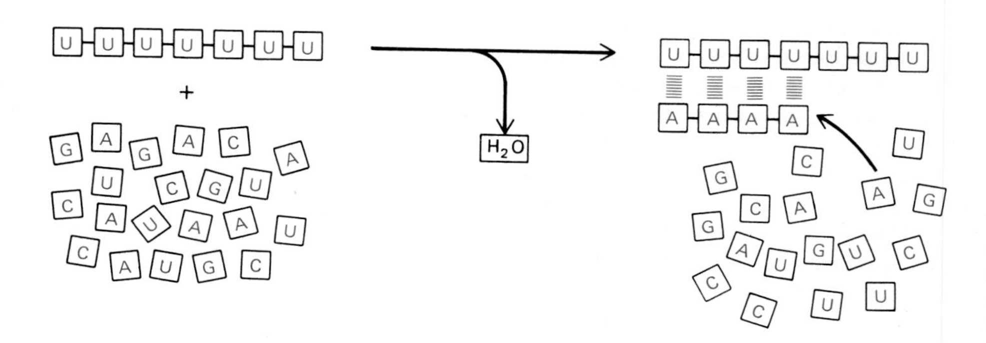

Figure 1–4 Preferential binding occurs between pairs of nucleotides (C with G and A with U) by relatively weak chemical bonds (*above*). This pairing enables one polynucleotide to act as a template for the synthesis of another (*left*).

Specific pairing between complementary nucleotides probably played a crucial part in the origin of life. Consider, for example, a polynucleotide such as RNA, made of a string of four nucleotides, containing the bases uracil (U), adenine (A), cytosine (C), and guanine (G). Because of complementary pairing between the bases A and U and between the bases G and C, when RNA is added to a mixture of activated nucleotides under conditions that favor polymerization, new RNA molecules are produced in which nucleotides are joined in a sequence that is complementary to the original. That is, the new molecules are rather like a mold of the original, with each A in the original corresponding to a U in the copy, and so on. The sequence of nucleotides in the original RNA strand contains information that is, in essence, preserved in the newly formed complementary strands: a second round of copying, with the complementary strand as a template, restores the original sequence (Figure 1–5).

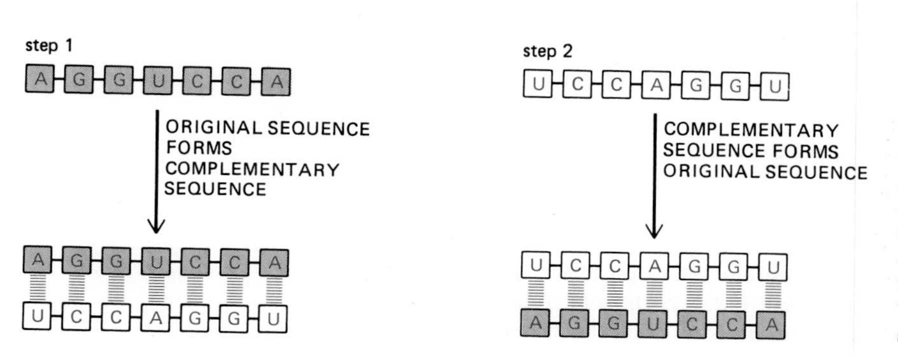

Figure 1–5 Replication of a polynucleotide sequence (here an RNA molecule). In step 1, the original RNA molecule acts as a template to form an RNA molecule of complementary sequence. In step 2, this complementary RNA molecule itself acts as a template, forming RNA molecules of the original sequence. Since each templating molecule can produce many copies of the complementary strand, these reactions can result in the "multiplication" of the original sequence.

Such *complementary templating* mechanisms are elegantly simple, and they lie at the heart of information transfer processes in biological systems. Genetic information contained in every cell is encoded in the sequences of nucleotides in its polynucleotide molecules, and this information is passed on (inherited) from generation to generation by means of complementary base-pairing interactions.

Templating mechanisms, however, require catalysts to promote them: without specific catalysis, templated polymerization is slow and inefficient and other, competing reactions hinder the formation of accurate replicas. Today, the catalytic functions that replicate polynucleotides are provided by highly specialized proteins called *enzymes,* which would not have been present in the "prebiotic soup." On the primitive earth, metal ions and minerals such as clays would have provided some catalytic help. But most important, RNA itself can act as a catalyst: RNA molecules have both the templating properties essential for replication and the potential to fold up to form complex surfaces that catalyze specific reactions. As we shall now discuss, this special versatility of RNA molecules is likely to have provided the basis for the evolution of the first living systems.

Self-replicating Molecules Undergo Natural Selection[2]

Errors inevitably occur in any copying process, and imperfect copies of the original will be propagated. With repeated replication, therefore, the sequence of nucleotides in an original polynucleotide molecule will undergo substantial changes, generating in this way a variety of different molecules. In the case of RNA, these molecules are likely to have different functional properties. RNA molecules are not just strings of symbols that carry information in an abstract way. They also have chemical personalities that affect their behavior. In particular, the specific sequence of nucleotides governs how the molecule folds up in solution. Just as the nucleotides in a polynucleotide can pair with free complementary nucleotides in their environment to form a new polymer, so they can pair with complementary nucleotide residues within the polymer itself. A sequence GGGG in one part of a polynucleotide chain can form a relatively strong association with a CCCC sequence in another region of the same molecule. Such associations produce various three-dimensional folds, and the molecule as a whole takes on a unique shape that depends entirely on the sequence of its nucleotides (Figure 1–6).

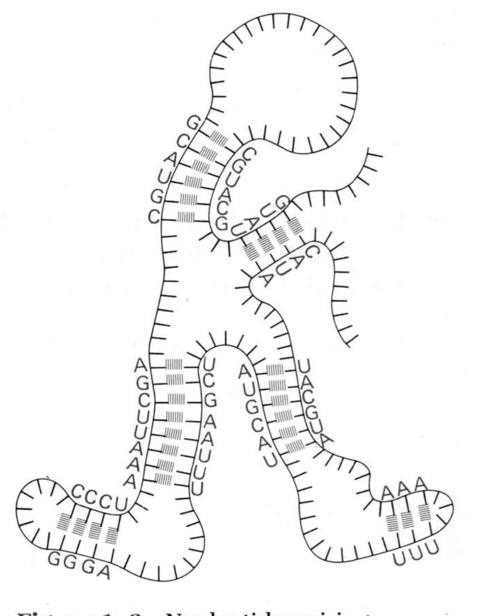

Figure 1–6 Nucleotide pairing between different regions of the same polynucleotide (RNA) chain causes the molecule to adopt a distinctive shape.

The three-dimensional folded structure of a polynucleotide affects its stability, its actions on other molecules, and its ability to replicate, so not all polynucleotide shapes will be equally successful in a replicating mixture. In laboratory studies, replicating systems of RNA molecules have been shown to undergo a form of natural selection in which different favorable sequences eventually predominate, depending on the exact conditions.

An RNA molecule therefore has two special characteristics: it carries information encoded in its nucleotide sequence that it can pass on by the process of replication, and it has a unique folded structure that determines how it will interact with other molecules and respond to the ambient conditions. These two features—one informational, the other functional—are the two properties essential for evolution. The nucleotide sequence of an RNA molecule is analogous to the *genotype*—the hereditary information—of an organism. The folded three-dimensional structure is analogous to the *phenotype*—the expression of the genetic information on which natural selection operates.

Specialized RNA Molecules Can Catalyze Biochemical Reactions[2,3]

Natural selection depends on the environment, and for a replicating RNA molecule a critical component of the environment is the set of other RNA molecules in the mixture. Besides acting as templates for their own replication, these can catalyze the breakage and formation of covalent bonds, including bonds between nucleotides. For example, some specialized RNA molecules can catalyze a change in other RNA molecules, cutting the nucleotide sequence at a particular point; and other types of RNA molecules spontaneously cut out a portion of their own nucleotide sequence and rejoin the cut ends (a process known as self-splicing). Each

RNA-catalyzed reaction depends on a specific arrangement of atoms that forms on the surface of the catalytic RNA molecule, causing particular chemical groups on one or more of its nucleotides to become highly reactive.

Certain catalytic activities would have had a cardinal importance in the primordial soup. Consider in particular an RNA molecule that catalyzes the process of templated polymerization, taking any given RNA molecule as template. This catalytic molecule, by acting on copies of itself, can replicate with heightened speed and efficiency (Figure 1–7A). At the same time, it can promote the replication of any other types of RNA molecules in its neighborhood. Some of these may have catalytic actions that help or hinder the survival or replication of RNA in other ways. If beneficial effects are reciprocated, the different types of RNA molecules, specialized for different activities, may evolve into a cooperative system that replicates with unusually great efficiency (Figure 1–7B).

Information Flows From Polynucleotides to Polypeptides[4]

It is suggested, therefore, that between 3.5 and 4 billion years ago, somewhere on earth, self-replicating systems of RNA molecules began the process of evolution. Systems with different sets of nucleotide sequences competed for the available precursor materials to construct copies of themselves, just as organisms now compete; success depended on the accuracy and the speed with which the copies were made and on the stability of those copies.

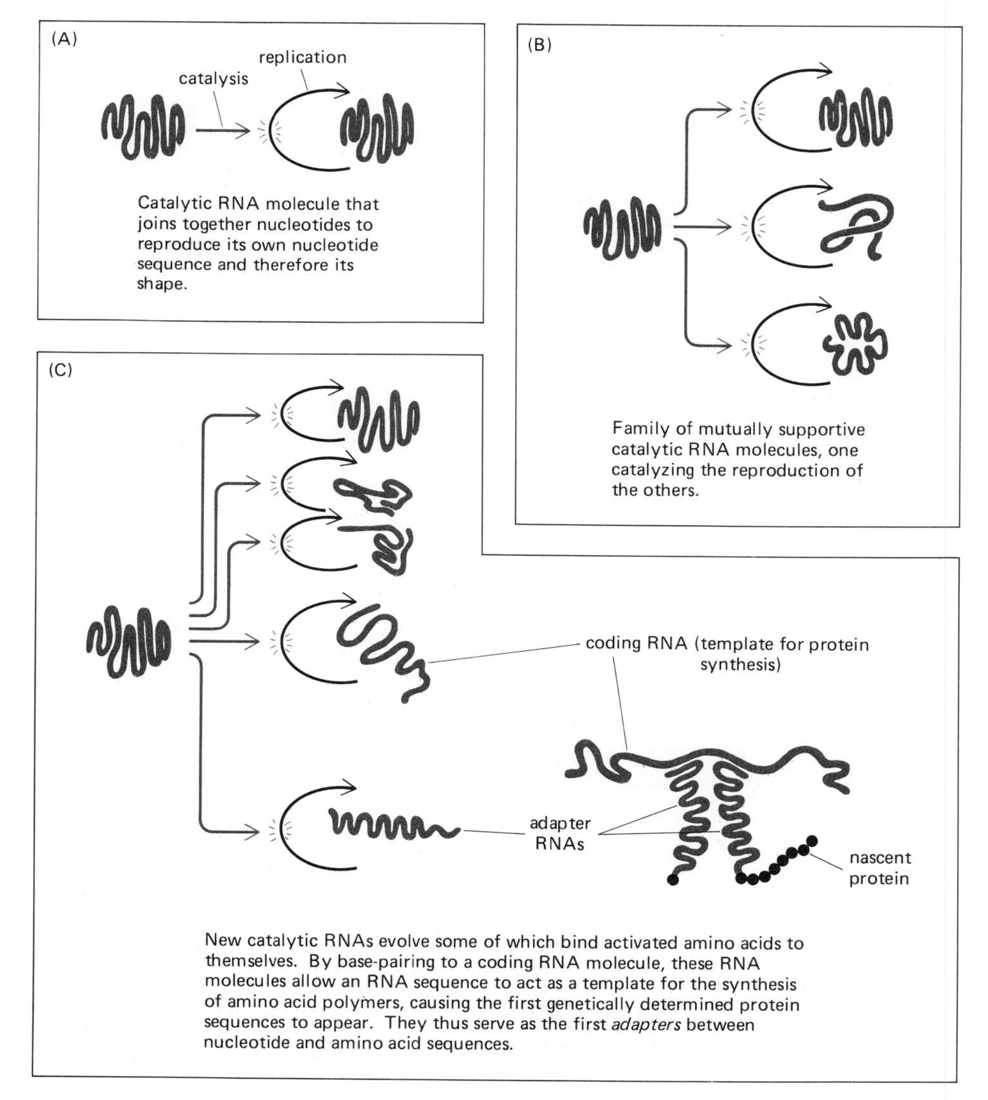

Figure 1–7 Diagram of three successive steps in the evolution of a self-replicating system of RNA molecules capable of directing protein synthesis.

However, while the structure of polynucleotides is well suited for information storage and replication, the catalytic capabilities of RNA molecules are apparently too limited to provide all the functions of a modern cell. Greater versatility is attained by polypeptides, which are composed of many different amino acids with chemically diverse side chains. As we shall discuss in Chapter 3, polypeptides are able to adopt diverse three-dimensional forms that bristle with reactive sites, and they are therefore ideally suited to a wide range of structural and chemical tasks. Even random polymers of amino acids produced by prebiotic synthetic mechanisms are likely to have had catalytic properties, some of which could have enhanced the replication of RNA molecules. Any polynucleotide that helped guide the synthesis of a useful polypeptide in its environment would have had a great advantage in the evolutionary struggle for survival.

But how could polynucleotides exert such control? How could the information encoded in their sequences specify the sequences of polymers of a different type? Clearly, the polynucleotides must act as catalysts to join selected amino acids together. In present-day organisms, a collaborative system of RNA molecules plays a central part in directing the synthesis of polypeptides—that is, **protein synthesis**—but the process is aided by other proteins synthesized previously. The biochemical machinery for protein synthesis is remarkably elaborate. One RNA molecule carries the genetic information for a particular polypeptide in the form of a code, while other RNA molecules act as adapters, each binding a specific amino acid. These two types of RNA molecules form complementary base pairs with one another to enable sequences of nucleotides in the coding RNA molecule to direct the incorporation of specific amino acids held on the adapter RNAs into a growing polypeptide chain. Precursors to these two types of RNA molecules presumably directed the first protein synthesis without the aid of proteins (Figure 1–7C).

Today, these events in the assembly of new proteins take place on the surface of *ribosomes*—complex particles composed of several large RNA molecules of yet another class, together with more than 50 different types of protein. In Chapter 5 we shall see that the ribosomal RNA in these particles plays a central catalytic role in the process of protein synthesis and forms more than 60% of the ribosome's mass. At least in evolutionary terms, it is the fundamental component of the ribosome.

It seems likely, then, that RNA guided the primordial synthesis of proteins, perhaps in a clumsy and primitive fashion. In this way, RNA was able to create tools—in the form of proteins—for more efficient biosynthesis, and some of these could have been put to use in the replication of RNA and in the process of tool production itself.

The synthesis of specific proteins under the guidance of RNA required the evolution of a code by which the polynucleotide sequence specifies the amino acid sequence that makes up the protein. This code—the *genetic code*—is spelled out in a "dictionary" of three-letter words: different triplets of nucleotides encode specific amino acids. The code seems to have been selected arbitrarily, and yet it is virtually the same in all living organisms. This strongly suggests that all present-day cells have descended from a single line of primitive cells that evolved the mechanism of protein synthesis.

Once the evolution of nucleic acids had advanced to the point of specifying powerful catalytic proteins, or *enzymes*, to aid in their own manufacture, the proliferation of the replicating system would have been greatly speeded up. The potentially explosive nature of such an autocatalytic process can be seen today in the life cycle of some bacterial viruses: after they have entered a bacterium, such viruses direct the synthesis of proteins that catalyze selectively their own replication, so that within a short time they take over the entire cell (Figure 1–8).

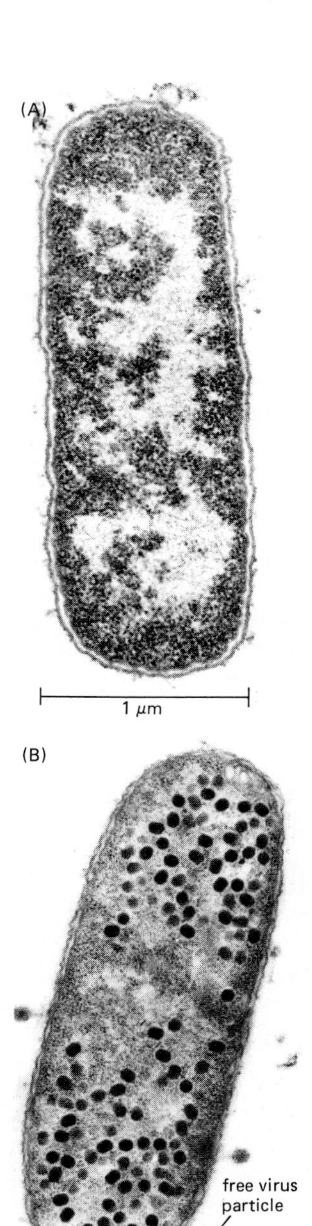

Figure 1–8 Micrograph of a bacterial cell (*Escherichia coli*) (A) in its normal healthy state and (B) just over an hour after being infected with a virus (bacteriophage T4). The original free virus particles, some of which can be seen still adhering to the outside of the cell, injected their DNA into the cell; this then directed the synthesis of specific viral proteins, some of which degraded the host cell DNA while others catalyzed the replication of the virus. At the stage shown, the newly made viral DNA has been packaged into protein coats, seen as dense particles in the cytoplasm. The cell is about to burst open, releasing several hundred new virus particles to the surroundings. (Courtesy of E. Kellenberger.)

Membranes Defined the First Cell

One of the crucial events leading to the formation of the first cell must have been the development of an outer membrane. For example, the proteins synthesized under the control of a certain species of RNA would not facilitate reproduction of that species of RNA unless they remained in the neighborhood of the RNA; more-

Figure 1–9 Schematic drawing showing the evolutionary significance of cell-like compartments. In a mixed population of self-replicating RNA molecules capable of protein synthesis (as illustrated in Figure 1–7), any improved form of RNA that is able to produce a more useful protein must share this protein with all of its competitors. However, if the RNA is enclosed within a compartment, such as a lipid membrane, then any protein it makes is retained for its own use; the RNA can therefore be selected on the basis of its making a better protein.

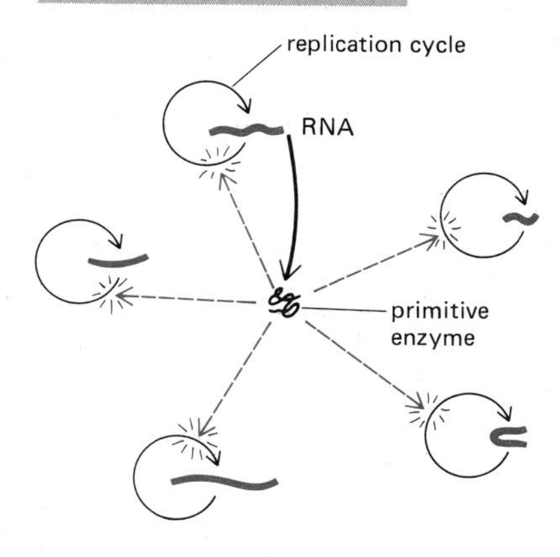

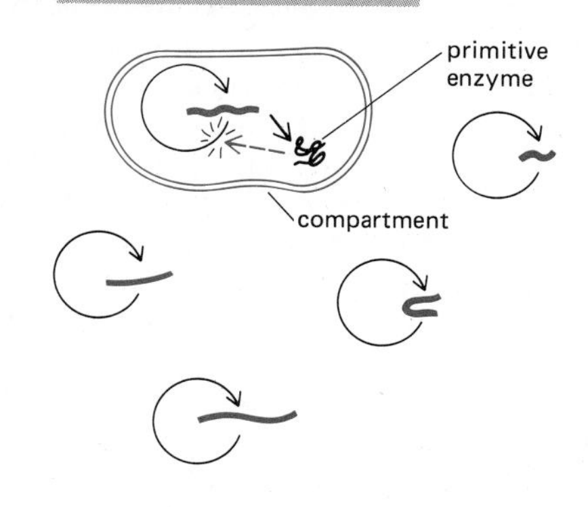

over, as long as these proteins were free to diffuse among the population of replicating RNA molecules, they could benefit equally any competing species of RNA that might be present. If a variant RNA arose that made a superior type of enzyme, the new enzyme could not contribute *selectively* to the survival of the variant RNA in its competition with its fellows. Selection of RNA molecules according to the quality of the proteins they generated could not begin until some form of compartment evolved to contain the proteins made by an RNA molecule and thereby make them available only to the RNA that had generated them (Figure 1–9).

The need for containment is easily fulfilled by another class of molecules, which has the simple physicochemical property of being *amphipathic,* that is, consisting of one part that is hydrophobic (water insoluble) and another part that is hydrophilic (water soluble). When such molecules are placed in water, they aggregate with their hydrophobic portions as much in contact with one another as possible and with their hydrophilic portions in contact with the water. Amphipathic molecules of appropriate shape spontaneously aggregate to form *bilayers,* creating small closed vesicles whose aqueous contents are isolated from the external medium (Figure 1–10). The phenomenon can be demonstrated in a test tube by simply mixing phospholipids and water together: under appropriate conditions, small vesicles will form. All present-day cells are surrounded by a **plasma membrane** consisting of amphipathic molecules—mainly phospholipids—in this configuration; in cell membranes, the lipid bilayer also contains amphipathic proteins. In the electron microscope such membranes appear as sheets about 5 nm thick, with a distinctive three-layered appearance due to the tail-to-tail packing of the phospholipid molecules.

It is not clear at what point in the evolution of biological catalysts the first cells were formed. They may have originated when phospholipid molecules in the prebiotic soup spontaneously assembled into membranous structures enclosing a self-replicating mixture of catalytic RNA molecules. It is more commonly assumed, however, that protein synthesis had to evolve before there were cells. In either case, once they were sealed within a closed membrane, RNA molecules could begin to evolve, not merely on the basis of their own structure, but also according to their effect on the other molecules in the same compartment: the nucleotide sequences of the RNA molecules could now be expressed in the character of the cell as a whole.

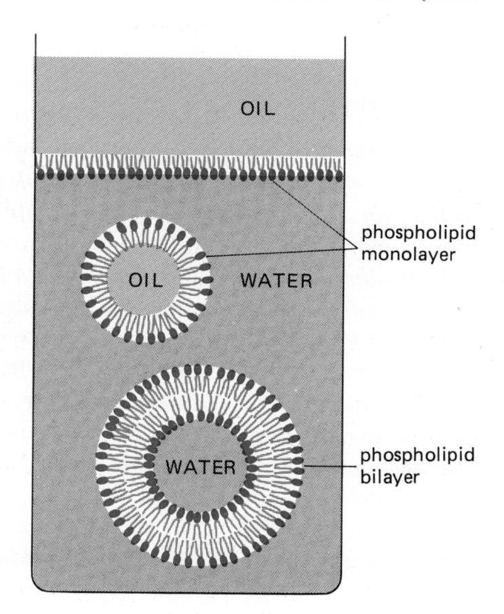

Figure 1–10 Cell membranes are formed of phospholipids. Because these molecules have hydrophilic heads and lipophilic tails, they will align themselves at an oil-water interface with their heads in the water and their tails in the oil. In water, they will associate to form closed bilayer vesicles in which the lipophilic tails are in contact with one another and the hydrophilic heads are exposed to the water.

All Present-Day Cells Use DNA as Their Hereditary Material[4,5]

The picture we have presented is, of course, speculative: there are no fossil records that trace the origins of the first cell. Nevertheless, there is persuasive evidence from present-day organisms and from experiments that the broad features of this evolutionary story are correct. The prebiotic synthesis of small molecules, the self-replication of catalytic RNA molecules, the translation of RNA sequences into amino acid sequences, and the assembly of lipid molecules to form membrane-bounded compartments—all presumably occurred to generate primitive cells 3.5 or 4 billion years ago.

It is useful to compare these early cells with the simplest present-day cells, the **mycoplasmas.** Mycoplasmas are small bacteriumlike organisms that normally lead a parasitic existence in close association with animal or plant cells (Figure 1–11). Some have a diameter of about 0.3 μm and contain only enough nucleic acid to direct the synthesis of about 750 different proteins. Some of these proteins are enzymes, some are structural; some lie in the cell's interior, others are embedded in its membrane. Together they synthesize essential small molecules that are not available in the environment, redistribute the energy needed to drive biosynthetic reactions, and maintain appropriate conditions inside the cell.

The first cells on the earth presumably contained many fewer components than a mycoplasma and divided much more slowly. There was, however, a more fundamental difference between these primitive cells and a mycoplasma, or indeed any other present-day cell: the hereditary information in all cells alive today is stored in DNA rather than in the RNA that is thought to have stored the hereditary information in primitive cells. Both types of polynucleotides are found in present-day cells, but they function in a collaborative manner, each having evolved to perform specialized tasks. Small chemical differences fit the two kinds of molecules for distinct functions. As the permanent repository of genetic information, DNA is more stable than RNA. This is partly because DNA lacks a sugar hydroxyl group that makes RNA more susceptible to hydrolysis. But it is also because DNA, unlike RNA, exists principally in a double-stranded form, composed of a pair of complementary polynucleotide molecules. Its double-stranded structure not only makes DNA relatively easy to replicate (as will be explained in Chapter 3), it also permits a repair mechanism to operate that uses the intact strand as a template for the correction or repair of the associated damaged strand. DNA guides the synthesis of specific RNA molecules, again by the principle of complementary base-pairing, though now this pairing is between slightly different types of nucleotides. The resulting single-stranded RNA molecules then perform two primeval functions: they direct protein synthesis both as coding RNA molecules (*messenger* RNAs) and as RNA catalysts (*ribosomal* and other nonmessenger RNAs).

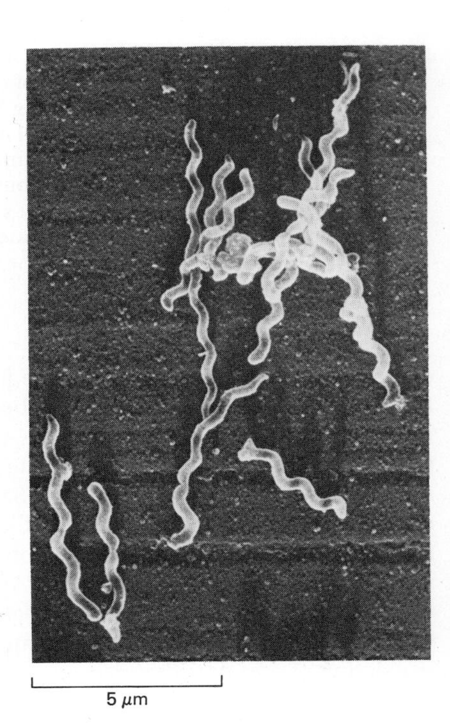

Figure 1–11 *Spiroplasma citrii,* a mycoplasma that grows in plant cells. (Courtesy of J. Burgess.)

The suggestion, in short, is that RNA came first in evolution, having both genetic and catalytic properties; after efficient protein synthesis evolved, DNA took over the primary genetic function and proteins became the major catalysts, while RNA remained primarily as the intermediary connecting the two (Figure 1–12). DNA would have become necessary only when cells became so complex that they required an amount of genetic information greater than that which could be stably maintained in RNA molecules.

Summary

Living cells probably arose on earth by the spontaneous aggregation of molecules about 3.5 billion years ago. From our knowledge of present-day organisms and the molecules they contain, it seems likely that the development of the autocatalytic mechanisms fundamental to living systems began with the evolution of families of RNA molecules that could catalyze their own replication. With time, one of these families of cooperating RNA catalysts developed the ability to direct synthesis of polypeptides. Early cells are likely to have relied heavily on both RNA and protein catalysis and to have contained only RNA as their hereditary material. Finally, as the accumulation of additional protein catalysts allowed more efficient and complex cells to evolve, the DNA double helix replaced RNA as a more stable molecule for storing the increased amounts of genetic information required by such cells.

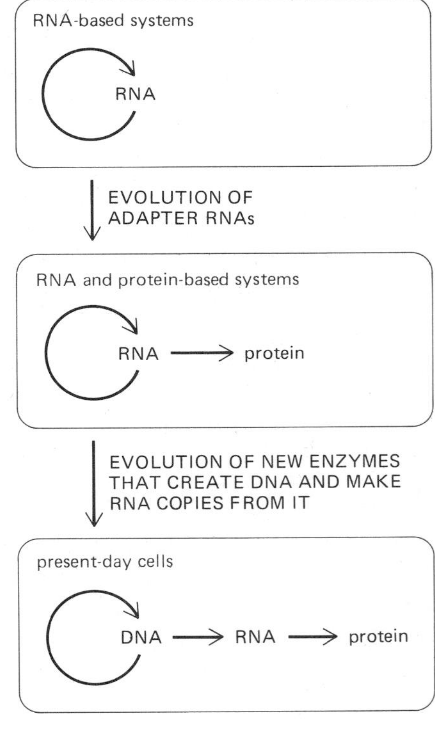

Figure 1–12 Suggested stages of evolution from simple self-replicating systems of RNA molecules to present-day cells, in which DNA is the repository of genetic information and RNA acts largely as a go-between to direct protein synthesis.

From Procaryotes to Eucaryotes[6]

It is thought that all organisms living now on earth derive from a single primordial cell born more than three billion years ago. This cell, outreproducing its competitors, took the lead in the process of cell division and evolution that eventually covered the earth with green, changed the composition of its atmosphere, and made it the home of intelligent life. The family resemblances among all organisms seem too strong to be explained in any other way. One important landmark along this evolutionary road occurred about 1.5 billion years ago, when there was a transition from small cells with relatively simple internal structures—the so-called **procaryotic** cells, which include the various types of bacteria—to a flourishing of larger and radically more complex *eucaryotic* cells such as are found in higher animals and plants.

Procaryotic Cells Are Structurally Simple but Biochemically Diverse[7]

Bacteria are the simplest organisms found in most natural environments. They are spherical or rod-shaped cells, commonly several micrometers in linear dimension (Figure 1–13). They often possess a tough protective coat, called a *cell wall,* beneath which a plasma membrane encloses a single cytoplasmic compartment containing DNA, RNA, proteins, and small molecules. In the electron microscope this cell interior appears as a matrix of varying texture without any obvious organized internal structure (see Figure 1–8A).

Bacteria are small and can replicate quickly by simply dividing in two by *binary fission.* When food is plentiful, "survival of the fittest" generally means survival of those that can divide the fastest. Under optimal conditions, a single procaryotic cell can divide every 20 minutes and thereby give rise to 5 billion cells (approximately equal to the present human population on earth) in less than 11 hours. The ability to divide quickly enables populations of bacteria to adapt rapidly to changes in their environment. Under laboratory conditions, for example, a population of bacteria maintained in a large vat will evolve within a few weeks by spontaneous mutation and natural selection to utilize new types of sugar molecules as carbon sources.

In nature, bacteria live in an enormous variety of ecological niches, and they show a corresponding richness in their underlying biochemical composition. Two distantly related groups can be recognized: the *eubacteria,* which are the commonly encountered forms that inhabit soil, water, and larger living organisms; and the *archaebacteria,* which are found in such incommodious environments as bogs, ocean depths, salt brines, and hot acid springs (Figure 1–14).

There exist species of bacteria that can utilize virtually any type of organic molecule as food, including sugars, amino acids, fats, hydrocarbons, polypeptides, and polysaccharides. Some are even able to obtain their carbon atoms from CO_2 and their nitrogen atoms from N_2. Despite their relative simplicity, bacteria have survived for longer than any other organisms and still are the most abundant type of cell on earth.

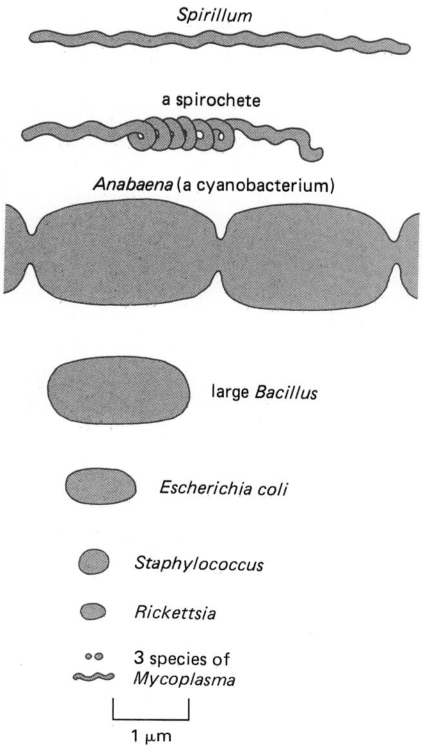

Figure 1–13 Some procaryotic cells drawn to scale.

Metabolic Reactions Evolve[7,8]

A bacterium growing in a salt solution containing a single type of carbon source, such as glucose, must carry out a large number of chemical reactions. Not only must it derive from the glucose the chemical energy needed for many vital processes, it must also use the carbon atoms of glucose to synthesize every type of organic molecule that the cell requires. These reactions are catalyzed by hundreds of enzymes working in reaction "chains" so that the product of one reaction is the substrate for the next; such enzymatic chains, called *metabolic pathways,* will be discussed in the following chapter.

Originally, when life began on earth, there was probably little need for such metabolic reactions. Cells could survive and grow on the molecules in their surroundings—a legacy from the prebiotic soup. As these natural resources were exhausted, organisms that had developed enzymes to manufacture organic mole-

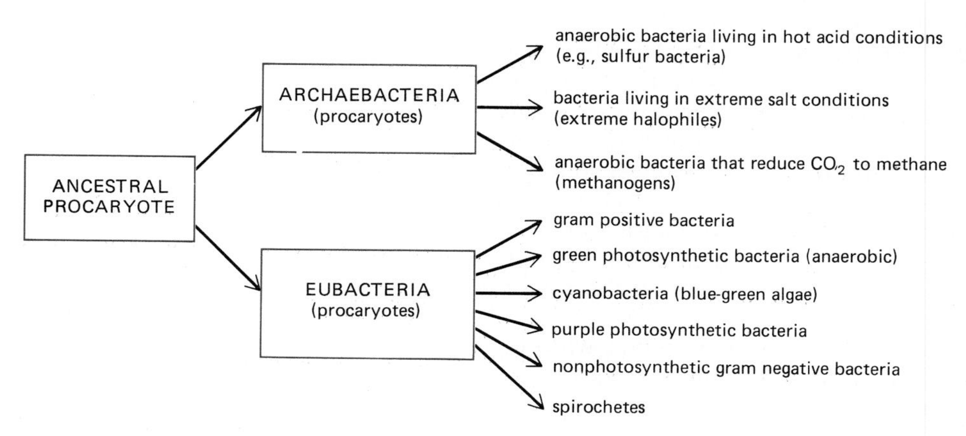

Figure 1–14 Family relationships between present-day bacteria; arrows indicate probable paths of evolution. The origin of eucaryotic cells is discussed later in the text.

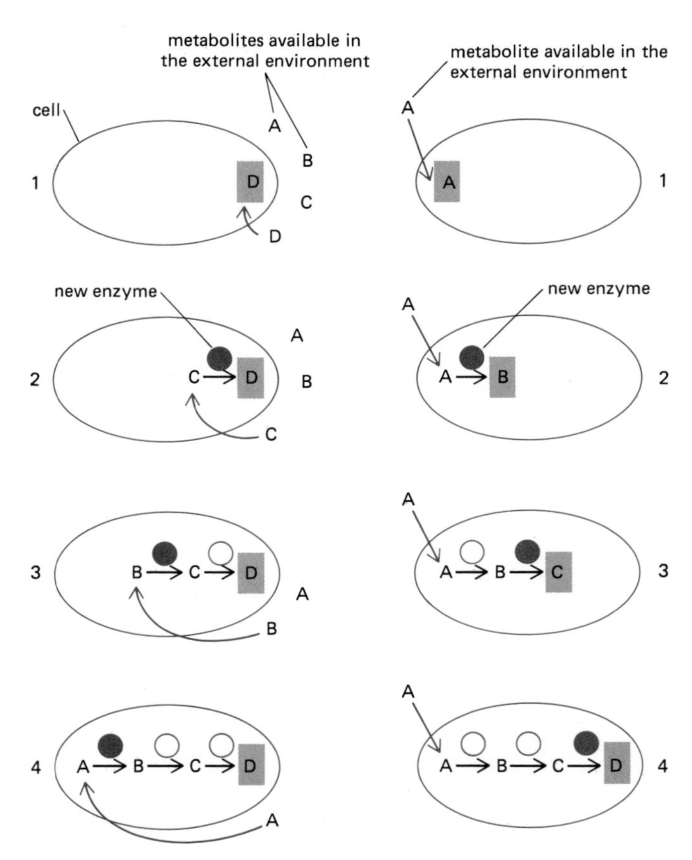

Figure 1–15 Schematic drawing showing two possible ways in which metabolic pathways might have evolved. The cell on the left is provided with a supply of related substances (A, B, C, and D) produced by prebiotic synthesis. One of these, substance D, is metabolically useful. As the cell exhausts the available supply of D, a selective advantage is obtained by the evolution of a new enzyme that is able to produce D from the closely related substance C. Fundamentally important metabolic pathways may have evolved by a series of similar steps. On the right, a metabolically useful compound A is available in abundance. An enzyme appears in the course of evolution that, by chance, has the ability to convert substance A to substance B. Other changes then occur within the cell that enable it to make use of the new substance. The appearance of further enzymes can build up a long chain of reactions.

cules had a strong selective advantage. In this way, the complement of enzymes possessed by cells is thought to have gradually increased, generating the metabolic pathways of present organisms. Two plausible ways in which a metabolic pathway could arise in evolution are illustrated in Figure 1–15.

If metabolic pathways evolved by the sequential addition of new enzymatic reactions to existing ones, the most ancient reactions should, like the oldest rings in a tree trunk, be closest to the center of the "metabolic tree," where the most fundamental of the basic molecular building blocks are synthesized. This position in metabolism is firmly occupied by the chemical processes that involve sugar phosphates, among which the most central of all is probably the sequence of reactions known as **glycolysis,** by which glucose can be degraded in the absence of oxygen (that is, *anaerobically*). The oldest metabolic pathways would have had to be anaerobic, because there was no free oxygen in the atmosphere of the primitive earth. Glycolysis occurs in virtually every living cell and drives the formation of the compound *adenosine triphosphate,* or *ATP,* which is used by all cells as a versatile source of chemical energy.

Linked to these core reactions of sugar phosphates are hundreds of other chemical processes. Some of these are responsible for the synthesis of small molecules, many of which in turn are utilized in further reactions to make the large polymers specific to the organism. Other reactions are used to degrade complex molecules, taken in as food, into simpler chemical units. One of the most striking features of these metabolic reactions is that they take place similarly in all kinds of organisms. Certainly differences exist: many specialized products of metabolism are restricted to certain genera or species. Even the common amino acid lysine is made in different ways in bacteria, in yeast, and in green plants and is not made at all in higher animals. But in broad terms the majority of reactions and most of the enzymes that catalyze them are found in all living things, from bacteria to people; for this reason they are believed to have been present in the primitive ancestral cells that gave rise to all these organisms.

The enzymes that catalyze the fundamental metabolic reactions, while continuing to serve the same essential functions, have undergone progressive modifications as organisms have evolved into divergent forms. For this reason, the

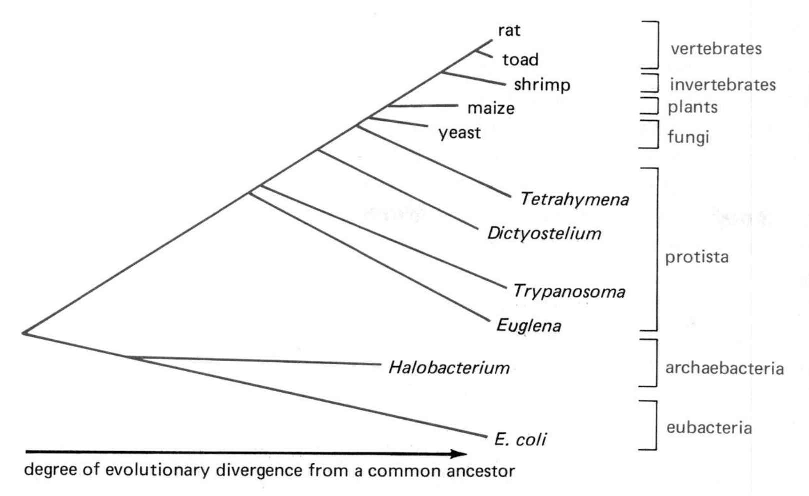

Figure 1–16 Evolutionary relationships of organisms deduced from the nucleotide sequences of their small-subunit ribosomal RNA genes. These genes contain highly conserved sequences, which change so slowly that they can be used to measure phylogenetic relationships spanning the entire range of living organisms. The data suggest that the plant, animal, and fungal lineages diverged from a common ancestor relatively late in the history of eucaryotic cells. (Adapted from M.L. Sogin, H.J. Elwood, and J.H. Gunderson, *Proc. Natl. Acad. Sci. USA* 83:1383–1387, 1986.)

amino acid sequence of the same type of enzyme in different living species provides a valuable indication of the evolutionary relationship between these species. The evidence obtained closely parallels that from other sources, such as the fossil record. An even richer source of information is locked in the living cell in the sequences of nucleotides in DNA, and modern methods of analysis allow these *DNA sequences* to be determined in large numbers and compared between species. Comparisons of highly conserved sequences, which have a central function and therefore change only slowly during evolution, can reveal relationships between organisms that diverged long ago (Figure 1–16), while very rapidly evolving sequences can be used to determine how more closely related species evolved (Figure 1–17). It is expected that continued application of these methods will enable the course of evolution to be followed with unprecedented accuracy.

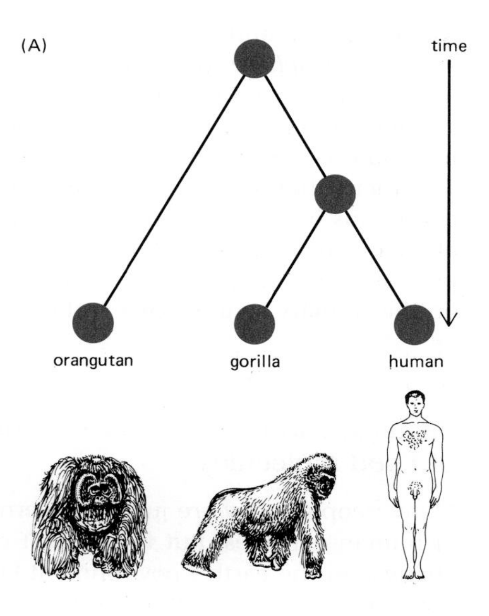

Figure 1–17 Is a human being more closely related to the gorilla or to the orangutan? The question can be answered by comparing their DNA sequences, which gives the pedigree shown in (A). Mitochondrial DNA is usually chosen for such comparisons between closely related organisms, because this DNA undergoes evolutionary change about 5 to 10 times more rapidly than the bulk of the DNA in the cell (the nuclear DNA). (B) The first 75 nucleotides of the same gene (a mitochondrial gene coding for a subunit of the enzyme NAD dehydrogenase) are shown for each species. A colored letter marks each point where the gorilla or orangutan differs from humans. The boxes beneath the nucleotide sequences symbolize the amino acids in the corresponding proteins. The apes' amino acids are identified in color where they differ from those of the human. Analyzing additional DNA sequences in this way, one finds that the gorilla sequence differs from that of humans by 10% of its nucleotides, whereas orangutans differ both from humans and from gorillas by 17%. Assuming that these differences are due to mutations occurring at random and with a similar frequency in the lineage of each species, one can deduce the pedigree shown in (A). (Data from W.M. Brown, E.M. Prager, A. Wang, and A.C. Wilson, *J. Mol. Biol.* 18:225–239, 1982.)

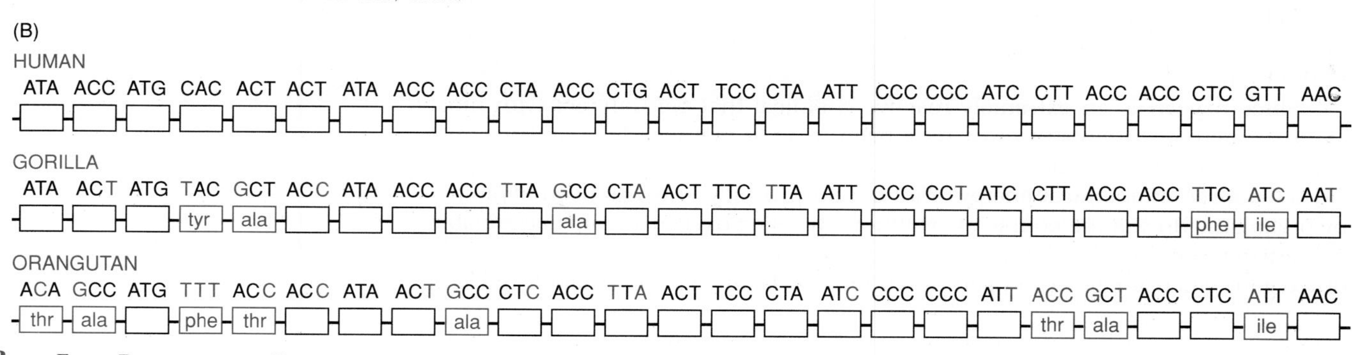

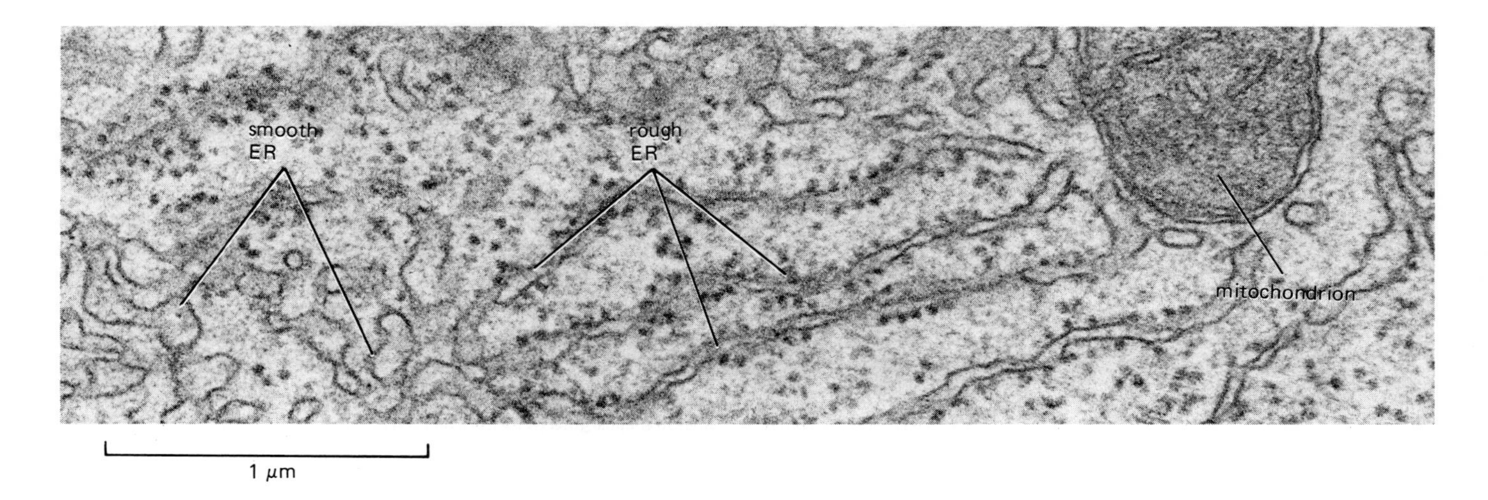

Eucaryotic Cells Contain a Rich Array of Internal Membranes

Eucaryotic cells are usually much larger in volume than procaryotic cells, commonly by a factor of a thousand or more, and they carry a proportionately larger quantity of most cellular materials; for example, a human cell contains about one thousand times as much DNA as a typical bacterium. This large size creates problems. Since all the raw materials for the biosynthetic reactions occurring in the interior of a cell must ultimately enter and leave by passing through the plasma membrane covering its surface, and since the membrane is also the site of many important reactions, an increase in cell volume requires an increase in cell surface. But it is a fact of geometry that simply scaling up a structure increases the volume as the cube of the linear dimension while the surface area increases only as the square. Therefore, if the large eucaryotic cell is to keep as high a ratio of surface to volume as the procaryotic cell, it must supplement its surface area by means of convolutions, infoldings, and other elaborations of its membrane.

This probably explains in part the complex profusion of **internal membranes** that is a basic feature of all eucaryotic cells. Membranes surround the nucleus, the mitochondria, and (in plant cells) the chloroplasts. They form a labyrinthine compartment called the **endoplasmic reticulum** (Figure 1–23), where lipids and proteins of cell membranes, as well as materials destined for export from the cell, are synthesized. They also form stacks of flattened sacs constituting the **Golgi apparatus** (Figure 1–24), which is involved in the modification and transport of the molecules made in the endoplasmic reticulum. Membranes surround **lysosomes,** which contain stores of enzymes required for intracellular digestion and so prevent them from attacking the proteins and nucleic acids elsewhere in the cell. In the same way, membranes surround **peroxisomes,** where dangerously reactive hydrogen peroxide is generated and degraded during the oxidation of various molecules by O_2. Membranes also form small vesicles and, in plants, a large liquid-filled *vacuole.* All these membrane-bounded structures correspond to distinct internal compartments within the cytoplasm. In a typical animal cell, these compartments (or organelles) occupy nearly half the total cell volume. The remaining compartment of the cytoplasm, which includes everything other than the membrane-bounded organelles, is usually referred to as the **cytosol.**

All of the aforementioned membranous structures lie in the interior of the cell. How, then, can they help to solve the problem we posed at the outset and provide the cell with a surface area that is adequate to its large volume? The answer is that there is a continual exchange between the internal membrane-bounded compartments and the outside of the cell, achieved by *endocytosis* and *exocytosis,* processes unique to eucaryotic cells. In endocytosis, portions of the external surface membrane invaginate and pinch off to form membrane-bounded cytoplasmic vesicles that contain both substances present in the external medium and molecules previously adsorbed on the cell surface. Very large particles or even entire foreign cells can be taken up by *phagocytosis*—a special form of endocytosis.

Figure 1–23 Electron micrograph of a thin section of a mammalian cell showing both smooth and rough regions of the endoplasmic reticulum (ER). The smooth regions are involved in lipid metabolism; the rough regions, studded with ribosomes, are sites of synthesis of proteins that are destined to leave the cytosol and enter certain other compartments of the cell. (Courtesy of George Palade.)

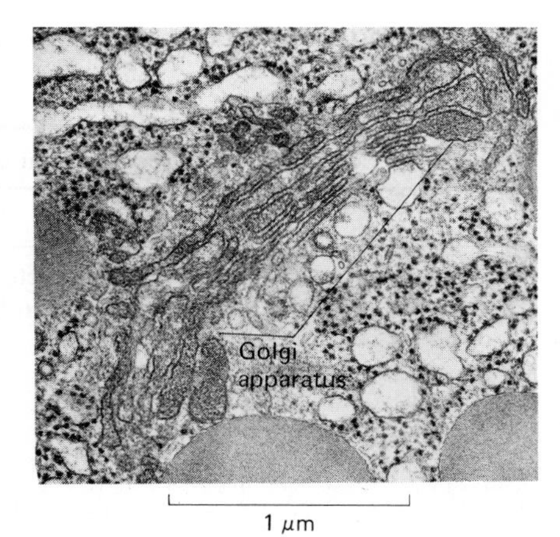

Figure 1–24 Electron micrograph of a thin section of a mammalian cell showing the Golgi apparatus, which is composed of flattened sacs of membrane arranged in multiple layers (see also Panel 1–1, pp. 16–17). The Golgi apparatus is involved in the synthesis and packaging of molecules destined to be secreted from the cell, as well as in the routing of newly synthesized proteins to the correct cellular compartment. (Courtesy of Daniel S. Friend.)

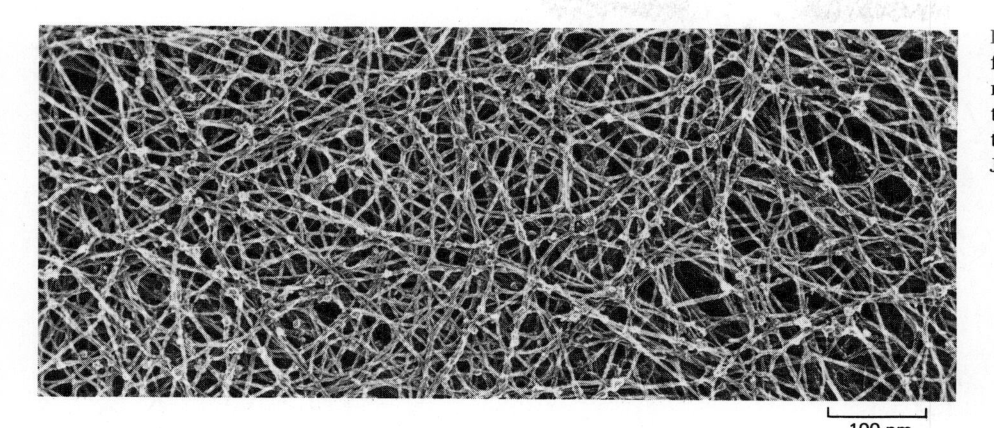

Figure 1–25 A network of actin filaments underlying the plasma membrane of an animal cell is seen in this electron micrograph prepared by the deep-etch technique. (Courtesy of John Heuser.)

Exocytosis is the reverse process, whereby membrane-bounded vesicles inside the cell fuse with the plasma membrane and release their contents into the external medium. In this way, membranes surrounding compartments deep inside the cell serve to increase the effective surface area of the cell for exchanges of matter with the external world.

As we shall see in later chapters, the various membranes and membrane-bounded compartments in eucaryotic cells have become highly specialized, some for secretion, some for absorption, some for specific biosynthetic processes, and so on.

Eucaryotic Cells Have a Cytoskeleton

The larger a cell is, and the more elaborate and specialized its internal structures, the greater is its need to keep these structures in their proper places and to control their movements. All eucaryotic cells have an internal skeleton, the **cytoskeleton,** that gives the cell its shape, its capacity to move, and its ability to arrange its organelles and transport them from one part of the cell to another. The cytoskeleton is composed of a network of protein filaments, two of the most important of which are *actin filaments* and *microtubules* (Figure 1–25). These two must date from a very early epoch in evolution since they are found almost unchanged in all eucaryotes. Both are involved in the generation of cellular movements; actin filaments, for example, participate in the contraction of muscle, whereas microtubules are the main structural and force-generating elements in *cilia* and *flagella*—the long projections on some cell surfaces that beat like whips and serve as instruments of propulsion.

Actin filaments and microtubules are also essential for the internal movements that occur in the cytoplasm of all eucaryotic cells. Thus microtubules in the form of a *mitotic spindle* are a vital part of the usual machinery for partitioning DNA equally between the two daughter cells when a eucaryotic cell divides. Without microtubules, therefore, the eucaryotic cell could not reproduce. In this and other examples, movement by free diffusion would be either too slow or too haphazard to be useful. In fact, the organelles in a eucaryotic cell often appear to be attached, directly or indirectly, to the cytoskeleton and, when they move, to be propelled along cytoskeletal tracks.

Protozoa Include the Most Complex Cells Known[12]

The complexity that can be achieved by a single eucaryotic cell is nowhere better illustrated than in *protists* (Figure 1–26). These are free-living, single-celled eucaryotes that are evolutionarily diverse (see Figure 1–16) and exhibit a bewildering variety of different forms and behaviors: they can be photosynthetic or carnivorous, motile or sedentary. Their anatomy is often complex and includes such structures as sensory bristles, photoreceptors, flagella, leglike appendages, mouth parts, stinging darts, and musclelike contractile bundles. Although they are single cells, they can

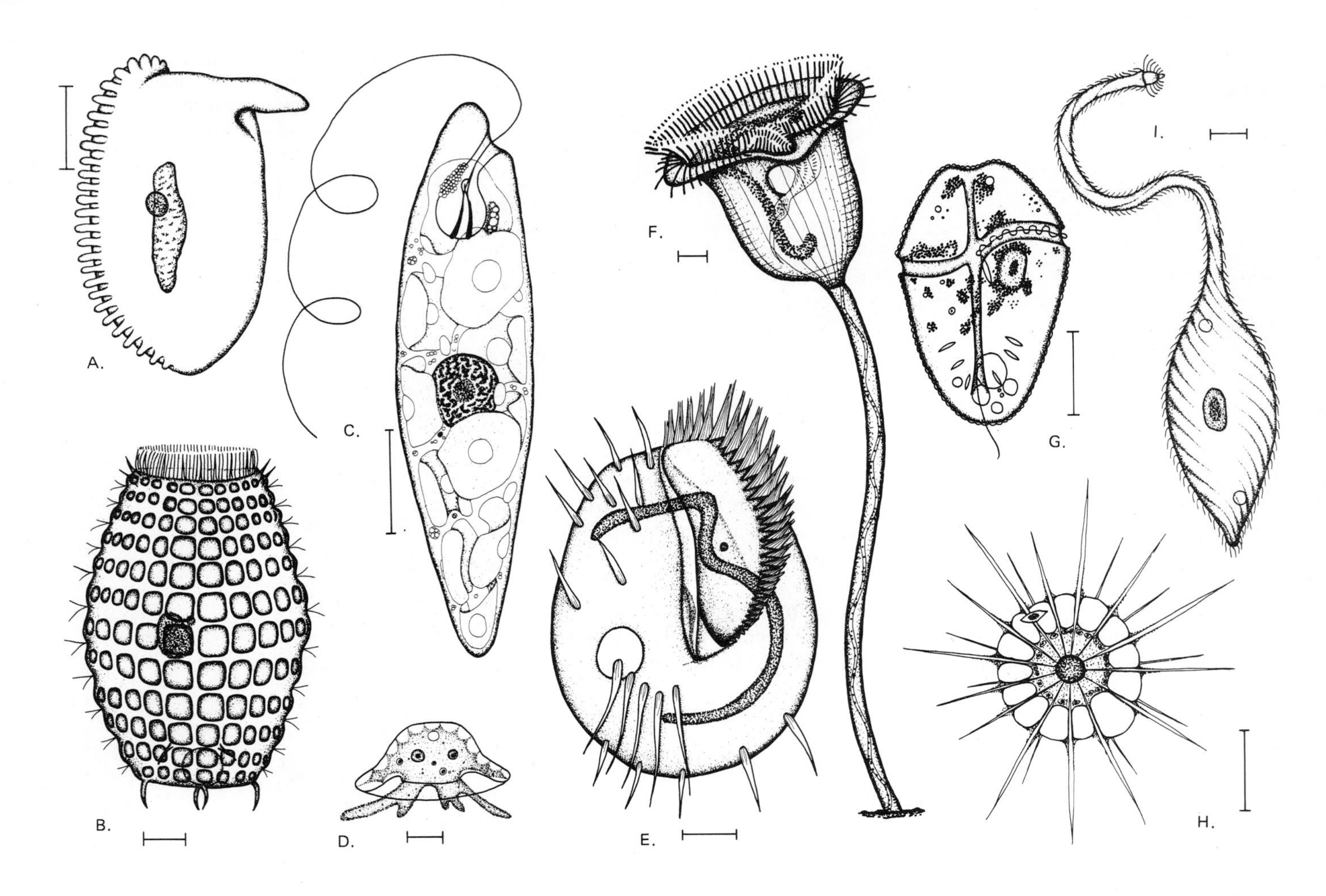

be as intricate and versatile as many multicellular organisms. This is particularly true of the group of protists known as **protozoa**—or "first animals."

Didinium is a carnivorous protozoan. It has a globular body, about 150 μm in diameter, encircled by two fringes of cilia; its front end is flattened except for a single protrusion rather like a snout (Figure 1–27). *Didinium* swims around in the water at high speed by means of the synchronous beating of its cilia. When it encounters a suitable prey, usually another type of protozoan, such as a *Paramecium,* it releases numerous small paralyzing darts from its snout region. Then the *Didinium* attaches to and devours the *Paramecium,* inverting like a hollow ball to engulf the other cell, which is as large as itself. Most of this complex behavior—swimming, and paralyzing and capturing its prey—is generated by the cytoskeletal structures lying just beneath the plasma membrane. Included in this *cell cortex,* for example, are the parallel bundles of microtubules that form the core of each cilium and enable it to beat.

Predatory behavior of this sort and the set of features on which it depends—large size, the capacity for phagocytosis, and the ability to move in pursuit of prey—are peculiar to eucaryotes. Indeed, it is probable that these features came very early in eucaryotic evolution, making possible the subsequent capture of bacteria for domestication as mitochondria and chloroplasts.

Figure 1–26 An assortment of protists, illustrating some of the enormous variety to be found among this class of single-celled organisms. These drawings are done to different scales, but in each case the bar denotes 10 μm. The organisms in (A), (B), (E), (F), and (I) are ciliates; (C) is an euglenoid; (D) is an amoeba; (G) is a dinoflagellate; (H) is a heliozoan. (From M.A. Sleigh, The Biology of Protozoa. London: Edward Arnold, 1973.)

Genes Can Be Switched On and Off

The protozoa, for all their marvels, do not represent the peak of eucaryotic evolution. Greater things were achieved, not by concentrating every sort of complexity in a single cell, but by dividing the labor among different types of cells. *Multicellular organisms* evolved in which cells closely related by ancestry became differentiated from one another, some developing one feature to a high degree, others another, so forming the specialized parts of one great cooperative enterprise.

The various specialized cell types in a single higher plant or animal often appear radically different (Panel 1–2, pp. 24–25). This seems paradoxical, since all of the cells in a multicellular organism are closely related, having recently descended from the same precursor cell—the fertilized egg. Common lineage implies similar genes; how then do the differences arise? In a few cases, cell specialization involves the loss of genetic material. An extreme example is the mammalian red blood cell, which loses its entire nucleus in the course of differentiation. But the overwhelming majority of cells in most plant and animal species retain all of the genetic information contained in the fertilized egg. Specialization depends on changes in *gene expression,* not on the loss or acquisition of genes.

Even bacteria do not make all of their types of protein all of the time but are able to adjust the level of synthesis according to external conditions. Proteins required specifically for the metabolism of lactose, for example, are made by many bacteria only when this sugar is available for use; and when conditions are unfavorable for cell proliferation, some bacteria arrest most of their normal metabolic processes and form *spores,* which have tough, impermeable outer walls and a cytoplasm of altered composition.

Eucaryotic cells have evolved far more sophisticated mechanisms for controlling gene expression, and these affect entire systems of interacting gene products. Groups of genes are activated or repressed in response to both external and internal signals. Membrane composition, cytoskeleton, secretory products, even metabolism—all these and other features must change in a coordinated manner when cells become differentiated. Compare, for example, a skeletal muscle cell specialized for contraction with an *osteoblast,* which secretes the hard matrix of bone in the same animal (Panel 1–2, pp. 24–25). Such radical transformations of cell character reflect stable changes in gene expression. The controls that bring about these changes have evolved in eucaryotes to a degree unmatched in procaryotes.

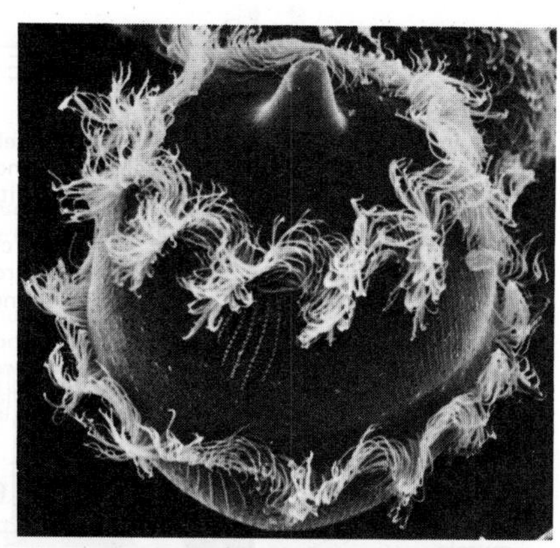

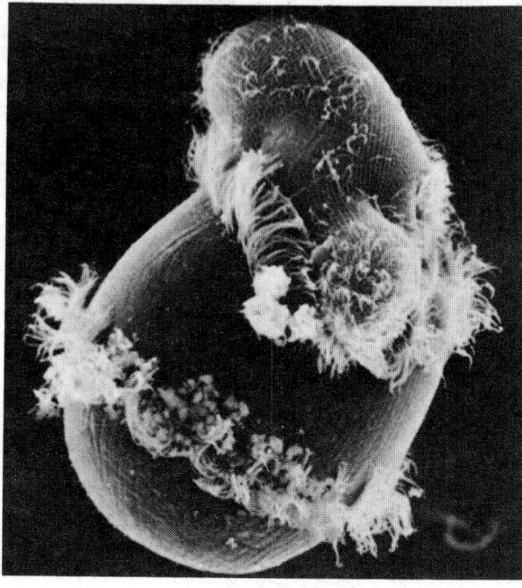

Figure 1–27 Scanning electron micrographs showing one protozoan eating another. Protozoans are single-cell animals that show an amazing diversity of form and behavior. *Didinium* (*above*), a ciliated protozoan, has two circumferential rings of motile cilia and a snoutlike protuberance at its leading end, with which it captures its prey. Below, *Didinium* is shown engulfing another protozoan, *Paramecium.* (Courtesy of D. Barlow.)

Eucaryotic Cells Have Vastly More DNA Than They Need for the Specification of Proteins

Eucaryotic cells contain a very large quantity of DNA. In human cells, as we have said, there is about a thousand times more DNA than in typical bacteria, while the cells of some amphibians have a DNA content more than 10 times that of humans (Figure 1–28). Yet it seems that only a small fraction of this DNA—perhaps 1% in human cells—carries the specifications for proteins that are actually made by the organism. Why then is the remaining 99% of the DNA there? One hypothesis is that much of it acts merely to increase the physical bulk of the nucleus. Another is that it is in large part parasitic—a collection of DNA sequences that have over the ages accumulated in the cell, exploiting the cell's machinery for their own reproduction and bringing no immediate benefit in return. Indeed, the DNA of many species has been shown to contain sequences called *transposable elements,* which have the ability to "jump" occasionally from one location to another in the DNA and even to insert additional copies of themselves at new sites. Transposable elements could thus proliferate like a slow infection, becoming an ever larger proportion of the genetic material.

But evolution is opportunistic. Whatever the origins of the DNA that does not code for protein, it is certain that it now has some important functions. Part of this DNA is structural, enabling portions of the genetic material to become condensed or "packaged" in specific ways, as described in the next section, and some of the DNA is regulatory and helps to switch on and off the genes that direct the synthesis of proteins, thus playing a crucial role in the sophisticated control of gene expression in eucaryotic cells.

In Eucaryotic Cells the Genetic Material Is Packaged in Complex Ways

The length of DNA in eucaryotic cells is so great that the risk of entanglement and breakage becomes severe. Probably for this reason, proteins unique to eucaryotes, the *histones,* have evolved to bind to the DNA and wrap it up into compact and them, and even small changes in their properties are likely to result in disaster. Fundamental steps have been "frozen" into developmental processes, just as the genetic code or protein synthesis mechanisms have become frozen into the basic biochemical organization of the cell. In contrast, cells produced near the end of development have more freedom to change. It is presumably for this reason that the embryos of different species so often resemble each other in their early stages and, as they develop, seem sometimes to replay the steps of evolution.

Eucaryotic Organisms Possess a Complex Machinery for Reproduction

Within the multicellular organism there must be some cells that serve as precursors for a new generation. In higher plants and animals these cells have a highly specialized character and are called *germ cells.* The propagation of the species depends on them, and there is a powerful selection pressure to adjust the structure of the organism as a whole to provide the germ cells with the best chance of survival. Other cells may die, but as long as germ cells survive, new organisms of the same sort will be produced. In this sense the most fundamental distinction to be drawn in a multicellular organism is the distinction between germ cells and the rest, that is, between germ cells and *somatic* cells.

Not all multicellular organisms reproduce by means of distinctive differentiated germ cells. Many simple animals, including sponges and coelenterates, can reproduce by budding off portions of their bodies, and many plants do likewise. Germ cells are, however, a necessity for **sexual reproduction.** This process is so familiar to us that we take it for granted, but it is by no means the obvious way to reproduce: it is far more complicated than asexual reproduction and requires a large diversion of resources. Two individuals of the same species but different sex produce germ cells of usually very different character—*eggs* from one, *sperm* from the other. An egg cell fuses with a sperm cell to form a *zygote*—the single precursor cell for the development of a new organism, whose genes represent a partly random reassortment of the genes of the two parents. While they may also reproduce in other ways, almost all eucaryotic species, unicellular as well as multicellular, are capable of reproducing sexually. Eucaryotic cells have evolved a complex machinery for sex; our lives revolve around it. Strong selective pressures must have operated to favor the evolution of sexual reproduction in preference to simpler strategies based on ordinary cell division. Although it is surprisingly difficult to say with certainty what those selection pressures were, it is at least plain that sexual reproduction brings new possibilities for manipulating and recombining the genes of a species. It may thus have played a crucial part in permitting the evolution of novel genes in novel combinations, and so in engendering the endless variety of forms and functions seen in plants and animals today.

The Cells of the Vertebrate Body Exhibit More Than 200 Different Modes of Specialization

The wealth of diverse specializations to be found among the cells of a higher animal is incomparably greater than any procaryote can show. In a vertebrate, more than 200 distinct **cell types** are plainly distinguishable, and many of these types of cells certainly include, under a single name, a large number of more subtly different varieties. Panel 1–2 (pp. 24–25) shows a small selection. In this profusion of specialized behaviors one can see displayed, in a single organism, the astonishing versatility of the eucaryotic cell. Much of our current knowledge of the general properties of eucaryotic cells has depended on the study of such specialized types of cells because they demonstrate to exceptionally good advantage particular features on which all cells depend in some measure. Each feature and each organelle of the prototype that we have outlined in Panel 1–1 (pp. 16–17) is developed to an unusual degree or revealed with special clarity in one cell type or another. To take one arbitrary example, consider the *neuromuscular junction,* where just three types of cells are involved: a muscle cell, a nerve cell, and a Schwann cell. Each has a very different role (Figure 1–37):

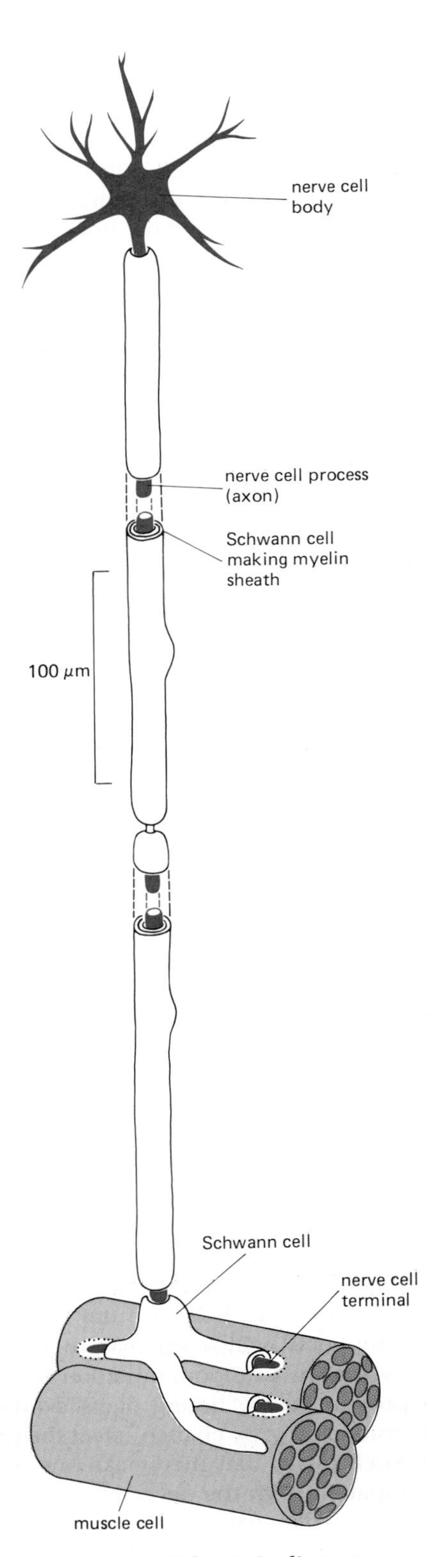

Figure 1–37 Schematic diagram showing a nerve cell, with its associated Schwann cells, contacting a muscle cell at a neuromuscular junction.

1. The muscle cell has made contraction its specialty. Its cytoplasm is packed with organized arrays of protein filaments, including vast numbers of actin filaments. There are also many mitochondria interspersed among the protein filaments, supplying ATP as fuel for the contractile apparatus.
2. The nerve cell stimulates the muscle to contract, conveying an excitatory signal to the muscle from the brain or spinal cord. The nerve cell is therefore extraordinarily elongated: its main body, containing the nucleus, may lie a meter or more from the junction with the muscle. The cytoskeleton is consequently well developed so as to maintain the unusual shape of the cell and to transport materials efficiently from one end of the cell to the other. The most crucial specialization of the nerve cell, however, is its plasma membrane, which contains proteins that act as ion *pumps* and ion *channels*, causing a movement of ions that is equivalent to a flow of electricity. Whereas all cells contain such pumps and channels in their plasma membranes, the nerve cell has exploited them in such a way that a pulse of electricity can propagate in a fraction of a second from one end of the cell to the other, conveying a signal for action.
3. Lastly, Schwann cells are specialists in the mass production of plasma membrane, which they wrap around the elongated portion of the nerve cell, laying down layer upon layer of membrane like a roll of tape, to form a *myelin sheath* that serves as insulation.

Cells of the Immune System Are Specialized for the Task of Chemical Recognition

Among all the cell systems that have evolved in higher animals, two stand out in different ways as pinnacles of complexity and sophistication: the *immune system* of the vertebrate is one, the *nervous system* is the other. Each far surpasses the performance of any artificial device—the vertebrate immune system in its capacity for chemical discrimination, the nervous system in its capacities for perception and control. Each system comprises a large number of different cell types and depends on complex interactions among them.

The protected and well-nourished environment in the interior of a multicellular animal is as inviting to foreign organisms as it is congenial to the animal's own cells. Hence there is a need for such animals to defend themselves against invading organisms—particularly viruses and bacteria. The primary task of the **immune system** is to destroy any such foreign microorganisms that may gain entry to the body.

As mentioned earlier, many eucaryotic cells are capable of phagocytosis: they can engulf and digest particles of matter from their surroundings. Among the differentiated cells in higher animals, there are professional phagocytic cells, such as *macrophages,* that specialize in this activity and can swallow up and destroy bacteria and other foreign cells (Figure 1–38). But there is a difficulty: it is good if the phagocytic cell attacks the foreign invader, but it would be disastrous if it were to attack also its own relatives and colleagues. The immune system therefore faces the problem of discriminating between the animal's own cells and those that are foreign—that is, of distinguishing between self and nonself.

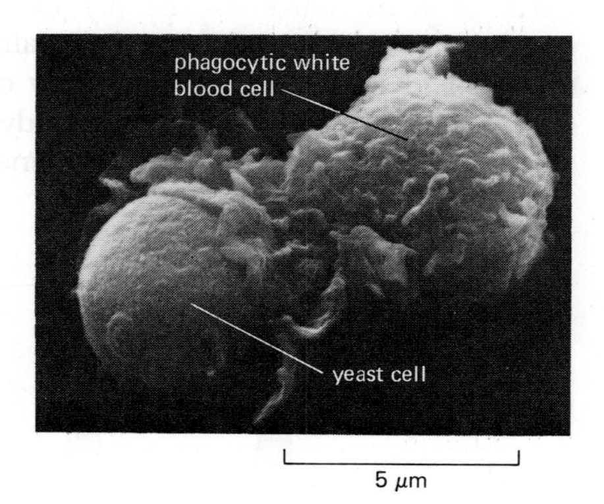

Figure 1–38 A scanning electron micrograph of a neutrophil—a type of professional phagocytic white blood cell—phagocytosing a yeast cell. (From J. Boyles and D.F. Bainton, *Cell* 24:905–914, 1981. © Cell Press.)

The vertebrates have consequently evolved a specialized class of discriminatory cells, the *lymphocytes.* These are not themselves phagocytic; instead, they collaborate to provide the phagocytic cells with cues that tell them whether to attack or let live. In particular, certain of the lymphocytes (the B lymphocytes) manufacture specific protein molecules, or *antibodies,* that bind selectively to particular arrangements of atoms on the surfaces of invading organisms or on the toxic molecules they produce. To brand a new type of invader as foreign, new types of antibody must be produced; and since the variety of possible invaders is vast and essentially unpredictable, the B lymphocytes must be capable of making an almost endless variety of antibodies. On the other hand, the system must not produce antibodies that bind to the animal's own cells and molecules.

The vast diversity of antibodies is generated by unique genetic mechanisms that create millions of genetically different lymphocytes, each able to proliferate

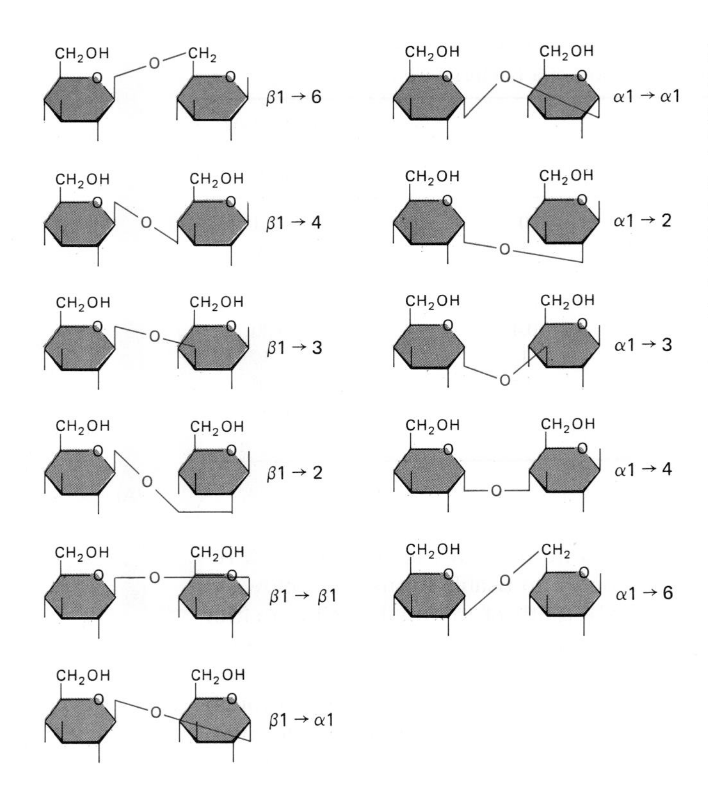

Figure 2–4 Eleven disaccharides consisting of two D-glucose units. Although these differ only in the type of linkage between the two glucose units, they are chemically distinct. Since the oligosaccharides associated with proteins and lipids may have six or more different kinds of sugar joined in both linear and branched arrangements through linkages such as those illustrated here, the number of possible distinct types of oligosaccharides is extremely large.

Glucose is the principal food compound of many cells. A series of oxidative reactions (see p. 61) leads from this hexose to various smaller sugar derivatives and eventually to CO_2 and H_2O. The net result can be written

$$C_6H_{12}O_6 + 6O_2 \rightarrow 6CO_2 + 6H_2O + \text{energy}$$

In the course of glucose breakdown, energy and "reducing power," both of which are essential in biosynthetic reactions, are salvaged and stored, mainly in the form of two crucial molecules, called **ATP** and **NADH,** respectively (see p. 66).

Simple polysaccharides composed only of glucose residues—principally *glycogen* in animal cells and *starch* in plant cells—are used to store energy for future use. But sugars do not function exclusively in the production and storage of energy. Important extracellular structural materials (such as cellulose) are composed of simple polysaccharides, and smaller but more complex chains of sugar molecules are often covalently linked to proteins in *glycoproteins* and to lipids in *glycolipids.*

Fatty Acids Are Components of Cell Membranes[4]

A fatty acid molecule, such as *palmitic acid* (Figure 2–5), has two distinct regions: a long hydrocarbon chain, which is hydrophobic (water insoluble) and not very reactive chemically, and a carboxylic acid group, which is ionized in solution (COO^-), extremely hydrophilic (water soluble), and readily forms esters and amides. In fact, almost all of the fatty acid molecules in a cell are covalently linked to other molecules by their carboxylic acid group. The many different fatty acids found in cells differ in such chemical features as the length of their hydrocarbon chains and the number and position of the carbon-carbon double bonds they contain (Panel 2–4, pp. 52–53).

Fatty acids are a valuable source of food since they can be broken down to produce more than twice as much usable energy, weight for weight, as glucose. They are stored in the cytoplasm of many cells in the form of droplets of *triglyceride* molecules, which consist of three fatty acid chains, each joined to a glycerol molecule (Panel 2–4, pp. 52–53); these molecules are the animal fats familiar from

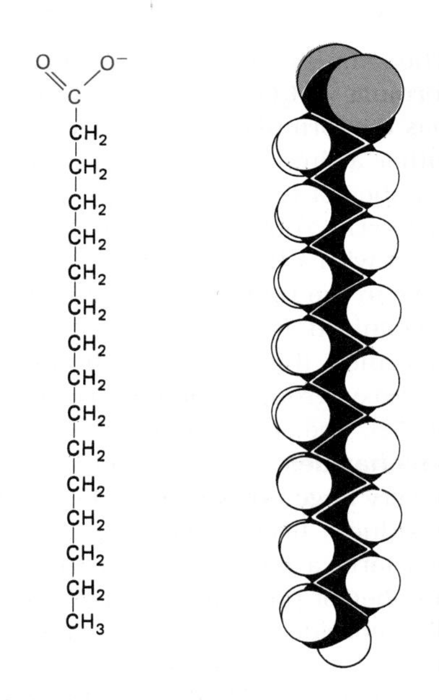

Figure 2–5 Palmitic acid. The carboxylic acid group *(color)* is shown in its ionized form. A space-filling model is presented on the right.

everyday experience. When required, the fatty acid chains can be released from triglycerides and broken down into two-carbon units. These two-carbon units, present as the acetyl group in a water-soluble molecule called *acetyl CoA*, are then further degraded in various energy-yielding reactions, which will be described below.

But the most important function of fatty acids is in the construction of cell membranes. These thin, impermeable sheets that enclose all cells and surround their internal organelles are composed largely of **phospholipids,** which are small molecules that resemble triglycerides in that they are constructed mostly from fatty acids and glycerol. However, in phospholipids the glycerol is joined to two rather than three fatty acid chains. The remaining site on the glycerol is coupled to a phosphate group, which is in turn attached to another small hydrophilic compound such as *ethanolamine, choline,* or *serine.*

Each phospholipid molecule, therefore, has a hydrophobic tail—composed of the two fatty acid chains—and a hydrophilic polar head group, where the phosphate is located. Thus phospholipid molecules are, in effect, detergents, and this is evident in their properties. A small amount of phospholipid will spread over the surface of water to form a *monolayer* of phospholipid molecules; in this thin film, the tail regions pack together very closely facing the air and the head groups are in contact with the water (Panel 2–4, pp. 52–53). Two such films can combine tail to tail to make a phospholipid sandwich, or **lipid bilayer,** which is the structural basis of all cell membranes.

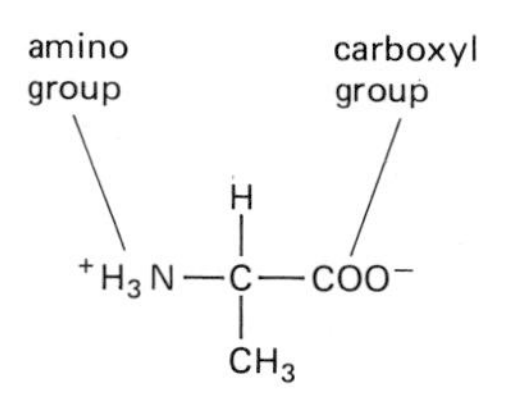

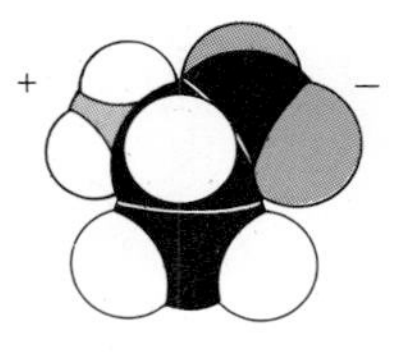

Figure 2–6 The amino acid alanine as it exists at pH 7 in its ionized form. When incorporated into a polypeptide chain, the charges on the amino and carboxyl groups of the free amino acid disappear. A space-filling model is shown below the structural formula.

Amino Acids Are the Subunits of Proteins

The common amino acids are chemically varied, but they all contain a carboxylic acid group and an amino group, both linked to a single carbon atom (Figure 2–6). They serve as subunits in the synthesis of **proteins,** which are long linear polymers of amino acids joined head to tail by a *peptide bond* between the carboxylic acid group of one amino acid and the amino group of the next (Figure 2–7). There are 20 common amino acids in proteins, each with a different *side chain* attached to the α-carbon atom (Panel 2–5, pp. 54–55). The same 20 amino acids occur over and over again in all proteins, including those made by bacteria, plants, and animals. Although the choice of precisely these 20 amino acids is probably an example of an evolutionary accident, the chemical versatility they provide is vitally important. For example, 5 of the 20 amino acids have side chains that can carry a charge (Figure 2–8), whereas the others are uncharged but reactive in specific ways (Panel 2–5, pp. 54–55). As we shall see, the properties of the amino acid side chains, in aggregate, determine the properties of the proteins they constitute and underlie all of the diverse and sophisticated functions of proteins.

Nucleotides Are the Subunits of DNA and RNA[5]

In nucleotides, one of several different nitrogen-containing ring compounds (often referred to as *bases* because they can combine with H^+ in acidic solutions) is linked to a five-carbon sugar (either *ribose* or *deoxyribose*) that carries a phosphate group. There is a strong family resemblance between the nitrogen-containing rings found in nucleotides. *Cytosine* (C), *thymine* (T), and *uracil* (U) are called **pyrimidine** compounds because they are all simple derivatives of a six-membered pyrimidine ring; *guanine* (G) and *adenine* (A) are **purine** compounds, with a second five-membered ring fused to the six-membered ring. Each nucleotide is named by reference to the unique base that it contains (Panel 2–6, pp. 56–57).

Nucleotides can act as carriers of chemical energy. The triphosphate ester of adenine, **ATP** (Figure 2–9), above all others, participates in the transfer of energy in hundreds of individual cellular reactions. Its terminal phosphate is added using energy from the oxidation of foodstuffs, and this phosphate can be readily split off by hydrolysis to release energy that drives energetically unfavorable biosynthetic reactions elsewhere in the cell. Other nucleotide derivatives serve as carriers for the transfer of particular chemical groups, such as hydrogen atoms or sugar residues, from one molecule to another. And a cyclic phosphate-containing ad-

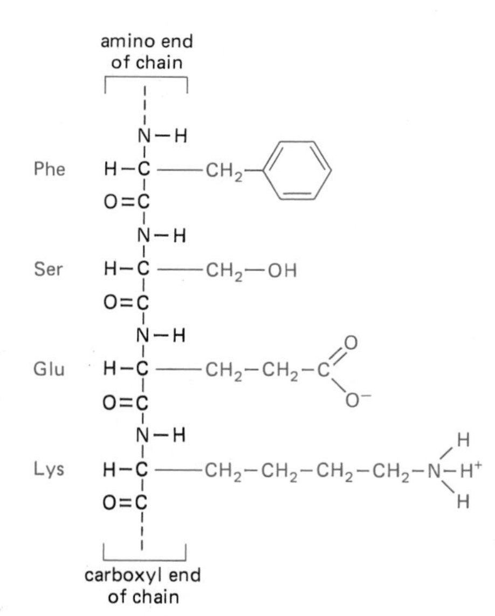

Figure 2–7 A small part of a protein molecule. The four amino acids shown are linked together by a type of covalent bond called a peptide bond. A protein is therefore also sometimes referred to as a polypeptide. The amino acid *side chains* are shown here in color.

HYDROPHILIC AND HYDROPHOBIC MOLECULES

Because of the polar nature of water molecules, they will cluster around ions and other polar molecules.

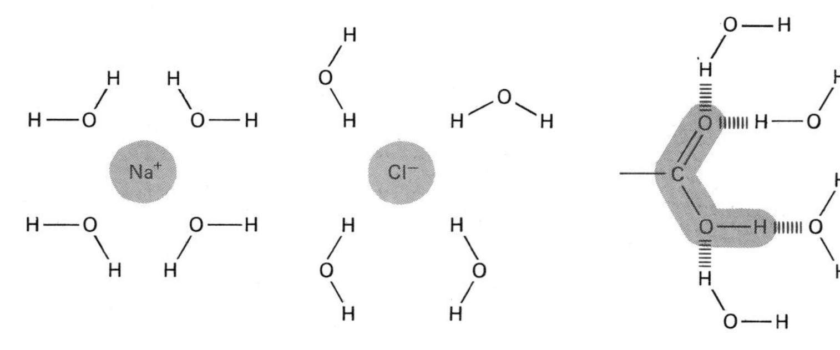

Molecules that can thereby be accommodated in water's hydrogen-bonded structures are hydrophilic and relatively water-soluble.

Nonpolar molecules interrupt the H-bonded structure of water without forming favorable interactions with water molecules. They are therefore hydrophobic and quite insoluble in water.

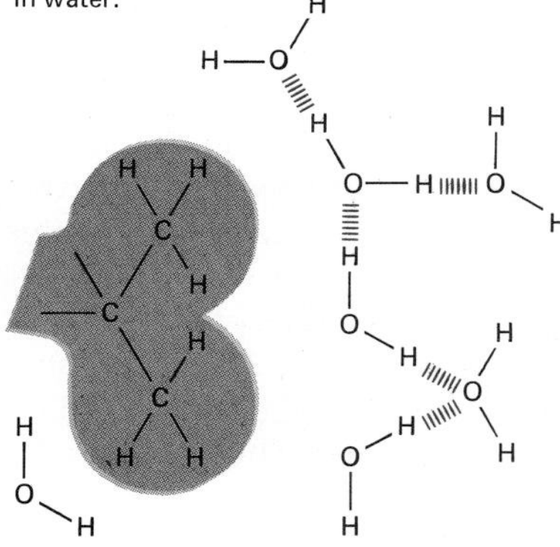

WATER

Although a water molecule has an overall neutral charge (having the same number of electrons and protons), the electrons are asymmetrically distributed, which makes the molecule polar.

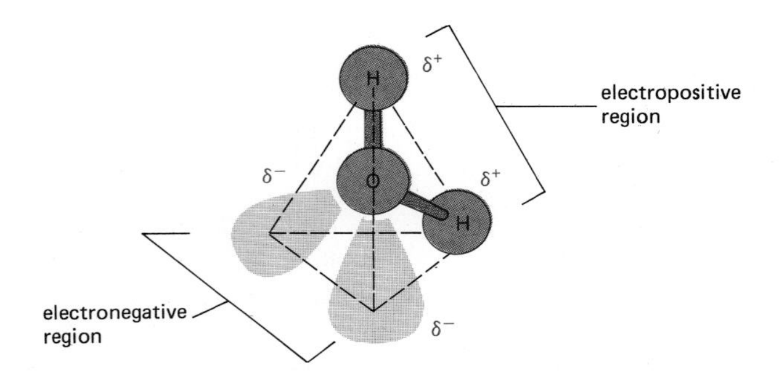

The oxygen nucleus draws electrons away from the hydrogen nuclei, leaving these nuclei with a small net positive charge. The excess of electron density on the oxygen atom creates weakly negative regions at the other two corners of an imaginary tetrahedron.

WATER STRUCTURE

Molecules of water join together transiently in a hydrogen-bonded lattice. Even at 37°C, 15% of the water molecules are joined to four others in a short-lived assembly known as a "flickering cluster."

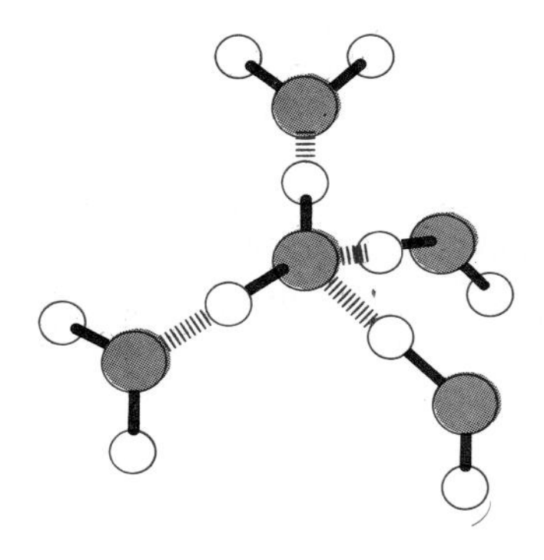

The cohesive nature of water is responsible for many of its unusual properties, such as high surface tension, specific heat, and heat of vaporization.

HYDROGEN BONDS

Because they are polarized, two adjacent H_2O molecules can form a linkage known as a hydrogen bond. Hydrogen bonds have only about 1/20 the strength of a covalent bond.

Hydrogen bonds are strongest when the three atoms lie in a straight line.

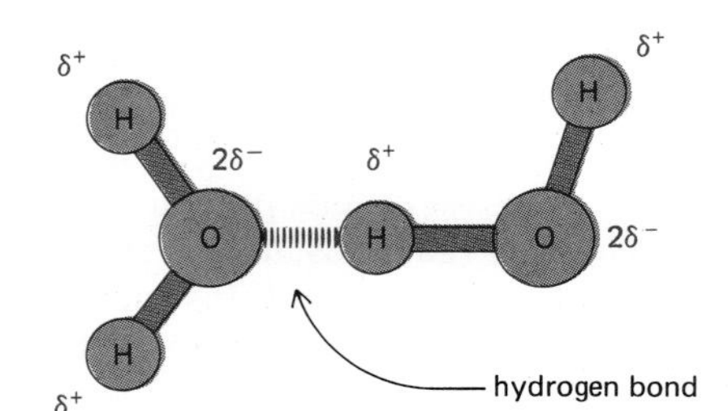

 Panel 2–1 The chemical properties of water and their influence on the behavior of biological molecules.

HYDROPHOBIC REPULSION CAN HOLD MOLECULES TOGETHER

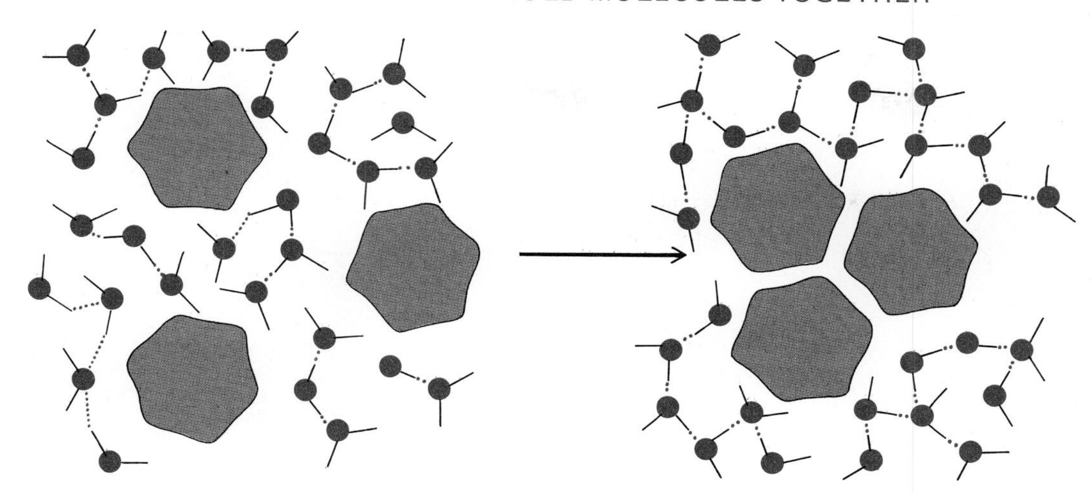

Small oil droplets coalesce into large oil drops in water since they thereby cause less disruption to the large hydrogen-bonded network of water molecules. For the same reason, hydrophobic molecules in aqueous solution will tend to be pushed together into larger aggregates by the water.

ACIDS AND BASES

An acid is a molecule that releases an H^+ ion (proton) in solution.

e.g.,

$$CH_3-C(=O)OH \rightleftharpoons CH_3-C(=O)O^- + H^+$$

acid — base — proton

A base is a molecule that accepts an H^+ ion (proton) in solution.

e.g.,

$$CH_3-NH_2 + H^+ \rightleftharpoons CH_3-NH_3^+$$

base — proton — acid

Water itself has a slight tendency to ionize and therefore can act as both a weak acid and as a weak base. When it acts as an acid, it releases a proton to form a hydroxyl ion. When it acts as a base, it accepts a proton to form a hydronium ion. Most protons in aqueous solutions exist as hydronium ions.

$$H-O-H \cdots OH_2 \rightleftharpoons OH^- + H_3O^+$$

hydroxyl ion — hydronium ion

pH

The acidity of a solution is defined by the concentration of H^+ ions it possesses. For convenience we use the pH scale where

$$pH = -\log_{10}[H^+]$$

For pure water

$[H^+] = 10^{-7}$ moles/liter

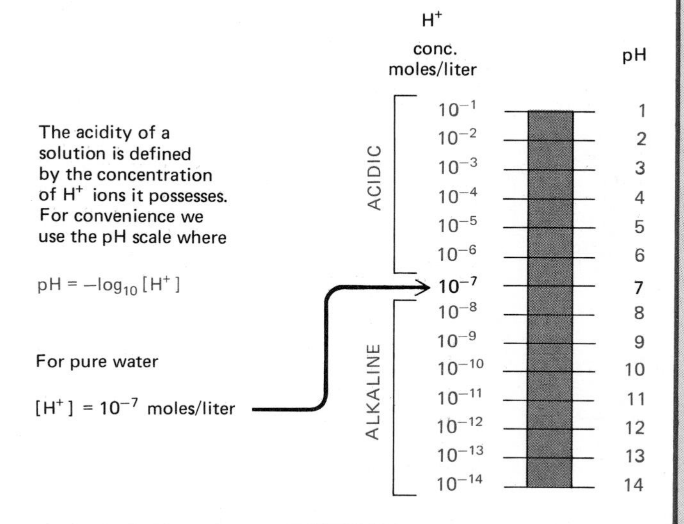

OSMOSIS

If two aqueous solutions are separated by a membrane that allows only water molecules to pass, water will move into the solution containing the greatest concentration of solute molecules by a process known as osmosis.

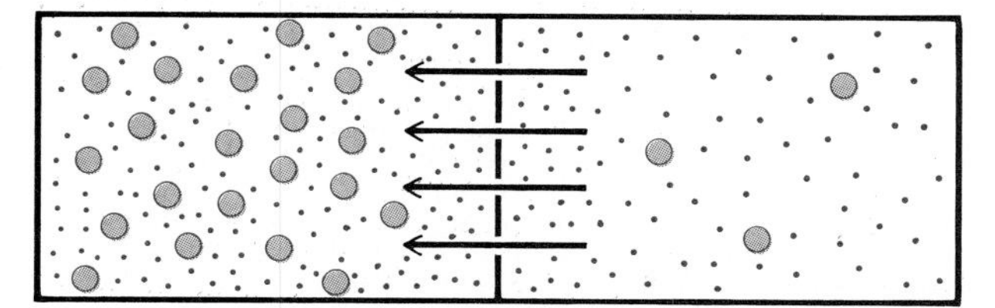

This movement of water from a hypotonic to a hypertonic solution can cause an increase in hydrostatic pressure in the hypertonic compartment. Two solutions that have identical solute concentrations and are therefore osmotically balanced are said to be isotonic.

CARBON SKELETONS

The unique role of carbon in the cell comes from its ability to form strong covalent bonds with other carbon atoms. Thus carbon atoms can join to form chains.

[also written as]

or branched trees

[also written as]

or rings

[also written as]

COVALENT BONDS

Atoms in biological molecules are usually joined by covalent bonds (formed by sharing pairs of electrons). Each atom can form a fixed number of such bonds in a definite spatial arrangement.

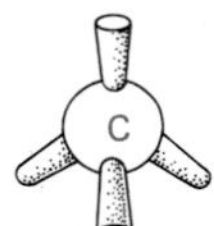

O

Double bonds exist and have a different spatial arrangement.

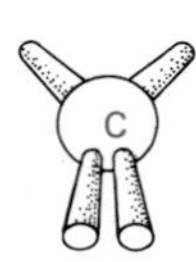

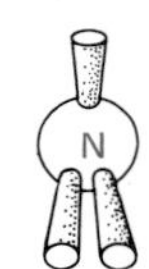

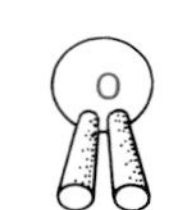

HYDROCARBONS

Carbon and hydrogen together make stable compounds called hydrocarbons. These are nonpolar, do not form hydrogen bonds, and are generally insoluble in water.

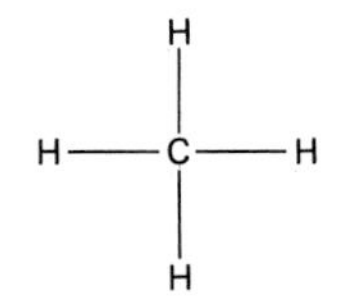

methane

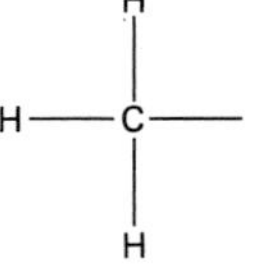

methyl group

Part of a fatty acid chain:

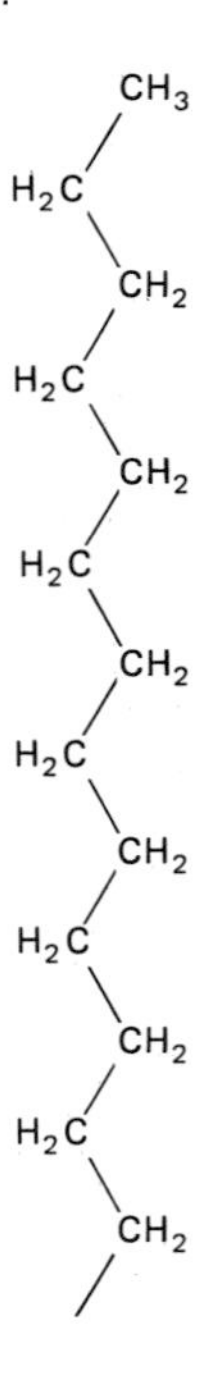

RESONANCE AND AROMATICITY

The carbon chain can include double bonds. If these are on alternate carbon atoms, the bonding electrons move within the molecule, stabilizing the structure by a phenomenon called resonance.

the truth is somewhere between these two structures

When resonance occurs throughout a ring compound, an aromatic ring is generated.

[often written as]

 Panel 2–2 Chemical bonds and groups commonly encountered in biological molecules.

C—O COMPOUNDS

Many biological compounds contain a carbon bonded to an oxygen. For example:

alcohol

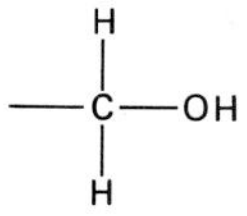

The —OH is called a hydroxyl group.

aldehyde

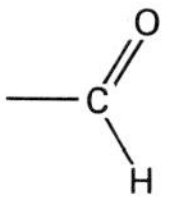

ketone

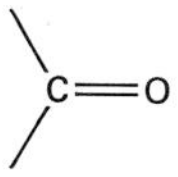

The C=O is called a carbonyl group.

carboxylic acid

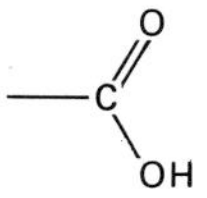

The —COOH is called a carboxyl group. In water this loses a H^+ ion to become $-COO^-$.

esters

Esters are formed by combining an acid and an alcohol:

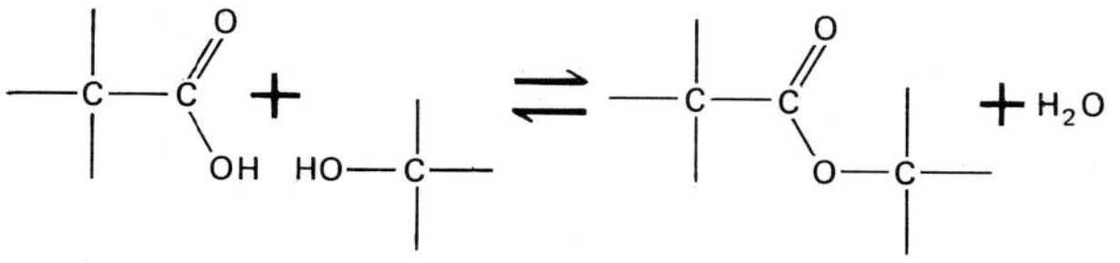

acid alcohol ester

C—N COMPOUNDS

Amines and amides are two important examples of compounds containing a carbon linked to a nitrogen.

Amines in water combine with a H^+ ion to become positively charged.

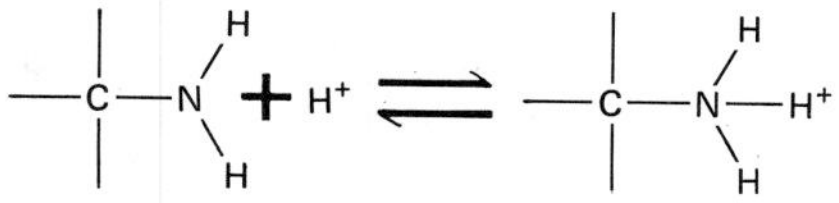

They are theretore basic.

Amides are formed by combining an acid and an amine. They are more stable than esters. Unlike amines, they are uncharged in water. An example is the peptide bond.

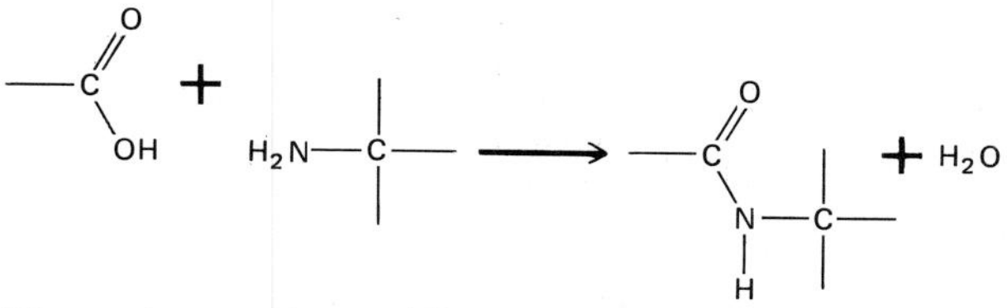

Nitrogen also occurs in several ring compounds, including important constituents of nucleic acids: purines and pyrimidines.

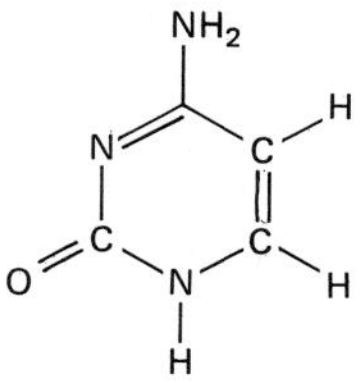

cytosine (a pyrimidine)

PHOSPHATES

Inorganic phosphate is a stable ion formed from phosphoric acid, H_3PO_4. It is often written as P_i.

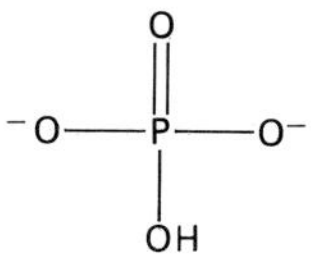

Phosphate esters can form between a phosphate and a free hydroxyl group.

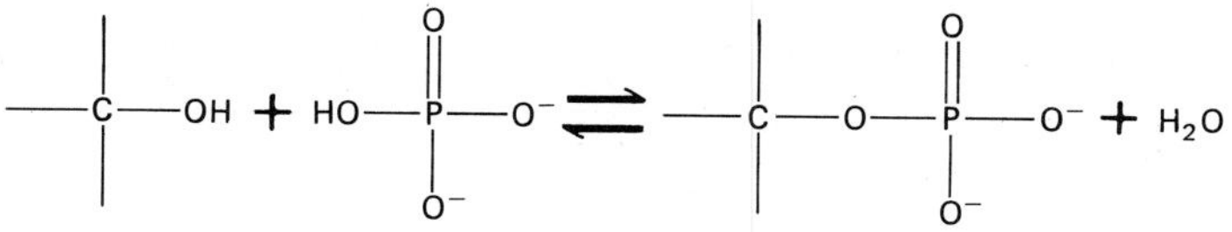

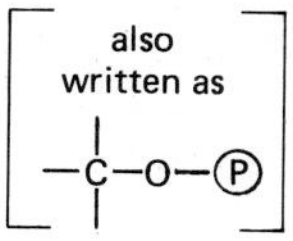

The combination of a phosphate and a carboxyl group, or two or more phosphate groups, gives an acid anhydride.

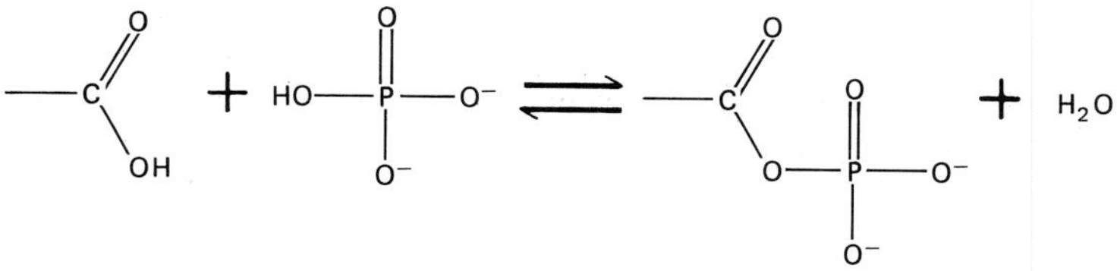

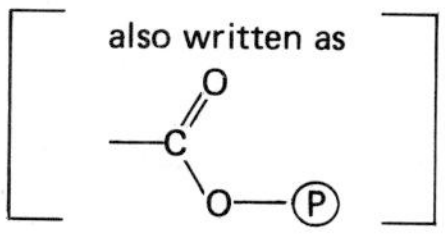

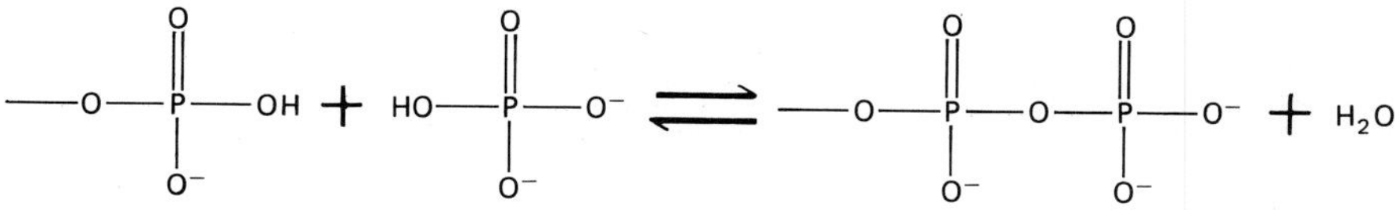

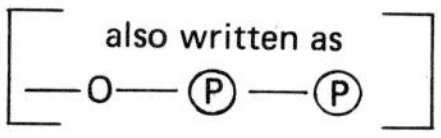

HEXOSES $n = 6$

Two common hexoses are

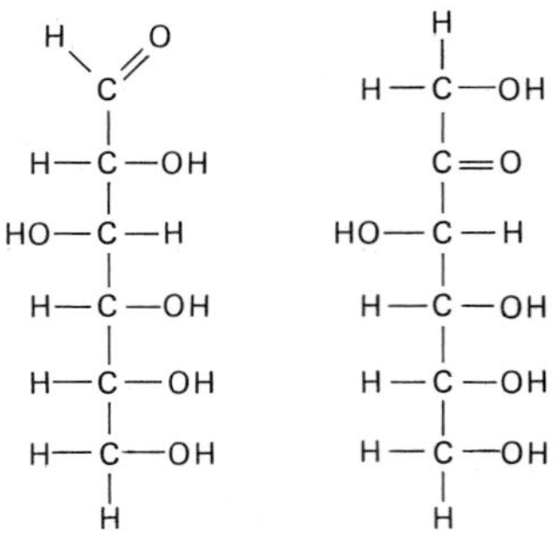

glucose fructose

MONOSACCHARIDES

Monosaccharides are aldehydes or ketones

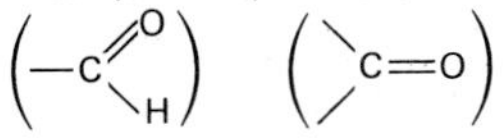

that also have two or more hydroxyl groups. Their general formula is $(CH_2O)_n$. The simplest are trioses ($n = 3$) such as

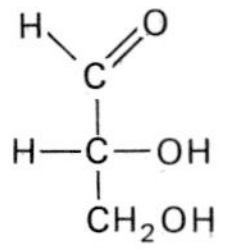

glyceraldehyde (an aldose)

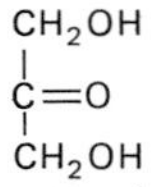

dihydroxyacetone (a ketose)

PENTOSES $n = 5$

A common pentose is

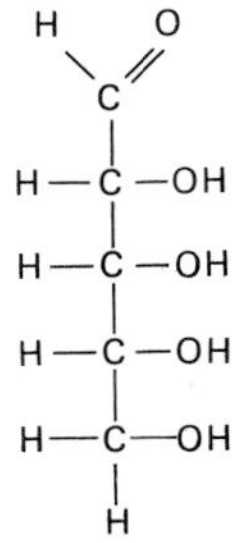

ribose

D-glucose (open-chain form)

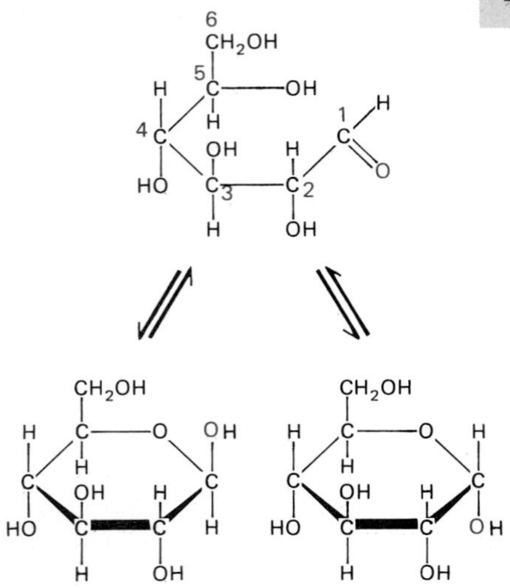

D-ribose (open-chain form)

RING FORMATION

The aldehyde or ketone group of a sugar can react with a hydroxyl group.

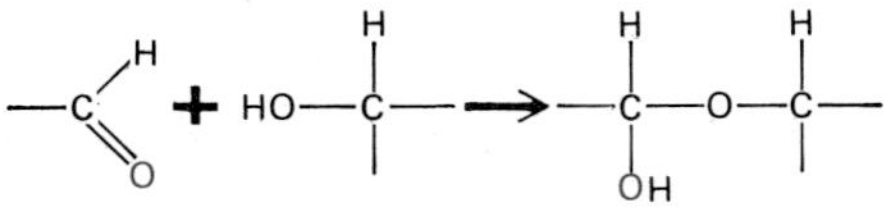

For the larger sugars ($n > 4$) this happens within the same molecule to form a 5- or 6-membered ring.

β-D-glucose α-D-glucose

STEREOISOMERS

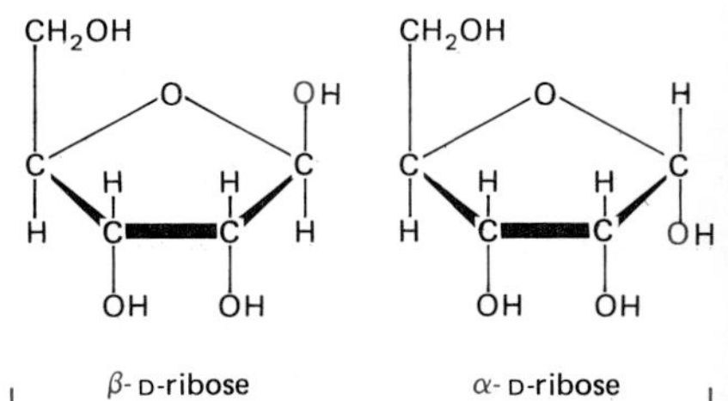

β-D-ribose α-D-ribose

STEREOISOMERS

NUMBERING

The carbon atoms of a sugar are numbered from the end closest to the aldehyde or ketone.

STEREOISOMERS

Monosaccharides have many isomers that differ only in the orientation of their hydroxyl groups – e.g., glucose, galactose, and mannose are isomers of each other.

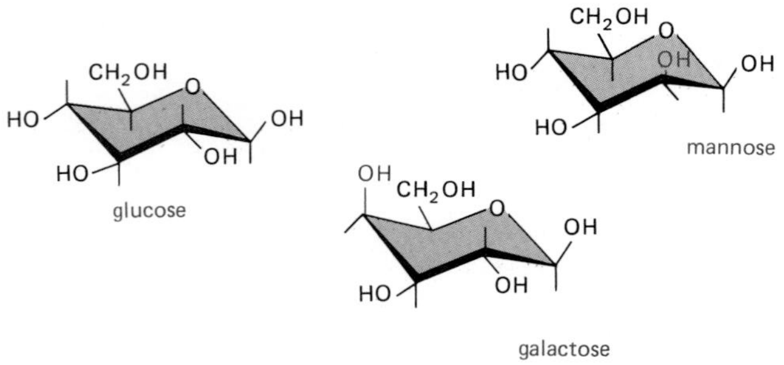

glucose

mannose

galactose

D AND L FORMS

Two isomers that are mirror images of each other have the same chemistry and therefore are given the same name and distinguished by the prefix D or L.

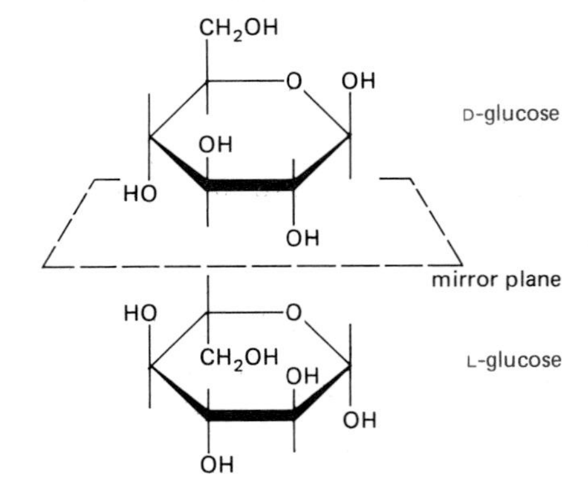

 Panel 2–3 An outline of some of the types of sugars commonly found in cells.

α- AND β-LINKS

The hydroxyl group on the carbon that carries the aldehyde or ketone can rapidly change from one position to another. These two positions are called α- and β-.

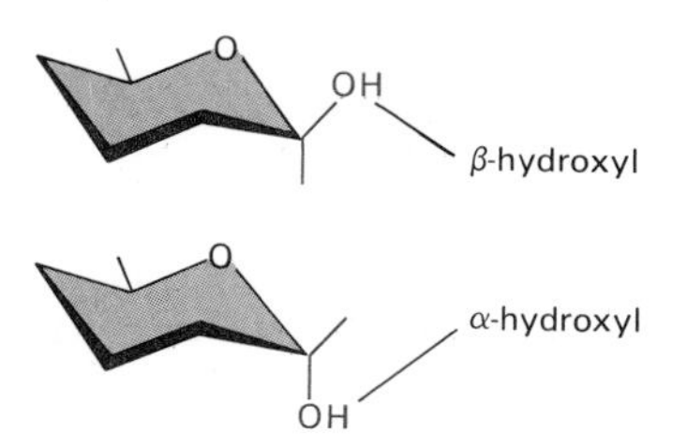

As soon as one sugar is linked to another, the α- or β-form is frozen.

SUGAR DERIVATIVES

The hydroxyl groups of a simple monosaccharide can be replaced by other groups. For example

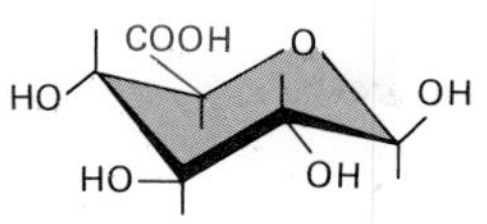

D-glucuronic acid

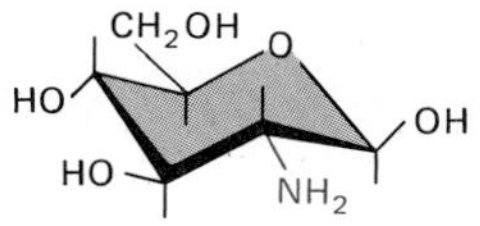

D-glucosamine

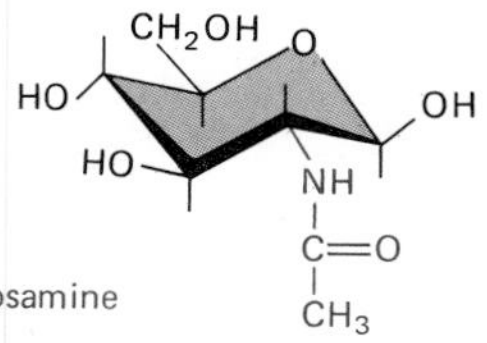

N-acetyl-D-glucosamine

DISACCHARIDES

The carbon that carries the aldehyde or the ketone can react with any hydroxyl group on a second sugar molecule to form a glycosidic bond. Three common disaccharides are maltose (glucose α1,4 glucose), lactose (galactose β1,4 glucose), and sucrose (glucose α1,2 fructose).

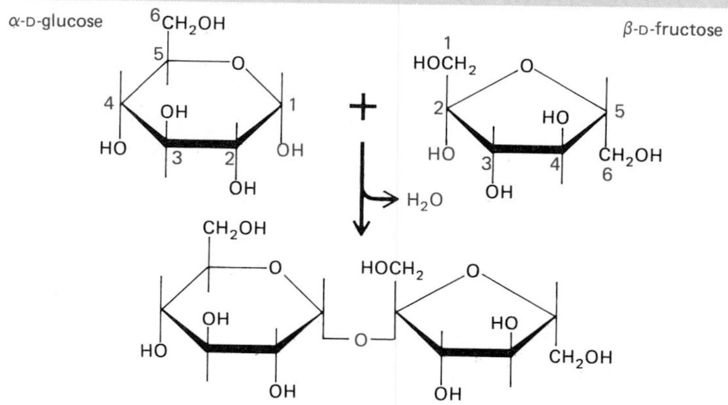

α-D-glucose + β-D-fructose → H_2O

sucrose (glucose α1,2 fructose)

OLIGOSACCHARIDES AND POLYSACCHARIDES

Large linear and branched molecules can be made from simple repeating units. Short chains are called oligosaccharides, while long chains are called polysaccharides. Glycogen, for example, is a polysaccharide made entirely of glucose units joined together.

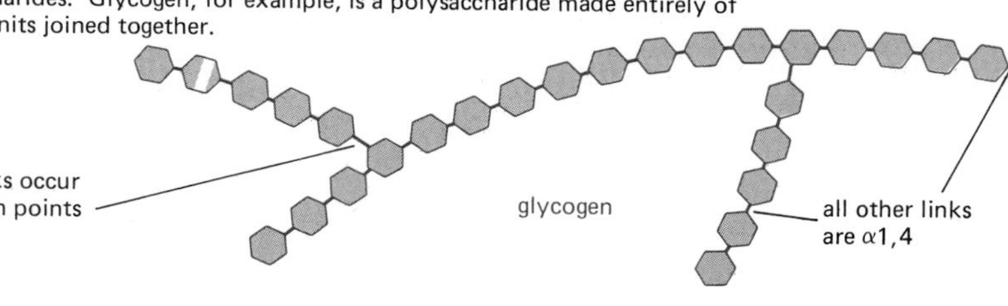

α1,6 links occur at branch points

COMPLEX OLIGOSACCHARIDES

In many cases a sugar sequence is nonrepetitive. Very many different molecules are possible. Such complex oligosaccharides are usually linked to proteins or to lipids

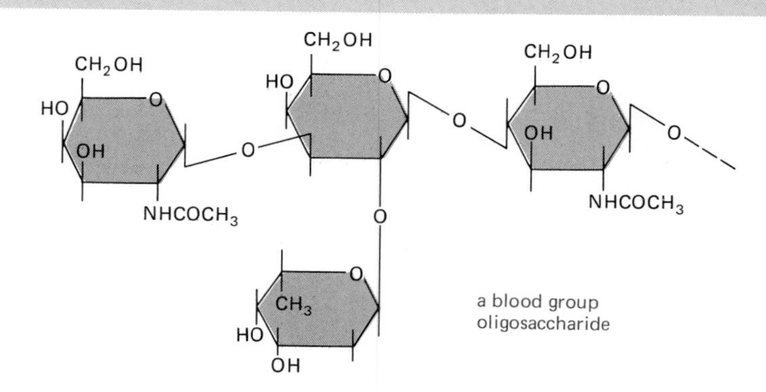

a blood group oligosaccharide

COMMON FATTY ACIDS

These are carboxylic acids with long hydrocarbon tails.

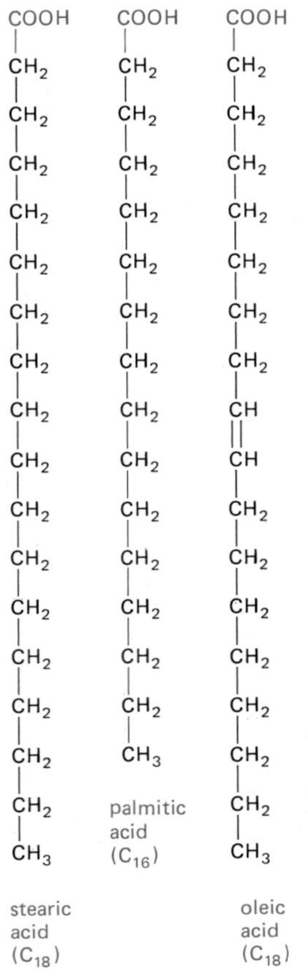

stearic acid (C_{18})	palmitic acid (C_{16})	oleic acid (C_{18})
COOH	COOH	COOH
CH_2	CH_2	CH_2
CH_2	CH_2	CH_2
CH_2	CH_2	CH_2
CH_2	CH_2	CH_2
CH_2	CH_2	CH_2
CH_2	CH_2	CH_2
CH_2	CH_2	CH_2
CH_2	CH_2	CH
CH_2	CH_2	CH (double bond to CH above)
CH_2	CH_2	CH_2
CH_2	CH_2	CH_2
CH_2	CH_2	CH_2
CH_2	CH_2	CH_2
CH_2	CH_2	CH_2
CH_2	CH_3	CH_2
CH_2		CH_2
CH_3		CH_3

Hundreds of different kinds of fatty acids exist. Some have one or more double bonds and are said to be unsaturated.

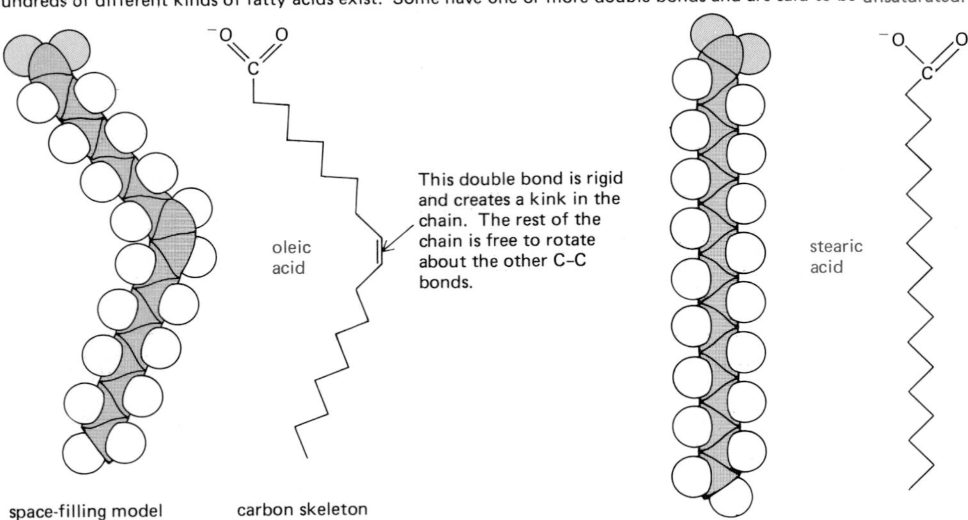

TRIGLYCERIDES

Fatty acids are stored as an energy reserve (fat) through an ester linkage to glycerol to form triglycerides.

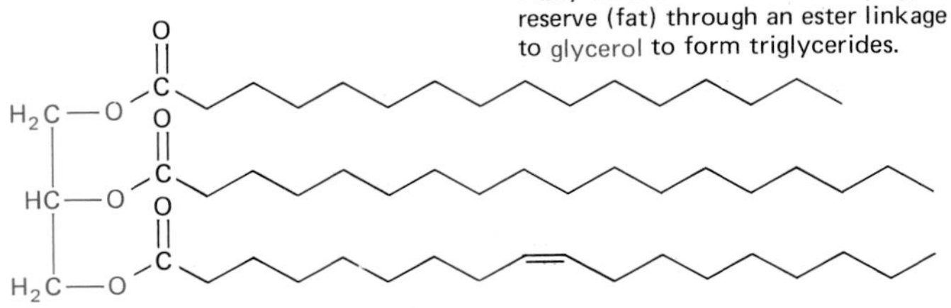

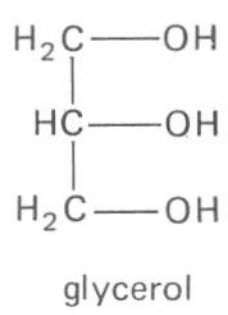

H_2C—O—C(=O)—
HC—O—C(=O)—
H_2C—O—C(=O)—

H_2C—OH
HC—OH
H_2C—OH

glycerol

CARBOXYL GROUP

If free, the carboxyl group of a fatty acid will be ionized.

C(=O)O^-

But more usually it is linked to other groups to form either esters

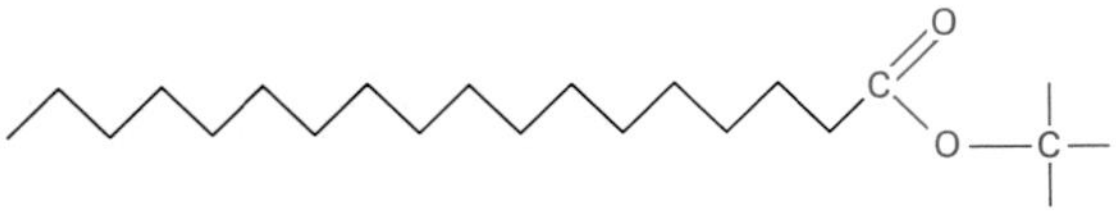

C(=O)—O—C—

or amides

C(=O)—N(H)—

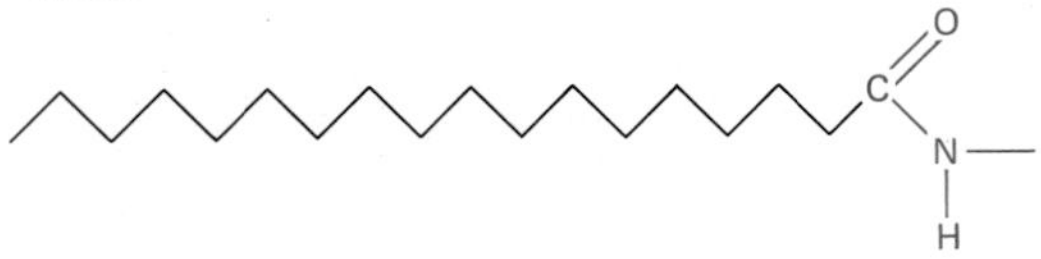

PHOSPHOLIPIDS

Phospholipids are the major constituent of cell membranes.

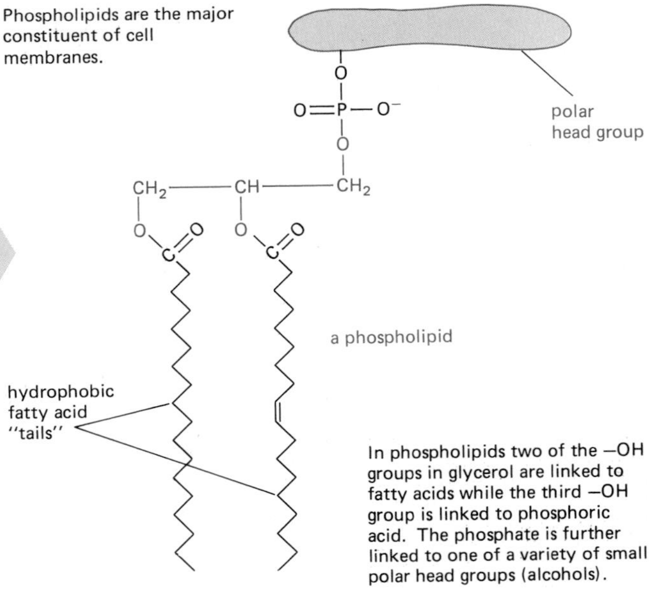

O
O=P—O^-
O
CH_2—CH—CH_2
O O
C=O C=O

polar head group

a phospholipid

hydrophobic fatty acid "tails"

In phospholipids two of the –OH groups in glycerol are linked to fatty acids while the third –OH group is linked to phosphoric acid. The phosphate is further linked to one of a variety of small polar head groups (alcohols).

Panel 2–4 An outline of some of the types of fatty acids commonly encountered in cells and the structures that they form.

LIPID AGGREGATES

Fatty acids have a hydrophilic head and a hydrophobic tail.

In water they can form a surface film or form small micelles.

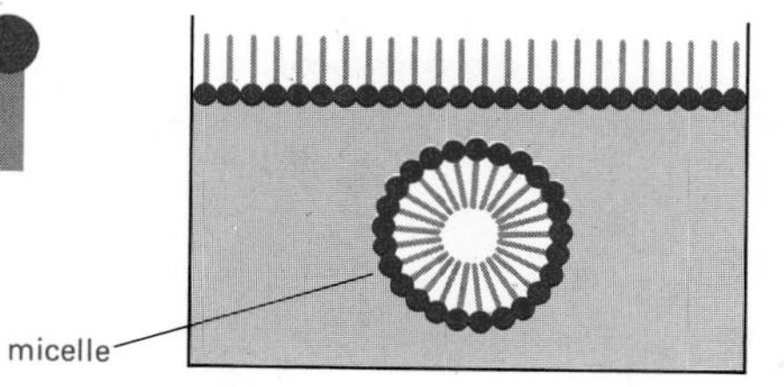

Their derivatives can form larger aggregates held together by hydrophobic forces:

Triglycerides form large spherical fat droplets in the cell cytoplasm.

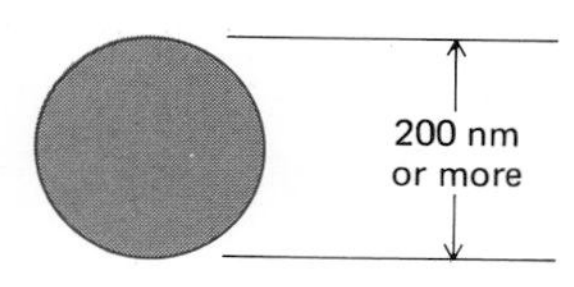

Phospholipids and glycolipids form self-sealing lipid bilayers that are the basis for all cellular membranes.

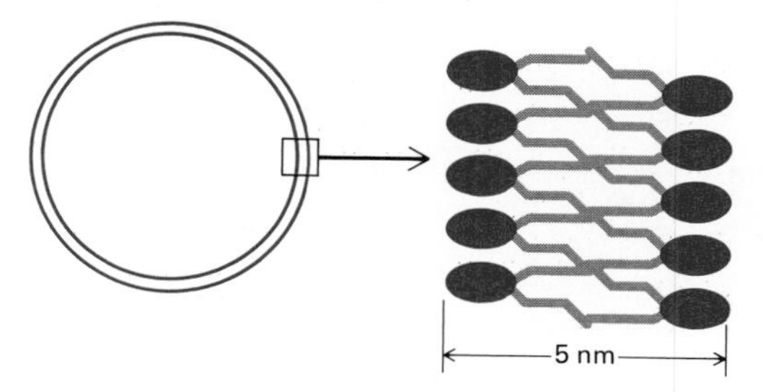

OTHER LIPIDS

Lipids are defined as the water-insoluble molecules in cells that are soluble in organic solvents. Two other common types of lipid are steroids and polyisoprenoids. Both are made from isoprene units.

CH_3—C(=CH_2)—CH=CH_2 isoprene

STEROIDS

Steroids have a common multiple-ring structure.

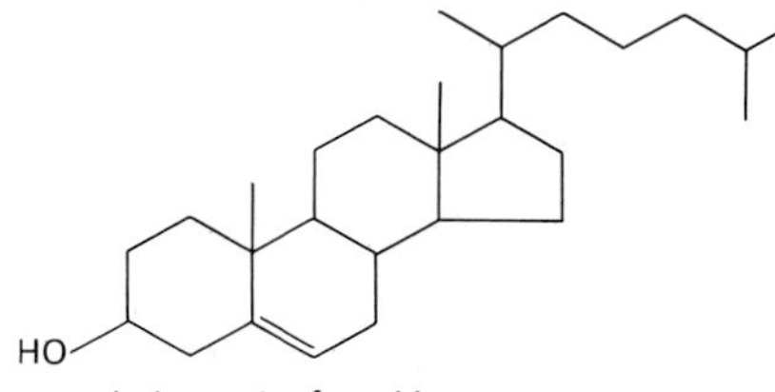

cholesterol— found in many membranes

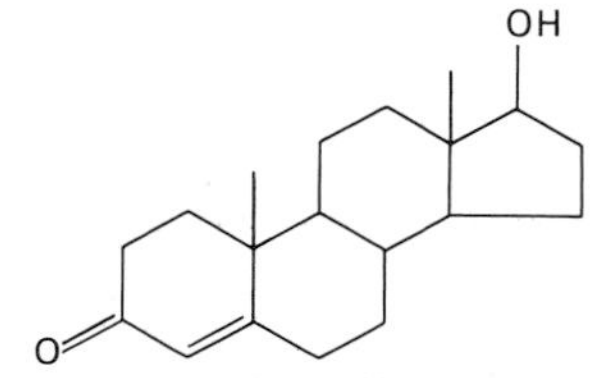

testosterone— male steroid hormone

GLYCOLIPIDS

Like phospholipids, these compounds are composed of a hydrophobic region, containing two long hydrocarbon tails, and a polar region, which now contains one or more sugar residues and no phosphate.

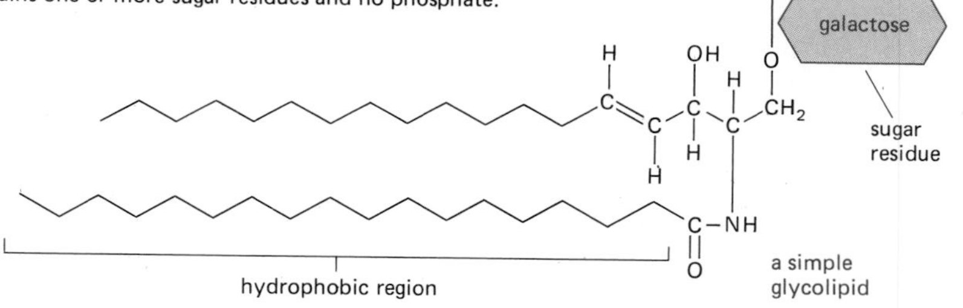

a simple glycolipid

POLYISOPRENOIDS

long chain polymers of isoprene

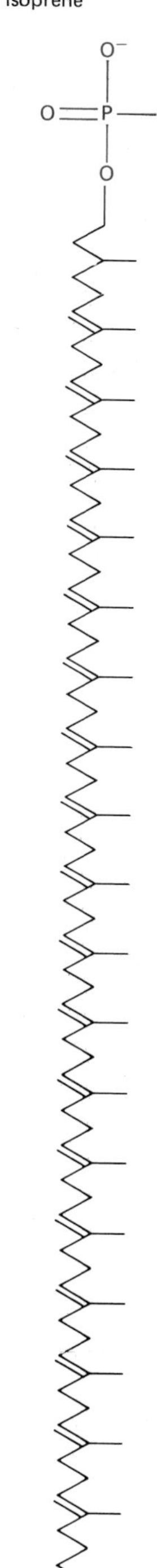

dolichol phosphate— used to carry activated sugars in the membrane-associated synthesis of glycoproteins and some polysaccharides.

THE AMINO ACID

The general formula of an amino acid is

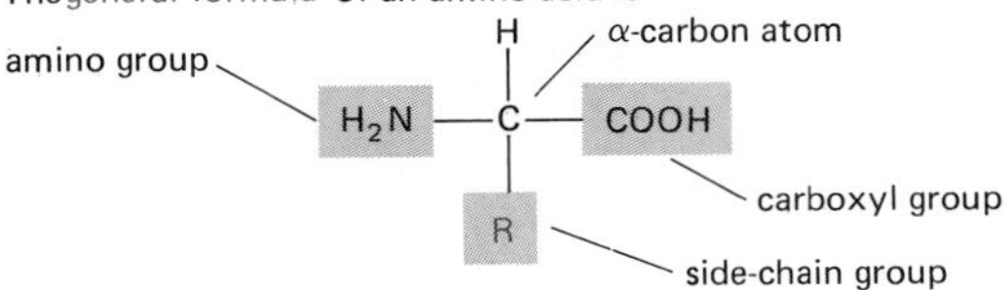

R is commonly one of 20 different side chains. At pH 7 both the amino and carboxyl groups are ionized.

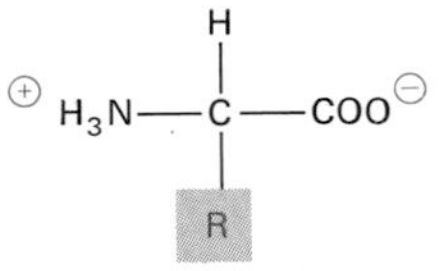

$$^{\oplus}H_3N—\underset{\underset{R}{|}}{\overset{\overset{H}{|}}{C}}—COO^{\ominus}$$

OPTICAL ISOMERS

The α-carbon atom is asymmetric, which allows for two mirror image (or stereo-) isomers, D and L.

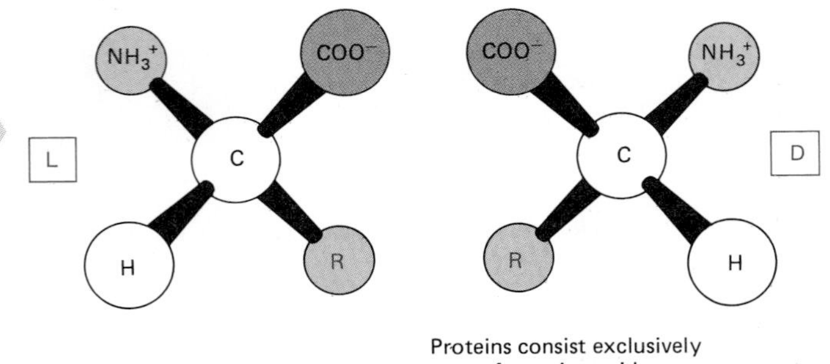

Proteins consist exclusively of L-amino acids.

PEPTIDE BONDS

Amino acids are commonly joined together by an amide linkage, called a peptide bond.

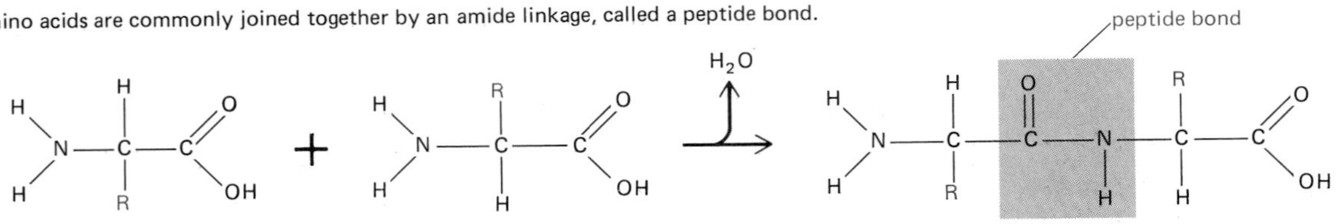

H_2N–CH(R)–COOH + H_2N–CH(R)–COOH → (−H_2O) H_2N–CH(R)–CO–NH–CH(R)–COOH

peptide bond

Proteins are long polymers of amino acids linked by peptide bonds, and they are always written with the *N*-terminus toward the left. The sequence of this tripeptide is His Cys Val.

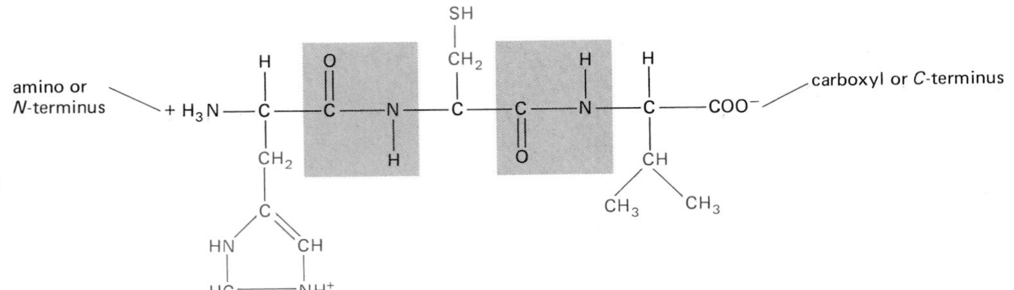

FAMILIES OF AMINO ACIDS

The common amino acids are grouped according to whether their side chains are

- acidic
- basic
- uncharged polar
- nonpolar

These 20 amino acids are given both three-letter and one-letter abbreviations.

Thus: alanine = Ala = A

BASIC SIDE CHAINS

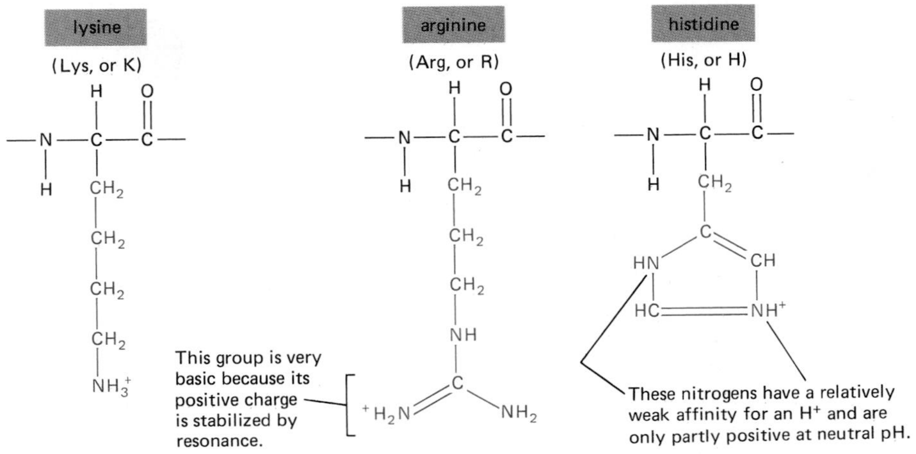

lysine

(Lys, or K)

—NH—CH(—C(=O)—)—CH_2—CH_2—CH_2—CH_2—NH_3^+

arginine

(Arg, or R)

—NH—CH(—C(=O)—)—CH_2—CH_2—CH_2—NH—C(=^+H_2N)—NH_2

This group is very basic because its positive charge is stabilized by resonance.

histidine

(His, or H)

—NH—CH(—C(=O)—)—CH_2—C (imidazole ring: HN, CH, HC, NH^+)

These nitrogens have a relatively weak affinity for an H^+ and are only partly positive at neutral pH.

 Panel 2–5 The 20 amino acids involved in the synthesis of proteins.

ACIDIC SIDE CHAINS

aspartic acid

(Asp, or D)

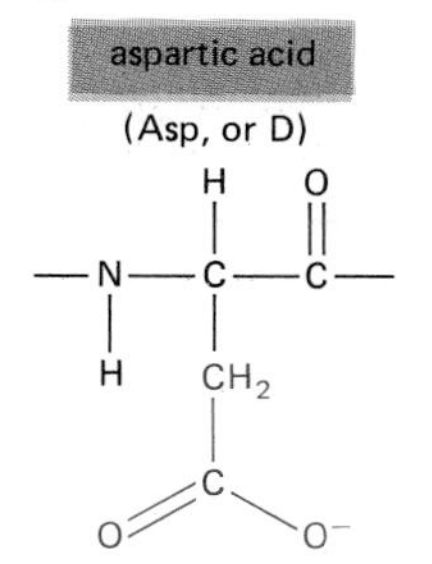

glutamic acid

(Glu, or E)

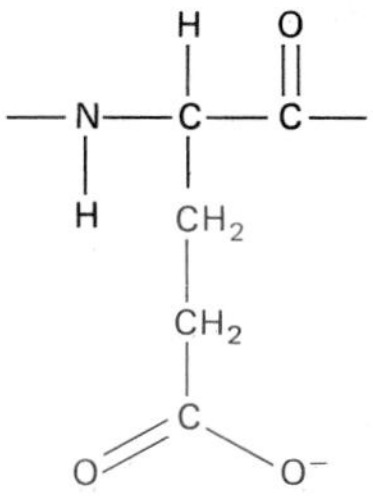

Amino acids with uncharged polar side chains are relatively hydrophilic and are usually on the outside of proteins, while the side chains on nonpolar amino acids tend to cluster together on the inside. Amino acids with basic and acidic side chains are very polar and they are nearly always found on the outside of protein molecules.

UNCHARGED POLAR SIDE CHAINS

asparagine

(Asn, or N)

glutamine

(Gln, or Q)

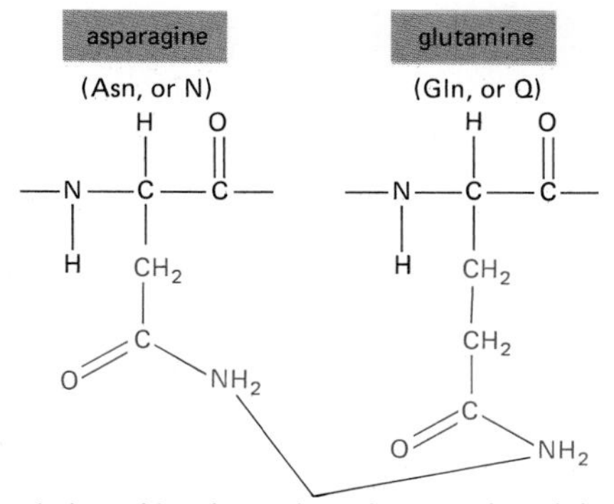

Although the amide N is not charged at neutral pH, it is polar.

serine

(Ser, or S)

threonine

(Thr, or T)

tyrosine

(Tyr, or Y)

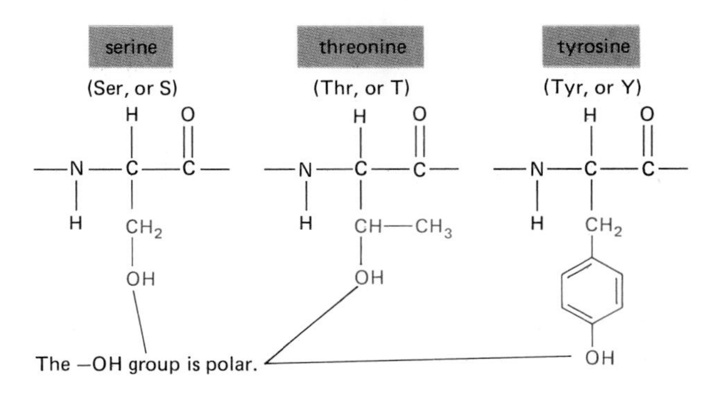

The –OH group is polar.

NONPOLAR SIDE CHAINS

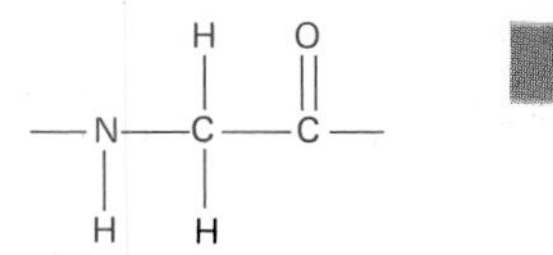

glycine

alanine

(Ala, or A)

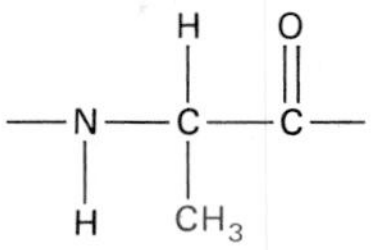

valine

(Val, or V)

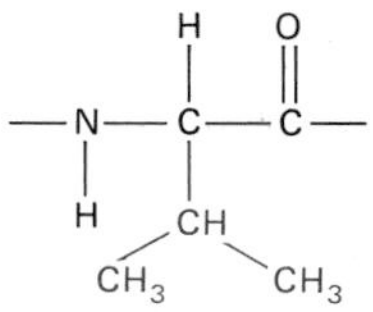

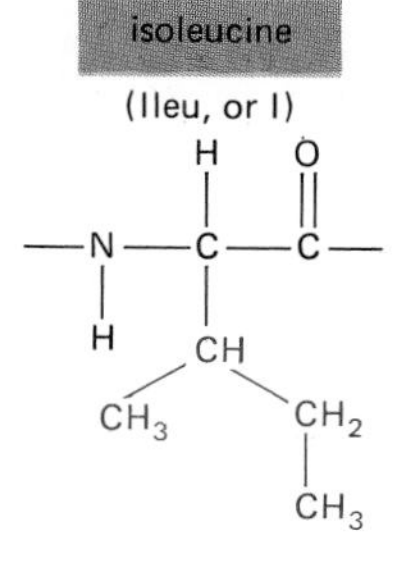

leucine

(Leu, or L)

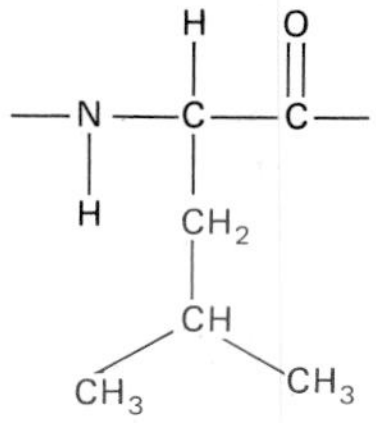

isoleucine

(Ileu, or I)

proline

(Pro, or P)

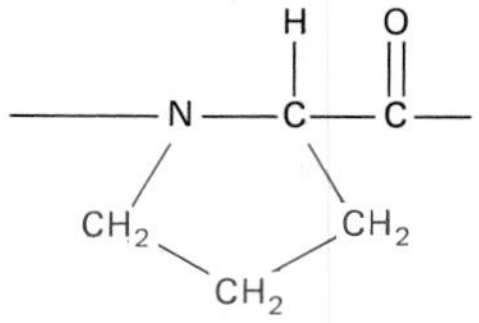

(actually an imino acid)

phenylalanine

(Phe, or F)

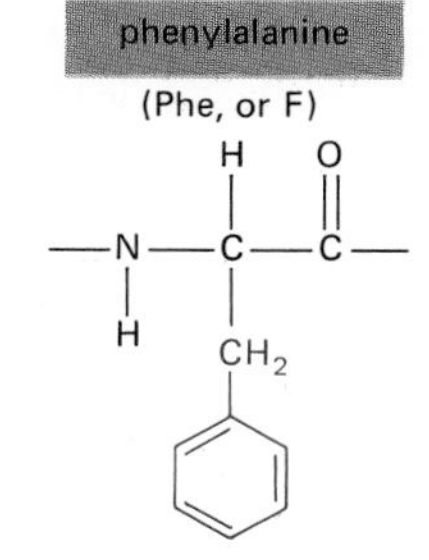

methionine

(Met, or M)

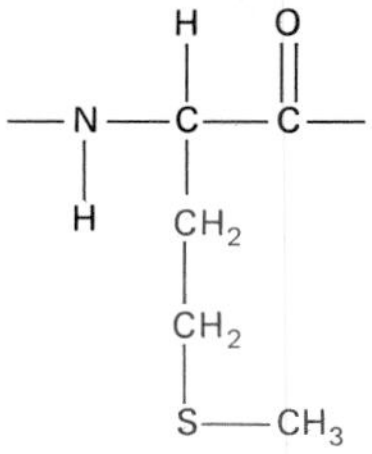

tryptophan

(Trp, or W)

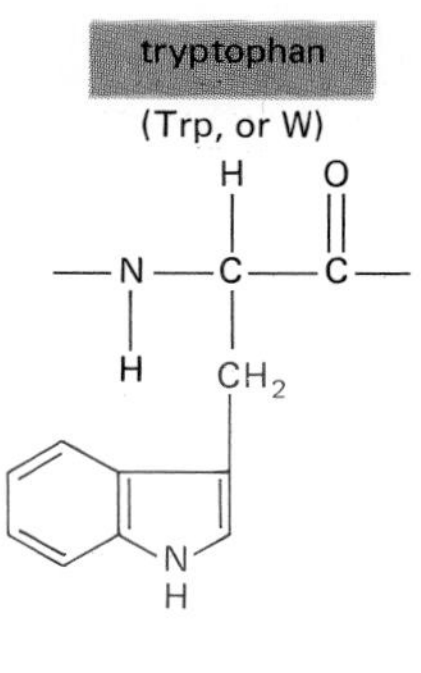

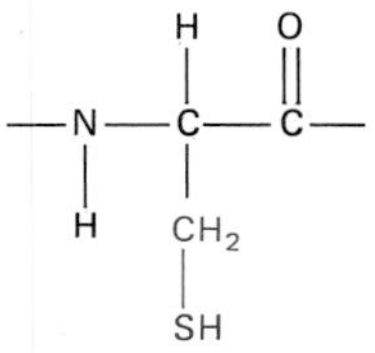

cysteine

(Cys, or C)

Paired cysteines allow disulfide bonds to form in proteins.

$—CH_2—S—S—CH_2—$

BASES

The bases are N-containing ring compounds, either purines or pyrimidines.

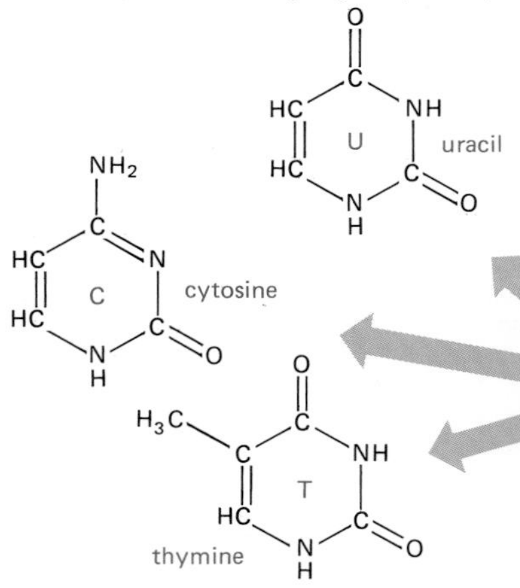

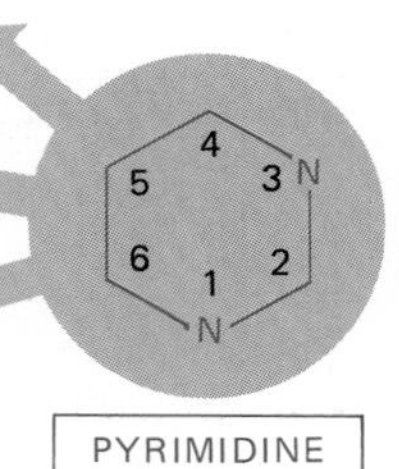

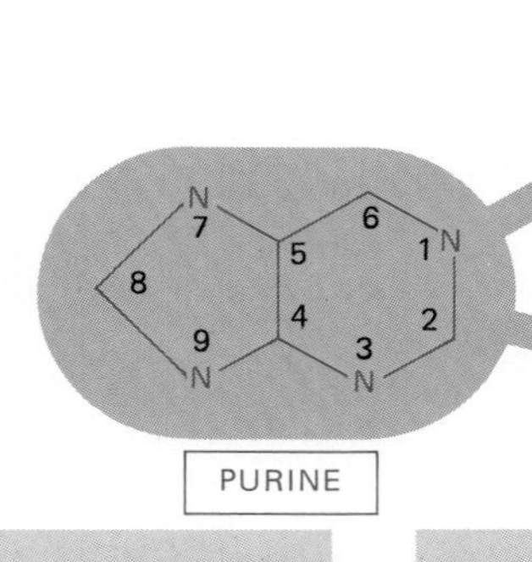

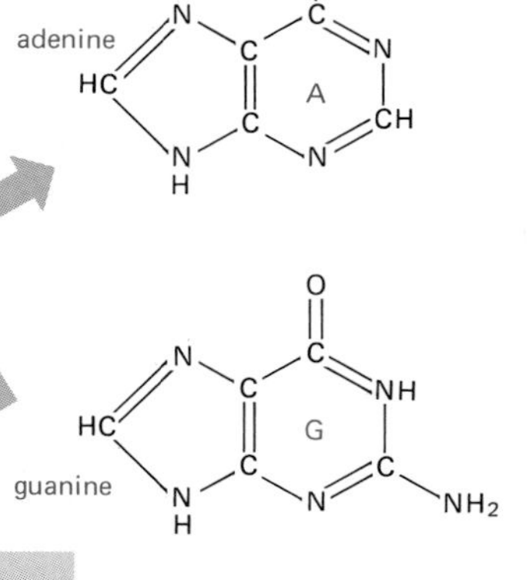

PHOSPHATES

The phosphates are normally joined to the C5 hydroxyl of the ribose or deoxyribose sugar. Mono-, di-, and triphosphates are common.

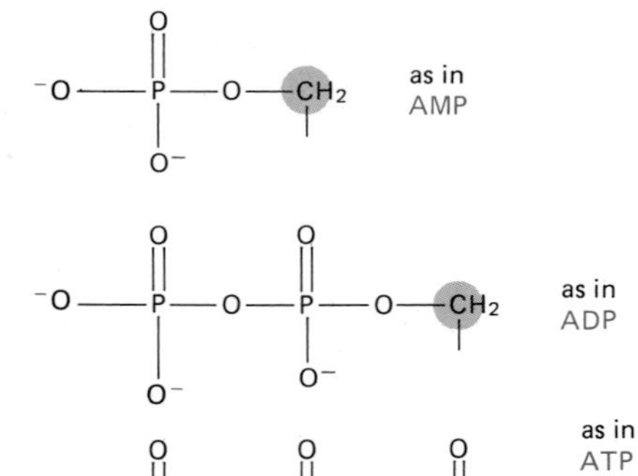

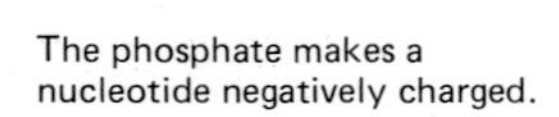

The phosphate makes a nucleotide negatively charged.

NUCLEOTIDES

A nucleotide consists of a nitrogen-containing base, a 5-carbon sugar, and one or more phosphate groups.

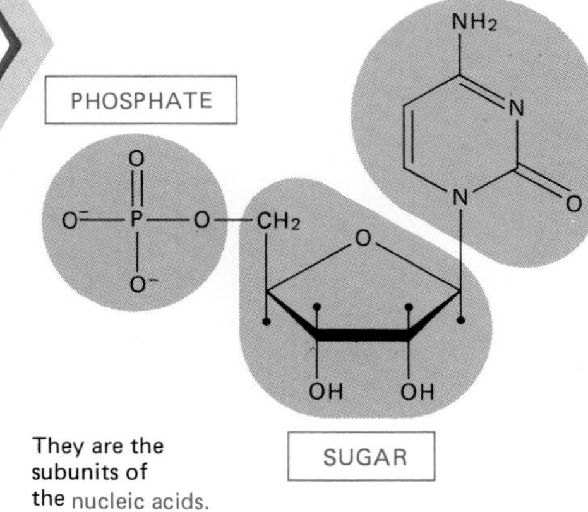

They are the subunits of the nucleic acids.

BASE-SUGAR LINKAGE

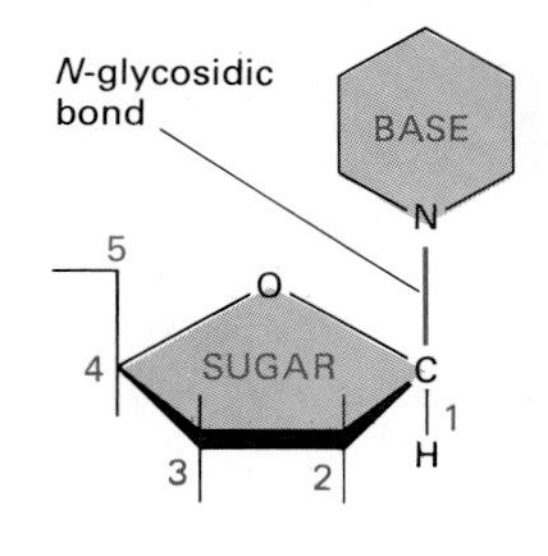

The base is linked to the same carbon (C1) used in sugar-sugar bonds.

SUGARS

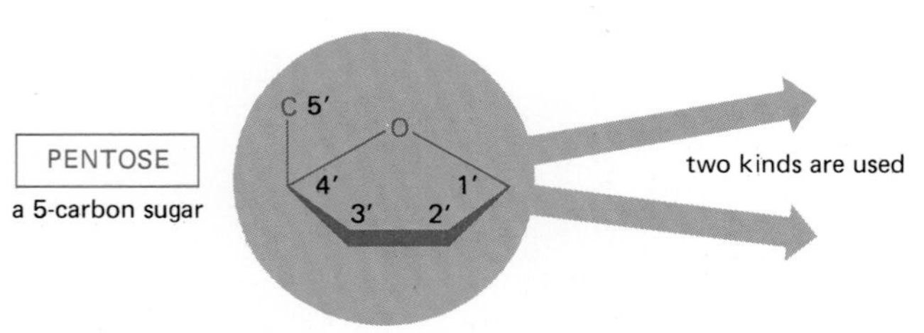

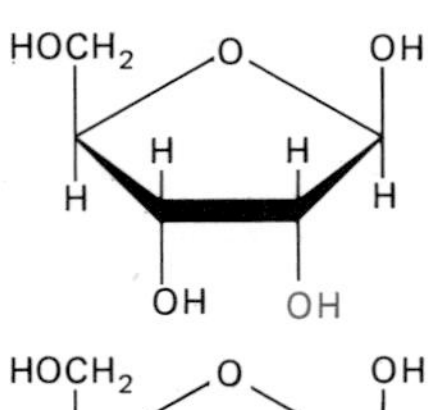

β-D-RIBOSE
used in ribonucleic acid

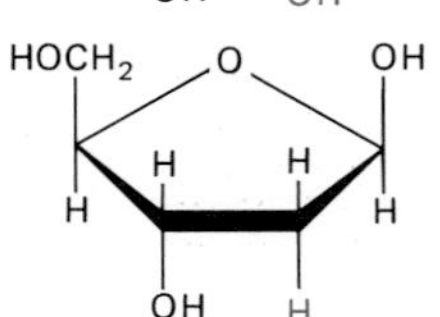

β-D-2-DEOXYRIBOSE
used in deoxyribonucleic acid

Each numbered carbon on the sugar of a nucleotide is followed by a prime mark; therefore, one speaks of the "5-prime carbon," etc.

Panel 2–6 A survey of the major types of nucleotides and their derivatives encountered in cells.

NOMENCLATURE

The names can be confusing, but the abbreviations are clear.

BASE + SUGAR = NUCLEOSIDE

BASE + SUGAR + PHOSPHATE = NUCLEOTIDE

BASE	NUCLEOSIDE	ABBR.
adenine	adenosine	A
guanine	guanosine	G
cytosine	cytidine	C
uracil	uridine	U
thymine	thymidine	T

Nucleotides are abbreviated by three capital letters as follows:

AMP = adenosine monophosphate
dAMP = deoxyadenosine monophosphate
UDP = uridine diphosphate
ATP = adenosine triphosphate
etc.

NUCLEIC ACIDS

Nucleotides are joined together by a phosphodiester linkage between 5′ and 3′ carbon atoms to form nucleic acids. The linear sequence of nucleotides in a nucleic acid chain is commonly abbreviated by a one-letter code, A–G–C–T–T–A–C–A, with the 5′ end of the chain written at the left.

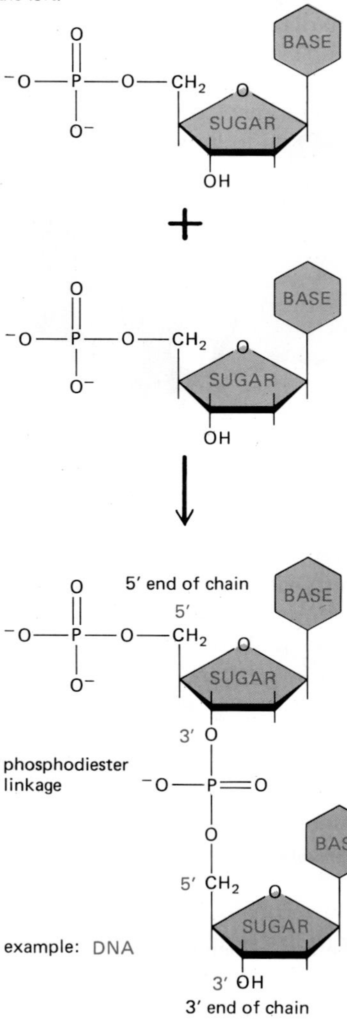

BASE
SUGAR
OH

+

BASE
SUGAR
OH

5′ end of chain

BASE
SUGAR

phosphodiester linkage

BASE
SUGAR

3′ end of chain

example: DNA

NUCLEOTIDES HAVE MANY OTHER FUNCTIONS

1. They carry chemical energy in their easily hydrolyzed acid-anhydride bonds.

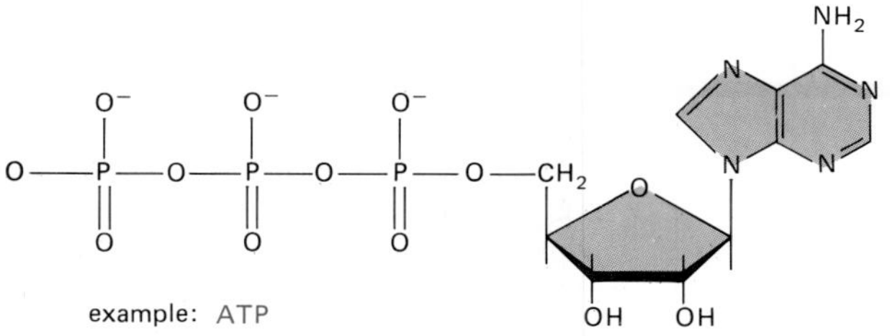

example: ATP

2. They combine with other groups to form coenzymes.

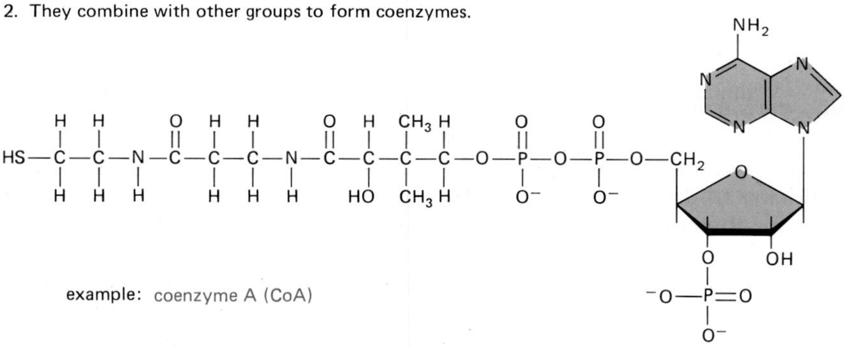

example: coenzyme A (CoA)

3. They are used as specific signaling molecules in the cell.

example: cyclic AMP

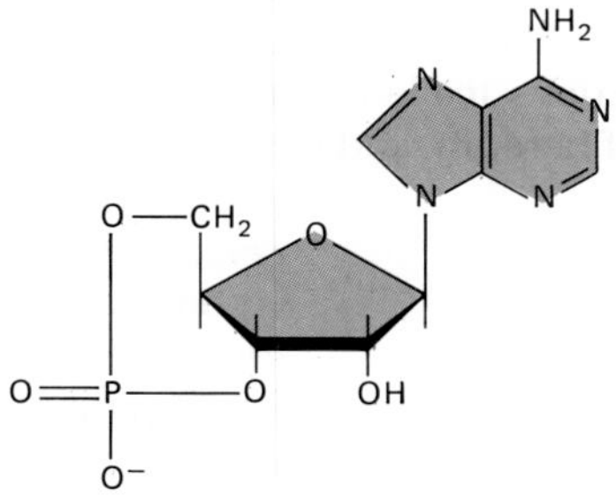

reactions are said to be *coupled,* as will be explained subsequently. It is the tight coupling of heat production to an increase in order that distinguishes the metabolism of the cell from the wasteful burning of fuel in a fire.

Figure 2–11 illustrates in a general way how coupled reactions release heat energy, which disorders the environment, thereby compensating for the increase in order they create in the cell. It is the resulting increase in the disorder of the universe that drives the coupled reactions in the direction of cell ordering.

Energy cannot be created or destroyed in chemical reactions. Therefore, in our example, the heat loss from the box that drives the production of biological order inside it requires an input of energy if it is to continue. This energy must be in a form other than heat. For plants, the energy is initially derived from the electromagnetic radiation of the sun; for animals, it is derived from the energy stored in the covalent bonds of the organic molecules they eat. However, since these organic nutrients are themselves produced by photosynthetic organisms, such as green plants, the sun is in fact the ultimate energy source for both types of organisms.

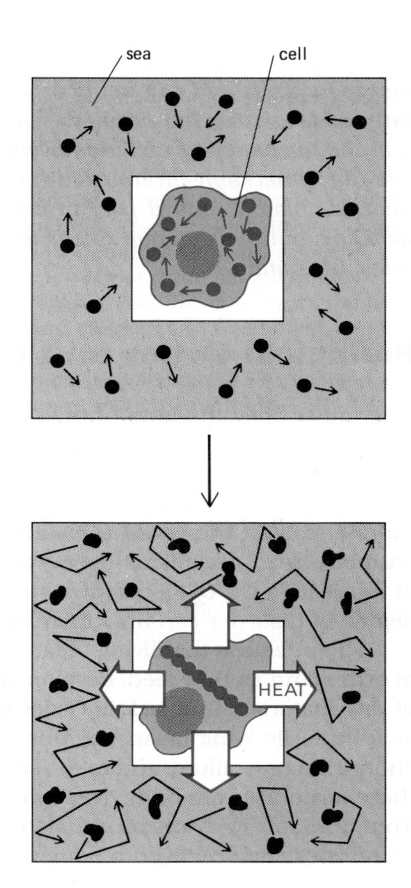

Figure 2–11 For a simple thermodynamic analysis of a living cell, it is useful to consider it and its immediate environment enclosed in a sealed box that allows heat but not molecules to be exchanged with the rest of the universe, which we have designated here as the "sea." In the upper diagram, the molecules of both the cell and the rest of the universe are depicted in a relatively disordered state. In the lower diagram, heat has been released from the cell by a reaction that orders the molecules it contains. The increase in random motion, including bond distortions, of the molecules in the rest of the universe creates a disorder that more than compensates for the increased order in the cell, as required by the laws of thermodynamics for spontaneous processes. In this way the release of heat by a cell to its surroundings allows it to become more highly ordered internally at the same time that the universe as a whole becomes more disordered.

Photosynthetic Organisms Use Sunlight to Synthesize Organic Compounds[8]

Solar energy enters the living world (the *biosphere*) by means of the **photosynthesis** carried out by photosynthetic organisms—either plants or bacteria. In photosynthesis, electromagnetic energy is converted into chemical bond energy. At the same time, however, part of the energy of sunlight is converted into heat energy, and the release of this heat to the environment increases the disorder of the universe and thereby drives the photosynthetic process.

The reactions of photosynthesis are described in detail in Chapter 7. In broad terms, they occur in two distinct stages. In the first (the *light reactions*), the visible radiation excites an electron in a pigment molecule, which, in returning to a lower energy state, provides the energy needed for the synthesis of ATP and NADPH. In the second (the *dark reactions*), the ATP and NADPH are used to drive a series of "carbon-fixation" reactions in which CO_2 from the air is used to form sugar molecules (Figure 2–12).

The *net* result of photosynthesis, so far as the green plant is concerned, can be summarized by the equation

$$\text{energy} + CO_2 + H_2O \rightarrow \text{sugar} + O_2$$

which is the reverse of the oxidative decomposition of a sugar. However, this simple equation hides the complex nature of the dark reactions, which involve many linked reaction steps. Furthermore, although the initial fixation of CO_2 results in sugars, subsequent metabolic reactions soon convert these into the other small and large molecules essential to the plant cell.

Chemical Energy Passes from Plants to Animals

Animals and other nonphotosynthetic organisms cannot capture energy from sunlight directly and so have to survive on "second-hand" energy obtained by eating plants or on "third-hand" energy obtained by eating other animals. The organic molecules made by plant cells provide both building blocks and fuel to the organisms that feed on them. All types of plant molecules can serve this purpose—sugars, proteins, polysaccharides, lipids, and many others.

The transactions between plants and animals are not all one-way. Plants, animals, and microorganisms have existed together on this planet for so long that many of them have become an essential part of the others' environment. The oxygen released by photosynthesis is consumed in the combustion of organic molecules by nearly all organisms, and some of the CO_2 molecules that are "fixed" today into larger organic molecules by photosynthesis in a green leaf were yesterday released into the atmosphere by the respiration of an animal. Thus, carbon utilization is a cyclic process that involves the biosphere as a whole and crosses boundaries between individual organisms (Figure 2–13). Similarly, atoms of nitrogen, phosphorus, and sulfur can, in principle, be traced from one biological molecule to another in a series of similar cycles.

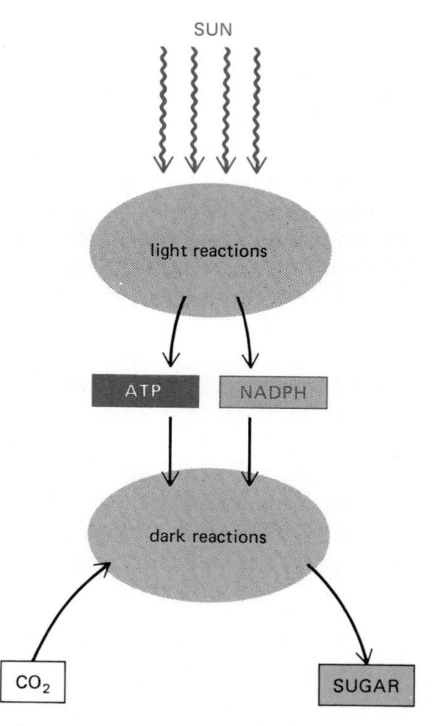

Figure 2–12 The two stages of photosynthesis in a green plant.

Cells Obtain Energy by the Oxidation of Biological Molecules[9]

The carbon and hydrogen atoms in a cell are not in their most stable form. Because the earth's atmosphere contains a great deal of oxygen, the most energetically stable form of carbon is as CO_2 and that of hydrogen is as H_2O. A cell is therefore able to obtain energy from sugar or protein molecules by allowing their carbon and hydrogen atoms to combine with oxygen to produce CO_2 and H_2O, respectively. However, the cell does not oxidize molecules in one step, as occurs in a fire. It takes them through a large number of reactions that only rarely involve the direct addition of oxygen. Before we can consider these reactions and the driving force behind them, we need to discuss what is meant by the process of oxidation.

Oxidation, in the sense used above, does not mean only the addition of oxygen atoms; rather, it applies more generally to any reaction in which electrons are transferred from one atom to another. **Oxidation** in this sense refers to the removal of electrons, and **reduction**—the converse of oxidation—means the addition of electrons. Thus, Fe^{2+} is oxidized if it loses an electron to become Fe^{3+}, and a chlorine atom is reduced if it gains an electron to become Cl^-. The same terms are used when there is only a partial shift of electrons between atoms linked by a covalent bond. For example, when a carbon atom becomes covalently bonded to an electronegative atom such as oxygen, chlorine, or sulfur, it gives up more than its equal share of electrons—acquiring a partial positive charge—and so is said to be oxidized. Conversely, a carbon atom in a C-H linkage has more than its share of electrons, and so it is said to be reduced (Figure 2–14).

The combustion of food materials in a cell converts the C and H atoms in organic molecules (where they are both in a relatively electron-rich, or reduced, state) to CO_2 and H_2O, where they have given up electrons and are therefore highly oxidized. The shift of electrons from carbon and hydrogen to oxygen allows all of these atoms to achieve a more stable state and hence is energetically favorable.

The Breakdown of Organic Molecules Takes Place in Sequences of Enzyme-catalyzed Reactions[10]

Although the most energetically favorable form of carbon is as CO_2 and that of hydrogen is as H_2O, a living organism does not disappear in a puff of smoke for the same reason that the book in your hands does not burst into flame: the

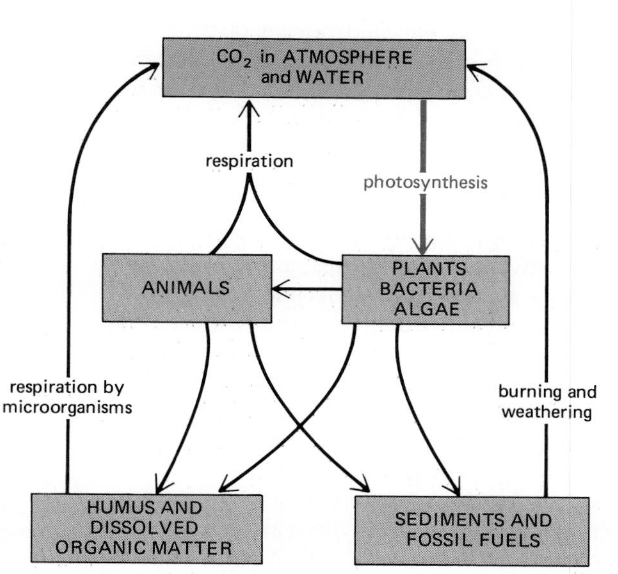

Figure 2–13 The carbon cycle. Individual carbon atoms are incorporated into organic molecules of the living world by the photosynthetic activity of plants, bacteria, and marine algae. They pass to animals, microorganisms, and organic material in soil and oceans in cyclic paths. CO_2 is restored to the atmosphere when organic molecules are oxidized by cells or burned by humans as fossil fuels.

molecules of both exist in metastable energy troughs (Figure 2–15) and require *activation energy* before they can pass to more stable configurations. In the case of the book, the activation energy can be provided by a lighted match. For a living cell, the combustion is achieved in a less sudden and destructive fashion. Highly specific protein catalysts, or **enzymes,** combine with biological molecules in such a way that they reduce the activation energy of particular reactions that the bound molecules can undergo. By selectively lowering the activation energy of one pathway or another, enzymes determine which of several alternative reaction paths is followed (Figure 2–16). In this way, each of the many different molecules in a cell is directed through specific reaction pathways, thereby determining the cell's chemistry.

The success of living organisms is attributable to the cell's ability to make enzymes of many different types, each with precisely specified properties. Each enzyme has a unique shape containing an *active site.* This active site binds a particular set of other molecules (called *substrates*) in such a way as to speed up a particular one of the many chemical reactions that the substrates can undergo, often by a factor of as much as 10^{14}. Like all other catalysts, enzyme molecules themselves are not changed after participating in a reaction and therefore can function over and over again.

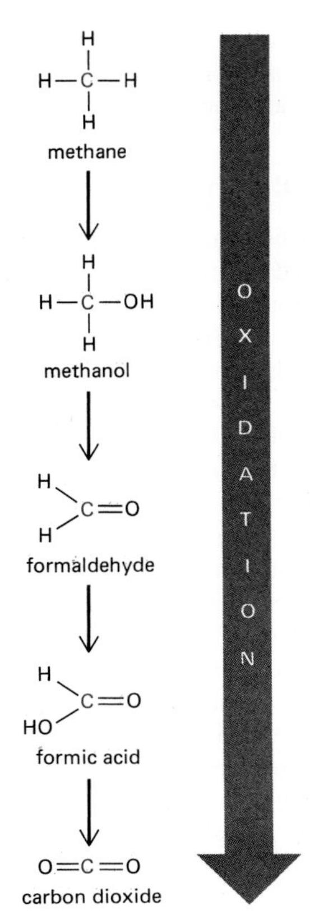

Figure 2–14 The carbon atom of methane may be converted to that of carbon dioxide by the successive removal of its hydrogen atoms. With each step, electrons are shifted away from the carbon as it passes to a more energetically stable state. Thus, in the process shown, the carbon atom becomes progressively more oxidized.

Part of the Energy Released in Oxidation Reactions Is Coupled to the Formation of ATP[11]

Cells derive useful energy from the "burning" of glucose only because they burn it in a very complex and controlled way. By means of enzyme-directed reaction paths, the synthetic, or *anabolic,* chemical reactions that create biological order are closely coupled to the degradative, or *catabolic,* reactions that provide the energy. The crucial difference between a **coupled reaction** and an uncoupled reaction is illustrated by the mechanical analogy in Figure 2–17, where an energetically favorable chemical reaction is represented by rocks falling from a cliff. The kinetic energy of falling rocks would normally be entirely wasted in the form of heat generated when they hit the ground (section A). But, by careful design, part of the kinetic energy could be used to drive a paddle wheel that lifts a bucket of water (section B). Because the rocks can reach the ground only by moving the paddle wheel, we say that the spontaneous reaction of rock falling has been directly coupled to the nonspontaneous reaction of lifting the bucket of water. Note that because part of the energy is now used to do work in section B, the rocks hit the ground with less velocity than in section A, and therefore correspondingly less energy is wasted as heat.

In cells, enzymes play the role of paddle wheels in our analogy and couple the spontaneous burning of foodstuffs to reactions that generate the nucleoside triphosphate ATP. Just as the energy stored in the elevated bucket of water in Figure 2–17 can be dispensed in small doses to drive a wide variety of different hydraulic machines (section C), ATP serves as a convenient and versatile store, or currency, of energy to drive many different chemical reactions that the cell needs.

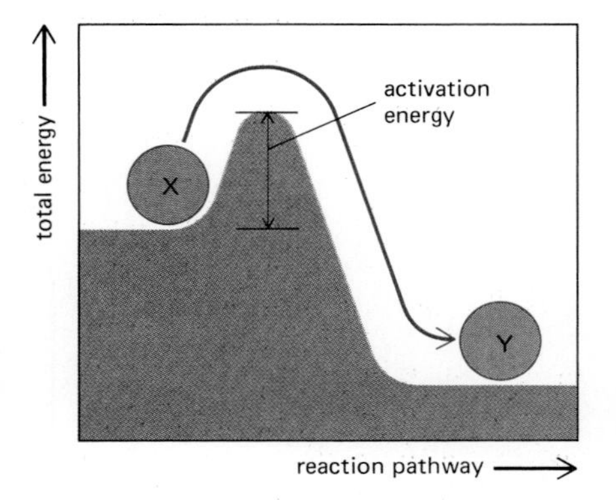

Figure 2–15 The principle of activation energy. Compound X can achieve a lower, more favorable energy state by being converted to compound Y. However, this transition will not take place unless X can acquire enough activation energy from its surroundings to undergo the reaction.

The Hydrolysis of ATP Generates Order in Cells[12]

How does ATP act as a carrier of chemical energy? Under the conditions existing in the cytoplasm, the breakdown of ATP by hydrolysis to release inorganic phosphate (P_i) occurs very readily and releases a great deal of usable energy (see p. 74). A chemical group that is linked by such a reactive bond is readily transferred to another molecule; for this reason, the terminal phosphate in ATP can be considered to exist in an *activated* state. The bond broken in this hydrolysis reaction is sometimes described as a *high-energy bond.* However, the important fact is that in aqueous solution the hydrolysis creates two molecules of much lower energy (ADP and P_i); there is nothing special about the covalent bond itself.

Other chemical reactions can be driven by the energy released by ATP hydrolysis through enzymes that couple the reactions to ATP hydrolysis. Among the many hundreds of these reactions are those involved in the synthesis of biological

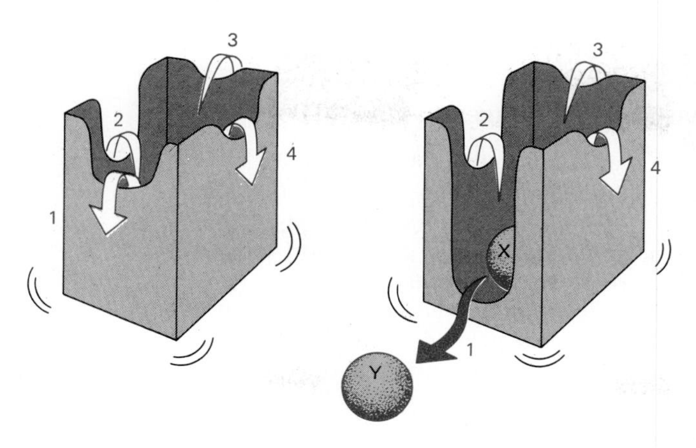

Figure 2–16 A "jiggling-box" model illustrating how enzymes direct molecules along desired reaction pathways. In this model the colored ball represents a potential enzyme substrate (compound X) that is bouncing up and down due to the constant bombardment of colliding water molecules. The four walls of the box represent the activation energy barriers for four different chemical reactions that are energetically favorable. In the left-hand box, none of these reactions occurs because the energy available from collisions is insufficient to surmount any of the energy barriers. In the right-hand box, enzyme catalysis lowers the activation energy for reaction number 1 only. It thereby allows this reaction to proceed with available energies, causing compound Y to form from compound X.

molecules, in the active transport of molecules across cell membranes, and in the generation of force and movement. These three types of processes play a vital part in establishing biological order. The macromolecules formed in biosynthetic reactions carry information, catalyze specific reactions, and are assembled into highly ordered structures within cells and in the extracellular space. Membrane-bound pumps maintain the special internal composition of cells and permit signals to pass within and between cells. Finally, the production of force and movement enables the cytoplasmic contents of cells to become organized and the cells themselves to move about and assemble into organized tissues.

Summary

Living cells are highly ordered and must create order within themselves in order to survive and grow. This is thermodynamically possible only because of a continual input of energy, part of which is released from the cells to their environment as heat. The energy comes ultimately from the electromagnetic radiation of the sun, which drives the formation of organic molecules in photosynthetic organisms such as green plants. Animals obtain their energy by taking up these organic molecules and oxidizing them in a series of enzyme-catalyzed reactions that are coupled to the formation of ATP. ATP is a common currency of energy in all cells, and its hydrolysis is coupled to other reactions to drive a variety of energetically unfavorable processes that create order.

Figure 2–17 A mechanical model illustrating the principle of coupled chemical reactions. The spontaneous reaction shown in (A) might serve as an analogy for the direct oxidation of glucose to CO_2 and H_2O, which produces heat only. In (B) the same reaction is coupled to a second reaction; the second reaction might serve as an analogy for the synthesis of ATP. The more versatile form of energy produced in (B) can be used to drive other cellular processes, as in (C). ATP is the most versatile form of energy in cells.

kinetic energy transformed into heat energy only

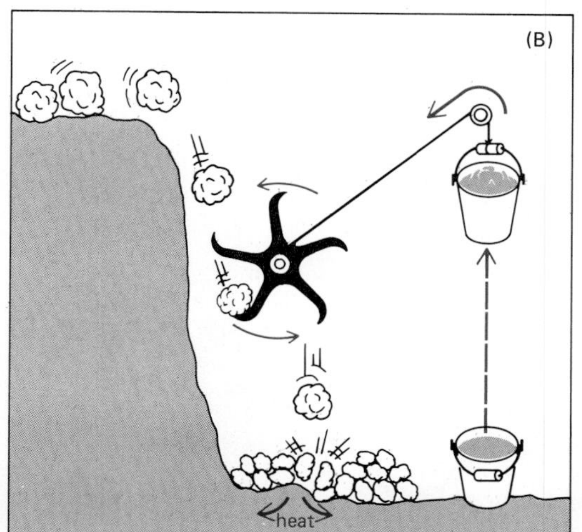

part of the kinetic energy is used to lift a bucket of water, and a correspondingly smaller amount is released as heat

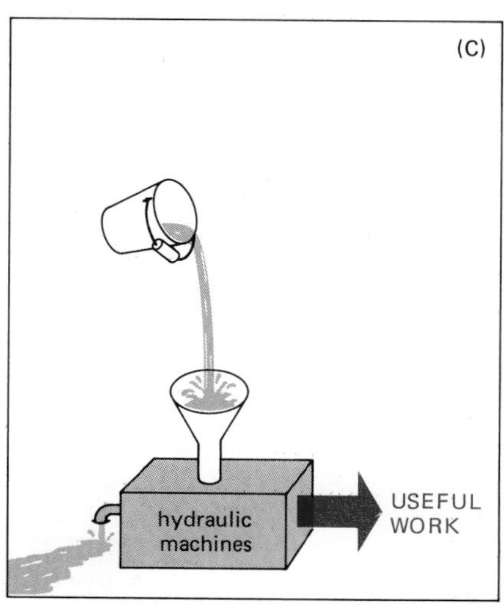

the potential kinetic energy stored in the elevated bucket of water can be used to drive a wide variety of different hydraulic machines

Food and the Derivation of Cellular Energy[6,13]

Food Molecules Are Broken Down in Three Stages to Give ATP

The proteins, lipids, and polysaccharides that make up the major part of the food we eat must be broken down into smaller molecules before our cells can use them. The enzymatic breakdown, or catabolism, of these molecules may be regarded as proceeding in three stages (Figure 2–18). We shall give a short outline of these stages before discussing the last two of them in more detail.

In stage 1, called *digestion*, large polymeric molecules are broken down into their monomeric subunits—proteins into amino acids, polysaccharides into sugars, and fats into fatty acids and glycerol. These processes occur mainly outside cells through the action of secreted enzymes. In stage 2, the resultant small molecules enter cells and are further degraded in the cytoplasm. Most of the carbon and hydrogen atoms of sugars are converted into *pyruvate*, which then enters mitochondria, where it is converted to the acetyl groups of the chemically reactive compound *acetyl coenzyme A* (*acetyl CoA*, Figure 2–19). Major amounts of acetyl CoA are also produced by the oxidation of fatty acids. In stage 3, the acetyl group

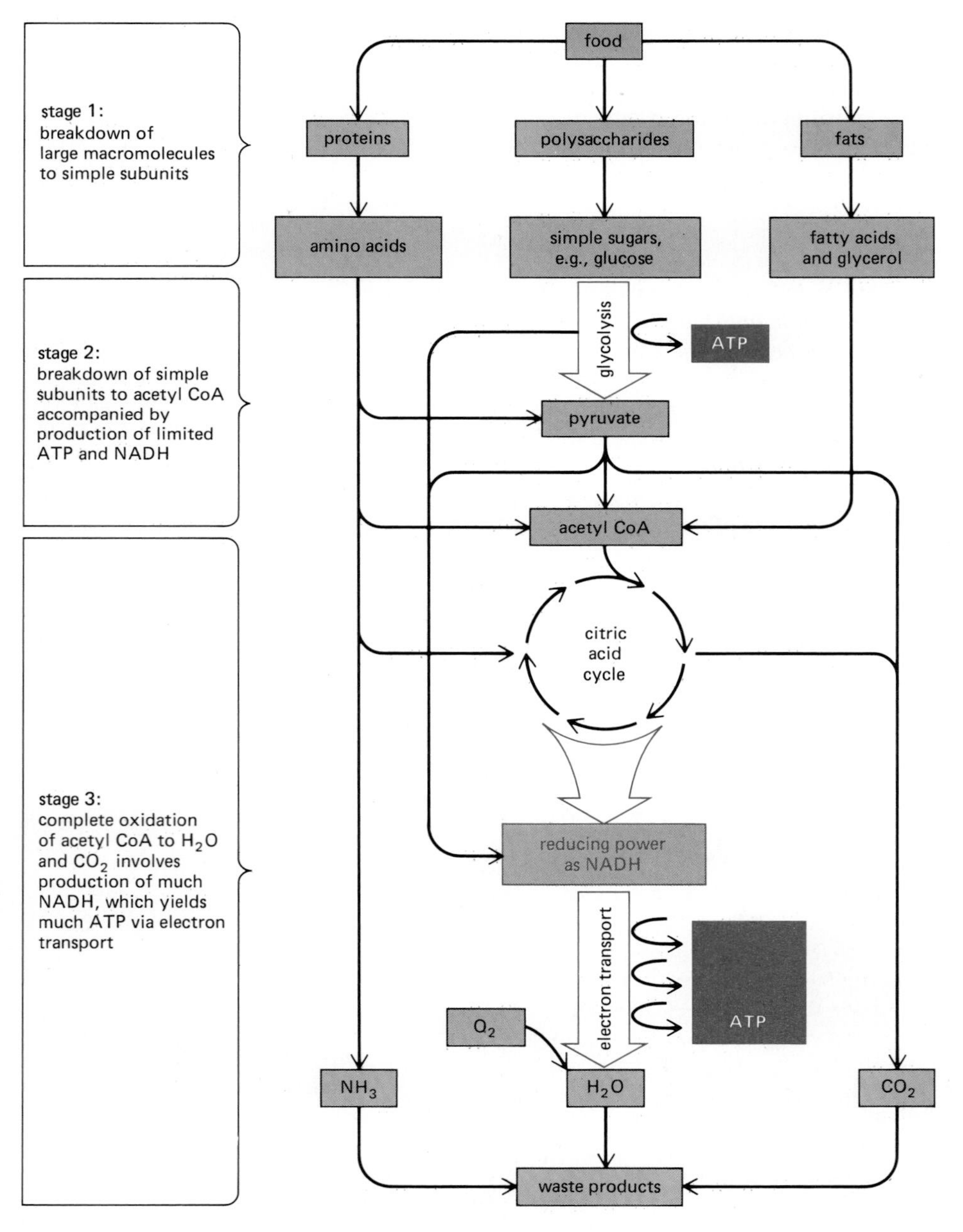

Figure 2–18 Simplified diagram of the three stages of catabolism that lead from food to waste products. This series of reactions produces ATP, which is then used to drive biosynthetic reactions and other energy-requiring processes in the cell.

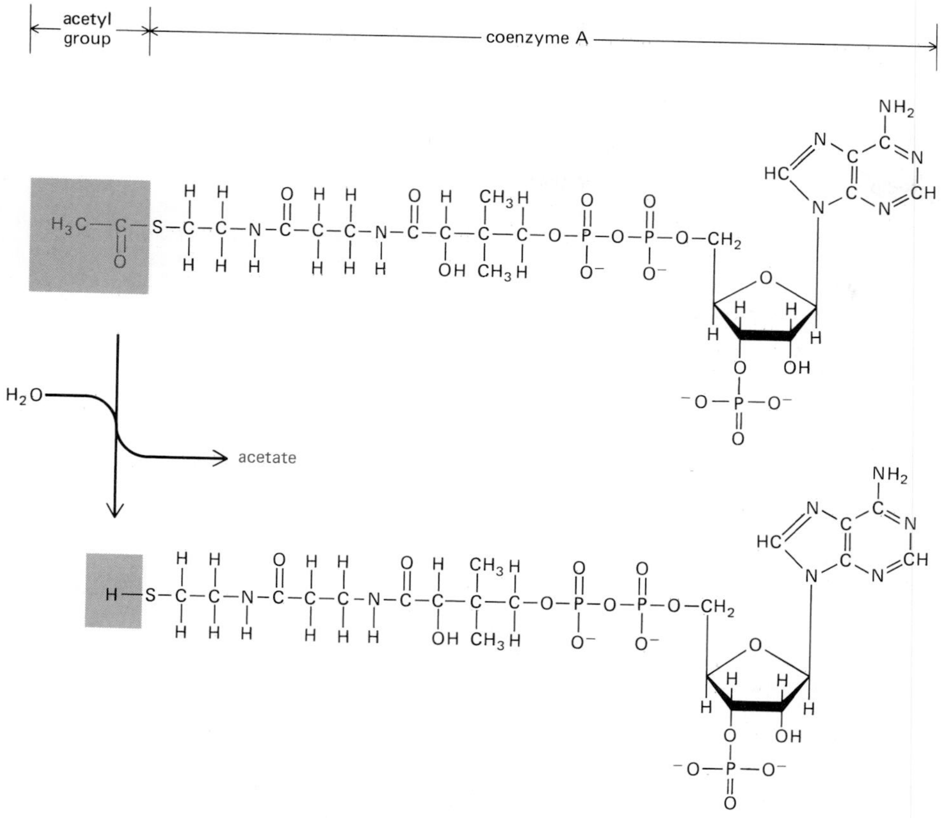

Figure 2–19 The structure of the crucial metabolic intermediate acetyl coenzyme A (acetyl CoA). Acetyl groups produced in stage 2 of catabolism (see Figure 2–18) are covalently linked to coenzyme A (CoA).

of acetyl CoA is completely degraded to CO_2 and H_2O. It is in this final stage that most of the ATP is generated. Through a series of coupled chemical reactions, more than half of the energy theoretically derivable from the combustion of carbohydrates and fats to H_2O and CO_2 is channeled into driving the energetically unfavorable reaction P_i + ADP → ATP. Because the rest of the combustion energy is released by the cell as heat, this generation of ATP creates net disorder in the universe, in conformity with the second law of thermodynamics.

Through the production of ATP, the energy originally derived from the combustion of carbohydrates and fats is redistributed as a conveniently packaged form of chemical energy. Roughly 10^9 molecules of ATP are in solution throughout the intracellular space in a typical cell, where their energetically favorable hydrolysis back to ADP and phosphate provides the driving energy for a large number of different coupled reactions that would otherwise be energetically unfavorable.

Glycolysis Can Produce ATP Even in the Absence of Oxygen

The most important part of stage 2 of catabolism is a sequence of reactions known as **glycolysis**—the lysis (splitting) of glucose. In glycolysis, a glucose molecule with six carbon atoms is converted into two molecules of pyruvate, each with three carbon atoms. This conversion involves a sequence of nine enzymatic steps that create phosphate-containing intermediates (Figure 2–20). Logically, the sequence can be divided into three parts: (1) in steps 1 to 4, glucose is converted to two molecules of the three-carbon aldehyde *glyceraldehyde 3-phosphate*—a conversion that requires an investment of energy in the form of ATP hydrolysis to provide the two phosphates; (2) in steps 5 and 6, the aldehyde group of each glyceraldehyde 3-phosphate molecule is oxidized to a carboxylic acid, and the energy from this reaction is coupled to the synthesis of ATP from ADP and inorganic phosphate; and (3) in steps 7, 8, and 9, the same two phosphate molecules that were added to sugars in the first reaction sequence are transferred back to ADP to form ATP, thereby repaying the original investment of two ATP molecules hydrolyzed in the first reaction sequence.

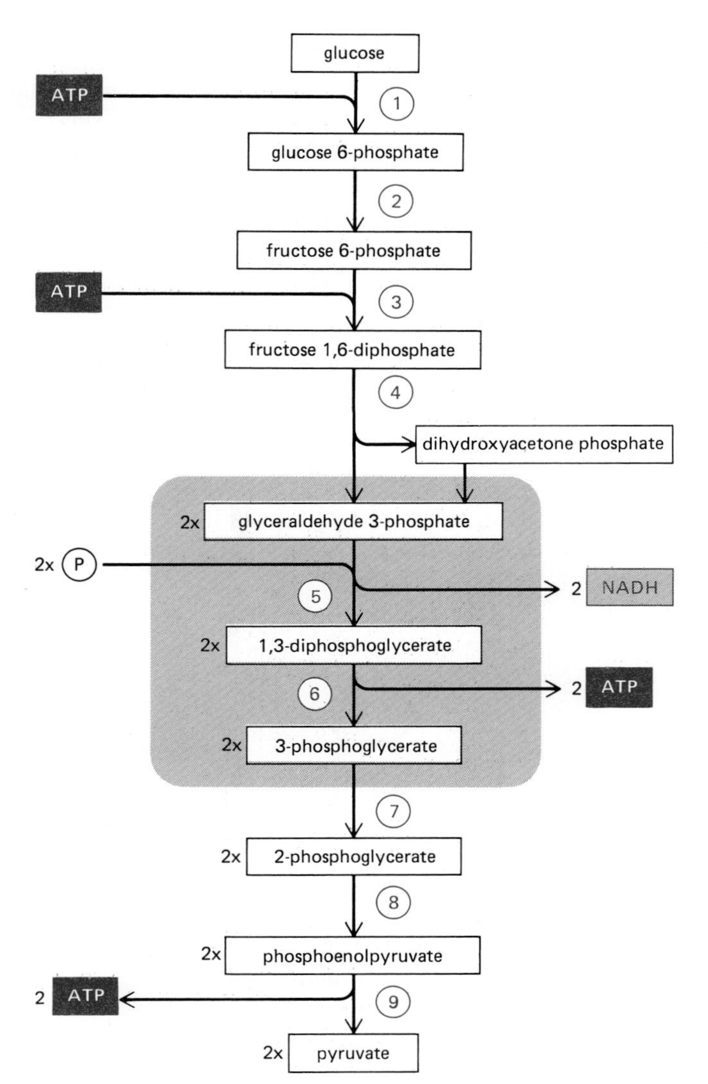

Figure 2–20 Intermediates of glycolysis. Each reaction shown is catalyzed by a different enzyme. In the series of reactions designated as step 4, a six-carbon sugar is cleaved to give two three-carbon sugars, so that the number of molecules at every step after this is doubled. Steps 5 and 6 in the colored box are the reactions responsible for the net synthesis of ATP and NADH molecules (see Figure 2–21).

At the end of glycolysis, the ATP balance sheet shows a net profit of the two molecules of ATP (per glucose molecule) that were produced in steps 5 and 6. As the only reactions in the sequence in which a high-energy phosphate linkage is created from inorganic phosphate, these two steps lie at the heart of glycolysis. They also provide an excellent illustration of the way in which reactions in the cell can be coupled together by enzymes to harvest the energy released by oxidations (Figure 2–21). The overall result is that an aldehyde group on a sugar is oxidized to a carboxylic acid and an inorganic phosphate group is transferred to a high-energy linkage on ATP; in addition, a molecule of NAD^+ is reduced to NADH (Figure 2–22). This elegant set of coupled reactions was probably among the earliest metabolic steps to appear in the evolving cell. Besides being central to glycolysis, these reactions are driven in reverse during photosynthesis by the large quantities of NADPH and ATP produced by the light-activated reactions; they thereby play a pivotal role in the photosynthetic carbon-fixation process by producing glyceraldehyde 3-phosphate (p. 370).

For most animal cells, glycolysis is only a prelude to stage 3 of catabolism, since the pyruvic acid that is formed quickly enters the mitochondria to be completely oxidized to CO_2 and H_2O. However, in the case of anaerobic organisms (those that do not utilize molecular oxygen) and for tissues, such as skeletal muscle, that can function under anaerobic conditions, glycolysis can become a major source of the cell's ATP. Here, instead of being degraded in mitochondria, the pyruvate molecules stay in the cytosol and, depending on the organism, can be

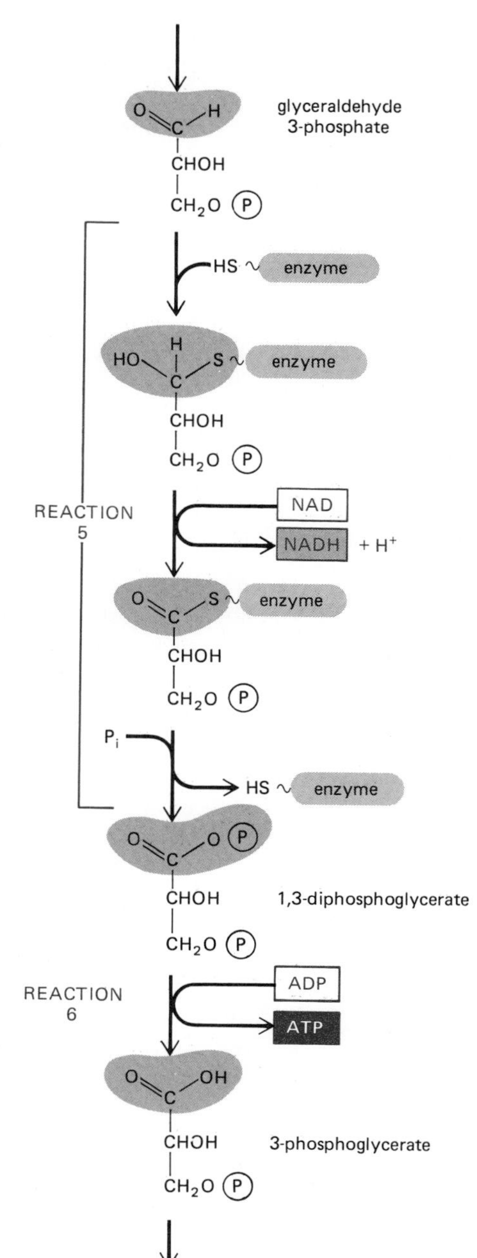

Figure 2–21 Steps 5 and 6 of glycolysis: the oxidation of an aldehyde to a carboxylic acid is coupled to the formation of ATP and NADH (see also Figure 2–20). As shown, step 5 begins when the enzyme *glyceraldehyde 3-phosphate dehydrogenase* forms a covalent bond to the carbon carrying the aldehyde group on glyceraldehyde 3-phosphate. Next, hydrogen (as a hydride ion—a proton plus two electrons) is removed from the enzyme-linked aldehyde group in glyceraldehyde 3-phosphate and transferred to the important hydrogen carrier NAD^+ (see Figure 2–22). This oxidation step creates a sugar carbonyl group attached to the enzyme in a high-energy linkage. This linkage is then broken by a phosphate ion from solution, creating a high-energy sugar-phosphate bond instead. In these last two reactions, the enzyme has *coupled* the energetically favorable process of oxidizing an aldehyde to the energetically unfavorable formation of a high-energy phosphate bond. Finally, in step 6 of glycolysis, the newly created reactive phosphate group is transferred to ADP to form ATP, leaving a free carboxylic acid group on the oxidized sugar.

converted into ethanol plus CO_2 (as in yeast) or into lactate (as in muscle), which is then excreted. In such anaerobic energy-yielding reactions, called **fermentations,** the further reaction of pyruvate is required in order to use up the reducing power produced in reaction 5 of glycolysis, thereby regenerating the NAD^+ required for glycolysis to continue (see p. 381).

Oxidative Catabolism Yields a Much Greater Amount of Usable Energy[14]

The anaerobic generation of ATP from glucose through the reactions of glycolysis is relatively inefficient. The end products of anaerobic glycolysis still contain a great deal of chemical energy that can be released by further oxidation. The evolution of *oxidative catabolism* (cellular respiration) became possible only after molecular oxygen had accumulated in the earth's atmosphere as a result of photosynthesis by the cyanobacteria. Earlier, anaerobic catabolic processes had presumably dominated life on earth. The addition of an oxygen-requiring stage to the catabolic process (stage 3 in Figure 2–18) provided cells with a much more powerful and efficient method for extracting energy from food molecules. This third stage begins with the *citric acid cycle* (also called the tricarboxylic acid cycle, or the Krebs cycle) and ends with *oxidative phosphorylation,* both of which occur in aerobic bacteria and the mitochondria of eucaryotic cells.

Metabolism Is Dominated by the Citric Acid Cycle[15]

The primary function of the **citric acid cycle** is to oxidize acetyl groups that enter the cycle in the form of acetyl CoA molecules. The reactions form a cycle because the acetyl group is not oxidized directly, but only after it has been covalently added to a larger molecule, *oxaloacetate,* which is regenerated at the end of one turn of the cycle. As illustrated in Figure 2–23, the cycle begins with the reaction between acetyl CoA and oxaloacetate to form the tricarboxylic acid molecule called *citric acid* (or *citrate*). A series of reactions then occurs in which two of the six carbons of citrate are oxidized to CO_2, forming another molecule of oxaloacetate to repeat the cycle. (Because the two carbons that are newly added in each cycle enter a different part of the citrate molecule from the part oxidized to CO_2, it is only after several cycles that their turn comes to be oxidized.) The CO_2 produced in these reactions then diffuses from the mitochondrion (or from the bacterium) and leaves the cell.

The energy made available when the C—H and C—C bonds in citrate are oxidized is captured in several different ways in the course of the citric acid cycle. At one step in the cycle (succinyl CoA to succinate), a high-energy phosphate linkage is created by a mechanism resembling that described for glycolysis above. (Although this reaction produces GTP rather than ATP, all nucleoside triphosphates are equivalent energetically because of exchange reactions such as ADP +

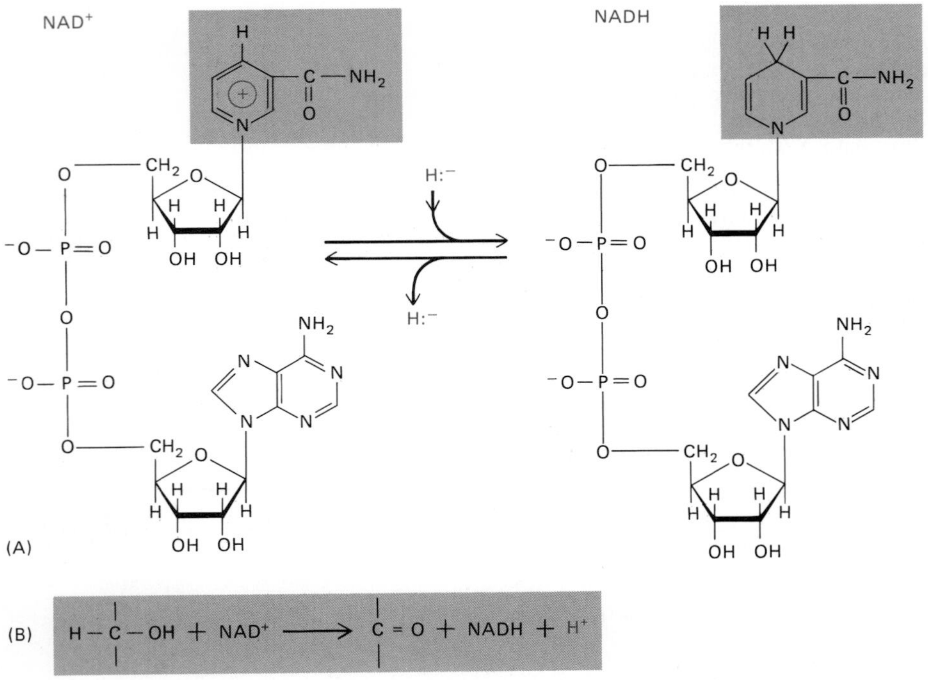

(B) $H-C-OH + NAD^+ \longrightarrow C=O + NADH + H^+$

Figure 2–22 NADH and NAD^+ are the most important carriers of hydrogen in catabolic reactions. Their structures are shown in (A). NAD is an abbreviation for *nicotinamide adenine dinucleotide,* reflecting the fact that the right half of the molecule, as drawn, is adenosine monophosphate (AMP). The part of the NAD^+ molecule known as the nicotinamide ring (in the colored box) is able to accept a hydrogen atom together with an additional electron (a hydride ion, H^-), forming NADH. In this reduced form, the nicotinamide ring has a reduced stability because it is no longer stabilized by resonance. As a result, the added hydride ion is easily transferred to other molecules.

(B) An example of a reaction involving NAD^+ and NADH. In the biological oxidation of a substrate molecule such as an alcohol, two hydrogen atoms are lost from the substrate. One of these is added as a hydride ion to NAD^+, producing NADH, while the other is released into solution as a proton (H^+). (See also Figure 7–18, p. 351.)

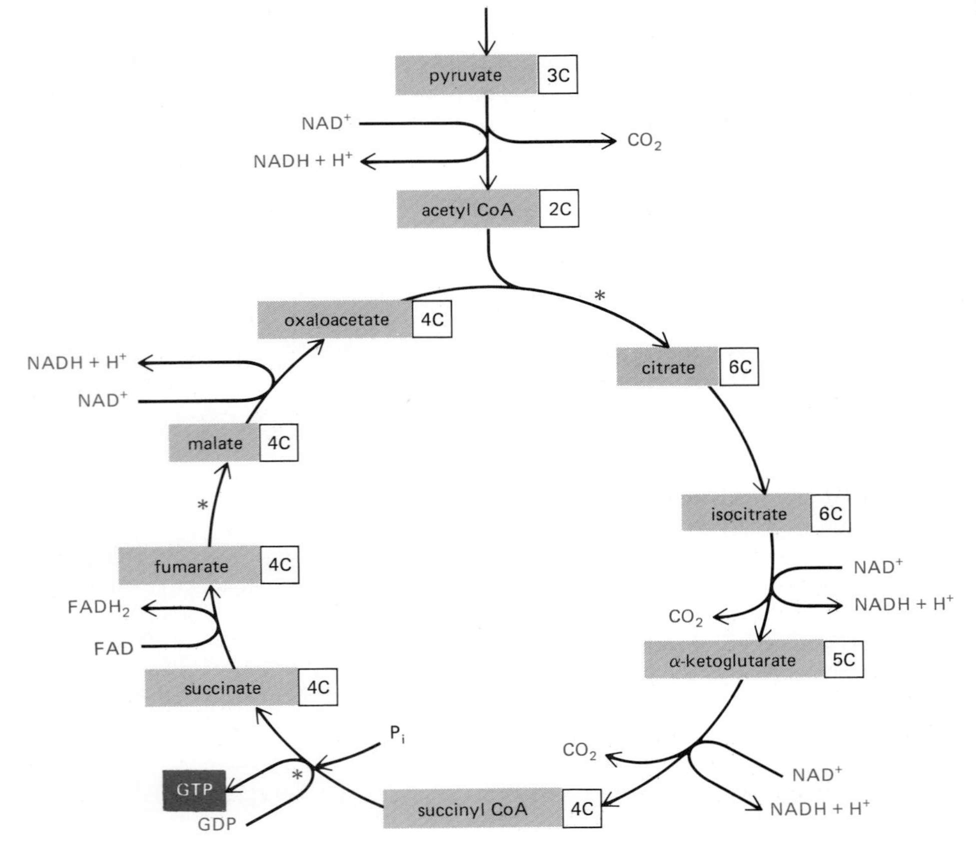

Figure 2–23 The citric acid cycle. In mitochondria and in aerobic bacteria, the acetyl groups produced from pyruvate are further oxidized. The carbon atoms of the acetyl groups are converted to CO_2, while the hydrogen atoms are transferred to the carrier molecules NAD^+ and FAD. Additional oxygen and hydrogen atoms enter the cycle in the form of water at the steps marked with an asterisk (*). For details, see Figure 7–14, p. 348.

GTP ⇌ ATP + GDP.) All of the remaining energy of oxidation that is captured is channeled into the conversion of hydrogen- (or hydride ion-) carrier molecules to their reduced forms; for each turn of the cycle, three molecules of NAD^+ are converted to NADH and one *flavin adenine nucleotide* (FAD) is converted to $FADH_2$. The energy carried by the activated hydrogen on these carrier molecules will subsequently be harnessed through the reactions of *oxidative phosphorylation* (to be considered in more detail below), which require molecular oxygen from the atmosphere.

The additional oxygen atoms required to make CO_2 from the acetyl groups entering the citric acid cycle are supplied not by molecular oxygen but by water. Three molecules of water are split in each cycle, and their oxygen atoms are used to make CO_2. Some of their hydrogen atoms enter substrate molecules and are raised to a higher energy and removed (together with the hydrogen atoms of the acetyl groups) to carrier molecules such as NADH.

In the eucaryotic cell, the mitochondrion is the center toward which all catabolic processes lead, whether they begin with sugars, fats, or proteins. For, in addition to pyruvate, fatty acids and some amino acids also pass from the cytosol into mitochondria, where they are converted into acetyl CoA or one of the other intermediates of the citric acid cycle. The mitochondrion also functions as the starting point for some biosynthetic reactions by producing vital carbon-containing intermediates, such as *oxaloacetate* and *α-ketoglutarate*. These substances are transferred back from the mitochondrion to the cytosol, where they serve as precursors for the synthesis of essential molecules, such as amino acids.

In Oxidative Phosphorylation, the Transfer of Electrons to Oxygen Drives ATP Formation[9,16]

Oxidative phosphorylation is the last step in catabolism and the point at which the major portion of metabolic energy is released. In this process, molecules of NADH and $FADH_2$ transfer the electrons that they have gained from the oxidation of food molecules to molecular oxygen, O_2. The reaction, which is formally equivalent to the burning of hydrogen in air to form water, releases a great deal of chemical energy. Part of this energy is used to make the major portion of the cell's ATP; the rest is liberated as heat.

Although the overall chemistry of NADH and $FADH_2$ oxidation involves a transfer of hydrogen to oxygen, complete hydrogen atoms are not transferred directly. It is the *electrons* from the hydrogen atoms that are important. This is because a hydrogen atom can be readily dissociated into its constituent electron and proton (H^+). The electron can then be transferred separately to a molecule that accepts only electrons, while the proton remains in aqueous solution. By the same reasoning, if an electron alone is donated to a molecule with a strong affinity for hydrogen, then a hydrogen atom will be reconstituted automatically by the capture of a proton from solution. In the course of oxidative phosphorylation, electrons from NADH and $FADH_2$ pass down a chain of carrier molecules, but the presence or absence of intact hydrogen atoms depends on the nature of the carrier.

In a eucaryotic cell, this series of electron transfers along the **electron-transport chain** takes place on the inner membrane of the mitochondrion, in which all of the carrier molecules are embedded. At each step of the transfer, the electrons fall to a lower energy state, until at the end they are transferred to oxygen molecules. Oxygen molecules have the highest affinity of all of the carriers for electrons, and electrons bound to oxygen are thus in their lowest energy state. The energy released as these electrons fall to lower energy states is harnessed, in a way that is not fully understood, to pump protons from the inner mitochondrial compartment to the outside (Figure 2–24). An *electrochemical proton gradient* is thereby generated across the inner mitochondrial membrane. This gradient, in turn, drives a flux of protons back through a special enzyme complex in the same membrane, causing the enzyme (*ATP synthetase*) to add a phosphate group to ADP, and thereby generating ATP inside the mitochondrion. Finally, the newly

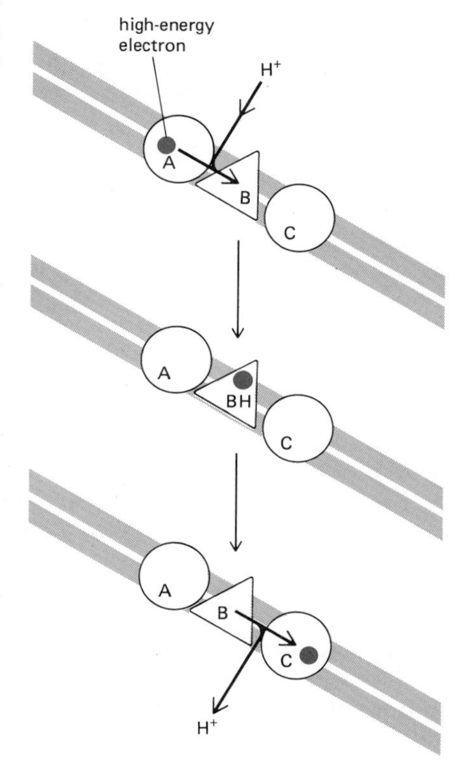

Figure 2–24 The generation of a H^+ gradient across a membrane by electron-transport reactions. A high-energy electron (derived, for example, from the oxidation of a metabolite) is passed sequentially by carriers A, B, and C to a lower energy state. In this diagram, carrier B is arranged in the membrane in such a way that it takes up H^+ from one side and releases it to the other as the electron passes. The resulting H^+ gradient represents a form of stored energy that is harnessed by other membrane proteins in the mitochondrion to drive the formation of ATP (see also Figure 7–35, p. 363).

made ATP is transferred from the mitochondrion to the rest of the cell, where it drives a variety of metabolic reactions.

The nature of the electron-transport chain and the mechanism of ATP synthesis will be described in detail in Chapter 7.

Amino Acids and Nucleotides Are Part of the Nitrogen Cycle

In our discussion so far, we have concentrated mainly on carbohydrate metabolism. We have not yet considered the metabolism of nitrogen or sulfur. These two elements are constituents of proteins and nucleic acids, which are the two most important classes of macromolecules in the cell and make up approximately two-thirds of its dry weight. Atoms of nitrogen and sulfur pass from compound to compound and between organisms and their environment in a series of reversible cycles.

Although molecular nitrogen is abundant in the earth's atmosphere, it is chemically unreactive. Only a few living species are able to incorporate it into organic molecules, a process called **nitrogen fixation.** Nitrogen fixation occurs in certain microorganisms and by some geophysical processes, such as lightning discharge. It is essential to the biosphere as a whole, for without it life would not exist on this planet. Only a small fraction of the nitrogenous compounds in today's organisms, however, represents fresh products of nitrogen fixation. Most organic nitrogen has been in circulation for some time, passing from one living organism to another. Thus nitrogen-fixing reactions can be said to perform a "topping-up" function for the total nitrogen supply.

Vertebrates receive virtually all of their nitrogen in their dietary intake of proteins and nucleic acids. In the body, these macromolecules are broken down to component amino acids and nucleotides, which are then repolymerized into new proteins and nucleic acids or utilized to make other molecules. About half of the 20 amino acids found in proteins are *essential amino acids* (Figure 2–25) for vertebrates: they cannot be synthesized from other ingredients of the diet. The others can be so synthesized, using a variety of raw materials, including intermediates of the citric acid cycle. The essential amino acids are made in other organisms, usually by long and energetically expensive pathways that have been lost in the course of vertebrate evolution.

The nucleotides needed to make RNA and DNA can be synthesized using specialized biosynthetic pathways: there are no "essential nucleotides" that must be provided in the diet. All of the nitrogens in the purine and pyrimidine bases (as well as some of the carbons) are derived from the plentiful amino acids glutamine, aspartic acid, and glycine, whereas the ribose and deoxyribose sugars are derived from glucose.

Amino acids that are not utilized in biosynthesis can be oxidized to generate metabolic energy. Most of their carbon and hydrogen atoms eventually form CO_2 or H_2O, whereas their nitrogen atoms are shuttled through various forms and eventually appear as urea, which is excreted. Each amino acid is processed differently, and a whole constellation of enzymatic reactions exists for their catabolism.

THE ESSENTIAL AMINO ACIDS
THREONINE
METHIONINE
LYSINE
VALINE
LEUCINE
ISOLEUCINE
HISTIDINE
PHENYLALANINE
TRYPTOPHAN

Figure 2–25 The nine essential amino acids, which cannot be synthesized by human cells and so must be supplied in the diet.

Summary

Animal cells can be considered to derive energy from food in three stages. In stage 1, proteins, polysaccharides, and fats are broken down by extracellular reactions to small molecules. In stage 2, these small molecules are degraded within cells to produce acetyl CoA and a limited amount of ATP and NADH. These are the only reactions that can yield energy in the absence of oxygen. In stage 3, the acetyl CoA molecules are degraded in mitochondria to give CO_2 and hydrogen atoms that are linked to carrier molecules such as NADH. Electrons from the hydrogen atoms are passed through a complex chain of membrane-bound carriers, finally being passed to molecular oxygen to form water. Driven by the energy released in these electron-transfer steps, hydrogen ions (H^+) are transported out of the mitochondria. The resulting electrochemical proton gradient across the inner mitochondrial membrane is harnessed to drive the synthesis of most of the cell's ATP.

Biosynthesis and the Creation of Order[17]

Thousands of different chemical reactions are occurring in a cell at any instant of time. The reactions are all linked together in chains and networks in which the product of one reaction becomes the substrate of the next. Most of the chemical reactions in cells can be roughly classified as being concerned either with catabolism or with biosynthesis. Having discussed the catabolic reactions, we now turn to the reactions of biosynthesis. These begin with the intermediate products of glycolysis and the citric acid cycle (and closely related compounds) and generate the larger and more complex molecules of the cell.

The Free-Energy Change for a Reaction Determines Whether It Can Occur[18]

Although enzymes speed up energetically favorable reactions, they cannot force energetically unfavorable reactions to occur. In terms of a water analogy, enzymes by themselves cannot make water run uphill. But cells must do just that in order to grow and divide; they must build highly ordered and energy-rich molecules from small and simple ones. We have seen that, in a general way, this is done through enzymes that couple energetically favorable reactions, which consume energy derived ultimately from the sun and produce heat, to energetically unfavorable reactions, which produce biological order. Let us examine in greater detail how such coupling is achieved.

First, we must consider more carefully the term "energetically favorable," which we have so far used loosely, without giving it a definition. As explained earlier, a chemical reaction can proceed spontaneously only if it results in a net increase in the disorder of the universe. Disorder increases when energy is dissipated as heat; and the criterion for an increase of disorder can conveniently be expressed in terms of a quantity called the **free energy, *G*.** This is defined in such a way that changes in its value, denoted by **ΔG,** measure the amount of disorder created in the universe when a reaction takes place, as explained in Panel 2–7 (pp. 72–73). "Energetically favorable" reactions, by definition, are those that release a large quantity of free energy, or in other words have a large *negative* ΔG and create much disorder. Such reactions have a strong tendency to occur spontaneously, although their rate will depend on other factors, such as the availability of specific enzymes (see below). Conversely, reactions with a *positive* ΔG, such as those in which two amino acids are joined together to form a peptide bond, by themselves create order in the universe and therefore cannot occur spontaneously. Energetically unfavorable reactions of this kind can take place only if they are coupled to a second reaction with a negative ΔG so large that the ΔG of the entire process is negative.

The course of most reactions can be predicted quantitatively. A large body of thermodynamic data has been collected that makes it possible to calculate the change in free energy for most of the important metabolic reactions of the cell. The overall free-energy change for a pathway is then simply the sum of the energy changes in each of its component steps. Consider two reactions

$$X \rightarrow Y \quad \text{and} \quad C \rightarrow D$$

where the ΔG values are $+1$ and -13 kcal/mole, respectively. (Recall that a mole is 6×10^{23} molecules of a substance.) If these two reactions can be coupled together, the ΔG for the coupled reaction will be -12 kcal/mole. Thus, the unfavorable reaction $X \rightarrow Y$, which will not occur spontaneously, can be driven by the favorable reaction $C \rightarrow D$, provided that a mechanism exists by which the two reactions can be coupled together.

Biosynthetic Reactions Are Often Directly Coupled to ATP Hydrolysis

Consider a typical biosynthetic reaction in which two monomers, A and B, are to be joined in a *dehydration* (also called *condensation*) reaction, in which water is released:

THE IMPORTANCE OF FREE ENERGY FOR CELLS

Life is possible because of the complex network of interacting chemical reactions occurring in every cell. In viewing the metabolic pathways that comprise this network, one might suspect that the cell has had the ability to evolve an enzyme to carry out any reaction that it needs. But this is not so. Although enzymes are powerful catalysts, they can speed up only those reactions that are thermodynamically possible; other reactions proceed in cells only because they are *coupled* to very favorable reactions that drive them. The question of whether a reaction can occur spontaneously, or instead needs to be coupled to another reaction, is central to cell biology. The answer is obtained by reference to a quantity called the *free energy*: the total change in free energy during a set of reactions determines whether or not the entire reaction sequence can occur. In this panel, we will explain some of the fundamental ideas—derived from a special branch of chemistry and physics called *thermodynamics*—that are required for understanding what free energy is and why it is so important to cells.

ENERGY RELEASED BY CHANGES IN CHEMICAL BONDING IS CONVERTED INTO HEAT

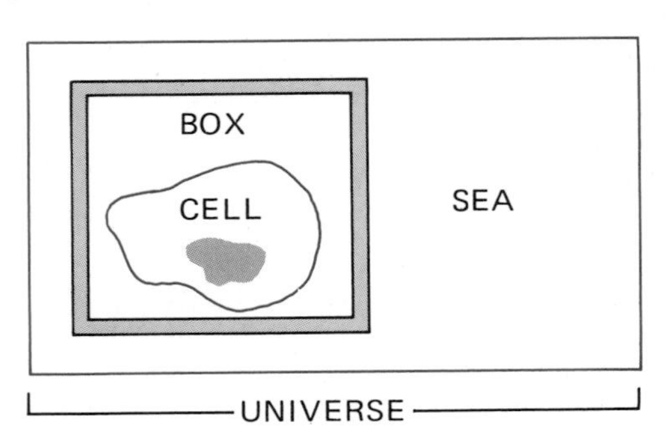

An *enclosed system* is defined as a collection of molecules that does not exchange matter with the rest of the universe (for example, the "cell in a box" described previously in Figure 2–11 and shown above). Any such system will contain molecules with a total energy E. This energy will be distributed in a variety of ways: some as the translational energy of the molecules, some as their vibrational and rotational energies, but most as the bonding energies between the individual atoms that make up the molecules. Suppose that a reaction occurs in the system. The first law of thermodynamics places a constraint on what types of reactions can occur in the system: it states that "in any process, the total energy of the universe remains constant." For example, suppose that reaction $A \rightarrow B$ occurs in the enclosed system and releases a great deal of chemical bond energy. This energy will initially increase the intensity of molecular motions (translational, vibrational, and rotational) in the system, which is equivalent to raising its temperature. However, these increased motions will soon be transferred out of the system by a series of molecular collisions that heat up first the walls of the box and then the outside world (represented by the sea in our example). In the end, the system returns to its initial temperature, by which time all the chemical bond energy released in the box has been converted into heat energy and transferred out of the box to the surroundings. According to the first law, the change in the energy in the box (ΔE_{box}, which we shall denote as ΔE) must be equal and opposite to the amount of heat energy transferred, which we shall designate as h: that is, $\Delta E = -h$. Thus, the energy in the box (E) decreases when heat leaves the system.

E also can change during a reaction due to work being done on the outside world. For example, suppose that there is a small increase in the volume (ΔV) of the box during a reaction. Since the walls of the box must push against the constant pressure (P) in the surroundings in order to expand, this does work on the outside world and requires energy. The energy used is $P(\Delta V)$, which according to the first law must decrease the energy in the box (E) by the same amount. In most reactions, chemical bond energy is converted into both work and heat. *Enthalpy* (H) is a composite function that includes both of these ($H = E + PV$). To be rigorous, it is the change in enthalpy (ΔH) in an enclosed system and not the change in energy that is equal to the heat transferred to the outside world during a reaction. Reactions in which H decreases release heat to the surroundings and are said to be "exothermic," while reactions in which H increases absorb heat from the surroundings and are said to be "endothermic." Thus, $-h = \Delta H$. However, the volume change is negligible in most biological reactions, so to a good approximation:

$$-h = \Delta H \cong \Delta E$$

THE SECOND LAW OF THERMODYNAMICS

Consider a container in which 1000 coins are all lying heads up. If the container is shaken vigorously, subjecting the coins to the types of random motions that all molecules experience due to their frequent collisions with other molecules, one will end up with about half of the coins oriented heads down. The reason for this reorientation is that there is only a single way in which the original state of the coins can be reinstated (every coin must lie heads up), whereas there are many different ways (about 10^{298}) to achieve a state in which there is an equal mixture of heads and tails; in fact, there are more ways to achieve a 50:50 state than to achieve any other state. Each state has a probability of occurrence that is proportional to the number of ways it can be realized. The second law of thermodynamics states that "systems will change spontaneously from states of lower probability to states of higher probability." Since states of lower probability are said to be more "ordered" than states of higher probability, the second law can be restated: "the universe constantly changes so as to become more disordered."

THE ENTROPY, S

The second law (but not the first law) allows one to predict the *direction* of a particular reaction. But to make it useful for this purpose, one needs a convenient measure of the probability, or equivalently, the degree of disorder of a state. The entropy (S) is such a measure. It is a logarithmic function of the probability such that the *change in entropy* (ΔS) that occurs when the reaction A → B converts one mole of A into one mole of B is

$$\Delta S = R \ln p_B/p_A$$

where p_A and p_B are the probabilities of the two states A and B, R is the gas constant (2 cal deg^{-1} mole^{-1}), and ΔS is measured in entropy units (eu). In our initial example of 1000 coins, the relative probability of all heads (state A) versus half heads and half tails (state B) is equal to the ratio of the number of different ways that the two results can be obtained. One can calculate that $p_A = 1$ and $p_B = 1000!(500! \times 500!) = 10^{298}$. Therefore, the entropy change for the reorientation of the coins when their container is vigorously shaken and an equal mixture of heads and tails is obtained is $R \ln (10^{298})$, or about 1370 eu per mole of such containers (6×10^{23} containers). We see that, because ΔS defined above is positive for the transition from state A to state B ($p_B/p_A > 1$), reactions with a large *increase* in S (that is, for which $\Delta S > 0$) are favored and will occur spontaneously.

As discussed in the text, heat energy causes the random commotion of molecules. Because the transfer of heat from an enclosed system to its surroundings increases the number of different arrangements that the molecules in the outside world can have, it increases their entropy. It can be shown that the release of a fixed quantity of heat energy has a greater disordering effect at low temperature than at high temperature and that the value of ΔS for the surroundings, as defined above (ΔS_{sea}), is precisely equal to the amount of heat transferred to the surroundings from the system (h) divided by the absolute temperature (T):

$$\Delta S_{sea} = h/T$$

THE GIBBS FREE ENERGY, G

When dealing with an enclosed biological system, one would like to have a simple way of predicting whether a given reaction will or will not occur spontaneously in the system. We have seen that the crucial question is whether the entropy change for the universe is positive or negative when that reaction occurs. In our idealized system, the cell in a box, there are two separate components to the entropy change of the universe—the entropy change for the system enclosed in the box and the entropy change for the surrounding "sea"—and both must be added together before any prediction can be made. For example, it is possible for a reaction to absorb heat and thereby decrease the entropy of the sea ($\Delta S_{sea} < 0$) and at the same time to cause such a large degree of disordering inside the box ($\Delta S_{box} > 0$) that the total $\Delta S_{universe} = \Delta S_{sea} + \Delta S_{box}$ is greater than 0. In this case, the reaction will occur spontaneously, even though the sea gives up heat to the box during the reaction. An example of such a reaction is the dissolving of sodium chloride in a beaker containing water (the "box"), which is a spontaneous process even though the temperature of the water drops as the salt goes into solution.

Chemists have found it useful to define a number of new "composite functions" that describe *combinations* of physical properties of a system. The properties that can be combined include the temperature (T), pressure (P), volume (V), energy (E), and entropy (S). The enthalpy (H) is one such composite function. But by far the most useful composite function for biologists is the *Gibbs free energy, G*. It serves as an accounting device that allows one to deduce the entropy change of the universe resulting from a chemical reaction in the box, while avoiding any separate consideration of the entropy change in the sea. For a box of volume V at pressure P, the definition of G is

$$G = H - TS$$

where H is the enthalpy described above ($E + PV$), T is the absolute temperature, and S is the entropy. Each of these quantities applies to the inside of the box only. The change in free energy during a reaction in the box (the G of the products minus the G of the starting materials) is denoted as ΔG and, as we shall now demonstrate, is a direct measure of the amount of disorder that is created in the universe when the reaction occurs.

At constant temperature, the change in free energy (ΔG) during a reaction equals $\Delta H - T\Delta S$. Remembering that $\Delta H = -h$, the heat absorbed from the sea, we have

$$-\Delta G = -\Delta H + T\Delta S$$
$$-\Delta G = h + T\Delta S, \text{ and } -\Delta G/T = h/T + \Delta S$$

But h/T is equal to the entropy change of the sea (ΔS_{sea}), and the ΔS in the above equation is ΔS_{box}. Therefore

$$-\Delta G/T = \Delta S_{sea} + \Delta S_{box} = \Delta S_{universe}$$

We conclude that a reaction will proceed in the direction that causes the change in free energy (ΔG) to be *less than zero,* because in this case there will be a positive entropy change in the universe when the reaction occurs. Thus, the **free energy** change is a **direct measure of the entropy change of the universe.**

For a complex set of coupled reactions involving many different molecules, the total free-energy change can be computed simply by adding up the free energies of all the different molecular species after the reaction and comparing this value to the sum of free energies before the reaction; for common substances, the required free-energy values can be found from published tables. In this way one can predict the direction of a reaction and thereby readily disprove any proposed mechanism. Thus, for example, from the observed values for the magnitude of the electrochemical proton gradient across the inner mitochondrial membrane and the ΔG for ATP hydrolysis inside the mitochondrion, one can be certain that the enzyme ATP synthetase requires the passage of more than one proton for each molecule of ATP that it synthesizes (see p. 358).

The value of ΔG for a reaction is a direct measure of the degree of displacement of the reaction from equilibrium. The large negative value for ATP hydrolysis in a cell merely reflects the fact that cells keep the ATP hydrolysis reaction as much as ten orders of magnitude away from equilibrium (see p. 354). If a reaction reaches equilibrium, $\Delta G = 0$, and the reaction then proceeds at precisely equal rates in the forward and backward direction. For ATP hydrolysis, equilibrium is reached when the vast majority of the ATP has been hydrolyzed, as occurs in a dead cell.

The Coordination of Catabolism and Biosynthesis [19]

Metabolism Is Organized and Regulated

Some idea of how intricate a cell is when viewed as a chemical machine can be obtained from Figure 2–35, which is a chart showing only some of the enzymatic pathways in a cell. All of these reactions occur in a cell that is less than 0.1 mm in diameter, and each requires an enzyme that is itself the product of a whole series of information-transfer and protein-synthesis reactions. For a typical small molecule—the amino acid *serine*, for example—there are half a dozen or more enzymes that can modify it chemically in different ways: it can be linked to AMP (adenylated) in preparation for protein synthesis, or degraded to glycine, or converted to pyruvate in preparation for oxidation; it can be acetylated by acetyl CoA or transferred to a fatty acid to make phosphatidyl serine. All of these different pathways compete for the same serine molecule, and similar competitions for thousands of other small molecules go on at the same time. One might think that the whole system would need to be so finely balanced that any minor upset, such as a temporary change in dietary intake, would be disastrous.

In fact, the cell is amazingly stable. Whenever it is perturbed, the cell reacts so as to restore its initial state. It can adapt and continue to function during starvation or disease. Mutations of many kinds can eliminate particular reaction pathways, and yet—provided that certain minimum requirements are met—the cell survives. It does so because an elaborate network of control mechanisms regulates and coordinates the rates of its reactions. Some of the higher levels of control will be considered in later chapters. Here we are concerned only with the simplest mechanisms that regulate the flow of small molecules through the various metabolic pathways.

Figure 2–35 Some of the chemical reactions occurring in a cell. (A) About 500 common metabolic reactions are shown diagrammatically, with each chemical species represented by a filled circle. The centrally placed reactions of the glycolytic pathway and the citric acid cycle are shown in solid color. A typical mammalian cell synthesizes more than 10,000 different proteins, a major proportion of which are enzymes. In the arbitrarily selected segment of this metabolic maze (color shaded), cholesterol is synthesized from acetyl CoA. To the right and below the maze, this segment is shown in detail in an enlargement (B). ▶

Metabolic Pathways Are Regulated by Changes in Enzyme Activity[20]

The concentrations of the various small molecules in a cell are buffered against major changes by a process known as **feedback regulation,** which fine-tunes the flux of metabolites through a particular pathway by temporarily increasing or decreasing the activity of crucial enzymes. For example, the first enzyme of a series of reactions is usually inhibited by a *negative feedback* effect of the final product of that pathway: if large quantities of the final product accumulate, further entry of precursors into the reaction pathway is automatically inhibited (Figure 2–36). Where pathways branch or intersect, as they often do, there are usually multiple points of control by different final products. The complexity of such feedback control processes is illustrated in Figure 2–37, which shows the pattern of enzyme regulation observed in a set of related amino acid pathways.

Feedback regulation can work almost instantaneously and is reversible; in addition, a given end product may activate enzymes leading along other pathways, as well as inhibit enzymes that cause its own synthesis. The molecular basis for this type of control in cells is well understood, but since an explanation requires some knowledge of protein structure, it will be deferred until Chapter 3 (p. 128).

Catabolic Reactions Can Be Reversed by an Input of Energy[21]

By regulating a few enzymes at key points in a metabolic network, large-scale changes that affect the metabolism of the entire cell can be achieved. For example, a special pattern of feedback regulation enables a cell to switch from glucose degradation to glucose biosynthesis (denoted *gluconeogenesis*). The need for gluconeogenesis is especially acute in periods of violent exercise, when the glucose needed for muscle contraction is generated from lactic acid by liver cells, and also in periods of starvation, when glucose must be formed from the glycerol portion of fats and from amino acids for survival.

The normal breakdown of glucose to pyruvate during glycolysis is catalyzed by a number of enzymes acting in series. The reactions catalyzed by most of these enzymes are readily reversible, but three reaction steps (numbers 1, 3, and 9 in

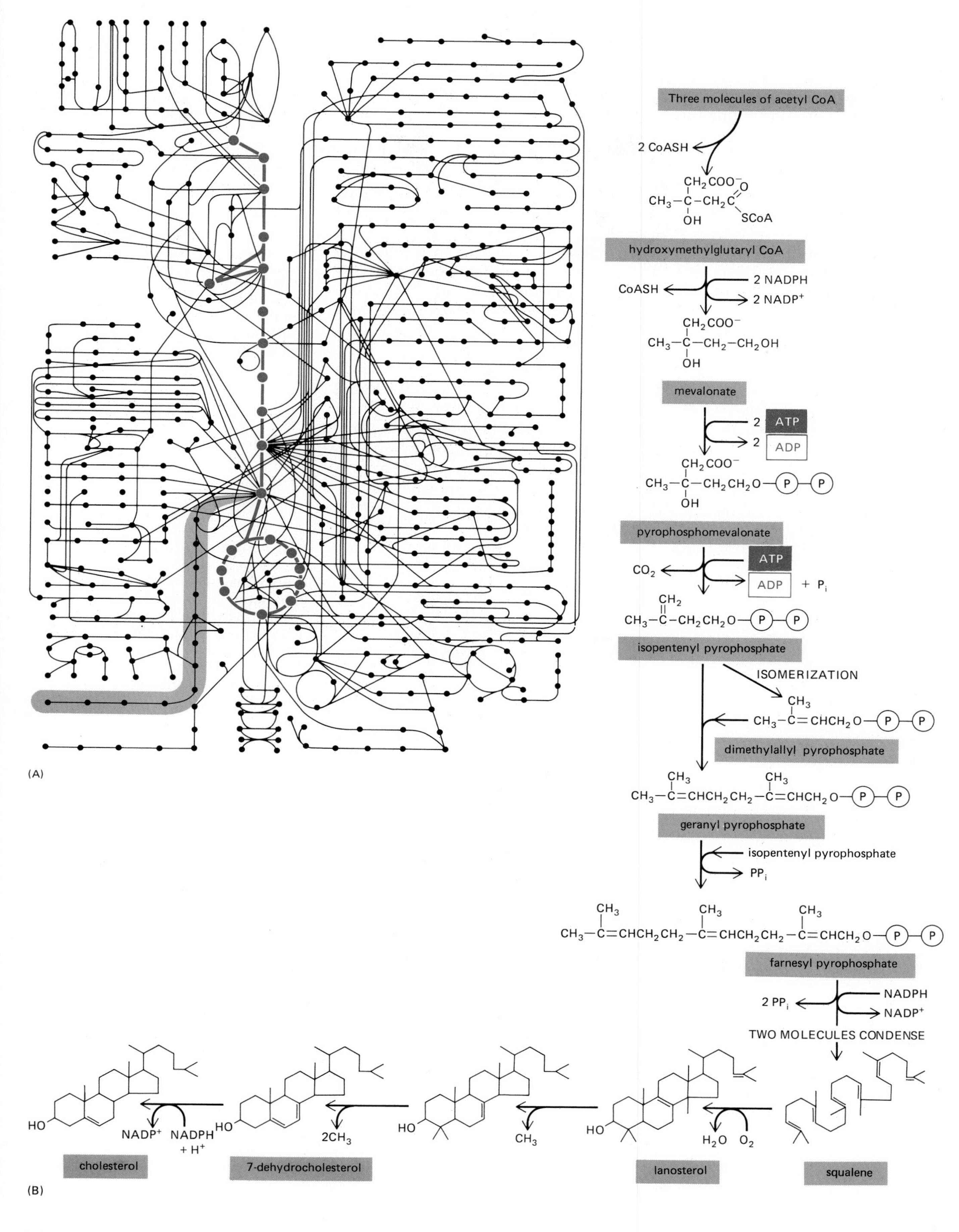

Three molecules of acetyl CoA
2 CoASH
hydroxymethylglutaryl CoA
CoASH
2 NADPH
2 NADP+
mevalonate
2 ATP
2 ADP
pyrophosphomevalonate
CO2
ATP
ADP
+ Pi
isopentenyl pyrophosphate
ISOMERIZATION
dimethylallyl pyrophosphate
geranyl pyrophosphate
isopentenyl pyrophosphate
PPi
farnesyl pyrophosphate
2 PPi
NADPH
NADP+
TWO MOLECULES CONDENSE
squalene
H2O
O2
lanosterol
CH3
2CH3
7-dehydrocholesterol
NADP+
NADPH + H+
cholesterol
HO
(A)
(B)

the sequence of Figure 2–20) are effectively irreversible. In fact, it is the large negative free-energy change that occurs in these reactions that normally drives the breakdown of glucose. For the reactions to proceed in the opposite direction and make glucose from pyruvate, each of these three reactions must be bypassed. This is achieved by substituting three alternative, enzyme-catalyzed bypass reactions that are driven in the uphill direction by an input of chemical energy (Figure 2–38). Thus, whereas two ATP molecules are generated as each molecule of glucose is degraded to two molecules of pyruvate, the reverse reaction during gluconeogenesis requires the hydrolysis of four ATP and two GTP molecules. This is equivalent, in total, to the hydrolysis of six molecules of ATP for every molecule of glucose synthesized.

The bypass reactions in Figure 2–38 must be controlled so that glucose is broken down rapidly when energy is needed but is synthesized when the cell is nutritionally replete. If both forward and reverse reactions were allowed to proceed without restraint, they would shuttle large quantities of metabolites backward and forward in futile cycles that would consume large amounts of ATP and generate heat for no purpose.

The elegance of the control mechanisms involved can be illustrated by a single example. Step 3 of glycolysis is one of the reactions that must be bypassed during glucose formation. Normally this step involves the addition of a second phosphate group to fructose 6-phosphate from ATP and is catalyzed by the enzyme *phosphofructokinase.* This enzyme is activated by AMP, ADP, and inorganic phosphate, whereas it is inhibited by ATP, citrate, and fatty acids. Therefore, the enzyme is activated by the accumulation of the products of ATP hydrolysis when energy supplies are low, and it is inactivated when energy (in the form of ATP) or food

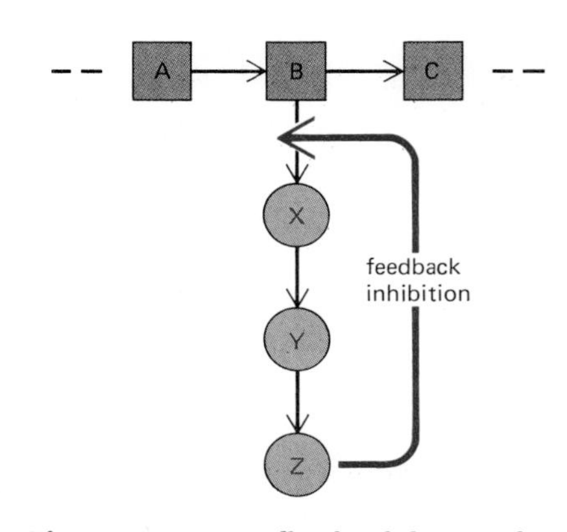

Figure 2–36 Feedback inhibition of a single biosynthetic pathway. The end product Z inhibits the first enzyme that is unique to its synthesis and thereby controls its own level in the cell.

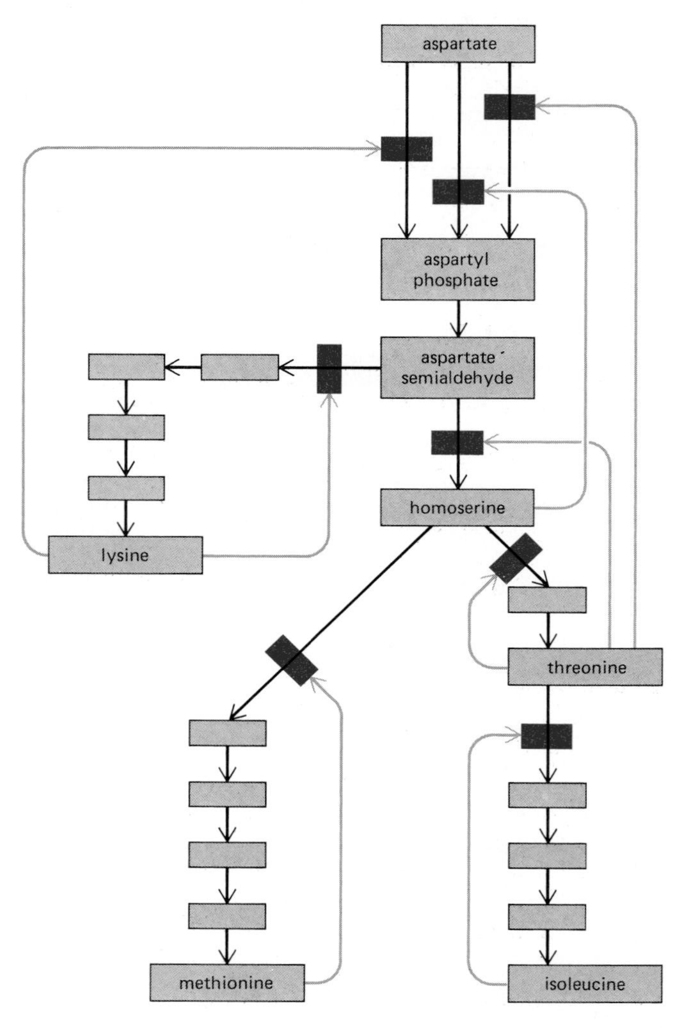

Figure 2–37 Feedback inhibition in the synthesis of the amino acids lysine, methionine, threonine, and isoleucine in bacteria. The colored arrows indicate positions at which products "feed back" to inhibit enzymes. Note that three different enzymes (called *isozymes*) catalyze the initial reaction, each inhibited by a different product.

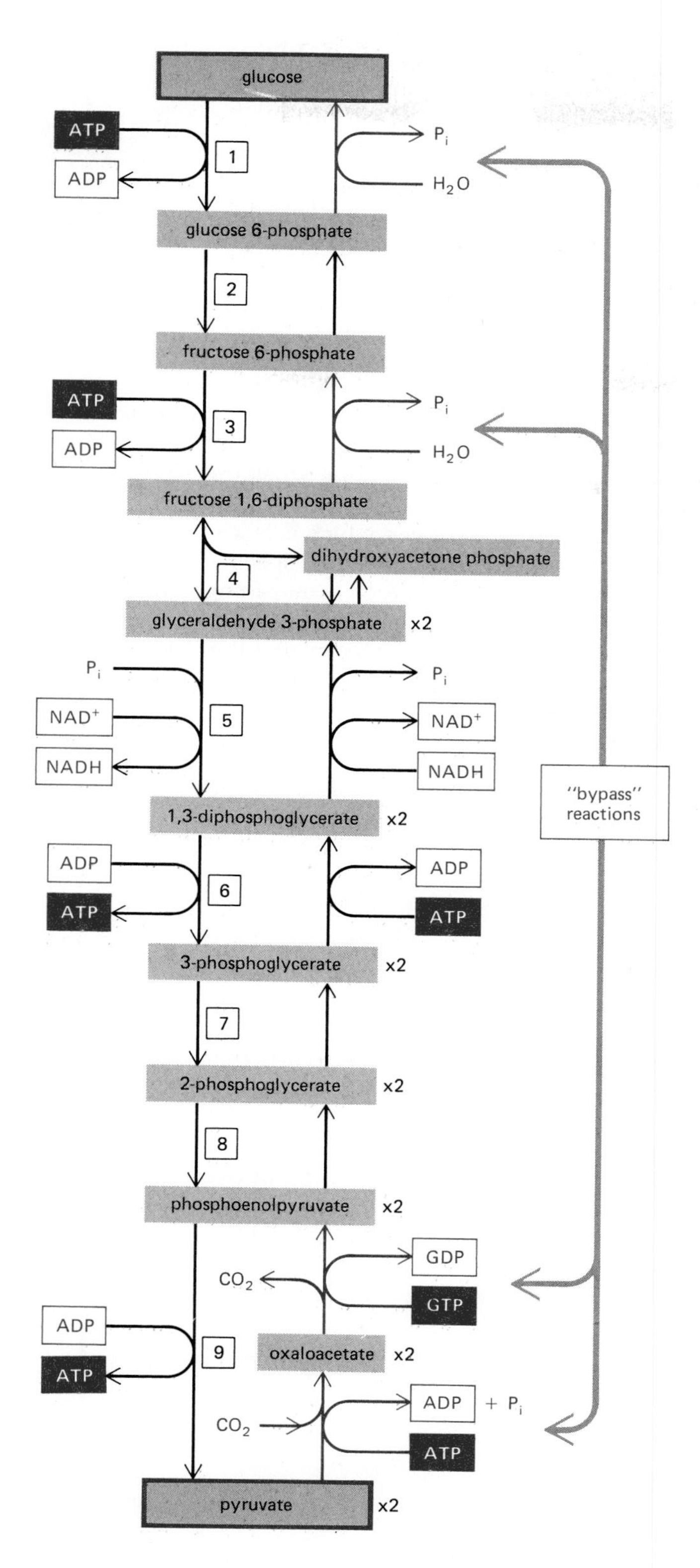

Figure 2–38 Comparison of the reactions that produce glucose during gluconeogenesis with those that degrade glucose during glycolysis. The degradative (glycolytic) reactions are energetically favorable (the free-energy change is less than zero), whereas the synthetic reactions require an input of energy. To synthesize glucose, different "bypass enzymes" are needed that bypass reactions 1, 3, and 9 of glycolysis. The overall flux of reactants between glucose and pyruvate is determined by feedback control mechanisms that operate to control the metabolism of the molecules that participate in these three glycolytic steps.

supplies such as fatty acids or citrate (derived from amino acids) are abundant. *Fructose diphosphatase* is the enzyme that catalyzes the reverse (bypass) reaction (the hydrolysis of fructose 1,6-diphosphate to fructose 6-phosphate, leading to the formation of glucose); this enzyme is regulated in the opposite way by the same feedback control molecules, so that it is stimulated when the phosphofructokinase is inhibited.

Note that phosphofructokinase is activated by ADP, which is a product of the reaction it catalyzes (ATP + fructose 6-phosphate → ADP + fructose 1,6-diphosphate), and is inhibited by ATP, which is one of its substrates. As a result, this enzyme can turn itself on, being subject to a complex form of positive feedback control. Under certain circumstances such feedback control gives rise to striking oscillations in the activity of the enzyme, causing corresponding oscillations in the concentrations of various glycolytic intermediates (Figure 2–39). Although the physiological significance of these particular oscillations is not known, they illustrate how a biological oscillator can be produced by a few enzymes. In principle, such oscillations could provide an internal clock, enabling a cell to "measure time" and, for example, to perform certain functions at fixed intervals.

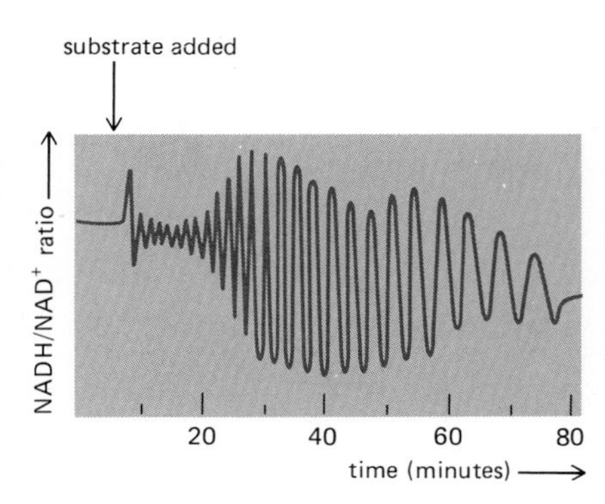

Figure 2–39 The abrupt addition of glucose to an extract containing the enzymes and cofactors required for glycolysis can produce large cyclic fluctuations in the levels of intermediates such as NADH. These metabolic oscillations arise, in part, from the positive feedback control of the glycolytic enzyme phosphofructokinase.

Enzymes Can Be Switched On and Off by Covalent Modification[22]

The types of feedback control just described permit the rates of reaction sequences to be continuously and automatically regulated in response to second-by-second fluctuations in metabolism. Cells have different devices for regulating enzymes when longer-lasting changes in activity, occurring over minutes or hours, are required. These involve reversible covalent modification of enzymes, which is often, but not always, accomplished by the addition of a phosphate group to a specific serine, threonine, or tyrosine residue in the enzyme. The phosphate comes from ATP, and its transfer is catalyzed by a family of enzymes known as *protein kinases.*

We shall describe in the following chapter how phosphorylation can alter the shape of an enzyme in such a way as to increase or inhibit its activity. The subsequent removal of the phosphate group, which reverses the effect of the phosphorylation, is achieved by a second type of enzyme, called a *phosphoprotein phosphatase.* Covalent modification of enzymes adds another dimension to metabolic control because it allows specific reaction pathways to be regulated by signals (such as hormones) that are unrelated to the metabolic intermediates themselves.

Reactions Are Compartmentalized Both Within Cells and Within Organisms[23]

Not all of a cell's metabolic reactions occur within the same subcellular compartment. Because different enzymes are found in different parts of the cell, the flow of chemical components is channeled physically as well as chemically.

The simplest form of such spatial segregation occurs when two enzymes that catalyze sequential reactions form an enzyme complex, and the product of the first enzyme does not have to diffuse through the cytoplasm to encounter the second enzyme. The second reaction begins as soon as the first is over. Some large enzyme aggregates carry out whole series of reactions without losing contact with the substrate. For example, the conversion of pyruvate to acetyl CoA proceeds in three chemical steps, all of which take place on the same large enzyme complex (Figure 2–40), and in fatty acid synthesis an even longer sequence of reactions is catalyzed by a single enzyme assembly. Not surprisingly, some of the largest en-

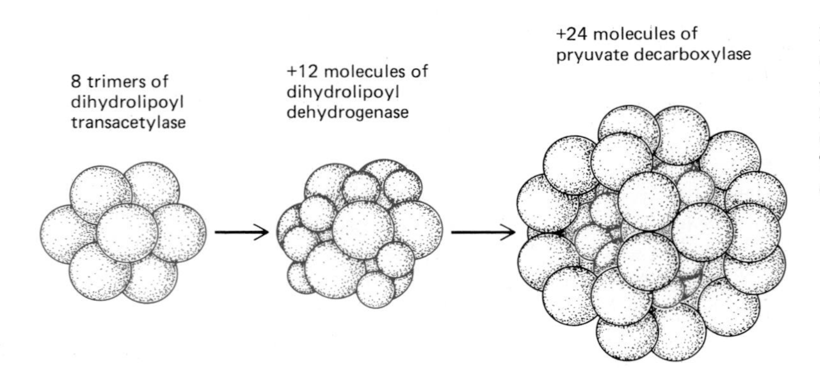

Figure 2–40 The structure of pyruvate dehydrogenase—an example of a large multienzyme complex in which reaction intermediates are passed directly from one enzyme to another. This enzyme complex catalyzes the conversion of pyruvate to acetyl CoA.

zyme complexes are concerned with the synthesis of macromolecules such as proteins and DNA.

The next level of spatial segregation in cells involves the confinement of functionally related enzymes within the same membrane or within the aqueous compartment of an organelle that is bounded by a membrane. The oxidative metabolism of glucose is a good example (Figure 2–41). After glycolysis, pyruvate is actively taken up from the cytosol into the inner compartment of the mitochondrion, which contains all of the enzymes and metabolites involved in the citric acid cycle. Moreover, the inner mitochondrial membrane itself contains all of the enzymes that catalyze the subsequent reactions of oxidative phosphorylation, including those involved in the transfer of electrons from NADH to O_2 and in the synthesis of ATP. The entire mitochondrion can therefore be regarded as a small ATP-producing factory. In the same way, other cellular organelles, such as the nucleus, the Golgi apparatus, and the lysosomes, can be viewed as specialized compartments where functionally related enzymes are confined to perform a specific task. In a sense, the living cell is like a city, with many specialized services concentrated in different areas that are extensively interconnected by various paths of communication.

Spatial organization in a multicellular organism extends beyond the individual cell. The different tissues of the body have different sets of enzymes and make distinct contributions to the chemistry of the organism as a whole. In addition to differences in specialized products such as hormones or antibodies, there are significant differences in the "common" metabolic pathways among various types of cells in the same organism. Although virtually all cells contain the enzymes of glycolysis, the citric acid cycle, lipid synthesis and breakdown, and amino acid metabolism, the levels of these processes in different tissues are differently regulated. Nerve cells, which are probably the most fastidious cells in the body, maintain almost no reserves of glycogen or fatty acids and rely almost entirely on a supply of glucose from the bloodstream. Liver cells supply glucose to actively contracting muscle cells and recycle the lactic acid produced by muscle cells back into glucose (Figure 2–42). All types of cells have their distinctive metabolic traits and cooperate extensively in the normal state, as well as in response to stress and starvation.

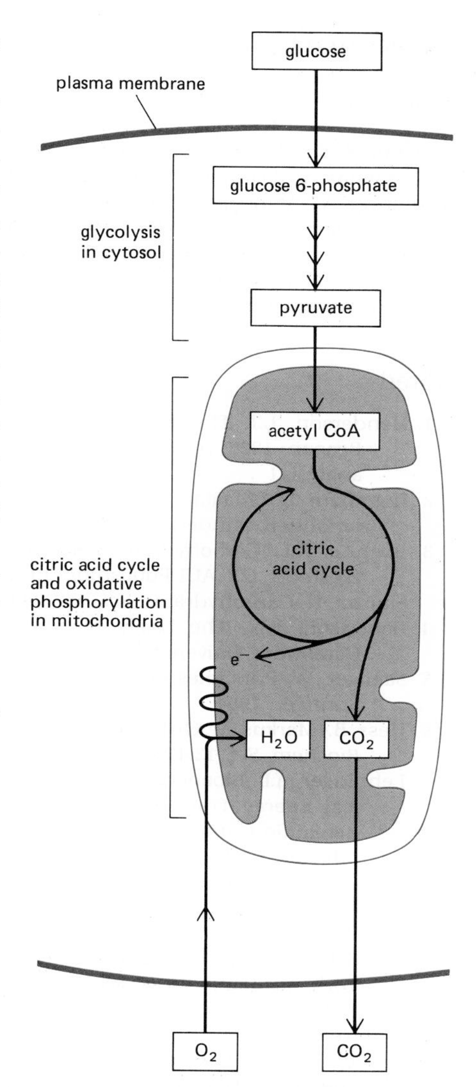

Figure 2–41 Segregation of the various steps in the breakdown of glucose in the eucaryotic cell. Glycolysis occurs in the cytosol, whereas the reactions of the citric acid cycle and oxidative phosphorylation take place only in mitochondria.

Summary

The many thousands of different chemical reactions carried out simultaneously by a cell are closely coordinated. A variety of control mechanisms regulate the activities of key enzymes in response to the changing conditions in the cell. One very common form of regulation is a rapidly reversible feedback inhibition exerted on the first enzyme of a pathway by the final product of that pathway. A longer-lasting form of regulation involves the chemical modification of one enzyme by another, often by phosphorylation. Combinations of regulatory mechanisms can produce major and long-lasting changes in the metabolism of the cell. Not all cellular reactions occur within the same intracellular compartment, and spatial segregation by internal membranes permits organelles to specialize in their biochemical tasks.

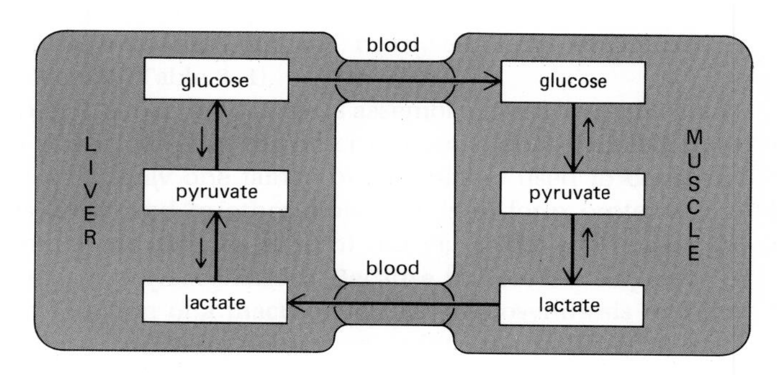

Figure 2–42 Schematic view of the metabolic cooperation between liver and muscle cells. The principal fuel of actively contracting muscle cells is glucose, much of which is supplied by liver cells. Lactic acid, the end product of anaerobic glucose breakdown by glycolysis in muscle, is converted back to glucose in the liver by the process of gluconeogenesis.

Table 3–1 Approximate Chemical Compositions of a Typical Bacterium and a Typical Mammalian Cell

	Percent of Total Cell Weight	
Component	E. Coli *Bacterium*	*Mammalian Cell*
H_2O	70	70
Inorganic ions (Na^+, K^+, Mg^{2+}, Ca^{2+}, Cl^-, etc.)	1	1
Miscellaneous small metabolites	3	3
Proteins	15	18
RNA	6	1.1
DNA	1	0.25
Phospholipids	2	3
Other lipids	—	2
Polysaccharides	2	2
Total cell volume:	2×10^{-12} cm^3	4×10^{-9} cm^3
Relative cell volume:	1	2000

Proteins, polysaccharides, DNA, and RNA are macromolecules. Lipids are not generally classed as macromolecules even though they share some of their features; for example, most are synthesized as linear polymers of a smaller molecule (the acetyl group on acetyl CoA) and self-assemble into larger structures (membranes). Note that water and protein comprise most of the mass of both mammalian and bacterial cells.

The Specific Interactions of a Macromolecule Depend on Weak Noncovalent Bonds[2]

A macromolecular chain is held together by *covalent* bonds, which are strong enough to preserve the sequence of subunits for long periods of time. Although the sequence of subunits determines the information content of a macromolecule, utilizing that information depends largely on much weaker, *noncovalent* bonds. These weak bonds form between different parts of the same macromolecule and between different macromolecules. They therefore play a major part in determining both the three-dimensional structure of macromolecular chains and how these structures interact with one another.

The noncovalent bonds encountered in biological molecules are usually classified into three types: **ionic bonds, hydrogen bonds,** and **van der Waals attractions.** Another important weak force is created by the three-dimensional structure of water, which tends to force hydrophobic groups together in order to minimize their disruptive effect on the hydrogen-bonded network of water molecules (see Panel 2–1, pp. 46–47). This expulsion from the aqueous solution generates what is sometimes thought of as a fourth kind of weak noncovalent bond. These four types of weak bonds are the subject of Panel 3–1, pp. 90–91.

In an aqueous environment each noncovalent bond is 30 to 300 times weaker than the typical covalent bonds that hold biological molecules together (Table 3–2) and only slightly stronger than the average energy of thermal collisions at 37°C. A single noncovalent bond—unlike a single covalent bond—is therefore too weak to withstand the thermal motions that tend to pull molecules apart, and large numbers of noncovalent bonds are needed to hold two molecular surfaces together. Large numbers of noncovalent bonds can form between two surfaces only when large numbers of atoms on the surfaces are precisely matched to each other (Figure 3–2), which accounts for the specificity of biological recognition, such as occurs between an enzyme and its substrates.

The weak noncovalent forces determine how different regions of the same macromolecule fit together, in addition to determining how that macromolecule will interact with other molecules. However, as explained at the top of Panel 3–1, atoms behave almost as if they were hard spheres with a definite radius (their

Table 3–2 Covalent and Noncovalent Chemical Bonds

Bond Type	Length (nm)	Strength (kcal/mole) In Vacuum	Strength (kcal/mole) In Water
Covalent	0.15	90	90
Ionic	0.25	80	3
Hydrogen	0.30	4	1
Van der Waals attraction (per atom)	0.35	0.1	0.1

The strength of a bond can be measured by the energy required to break it, here given in kilocalories per mole (kcal/mole). (*One kilocalorie* is the quantity of energy needed to raise the temperature of 1000 g of water by 1°C. An alternative unit in wide use is the kilojoule, kJ, equal to 0.24 kcal.) Individual bonds vary a great deal in strength, depending on the atoms involved and their precise environment, so that the above values are only a rough guide. Note that the aqueous environment in a cell will greatly weaken both the ionic and the hydrogen bonds between nonwater molecules (Panel 3–1, pp. 90–91). The bond length is the center-to-center distance between the two interacting atoms; the length given here for a hydrogen bond is that between its two nonhydrogen atoms.

"van der Waals radius"). The requirement that no two atoms overlap severely limits the number of three-dimensional arrangements of atoms (or **conformations**) that are possible for each polypeptide chain. Nevertheless, a long flexible chain such as a protein can still fold in an enormous number of ways, each conformation having a different set of weak intrachain interactions. In practice, however, most proteins in a cell fold stably in only one way: during the course of evolution, the sequence of amino acid subunits has been selected so that one conformation is able to form many more favorable intrachain interactions than any other.

A Helix Is a Common Structural Motif in Biological Structures Made from Repeated Subunits[3]

Biological structures are often formed by linking subunits that are very similar to each other—such as amino acids or nucleotides—into a long, repetitive chain (see p. 78). If all the subunits are identical, neighboring subunits in the chain will often fit together in only one way, adjusting their relative positions so as to minimize the free energy of the contact between them. Each subunit will be positioned in exactly the same way in relation to its neighboring subunits, so that subunit 3 will fit onto subunit 2 in the same way that subunit 2 fits onto subunit 1, and so on. Because it is very rare for subunits to join up in a straight line, this arrangement will generally result in a **helix**—a regular structure that resembles a spiral stair-

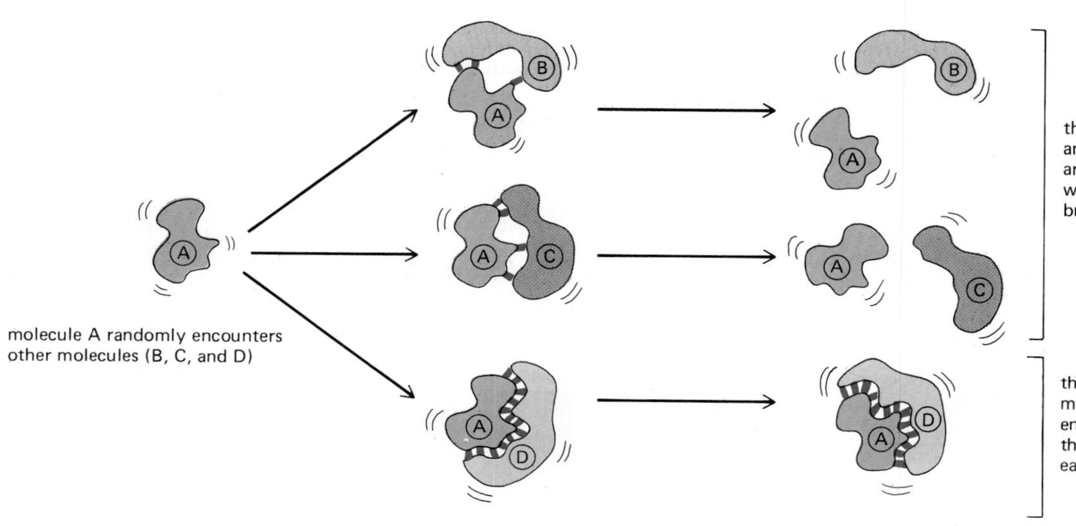

Figure 3–2 How weak bonds mediate recognition between macromolecules.

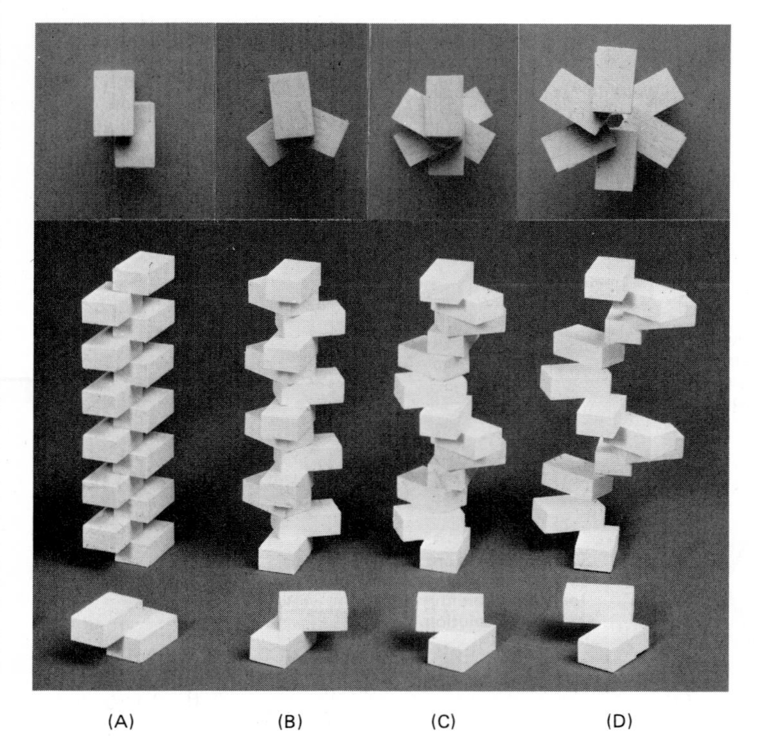

Figure 3–3 A helix will form when a series of subunits bind to each other in a regular way. In the foreground the interaction between two subunits is shown; behind it are the helices that result. These helices have (A) two, (B) three, and (C) and (D) six subunits per turn. At the top, the arrangement of subunits has been photographed from directly above the helix. Note that the helix in (D) has a wider path than that in (C).

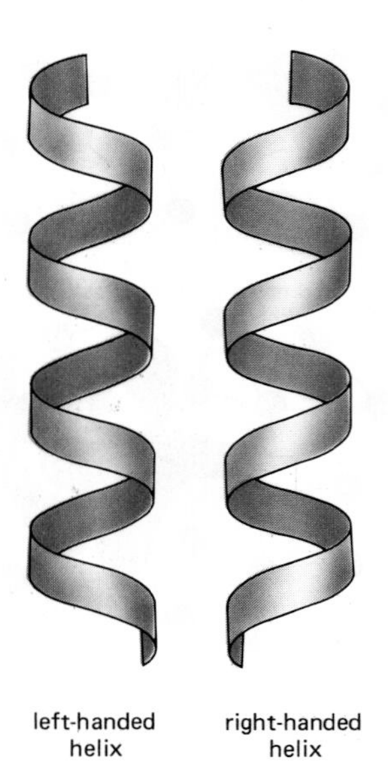

Figure 3–4 Comparison of a left-handed and a right-handed helix. As a reference, it is useful to remember that standard screws, which insert when turned clockwise, are right-handed. Note that a helix preserves the same handedness when it is turned upside down.

case, as illustrated in Figure 3–3. Depending on the twist of the staircase, a helix is said to be either right-handed or left-handed (Figure 3–4). Handedness is not affected by turning the helix upside down, but it is reversed if the helix is reflected in a mirror.

Helices occur commonly in biological structures, whether the subunits are small molecules that are covalently linked together (as in DNA) or large protein molecules that are linked by noncovalent forces (as in actin filaments). This is not surprising. A helix is an unexceptional structure, generated simply by stacking many similar subunits next to each other, each in the same strictly repeated relationship to the one before.

Diffusion Is the First Step to Molecular Recognition[4]

Before two molecules can bind to each other, they must come into close contact. This is achieved by the thermal motions that cause molecules to wander, or *diffuse,* from their starting positions. As the molecules in a liquid rapidly collide and bounce off one another, an individual molecule moves first one way and then another, its path constituting a "random walk" (Figure 3–5). The average distance that each type of molecule travels from its starting point is proportional to the square root of the time involved: that is, if it takes a particular molecule 1 second on average to go 1 μm, it will go 2 μm in 4 seconds, 10 μm in 100 seconds, and so on. Diffusion is therefore an efficient way for molecules to move limited distances but an inefficient way for molecules to move long distances.

Figure 3–5 A random walk. Molecules in solution move in a random fashion due to the continual buffeting they receive in collisions with other molecules. This movement allows small molecules to diffuse from one part of the cell to another in a surprisingly short time: such molecules will generally diffuse across a typical animal cell in less than a second.

Experiments performed by injecting fluorescent dyes and other labeled molecules into cells show that the diffusion of small molecules through the cytoplasm is nearly as rapid as it is in water. A molecule the size of ATP, for example, requires only about 0.2 second to diffuse an average distance of 10 μm—the diameter of a small animal cell. Large macromolecules, however, move much more slowly. Not only is their diffusion rate intrinsically slower, but their movement is retarded by frequent collisions with many other macromolecules that are held in place by molecular associations in the cytoplasm (Figure 3–6).

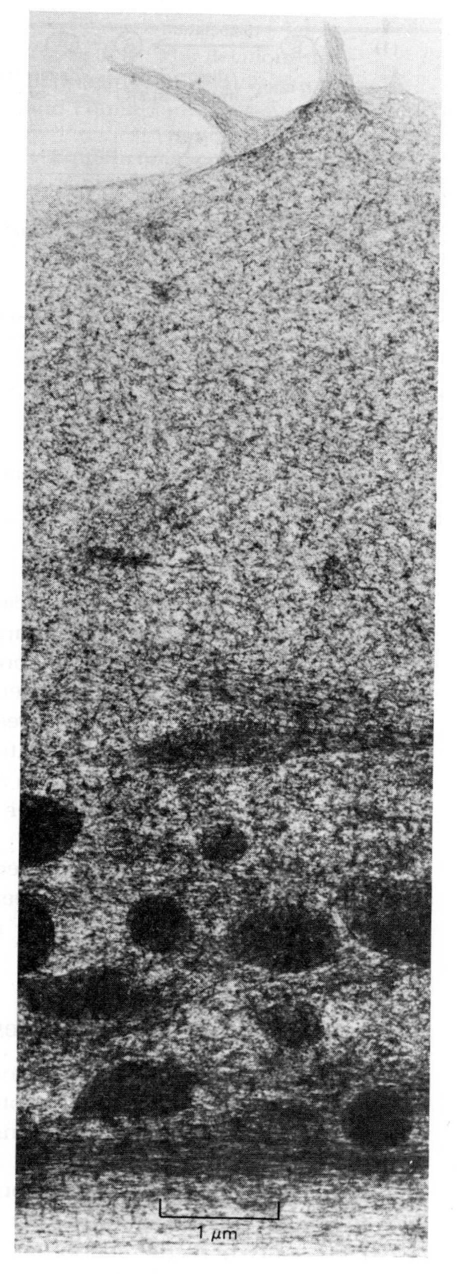

Figure 3–6 An electron micrograph of a region of the cytoplasm of an animal cell illustrating the high concentration of proteins it contains. Macromolecules diffuse relatively slowly in the cytoplasm because they interact with many other macromolecules; small molecules diffuse nearly as rapidly as they do in water. The cell shown in this micrograph was prepared by a special rapid freezing method that optimally preserves cytoplasmic structures. (Reproduced from P.C. Bridgman and T.S. Reese, *J. Cell Biol.* 99:1655–1668, 1984. By copyright permission of the Rockefeller University Press.)

Thermal Motions Bring Molecules Together and Then Pull Them Apart[5]

Encounters between two macromolecules or between a macromolecule and a small molecule occur randomly through simple diffusion. An encounter may lead immediately to the formation of a complex, in which case the rate of complex formation is said to be *diffusion-limited.* Alternatively, the rate of complex formation may be slower, requiring some adjustment of the structure of one or both molecules before the interacting surfaces can fit together, so that most often the two colliding molecules will bounce off each other without sticking. In either case, once the two interacting surfaces have come sufficiently close together, they form multiple weak bonds with each other that persist until random thermal motion causes the molecules to dissociate again.

In general, the stronger the binding of the molecules in the complex, the slower their rate of dissociation. At one extreme the total energy of the bonds formed is negligible compared with that of thermal motion, and the two molecules dissociate as rapidly as they came together. At the other extreme the total bond energy is so high that dissociation rarely occurs. The precise strength of the bonding between two molecules is a useful index of the specificity of the recognition process.

To illustrate how the binding strength is measured, let us consider a reaction in which molecule A binds to molecule B. The reaction will proceed until it reaches an *equilibrium point,* at which the rates of formation and dissociation are equal. The concentrations of A, B, and the complex AB at this point can be used to determine an **equilibrium constant (*K*)** for the reaction, as explained in Figure 3–7. This constant is sometimes termed the **affinity constant** and is commonly employed as a measure of the strength of binding between two molecules: the *stronger* the binding, the *larger* is the value of the affinity constant.

The equilibrium constant of a reaction in which two molecules bind to each other is related directly to the standard free-energy change for the binding (ΔG°) by the equation described in Table 3–3. The table also lists the ΔG° values corresponding to a range of K values. Affinity constants for simple binding interactions in biological systems often range between 10^3 and 10^{12} liters/mole; this corresponds to binding energies in the range 4–17 kcal/mole, which could arise from 4 to 17 average hydrogen bonds.

The strongest interactions occur whenever a biological function requires that two macromolecules remain tightly associated for a long period of time—for example, when a gene regulatory protein binds to DNA to turn off a gene (see p. 558). Weaker interactions occur when the function demands a rapid change in the structure of a complex—for example, when two interacting proteins change partners during the movements of a protein machine (see p. 131).

Atoms and Molecules Are in Constant Motion[6]

The chemical reactions in a cell occur at amazingly fast rates. A typical enzyme molecule, for example, will catalyze on the order of 1000 reactions per second, and rates of more than 10^6 reactions per second are achieved by some enzymes, such as catalase. Since each reaction requires a separate encounter between an enzyme and a substrate molecule, such rates are possible only because the molecules are moving so rapidly. Molecular motions can be classified broadly into three kinds: (1) the movement of a molecule from one place to another (translational motion), (2) the rapid back-and-forth movement of covalently linked atoms with respect to one another (vibrations), and (3) rotations. All of these motions are important in bringing the surfaces of interacting molecules together.

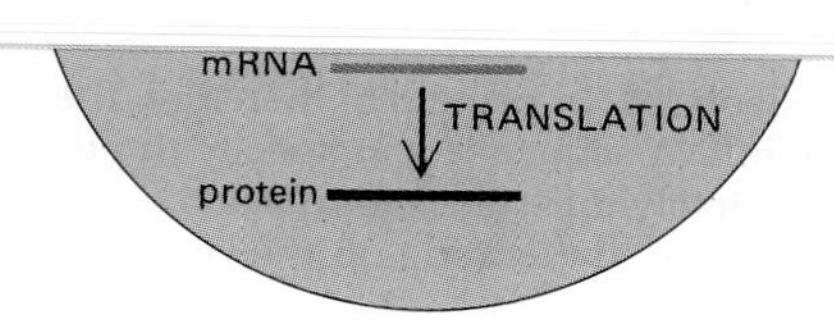

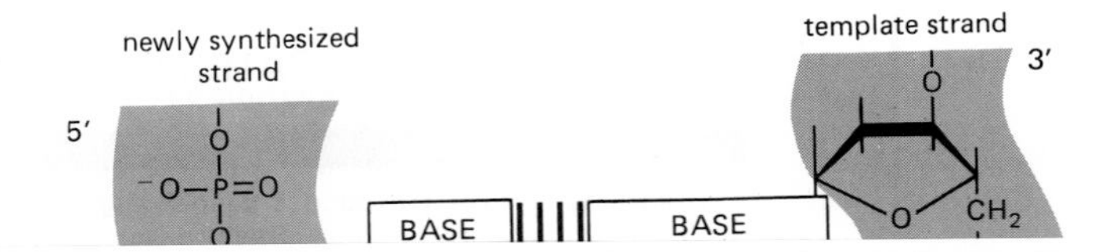

Figure 3–10 Addition of a deoxyribonucleotide to the 3′ end of a polynucleotide chain is the fundamental reaction by which DNA is synthesized. As shown, base-pairing between this incoming

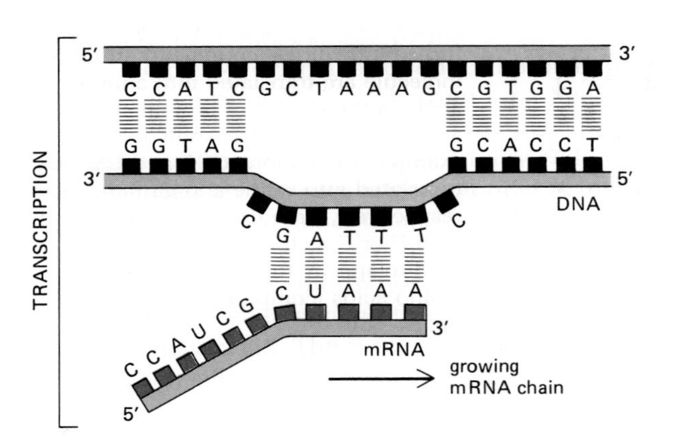

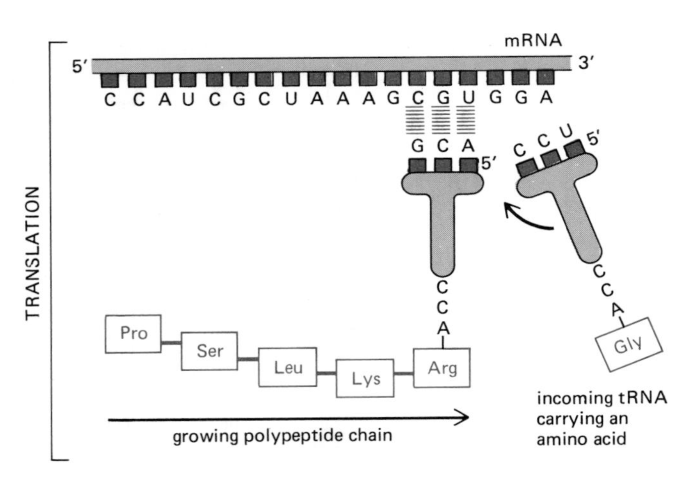

Figure 3–17 Information flow in protein synthesis. The nucleotides in an mRNA molecule are joined together to form a complementary copy of a segment of one strand of DNA. They are then matched three at a time to complementary sets of three nucleotides in the anticodon regions of tRNA molecules. At the other end of each type of tRNA molecule, a specific amino acid is held in a high-energy linkage, and when matching occurs, this amino acid is added to the end of the growing polypeptide chain. Thus, translation of the mRNA nucleotide sequence into an amino acid sequence depends on complementary base-pairing between codons in the mRNA and corresponding tRNA anticodons. The molecular basis of information transfer in translation is therefore very similar to that in DNA replication and transcription. Note that the mRNA is both synthesized and translated starting from its 5′ end.

dues at either end of the "L" are especially important for the function of the tRNA molecule in protein synthesis: one forms the *anticodon* that base-pairs to a complementary triplet in an mRNA molecule (the codon), while the *CCA sequence* at the 3′ end of the molecule is attached covalently to a specific amino acid (Figure 3–16A).

The RNA Message Is Read from One End to the Other by a Ribosome[18]

The codon recognition process by which genetic information is transferred from mRNA via tRNA to protein depends on the same type of base-pair interactions that mediate the transfer of genetic information from DNA to DNA and from DNA to RNA (Figure 3–17). But the mechanics of ordering the tRNA molecules on the mRNA are complicated and require a **ribosome,** a complex of more than 50 different proteins associated with several structural RNA molecules (rRNAs). Each ribosome is a large protein-synthesizing machine on which tRNA molecules position themselves so as to read the genetic message encoded in an mRNA molecule. The ribosome first finds a specific start site on the mRNA that sets the reading frame and determines the amino-terminal end of the protein. Then, as the ribosome moves along the mRNA molecule, it translates the nucleotide sequence into an amino acid sequence one codon at a time, using tRNA molecules to add amino acids to the growing end of the polypeptide chain (Figure 3–18). When a ribosome reaches the end of the message, both it and the freshly made carboxyl end of the protein are released from the 3′ end of the mRNA molecule into the cytoplasm.

Ribosomes operate with remarkable efficiency: in 1 second a single bacterial ribosome adds about 20 amino acids to a growing polypeptide chain. Ribosome structure and the mechanism of protein synthesis are discussed in Chapter 5.

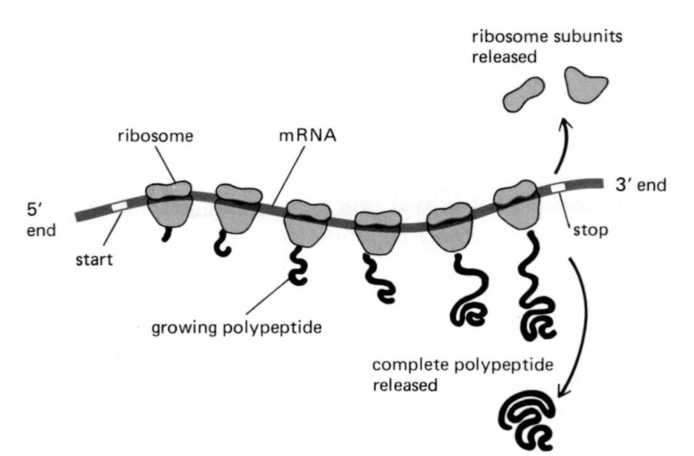

Figure 3–18 Synthesis of a protein by ribosomes attached to an mRNA molecule. Ribosomes become attached to a start signal near the 5′ end of the mRNA molecule and then move toward the 3′ end, synthesizing protein as they go. A single mRNA will usually have a number of ribosomes traveling along it at the same time, each making a separate but identical polypeptide chain; the entire structure is known as a *polyribosome.*

Some RNA Molecules Function as Catalysts[19]

RNA molecules have commonly been viewed as strings of nucleotides with a relatively uninteresting chemistry. In 1981 this view was shattered by the discovery of a catalytic RNA molecule with the type of sophisticated chemical reactivity that biochemists had previously associated only with proteins. The ribosomal RNA molecules of the ciliated protozoon *Tetrahymena* are initially synthesized as a large precursor from which one of the rRNAs had been shown to be produced by an RNA splicing reaction. The surprise came with the discovery that this splicing can occur *in vitro* in the absence of protein. It was subsequently shown that the intron sequence itself has an enzymelike catalytic activity that carries out the two-step reaction illustrated in Figure 3–19. The 400-nucleotide-long intron sequence was then synthesized in a test tube and shown to fold up to form a complex surface that can function like an enzyme in reactions with other RNA molecules. For example, it can bind two specific substrates tightly—a guanine nucleotide and an RNA chain—and catalyze their covalent attachment so as to sever the RNA chain at a specific site (Figure 3–20).

In this model reaction, which mimics the first step in Figure 3–19, the same intron sequence acts repeatedly to cut many RNA chains. Although RNA splicing is most commonly achieved by means that are not autocatalytic (see p. 533), self-splicing RNAs with intron sequences related to that in *Tetrahymena* have been discovered in other types of cells, including fungi and bacteria. This suggests that these RNA sequences may have arisen before the eucaryotic and procaryotic lineages diverged about 1.5 billion years ago.

Several other families of catalytic RNAs have recently been discovered. Most tRNAs, for example, are initially synthesized as a larger precursor RNA, and an RNA molecule has been shown to play the major catalytic role in an RNA-protein complex that recognizes these precursors and cleaves them at specific sites. A catalytic RNA sequence also plays an important part in the life cycle of many plant

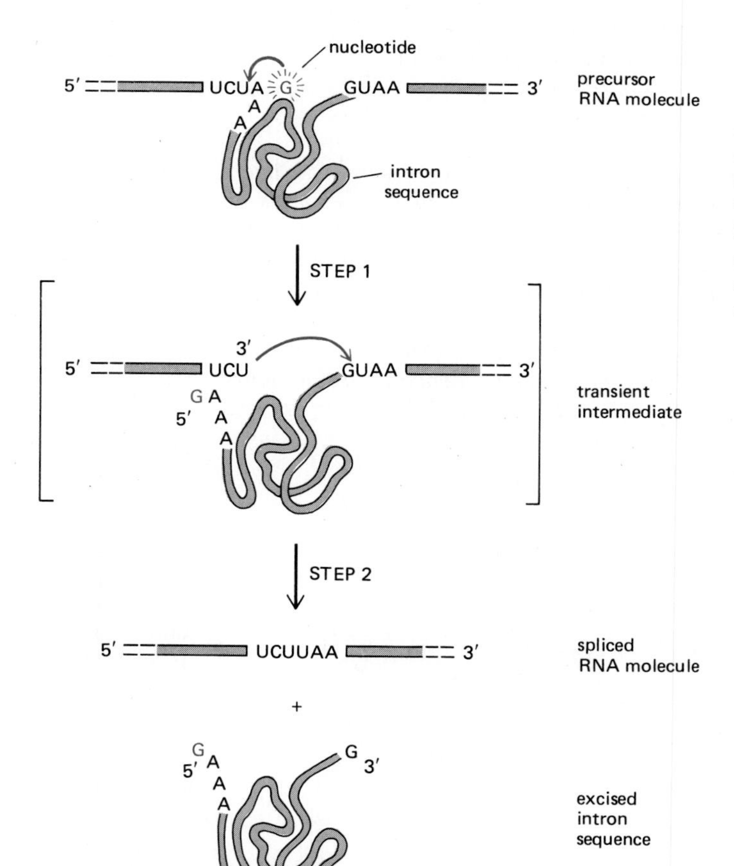

Figure 3–19 Diagram of the self-splicing reaction in which an intron sequence catalyzes its own excision from a *Tetrahymena* ribosomal RNA molecule. As shown, the reaction is initiated when a G nucleotide is added to the intron sequence, cleaving the RNA chain in the process; the newly created 3′ end of the RNA chain then attacks the other side of the intron to complete the reaction.

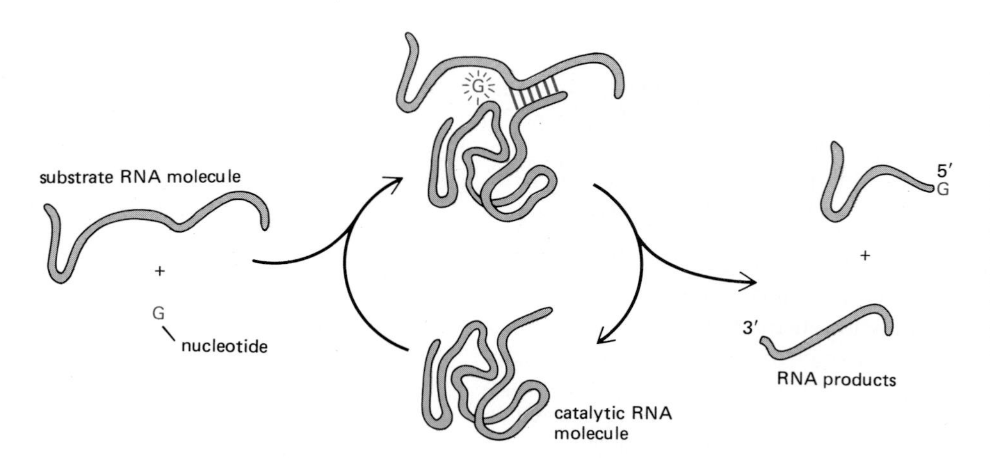

Figure 3–20 An enzymelike reaction catalyzed by the purified *Tetrahymena* intron sequence. In this reaction, which corresponds to the first step in Figure 3–19, both a specific substrate RNA molecule and a G nucleotide become tightly bound to the surface of the catalytic RNA molecule. The nucleotide is then covalently attached to the substrate RNA molecule, cleaving it at a specific site. The release of the resulting two RNA chains frees the intron sequence for further cycles of reaction.

viroids, and a similar sequence appears in a frog RNA molecule, although its role in the frog is unknown. Most remarkably, ribosomes are now suspected to function largely by RNA-based catalysis, with the ribosomal proteins playing a supporting role to the ribosomal RNAs, which make up more than half the mass of the ribosome.

How is it possible for an RNA molecule to act like an enzyme? The example of tRNA indicates that RNA molecules can fold up in highly specific ways. A proposed two-dimensional structure for the core of the self-splicing *Tetrahymena* intron sequence is shown in Figure 3–21. Interactions between different parts of this RNA molecule (analogous to the unusual hydrogen bonds in tRNA molecules—see Figure 3–16) are responsible for folding it further to create a complex three-dimensional surface with catalytic activity. An unusual juxtaposition of atoms can strain covalent bonds and thereby make selected atoms in the folded RNA chain unusually reactive.

As explained in Chapter 1, the discovery of catalytic RNA molecules has profoundly changed our views of how the first living cells arose (see p. 7).

Summary

Genetic information is carried in the linear sequence of nucleotides in DNA. Each molecule of DNA is a double helix formed from two complementary strands of nucleotides held together by hydrogen bonds between G-C and A-T base pairs. Duplication of the genetic information occurs by the polymerization of a new complementary strand onto each of the old strands of the double helix during DNA replication.

The expression of the genetic information stored in DNA involves the translation of a linear sequence of nucleotides into a co-linear sequence of amino acids in proteins. A limited segment of DNA is first copied into a complementary strand of RNA. This primary RNA transcript is spliced to remove intron sequences, pro-

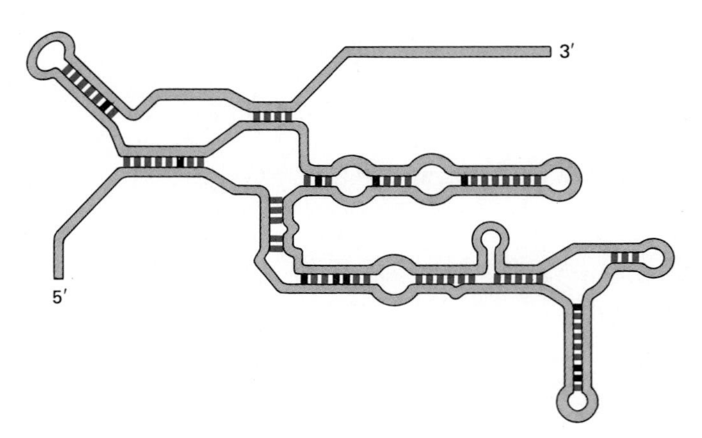

Figure 3–21 A two-dimensional view of the catalytic core of the intron RNA sequence illustrated in Figures 3–19 and 3–20. Normal complementary base pairs are shown in color, while weaker base-pair interactions, such as G-U pairs, are shown in black. This molecule is about 240 nucleotides long; it is normally tightly folded in three dimensions, but its precise conformation is unknown. Self-splicing RNAs with related structures have been discovered in the mitochondria of fungi and in a bacterial virus (bacteriophage T4).

ducing an mRNA molecule. Finally, the mRNA is translated into protein in a complex set of reactions that occur on a ribosome. The amino acids used for protein synthesis are first attached to a family of tRNA molecules, each of which recognizes, by complementary base-pairing interactions, particular sets of three nucleotides in the mRNA. The sequence of nucleotides in the mRNA is then read from one end to the other in sets of three, according to a universal genetic code.

Other RNA molecules in cells function as enzymelike catalysts. These RNA molecules fold up to create a surface containing nucleotides that have become unusually reactive.

Protein Structure[20]

To a large extent cells are made of protein, which constitutes more than half of their dry weight (see Table 3–1). Proteins determine the shape and structure of the cell and also serve as the main instruments of molecular recognition and catalysis. Although DNA stores the information required to make a cell, it has little direct influence on cellular processes. The gene for hemoglobin, for example, cannot carry oxygen: that is a property of the protein specified by the genes. In computer terminology, the DNA and the mRNA represent the "software"—instructions that a cell receives from its parent. Proteins and catalytic RNA molecules constitute the "hardware"—the machinery that executes the program stored in the memory.

DNA and RNA are chains of nucleotides that are chemically very similar to one another. In contrast, proteins are made from an assortment of 20 very different amino acids, each with a distinct chemical personality (see Panel 2–5, pp. 54–55). This variety allows for enormous versatility in the chemical properties of different proteins, and it presumably explains why evolution has selected proteins rather than RNA molecules to catalyze most cellular reactions.

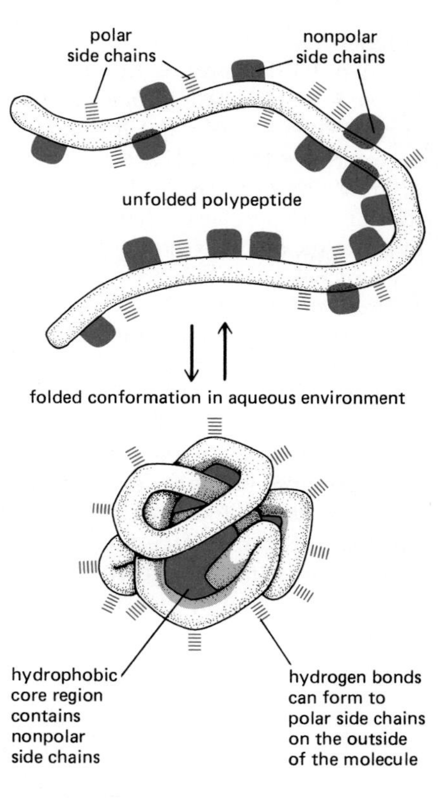

Figure 3–22 How a protein folds into a globular conformation. The polar amino acid side chains tend to gather on the outside of the protein, where they can interact with water; the nonpolar amino acid side chains are buried on the inside to form a hydrophobic core that is "hidden" from water.

The Shape of a Protein Molecule Is Determined by Its Amino Acid Sequence[21]

Many of the bonds in a long polypeptide chain allow free rotation of the atoms they join, giving the protein backbone great flexibility. In principle, then, any protein molecule could adopt an almost unlimited number of shapes (*conformations*). Most polypeptide chains, however, fold into only one particular conformation determined by their amino acid sequence. This is because the side chains of the amino acids associate with one another and with water to form various weak noncovalent bonds (see Panel 3–1, pp. 90–91). Provided that the appropriate side chains are present at crucial positions in the chain, large forces are developed that make one particular conformation especially stable.

Most proteins fold spontaneously into their correct shape. By treatment with certain solvents, a protein can be unfolded, or *denatured,* to give a flexible polypeptide chain that has lost its native conformation. When the denaturing solvent is removed, the protein will usually refold spontaneously into its original conformation, indicating that all the information necessary to specify the shape of a protein is contained in the amino acid sequence itself.

One of the most important factors governing the folding of a protein is the distribution of its polar and nonpolar side chains. The many hydrophobic side chains in a protein tend to be pushed together in the interior of the molecule, which enables them to avoid contact with the aqueous environment (just as oil droplets coalesce after being mechanically dispersed in water). By contrast, the polar side chains tend to arrange themselves near the outside of the protein molecule, where they can interact with water and with other polar molecules (Figure 3–22). Since the peptide bonds are themselves polar, they tend to interact both with one another and with polar side chains to form hydrogen bonds (Figure 3–23); nearly all polar residues buried within the protein are paired in this way. Hydrogen bonds thus play a major part in holding together different regions of

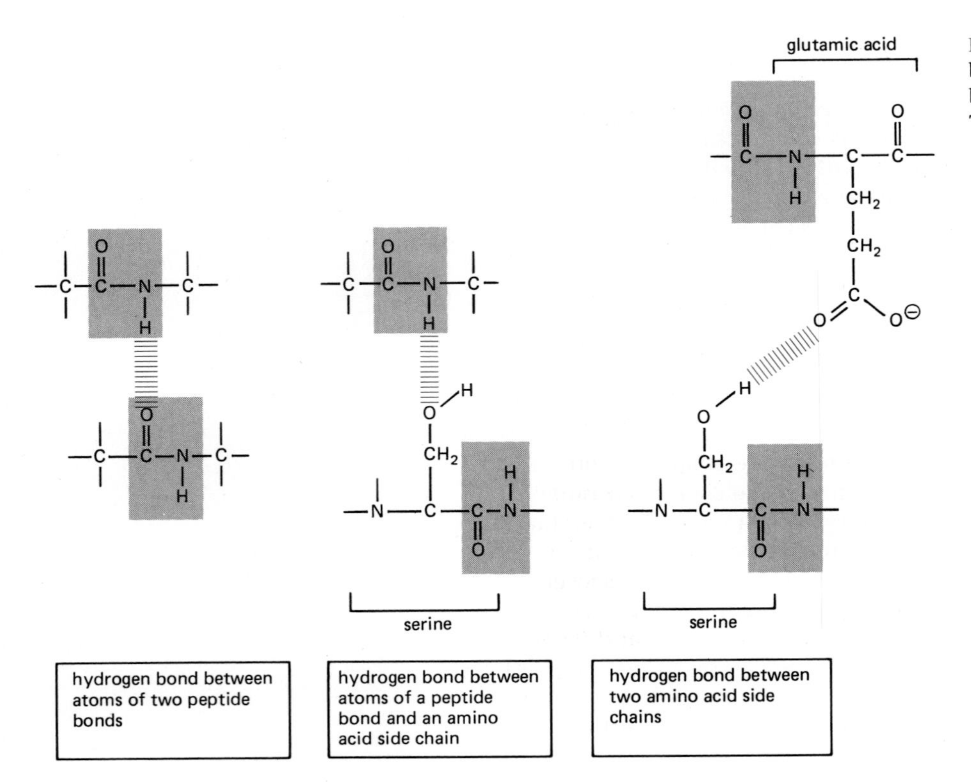

Figure 3–23 Some of the hydrogen bonds (shown in color) that can form between the amino acids in a protein. The peptide bonds are shaded in gray.

polypeptide chain in a folded protein molecule, and they are crucially important for many of the binding interactions that occur on protein surfaces.

Secreted or cell-surface proteins often form additional *covalent* intrachain bonds. For example, the formation of **disulfide bonds** (also called S—S bonds) between the two —SH groups of neighboring cysteine residues in a folded polypeptide chain (Figure 3–24) often serves to stabilize the three-dimensional structure of extracellular proteins. These bonds are not required for the specific folding of proteins, since folding occurs normally in the presence of reducing agents that prevent S—S bond formation. In fact, S—S bonds are rarely, if ever, formed in protein molecules in the cytosol because the high cytosolic concentration of —SH reducing agents breaks such bonds (see p. 445).

The net result of all the individual amino acid interactions is that most protein molecules fold up spontaneously into precisely defined conformations, usually compact and globular but sometimes long and fibrous. The inner core is composed of clustered hydrophobic side chains—packed into a tight, nearly crystalline arrangement—while a very complex and irregular exterior surface is formed by the more polar side chains. The positioning and chemistry of the different atoms on this intricate surface make each protein unique and enable it to bind specifically to other macromolecular surfaces and to certain small molecules (see below). From both a chemical and a structural standpoint, proteins are the most sophisticated molecules known.

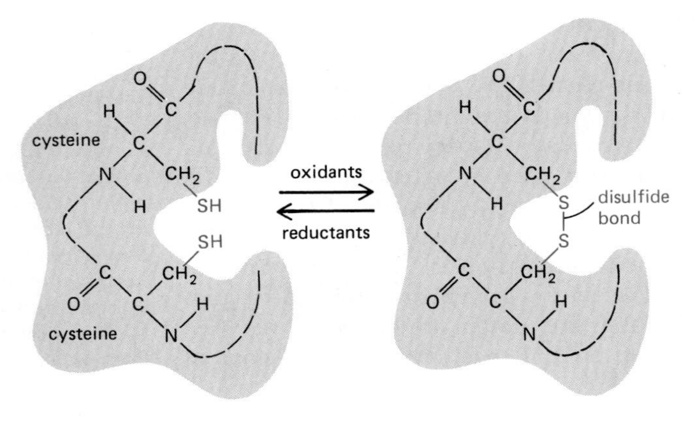

Figure 3–24 The formation of a covalent disulfide bond between the side chains of neighboring cysteine residues in a protein.

Common Folding Patterns Recur in Different Protein Chains[22]

Although all the information required for the folding of a protein chain is contained in its amino acid sequence, we have not yet learned how to "read" this information so as to predict the detailed three-dimensional structure of a protein whose sequence is known. Consequently, the folded conformation can be determined only by an elaborate *x-ray diffraction analysis* performed on crystals of the protein. So far, more than 100 types of proteins have been completely analyzed by this technique. Each of these proteins has a specific conformation so intricate and irregular that it would require a chapter to describe any one of them in full three-dimensional detail.

When the three-dimensional structures of different protein molecules are compared, it becomes clear that, although the overall conformation of each protein is unique, several folding patterns recur repeatedly in parts of these macromolecules. Two patterns are particularly common because they result from regular hydrogen-bonding interactions between the peptide bonds themselves rather than between the side chains of particular amino acids. Both patterns were correctly predicted in 1951 from model-building studies based on the different x-ray diffraction patterns of silk and hair. The two regular folding patterns discovered are now known as the *β sheet,* which occurs in the protein fibroin, found in silk; and the *α helix,* which occurs in the protein α-keratin, found in skin and its appendages, such as hair, nails, and feathers.

The core of most (but not all) globular proteins contains extensive regions of **β sheet.** In the example illustrated in Figure 3–25, which shows part of an antibody molecule, an *antiparallel β sheet* is formed when an extended polypeptide chain folds back and forth upon itself, with each section of the chain running in the direction opposite to that of its immediate neighbors. This gives a very rigid structure held together by hydrogen bonds that connect the peptide bonds in neighboring chains. The antiparallel β sheet and the closely related *parallel β sheet* (which is formed by regions of polypeptide chain that run in the same direction) frequently serve as the framework around which globular proteins are constructed.

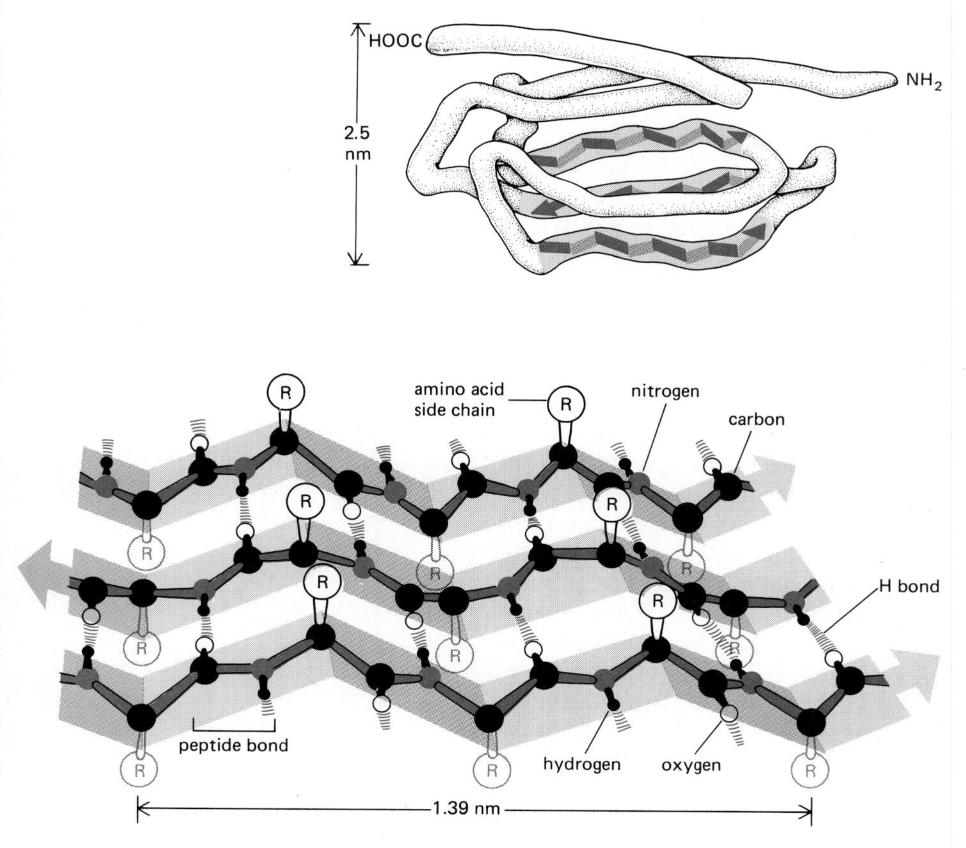

Figure 3–25 A β sheet is a common structure formed by parts of the polypeptide chain in globular proteins. At the top, a domain of 115 amino acids from an immunoglobulin molecule is shown; it consists of a sandwichlike structure of two β sheets, one of which is drawn in color. At the bottom, a perfect antiparallel β sheet is shown in detail. Note that every peptide bond is hydrogen-bonded to a neighboring peptide bond. The actual sheet structures in globular proteins are usually less regular than the β sheet shown here, and many sheets are slightly twisted (see Figure 3–27).

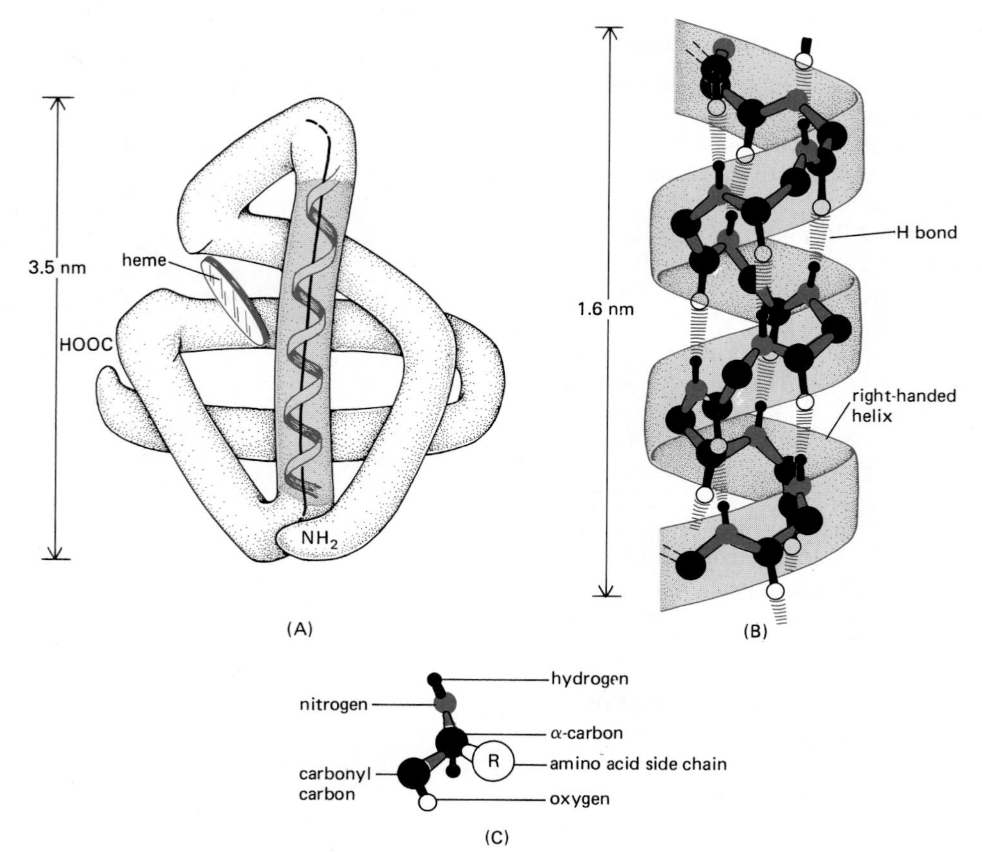

Figure 3–26 An α helix is another common structure formed by parts of the polypeptide chain in proteins. (A) The oxygen-carrying molecule myoglobin (153 amino acids long) is shown, with one region of α helix outlined in color. (B) A perfect α helix is shown in detail. As in the β sheet, every peptide bond is hydrogen-bonded to a neighboring peptide bond. (C) The atoms in an amino acid residue. Note that for clarity in (B) both the side chains [which protrude radially along the outside of the helix and are denoted by R in (C)] and the hydrogen atom are omitted on the α-carbon atom of each amino acid (see also Figure 3–27).

An **α helix** is generated when a single polypeptide chain turns regularly about itself to make a rigid cylinder in which each peptide bond is regularly hydrogen-bonded to other peptide bonds nearby in the chain. Many globular proteins contain short regions of such α helices (Figure 3–26) and those portions of a transmembrane protein that cross the lipid bilayer are nearly always α helices because of the constraints imposed by the hydrophobic lipid environment (see p. 285). In aqueous environments an isolated α helix is usually not stable on its own. However, two identical α helices that have a repeating arrangement of nonpolar side chains will twist around each other gradually to form a particularly stable structure known as a *coiled-coil* (see p. 617). Long rodlike coiled-coils are found in many fibrous proteins, such as the intracellular α-keratin fibers that reinforce skin and its appendages. Space-filling representations of an α helix and a β sheet from actual proteins are shown with and without their side chains in Figure 3–27.

Proteins Are Amazingly Versatile Molecules[23]

Because of the variety of their amino acid side chains, proteins are remarkably versatile with respect to the types of structures they can form. Contrast, for example, two abundant proteins secreted by cells in connective tissue—collagen and elastin—both present in the extracellular matrix. In **collagen** molecules three separate polypeptide chains, each rich in the amino acid proline and containing the amino acid glycine at every third residue, are wound around one another to generate a regular triple helix (see p. 809). These collagen molecules are, in turn, packed together into fibrils in which adjacent molecules are tied together by covalent cross-links between neighboring lysine residues, giving the fibril enormous tensile strength (Figure 3–28).

Elastin is at the opposite extreme. Its relatively loose and unstructured polypeptide chains are covalently cross-linked to generate a rubberlike elastic meshwork that enables tissues such as arteries and lungs to deform and stretch without damage (see p. 815). As illustrated in Figure 3–29, the elasticity is due to the ability

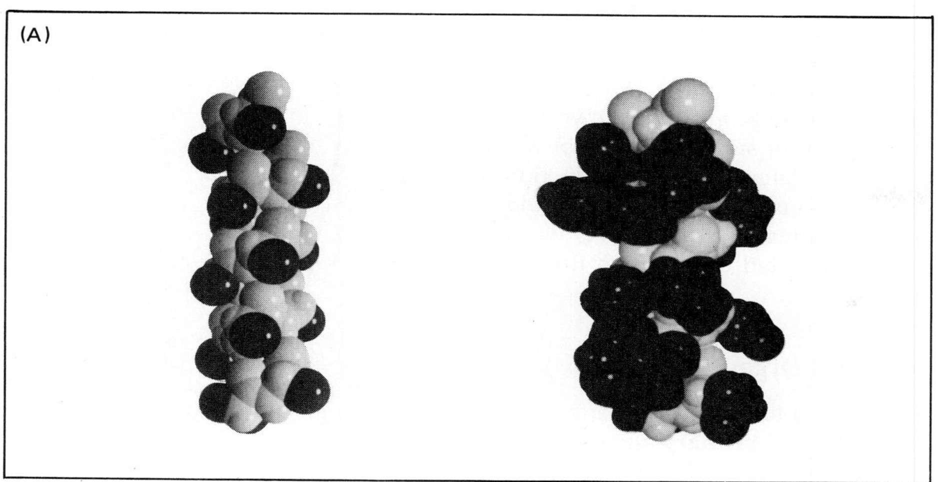

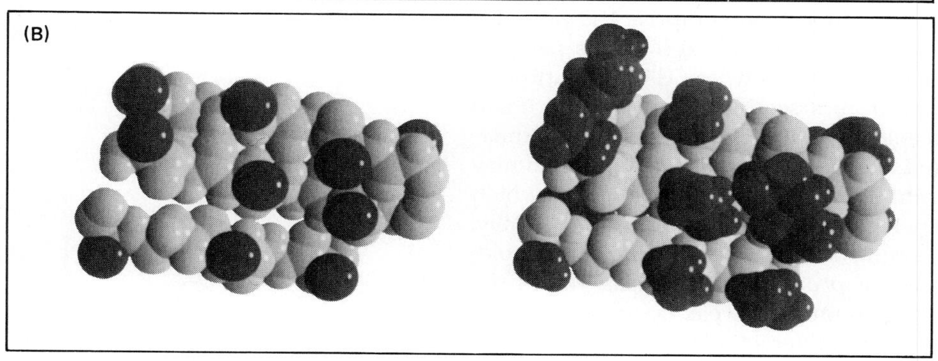

Figure 3–27 Space-filling models of an α helix and a β sheet with (*right*) and without (*left*) their amino acid side chains. (A) An α helix (part of the structure of myoglobin). (B) A region of β sheet (part of the structure of an immunoglobulin domain). In the photographs on the left, each side chain is represented by a single darkly shaded atom (the R groups in Figures 3–25 and 3–26); the entire side chain is shown on the right. (Courtesy of Richard J. Feldmann.)

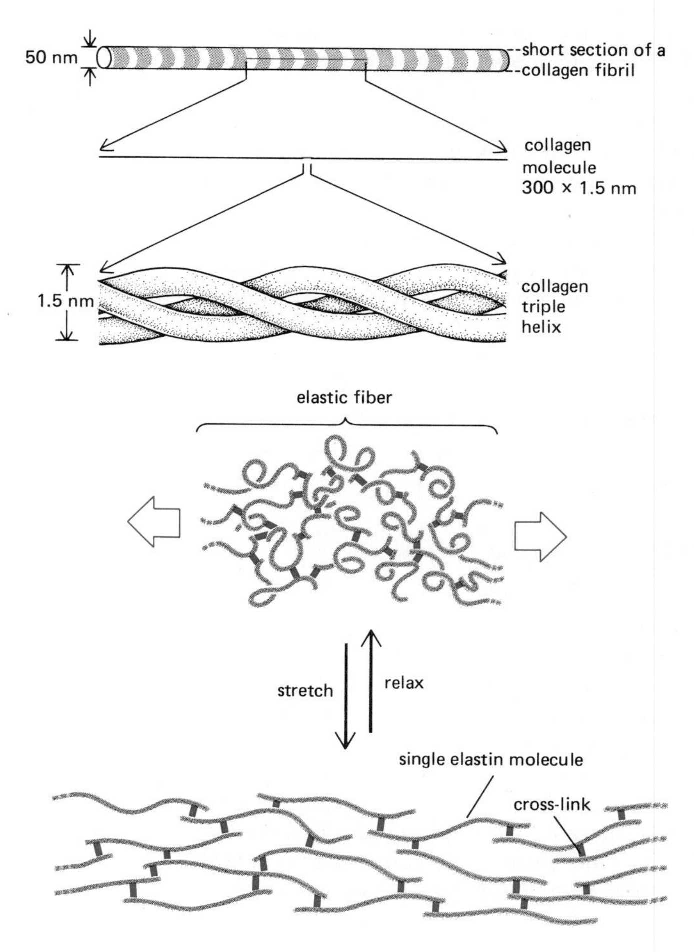

Figure 3–28 Collagen is a triple helix formed by three extended protein chains that wrap around each other. Many rodlike collagen molecules are cross-linked together in the extracellular space to form inextensible collagen fibrils (*top*) that have the tensile strength of steel.

Figure 3–29 Elastin polypeptide chains are cross-linked together to form elastic fibers. Each elastin molecule uncoils into a more extended conformation when the fiber is stretched. The striking contrast between the physical properties of elastin and collagen is due entirely to their very different amino acid sequences.

of individual protein molecules to uncoil reversibly whenever a stretching force is applied.

It is remarkable that the same basic chemical structure—a chain of amino acids—can form so many different structures: a rubberlike elastic meshwork (elastin), an inextensible cable with the tensile strength of steel (collagen), or any of the wide variety of catalytic surfaces on the globular proteins that function as enzymes. Figure 3–30 illustrates and compares the range of shapes that could, in theory, be adopted by a polypeptide chain 300 amino acids long. As we have already emphasized, the conformation actually adopted depends on the amino acid sequence.

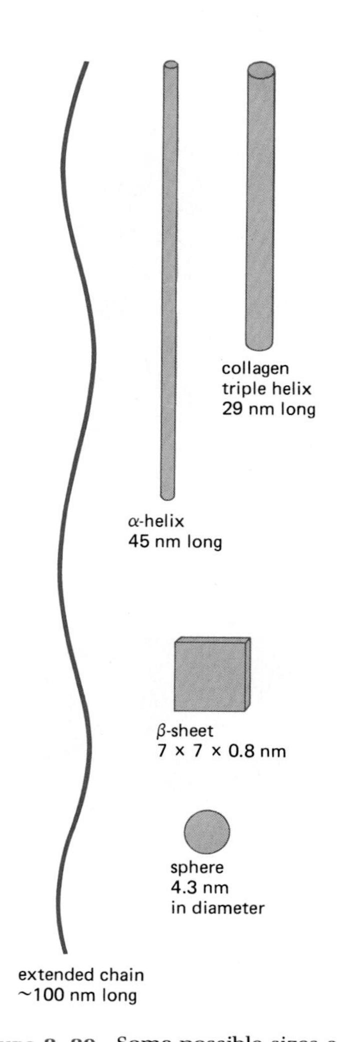

Figure 3–30 Some possible sizes and shapes of a protein molecule 300 amino acid residues long. The structure formed is determined by the amino acid sequence. (Adapted from D.E. Metzler, Biochemistry. New York: Academic Press, 1977.)

Proteins Have Different Levels of Structural Organization[24]

In describing the structure of a protein, it is helpful to distinguish various levels of organization. The amino acid sequence is called the **primary structure** of the protein. Regular hydrogen-bond interactions within contiguous stretches of polypeptide chain give rise to α helices and β sheets, which comprise the protein's **secondary structure.** Certain combinations of α helices and β sheets pack together to form compactly folded globular units, each of which is called a protein **domain.** Domains are usually constructed from a section of polypeptide chain that contains between 50 and 350 amino acids, and they seem to be the modular units from which proteins are constructed (see below). While small proteins may contain only a single domain, larger proteins contain a number of domains, which are usually connected by relatively open lengths of polypeptide chain. Finally, individual polypeptides often serve as subunits for the formation of larger molecules, sometimes called *protein assemblies* or *protein complexes,* in which the subunits are bound to one another by a large number of weak noncovalent interactions (see p. 88); in extracellular proteins these interactions are often stabilized by disulfide bonds.

The three-dimensional structure of a protein can be illustrated in various ways. Consider the unusually small protein basic pancreatic trypsin inhibitor (BPTI), which contains 58 amino acid residues folded into one domain. BPTI can be shown as a stereo pair displaying all of its nonhydrogen atoms (Figure 3–31A) or as an accurate space-filling model, where most of the details are obscured (Figure 3–31B). Alternatively, it can be shown more schematically, with all of the side chains and actual atoms omitted so that it is easier to follow the course of the main polypeptide chain (Figures 3–31C, D, and E). Such schematic drawings are essential for displaying the structure of proteins, which are usually much larger than BPTI, making it possible to trace the irregular course of the polypeptide chain within each domain (Figure 3–32).

Figure 3–33 shows how the structure of a large protein can be resolved into several levels of organization, each level constructed from the one below it in a hierarchical fashion. These levels of increased organizational complexity may correspond to the steps by which a newly synthesized protein folds into its final native structure inside the cell.

Relatively Few of the Many Possible Polypeptide Chains Would Be Useful

Since each of the 20 amino acids is chemically distinct and each can, in principle, occur at any position in a protein chain, there are $20 \times 20 \times 20 \times 20 = 160{,}000$ different possible polypeptide chains 4 amino acids long, or 20^n different possible polypeptide chains n amino acids long. For a typical protein length of about 300 amino acids, more than 10^{390} different proteins can be made.

We know, however, that only a very small fraction of these possible proteins would adopt a stable three-dimensional conformation. The vast majority would have many different conformations of roughly equal energy, each with different chemical properties. Proteins with such variable properties would not be useful and would therefore be eliminated by natural selection in the course of evolution.

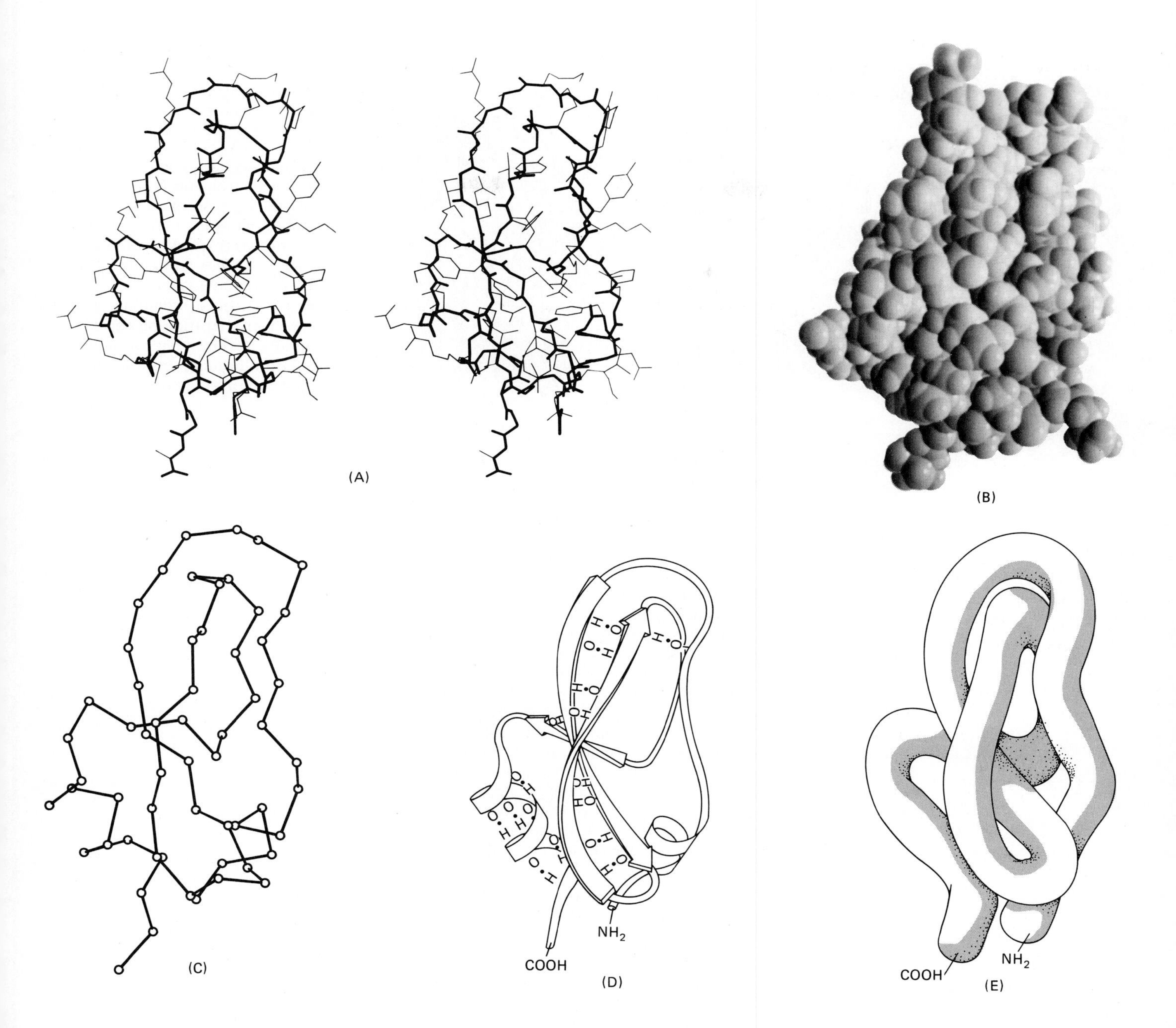

Figure 3–31 The three-dimensional conformation of the small protein, basic pancreatic trypsin inhibitor (BPTI) as seen in five commonly used representations. (A) A stereo pair illustrating the positions of all nonhydrogen atoms. The main chain is shown with heavy lines and the side chains with thin lines. (B) Space-filling model showing the van der Waals radii of all atoms (see Panel 3–1, pp. 90–91). (C) Backbone wire model composed of lines that connect each α-carbon along the polypeptide backbone. (D) "Ribbon model," which represents all regions of regular hydrogen-bonded interactions as either helices (α helices) or sets of arrows (β sheets) pointing toward the carboxyl-terminal end of the chain; in this example the hydrogen bonds are shown. (E) "Sausage model," which shows the course of the polypeptide chain but omits all detail. Note that the core of all globular proteins is densely packed with atoms. Thus the impression of an open structure produced by models (C), (D), and (E) is misleading. (B and C, courtesy of Richard J. Feldmann; A and D, courtesy of Jane Richardson.)

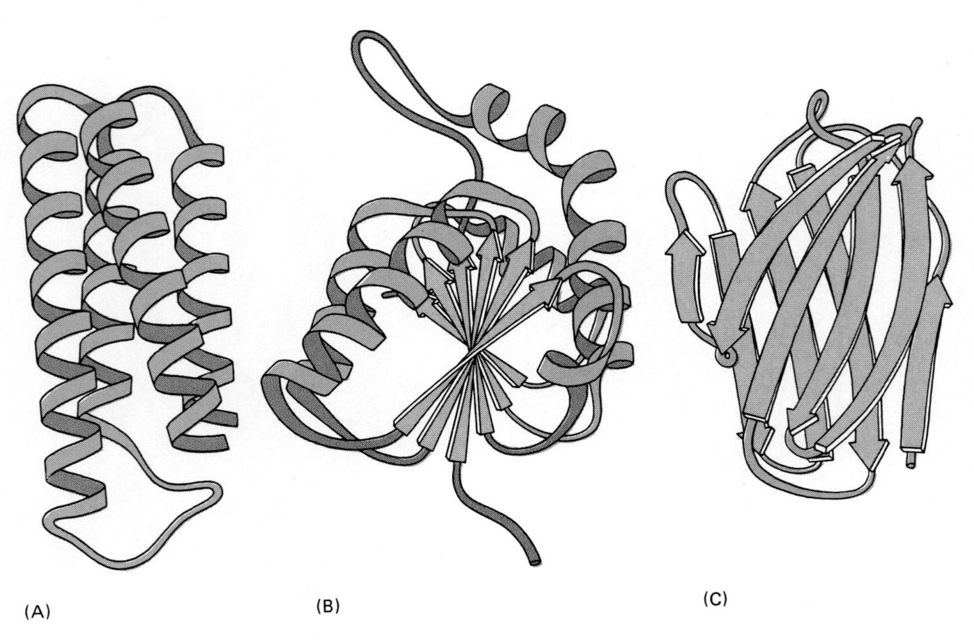

Figure 3–32 Ribbon models of the three-dimensional structure of several differently organized protein domains. (A) Cytochrome b_{562}, a single-domain protein composed almost entirely of α helices. (B) The NAD-binding domain of lactic dehydrogenase composed of a mixture of α helices and β sheets. (C) The variable domain of an immunoglobin light chain composed of a sandwich of two β sheets. In these examples the α helices and connecting strands are colored, while strands organized as β sheets are denoted by gray arrows. Note that the polypeptide chain generally traverses back and forth across the entire domain, making sharp turns only at the protein surface. (Drawings courtesy of Jane Richardson.)

Figure 3–33 The three-dimensional structure of a protein can be described in terms of different levels of folding, each of which is constructed from the preceding one in hierarchical fashion. These levels are illustrated here using the catabolite activator protein (CAP), a bacterial gene regulatory protein with two domains. When the large domain binds cyclic AMP, it causes a conformational change in the protein that enables the small domain to bind to a specific DNA sequence (see p. 560). The amino acid sequence is termed the *primary structure* and the first folding level the *secondary structure*. As indicated under the brackets at the bottom of this figure, the combination of the second and third folding levels shown here is commonly termed the *tertiary structure*, and the fourth level (the assembly of subunits) the *quaternary structure* of a protein. (Modified from a drawing by Jane Richardson.)

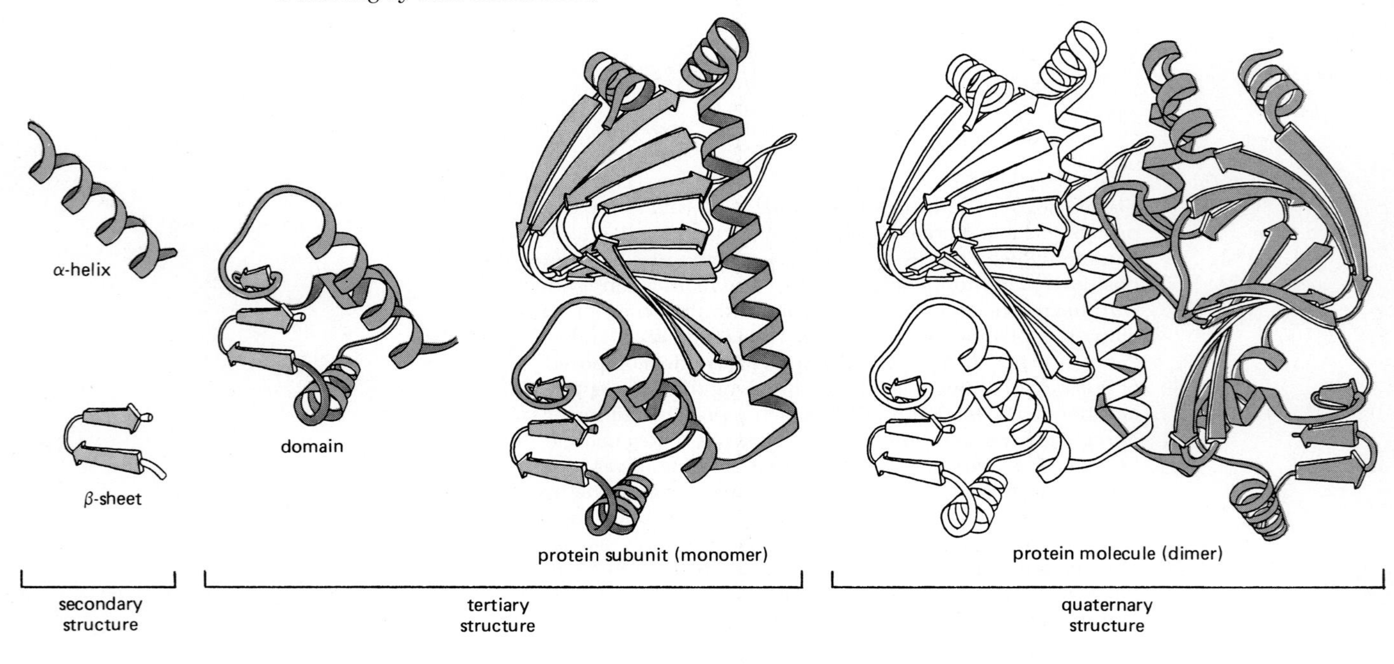

Present-day proteins have an amazingly sophisticated structure and chemistry because of their unique folding properties. Not only is the amino acid sequence such that a single conformation is extremely stable, but this conformation has the precise chemical properties that enable the protein to perform a specific catalytic or structural function in the cell. Proteins are so precisely built that the change of even a few atoms in one amino acid can sometimes disrupt the structure and cause a catastrophic change in function.

New Proteins Usually Evolve by Minor Alterations of Old Ones[25]

Cells have genetic mechanisms that allow genes to be duplicated, modified, and recombined in the course of evolution (see p. 599). Consequently, once a protein with useful surface properties has evolved, its basic structure can be incorporated in many other proteins. Proteins of different but related function in present-day organisms often have similar amino acid sequences. Such families of proteins are believed to have evolved from a single ancestral gene that duplicated in the course of evolution to give rise to other genes in which mutations gradually accumulated to produce related proteins with new functions.

Consider the family of protein-cleaving (proteolytic) enzymes, the **serine proteases,** which includes the digestive enzymes chymotrypsin, trypsin, and elastase and some of the proteases in the blood-clotting and complement (see p. 1031) enzymatic cascades. When two of these enzymes are compared, about 40% of the positions in their amino acid sequences are found to be occupied by the same amino acid (Figure 3–34). The similarity of their three-dimensional conformations as determined by x-ray crystallography is even more striking: most of the detailed twists and turns in their polypeptide chains, which are several hundred amino acids long, are identical (Figure 3–35).

The various serine proteases nonetheless have quite distinct functions. Some of the amino acid changes that make these enzymes different were presumably selected in the course of evolution because they resulted in changes in substrate specificity and regulatory properties, giving them the different functional properties they have today. Other amino acid changes are likely to be "neutral," having neither a beneficial nor a damaging effect on the basic structure and function of the enzyme. Since mutation is a random process, there must also have been many deleterious changes that altered the three-dimensional structure of these enzymes sufficiently to inactivate them. Such inactive proteins would have been lost whenever the individual organisms making them were at enough of a disadvantage to be eliminated by natural selection. It is not surprising, then, that cells contain whole sets of structurally related polypeptide chains that have a common ancestry but different functions.

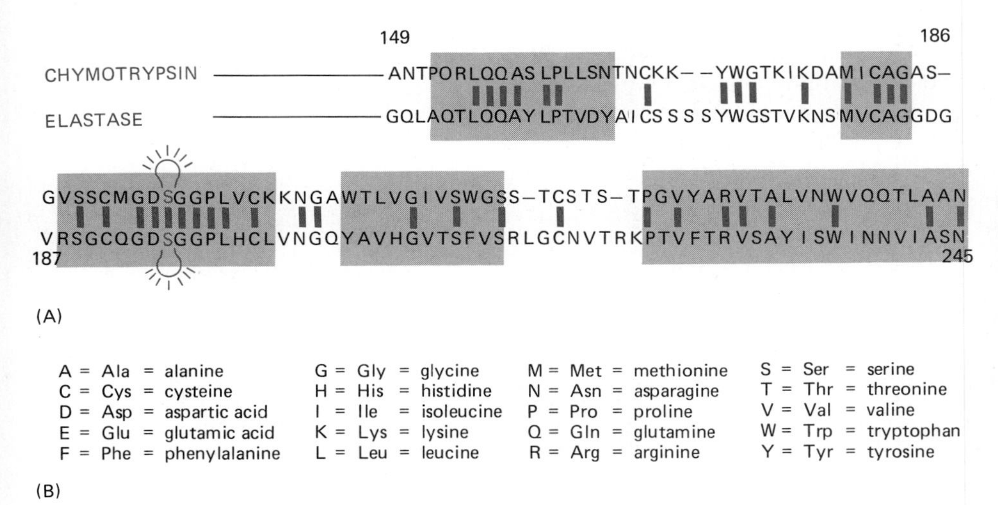

Figure 3–34 (A) Comparison of the amino acid sequences of two members of the serine protease family of enzymes. The carboxyl-terminal portions of the two proteins are shown (amino acids 149 to 245). Identical amino acids are connected by colored bars, and the serine residue in the active site at position 195 is highlighted. In the boxed sections of the polypeptide chains, each amino acid occupies a closely equivalent position in the three-dimensional structures of the two enzymes (see Figure 3–35). (B) The standard one-letter and three-letter codes for amino acids. (Modified from J. Greer, *Proc. Natl. Acad. Sci. USA* 77:3393–3397, 1980.)

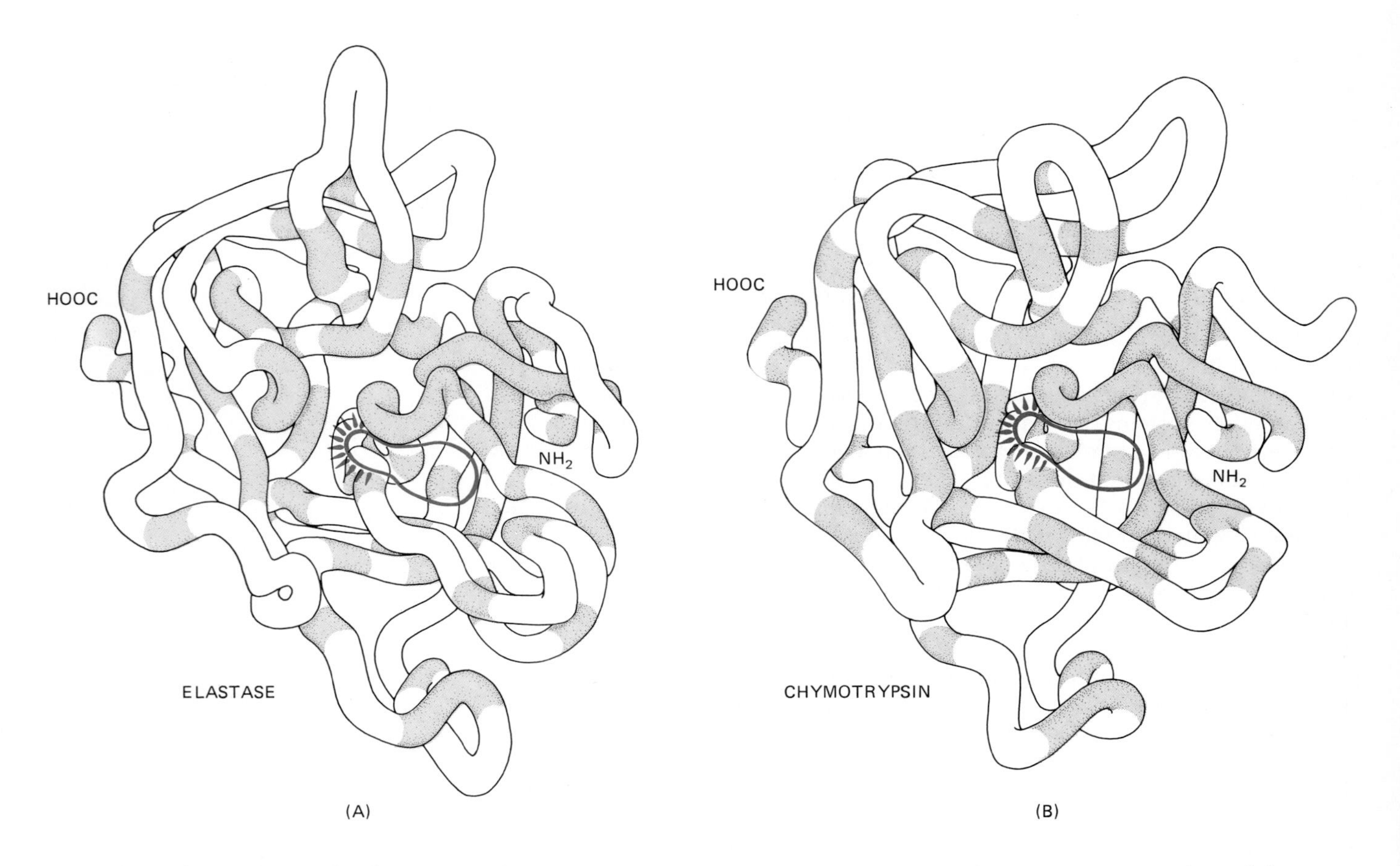

Figure 3–35 Comparison of the conformations of the two evolutionarily related proteases shown in Figure 3–34: elastase (A) and chymotrypsin (B). Although only those amino acid residues in the polypeptide chain shaded in color are the same in the two proteins, their conformations are very similar everywhere. The active site, which is circled, contains an activated serine residue (see Figure 3–47). Chymotrypsin contains more than two chain termini because it is formed by the proteolytic cleavage of chymotrypsinogen, an inactive precursor.

New Proteins Can Evolve by Recombining Preexisting Polypeptide Domains[26]

Once a number of stable protein surfaces have been made in a cell, new surfaces with different binding properties can be generated by joining two or more individual proteins together by noncovalent interactions between them. This combining of globular proteins to make larger, functional protein assemblies is common; although a typical polypeptide chain has a molecular weight of 40,000 to 50,000 (about 300 to 400 amino acids), and relatively few polypeptide chains are more than three times this size, many protein complexes have molecular weights of a million or more.

A related but distinct way of making a new protein from existing chains is to join the corresponding DNA sequences to make a gene that encodes a single large polypeptide chain (see p. 602). Proteins in which different parts of the polypeptide chain fold independently into separate globular domains are believed to have evolved in this way. Many proteins have such "multidomain" structures, and, as might be expected from the evolutionary considerations discussed above, an important binding site for another molecule frequently lies at the site where the separate domains are juxtaposed (Figure 3–36). The structure of one multidomain protein of this type is shown in Figure 3–37.

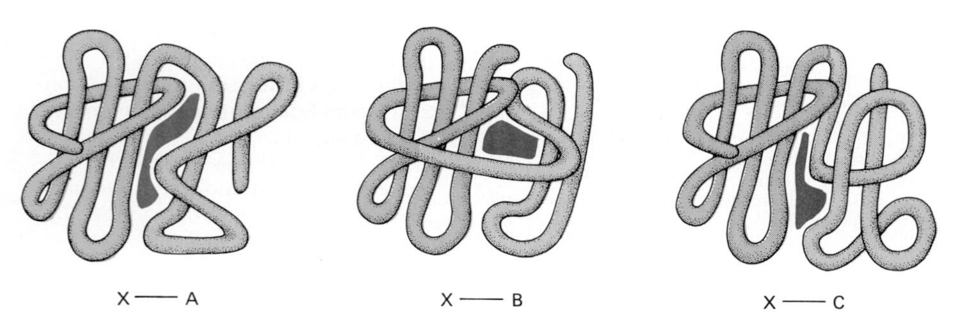

Figure 3–36 The general principle by which the juxtaposition of separate protein surfaces in the course of evolution has given rise to proteins that contain new binding sites for other molecules (*ligands*—see p. 122). As indicated here, the ligand-binding sites often lie at the interface between two protein domains.

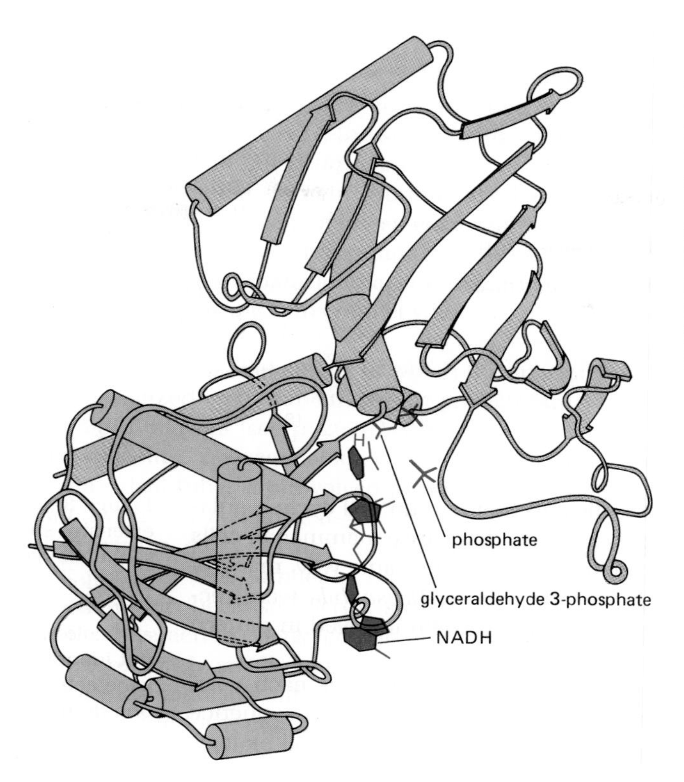

Figure 3–37 The structure of the glycolytic enzyme glyceraldehyde 3-phosphate dehydrogenase. The protein is composed of two domains, each shown differently shaded, with regions of α helix represented by cylinders and regions of β sheet indicated by arrows. The details of the reaction catalyzed by the enzyme are shown in Figure 2–21. Note that the three bound substrates lie at an interface between the two domains. (Courtesy of Alan J. Wonacott.)

Another way of reutilizing an amino acid sequence is especially widespread among long fibrous proteins such as collagen (see Figure 3–28). In these cases a structure is formed from multiple internal repeats of an ancestral amino acid sequence (see p. 602). Putting together amino acid sequences by joining preexisting coding DNA sequences is clearly a much more efficient strategy for a cell than the alternative of deriving new protein sequences from scratch by random DNA mutation.

Structural Homologies Can Help Assign Functions to Newly Discovered Proteins[27]

The development of techniques for rapidly sequencing DNA molecules has made it possible to determine the amino acid sequences of many proteins from the nucleotide sequences of their genes (see p. 186). An ever-enlarging "protein data base" is therefore available that is routinely scanned by computers to search for possible sequence homologies between a newly sequenced protein and previously studied ones. Although sequences have so far been determined for only a few percent of the proteins in eucaryotic organisms, it is common to find that a newly sequenced protein is homologous to some other, known protein over part of its length, indicating that most proteins may have descended from relatively few ancestral types. As expected, the sequences of many large proteins often show signs of having evolved by the joining of preexisting domains in new combinations—a process called "domain shuffling" (Figure 3–38).

The discovery of domain homologies can also be useful in another way. It is much more difficult to determine the three-dimensional structure of a protein than to determine its amino acid sequence. But the conformation of a newly sequenced protein domain can be guessed if it is homologous to a domain of a protein whose conformation has already been determined by x-ray diffraction analysis. By assuming that the twists and turns of the polypeptide chain will be conserved in the two proteins despite the presence of discrepancies in amino acid sequence, one can often sketch the structure of the new protein with reasonable accuracy.

These protein comparisons are also important because related structures often imply related functions. Many years of experimentation can be saved by discovering an amino acid sequence homology with a protein of known function. For

If an enzyme speeds up the rate of the forward reaction, A + B → AB, by a factor of 10^8, it must speed up the rate of the backward reaction, AB → A + B, by a factor of 10^8 as well. The *ratio* of the forward to the backward rates of reaction depends only on the concentrations of A, B, and AB. The equilibrium point remains precisely the same whether or not the reaction is catalyzed by an enzyme.

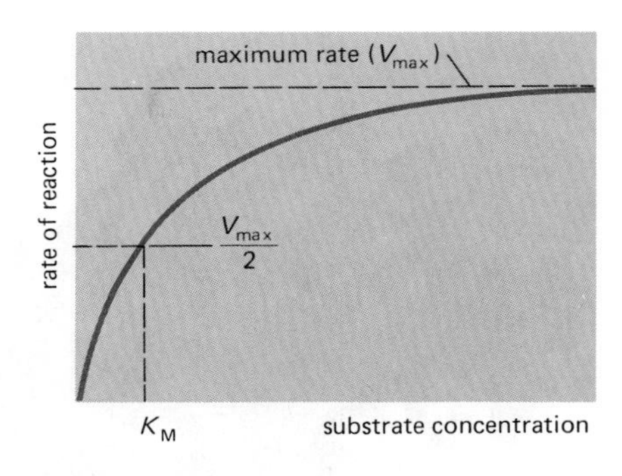

Figure 3–54 The rate of an enzyme reaction (V) increases as the substrate concentration increases until a maximum value (V_{max}) is reached. At this point all substrate-binding sites on the enzyme molecules are fully occupied, and the rate of reaction is limited by the rate of the catalytic process on the enzyme surface. For most enzymes the concentration of substrate at which the reaction rate is half-maximal (K_M) is a measure of how tightly the substrate is bound, with a large value of K_M corresponding to weak binding, and vice versa.

Many Enzymes Make Reactions Proceed Preferentially in One Direction by Coupling Them to ATP Hydrolysis[35]

The living cell is a chemical system that is far from equilibrium. The product of each enzyme usually serves as a substrate for another enzyme in the metabolic pathway and is rapidly consumed. More important, by means of enzyme-catalyzed pathways, previously described in Chapter 2, many reactions are driven in one direction by being *coupled* to the energetically favorable hydrolysis of ATP to ADP and inorganic phosphate (see p. 74). To make this strategy effective, the ATP pool is itself maintained at a level far from its equilibrium point, with a high ratio of ATP to its hydrolysis products (see Chapter 7, p. 354). This ATP pool thereby serves as a "storage battery" that keeps energy and atoms continually passing through the cell, directed along pathways determined by the enzymes present. For a living system, approaching chemical equilibrium means decay and death.

Multienzyme Complexes Help to Increase the Rate of Cell Metabolism[36]

The efficiency of enzymes in accelerating chemical reactions is crucial to the maintenance of life. Cells, in effect, must race against the unavoidable processes of decay, which run downhill toward chemical equilibrium. If the rates of desirable reactions were not greater than the rates of competing side reactions, a cell would soon die. Some idea of the rate at which cellular metabolism proceeds can be obtained by measuring the rate of ATP utilization. A typical mammalian cell turns over (that is, completely degrades and replaces) its entire ATP pool once every one or two minutes. For each cell this turnover represents the utilization of roughly 10^7 molecules of ATP per second (or, for the human body, about a gram of ATP every minute).

The rates of cellular reactions are rapid because of the effectiveness of enzyme catalysis. Many important enzymes have become so efficient that there is no possibility of further useful improvement: the factor limiting the reaction rate is no longer the intrinsic speed of action of the enzyme but the frequency with which the enzyme collides with its substrate. Such a reaction is said to be *diffusion-limited* (see pp. 92–94).

If a reaction is diffusion-limited, its rate will depend on the concentration of both the enzyme and its substrate. For a sequence of reactions to occur very rapidly, each metabolic intermediate and enzyme involved must therefore be present in high concentration. Given the enormous number of different reactions carried out by a cell, there are limits to the concentrations of substrates that can be achieved. In fact, most metabolites are present in micromolar (10^{-6} M) concentrations, and most enzyme concentrations are much less. How is it possible, therefore, to maintain very fast metabolic rates?

The answer lies in the spatial organization of cell components. Reaction rates can be increased without raising substrate concentrations by bringing the various enzymes involved in a reaction sequence together to form a large protein assembly known as a **multienzyme complex.** In this way the product of enzyme A is passed directly to enzyme B, and so on to the final product, and diffusion rates need not be limiting even when the concentration of substrate in the cell as a whole is very low. Such enzyme complexes are very common (the structure of one, pyruvate dehydrogenase, was shown in Figure 2–40), and they are involved in nearly all aspects of metabolism, including the central genetic processes of DNA, RNA, and protein synthesis. In fact, it may be that few enzymes in eucaryotic cells diffuse freely in solution; instead, most may have evolved binding sites that concentrate them with other proteins of related function in particular regions of the cell, thereby increasing the rate and efficiency of the reactions they catalyze (see p. 668).

Cells have another way of increasing the rate of metabolic reactions, which depends on intracellular membranes.

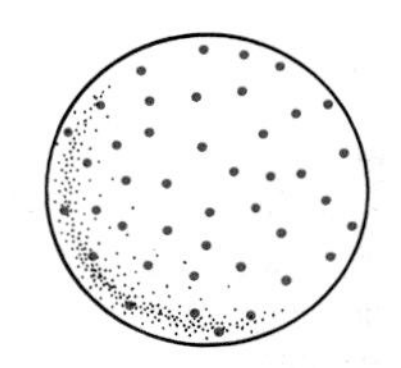

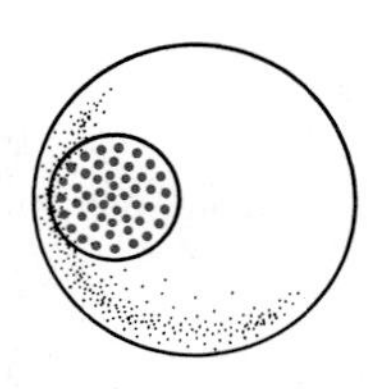

Figure 3–55 A large increase in the concentration of interacting molecules can be achieved by confining them to the same membrane-bounded compartment in a eucaryotic cell.

Intracellular Membranes Increase the Rates of Diffusion-limited Reactions[37]

The extensive intracellular membranes of eucaryotic cells act in at least two ways to increase the rates of reactions that would otherwise be limited by the speed of diffusion. First, membranes can segregate certain substrates and the enzymes that act on them into the same membrane-bounded compartment, such as the endoplasmic reticulum or the cell nucleus. If, for example, the compartment occupies a total of 10% of the volume of the cell, the concentration of reactants in the compartment can be 10 times greater than in a similar cell with no compartmentalization (Figure 3–55).

A second way membranes can increase the rate of a reaction is by restricting the diffusion of reactants to the two dimensions of the membrane itself. Enzymes and their substrates confined to two dimensions will collide with one another much more frequently than if they were diffusing in three dimensions, even though the rate of diffusion of molecules in a membrane is about 100-fold slower than in aqueous solution (Figure 3–56). This process seems certain to operate in the case of the enzymes and substrates involved in the synthesis of lipid molecules, where the substrates dissolve directly in the lipid bilayer; it probably also operates to accelerate many other reactions that utilize membrane-bound enzymes.

A similar "diffusion to capture" mechanism has been shown to increase the rate at which some gene regulatory proteins find the specific DNA sequences they bind to on chromosomes. These proteins have a weak affinity for any region of DNA. Each time they encounter a chromosome, they "bind and slide," thereby scanning long lengths of DNA for the presence of their specific binding sites.

Protein Molecules Can Reversibly Change Their Shape[38]

Natural selection has generally favored the evolution of polypeptides that adopt specific stable conformations. However, many protein molecules—perhaps most—have two or more slightly different conformations available to them, and by shifting reversibly from one to another, they can alter their function. Such an **allosteric protein** may, for example, be able to form within itself several alternative sets of hydrogen bonds of roughly equal energy, each alternative set requiring different spatial relationships between two domains of the polypeptide chain. The alternative stabilized conformations will generally be separated from one another by unstable intermediate states, so that the molecule will "flip-flop" between the stabilized conformations.

Each distinct conformation of an allosteric protein has a somewhat different surface and thus a different ability to interact with other molecules. Often only

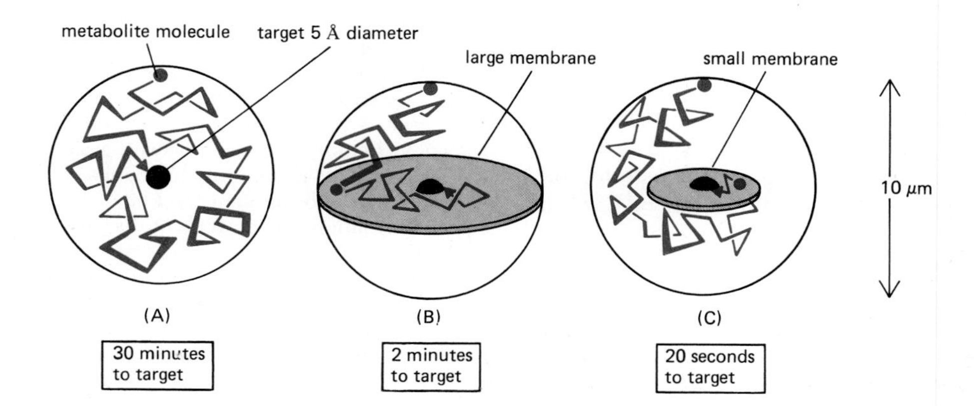

Figure 3–56 Reaction rates can increase when membranes convert diffusion in three dimensions to diffusion in two dimensions. The results of a series of theoretical calculations are shown here. (A) An average molecule diffusing in three dimensions will find a single "target" inside a sphere 10 μm in diameter within 30 minutes. (B) The diffusion time is greatly reduced if the target is fixed in a membrane; here it takes about 1 second for an average molecule to hit a large internal membrane, followed by a mean time of 2 minutes of diffusion in the membrane to find the target. (C) With an internal membrane that has a tenfold reduced surface area, an average molecule will require 10 seconds to find the membrane but will now hit the target by diffusion in the membrane about 10 times more quickly than in (B). Thus the efficiency of collision is nearly 100-fold greater in (C) than in (A).

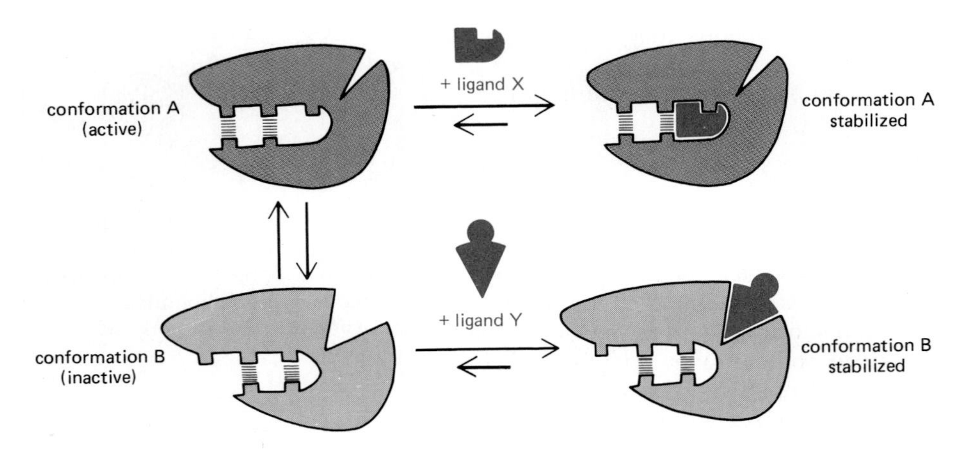

Figure 3–57 Each conformation of an allosteric protein will be stabilized by the binding of a ligand that it binds preferentially. The tight binding of a ligand to only one conformation of an allosteric protein will shift the protein into the conformation that best binds the ligand. Therefore, a high concentration of ligand X will activate the protein shown, whereas a high concentration of ligand Y will inactivate it.

one of two conformations has a high affinity for a particular ligand; in this case the presence or absence of the ligand will determine the conformation that the protein adopts (Figure 3–57). When there are two distinct ligands, each specific to a particular surface of the same protein, the concentration of one molecule will change the affinity of the protein for the other. Allosteric changes of this type are fundamental to the regulation of most biological processes.

Allosteric Proteins Help Regulate Metabolism[39]

Allosteric proteins are essential to the **feedback regulation** that controls the flux through a metabolic pathway (see p. 80). Enzymes that act early in a pathway, for example, are almost always allosteric proteins that can exist in two different conformations. One is an active conformation that binds substrate at its **active site** and catalyzes its conversion to the next substance in the pathway. The other is an inactive conformation that tightly binds the final product of the same pathway at a different place on the protein surface (the **regulatory site**). As the final product accumulates, it binds and converts the enzyme to its inactive conformation (*negative feedback*), which is stabilized because the product can bind to the enzyme only in this form. In other cases an enzyme involved in a metabolic pathway is *activated* by an allosteric transition that occurs when it binds a ligand that accumulates when the cell is deficient in a product of the pathway. Here the ligand binds preferentially to the active form of the enzyme (*positive feedback*), and this binding is required to flip the enzyme from an inactive to an active conformation (see Figure 3–57). As a result of both negative and positive feedback regulation, a cell makes a given product only when that product is needed, and a relatively constant concentration of each metabolite is thereby maintained.

Allosteric Proteins Are Vital for Cell Signaling[40]

As we have noted, allosteric proteins such as those involved in feedback regulation have at least two binding sites, one for the enzyme substrate and one or more for regulatory ligands. These sites occupy different regions of the protein surface, and the ligands they recognize can be totally different. Because the binding of a ligand to one site can affect another site by changing the protein's conformation, any metabolic process can be regulated by any other in the cell, regardless of its chemical nature. For example, the production and breakdown of glycogen in muscle cells is linked to the concentration of Ca^{2+} by means of allosteric enzymes whose activities are changed by changes in the concentration of Ca^{2+} in the cytosol (see p. 712).

Allosteric proteins mediate especially sensitive responses to signals when, as is often the case, they behave cooperatively as identical subunits in a symmetrical assembly. In these proteins the change in the conformation of one subunit caused by ligand binding can help the neighboring subunits to bind the same ligand (Figure 3–58), with the result that a relatively small change in ligand concentration

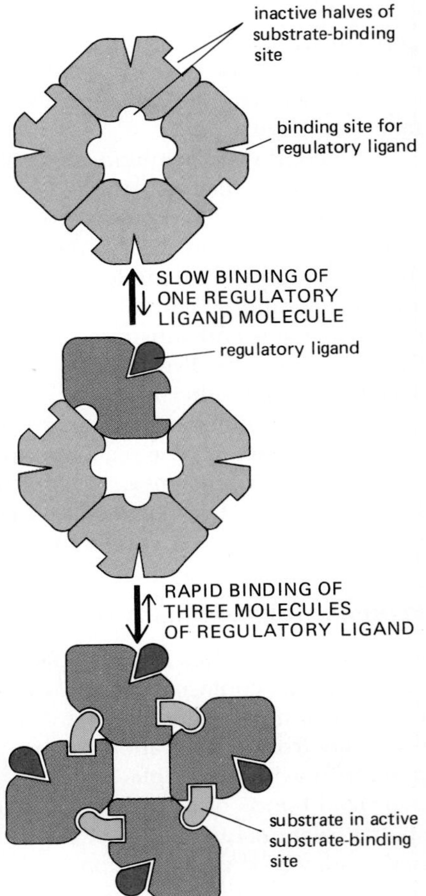

Figure 3–58 Schematic diagram illustrating how the conformation of one subunit influences that of its neighbors in a symmetrical protein composed of identical allosteric subunits. The binding of a single regulatory ligand molecule to one subunit of the enzyme changes the conformation of this subunit, as in Figure 3–57. Because this conformational change makes it easier for neighboring subunits to change their shape to the tight-binding conformation, the binding of the first molecule of ligand increases the affinity with which the other subunits bind the same ligand. The enzyme will therefore be activated by a relatively small increase in the concentration of the regulatory ligand (see Figure 3–59).

in the surrounding medium switches the whole assembly from an active to an inactive conformation, or vice versa. If the ligand binds preferentially to the active conformation of each enzyme subunit, this behavior will produce a sharp increase in enzyme activity as the ligand concentration is raised (Figure 3–59). The structure of one well-studied allosteric enzyme, aspartate transcarbamoylase, is shown in Figure 3–60.

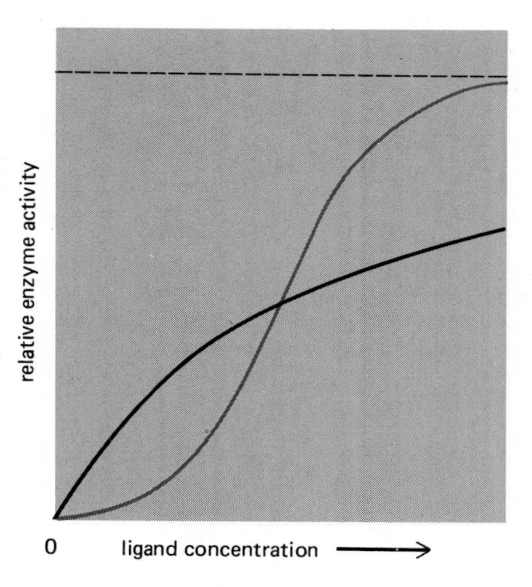

Figure 3–59 The activity of the multisubunit allosteric enzyme illustrated in Figure 3–58 would be expected to increase steeply with ligand concentration (as shown in color) due to the "cooperative" binding of the ligand molecules. In contrast, the activation of an allosteric enzyme containing only one subunit would be expected to display the simple kinetics shown in black. The dotted line indicates the maximum level of activity attained at very high concentrations of the ligand, which is assumed to be the same in both cases.

Proteins Can Be Pushed or Pulled into Different Shapes[40,41]

All of the directed movements in cells depend on forces generated by proteins. But how can a protein molecule be made to move in a controlled fashion? Before we answer this question, we must discuss some of the ways in which cells regulate the conformation of allosteric proteins. Consider an allosteric protein that can adopt two alternative conformations—a low-energy form C (inactive) and a high-energy form C* (active)—that differ in energy by about 4.3 kcal/mole (the energy available from forming about four hydrogen bonds on a protein surface). Given this energy difference, conformation C will be favored by about 1000 to 1 (see Table 3–3, p. 95), and the protein will almost always be in its inactive conformation. However, there are two ways the protein can be forced to adopt the active conformation.

The molecule can, in a sense, be "pulled" into the active C* conformation by binding to a low-molecular-weight ligand. Provided this ligand binds only to C*, the energy of this conformation will be selectively reduced without affecting that of C. Since the ligand binds relatively weakly to the protein (most of its binding energy having been used up in pulling the shape of the protein to fit the ligand), it can readily dissociate, and so this conformational change in the protein is perfectly reversible.

Alternatively, an input of chemical energy can be used to convert conformation C to the active form C* in a less readily reversible manner. A common mechanism involves the transfer of a phosphate group from ATP to a serine, threonine, or tyrosine residue in the protein, forming a covalent linkage. Suppose that this phosphorylation reaction, driven by the highly favorable hydrolysis of ATP to ADP, creates a charge repulsion unfavorable to conformation C. If this repulsion is reduced in the active C* form, the change from C to C* will be greatly encouraged by the phosphorylation (Figure 3–61). Controlled protein phosphorylation is commonly observed to activate or inhibit the function of specific proteins in eucaryotic cells (see p. 417); in fact, about one-tenth of all the proteins made in a mammalian cell contain covalently bound phosphate.

It is sometimes observed that ATP can be made *in vitro* by adding ADP to such a phosphorylated protein, demonstrating directly that a great deal of the energy released by ATP hydrolysis has been stored in straining the protein's conformation during the initial phosphorylation event. But how do such energy-driven changes in protein conformation produce movement and, thereby, do useful work in a cell?

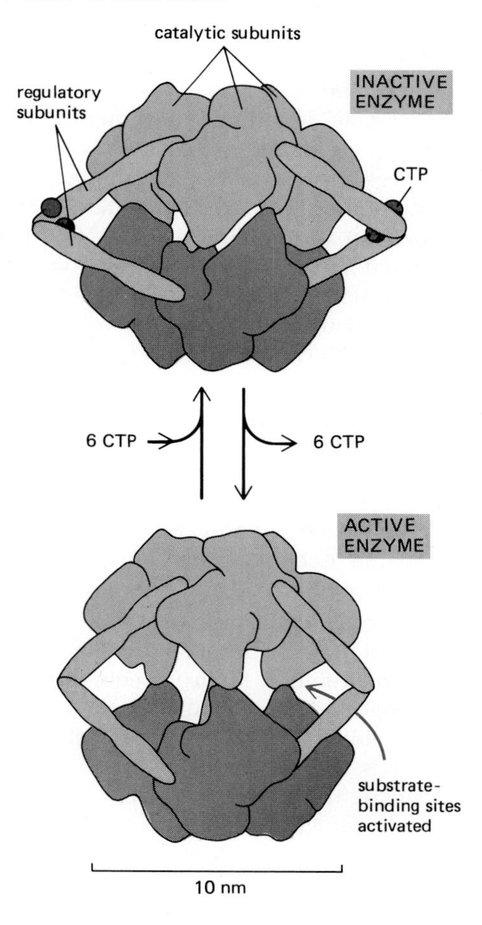

Figure 3–60 The enzyme *aspartate transcarbamoylase* is turned off in response to the binding of cytosine triphosphate (CTP). The enzyme complex contains six catalytic subunits and six regulatory subunits, and the structures of its inactive and active forms have been determined by x-ray diffraction analyses. Each of the regulatory subunits can bind one molecule of CTP, one of the final products in the pathway that begins when this enzyme catalyzes the formation of carbamoyl aspartate from carbamoyl phosphate and aspartic acid. By means of this negative feedback regulation, the enzyme is prevented from producing more CTP than the cell needs. (Based on K.L. Krause, K.W. Volz, and W.N. Lipscomb, *Proc. Natl. Acad. Sci. USA* 82:1643–1647, 1985.)

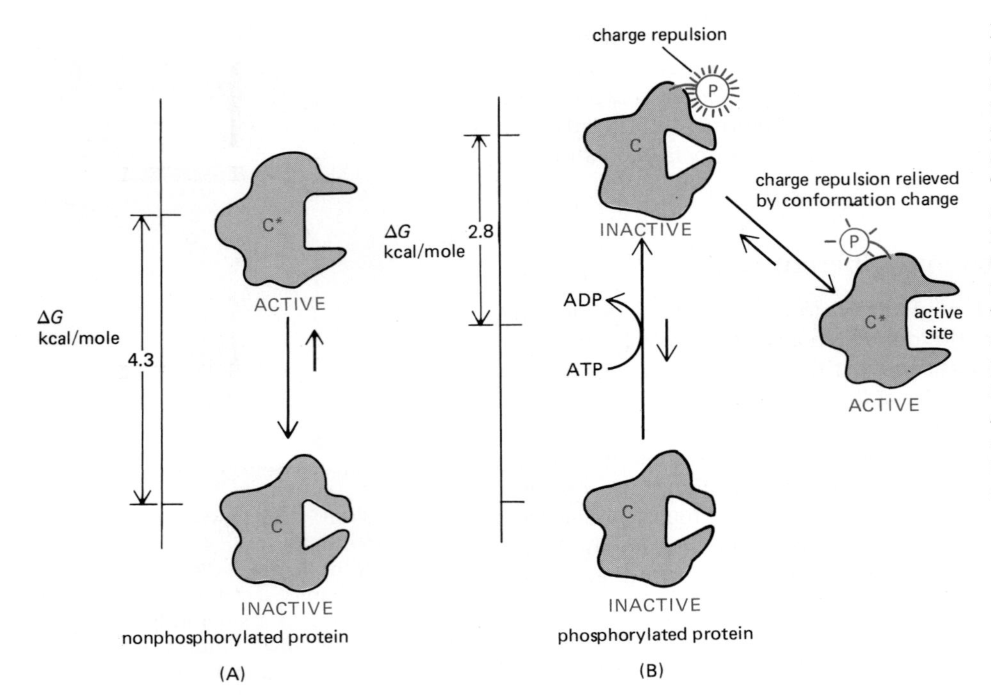

Figure 3–61 Phosphorylation by ATP can activate an allosteric protein. In this example the inactive conformation of the nonphosphorylated protein (A) is favored 1000 to 1 because of a free energy difference of 4.3 kcal/mole (see Table 3–3). When phosphorylated, the active conformation of the protein (B) is favored 100 to 1 (2.8 kcal/mole) because the phosphorylation produces an unfavorable charge repulsion, part of which is relieved by a shift to the active conformation C*. In this way phosphorylation "pushes" the enzyme into the active conformation. Alternatively, phosphorylation could create a charge attraction that brings together two separated parts of an allosteric protein.

Energy-driven Changes in Protein Conformations Can Do Useful Work[42]

Suppose that a protein is required to "walk" along a narrow thread, such as an actin filament or a DNA molecule. Figure 3–62 shows schematically how an allosteric protein might do this by adopting different conformations. With nothing to drive these conformational changes in an orderly way, the shape changes will be perfectly reversible and the protein will wander randomly back and forth along the thread.

Since directional movement of a protein does net work, the laws of thermodynamics demand that such movement depletes free energy from some other source (otherwise the protein could be used to make a perpetual motion machine). Therefore, no matter what modifications we make to the model in Figure 3–62, such as adding ligands that favor certain conformations, the protein molecule shown cannot go anywhere without an input of energy.

What is needed is some means of making the series of protein conformation changes unidirectional. For example, the entire cycle would proceed in one direction if any one of the steps could be made irreversible. One way to do this is through the mechanism just discussed for driving allosteric changes in a protein molecule by a phosphorylation-dephosphorylation cycle. Allosteric changes in proteins can also be driven by ATP hydrolysis without involving a phosphorylation of the protein. In the modified walking scheme shown in Figure 3–63, for example, ATP binding "pulls" the protein from conformation 1 to conformation 2; the bound ATP is then hydrolyzed to produce bound ADP and inorganic phosphate (P_i), causing a change from conformation 2 to conformation 3; and finally, the release of the bound ADP and P_i drives the protein back to conformation 1 again.

Because the transitions $1 \rightarrow 2 \rightarrow 3 \rightarrow 1$ are driven by the energy of ATP hydrolysis, and because the intracellular pools of ATP and its hydrolysis products are maintained far from equilibrium (see p. 354), this series of conformational changes will be effectively irreversible under physiological conditions (that is, the probability that ADP will recombine with P_i to form ATP by the route $1 \rightarrow 3 \rightarrow 2 \rightarrow 1$ is extremely low). Therefore, the entire cycle will go in only one direction, causing the protein molecule to move continuously to the right in this schematic example. Examples of proteins that generate directional movement in this way include *myosin* (which moves along actin filaments), *dynein* (which moves along microtubules), and *DNA helicase* (which moves along DNA).

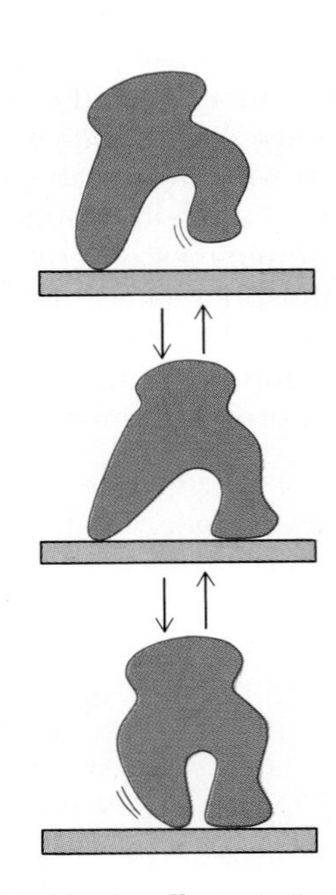

Figure 3–62 An allosteric "walking" protein. Although its three different conformations allow it to wander randomly back and forth while bound to a filament, the protein cannot move uniformly in a single direction.

Many proteins use similar mechanisms to create coherent movement. They all have the ability to go through cyclic changes in shape that are coupled to the hydrolysis of ATP; some are transiently phosphorylated in the process, whereas others are not.

ATP-driven Membrane-bound Allosteric Proteins Can Act as Pumps[43]

Besides generating mechanical force, allosteric proteins can use the energy of ATP hydrolysis to do other forms of work, such as pumping specific ions into or out of the cell. An important example is the **Na^+-K^+ ATPase** found in the plasma membrane of all animal cells, which pumps 3 Na^+ out of the cell and 2 K^+ in during each cycle of conformational change driven by ATP-mediated phosphorylation (see p. 305). This ATP-driven pump consumes more than 30% of the total energy requirement of most cells. By continuously pumping Na^+ out and K^+ in, it creates a cell interior in which the Na^+ concentration is low and the K^+ concentration is high with respect to the cell exterior, thereby generating two ion gradients across the plasma membrane, each the reverse of the other. The energy stored in these and other ion gradients is, in turn, used to drive conformational changes in a variety of other membrane-bound allosteric proteins, enabling them to do useful work for the cell.

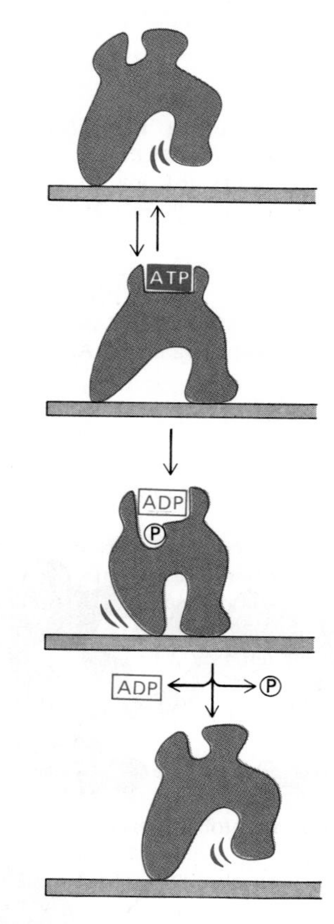

Figure 3–63 An allosteric "walking" protein in which an orderly transition among three conformations is driven by the hydrolysis of a bound ATP molecule. Because one of these transitions is coupled to the hydrolysis of ATP, the cycle is essentially irreversible. By repeated cycles, the protein moves consistently to the right along the filament.

Membrane-bound Allosteric Proteins Can Harness the Energy Stored in Ion Gradients to Do Useful Work[43,44]

Although critically important, ATP and the other nucleoside triphosphates are not the only sources of readily available energy that proteins can use to do useful work. Ion gradients across various cell membranes can store and release energy in a fashion analogous to differences of water pressure on either side of a dam. The large Na^+ gradient across the plasma membrane generated by the Na^+-K^+ ATPase, for example, is used to drive many other plasma membrane-bound protein pumps that transport glucose or specific amino acids into the cell (see p. 309).

Membrane-bound allosteric pumps driven by the hydrolysis of ATP can also work in reverse and employ the energy in the ion gradient to synthesize ATP. In fact, we shall see in Chapter 7 that the energy available in the H^+ (proton) gradient across the inner mitochondrial membrane is used in just such a fashion to synthesize most of the ATP required by animal cells.

Protein Machines Play Central Roles in Many Biological Processes[45]

Complex cellular processes such as DNA replication or protein synthesis are mediated by multienzyme complexes that function as sophisticated "protein machines." For example, the numerous protein parts of the DNA replication apparatus move relative to one another without disassembling, allowing the entire complex to travel rapidly in a zipperlike fashion along the DNA (see p. 232).

In such protein machines, the hydrolysis of bound nucleoside triphosphate molecules drives ordered conformational changes in the individual proteins, enabling groups of these proteins to move coordinately. In this way the appropriate enzymes are moved directly into the positions where they are needed to carry out each reaction in a series, instead of waiting for the random collision of each separate component that would otherwise be required. A simple mechanical analogy that captures the sense of such "high-technology" solutions to the cell's needs is illustrated in Figure 3–64. It seems likely that most of the major processes occurring in cells are mediated by such highly sophisticated, multicomponent protein machines.

Summary

The biological function of a protein depends on the detailed chemical properties of its surface. Binding sites are formed as surface cavities in which precisely positioned amino acid side chains are brought together by protein folding. By binding

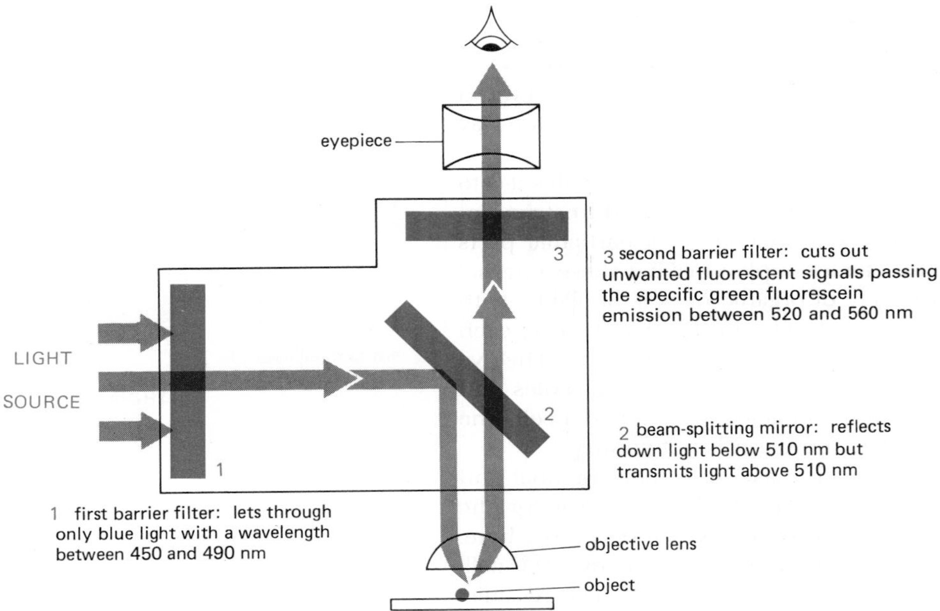

Figure 4–6 The optical system of a modern fluorescence microscope. A filter set consists of two barrier filters and a dichroic (beam-splitting) mirror. In this example the filter set for detection of fluorescein fluorescence is shown. High-numerical-aperture objective lenses are especially important in this type of microscopy since, for a given magnification, the brightness of the fluorescent image is proportional to the fourth power of the numerical aperture (see also Figure 4–4).

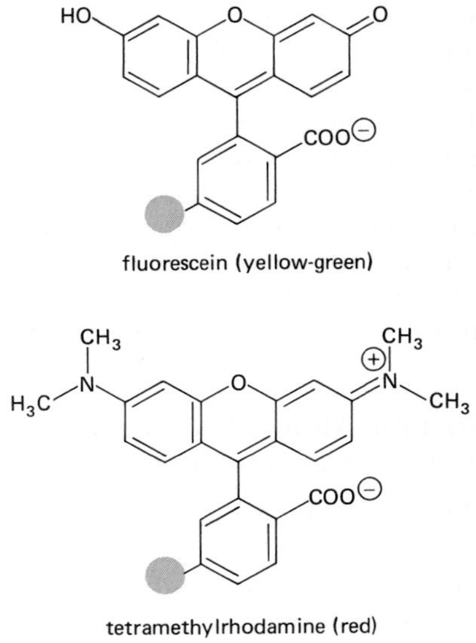

fluorescein (yellow-green)

tetramethylrhodamine (red)

Figure 4–7 The structures of fluorescein and tetramethylrhodamine, two dyes that are commonly used for fluorescence microscopy. Fluorescein emits yellow-green light when activated by light of the appropriate wavelength, whereas the rhodamine dye emits red light. The portion of each molecule shown in color denotes the position of a chemically reactive group; at this position a covalent bond is commonly formed between the dye and a protein (or other molecule). Commercially available versions of these dyes with different types of reactive groups allow the dye to be targeted either to an —SH group or to an $—NH_2$ group on a protein.

Figure 4–8 Fluorescence micrographs of a portion of the surface of an early *Drosophila* embryo in which the microtubules have been labeled with an antibody coupled to fluorescein (*left panel*) and the actin filaments have been labeled with an antibody coupled to rhodamine (*middle panel*). In addition, the chromosomes have been labeled with a third dye that fluoresces only when it binds to DNA (*right panel*). At this stage, all the nuclei of the embryo share a common cytoplasm, and they are in the metaphase stage of mitosis. The three micrographs were taken of the same region of a fixed embryo using three different filter sets in the fluorescence microscope (see also Figure 4–6). (Courtesy of Tim Karr.)

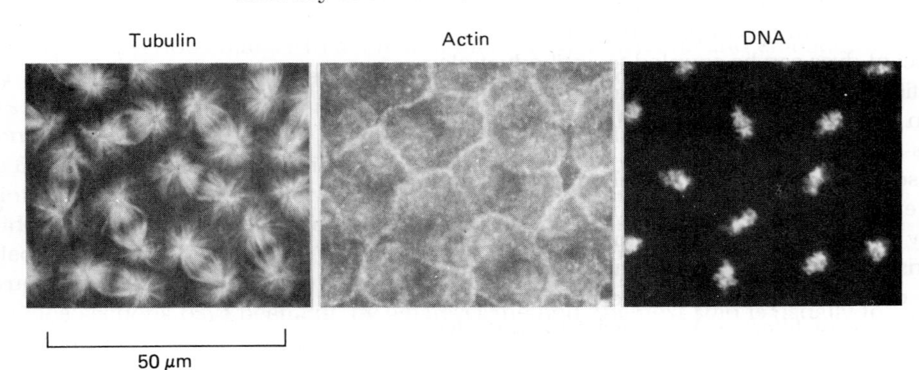

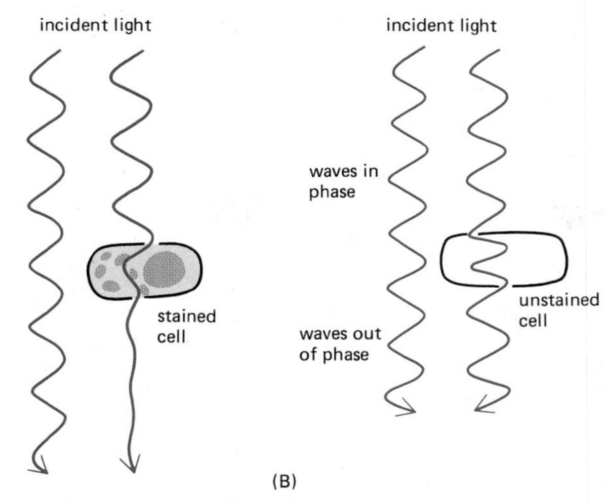

Figure 4–9 Two ways of increasing contrast in light microscopy. The stained portions of the cell in (A) reduce the amplitude of light waves of particular wavelengths passing through them. A colored image of the cell is thereby obtained that is visible in the ordinary way. Light passing through the unstained, living cell (B) undergoes scarcely any change in amplitude, and the structural details cannot be seen even if the image is highly magnified. However, the phase of the light is altered by its passage through the cell, and small phase differences can be made visible by exploiting interference effects using a phase-contrast or a differential-interference-contrast microscope.

Important new methods, to be discussed later, enable fluorescence microscopy to be used to monitor changes in the concentration and location of specific molecules inside *living* cells (see p. 142 and p. 156).

Living Cells Are Seen Clearly in a Phase-Contrast or a Differential-Interference-Contrast Microscope[2]

The possibility that some components of the cell may be lost or distorted during specimen preparation has always worried microscopists. The only certain way to avoid the problem is to examine cells while they are alive, without fixing or freezing. For this purpose, microscopes with special optical systems are useful.

When light passes through a living cell, the phase of the light wave is changed according to the cell's refractive index: light passing through a relatively thick or dense part of the cell, such as the nucleus, is retarded and its phase consequently shifted relative to light that has passed through an adjacent thinner region of the cytoplasm. Both the **phase-contrast microscope** and the **differential-interference-contrast microscope** exploit the interference effects produced when these two sets of waves recombine, thereby creating an image of the cell's structure (Figure 4–9). Both types of light microscopy are widely used to visualize living cells.

A simpler way to see some of the features of a living cell is to observe the light that is scattered by its various components. In the **dark-field microscope,** the illuminating rays of light are directed from the side, so that only scattered light enters the microscope lenses. Consequently, the cell appears as an illuminated object against a black background. Images of the same cell obtained by four different kinds of light microscopy are shown in Figure 4–10.

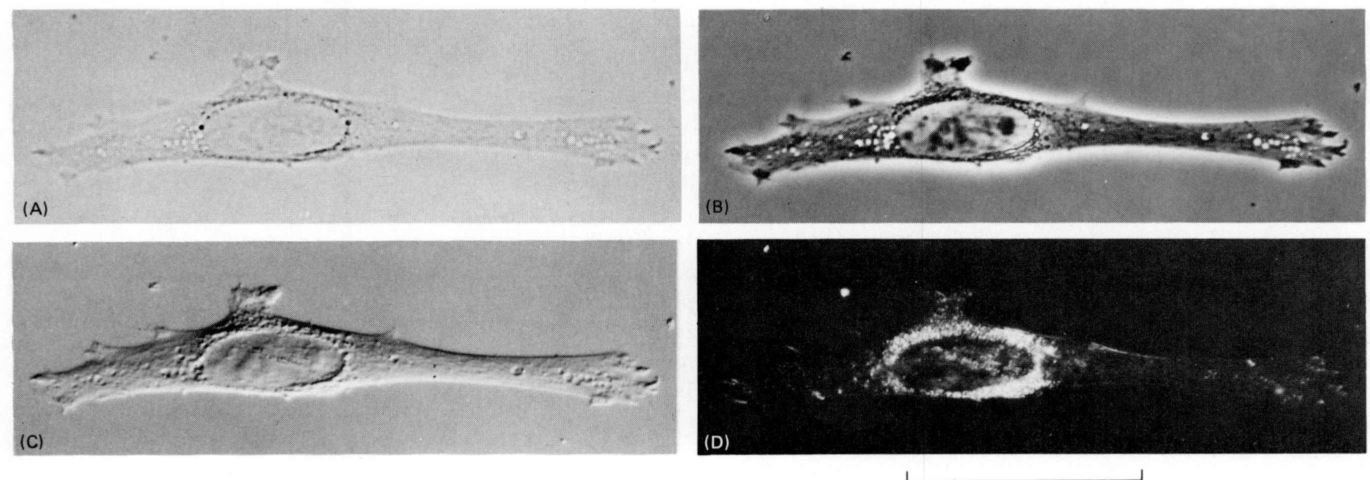

Figure 4–10 A fibroblast in tissue culture visualized with four types of light microscopy. The image in (A) was obtained by the simple transmission of light through the cell, a technique known as *bright-field microscopy.* The other images were obtained by techniques discussed in the text: (B) phase-contrast microscopy, (C) Nomarski differential-interference-contrast microscopy, and (D) dark-field microscopy. All four types of image can be obtained with most modern microscopes simply by interchanging optical components.

One of the great advantages of phase-contrast, interference, and dark-field microscopy is that they make it possible to watch the movements involved in such processes as mitosis and cell migration. Since many cellular motions are too slow to be seen in real time, it is often helpful to take time-lapse motion pictures (*microcinematography*) or video recordings. Here, successive frames separated by a short time delay are recorded, so that when the resulting film or video tape is projected or played at normal speed, events appear greatly speeded up.

Images Can Be Enhanced and Analyzed by Electronic Techniques[5]

In recent years, video cameras and the associated technology of **image processing** have had a major impact on light microscopy. They have not only enabled certain practical limitations of microscopes—due to imperfections in the optical system—to be overcome, but they have also circumvented two fundamental limitations of the human eye. These limitations are that (1) the eye cannot see in extremely dim light and (2) it cannot perceive small differences in light intensity against a bright background. The first of these problems can be overcome by attaching highly light-sensitive video cameras (of the kind used in night surveillance) to a microscope. It is then possible to observe cells for long periods at very low light levels, thereby avoiding the damaging effects of prolonged bright light (and heat). Such *image-intensification systems* are especially important for viewing fluorescent molecules in living cells.

Figure 4–11 Light-microscope images of unstained microtubules that have been visualized by differential-interference-contrast microscopy followed by electronic image processing. (A) shows the original unprocessed image; (B) shows the final result of an electronic process that greatly enhances contrast and reduces "noise." Although the microtubules are only 0.025 μm in diameter, they appear as much wider filaments because of diffraction effects. (Courtesy of Bruce Schnapp.)

Because images produced by video cameras are in electronic form, they can be readily digitized, fed to a computer and processed in various ways to extract latent information. Such image processing makes it possible to compensate for various optical faults in microscopes so as to attain the theoretical limit of resolution. Moreover, by using video systems linked to image processors, contrast can be greatly enhanced so that the eye's limitations in detecting small differences are overcome. Although this processing also enhances the effects of random background irregularities in the optical system, this "noise" can be removed by electronically subtracting an image of a blank area of the field. Small transparent objects then become visible that were previously impossible to distinguish from the background.

The high contrast attainable by computer-assisted interference microscopy makes it possible to see even very small objects such as single microtubules (Figure 4–11), which have a diameter less than one-tenth the wavelength of light (0.025 μm). Individual microtubules can also be seen in a fluorescence microscope if they are fluorescently labeled (see Figure 4–56). In both cases, however, the unavoidable diffraction effects badly blur the image, so that the microtubules appear at least 0.2 μm wide, making it impossible to distinguish a single microtubule from a bundle of several microtubules. Such a distinction requires electron microscopes, which have pushed the limit of resolution well below the wavelength of visible light.

The Electron Microscope Resolves the Fine Structure of the Cell[6]

The relationship between the limit of resolution and the wavelength of the illuminating light (see Figure 4–4) holds true for any form of radiation, whether it is a beam of light or a beam of electrons; but with electrons the limit of resolution can be made very small. The wavelength of an electron decreases as its velocity increases. In an electron microscope with an accelerating voltage of 100,000 V, the wavelength of an electron is 0.004 nm, and in theory the resolution of such a microscope should be about 0.002 nm. However, because the aberrations of an electron lens are considerably harder to correct than those of a glass lens, the practical resolving power of most modern electron microscopes is, at best, 0.1 nm (1 Å) (Figure 4–12). Furthermore, problems of specimen preparation, contrast, and radiation damage effectively limit the normal resolution for biological objects to 2 nm (20 Å)—about 100 times better than the resolution of the light microscope. Some of the landmarks in the development of electron microscopy are outlined in Table 4–2.

Figure 4–12 Electron micrograph of a thin layer of gold showing the individual atoms as bright spots. The distance between adjacent gold atoms is about 0.2 nm (2 Å). (Courtesy of Graham Hills.)

Table 4–2 Major Events in the Development of the Electron Microscope and Its Applications to Cell Biology

Year	Event
1897	**J. J. Thomson** announced the existence of negatively charged particles, later termed *electrons.*
1924	**de Broglie** proposed that a moving electron has wavelike properties.
1926	**Busch** proved that is was possible to focus a beam of electrons with a cylindrical magnetic lens, laying the foundations of electron optics.
1931	**Ruska** and colleagues built the first transmission electron microscope.
1935	**Knoll** demonstrated the feasibility of the scanning electron microscope; three years later a prototype instrument was built by **Von Ardenne.**
1939	**Siemens** produced the first commercial transmission electron microscope.
1944	**Williams** and **Wyckoff** introduced the metal shadowing technique.
1945	**Porter, Claude,** and **Fullam** used the electron microscope to examine cells in tissue culture after fixing and staining them with OsO_4.
1948	**Pease** and **Baker** reliably prepared thin sections (0.1 to 0.2 μm thick) of biological material.
1952	**Palade, Porter,** and **Sjöstrand** developed methods of fixation and thin sectioning that enabled many intracellular structures to be seen for the first time. In one of the first applications of these techniques, **H. E. Huxley** showed that skeletal muscle contains overlapping arrays or protein filaments, supporting the "sliding filament" hypothesis of muscle contraction.
1953	**Porter and Blum** developed the first widely accepted ultramicrotome, incorporating many features introduced by **Claude** and **Sjöstrand** previously.
1956	**Glauert** and associates showed that the epoxy resin Araldite was a highly effective embedding agent for electron microscopy. **Luft** introduced another embedding resin, Epon, five years later.
1957	**Robertson** described the trilaminar structure of the cell membrane, seen for the first time in the electron microscope.
1957	Freeze-fracture techniques, initially developed by **Steere**, were perfected by **Moor and Mühlethaler**. Later (1966), **Branton** demonstrated that freeze-fracture allows the interior of the membrane to be visualized.
1959	**Singer** used antibodies coupled to ferritin to detect cellular molecules in the electron microscope.
1959	**Brenner and Horne** developed the negative staining technique, invented four years previously by **Hall,** into a generally useful technique for visualizing viruses, bacteria, and protein filaments.
1963	**Sabatini, Bensch, and Barrnett** introduced glutaraldehyde (usually followed by OsO_4) as a fixative for electron microscopy.
1965	**Cambridge Instruments** produced the first commercial scanning electron microscope.
1968	**de Rosier and Klug** described techniques for the reconstruction of three-dimensional structures from electron micrographs.
1975	**Henderson and Unwin** determined the first structure of a membrane protein by computer-based reconstruction from electron micrographs of unstained samples.
1979	**Heuser, Reese,** and colleagues developed a high-resolution, deep-etching technique based on very rapid freezing.

In overall design, the **transmission electron microscope (TEM)** is similar to a light microscope, although it is much larger and upside down (Figure 4–13). The source of illumination is a filament or cathode that emits electrons at the top of a cylindrical column about two meters high. Since electrons are scattered by collisions with air molecules, air must first be pumped out of the column to create a vacuum. The electrons are then accelerated from the filament by a nearby anode and allowed to pass through a tiny hole to form an electron beam that travels down the column. Magnetic coils placed at intervals along the column focus the electron beam, just as glass lenses focus the light in a light microscope. The specimen is put into the vacuum, through an airlock, into the path of the electron beam. Some of the electrons passing through the specimen are scattered according

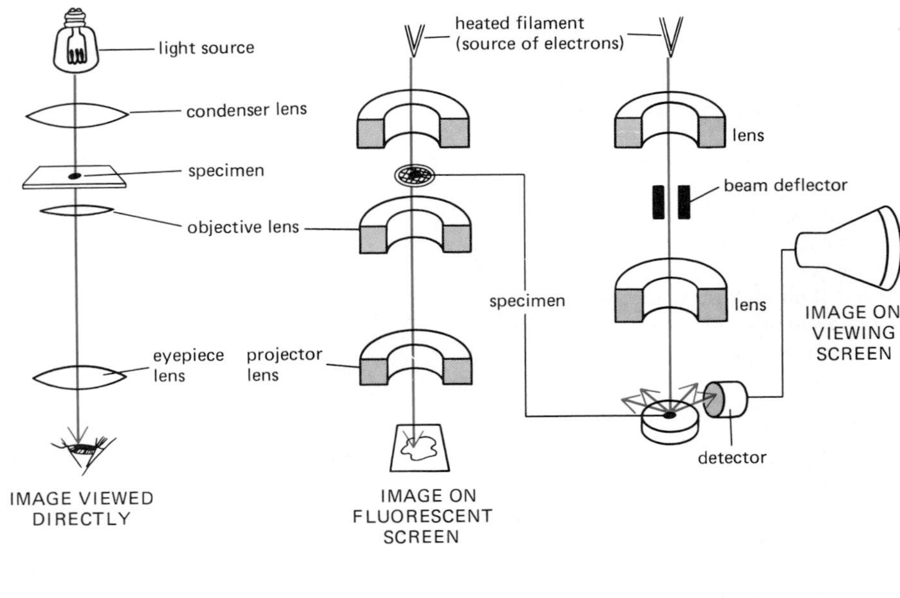

Figure 4–13 Principal features of a light microscope, a transmission electron microscope, and a scanning electron microscope, drawn to emphasize the similarities of overall design. The two types of electron microscopes require that the specimen be placed in a vacuum.

to the local density of the material; the remainder are focused to form an image—in a manner analogous to the way an image is formed in a light microscope—either on a photographic plate or on a phosphorescent screen. Because the scattered electrons are lost from the beam, the dense regions of the specimen show up in the image as areas of reduced electron flux.

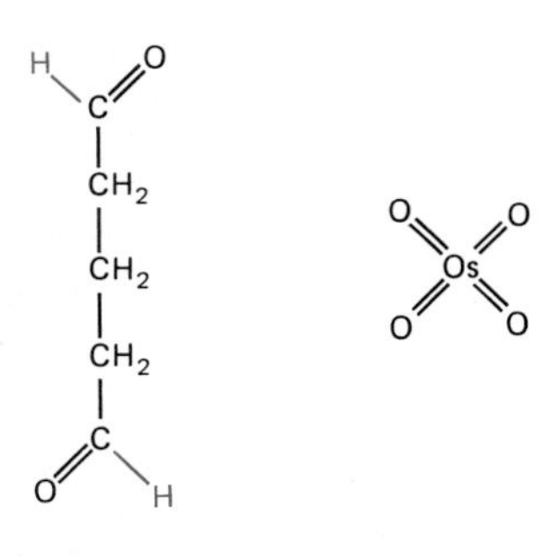

Figure 4–14 Glutaraldehyde and osmium tetroxide are common fixatives used for electron microscopy. The two reactive aldehyde groups of glutaraldehyde enable it to cross-link various types of molecules, forming covalent bonds between them. Osmium tetroxide is reduced by many organic compounds with which it forms cross-linked complexes. It is especially useful for fixing cell membranes since it reacts with the C=C double bonds present in many fatty acids.

Biological Specimens Require Special Preparation for the Electron Microscope[7]

In the early days of its application to biological materials, the electron microscope revealed many previously unimagined structures in cells. But before these discoveries could be made, electron microscopists had to develop new procedures for embedding, cutting, and staining tissues.

Since any specimen will be exposed to a very high vacuum in the electron microscope, there is no possibility of viewing it in the living, wet state. Tissues are usually preserved by fixation—first with *glutaraldehyde,* which covalently cross-links protein molecules to their neighbors, and then with *osmium tetroxide,* which binds to and stabilizes lipid bilayers as well as proteins (Figure 4–14). Since electrons have very limited penetrating power, the fixed tissues normally have to be cut into extremely thin sections (50 to 100 nm thick—about 1/200 of the thickness of a single cell) before they are viewed. This is achieved by dehydrating the specimen and permeating it with a monomeric resin that polymerizes to form a solid block of plastic; the block is then cut with a fine glass or diamond knife on a special microtome. These *thin sections,* free of water and other volatile solvents, are placed on a small circular metal grid for viewing in the microscope (Figure 4–15).

Contrast in the electron microscope depends on the atomic number of the atoms in the specimen: the higher the atomic number, the more electrons are scattered and the greater is the contrast. Biological molecules are composed of atoms of very low atomic number (mainly carbon, oxygen, nitrogen, and hydrogen). To make them visible, they are usually impregnated (before or after sectioning) with the salts of heavy metals such as osmium, uranium, and lead. Different cellular constituents are revealed with various degrees of contrast according to their degree of impregnation, or "staining," with these salts. For example, lipids tend to stain darkly with osmium, revealing the location of cell membranes (Figure 4–16).

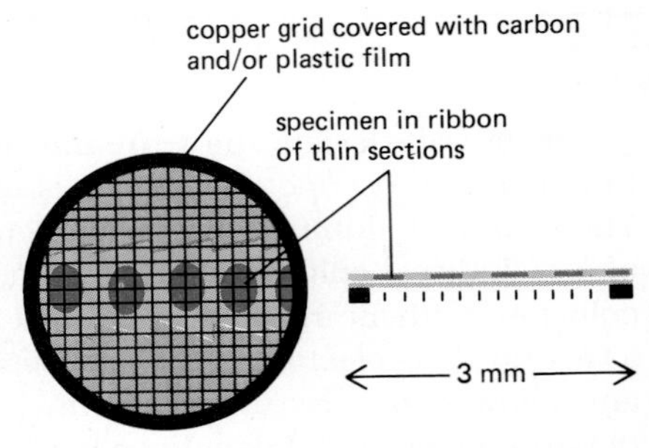

Figure 4–15 Diagram of the copper grid used to support the thin sections of a specimen in the transmission electron microscope.

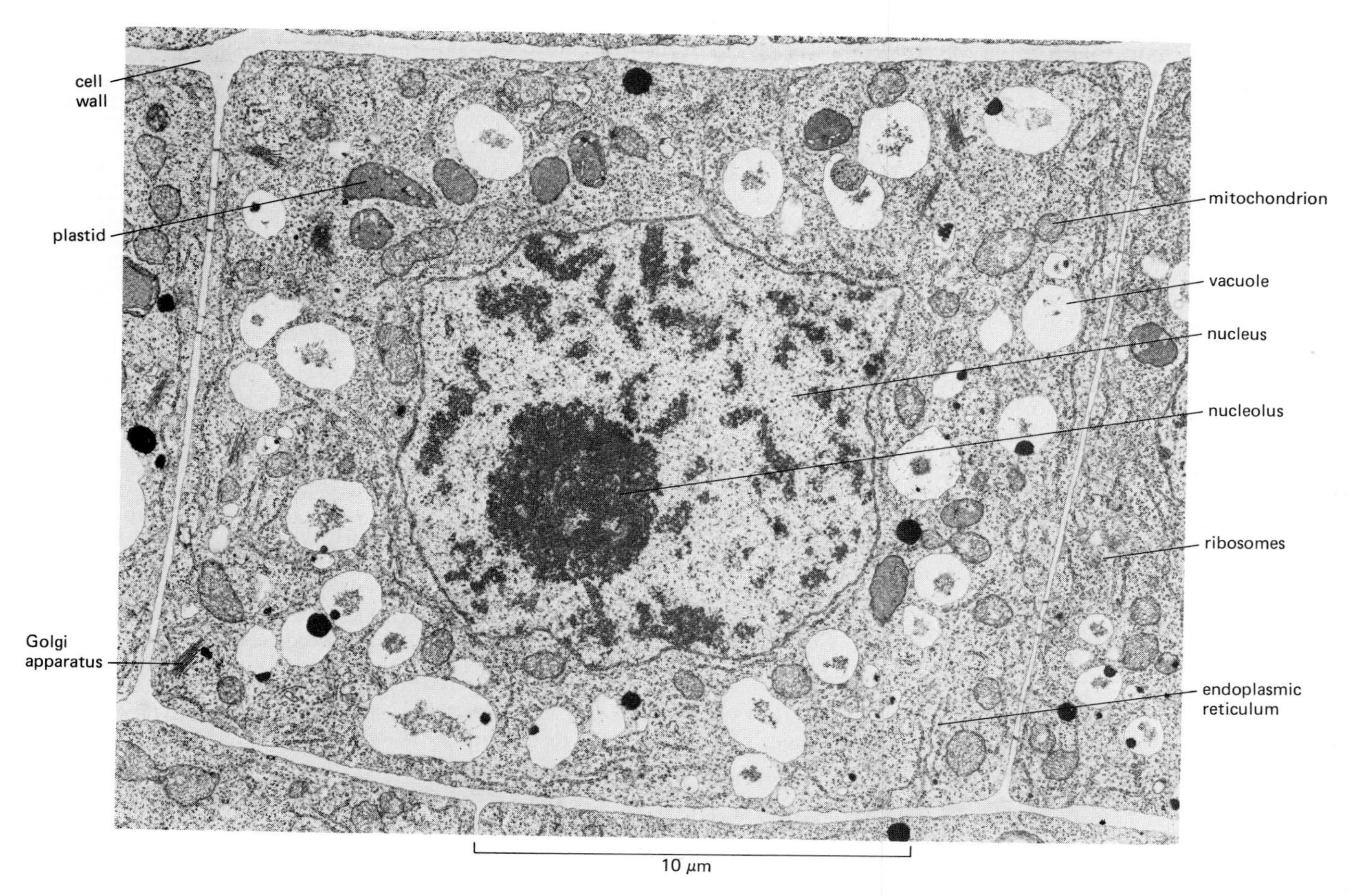

Figure 4–16 Thin section of a root tip cell from a grass, stained with osmium and other heavy metal ions. The cell wall, nucleus, vacuoles, mitochondria, endoplasmic reticulum, Golgi apparatus, and ribosomes are easily seen. (Courtesy of Brian Gunning.)

In some cases specific macromolecules can be located in thin sections by techniques adapted from light microscopy. Certain enzymes in cells can be detected by incubating the specimen with a substrate whose reaction leads to the local deposition of an electron-dense precipitate (Figure 4–17). Alternatively, as discussed on page 177, antibodies can be coupled to an indicator enzyme (usually peroxidase) or to an electron-dense marker (usually tiny spheres of metallic gold, which are referred to as *colloidal gold* particles) and then used to locate the macromolecules that the antibodies recognize.

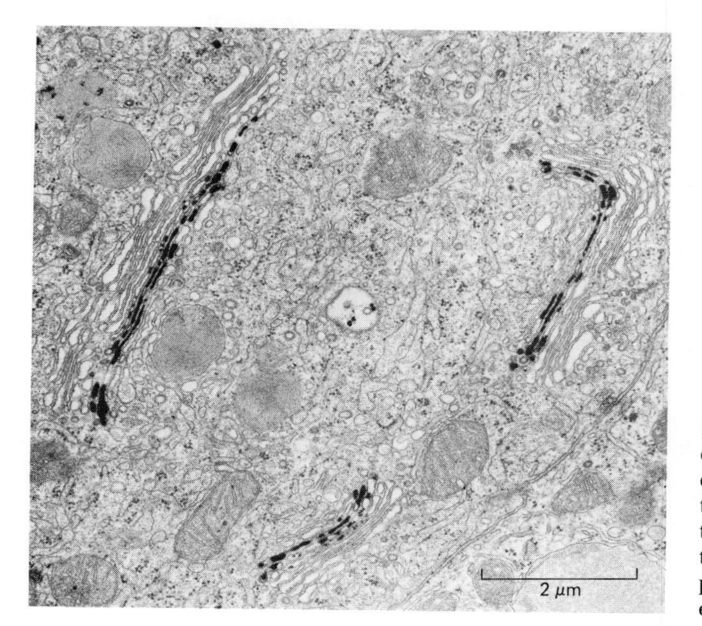

Figure 4–17 Electron micrograph of a cell showing the location of a particular enzyme (nucleotide diphosphatase) in the Golgi apparatus. A thin section of the cell was incubated with a substrate that formed an electron-dense precipitate upon reaction with the enzyme. (Courtesy of Daniel S. Friend.)

Three-dimensional Images of Surfaces Can Be Obtained by Scanning Electron Microscopy[8]

Thin sections are effectively two-dimensional slices of tissue and fail to convey the three-dimensional arrangement of cellular components. Although the third dimension can be reconstructed from hundreds of serial sections (Figure 4–18), this is a lengthy and tedious process.

Fortunately, there are more direct means to obtain a three-dimensional image. One is to examine a specimen in a **scanning electron microscope (SEM),** which is usually a smaller and simpler device than a transmission electron microscope. Whereas the transmission electron microscope uses the electrons that have passed through the specimen to form an image, the scanning electron microscope uses electrons that are scattered or emitted from the specimen's surface. The specimen to be examined is fixed, dried, and coated with a thin layer of heavy metal. The specimen is then scanned with a very narrow beam of electrons. The quantity of electrons scattered or emitted as this primary beam bombards each successive point of the metallic surface is measured and used to control the intensity of a second beam, which moves in synchrony with the primary beam and forms an image on a television screen. In this way, a highly enlarged image of the surface as a whole is built up.

The SEM technique provides a tremendous depth of focus; moreover, since the amount of electron scattering depends on the angle of the surface relative to the beam, the image has highlights and shadows that give it a three-dimensional appearance (Figure 4–19). Only surface features can be examined, however, and in most forms of SEM the resolution attainable is not very high (about 10 nm, with an effective magnification of up to 20,000 times). As a result, the technique is usually used to study whole cells and tissues rather than subcellular organelles.

To a limited degree, a three-dimensional image of the interior of a cell can be obtained from conventionally stained thin sections by tilting the specimen in the electron beam of a transmission electron microscope and photographing it from two different angles. When the resulting pair of micrographs is examined through

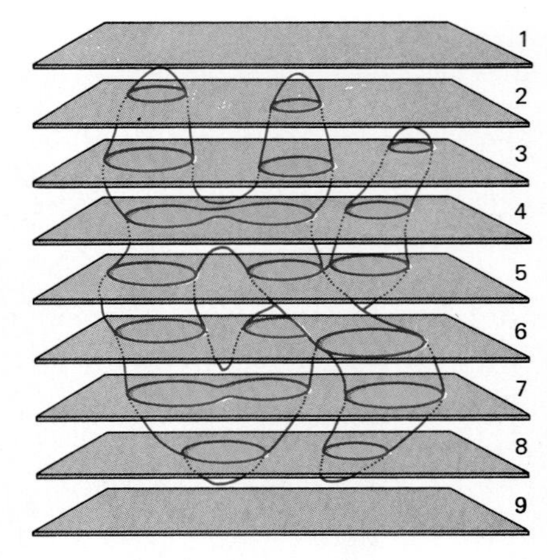

Figure 4–18 Single thin sections sometimes give misleading impressions. In this example, most sections through a cell containing a branched mitochondrion will appear to contain two or three separate mitochondria. Sections 4 and 7, moreover, might be interpreted as showing a mitochondrion in the process of dividing. However, the true three-dimensional shape can be reconstructed from serial sections.

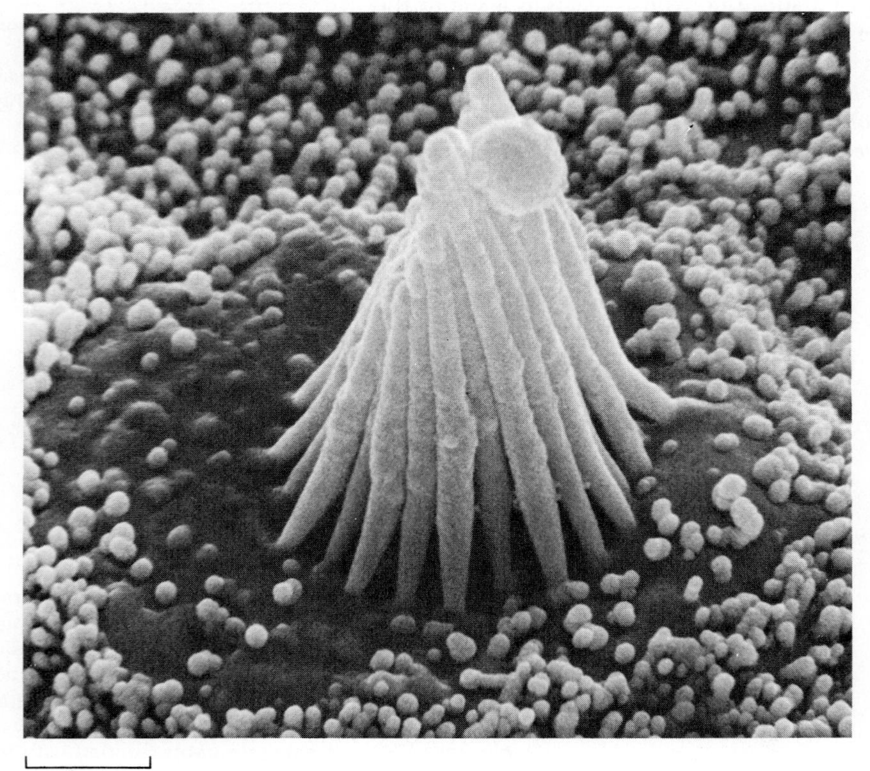

Figure 4–19 Scanning electron micrograph of the organ pipe-like arrangement of stereocilia projecting from the surface of hair cells in the inner ear. (Courtesy of R. Jacobs and A.J. Hudspeth.)

stereo glasses, a three-dimensional image is simulated. The depth of specimen that can be examined in this way depends on the penetrating power of the electrons and hence on their energy. For this reason, **high-voltage electron microscopes** have been built that accelerate the electron beam through 1,000,000 V rather than 100,000 V. These giant machines allow one to examine sections that are several micrometers thick by transmission electron microscopy.

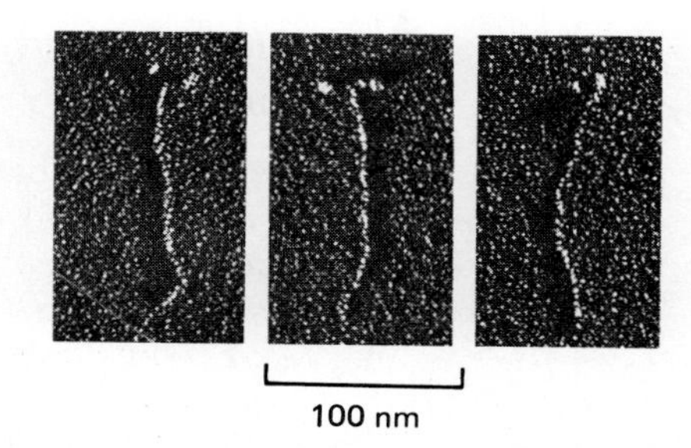

Figure 4–20 Electron micrographs of individual myosin protein molecules that have been shadowed with platinum. Myosin is a major component of the contractile apparatus of muscle; as shown here, it is composed of two globular head regions linked to a common rodlike tail. (Courtesy of Arthur Elliot.)

Metal Shadowing Allows Surface Features to Be Examined at High Resolution by Transmission Electron Microscopy[9]

The transmission electron microscope can be used to study the surface of a specimen at very high magnification, allowing individual macromolecules to be seen. As for scanning electron microscopy, a thin film of a heavy metal such as platinum is evaporated onto the dried specimen. The metal is sprayed on from an oblique angle, so as to deposit a coating that is thicker in some places than others—a process known as **shadowing** because a shadow effect is created that gives the image a three-dimensional appearance.

Some specimens coated in this way are thin enough or small enough for the electron beam to penetrate them directly; this is the case for individual molecules, viruses, and cell walls (Figure 4–20). But for thicker specimens the organic material of the cell must be dissolved away after shadowing so that only the thin metal *replica* of the surface of the specimen is left. The replica is reinforced with a film of carbon so that it can be placed on a grid and examined in the transmission electron microscope in the ordinary way (Figure 4–21).

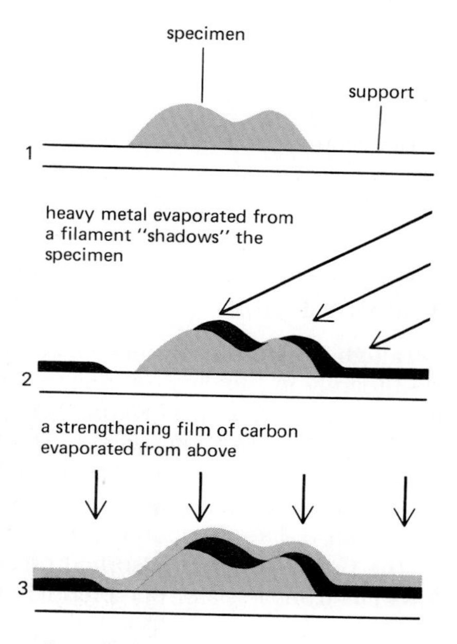

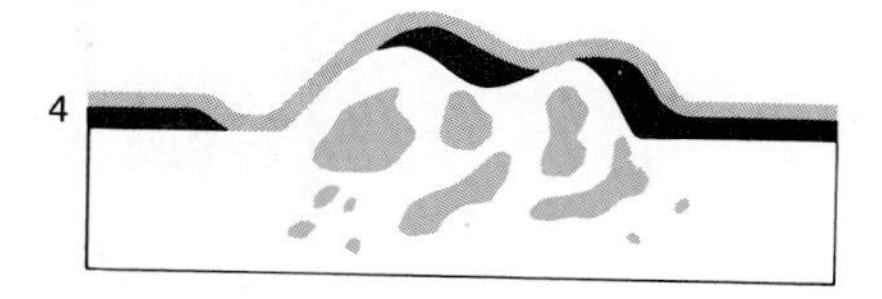

Figure 4–21 Preparation of a metal-shadowed replica of the surface of a specimen. Note that the thickness of the metal reflects the surface contours of the original specimen.

Freeze-Fracture and Freeze-Etch Electron Microscopy Provide Unique Views of the Cell Interior[10]

Two methods that use metal replicas have been particularly useful in cell biology. One of these, **freeze-fracture** electron microscopy, provides a way of visualizing the interior of cell membranes. Cells are frozen at the temperature of liquid nitrogen (−196°C) in the presence of a *cryoprotectant* (antifreeze) to prevent distortion from ice crystal formation, and then the frozen block is cracked with a knife blade. The fracture plane often passes through the hydrophobic middle of lipid bilayers, thereby exposing the interior of cell membranes. The resulting fracture faces are shadowed with platinum, the organic material is dissolved away, and the replicas are floated off and viewed in the electron microscope (as in Figure 4–21). Such replicas are studded with small bumps, called *intramembrane particles,* which represent large membrane proteins. The technique provides a convenient and dramatic way to visualize the distribution of such proteins in the plane of a membrane (Figure 4–22).

Another important and related replica method is **freeze-etch** electron microscopy, which can be used to examine either the exterior or interior of cells. In this technique the cells are again frozen at a very low temperature, and the frozen block is cracked with a knife blade. But now the ice level is lowered around the cells (and to a lesser extent within the cells) by the sublimation of water in a vacuum as the temperature is raised (a process called *freeze-drying*) (Figure 4–23). The parts of the cell exposed by this *etching* process are then shadowed as before to make a platinum replica.

Cryoprotectants cannot be used in freeze-etching since they are nonvolatile and remain in the specimen as the water sublimes. Therefore, to obtain high-resolution images by this technique, one must prevent the formation of large ice crystals by freezing the specimen extremely rapidly (at a cooling rate greater than 20°C per millisecond). One way of achieving such *rapid freezing* is by using a special device to slam the sample against a copper block cooled to −269°C with liquid helium. Particularly impressive results are achieved if rapidly frozen cells are *deep-etched* by prolonged freeze-drying. This technique exposes structures in the interior of the cell, revealing their three-dimensional organization with exceptional clarity (Figure 4–24).

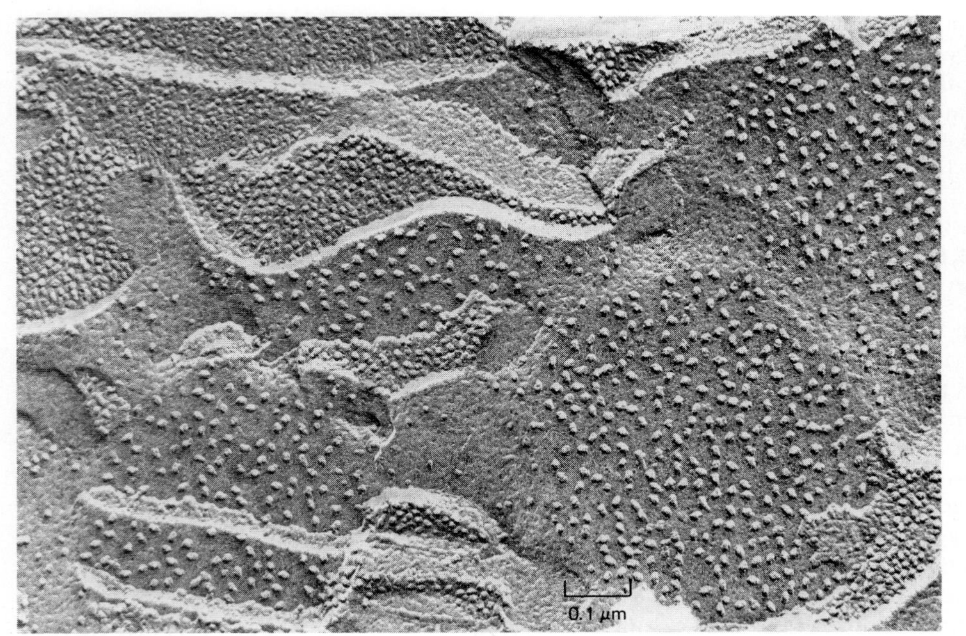

Figure 4–22 Freeze-fracture electron micrograph of the thylakoid membranes from the chloroplast of a plant cell. The thylakoid membranes, which carry out photosynthesis, are stacked up in multiple layers (see p. 367). The plane of the fracture has moved from layer to layer, passing through the middle of each lipid bilayer and exposing transmembrane proteins that have sufficient bulk in the interior of the bilayer to cast a shadow and show up as intramembrane particles in this platinum replica. The largest particles seen in the membrane are the complete photosystem II—a complex of multiple proteins. (Courtesy of L.A. Staehelin.)

Because a metal shadowed replica rather than the sample itself is viewed under vacuum in the microscope, both freeze-fracture and freeze-etch microscopy can be used to study frozen unfixed cells, thereby avoiding the risk of artifacts caused by fixation.

Negative Staining and Cryoelectron Microscopy Allow Macromolecules to Be Viewed at High Resolution[11]

Although isolated macromolecules, such as DNA or large proteins, can be visualized readily in the electron microscope if they are shadowed with a heavy metal to provide contrast (see Figure 4–20), finer detail can be seen by using **negative staining.** Here the molecules, supported on a thin film of carbon (which is nearly transparent to electrons), are washed with a concentrated solution of a heavy metal salt such as uranyl acetate. After the sample has dried, a very thin film of metal salt covers the carbon film everywhere except where it has been excluded by the presence of an adsorbed macromolecule. Because the macromolecule allows electrons to pass much more readily than does the surrounding heavy metal

Figure 4–23 Freeze-etch electron microscopy begins when the frozen specimen is fractured with a knife (A). The ice level is then lowered by the sublimation of water in a vacuum, exposing structures in the cell that were near the fracture plane (B). Following these steps, a replica of the still-frozen surface is prepared (as described in Figure 4–21) and examined in the transmission electron microscope.

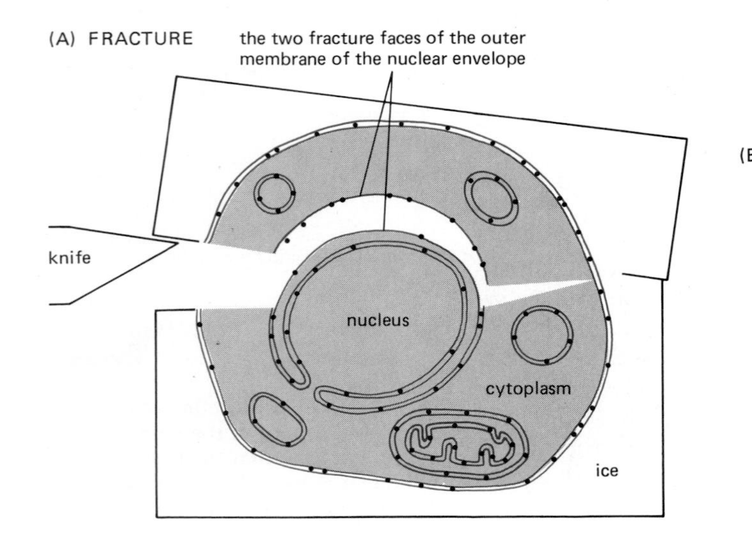

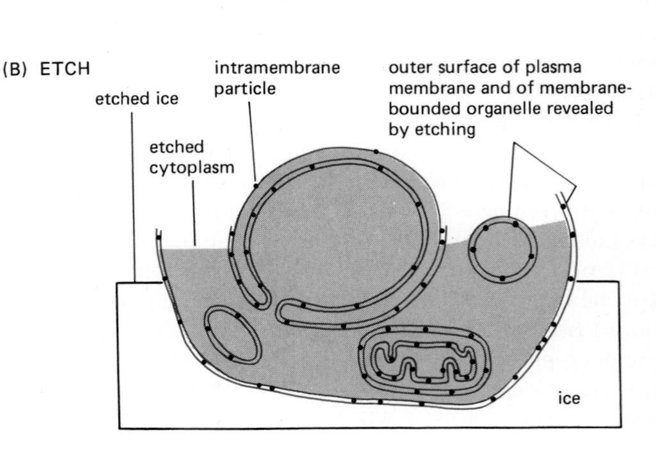

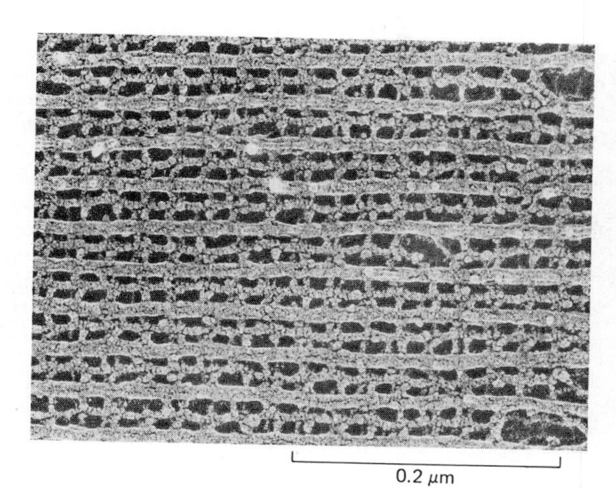

Figure 4–24 Regular array of protein filaments in an insect muscle. To obtain this image, the muscle cells were rapidly frozen in liquid helium, fractured through the cytoplasm, and subjected to deep etching. A metal replica was then prepared and examined at high magnification. (Courtesy of Roger Cooke and John Heuser.)

stain, a reversed or negative image of the molecule is created. Negative staining is especially useful for viewing large macromolecular aggregates such as viruses or ribosomes and for seeing the subunit structure of protein filaments (Figure 4–25).

Both negative staining and shadowing are capable of providing high-contrast surface views of small macromolecular assemblies; but both are limited in resolution by the size of the metal particles in the shadow or stain, which only roughly outline the surface of a molecule or macromolecular assembly. Recent methods provide an alternative that has allowed even the interior features of three-dimensional structures such as viruses to be visualized directly at high resolution. In this technique, called **cryoelectron microscopy,** a very thin (~100 nm) layer of rapidly frozen hydrated sample is prepared on a microscope grid. A special sample holder is required to keep this hydrated specimen at −160°C in the vacuum of the microscope, where it can be viewed directly without fixation, staining, or drying. The homogeneity of the vitrified water layer and the use of underfocus phase contrast allow surprisingly clear images to be obtained of these unstained specimens (Figure 4–26).

Regardless of the method used, a single protein molecule gives only a weak and ill-defined image in the electron microscope. Efforts to get better information by prolonging the time of inspection or by increasing the intensity of the illuminating beam are self-defeating because they damage and disrupt the object under examination. To discover the details of molecular structure, therefore, it is necessary to combine the information obtained from many molecules in such a way as to average out the random errors in the individual images. This is possible for viruses or protein filaments, in which the individual subunits are present in regular repeating arrays (see p. 118); it is also possible for any substance that can be made to form a crystalline array in two dimensions in which large numbers of molecules are held in identical orientation and in regularly spaced positions. Given an electron micrograph of either type of array, one can use image-processing techniques to compute the average image of an individual molecule, revealing details obscured by the random "noise" in the original picture.

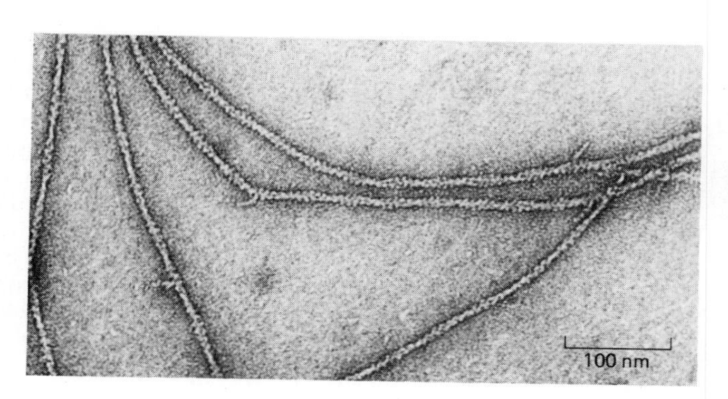

Figure 4–25 Electron micrograph of negatively stained actin filaments. Each filament is about 8 nm in diameter and is seen, on close inspection, to be composed of a helical chain of globular actin molecules. (Courtesy of Roger Craig.)

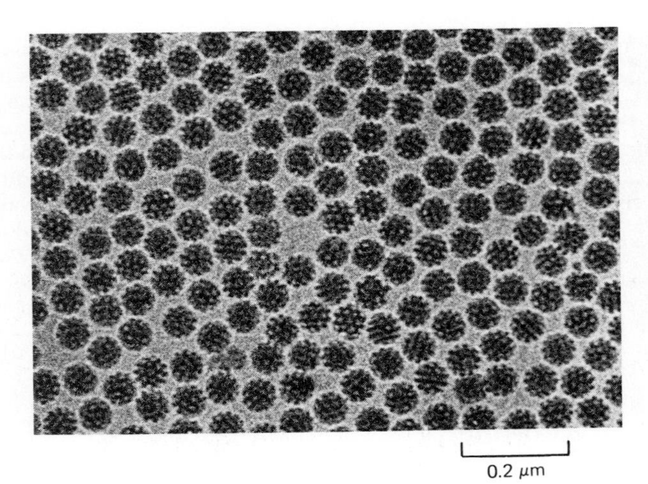

Figure 4–26 Semliki forest virus in a thin layer of unstained vitrified water viewed by cryoelectron microscopy at −160°C. A series of such micrographs are taken at different known degrees of underfocus to make use of phase contrast. A large number of these images can then be combined by image processing methods to produce a three-dimensional image at high resolution. (Courtesy of Jacques Dubochet; see also S.D. Fuller, *Cell* 48:923–934, 1987.)

Image reconstructions of this type have allowed the interior structure of an enveloped virus to be obtained to a resolution of 3.5 nm and have given the shape of an individual protein molecule to a resolution somewhat less than 0.5 nm (5 Å). But even in its most sophisticated forms, electron microscopy falls short of providing a full description of molecular structure, because the atoms in a molecule are separated by distances of only 0.1 or 0.2 nm. To resolve molecular structure in atomic detail, another technique is needed, using x-rays rather than electrons.

The Structure of an Object in a Crystalline Array Can Be Deduced from the Diffraction Pattern It Creates[12]

Like light, x-rays are a form of electromagnetic radiation, but because of their much shorter wavelength, they allow resolution of much finer detail. Unlike visible light or beams of electrons, however, x-rays cannot be focused to form an image of the usual sort after passing through a specimen. Instead, the structure of the specimen is deduced by the technique of **x-ray diffraction.**

Consider a single object (such as a single molecule) placed in a beam of radiation of any sort whose wavelength is small compared to the dimensions of the object. The object will scatter some of the radiation. The scattered radiation can be thought of as consisting of a family of overlapping waves, each emanating from a different part of the object. As the waves overlap, they undergo interference, producing a distribution of radiation known as a **diffraction pattern.** The diffraction pattern could be recorded on a photographic plate placed at some distance from the object and described in terms of the amounts of scattered radiation sent out by the object in different directions (Figure 4–27). The structure of the object determines the form of the diffraction pattern. Conversely, given a full description of the diffraction pattern, it is possible, in theory, to calculate the

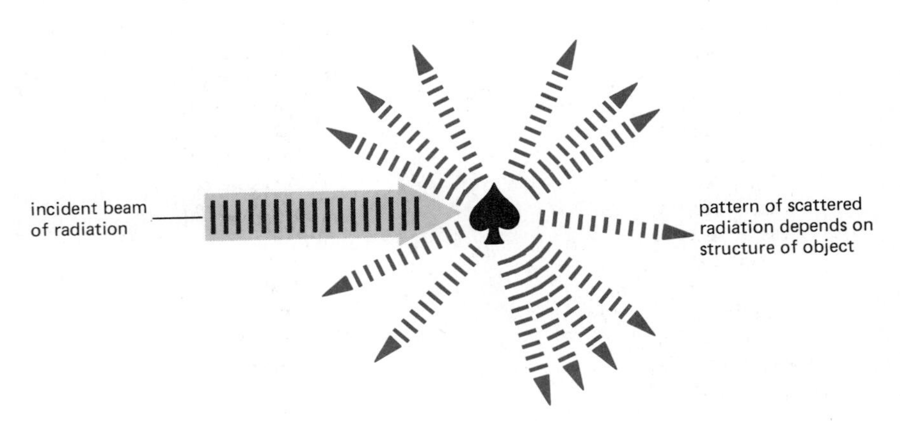

Figure 4–27 The scattering of radiation by a single object whose dimensions are comparable to the wavelength of the radiation. Radiation falling on the object is scattered with different intensities in different directions. The intensity of the scattered beam in a given direction depends on the way in which radiation scattered from one part of the object interferes with that scattered from another part. In the diagram, the resultant intensity of scattering in the various possible directions is indicated by the number of colored arrows radiating from the object.

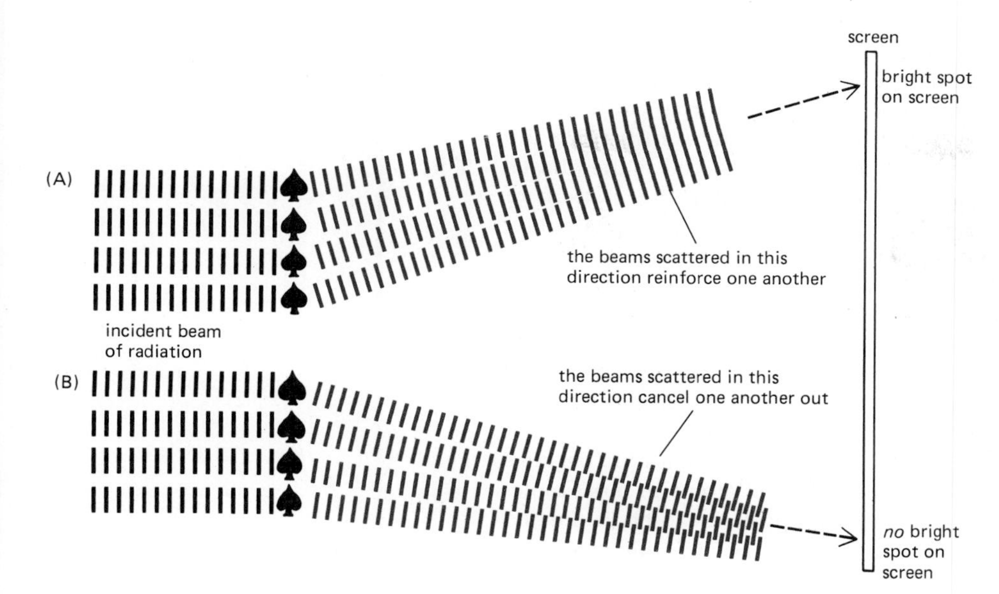

Figure 4–28 The scattering of radiation by a crystal. When many identical objects are arranged in a crystalline array, the radiation scattered by each object interferes with that scattered by the others. Only in certain directions (depending on the spacing of the objects in the array) do the individual scattered beams reinforce, producing bright spots in the diffraction pattern of the array. The intensity of each bright spot varies according to the intensity with which an individual object in the array would scatter radiation in that direction if the object were examined in isolation (see Figure 4–27).

structure of the object that produced it. In practice, the diffraction pattern due to a single molecule would be far too faint and erratic for such a purpose.

Suppose now that many identical objects are arranged in a crystalline array and again illuminated with a beam of radiation (Figure 4–28). Now the total amount of scattered radiation is much greater. However, the radiation scattered by each object will interfere with that scattered by the others. Only in certain directions, depending on the spacing of the objects in the array, will the individual scattered beams reinforce, producing a bright spot in the diffraction pattern. The complete diffraction pattern of the crystalline array will thus consist of many such discrete bright spots of differing intensities (Figure 4–29).

The relative intensities of the various spots in the diffraction pattern will depend on the scattering properties of the individual objects in the array. In fact, the intensity of a given spot is proportional to the average intensity of the radiation that would be scattered in the same direction from a representative single object standing alone. Thus, while the *positions* of the spots in the diffraction pattern depend on the arrangement of objects in the array, the *intensities* of the spots give information as to the internal structure of a representative single object. This information, moreover, is precise and plentiful because it is obtained by combining contributions from a very large number of equivalent sources. In fact, from a full description of the diffraction pattern of the crystalline array, it is possible to calculate the structure of the individual objects from which it is built.

X-ray Diffraction Reveals the Three-dimensional Arrangement of the Atoms in a Molecule[13]

If diffraction patterns are to be used to analyze molecular structure, the diffracted radiation must have a wavelength shorter than the distance between the atoms in a molecule. Since x-rays can have wavelengths of around 0.1 nm (which is the diameter of a hydrogen atom), they are ideally suited for investigating the arrangement of individual atoms in molecules—something that cannot be done with even the highest-resolution electron microscope. A further virtue of x-rays is that they have greater penetrating power than electrons; thus much thicker specimens can be used. Finally, because a vacuum is not required, thick hydrated specimens can be examined, and the distortions induced by the preparative procedures necessary for most forms of electron microscopy are thereby avoided.

For high-resolution studies, large, highly ordered crystals are required (Figure 4–30). The x-rays that penetrate the crystal are scattered chiefly by the electrons in its constituent atoms. Large atoms with many electrons scatter x-rays more than do small atoms, so atoms of C, N, O, and P are much more readily detected

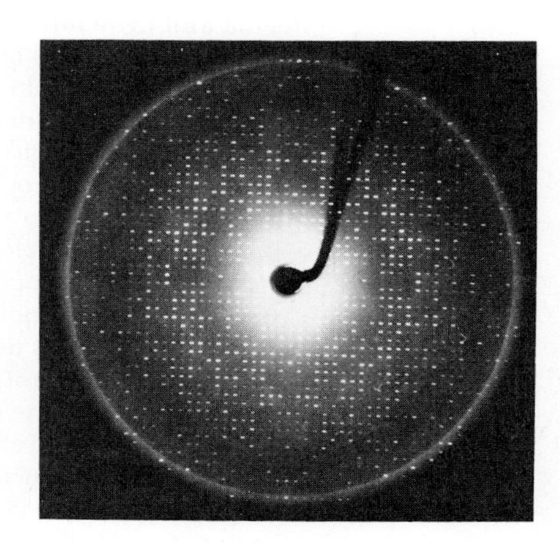

Figure 4–29 Part of the x-ray diffraction pattern obtained from a protein crystal. This particular crystal was used to help determine the atomic structure of the proteolytic enzyme trypsin. (Courtesy of Robert Stroud.)

than atoms of H; for the same reason, very heavy atoms create an intense beacon of scattered x-rays. Deducing the three-dimensional structure of a large molecule from the diffraction pattern of its crystal is a complex task and was not achieved for a protein molecule until 1960 (Table 4–3). It frequently requires measuring the positions and intensities of hundreds of thousands of spots, as well as the phases of the waves at each spot (which requires comparing the diffraction pattern of crystals with and without heavy atoms bound to specific sites on the molecule).

In recent years, x-ray diffraction analysis has become increasingly automated. The diffracted x-rays are measured electronically by sophisticated detectors that greatly speed the process of data collection, and powerful computers perform the huge numbers of calculations required. The slowest step now is obtaining suitable crystals of the macromolecule of interest; this requires large amounts of the pure macromolecule and often involves years of trial-and-error searching for the proper crystallization conditions.

Despite the difficulties, x-ray diffraction is a widely used technique because it is the only way to determine the detailed arrangement of atoms in most molecules. With reasonably good crystals, the structure of a protein can be calculated to a resolution of 0.3 nm, revealing the main course of the polypeptide chain but few other details. With very high-quality crystals (and much more work), a resolution of 0.15 nm is obtainable, revealing the position of almost all of the nonhydrogen atoms in the protein. The structures of more than a hundred proteins and of several small RNA and DNA molecules have now been determined in this way.

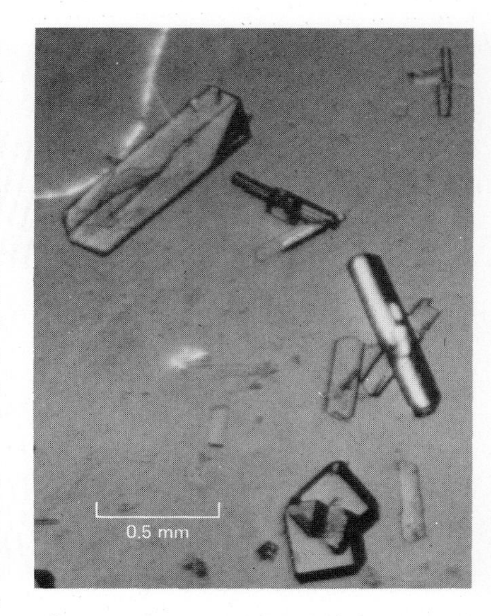

Figure 4–30 Crystals of the enzyme glycogen phosphorylase viewed in a light microscope. (Courtesy of Robert Fletterick.)

Table 4–3 Landmarks in the Development of X-ray Crystallography and Its Application to Biological Molecules

1864	**Hoppe-Seyler** crystallized, and named, the protein hemoglobin.
1895	**Röntgen** observed that a new form of penetrating radiation, which he named x-rays, was produced when cathode rays (electrons) hit a metal target.
1912	**Von Laue** obtained the first x-ray diffraction patterns by passing x-rays through a crystal of zinc sulfide.
	W. L. Bragg proposed a simple relationship between an x-ray diffraction pattern and the arrangement of atoms in a crystal that produces the pattern.
1926	**Summer** obtained crystals of the enzyme urease from extracts of jack beans and demonstrated that proteins possess catalytic activity.
1931	**Pauling** published his first essays on "The Nature of the Chemical Bond," detailing the rules of covalent bonding.
1934	**Bernal and Crowfoot** presented the first detailed x-ray diffraction patterns of a protein obtained from crystals of the enzyme pepsin.
1935	**Patterson** developed an analytical method for determining interatomic spacings from x-ray data.
1941	**Astbury** obtained the first x-ray diffraction pattern of DNA.
1951	**Pauling and Corey** proposed the structure of a helical conformation of a chain of L-amino acids—the α helix—and the structure of the β sheet, both of which were later found in many proteins.
1953	**Watson and Crick** proposed the double-helix model of DNA, based on x-ray diffraction patterns obtained by **Franklin and Wilkins.**
1954	**Perutz** and colleagues developed heavy-atom methods to solve the phase problem in protein crystallography.
1960	**Kendrew** described the first detailed structure of a protein (sperm whale myoglobin) to a resolution of 0.2 nm, and **Perutz** proposed a lower-resolution structure of the larger protein hemoglobin.
1966	**Phillips** described the structure of lysozyme, the first enzyme to be analyzed in detail.
1976	**Kim and Rich** and **Klug** and colleagues described the detailed three-dimensional structure of tRNA determined by x-ray diffraction.
1977–1978	**Holmes** and **Klug** determined the structure of tobacco mosaic virus (TMV), and **Harrison** and **Rossman** determined the structure of two small spherical viruses.

Summary

Many light microscopic techniques are available for observing cells. Stained fixed cells can be studied with conventional optics, while labeled antibodies can be used to locate specific molecules in cells with the fluorescence microscope. Cells in their natural living state can be seen with phase-contrast, interference, or dark-field optics. Such studies of living cells by light microscopy are facilitated by electronic image-processing techniques, which greatly enhance sensitivity and increase resolution.

Determining the detailed structure of the membranes and organelles in cells requires the higher resolution attainable in the transmission electron microscope. The shapes of isolated macromolecules that have been shadowed with a heavy metal or outlined by negative staining can also be visualized readily by electron microscopy. But the precise position of each atom in a molecule can be determined only if the molecule will assemble into large crystals, in which case the complete three-dimensional structure of the molecule can be deduced by x-ray diffraction analysis.

Probing Chemical Conditions in the Interior of Living Cells

The classical methods of microscopy give good views of cell architecture but not much information about cell chemistry. We have seen that antibodies can be used to locate specific macromolecules in cells, but it is also important to be able to investigate the distribution and concentration of the small molecules. The concentrations of fundamental metabolites such as ATP or glucose and of inorganic ions must be regulated precisely and rapidly in order to maintain life, and they can be very different in different regions of cells and tissues. Moreover, small molecules such as cyclic AMP, Ca^{2+}, and H^{+} function as intracellular "messengers," and it is important to be able to monitor changes in their concentrations in response to extracellular signals. In this section we discuss some methods adapted from conventional chemistry that allow the chemical conditions inside a cell to be determined while the cell is alive.

Nuclear Magnetic Resonance (NMR) Can Be Used to Assay the Chemistry of Populations of Living Cells[14]

The nuclei of many atoms have a magnetic moment: that is, they have an intrinsic magnetization, like bar magnets. Because their magnetic behavior is influenced by surrounding atoms, these nuclei can be made to reveal the chemical nature of their environment by a method called **nuclear magnetic resonance (NMR)** spectroscopy, which is harmless to living cells. When atomic nuclei with a magnetic moment are placed in a strong magnetic field, they will adopt one of a limited number of permitted orientations. Each orientation will have an energy that depends on the strength of the field and on the chemical environment. If a set of atoms in identical chemical environments are irradiated with radio waves, the energy of these waves will be strongly absorbed when the waves have a precisely defined frequency that corresponds to the energy difference between two such nuclear orientations. This is the so-called *resonance frequency*. A sample of tissue, containing atoms in a variety of different molecules and environments, will absorb energy at a variety of different resonance frequencies. The graph of absorbance against resonance frequency for a given sample constitutes its *NMR spectrum*. This spectrum reflects the structure and relative quantities of each of the different molecules present that contain magnetic nuclei.

In chemistry laboratories, NMR is a routine analytical technique for determining the molecular structure of small molecules in solution. Improvements in instrumentation have made it possible to apply NMR methods to problems of biological interest. For example, the NMR signal from protons (hydrogen nuclei) is widely used to investigate proteins, nucleic acids, and other macromolecules in

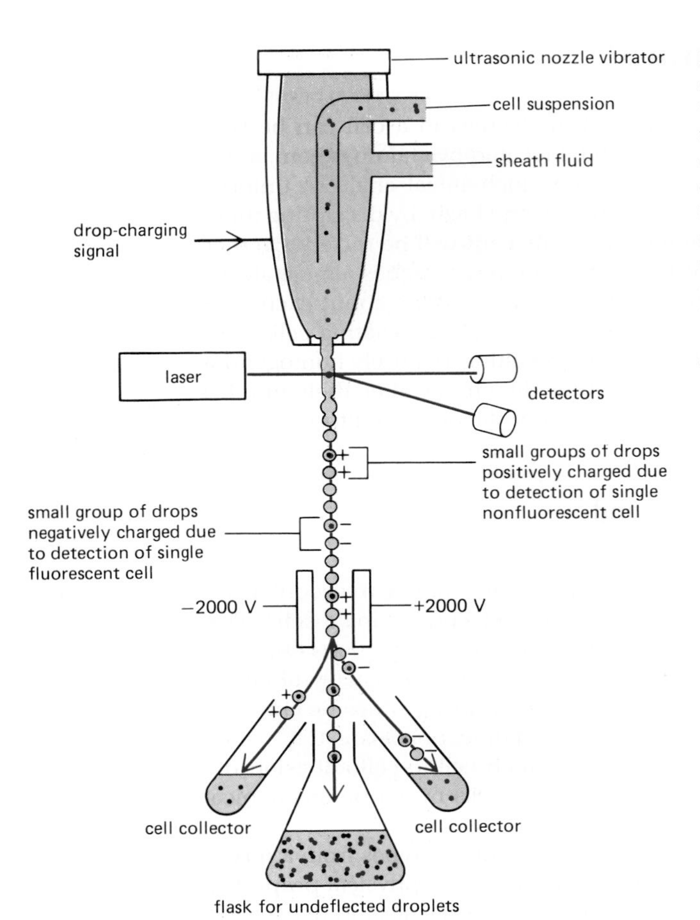

Figure 4–38 A fluorescence-activated cell sorter. When a cell passes through the laser beam, it is monitored for fluorescence. Droplets containing single cells are given a negative or positive charge, depending on whether the cell is fluorescent or not. Droplets are then deflected by an electric field into collection tubes according to their charge. Note that the cell concentration must be adjusted so that most droplets contain no cell; most of these flow to a waste container together with any cell clumps.

Cells Can Be Grown in a Culture Dish[20]

Given the appropriate conditions, most kinds of plant and animal cells will live, multiply, and even express differentiated properties in a tissue-culture dish. The cells can be watched under the microscope or analyzed biochemically, and the effects of adding or removing specific molecules such as hormones or growth factors can be explored. In addition, in a mixed culture the interactions between one cell type and another can be studied. Experiments on cultured cells are sometimes said to be carried out *in vitro* (literally, "in glass"); by contrast, experiments on intact organisms are said to be carried out *in vivo.* These two terms are used in a very different sense by most biochemists and cell biologists—for whom *in vitro* refers to biochemical reactions occurring outside living cells, while *in vivo* refers to any reaction taking place inside a living cell.

Tissue culture began in 1907 with an experiment designed to settle a controversy in neurobiology. The hypothesis under examination was known as the *neuronal doctrine,* which states that each nerve fiber is the outgrowth of a single nerve cell and not the product of the fusion of many cells. To test this contention, small pieces of spinal cord were placed on clotted tissue fluid in a warm, moist chamber and observed at regular intervals under the microscope. After a day or so, individual nerve cells could be seen extending long, thin processes into the clot. Thus the neuronal doctrine was validated, and the foundations for the cell-culture revolution were laid.

The original experiments in 1907 involved the culture of small tissue fragments, or **explants.** Today, cultures are more commonly made from suspensions of cells dissociated from tissues as described above. Unlike bacteria, most tissue cells are not adapted to living in suspension and require a solid surface on which to grow and divide. This mechanical support was originally provided by a clot of tissue fluid but is now usually the surface of a plastic tissue-culture dish (Figure

4–39). Cells vary in their requirements, however, and some will not grow or differentiate unless the culture dish is coated with specific extracellular matrix components, such as collagen.

Cultures prepared directly from the tissues of an organism, either with or without an initial cell fractionation step, are called **primary cultures.** In most cases, cells in primary cultures can be removed from the culture dish and used to form a large number of **secondary cultures;** they may be repeatedly subcultured in this way for weeks or months. Such cells often display many of the differentiated properties appropriate to their origin: fibroblasts continue to secrete collagen; cells derived from embryonic skeletal muscle fuse to form giant muscle fibers that spontaneously contract in the culture dish; nerve cells extend axons that are electrically excitable and make synapses with other nerve cells; and epithelial cells form extensive sheets with many of the properties of an intact epithelium. Since these phenomena occur in culture, they are accessible to study in ways that are not possible in intact tissues.

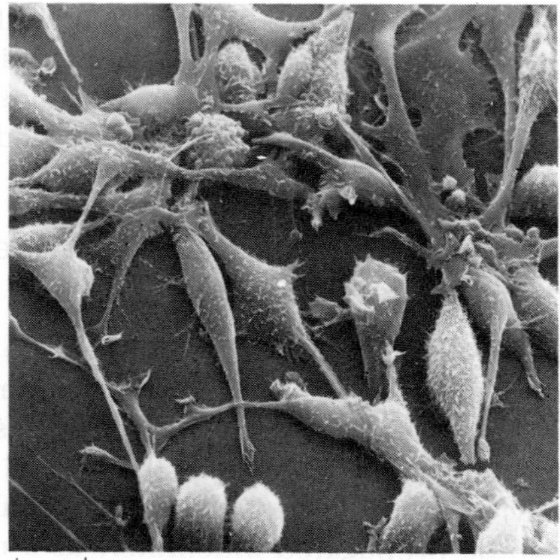

Figure 4–39 Scanning electron micrograph of rat fibroblasts growing on a plastic surface in tissue culture. (Courtesy of Guenter Albrecht-Buehler.)

Chemically Defined Media Permit Identification of Specific Growth Factors[21]

Until the early 1970s, tissue culture was something of a blend of science and witchcraft. Although tissue fluid clots were replaced by dishes of liquid media containing well-defined mixtures of salts, amino acids, and vitamins, most media also included a proportion of some poorly defined biological material, such as horse serum or fetal calf serum or a crude extract made from chick embryos. Such media are still used today for most routine tissue culture (Table 4–4), but they are inappropriate for determining the specific requirements for the growth and differentiation of a particular type of cell.

This difficulty led to the development of various chemically defined media that support the growth of different types of cells. In these media, each component is a known molecule. In addition to small molecules, these media usually contain one or more of the various protein **growth factors** that most cells require in order to survive and proliferate in culture: for example, certain nerve cells need trace amounts of *nerve growth factor (NGF)* to differentiate and survive in culture as

Table 4–4 Composition of a Typical Medium Suitable for the Cultivation of Mammalian Cells

Amino Acids	Vitamins	Salts	Miscellaneous
Arginine	Biotin	NaCl	Glucose
Cystine	Choline	KCl	Penicillin
Glutamine	Folate	NaH_2PO_4	Streptomycin
Histidine	Nicotinamide	$NaHCO_3$	Phenol red
Isoleucine	Pantothenate	$CaCl_2$	Whole serum
Leucine	Pyridoxal	$MgCl_2$	
Lysine	Thiamine		
Methionine	Riboflavin		
Phenylalanine			
Threonine			
Tryptophan			
Tyrosine			
Valine			

Glucose is used at a concentration of 5 to 10 mM. The amino acids are all in the L form and, with one or two exceptions, are used at concentrations of 0.1 or 0.2 mM; vitamins are used at a 100-fold lower concentration, that is, about 1 μM. Serum, which is usually from horse or calf, is added to make up 10% of the total volume. Penicillin and streptomycin are antibiotics added to suppress the growth of bacteria. Phenol red is a pH indicator dye whose color is monitored to assure a pH of about 7.4.

Cultures are usually grown in a plastic or glass container with a suitably prepared surface that allows the attachment of cells. The containers are kept in a incubator at 37°C in an atmosphere of 5% CO_2, 95% air.

they occur in the intact polypeptide chain. This was traditionally achieved by comparing the sequences of different sets of overlapping peptide fragments obtained by cleaving the same protein with different proteolytic enzymes.

Improvements in protein sequencing technology have greatly increased its speed and sensitivity, allowing analysis of minute samples; the sequence of several dozen amino acids at the amino-terminal end of a peptide can be obtained overnight from a few micrograms of protein—the amount available from a single band on an SDS polyacrylamide gel. This has been important for characterizing many minor cell proteins, such as the receptors for steroid and polypeptide hormones. Knowing the sequence of as few as 20 amino acids of a protein is frequently enough to allow a DNA probe to be designed that allows cloning of its gene (see p. 262). Once the gene has been isolated, the rest of the protein's amino acid sequence can be deduced from the DNA sequence by reference to the genetic code. This is a major advantage because, even with automation, the direct determination of the entire amino acid sequence of a protein is a major undertaking. A protein of 100 residues can often be sequenced in a month of hard work, but the difficulty increases steeply with the length of the polypeptide chain, and the chemical peculiarities of individual peptide fragments prevent the process from being routine. Since DNA sequencing can be done so quickly and simply (see below), the sequences of most proteins are now determined largely from the nucleotide sequences of their genes.

Summary

Populations of cells can be analyzed biochemically by disrupting them and fractionating their contents by ultracentrifugation. Further fractionations allow functional cell-free systems to be developed; such systems are required to determine the molecular details of complex cellular processes. For example, protein synthesis, DNA replication, RNA splicing, and various types of intracellular transport are all currently being studied in this way.

The major proteins in soluble cell extracts can be purified by column chromatography; depending on the type of column matrix, biologically active proteins can be separated according to their molecular weight, hydrophobicity, charge characteristics, or affinity for other molecules. In a typical purification the sample is passed through several different columns in turn—the enriched fractions obtained from one column being applied to the next. Once a protein has been purified to homogeneity, its biological activities can be examined in detail. In addition, a small part of the protein's amino acid sequence can be determined and its gene can be cloned; the remaining amino acid sequence is then obtained from the nucleotide sequence of the gene.

The molecular weight and subunit composition of even very small amounts of a protein can be determined by SDS polyacrylamide-gel electrophoresis. In two-dimensional gel electrophoresis, proteins are resolved as separate spots by isoelectric focusing in one dimension followed by SDS polyacrylamide-gel electrophoresis in a second dimension. These electrophoretic separations can be applied even to proteins that are normally insoluble in water.

Tracing Cellular Molecules with Radioactive Isotopes and Antibodies

Almost any property of a molecule—physical, chemical, or biological—can in principle be used as a means of detecting it. In cell biological studies, molecules are often monitored either by their optical properties—whether in their native state or after staining them with a dye—or by their biochemical activity. In this section we consider two other detection methods that have been particularly useful: those involving *radioisotopes* and those utilizing *antibodies*. Each of these methods is capable of detecting specific molecules in a complex mixture with great sensitivity: under optimal conditions they can detect fewer than 1000 molecules in a sample.

Radioactive Atoms Can Be Detected with Great Sensitivity[31]

Most naturally occurring elements are a mixture of slightly different *isotopes.* These differ from each other in the mass of their atomic nuclei, but because they have the same number of electrons, they have the same chemical properties. The nuclei of radioactive isotopes, or **radioisotopes,** are unstable and undergo random disintegration to produce different atoms. In the course of these disintegrations, energetic subatomic particles such as electrons or radiations such as γ-rays are given off.

Although naturally occurring radioisotopes are rare (because of their instability), radioactive atoms can be produced in large amounts in nuclear reactors in which stable atoms are bombarded with high-energy particles. As a result, many biologically important elements are readily available in radioisotopically labeled form (Table 4–11). The radiation they emit is detected in various ways. Electrons (β particles) can be detected in a *Geiger counter* by the ionization they produce in a gas. They can also be measured in a *scintillation counter* by the small flashes of light they induce in a scintillation fluid. These methods make it possible to measure the quantity of a particular radioisotope present in a biological specimen. It is also possible to localize the isotope by using *autoradiography* to detect its effect on the grains of silver halide in a photographic emulsion (see p. 176). All these methods of detection are capable of extreme sensitivity; in favorable circumstances, nearly every disintegration—and therefore every radioactive atom that decays—can be detected.

Table 4–11 Some Radioisotopes in Common Use in Biological Research

Isotope	Half-Life
^{32}P	14 days
^{131}I	8.1 days
^{35}S	87 days
^{14}C	5570 years
^{45}Ca	164 days
^{3}H	12.3 years

The isotopes are arranged in decreasing order of the energy of the β radiation (electrons) they emit. ^{131}I also emits γ radiation. The *half-life* is the time required for 50% of the atoms of an isotope to disintegrate.

Radioisotopes Are Used to Trace Molecules in Cells and Organisms[32]

One of the earliest uses of radioactivity in biology was to trace the chemical pathway of carbon during photosynthesis. Unicellular green algae were maintained in an atmosphere containing radioactively labeled CO_2 ($^{14}CO_2$), and at various times after they had been exposed to sunlight their soluble contents were separated by paper chromatography. Small molecules containing ^{14}C atoms derived from CO_2 were detected by a sheet of photographic film placed over the dried paper chromatogram. In this way most of the principal components in the photosynthetic pathway from CO_2 to sugar were identified.

Radioactive molecules can be used to follow the course of almost any process in cells. In a typical experiment a precursor in radioactive form is added to cells so that the radioactive molecules mix with the preexisting unlabeled ones; both are treated identically by the cell, since they differ only in the weight of their atomic nuclei. Changes in the location or chemical form of the radioactive molecules can be followed as a function of time. The resolution of such experiments is often sharpened by using a **pulse-chase** labeling protocol, in which the radioactive material (the *pulse*) is added for only a very brief period and then washed away and replaced by nonradioactive molecules (the *chase*). Samples are taken at regular intervals, and the chemical form or location of the radioactivity is identified for each sample (Figure 4–54).

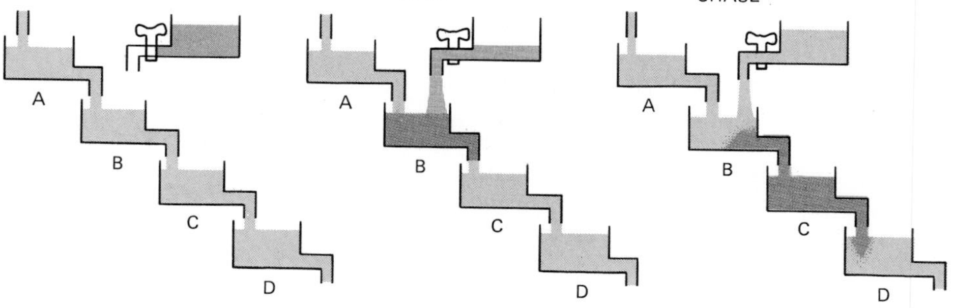

Figure 4–54 The logic of a typical pulse-chase experiment using radioisotopes. The chambers labeled A, B, C, and D represent either different compartments in the cell (detected by autoradiography or by cell-fractionation experiments) or different chemical compounds (detected by chromatography or other chemical methods).

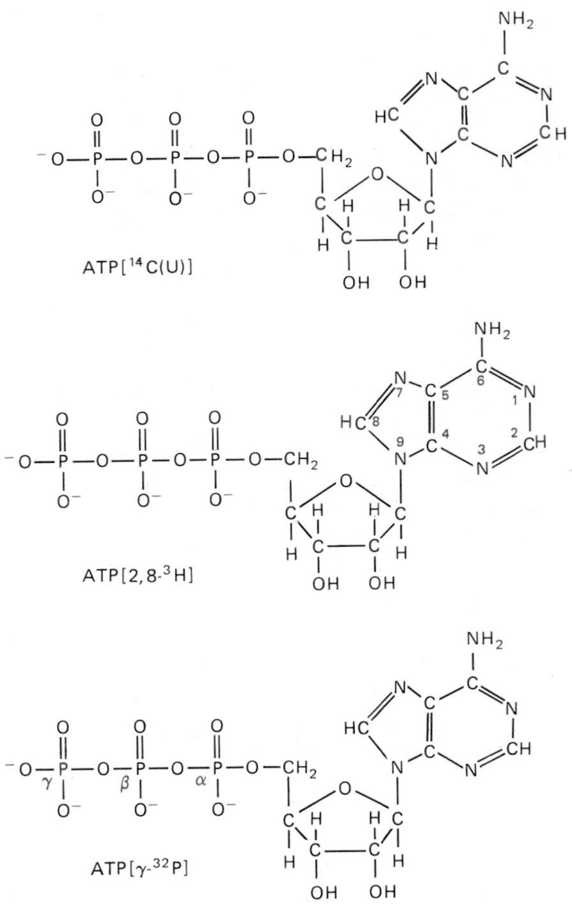

ATP[$^{14}C(U)$]

ATP[2,8-^{3}H]

ATP[γ-^{32}P]

Figure 4–55 Three commercially available radioactive forms of ATP, with the radioactive atoms shown in color. The nomenclature used to identify the position and type of the radioactive atoms is also shown.

Radioisotopic labeling is uniquely valuable as a way of distinguishing between molecules that are chemically identical but have different histories—for example, those that differ in their time of synthesis. The use of radioactive tracers, in fact, has shown that almost all the molecules in a living cell are continually being degraded and replaced, even when the cell is not growing and is apparently in a steady state. Such "turnover" processes, which sometimes take place very slowly, would be almost impossible to detect without radioisotopes.

Today nearly all common small molecules are available in radioactive form from commercial sources, and virtually any biological molecule, no matter how complicated, can be radioactively labeled. Compounds are often made with radioactive atoms incorporated at particular positions in their structure, enabling the separate fates of different parts of the same molecule to be followed during biological reactions (Figure 4–55).

One of the important uses of radioactivity in cell biology is in the localization by **autoradiography** of radioactive compounds in sections of whole cells or tissues. In this procedure living cells are briefly exposed to a "pulse" of a specific radioactive compound and incubated for a variable period before being fixed and processed for light or electron microscopy. Each preparation is then overlaid with a thin film of photographic emulsion. After remaining in the dark for a number of days—during which time the radioisotope decays—the emulsion is developed. The position of the radioactivity in each cell can be determined by the position of the developed silver grains. For example, incubation of cells with a radioactive DNA precursor (*^{3}H-thymidine*) shows that DNA is made in the nucleus and remains there. By contrast, labeling cells with a radioactive RNA precursor (*^{3}H-uridine*) reveals that RNA is initially made in the cell nucleus and then moves rapidly into the cytoplasm.

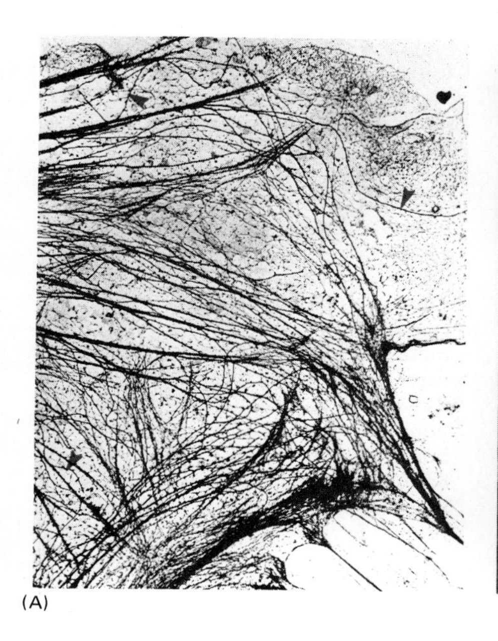
(A)

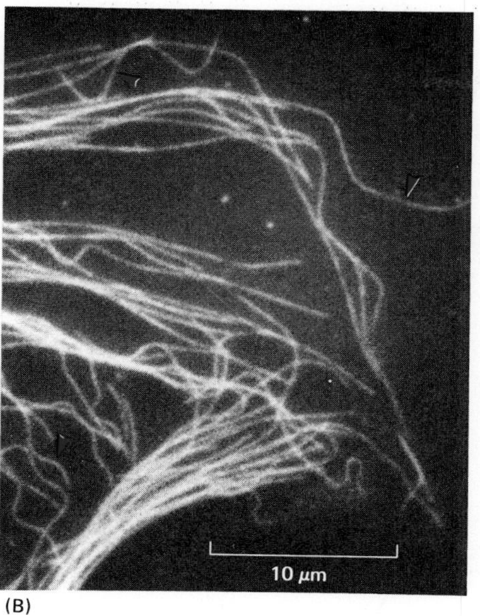

(B)

Figure 4–56 (A) An electron micrograph of the periphery of a cultured epithelial cell showing the distribution of microtubules and other filaments. (B) The same area stained with a fluorescent antibody to tubulin, the protein subunit of microtubules, using the technique of indirect immunocytochemistry (see Figure 4–58). Arrows indicate individual microtubules that are readily recognizable in the two figures. (From M. Osborn, R. Webster, and K. Weber, *J. Cell Biol.* 77:R27–R34, 1978. Reproduced by copyright permission of the Rockefeller University Press.)

Antibodies Can Be Used to Detect and Isolate Specific Molecules[33]

Antibodies are proteins produced by vertebrates as a defense against infection (see Chapter 18). They are unique among proteins because they are made in millions of different forms, each with a different binding site that specifically recognizes the molecule (called an *antigen*) that induced its production. The precise antigen specificity of antibodies makes them powerful tools for the cell biologist. Labeled with fluorescent dyes, they are invaluable for locating specific molecules in cells by fluorescence microscopy (Figure 4–56); labeled with electron-dense particles such as colloidal gold spheres, they are used to locate particular molecules at high resolution in the electron microscope (Figure 4–57). As biochemical tools they are used to detect and quantify molecules in cell extracts and to identify specific proteins after they have been fractionated by electrophoresis in polyacrylamide gels. On a preparative scale, antibodies can be coupled to an inert matrix to produce an affinity column that is then used either to purify a specific molecule from a crude cell extract or, if the molecule is on the cell surface, to pick out specific types of living cells from a heterogeneous population.

The sensitivity of antibodies as probes for detecting specific molecules in cells and tissues is frequently enhanced by a signal-amplification method. For example, although a marker molecule such as a fluorescent dye can be linked directly to an antibody used for specific recognition (the *primary antibody*), a stronger signal is achieved by using an unlabeled primary antibody and then detecting it with a group of labeled *secondary antibodies* that bind to it (Figure 4–58A).

An alternative amplification system exploits the exceptionally high binding affinity of *biotin* (a small water-soluble vitamin) for *streptavidin* (a bacterial protein). If the primary antibody has been covalently coupled to biotin, streptavidin can be directly labeled with a marker and used in place of a secondary antibody. The streptavidin can also be used to link a single biotin-coupled antibody molecule to a whole network of biotin-coupled marker molecules (Figure 4–58B). Similar networks can be formed by extending the procedure in Figure 4–58A with a third layer of antibodies.

The most sensitive amplification methods use an enzyme as the marker molecule. For example, alkaline phosphatase produces inorganic phosphate, and coupling the enzyme to a secondary antibody permits a sensitive chemical test for phosphate to be employed to reveal the presence of an antibody-antigen complex. Since each enzyme molecule acts catalytically to generate many thousands of molecules of product, such an *enzyme-linked immunoassay (ELIZA)* allows even tiny amounts of antigen to be detected. These assays are frequently used in medicine as a sensitive test for various types of infections.

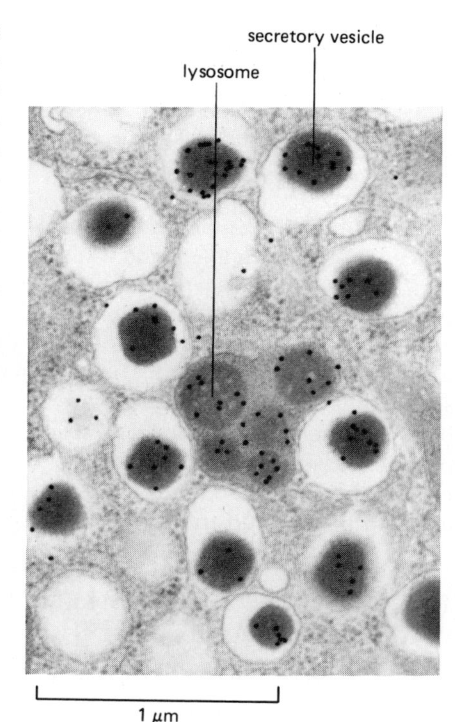

Figure 4–57 Immunocytochemical localization of specific protein molecules in electron micrographs by labeling with antibodies coupled to colloidal gold particles. A thin section of an insulin-secreting cell is shown in which insulin molecules have been labeled with anti-insulin antibodies bound to tiny gold spheres (each seen as a black dot). Most of the insulin is stored in the dense cores of secretory vesicles; in addition, some cores are being degraded in lysosomes. (From L. Orci, *Diabetologia* 28:528–546, 1985.)

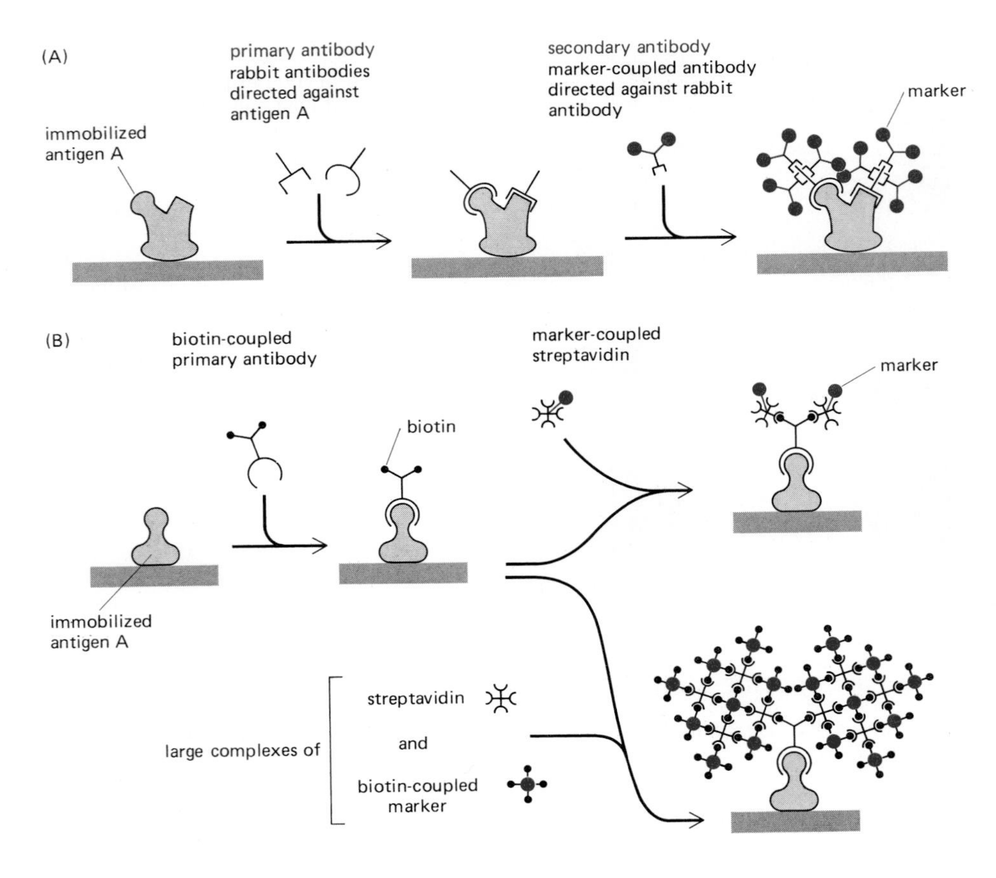

Figure 4–58 How antibodies are used to detect a particular molecule with great sensitivity. (A) The method illustrated here (known as *indirect immunocytochemistry*) has an enhanced sensitivity because the primary antibody (defined as the antibody molecule that binds to a recognized molecule of antigen) is itself recognized by many molecules of a second type of antibody. This secondary antibody is covalently coupled to a marker molecule that makes it readily detectable. Commonly used marker molecules include fluorescein or rhodamine dyes (for fluorescence microscopy), the enzyme horseradish peroxidase (for either bright-field light microscopy or electron microscopy), the iron-containing protein ferritin or colloidal gold spheres (for electron microscopy), and the enzyme alkaline phosphatase (for biochemical detection). (B) Modifications of the method in (A) that replace the secondary antibody by exploiting the high-affinity interaction between streptavidin and biotin. Because each streptavidin molecule can bind four biotin molecules, it can cross-link many biotinylated marker molecules into a large three-dimensional network. At the bottom an especially sensitive "sandwich technique" is shown that uses such networks to produce very heavy labeling of each primary antibody molecule.

Antibodies are made most simply by injecting a sample of the antigen several times into an animal such as a rabbit or a goat and then collecting the antibody-rich serum. This *antiserum* contains a heterogeneous mixture of antibodies, each produced by a different antibody-secreting cell (a B lymphocyte). The different antibodies recognize various parts of the antigen molecule as well as impurities in the antigen preparation. The specificity of an antiserum for a particular antigen sometimes can be sharpened by removing the unwanted antibody molecules that bind to other molecules; for example, an antiserum produced against protein X can be passed through an affinity column of antigens Y and Z to remove any contaminating anti-Y and anti-Z antibodies. Even so, the heterogeneity of such antisera has limited their usefulness.

Hybridoma Cell Lines Provide a Permanent Source of Monoclonal Antibodies[34]

In 1976 the problem of antiserum heterogeneity was overcome by the development of a technique that revolutionized the use of antibodies as tools in cell biology. The technique involves propagating a clone of cells from a single antibody-secreting B lymphocyte so that a homogeneous preparation of antibodies can be obtained in large quantities. But B lymphocytes normally have a limited life-span in culture. To overcome this limitation, individual antibody-producing B lymphocytes from an immunized mouse are fused with cells derived from an "immortal" B lymphocyte tumor. From the resulting heterogeneous mixture of hybrid cells, those hybrids that have both the ability to make a particular antibody and the ability to multiply indefinitely in tissue culture are selected. These **hybridomas** are propagated as individual clones, each of which provides a permanent and stable source of a single type of **monoclonal antibody** (Figure 4–59).

Since they are the product of a single B lymphocyte clone, monoclonal antibodies are made as a population of identical antibody molecules, each with an identical antigen-binding site. This site will recognize, for example, a particular

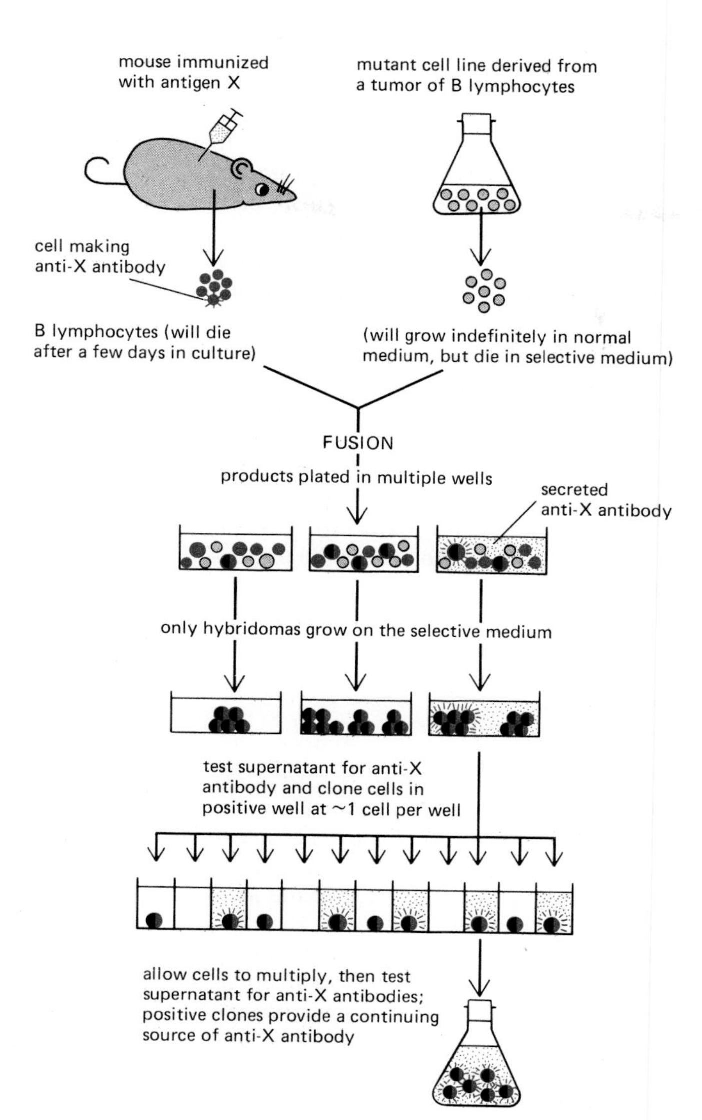

Figure 4–59 Preparation of hybrid cells, or hybridomas, that secrete homogeneous monoclonal antibodies against a particular antigen (X). The selective growth medium used contains an inhibitor (aminopterin) that blocks the normal biosynthetic pathways by which nucleotides are made. The cells must therefore use a bypass pathway to synthesize their nucleic acids, and this pathway is defective in the mutant cell line to which the normal B lymphocytes are fused. Because neither cell type used for the initial fusion can grow on its own, only the cell hybrids survive.

conformation of a defined sequence of five or six amino acid side chains on a protein. Their uniform specificity alone makes monoclonal antibodies much more useful for most purposes than conventional antisera, which usually contain a mixture of antibodies that recognize a variety of different antigenic sites on even a small macromolecule.

But the most important advantage of the hybridoma technique is that monoclonal antibodies can be made against molecules that constitute only a minor component of a complex mixture. In an ordinary antiserum made against such a mixture, the number of antibody molecules that recognize the minor component would be too small to be useful. But if the B lymphocytes that produce all these antibodies are made into hybridomas and a desired hybridoma clone is selected from the large mixture of clones, it can be propagated indefinitely to produce the desired antibody in large quantities. In principle, therefore, a monoclonal antibody can be made against any protein in a biological sample; once the antibody is made, it can be used as a specific probe—both to track down and localize the protein that induced its formation and to purify the protein in order to study its structure and function. Since fewer than 5% of the estimated 10,000 proteins in a typical mammalian cell have thus far been isolated, many monoclonal antibodies made against impure protein mixtures in fractionated cell extracts identify new proteins. Using monoclonal antibodies and gene-cloning technology (see below), it is no longer difficult to identify and characterize novel proteins and genes; the problem is to determine their function.

Antibodies and Other Macromolecules Can Be Injected into Living Cells[35]

Antibody molecules can be used to determine the function of the molecules to which they bind. Antibodies to nerve growth factor, for example, when injected into a newborn mouse, prevent the development of certain classes of nerve cells that depend on the growth factor for their survival. Similarly, antibodies that react with molecules on the surface of a particular type of cell can be used to kill the cells; by specifically eliminating one cell type from a mixed population, one can establish the importance of that cell type for different biological functions.

Proteins inside living cells cannot be reached by antibodies added externally because the plasma membrane is impermeable to large molecules. Because the plasma membrane is self-sealing, however, it is possible to introduce antibodies and other macromolecules into the cytoplasm of eucaryotic cells in other ways—most commonly by injecting them through a fine glass needle (see p. 157). For example, anti-myosin antibodies injected into a sea urchin egg prevent the egg cell from dividing in two, even though nuclear division occurs normally. This observation demonstrates that myosin plays a crucial part in the contractile process that divides the cytoplasm during mitosis but suggests that it is not required for nuclear division. The high specificity of monoclonal antibodies and the ease of producing them in concentrated form make them particularly suitable for this type of application.

Summary

Any molecule in the cell can be "labeled" by the incorporation of one or more radioactive atoms. The unstable nuclei of these atoms disintegrate, emitting radiation that allows the molecule to be detected and its movements and metabolism traced in the cell. Applications of radioisotopes in cell biology include the analysis of metabolic pathways by pulse-chase methods and the determination of the location of individual molecules in a cell by autoradiography.

Antibodies are also versatile and sensitive tools for detecting and localizing specific biological molecules. Vertebrates make millions of different antibody molecules, each with a binding site that recognizes a specific region of a macromolecule. The hybridoma technique allows monoclonal antibodies of a single specificity to be obtained in virtually unlimited amounts. In principle, monoclonal antibodies can be made against any cell macromolecule; these can be used to locate and purify the molecule and, in some cases, to analyze its function.

Recombinant DNA Technology[36]

The central challenge in modern cell biology is to understand the workings of the cell in molecular detail. We have discussed a number of powerful techniques for purifying, analyzing, and tracing the proteins of a cell. In this final section we shall discuss how one examines the structure and function of the cell's DNA. Classically, the only way to investigate the information content of the DNA was through genetics, which allows gene functions to be deduced from the phenotypes of mutant organisms and their progeny. This approach continues to provide unique insights, but it has been supplemented in recent years by a set of techniques known collectively as "recombinant DNA technology." These techniques greatly facilitate genetic studies by allowing both direct control and detailed chemical analysis of the genetic material. In addition, by making even minor cell proteins available in large quantities, the same methods have revolutionized biochemical studies of protein structure and function.

Recombinant DNA Technology Has Revolutionized Cell Biology[37]

Until the early 1970s, DNA was the most difficult cellular molecule for the biochemist to analyze. Enormously long and chemically monotonous, the nucleotide sequence of DNA could be approached only by indirect means—such as through protein or RNA sequencing or by genetic analysis. Today the situation has changed entirely. From being the most difficult macromolecule of the cell to analyze, DNA has become the easiest. It is now possible to excise specific regions of DNA, to obtain them in virtually unlimited quantities, and to determine the sequence of their nucleotides at a rate of hundreds of nucleotides a day. By variations of the same techniques, an isolated gene can be altered (engineered) at will and transferred back into cells in culture or (with rather more difficulty) into the germ line of animals, where the modified gene becomes incorporated as a permanent functional part of the genome.

These technical breakthroughs have had a dramatic impact on cell biology by allowing the study of cells and their macromolecules in previously unimagined ways. For example, they have provided the means to determine the functions of many newly discovered proteins and their individual domains and to unravel the complex mechanisms by which eucaryotic gene expression is regulated. They have also made available large amounts of the rare proteins that regulate cell proliferation and development. In commercial laboratories, similar techniques offer great promise for the large-scale economical production of protein hormones and vaccines, previously available only with great labor and cost.

Recombinant DNA technology comprises a mixture of techniques, some new and some borrowed from other fields such as microbial genetics (Table 4–12). The most important of these techniques are (1) the specific cleavage of DNA by *restriction nucleases*, which greatly facilitates the isolation and manipulation of individual genes; (2) rapid *sequencing* of all the nucleotides in a purified DNA fragment, which makes it possible to determine the precise boundaries of a gene and the amino acid sequence it encodes; (3) *nucleic acid hybridization*, which makes it possible to find specific sequences of DNA or RNA with great accuracy and sensitivity on the basis of their ability to bind a complementary nucleic acid sequence; (4) *DNA cloning*, whereby a specific DNA fragment is integrated into a self-replicating genetic element (plasmid or virus) that inhabits bacteria so that a single DNA molecule can be reproduced to generate many billions of identical copies; and (5) *genetic engineering*, by which DNA sequences are altered to make modified versions of genes, which are reinserted back into cells or organisms.

In the sections that follow, we shall explain how recombinant DNA technology has generated new experimental approaches that have revolutionized cell biology. However, a proper appreciation of DNA cloning and genetic engineering requires an understanding of the natural mechanisms by which cells replicate and decode

Table 4–12 Some Major Steps in the Development of Recombinant DNA Technology

1869	**Miescher** isolated DNA for the first time.
1944	**Avery** provided evidence that DNA, rather than protein, carries the genetic information during bacterial transformation.
1953	**Watson and Crick** proposed the double-helix model for DNA structure based on x-ray results of **Franklin and Wilkins.**
1957	**Kornberg** discovered DNA polymerase, the enzyme now used to produce highly radioactive DNA probes.
1961	**Marmur and Doty** discovered DNA renaturation, establishing the specificity and feasibility of nucleic acid hybridization reactions.
1962	**Arber** provided the first evidence for the existence of DNA restriction nucleases, leading to their later purification and use in DNA sequence characterization by **Nathans and H. Smith.**
1966	**Nirenberg, Ochoa,** and **Khorana** elucidated the genetic code.
1967	**Gellert** discovered DNA ligase, the enzyme used to join DNA fragments together.
1972–1973	DNA cloning techniques were developed by the laboratories of **Boyer, Cohen, Berg,** and their colleagues at Stanford University and the University of California at San Francisco.
1975–1977	**Sanger and Barrell** and **Maxam and Gilbert** developed rapid DNA-sequencing methods.
1981–1982	**Palmiter and Brinster** produced transgenic mice; **Spradling and Rubin** produced transgenic fruit flies.

DNA. We shall therefore postpone our main discussion of the details of gene cloning and engineering until Chapter 5, where they will be explained after an introduction to the basic genetic mechanisms.

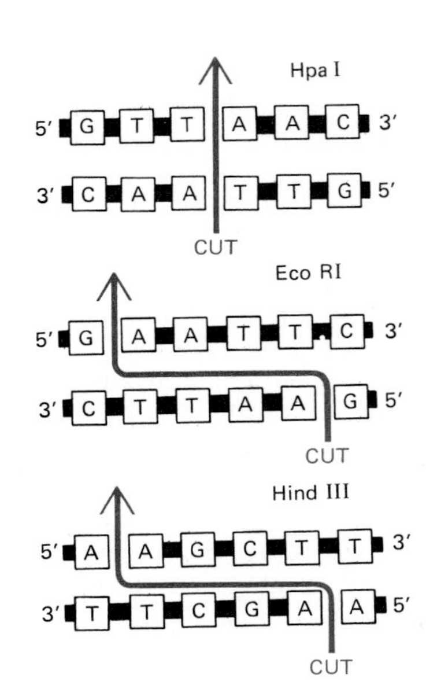

Figure 4–60 The DNA nucleotide sequences recognized by three widely used restriction nucleases. As in these examples, such sequences are often six base pairs long and "palindromic"—that is, the nucleotide sequences of the two strands are the same in the short region of helix that is recognized. The two strands of DNA are cut at or near the recognition sequence, often with a staggered cleavage that creates a cohesive end—as for Eco RI and Hind III. Restriction nucleases are obtained from various species of bacteria: Hpa I is from *Hemophilus parainfluenzae;* Eco RI is from *Escherichia coli;* and Hind III is from *Hemophilus influenzae.*

Restriction Nucleases Hydrolyze DNA Molecules at Specific Nucleotide Sequences[38]

Many bacteria make enzymes called **restriction nucleases,** which protect the bacteria by degrading the DNA molecules carried into the cell by viruses. Each enzyme recognizes a specific sequence of four to eight nucleotides in DNA. These sequences, where they occur in the genome of the bacterium itself, are "camouflaged" by methylation at an A or a C residue; where the sequences occur in foreign DNA, they are generally not methylated and so are cleaved by the restriction nuclease (Figure 4–60). Many restriction nucleases have been purified from different species of bacteria; more than 100, most of which recognize different nucleotide sequences, are now available commercially.

A particular restriction nuclease will cut any double-helical DNA molecule extracted from a cell into a series of specific DNA fragments known as **restriction fragments.** By comparing the sizes of the restriction fragments produced from a particular genetic region after treatment with a combination of different restriction nucleases, a **restriction map** of that region can be constructed showing the location of each cutting (restriction) site in relation to its neighboring restriction sites (Figure 4–61). Because each restriction nuclease recognizes a different short DNA sequence, a restriction map reflects the arrangement of specific nucleotide sequences in the region. This means that different regions of DNA can be compared (by comparing their restriction maps) without having to determine their nucleotide sequences. For example, by comparing the restriction maps illustrated in Figure 4–62, we know that the chromosomal regions that code for hemoglobin chains in humans, orangutans, and chimpanzees have remained largely unchanged during the 5 to 10 million years since these species first diverged. Restriction maps are also important in DNA cloning and genetic engineering, enabling one to locate a gene of interest on a particular restriction fragment, as will be explained later.

Any DNA Sequence Can Be Produced in Large Amounts by DNA Cloning[39]

Many restriction nucleases produce staggered cuts, which leave short single-stranded tails at the two ends of each fragment. These are known as *cohesive ends,* since each tail can form complementary base pairs with the tail at any other end produced by the same enzyme (Figure 4–63). The cohesive ends generated by restriction enzymes allow any two DNA fragments to be easily joined together, as long as the fragments were generated with the same restriction nuclease (or with another nuclease that produces the same cohesive ends). In this way a fragment of DNA from any source can be inserted into the purified DNA genome of a self-

Figure 4–61 A simple example illustrating how the cutting sites for different restriction nucleases (known as *restriction sites*) are positioned relative to each other on double helical DNA molecules to create a *restriction map.* (kb, kilobases; an abbreviation designating either 1000 nucleotides or 1000 nucleotide pairs.)

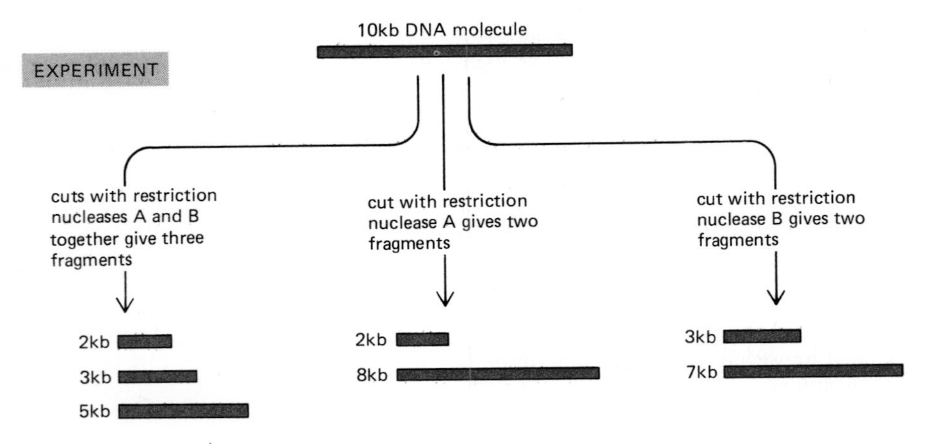

CONCLUSION: enzyme A cuts near one end of the molecule. Enzyme B must cut either near the same end or near the other end. The size of the fragments produced by both enzymes acting together rules out the first alternative and leads to the unambiguous order of restriction nuclease cutting sites shown below.

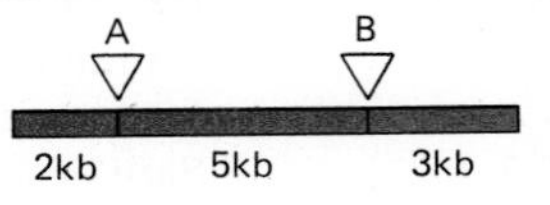

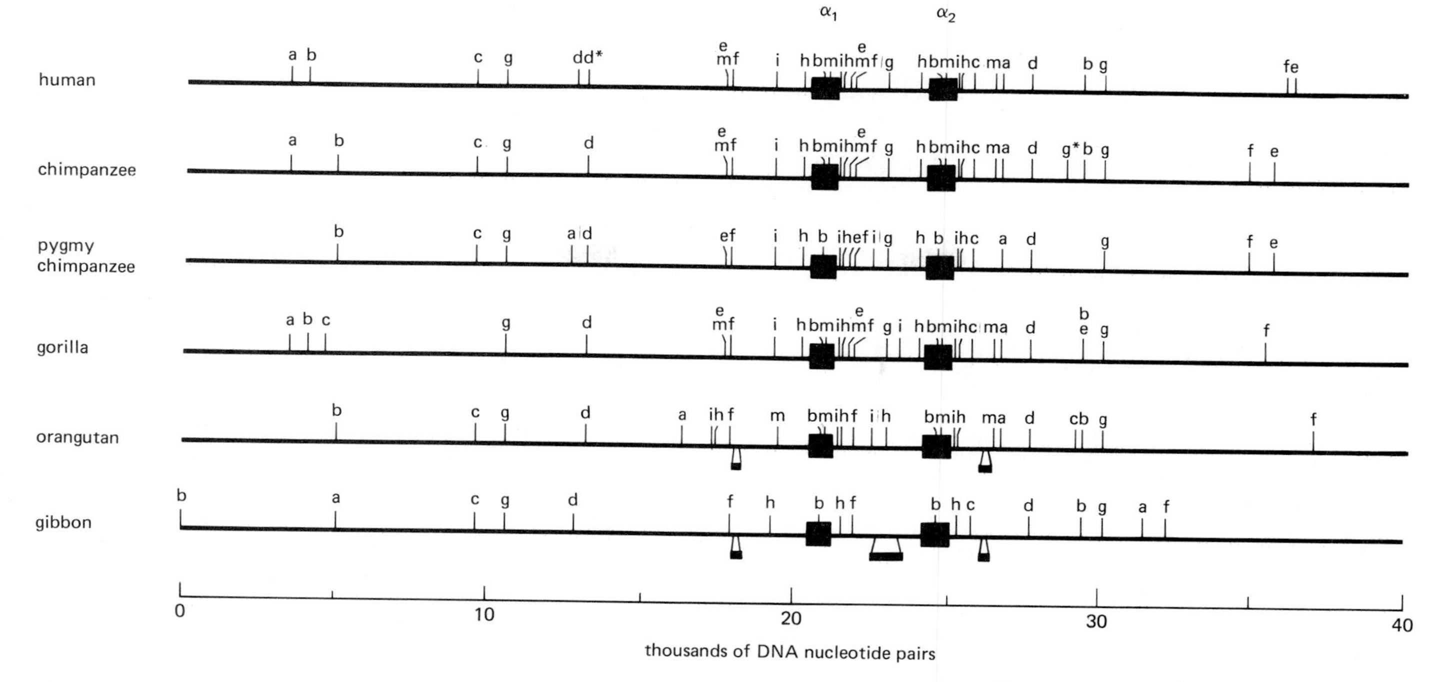

Figure 4–62 Restriction maps of human and various primate DNAs in a cluster of genes coding for hemoglobin. The two squares in each map indicate the positions of the DNA corresponding to the α-globin genes. Each letter stands for a site cut by a different restriction nuclease. As in Figure 4–61, the location of each cut was determined by comparing the sizes of the DNA fragments generated by treating the DNAs with the various restriction nucleases, individually and in combinations. (Courtesy of Elizabeth Zimmer and Alan Wilson.)

replicating genetic element, which is generally either a plasmid or a bacterial virus (see p. 259). A clone of bacteria containing the resultant plasmid or virus can then serve as a factory for production of unlimited amounts of the DNA fragment—a process called **DNA cloning.** The initial DNA fragment may be derived directly from genomic DNA; or it may be made from *cDNA,* which is DNA that has been made by copying messenger RNA. The procedures will be described in detail in Chapter 5 and will be only briefly outlined here.

The first step in obtaining **genomic DNA clones** generally involves purifying DNA from the entire cellular genome and cleaving it with a restriction nuclease. This produces an enormous number of different DNA fragments: for example, between 10^5 and 10^7 fragments are generated from a mammalian genome. The cloning process will therefore produce millions of cell colonies (*clones*), most of which will carry a *different* DNA fragment. The most difficult part of genomic cloning is finding the one clone in a million that contains the DNA fragment of interest (see p. 262).

In order to obtain **cDNA clones,** one begins by purifying mRNA (or a subfraction of the mRNA) from cells. This mRNA is then used as a template for *reverse transcriptase,* an enzyme produced by certain viruses that synthesize DNA by copying an RNA sequence (the reverse of the usual transcription process, in which RNA is copied from a DNA sequence). The enzyme produces a complementary DNA copy (hence "cDNA") of each mRNA molecule present. These single-stranded cDNA molecules are then converted into double-stranded DNA molecules (see p. 260), which are then cloned using methods similar to those used to clone genomic DNA fragments.

As explained in Chapter 5 (p. 260), there are important differences between genomic and cDNA clones. In particular, because of the extensive RNA splicing that occurs in higher eucaryotes (p. 531), only cDNA clones are likely to contain an uninterrupted form of the nucleotide sequences that code for proteins.

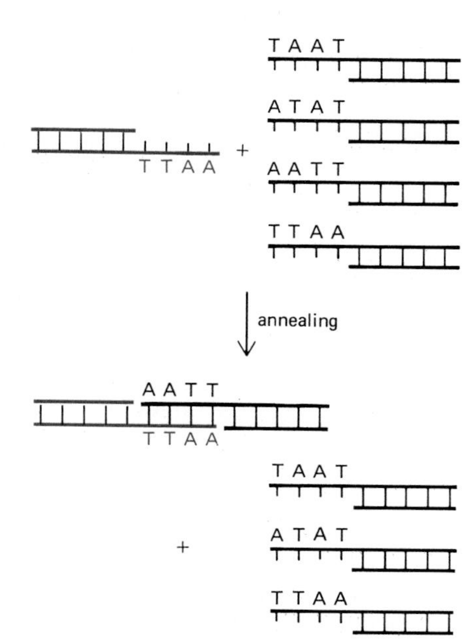

Figure 4–63 Many kinds of restriction nucleases produce DNA fragments with cohesive ends. DNA fragments with the same cohesive ends can readily join by complementary base-pairing between their cohesive ends as illustrated. The two DNA fragments that join in this example were both produced by the Eco RI restriction nuclease (see Figure 4–60).

Gel Electrophoresis Rapidly Separates DNA Molecules of Different Sizes[40]

In the early 1970s it was found that the length and purity of DNA molecules could be accurately determined by the same types of gel electrophoresis methods that had proved so useful in the analysis of protein chains (p. 170). The procedure is actually simpler than for proteins; because each nucleotide in a nucleic acid molecule already carries a single negative charge, there is no need for the negatively charged detergent SDS that is required to make protein molecules move in a uniform manner toward the positive electrode. For DNA fragments less than 500 nucleotides long, specially designed polyacrylamide gels allow molecules that differ in length by as little as a single nucleotide to be separated from each other (Figure 4–64A). However, the pores in polyacrylamide gels are too small to permit larger DNA molecules to pass; to separate these by size, the much more porous gels formed by dilute solutions of agarose (a polysaccharide isolated from seaweed) are used (Figure 4–64B). Both these DNA separation methods are widely used for analytical and preparative purposes.

A recent variation of agarose gel electrophoresis, called *pulsed-field gel electrophoresis,* makes it possible to separate even huge DNA molecules. Ordinary gel electrophoresis fails to separate these molecules because the steady electric field stretches them out into snakelike configurations that travel end-first through the gel at a rate that is independent of their length. But frequent alterations in the direction of the electric field force the molecules to reorient in order to move—a process that takes more time for larger molecules. Because entire bacterial or yeast chromosomes appear as discrete bands on such pulsed-field gels (Figure 4–64C), chromosome rearrangements can be detected directly. In addition, genes can be mapped to specific yeast chromosomes by using the hybridization of cloned DNA molecules from a gene to detect complementary DNA sequences in the gel (see p. 190).

The DNA bands on agarose or polyacrylamide gels will of course be invisible unless the DNA is labeled or stained in some way. One sensitive method of staining DNA is to soak the gel after electrophoresis in the dye *ethidium bromide,* which fluoresces under ultraviolet light when it is bound to DNA (Figure 4–64B and C).

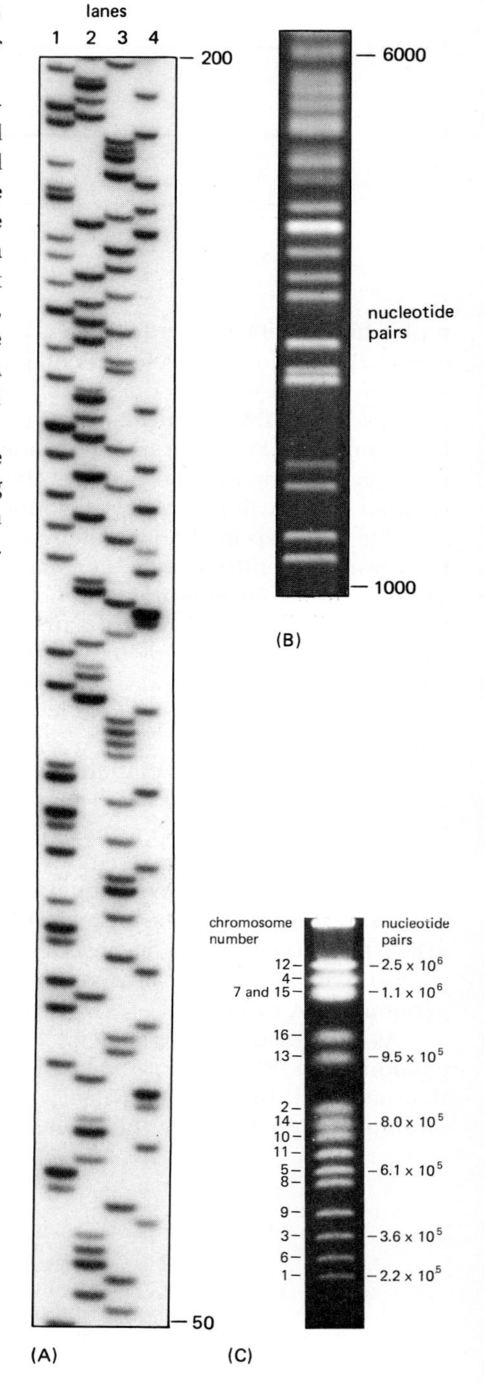

Figure 4–64 Gel electrophoresis is a powerful technique for separating DNA molecules according to their size. In the three examples shown, electrophoresis is from top to bottom, so that the largest DNA molecules are near the top of the gel. In (A) a polyacrylamide gel with small pores is used to fractionate single strands of DNA. In the size range 10 to 500 nucleotides, DNA molecules that differ in size by only a single nucleotide can be separated from each other. In this example, lanes 1 through 4 represent the products of four separate DNA sequencing reactions in which the chain-terminating dideoxyribonucleotides G, A, T, and C have been included, respectively (see Figure 4–68); since the DNA molecules used in these reactions are radiolabeled, their positions can be determined by autoradiography, as shown. In (B) an agarose gel with medium-sized pores is used to separate double-stranded DNA molecules. This method is most useful in the size range 300 to 10,000 nucleotide pairs. These DNA molecules are restriction fragments produced from the genome of a bacterial virus, and they have been detected by their fluorescence when stained with the dye ethidium bromide. In (C) the technique of pulsed-field agarose gel electrophoresis has been used to separate 16 different yeast (*Saccharomyces cerevisiae*) chromosomes that range in size from 220,000 to 2,500,000 DNA nucleotide pairs. DNA molecules as large as 10^7 nucleotide pairs can be separated on these gels. (A, courtesy of Leander Lauffer and Peter Walter; B, courtesy of Ken Kreuzer; C, from D. Vollrath and R.W. Davis, *Nucleic Acids Res.* 15:7876, 1987.)

An even more sensitive detection method involves incorporating a radioisotope into the DNA molecules before electrophoresis; ^{32}P is usually used since it can be incorporated into DNA phosphates and emits a very energetic β particle that is easy to detect by autoradiography (Figure 4–64A).

Purified DNA Molecules Can Be Labeled with Radioisotopes *in Vitro*[41]

Two procedures are widely used to radiolabel isolated DNA molecules. The first uses an *E. coli* enzyme, *DNA polymerase I*, to insert a large number of radioactive nucleotides (usually labeled with ^{32}P) into each DNA molecule (Figure 4–65A), thereby producing very radioactive "DNA probes" for nucleic acid hybridization reactions (see below). The second procedure uses the bacteriophage enzyme *polynucleotide kinase* to transfer a single ^{32}P-labeled phosphate from ATP to the 5′ end of each DNA chain (Figure 4–65B). Because only one ^{32}P atom is incorporated by the kinase into each DNA strand, the DNA molecules are usually not radioactive enough to be used as DNA probes; but because they are labeled only at one end, they are invaluable for DNA sequencing and DNA footprinting, as we shall now discuss.

Isolated DNA Fragments Can Be Rapidly Sequenced[42]

Methods were developed in the late 1970s that allow the nucleotide sequence of any purified DNA fragment to be determined simply and quickly. As a result, the complete DNA sequences of many hundreds of mammalian genes have been obtained, including those coding for insulin, hemoglobin, interferon, and cytochrome c. The volume of DNA sequence information is already so large (many millions of nucleotides) that computers must be used to store and analyze it. Several continuous stretches of DNA sequence have been determined that contain

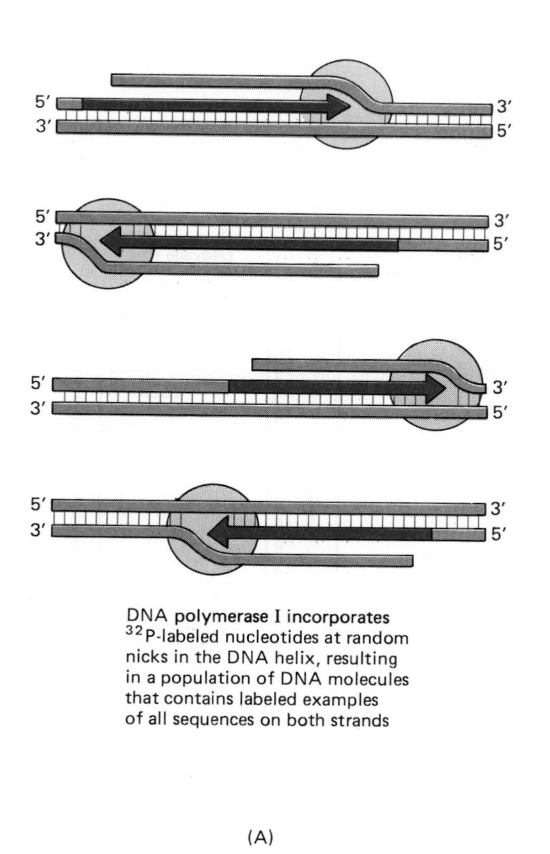

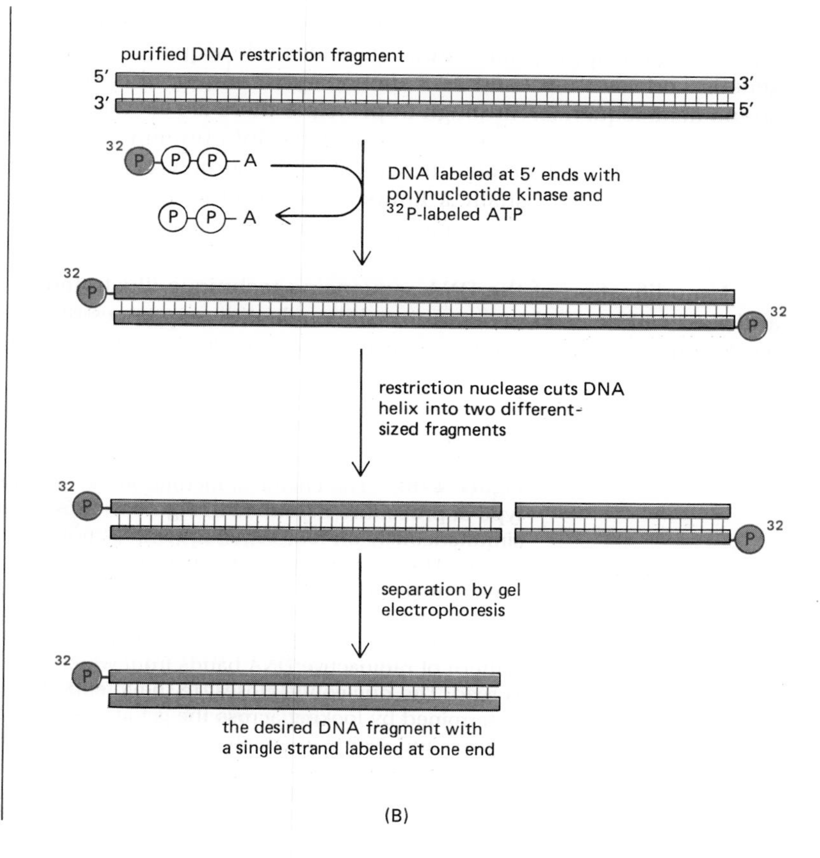

Figure 4–65 Two enzymatic procedures are used routinely for making DNA molecules radioactive. (A) DNA polymerase I labels all the nucleotides in a DNA molecule and can thereby produce highly radioactive DNA probes. (B) Polynucleotide kinase labels only the 5′ ends of DNA strands; therefore, when labeling is followed by restriction nuclease cleavage, as shown, DNA molecules containing a single 5′ end-labeled strand can be readily obtained.

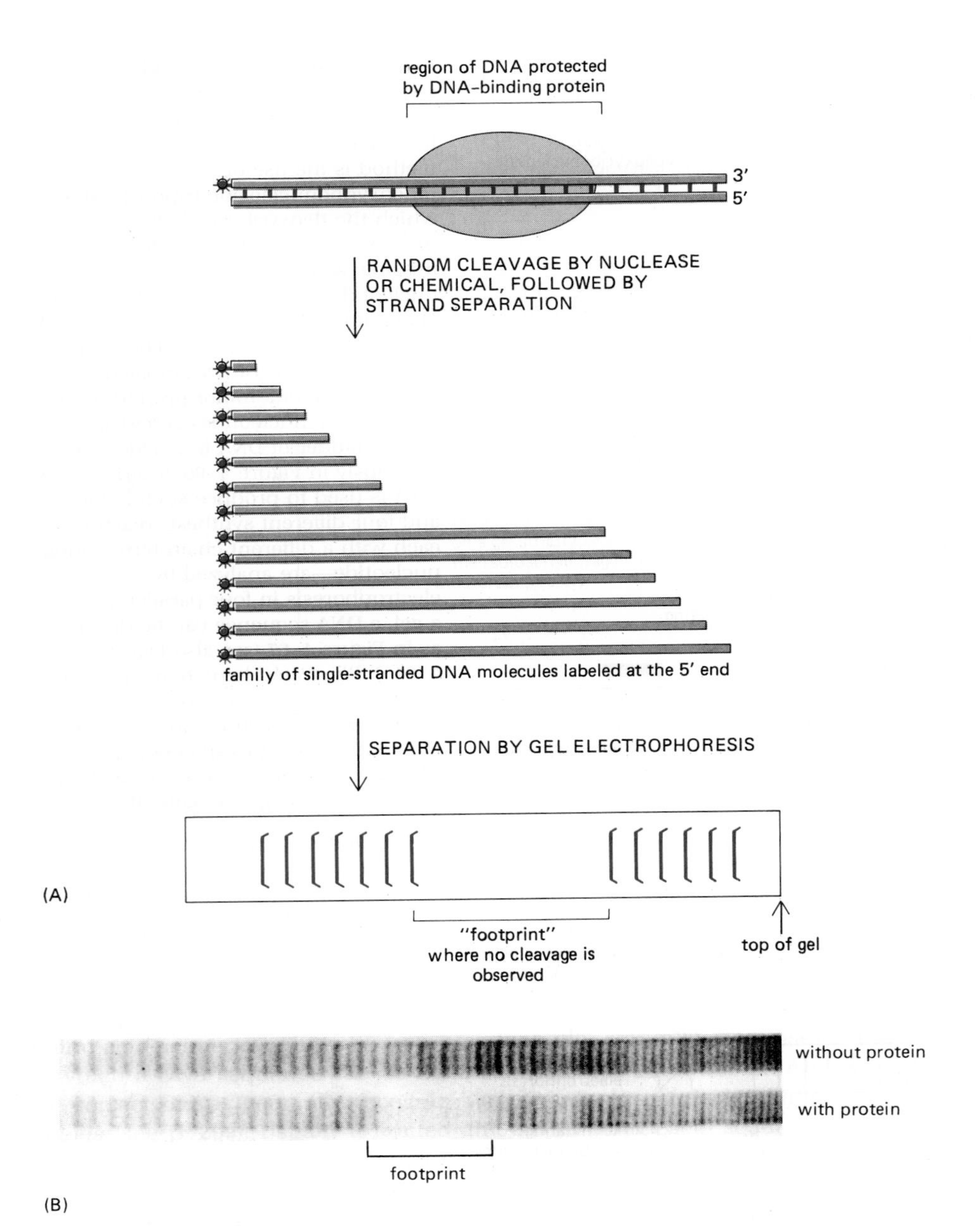

Figure 4–69 The DNA footprinting technique. (A) A protein binds tightly to a specific DNA sequence that is eight nucleotides long, thereby protecting these eight nucleotides from the cleaving agent. If the same reaction were carried out without the DNA-binding protein, a complete ladder of bands would be seen on the gel (not shown). (B) An actual footprint used to determine the binding site for a human protein that stimulates the transcription of specific eucaryotic genes. These results locate the binding site about 60 nucleotides upstream from the start site for RNA synthesis. The cleaving agent was a small, iron-containing organic molecule that normally cuts at every phosphodiester bond with nearly equal frequency. (B, courtesy of Michele Sawadogo and Robert Roeder.)

Nucleic Acid Hybridization Reactions Provide a Sensitive Way of Detecting Specific Nucleotide Sequences[43]

When an aqueous solution of DNA is heated at 100°C or exposed to a very high pH (pH ⩾ 13), the complementary base pairs that normally hold the two strands of the double helix together are disrupted and the double helix rapidly dissociates into two single strands. This process, called *DNA denaturation,* was for many years thought to be irreversible. However, in 1961 it was discovered that complementary single strands of DNA will readily re-form double helices (a process called **DNA renaturation** or **hybridization**) if they are kept for a prolonged period at 65°C. Similar hybridization reactions will occur between any two single-stranded nucleic acid chains (DNA:DNA, RNA:RNA, or RNA:DNA), provided that they have a complementary nucleotide sequence.

The rate of double-helix formation during hybridization reactions is limited by the rate at which two complementary nucleic acid chains happen to collide, which depends on their concentration in the solution. Hybridization rates can therefore be used to determine the concentration of any desired DNA or RNA sequence in a mixture of other sequences. This assay requires a pure single-

Figure 4–70 Measurement of the number of copies of a specific gene in a sample of DNA by means of DNA hybridization. The radioactive single-stranded DNA fragment used in such experiments is commonly referred to as a *DNA probe;* the chromosomal DNA is not radioactively labeled here.

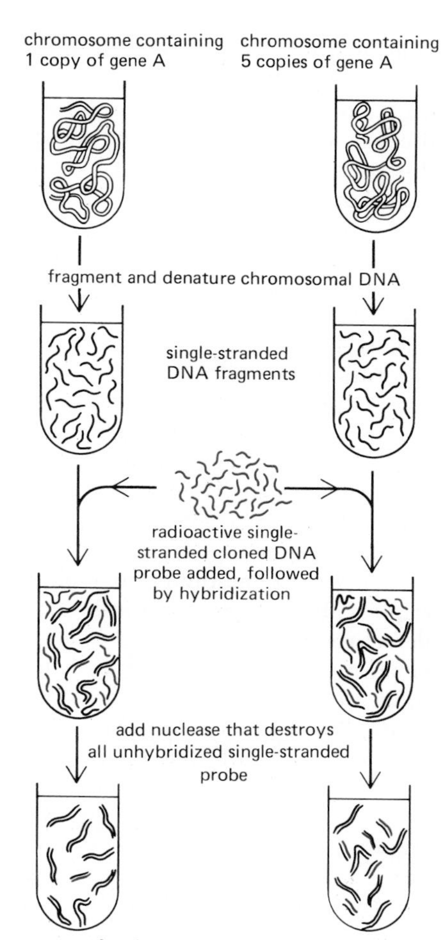

stranded DNA fragment that is complementary in sequence to the nucleic acid (DNA or RNA) one wishes to detect; the DNA can be obtained by cloning, or if the sequence is short, it can be synthesized by chemical means. In either case the DNA fragment is heavily radiolabeled with ^{32}P (see Figure 4–65) so that its incorporation into double-stranded molecules can be followed during the course of a hybridization reaction. A single-stranded DNA molecule used as an indicator in this way is known as a **DNA probe;** it can be anywhere from 15 to thousands of nucleotides long.

Hybridization reactions using DNA probes are so sensitive and selective that complementary sequences present at a concentration as low as one molecule per cell can be detected (Figure 4–70). It is thus possible to determine how many copies of a particular DNA sequence (contained in the probe) are present in a cell's genome. The same technique can be used to search for related but nonidentical genes; for example, once an interesting gene has been cloned from a mouse or a chicken, part of its sequence can be used as a probe to find the corresponding gene in a human.

Alternatively, DNA probes can be used in hybridization reactions with RNA rather than DNA to find out whether a cell is expressing a given gene. In this case a DNA probe that contains part of the gene's sequence is hybridized with RNA purified from the cell in question to see whether the RNA includes molecules matching the probe DNA and, if so, in what quantities. In somewhat more elaborate procedures the DNA probe is treated with specific nucleases after the hybridization is complete to determine the exact regions of the DNA probe that have paired with cellular RNA molecules. One can thereby determine the start and stop sites for RNA transcription (Figure 4–71); in the same way, one can identify the precise boundaries of the regions that are cut out of the RNA transcripts by *RNA splicing* (the intron sequences—see p. 533).

Large numbers of genes are switched on and off in elaborate patterns as an embryo develops. The hybridization of DNA probes to cellular RNAs allows one to determine whether a particular gene is off or on; moreover, when the expression of a gene changes, one can determine whether the change is due to controls that act on the transcription of DNA, the splicing of the gene's RNA, or the translation of its mature mRNA molecules into protein. Hybridization methods are in such wide use in cell biology today that it is difficult to imagine what it would be like to study gene structure and expression without them.

Northern and Southern Blotting Facilitate Hybridization with Electrophoretically Separated Nucleic Acid Molecules[44]

DNA probes are often used in conjunction with gel electrophoresis to detect the nucleic acid molecules with sequences that are complementary to all or part of the probe. The electrophoresis fractionates the many different RNA or DNA mole-

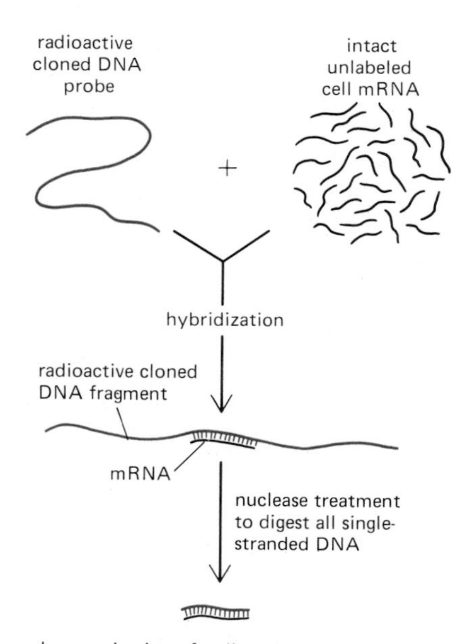

Figure 4–71 The use of nucleic acid hybridization to determine the region of a cloned DNA fragment that is transcribed into mRNA. The method shown requires a nuclease that cuts the DNA chain only where it is not base-paired to a complementary RNA chain. Both the beginning and the end of an RNA molecule can be exactly mapped in this way; in addition, the positions of introns (intervening sequences) in eucaryotic genes are mapped by similar procedures.

cules in a crude mixture according to their size before the hybridization reaction is carried out; if molecules of only one or a few sizes become labeled with the probe, one can be certain that the hybridization was indeed specific. Moreover, the size information obtained can be invaluable in itself. An example will illustrate this point.

Suppose that one wishes to determine the nature of the defect in mutant mice that produces abnormally low amounts of *albumin*, a protein that liver cells normally secrete into the blood in large amounts. First one collects identical samples of liver tissue from defective and normal mice (the latter serving as controls) and disrupts the cells in a strong detergent to inactivate cellular nucleases that might otherwise degrade the nucleic acids. Next one separates the RNA and DNA from all of the other cell components: the proteins present are completely denatured and removed by repeated extractions with phenol—a potent organic solvent that is partly miscible with water; the nucleic acids, which remain in the aqueous phase, are then precipitated with alcohol to separate them from the small molecules of the cell. Then one separates the DNA from the RNA by their different solubilities in alcohols and degrades any contaminating nucleic acid of the unwanted type by treatment with highly specific enzymes—either RNase or DNase.

To analyze albumin-encoding RNAs with a DNA probe for such RNA, a technique called **Northern blotting** is used. First, the intact RNA molecules from defective and control liver cells are fractionated into a series of bands by gel electrophoresis. Then, to make the RNA molecules accessible to DNA probes, a replica of the gel is made by transferring ("blotting") the fractionated RNA molecules onto a sheet of nitrocellulose or nylon paper. The RNA molecules that hybridize to the radioactive DNA probe (because they contain part of the normal albumin gene sequence) are then located by incubating the paper with a solution containing the probe and detecting the hybridized probe by autoradiography (Figure 4–72). Because small nucleic acid molecules move more rapidly through the gel than large ones, the size of each RNA band that binds the probe can be determined by reference to the rates of migration of RNA molecules of known size

Figure 4–72 Northern and Southern blotting analyses. After the indicated mixture of RNA or DNA molecules is fractionated by electrophoresis through an agarose gel, the many different RNA or DNA molecules present are transferred to nitrocellulose or nylon paper by blotting. The paper sheet is then exposed to a radioactive DNA probe for a prolonged period under hybridization conditions. The sheet is washed thoroughly afterward, so that only those immobilized RNA or DNA molecules that hybridize to the probe become radioactively labeled and show up as bands on autoradiographs of the paper sheet.

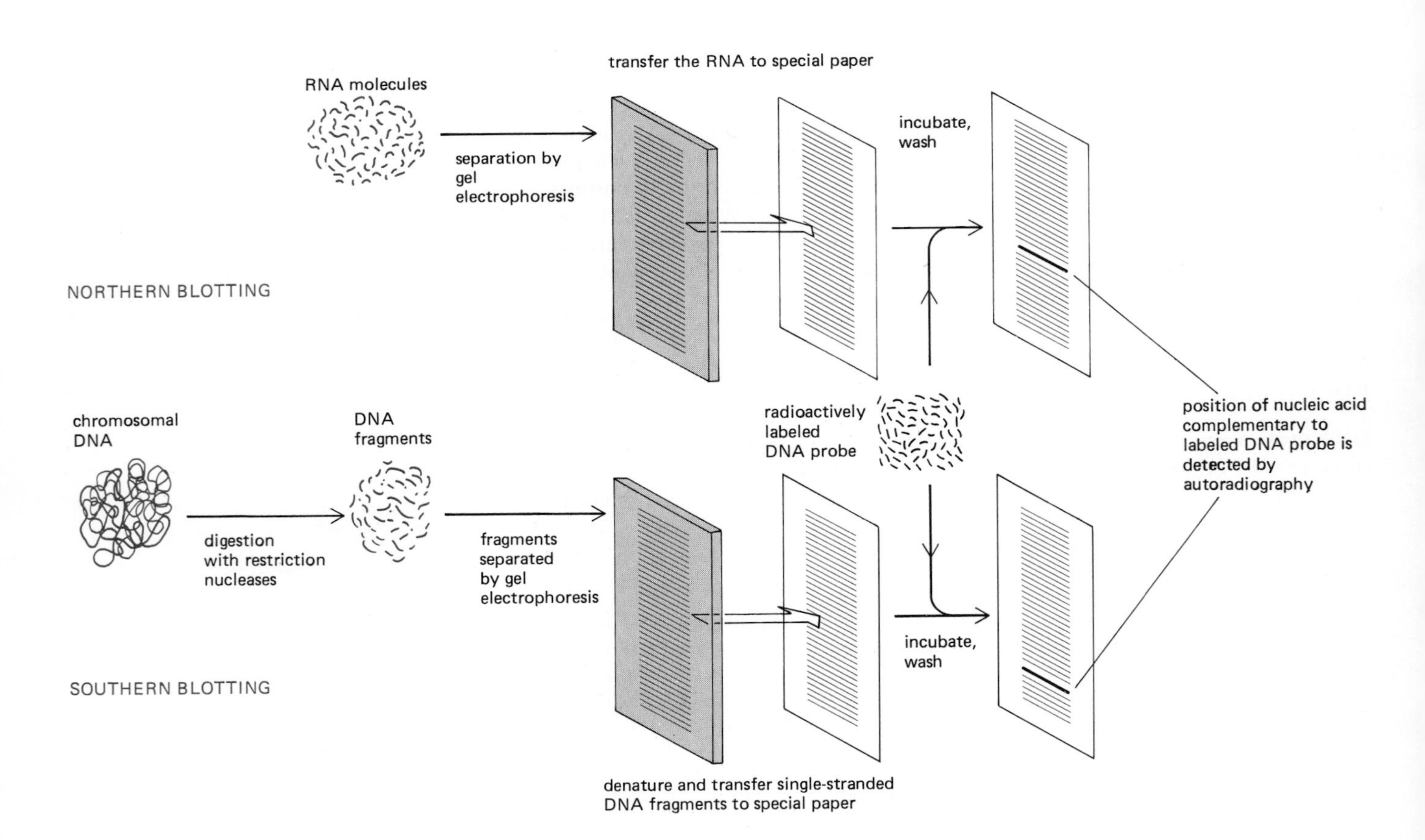

(*RNA standards*). In this way one might discover that liver cells from defective mice make albumin RNA in normal amounts and of normal size; alternatively, normal albumin RNA might be detected in greatly reduced amounts. Another possibility is that the mutant albumin RNA might be abnormally short and therefore move unusually quickly through the gel, in which case the gel blot could be retested with more selective DNA probes to reveal what part of the normal RNA is missing.

To characterize the structure of the albumin gene in the defective mice, an analogous method, called **Southern blotting,** which analyzes DNA rather than RNA, is used. Isolated DNA is first cut into readily separable fragments with restriction nucleases. The fragments are then separated according to their size by gel electrophoresis, and those complementary to the albumin DNA probe are identified by blotting and hybridization, as just described for RNA (see Figure 4–72). By repeating this procedure using different restriction nucleases, a detailed *restriction map* could be constructed for the genome in the region of the albumin gene (see p. 182). From this map one could determine if the albumin gene has been rearranged in the defective animals—for example, by the deletion or the insertion of a short DNA sequence.

Synthetic DNA Molecules Facilitate the Prenatal Diagnosis of Genetic Diseases[45]

At the same time that microbiologists were developing DNA cloning techniques, organic chemists were improving the methods for synthesizing short DNA chains. Today such *DNA oligonucleotides* are routinely produced by machines that can automatically synthesize any sequence up to 80 nucleotides long overnight. This ability to produce DNA molecules of a desired sequence makes it possible to redesign genes at will, an important aspect of genetic engineering, as will be explained in Chapter 5 (p. 266).

Another important use for DNA oligonucleotides is in the prenatal diagnosis of genetic diseases. More than 500 human genetic diseases are attributable to single-gene defects. In most of these the mutation is recessive: that is, it is harmful only when an individual inherits a defective copy of the gene from both parents. One goal of modern medicine is to identify those fetuses that carry two bad copies of the affected gene long before birth so that the mother, if she wishes, can have the pregnancy terminated. For example, in sickle-cell anemia the exact nucleotide change in the mutant gene is known (the sequence GAG is changed to GTG in the DNA strand that codes for the β chain of hemoglobin). For prenatal diagnosis, two DNA oligonucleotides are synthesized—one corresponding to the normal gene sequence in the region of the mutation and the other corresponding to the mutated sequence. By keeping these sequences short (about 20 nucleotides) and selecting a hybridization temperature where only the perfectly matched helix is stable, they can be used as radioactive probes to distinguish between the two forms of the gene. The diagnosis involves isolating DNA from fetal cells collected by amniocentesis and then using the oligonucleotide probes for Southern blotting (see Figure 4–72). A defective fetus can be readily recognized because its DNA will hybridize *only* with the oligonucleotide that is complementary to the mutant DNA sequence. For many genetic abnormalities, the exact nucleotide sequence change is not known. For an increasing number of these, prenatal diagnosis is still possible by using Southern blotting to assay for specific variations in the human genome (called *restriction fragment length polymorphisms*, or *RFLPs*) that are known to be closely linked to the defective gene.

Hybridization at Reduced Stringency Allows Distantly Related Genes to Be Identified[46]

New genes arise during evolution by the duplication and divergence of old genes and by the reutilization of portions of old genes in new combinations (p. 599). For this reason, most genes have a family of close relatives elsewhere in the genome, and some of them are likely to have a related function. Laborious methods are usually required to isolate a DNA clone corresponding to the first member of such

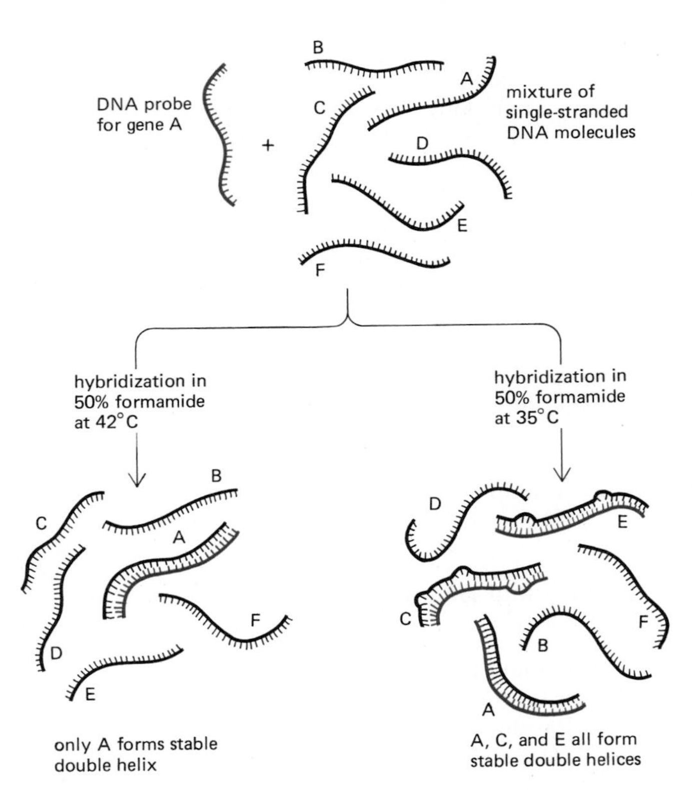

Figure 4–73 Comparison of stringent and reduced-stringency hybridization conditions. In the reaction on the left (stringent conditions), the solution is kept only a few degrees below the temperature at which a perfect DNA helix denatures (its *melting temperature*), so that the imperfect helices that can form under the conditions of reduced stringency on the right are unstable. Only the hybridization conditions on the right can be used to find genes that are nonidentical but related to gene A.

a *gene family* (see p. 262). However, additional genes that produce proteins with related functions can often be isolated relatively easily by using sequences from the first gene as DNA probes. Because the new genes are unlikely to have *identical* sequences, hybridizations with the DNA probe are usually carried out under conditions of "reduced stringency"—defined as conditions that allow even an imperfect match with the probe sequence to form a stable double helix (Figure 4–73).

Although using reduced stringency for hybridization carries the risk of obtaining a false signal from a chance region of short sequence homology in an unrelated DNA sequence, such hybridization represents one of the most powerful uses of recombinant DNA technology. For example, this approach has led to the isolation of a whole family of DNA-binding proteins that function as master regulators of gene expression during embryonic development in *Drosophila* (p. 937). It has also made it possible to isolate members of this same gene family from a variety of other organisms, including humans.

In Situ Hybridization Techniques Locate Specific Nucleic Acid Sequences in Chromosomes and Cells[47]

Nucleic acids, no less than other macromolecules, occupy precise positions in cells and tissues, and a great deal of potential information is lost when these molecules are extracted by homogenization. For this reason, techniques have been developed in which nucleic acid probes are used in much the same way as labeled antibodies to locate specific nucleic acid sequences *in situ,* a procedure called ***in situ* hybridization.** This can now be done both for DNA in chromosomes and for RNA in cells. Highly radioactive nucleic acid probes can be hybridized to chromosomes that have been exposed briefly to a very high pH to disrupt their DNA base pairs. The chromosomal regions that bind the radioactive probe during the hybridization step are visualized by autoradiography. The spatial resolution of this technique can be improved by labeling the DNA probes chemically instead of radioactively. Most commonly, the probes are synthesized with nucleotides that contain a biotin side chain (see p. 177), and the hybridized probes are detected by staining with a network of streptavidin and some type of marker molecule (Figure 4–74).

In situ hybridization methods have also been developed that reveal the distribution of specific RNA molecules in cells within tissues. In this case the tissues are not exposed to a high pH, so the chromosomal DNA remains double-stranded and cannot bind the probe. Instead the tissue is gently fixed so that its RNA is retained in an exposed form that will hybridize when the tissue is incubated with a complementary DNA probe. In this way striking patterns of differential gene expression have been observed in developing *Drosophila* embryos (Figure 4–75); these patterns have provided new insights into the mechanisms that distinguish between cells in different positions during development (p. 927).

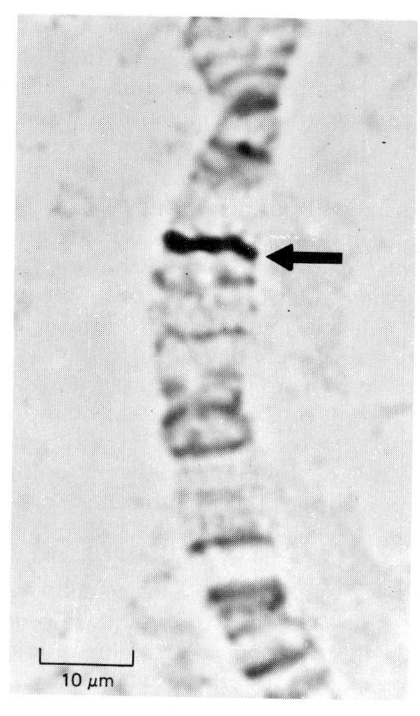

Figure 4–74 Localization of a *Drosophila* gene by *in situ* hybridization of a biotin-labeled cloned DNA probe with *Drosophila* polytene chromosomes. The DNA in this unusually large chromosome (see p. 508) has been partially denatured so that the probe can hybridize to it. After hybridization and washing, the location of the bound probe is detected by treating the chromosome with a network of horseradish peroxidase enzymes conjugated to streptavidin, as in Figure 4–58B. The peroxidase is made visible by the insoluble dark product deposited (*arrow*) when the treated chromosomes are incubated with hydrogen peroxide and a reactive dye molecule. (Courtesy of Todd Laverty and Gerald Rubin.)

Recombinant DNA Techniques Allow Even the Minor Proteins of a Cell to Be Studied[48]

Until very recently, the only proteins in a cell that could be studied easily were the relatively abundant ones. Starting with several hundred grams of cells, a major protein—one that constitutes 1% or more of the total cellular protein—can be purified by sequential chromatography steps to yield perhaps 0.1 g (100 mg) of pure protein. This quantity is sufficient for conventional amino acid sequencing, for detailed analysis of biological activity (if known), and for the production of antibodies, which can then be used to localize the protein in the cell. Moreover, if suitable crystals can be grown (usually a difficult task), the three-dimensional structure of the protein can be determined by x-ray diffraction techniques. In this way the structure and function of many abundant proteins—including hemoglobin, trypsin, immunoglobulin, and lysozyme—have been analyzed.

The vast majority of the thousands of different proteins in a eucaryotic cell, however, including many with crucially important functions, are present in only very small amounts. For most of them it is extremely difficult, if not impossible, to obtain more than a few micrograms of pure material. One of the most important contributions of DNA cloning and genetic engineering to cell biology is that they have made it possible to manufacture any of the cell's proteins, including the minor ones, in large amounts. This is done by first cloning the gene for the protein of interest and then inserting it into a special plasmid called an *expression vector*. The vector is engineered in such a way that when it is introduced into an appropriate type of bacterium, yeast, or mammalian cell, the inserted gene directs the synthesis of very large amounts of the protein of interest (see p. 265). In principle, therefore, any protein can now be made available for the kinds of detailed structural and functional studies that were previously possible only for a rare few.

Mutant Organisms Best Reveal the Function of a Gene[49]

Suppose that one has cloned a gene that codes for a newly discovered protein. How can one discover what the protein does in the cell? This has become a common problem in cell biology and one that is surprisingly difficult to solve, since neither the three-dimensional structure of the protein nor the complete nucleotide sequence of its gene is usually sufficient to deduce the protein's function. Moreover, many proteins—such as those that have a structural role in the

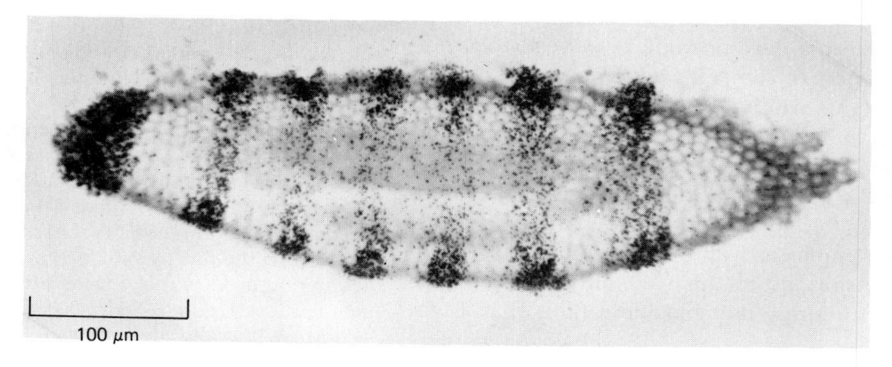

Figure 4–75 Autoradiograph of a section of a very young *Drosophila* embryo that has been subjected to *in situ* hybridization using a radioactive DNA probe complementary to a gene involved in segment development. The probe has been hybridized to RNA in the embryo, and the pattern of exposed silver grains reveals that the RNA made by the gene (called *ftz*) is localized in alternating stripes that are three or four cells wide across the embryo. At this stage of development (cellular blastoderm), the embryo contains about 6000 cells. (From E. Hafen, A. Kuriowa, and W.J. Gehring, *Cell* 37:833–841, 1984.)

cell or normally form part of a large multienzyme complex—will have no obvious activity when isolated from the other components of the functional unit.

One approach, mentioned earlier (see p. 180), is to inactivate the particular protein with a specific antibody and observe how cells are affected. Although this provides a powerful way to test protein function, for an intracellular protein the effect is transitory because the microinjected antibody is eventually diluted out during cell proliferation or destroyed by intracellular degradation.

Genetic approaches provide a more powerful solution to this problem. Mutants that lack a particular protein or, more usefully, synthesize a temperature-sensitive version of the protein (one that is inactivated by a small increase or decrease in temperature) may quickly reveal the function of the normal molecule. This approach has been immensely useful, for example, in demonstrating the function of enzymes involved in the principal metabolic pathways of bacteria and in discovering many of the gene products responsible for the orderly development of the *Drosophila* embryo. Traditionally, it has been most generally applicable to organisms that reproduce very rapidly—such as bacteria, yeasts, nematode worms, and fruit flies. By treating these organisms with agents that alter their DNA (*mutagens*), very large numbers of mutants can be isolated quickly and then screened for a particular defect of interest. For example, by screening populations of mutagen-treated bacteria for cells that stop making DNA when they are shifted from 30°C to 42°C, many temperature-sensitive mutants were isolated in the genes that encode the bacterial proteins required for DNA replication. These mutants were later used to identify and characterize the corresponding DNA replication proteins (see p. 238).

Humans do not reproduce rapidly, nor are they ever intentionally treated with mutagens. Moreover, any human with a serious defect in an essential process, such as DNA replication, would die long before birth. However, many mutations that are compatible with life—for example, tissue-specific defects in lysosomes or in cell-surface receptors—have arisen spontaneously in the human population. Analysis of the phenotypes of the affected individuals, together with studies of their cultured cells, have provided many unique insights into important cell functions (for example, see p. 329 and p. 464). Although such mutants are extremely rare, they are very efficiently discovered because of a unique human property: the mutant individuals call attention to themselves by seeking special medical care.

Cells and Organisms Containing Altered Genes Can Be Made to Order[50]

Although it is often not difficult to obtain mutants that are deficient in a particular process, such as DNA replication or eye development, it can take many years to trace the defect to a particular altered protein. Recently, recombinant DNA technology has made possible a different type of genetic approach in which one starts with a protein and creates a mutant cell or organism in which that protein (or its expression) is abnormal. Because the new approach reverses the traditional gene-to-protein direction of genetic analysis, it is commonly referred to as *reverse genetics*.

Reverse genetics begins with a protein with interesting properties that has been isolated from a cell. By methods that will be described in Chapter 5 (p. 258), the gene encoding the protein is cloned and its nucleotide sequence is determined; this sequence is then altered by biochemical means to create a mutant gene that codes for an altered version of the protein. The mutant gene is transferred into a cell, where it can integrate into a chromosome by genetic recombination (p. 239) to become a permanent part of the cell's genome. If the gene is expressed, the cell and all of its descendants will now synthesize an altered protein. If the original cell used for the gene transfer is a fertilized egg, whole multicellular organisms can be obtained that contain the mutant gene, and some of these **transgenic organisms** will pass the gene on to their progeny as a permanent part of their germ line (Figure 4–76). Such *genetic transformations* are now routinely performed with organisms as complex as fruit flies and mammals. Technically, even humans could now be transformed in this way, although such experiments are not performed for fear of the unpredictable aberrations that might occur in such individuals.

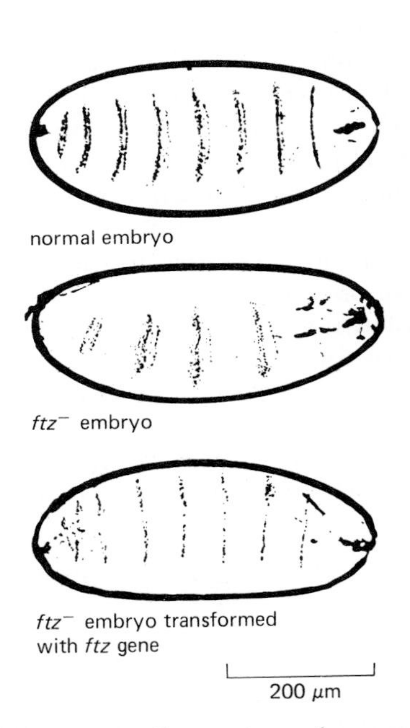

Figure 4–76 Comparison of a normal *Drosophila* larva and two mutant larvae that contain defective *ftz* genes. One of the defective (ftz^-) larvae has been transformed by the injection of a DNA clone containing the normal *ftz* gene sequence into the egg of one of its ancestors. This added DNA sequence has become permanently integrated into one of the fly's chromosomes and is therefore faithfully inherited and expressed. The *ftz* gene is required for normal development, and the addition of this gene to the genome is seen to restore the larval segments that are missing in the ftz^- organism. For the technique used to make transgenic animals, see Figure 5–88, p. 268. (Courtesy of Walter Gehring.)

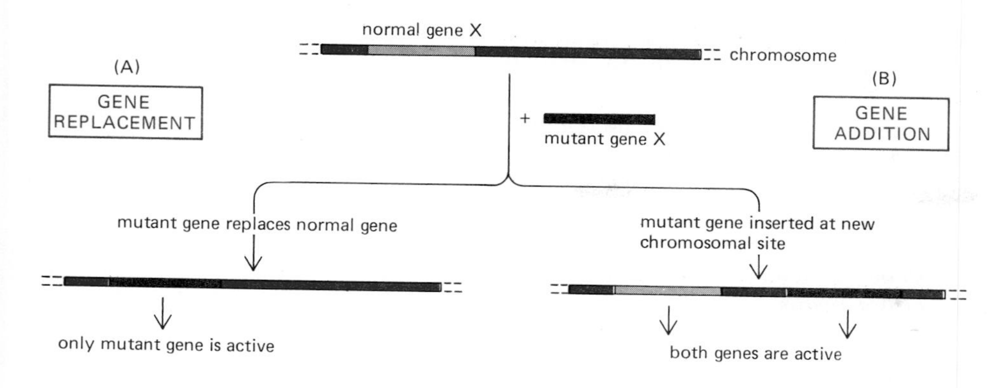

Figure 4–77 A gene whose nucleotide sequence has been altered can be inserted back into the chromosomes of an organism. In bacteria and yeast, it is possible to select for mutants that have undergone a gene *replacement* (A), in which a genetic recombination event inserts the mutant gene in place of the normal gene; in these cases, only the mutant gene remains in the cell. In higher eucaryotes, gene *addition* (B) generally occurs instead of gene replacement. The transformed cell or organism now contains the mutated gene in addition to the normal gene. Gene replacements are expected to be rare in organisms with large amounts of DNA, since the mutant gene must pair with the normal gene and replace it by genetic recombination despite a huge excess of other DNA sequences.

Engineered Genes Encoding Antisense RNA Can Create Specific Dominant Mutations[51]

In bacterial and yeast cells, which are generally haploid, an artificially introduced mutant gene will recombine with the normal gene often enough to make it possible to select cells in which the mutant gene has replaced the single copy of the normal gene (Figure 4–77A). In this way cells can be made to order that produce a specific protein only in its mutant form. The function of the normal protein can usually be determined from the phenotype of such cells. In higher eucaryotes, such as mammals or fruit flies, methods are not yet available for the easy replacement of a normal gene by the insertion of a cloned mutant gene. Genetic transformation in these organisms generally leads instead to insertion of the cloned gene at a random place in the genome, so that the cell (or the organism) ends up with the mutated gene in addition to its normal gene copies (Figure 4–77B).

It would be enormously useful to be able to create specific *dominant* mutations in higher eucaryotic cells by artificially inserting a mutant gene that would eliminate the activity of its normal counterparts in the cell. An ingenious and promising approach exploits the specificity of hybridization reactions between two complementary nucleic acid chains to accomplish this goal. Normally only one of the two DNA strands in a given portion of double helix is transcribed into RNA, and it is always the same strand for a given gene. If a cloned gene is engineered so that only the opposite DNA strand is transcribed, the result will be **antisense RNA** with a sequence complementary to the normal RNA transcripts. Antisense RNA, when synthesized in large enough amounts, will often hybridize with the "sense" RNA made by the normal genes and thereby inhibit the synthesis of the corresponding protein (Figure 4–78).

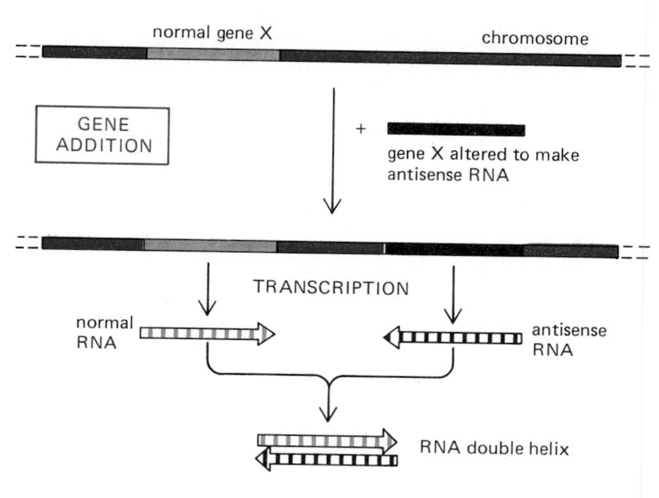

Figure 4–78 The antisense RNA strategy for generating dominant mutations. As illustrated, mutant genes that have been engineered to produce RNA that is complementary in sequence to the RNA made by the normal gene can cause double-stranded RNA to form inside cells. If a large excess of the antisense RNA is produced, it can hybridize with—and thereby inactivate—most of the normal RNA produced by gene X. In the future it may become possible to inactivate any gene in this way. At present the technique seems to work for some genes but not others.

If a protein is required for the survival of the cell (or the organism), the dominant mutants just described will die, making it impossible to test the function of the protein. To avoid this problem, one can engineer a gene so that it produces antisense RNA only on command—for example, in response to an increase in temperature or to the presence of a specific signaling molecule. Cells or organisms containing such an *inducible* antisense gene can be deprived of the specific protein at a particular time, and the effect can then be followed. Although some technical problems remain to be worked out, in the future this technique for producing dominant mutations to inactivate specific genes is likely to be widely used to determine the functions of proteins in higher organisms.

Summary

Recombinant DNA technology has revolutionized the study of the cell. Any region of a cell's DNA can now be excised with restriction nucleases and inserted into a self-replicating genetic element (a plasmid or a virus) to produce a "genomic DNA clone." Alternatively, a DNA copy of any mRNA molecule can be used to produce a "cDNA clone." Unlimited amounts of a highly purified DNA molecule can thereby be obtained and its nucleotide sequence determined at a rate of hundreds of nucleotides per day, revealing the amino acid sequence of the protein it encodes. Genetic engineering techniques can then be used to create a mutant gene and insert it into a cell's chromosome so that it becomes a permanent part of the genome. If the cell used for this gene transfer is a fertilized egg, transgenic organisms can be produced that express the mutant gene and will pass it on to their progeny. Especially important for cell biology is the power that such technology gives the experimenter to alter cells in highly specific ways—allowing one to discern the effect on the cell of a change in a single protein that has been intentionally mutated by "reverse genetic" techniques.

The consequences of recombinant DNA technology are far-reaching in other ways as well. Bacteria, yeasts, or mammalian cells can be engineered to synthesize any desired protein in large quantities, making it possible to analyze the structure and function of the protein in detail, or to use the protein as a drug or a vaccine for medical purposes. In addition, highly specific DNA probes are readily obtained that—by means of nucleic acid hybridization reactions—allow determination of the spatial pattern of gene expression in a tissue, the localization of genes on chromosomes, and the discovery of genes with related functions in a gene family.

References

General

Cantor, C.R.; Schimmel, P.R. Biophysical Chemistry, 3 vols. New York: W.H. Freeman, 1980. (A comprehensive account of the physical principles underlying many biochemical and biophysical techniques.)

Freifelder, D. Physical Biochemistry, 2nd ed. New York: W.H. Freeman, 1982.

Van Holde, K.E. Physical Biochemistry, 2nd ed. Englewood, NJ: Prentice-Hall, 1985.

Cited

1. Bradbury, S. An Introduction to the Optical Microscope. Oxford, U.K.: Oxford University Press, 1984.

   Fawcett, D.W. A Textbook of Histology, 11th ed. Philadelphia: Saunders, 1986.

2. Spencer, M. Fundamentals of Light Microscopy. Cambridge, U.K.: Cambridge University Press, 1982.

3. Boon, M.E.; Drijver, J.S. Routine Cytological Staining Methods. London: Macmillan, 1986.

4. Ploem, J.S.; Tanke, H.J. Introduction to Fluorescence Microscopy. Royal Microscopical Society Microscopy Handbook No. 10. Oxford, U.K.: Oxford Scientific Publications, 1987.

   Willingham, M.C.; Pastan, I. An Atlas of Immunofluorescence in Cultured Cells. Orlando, FL: Academic, 1985.

5. Allen, R.D. New observations on cell architecture and dynamics by video-enhanced contrast optical microscopy. *Annu. Rev. Biophys. Biophys. Chem.* 14:265–290, 1985.

   Inoué, S. Video Microscopy. New York: Plenum Press, 1986.

6. Pease, D.C.; Porter, K.R. Electron microscopy and ultramicrotomy. *J. Cell Biol.* 91:287s–292s, 1981.

7. Wischnitzer, S. Introduction to Electron Microscopy. 3rd ed. Elmsford, NY: Pergammon, 1981.

8. Everhart, T.E.; Hayes, T.L. The scanning electron microscope. *Sci. Am.* 226(1):54–69, 1972.

   Hayat, M.A. Introduction to Biological Scanning Electron Microscopy. Baltimore: University Park Press, 1978.

   Kessel, R.G.; Kardon, R.H. Tissues and Organs. New York: W.H. Freeman, 1979. (An atlas of vertebrate tissues seen by scanning electron microscopy.)

9. Sommerville, J.; Scheer, U., eds. Electron Microscopy in Molecular Biology: A Practical Approach. Washington, D.C.: IRL Press, 1987.

10. Heuser, J. Quick-freeze, deep-etch preparation of samples for 3-D electron microscopy. *Trends Biochem. Sci.* 6:64–68, 1981.

Pinto da Silva, P.; Branton, D. Membrane splitting in freeze-etching. *J. Cell Biol.* 45:598–605, 1970.

11. Chiu, W. Electron microscopy of frozen, hydrated specimens. *Annu. Rev. Biophys. Biophys. Chem.* 15:237–257, 1986.

Unwin, P.N.T.; Henderson, R. Molecular structure determination by electron microscopy of unstained crystalline specimens. *J. Mol. Biol.* 94:425–440, 1975.

12. Glusker, J.P.; Trueblood, K.N. Crystal Structure Analysis: A Primer. Oxford, U.K.: Oxford University Press, 1985.

13. Alzari, P.M.; Lascombe, M.B.; Poljak, R.J. Three-dimensional structure of antibodies. *Annu. Rev. Immunol.* 6:555–580, 1988.

Kendrew, J.C. The three-dimensional structure of a protein molecule. *Sci. Am.* 205(6):96-111, 1961.

Perutz, M.F. The hemoglobin molecule. *Sci. Am.* 211(5):64–76, 1964.

14. Cooke, R.M.; Cambell, I.D. Protein structure determination by NMR. *Bioessays* 8:52–56, 1988.

Shulman, R.G. NMR spectroscopy of living cells. *Sci. Am.* 248(1): 86–93, 1983.

Wüthrich, K.; Wagner, G. Internal dynamics of proteins. *Trends Biochem. Sci.* 9:152–154, 1984.

15. Ammann, D. Ion Selective Microelectrodes: Principles, Design and Application. Berlin: Springer-Verlag, 1986.

Auerbach, A.; Sachs, F. Patch clamp studies of single ionic channels. *Annu. Rev. Biophys. Bioeng.* 13:269–302, 1984.

16. Grynkiewicz, G.; Poenie, M.; Tsien, R.Y. A new generation of Ca^{2+} indicators with greatly improved fluorescence properties. *J. Biol. Chem.* 260:3440–3450, 1985.

Tsien, R.Y.; Poenie, M. Fluorescence ratio imaging: a new window into intracellular ionic signalling. *Trends Biochem. Sci.* 11:450, 1986.

17. Celis, J.E.; Graessmann, A.; Logter, A., eds. Microinjection and Organelle Transplantation Techniques. London: Academic Press, 1986.

Gomperts, B.D.; Fernandez, J.M. Techniques for membrane permeabilization. *Trends Biochem. Sci.* 10:414–417, 1985.

Ostro, Marc J. Liposomes. *Sci. Am.* 256(1):102–111, 1987.

Ureta, T.; Radojkovic, J. Microinjected frog oocytes. *Bioessays* 2:221–225, 1985.

18. Freshney, R.I. Culture of Animal Cells: A Manual of Basic Technique. New York: Liss, 1987.

19. Herzenberg, L.A.; Sweet, R.G.; Herzenberg, L.A. Fluorescence-activated cell sorting. *Sci. Am.* 234(3):108–116, 1976.

Kamarck, M.E. Fluorescence-activated cell sorting of hybrid and transfected cells. *Methods Enzymol.* 151:150–165, 1987.

Nolan, G.P; Fiering, S.; Nicolas, J.F.; Herzenberg, L.A. Fluorescence- activated cell analysis and sorting of viable mammalian cells based on beta-D-galactosidase activity after transduction of *E. coli lac* Z. *Proc. Nat. Acad. Sci. USA.* 85:2603–2607, 1988.

20. Harrison, R.G. The outgrowth of the nerve fiber as a mode of protoplasmic movement. *J. Exp. Zool.* 9:787–848, 1910. (Possibly the first use of tissue culture.)

21. Ham, R.G. Clonal growth of mammalian cells in a chemically defined, synthetic medium. *PNAS* 53:288–293, 1965.

Loo, D.T.; Fuquay, J.I.; Rawson, C.L.; Barnes, D.W. Extended culture of mouse embryo cells without senescence: inhibition by serum. *Science* 236:200–202, 1987.

Sirabasku, D.A.; Pardee, A.B.; Sato, G.H., eds. Growth of Cells in Hormonally Defined Media. Cold Spring Harbor, NY: Cold Spring Harbor Laboratory, 1982.

22. Ruddle, F.H.; Creagan, R.P. Parasexual approaches to the genetics of man. *Annu. Rev. Genet.* 9:407–486, 1975.

23. Colowick, S.P.; Kaplan, N.O., eds. Methods in Enzymology, Vols. 1–.... San Diego, CA: Academic Press, 1955–1988. (A multivolume series containing general and specific articles on many procedures.)

Cooper, T.G. The Tools of Biochemistry. New York: Wiley, 1977.

Scopes, R.K. Protein Purification Principles and Practice, 2nd ed. New York: Springer-Verlag, 1987.

24. Claude, A. The coming of age of the cell. *Science* 189:433–435, 1975.

deDuve, C.; Beaufay, H. A short history of tissue fractionation. *J. Cell Biol.* 91:293s–299s, 1981.

Meselson, M.; Stahl, F.W. The replication of DNA in *Escherichia coli. Proc. Natl. Acad. Sci. USA* 47:671–682, 1958. (Density gradient centrifugation was used to show the semiconservative replication of DNA.)

Palade, G. Intracellular aspects of the process of protein synthesis *Science* 189:347–358, 1975.

Sheeler, P. Centrifugation in Biology and Medical Science. New York: Wiley, 1981.

25. Morré, D.J; Howell, K.E.; Cook, G.M.W.; Evans, W.H., eds. Cell Free Analysis of Membrane Traffic. New York: Liss, 1986.

Nirenberg, N.W.; Matthaei, J.H. The dependence of cell free protein synthesis in *E. coli* on naturally occurring or synethetic polyribonucleotides. *Proc. Natl. Acad. Sci. USA* 47:1588–1602, 1961.

Racker, E. A New Look at Mechanisms in Bioenergetics. New York: Academic Press, 1976. (Cell-free systems in the working out of energy metabolism.)

Zamecnik, P.C. An historical account of protein synthesis with current overtones—a personalized view. *Cold Spring Harbor Symp. Quant. Biol.* 34:1–16, 1969.

26. Dean, P.D.G.; Johnson, W.S.; Middle, F.A. Affinity Chromatography: A Practical Approach. Arlington, VA: IRL Press, 1985.

Gilbert, M.T. High Performance Liquid Chromatography. Littleton, MA.: John Wright-PSG, 1987.

27. Andrews, A.T. Electrophoresis, 2nd ed. Oxford, U.K.: Clarendon Press, 1986.

Laemmli, U.K. Cleavage of the structural proteins during the assembly of the head of bacteriophage T4. *Nature* 227: 680–685, 1970.

28. Celis, J.E.; Bravo, R., eds. Two-Dimensional Gel Electrophoresis of Proteins. New York: Academic Press, 1983.

O'Farrell, P.H. High-resolution two-dimensional electrophoresis of proteins. *J. Biol. Chem.* 250:4007–4021, 1975.

29. Cleveland, D.W.; Fisher, S.G.; Kirschner, M.W.; Laemmli, U.K. Peptide mapping by limited proteolysis in sodium dodecyl sulfate and analysis by gel electrophoresis. *J. Biol. Chem.* 252:1102–1106, 1977.

Ingram, V.M. A specific chemical difference between the globins of normal human and sickle-cell anemia hemoglobin. *Nature* 178:792–794, 1956.

30. Edman, P.; Begg, G. A protein sequenator. *Eur. J. Biochem.* 1: 80–91, 1967. (First automated determination of sequencer.)

Hewick, R.M.; Hunkapiller, M.W.; Hood, L.E.; Dreyer, W.J. A gas-liquid-solid phase peptide and protein sequenator. *J. Biol. Chem.* 256:7990–7997, 1981.

Sanger, F. The arrangement of amino acids in proteins. *Adv. Protein Chem.* 7:1–67, 1952.

Walsh, K.A.; Ericsson, L.H.; Parmelee, D.C.; Titani, K. Advances in protein sequencing. *Annu. Rev. Biochem.* 50:261–284, 1981.

31. Chase, G.D.; Rabinowitz, J.L. Principles of Radio Isotope Methodology, 2nd ed. Minneapolis: Burgess, 1962.

Dyson, N.A. An Introduction to Nuclear Physics with Applications in Medicine and Biology. Chichester, U.K.: Horwood, 1981.

32. Calvin, M. The path of carbon in photosynthesis. *Science* 135:879–889, 1962. (One of the earliest uses of radio isotopes in biology.)

Rogers, A.W. Techniques of Autoradiography, 3rd ed. New York: Elsevier/ North Holland, 1979.

33. Anderton, B.H.; Thorpe, R.C. New methods of analyzing for antigens and glycoproteins in complex mixtures. *Immunol. Today* 2:122–127, 1980.

Coons, A.H. Histochemistry with labeled antibody. *Int. Rev. Cytol.* 5:1–23, 1956.

34. Milstein, C. Monoclonal antibodies. *Sci. Am.* 243(4):66–74, 1980.

 Yelton, D.E.; Scharff, M.D. Monoclonal antibodies: a powerful new tool in biology and medicine. *Annu. Rev. Biochem.* 50:657–680, 1981.

35. Mabuchi, I.; Okuno, M. The effect of myosin antibody on the division of starfish blastomeres. *J. Cell Biol.* 74:251–263, 1977.

 Mulcahy, L.S.; Smith, M.R.; Stacey, D.W. Requirement for ras proto-oncogene function during serum stimulated growth of NIH 3T3 cells. *Nature* 313:241–243, 1985.

36. Drlica, K. Understanding DNA and Gene Cloning. New York: Wiley, 1984.

 Sambrook, J.; Fritsch, E.F.; Maniatis, T. Molecular Cloning: A Laboratory Manual, 2nd ed. Cold Spring Harbor, NY: Cold Spring Harbor Laboratory, 1989.

 Watson, J.D.; Tooze, J. The DNA Story: A Documentary History of Gene Cloning. New York: W.H. Freeman, 1981.

 Watson, J.D.; Tooze, J.; Kurtz, D.T. Recombinant DNA: A Short Course. New York: W.H. Freeman, 1983.

37. Garoff, H. Using recombinant DNA techniques to study protein targeting in the eucaryotic cell. *Annu. Rev. Cell Biol.* 1:403–445, 1985.

 Jackson, Ian J. The real reverse genetics: targeted mutagenesis in the mouse. *Trends Genet.* 3:119, 1987.

 Kelly, J.H.; Darlington, G.J. Hybrid genes: molecular approaches to tissue-specific gene regulation. *Annu. Rev. Genet.* 19:273–296, 1985.

38. Nathans, D.; Smith, H.O. Restriction endonucleases in the analysis and restructuring of DNA molecules. *Annu. Rev. Biochem.* 44:273–293, 1975.

 Smith, H.O. Nucleotide sequence specificity of restriction endonucleases. *Science* 205:455–462, 1979.

39. Cohen, S.N. The manipulation of genes. *Sci. Am.* 233(1):24–33, 1975.

 Maniatis, T.; et al. The isolation of structural genes from libraries of eucaryotic DNA. *Cell* 15:687–701, 1978.

 Novick, R.P. Plasmids. *Sci. Am.* 243(6):102–127, 1980.

40. Cantor, C.R.; Smith, C.L.; Mathew, M.K. Pulsed-field gel electrophoresis of very large DNA molecules. *Annu. Rev. Biophys. Biophys. Chem.* 17:287–304, 1988.

 Maxam, A.M.; Gilbert, W. A new method of sequencing DNA. *Proc. Natl. Acad. Sci. USA* 74:560–564, 1977.

 Southern, E.M. Gel electrophoresis of restriction fragments. *Methods Enzymol.* 68:152–176, 1979.

41. Rigby, P.W.; Dieckmann, M.; Rhodes, C.; Berg, P. Labeling deoxyribonucleic acid to high specific acitivity *in vitro* by nick translation with DNA polymerase I. *J. Mol. Biol.* 113:237–251, 1977.

42. Galas, D.J.; Schmitz, A. DNAse footprinting: a simple method for the detection of protein-DNA binding specificity. *Nucleic Acids Res.* 5:3157–3170, 1978.

 Gilbert, W. DNA sequencing and gene structure. *Science* 214:1305–1312, 1981.

 Prober, J.M.; et al. A system for rapid DNA sequencing with fluorescent chain terminating dideoxynucleotides. *Science* 238:336–341, 1987.

 Sanger, F. Determinating of nucleotide sequences in DNA. *Science* 214:1205–1210, 1981.

 Tullius, T.D. Chemical "snapshots" of DNA: using the hydroxyl radical to study the structure of DNA and DNA protein complexes. *Trends Biochem. Sci.* 12:297–300, 1987.

43. Berk, A.J.; Sharp, P.A. Sizing and mapping of early adenovirus mRNAS by gel electrophoresis of S1 endonuclease digested hybrids. *Cell* 12:721–732, 1977.

 Hood, L.E.; Wilson, J.H.; Wood, W.B. Molecular Biology of Eucaryotic Cells: A Problems Approach, pp. 56–61, 192–210. Menlo Park, CA: Benjamin-Cummings, 1975.

 Wetmer, J.G. Hybridization and renaturation kinetics of nucleic acids. *Annu. Rev. Biophys. Bioeng.* 5:337–361, 1976.

44. Alwine, J.C.; Kemp, D.J.; Stark, G.R. A method for detection of specific RNAs in agarose gels by transfer to diazobenzyloxymethyl-paper and hybridization with DNA probes. *Proc. Natl. Acad. Sci. USA* 74:5350–5354, 1977.

 Southern, E.M. Detection of specific sequences among DNA fragments separated by gel electrophoresis. *J. Mol. Biol.* 98:503–517, 1975.

45. Itakura, K.; Rossi, J.J., Wallace, R.B. Synthesis and use of synthetic oligonucleotides. *Annu. Rev. Biochem.* 53:323–356, 1984.

 Ruddle, F.H. A new era in mammalian gene mapping: somatic cell genetics and recombinant DNA methodologies *Nature* 294:115–119, 1981.

 White, R.; Lalovel, J.M. Chromosome mapping with DNA markers. *Sci. Am.* 258(2):40–48, 1988.

46. McGinnis, W.; Garber, R.L.; Wirz, J.; Kuroiwa, A.; Ghering, W.J. A homologous protein-coding sequence in *Drosophila* homeotic genes and its conservation in other metazoans. *Cell* 37:403–408, 1984.

47. Gerhard, D.S.; Kawasaki, E.S.; Bancroft, F.C.; Szabo, P. Localization of a unique gene by direct hybridization. *Proc. Natl. Acad. Sci. USA* 78:3755–3759, 1981.

 Huten, E.; Kuroiwa, A.; Gehring, W.J. Spatial distribution of transcripts from the segmentation gene *fushi tarazu* during *Drosophila* embryonic development. *Cell* 37:833–841, 1984.

 Pardue, M.L.; Gall, J.G. Molecular hybridization of radioactive DNA to the DNA of cytological preparations. *Proc. Natl. Acad. Sci. USA* 64:600–604, 1969.

48. Abelson, J.; Butz, E., eds. Recombinant DNA. *Science* 209:1317–1438, 1980.

 Gilbert, W.; Villa-Komaroff, L. Useful proteins from recombinant bacteria. *Sci. Am.* 242(4):74–94, 1980.

49. Lederberg, J.; Lederberg, E.M. Replica plating and indirect selection of bacterial mutants. *J. Bacteriol.* 63:399–406, 1952.

 Nüsslein-Volhard, C.; Wieschaus, E. Mutations affecting segment number and polarity in *Drosophila. Nature* 287:795–801, 1980.

50. Palmiter, R.D.; Brinster, R.L. Germ line transformation of mice. *Annu. Rev. Genet.* 20:465–499, 1986.

 Pellicer, A.; et al. Altering genotype and phenotype by DNA-mediated gene transfer. *Science* 209:1414–1422, 1980.

 Rubin, G.M.; Spradling, A.C. Genetic transformation of *Drosophila* with transposable element vectors. *Science* 218:348–353, 1982.

 Shortle, D.; Nathans, D. Local mutagenesis: a method for generating viral mutants with base substitutions in preselected regions of the viral genome. *Proc. Natl. Acad. Sci. USA* 75:2170–2174, 1978.

 Struhl, K. The new yeast genetics. *Nature* 305:391–397, 1983.

51. Melton, D.A.; Rebagliati, M.R. Antisense RNA injections in fertilized eggs as a test for the function of localized mRNAs. *J. Embryol. Exp. Morphol.* 97:211–221, 1986.

 Rosenberg, V.B.; Preiss, A.; Seifert, E.; Jäckle, H.; Knipple, D.C. Production of phenocopies by Krüppel anti-sense RNA injection into *Drosophila* embryos. *Nature* 313:703–706, 1985.

 Weintraub, H.; Izant, J.G.; Harland, R.M. Anti-sense RNA as a molecular tool for genetic analysis. *Trends Genet.* 1:22–25, 1985.

The Molecular Organization of Cells

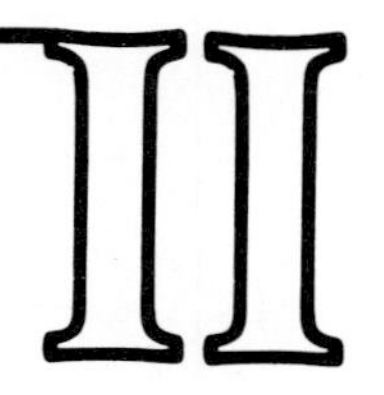

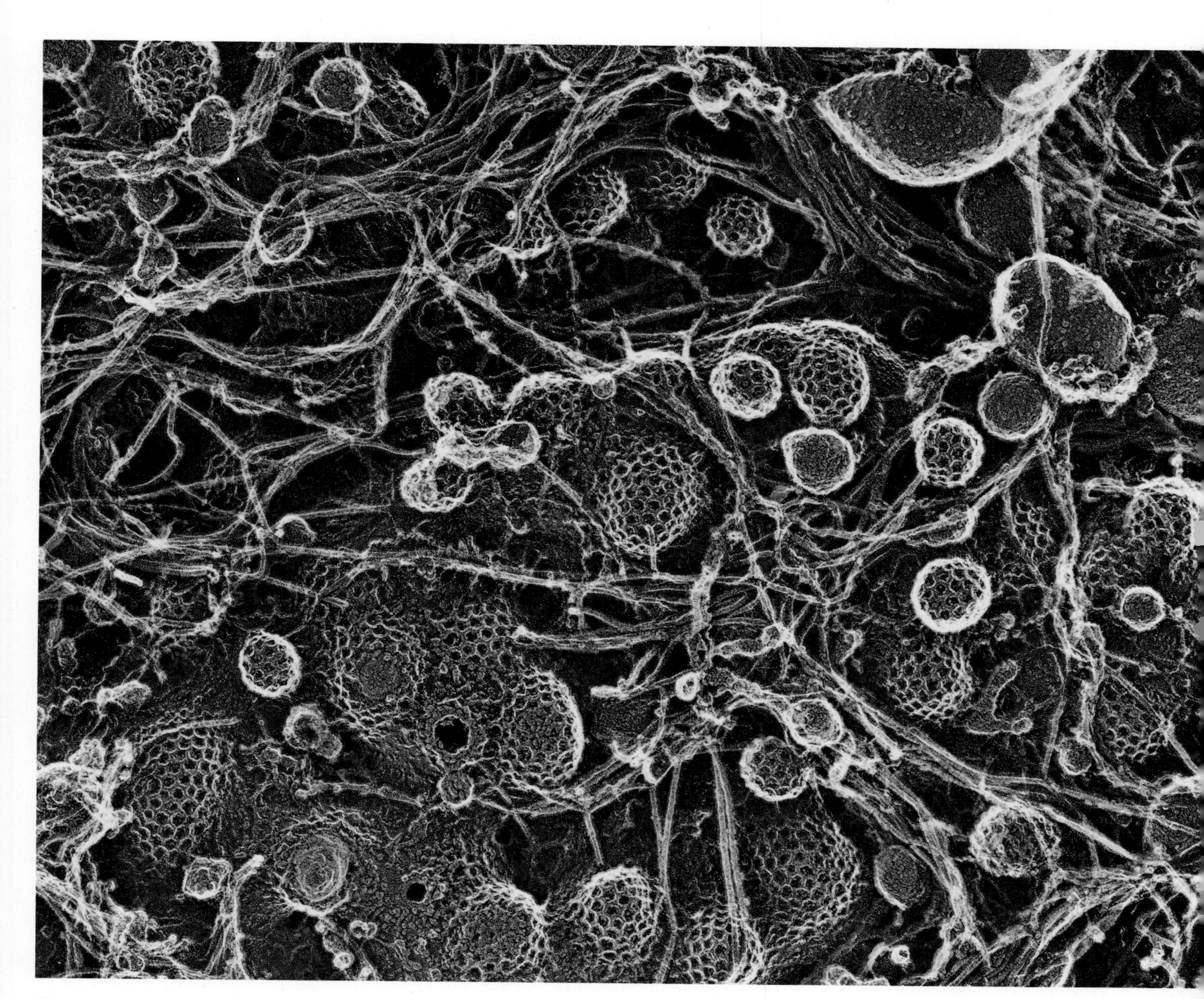

Electron micrograph of coated pits and vesicles on the inner face of the plasma membrane of a mouse liver cell. Numerous keratin filaments are also seen. (From N. Hirokawa and J. Heuser, *Cell* 30:395–406, 1982.)

Electron micrograph of DNA spilling out of a single disrupted human chromosome. Only about half of the DNA in one chromosome is shown. (Courtesy of James Paulson and Ulrich Laemmli.)

Basic Genetic Mechanisms

5

The ability of cells to maintain a high degree of order in a chaotic universe stems from the genetic information that is expressed, maintained, replicated, and occasionally improved by the basic genetic processes—*RNA and protein synthesis, DNA repair, DNA replication,* and *genetic recombination.* These processes, which produce and maintain the proteins and nucleic acids of a cell, are all one-dimensional: the information in a linear sequence of nucleotides is used to specify either another linear chain of nucleotides (a DNA or RNA molecule) or a linear chain of amino acids (a protein molecule). Genetic events are therefore conceptually simple compared with most other cellular processes, which are largely the result of information expressed in the complex three-dimensional surfaces of protein molecules. Perhaps that is why we understand more about genetic mechanisms than about most other cellular processes.

In this chapter we examine the molecular machinery that repairs, replicates, and alters the DNA of the cell. We shall see that the machinery depends on enzymes that cut, copy, and recombine nucleotide sequences. We explain how these and other enzymes can be parasitized by viruses, plasmids, and transposable genetic elements, which not only direct their own replication but can alter the cell genome by genetic recombination events. Finally, we discuss how the basic genetic mechanisms can be exploited in the laboratory to isolate genes and their products and to make them to order in virtually unlimited quantities.

First, however, we reconsider a central topic mentioned briefly in Chapter 3: the mechanisms of RNA and protein synthesis.

RNA and Protein Synthesis

Proteins constitute more than half the total dry mass of a cell, and their synthesis is central to cell maintenance, growth, and development. Protein synthesis depends on the collaboration of several classes of RNA molecules and requires a series of preparatory steps. First, a molecule of *messenger RNA (mRNA)* must be copied from the DNA that encodes the protein to be synthesized. Meanwhile, in the cytoplasm, each of the 20 amino acids from which the protein is to be built must be attached to its specific *transfer RNA (tRNA)* molecule, and the subunits of the ribosome on which the new protein is to be made must be preloaded with auxiliary protein factors. Protein synthesis begins when all of these components come together in the cytoplasm to form a functional ribosome. As a single molecule of mRNA moves stepwise through each ribosome, the sequence of nucleotides in the mRNA molecule is translated into a corresponding sequence of amino acids to produce a distinctive protein chain. We shall begin by considering how the many different RNA molecules in a cell are made.

RNA Polymerase Copies DNA into RNA: the Process of DNA Transcription[1]

RNA is synthesized on a DNA template in a process known as **DNA transcription.** Transcription generates the mRNAs that carry the information for protein synthesis, as well as the transfer, ribosomal, and other RNA molecules that have structural and catalytic functions. These RNA molecules are synthesized by **RNA polymerase** enzymes, which make an RNA copy of a DNA sequence. In eucaryotes, different RNA polymerase molecules synthesize different types of RNA, but most of what we know about RNA polymerase comes from studies on bacteria, where a single species of the enzyme mediates all RNA synthesis.

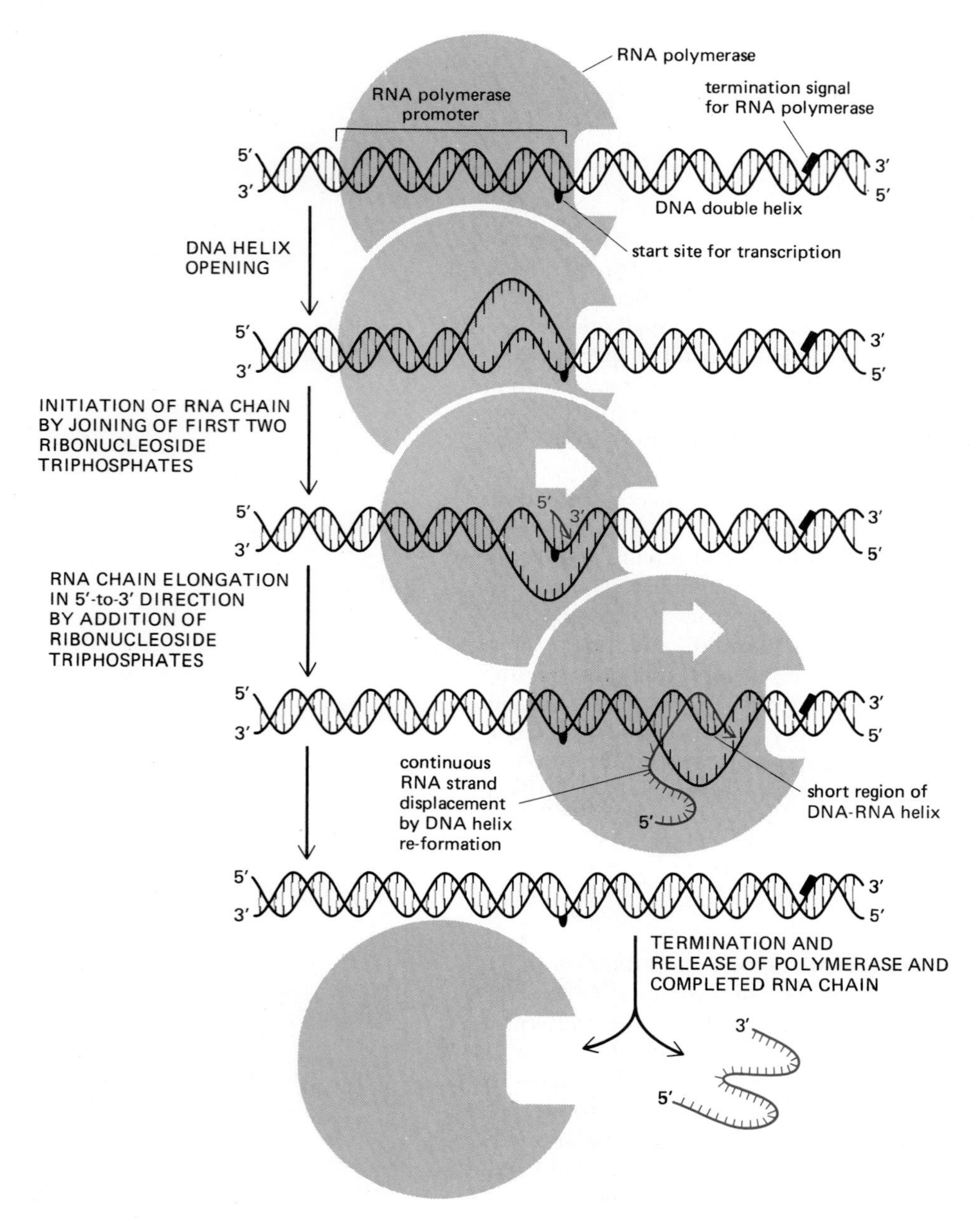

Figure 5–1 The synthesis of an RNA molecule by RNA polymerase. The enzyme begins its synthesis at a special start signal on the DNA called a promoter and completes its synthesis at a termination (stop) signal, whereupon both the polymerase and its completed RNA chain are released. During RNA chain elongation, polymerization rates average about 30 nucleotides per second at 37°C. Therefore, an RNA chain of 5000 nucleotides takes about 3 minutes to complete.

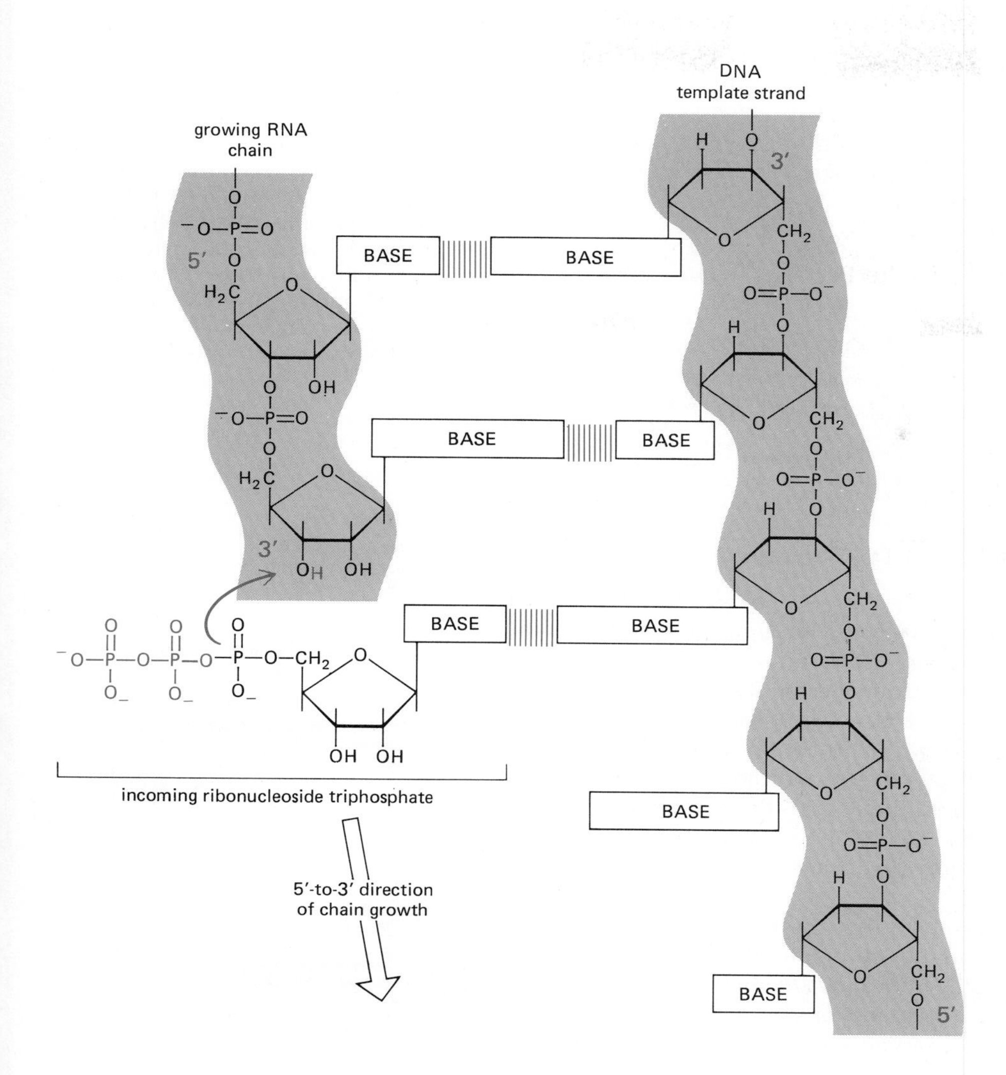

Figure 5–2 The chain elongation reaction catalyzed by an RNA polymerase enzyme. In each step an incoming ribonucleoside triphosphate is selected for its ability to base-pair with the exposed DNA template strand; a ribonucleoside monophosphate is then added to the growing 3′-OH end of the RNA chain (*colored arrow*), and pyrophosphate is released (*colored atoms*). The new RNA chain therefore grows by one nucleotide at a time in the 5′-to-3′ direction, and it is complementary in sequence to the DNA template strand. The reaction is driven both by the favorable free-energy change that accompanies the release of pyrophosphate and by the subsequent hydrolysis of the pyrophosphate to inorganic phosphate (see p. 74).

The bacterial RNA polymerase is a large multisubunit enzyme associated with several additional protein subunits that enter and leave the polymerase-DNA complex at different stages of transcription (see p. 524). Free RNA polymerase molecules collide randomly with the chromosome, sticking only weakly to most DNA. However, the polymerase binds very tightly when it collides with a specific DNA sequence, called the **promoter,** that contains the *start site* for RNA synthesis and signals where RNA synthesis should begin. The reactions that ensue are outlined in Figure 5–1. After binding to the promoter, the RNA polymerase opens up a local region of the double helix to expose the nucleotides on a short stretch of DNA on each strand. One of the two exposed DNA strands acts as a template for complementary base-pairing with incoming ribonucleoside triphosphate monomers, two of which are joined together by the polymerase to begin an RNA chain. The RNA polymerase molecule then moves stepwise along the DNA, unwinding the DNA helix just ahead to expose a new region of template for complementary base-pairing. In this way the growing RNA chain is extended by one nucleotide at a time in the 5′-to-3′ direction (Figure 5–2). The chain elongation process continues until the enzyme encounters a second special sequence in the DNA, the **termination signal,** at which point the polymerase releases both the DNA template and the newly made RNA chain (see also Figure 5–6B).

As the enzyme moves, an RNA-DNA double helix is formed at the enzyme's active site. This helix is a very short one because the DNA-DNA helix immediately rewinds at the rear of the polymerase, displacing the RNA just made (Figure 5–3). As a result, each completed RNA chain is released from the DNA template as a free, single-stranded RNA molecule, typically between 70 and 10,000 nucleotides long.

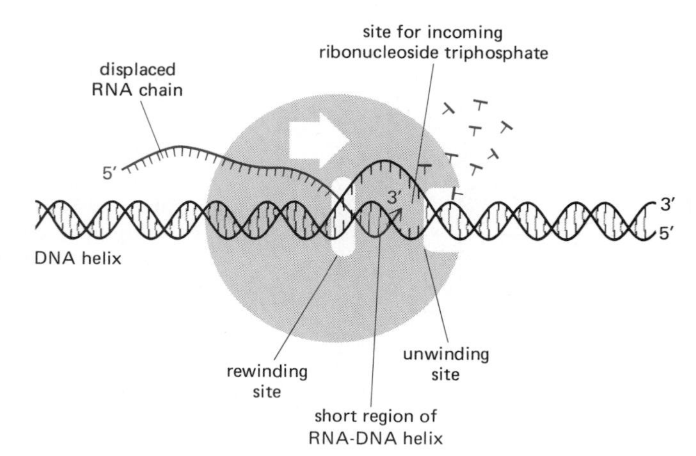

Figure 5–3 A moving RNA polymerase molecule is continuously unwinding the DNA helix ahead of the polymerization site while rewinding the two DNA strands behind this site to displace the newly formed RNA chain. A short region of RNA-DNA helix (about 17 nucleotide pairs for the bacterial enzyme) is therefore formed only transiently, and the final RNA product is released as a single-stranded copy of one of the two DNA strands.

The Promoter Sequence Defines Which DNA Strand Is to Be Transcribed[2]

In principle, any region of the DNA double helix could be copied into two different mRNA molecules—one from each of the two DNA strands. In reality, only one DNA strand is used as a template in each region, and the RNA made is therefore equivalent in nucleotide sequence to the opposite, nontemplate DNA strand. Which of the two strands is copied is determined by the promoter, which is an oriented DNA sequence that points the RNA polymerase in one direction or the other. Because RNA chains are synthesized only in the 5′-to-3′ direction, the promoter determines which DNA strand is copied (Figure 5–4). The DNA strand that is copied into RNA is often different for neighboring genes, as illustrated for a small region of a chromosome in Figure 5–5.

DNA footprinting studies (p. 187) indicate that the *E. coli* RNA polymerase covers a large region of DNA when it binds to a promoter; the polymerase "footprint" is about 60 base pairs long, extending from about 40 nucleotides "upstream" to 20 nucleotides "downstream" from the start site for transcription. Comparison of the DNA sequences of many strong promoters reveals that the polymerase primarily recognizes two highly conserved DNA sequences, each six nucleotides long, which are located upstream from the start site and separated from each other by about 17 nucleotides of unrecognized DNA (Figure 5–6A). Conserved sequences found in all examples of a particular type of regulatory region in DNA, such as promoters, are called **consensus sequences.** For example, when large numbers of *E. coli* promoters are compared, the two consensus hexanucleotide sequences are T_{82} T_{84} G_{78} A_{65} C_{54} A_{45} and T_{80} A_{95} T_{45} A_{60} A_{50} T_{96} (where the number is the percentage of cases where the nucleotide is present at that position in the sequence; 100 means that the nucleotide is always present, whereas 25 means that it is present in one of four promoters—the frequency expected by chance). In general, strong promoters have sequences that match the two promoter consensus sequences closely, whereas weak promoters (those associated with genes that produce relatively small amounts of mRNA) match these sequences less well.

Eucaryotic cells have three different RNA polymerases. One of these makes all of the RNAs that code for proteins (that is, the mRNAs); the other two make RNA molecules with structural and catalytic roles (such as ribosomal RNAs and transfer

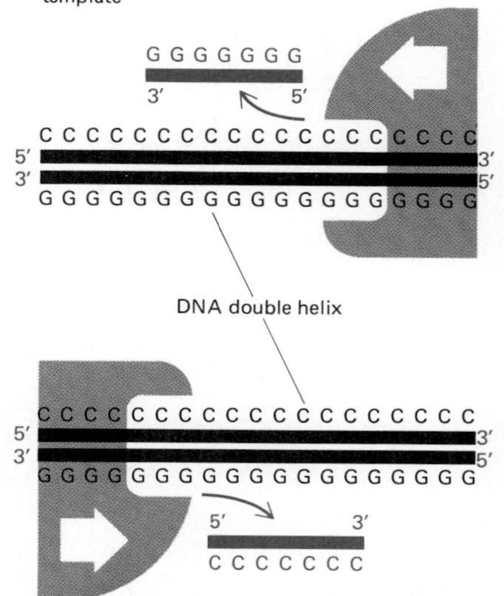

Figure 5–4 Because the DNA strand serving as template must be traversed from its 3′ end to its 5′ end (see Figure 5–2), the direction of RNA polymerase movement determines which of the two DNA strands will serve as a template for the synthesis of RNA, as shown in the diagram. The direction of polymerase movement is, in turn, determined by the orientation of the promoter sequence at which the RNA polymerase starts.

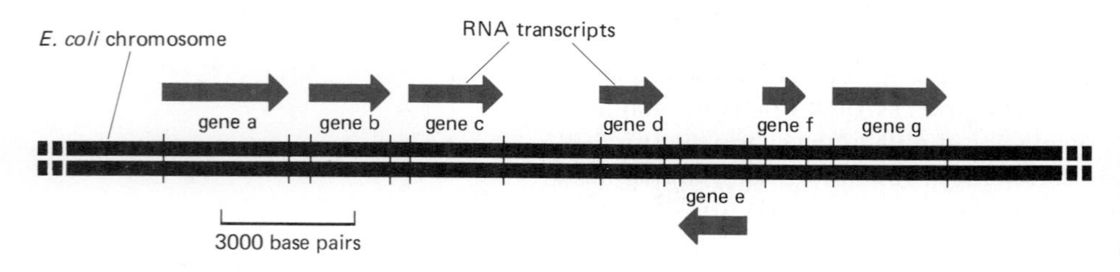

Figure 5–5 A short portion of a typical bacterial chromosome, illustrating the pattern of DNA transcription involved in the expression of several neighboring genes.

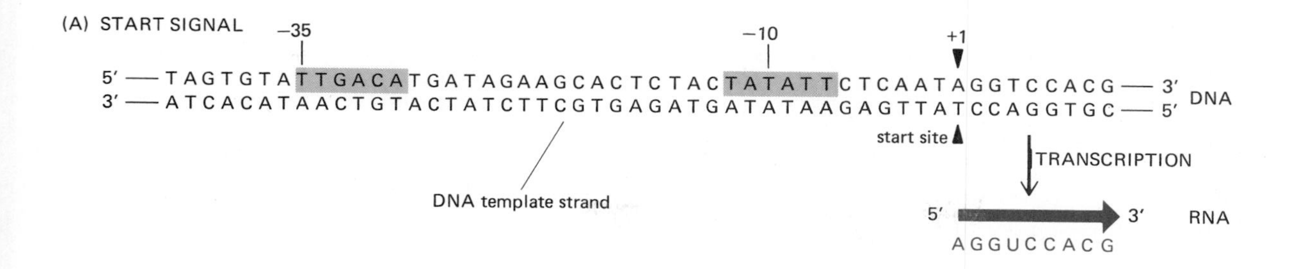

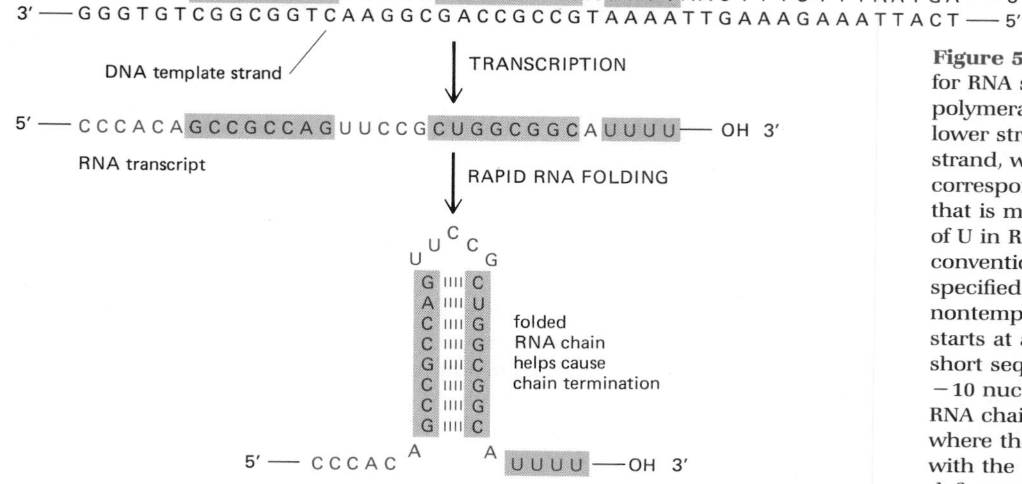

Figure 5–6 The start and stop signals for RNA synthesis by bacterial RNA polymerase of *E. coli.* Note that the lower strand of DNA is the template strand, whereas the upper strand corresponds in sequence to the RNA that is made (except for the substitution of U in RNA for T in DNA—see p. 98). By convention, when a DNA sequence is specified, it is presented for the nontemplate strand. (A) The polymerase starts at a promoter sequence. Two short sequences (*color*), about −35 and −10 nucleotides from the start of the RNA chain, are thought to determine where the polymerase binds; together with the start site, these sequences define a promoter. Major modifications in either of these two consensus sequences eliminate promoter activity, while modifications elsewhere do not. (B) The polymerase stops when it synthesizes a run of U residues from a complementary run of A residues on the template strand, provided that it has just synthesized a self-complementary RNA nucleotide sequence (shown in gray). This combination of DNA sequences therefore constitutes a termination signal. The precise sequence of nucleotides in the self-complementary region can vary; it is the hairpin helix that rapidly forms in this region of the newly synthesized RNA chain that is crucial for stopping transcription.

RNAs). All three are large multisubunit enzymes that resemble the bacterial enzyme, but the promoters each enzyme recognizes are more complex and not as well characterized (see p. 526). It is unclear why both bacterial and eucaryotic RNA polymerases are such complicated molecules, with multiple subunits and a total mass of more than 500,000 daltons, when some bacterial viruses encode single-chain RNA polymerases of one-fifth this mass that catalyze RNA synthesis at least as well as the host cell enzyme. Presumably the multiple subunit composition of the cellular RNA polymerases is important for regulatory aspects of cellular RNA synthesis that have not yet been well defined.

This outline of DNA transcription omits many details; usually other complex steps must occur before an mRNA molecule is produced. For example, *gene regulatory proteins* help to determine which regions of DNA are transcribed by the RNA polymerase and thereby play a major part in determining which proteins are made by a cell. Moreover, although mRNA molecules are produced directly by DNA transcription in procaryotes, in higher eucaryotic cells most RNA transcripts are altered extensively—by a process called *RNA splicing*—before they leave the cell nucleus and enter the cytoplasm as mRNA molecules (see p. 531). These aspects of mRNA production will be discussed in Chapters 9 and 10, when we consider the cell nucleus and the control of gene expression. For now, let us assume that functional mRNA molecules have been produced and proceed to examine how they direct protein synthesis.

Transfer RNA Molecules Act as Adaptors That Translate Nucleotide Sequences into Protein Sequences[3]

All cells contain a set of **transfer RNAs (tRNAs),** each of which is a small RNA molecule (most have a length between 70 and 90 nucleotides). The tRNAs, by binding at one end to a specific codon in the mRNA and at their other end to the

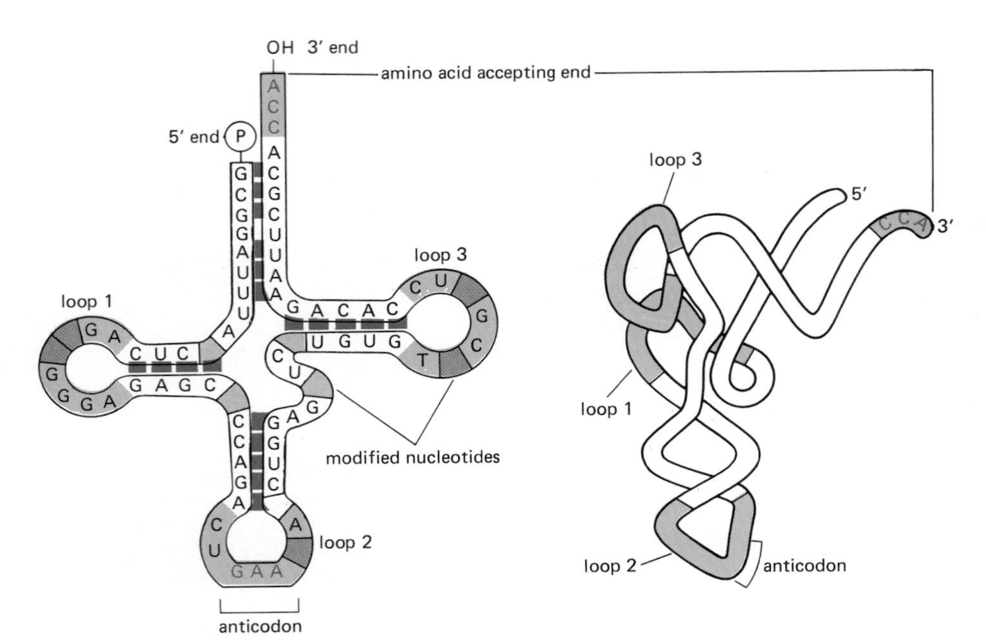

Figure 5–7 The structure of a typical tRNA molecule. The base-paired regions in the molecule are shown schematically at the left in a "cloverleaf" structure, and an outline of the overall three-dimensional conformation determined by x-ray diffraction is shown at the right. Note that the molecule is L-shaped, with one end designed to accept the amino acid while the other contains the three nucleotides of the anticodon. The amino acid is attached to the A residue of a CCA sequence at the 3′ end of the molecule (see Figure 5–11).

amino acid specified by that codon, enable amino acids to line up according to the sequence of nucleotides in the mRNA. Each tRNA is designed to carry only one of the 20 amino acids used for protein synthesis: a tRNA that carries glycine is designated $tRNA^{Gly}$, and so on. Each of the 20 amino acids has at least one type of tRNA assigned to it, and most have several. Before an amino acid is incorporated into a protein chain, it is attached by its carboxyl end to the 3′ end of an appropriate tRNA molecule. This attachment serves two purposes. First, and most important, it covalently links the amino acid to a tRNA containing the correct **anticodon**—the sequence of three nucleotides that is complementary to the three-nucleotide *codon* that specifies that amino acid on an mRNA molecule. Codon-anticodon pairings enable each amino acid to be inserted into a growing protein chain according to the dictates of the sequence of nucleotides in the mRNA, thereby allowing the genetic code to be used to translate nucleotide sequences into protein sequences. This is the essential "adaptor" function of the tRNA molecule: with one end attached to an amino acid and the other paired to a codon, the tRNA converts sequences of nucleotides into sequences of amino acids.

The second function of the amino acid attachment is to activate the amino acid by generating a high-energy linkage at its carboxyl end so that it can react with the amino group of the next amino acid in the sequence to form a *peptide bond.* The activation process is necessary for protein synthesis because nonactivated amino acids cannot be added directly to a growing polypeptide chain. (By contrast, the reverse process, in which a peptide bond is hydrolyzed by the addition of water, can occur spontaneously.)

The function of a tRNA molecule depends on its precisely folded three-dimensional structure. A few tRNAs have been crystallized and their complete structures determined by x-ray diffraction analyses. Both complementary base-pairings and unusual base interactions are required to fold a tRNA molecule (see Figure 3–16, p. 103). The nucleotide sequences of tRNA molecules from many different organisms reveal that tRNAs can form the loops and base-paired stems of a "cloverleaf" structure, and all are thought to fold further to adopt the L-shaped conformation detected in crystallographic analyses (Figure 5–7). In the native structure, the amino acid is attached to one end of the "L," while the anticodon is located at the other (Figure 5–8).

The nucleotides in a completed nucleic acid chain (like the amino acids in proteins, p. 416) can be covalently modified to modulate the biological activity of the nucleic acid molecule. Such posttranscriptional modifications are especially common in tRNA molecules, which contain many different modified nucleotides

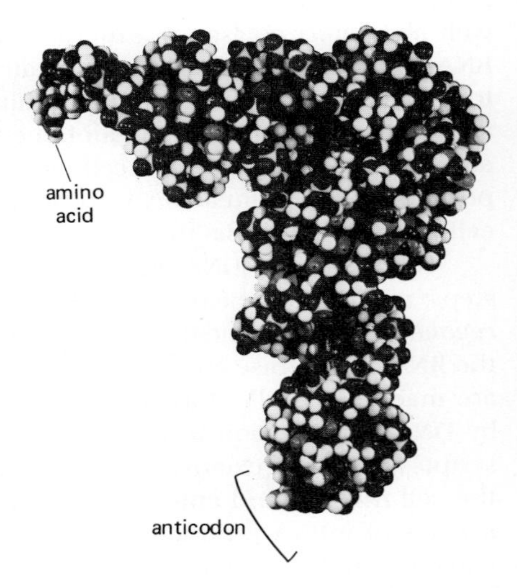

Figure 5–8 A space-filling model of a tRNA molecule with its bound amino acid. There are many different tRNA molecules, including at least one for each different amino acid. Although they differ in nucleotide sequence, they are all folded in a similar way. The particular tRNA molecule shown binds phenylalanine and is therefore denoted $tRNA^{Phe}$. (Courtesy of Sung-Hou Kim.)

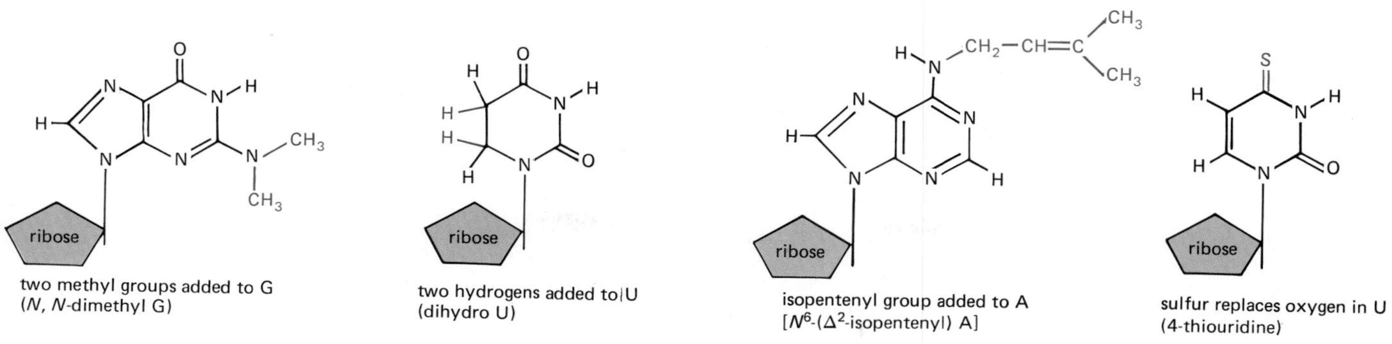
ribose
two methyl groups added to G
(N, N-dimethyl G)
two hydrogens added to U
(dihydro U)
isopentenyl group added to A
[N^6-(Δ^2-isopentenyl) A]
sulfur replaces oxygen in U
(4-thiouridine)

Figure 5–9 A few of the unusual nucleotides found in tRNA molecules. These nucleotides are produced by covalent modification of a normal nucleotide after it has been incorporated into a polynucleotide chain. In most tRNA molecules about 10% of the nucleotides are modified (see Figure 5–7).

(Figure 5–9). Some of the modified nucleotides affect the conformation and base-pairing of the anticodon and thereby facilitate the recognition of the appropriate mRNA codon by the tRNA molecule.

Specific Enzymes Couple Each Amino Acid to Its Appropriate tRNA Molecule[4]

Only the tRNA molecule, and not its attached amino acid, determines where the amino acid is added during protein synthesis. This was established by an ingenious experiment in which an amino acid (cysteine) was chemically converted into a different amino acid (alanine) after it was already attached to its specific tRNA. When such "hybrid" tRNA molecules were used for protein synthesis in a cell-free system, the wrong amino acid was inserted at every point in the protein chain where that tRNA was used. Thus the success of decoding is crucially dependent on the accuracy of the mechanism that normally links each activated amino acid specifically to its corresponding tRNA molecules.

How does a tRNA molecule become covalently linked to the one amino acid out of 20 that is its appropriate partner? The mechanism depends on enzymes called **aminoacyl-tRNA synthetases** that couple each amino acid to its appropriate set of tRNA molecules. There is a different synthetase enzyme for every amino acid (20 synthetases in all): one attaches glycine to all $tRNA^{Gly}$ molecules, another attaches alanine to all $tRNA^{Ala}$ molecules, and so on. The coupling reaction that creates an **aminoacyl-tRNA** molecule is catalyzed in two steps, as illustrated in Figure 5–10. The structure of the amino acid-RNA linkage is shown in Figure 5–11.

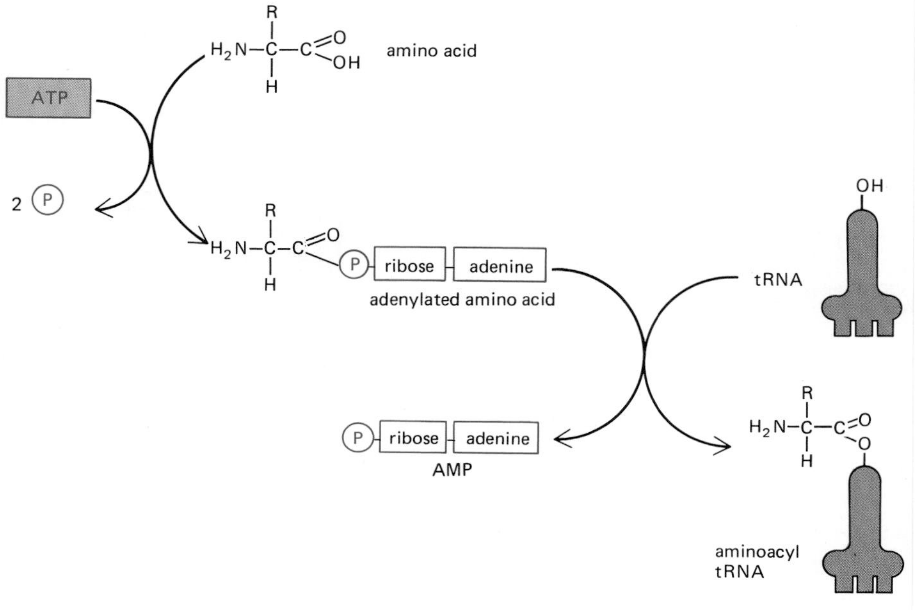
amino acid
ATP
2 P
ribose
adenine
adenylated amino acid
tRNA
OH
AMP
aminoacyl
tRNA

Figure 5–10 The two-step process in which an amino acid is activated for protein synthesis by an aminoacyl-tRNA synthetase enzyme. As indicated, the energy of ATP hydrolysis is used to attach each amino acid to its tRNA molecule in a high-energy linkage. The amino acid is first activated through the linkage of its carboxyl group directly to an AMP moiety, forming an *adenylated amino acid*; the linkage of the AMP, normally an unfavorable reaction, is driven by the hydrolysis of the ATP molecule that donates the AMP. Without leaving the synthetase enzyme, the AMP-linked carboxyl group on the amino acid is then transferred to a hydroxyl group on the sugar at the 3′ end of the tRNA molecule. This transfer joins the amino acid by an activated ester linkage to the tRNA and forms the final aminoacyl-tRNA molecule. The synthetase enzyme is not shown in these diagrams.

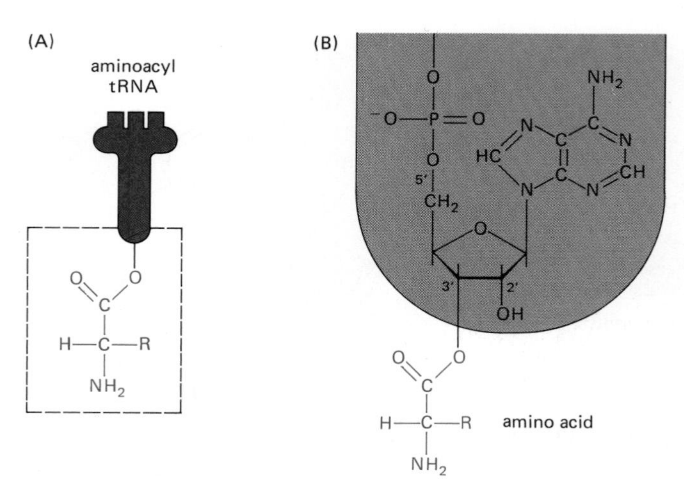

Figure 5–11 The structure of the aminoacyl-tRNA linkage. The carboxyl end of the amino acid forms an ester bond to ribose. Because the hydrolysis of this ester bond is associated with a large favorable change in free energy, an amino acid held in this way is said to be *activated.* (A) Schematic of the structure. (B) Actual structure corresponding to boxed region in (A). As in Figure 5–10, the "R-group" on the amino acid indicates one of the 20 side chains (see pp. 54–55).

Although the tRNA molecules serve as the final adaptors in converting nucleotide sequences into amino acid sequences, the aminoacyl-tRNA synthetase enzymes are adaptors of equal importance to the decoding process. Thus the genetic code is translated by two sets of adaptors that act sequentially, each matching one molecular surface to another with great specificity; it is their combined action that associates each sequence of three nucleotides in the mRNA molecule—each codon—with its particular amino acid (Figure 5–12).

5-6 Amino Acids Are Added to the Carboxyl-Terminal End of a Growing Polypeptide Chain

The fundamental reaction of protein synthesis is the formation of a peptide bond between the carboxyl group at the end of a growing polypeptide chain and a free amino group on an amino acid. Consequently, a protein is synthesized stepwise from its amino-terminal end to its carboxyl-terminal end. Throughout the entire process, the growing carboxyl end of the polypeptide chain remains activated by its covalent attachment to a tRNA molecule (a *peptidyl-tRNA* molecule). This high-energy covalent linkage is disrupted in each cycle but is immediately replaced by the identical linkage on the most recently added amino acid (Figure 5–13). In this way, each amino acid added carries with it the activation energy for the addition of the *next* amino acid rather than the energy for its own addition—an example of the "head-growth" type of polymerization described in Chapter 2 (Figure 2–34, p. 79).

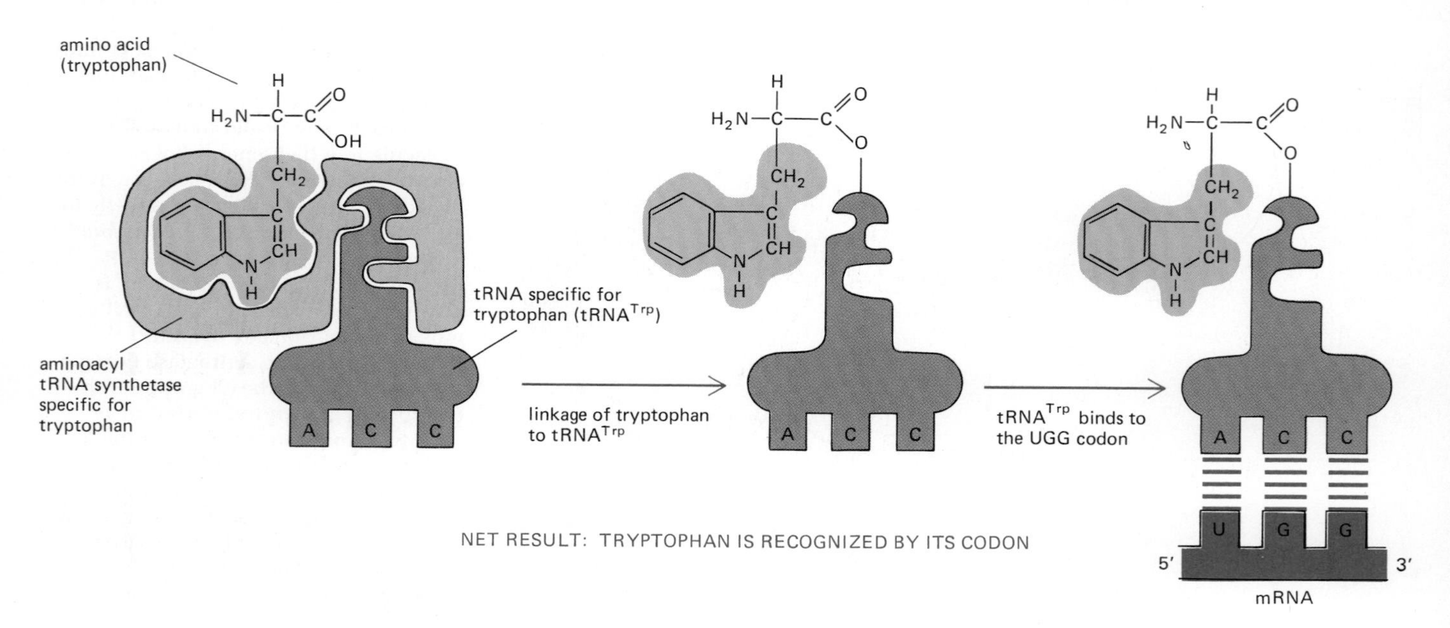

Figure 5–12 How the genetic code is translated by means of two linked "adaptors": the aminoacyl-tRNA synthetase enzyme, which couples a particular amino acid to its corresponding tRNA, and the tRNA molecule, which then binds to the appropriate nucleotide sequence on the mRNA.

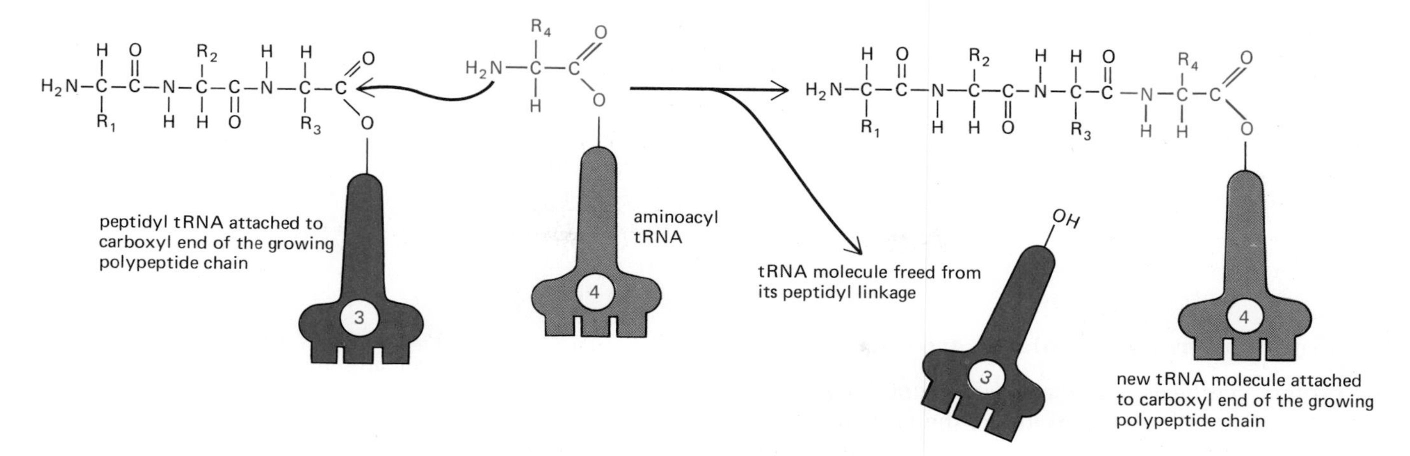

Figure 5–13 A polypeptide chain grows by the stepwise addition of amino acids to its carboxyl-terminal end. The formation of each peptide bond is energetically favorable because the growing carboxyl terminus has been activated by the covalent attachment of a tRNA molecule. The peptidyl-tRNA linkage that activates the growing end is regenerated in each cycle.

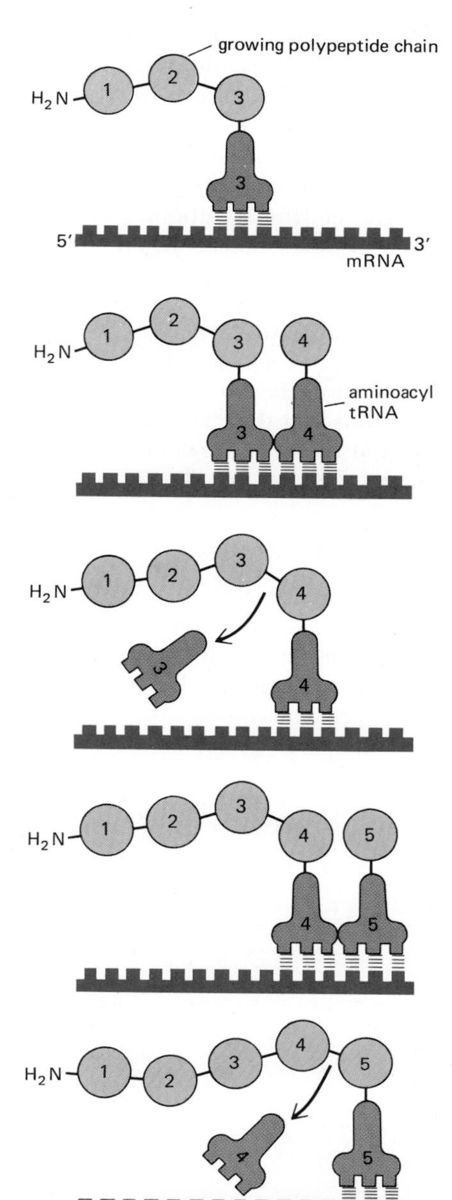

5-3 The Genetic Code Is Degenerate[5]

In the course of protein synthesis, the translation machinery moves in the 5′-to-3′ direction along an mRNA molecule and the mRNA sequence is read three nucleotides at a time. As we have seen, each amino acid is specified by the triplet of nucleotides (**codon**) in the mRNA molecule that pairs with a sequence of three complementary nucleotides at the anticodon tip of a particular tRNA. Because only one of the many types of tRNA molecules in a cell can base-pair with each codon, the codon determines the specific amino acid residue to be added to the growing polypeptide chain end (Figure 5–14).

Since RNA is constructed from four types of nucleotides, there are 64 possible sequences composed of three nucleotides (4 × 4 × 4), and most of them occur somewhere in most mRNA molecules. Three of these 64 sequences do not code for amino acids but instead specify the termination of a polypeptide chain; they are known as *stop codons.* That leaves 61 codons to specify only 20 different amino acids. For this reason, most of the amino acids are represented by more than one codon (Figure 5–15), and the genetic code is said to be *degenerate.* Two amino acids, methionine and tryptophan, have only one codon each, and they are the least abundant amino acids in proteins.

The degeneracy of the genetic code implies that either there is more than one tRNA for each amino acid or a single tRNA molecule can base-pair with more than one codon. In fact, both situations occur. For some amino acids there is more than one tRNA molecule, and some tRNA molecules are constructed so that they require accurate base-pairing only at the first two positions of the codon and can tolerate a mismatch (or wobble) at the third. This *wobble base-pairing* explains why so many of the alternative codons for an amino acid differ only in their third nucleotide (see Figure 5–15). The standard wobble pairings make it possible to fit the 20 amino acids to 61 codons with as few as 31 different tRNA molecules; in animal mitochondria a more extreme "wobble" allows protein synthesis with only 22 tRNAs (see p. 392).

Figure 5–14 Each amino acid added to the growing end of a polypeptide chain is selected by complementary base-pairing between the anticodon on its attached tRNA molecule and the next codon on the mRNA chain.

GCA GCC GCG GCU	AGA AGG CGA CGC CGG CGU	GAC GAU	AAC AAU	UGC UGU	GAA GAG	CAA CAG	GGA GGC GGG GGU	CAC CAU	AUA AUC AUU	UUA UUG CUA CUC CUG CUU	AAA AAG	AUG	UUC UUU	CCA CCC CCG CCU	AGC AGU UCA UCC UCG UCU	ACA ACC ACG ACU	UGG	UAC UAU	GUA GUC GUG GUU	UAA UAG UGA
Ala	Arg	Asp	Asn	Cys	Glu	Gln	Gly	His	Ile	Leu	Lys	Met	Phe	Pro	Ser	Thr	Trp	Tyr	Val	stop
A	R	D	N	C	E	Q	G	H	I	L	K	M	F	P	S	T	W	Y	V	

Figure 5–15 The genetic code. The standard one-letter abbreviation for each amino acid is presented below its three-letter abbreviation. Codons are written with the 5′-terminal nucleotide on the left. Note that most amino acids are represented by more than one codon and that variation is common at the third nucleotide (see also Figure 3–15, p. 103).

The Events in Protein Synthesis Are Catalyzed on the Ribosome[6]

The protein synthesis reactions just described require a complex catalytic machinery to guide them. For example, the growing end of the polypeptide chain must be kept in register with the mRNA molecule to ensure that each successive codon in the mRNA engages precisely with the anticodon of a tRNA molecule and does not slip by one nucleotide, thereby changing the reading frame (see p. 102). This and the other events in protein synthesis are catalyzed by **ribosomes,** which are large complexes of RNA and protein molecules. Eucaryotic and procaryotic ribosomes are very similar in design and function. Each is composed of one large and one small subunit, which fit together to form a complex with a mass of several million daltons (Figure 5–16). The small subunit binds the mRNA and tRNAs, while the large subunit catalyzes peptide bond formation.

More than half of the weight of a ribosome is RNA, and there is increasing evidence that the **ribosomal RNA (rRNA)** molecules play a central part in its catalytic activities. Although the rRNA molecule in the small ribosomal subunit varies in size depending on the organism, its complicated folded structure is highly conserved (Figure 5–17); there are also close homologies between the rRNAs of the large ribosomal subunits in different organisms. Ribosomes contain a large number of proteins (Figure 5–18), but many of these have been relatively poorly conserved in sequence during evolution, and a surprising number seem not to be essential for ribosome function. Therefore, as discussed below (p. 219), it has been suggested that the ribosomal proteins mainly enhance the function of the rRNAs and that the RNA molecules rather than the protein molecules catalyze many of the reactions on the ribosome.

A Ribosome Moves Stepwise Along the mRNA Chain[6,7]

A ribosome contains three binding sites for RNA molecules: one for mRNA and two for tRNAs. One site, called the **peptidyl-tRNA binding site,** or **P-site,** holds the tRNA molecule that is linked to the growing end of the polypeptide chain. Another site, called the **aminoacyl-tRNA binding site,** or **A-site,** holds the

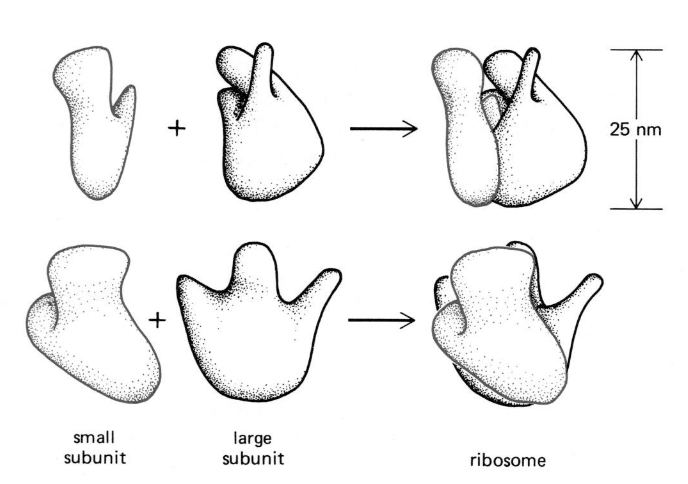

Figure 5–16 A three-dimensional model of the bacterial ribosome as viewed from two different angles. The positions of many ribosomal proteins in this structure have been determined by using an electron microscope to visualize the positions where specific antibodies bind, as well as by measuring the neutron scattering from ribosomes containing one or more deuterated proteins. (After J. A. Lake, *Ann. Rev. Biochem.* 54:507–530, 1985.)

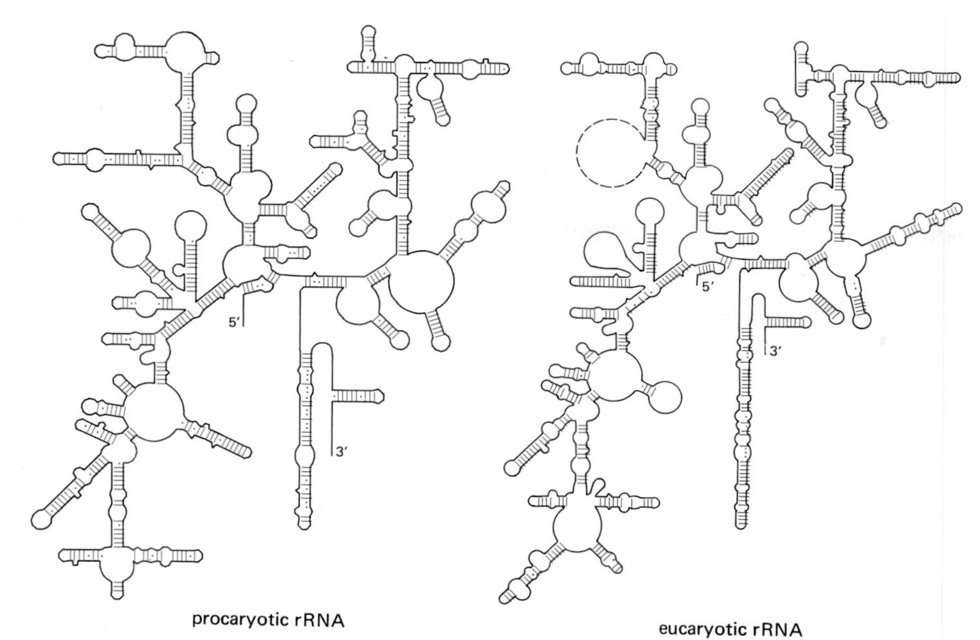

Figure 5–17 The complex array of loops and base-paired stems in the folded structure of *E. coli* 16S rRNA (*left*) and yeast (*S. cerevisiae*) 18S rRNA (*right*). The major structural features appear to be shared by all known 16S-like rRNAs, including the rRNA of archaebacteria. Dots indicate postulated weak interactions between bases, such as G-U base pairs. (After R.R. Gutell, B. Weiser, C.R. Woese, and H.F. Noller, *Prog. Nucleic Acid Res. Mol. Biol.* 32:155–216, 1985.)

Figure 5–18 A comparison of the structures of procaryotic and eucaryotic ribosomes. Ribosomal components are commonly designated by their "S values," which indicate their rate of sedimentation in an ultracentrifuge (see p. 165). Note that, despite the differences in the number and size of their rRNA and protein components, both ribosomes have nearly the same structure and function in very similar ways. For example, although the 18S and 28S rRNAs of the eucaryotic ribosome contain many extra nucleotides not present in their bacterial counterparts, these nucleotides are present as multiple insertions that leave the basic structure of each rRNA largely unchanged (see Figure 5–17).

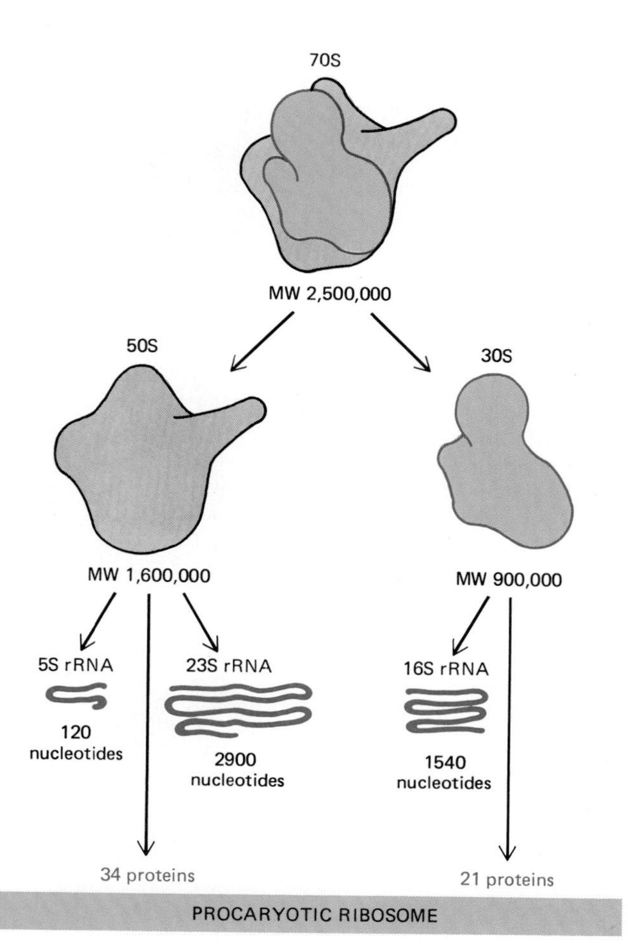

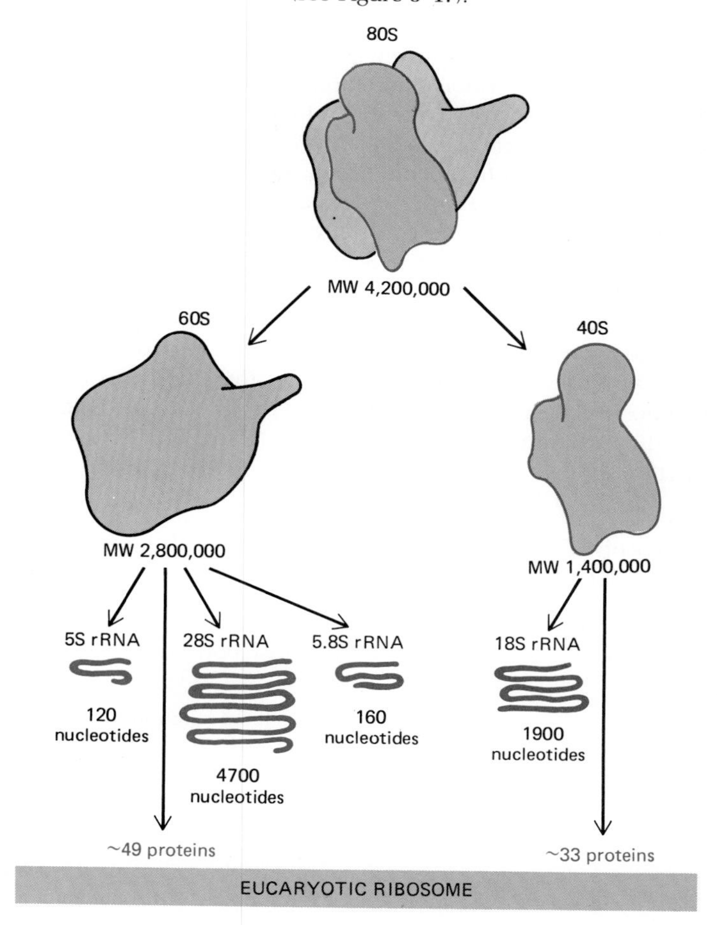

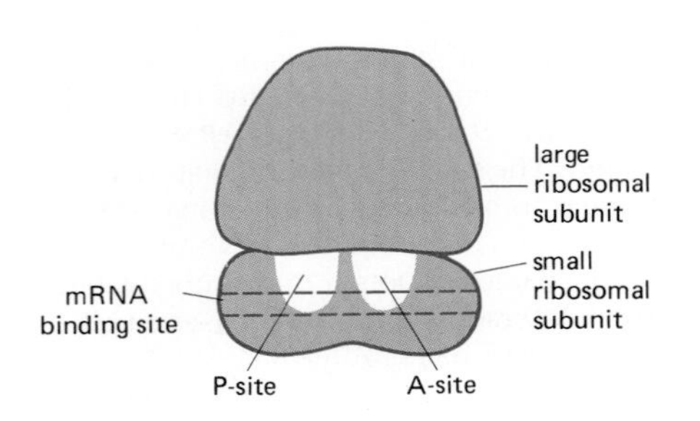

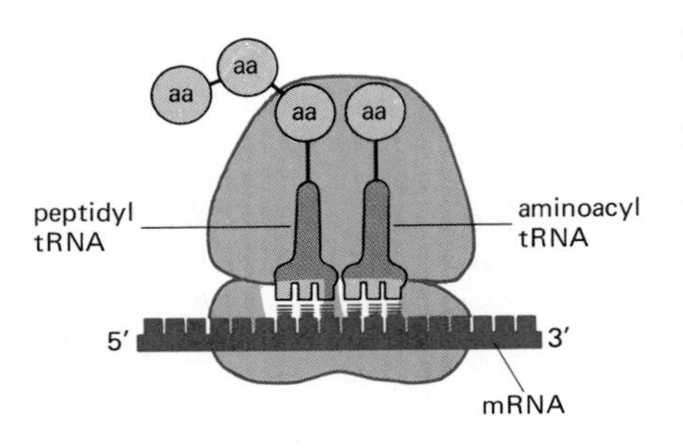

Figure 5–19 The three major RNA binding sites on a ribosome. An empty ribosome is shown at the left and a loaded ribosome at the right. The representation of a ribosome used here and in the next three figures is highly schematic; for a more accurate view, see Figures 5–16 and 5–23.

incoming tRNA molecule charged with an amino acid. A tRNA molecule is held tightly at either site only if its anticodon forms base pairs with a complementary codon on the mRNA molecule that is bound to the ribosome. The A- and P-sites are so close together that the two tRNA molecules are forced to form base pairs with adjacent codons in the mRNA molecule (Figure 5–19).

The process of polypeptide chain elongation on a ribosome can be considered as a cycle with three discrete steps (Figure 5–20). In step 1, an aminoacyl-tRNA molecule becomes bound to a vacant ribosomal A-site (adjacent to an occupied P-site) by forming base pairs with the three mRNA nucleotides exposed at the A-site. In step 2, the carboxyl end of the polypeptide chain is uncoupled from the tRNA molecule in the P-site and joined by a peptide bond to the amino acid linked to the tRNA molecule in the A-site. This reaction is catalyzed by **peptidyl transferase.** Its enzymelike activity requires the integrity of the ribosome and is thought to be mediated by a specific region of the major rRNA molecule in the large subunit. In step 3, the new peptidyl-tRNA in the A-site is translocated to the P-site as the ribosome moves exactly three nucleotides along the mRNA molecule. This step requires energy and is driven by a series of conformational changes induced in one of the ribosomal components by the hydrolysis of a GTP molecule (see p. 130).

As part of the translocation process of step 3, the free tRNA molecule that was generated in the P-site during step 2 is released from the ribosome to reenter the cytoplasmic tRNA pool. Therefore, upon completion of step 3, the unoccupied A-site is free to accept a new tRNA molecule linked to the next amino acid, which starts the cycle again. In a bacterium each cycle requires about one-twentieth of a second under optimal conditions, so that the complete synthesis of an average-sized protein of 400 amino acids is accomplished in about 20 seconds. Ribosomes move along an mRNA molecule in the 5′ to 3′ direction, which is also the direction of RNA synthesis (see Figure 5–2).

In most cells, protein synthesis consumes more energy than any other biosynthetic process. At least four high-energy phosphate bonds are split to make each new peptide bond: Two of these are required to charge each tRNA molecule with an amino acid (see Figure 5–10), and two more drive steps in the cycle of reactions occurring on the ribosome during synthesis itself—one for the aminoacyl-tRNA binding in step 1 (see p. 217) and one for the ribosome translocation in step 3.

5-5 5-8 A Protein Chain Is Released from the Ribosome When One of Three Different Stop Codons Is Reached[6,8]

Three of the 64 possible codons (UAA, UAG, and UGA) in an mRNA molecule are **stop codons,** which terminate the translation process. Cytoplasmic proteins called

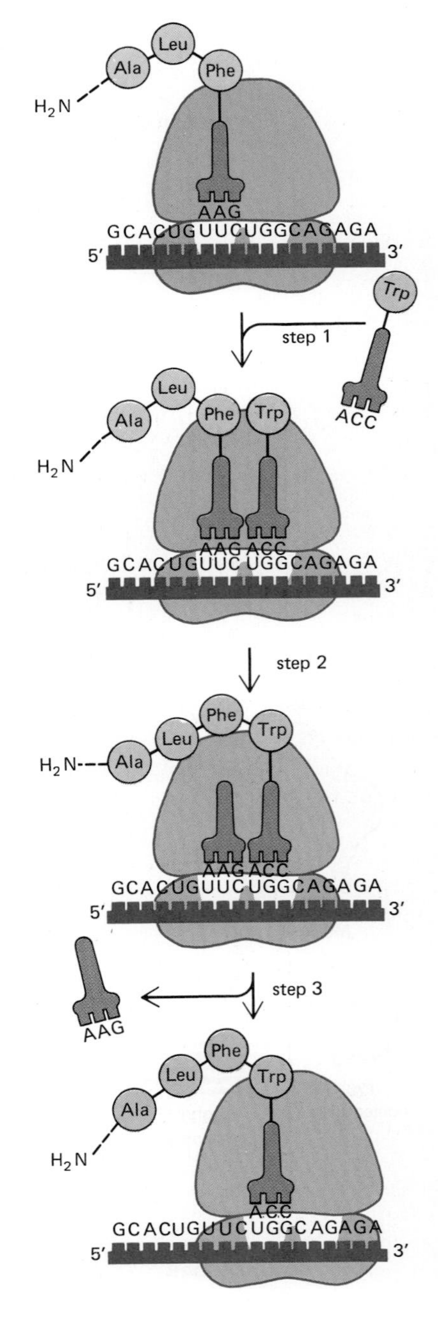

Figure 5–20 The elongation phase of protein synthesis on a ribosome. The three-step cycle shown is repeated over and over during the synthesis of a protein chain. An aminoacyl-tRNA molecule binds to the A-site on the ribosome in step 1; a new peptide bond is formed in step 2; and the ribosome moves a distance of three nucleotides along the mRNA chain in step 3, ejecting an old tRNA molecule and "resetting" the ribosome so that the next aminoacyl-tRNA molecule can bind.

release factors bind directly to any stop codon that reaches the A-site on the ribosome. This binding alters the activity of the peptidyl transferase, causing it to catalyze the addition of a water molecule instead of an amino acid to the peptidyl-tRNA. This reaction frees the carboxyl end of the growing polypeptide chain from its attachment to a tRNA molecule. Since only this attachment normally holds the growing polypeptide to the ribosome, the completed protein chain is immediately released into the cytoplasm (Figure 5–21). The ribosome then releases the mRNA and dissociates into its two separate subunits, which can assemble on another mRNA molecule to begin a new round of protein synthesis by the process to be described next.

5-4 The Initiation Process Sets the Reading Frame for Protein Synthesis[6,9]

In principle, an RNA sequence can be decoded in one of three different **reading frames,** each of which will specify a completely different polypeptide chain (see Figure 3–14, p. 102). Which of the three frames is actually read is determined by the way that a ribosome assembles on an mRNA molecule. During the *initiation* phase of protein synthesis, the two subunits of the ribosome are brought together at the exact spot on the mRNA where the polypeptide chain is to begin.

The initiation process is complicated, involving a number of steps catalyzed by proteins called **initiation factors (IF),** many of which are themselves composed of several polypeptide chains. Because the process is so complex, many of the details of initiation are still uncertain. It is clear, however, that each ribosome is assembled onto an mRNA chain in two steps; only after the small ribosomal subunit loaded with initiation factors finds the start codon does the large subunit bind.

Before a ribosome can begin a new protein chain, it must bind an aminoacyl tRNA molecule in its P-site, where normally only peptidyl tRNA molecules are bound (Figure 5–22). A special tRNA molecule is required for this purpose. This **initiator tRNA** provides the amino acid that starts a protein chain, and it always carries methionine (aminoformyl methionine in bacteria). In eucaryotes the initiator tRNA molecule is loaded onto the small ribosomal subunit before the mRNA is bound. An important initiation factor called *eucaryotic initiation factor 2 (eIF-2)* binds tightly to the initiator tRNA, and it is required to position the tRNA on the small subunit. In some cells the overall rate of protein synthesis is controlled by this factor (see below).

As described in the next section, the small ribosomal subunit helps its bound initiator tRNA molecule to find a special AUG codon (the *start codon*) on an mRNA molecule. Once this has occurred, the several initiation factors that were previously associated with the small ribosomal subunit are discharged to make way for the binding of a large ribosomal subunit to the small one. Because the initiator tRNA molecule is bound to the P-site of the ribosome, the synthesis of a protein chain can begin directly with the binding of a second aminoacyl-tRNA molecule to the A-site of the ribosome (Figure 5–22). Thus a complete functional ribosome is assembled, with the mRNA molecule threaded through it (Figure 5–23). Further steps in the elongation phase of protein synthesis then proceed as described previously (see step 2 of Figure 5–20). Because an initiator tRNA molecule has begun each polypeptide chain, all newly made proteins have a methionine (or the aminoformyl derivative of methionine in bacteria) as their amino-terminal residue. The methionine is often removed shortly after its incorporation by a specific aminopeptidase; this trimming process is important because the amino acid left at the amino terminus can determine the protein's lifetime in the cell by its effects on a ubiquitin-dependent degradation pathway (see p. 419).

Evidently the correct initiation site on the mRNA molecule has to be selected by the small subunit acting in concert with initiation factors but in the absence of the large subunit, which is presumably why all ribosomes are formed from two separate subunits. We shall now consider how this selection occurs.

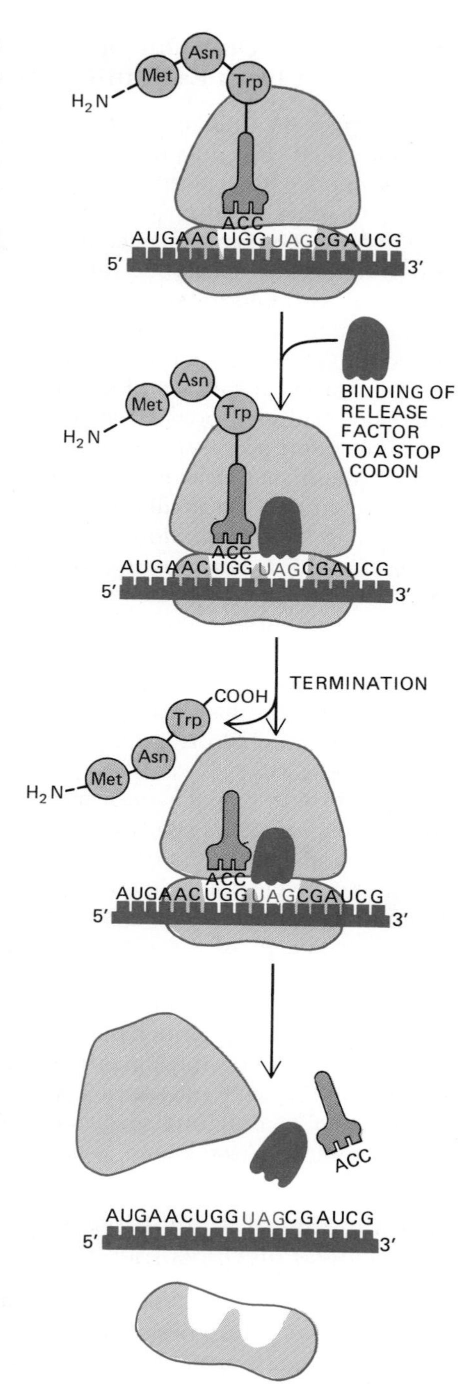

Figure 5–21 The final phase of protein synthesis. The binding of release factor to a stop codon terminates translation; the completed polypeptide is released, and the ribosome dissociates into its two separate subunits.

5-45

In Eucaryotes Only One Species of Polypeptide Chain Is Usually Synthesized from Each mRNA Molecule[10]

A messenger RNA molecule will typically contain many different AUG sequences, each of which can code for methionine. In eucaryotes, however, only one of these AUG sequences will normally be recognized by the initiator tRNA and thereby serve as a **start codon.** How is this start codon distinguished by the ribosome?

The mechanism for selecting a start codon is different in eucaryotes and procaryotes. The eucaryotic mRNAs (except those that are synthesized in mitochondria and chloroplasts) are extensively modified in the nucleus immediately after their transcription (see p. 531). Two general modifications are the addition of a unique "cap" structure, composed of a 7-methylguanosine residue linked to a triphosphate at the 5′ end (Figure 5–24) and the addition of a run of about 200 adenylic residues ("poly A") at the 3′ end. What part, if any, the poly A plays in the translation process is uncertain. But the 5′ cap structure is essential for efficient protein synthesis. Experiments carried out with extracts of eucaryotic cells have shown that the small ribosomal subunit first binds at the 5′ end of an mRNA chain, aided by recognition of the 5′ cap (Figure 5–22). This subunit then moves along the mRNA chain, carrying its bound initiator tRNA in a search of an AUG start codon. The requirements for a start codon are apparently not very stringent, since it seems that only a few additional nucleotides besides the AUG itself are important. In most RNAs, only the first suitable AUG will be used; once a start codon near the 5′ end has been selected, none of the many other AUG codons farther down the chain can serve as initiation sites. As a result, only a single species of polypeptide chain is usually synthesized from an mRNA molecule (for exceptions see p. 594).

In all these respects, procaryotic mRNAs are quite different from eucaryotic mRNAs (Figure 5–25). Bacterial mRNAs have no 5′ cap structure. Instead they contain a specific initiation-site sequence, up to six nucleotides long, which can occur at multiple points in the same mRNA molecule. These sequences are located four to seven nucleotides upstream from an AUG, and they form base pairs with a specific region of the rRNA in a ribosome to signal the initiation of protein synthesis at this nearby start codon. Bacterial ribosomes, unlike eucaryotic ribosomes, bind directly to the middle of an mRNA molecule to recognize a start codon and initiate a polypeptide chain. As a result, bacterial messenger RNAs are commonly *polycistronic,* meaning that they encode multiple proteins separately translated from the same mRNA molecule. Eucaryotic mRNAs, in contrast, are typically *monocistronic,* only one species of polypeptide chain being translated per messenger molecule (see Figure 5–25).

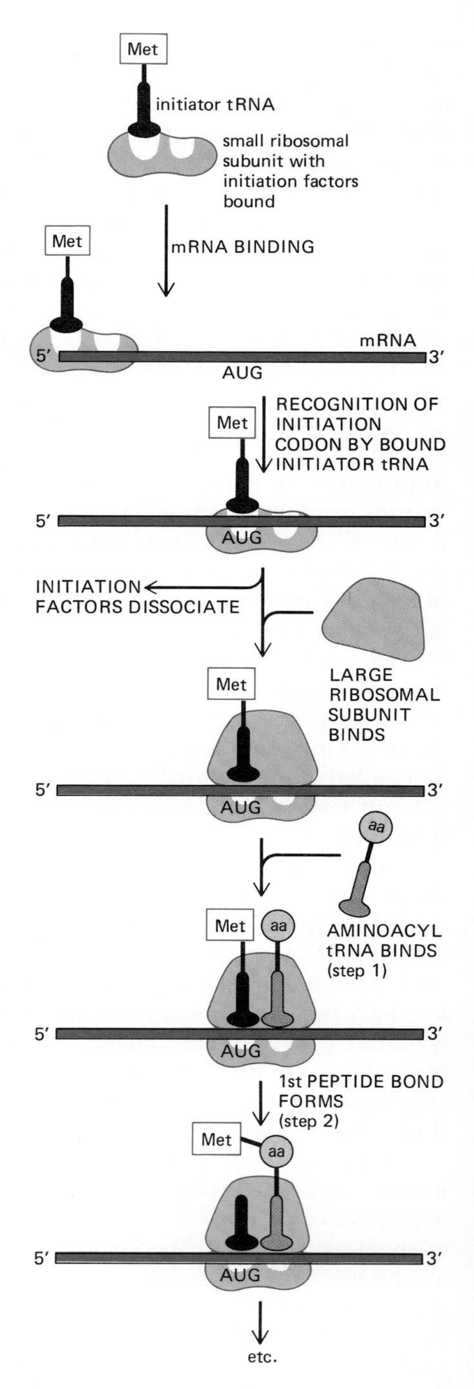

Figure 5–22 The initiation phase of protein synthesis. Events are illustrated as they occur in eucaryotes, but a very similar process occurs in bacteria. Step 1 and step 2 refer to steps in the elongation reaction shown in Figure 5–20.

The Binding of Many Ribosomes to an Individual mRNA Molecule Generates Polyribosomes[11]

The complete synthesis of a protein takes 20 to 60 seconds on average. But even during this very short period, multiple initiations usually take place on each mRNA molecule being translated. A new ribosome hops onto the 5′ end of the mRNA molecule almost as soon as the preceding ribosome has translated enough of the amino acid sequence to get out of the way. Such mRNA molecules are thus present in **polyribosomes,** or **polysomes,** formed by several ribosomes spaced as close as 80 nucleotides apart along a single messenger molecule (Figures 5–26 and 5–27). In procaryotes (but *not* eucaryotes), RNA is accessible to ribosomes as soon as it is made. Thus, ribosomes will begin their synthesis at the 5′ end of a new mRNA molecule and then follow behind the RNA polymerase as it completes an mRNA chain.

Polyribosomes are a common feature of cells. They can be isolated and separated from single ribosomes in the cytosol by ultracentrifugation after cell lysis (Figure 5–28). The mRNA purified from these polyribosomes can be used to determine if the protein encoded by a particular DNA sequence is being actively synthesized in the cells used to prepare the polyribosomes. These mRNA molecules can also serve as the starting material for the preparation of specialized cDNA libraries (see p. 261).

0-25 The Overall Rate of Protein Synthesis in Eucaryotes Is Controlled by Initiation Factors[12]

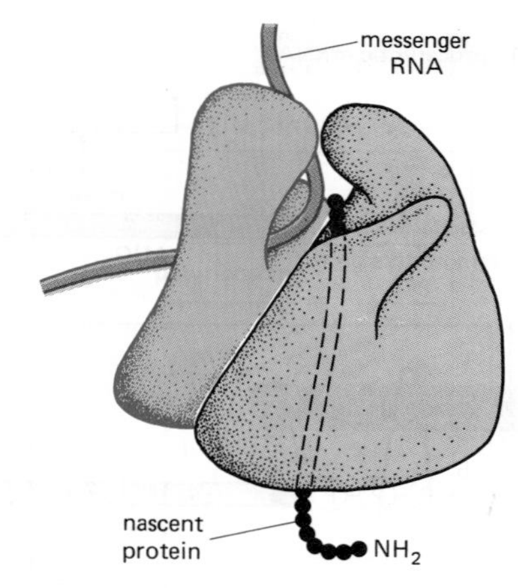

Figure 5–23 A three-dimensional model of a functioning bacterial ribosome. The small (*colored*) subunit and the large (*black*) subunit form a complex through which the messenger RNA is threaded. Although the exact paths of the mRNA and the nascent protein chain are unknown, the addition of amino acids occurs in the general region shown. (Modified from J.A. Lake, *Annu. Rev. Biochem.* 54:507–530, 1985.)

As we shall discuss in Chapter 13, the cells in a multicellular organism multiply only when they are in an environment where they are stimulated by specific growth factors. Although the mechanisms by which growth factors act are incompletely understood, one of their major effects must be to increase the overall rate of protein synthesis (p. 746). What determines this rate? Direct studies in tissues are difficult; but when eucaryotic cells in culture are starved of nutrients, there is a marked reduction of the rate of polypeptide-chain *initiation,* which can be shown to result from inactivation of the protein synthesis initiation factor eIF-2. In at least one type of cell (immature red blood cells), the activity of eIF-2 is reduced in a controlled way by the phosphorylation of one of its three protein subunits. This finding suggests that the rate of eucaryotic protein synthesis may be controlled by specific protein kinases that, when activated, reduce the frequency of initiation events. The action of growth factors may be mediated in part by intracellular regulatory molecules that inhibit or counteract the effect of such protein kinases.

The initiation factors required for protein synthesis are much more numerous and complex in eucaryotes than in procaryotes, even though they perform the same basic functions. Many of the extra components could be regulatory proteins that respond to different growth factors and help coordinate cell growth and proliferation in multicellular organisms. Such controls are not needed in bacteria, which generally grow as fast as the nutrients in their environment allow.

5-9 The Fidelity of Protein Synthesis Is Improved by Two Separate Proofreading Processes[13]

The error rate in protein synthesis can be estimated by monitoring the frequency of incorporation of an amino acid into a protein that normally lacks that amino acid. Error rates of about one amino acid misincorporated for every 10^4 amino acids polymerized are observed, which means that only about one in every 25 protein molecules of average size (400 amino acids) should contain an error. The fidelity of the decoding process depends on the accuracy of the two adaptor mechanisms previously discussed: the linking of each amino acid to its corresponding tRNA molecule and the base-pairing of the codons in mRNA to the anticodons in tRNA (see Figure 5–12). Not surprisingly, cells have evolved "proofreading" mechanisms to reduce the number of errors in both these crucial steps of protein synthesis.

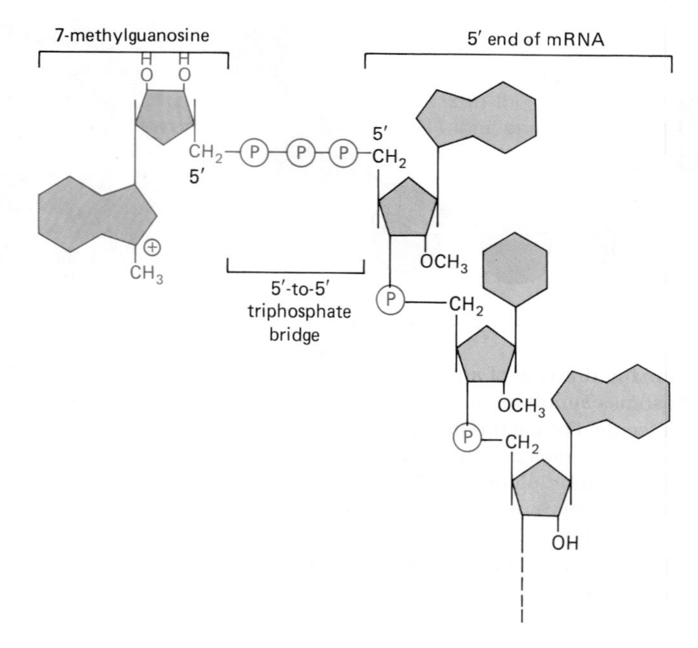

Figure 5–24 The structure of the cap at the 5′ end of eucaryotic mRNA molecules. Note the unusual 5′-to-5′ linkage to the positively charged 7-methylguanosine and the methylation of the 2′ hydroxyl group on the first ribose sugar in the RNA. (The second sugar is not always methylated.)

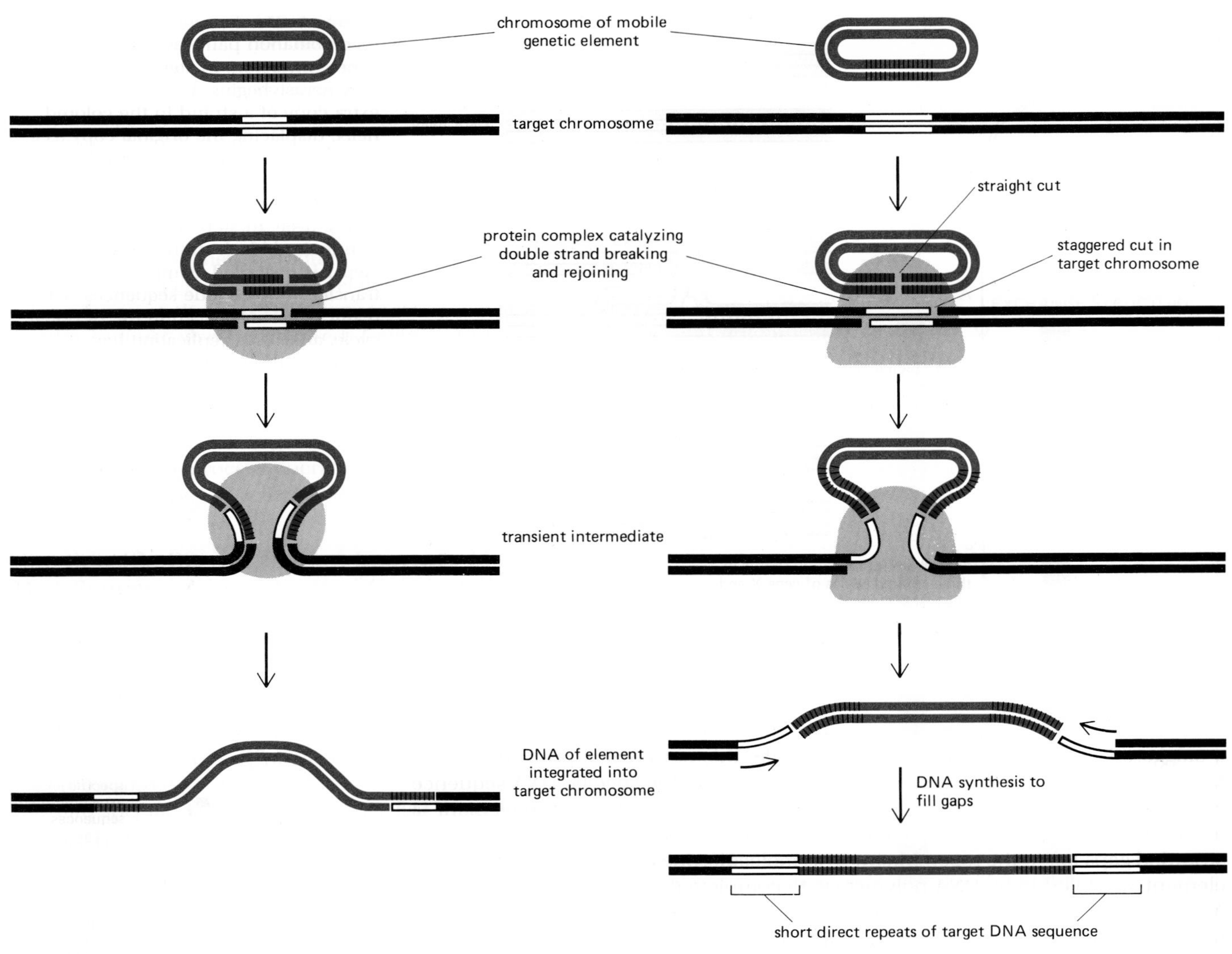

Figure 5–67 Comparison of the mechanisms used by two different classes of site-specific recombination enzymes. In each case a specific enzyme (*gray*) binds to the cross-hatched DNA sequence on the mobile genetic element and holds this sequence close to a site on the target chromosome. In (A) the enzyme makes a staggered cut on either side of a very short homologous DNA sequence on both chromosomes (12 nucleotides in the case of lambda integrase) and then switches the partner strands so as to form a short heteroduplex joint. In (B) the enzyme makes a staggered cut in the target chromosome and joins the protruding ends directly to the evenly cut ends of the mobile element DNA. This type of enzyme leaves short repeats of target DNA sequence (3 to 12 nucleotides in length, depending on the enzyme) on either side of the integrated DNA segment.

Viruses, Plasmids, and Transposable Genetic Elements[42]

In our description of the basic genetic mechanisms, we have so far focused on their selective advantage for the cell. We saw that the short-term survival of the cell depends absolutely on the maintenance of genetic information by DNA repair, while the multiplication of the cell requires rapid and accurate DNA replication. On a longer time scale the appearance of genetic variants, on which evolution of the species depends, is greatly facilitated by the reassortment of genes and the occasional rearrangement of DNA sequences caused by genetic recombination. We shall now examine a group of genetic elements that seem to act as parasites, subverting the genetic mechanisms of the cell for their own advantage.

Under certain circumstances, DNA sequences can replicate independently of the rest of the genome. Such sequences have widely different degrees of independence from their host cells. Of these, virus chromosomes are the most independent because they have a protein coat that allows them to move freely from cell to cell. To varying degrees the viruses are closely related to more host cell-dependent DNA sequences known as plasmids and transposable elements, which lack a coat and are therefore confined to replicate with a single cell and its progeny. More primitive still are some DNA sequences that are suspected of being mobile because they are repeated many times in a cell's chromosome—but that move or multiply so rarely that it is not clear if they should be considered as separate genetic elements at all.

All these quasi-independent genetic elements must heavily exploit the metabolism of the host cell in order to multiply, and they have thereby served as important tools for investigating the normal cell machinery. We shall begin our discussion with the viruses, which are the best understood of the mobile genetic elements. Then we shall describe the properties of plasmids and transposable elements, some of which bear a remarkable resemblance to viruses and may in fact have been their ancestors.

5-39 Viruses Are Mobile Genetic Elements[43]

Viruses were first described as disease-causing agents that can multiply only in cells and that by virtue of their tiny size pass through ultrafine filters that hold back even the smallest bacteria. Before the advent of the electron microscope, their nature was obscure, although it was suspected that they might be naked genes that had somehow acquired the ability to move from one cell to another. The use of ultracentrifuges in the 1930s made it possible to separate viruses from host cell components, and by the early 1940s the generalization emerged that all viruses contain nucleic acids. The idea that viruses and genes carry out similar functions was confirmed by studies on the bacterial viruses (*bacteriophages*). In 1952 it was shown for the bacteriophage T4 that only the phage DNA, and not the phage protein, enters the bacterial host cell and initiates the replication events that lead to the production of several hundred progeny viruses in every infected cell.

These observations led to the notion of viruses as genetic elements enclosed by a protective coat that enables them to move from one cell to another. Virus multiplication per se is often lethal to the cells in which it occurs; in many cases the infected cell breaks open (*lyses*) and thereby allows the progeny viruses access to nearby cells. Many of the clinical manifestations of viral infection reflect this cytolytic effect of the virus. For example, both the cold sores formed by herpes simplex virus and the lesions caused by smallpox reflect the killing of the epithelial cells in a local area of the skin.

The type of nucleic acid in a virus, the structure of its coat, its mode of entry into the host cell, and its mechanism of replication once inside all vary considerably from one type of virus to another. Electron micrographs illustrating some structural differences among viruses are presented in Figure 5–68.

The Outer Coat of a Virus May Be a Protein Capsid or a Membrane Envelope[44]

Initially it was thought that the outer coat of a virus might be constructed from a single type of protein molecule. Viral infections were believed to start with the dissociation of the viral chromosome (its nucleic acid) from its protein coat inside the host cell, followed by replication of the chromosome to form many identical

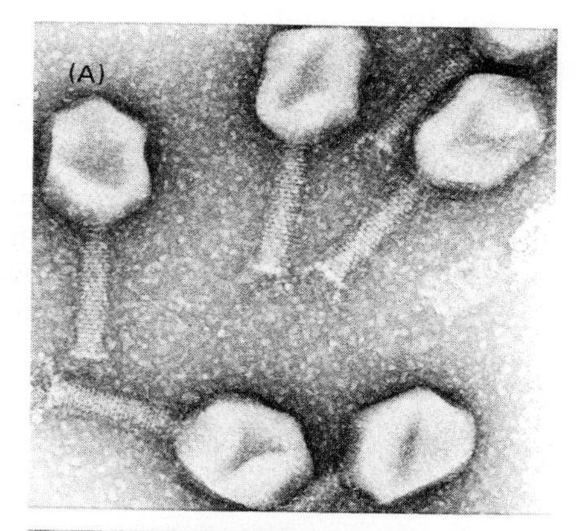

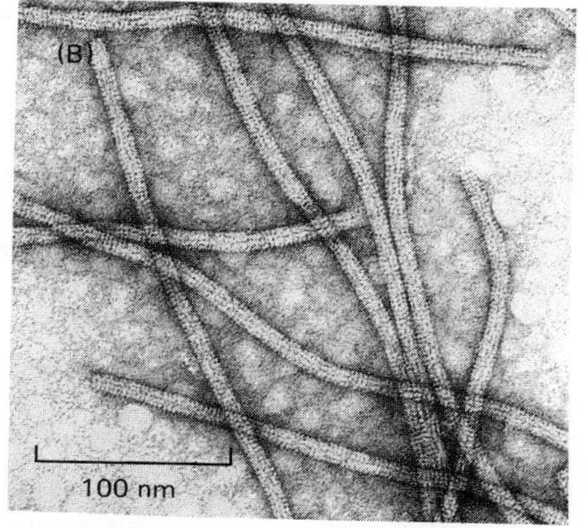

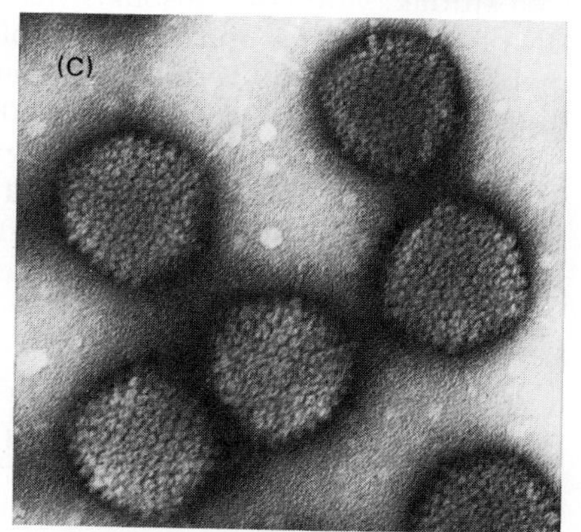

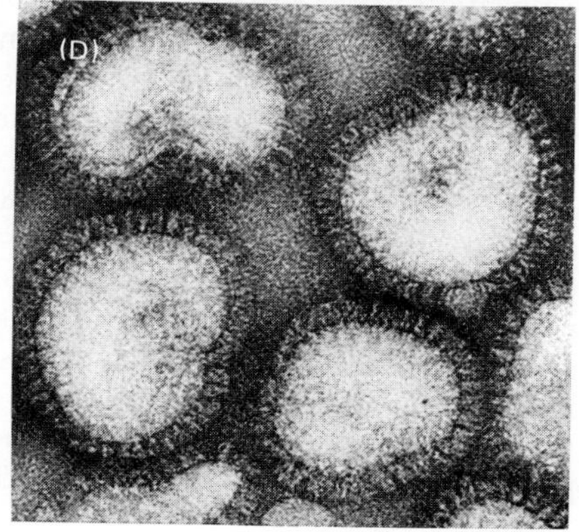

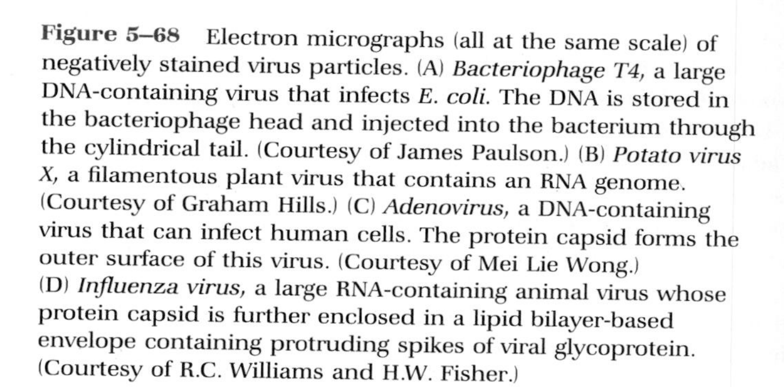
Figure 5–68 Electron micrographs (all at the same scale) of negatively stained virus particles. (A) *Bacteriophage T4,* a large DNA-containing virus that infects *E. coli.* The DNA is stored in the bacteriophage head and injected into the bacterium through the cylindrical tail. (Courtesy of James Paulson.) (B) *Potato virus X,* a filamentous plant virus that contains an RNA genome. (Courtesy of Graham Hills.) (C) *Adenovirus,* a DNA-containing virus that can infect human cells. The protein capsid forms the outer surface of this virus. (Courtesy of Mei Lie Wong.) (D) *Influenza virus,* a large RNA-containing animal virus whose protein capsid is further enclosed in a lipid bilayer-based envelope containing protruding spikes of viral glycoprotein. (Courtesy of R.C. Williams and H.W. Fisher.)

copies. After the synthesis of new copies of the virus-specific coat protein from virally encoded messenger RNA molecules, formation of the progeny virus particles would occur by the spontaneous assembly of these coat protein molecules around the progeny viral chromosomes (Figure 5–69).

It is now known that these ideas vastly oversimplify the diversity of virus life cycles. For example, the protein shell that surrounds the nucleic acid of most viruses (the **capsid**) contains more than one type of polypeptide chain, often arranged in several layers. In many viruses, moreover, the protein capsid is further enclosed by a lipid bilayer membrane that contains proteins. Many of these *enveloped viruses* acquire their envelope in the process of budding from the plasma membrane (Figure 5–70). This budding process allows the virus particles to leave the cell without disrupting the plasma membrane and, therefore, without killing the cell. While their lipid components are identical to those found in the plasma membrane of the host cell, the proteins in the lipid bilayer are virus-specific. The assembly of viral envelopes at the plasma membrane is discussed in Chapter 8 (p. 469), and the assembly of a viral capsid is illustrated in Figure 3–43 (p. 120).

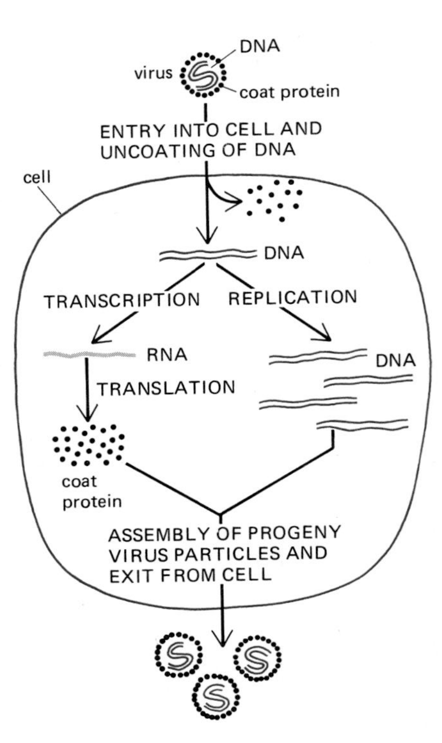

Figure 5–69 The simplest of all viral life cycles. The hypothetical virus shown consists of a small double-stranded DNA molecule that codes for only a single viral capsid protein. No known virus is this simple.

5-45 Viral Genomes Come in a Variety of Forms and Can Be Either RNA or DNA[45]

When the DNA double helix was discovered, it seemed logical that genetic information should be stored only in this form since, as we have seen, it has obvious advantages for DNA stability and repair. If one polynucleotide chain is accidentally damaged, its complementary chain permits the damage to be readily corrected. This concern with repair, however, need not bother small viral chromosomes that contain only several thousand nucleotides—for the chance of accidental damage is very small compared with the risk to a cell genome containing millions of nucleotides.

The genetic information of a virus can, therefore, be carried in a variety of unusual forms, including RNA instead of DNA. A viral chromosome may be a single-stranded RNA chain (tobacco mosaic virus), a double-stranded RNA helix (reovirus), a circular single-stranded DNA chain (M13 and ϕX174 bacteriophages), or a linear single-stranded DNA chain (parvoviruses). Moreover, although the first well-studied viral chromosomes were simple linear DNA double helices, circular DNA double helices and more complex linear DNA double helices are also common. For example, several viruses have protein molecules covalently attached to the 5′ ends of their DNA strands, and the DNA double helices from the very large poxviruses have their opposite strands at each end covalently joined through phosphodiester linkages (Figure 5–71). Each type of genome requires unique enzymatic tricks for its replication and thus must encode not only the viral coat protein but also one or more of the enzymes needed to replicate the viral nucleic acid.

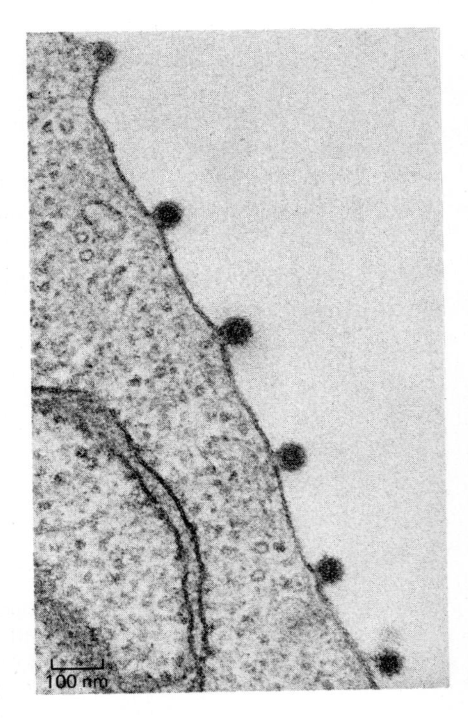

Figure 5–70 Electron micrograph of a thin section of an animal cell from which several copies of an enveloped virus (Semliki forest virus) are budding. This virus contains a single-stranded RNA genome. (Courtesy of M. Olsen and G. Griffiths.)

5-40 A Viral Chromosome Codes for Enzymes Involved in the Replication of Its Nucleic Acid[46]

The amount of information that a virus brings into a cell to ensure its own selective replication varies greatly. For example, the DNA of the relatively large bacteriophage T4 codes for at least 30 different enzymes that ensure the rapid replication of the T4 chromosome in preference to the DNA of its *E. coli* host cell (Figure 5–72). Some of these proteins mediate continuous rounds of T4 DNA replication, with the unusual feature that 5-hydroxymethylcytosine is incorporated in place of cytosine in the DNA. The unusual base composition of the T4 DNA makes it distinguishable from host DNA and selectively protects it from nucleases also encoded in the T4 genome that thus degrade only the *E. coli* DNA. Still other proteins alter host cell RNA polymerase molecules so that they transcribe different sets of bacteriophage genes at different stages of infection, appropriately to the needs of the phage.

Smaller DNA viruses, such as the monkey virus SV40 and the tiny bacteriophage ϕX174, carry much less genetic information. They rely more on host cell

RNA VIRUSES

single-stranded RNA, e.g., tobacco mosaic virus, bacteriophage R17, and poliovirus

double-stranded RNA, e.g., reovirus

DNA VIRUSES

single-stranded DNA, e.g., parvovirus

single-stranded circular DNA, e.g., bacteriophages ϕX174 and M13

double-stranded DNA, e.g., bacteriophage T4 and herpesvirus

double-stranded circular DNA, e.g., SV40 and polyoma virus

double-stranded DNA with covalently linked terminal protein, e.g., adenovirus

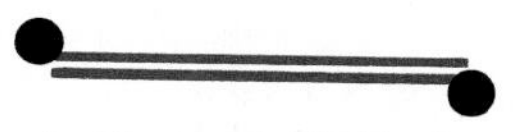

double-stranded DNA with each end covalently sealed, e.g., poxvirus

Figure 5–71 Schematic drawings (not to scale) of several types of viral genomes. The smallest viruses contain only a few genes and can have an RNA or a DNA genome; the largest viruses contain hundreds of genes and have a double-stranded DNA genome. The peculiar ends on some of these DNA molecules (as well as the circular forms) overcome the difficulty of replicating the last few nucleotides at the end of a DNA chain (see p. 519).

enzymes to carry out their protein and DNA synthesis, parasitizing host cell DNA replication proteins, including the DNA polymerase enzyme. Even the smallest DNA viruses, however, code for proteins that selectively initiate the synthesis of their own DNA, recognizing a particular nucleotide sequence in the virus that serves as a *replication origin*. This is essential because a virus must override the cellular control signals that would otherwise cause the viral DNA to replicate in pace with the host cell DNA, doubling only once in each cell cycle. We do not yet understand how eucaryotic cells regulate their own DNA synthesis (see p. 732), and the mechanisms used by viruses to escape from this regulation—which are much more accessible to study—should provide insights into the host regulatory mechanisms.

RNA viruses have very specialized requirements for replication since, to reproduce their genomes, they must copy RNA molecules, which means polymerizing nucleoside triphosphates on an RNA template. Cells normally do not have enzymes to carry out this reaction, so even the smallest RNA viruses must encode their own RNA-dependent nucleic acid polymerase enzymes in order to replicate.

We shall now look in more detail at the replication mechanisms of the various types of viruses.

5-42 Both RNA Viruses and DNA Viruses Replicate Through the Formation of Complementary Strands[47]

The replication of the genomes of RNA viruses, like DNA replication, occurs through the formation of complementary strands. For most RNA viruses this process is catalyzed by specific RNA-dependent RNA polymerase enzymes (*replicases*). These enzymes are encoded by the viral RNA chromosome and are often incorporated into the progeny virus particles, so that upon entry of the virus into a cell, they immediately begin replicating the viral RNA. Replicases are always packaged into the capsid of the so-called *negative-strand RNA viruses*, such as influenza or vesicular stomatitis virus. Negative-strand viruses are so called because the infecting strand does not code for protein; only its complementary strand carries the coding sequences. Thus the infecting strand remains impotent without a preformed replicase. In contrast, the viral RNA of *positive-strand RNA viruses*, such as poliovirus, can serve as mRNA and therefore the naked genome itself is infectious.

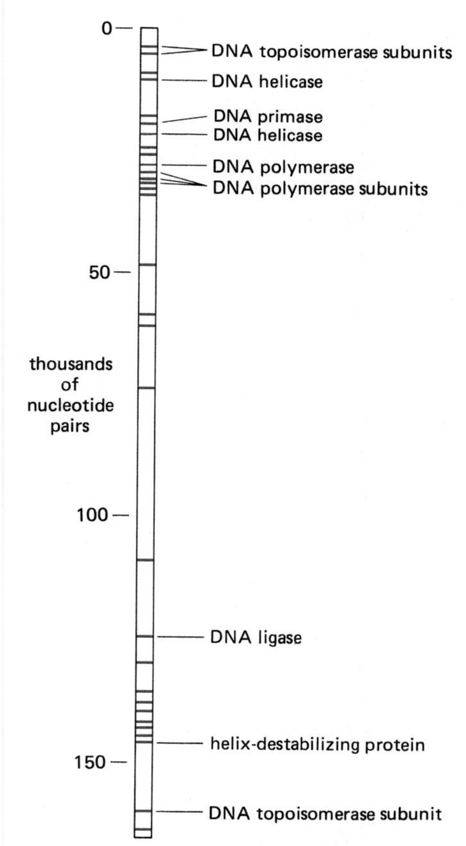

Figure 5–72 The T4 bacteriophage chromosome, showing the positions of the more than 30 genes involved in T4 DNA replication. The genome of bacteriophage T4 consists of more than 160,000 nucleotide pairs and encodes more than 200 different proteins, including those involved in DNA replication (some of which are labeled). The remaining proteins include many that are involved in the bacteriophage head and tail assemblies (see Figure 5–68A).

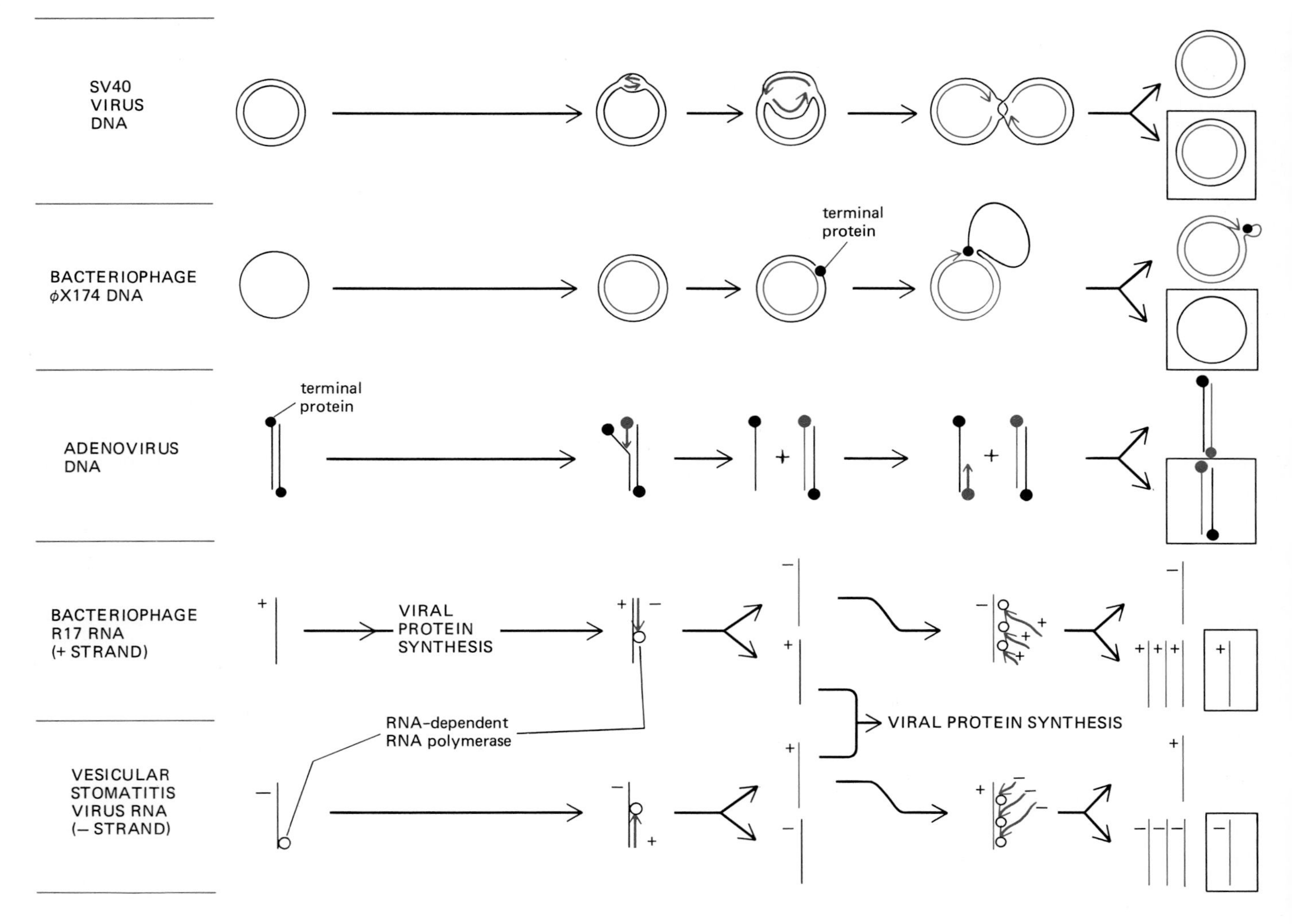

Figure 5–73 Some of the diverse strategies used by different viruses to replicate their genomes. Where indicated, *terminal proteins* are covalently attached to the ends of DNA chains; these proteins play an important role in the replication process. A major difference between the life cycles of *positive-strand* and *negative-strand* RNA viruses is that the latter must synthesize a positive RNA strand before making viral proteins. A negative-strand virus, therefore, must carry within its capsid one or more molecules of the viral RNA-dependent RNA polymerase (replicase). The final RNA or DNA product, which is identical to the infecting viral genome at the left, is boxed.

The synthesis of viral RNA always begins at the 3′ end of the RNA template (starting at the 5′ end of the new viral RNA molecule) and progresses until the 5′ end of the template is reached. There are no error-correcting mechanisms for viral RNA synthesis, and error rates are similar to those in DNA transcription (about one error in 10^4 nucleotides synthesized). This is not a serious deficiency as long as the RNA chromosome is relatively short; for this reason the genomes of all RNA viruses are small relative to those of the large DNA viruses.

All DNA viruses begin their replication at a replication origin, which binds special initiator proteins that attract the replication enzymes of the host cell (see p. 235). However, there are many different replication pathways. The complexity of these diverse replication schemes reflects, in part, the problem of replicating the ends of a simple linear DNA molecule, given a DNA polymerase enzyme that cannot begin synthesis without a primer (p. 229). DNA viruses have solved this problem in a variety of ways: some have circular DNA genomes and thus no ends; others have linear DNA genomes that repeat their terminal sequences or end in loops; while still others have special terminal proteins that serve to prime the DNA polymerase directly.

Some of the ways in which viruses are known to replicate their genomes are outlined in Figure 5–73.

5-43 Viral Chromosomes Can Integrate into Host Chromosomes[48]

The end result of the entry of a viral chromosome into a cell is not always its immediate multiplication to produce large numbers of progeny. Many viruses enter a *latent* state, in which their genomes are present but inactive in the cell and no progeny are produced. Viral latency was discovered when it was found that exposure to ultraviolet light induced many apparently uninfected bacteria to produce progeny bacteriophages. Subsequent experiments showed that these *lysogenic bacteria* carry in their chromosomes a dormant but complete viral chromosome. Such integrated viral chromosomes are called **proviruses.**

Bacteriophages that can integrate their DNA into bacterial chromosomes are known as *lysogenic bacteriophages.* The best-known example is the bacteriophage lambda, whose integrase protein we have already discussed. When lambda infects a suitable *E. coli* host cell, it normally multiplies to produce several hundred progeny particles that are released when the bacterial cell lyses; this is called a *lytic infection.* More rarely, the free ends of the linear infecting DNA molecules join to form a DNA circle that becomes integrated into the circular host *E. coli* chromosome by a site-specific recombination event (see p. 246). The resulting lysogenic bacterium, carrying the proviral lambda chromosome, multiples normally until it is subjected to an environmental insult, such as exposure to ultraviolet light or ionizing radiation. The resulting cell debilitation induces the integrated provirus to leave the host chromosome and begin a normal cycle of viral replication. In this way the integrated provirus need not perish with its damaged host cell but has a chance to escape to a nearby normal *E. coli* cell (Figure 5–74).

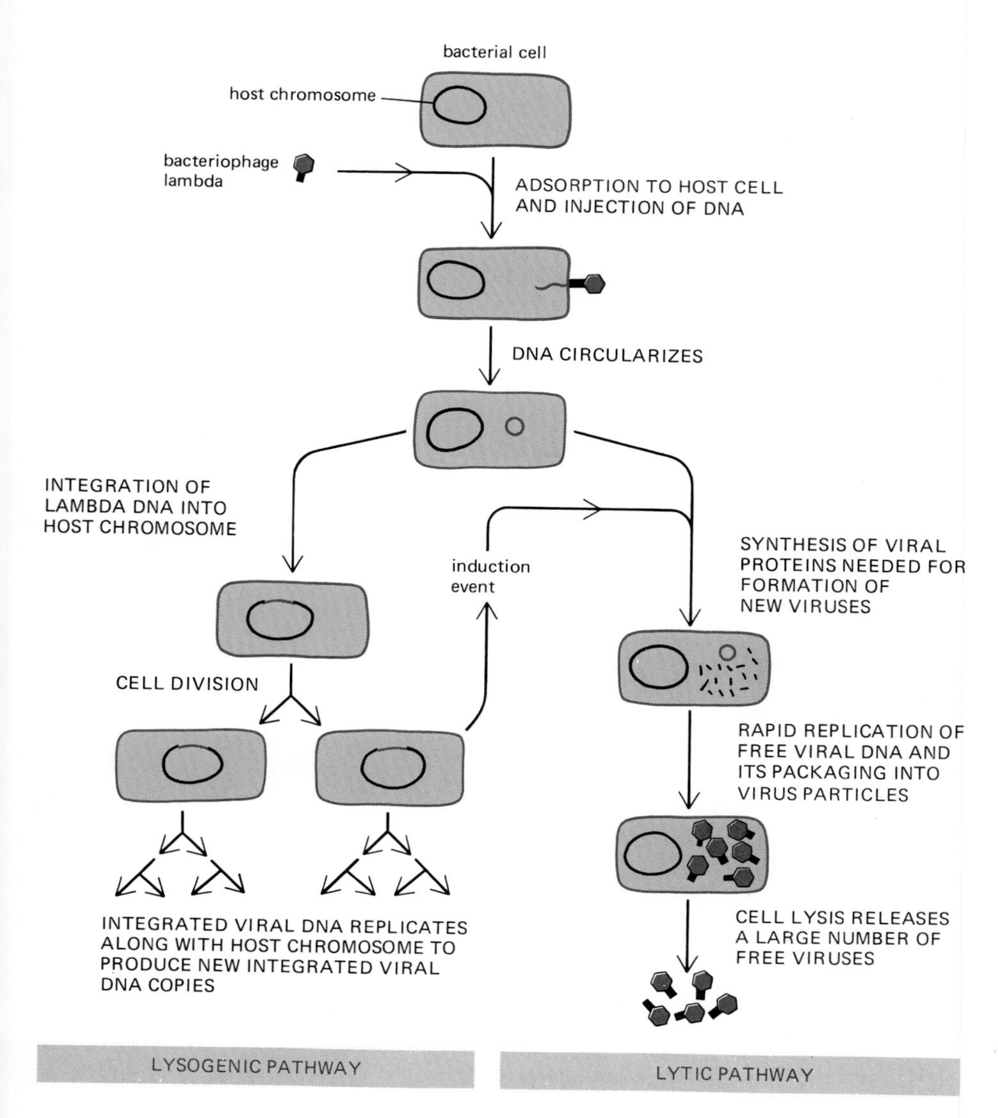

Figure 5–74 The life cycle of bacteriophage lambda. The lambda genome contains about 50,000 nucleotide pairs and encodes about 50 proteins. Its double-stranded DNA can exist in both linear and circular forms. As shown, the bacteriophage can multiply by either a lytic or a lysogenic pathway in the *E. coli* bacterium. When the bacteriophage is growing in the lysogenic state, damage to the cell causes the integrated viral DNA (provirus) to exit from the host chromosome and shift to lytic growth. The entrance and exit of the DNA from the chromosome are site-specific genetic recombination events catalyzed by the lambda *integrase* protein (see p. 495).

The Continuous Synthesis of Viral Proteins Can Make Cells Cancerous[49]

Animal cells, like bacteria, can offer viruses an alternative to lytic growth. *Permissive* cells permit DNA viruses to multiply lytically and kill the cell. *Nonpermissive* cells may allow the DNA virus to enter but not to replicate lytically; in a small percentage of such cells, the viral chromosome either becomes integrated into the host cell genome, where it is replicated along with the host chromosomes, or forms a plasmid—a circular DNA molecule—that replicates in a controlled fashion without killing the cell. This sometimes results in a genetic change in the non-permissive cells, causing them to proliferate in an ill-controlled way and thus transforming them into their cancerous equivalents. In this case the DNA virus is called a *DNA tumor virus* and the process is called virus-mediated *neoplastic transformation.* The most extensively studied DNA tumor viruses are two papovaviruses, SV40 and polyoma. Their transforming ability has been traced to several viral proteins that cooperate to drive quiescent cells from G_0 to S phase (see p. 517). In permissive cells the shift to S phase provides the virus with all of the host cell replication enzymes required for viral DNA synthesis. The synthesis of these viral proteins by a provirus in a nonpermissive cell overrides some of the normal growth control mechanisms in the cell and in all of its progeny.

RNA Tumor Viruses Are Retroviruses[50]

For one group of RNA viruses, the so-called *RNA tumor viruses,* the infection of a permissive cell often leads simultaneously to a nonlethal release of progeny virus from the cell surface by budding and a permanent genetic change in the infected cell that makes it cancerous. How RNA virus infection could lead to a permanent genetic alteration was unclear until the discovery of the enzyme *reverse transcriptase,* which transcribes the infecting RNA chains of these viruses into complementary DNA molecules that integrate into the host cell genome. RNA tumor viruses—which include the first well-known tumor virus, the Rous sarcoma virus—are members of a large class of viruses known as **retroviruses.** These viruses are so named because they reverse the normal process in which DNA is transcribed into RNA as part of their normal life cycle. The virus that causes acquired immune deficiency syndrome (AIDS) is also a retrovirus.

The life cycle of a retrovirus is outlined in Figure 5–75. The enzyme **reverse transcriptase** is an unusual DNA polymerase that uses either RNA or DNA as a template; it is encoded by the retrovirus RNA and is packaged inside the viral capsid. When the single-stranded RNA of the retrovirus enters a cell, the reverse transcriptase first makes a DNA copy of the RNA strand to form a DNA-RNA hybrid helix, which is then used by the same enzyme to make a double helix with two DNA strands. This DNA copy of the RNA genome then circularizes and integrates into a host cell chromosome. The integration is aided by a virus-encoded site-specific recombination enzyme that recognizes a particular viral DNA sequence and helps to catalyze the insertion of the viral DNA into virtually any site on a host cell chromosome (see Figure 5–67B). The next step in the infectious process is transcription of the integrated viral DNA by host cell RNA polymerase, producing large numbers of viral RNA molecules identical to the original infecting genome. Finally, these RNA molecules are translated to produce the capsid, envelope, and reverse transcriptase proteins that are assembled with the RNA into new enveloped virus particles, which bud from the plasma membrane (see Figure 5–75).

Both RNA and DNA tumor viruses transform cells because the permanent presence of the viral DNA in the cell causes the synthesis of new proteins that alter the control of host cell proliferation. The genes that code for such proteins are called *oncogenes.* Unlike DNA tumor viruses, whose oncogenes typically encode normal viral proteins essential for viral multiplication, the oncogenes carried by RNA tumor viruses are modified versions of normal host cell genes that are not required for viral replication. Since only a limited amount of RNA can be packed into the capsid of a retrovirus, the acquired oncogene sequences often replace an essential part of the retroviral genome, making the virus defective. We shall discuss

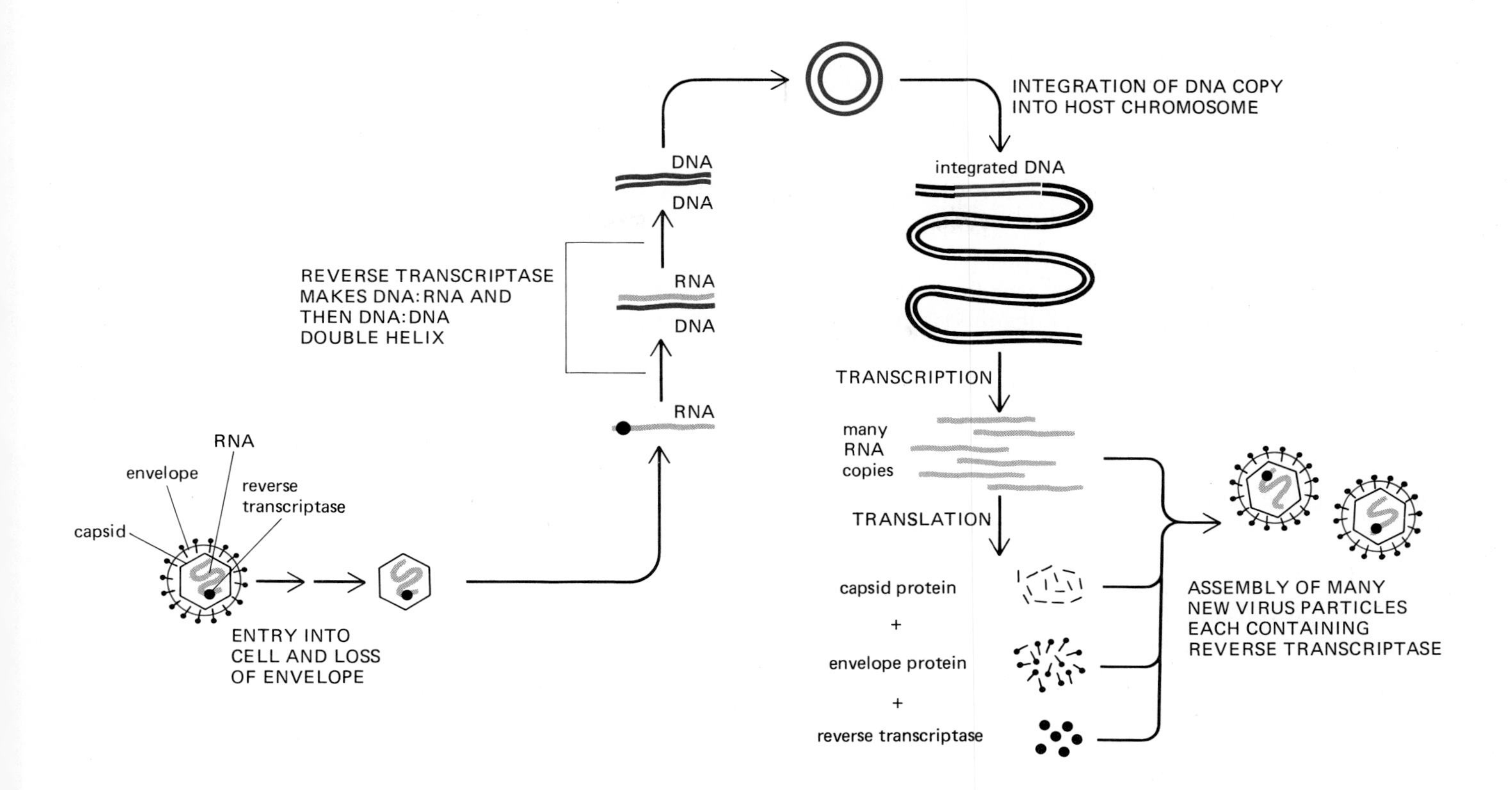

Figure 5–75 The life cycle of a retrovirus. The retrovirus genome consists of an RNA molecule of about 8500 nucleotides; two such molecules are packaged into each viral particle. The enzyme *reverse transcriptase* is a DNA polymerase that first makes a DNA copy of the viral RNA molecule and then a second DNA strand, generating a double-stranded DNA copy of the RNA genome. The integration of this DNA double helix into the host chromosome, catalyzed by a viral protein, is required for the synthesis of new viral RNA molecules by the host cell RNA polymerase.

elsewhere how viral oncogenes have provided important clues to the causes and nature of cancer, as well as to the normal mechanisms that control cell growth and division in multicellular animals (see p. 754 and p. 1205).

Some Transposable Elements Are Close Relatives of Retroviruses[51]

Because many viruses can move in and out of their host chromosomes, any large genome is likely to contain a number of different proviruses. These genomes are also likely to house a variety of mobile DNA sequences that do not form viral particles and cannot leave the cell. Such **transposable elements** range in length from a few hundred to tens of thousands of base pairs, and they are usually present in multiple copies per cell. One can consider these elements as tiny parasites hidden in chromosomes. Each transposable element is occasionally activated to move to another DNA site in the same cell by a process called *transposition,* catalyzed by its own site-specific recombination enzyme. These enzymes, referred to as *transposases,* are often encoded in the DNA of the element itself. Since most transposable elements move only very rarely (once in 10^5 cell generations for many elements in bacteria), it is often difficult to discriminate them from nonmobile parts of the chromosome. It is not known what suddenly triggers their movement.

Transposition can occur by a variety of mechanisms. One large family of transposable elements uses a mechanism that is indistinguishable from part of a retrovirus life cycle. These elements, called *retrotransposons,* are present in organisms as diverse as yeast, *Drosophila,* and mammals. One of the best-understood retrotransposons is the so-called Ty1 element of yeast. The first step in its transposition is the transcription of the entire transposable element, producing an RNA copy of the element that is more than 5000 nucleotides long. This transcript encodes a reverse transcriptase enzyme that makes a circular double-stranded DNA copy of the RNA molecule via a RNA-DNA hybrid intermediate, precisely mimicking the early stages of infection by a retrovirus (see Figure 5–75). The analogy continues as the DNA circle integrates into a randomly selected site on the chromosome. As for a retrovirus, integration proceeds by the type of site-specific mechanism shown in Figure 5–67B, presumably utilizing a transposase that is also encoded by the long RNA transcript. Although the resemblance to a retrovirus is striking, without a functional protein coat the Ty1 element is constrained to move within a single cell and its progeny.

5-44 Other Transposable Elements Transfer Themselves Directly from One Site in the Genome to Another[52]

Unlike retrotransposons, many transposable elements seem never to exist free of the host chromosome; the transposases that catalyze their movement act on the DNA of the element while it is still integrated in the host genome. The transposase is thought to bind to a short sequence that is repeated in reverse orientation at each end of the element, thereby holding these two ends close together while catalyzing the subsequent recombination event (Figure 5–76). For some transposable elements the transposition mechanism involves only the breaking and rejoining of DNA, with the two ends of the element being inserted into a staggered nick made elsewhere on a chromosome (see Figure 5–67B). Such a transposable element moves directly from one chromosomal site to another without DNA replication. However, the DNA sequence is often altered when the break in the vacated chromosome reseals, causing a mutation at the old chromosomal site.

Other transposable elements replicate when they move. In the best-studied example, site-specific recombination triggers a localized synthesis of DNA; one copy of the replicated transposable element is inserted at a randomly selected new chromosomal site, while the other copy remains at the old one (Figure 5–77). The mechanism is closely related to the nonreplicative mechanism just described; indeed, some transposable elements can move by either pathway.

In addition to moving themselves, all types of transposable elements occasionally move or rearrange neighboring DNA sequences of the host genome. For example, they frequently cause deletions of adjacent nucleotide sequences or carry them to another site. The presence of transposable elements makes the arrangement of the DNA sequences in chromosomes much less stable than previously thought, and it is likely that they have been responsible for many important evolutionary changes in genomes (see pp. 605–609).

Are the transposable elements also of evolutionary importance as the most ancient ancestors of viruses? Although the precursors of retroviruses were almost certainly retrotransposons, all present-day transposable elements rely heavily on

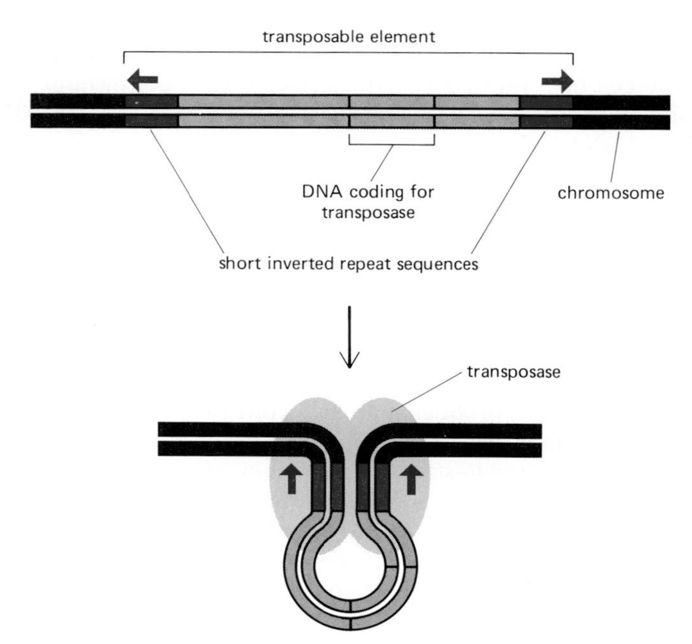

Figure 5–76 The structure of a transposable element that moves directly from one chromosomal site to another. Transposable elements of this type can be recognized by the "inverted repeat DNA sequences" at their ends. Experiments show that the repetition of these sequences, which can be as short as 20 nucleotides, is all that is necessary for the DNA between them to be transposed by the particular transposase enzyme associated with the element. The protein complex that forms in the first stage of movement is shown. Subsequent events involve cleavage of the DNA at the ends of the inverted repeat sequences and occur by modifications of the recombination scheme shown previously in Figure 5–67B (see also Figure 5–77).

DNA metabolism. But very early cells are thought to have had RNA rather than DNA genomes (see p. 10), so we must look to RNA metabolism for the ultimate origin of viruses.

Most Viruses Probably Evolved from Plasmids[53]

Even the largest viruses depend heavily on their host cells for biosynthesis; for example, no known virus makes its own ribosomes or generates the ATP it requires. Clearly, therefore, cells must have evolved before viruses. The precursors of the first viruses were probably small nucleic acid fragments that developed the ability to multiply independent of the chromosomes of their host cells. Such independently replicating elements, called **plasmids,** can replicate indefinitely outside the host chromosome. Plasmids occur in both DNA and RNA forms (see p. 259), and, like viruses, they contain a special nucleotide sequence that serves as an origin of replication. Unlike viruses, however, they cannot make a protein coat and therefore cannot move from cell to cell readily. Many also cannot integrate into chromosomes.

The first RNA plasmids may have resembled the *viroids* found in some plant cells. These small RNA circles, only 300 to 400 nucleotides long, are replicated despite the fact that they do not code for any protein (see Fig. 10–61, p. 598). Having no protein coat, viroids exist as naked RNA molecules and pass from plant to plant only when the surfaces of both donor and recipient cells are damaged so that there is no membrane barrier for the viroid to pass. Under the pressure of natural selection, such independently replicating elements could be expected to acquire nucleotide sequences from the host cell that would facilitate their own multiplication, including sequences that code for proteins. Some present-day plasmids are indeed quite complex, encoding proteins and RNA molecules that regulate their replication, as well as proteins that control their partitioning into daughter cells. The largest known plasmids are double-stranded DNA circles more than 100,000 base pairs long.

The first virus probably appeared when an RNA plasmid acquired a gene coding for a capsid protein. But a capsid can enclose only a limited amount of nucleic acid; therefore a virus is limited in the number of genes it can contain. Forced to make optimal use of their limited genomes, some small viruses (like ϕX174) evolved *overlapping genes,* in which part of the nucleotide sequence encoding one protein is used (in the same or a different reading frame) to encode a second protein. Other viruses evolved larger capsids and consequently could accommodate more genes.

With their unique ability to transfer nucleic acid sequences across species barriers, viruses have almost certainly played an important part in the evolution of the organisms they infect. Many recombine frequently with their host cell genome and with one another and in this way can pick up small pieces of host chromosome at random and carry them to different cells or organisms. Moreover, integrated copies of viral DNA (proviruses) have become a normal part of the genome of most organisms. Examples of such proviruses include the lambda family of bacteriophages and the so-called endogenous retroviruses found in numerous copies in vertebrate genomes. The integrated viral DNA can become altered so that it cannot produce a complete virus but can still encode proteins, some of which may be useful to the host cell. Therefore, viruses, like sexual reproduction, can speed up evolution by promoting the mixing of gene pools.

The process in which DNA sequences are transferred between different host cell genomes by means of a virus is called *DNA transduction,* and several viruses that transduce DNA with particularly high frequencies are commonly used by researchers to move genes from one cell to another. Viruses and their close relatives, plasmids and transposable elements, have also been important to cell biology in many other ways. Because of their relative simplicity, for example, studies of their reproduction have progressed unusually rapidly and have illuminated many of the basic genetic mechanisms in cells. In addition, both viruses and plasmids have been crucial elements in the development of the recombinant DNA technologies that we shall consider next.

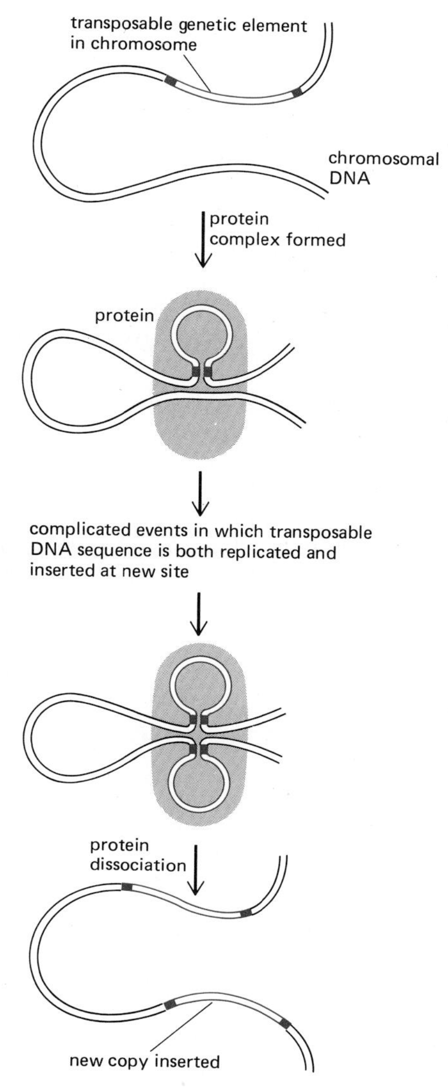

Figure 5–77 Schematic diagram illustrating the movement of one type of transposable element within a chromosome. The element shown replicates during transposition, its movement occurring without it being excised from its original site. The two inverted repeat DNA sequences that commonly flank the two ends of transposable elements are shown as colored squares. At the start of transposition, the transposase cuts one of the two DNA strands at each end of the element, and the element then serves as a template for DNA synthesis that begins by the the addition of nucleotides to the 3′ ends of chromosomal DNA sequences. Further details are known, but the process is too complex to be illustrated here.

Summary

Viruses are infectious particles that consist of a DNA or RNA molecule (the viral genome) packaged in a protein capsid, which in the enveloped viruses is surrounded by a lipid bilayer-based membrane. Both the structure of the viral genome and its mode of replication vary widely among viruses. A virus can multiply only inside a host cell, whose genetic mechanisms it subverts for its own reproduction. A common outcome of a viral infection is the lysis of the infected cell and release of infectious viral particles. In some cases, however, the viral chromosome instead integrates into a host cell chromosome, where it is replicated as a provirus along with the host genome. Many viruses are thought to have evolved from plasmids, which are self-replicating DNA or RNA molecules that lack the ability to wrap themselves in a protein coat.

Transposable elements are DNA sequences that differ from viruses in being able to multiply only in their host cell and its progeny; like plasmids, they cannot leave the cell. Unlike plasmids, they normally replicate only as an integral part of a chromosome. However, some transposable elements are closely related to retroviruses and can move from place to place in the genome by the reverse transcription of an RNA intermediate. Other transposable elements can move without ever detaching from the chromosomes. Although both viruses and transposable elements can be viewed as parasites, many of the DNA sequence rearrangements they cause are important for the evolution of cells and organisms.

DNA Cloning and Genetic Engineering[54]

The discoveries we have discussed in this chapter resulted from curiosity about the cell and the fundamental mechanisms of heredity. In recent years, however, this basic knowledge has been put to practical use. The techniques of *DNA cloning* and *genetic engineering* enable specific genes to be isolated in quantity, redesigned, and then inserted back into cells and organisms. They form part of the panel of methods previously introduced in Chapter 4, known collectively as *recombinant DNA technology*. Recombinant DNA technology has revolutionized the study of living cells. It has also provided medicine and industry with an efficient means of producing specific proteins in large amounts that previously were available only in extremely small quantities, if at all.

5-48
5-49
5-50

Restriction Nucleases Facilitate the Cloning of Genes[55]

In the 1960s the goal of isolating a single gene from a large chromosome seemed impossibly remote. Unlike a protein, a gene does not exist as a discrete entity in cells, but rather as a small region of a much larger DNA molecule. Although the DNA molecules in a cell can be randomly broken into small pieces by mechanical force, a fragment containing an individual gene in a mammalian genome would still be only one part in a million of the total DNA fragments and indistinguishable in size from other genes. How could such a gene be purified? Since all DNA molecules consist of an approximately equal mixture of the same four nucleotides, they cannot be readily separated, as proteins can, on the basis of their different charges and binding properties (see p. 167). Moreover, even if a purification scheme could be devised, vast amounts of DNA would be needed to yield enough of any particular gene to be useful for further experiments.

The solution to all of these problems began to appear with the discovery of **restriction nucleases.** These enzymes, which can be purified from bacteria, cut the DNA double helix at specific sequences of four to eight nucleotides, producing DNA fragments of strictly defined sizes that are known as **restriction fragments** (see p. 182). Different species of bacteria make restriction nucleases with different sequence specificities, and it is relatively simple to find a restriction nuclease that will create a small DNA fragment that includes a particular gene. The size of the restriction fragment can then be used as a basis for partially purifying the gene from a mixture.

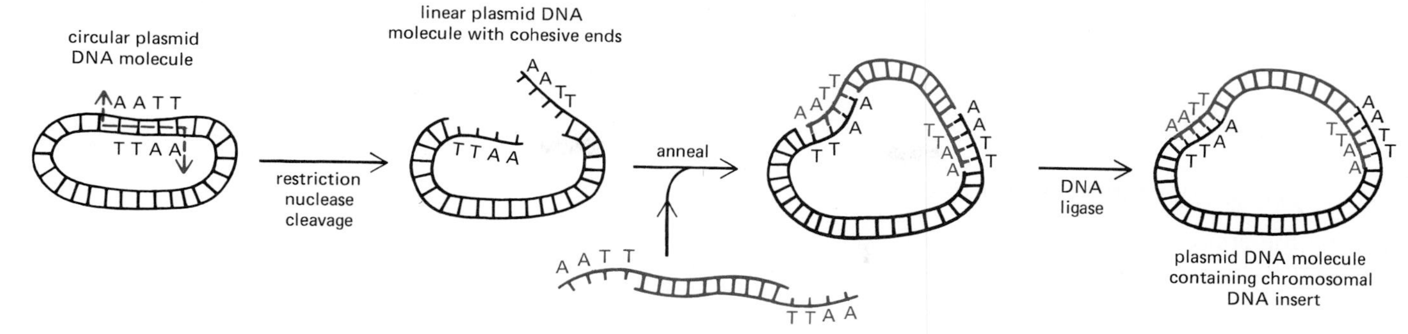

Figure 5–78 The cohesive ends produced by many kinds of restriction nucleases (see Figure 4–63) allow two DNA fragments to be joined by complementary base-pair interactions. DNA fragments joined in this way can be covalently linked in a highly efficient reaction catalyzed by the enzyme DNA ligase. In this example, a recombinant plasmid DNA molecule containing a chromosomal DNA insert is formed.

Another property of restriction nucleases that is helpful for gene cloning is that many of them produce staggered cuts that leave short, single-stranded tails on both ends of the DNA fragment. These are known as *cohesive ends* since they can form complementary base pairs with any other end produced by the same enzyme (see p. 182). The cohesive ends generated by a restriction nuclease make it possible to join two double-helical DNA fragments from different genomes by complementary base-pairing (Figure 5–78). For example, a DNA fragment containing a human gene can be joined in a test tube to the chromosome of a bacterial virus and the new *recombinant DNA molecule* then introduced into a bacterial cell. Starting with only one such recombinant DNA molecule that infects a single cell, the normal replication mechanism of the virus can produce more than 10^{12} identical virus DNA molecules in less than a day, thereby amplifying the amount of the attached human DNA fragment by the same factor. A virus used in this way is known as a *cloning vector*.

5-51 A DNA Library Can Be Made Using Either Viral or Plasmid Vectors[56]

In order to clone a gene, one begins by constructing a *DNA library* in either a virus or a plasmid vector. The principles underlying the methods used for cloning genes are the same for either cloning vector, although the details may be different. For simplicity, in this chapter we shall ignore these differences and illustrate the methods discussed in reference to plasmid vectors.

The **plasmid vectors** used for gene cloning are small circular molecules of double-stranded DNA derived from larger plasmids that occur naturally in bacteria, yeast, and mammalian cells (see p. 257). They generally account for only a minor fraction of the total host cell DNA, but they can easily be separated on the basis of their small size from chromosomal DNA molecules, which are large and form a pellet upon centrifugation. For use as cloning vectors, the purified plasmid DNA circles are first cut with a restriction nuclease to create linear DNA molecules. The cellular DNA to be used in constructing the library is cut with the same restriction nuclease, and the resulting restriction fragments (including those containing the gene to be cloned) are then added to the cut plasmids and annealed to form recombinant DNA circles. These recombinant molecules containing foreign DNA inserts are then covalently sealed with the DNA ligase enzyme described previously (see p. 223) to form intact DNA circles (see Figure 5–78).

In the next step in preparing the library, the recombinant DNA circles are introduced into cells (usually bacterial or yeast cells) that have been made transiently permeable to DNA; such cells are said to be *transfected* with the plasmids. As these cells grow and divide, the recombinant plasmids also replicate to produce an enormous number of copies of DNA circles containing the foreign DNA (Figure 5–79). Many bacterial plasmids carry genes for antibiotic resistance, a property that can be exploited to select those cells that have been successfully transfected; if the bacteria are grown in the presence of the antibiotic, only cells containing plasmids will survive. These surviving bacteria are said to contain a DNA library. However, only a few of these bacteria will harbor the particular recombinant plas-

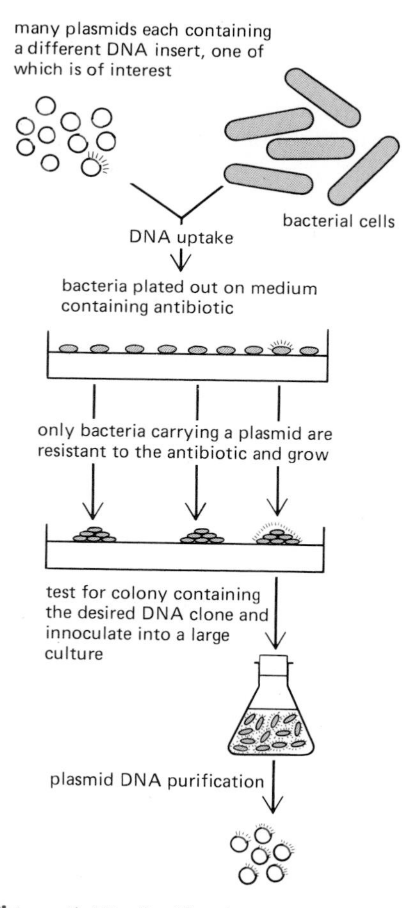

Figure 5–79 Purification and amplification of a specific DNA sequence by DNA cloning in a bacterium.

mids that contain the gene to be isolated. One needs to be able to identify these cells in order to recover the DNA of interest in pure form and in useful quantities. Before discussing how this is achieved, we need to describe a second way of generating a DNA library that is commonly used in gene cloning.

Two Types of DNA Libraries Serve Different Purposes[57]

Cleaving the entire genome of a cell with a specific restriction nuclease as just described is sometimes called the "shotgun" approach to gene cloning. It produces a very large number of DNA fragments—on the order of a million for a mammalian genome—which will generate millions of different colonies of transfected cells. Each of these colonies will be composed of a *clone* derived from a single ancestor cell and therefore harbor a recombinant plasmid with the same inserted genomic DNA sequence. Such a plasmid is said to contain a **genomic DNA clone,** and the entire collection of plasmids is said to comprise a **genomic DNA library.** But because the genomic DNA is cut into fragments at random, only some fragments will contain genes; many will contain only a portion of a gene, while most of the genomic DNA clones obtained from the DNA of a higher eucaryotic cell will contain only noncoding DNA, which comprises most of the DNA in such genomes (see p. 485).

An alternative strategy begins the cloning process by selecting only those DNA sequences that are transcribed into RNA and thus are presumed to correspond to genes. This is done by extracting the mRNA (or a purified subfraction of the mRNA) from cells and then making a **complementary DNA (cDNA)** copy of each mRNA molecule present; this reaction is catalyzed by the *reverse transcriptase* enzyme of retroviruses, which synthesizes a DNA chain on an RNA template (see p. 254). The single-stranded DNA molecules synthesized by the reverse transcriptase are converted into double-stranded DNA molecules by DNA polymerase, and these molecules are inserted into plasmids and cloned (Figure 5–80). Each clone obtained in this way is called a **cDNA clone,** and the entire collection of clones derived from one mRNA preparation constitutes a **cDNA library.**

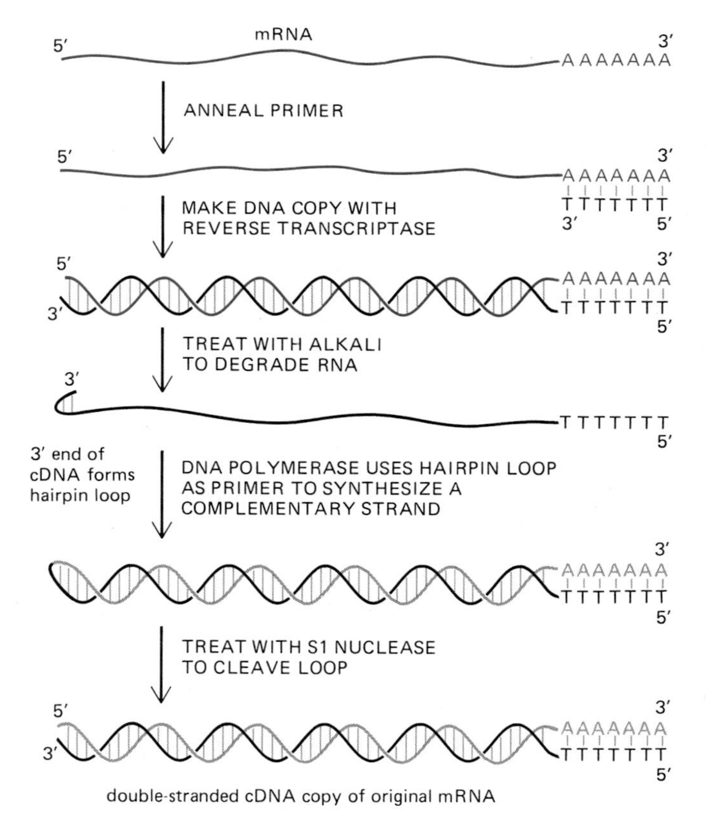

Figure 5–80 The synthesis of cDNA. A DNA copy (cDNA) of an mRNA molecule is produced by the enzyme reverse transcriptase (see p. 254), thereby forming a DNA/RNA hybrid helix. Treating the DNA/RNA hybrid with alkali selectively degrades the RNA strand into nucleotides. The remaining single-stranded cDNA is then copied into double-stranded cDNA by the enzyme DNA polymerase. As indicated, both reverse transcriptase and DNA polymerase require a primer to begin their synthesis. For reverse transcriptase a small oligonucleotide is used; in this example oligo(dT) has been annealed with the long poly A tract at the 3′ end of most mRNAs (see p. 528).

Note that the double-stranded cDNA molecule produced here lacks cohesive ends; such "blunt-ended" DNA molecules can be cloned by one of several procedures analogous to that shown in Figure 5–78 but less efficient. For example, synthetic oligonucleotides containing restriction-enzyme cutting sites can be ligated onto the DNA ends, or single-stranded DNA "tails" can be added enzymatically to facilitate the insertion of the cDNA molecule into a cloning vector.

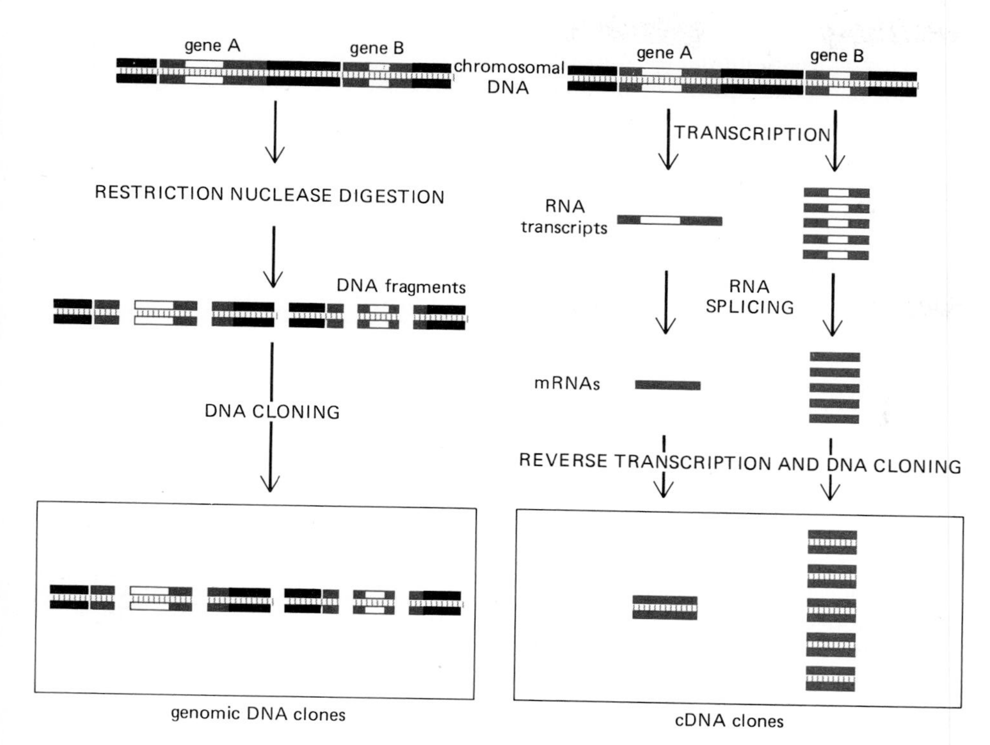

Figure 5–81 Diagrammatic representation of the differences between cDNA clones and genomic DNA clones. In this example gene A is infrequently transcribed while gene B is frequently transcribed. Both types of RNA transcripts are processed by RNA splicing to remove a single intron sequence during the formation of the mRNA. Most genes contain many introns (see Table 9–1, p. 486).

There are important differences between genomic DNA clones and cDNA clones, as illustrated in Figure 5–81. Genomic clones represent a random sample of all of the DNA sequences in an organism and with very rare exceptions will be the same regardless of the cell type used to prepare them. By contrast, cDNA clones contain only those regions of the genome that have been transcribed into mRNA, and—since the cells of different tissues produce distinct sets of mRNA molecules (see p. 553)—a different cDNA library will be obtained for each type of cell used to prepare the library.

The use of a cDNA library for gene cloning has several advantages. First, many proteins are produced in very large quantities by specialized cells, in which case the mRNA encoding the protein is produced in such excess that a cDNA library prepared from the cells will be highly enriched for the cDNA molecules encoding the protein (Figure 5–81). The abundance of the cDNA of interest greatly reduces the problem of identifying the desired clone in the library. Hemoglobin, for example, is made in large amounts by developing erythrocytes, and, for this reason, the globin genes were among the first to be cloned.

A second advantage of cDNA clones is that they contain the uninterrupted coding sequence of a gene. Eucaryotic genes, as we have seen (p. 102), usually consist of coding sequences of DNA separated by noncoding sequences, so that the production of mRNA from these genes entails the removal of the noncoding sequences (intron sequences) from the initial RNA transcript and the splicing together of the remaining coding sequences. Neither bacterial nor yeast cells will make these modifications to the RNA produced from a gene of a higher eucaryotic cell. Thus, if the aim of the cloning is either to deduce the amino acid sequence of the protein from the DNA or to produce the protein in bulk by expressing the cloned gene in a bacterial or yeast cell, it is preferable to start with cDNA.

Genomic and cDNA libraries are inexhaustible resources that are widely shared among investigators, and an increasing number of such libraries are available from commercial sources.

cDNA Libraries Can Be Prepared from Selected Populations of mRNA Molecules[58]

When cDNAs are prepared from cells that express the gene of interest at extremely high levels, the majority of cDNA clones may contain the gene sequence, which can therefore be selected with minimal effort. For less abundantly transcribed genes, various methods can be used to enrich for particular mRNAs before making the cDNA library. For example, if an antibody against the protein is available, it can be used to precipitate selectively those isolated polyribosomes (see p. 214) that contain the appropriate growing polypeptide chains. Since these ribosomes will also contain the mRNA coding for the protein, the precipitate may be enriched in the desired mRNA by as much as 1000-fold.

Substrative hybridization provides a powerful alternative way of enriching for particular nucleotide sequences prior to cDNA cloning. This selection procedure can be used, for example, if two closely related cell types are available from the same species of organism but only one type produces the protein or proteins of interest. An early use was to identify cell-surface receptor proteins present on T lymphocytes but not on B lymphocytes (see p. 1037). It can also be used where a cell that expresses the protein has a mutant counterpart that does not. The first step is to synthesize cDNA molecules using the mRNA from the cell type that makes the protein of interest. These cDNAs are then hybridized with a large excess of mRNA molecules from the second cell type. Those rare cDNA sequences that fail to find a complementary mRNA partner, and therefore are likely to represent mRNA sequences present only in the first cell type, are then purified by a simple biochemical procedure that separates single-stranded from double-stranded nucleic acids (Figure 5–82). Besides providing a powerful way to clone genes whose products are known to be restricted to a specific differentiated cell type, cDNA libraries prepared after subtractive hybridization are useful for defining the differences in gene expression between any two related cells.

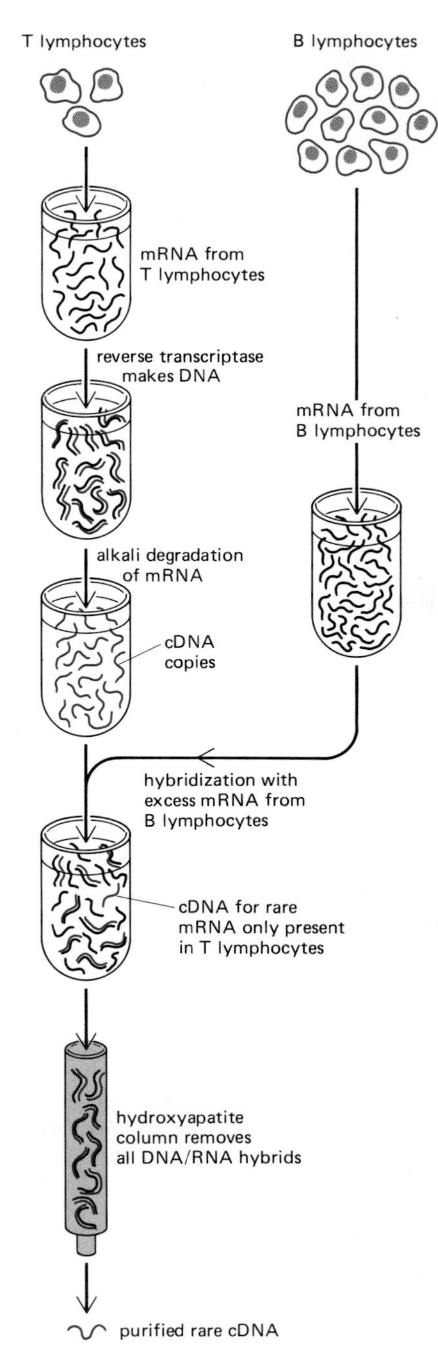

Figure 5–82 Subtractive hybridization is used in this example to purify rare cDNA clones corresponding to mRNA molecules present in T lymphocytes but not in B lymphocytes. Because the two cell types are very closely related, most of the mRNAs will be common to both cell types, and this technique is thus a powerful way to enrich for those specialized molecules that distinguish the two cells.

5-52 Hybridization with a Radioactive DNA Probe Can Be Used to Identify the Clones of Interest in a DNA Library[59]

The most difficult part of gene cloning is often the identification of the rare colonies in the library that contain the DNA fragment of interest. This is especially true in the case of a genomic library, where one has to identify one bacterial cell in a million to select a specific mammalian gene. The technique most frequently used is a form of *in situ* hybridization that takes advantage of the exquisite specificity of the base-pairing interactions between two complementary nucleic acid molecules (see p. 192). Culture dishes containing the growing bacterial colonies are blotted with a piece of filter paper, to which some members of each bacterial colony adhere. The adhering colonies, known as *replicas,* are treated with alkali and then incubated with a radioactive DNA probe containing part of the sequence of the gene being sought (Figure 5–83). If necessary, millions of bacterial clones can be screened in this way to find the one clone that hybridizes with the probe.

The way the specific DNA probes are made will depend on the information available about the gene to be cloned. In many cases the protein of interest has been identified by biochemical studies and purified in small amounts. Even a few micrograms of pure protein are enough to determine the sequence of its first 30 or so amino acids. From this amino acid sequence, the corresponding nucleotide sequence can be deduced using the genetic code. Two sets of DNA oligonucleotides, each about 20 nucleotides long, are then synthesized by chemical methods and radioactively labeled. The two sets of oligonucleotides are chosen to match different parts of the predicted nucleotide sequence of the gene (Figure 5–84). Colonies of cells that hybridize with both sets of DNA probes are strong candidates for containing the desired gene and are saved for further characterization (see below).

The Selection of Overlapping DNA Clones Allows One to "Walk" Along the Chromosome to a Nearby Gene of Interest[60]

Many of the most interesting genes—for example, those that control development—are known only from genetic analysis of mutants in such organisms as the fruit fly *Drosophila* and the nematode *C. elegans.* The protein products of these genes are unknown and may be present in very small quantities in a few cells or

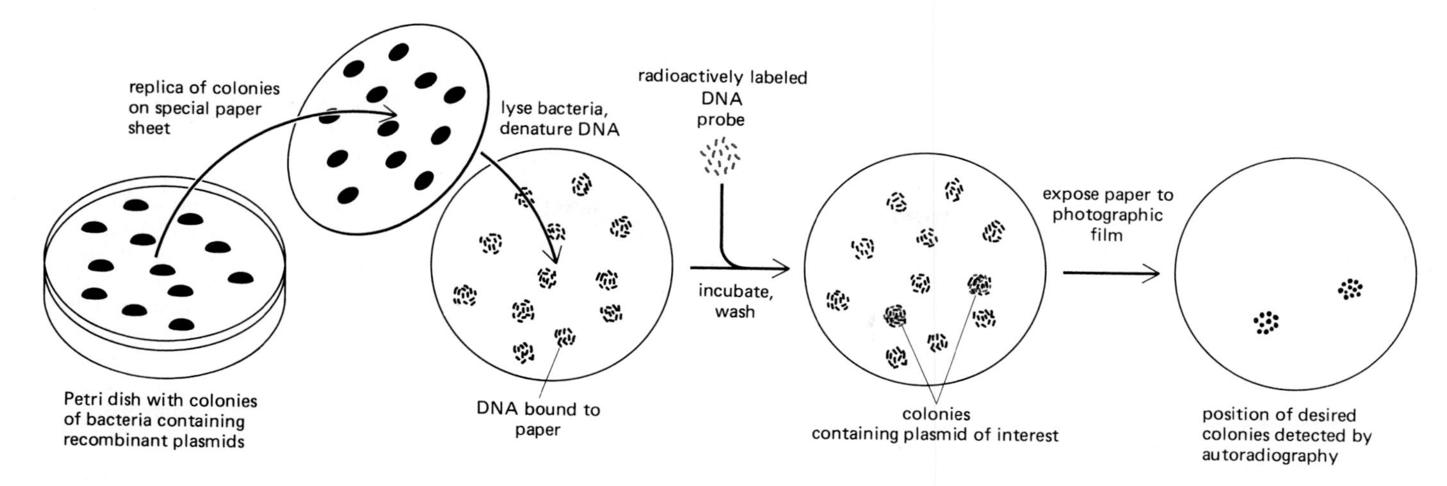

Figure 5–83 An efficient technique commonly used to detect a bacterial colony carrying a particular DNA clone (see also Figure 5–79). Each bacterial cell carrying a recombinant plasmid develops into a colony of identical cells, visible as a white spot on the nutrient agar. A replica of the culture is then made by pressing a piece of absorbent paper against the surface. This replica is treated with alkali (to break open the adherent cells and denature the plasmid DNA) and then hybridized to a highly radioactive DNA probe. Those bacterial colonies that have bound the probe are identified by autoradiography.

produced only at one stage of development. However, a study of the genetic linkage between different mutations can be used to generate *chromosome maps,* which give the relative locations of the genes. Once one mapped gene has been cloned, the clones in a genomic DNA library that correspond to neighboring genes can be identified using a technique known as **chromosome walking.** For this method, two different genomic DNA libraries can be used, each prepared by fragmenting the same DNA with a different restriction nuclease. Figure 5–85 illustrates how a clone in one library can be used as a DNA probe to find an overlapping clone in the other library. This new clone is then used to prepare a DNA probe, which is used to find another overlapping clone in the first library, and so on. In this way one can walk along a chromosome one clone at a time, in steps of 30,000 base pairs or more. But how does one know when the gene of interest (identified originally by a deleterious mutation) has been reached? The standard way is continually to compare the size of the DNA restriction fragments produced from mutant and normal chromosomes by Southern blot analysis (see p. 191), using each new clone as a probe as the walk proceeds. Some of the mutants will have

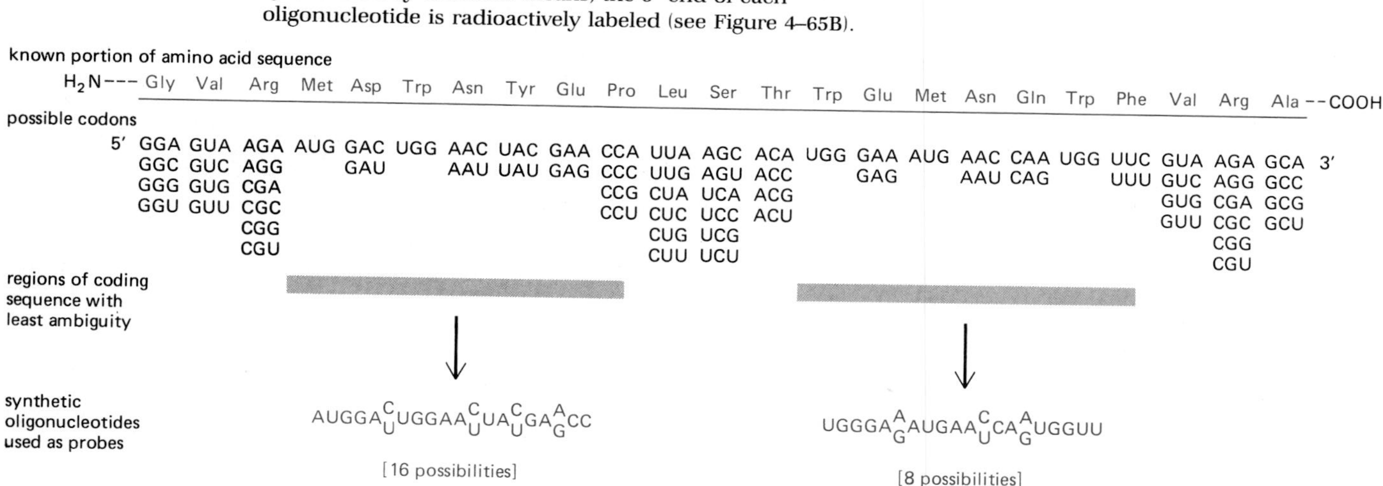

Figure 5–84 Selecting regions of a known amino acid sequence to make synthetic oligonucleotide probes. Only one nucleotide sequence will actually code for the protein. But the degeneracy of the genetic code means that several different nucleotide sequences will give the same amino acid sequence, and it is impossible to tell in advance which is the correct one. Because it is desirable to have as large a fraction of the correct nucleotide sequence as possible in the mixture of oligonucleotides to be used as a probe, those regions with the fewest possibilities are chosen, as illustrated. After the oligonucleotide mixture is synthesized by chemical means, the 5′ end of each oligonucleotide is radioactively labeled (see Figure 4–65B).

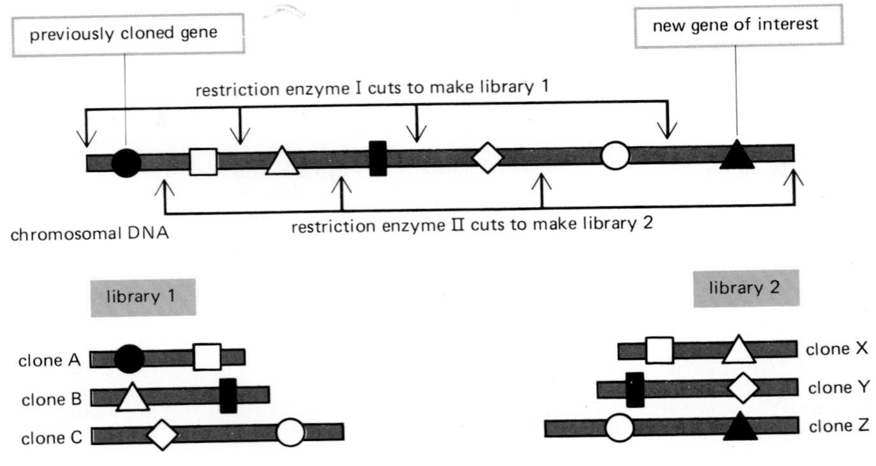

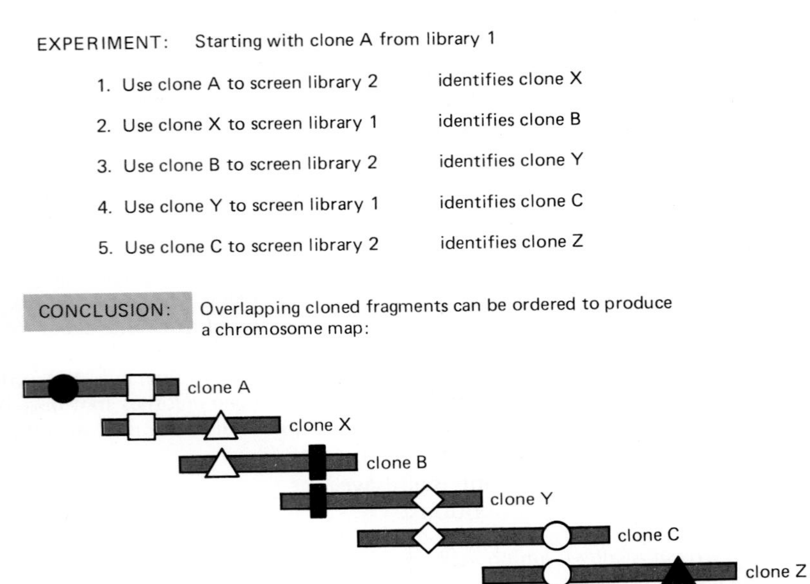

Figure 5–85 The use of overlapping DNA clones to find a new gene by "chromosome walking." To speed up the walk, genomic libraries containing very large cloned DNA molecules are optimal. To probe for the next clone in the walk, one uses a short ^{32}P-labeled DNA fragment from one end of the previously identified clone: If a "right-handed" end is used, for example, the walk will go in the "rightward" direction, as shown in this example. By using a small end fragment as a probe, one also reduces the probability that the probe will contain a repeated DNA sequence that would hybridize with many clones from different parts of the genome and thereby interrupt the walk.

been caused by small deletions or insertions of DNA sequences in the relevant gene, and these can be readily identified. When the deleterious mutations in a typical human gene are analyzed, for example, about 1 in 10 turns out to be a deletion that is easily detected by Southern blotting, and the precise molecular defects responsible for human genetic diseases are increasingly being identified in this way.

By using related methods, it has been possible to order (map) a nearly complete set of large genomic clones along the chromosomes of the nematode *C. elegans*. Such large clones, each about 30,000 base pairs in length, are prepared in bacteriophage lambda vectors called *cosmids*, which are specially designed to accept only large DNA inserts. It takes a few thousand cosmid clones to cover the entire genome of an organism such as *C. elegans* or *Drosophila*. To map the entire human genome in this way would require ordering more than 100,000 such cosmid clones, which is time-consuming but technically feasible. Moreover, human DNA fragments that are 10 times larger than cosmid clones (300,000 base pairs) can be cloned as artificial chromosomes in yeast cells; in principle, the human genome could be mapped as 10,000 clones of this type (see Fig. 9–5, p. 485).

In the near future, ordered sets of genomic clones will no doubt be available from centralized DNA libraries for use by all research workers. A complete library will be available for each commonly studied organism, with each DNA insert catalogued according to its chromosome of origin and numbered sequentially with respect to the positions of all other DNA fragments derived from the same chromosome. One will then begin a "chromosome walk" simply by obtaining from the library all the clones covering the region of the genome that contains the mutant gene of interest. These clones would then be used to make DNA probes to locate the altered gene precisely. Many of the mutant genes that cause genetic diseases in humans will eventually be isolated in this way.

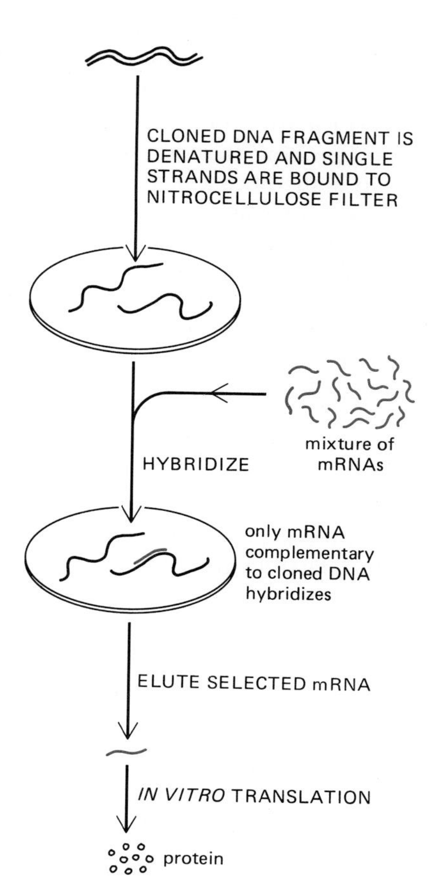

Figure 5–86 The technique of hybrid selection. The purified mRNA molecules are eluted from the filter by subjecting it to conditions that separate the two strands of an RNA-DNA helix.

In Vitro Translation Facilitates Identification of the Correct DNA Clone[61]

Despite the power of *in situ* hybridization methods to find specific cDNA or genomic clones from a DNA library, they usually pick out many "false positive" clones. Further ingenuity is required to discriminate between these and authentic positive clones. The task is easiest when the desired clone encodes a protein that has already been characterized by other means. In this case, each candidate cloned DNA fragment is used to purify its complementary mRNA molecules from a mixture of cellular mRNAs by a process called **hybrid selection,** in which an excess of the DNA fragment is separated into single strands and immobilized on a filter that is then used to select complementary mRNA molecules by RNA-DNA hybridization (Figure 5–86). The mRNA purified this way is then allowed to direct protein synthesis in a cell-free system using radioactive amino acids. Finally, the radioactive protein produced is characterized and compared with the expected protein product of the desired clone. A match in such a test is normally a prerequisite for concluding that a cloned DNA fragment encodes the given protein.

Expression Vectors Allow cDNA Clones to Be Used to Overproduce Proteins[62]

Very often there is no biochemical information about the protein encoded by a cloned DNA fragment. This is usually the case, for example, when the clone has been identified by subtractive hybridization or by chromosome walking to a mutant gene. In these cases, moreover, the mRNA for the protein in question is often present in such low abundance and in such a limited group of cells that hybrid selection of the complementary mRNA is not feasible. Other methods must then be used to characterize the protein product of the cloned gene. One method is to synthesize a short protein fragment (an *oligopeptide*) corresponding to the deduced amino acid sequence of the protein product of the sequenced cDNA molecule, and then to raise antibodies against the oligopeptide. In many cases the antibodies will recognize the same amino acid sequence when it occurs as part of the natural protein molecule, and these then provide a means to detect, locate, and purify the protein encoded by the original cDNA. In conjunction with cloning by subtractive hybridization, this immunological approach is a powerful way to identify cell type-specific proteins and investigate the development, properties, and functions of each type of cell in a multicellular organism.

The most direct way to characterize the protein encoded by a cloned cDNA, however, is to allow the cDNA itself to direct protein synthesis in a host cell. Plasmids or viruses used for this purpose are called **expression vectors,** and they are constructed so that the cDNA clone is connected directly to a DNA sequence that acts as a strong promoter for DNA transcription. A variety of expression vectors are available, each engineered to function in the type of cell in which the protein is to be made. By means of such *genetic engineering,* bacteria, yeast, or mammalian cells can be induced to make vast quantities of useful proteins—such as human growth hormone, interferon, and viral antigens for vaccines. Bacterial cells with plasmids or viral vectors that have been engineered in this way are especially adept at protein production, and it is common for an engineered gene to produce more than 10% of the total cell protein. Because the production of such a large quantity of a single protein frequently kills the cell, special "inducible" promoters have been designed to allow transcription of the gene to be delayed until a few hours before the cells are collected for protein isolation. Some plasmid expression vectors, for example, contain a bacteriophage lambda promoter that is kept silent by a heat-sensitive gene-repressor protein; the promoter can be turned on at the desired time by increasing the temperature of the bacteria to 42°C, and large amounts of the protein of interest will suddenly be made.

If a cDNA library is prepared in an expression vector, each clone will in general produce a different protein, and it is then possible to identify the clones of interest by their protein product instead of by their nucleic acid sequence. This is usually done by probing the clones with radioactively labeled antibodies. Alternatively, one can sometimes test directly for the biological activity of the gene product. This is an especially powerful way to search for eucaryotic genes coding for secreted growth factors. For example, cDNA clones from cells that produce a growth factor can be prepared in an expression vector that replicates in mammalian cells. A mixture of transfected cells containing many different such clones is grown in a small dish, where any cell expressing the gene will secrete the growth factor into the culture medium. Samples of the medium are then tested for the growth factor by adding the medium to cultures of other cells known to respond to that particular factor. The test is so sensitive that it will usually be positive even if only one cell in a thousand in the original culture contains a gene encoding the growth factor. Repeated rounds of testing allow the search to be narrowed down to the single clone in the mixture that is producing the growth factor. In this way previously unrecognized growth factors have been discovered and isolated in a few months; isolating growth factors using standard biochemical techniques can require tedious purifications of more than a millionfold that take years.

5-53 Genes Can Be Redesigned to Produce Proteins of Any Desired Sequence[63]

Both the coding sequence of a gene and its regulatory regions can be redesigned to change the functional properties of the protein product, the amount of protein made, or the particular cell type in which the protein is produced.

The coding sequence of a gene can be extensively altered—for example, by fusing part of it to the coding sequence of a different gene to produce a novel hybrid gene that encodes a *fusion protein.* Such proteins are frequently used to test the function of different domains of a protein molecule. For example, most nuclear proteins contain specific short sequences of amino acids that are recognized as signals for their direct import into the cell nucleus. By artificially attaching different parts of nuclear proteins to a cytoplasmic protein using gene fusion techniques, the "signal peptides" responsible for nuclear import have been identified (see p. 424).

Special techniques are required to alter a gene in more subtle ways, so that the protein it encodes differs from the original by one or a few amino acids. The first step is the chemical synthesis of a short DNA molecule containing the altered portion of the gene's nucleotide sequence. This synthetic DNA oligonucleotide is hybridized with single-stranded plasmid DNA that contains the DNA sequence to be altered, using conditions that allow imperfectly matched DNA strands to pair (Figure 5–87). The synthetic oligonucleotide will now serve as a primer for DNA synthesis by DNA polymerase, thereby generating a DNA double helix that incorporates the altered sequence into the gene. The modified gene is then inserted into an expression vector so that the redesigned protein can be produced in large enough quantities for detailed studies. By changing selected amino acids in a protein in this way, one can analyze which parts of the polypeptide chain are important in such fundamental processes as protein folding, protein-ligand interactions, and enzymatic catalysis.

To determine which parts of a eucaryotic gene are responsible for regulating its expression, suspected regulatory DNA sequences of the gene (see p. 564) can be joined to the coding sequence of a gene that encodes a readily detected marker enzyme not normally present in eucaryotic cells. One of the most commonly used marker enzymes is the bacterial protein *chloramphenicol acetyltransferase (CAT).* The resulting recombinant DNA molecule is then inserted into a eucaryotic cell to test whether the suspected regulatory sequences promote expression of the CAT gene. This is most conveniently done by transfecting cells in culture and testing for CAT enzyme activity after an overnight incubation. Although many of the transfected cells will transiently express the foreign gene, very few will retain the gene permanently. To obtain *permanent transformants* it is necessary to select

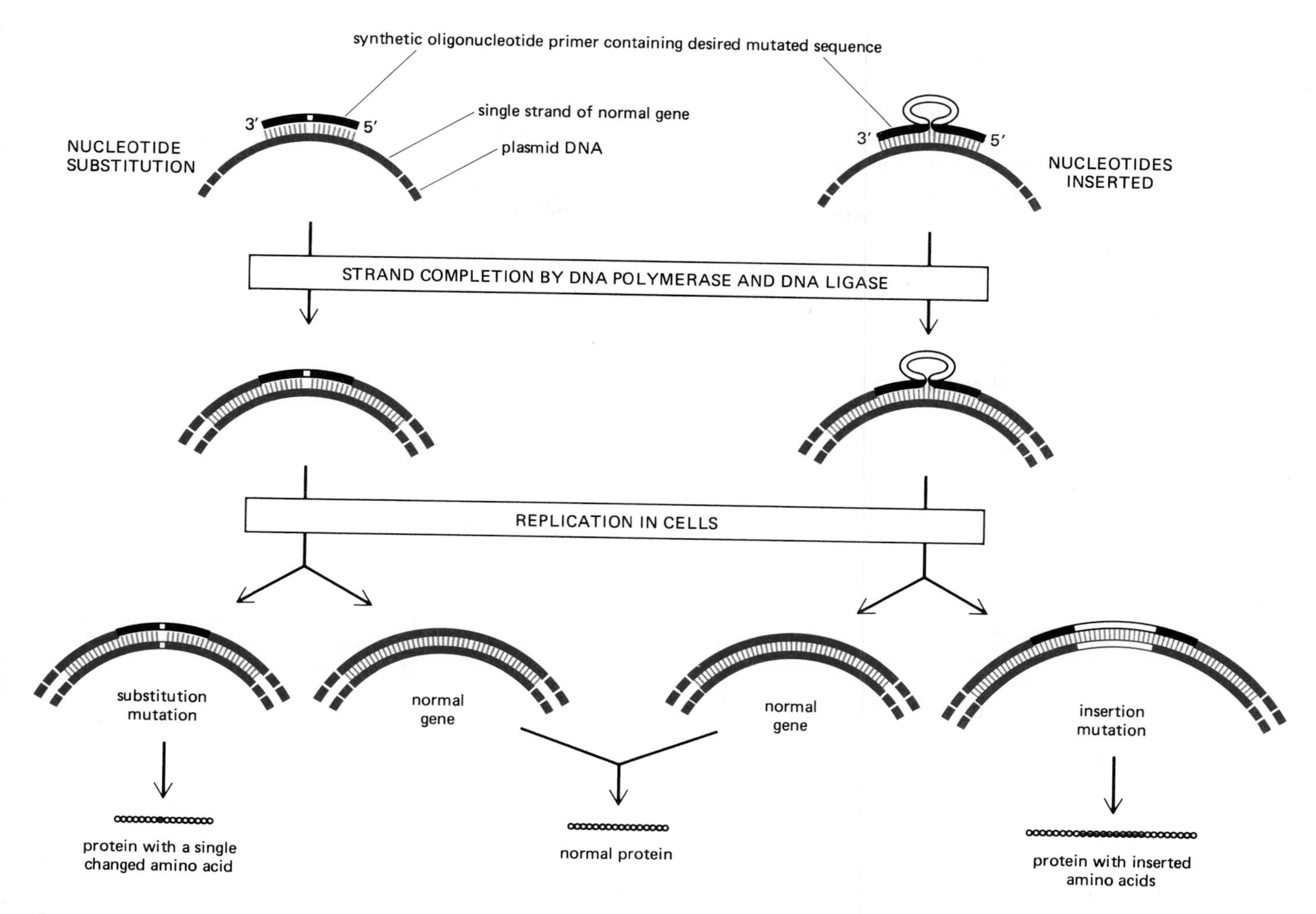

the rare cell clones that have stably incorporated the recombinant DNA molecules into their chromosomes. By transfecting different cell types with such recombinant DNA molecules, it has been possible to identify regulatory DNA sequences that enable a gene to be expressed in one specific cell type but not in others (see p. 565).

Figure 5–87 The use of synthetic oligonucleotides to modify the protein-coding regions of genes. Only two of the many types of changes that can be engineered in this way are shown. For example, with an appropriate oligonucleotide, more than one amino acid substitution can be made at a time, or one or more amino acids can be deleted. As indicated, because only one of the two DNA strands in the original recombinant plasmid is altered by this procedure, only half of the transfected cells will end up with a plasmid that contains the desired mutant gene. Note that most of the plasmid sequence is not illustrated here.

Engineered Genes Can Be Permanently Inserted into the Germ Line of Mice or Fruit Flies to Produce Transgenic Animals[64]

The ultimate test of the function of an altered gene is to reinsert it into an organism and see what effect it has. Ideally one would like to be able to replace the normal gene with the altered one so that the function of the mutant protein can be analyzed in the absence of the normal protein. The only eucaryotic organism in which this can be done routinely is yeast. DNA fragments transferred into growing yeast cells are efficiently integrated as single copies into the homologous chromosomal sites by general recombination events (see p. 239), so that a redesigned gene can be inserted in place of the endogenous copy of the gene. This provides a powerful way to study the function of yeast genes (see p. 195).

When DNA is transferred into mammalian cells, by contrast, there is presently no way to control how and where it becomes integrated into a chromosome. Only about one in a thousand integrative events leads to gene replacement. Instead linear DNA fragments are rapidly ligated end to end by enzymes in the cell to form long tandem arrays, which usually become integrated into a chromosome at an apparently randomly selected site. Fertilized mammalian eggs behave like other mammalian cells in this respect. A mouse egg injected with 200 copies of a linear DNA molecule will often develop into a mouse containing a tandem array of copies of the injected gene integrated at a single random site in one of its chromosomes

The PCR method is extremely sensitive: it can detect a single DNA molecule in a sample. Trace amounts of RNA can be analyzed in a similar way by converting them to DNA sequences with reverse transcriptase (see p. 260). The PCR cloning technique is rapidly replacing Southern blotting for prenatal diagnosis of genetic diseases and for the detection of low levels of viral infection. It also has great promise for forensic medicine, as it allows the unambiguous identification of the human source of a single cell.

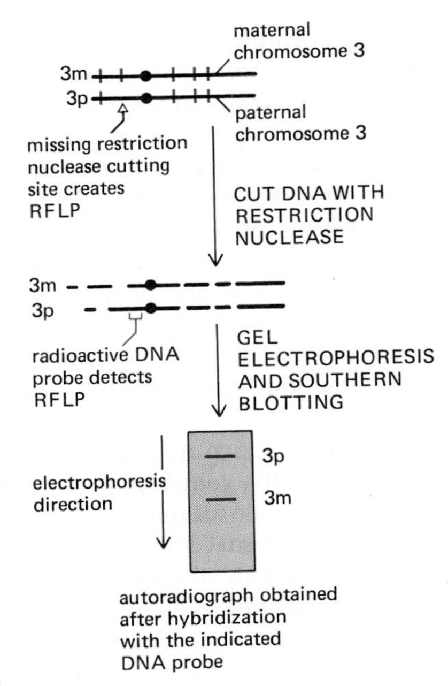

Figure 5–90 Detection of a restriction fragment length polymorphism (RFLP) by Southern blotting. For simplicity the chromosomes are shown to contain only a few restriction-enzyme cutting sites, whereas in reality they contain many thousands of such sites. If the PCR technique (see p. 269) is used to amplify the appropriate DNA region before the restriction nuclease treatment, the same test can be applied without radioisotopes and the blotting step can be omitted.

Recombinant DNA Methods Greatly Facilitate Genetic Approaches to the Mapping and Analysis of Large Genomes[66]

10-31

Recently developed methods make it possible to prepare detailed maps of very large genomes. There are two categories of maps: (1) *Physical maps* are based on the DNA molecules that comprise each chromosome. They include restriction maps (see p. 182) and ordered libraries of genomic DNA clones (see p. 264). (2) *Genetic linkage maps* are based on the frequency of coinheritance of two or more features that serve as *genetic markers,* each different in the mother and father and attributable to a specific chromosomal site. Traditionally these markers were genes whose expression is detected by means of their effects (as with genes that cause genetic diseases such as muscular dystrophy). More recently, recombinant DNA methods have made it possible to use as genetic markers short DNA sequences that contain a restriction nuclease cutting site and differ between individuals; such a sequence is particularly useful for genetic mapping because it creates a **restriction fragment length polymorphism,** or **RFLP**, which can be readily detected by Southern blotting (see p. 191) with a nearby DNA probe (Figure 5–90).

If two genetic markers are on different chromosomes, their inheritance will be *unlinked*—that is, they will have only a 50:50 chance of being inherited together. The same is true for markers at opposite ends of a single chromosome because of the high probability that they will be separated during the extensive crossing over that occurs during meiosis in the development of eggs and sperm (see p. 847). The closer together two markers are on the same chromosome, the larger is the chance that they will not be separated by crossover events and will therefore be coinherited. By screening large family groups for the coinheritance of a gene of interest (such as one associated with a disease) and a large number of individual RFLPs, a few RFLP markers can be identified that surround the gene. In this way DNA sequences in the vicinity of the gene can be located and eventually the DNA corresponding to the gene itself can be found (Figure 5–91). Many genes that cause

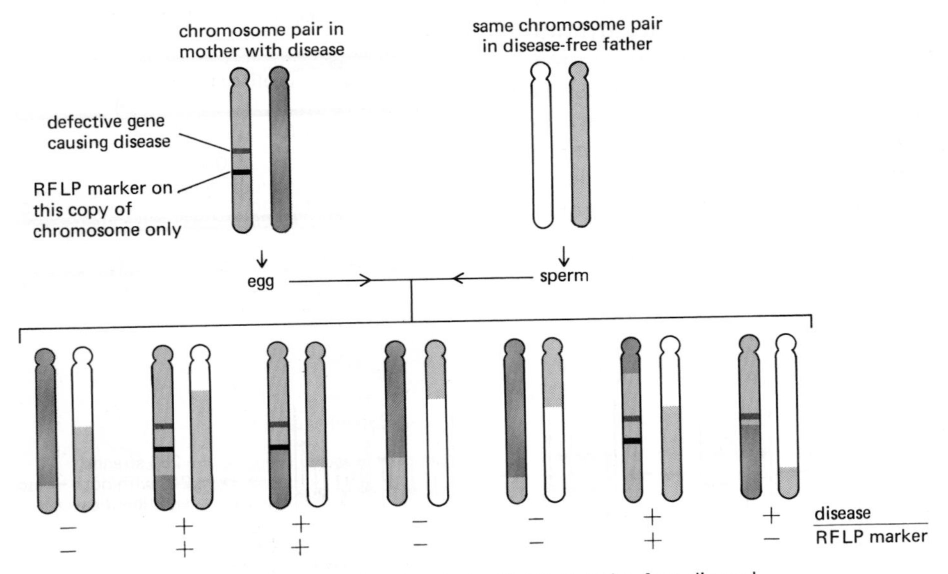

Figure 5–91 Genetic linkage analysis detects the coinheritance of a gene that causes a specific human phenotype (here a genetic disease) with a nearby RFLP marker. Analyses of this type are allowing many human genes to be cloned and their gene products to be analyzed.

human diseases are being located in this way. Once the gene is isolated, the protein it encodes can be analyzed in detail (see p. 193).

To facilitate these and other studies, an intensive effort is underway to prepare a high-resolution RFLP map of the human genome, with thousands of RFLP markers spaced an average of 10^6 nucleotide pairs apart (on average, two markers separated by this distance will be coinherited by 99 of every 100 progeny). With such a map it will become possible to use genetic linkage studies to locate genes that have been identified only by their effect on humans to one or a few large DNA clones in an ordered genomic DNA library. Isolating the gene could then be accomplished in most cases in a matter of months.

Given the rapid advances in recombinant DNA technology, it should soon be possible to sequence very long stretches of DNA at reasonable cost. This will make it possible to sequence the entire genomes of bacteria, yeasts, nematodes, and *Drosophila,* as well as major portions of the genomes of various higher plants and vertebrates, including humans.

Summary

In DNA cloning the DNA fragment containing a gene of interest is typically identified by using a radiolabeled DNA probe or, following expression of the gene in host cells, by using an antibody to detect the protein the gene encodes. The cells carrying this DNA fragment are then allowed to proliferate, producing large amounts of the gene or gene product. In genetic engineering, the nucleotide sequence of such a cloned DNA fragment is altered, joined to other DNA sequences, and then inserted back into cells. Together, DNA cloning and genetic engineering provide powerful research tools for the cell biologist. In principle, a gene encoding a protein of any desired amino acid sequence can be constructed and attached to a promoter DNA sequence that causes the timing and pattern of the gene's expression to be controlled in a desired way. The new gene can then be inserted either into cultured cells or into the germ line of a mouse or a fruit fly. In transgenic animals one can examine the effect of the expression of the gene on many different cells and tissues.

References

General

Judson, H.F. The Eighth Day of Creation: Makers of the Revolution in Biology. New York: Simon & Schuster, 1979. (History derived from personal interviews.)

Lewin, B. Genes, 3rd ed. New York: Wiley, 1987.

Stent, G.S. Molecular Genetics: An Introductory Narrative. San Francisco: Freeman, 1971. (Clear descriptions of the classical experiments.)

Watson, J.D.; Hopkins, N.H.; Roberts, J.W.; Steitz, J.A.; Weiner, A.M. Molecular Biology of the Gene, 4th ed. Menlo Park, CA: Benjamin-Cummings, 1987.

Cited

1. Chamberlin, M. Bacterial DNA-dependent RNA polymerases. In The Enzymes, 3rd ed., Vol. 15B (P. Boyer, ed.), pp. 61–108. New York: Academic Press, 1982.

   McClure, W. Mechanism and control of transcription initiation in prokaryotes. *Annu. Rev. Biochem.* 54:171–204, 1985.

   Yager, T.D.; von Hippel, P.H. Transcript elongation and termination in *E. coli.* In *Escherichia coli* and *Salmonella typhimurium:* Cellular and Molecular Biology (F.C. Niedhardt, ed.), pp. 1241–1275. Washington, DC: American Society for Microbiology, 1987.

2. Hawley, D.K.; McClure, W.R. Compilation and Analysis of *Escherichia coli* promoter DNA sequences. *Nucleic Acids Res.* 11:2237–2255, 1983.

   Siebenlist, U.; Simpson, R.; Gilbert, W. *E. coli* RNA polymerase interacts homologously with two different promoters. *Cell* 20:269–281, 1980.

3. Rich, A.; Kim, S.H. The three-dimensional structure of transfer RNA. *Sci. Am.* 238(1):52–62, 1978.

   Schimmel, P.R.; Söll, D.; Abelson, J.N., eds. Transfer RNA: Structure, Properties and Recognition, and Biological Aspects (2 volumes). Cold Spring Harbor, NY: Cold Spring Harbor Laboratory, 1980.

4. Schimmel, P. Aminoacyl tRNA synthetases: general scheme of structure-function relationships in the polypeptides and recognition of transfer RNAs. *Annu. Rev. Biochem.* 56:125–158, 1987.

5. Crick, F.H.C. The genetic code. III. *Sci. Am.* 215(4):55–62, 1966.

   The Genetic Code. *Cold Spring Harbor Symp. Quant. Biol.,* Vol. 31, 1966. (The original experiments that defined the code.)

6. Moore, P.B. The ribosome returns. *Nature* 331:223–227, 1988.

   Noller, H.F. Structure of ribosomal RNA. *Annu. Rev. Biochem.* 53:119–162, 1984.

   Spirin, A.S. Ribosome Structure and Protein Synthesis. Menlo Park, CA: Benjamin-Cummings, 1986.

7. Clark, B. The elongation step of protein biosynthesis. *Trends Biochem. Sci.* 5:207–210, 1980.

   Watson, J.D. The involvement of RNA in the synthesis of proteins. *Science* 140:17–26, 1963. (A description of how ribosome function was initially deciphered.)

8. Caskey, C.T. Peptide chain termination. *Trends Biochem. Sci.* 5:234–237, 1980.

6-6 6-7

The Lipid Bilayer Is a Two-dimensional Fluid[3]

Surprisingly, it was only in the early 1970s that researchers first recognized that individual lipid molecules are able to diffuse freely within lipid bilayers. The initial demonstration came from studies of synthetic lipid bilayers. Two types of synthetic bilayers have been very useful in experimental studies: (1) bilayers made in the form of spherical vesicles, called **liposomes,** which can vary in size from about 25 nm to 1 μm in diameter depending on how they are produced (Figure 6–4); and (2) planar bilayers, called **black membranes,** formed across a hole in a partition between two aqueous compartments (Figure 6–5).

A variety of techniques have been used to measure the motion of individual lipid molecules and of their different parts. For example, one can construct a lipid molecule whose polar head group carries a "spin label," such as a nitroxyl group (⪢N—O); this contains an unpaired electron whose spin creates a paramagnetic signal that can be detected by electron spin resonance (ESR) spectroscopy. (The principles of this technique are similar to those of nuclear magnetic resonance; see p. 153.) The motion and orientation of a spin-labeled lipid in a bilayer can be deduced from the ESR spectrum. Such studies show that lipid molecules in synthetic bilayers very rarely migrate from the monolayer on one side to that on the other; this process, called "flip-flop," occurs less often than once a month for any individual lipid molecule (Figure 6–6). On the other hand, lipid molecules readily exchange places with their neighbors *within* a monolayer ($\sim 10^7$ times a second). This gives rise to a rapid lateral diffusion, with a diffusion coefficient (D) of about 10^{-8} cm^2/sec, which means that an average lipid molecule diffuses the length of a large bacterial cell (~2 μm) in about 1 second. These studies have also shown that individual lipid molecules rotate very rapidly about their long axes and that their hydrocarbon chains are flexible, the greatest degree of flexion occurring near the center of the bilayer and the smallest adjacent to the polar head group (see Figure 6–6).

Similar studies have been carried out with labeled lipid molecules in isolated biological membranes and in relatively simple whole cells such as mycoplasma, bacteria, and nonnucleated red blood cells. The results are generally the same as for synthetic bilayers, and they demonstrate that the lipid component of a biological membrane is a two-dimensional liquid in which the constituent molecules are free to move laterally. As in synthetic bilayers, individual lipid molecules are normally confined to their own monolayer. There is an exception, however: in membranes where lipids are actively synthesized, such as the endoplasmic reticulum, there must be a rapid flip-flop of specific lipids across the bilayer, and membrane-bound enzymes called *phospholipid translocators* (see p. 449) are present to catalyze it.

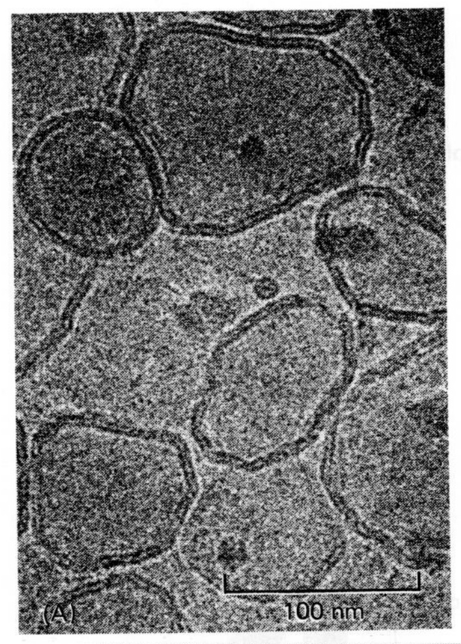

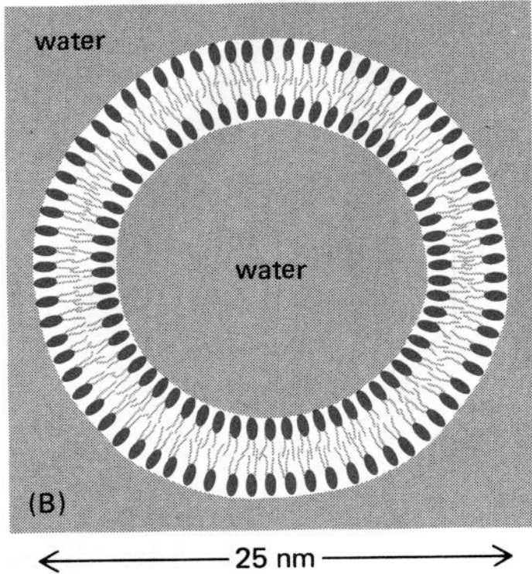

Figure 6–4 (A) An electron micrograph of unfixed, unstained phospholipid vesicles (liposomes) in water. Note that the bilayer structure of the vesicles is readily apparent. (B) A drawing of a small spherical liposome seen in cross-section. Liposomes are commonly used as model membranes in experimental studies. (A, courtesy of Jean Lepault.)

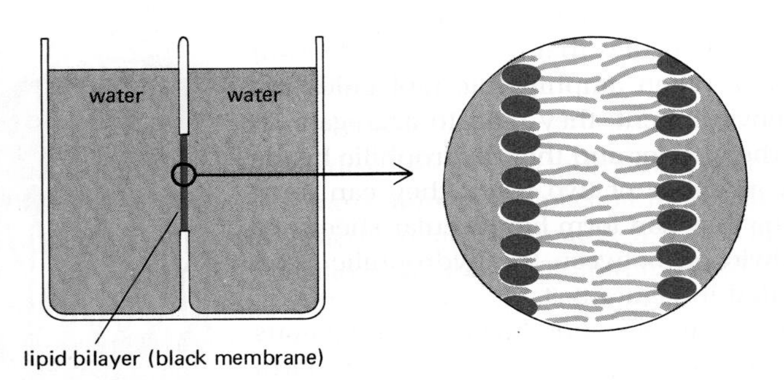

Figure 6–5 A cross-sectional view of a synthetic lipid bilayer, called a black membrane, formed across a small hole in a partition separating two aqueous compartments. Black membranes are used to measure the permeability properties of artificial membranes.

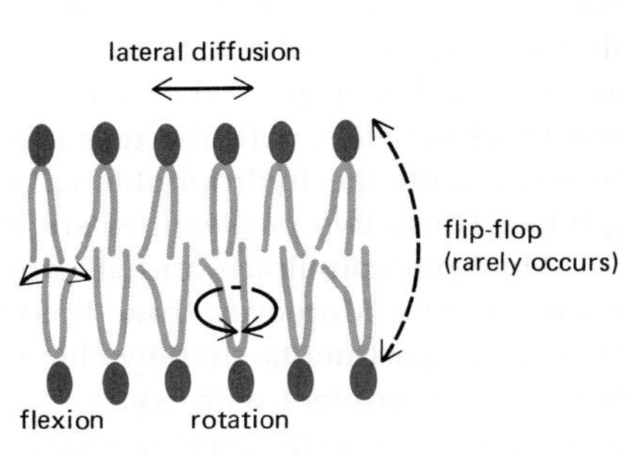

Figure 6–6 The types of movement possible for phospholipid molecules in a lipid bilayer.

The Fluidity of a Lipid Bilayer Depends on Its Composition[4]

A synthetic lipid bilayer made from a single type of phospholipid changes from a liquid state to a rigid crystalline (or gel) state at a characteristic freezing point. This change of state is called a *phase transition,* and the temperature at which it occurs is lower (that is, the membrane becomes more difficult to freeze) if the hydrocarbon chains are short or have double bonds. A shorter chain length reduces the tendency of the hydrocarbon tails to interact with one another, and *cis*-double bonds produce kinks in the hydrocarbon chains that make them more difficult to pack together (Figure 6–7).

In synthetic bilayers containing a mixture of phospholipids with varying degrees of saturation (and therefore different phase-transition points), *phase separations* can occur: when their freezing points are reached, individual phospholipid molecules of the same type aggregate spontaneously within the bilayer to form frozen patches. In biological membranes, saturated and unsaturated fatty acid chains are usually bonded together in the same lipid molecule (that is, one chain is unsaturated while the other is not), so that phase separations of this kind are unlikely to occur.

Another determinant of membrane fluidity is **cholesterol.** Eucaryotic plasma membranes contain large amounts of cholesterol, up to one molecule for every phospholipid molecule. Cholesterol molecules orient themselves in the bilayer with their hydroxyl groups close to the polar head groups of the phospholipid molecules; their rigid, platelike steroid rings interact with—and partly immobilize—those regions of the hydrocarbon chains that are closest to the polar head groups, leaving the rest of the chain flexible (Figure 6–8). Although cholesterol tends in this way to make lipid bilayers less fluid, at the high concentrations found in most eucaryotic plasma membranes it also prevents the hydrocarbon chains from coming together and crystallizing. In this way cholesterol inhibits possible phase transitions.

In addition to affecting fluidity, cholesterol decreases the permeability of lipid bilayers to small water-soluble molecules and is thought to enhance both the flexibility and the mechanical stability of the bilayer. Changes in membrane shape that would require the two sides of the lipid bilayer to compress or expand to very different degrees are facilitated because cholesterol, unlike phospholipids, can readily redistribute (flip-flop) between the two monolayers in response to such forces. The flip-flop of cholesterol molecules occurs with a low energy barrier, and is therefore rapid, because its small polar head group (a hydroxyl group) can pass relatively easily through the center of the bilayer.

The importance of cholesterol in maintaining the mechanical stability of higher eucaryotic cell membranes is suggested by mutant animal cell lines that are unable to synthesize cholesterol. Such cells rapidly lyse (break open and release their

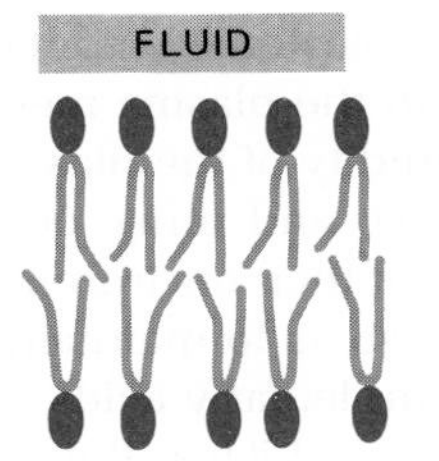

Figure 6–7 Double bonds in unsaturated hydrocarbon chains increase the fluidity of a phospholipid bilayer by making it more difficult to pack the chains together.

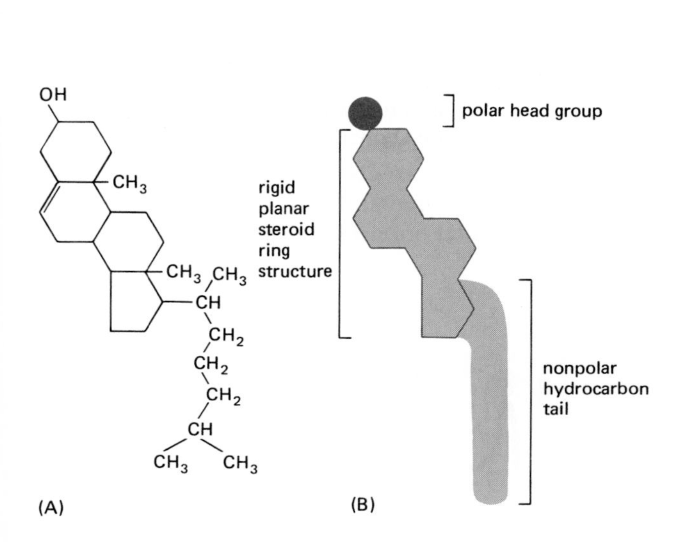

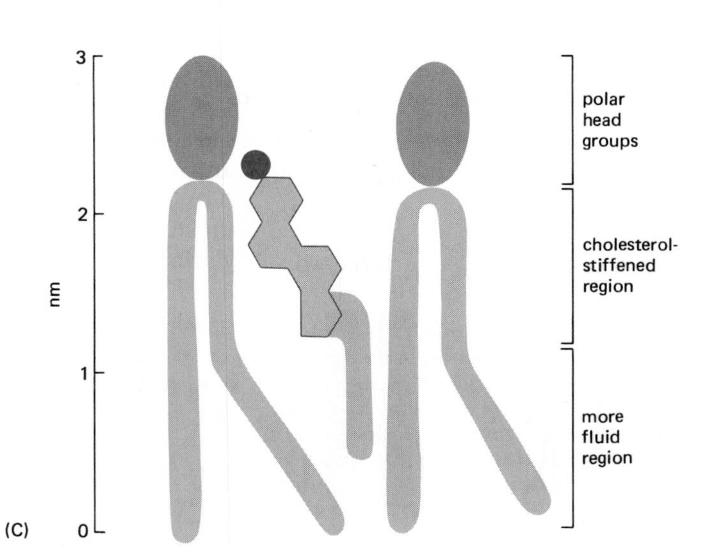

Figure 6–8 Cholesterol represented by a formula (A) and a schematic drawing (B), and depicted interacting with two phospholipid molecules in a monolayer (C).

charged phospholipid in order to act (see p. 704). The importance of having specific inositol phospholipids concentrated in the cytoplasmic leaflet of the plasma membrane lipid bilayer, where they can be used to generate intracellular mediators in response to extracellular signals, is discussed in Chapter 12 (see p. 702).

Glycolipids Are Found on the Surface of All Plasma Membranes, but Their Function Is Unknown[7]

The lipid molecules that show the most striking and consistent asymmetry in distribution in the plasma membranes of animal cells are the oligosaccharide-containing lipid molecules called **glycolipids.** These intriguing molecules are found only in the outer half of the bilayer, and their sugar groups are exposed at the cell surface (see Figure 6–10), suggesting some role in interactions of the cell with its surroundings. The asymmetric distribution of glycolipids in the bilayer derives from the addition of sugar groups to the lipid molecules in the lumen of the Golgi apparatus, which is topologically equivalent to the exterior of the cell (see p. 408).

Glycolipids probably occur in all animal cell plasma membranes, and they generally constitute about 5% of the lipid molecules in the outer monolayer. They differ remarkably from one animal species to another and even among tissues in the same species. In bacteria and plants almost all glycolipids are derived from glycerol-based lipids, as is the common phospholipid phosphatidylcholine. In animal cells, however, they are almost always produced from *ceramide,* as is the phospholipid sphingomyelin (see Figure 6–9). These *glycosphingolipids* have a general structure that is similar to that of the glycerol-based lipids, having a polar head group and two hydrophobic fatty acid chains. However, one of the fatty acid chains is initially coupled to serine to form the amino alcohol *sphingosine,* to which the second fatty acid chain is then linked to form ceramide (Figure 6–11).

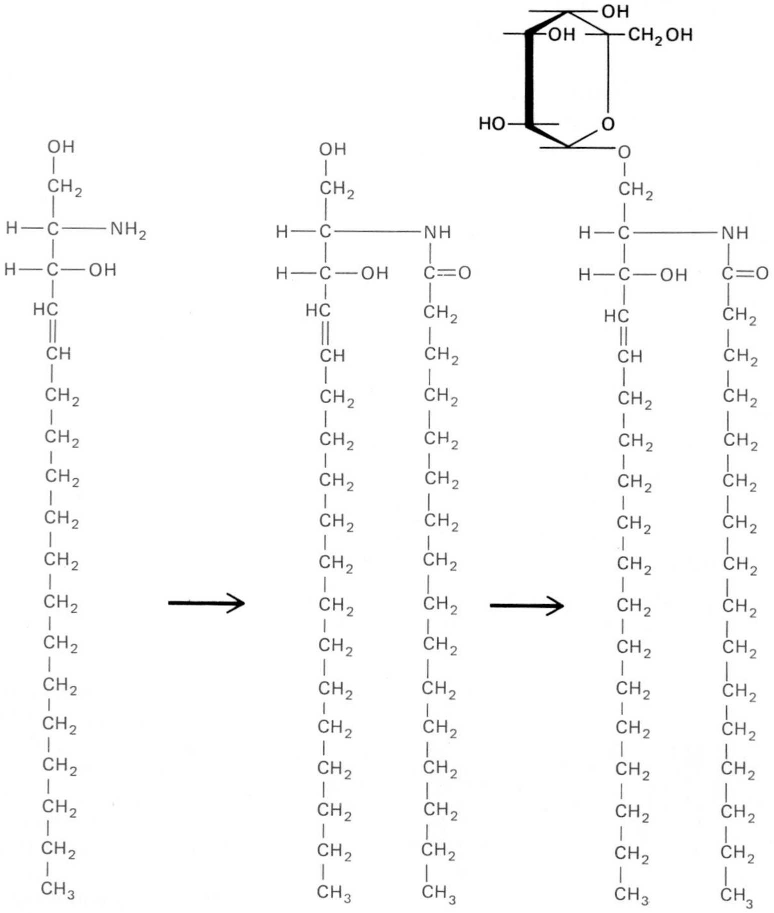

sphingosine → ceramide → galactocerebroside

Figure 6–11 Final steps in the synthesis of the simple glycosphingolipid galactocerebroside. Sphingosine is formed by condensing the amino acid serine with one fatty acid; a second fatty acid is then added to form ceramide, as shown. Ceramide is made in the endoplasmic reticulum, and the carbohydrate is added in the Golgi apparatus (see p. 449). Ceramide is also used to form the major phospholipid sphingomyelin (see Figure 6–9).

Glycolipids are distinguished from one another by their polar head group, which consists of one or more sugar residues. Among the most widely distributed glycolipids in the plasma membranes of both eucaryotic and procaryotic cells are the **neutral glycolipids,** whose polar head groups consist of anywhere from 1 to 15 or more neutral (uncharged) sugars, depending on the organism and cell type. One example is *galactocerebroside,* one of the simplest glycolipids, which has only galactose as its polar head group (see Figure 6–11). It is the main glycolipid in *myelin,* which consists of many concentric layers of plasma membrane wound around a nerve cell process (an axon) by a specialized myelinating cell (see p. 1073). The myelinating cells can be distinguished by the large amount of galactocerebroside in their plasma membrane, where it constitutes almost 40% of the outer monolayer. Galactocerebroside is largely absent from the membranes of most other cells, and it is thought to play an important part in the specific interaction between the myelinating cell and the axon.

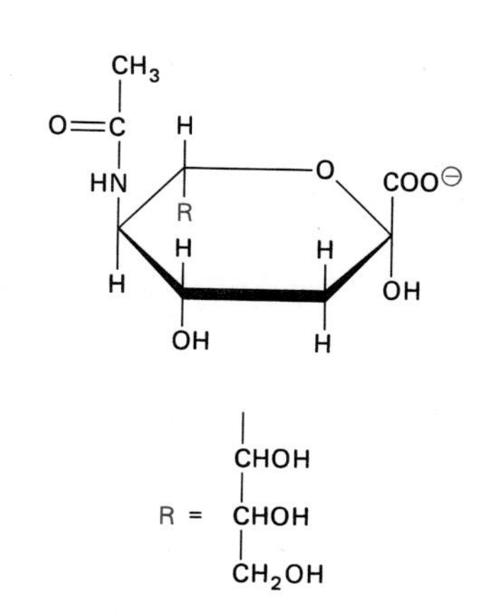

Figure 6–12 The structure of sialic acid (*N*-acetylneuraminic acid, or NANA). In cells this acid exists in its ionized form (—COO^-), as shown.

The most complex of the glycolipids, the **gangliosides,** contain one or more sialic acid residues (also known as *N*-acetylneuraminic acid, or NANA), which gives them a net negative charge (Figure 6–12). Gangliosides are most abundant in the plasma membrane of nerve cells, where they constitute 5–10% of the total lipid mass, although they are found in much smaller quantities in most cell types. So far more than 40 different gangliosides have been identified. Some common examples are shown in Figure 6–13, where the nomenclature used to describe them is also introduced.

There are only hints as to what the functions of glycolipids might be. For example, the ganglioside G_{M1} (Figure 6–13) acts as a cell-surface receptor for the bacterial toxin that causes the debilitating diarrhea of cholera; cholera toxin binds to and enters only those cells with G_{M1} on their surface, including intestinal epithelial cells. The entry of cholera toxin into the cell leads to a prolonged increase in the concentration of intracellular cyclic AMP, which in turn causes a large efflux of Na^+ and water into the intestine (see p. 697). Although binding bacterial toxins cannot be the *normal* function of gangliosides, such observations suggest that these glycolipids may also serve as receptors for normal signaling between cells.

Summary

Biological membranes consist of a continuous double layer of lipid molecules in which various membrane proteins are embedded. This lipid bilayer is fluid, with individual lipid molecules able to diffuse rapidly within their own monolayer. However, most types of lipid molecules very rarely flip-flop spontaneously from one

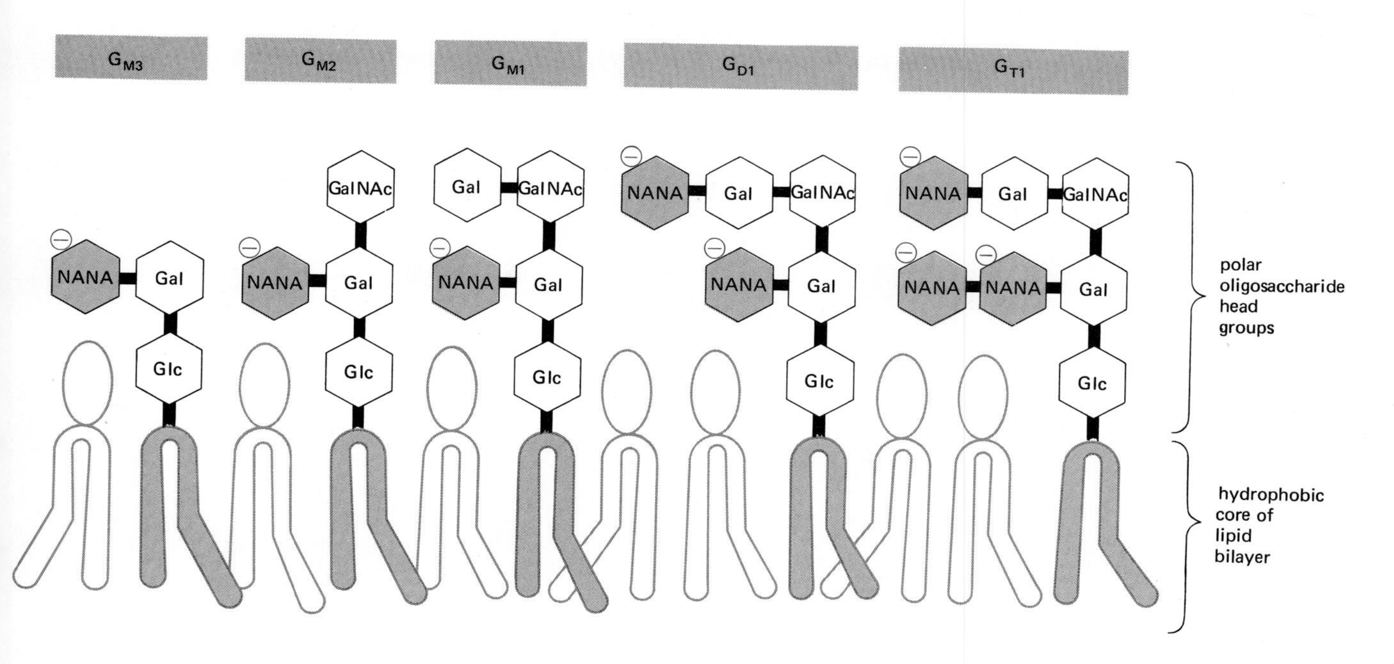

Figure 6–13 Some representative gangliosides with their standard designations. In G_{M1}, G_{M2}, G_{M3}, G_{D1}, and G_{T1}, the letters M, D, and T refer to the number of sialic acid residues (mono, di, and tri, respectively), while the number that follows the letter is determined by subtracting the number of uncharged sugar residues from 5. NANA = *N*-acetylneuraminic (sialic) acid; Gal = galactose; Glc = glucose; GalNAc = *N*-acetylgalactosamine. Gal, Glc, and GalNAc are all uncharged, whereas NANA carries a negative charge (see Figure 6–12).

monolayer to the other. Membrane lipid molecules are amphipathic and assemble spontaneously into bilayers when placed in water; the bilayers form sealed compartments that reseal if torn. There are three major classes of lipid molecules in the plasma membrane bilayer—phospholipids, cholesterol, and glycolipids—and the lipid compositions of the inner and outer monolayers are different. Different mixtures of lipids are found in the plasma membranes of cells of different types, as well as in the various internal membranes of a single eucaryotic cell. For the most part the functional significance of the different lipid compositions of different membranes is unknown.

Membrane Proteins

Although the basic structure of biological membranes is provided by the lipid bilayer, most of the specific functions are carried out by proteins. Accordingly, the amounts and types of proteins in a membrane are highly variable: in the myelin membrane, which serves mainly to insulate nerve cell axons, less than 25% of the membrane mass is protein, whereas in the membranes involved in energy transduction (such as the internal membranes of mitochondria and chloroplasts), approximately 75% is protein; the usual plasma membrane is somewhere in between, with about 50% of the mass being protein. Because lipid molecules are small in comparison to protein molecules, there are always many more lipid molecules than protein molecules in membranes—about 50 lipid molecules for each protein molecule in a membrane that is 50% protein by mass.

6-10 6-11 The Polypeptide Chain of Many Membrane Proteins Crosses the Lipid Bilayer One or More Times[8]

Many membrane proteins extend across the lipid bilayer (examples 1 and 2 in Figure 6–14). Like their lipid neighbors, these so-called **transmembrane proteins** are amphipathic: they have hydrophobic regions that pass through the membrane and interact with the hydrophobic tails of the lipid molecules in the interior of the bilayer and hydrophilic regions that are exposed to water on both sides of the membrane. The hydrophobicity of some of these membrane proteins is increased by the covalent attachment of a fatty acid chain that is inserted in the cytoplasmic leaflet of the bilayer (see example 1 in Figure 6–14). Some intracellular membrane proteins are associated with the bilayer only by means of such a fatty acid chain (see example 3 in Figure 6–14 and p. 417), while some cell-surface proteins are attached to the bilayer only by a covalent linkage (via a specific oligosaccharide) to phosphatidylinositol, a minor phospholipid, in the outer lipid monolayer of the plasma membrane (see example 4 in Figure 6–14 and p. 448).

Figure 6–14 Five ways in which membrane proteins can be associated with the lipid bilayer. Transmembrane proteins extend across the bilayer as a single α helix (1) or as multiple α helices (2); some of these "single-pass" and "multipass" proteins have a covalently attached fatty acid chain inserted in the cytoplasmic monolayer (1). Other membrane proteins are attached to the bilayer solely by a covalently attached lipid—either a fatty acid chain in the cytoplasmic monolayer (3) or, less often, via an oligosaccharide, to a minor phospholipid, phosphatidylinositol, in the noncytoplasmic monolayer (4). Finally, many proteins are attached to the membrane only by noncovalent interactions with other membrane proteins (5). The details of how membrane proteins become associated with the lipid bilayer in these ways are discussed in Chapter 8.

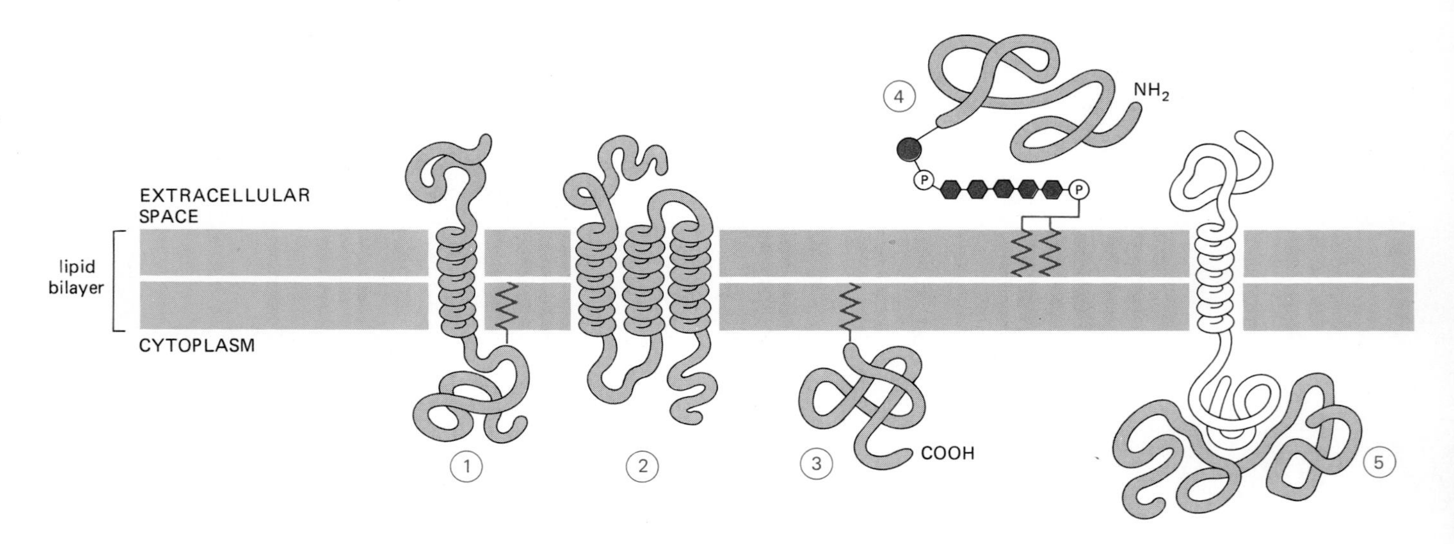

Figure 6–15 A segment of a transmembrane polypeptide chain crossing the lipid bilayer as an α helix, as determined by x-ray diffraction analyses of crystals of a membrane protein. Only the α-carbon backbone of the polypeptide chain is shown, and the hydrophobic amino acids are indicated in color. The protruding nonpolar side chains of the amino acid residues (not shown) interact with the hydrophobic fatty acid chains in the interior of the lipid bilayer, while the polar peptide bonds shield themselves from the hydrophobic environment of the bilayer by forming hydrogen bonds with one another (not shown). The polypeptide segment shown is part of the bacterial photosynthetic reaction center illustrated in Figure 6–32, p. 294. (Based on data from J. Deisenhofer *et al.*, *Nature* 318:618–624, 1985, and H. Michel *et al.*, *EMBO J.* 5:1149–1158, 1986.)

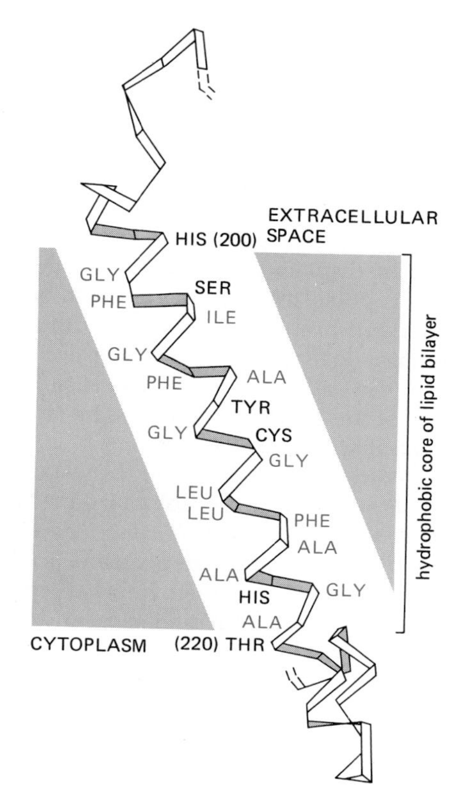

Other proteins associated with membranes do not extend into the hydrophobic interior of the lipid bilayer at all but are bound to one or other face of the membrane by noncovalent interactions with other membrane proteins (see example 5 in Figure 6–14). Many of these can be released from the membrane by relatively gentle extraction procedures, such as exposure to solutions of very high or low ionic strength or extreme pH, which interfere with protein-protein interactions but leave the lipid bilayer intact; these proteins are referred to operationally as **peripheral membrane proteins.** By contrast, transmembrane proteins, proteins linked to phosphatidylinositol, and some proteins held in the bilayer by a fatty acid chain, as well as some other tightly bound proteins, can be released only by disrupting the bilayer with detergents or organic solvents and are called **integral membrane proteins.**

In transmembrane proteins the parts of the polypeptide chain that are buried in the hydrophobic environment of the lipid bilayer are composed largely of amino acid residues with nonpolar side chains. But the peptide bonds themselves are polar, and because water is absent, all peptide bonds in the bilayer are driven to form hydrogen bonds with one another (see Figure 3–26, p. 110). The hydrogen bonding between peptide bonds is maximized if the polypeptide chain forms a regular α helix as it crosses the bilayer, and this is how most polypeptide chains traverse the membrane (Figure 6–15). In those cases where multiple strands of the polypeptide chain cross the bilayer, peptide bonds could, in principle, satisfy their hydrogen-bonding requirements if the strands were arranged as a β sheet (see p. 109). However, in such "multipass" membrane proteins the polypeptide chain usually crosses the bilayer as a series of α helices rather than as a β sheet (see example 2 in Figure 6–14). The strong drive to maximize hydrogen bonding in the absence of water also means that a polypeptide chain that enters the bilayer is likely to pass entirely through it before changing direction, since chain bending requires a loss of regular hydrogen-bonding interactions. Probably for this reason, there is still no established example of a membrane protein in which the polypeptide chain extends only partway across the lipid bilayer.

A transmembrane protein always has a unique orientation in the membrane. This reflects both the asymmetrical manner in which it is synthesized and inserted into the lipid bilayer in the endoplasmic reticulum (see p. 441) and the different functions of its cytoplasmic and extracellular domains. The great majority of transmembrane proteins are glycosylated. As in the case of glycolipids, the oligosaccharide chains are always present on the extracellular side of the membrane, since the sugar residues are added in the lumen of the endoplasmic reticulum and Golgi apparatus (see p. 454). A further asymmetry involves protein sulfhydryl (SH) groups, which remain reduced as cysteines in cytoplasmic domains but often participate in the formation of inter- or intrachain disulfide (S—S) bonds in extracellular domains (Figure 6–16).

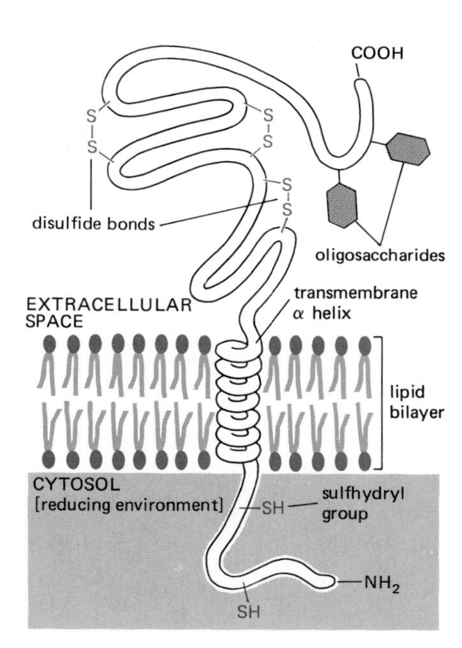

Figure 6–16 A typical "single-pass" transmembrane glycoprotein. Note that the polypeptide chain traverses the lipid bilayer as a right-handed α helix and that the oligosaccharide chains and disulfide bonds are all on the outside of the cell. Disulfide bonds do not form between the sulfhydryl groups on the cytoplasmic domain of the protein because of the reducing environment in the cytosol (see p. 445).

Membrane Proteins Can Be Solubilized and Purified in Detergents[9]

In general, transmembrane proteins (and some other tightly bound membrane proteins) can be solubilized only by agents that disrupt hydrophobic associations and destroy the bilayer. The most useful among these for the membrane biochemist are **detergents,** which are small amphipathic molecules that tend to form micelles in water (Figure 6–17). When mixed with membranes, the hydrophobic ends of detergents bind to the hydrophobic regions of the membrane proteins, thereby displacing the lipid molecules. Since the other end of the detergent molecule is polar, this binding tends to bring the membrane proteins into solution as detergent-protein complexes (although some tightly bound lipid molecules also remain) (Figure 6–18). The polar ends of detergents can be either charged (ionic), as in the case of *sodium dodecyl sulfate (SDS),* or uncharged (nonionic), as in the case of the *Triton* detergents. The structures of these two common detergents are illustrated in Figure 6–19.

With strong ionic detergents such as SDS, even the most hydrophobic membrane proteins can be solubilized. This allows them to be analyzed by *SDS polyacrylamide-gel electrophoresis* (see p. 169), a procedure that has revolutionized the study of membrane proteins. Such strong detergents unfold (denature) proteins by binding to their internal "hydrophobic cores" (see Figure 3–22, p. 107), thereby rendering them inactive and unusable for functional studies. Nonetheless, proteins can be readily purified in their SDS-denatured form, and in some cases the purified protein can be renatured, with recovery of functional activity, by removing the detergent.

Less hydrophobic membrane proteins can be solubilized by a low concentration of a mild detergent, enabling the solubilized protein to be purified in an active, if not entirely normal, form. If the detergent is then removed in the absence of phospholipid, the membrane protein molecules usually aggregate and precipitate

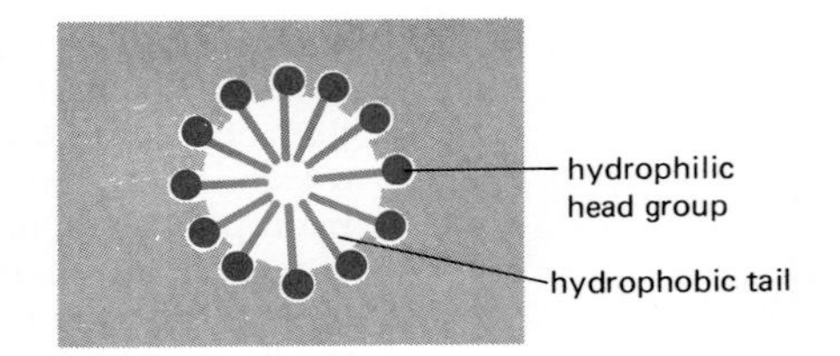

Figure 6–17 A detergent micelle in water, shown in cross-section. Because they have both polar and nonpolar ends, detergent molecules are amphipathic.

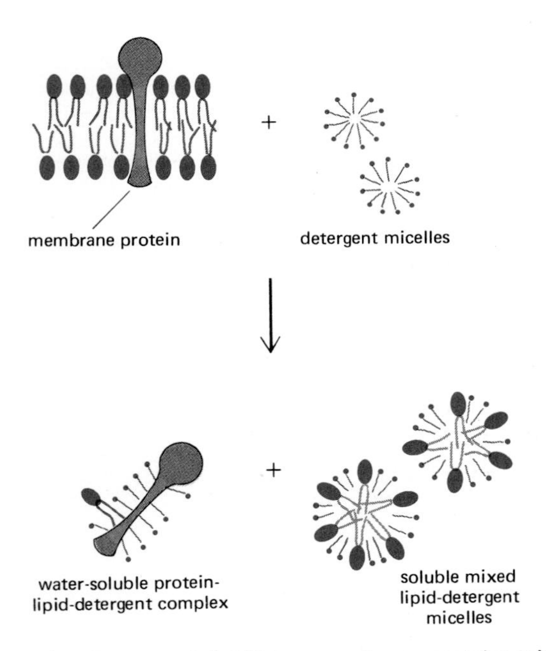

Figure 6–18 Solubilizing membrane proteins with detergent. The detergent disrupts the lipid bilayer and brings the proteins into solution as protein-lipid-detergent complexes. The phospholipids in the membrane are also solubilized by the detergent.

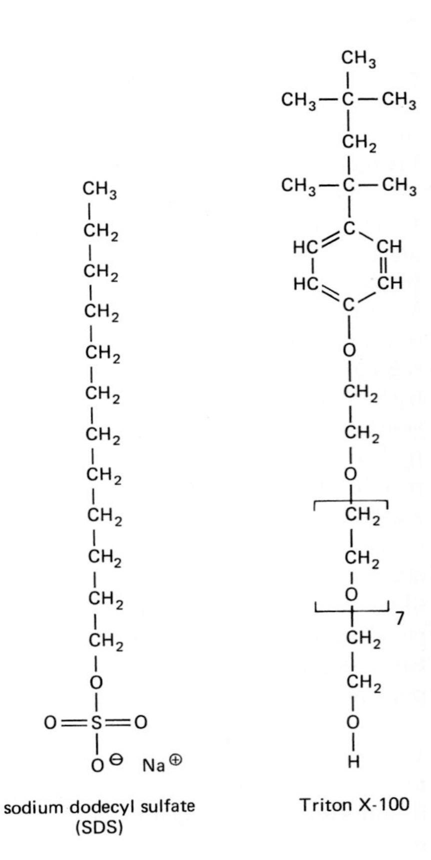

Figure 6–19 The structure of two commonly used detergents: sodium dodecyl sulfate (SDS), an anionic detergent, and Triton X-100, a nonionic detergent. Note that the bracketed portion of Triton X-100 is repeated seven times.

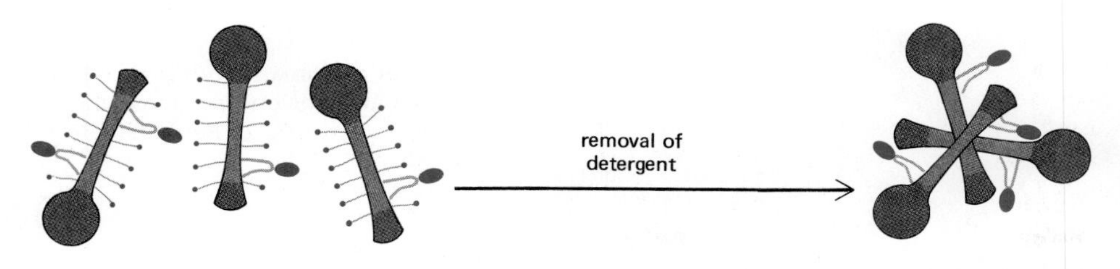

Figure 6–20 When the detergent is removed from detergent-solubilized membrane proteins, the relatively naked protein molecules (with only a few attached lipid molecules) tend to bury their hydrophobic regions by clustering together, forming large aggregates that precipitate from solution.

out of solution (Figure 6–20). If, however, the purified protein is mixed with phospholipids before the detergent is removed, the active protein will usually insert into the lipid bilayer formed by the phospholipids (Figure 6–21). In this way functionally active membrane protein systems can be reconstituted from purified components, providing a powerful means to analyze their activities. For example, if a purified protein can be shown to pump ions across a synthetic lipid bilayer in the absence of other proteins, then that protein is unequivocally an ion pump; moreover, by controlling the access of the protein to ATP and ions, its detailed mechanism of action can be dissected (see p. 304).

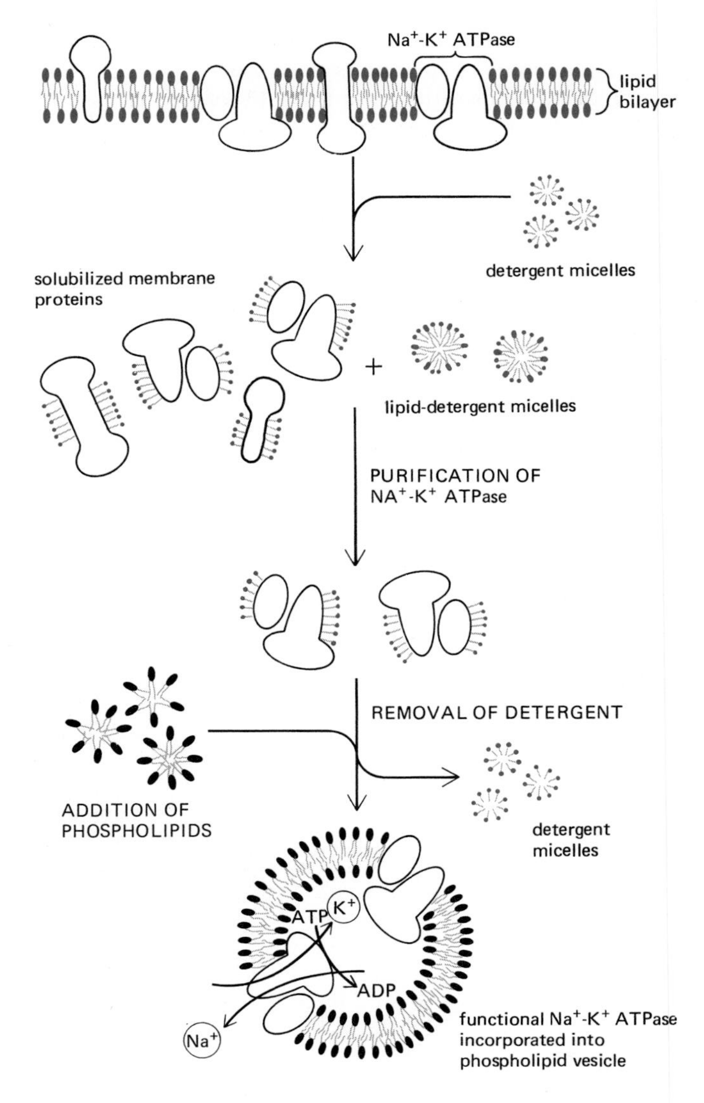

Figure 6–21 Solubilizing, purifying, and reconstituting functional Na^+-K^+ ATPase molecules into phospholipid vesicles. The Na^+-K^+ ATPase is an ion pump that is present in the plasma membrane of most animal cells; it uses the energy of ATP hydrolysis to pump Na^+ out of the cell and K^+ in. The reconstitution is carried out in the presence of high Na^+ and ATP so that these molecules end up in sufficiently high concentration inside the vesicles to allow the ATPase to function as a pump. The detergent is generally removed either by prolonged dialysis or by any of various forms of chromatography.

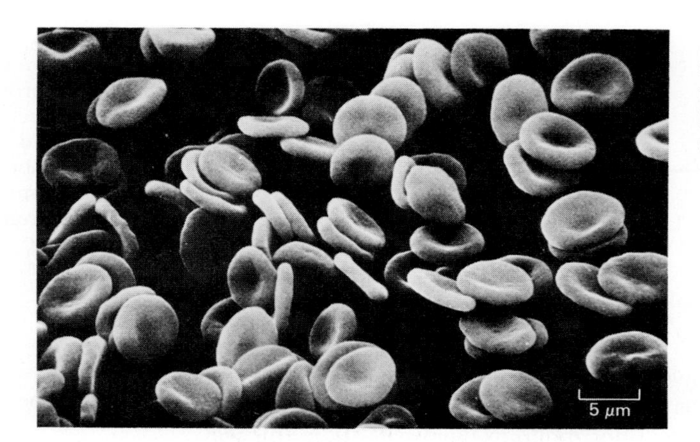

Figure 6–22 A scanning electron micrograph of human red blood cells. The cells have a biconcave shape and lack nuclei. (Courtesy of Bernadette Chailley.)

6-12 The Cytoplasmic Side of Membrane Proteins Can Be Studied in Red Blood Cell Ghosts[10]

More is known about the plasma membrane of the human red blood cell (Figure 6–22) than about any other eucaryotic membrane. There are a number of reasons for this: (1) Red blood cells are available in large numbers (from blood banks, for example) relatively uncontaminated by other cell types. (2) Since these cells have no nucleus or internal organelles, the plasma membrane is their only membrane, and it can be isolated without contamination by internal membranes (thus avoiding a serious problem encountered in plasma membrane preparations from other cell types, in which the plasma membrane typically constitutes less than 5% of the total membrane [see Table 8–2, p. 408]). (3) It is easy to prepare empty red blood cell membranes, or "ghosts," by exposing the cells to a hypotonic salt solution. Because the solution has a lower salt concentration than the cell interior, water flows into the red cells (see p. 308), causing them to swell and burst (lyse) and release their hemoglobin (the major nonmembrane protein). (4) Membrane ghosts can be studied while they are still leaky (in which case any reagent can interact with molecules on both faces of the membrane), or they can be allowed to reseal so that water-soluble reagents cannot reach the internal face. Moreover, since sealed *inside-out* vesicles can also be prepared from red blood cell ghosts (Figure 6–23), the external side and internal (cytoplasmic) side of the membrane can be studied separately. The use of sealed and unsealed red cell ghosts first made it possible to demonstrate that some membrane proteins extend all the way through the lipid bilayer (see below) and that the lipid compositions of the two halves of the bilayer are different. Like most of the basic principles initially demonstrated in red blood cell membranes, these findings have since been extended to the membranes of nucleated cells.

The "sidedness" of a membrane protein can be determined in several ways. One is to use a covalent labeling reagent (for example, one carrying a radioactive or fluorescent marker) that is water soluble and therefore cannot penetrate the lipid bilayer; it attaches covalently to specific groups only on the exposed side of

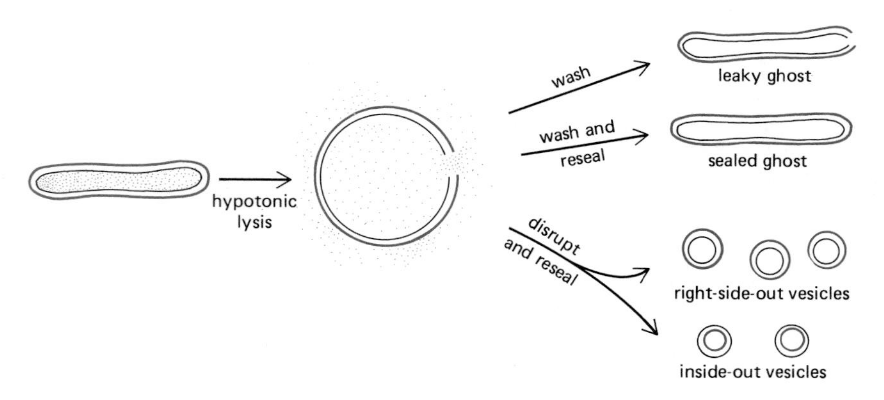

Figure 6–23 The preparation of sealed and unsealed red blood cell ghosts and of right-side-out and inside-out vesicles. As indicated, there is evidence that the red cells rupture in only one place, giving rise to ghosts with a single hole in them. The smaller vesicles are produced by mechanically disrupting the ghosts; the orientation of the membrane in these vesicles can be either right-side-out or inside-out, depending on the ionic conditions used during the disruption procedure.

the membrane. The membranes are then solubilized, the proteins separated by SDS polyacrylamide-gel electrophoresis, and the labeled proteins detected either by their radioactivity (by autoradiography of the gel) or by their fluorescence (by exposing the gel to ultraviolet light). By using such *vectorial labeling* it is possible to determine how a particular protein, detected as a band on a gel, is oriented in the membrane: if it is labeled from both the external side (when intact cells or sealed ghosts are labeled) and the internal (cytoplasmic) side (when sealed inside-out vesicles are labeled), then it must be a transmembrane protein. An alternative approach is to expose either the external or internal surface to membrane-impermeant proteolytic enzymes: if a protein is partially digested from both surfaces, it must be a transmembrane protein. In addition, labeled antibodies can be used to determine if a specific part of a transmembrane protein is exposed on one side of the membrane or the other.

When the plasma membrane proteins of the human red blood cell are studied by SDS polyacrylamide-gel electrophoresis, approximately 15 major protein bands are detected, varying in molecular weight from 15,000 to 250,000. Three of these proteins—*spectrin, glycophorin,* and *band 3*—account for more than 60% (by weight) of the total membrane protein (Figure 6–24). Each of these proteins is arranged in the membrane in a different manner. We shall, therefore, use them as examples of three major ways that proteins are associated with membranes.

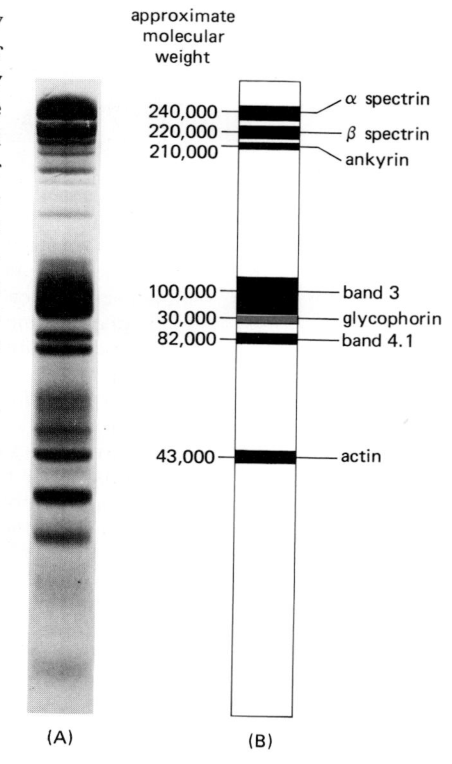

Figure 6–24 SDS polyacrylamide-gel electrophoresis pattern of the proteins in the human red blood cell membrane stained with Coomassie blue (A). The positions of some of the major proteins in the gel are indicated in the drawing in (B); glycophorin is shown in color to distinguish it from band 3. Other bands in the gel are omitted from the drawing. The large amount of carbohydrate in glycophorin molecules slows their migration so that they run almost as slowly as the much larger band 3 molecules. (A, courtesy of Ted Steck.)

6-14 Spectrin Is a Cytoskeletal Protein Noncovalently Associated with the Cytoplasmic Side of the Red Blood Cell Membrane[11]

Most of the protein molecules associated with the human red blood cell membrane are peripheral membrane proteins associated with the cytoplasmic side of the lipid bilayer. The most abundant of these proteins is **spectrin,** a long, thin, flexible rod about 100 nm in length that constitutes about 25% of the membrane-associated protein mass (about 2.5×10^5 copies per cell). It is the principal component of the protein meshwork (the *cytoskeleton*) that underlies the red blood cell membrane, maintaining the structural integrity and biconcave shape of this membrane (see Figure 6–22); if the cytoskeleton is extracted from red blood cell ghosts in low-ionic-strength solutions, the membrane fragments into small vesicles.

Spectrin is composed of two very large polypeptide chains, α spectrin (~240,000 daltons) and β spectrin (~220,000 daltons). Each chain is thought to be made up of many α-helical segments interwound in groups of three and connected by nonhelical regions (Figure 6–25). The spectrin heterodimers self-associate head-to-head to form 200-nm-long tetramers. The tail ends of five or six tetramers are linked together by binding to short actin filaments and to another protein (*band 4.1*) in a "junctional complex," forming a deformable, netlike meshwork that underlies the entire cytoplasmic surface of the membrane (Figure 6–26). It is this spectrin-based cytoskeleton that enables the red cell to withstand the stress on its membrane as it is forced through narrow capillaries. Anemic mice and humans with a genetic abnormality of spectrin have red cells that are spherical (instead of concave) and abnormally fragile; the severity of the anemia increases with the degree of the spectrin deficiency.

The protein mainly responsible for attaching the spectrin cytoskeleton to the red cell plasma membrane was identified by binding radiolabeled spectrin to red cell membranes from which spectrin and various other peripheral proteins had been removed. These experiments showed that the rebinding of spectrin depends on a large intracellular attachment protein called **ankyrin,** which binds both to β spectrin and to the cytoplasmic domain of the transmembrane protein band 3 (see Figure 6–26). By connecting band 3 protein to spectrin, ankyrin links the spectrin network to the membrane; it also greatly reduces the rate of diffusion of the band 3 molecules in the lipid bilayer. The spectrin-based cytoskeleton may also be attached to the membrane by a second mechanism: the cytoskeletal protein band 4.1—which binds to spectrin and actin—has been shown to bind to the cytoplasmic domain of glycophorin, the other major transmembrane protein in red blood cells.

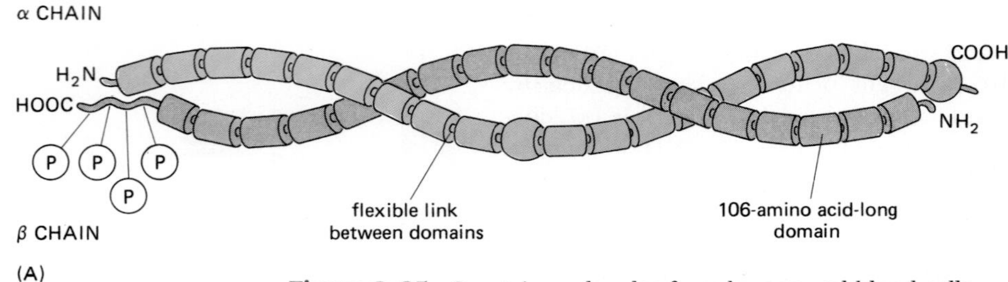

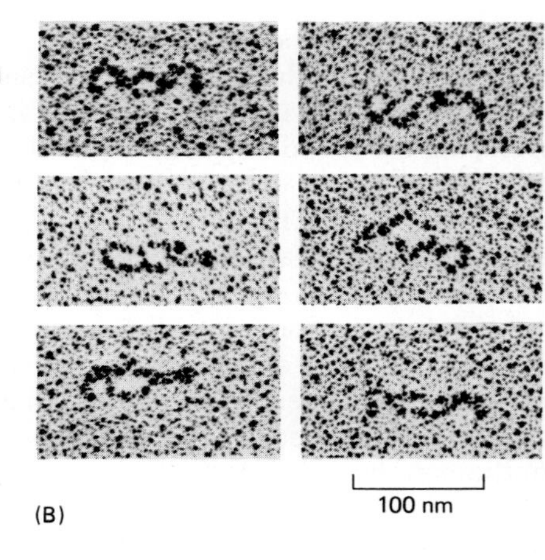

Figure 6–25 Spectrin molecules from human red blood cells shown schematically in (A) and in electron micrographs in (B). Each spectrin heterodimer consists of two antiparallel, loosely intertwined, flexible polypeptide chains; these are attached noncovalently to each other at multiple points, including both ends. The phosphorylated "head" end, where two dimers associate to form a tetramer, is on the left. Both the α and β chains are composed largely of repeating domains 106 amino acids long. It has been proposed that each domain is organized into groups of three α helices (not shown) that are linked by nonhelical regions. In (B) the spectrin molecules have been shadowed with platinum. (A, adapted from D.W. Speicher and V.T. Marchesi, *Nature* 311:177–180, 1984; B, courtesy of D.M. Shotton, with permission from D.M. Shotton, B.E. Burke, and D. Branton, *J. Mol. Biol.* 131:303–329, 1979, © Academic Press Inc. [London] Ltd.)

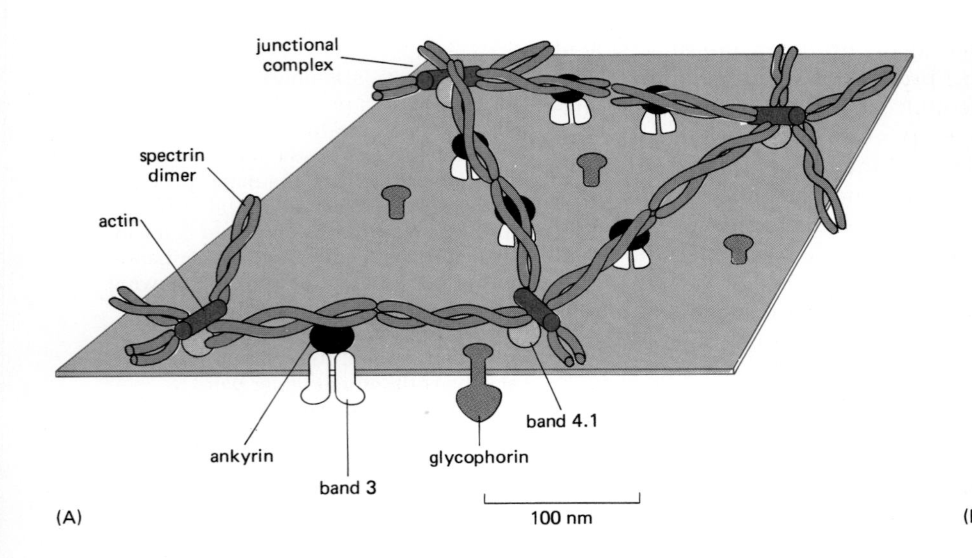

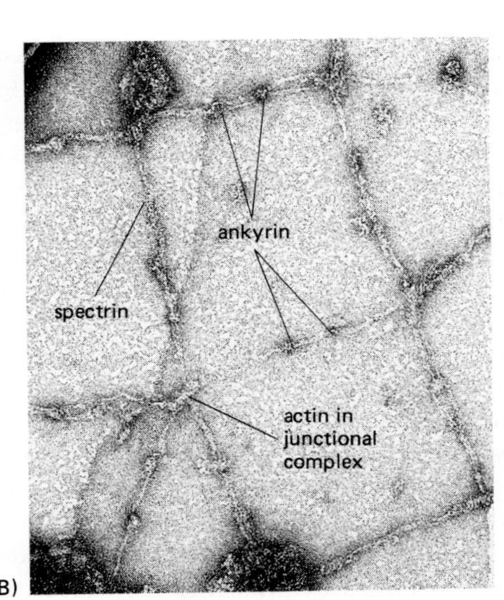

Figure 6–26 Schematic drawing (A) and electron micrograph (B) of the spectrin-based cytoskeleton on the cytoplasmic side of the human red blood cell membrane. The arrangement shown in (A) has been deduced mainly from studies on the interactions of purified proteins *in vitro*. Spectrin dimers associate head-to-head to form tetramers that are linked together into a netlike meshwork by junctional complexes composed of short actin filaments (containing about 15 actin monomers) and band 4.1 (and two or three other proteins that are not shown). This cytoskeleton is linked to the membrane by the indirect binding of spectrin tetramers to band 3 proteins via ankyrin molecules and may also be linked by the binding of band 4.1 proteins to glycophorin (not shown). The electron micrograph in (B) shows the cytoskeleton on the cytoplasmic side of a red blood cell membrane after fixation and negative staining. The spectrin meshwork has been purposely stretched out to allow the details of its structure to be seen; in the normal cell the meshwork shown would occupy only about one-tenth of this area. (Courtesy of T. Byers and D. Branton, *Proc. Natl. Acad. Sci. USA* 82:6153–6157, 1985.)

An analogous but more elaborate and complicated cytoskeletal network exists beneath the plasma membrane of nucleated cells. This network, which constitutes the cortical region (or *cortex*) of the cytoplasm, is rich in actin filaments that are thought to be attached to the plasma membrane in numerous ways (see p. 632). Proteins that are structurally homologous to spectrin, ankyrin, and band 4.1 are present in the cortex, but their organization and functions are as yet poorly understood (see p. 641).

Glycophorin Extends Through the Red Blood Cell Lipid Bilayer as a Single α Helix[12]

Glycophorin is one of the two major proteins exposed on the outer surface of the human red blood cell and was the first membrane protein for which the complete amino acid sequence was determined. It is a small transmembrane glycoprotein (131 amino acid residues) with most of its mass on the external surface of the membrane, where its hydrophilic amino-terminal end is located. This part of the protein carries all of the carbohydrate (about 100 sugar residues on 16 separate oligosaccharide side chains), which accounts for 60% of the molecule's mass. In fact, the great majority of the total red blood cell surface carbohydrate (including more than 90% of the sialic acid and, therefore, most of the negative charge of the surface) is carried by glycophorin molecules. The hydrophilic carboxyl-terminal tail of glycophorin is exposed to the cytosol, while a hydrophobic α-helical segment 23 amino acids long spans the lipid bilayer.

Despite there being more than 6×10^5 glycophorin molecules per cell, their function remains unknown. Indeed, individuals whose red cells lack a major subset of these molecules appear to be perfectly healthy. Although glycophorin itself is found only in red blood cells, its structure is representative of a common class of transmembrane glycoproteins that traverse the lipid bilayer as a single α helix—so-called *single-pass membrane proteins* (see example 1 in Figure 6–14 and Figure 6–16). A variety of cell-surface receptors (see p. 706), for example, belongs to this class.

6-13 Band 3 of the Human Red Blood Cell Membrane Is an Anion Transport Protein[13]

Unlike glycophorin, the **band 3 protein** is known to play an important part in cell function. It derives its name from its position relative to the other membrane proteins after electrophoresis in SDS polyacrylamide gels (see Figure 6–24). Like glycophorin, band 3 is a transmembrane protein, but it is a *multipass membrane protein,* traversing the membrane in a highly folded conformation: the polypeptide chain (about 930 amino acids long) extends across the bilayer at least 10 times. Each red blood cell contains about 10^6 band 3 polypeptide chains, which are thought to form dimers and possibly tetramers in the membrane.

The main function of red blood cells is to carry O_2 from the lungs to the tissues and CO_2 from the tissues to the lungs. Band 3 protein is instrumental in this exchange. Red cells dispose of the CO_2 they accumulate in the tissues by ejecting HCO_3^- in exchange for Cl^- as they move through the lungs. This exchange takes place through an anion transport protein and can be blocked by specific inhibitors that bind to the transport protein. By modifying the inhibitors so that they radioactively label the protein to which they bind, the anion transport protein has been identified as band 3. More recently, anion transport has been reconstituted *in vitro* using purified band 3 protein incorporated into phospholipid vesicles. A closely related anion transporter is found in many nucleated cells, where it helps to control intracellular pH (see p. 309).

Band 3 proteins can be seen as distinct *intramembrane particles* by the technique of **freeze-fracture electron microscopy,** in which cells are frozen in liquid nitrogen and the resulting block of ice is fractured with a knife (see p. 147). The fracture plane tends to pass through the hydrophobic middle of membrane lipid bilayers, separating them into their two monolayers. The exposed *fracture faces* are then shadowed with platinum, and the resulting platinum replica is examined with an electron microscope. As illustrated in Figure 6–27, two different fracture

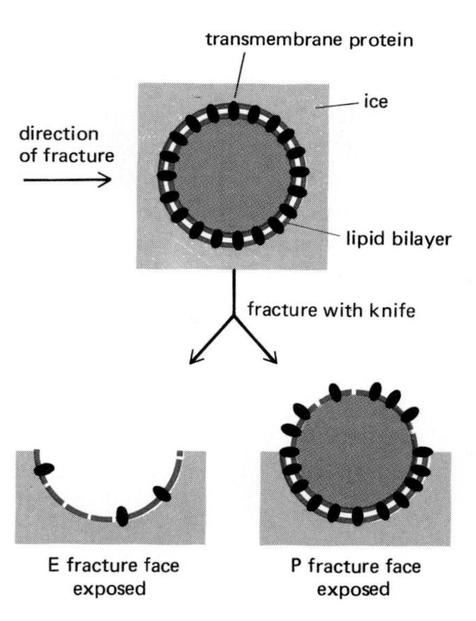

Figure 6–27 How freeze-fracture electron microscopy provides images of the hydrophobic interior of the cytoplasmic (or protoplasmic) half of the bilayer (called the P face) and the external half of the bilayer (called the E face). After the fracturing process shown here, the exposed fracture faces are shadowed with platinum and carbon, the organic material is digested away, and the resulting platinum replica is examined in the electron microscope (see also Figure 4–23, p. 148).

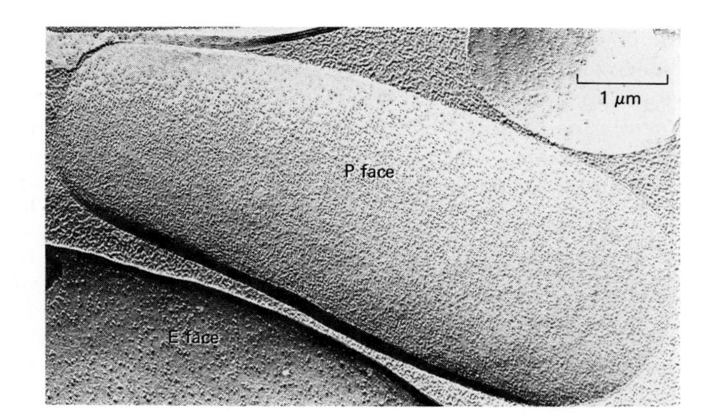

Figure 6–28 Freeze-fracture electron micrograph of human red blood cells. Note that the density of intramembrane particles on the protoplasmic (P) face is higher than on the external (E) face. (Courtesy of L. Engstrom and D. Branton.)

faces are exposed and replicated in this technique—the face representing the hydrophobic interior of the cytoplasmic (or protoplasmic) half of the bilayer (called the **P face**) and the face representing the hydrophobic interior of the external half of the bilayer (called the **E face**). When examined in this way, human red blood cell membranes are studded with intramembrane particles that are relatively homogeneous in size (7.5 nm in diameter), randomly distributed, and more concentrated on the P face than on the E face (Figure 6–28). These are thought to be principally band 3 molecules: when synthetic lipid bilayers are reconstituted with purified band 3 protein molecules, typical 7.5-nm intramembrane particles are observed when the bilayers are fractured. Figure 6–29 illustrates why band 3 molecules are seen in freeze-fracture electron microscopy of red blood cell membranes but glycophorin molecules probably are not.

In a general way, it is not difficult to imagine how a transmembrane protein such as band 3, with much of its mass in the lipid bilayer, could mediate the passive transport of polar molecules across the nonpolar bilayer. The band 3 protein (or its dimer or tetramer) could provide a transmembrane hydrophilic pathway along which the Cl^- and HCO_3^- ions are transported without having to make contact with the hydrophobic interior of the lipid bilayer (see Figure 6–43A, p. 302). It is difficult to see how a molecule such as glycophorin, which spans the bilayer as a single α helix, could mediate this type of transport on its own.

An understanding of how membrane transport proteins work requires precise information about their three-dimensional structure in the bilayer. The first plasma membrane transport protein for which such detail became known was *bacteriorhodopsin,* a protein that serves as a light-activated proton (H^+) pump in the

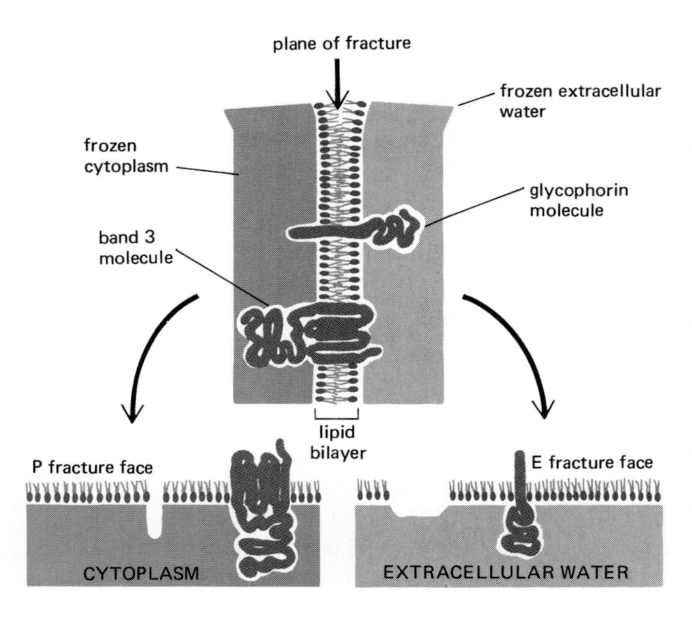

Figure 6–29 Probable fates of band 3 and glycophorin molecules in the human red blood cell membrane during freeze-fracture. When the lipid bilayer is split, either the inside or outside half of each transmembrane protein is pulled out of the frozen monolayer with which it is associated; the protein tends to remain with the monolayer that contains the main bulk of the protein. For this reason band 3 molecules usually remain with the inner (P) fracture face; since they have sufficient mass above the fracture plane, they are readily seen as intramembrane particles. Glycophorin molecules usually remain with the outer (E) fracture face, but it is thought that their cytoplasmic tails have insufficient mass to be seen.

plasma membrane of certain bacteria. The structure of bacteriorhodopsin is similar to that of various other membrane proteins (see p. 705), and it merits a brief digression here.

Figure 6–30 Schematic drawing of the bacterium *Halobacterium halobium* showing the patches of purple membrane that contain bacteriorhodopsin molecules. These bacteria, which live in saltwater pools where they are exposed to a large amount of sunlight, have evolved a variety of light-activated proteins, including bacteriorhodopsin—a light-activated proton pump in their plasma membrane.

Bacteriorhodopsin Is a Proton Pump That Traverses the Bilayer as Seven α Helices[14]

The "purple membrane" of the bacterium *Halobacterium halobium* is a specialized patch in the plasma membrane (Figure 6–30) that contains a single species of protein molecule, **bacteriorhodopsin.** Each bacteriorhodopsin molecule contains a single light-absorbing prosthetic group, or chromophore (called *retinal*), which is related to vitamin A and is identical to the chromophore found in rhodopsin of the vertebrate retinal rod cell (see p. 1106). Retinal is covalently linked to a lysine side chain of the protein; when it is activated by a single photon of light, the excited chromophore causes a conformational change in the protein that results in the transfer of one or two H^+ from the inside to the outside of the cell. This transfer establishes a H^+ and voltage gradient across the plasma membrane, which in turn drives the production of ATP by a second protein in the cell's plasma membrane.

Because the numerous bacteriorhodopsin molecules in the cell membrane are arranged in a planar crystalline lattice (like a two-dimensional crystal), it has been possible to determine the three-dimensional structure and orientation of bacteriorhodopsin in the membrane to a resolution of 0.7 nm by a combination of low-intensity electron microscopy and low-angle electron diffraction analysis. The latter procedure is analogous to the study of three-dimensional crystals of soluble proteins by x-ray diffraction analysis, although less structural detail can be obtained. As illustrated in Figure 6–31, these studies have shown that each bacteriorhodopsin molecule is folded into seven closely packed α helices (each containing about 25 amino acids), which pass roughly at right angles through the lipid bilayer. It seems likely that the protons are passed by the chromophore along a relay system set up by the side chains of the α helices; however, the molecular details are unknown.

Bacteriorhodopsin is a member of a family of membrane proteins with similar structures but different functions. For example, the light receptor protein *rhodopsin* in rod cells of the vertebrate retina and a number of cell-surface receptor proteins that bind specific hormones are also folded into seven transmembrane α helices (see p. 705). These proteins function as signal transducers rather than as transporters; each responds to an extracellular signal by activating another plasma membrane protein, which generates a chemical signal in the cytosol (see p. 685).

To understand the function of bacteriorhodopsin in molecular detail, it will be necessary to locate each of its atoms precisely, which will require x-ray diffraction studies of crystals of the protein. Because of their amphipathic nature, membrane proteins are extremely difficult to crystallize, and it was only in 1985 that the first one—a bacterial photosynthetic reaction center—was successfully studied by x-ray diffraction techniques. The results of these studies are of general importance to membrane biology because they show for the first time how multiple polypeptides can associate in a membrane to form a complex protein machine.

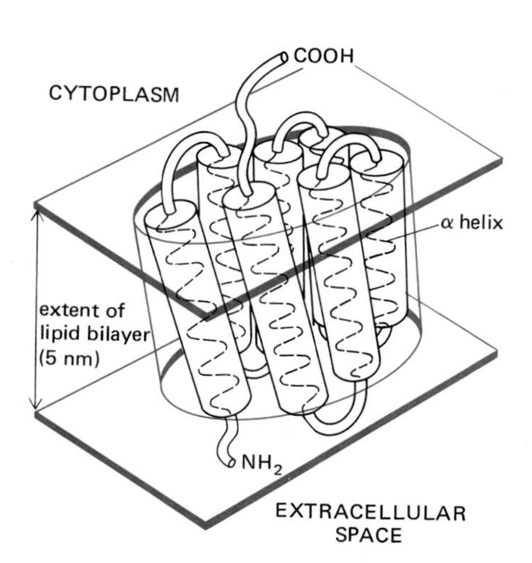

Figure 6–31 The structure of one bacteriorhodopsin molecule and its relationship to the lipid bilayer. The polypeptide chain crosses the bilayer as seven α helices. (Based on data from R. Henderson and P.N.T. Unwin, *Nature* 257:28–32, 1975.)

Four Different Polypeptide Chains in a Membrane-bound Complex Form a Bacterial Photosynthetic Reaction Center[15]

In Chapter 3 we discussed how different polypeptides associate to form large multienzyme assemblies that can carry out complex reactions with great efficiency because the subunits cooperate (see p. 126). Similar protein complexes are found in membranes, and the first of these to be understood in detail is a bacterial **photosynthetic reaction center.** This protein complex is located in the plasma membrane of the purple photosynthetic bacterium *Rhodopseudomonas viridis*, where it uses captured light energy to create a high-energy electron that it transfers across the membrane in less than a nanosecond. The electron is then passed to other electron carriers in the membrane, which use some of the energy released by the electron-transport process to synthesize ATP in the cytosol. The reaction center is a tetramer composed of four different polypeptides: L, M, H, and a cytochrome. It has been solubilized in detergent, crystallized as a protein-detergent complex, and analyzed by x-ray diffraction to obtain its complete three-dimensional structure.

The protein complex contains four molecules of chlorophyll and eight molecules of other coenzymes that carry electrons. The determination of the precise location of each of these coenzymes was a major advance in our understanding of photosynthesis, as discussed in Chapter 7 (see p. 375). Equally important, and more relevant to our present discussion, was the determination of the precise organization of the four protein subunits in the transmembrane protein complex. The L and M subunits are homologous, each containing five α helices that traverse the lipid bilayer of the plasma membrane (Figure 6–32). These two subunits form a heterodimer that is the core of the reaction center, its 10 α helices surrounding the electron carriers. The H subunit has a single transmembrane α helix, while the rest of the polypeptide chain is folded into a globular domain that protrudes on the cytoplasmic face of the membrane where it binds to the L-M heterodimer. The cytochrome is a peripheral membrane protein that is bound to the L-M heterodimer on the noncytoplasmic side of the membrane (see Figure 6–32).

The two "extra" subunits greatly increase the efficiency of the photosynthetic reaction catalyzed by the L-M heterodimer: the cytochrome keeps the heterodimer supplied with electrons, while the H subunit is believed to couple the reaction center to an array of light-harvesting proteins in the cell interior. The L-M heterodimer has been highly conserved during evolution, and a closely related pair of proteins is thought to form the core of one of the photosynthetic reaction centers in green plants (see p. 377).

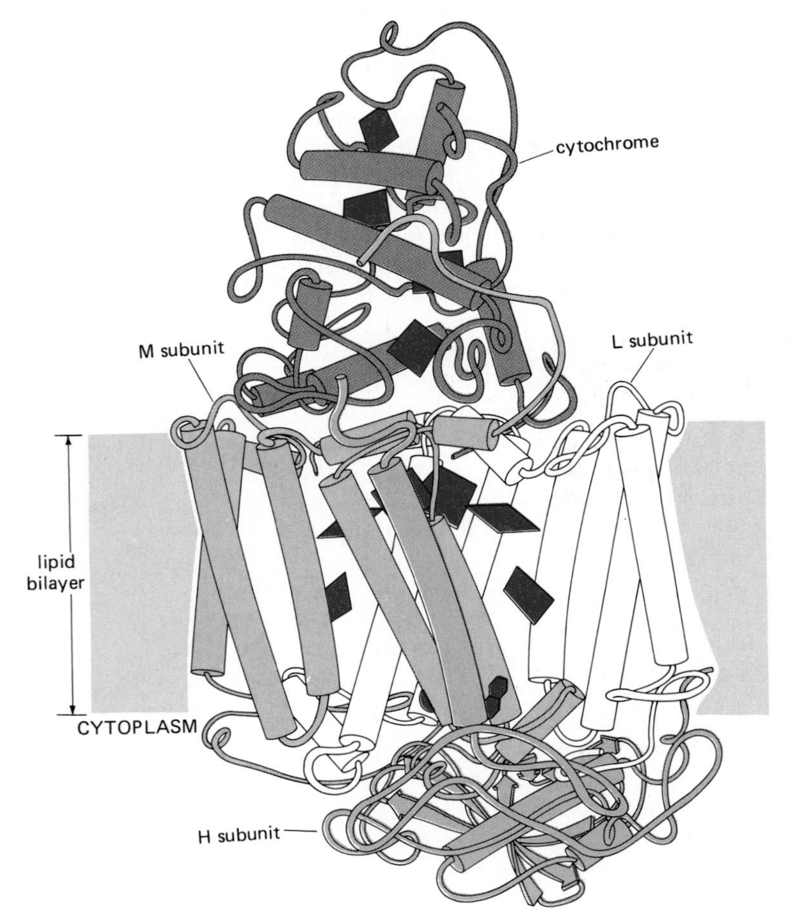

Figure 6–32 The structure of the photosynthetic reaction center of the bacterium *Rhodopseudomonas viridis* as determined by x-ray diffraction analysis of crystals of this transmembrane protein complex. The protein complex consists of four subunits, L, M, H, and a cytochrome: the L and M subunits form the core of the reaction center, and each contains five α helices that span the lipid bilayer. The locations of the electron carrier coenzymes are shown in solid color. (Adapted from a drawing by J. Richardson based on data from J. Deisenhofer, O. Epp, K. Miki, R. Huber, and H. Michel, *Nature* 318:618–624, 1985.)

6-15 Many Membrane Proteins Diffuse in the Plane of the Membrane[16]

Like membrane lipids, membrane proteins do not tumble (*flip-flop*) across the bilayer, and they rotate about an axis perpendicular to the plane of the bilayer (*rotational diffusion*). In addition, many membrane proteins are able to move laterally within the membrane (*lateral diffusion*). The first direct evidence that some plasma membrane proteins are mobile in the plane of the membrane was provided in 1970 by an experiment using hybrid cells (*heterocaryons*, see p. 162) that were artificially produced by fusing mouse cells with human cells. Two differently labeled antibodies were used to distinguish selected mouse and human plasma membrane proteins. Although at first the mouse and human proteins were confined to their own halves of the newly formed heterocaryon, the two sets of proteins diffused and mixed over the entire cell surface within half an hour or so (Figure 6–33). Further evidence for membrane protein mobility was soon provided by the discovery of a process called *patching* (see p. 333): when antibodies bind to specific proteins on the surface of cells, the proteins tend to become cross-linked into large clusters, indicating that the proteins are able to move laterally in the lipid bilayer.

The lateral diffusion rates of membrane proteins can be quantitated using the technique of *fluorescence recovery after photobleaching (FRAP)*. This method was first used to study the diffusion of individual rhodopsin molecules in the disc membranes of vertebrate rod cells. As we have seen, rhodopsin has a structure similar to bacteriorhodopsin, and it contains the same chromophore (retinal). The

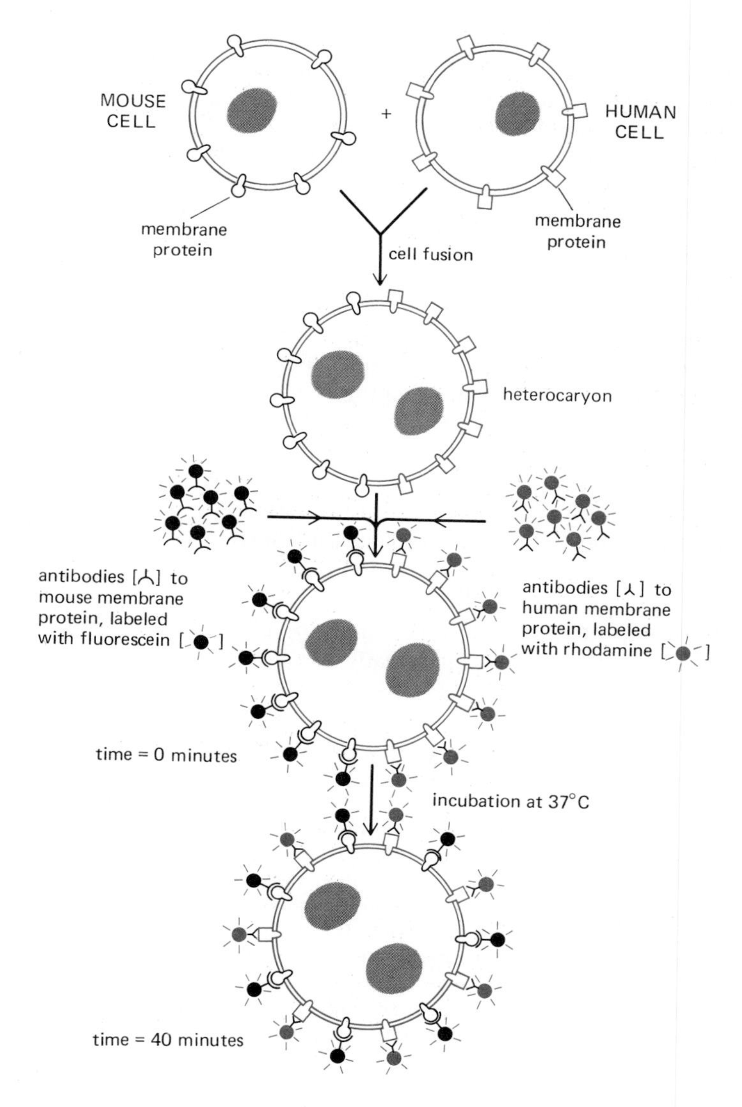

Figure 6–33 The experiment demonstrating the mixing of plasma membrane proteins on mouse-human hybrid cells. The mouse and human proteins are initially confined to their own halves of the newly formed heterocaryon plasma membrane, but they intermix with time. The two antibodies used to visualize the proteins can be distinguished in a fluorescence microscope because fluorescein is green whereas rhodamine is red. (Based on observations of L.D. Frye and M. Edidin, *J. Cell Sci.* 7:319–335, 1970, by permission of The Company of Biologists.)

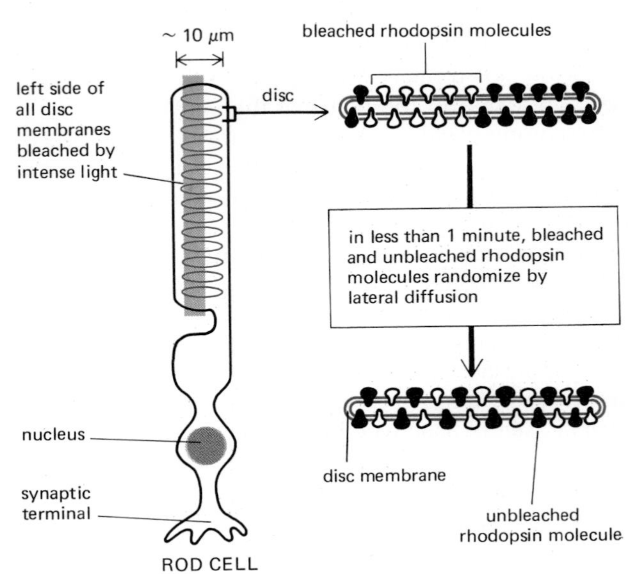

Figure 6–34 Measuring the rate of lateral diffusion of rhodopsin molecules in the disc membranes of a retinal photoreceptor (rod) cell by the FRAP technique. After the retinal chromophores in the rhodopsin molecules are bleached on one side of the cell, the rate at which the bleached and unbleached rhodopsin molecules mix by diffusion is measured. (Based on observations of M. Poo and R.A. Cone, *Nature* 247:438–441, 1974.)

diffusion of the rhodopsin molecules can be measured by bleaching the chromophore in the rhodopsin molecules on one side of a photoreceptor cell with an intense, highly focused beam of light and measuring the time taken for the bleached and unbleached rhodopsin molecules to mix by diffusion (Figure 6–34). The rate of diffusion, measured as a *diffusion coefficient (D)*, is about 5×10^{-9} cm^2/sec; this is only half the diffusion coefficient of a phospholipid molecule in a membrane (see p. 278) and is the highest diffusion coefficient of any membrane protein known.

The same technique can be used to study membrane proteins that do not contain chromophores by first attaching fluorescent ligands to them. Most commonly these are fluorescent monovalent antibodies (fragments of antibodies that have only one antigen-binding site and therefore cannot cross-link neighboring molecules—see p. 1013). The tightly bound ligands are then bleached in a small area by a laser beam, and the time taken for adjacent membrane proteins carrying unbleached fluorescent antibody molecules to diffuse into the bleached area is measured (Figure 6–35). The diffusion rates of various plasma membrane glycoproteins measured in this way have usually been found to be at least 5 to 50 times slower than those of rhodopsin molecules. These relatively slow diffusion rates are not an intrinsic property of the individual glycoprotein molecules themselves, as the same glycoproteins diffuse much more rapidly when inserted into synthetic lipid bilayers. The reasons for the slow diffusion rates measured by the FRAP technique for glycoproteins in plasma membranes are uncertain. One possibility is that the bulky oligosaccharide chains on the extracellular domains of such molecules interact with those on other glycoproteins in the membrane to slow diffusion: in at least some cases the removal of the carbohydrate greatly increases the diffusion rate of the protein.

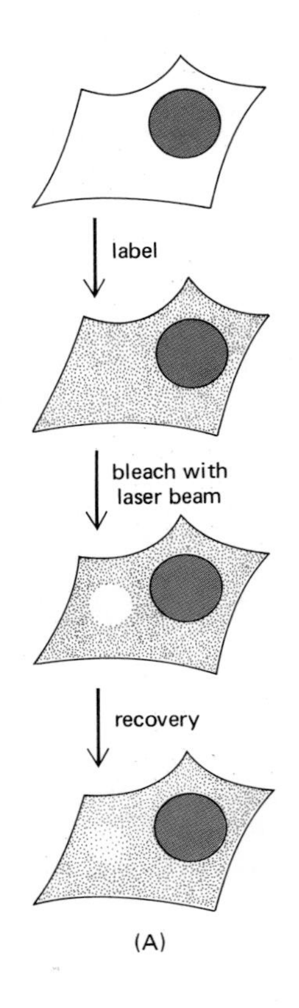

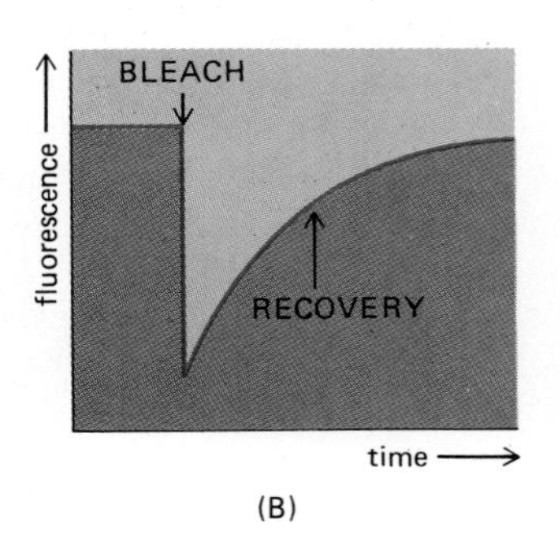

Figure 6–35 Measuring the rate of lateral diffusion of a plasma membrane glycoprotein by the FRAP technique. (A) A specific glycoprotein is labeled on the cell surface with a fluorescent monovalent antibody that binds only to that protein. After the antibodies are bleached in a small area using a laser beam, the fluorescence intensity recovers as the bleached molecules diffuse away and unbleached molecules diffuse into the irradiated area. (B) A graph showing the rate of recovery. The greater the diffusion coefficient of the membrane glycoprotein, the faster is the recovery.

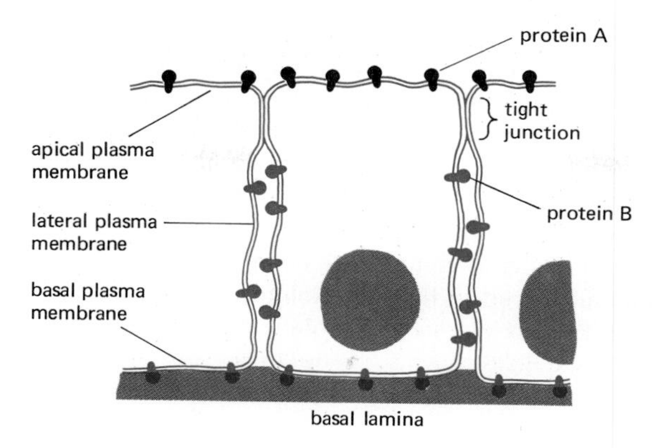

Figure 6–36 Diagram of an epithelial cell showing how a plasma membrane protein is restricted to a particular domain of the membrane. Protein A (in the apical membrane) and protein B (in the basal and lateral membranes) can diffuse laterally in their own domains but are prevented from entering the other domain, probably by the specialized cell junction called a *tight junction.* Lipid molecules in the outer (noncytoplasmic) monolayer of the plasma membrane are also unable to diffuse between the two domains, whereas those in the inner (cytoplasmic) monolayer are able to do so.

Cells Can Confine Proteins and Lipids to Specific Domains Within a Membrane[17]

The recognition that biological membranes are two-dimensional fluids was a major advance in understanding membrane structure and function, but it has become clear that the picture of a membrane as a lipid sea in which all proteins float freely is greatly oversimplified. Many cells have ways of confining membrane proteins to specific domains in a continuous lipid bilayer. For example, in epithelial cells, such as those that line the gut or the tubules of the kidney, certain plasma membrane enzymes and transport proteins are confined to the apical surface of the cells, whereas others are confined to the basal and lateral surfaces (Figure 6–36). As we shall discuss later, this asymmetric distribution of membrane proteins is often essential for the function of the epithelium (see p. 310). The lipid compositions of these two membrane domains are also different, demonstrating that epithelial cells can prevent the diffusion of lipid as well as protein molecules between the domains, although experiments with labeled lipids suggest that only lipid molecules in the outer monolayer of the membrane are confined in this way. The separation of both protein and lipid molecules is thought to be maintained, at least in part, by the barriers set up by a specific type of intercellular junction (called a *tight junction,* see p. 793). As discussed below, the membrane proteins that form intercellular junctions do not diffuse laterally in the interacting membranes (see Figure 6–38D).

A cell can also create membrane domains without using intercellular junctions. The mammalian spermatozoon, for instance, is a single cell that consists of two structurally and functionally distinct parts—a head and a tail (see p. 864)—covered by a continuous plasma membrane. When a sperm cell is examined by immunofluorescence microscopy using a variety of antibodies that react with cell-surface antigens, the plasma membrane is found to consist of at least three distinct domains (Figure 6–37). In some cases at least, the antigens are able to diffuse within the confines of their own domain; it is not known how they are confined.

In the two examples just considered, the diffusion of protein and lipid molecules is confined to specialized domains within a continuous plasma membrane.

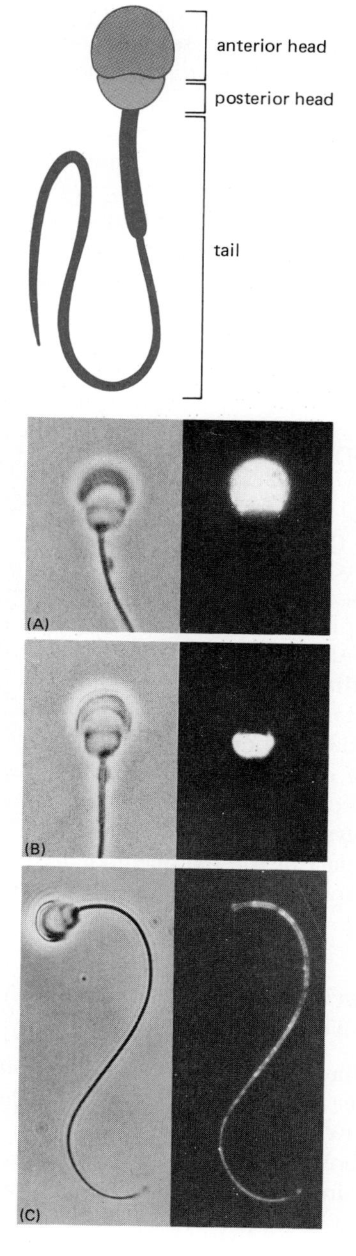

Figure 6–37 Three domains in the plasma membrane of guinea pig sperm defined with monoclonal antibodies (see p. 178). A guinea pig sperm is shown schematically in the upper drawing, while each of the three pairs of micrographs shown in (A), (B), and (C) shows cell-surface immunofluorescence staining with a different monoclonal antibody next to a phase-contrast micrograph of the same cell. The antibody shown in (A) labels only the anterior head, that in (B) only the posterior head, whereas that in (C) labels only the tail. (A and B, from D.G. Myles, P. Primakoff, and A.R. Bellvé, *Cell* 23:434–439, 1981, by copyright permission of Cell Press. C, from P. Primakoff and D.G. Myles, *Dev. Biol.* 98:417–428, 1983.)

Cells also have more drastic ways of immobilizing certain membrane proteins. One is exemplified by the purple membrane of *Halobacterium*, where the bacteriorhodopsin molecules assemble into large two-dimensional crystals in which the individual protein molecules are relatively fixed in relationship to one another; large aggregates of this kind diffuse very slowly. A more common way of restricting the lateral mobility of specific membrane proteins is to tether them to macromolecular assemblies inside or outside the cell. We have seen that some red blood cell membrane proteins are anchored to the cytoskeleton inside; in other cell types, plasma membrane proteins can be anchored to the cytoskeleton or to the extracellular matrix or to both. The four known ways of immobilizing specific membrane proteins are summarized in Figure 6–38.

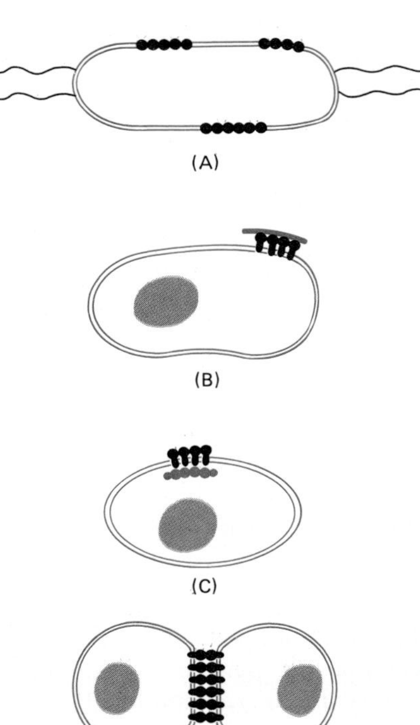

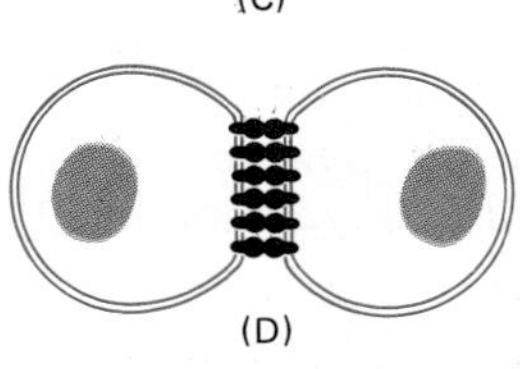

Figure 6–38 Four ways in which the lateral mobility of specific plasma membrane proteins can be restricted. The proteins can self-assemble into large aggregates (such as bacteriorhodopsin in the purple membrane of *Halobacterium*) (A), they can be tethered by interactions with assemblies of macromolecules outside (B) or inside (C) the cell, or they can interact with proteins on the surface of another cell (D).

Summary

Whereas the lipid bilayer determines the basic structure of biological membranes, proteins are responsible for most membrane functions, serving as specific receptors, enzymes, transport proteins, and so on. Most membrane proteins extend across the lipid bilayer: in some of these transmembrane proteins the polypeptide chain crosses the bilayer as a single α helix (single-pass proteins); in others the polypeptide chain crosses the bilayer multiple times as a series of α helices (multipass proteins). Other membrane-associated proteins do not span the bilayer but instead are attached to one or the other side of the membrane. Many of these are bound by noncovalent interactions with transmembrane proteins, but others are covalently attached to lipid molecules. Like the lipid molecules in the bilayer, many membrane proteins are able to diffuse in the plane of the membrane. On the other hand, cells have ways of immobilizing specific membrane proteins and of confining both membrane protein and lipid molecules to particular domains in a continuous lipid bilayer.

Membrane Carbohydrate

All eucaryotic cells have carbohydrate on their surface, both as oligosaccharide and polysaccharide chains covalently bound to membrane proteins and as oligosaccharide chains covalently bound to lipids (glycolipids). The total carbohydrate in plasma membranes constitutes between 2% and 10% of the membrane's total weight. Most plasma membrane protein molecules exposed at the cell surface carry sugar residues, whereas somewhat fewer than 1 in 10 lipid molecules in the outer lipid monolayer of most plasma membranes contain carbohydrate (see p. 282). Although the fiftyfold excess of lipid to protein molecules in membranes means that there are more lipid than protein molecules carrying carbohydrate in a typical membrane, there is more total carbohydrate attached to proteins: a single glycoprotein, such as glycophorin, can have many oligosaccharide side chains, whereas each glycolipid molecule has only one. Moreover, many plasma membranes contain integral *proteoglycan* molecules. Proteoglycans, which consist of long polysaccharide chains linked to a protein core, are found mainly outside the cell as part of the extracellular matrix (see p. 806); in the case of some integral membrane proteoglycans, however, the core is thought to extend across the lipid bilayer.

6-18 The Carbohydrate in Biological Membranes Is Confined Mainly to the Noncytosolic Surface[18]

As we have seen, biological membranes are strikingly asymmetrical: the lipids of the outer and inner lipid monolayers are different, as are the exposed polypeptides on the two surfaces. The distribution of carbohydrate is also asymmetrical, since the carbohydrate chains of the major glycolipids, glycoproteins, and proteoglycans of both internal and plasma membranes are located exclusively on the noncytosolic surface: in plasma membranes the sugar residues are exposed on the outside

Table 6–2 Commonly Used, Commercially Available Plant Lectins and the Specific Sugar Residues They Recognize

Lectin	Sugar Specificity
Concanavalin A (from jack beans)	α-D-glucose and α-D-mannose
Soybean lectin	α-galactose and N-acetyl-D-galactosamine
Wheat germ lectin	N-acetylglucosamine
Lotus seed lectin	fucose

of the cell, whereas in internal membranes they face inward toward the lumen of the membrane-bounded compartment (see p. 454).

There are two distinct ways in which oligosaccharides are attached to membrane glycoproteins: they may be *N-linked* to an asparagine residue in the polypeptide chain or *O-linked* to a serine or threonine residue (see p. 446). The *N*-linked oligosaccharides usually contain about 12 sugars and are constructed around a common core of mannose residues (see p. 452), whereas the *O*-linked oligosaccharides tend to be shorter (about 4 sugars long).

One convenient way to demonstrate the presence of cell-surface sugars is to use carbohydrate-binding proteins called **lectins.** These are proteins with binding sites that recognize a specific sequence of sugar residues. They were originally isolated from plants, where they are found in large quantities in many seeds; some of them are highly toxic and serve to deter animals from eating the seeds. More recently, lectins have been demonstrated in many other organisms, including mammals; some of them occur on cell surfaces and are thought to be involved in cell-cell recognition (for example, see Figure 15–42, p. 870). Since lectins bind to cell-surface glycoproteins, proteoglycans, and glycolipids, they are widely used as biochemical tools in cell biology to localize and isolate sugar-containing plasma membrane molecules. Some commonly used plant lectins and their sugar specificities are listed in Table 6–2.

The term *cell coat* or *glycocalyx* is often used to describe the carbohydrate-rich peripheral zone on the outside surface of most eucaryotic cells. This zone can be visualized by a variety of stains, such as ruthenium red (Figure 6–39), as well as by labeled lectins. Although the carbohydrate is attached mainly to intrinsic plasma membrane molecules, the glycocalyx can also contain both glycoproteins and proteoglycans that have been secreted and then adsorbed on the cell surface (Figure 6–40). Some of these adsorbed macromolecules are components of the extracellular matrix, so that where the plasma membrane ends and the extracellular matrix begins is largely a matter of semantics.

Although the high concentration of cell-surface carbohydrate must have important influences on many functions of the plasma membrane, the nature of these influences is not yet understood. The complexity of some of the oligosaccharides, taken together with their exposed position on the cell surface, suggests

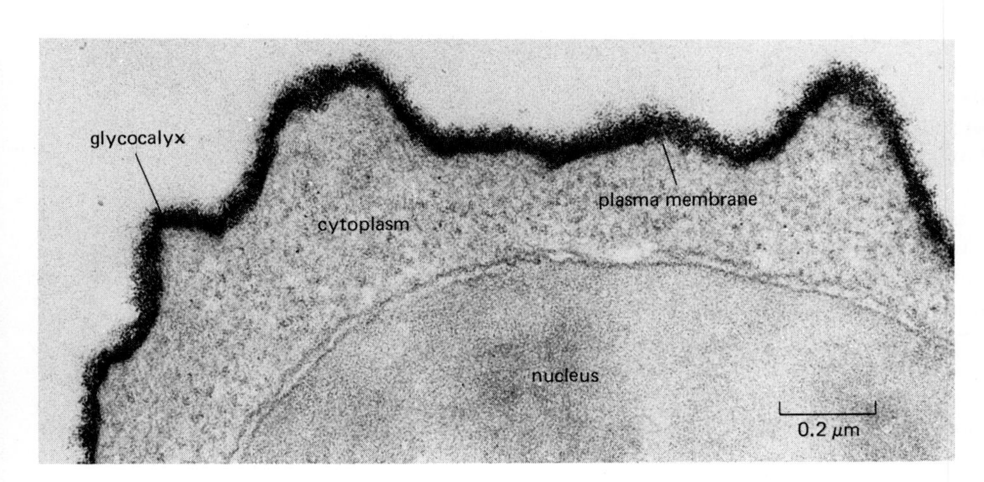

Figure 6–39 Electron micrograph of the surface of a lymphocyte stained with ruthenium red to show the cell coat (glycocalyx). (Courtesy of A.M. Glauert and G.M.W. Cook.)

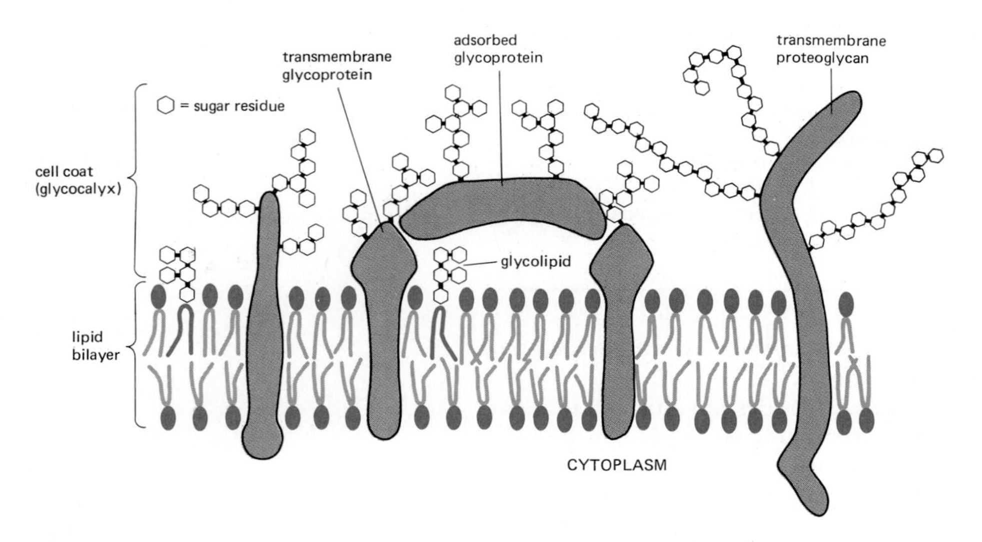

Figure 6–40 Diagram of the cell coat (glycocalyx), which is made up of the oligosaccharide side chains of glycolipids and integral membrane glycoproteins, and polysaccharide chains on integral proteoglycans. In addition, adsorbed glycoproteins and adsorbed proteoglycans (not shown) contribute to the glycocalyx in some cells. Note that all of the carbohydrate is on the outside of the membrane. Some integral glycoproteins and proteoglycans are covalently attached to phosphatidylinositol in the outer monolayers of the plasma membrane via a specific oligosaccharide (not shown, but see Figure 6–14–4, p. 284).

that they may play an important part in cell-cell and cell-matrix recognition processes. Although there is strong indirect evidence for this in a few cases (see pp. 869 and 875), for the most part this function of cell-surface carbohydrate has been difficult to demonstrate unambiguously.

Summary

In the plasma membrane of all eucaryotic cells, most of the proteins exposed on the cell surface and some of the lipid molecules in the outer lipid monolayer have oligosaccharide chains covalently attached to them. Some plasma membranes also contain integral proteoglycan molecules in which several polysaccharide chains are covalently linked to a transmembrane or lipid-linked core protein. Although the function of cell-surface carbohydrate remains uncertain, it is likely that at least some of it plays a part in cell-cell and cell-matrix recognition processes.

Membrane Transport of Small Molecules[19]

Because of its hydrophobic interior, the lipid bilayer is a highly impermeable barrier to most polar molecules and therefore prevents most of the water-soluble contents of the cell from escaping. For this very reason, however, cells have evolved special ways of transferring water-soluble molecules across their membranes. Cells must ingest essential nutrients and excrete metabolic waste products. They must also regulate intracellular ion concentrations, which means transporting specific ions into or out of the cell. Transport of small water-soluble molecules across the lipid bilayer is achieved by specialized transmembrane proteins, each of which is responsible for the transfer of a specific molecule or a group of closely related molecules. Cells have also evolved the means to transport macromolecules such as proteins and even large particles across their plasma membrane, but the mechanisms involved are very different from those used for transferring small molecules and will be discussed in a later section (see p. 323).

In this section we shall see that a combination of selective permeability and active transport across the plasma membrane creates large differences in the ionic composition of the cytosol compared with the extracellular fluid (Table 6–3). This enables cell membranes to store potential energy in the form of ion gradients. These transmembrane ion gradients are used to drive various transport processes, to convey electrical signals, and (in mitochondria, chloroplasts, and bacteria) to make ATP. Before discussing membrane transport proteins and the ion gradients that some of them generate, it is important to know something of the permeability properties of protein-free, synthetic lipid bilayers.

Table 6–3 Comparison of Ion Concentrations Inside and Outside a Typical Mammalian Cell

Component	Intracellular Concentration (mM)	Extracellular Concentration (mM)
Cations		
Na^+	5–15	145
K^+	140	5
Mg^{2+}	0.5	1–2
Ca^{2+}	10^{-4}	1–2
H^+	8×10^{-5} ($10^{-7.1}$ M or pH 7.1)	4×10^{-5} ($10^{-7.4}$ M or pH 7.4)
Anions*		
Cl^-	5–15	110

*Because the cell must contain equal + and − charges (that is, be electrically neutral), the large deficit in intracellular anions reflects the fact that most cellular constituents are negatively charged (HCO_3^-, PO_4^{3-}, proteins, nucleic acids, metabolites carrying phosphate and carboxyl groups, etc.). The concentrations of Ca^{2+} and Mg^{2+} given are for the free ions. There is a total of about 20 mM Mg^{2+} and 1−2 mM Ca^{2+} in cells, but this is mostly bound to proteins and other substances and, in the case of Ca^{2+}, stored within various organelles.

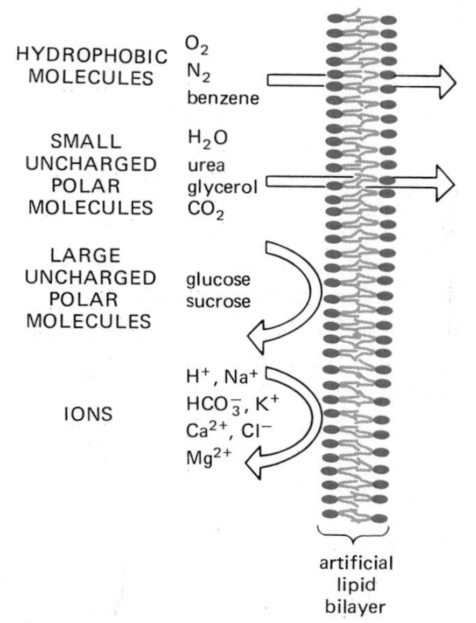

Figure 6–41 The relative permeability of a synthetic lipid bilayer to different classes of molecules. The smaller the molecule, and, more important, the fewer hydrogen bonds it makes with water, the more rapidly the molecule diffuses across the bilayer.

Protein-free Lipid Bilayers Are Impermeable to Ions but Freely Permeable to Water[20]

Given enough time, virtually any molecule will diffuse across a protein-free, synthetic lipid bilayer down its concentration gradient. The rate at which a molecule diffuses across such a lipid bilayer, however, varies enormously, depending largely on the size of the molecule and its relative solubility in oil. In general, the smaller the molecule and the more soluble it is in oil (that is, the more hydrophobic or nonpolar it is), the more rapidly it will diffuse across a bilayer. *Small nonpolar* molecules, such as O_2 (32 daltons), readily dissolve in lipid bilayers and therefore rapidly diffuse across them. *Uncharged polar* molecules also diffuse rapidly across a bilayer if they are small enough. For example, CO_2 (44 daltons), ethanol (46 daltons), and urea (60 daltons) cross rapidly; glycerol (92 daltons) less rapidly; and glucose (180 daltons) hardly at all (Figure 6–41). Importantly, water (18 daltons) diffuses very rapidly across lipid bilayers even though water molecules are relatively insoluble in oil. This is because water molecules have a very small volume and are uncharged.

In contrast, lipid bilayers are highly impermeable to all *charged* molecules (ions), no matter how small: the charge and high degree of hydration of such molecules prevents them from entering the hydrocarbon phase of the bilayer. Consequently, synthetic bilayers are 10^9 times more permeable to water than to even such small ions as Na^+ or K^+ (Figure 6–42).

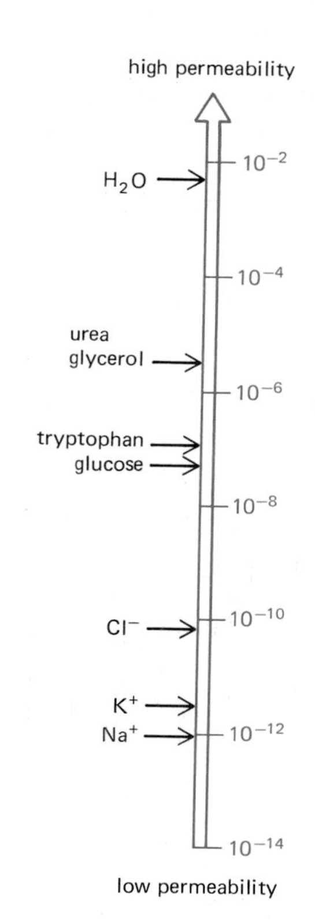

Figure 6–42 Permeability coefficients (cm/sec) for the passage of various molecules through synthetic lipid bilayers. The rate of flow of a solute across the bilayer is directly proportional to the difference in its concentration on the two sides of the membrane. Multiplying this concentration difference (in mol/cm³) by the permeability coefficient (cm/sec) gives the flow of solute in moles per second per square centimeter of membrane. For example, a concentration difference of tryptophan of 10^{-4} mol/cm³ ($10^{-4}/10^{-3}$ L = 0.1 M) would cause a flow of 10^{-4} mol/cm³ × 10^{-7} cm/sec = 10^{-11} mol/sec through 1 cm² of membrane, or 6×10^4 molecules/sec through 1 μm² of membrane.

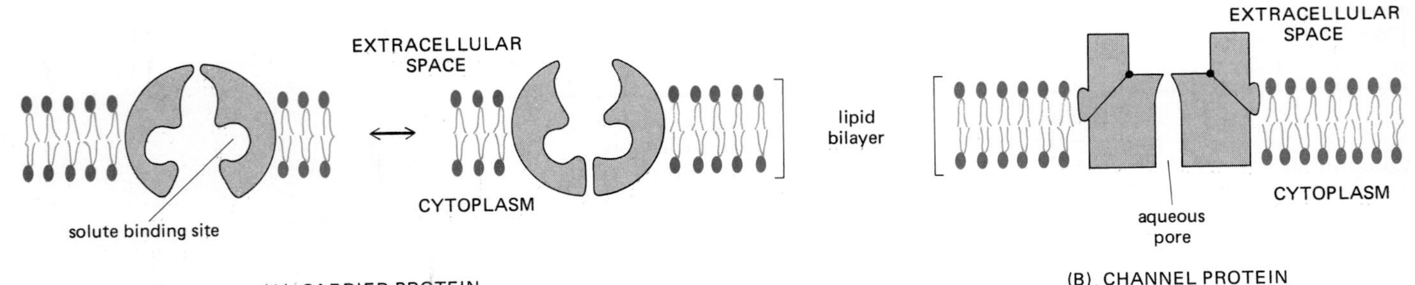

Figure 6–43 A simplified schematic view of the two classes of membrane transport proteins. A *carrier protein* alternates between two conformations, so that the solute binding site is sequentially accessible on one side of the bilayer and then on the other. In contrast, a *channel protein* forms a water-filled pore across the bilayer through which specific ions can diffuse.

6-19 Membrane Transport Proteins Can Act as Carriers or Channels[19]

Like synthetic lipid bilayers, cell membranes allow water and nonpolar molecules to permeate by simple diffusion. Cell membranes, however, are also permeable to various polar molecules such as ions, sugars, amino acids, nucleotides, and many cell metabolites that pass across synthetic lipid bilayers only very slowly. Special membrane proteins are responsible for transferring such solutes across cell membranes. These proteins, referred to as **membrane transport proteins,** occur in many forms and in all types of biological membranes. Each protein is designed to transport a particular class of molecule (such as ions, sugars, or amino acids) and often only certain molecular species of the class. The specificity of transport proteins was first indicated by studies in which single gene mutations were found to abolish the ability of bacteria to transport specific sugars across their plasma membrane. Similar mutations have now been discovered in humans suffering from a variety of inherited diseases that affect the transport of a specific solute in the kidney or intestine or both. For example, individuals with the inherited disease *cystinuria* are unable to transport certain amino acids (including cystine, the disulfide-linked dimer of cysteine) from either the urine or the intestine into the blood; the resulting accumulation of cystine in the urine leads to the formation of cystine "stones" in the kidneys.

All membrane transport proteins that have been studied in sufficient detail to establish their orientation in the membrane have been found to be multipass transmembrane proteins—that is, their polypeptide chain traverses the lipid bilayer multiple times. By forming a continuous protein pathway across the membrane, these proteins enable the specific solutes they transport to pass across the membrane without coming into direct contact with the hydrophobic interior of the lipid bilayer.

There are two major classes of membrane transport proteins: carrier proteins and channel proteins. **Carrier proteins** (also called *carriers* or *transporters*) bind the specific solute to be transported and undergo a conformational change in order to transfer the solute across the membrane. **Channel proteins,** on the other hand, form water-filled pores that extend across the lipid bilayer; when these pores are open, they allow specific solutes (usually inorganic ions of appropriate size and charge) to pass through them and thereby cross the membrane (Figure 6–43).

Active Transport Is Mediated by Carrier Proteins Coupled to an Energy Source[21]

All channel proteins and many carrier proteins allow solutes to cross the membrane only passively ("downhill")—a process called **passive transport** (or **facilitated diffusion**). If the transported molecule is uncharged, only the difference in its concentration on the two sides of the membrane (its *concentration gradient*) determines the direction of passive transport. If the solute carries a net charge, however, both its concentration gradient and the electrical potential difference across the membrane (the *membrane potential*) influence its transport. The concentration gradient and the electrical gradient together constitute the **electrochemical gradient** for each solute (Panel 6–2, p. 315). All plasma membranes have an electrical potential difference (voltage gradient) across them, with the inside negative compared to the outside. This potential favors the entry of positively charged ions into cells but opposes the entry of negatively charged ions.

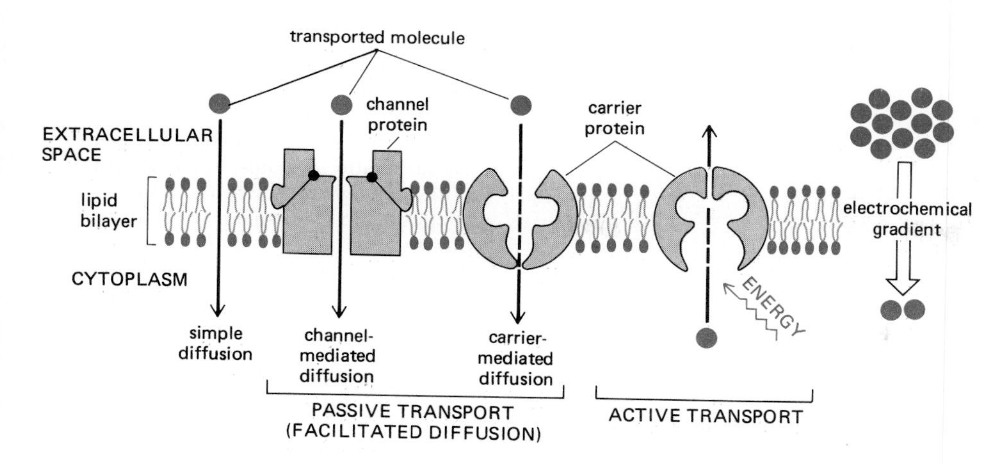

Figure 6–44 Schematic diagram of *passive transport* down an electrochemical gradient and *active transport* against an electrochemical gradient. Whereas simple diffusion and passive transport by membrane transport proteins (facilitated diffusion) occur spontaneously, active transport requires an input of metabolic energy. Only nonpolar molecules and small, uncharged polar molecules can cross the lipid bilayer directly by simple diffusion; the transfer of other polar molecules occurs at significant rates only through specific carrier proteins or channel proteins.

Cells also require transport proteins that will actively pump certain solutes across the membrane against their electrochemical gradient ("uphill"); this process, known as **active transport,** is always mediated by carrier proteins. In active transport the pumping activity of the carrier protein is directional because it is tightly coupled to a source of metabolic energy, such as ATP hydrolysis or an ion gradient, as will be discussed below. Thus transport by carrier proteins can be either active or passive, whereas transport by channel proteins is always passive (Figure 6–44).

6-20 Carrier Proteins Behave like Membrane-bound Enzymes[19]

The process by which a carrier protein specifically binds and transfers a solute molecule across the lipid bilayer resembles an enzyme-substrate reaction, and the carriers involved behave like specialized membrane-bound enzymes. Each type of carrier protein has a specific binding site for its solute (substrate). When the carrier is saturated (that is, when all these binding sites are occupied), the rate of transport is maximal. This rate, referred to as V_{max}, is characteristic of the specific carrier. In addition, each carrier protein has a characteristic binding constant for its solute, K_M, equal to the concentration of solute when the transport rate is half its maximum value (Figure 6–45). As with enzymes, the solute binding can be blocked specifically by competitive inhibitors (which compete for the same binding site and may or may not be transported by the carrier) or by noncompetitive inhibitors (which bind elsewhere and specifically alter the structure of the carrier). The analogy with an enzyme-substrate reaction is limited, however, since the transported solute is usually not covalently modified by the carrier protein.

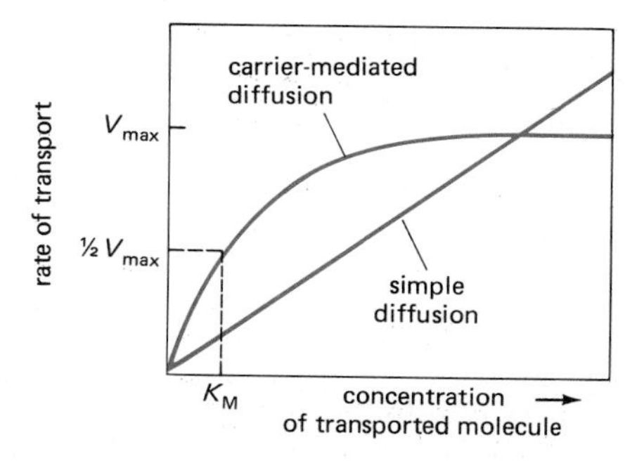

Figure 6–45 Kinetics of simple diffusion compared to carrier-mediated diffusion. Whereas the rate of the former is always proportional to the solute concentration, the rate of the latter reaches a maximum (V_{max}) when the carrier protein is saturated. The solute concentration when transport is at half its maximal value approximates the binding constant (K_M) of the carrier for the solute and is analogous to the K_M of an enzyme for its substrate (see p. 94).

Some carrier proteins simply transport a single solute from one side of the membrane to the other; they are called **uniports.** Others function as **coupled transporters,** in which the transfer of one solute depends on the simultaneous or sequential transfer of a second solute, either in the same direction (**symport**) or in the opposite direction (**antiport**) (Figure 6–46). Most animal cells, for ex-

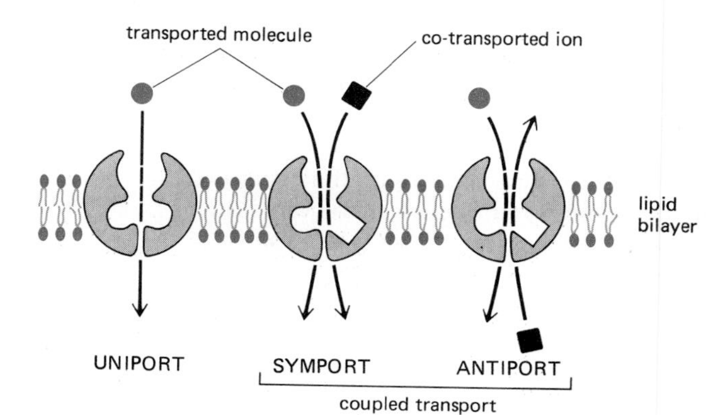

Figure 6–46 Schematic diagram of carrier proteins functioning as uniports, symports, and antiports.

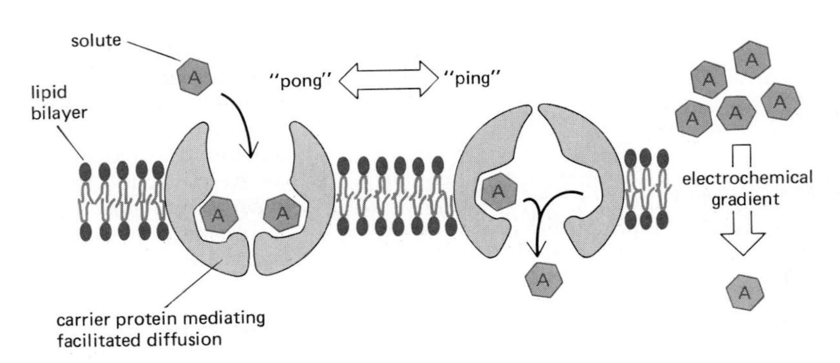

Figure 6–47 A hypothetical model showing how a conformational change in a carrier protein could mediate the facilitated diffusion of a solute A. The carrier protein shown can exist in two conformational states: in state "pong" the binding sites for solute A are exposed on the outside of the bilayer; in state "ping" the same sites are exposed on the other side of the bilayer. The transition between the two states is proposed to occur randomly and to be completely reversible. Therefore, if the concentration of A is higher on the outside of the bilayer, more A will bind to the carrier protein in the pong conformation than in the ping conformation, and there will be a net transport of A down its electrochemical gradient.

ample, must take up glucose from the extracellular fluid, where the concentration of the sugar is relatively high, by *passive* transport through glucose carriers that operate as uniports. By contrast, intestinal and kidney cells must take up glucose from the lumen of the intestine and kidney tubules, respectively, where the concentration of the sugar is low. These cells actively transport glucose by symport with Na^+, whose extracellular concentration is very high (see p. 304). As previously discussed, the anion carrier (band 3 protein) of the human red blood cell operates as an antiport to exchange Cl^- for HCO_3^- (see p. 291).

Although the molecular details are unknown, carrier proteins are thought to transfer the solute across the bilayer by undergoing a reversible conformational change that alternately exposes the solute-binding site first on one side of the membrane and then on the other. A schematic model of how such a carrier protein might operate is shown in Figure 6–47. Because carriers are now known to be multipass transmembrane proteins, it is highly unlikely that they ever tumble in the membrane or shuttle back and forth across the lipid bilayer as was once believed.

As we shall discuss below, it requires only a relatively minor modification of the model shown in Figure 6–47 to link the carrier protein to a source of energy, such as ATP hydrolysis (see Figure 6–49 below), or to an ion gradient (see Figure 6–51 below) to pump a solute uphill against its electrochemical gradient. An important example of a carrier protein that uses the energy of ATP hydrolysis to pump ions is the *Na^+-K^+ pump,* which plays a crucial part in generating and maintaining the Na^+ and K^+ gradients across the plasma membranes of animal cells.

6-23 The Plasma Membrane Na^+-K^+ Pump Is an ATPase[22]

The concentration of K^+ is typically 10 to 20 times higher inside cells than outside, whereas the reverse is true of Na^+ (see Table 6–3, p. 301). These concentration differences are maintained by a **Na^+-K^+ pump** that is found in the plasma membrane of virtually all animal cells. The pump operates as an antiport, actively pumping Na^+ out of the cell against its steep electrochemical gradient and pumping K^+ in. As explained below, the Na^+ gradient due to the pump regulates cell volume through its osmotic effects and is also exploited to drive transport of sugars and amino acids into the cell. Almost one-third of the energy requirement of a typical animal cell is consumed in fueling this pump; in electrically active nerve cells, which are repeatedly gaining small amounts of Na^+ and losing small amounts of K^+ during the propagation of action potentials (see below), this figure approaches two-thirds of the cell's energy requirement.

A major advance in understanding the Na^+-K^+ pump came with the discovery in 1957 that an enzyme that hydrolyzes ATP to ADP and phosphate requires Na^+ and K^+ for optimal activity. An important clue linking this **Na^+-K^+ ATPase** with the Na^+-K^+ pump was the observation that a known inhibitor of the pump, *ouabain,* also inhibits the ATPase. But the crucial evidence that ATP hydrolysis provides the energy for driving the pump came from studies of resealed red blood

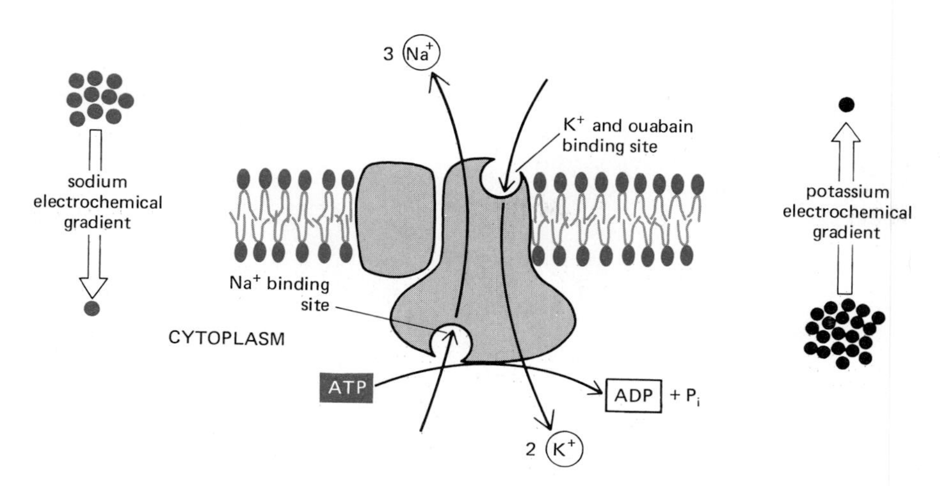

Figure 6–48 The Na^+-K^+ ATPase actively pumps Na^+ out and K^+ into a cell against their electrochemical gradients. For every molecule of ATP hydrolyzed inside the cell, three Na^+ are pumped out and two K^+ are pumped in. The specific pump inhibitor ouabain and K^+ compete for the same site on the external side of the ATPase.

cell ghosts (see p. 288), in which the concentrations of ions, ATP, and drugs on either side of the membrane could be varied and the effects on ion transport and ATP hydrolysis observed. It was found that (1) the transport of Na^+ and K^+ is tightly coupled to ATP hydrolysis, so that one cannot occur without the other; (2) ion transport and ATP hydrolysis can occur only when Na^+ and ATP are present inside the ghosts and K^+ is present on the outside; (3) ouabain is inhibitory only when present outside the ghosts, where it competes for the K^+ binding site; and (4) for every molecule of ATP hydrolyzed (100 ATP molecules can be hydrolyzed by each ATPase molecule each second), three Na^+ are pumped out and two K^+ are pumped in (Figure 6–48).

Although these experiments provided compelling evidence that ATP supplies the energy for pumping Na^+ and K^+ ions across the plasma membrane, they did not explain how ATP hydrolysis is coupled to ion transport. A partial explanation was provided by the finding that the terminal phosphate group of the ATP is transferred to an aspartic acid residue of the ATPase in the presence of Na^+. This phosphate group is subsequently hydrolyzed in the presence of K^+, and it is this last step that is inhibited by ouabain. The Na^+-dependent phosphorylation is coupled to a change in the conformation of the ATPase, which results in the transport of Na^+ out of the cell, whereas the K^+-dependent dephosphorylation, which occurs subsequently, results in the transport of K^+ into the cell during the return of the ATPase to its original conformation (Figure 6–49).

The Na^+-K^+ pump in red blood cell ghosts can be driven in reverse to produce ATP. When the Na^+ and K^+ gradients are experimentally increased to such an extent that the energy stored in their electrochemical gradients is greater than the chemical energy of ATP hydrolysis, these ions move down their electrochemical gradients and ATP is synthesized from ADP and phosphate by the Na^+-K^+ ATPase. Thus the phosphorylated form of the ATPase (step 2 in Figure 6–49) can relax either by donating its phosphate to ADP (step 2 to step 1) or by changing its conformation (step 2 to step 3). Whether the overall change in free energy is used to synthesize ATP (see p. 129) or to pump Na^+ out of the ghost depends on the relative concentrations of ATP, ADP and phosphate, and on the electrochemical gradients for Na^+ and K^+.

The Na^+-K^+ ATPase has been purified and found to consist of a large, multipass, transmembrane catalytic subunit (about 1000 amino acid residues long) and an associated smaller glycoprotein. The former has binding sites for Na^+ and ATP on its cytoplasmic surface and binding sites for K^+ and ouabain on its external surface, and is reversibly phosphorylated and dephosphorylated. The function of the glycoprotein is unknown. A functional Na^+-K^+ pump can be reconstituted from the purified complex: the ATPase is solubilized in detergent, purified, and mixed with appropriate phospholipids; when the detergent is removed, membrane vesicles are formed that pump Na^+ and K^+ in opposite directions in the presence of ATP (see Figure 6–21, p. 287).

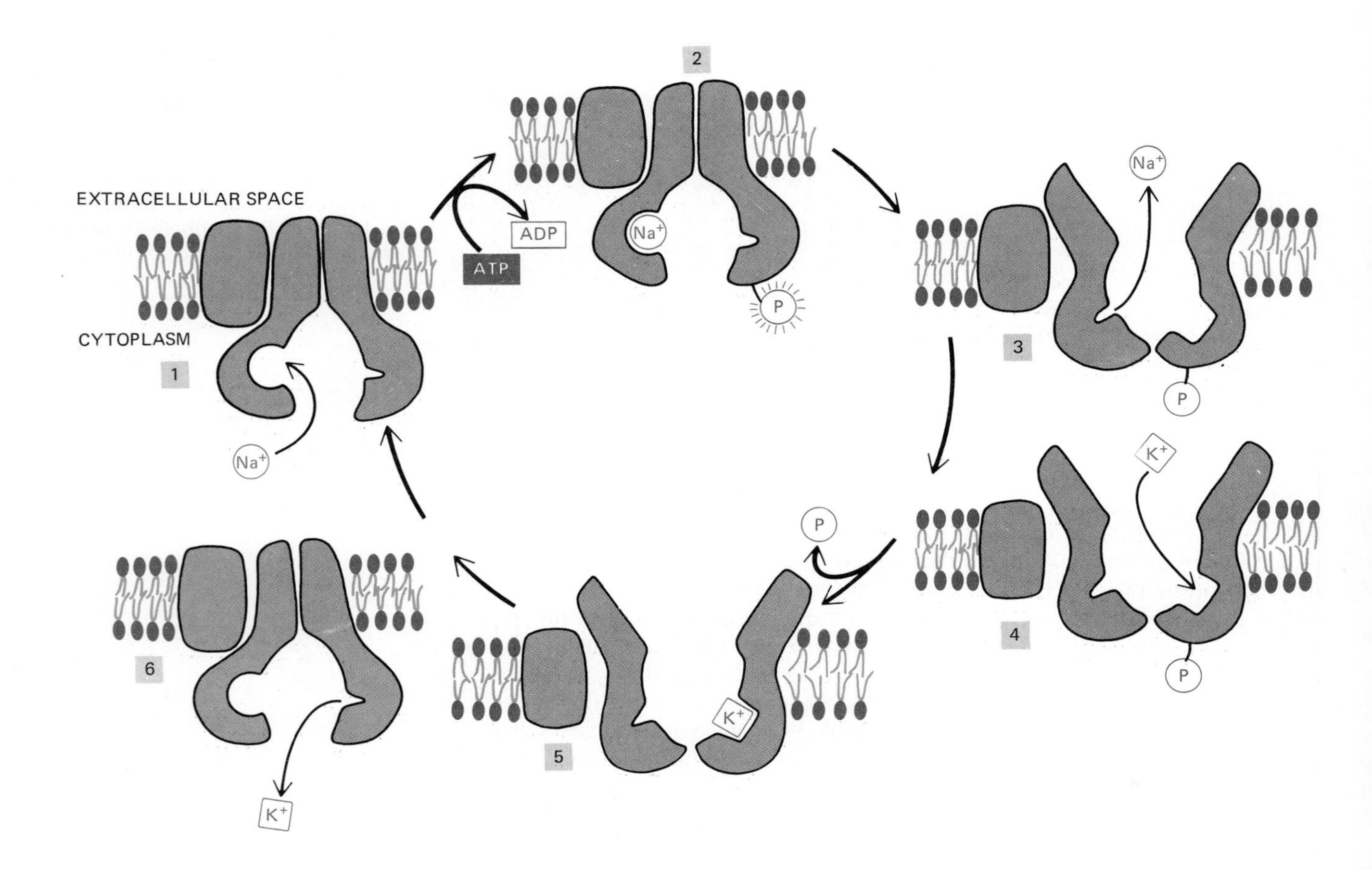

Figure 6–49 A schematic model of the Na^+-K^+ ATPase. The binding of Na^+ (1) and the subsequent phosphorylation by ATP (2) of the cytoplasmic face of the ATPase induce the protein to undergo a conformational change that transfers the Na^+ across the membrane and releases it on the outside (3). Then the binding of K^+ on the external surface (4) and the subsequent dephosphorylation (5) return the protein to its original conformation, which transfers the K^+ across the membrane and releases it into the cytosol (6). These changes in conformation are analogous to the ping ⇌ pong transitions shown in Figure 6–47 except that here the Na^+-dependent phosphorylation and the K^+-dependent dephosphorylation of the protein cause the conformational transitions to occur in an orderly manner, enabling the protein to do useful work. Although for simplicity only one Na^+ and one K^+ binding site are shown, in the real pump there are thought to be three Na^+ and two K^+ binding sites.

The Na^+-K^+ ATPase Is Required to Maintain Osmotic Balance and Stabilize Cell Volume[23]

Since the Na^+-K^+ ATPase drives three positively charged ions out of the cell for every two it pumps in, it is "electrogenic"; that is, it drives a net current across the membrane, tending to create an electrical potential, with the inside negative relative to the outside. This effect of the pump, however, seldom contributes more than 10% to the membrane potential. The remaining 90%, as we shall see, depends on the pump only indirectly; its immediate cause lies in an electrical force related to the unequal concentrations of K^+ on either side of the membrane. These K^+ concentrations are governed in turn by the need for K^+ inside the cell to balance the large negative charge carried by the cell's *fixed anions*—the many negatively charged organic molecules that are confined inside the cell, unable to cross the plasma membrane.

The Na^+-K^+ ATPase has a more direct role in regulating cell volume: it controls the solute concentration inside the cell, thereby regulating the osmotic forces that can make a cell swell or shrink (Figure 6–50). As explained in Panel 6–1 (see p. 308), the solutes inside a cell—including its fixed anions and the accompanying cations required for charge balance—create a large osmotic gradient that tends to "pull" water in. For animal cells this effect is counteracted by an opposite osmotic gradient due to a high concentration of inorganic ions—chiefly Na^+ and Cl^-—in the extracellular fluid. The Na^+-K^+ ATPase maintains osmotic balance by pumping out the Na^+ that leaks in down its steep electrochemical gradient; the Cl^- is kept out by the membrane potential, as will be explained below (see p. 314).

The importance of the Na^+-K^+ ATPase in controlling cell volume is indicated by the observation that many animal cells swell, and may burst, if they are treated with ouabain, which inhibits the Na^+-K^+ ATPase. There are, of course, other ways for a cell to cope with its osmotic problems. Plant cells and many bacteria are prevented from bursting by the semirigid cell wall that surrounds their plasma membrane; in amoebae the excess water that flows in osmotically is collected in

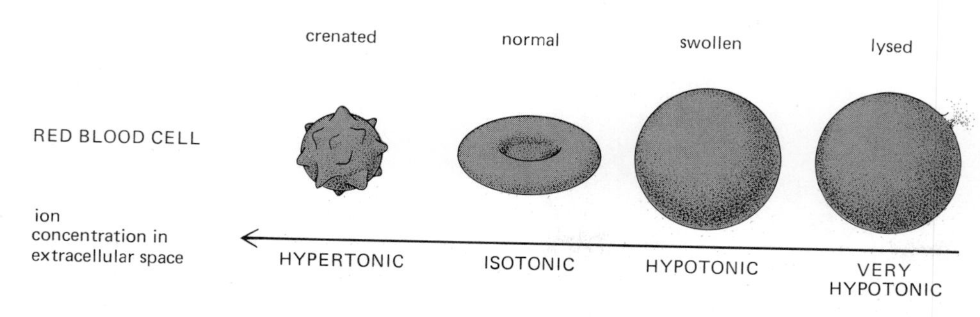

Figure 6–50 Response of a human red blood cell to changes in osmolarity (also called *tonicity*) of the extracellular fluid. Because the plasma membrane is freely permeable to water, water will move into or out of cells down its concentration gradient, a process called *osmosis.* If cells are placed in a *hypotonic solution* (i.e., a solution having a low solute concentration and therefore a high water concentration), there will be a net movement of water into the cells, causing them to swell and burst (lyse). Conversely, if cells are placed in a *hypertonic solution,* they will shrink (see also Panel 2–1, p. 47).

contractile vacuoles, which periodically discharge their contents to the exterior (see Panel 6–1, p. 308). But for most cells in multicellular animals, the Na^+-K^+ ATPase is crucial.

Some Ca^{2+} Pumps Are Also Membrane-bound ATPases[24]

Eucaryotic cells maintain very low concentrations of free Ca^{2+} in their cytosol ($\sim 10^{-7}$ M) in the face of very much higher extracellular Ca^{2+} concentrations ($\sim 10^{-3}$ M). Even a small influx of Ca^{2+} increases significantly the concentration of free Ca^{2+} in the cytosol, and the flow of Ca^{2+} down its steep concentration gradient in response to extracellular signals is one means of transmitting these signals rapidly across the plasma membrane (see p. 701). The Ca^{2+} gradient is in part maintained by Ca^{2+} pumps in the plasma membrane that actively transport Ca^{2+} out of the cell. One of these is known to be an ATPase, while the other is an antiporter that is driven by the Na^+ electrochemical gradient (see below).

The best-understood Ca^{2+} pump is a membrane-bound ATPase in the *sarcoplasmic reticulum* of muscle cells. The sarcoplasmic reticulum forms a network of tubular sacs in the cytoplasm of muscle cells and serves as an intracellular store of Ca^{2+}. (When an action potential depolarizes the muscle cell membrane, Ca^{2+} is released from the sarcoplasmic reticulum into the cytosol, stimulating the muscle to contract—see p. 624.) The Ca^{2+} pump is responsible for pumping Ca^{2+} from the cytosol into the sarcoplasmic reticulum. Like the Na^+-K^+ pump, it is an ATPase that is phosphorylated and dephosphorylated during its pumping cycle, and it pumps two Ca^{2+} ions into the sarcoplasmic reticulum for every ATP molecule hydrolyzed. Since the Ca^{2+} ATPase accounts for about 90% of the membrane protein of the sarcoplasmic reticulum, it has been relatively easy to purify. It is a single, large, multipass transmembrane polypeptide chain (containing about 1000 amino acid residues); and when it is incorporated into phospholipid vesicles (as described above for the Na^+-K^+ ATPase), it pumps Ca^{2+} as it hydrolyzes ATP. DNA cloning and sequencing experiments indicate that the Ca^{2+} ATPase is homologous in amino acid sequence to the large catalytic subunit of the Na^+-K^+ ATPase, revealing that these two ion pumps are evolutionarily related.

In nonmuscle cells an organelle equivalent to the sarcoplasmic reticulum likewise contains a Ca^{2+} ATPase that enables it to take up Ca^{2+} from the cytosol; the Ca^{2+} sequestered in this way is released back into the cytosol in response to specific extracellular signals, as discussed in Chapter 12 (see p. 701).

Membrane-bound Enzymes That Synthesize ATP Are Transport ATPases Working in Reverse[25]

The plasma membrane of bacteria, the inner membrane of mitochondria, and the thylakoid membrane of chloroplasts all contain an enzyme that is analogous to the two transport ATPases discussed above, but it normally works in reverse. Instead of ATP hydrolysis driving ion transport (as is usually the case for the Na^+-K^+ ATPase and Ca^{2+} ATPase), H^+ gradients across these membranes drive the synthesis of ATP from ADP and phosphate. The H^+ gradients are generated during the electron-transport steps of oxidative phosphorylation (in aerobic bacteria and mitochondria) or photosynthesis (in chloroplasts) or by the light-activated H^+

SOURCES OF INTRACELLULAR OSMOLARITY

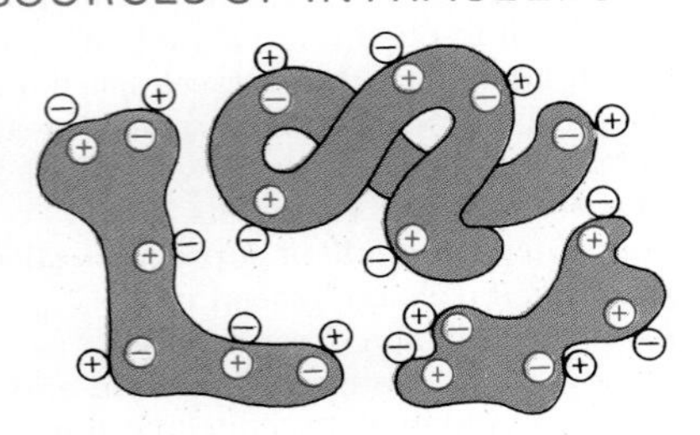

Macromolecules themselves contribute very little to the osmolarity of the cell interior since, despite their large size, each one counts only as a single molecule and there are relatively few of them compared to the number of small molecules in the cell. However, most biological macromolecules are highly charged, and they attract many inorganic ions of opposite charge. Because of their large numbers, these counterions make a major contribution to intracellular osmolarity.

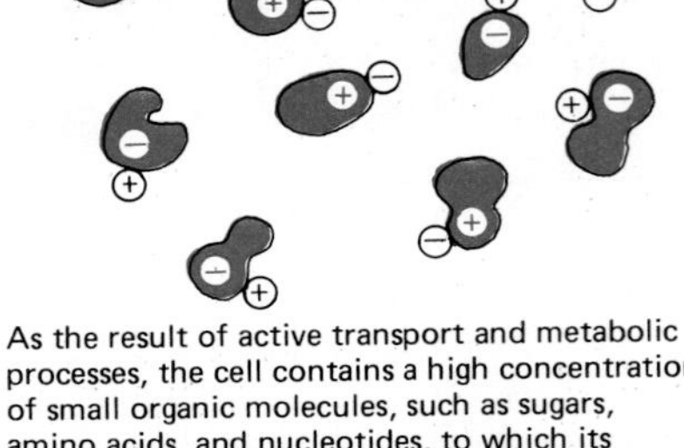

As the result of active transport and metabolic processes, the cell contains a high concentration of small organic molecules, such as sugars, amino acids, and nucleotides, to which its plasma membrane is impermeable. Because most of these metabolites are charged, they also attract counterions. Both the small metabolites and their counterions make a further major contribution to intracellular osmolarity.

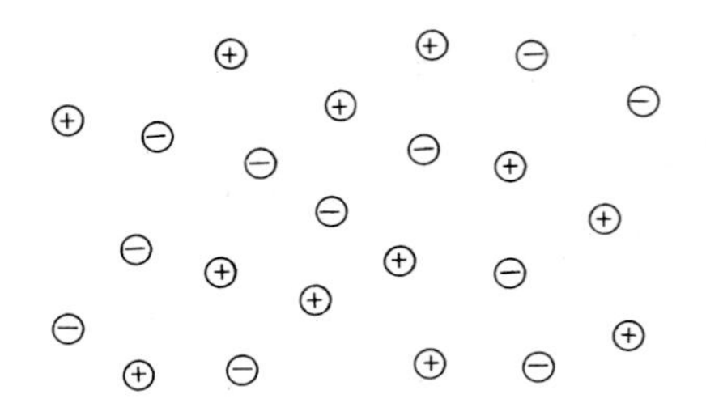

The osmolarity of the extracellular fluid is usually due mainly to small inorganic ions. These leak slowly across the plasma membrane into the cell. If they were not pumped out, and if there were no other molecules inside the cell that interacted with them so as to influence their distribution, they would eventually come to equilibrium with equal concentrations inside and outside the cell. However, the presence of charged macromolecules and metabolites in the cell that attract these ions gives rise to the Donnan effect: it causes the total concentration of inorganic ions (and therefore their contribution to the osmolarity) to be greater inside than outside the cell at equilibrium.

THE PROBLEM

Because of the above factors, a cell that does nothing to control its osmolarity will have a higher total concentration of solutes inside than outside. As a result, water will be higher in concentration outside the cell than inside. This difference in water concentration across the plasma membrane will cause water to move continuously into the cell by osmosis, causing it to rupture.

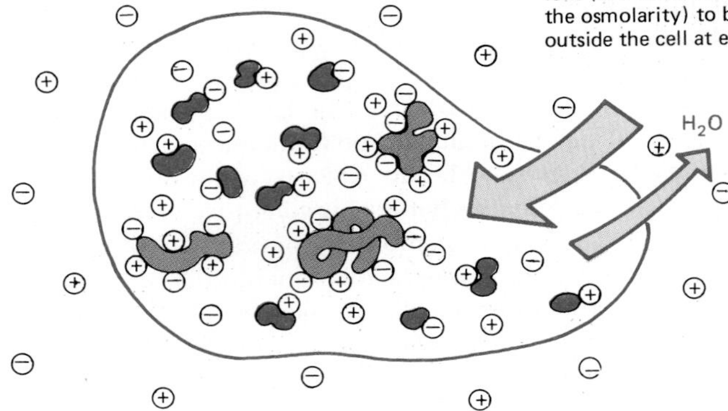

THE SOLUTION

Animal cells and bacteria control their intracellular osmolarity by actively pumping out inorganic ions, such as Na^+, so that their cytoplasm contains a lower total concentration of inorganic ions than the extracellular fluid, thereby compensating for their excess of organic solutes.

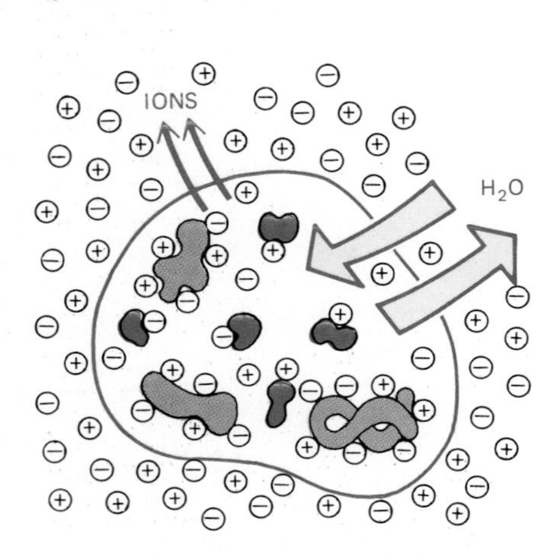

Plant cells are prevented from swelling by their rigid walls and so can tolerate an osmotic difference across their plasma membranes: an internal turgor pressure is built up, which at equilibrium forces out as much water as enters.

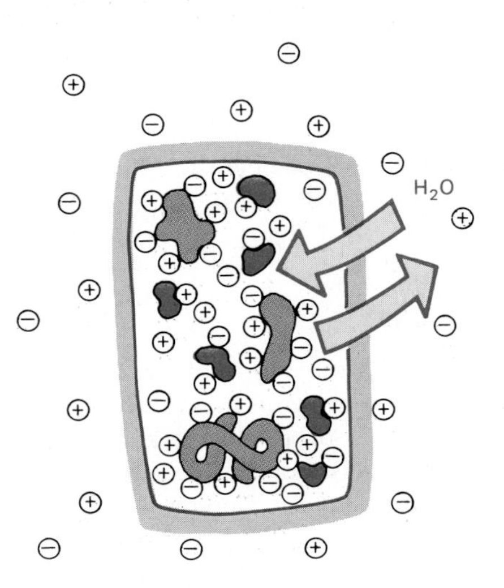

Many protozoa avoid becoming swollen with water, despite an osmotic difference across the plasma membrane, by periodically extruding water from special contractile vacuoles.

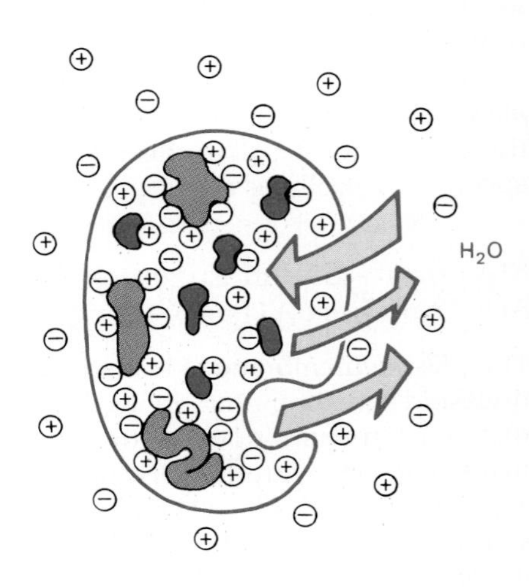

 Panel 6–1 Intracellular water balance: the problem and its solution.

pump (bacteriorhodopsin) in *Halobacterium*. The enzyme that normally synthesizes ATP, called *ATP synthetase*, can, like the transport ATPases, work in either direction, depending on the conditions: it can hydrolyze ATP and pump H^+ across the membrane, or it can synthesize ATP when H^+ flows through the enzyme in the reverse direction. ATP synthetase is responsible for producing nearly all of the ATP in most cells, and it is discussed in detail in Chapter 7 (see p. 356).

Active Transport Can Be Driven by Ion Gradients[26]

Many active transport systems are driven by the energy stored in ion gradients rather than by ATP hydrolysis. All of these function as coupled transporters—some as symports, others as antiports. In animal cells, Na^+ is the usual co-transported ion whose electrochemical gradient provides the driving force for the active transport of a second molecule. The Na^+ that enters the cell with the solute is pumped out by the Na^+-K^+ ATPase, which, by maintaining the Na^+ gradient, indirectly drives the transport. Intestinal and kidney epithelial cells, for instance, contain a variety of symport systems that are driven by the Na^+ gradient across the plasma membrane, each system specific for importing a small group of related sugars or amino acids into the cell. In these systems the solute and Na^+ bind to different sites on a carrier protein; because the Na^+ tends to move into the cell down its electrochemical gradient, the sugar or amino acid is, in a sense, "dragged" into the cell with it. The greater the Na^+ gradient, the greater the rate of solute entry; conversely, solute transport stops if the Na^+ concentration in the extracellular fluid is markedly reduced. A hypothetical (and highly simplified) model of how such a symport system could work is shown in Figure 6–51.

In bacteria and plants, most active transport systems driven by ion gradients depend on H^+ rather than Na^+ gradients. The active transport of many sugars and amino acids into bacterial cells, for example, is driven by the H^+ gradient across the plasma membrane. The *lactose carrier* (permease) is the most extensively studied example. It is a single transmembrane polypeptide (composed of about 400 amino acid residues) that is thought to traverse the lipid bilayer at least nine times. It functions as a H^+ symporter: one proton is co-transferred for every lactose molecule transported into the cell.

Ion gradients also can be used to drive antiport systems. Two important examples are the antiports that function together to regulate intracellular pH in many animal cells.

Antiports in the Plasma Membrane Regulate Intracellular pH[27]

Almost all vertebrate cells have a Na^+-driven antiport, called a **Na^+-H^+ exchange carrier,** in their plasma membrane that plays a crucial part in maintaining intracellular pH (pH_i, usually around 7.1 or 7.2). This carrier couples the efflux of H^+ to the influx of Na^+, and it thereby removes excess H^+ ions produced as a result of acid-forming reactions in the cell. The Na^+-H^+ exchanger is regulated by pH_i: when pH_i rises above 7.7 in chick muscle cells, for example, the exchanger is inactive; as pH_i falls, the activity of the exchanger increases, reaching half-maximal activity at around pH 7.4. This regulation is mediated by the binding of H^+ to a regulatory site on the cytoplasmic surface of the exchanger. The importance of the Na^+-H^+ exchanger in pH_i control is demonstrated by the fate of mutant fibroblasts that lack the exchanger: they rapidly die when exposed to a large acid load that has little effect on the survival of normal fibroblasts.

A **Cl^--HCO_3^- exchanger,** similar to the band 3 protein in the membrane of red blood cells (see p. 291), is also thought to play an important part in pH_i regulation in many nucleated cells. Like the Na^+-H^+ exchanger, the Cl^--HCO_3^- exchanger is regulated by pH_i, but in the opposite direction. Its activity increases as pH_i rises, increasing the rate at which HCO_3^- is ejected from the cell in exchange for Cl^-, thereby decreasing pH_i whenever the cytosol becomes too alkaline.

There is evidence that the Na^+-H^+ exchanger may be involved in transducing extracellular signals into intracellular ones, as well as in pH_i regulation. For example, most protein growth factors activate this antiport system in the course of

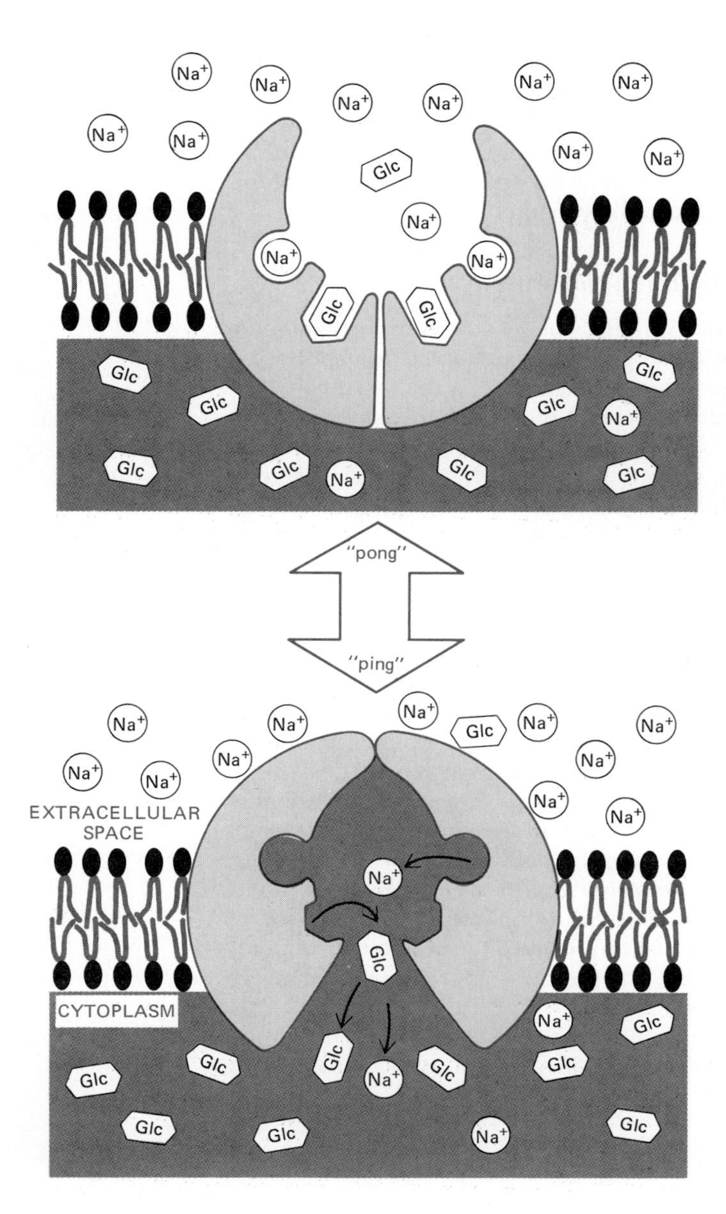

Figure 6–51 How a glucose pump could, in principle, be driven by a Na^+ gradient. The pump oscillates randomly between two states, "pong" and "ping," as in Figure 6–47. Although Na^+ binds equally well to the protein in either state, the binding of Na^+ induces an allosteric transition in the protein that greatly increases its affinity for glucose. Since the Na^+ concentration is higher in the extracellular space than in the cytosol, glucose is much more likely to bind to the pump in the "pong" state; therefore, both Na^+ and glucose enter the cell (via a pong → ping transition) much more often than they leave it (via a ping → pong transition). As a result, the system carries both glucose and Na^+ into the cell. By generating and maintaining the Na^+ gradient, the Na^+-K^+ ATPase indirectly provides the energy for such a symport system. For this reason, ion-driven carriers are said to mediate *secondary active transport* while transport ATPases are said to mediate *primary active transport*.

stimulating cell proliferation, increasing the pH_i from 7.1 or 7.2 to about 7.3. In some cases at least, they do so indirectly through the activation of a specific protein kinase (protein kinase C—see p. 704), which is thought to phosphorylate the exchanger, thereby increasing the affinity of its regulatory binding site for H^+ so that it remains active at a higher pH. Mutant cells that are deficient in the Na^+-H^+ exchanger, as well as cells that are treated with the drug *amiloride,* which inhibits the exchanger, fail to respond to these growth factors. These findings suggest that the activation of the exchanger and the resulting increase in pH_i play an important part in initiating cell proliferation. Similarly, following the fertilization of sea urchin eggs, an increase in pH_i caused by activation of the Na^+-H^+ exchanger activates both protein and DNA synthesis in the egg (see p. 874); it is not known which intracellular proteins respond to the increase in pH_i to cause these activations.

An Asymmetrical Distribution of Carrier Proteins in Epithelial Cells Underlies the Transcellular Transport of Solutes[28]

In some epithelial cells, such as those involved in absorbing nutrients from the gut, carrier proteins are distributed asymmetrically in the plasma membrane and thereby contribute to the **transcellular transport** of absorbed solutes. As shown

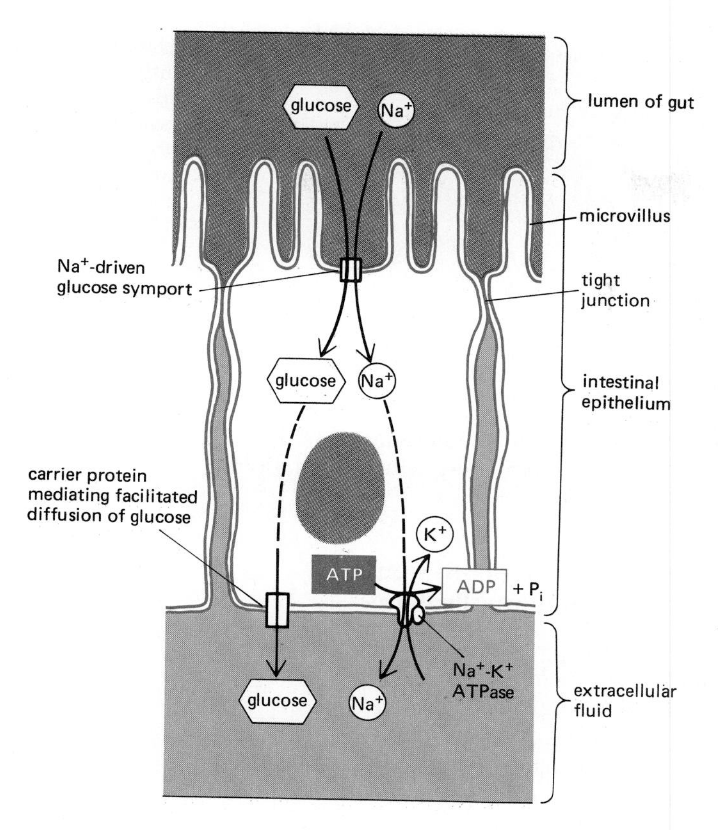

Figure 6–52 The asymmetrical distribution of transport proteins in the plasma membrane of an intestinal epithelial cell results in the transcellular transport of glucose from the gut lumen to the extracellular fluid (from where it passes into the blood). Glucose is pumped into the cell through the apical domain of the membrane by a Na^+-powered glucose symport, and glucose passes out of the cell (down its concentration gradient) by facilitated diffusion mediated by a different glucose carrier protein in the basal and lateral domain. The Na^+ gradient driving the glucose symport is maintained by the Na^+-K^+ ATPase in the basolateral plasma membrane domain, which keeps the internal concentration of Na^+ low.

Adjacent cells are connected by impermeable junctions (called *tight junctions*—see p. 793), which have a dual function in the transport process illustrated. These junctions prevent solutes from crossing the epithelium between cells, allowing a concentration gradient of glucose to be maintained across the cell sheet. The tight junctions are thought to serve also as diffusion barriers within the plasma membrane, confining the various carrier proteins to their respective membrane domains (see Figure 6–36, p. 297).

in Figure 6–52, Na^+-linked symports, located in the apical (absorptive) domain of the plasma membrane, actively transport nutrients into the cell, building up substantial concentration gradients, while Na^+-independent transport proteins in the basal and lateral (basolateral) domain allow nutrients to leave the cell passively down these concentration gradients. The Na^+-K^+ ATPase that maintains the Na^+ gradient across the plasma membrane of these cells is located in the basolateral domain. Related mechanisms are thought to be used by kidney and intestinal epithelial cells to pump water from one extracellular space to another.

In many of these epithelial cells, the plasma membrane area is greatly increased by the formation of thousands of **microvilli,** which extend as thin, fingerlike projections from the apical surface (see Figure 6–52). Such microvilli can increase the total absorptive area of a cell by as much as 25-fold, thereby greatly increasing its transport capabilities. Because the apical surface of gut epithelial cells is also a site where immobilized hydrolytic enzymes involved in the final stages of food digestion are located, both digestion and absorption are greatly enhanced by the increase in surface area afforded by the microvilli in this epithelium.

Active Transport in Bacteria Can Occur by "Group Translocation"[29]

Thus far we have seen that active transport can be driven by light (as in bacteriorhodopsin), by ATP hydrolysis, or by ion gradients. A fourth strategy, which operates in many bacteria, is to "trap" a molecule that has entered the cell passively by modifying it in such a way that it cannot escape by the same route. In some bacteria, for example, sugars are phosphorylated after their transfer across the plasma membrane. Because they are ionized and cannot leak out, the resulting sugar phosphates accumulate in the cell. Moreover, sugars entering the cell are phosphorylated immediately, so that the concentration of unphosphorylated sugars inside the cell is kept very low and sugar from outside can continue to enter down its concentration gradient. Because a phosphate group is transferred to the solute after its transport, this type of active transport is called **group translocation.** In the most extensively studied example, the phosphorylation mechanism is complex

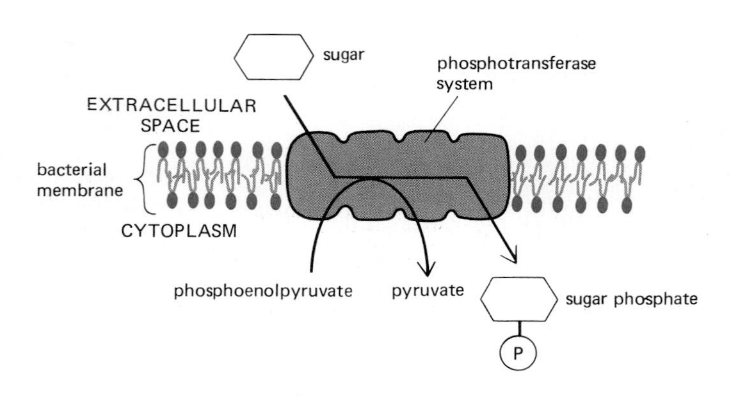

Figure 6–53 Active transport of sugars into bacteria by group translocation. A special "phosphotransferase system" of proteins in the bacterial membrane phosphorylates the sugar after the sugar's transport through the membrane. Phosphoenolpyruvate rather than ATP is the phosphate donor.

and highly regulated, involving at least four separate membrane proteins and phosphoenolpyruvate (rather than ATP) as the high-energy-phosphate donor (Figure 6–53).

Bacteria with Double Membranes Have Transport Systems That Depend on Water-soluble Substrate-binding Proteins[30]

As mentioned previously, the plasma membranes of all bacteria contain carrier proteins that use the H^+ gradient across the membrane to pump a variety of nutrients into the cell. But many bacteria, including *E. coli,* also have a surrounding *outer membrane,* through which solutes of up to 600 daltons can diffuse relatively freely through a variety of channel-forming proteins (known collectively as *porins*) (Figure 6–54). In these bacteria some sugars, amino acids, and small peptides are transported across the *inner (plasma) membrane* via a two-component transport system that utilizes water-soluble proteins located in the *periplasmic space* between the two membranes. These **periplasmic substrate-binding proteins** bind the specific molecule to be transported and, as a consequence, undergo a conformational change that enables them to bind to the second component in the transport system, which is a transmembrane carrier protein located in the inner membrane (Figure 6–55). It is thought that the substrate-binding proteins pass the bound solute to the carrier, which then uses the energy of ATP hydrolysis to transfer the solute across the inner membrane into the cell. The same periplasmic substrate-binding proteins serve as receptors in *chemotaxis,* an adaptive response that enables bacteria to swim toward an increasing concentration of a specific nutrient (see p. 720).

We now turn from carrier proteins to channel proteins.

Channel Proteins Form Aqueous Pores in the Plasma Membrane[31]

Unlike carrier proteins, **channel proteins** form water-filled pores across membranes. But whereas the channel-forming proteins of the outer membranes of bacteria (and of mitochondria and chloroplasts) have large, relatively unselective pores, channel proteins in the plasma membranes of animal and plant cells have

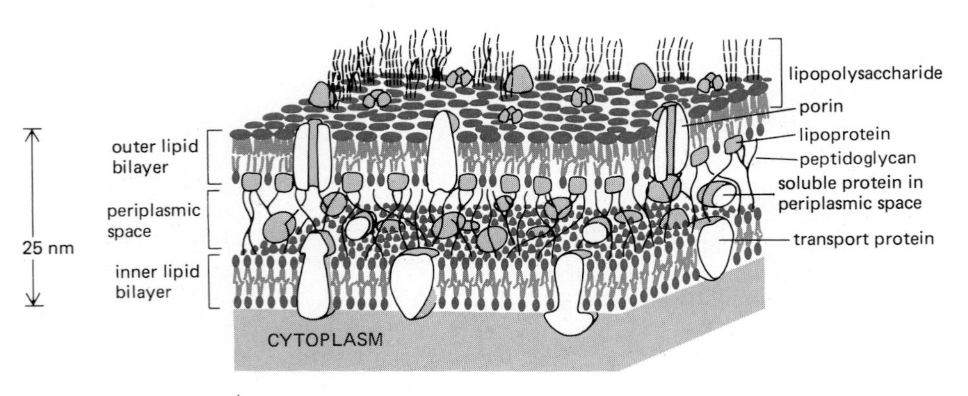

Figure 6–54 Schematic view of a small section of the double membrane of an *E. coli* bacterium. The inner membrane is the cell's plasma membrane. Between the inner and outer lipid bilayer membranes there is a highly porous, rigid peptidoglycan composed of protein and polysaccharide that constitutes the bacterial cell wall; it is attached to lipoprotein molecules in the outer membrane and fills the *periplasmic space.* This space also contains a variety of soluble protein molecules. The dashed black threads at the top represent the polysaccharide chains of the special lipopolysaccharide molecules that form the external monolayer of the outer membrane; for clarity, only a few of these chains are shown. Bacteria with double membranes are called *gram negative* because they do not retain the dark blue dye used in the gram staining procedure. Bacteria with single membranes (but thicker cell walls), such as staphylococci and streptococci, retain the blue dye and therefore are called *gram positive;* their single membrane is analogous to the inner (plasma) membrane of gram-negative bacteria.

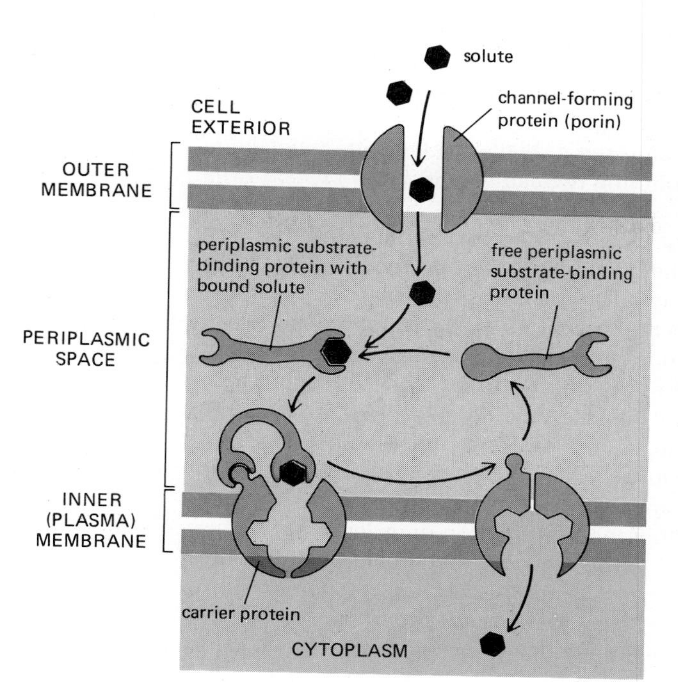

Figure 6–55 The transport system that depends on *periplasmic substrate-binding proteins* in bacteria with double membranes. The solute diffuses through channel-forming proteins (called *porins*) in the outer membrane and binds to a periplasmic substrate-binding protein. As a result, the substrate-binding protein undergoes a conformational change that enables it to bind to a carrier protein in the plasma membrane, which then picks up the solute and actively transfers it across the bilayer in a reaction driven by ATP hydrolysis. The peptidoglycan is omitted for simplicity; its porous structure allows the substrate-binding proteins and water-soluble solutes to move through it by simple diffusion.

small, highly selective pores. Almost all of the latter proteins are concerned specifically with ion transport and so are referred to as **ion channels.** More than 10^6 ions can pass through such a channel each second, which is a rate more than 100 times greater than the transport mediated by any known carrier protein. On the other hand, ion channels cannot be coupled to an energy source, so the transport they mediate is always passive ("downhill"), allowing specific ions, mainly Na^+, K^+, Ca^{2+}, or Cl^-, to diffuse down their electrochemical gradients across the lipid bilayer.

The channel proteins in plasma membranes show *ion selectivity,* permitting some ions to pass but not others. This suggests that their pores must be narrow enough in places to force permeating ions into intimate contact with the walls of the channel so that only ions of appropriate size and charge can pass. It is thought that the permeating ions have to shed most or all of their associated water molecules in order to get through the narrowest part of the channel. This both limits their maximum rate of passage and acts as a selective filter, letting only certain ions pass through. Thus, as ion concentrations are increased, the flux of ions through a channel increases proportionally but then levels off (saturates) at a certain maximum rate.

Another way in which ion channels differ from simple aqueous pores is that they are not continuously open. Instead they have "gates," which open briefly and then close again, as shown schematically in Figure 6–56. In most cases the gates open in response to a specific perturbation of the membrane. The main types of perturbations that are known to cause ion channels to open are a change in the voltage across the membrane (*voltage-gated channels*), mechanical stimulation

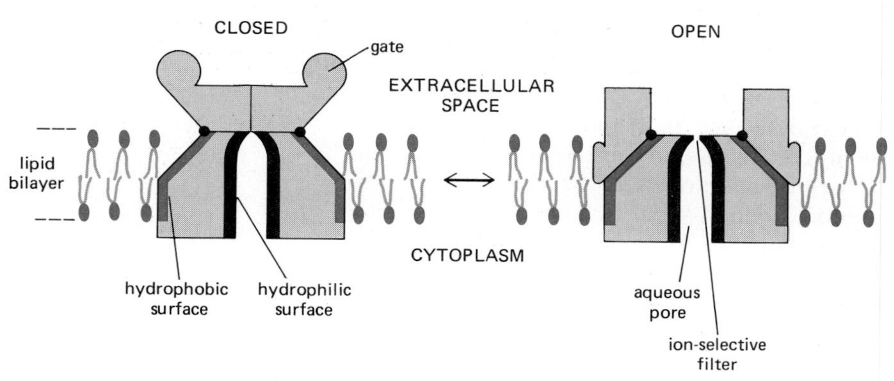

Figure 6–56 Schematic drawing of a gated ion channel in its closed and open conformations. A transmembrane protein, seen in cross-section, forms an aqueous pore across the lipid bilayer only when the gate is open. Hydrophilic amino acid side chains are thought to line the wall of the pore; hydrophobic side chains interact with the lipid bilayer. The pore narrows to atomic dimensions in one region (the "ion-selective filter"), where the ion selectivity of the channel is determined. A transient opening of the gate is caused by a specific perturbation of the membrane, which is different for different channels, as discussed in the text. The location of the gate and ion-selective filter, shown here on the external side of the membrane, is unknown for most channels.

(*mechanically gated channels*—see p. 1103), or the binding of a signaling molecule (*ligand-gated channels*). The signaling ligand can be either an extracellular mediator, called a *neurotransmitter* (*transmitter-gated channels*), or an intracellular mediator, such as an ion (*ion-gated channels*—see p. 1189), a nucleotide (*nucleotide-gated channels*—see p. 713), or a GTP-binding regulatory protein (*G-protein-gated channels*—see p. 706).

Approximately 50 types of ion channels have been described thus far, and new ones are still being discovered. They are responsible for the electrical excitability of nerve and muscle cells and mediate most forms of electrical signaling in the nervous system. A single nerve cell typically contains more than five kinds of ion channels. But these channels are not restricted to electrically excitable cells. They are present in all animal cells and are found in plant cells and microorganisms: they propagate the leaf-closing response of the mimosa plant, for example, and allow the single-celled paramecium to reverse direction after a collision.

Perhaps the most common ion channels are those that are permeable mainly to K^+ and are found in the plasma membrane of almost all animal cells. Because they seem not to require a specific membrane perturbation in order to open, they are sometimes called *K^+ leak channels.* Although poorly characterized and heterogeneous, they are the reason that most plasma membranes are much more permeable to K^+ than to other ions, and they play a critical part in maintaining the *membrane potential*—the voltage difference that is present across all plasma membranes.

6-24
6-26
The Membrane Potential Depends Largely on K^+ Leak Channels and the K^+ Gradient Across the Membrane

A **membrane potential** arises when there is a difference in the electric charge on the two sides of a membrane, due to a slight excess of positive ions over negative on one side and a slight deficit on the other. Such charge differences can result both from passive ion diffusion and from active electrogenic pumping. We shall see in Chapter 7 (p. 351) that most of the membrane potential of the mitochondrion is generated by electrogenic proton pumps in the mitochondrial inner membrane. Electrogenic pumps also generate most of the plasma membrane potential in plants and fungi. In typical animal cells, however, ion diffusion makes the largest contribution to the potential across the plasma membrane.

As explained earlier, the Na^+-K^+ ATPase helps to maintain osmotic balance across the animal cell membrane by keeping the intracellular concentration of Na^+ low. Because there is little Na^+ inside the cell, other cations have to be plentiful there to balance the charge carried by the cell's fixed anions—the negatively charged organic molecules that are confined inside the cell. This balancing role is performed largely by K^+, which is actively pumped into the cell by the Na^+-K^+ ATPase but can also move freely in or out through the **K^+ leak channels.** Thanks to the K^+ leak channels, K^+ comes very nearly to an equilibrium in which the electrical force due to negative charges attracting K^+ into the cell balances the tendency of K^+ to leak out down its concentration gradient. The membrane potential is the manifestation of this electrical force, and its size can be calculated from the steepness of the K^+ concentration gradient. The following argument may help to make this clear.

Suppose that initially there is no voltage gradient across the plasma membrane (the membrane potential is zero) but the concentration of K^+ is high inside the cell and low outside. K^+ will tend to leave the cell through the K^+ leak channels, driven by its concentration gradient. As K^+ moves out of the cell, it will leave behind negative charge, thereby creating an electrical field, or membrane potential, that will tend to oppose the further efflux of K^+. The net efflux of K^+ will halt if the membrane potential reaches a value where this electrical driving force on K^+ exactly balances the effect of its concentration gradient—that is, when the electrochemical gradient for K^+ is zero. Cl^- ions also equilibrate across the membrane, but because their charge is negative, the membrane potential keeps most of these diffusible ions out of the cell. This equilibrium condition, in which there is no net

6-24 6-25 6-26

THE NERNST EQUATION AND ION FLOW

The flow of any ion through a membrane channel protein is driven by the **electrochemical gradient** for that ion. This gradient represents the combination of two influences: the voltage gradient and the concentration gradient of the ion across the membrane. When these two influences just balance each other, the electrochemical gradient for the ion is zero and there is no *net* flow of the ion through the channel. The voltage gradient (membrane potential) at which this equilibrium is reached is called the **equilibrium potential** for the ion. It can be calculated from an equation that will be derived below, called the **Nernst equation**.

The Nernst equation is

$$V = \frac{RT}{zF} \ln \frac{C_o}{C_i}$$

where

V = the equilibrium potential in volts (internal potential minus the external potential)

C_o and C_i = outside and inside concentrations of the ion, respectively

R = the gas constant (2 cal mol^{-1} °K^{-1})

T = the absolute temperature (°K)

F = Faraday's constant (2.3×10^4 cal V^{-1} mol^{-1})

z = the valence (charge) of the ion

The Nernst equation is derived as follows:

A molecule in solution (a solute) always moves from a region of high concentration to a region of low concentration simply due to the pressure of numbers. Consequently, movement down a concentration gradient is accompanied by a favorable free-energy change ($\Delta G < 0$), whereas movement up a concentration gradient is accompanied by an unfavorable free-energy change ($\Delta G > 0$). (Free energy is introduced and discussed in Panel 2–7, pp. 72–73.) The free-energy change per mole of solute moved across the plasma membrane (ΔG_{conc}) is equal to $-RT \ln C_o/C_i$. If the solute is an ion, moving it into a cell across a membrane whose inside is at a voltage V relative to the outside will cause an additional free-energy change (per mole of solute moved) of $\Delta G_{volt} = zFV$. At the point where the concentration and voltage gradients just balance, $\Delta G_{conc} + \Delta G_{volt} = 0$ and the ion distribution is at equilibrium across the membrane. Thus,

$$zFV - RT \ln \frac{C_o}{C_i} = 0$$

and therefore,

$$V = \frac{RT}{zF} \ln \frac{C_o}{C_i} = 2.3 \frac{RT}{zF} \log_{10} \frac{C_o}{C_i}$$

For a univalent ion,

$$2.3 \frac{RT}{F} = 58 \text{ mV at } 20\text{ °C} \quad \text{and} \quad 61.5 \text{ mV at } 37\text{°C}$$

Therefore, when the membrane potential has a value of $61.5 \log_{10}([K^+]_o/[K^+]_i)$ millivolts (-89 mV when $[K^+]_o = 5$ mM and $[K^+]_i = 140$ mM), which is the K^+ equilibrium potential (V_K), there is no net flow of K^+ across the membrane. Similarly, when the membrane potential has a value of $61.5 \log_{10}([Na^+]_o/[Na^+]_i)$, the Na^+ equilibrium potential (V_{Na}), there is no net flow of Na^+. For a typical cell, V_K is between -70 mV and -100 mV and V_{Na} is between $+50$ mV and $+65$ mV.

For any particular membrane potential V_M, the net force tending to drive a particular type of ion out of the cell is proportional to the difference between V_M and the equilibrium potential for the ion: hence, for K^+ it is $V_M - V_K$ and for Na^+ it is $V_M - V_{Na}$. The actual current carried by each type of ion depends not only on this driving force but also on the ease with which that ion passes through its membrane channels, which is a function of the *conductance* of the channels. If the conductances of the sets of channels for K^+ and Na^+ are, respectively, g_K and g_{Na}, from Ohm's law the K^+ and Na^+ currents are, respectively, $g_K(V_M - V_K)$ and $g_{Na}(V_M - V_{Na})$. (**Conductance** is the reciprocal of resistance, and the unit of measurement is the reciprocal of ohm, or siemens, S. Most individual ion channels have conductances in the range of 1 to 150×10^{-12} S, or 1 to 150 pS.) At the resting membrane potential of most cells, the K^+ leak channels are the main type of channel open and, therefore, g_K dominates the total conductance of the membrane. During an action potential in nerve and muscle cells, large numbers of voltage-gated Na^+ channels briefly open, and g_{Na} comes to dominate the total membrane conductance for a moment.

Panel 6–2 The derivation of the Nernst equation.

exact balance of charges on each side of the membrane; membrane potential = 0

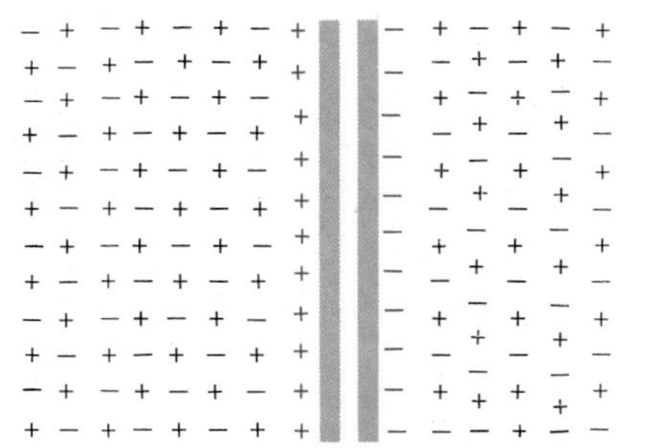

a few of the positive ions (*color*) cross the membrane from right to left, leaving their negative counterions behind (*color*); this sets up a nonzero membrane potential

Figure 6–57 A small flow of ions carries sufficient charge to cause a large change in the membrane potential. The ions that give rise to the membrane potential lie in a surface layer close to the membrane, held there by their electrical attraction to their oppositely charged counterparts (counterions) on the other side of the membrane. For a typical cell, 1 microcoulomb of charge (6×10^{12} univalent ions) per square centimeter of membrane, transferred from one side of the membrane to the other, will change the membrane potential by roughly 1 V. This means, for example, that in a spherical cell of diameter 10 μm, the number of K^+ ions that have to flow out to alter the membrane potential by 100 mV is only about 1/100,000 of the total number of K^+ ions in the cytosol.

flow of electric current across the membrane, defines the **resting membrane potential** for the cell. A simple but very important formula, the **Nernst equation,** expresses the equilibrium condition quantitatively and, as explained in Panel 6–2, makes it possible to calculate the resting membrane potential if the ratio of internal and external ion concentrations is known.

The number of ions that must move across the membrane to set up the membrane potential is minute. Thus one can think of the membrane potential as arising from movements of charge that leave ion *concentrations* practically unaffected and result in only a very slight discrepancy in the number of positive and negative ions on the two sides of the membrane (Figure 6-57). Moreover, these movements of charge are generally rapid, taking only a few milliseconds or less.

It is illuminating to consider what happens if the Na^+-K^+ ATPase is suddenly inactivated. First, there is an immediate slight drop in the membrane potential. This is because the pump is electrogenic and, when active, makes a small direct contribution to the membrane potential by driving a current out of the cell (see p. 306). Switching off the pump, however, does not abolish the major component of the resting potential, which is generated by the K^+ equilibrium mechanism outlined above. This persists as long as the K^+ concentration inside the cell stays high—typically for many minutes. The plasma membrane, however, is somewhat permeable to all small ions (including Na^+). Therefore, the ion gradients set up by pumping will eventually run down without the Na^+-K^+ ATPase, and the membrane potential established by diffusion through K^+ channels will fall as well. At the same time, the osmotic balance is upset (see p. 306); but if the cell does not burst, it eventually comes to a new resting state where Na^+, K^+, and Cl^- are all at equilibrium across the membrane. The membrane potential in this state is much less than it was in the normal cell with an active Na^+-K^+ pump.

The potential difference across the plasma membrane of a cell at rest varies between −20 mV and −200 mV, depending on the organism and cell type. Although the K^+ gradient always has a major influence on this potential, the gradients of other ions (and the disequilibrating effects of ion pumps) also have a significant effect: the more permeable the membrane for a given ion, the more strongly the membrane potential tends to be driven toward the equilibrium value for that ion. Consequently, almost any change of a membrane's permeability to ions causes a change in the membrane potential. This is the key principle relating the electrical excitability of cells to the activities of ion channels.

6-25 Voltage-gated Ion Channels Are Responsible for the Electrical Excitability of Nerve and Muscle Cells[33]

The plasma membranes of electrically excitable cells (mainly nerve and muscle cells) contain **voltage-gated ion channels,** which are responsible for generating **action potentials**—rapid, transient, self-propagating electrical excitations of the membrane. An action potential is triggered by a *depolarization* of the membrane—that is, by a shift in the membrane potential to a less negative value. A stimulus

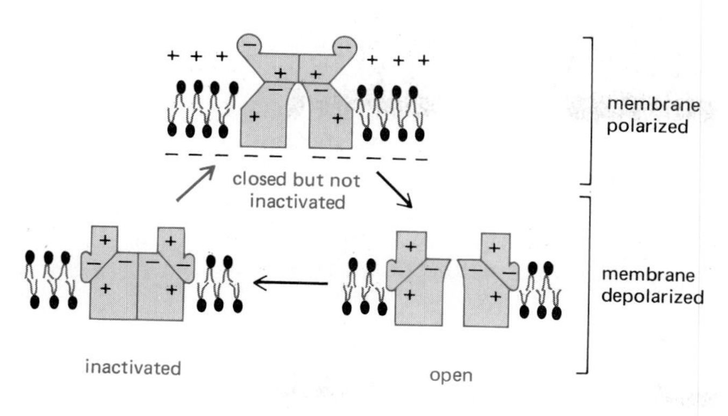

Figure 6–58 The voltage-gated Na^+ channel can adopt at least three conformations (states). Internal forces, represented here by attractions between charges on different parts of the channel, stabilize each state against small disturbances, but a sufficiently violent collision with other molecules can cause the channel to flip from one of these states to another. The state of lowest energy depends on the membrane potential because the different conformations have different charge distributions. When the membrane is at rest (highly polarized), the *closed but not inactivated* conformation has the lowest free energy and is therefore most stable; when the membrane is depolarized, the energy of the *open* conformation is lower and so the channel opens. But the free energy of the *inactive* conformation is lower still, and so, after a randomly variable period spent in the open state, the channel becomes inactivated. Thus the open conformation corresponds to a metastable state that can exist only transiently. The black arrows indicate the sequence that follows a sudden depolarization, while the red arrow indicates the return to the original conformation as the lowest energy state after the membrane is repolarized.

that causes a momentary partial depolarization promptly causes **voltage-gated Na^+ channels** to open, allowing a small amount of Na^+ to enter the cell. The influx of positive charge depolarizes the membrane further, thereby opening more Na^+ channels, which admit more Na^+ ions, causing still further depolarization. This process continues in a self-amplifying fashion until the potential in the local region of membrane has shifted from its resting value of about -70 mV all the way to the Na^+ equilibrium potential of about $+50$ mV (see Panel 6–2, p. 311). At this point, when the net electrochemical driving force for the flow of Na^+ is zero, the cell would come to a new resting state with all of its Na^+ channels permanently open if the open conformation of the channel were stable.

The cell is saved from such a permanent electrical spasm because the Na^+ channels have an automatic inactivating mechanism, and they rapidly reclose even though the membrane is still depolarized. In this *inactivated* state, the channel cannot open again until a few milliseconds after the membrane potential returns to its initial negative value. A schematic illustration of these three distinct states of the voltage-gated Na^+ channel—closed but not inactivated, open, and inactivated—is shown in Figure 6–58. How they contribute to the rise and fall of the action potential is shown in Figure 6–59.

The description just given of an action potential concerns only a small patch of plasma membrane. However, the self-amplifying depolarization of the patch is sufficient to depolarize neighboring regions of membrane, which then go through the same cycle. In this way the action potential spreads from the initial site of depolarization to involve the entire plasma membrane. A more detailed discussion

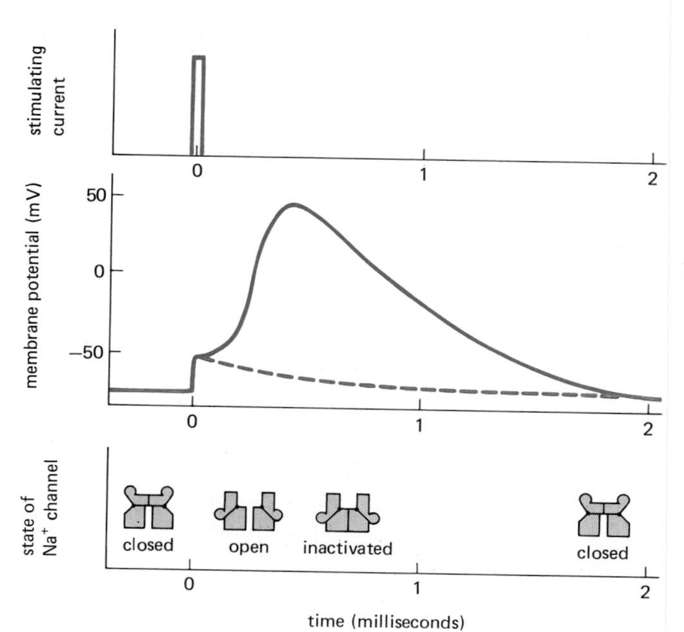

Figure 6–59 The triggering of an action potential by a brief pulse of current (shown in the upper graph), which partially depolarizes the membrane (middle graph). In the plot of membrane potential, the solid curve shows the course of an action potential due to the opening and subsequent inactivation of voltage-gated Na^+ channels. The return of the membrane potential to its initial value of -70 mV is automatic once the Na^+ channels close because of the continuous efflux of K^+ through K^+ channels (see p. 314). The membrane cannot fire a second action potential until the Na^+ channels, whose state is shown at the bottom, have returned to the closed but not inactivated conformation (see Figure 6–58); until then the membrane is refractory to stimulation. The dashed curve shows how the membrane potential would have simply relaxed back to the resting value after the initial depolarizing stimulus if there had been no voltage-gated ion channels in the membrane.

of the properties and functions of action potentials is given in Chapter 19 (see pp. 1065–1074).

Neurons and muscle cell membranes contain many thousands of voltage-gated Na^+ channels, and the current crossing the membrane is the sum of the currents flowing through all of these. This aggregate current can be recorded with a microelectrode; as described in Chapter 19 (see p. 1071). However, it is also possible to record current flowing through individual channels by means of *patch-clamp recording*, which provides a much more detailed picture of the properties of these channels.

6-27 Patch-Clamp Recording Indicates That Individual Na^+ Channels Open in an All-or-None Fashion[34]

Patch-clamp recording has revolutionized the study of ion channels. With this technique, transport through a single molecule of channel protein can be studied in a small patch of membrane covering the mouth of a micropipette (Figure 6–60). This makes it possible to record from ion channels in all sorts of cell types, including those that are not electrically excitable. Many of these cells, such as yeasts, are too small to be investigated by the traditional electrophysiologist's method of impalement with an intracellular electrode.

Patch-clamp recording indicates that individual Na^+ channels open in an all-or-nothing fashion: when open, the channel always has the same conductance, although the times of its opening and closing are random. Therefore, the aggregate current crossing the membrane of an entire cell through a large population of Na^+ channels does not indicate the *degree* to which a typical individual channel is open, but rather the average *probability* that it will open (Figure 6–61).

The phenomenon of voltage gating can be understood in terms of simple physical principles. The interior of the resting nerve or muscle cell is at an electric potential about 50–100 mV more negative than the external medium. This potential difference may seem small, but since it exists across a plasma membrane only about 5 nm thick, the resulting voltage gradient is about 100,000 V/cm. Proteins in the membrane are thus subjected to a very large electrical field. Membrane proteins, like all others, have a number of charged groups on their surface and polarized bonds between their various atoms. The electrical field therefore exerts forces on the molecular structure. For many membrane proteins the effects of changes in the membrane electrical field are probably insignificant. But voltage-gated ion channels have evolved a delicately poised sensitivity to the field: they can adopt a number of alternative conformations, whose stabilities depend on the strength of the field. Each conformation is stable against small disturbances, but it can "flip" to another conformation if given a sufficient jolt by the random thermal movements of the surroundings (see Figure 6–58).

The function of the voltage-gated Na^+ channel is blocked specifically by two paralytic poisons, *tetrodotoxin (TTX)*, made by puffer fish, and *saxitoxin*, made by certain marine dinoflagellates. Because of their high affinity and specificity, these toxins have been invaluable for pharmacological studies, for counting the number of Na^+ channels in a membrane, and for purifying them. In a skeletal muscle plasma membrane, for example, it can be shown that there are only a few hundred Na^+ channels per square micrometer, which is about one for every 10,000 phospholipid molecules. Despite this low density, the membranes are electrically excitable because each channel has a very large conductance, allowing more than 8000 ions to pass per millisecond.

The DNA sequence that encodes a voltage-gated Na^+ channel (in eels) was determined in 1984. It was found to code for a single very large polypeptide chain (about 1800 amino acid residues) containing four homologous transmembrane domains (each containing six putative membrane-spanning α helices), which are thought to cluster together to form the walls of an aqueous pore. More recently, the DNA sequence that codes for a voltage-gated Ca^{2+} channel was determined and shown to encode a large polypeptide with a structure very similar to that of the Na^+ channel, suggesting that voltage-gated ion channels belong to a family of evolutionarily and structurally related proteins. In each of these channels, one of the putative transmembrane segments contains regularly spaced, positively charged

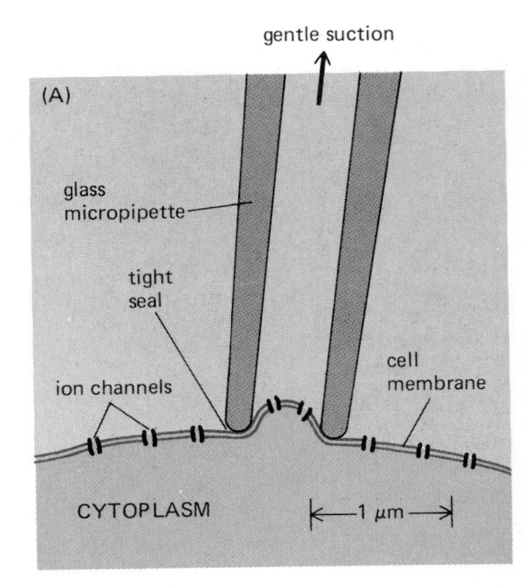

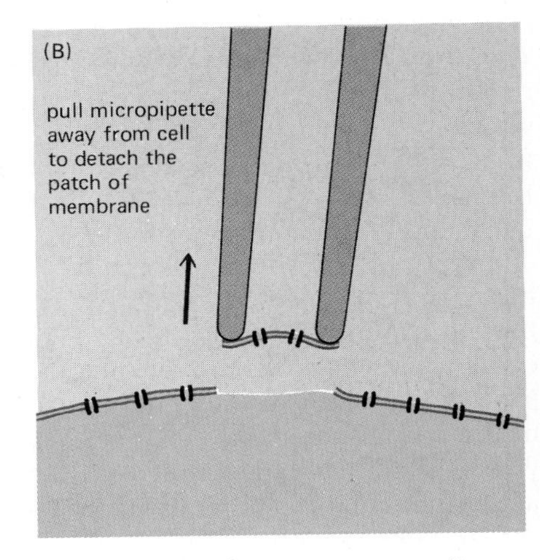

Figure 6–60 Patch-clamp recording. Because of the extremely tight seal between the micropipette and the membrane, current can enter or leave the micropipette only by passing through the channels in the patch of membrane covering its tip. Recordings of the current through these channels can be made with the patch still attached to the rest of the cell, as in (A), or detached, as in (B). The advantage of the detached patch is that it is easy to alter the composition of the solution on either side of the membrane to test the effect of various solutes on channel behavior. The detached patch can also be in the opposite orientation, with the cytoplasmic surface of the membrane facing the lumen of the pipette. (See also Figures 4–33 and 4–34, p. 156).

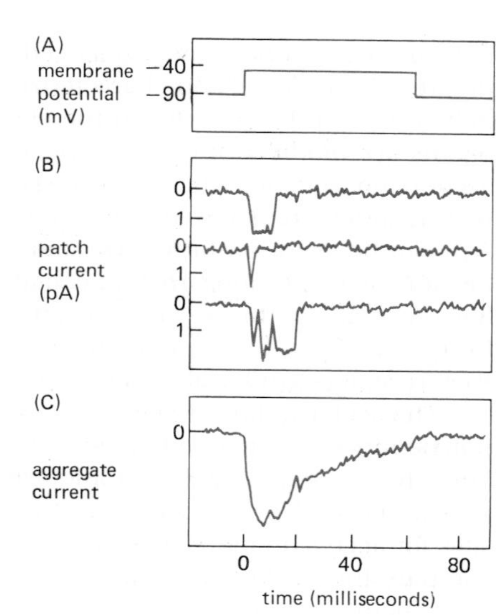

Figure 6–61 Recordings of current through a single voltage-gated Na^+ channel in a tiny patch of plasma membrane detached from an embryonic rat muscle cell (see Figure 6–60). The membrane was depolarized by an abrupt shift of potential, as indicated in (A). The three current records shown in (B) are from three experiments performed on the same patch of membrane. Each major current step in (B) represents the opening and closing of a single channel. Comparison of the three records shows that, whereas the times of channel opening and closing vary greatly, the rate at which current flows through an open channel is practically constant. The minor fluctuations in the current record arise largely from electrical noise in the recording apparatus. The sum of the currents measured in 144 repetitions of the same experiment is shown in (C). This aggregate current is equivalent to the usual Na^+ current that would be observed flowing through a relatively large region of membrane containing 144 channels. Comparison of (B) and (C) reveals that the time course of the aggregate current reflects the probability that any individual channel will be in the open state; this probability decreases with time as the channels in the depolarized membrane adopt their inactived conformation. The kinetics of channel opening and inactivation are much slower for this embryonic muscle cell than for a typical nerve cell. (Based on data from J. Patlak and R. Horn, *J. Gen. Physiol.* 79:333–351, 1982, by copyright permission of the Rockefeller University Press.)

amino acid residues, which together are thought to function as a voltage sensor, ensuring that the channel opens when the membrane is sufficiently depolarized (see Figure 6–58).

But more is known about the structure of another class of ion channels, which open in response to the binding of a specific neurotransmitter rather than to a change in membrane potential. These *transmitter-gated ion channels* also belong to a family of closely related proteins; unlike the voltage-gated Na^+ and Ca^{2+} channels, however, each of which is constructed from a single large polypeptide, all of the transmitter-gated ion channels studied so far are constructed from multiple homologous subunits.

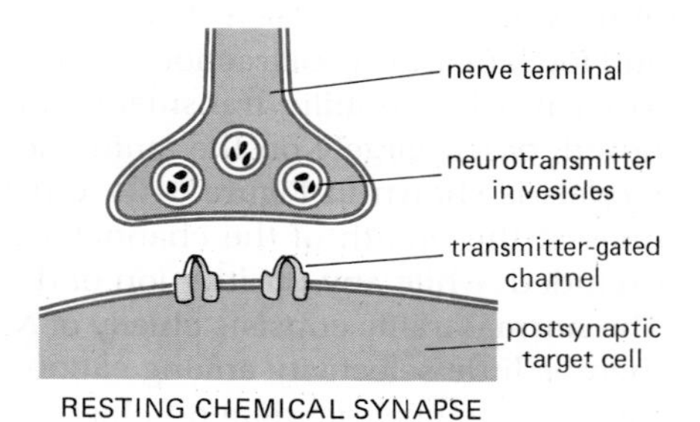

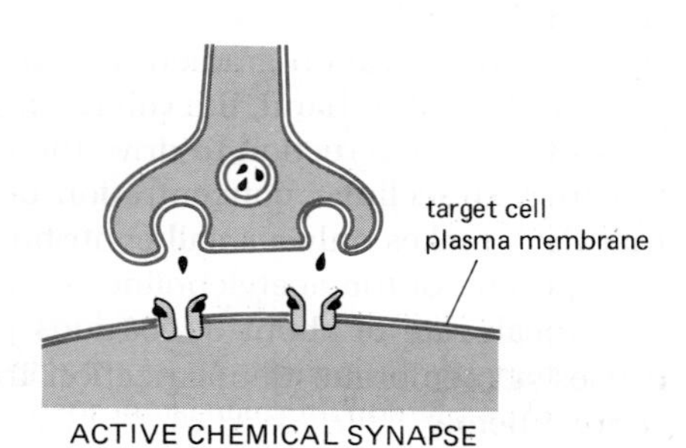

Figure 6–62 A chemical synapse. When an action potential reaches the nerve terminal, it stimulates the terminal to release its neurotransmitter; the neurotransmitter is contained in secretory vesicles and is released to the cell exterior when the vesicles fuse with the plasma membrane of the nerve terminal. The released neurotransmitter binds to and opens the transmitter-gated ion channels concentrated in the plasma membrane of the target cell at the synapse. The resulting ion flows alter the membrane potential of the target cell, thereby transmitting a signal from the excited nerve.

The Acetylcholine Receptor Is a Transmitter-gated Cation Channel[35]

Transmitter-gated ion channels are specialized for converting extracellular chemical signals into electrical signals and are located at specialized junctions (called *chemical synapses*) between nerve cells and their target cells. The channels are concentrated in the target cell plasma membrane in the region of the synapse. They open transiently in response to the binding of a *neurotransmitter* released by the nerve terminal, thereby producing a permeability change in the postsynaptic membrane of the target cell (Figure 6–62). Unlike the voltage-gated channels responsible for action potentials, transmitter-gated channels are relatively insensitive to the membrane potential and, therefore, cannot by themselves produce a self-amplifying excitation. Instead, they produce permeability changes, and hence changes of membrane potential, that are graded according to how much neurotransmitter is released at the synapse and how long it persists there. An action potential can be triggered only if voltage-gated channels are also present in the same target cell membrane.

In addition to having a characteristic ion selectivity, each transmitter-gated channel has a highly selective binding site for its particular neurotransmitter. The best-studied example of such a transmitter-gated channel is the **acetylcholine receptor** of skeletal muscle cells. This channel is opened transiently by *acetylcholine*, the neurotransmitter released from the nerve terminal at a *neuromuscular junction* (see p. 1075). The acetylcholine receptor has a special place in the history of ion channels. It was the first ion channel to be purified, the first to have its complete amino acid sequence determined, the first to be functionally reconstituted into synthetic lipid bilayers, and the first for which the electrical signal of a single open channel was recorded. Its gene was also the first channel protein gene

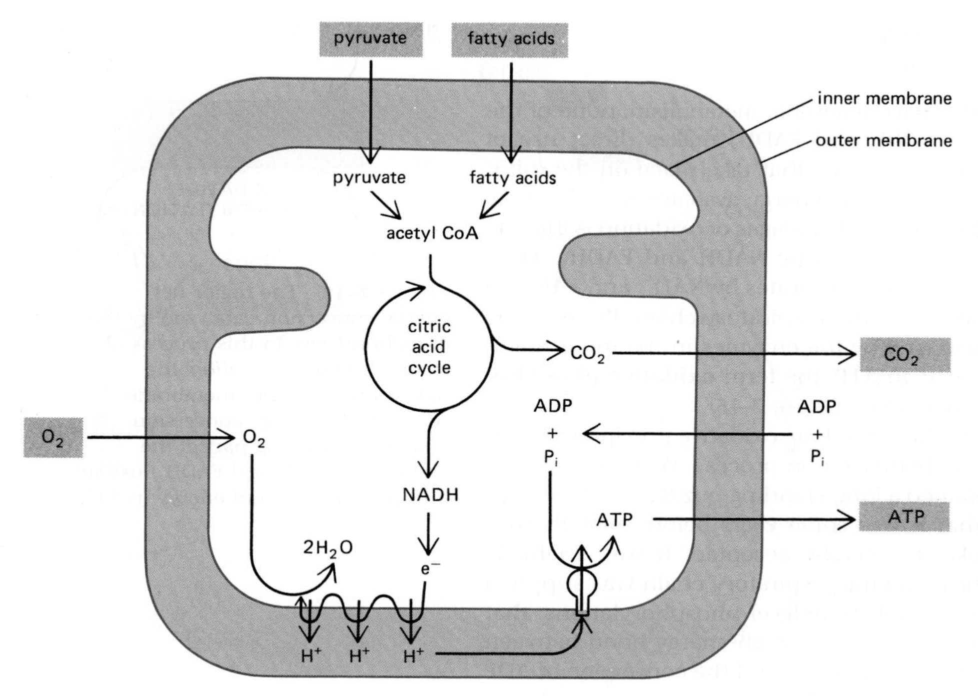

Figure 7–16 A summary of the flow of major reactants into and out of the mitochondrion. Pyruvate and fatty acids enter the mitochondrion and are metabolized by the citric acid cycle, which produces NADH. In the process of oxidative phosphorylation, high-energy electrons from NADH are then passed to oxygen by means of the respiratory chain in the inner membrane, producing ATP by a chemiosmotic mechanism.

NADH generated by glycolysis in the cytosol also passes electrons to the respiratory chain (not shown). Since NADH cannot pass across the mitochondrial inner membrane, the electron transfer from cytosolic NADH must be accomplished indirectly by means of one of several "shuttle" systems that transport another reduced compound into the mitochondrion; after being oxidized, this compound is returned to the cytosol, where it is reduced by NADH again.

Electrons Are Transferred from NADH to Oxygen Through Three Large Respiratory Enzyme Complexes[7]

Although the mechanism by which energy is harvested by the respiratory chain differs from that in other catabolic reactions, the principle is the same. The reaction $H_2 + \frac{1}{2}O_2 \rightarrow H_2O$ is made to occur in many small steps, so that most of the energy released can be converted into a storage form instead of being lost to the environment as heat. As in the formation of ATP and NADH in glycolysis or the citric acid cycle, this involves employing an indirect pathway for the reaction. The respiratory chain is unique in that the hydrogen atoms are first separated into protons and electrons. The electrons pass through a series of electron carriers in the mitochondrial inner membrane. When the electrons reach the end of this electron-transport chain, the protons are returned to neutralize the negative charges created by the final addition of the electrons to the oxygen molecule (Figure 7–17).

We shall outline the oxidation process starting from NADH, the major collector of reactive electrons derived from the oxidation of food molecules. Each hydrogen atom (which we shall denote as H·) consists of one electron (e^-) and one proton (H^+). The mechanism by which electrons are acquired by NADH was discussed earlier (p. 68) and is shown in greater detail in Figure 7–18. As this example makes clear, each molecule of NADH carries a *hydride ion* (a hydrogen atom plus an extra electron, $H:^-$) rather than a single hydrogen atom. However, because protons are freely available in aqueous solutions, carrying the hydride ion on NADH is equivalent to carrying two hydrogen atoms, or a hydrogen molecule ($H:^- + H^+ \rightarrow H_2$).

Electron transport begins when the hydride ion is removed from NADH to regenerate NAD^+ and is converted into a proton and two electrons ($H:^- \rightarrow H^+ + 2e^-$). The two electrons are passed to the first of the more than 15 different electron carriers in the respiratory chain. The electrons start with very high energy and gradually lose it as they pass along the chain. For the most part, the electrons pass from one metal atom to another, each metal atom being tightly bound to a protein molecule, which alters the electron affinity of the metal atom. The various types of electron carriers in the respiratory chain will be discussed in detail later (p. 359). Most important, the many proteins involved are grouped into three large *respiratory enzyme complexes*, each containing transmembrane proteins that hold the complex firmly in the mitochondrial inner membrane (see p. 360). Each complex in the chain has a greater affinity for electrons than its predecessor, and electrons pass sequentially from one complex to another until they are finally transferred to oxygen, which has the greatest affinity of all for electrons.

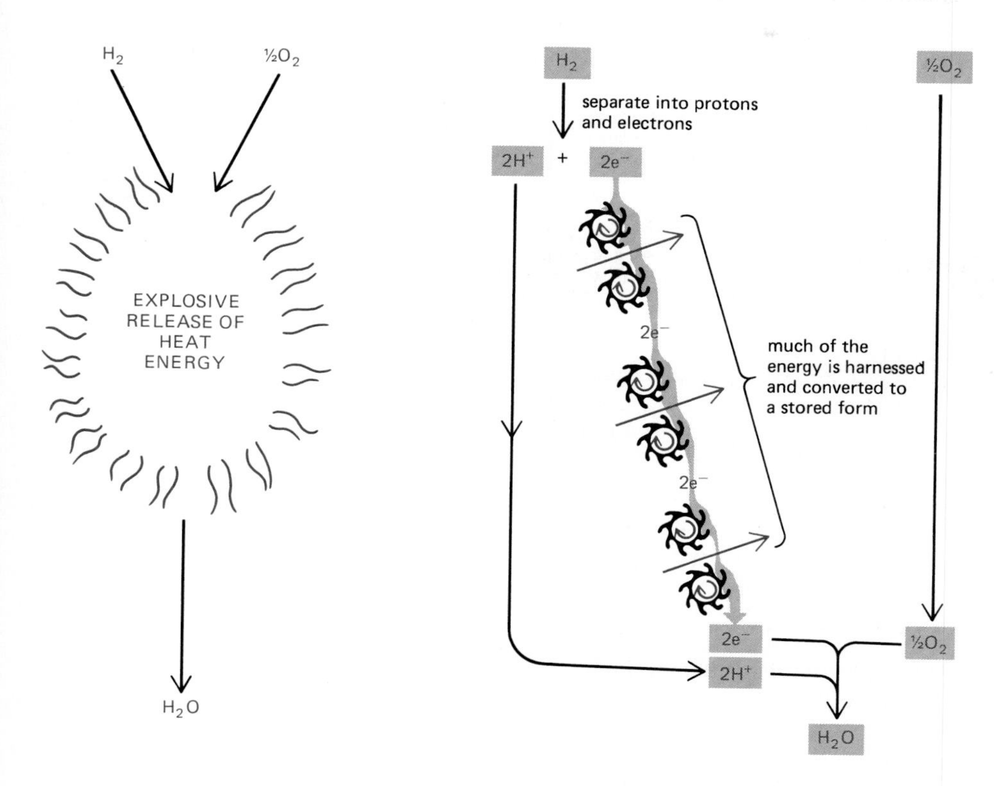

Figure 7–17 Illustration showing how most of the energy that would be released as heat if hydrogen were burned (*left panel*) is instead harnessed and stored in a form useful to the cell by means of the electron-transport chain in the mitochondrial inner membrane (*right panel*). The rest of the oxidation energy is released as heat by the mitochondrion. In reality the protons and electrons shown are removed from hydrogen atoms that are covalently linked to NADH or $FADH_2$ molecules (see Figure 7–18).

Energy Released by the Passage of Electrons Along the Respiratory Chain Is Stored as an Electrochemical Proton Gradient Across the Inner Membrane[8]

The close association of the electron carriers with protein molecules makes oxidative phosphorylation possible. The proteins guide the electrons along the respiratory chain so that the electrons move sequentially from one enzyme complex to another—with no short circuits. Most important, the transfer of electrons is coupled to allosteric changes in selected protein molecules, so that the energetically favorable flow of electrons pumps protons (H^+) across the inner membrane, from the matrix space to the intermembrane space and from there to the outside of the mitochondrion. This movement of protons has two major consequences.

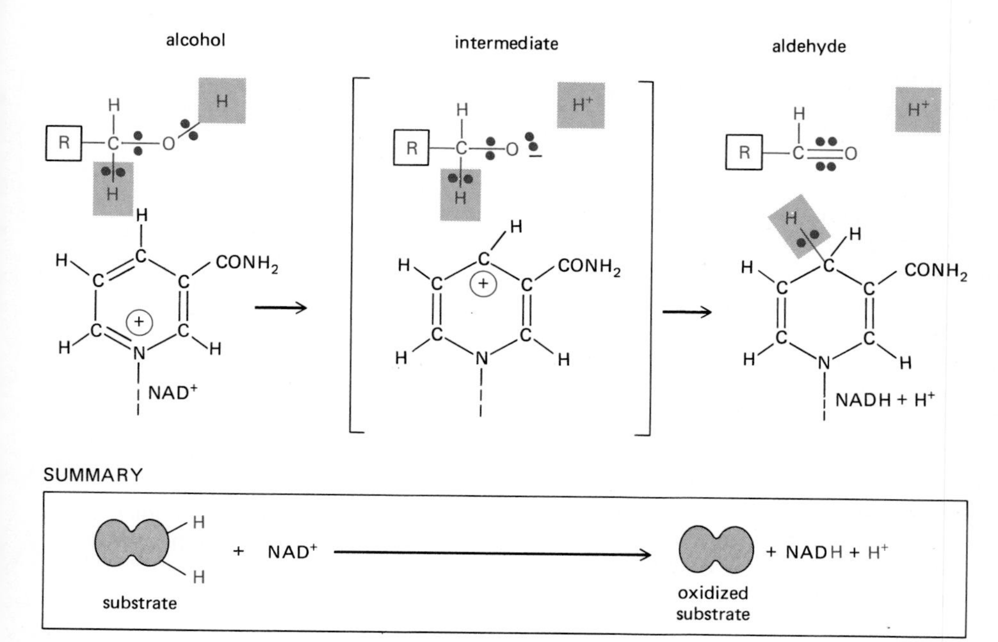

Figure 7–18 The biological oxidation of an alcohol to an aldehyde is thought to proceed in the manner shown. The components of two complete hydrogen atoms are lost from the alcohol: a hydride ion is transferred to NAD^+ and a proton escapes to the aqueous solution. Only the nicotinamide ring portion of the NAD^+ and NADH molecules is shown here (see Figure 2–22). The steps illustrated occur on a protein surface, being catalyzed by specific chemical groups on the enzyme alcohol dehydrogenase (not shown). (Modified with permission from P.F. Cook, N.J. Oppenheimer, and W.W. Cleland, *Biochemistry* 20:1817–1825, 1981. Copyright 1981 American Chemical Society.)

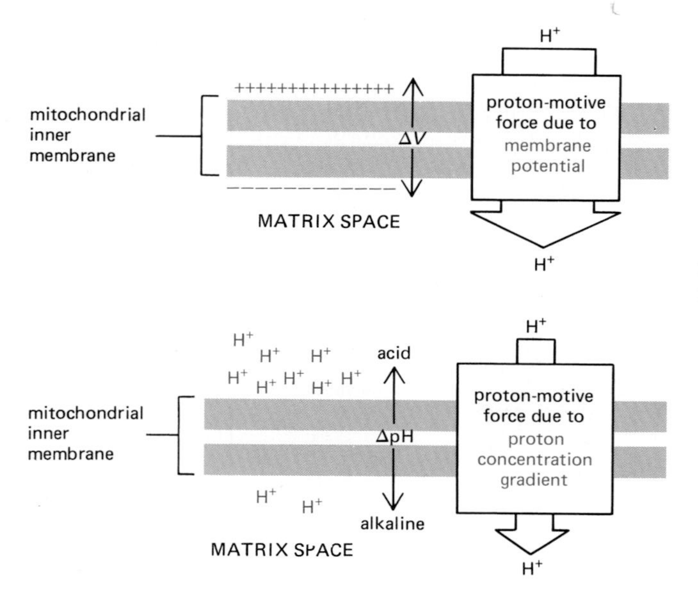

Figure 7–19 The two components of the electrochemical proton gradient. The total proton-motive force across the mitochondrial inner membrane consists of a large force due to the membrane potential (traditionally designated $\Delta\psi$ by experts, but designated ΔV in this text) and a smaller force due to the proton concentration gradient (ΔpH). Both forces act to drive protons into the matrix space.

(1) It generates a pH gradient across the inner mitochondrial membrane, with the pH higher in the matrix than in the cytosol, where the pH is generally close to 7. (Since small molecules equilibrate freely across the outer membrane of the mitochondrion, the pH in the intermembrane space is the same as in the cytosol.) (2) It generates a voltage gradient (membrane potential) across the inner mitochondrial membrane, with the inside negative and the outside positive (as a result of the net outflow of positive ions).

The pH gradient (ΔpH) acts to drive H^+ back into the matrix and OH^- out of the matrix and thus reinforces the effect of the membrane potential (ΔV), which acts to attract any positive ion into the matrix and to push any negative ion out. Together, these two forces are said to constitute an **electrochemical proton gradient** (Figure 7–19).

The electrochemical proton gradient exerts a **proton-motive force,** which can be measured in units of millivolts (mV). Since each ΔpH of 1 pH unit has an effect equivalent to a membrane potential of about 60 mV, the total proton-motive force equals $\Delta V - 60(\Delta pH)$. In a typical cell the proton-motive force across the inner membrane of a respiring mitochondrion is about 220 mV and is made up of a membrane potential of about 160 mV and a pH gradient of about −1 pH unit.

7-6 The Energy Stored in the Electrochemical Proton Gradient Is Used to Produce ATP and to Transport Metabolites and Inorganic Ions into the Matrix Space[9]

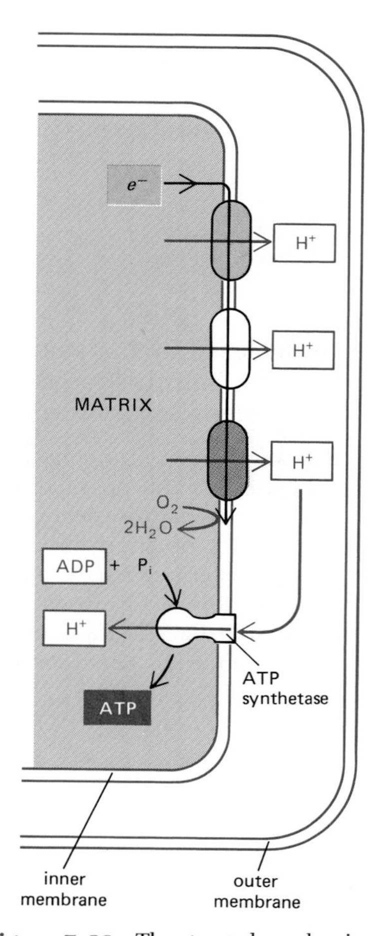

Figure 7–20 The general mechanism of oxidative phosphorylation. As a high-energy electron is passed along the electron-transport chain, some of the energy released is used to drive three respiratory enzyme complexes that pump protons out of the matrix space. These protons create an electrochemical proton gradient across the inner membrane that drives protons back through the ATP synthetase, a transmembrane protein complex that uses the energy of the proton flow to synthesize ATP from ADP and P_i in the matrix.

The mitochondrial inner membrane contains an unusually high proportion of protein, being approximately 70% protein and 30% phospholipid by weight. Many of the proteins belong to the electron-transport chain, which establishes the electrochemical proton gradient across the membrane. Another major component is the enzyme *ATP synthetase,* which catalyzes the synthesis of ATP. This is a large protein complex through which protons flow down their electrochemical gradient into the matrix. Like a turbine, ATP synthetase converts one form of energy to another, synthesizing ATP from ADP and P_i in the mitochondrial matrix in a reaction that is coupled to the inward flow of protons (Figure 7–20).

ATP synthesis is not the only process that is driven by the electrochemical proton gradient. The enzymes in the mitochondrial matrix, where the citric acid cycle and other metabolic reactions take place, must be supplied with high concentrations of substrates, and ATP synthetase must be supplied with ADP and phosphate. Thus many charged substrates must be transported across the inner membrane. This is achieved by various membrane carrier proteins (see p. 303), many of which actively transport specific molecules against their electrochemical

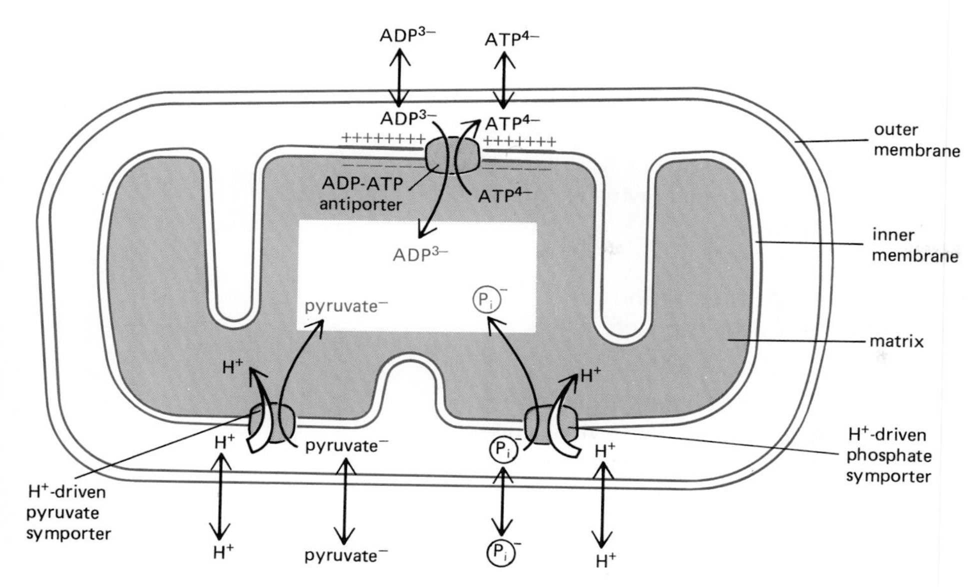

Figure 7–21 Some of the active transport processes driven by the electrochemical proton gradient across the mitochondrial inner membrane. The charge on each of the transported molecules is indicated for reference to the membrane potential, which is negative inside, as shown. The outer membrane is freely permeable to all of these compounds. See Chapter 6 for a discussion of symport and antiport transport mechanisms.

gradients, a process that requires an input of energy. For most metabolites the energy comes from *co-transporting* another molecule down its electrochemical gradient (see p. 309). The transport of ADP into the matrix space, for example, is mediated by an ADP-ATP antiport system: for each ADP molecule that moves in, an ATP molecule moves out down its electrochemical gradient. The transport of phosphate into the matrix space is mediated by a symport system that couples the inward movement of phosphate to the inward flow of protons down their electrochemical gradient so that the phosphate is dragged with them. Pyruvate is transported into the matrix in the same way (Figure 7–21). The electrochemical proton gradient is also used to import Ca^{2+}, which is thought to be important in regulating the activity of selected mitochondrial enzymes; the import of Ca^{2+} into mitochondria may also be important for removing Ca^{2+} from the cytosol when cytosolic Ca^{2+} levels become dangerously high (see p. 701).

The more energy from the electrochemical proton gradient is used to transport molecules and ions into the mitochondrion, the less there is to drive the ATP synthetase. If isolated mitochondria are incubated in a high concentration of Ca^{2+}, for example, they cease ATP production completely; all the energy in their electrochemical proton gradient is diverted to pumping Ca^{2+} into the matrix. Similarly, in certain specialized cells the electrochemical proton gradient is short-circuited so that their mitochondria produce heat instead of ATP (see p. 365). The use of the energy stored in the electrochemical proton gradient can be regulated by cells so that it is directed toward those activities most needed at the time.

7-8
7-38
The Rapid Conversion of ADP to ATP in Mitochondria Maintains a High Ratio of ATP to ADP in Cells[10]

Because of the antiporter in the inner membrane that pumps ADP into the matrix space in exchange for ATP (see Figure 7–21), ADP molecules produced by ATP hydrolysis in the cytosol rapidly enter mitochondria for recharging, while the ATP molecules formed in the mitochondrial matrix by oxidative phosphorylation are rapidly pumped into the cytosol, where they are needed. A typical ATP molecule in the human body shuttles in and out of a mitochondrion for recharging (as ADP) thousands of times a day, keeping the concentration of ATP in a cell about 10 times higher than that of ADP.

As discussed in Chapter 2, biosynthetic enzymes in cells guide their substrates along specific reaction paths, often driving energetically unfavorable reactions by coupling them to the energetically favorable hydrolysis of ATP (see Figure 2–27, p. 74). The highly charged ATP pool is thereby used to drive cellular processes in much the same way that a battery can be used to drive electric engines: if the activity of the mitochondria is halted, ATP levels fall and the cell's battery runs down, so that, eventually, energetically unfavorable reactions can no longer be driven by ATP hydrolysis.

It might seem that this state would not be reached until the concentration of ATP is zero, but in fact it is reached much sooner than that, at a concentration of ATP that depends on the concentrations of ADP and P_i. To explain why, we must consider some elementary principles of thermodynamics.

7-5 The Difference Between ΔG° and ΔG: A Large Negative Value of ΔG is Required for ATP Hydrolysis to Be Useful to the Cell[11]

The second law of thermodynamics states that chemical reactions proceed spontaneously in the direction that corresponds to an increase in the disorder of the *universe.* In Chapter 2 we noted that reactions that release energy to their surroundings as heat (such as the hydrolysis of ATP) tend to increase the disorder of the universe by increasing random molecular motions. Reactions will also affect the degree of disorder by changing the concentrations of the reactant and product molecules. The net change of disorder in the universe due to a reaction is proportional to the **change in free energy, ΔG,** that is associated with the reaction: reactions that bring about a large *decrease* in free energy (so that ΔG is very negative) create the most disorder in the universe and proceed most readily (see Panel 2–7, pp. 72–73).

When ATP is hydrolyzed to ADP and P_i under the conditions that normally exist in a cell, the free-energy change is roughly -11 to -13 kcal/mole. This extremely favorable ΔG requires that the concentration of ATP in the cell be kept high compared to the concentration of ADP and P_i. At "standard conditions," where ATP, ADP, and P_i are all present at the same concentration of 1 mole/liter, the ΔG for ATP hydrolysis—called the **standard free-energy change,** or **ΔG°**, of the reaction—is only -7.3 kcal/mole. At still lower concentrations of ATP relative to ADP and P_i, ΔG will become equal to zero. At this point the rate at which ADP and P_i will join to form ATP will be equal to the rate at which ATP hydrolyzes to form ADP and P_i. In other words, when $\Delta G = 0$, the reaction is at *equilibrium* (Figure 7–22).

At a fixed temperature, the value of ΔG° is a constant that depends only on the *nature* of the reactants. The value of ΔG, in contrast, is a variable that depends on the *concentrations* of the reactants, and its value indicates how far a reaction is from equilibrium. Therefore it is ΔG, not ΔG°, that determines if a reaction can be used to drive other reactions. Because the efficient conversion of ADP to ATP in mitochondria maintains such a high concentration of ATP relative to ADP and P_i, the ATP-hydrolysis reaction in cells is kept very far from equilibrium and ΔG is correspondingly very negative. Without this disequilibrium, ATP hydrolysis could not be used to direct the reactions of the cell, and many biosynthetic reactions would run backward rather than forward.

7-6 7-7 Cellular Respiration Is Remarkably Efficient

By means of oxidative phosphorylation, each pair of electrons in NADH provides energy for the formation of about 3 molecules of ATP. The pair of electrons in $FADH_2$, being at a lower energy, generates only about 2 ATP molecules. In all, about 12 molecules of ATP can be formed from each molecule of acetyl CoA that enters the citric acid cycle, which means that 24 ATP molecules are produced from 1 molecule of glucose and 96 ATP molecules from 1 molecule of palmitate, a 16-carbon fatty acid. If one includes the energy-yielding reactions that occur before acetyl CoA is formed, the complete oxidation of 1 molecule of glucose gives a net yield of about 36 ATPs, while about 129 ATPs are obtained from the complete oxidation of 1 molecule of palmitate. These numbers are approximate maximal values, since the actual amount of ATP made in the mitochondrion depends on what fraction of the electrochemical gradient energy is used for purposes other than ATP synthesis.

1. ATP $\xrightarrow{\text{hydrolysis}}$ ADP + P_i

hydrolysis rate = (hydrolysis rate constant) x (concentration of ATP)

2. ADP + P_i $\xrightarrow{\text{synthesis}}$ ATP

synthesis rate = (synthesis rate constant) x (conc. of phosphate) x (conc. of ADP)

3. AT EQUILIBRIUM: synthesis rate = hydrolysis rate

(synthesis rate constant) x (conc. of phosphate) x (conc. of ADP) = (hydrolysis rate constant) x (conc. of ATP)

$$\frac{(\text{conc. of ADP}) \times (\text{conc. of phosphate})}{(\text{conc. of ATP})} = \frac{(\text{hydrolysis rate constant})}{(\text{synthesis rate constant})} = \text{equilibrium constant } K$$

$$\frac{[\text{ADP}]\,[P_i]}{[\text{ATP}]} = K$$

4. For the reaction

ATP ⟶ ADP + P_i

the following equation applies:

$$\Delta G = \Delta G° + RT \ln \frac{[\text{ADP}]\,[P_i]}{[\text{ATP}]}$$

Where ΔG and $\Delta G°$ are in kilocalories per mole, R is the gas constant (2×10^{-3} kcal/mole °K), T is the absolute temperature (°K), and all the concentrations are in moles per liter.

When the concentrations of all reactants are at 1 M, $\Delta G = \Delta G°$ (since $RT \ln 1 = 0$). $\Delta G°$ is thus a constant defined as the standard free-energy change for the reaction.

At equilibrium the reaction has no net effect on the disorder of the universe, so $\Delta G = 0$. Therefore, at equilibrium,

$$-RT \ln \frac{[\text{ADP}]\,[P_i]}{[\text{ATP}]} = \Delta G°$$

But the concentrations of reactants at equilibrium must satisfy the equilibrium equation:

$$\frac{[\text{ADP}]\,[P_i]}{[\text{ATP}]} = K$$

Therefore, at equilibrium,

$$\Delta G° = -RT \ln K$$

We thus see that whereas $\Delta G°$ indicates the equilibrium point for a reaction, ΔG reveals *how far* the reaction is from equilibrium

Figure 7–22 The basic relationship between free-energy changes and equilibrium, as illustrated by the ATP hydrolysis reaction. The equilibrium constant shown here, K, is in units of moles per liter. (See Panel 2–7, pp. 72–73, for a discussion of free energy, and Figure 3–7, p. 94, for a definition of the equilibrium constant.)

When the free-energy changes for burning fats and carbohydrates directly to CO_2 and H_2O are compared to the total amount of energy generated and stored in the phosphate bonds of ATP during the corresponding biological oxidations, it is seen that the efficiency with which oxidation energy is converted into ATP bond energy is often greater than 50%. This is considerably better than the efficiency of most nonbiological energy-conversion devices. If cells worked with the efficiency of an electric motor or a gasoline engine (10–20%), an organism would have to eat voraciously in order to maintain itself. Moreover, since wasted energy is liberated as heat, large organisms would need much more efficient mechanisms for giving up heat to the environment.

Students sometimes wonder why the chemical interconversions in cells follow such complex pathways. The oxidation of sugars to CO_2 plus H_2O could certainly be accomplished more directly, eliminating the citric acid cycle and many of the steps in the respiratory chain. Although this would have made respiration easier to learn, it would have been a disaster for the cell. Oxidation produces huge amounts of free energy, which can be utilized efficiently only in small bits. The complex oxidative pathways involve many intermediates, each differing only slightly from its predecessor. As a result, the energy released is parceled out into small packets that can be efficiently converted to high-energy bonds in useful molecules such as ATP and NADH by means of coupled reactions (see Figure 2–17, p. 63).

Summary

The mitochondrion carries out most cellular oxidations and produces the bulk of the animal cell's ATP. The mitochondrial matrix space contains a large variety of enzymes, including those that oxidize pyruvate and fatty acids to acetyl CoA and those that oxidize this acetyl CoA to CO_2 through the citric acid cycle. Large amounts of NADH (and $FADH_2$) are produced by these oxidation reactions. The energy available from combining oxygen with the reactive electrons carried by NADH and $FADH_2$

is harnessed by an electron-transport chain in the mitochondrial inner membrane called the respiratory chain. The respiratory chain pumps protons out of the matrix to create a transmembrane electrochemical proton gradient, which includes contributions from both a membrane potential and a pH difference. The transmembrane gradient is in turn used both to synthesize ATP and to drive the active transport of selected metabolites across the mitochondrial inner membrane. The combination of these reactions is responsible for an efficient ATP-ADP exchange between the mitochondrion and the cytosol that keeps the cell's ATP pool highly charged.

The Respiratory Chain and ATP Synthetase[12]

Having considered in general terms how mitochondria function, let us now look in more detail at the respiratory chain—the electron-transport chain that is so crucial to all oxidative metabolism. Most of the elements of the chain are intrinsic components of the inner mitochondrial membrane, and they provide some of the clearest examples of the complex interactions that can occur among the individual proteins in a biological membrane.

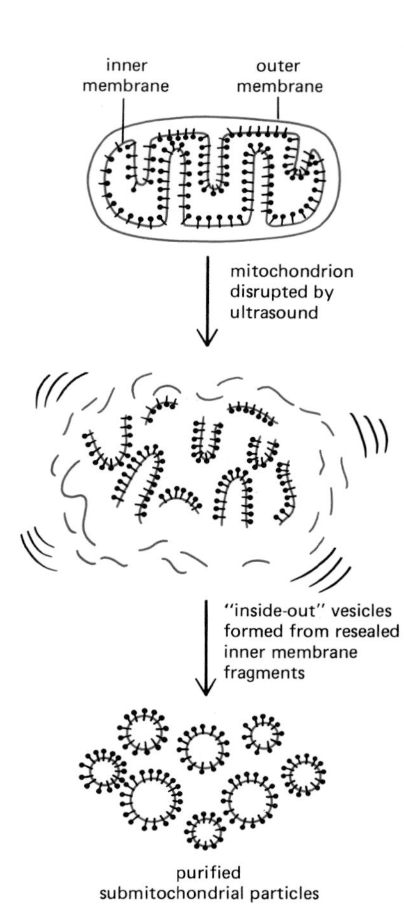

Figure 7–23 Outline of the procedure used to prepare submitochondrial particles from purified mitochondria. The particles are pieces of broken-off cristae that form closed vesicles.

Functional Inside-out Particles Can Be Isolated from Mitochondria[13]

The respiratory chain is relatively inaccessible to experimental manipulation in intact mitochondria. By disrupting mitochondria with ultrasound, however, it is possible to isolate functional *submitochondrial particles,* which consist of broken cristae that have resealed into small closed vesicles about 100 nm in diameter (Figure 7–23). When negatively stained submitochondrial particles are examined in an electron microscope, their outside surfaces are seen to be studded with tiny spheres attached to the membrane by stalks (Figure 7–24). In intact mitochondria these lollipoplike structures are located on the *inner* (matrix) side of the inner membrane. Thus the submitochondrial particles are inside-out vesicles of inner membrane, with what was previously their matrix-facing surface exposed to the surrounding medium. As a result, they can readily be provided with the membrane-impermeable metabolites that would normally be present in the matrix space. When NADH, ADP, and inorganic phosphate are added, such particles transport electrons from NADH to O_2 and couple this oxidation to ATP synthesis. This cell-free system provides an assay that makes it possible to purify the many proteins responsible for oxidative phosphorylation in a functional form.

7-11 ATP Synthetase Can Be Purified and Added Back to Membranes[14]

The first experiments to show that the various membrane proteins that catalyze oxidative phosphorylation can be separated without destroying their activity were performed in 1960. The tiny protein spheres studding the surface of submitochondrial particles were stripped from the particles and purified in soluble form. The stripped particles could still oxidize NADH in the presence of oxygen, but they could no longer synthesize ATP. On the other hand, the purified spheres on their own acted as ATPases, hydrolyzing ATP to ADP and P_i. When purified spheres (referred to as *F_1ATPases*) were added back to stripped submitochondrial particles, the reconstituted particles once again made ATP from ADP and P_i.

Subsequent work showed that the F_1ATPase is part of a larger transmembrane complex (about 500,000 daltons) containing at least nine different polypeptide chains, which is now known as **ATP synthetase** (also called *F_0F_1ATPase*). ATP synthetase comprises about 15% of the total inner membrane protein, and very similar enzyme complexes are present in both chloroplast and bacterial membranes. This protein complex contains a transmembrane proton carrier, and it synthesizes ATP when protons pass through it down their electrochemical gradient.

One of the most convincing demonstrations of the function of ATP synthetase came from an experiment performed in 1974. By that time, methods had been developed for transferring detergent-solubilized integral membrane proteins into

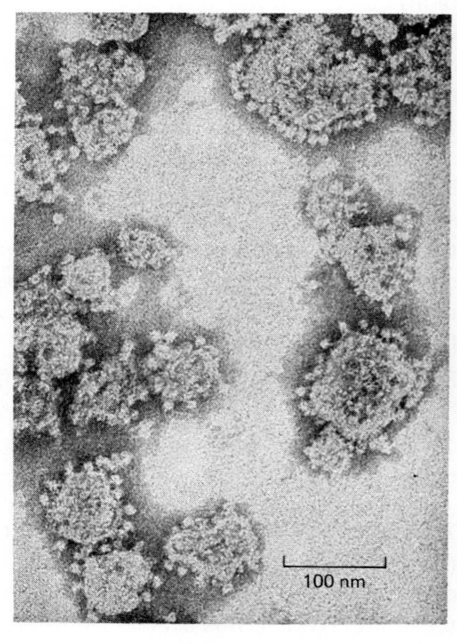

Figure 7–24 Electron micrograph of submitochondrial particles. (Courtesy of Efraim Racker.)

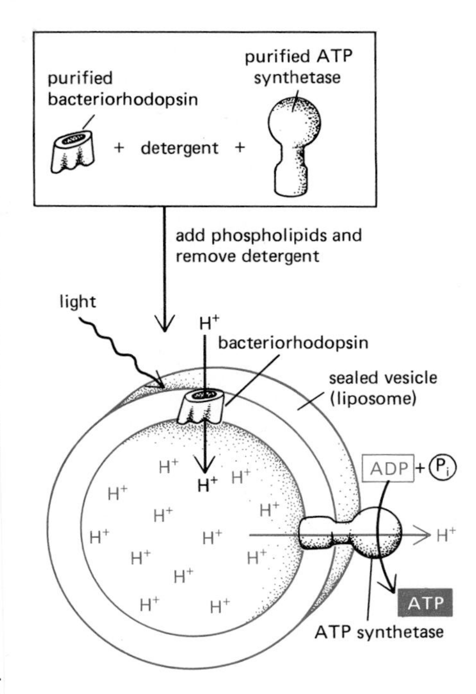

Figure 7–25 Outline of an important experiment demonstrating that the ATP synthetase can be driven by a simple proton flow. By combining a light-driven bacterial proton pump (bacteriorhodopsin), an ATP synthetase purified from ox heart mitochondria, and phospholipids, vesicles were produced that synthesized ATP in response to light.

lipid vesicles (liposomes) formed from purified phospholipids (see p. 278). It thus became possible to form a hybrid membrane that contained both a purified mitochondrial ATP synthetase and bacteriorhodopsin—a bacterial light-driven proton pump (see p. 293). When these vesicles were exposed to light, the protons pumped into the vesicle lumen by the bacteriorhodopsin flowed back out through the ATP synthetase, causing ATP to be made in the medium outside (Figure 7–25). Because a direct interaction between a bacterial proton pump and a mammalian ATP synthetase seems highly unlikely, this experiment strongly suggests that in mitochondria the proton translocation driven by electron transport and the ATP synthesis are separate events.

ATP Synthetase Can Function in Reverse to Hydrolyze ATP and Pump H^+ [15]

ATP synthetase is a reversible enzyme complex: it can either use the energy of ATP hydrolysis to pump protons across the inner mitochondrial membrane, or it can harness the flow of protons down an electrochemical proton gradient to make ATP (Figure 7–26). It thereby acts as a *reversible coupling device,* interconverting electrochemical-proton-gradient and chemical-bond energies. Its direction of action depends on the balance between the steepness of the electrochemical proton gradient and the local ΔG for ATP hydrolysis.

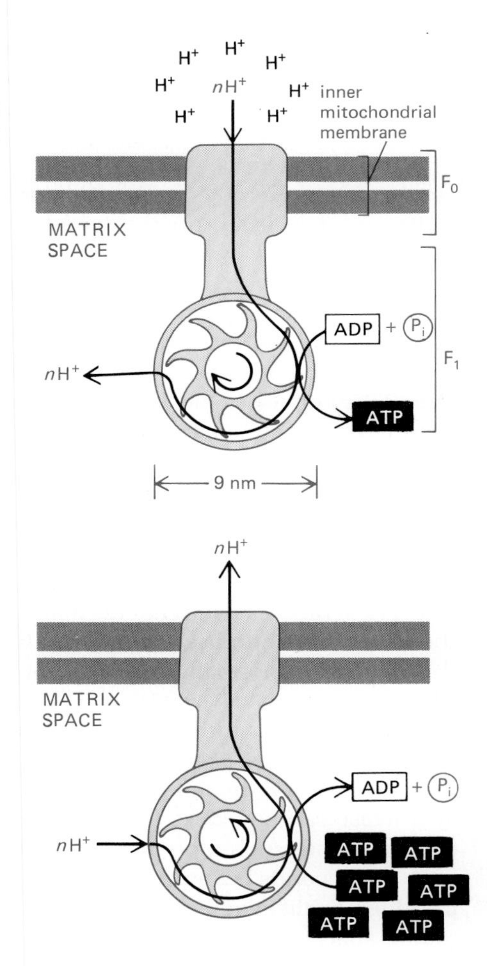

Figure 7–26 ATP synthetase is a reversible coupling device that interconverts the energies of the electrochemical proton gradient and chemical bonds. Composed of at least nine different polypeptide chains, it is also known as the F_0F_1ATPase. Five of its polypeptide chains make up the spherical head of the complex, which is called the F_1ATPase. The ATP synthetase can either synthesize ATP by harnessing the proton-motive force (*top*) or pump protons against their electrochemical gradient by hydrolyzing ATP (*bottom*). As explained in the text, the direction of operation at any given instant depends on the net free-energy change for the coupled processes of proton translocation across the membrane and the synthesis of ATP from ADP and P_i.

We have previously shown how the free-energy change (ΔG) for ATP hydrolysis depends on the concentrations of the three reactants ATP, ADP, and P_i (Figure 7–22); the ΔG for ATP synthesis is the negative of this value. The ΔG for proton translocation across the membrane is the sum of (1) the ΔG for moving a mole of any ion through a difference in membrane potential ΔV and (2) the ΔG for moving a mole of a molecule between any two compartments in which its concentration differs. The equation for the proton-motive force on p. 352 combines the same two components, replacing the concentration difference by an equivalent increment in the membrane potential to produce an "electrochemical potential" for the proton. Thus the ΔG for proton translocation and the proton-motive force measure the same potential, one in kilocalories and the other in millivolts. The conversion factor between them is the faraday. Thus,

$$\Delta G_{H^+} = -0.023 \text{ (proton-motive force)}$$

where ΔG_{H^+} is in kilocalories per mole (kcal/mole) and the proton-motive force is in millivolts (mV). For an electrochemical proton gradient of 220 mV, $\Delta G_{H^+} = -5.06$ kcal/mole.

ATP synthetase is so-called because it is normally driven by the large electrochemical proton gradient maintained by the respiratory chain (see Figure 7–20) to make most of the cell's ATP. The exact number of protons needed to make each ATP molecule is not known with certainty. To facilitate the calculations to be described below, however, we shall assume that one molecule of ATP is made by the ATP synthetase for every three protons driven through it.

Whether the ATP synthetase works in its ATP-synthesizing or its ATP-hydrolyzing direction at any instant depends on the exact balance between the favorable free-energy change for moving the three protons across the membrane into the matrix space (ΔG_{3H^+}, which is less than zero) and the unfavorable free-energy change for ATP *synthesis* in the matrix ($\Delta G_{\text{ATP synthesis}}$, which is greater than zero). As previously discussed, the value of $\Delta G_{\text{ATP synthesis}}$ depends on the exact concentrations of the three reactants ATP, ADP, and P_i in the mitochondrial matrix space (see Figure 7–22). The value of ΔG_{3H^+}, on the other hand, is proportional to the value of the proton-motive force across the inner mitochondrial membrane. The following example will help to explain how the balance between these two free-energy changes affects the ATP synthetase.

As explained in the legend to Figure 7–26, a single proton moving into the matrix down an electrochemical gradient of 220 mV liberates 5.06 kcal/mole of free energy, while the movement of three protons liberates three times this much free energy ($\Delta G_{3H^+} = -15.2$ kcal/mole). Thus, if the proton-motive force remains constant at 220 mV, the ATP synthetase will synthesize ATP until a ratio of ATP to ADP and P_i is reached where $\Delta G_{\text{ATP synthesis}}$ is just equal to $+15.2$ kcal/mole (here $\Delta G_{\text{ATP synthesis}} + \Delta G_{3H^+} = 0$). At this point there will be no further net ATP synthesis or hydrolysis by the ATP synthetase.

Suppose that a great deal of ATP is suddenly hydrolyzed by energy-requiring reactions in the cytosol—causing the ATP:ADP ratio in the matrix to fall. Now the value of $\Delta G_{\text{ATP synthesis}}$ will decrease (see Figure 7–22), and ATP synthetase will begin to synthesize ATP again to restore the original ATP:ADP ratio. Alternatively, if the proton-motive force drops suddenly and is then maintained at a constant 200 mV, ΔG_{3H^+} will change to -13.8 kcal/mole. As a result, ATP synthetase will start hydrolyzing some of the ATP in the matrix—until a new balance of ATP to ADP and P_i is reached (where $\Delta G_{\text{ATP synthesis}} = +13.8$ kcal/mole)—and so on.

In many bacteria, ATP synthetase is routinely reversed in a transition between aerobic and anaerobic metabolism, as we shall see later. The reversibility of the ATP synthetase is a property shared by other membrane proteins that couple ion movement to ATP synthesis or hydrolysis. For example, both the Na^+-K^+ pump and the Ca^{2+} pump, described in Chapter 6, hydrolyze ATP and use the energy released to pump specific ions across a membrane (see p. 305). If either of these pumps is exposed to an abnormally steep gradient of the ions it transports, it will act in reverse—synthesizing ATP from ADP and P_i instead of hydrolyzing it. Thus, like ATP synthetase, such pumps are able to convert the electrochemical energy stored in a transmembrane ion gradient directly into phosphate bond energy in ATP.

The Respiratory Chain Pumps H^+ Across the Inner Mitochondrial Membrane[16]

The ATP synthetase does not normally transport H^+ out of the matrix space across the inner mitochondrial membrane. Instead, the respiratory chain embedded in this membrane normally does, thereby generating the electrochemical proton gradient that drives ATP synthesis. The ability of the respiratory chain to translocate H^+ outward from the matrix space can be demonstrated experimentally under special conditions. For example, a suspension of isolated mitochondria can be provided with a suitable substrate for oxidation, and the H^+ flow through ATP synthetase can be blocked. In the absence of air, the injection of a small amount of oxygen into such a preparation causes a brief burst of respiration, which lasts for 1 to 2 seconds before all the oxygen is consumed. During this respiratory burst, a sudden acidification of the medium resulting from the extrusion of H^+ from the matrix space can be measured with a sensitive pH electrode.

A similar experiment can be carried out with a suspension of submitochondrial particles. In this case the medium becomes more basic when oxygen is injected, since protons are pumped *into* each vesicle because of its inside-out orientation.

7-12 Spectroscopic Methods Have Been Used to Identify Many Electron Carriers[17]

Many of the electron carriers in the respiratory chain absorb visible light and change color when they are oxidized or reduced. In general, each has an absorption spectrum and reactivity that is distinct enough to allow its behavior to be traced spectroscopically even in crude mixtures. It was therefore possible to purify these components long before their exact functions were known. Thus the *cytochromes* were discovered in 1925 as compounds that undergo rapid oxidation and reduction in living organisms as disparate as bacteria, yeasts, and insects. By observing cells and tissues with a spectroscope, three types of cytochromes were identified by their distinctive absorption spectra and designated cytochromes a, b, and c. This nomenclature has survived even though cells are now known to contain several cytochromes of each type, and the classification into types is not functionally important.

The **cytochromes** constitute a family of colored proteins that are related by the presence of a bound *heme group,* whose iron atom changes from the ferric (Fe III) to the ferrous (Fe II) state whenever it accepts an electron. The heme group consists of a *porphyrin* ring with a tightly bound iron atom held by four nitrogen atoms at the corners of a square (Figure 7–27). A related porphyrin ring is responsible for the red color of blood and the green color of leaves, being bound to iron in hemoglobin (see p. 601) and to magnesium in chlorophyll (see p. 373). The best understood of the many proteins in the respiratory chain is *cytochrome c,* whose three-dimensional structure has been determined by x-ray crystallography (Figure 7–28).

Iron-sulfur proteins are a second major family of electron carriers. In these proteins either two or four iron atoms are bound to an equal number of sulfur atoms and to cysteine side chains, forming an **iron-sulfur center** on the protein (Figure 7–29). There are more iron-sulfur centers than cytochromes in the respi-

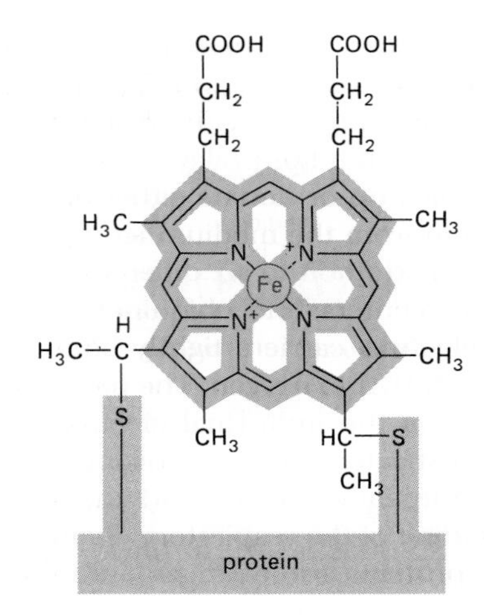

Figure 7–27 The structure of the heme group attached covalently to cytochrome c. Four of the six coordination positions on the iron are occupied by the porphyrin ring. The fifth and sixth coordination positions on the iron are perpendicular to the plane of the ring. In nearly all cytochromes these two positions are occupied by amino acid side chains, so that no other ligand can be bound. The exception is cytochrome a_3; as for hemoglobin and myoglobin, the sixth coordination position of the iron in this cytochrome oxidase component is free, so that it can bind oxygen.

There are five different cytochromes in the respiratory chain. Because the hemes in different cytochromes have slightly different structures and are held by their respective proteins in different ways, each of the cytochromes has a different affinity for an electron.

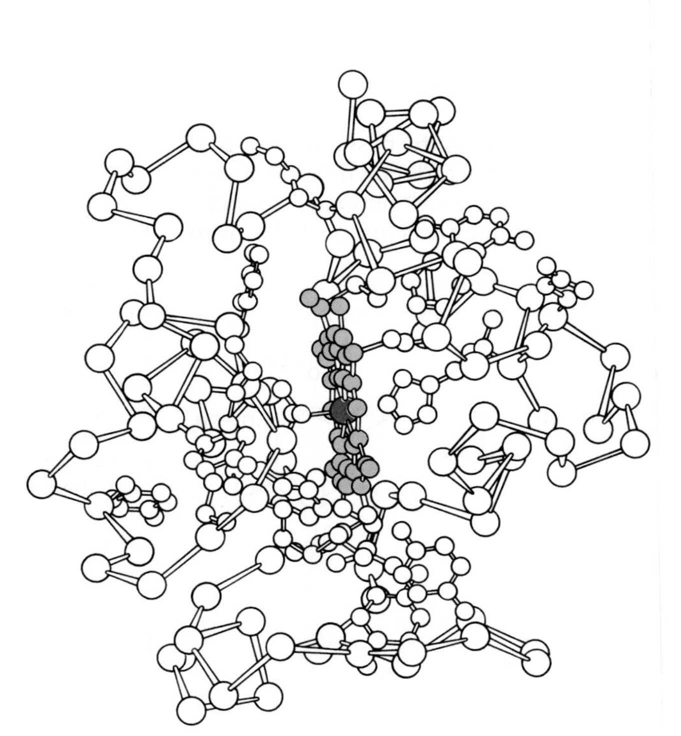

Figure 7–28 The three-dimensional structure of cytochrome c, an electron carrier in the electron-transport chain. This small protein contains just over 100 amino acids and is held loosely on the membrane by ionic interactions (see Figure 7–35). The iron atom (*dark color*) on the bound heme (*light color*) can carry a single electron (see also Figure 3–52A).

Summary

The respiratory chain in the inner mitochondrial membrane contains three major enzyme complexes that are involved in transferring electrons from NADH to O_2. Each of these can be purified, inserted into synthetic lipid vesicles, and then shown to pump protons when electrons are transported through it. In the native membrane the mobile electron carriers ubiquinone and cytochrome c complete the electron-transport chain by shuttling between the enzyme complexes. The path of electron flow is NADH → NADH dehydrogenase complex → ubiquinone → b-c_1 complex → cytochrome c → cytochrome oxidase complex → molecular oxygen (O_2).

The respiratory enzyme complexes couple the energetically favorable transport of electrons to the pumping of protons out of the matrix. The resulting electrochemical proton gradient is harnessed to make ATP by another transmembrane protein complex, ATP synthetase, through which the protons flow back into the matrix. The ATP synthetase is a reversible coupling device that normally converts a backflow of protons into ATP phosphate-bond energy, but it can also hydrolyze ATP to pump protons in the opposite direction if the electrochemical proton gradient is reduced. Its universal presence in mitochondria, chloroplasts, and bacteria testifies to the central importance of chemiosmotic mechanisms in cells.

Chloroplasts and Photosynthesis[26]

All animals and most microorganisms rely on the continual uptake of large amounts of organic compounds from their environment. These compounds provide both the carbon skeletons for biosynthesis and the metabolic energy that drives all cellular processes. It is believed that the first organisms on the primitive earth had access to an abundance of organic compounds produced by geochemical processes (see p. 4) but that most of these original compounds were used up billions of years ago. Since that time, virtually all of the organic materials required by living cells have been produced by *photosynthetic organisms,* including many types of photosynthetic bacteria. The most advanced photosynthetic bacteria are the cyanobacteria, which have minimal nutrient requirements. They use electrons from water and the energy of sunlight to convert atmospheric CO_2 into organic compounds. Moreover, in the course of splitting water [in the reaction $nH_2O + nCO_2 \xrightarrow{\text{light}} (CH_2O)_n + nO_2$], they liberate into the atmosphere the oxygen required for oxidative phosphorylation. As we shall explain later, it is thought that the evolution of cyanobacteria from more primitive photosynthetic bacteria first made possible the development of aerobic life forms.

In plants, which developed later, photosynthesis is carried out in a specialized intracellular organelle—the chloroplast. Chloroplasts carry out photosynthesis during the daylight hours. At night, when the production of energy-rich metabolites by photosynthesis ceases, the plant cell relies on its mitochondria (which closely resemble their counterparts in animal cells) to generate ATP.

Biochemical evidence suggests that chloroplasts are descendants of oxygen-producing photosynthetic bacteria that were endocytosed and lived in symbiosis with primitive eucaryotic cells. Mitochondria are also generally believed to be descended from endocytosed bacteria. The many differences between chloroplasts and mitochondria are thought to reflect their different bacterial ancestors as well as subsequent evolutionary divergence. Nevertheless, the fundamental mechanisms involved in light-driven ATP synthesis in chloroplasts and in respiration-driven ATP synthesis in mitochondria are very similar.

Chloroplasts Resemble Mitochondria but Have an Extra Compartment[27]

Chloroplasts carry out their energy interconversions by chemiosmotic mechanisms in much the same way that mitochondria do, and they are organized on the same principles (Figures 7–38 and 7–39). They have a highly permeable outer

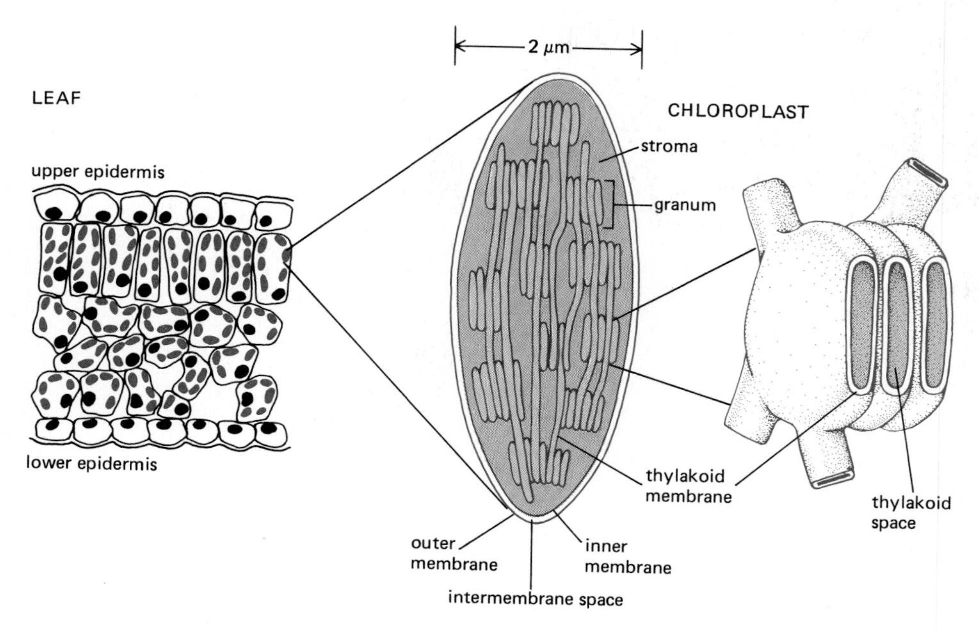

Figure 7–38 The chloroplast contains three distinct membranes (the outer membrane, the inner membrane, and the thylakoid membrane) that define three separate internal compartments—the intermembrane space, the stroma, and the thylakoid space. The thylakoid membrane contains all the energy-generating systems of the chloroplast. In electron micrographs this membrane appears to be broken up into separate units that enclose individual flattened vesicles (see Figure 7–39), but these are probably joined into a single, highly folded membrane in each chloroplast. As indicated, the individual thylakoids are interconnected, and they tend to stack to form aggregates called grana.

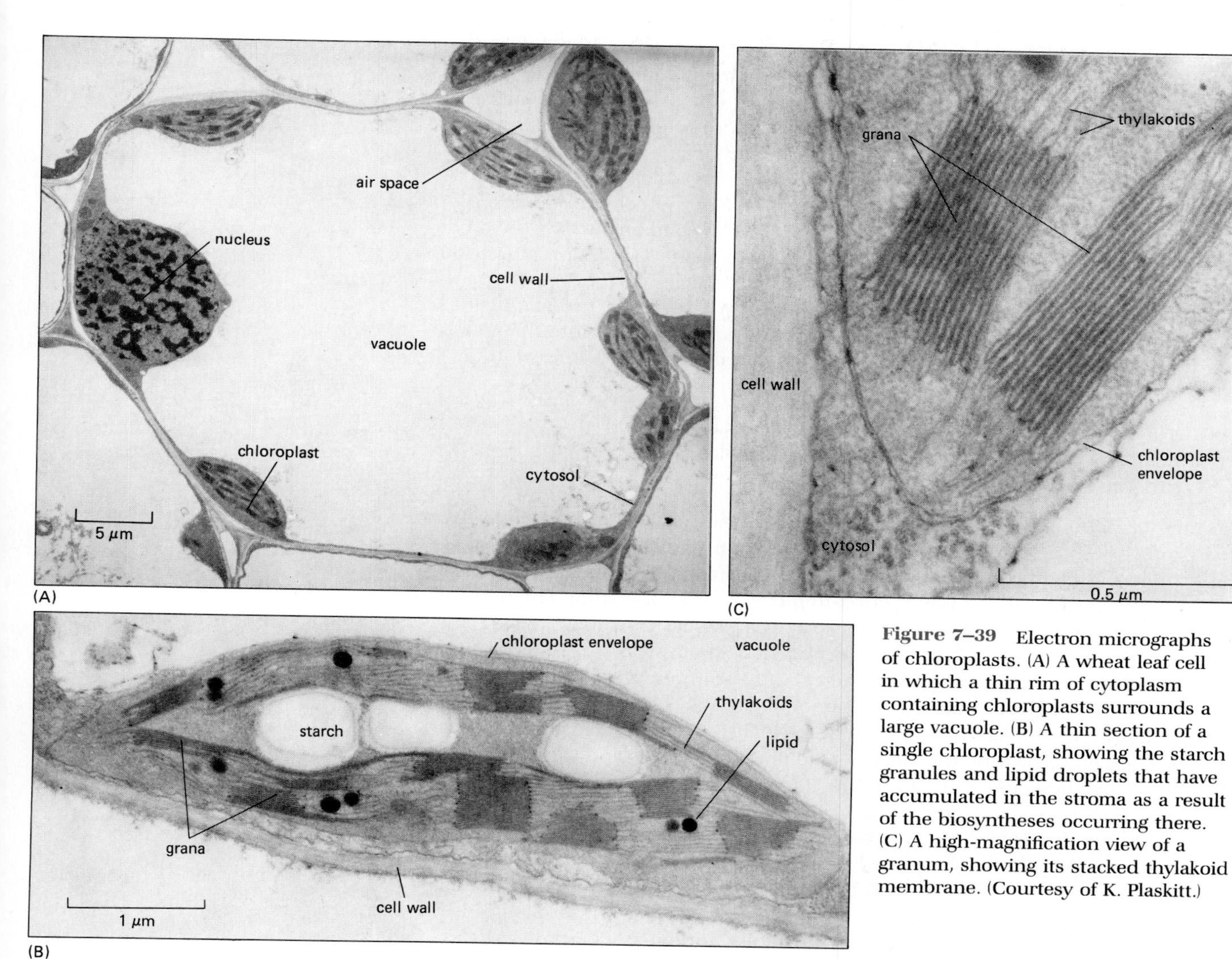

Figure 7–39 Electron micrographs of chloroplasts. (A) A wheat leaf cell in which a thin rim of cytoplasm containing chloroplasts surrounds a large vacuole. (B) A thin section of a single chloroplast, showing the starch granules and lipid droplets that have accumulated in the stroma as a result of the biosyntheses occurring there. (C) A high-magnification view of a granum, showing its stacked thylakoid membrane. (Courtesy of K. Plaskitt.)

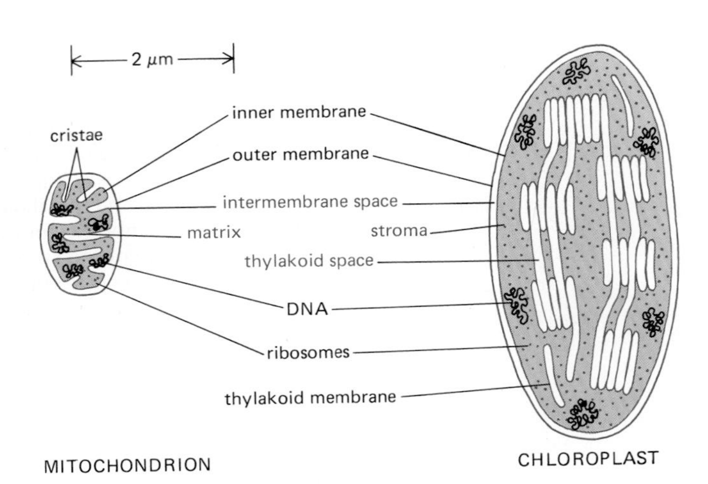

Figure 7–40 Comparison of a mitochondrion and a chloroplast. The chloroplast is generally much larger and contains a thylakoid membrane and thylakoid space. The mitochondrial inner membrane is folded into cristae.

membrane; a much less permeable inner membrane, in which special carrier proteins are embedded; and a narrow intermembrane space between. The inner membrane surrounds a large space called the **stroma,** which is analogous to the mitochondrial matrix and contains various enzymes, ribosomes, RNA, and DNA.

There is, however, an important difference between the organization of mitochondria and that of chloroplasts. The inner membrane of the chloroplast is not folded into cristae and does not contain an electron-transport chain. Instead, the photosynthetic light-absorbing system, the electron-transport chain, and an ATP synthetase are all contained in a third distinct membrane that forms a set of flattened disclike sacs, the **thylakoids** (see Figure 7–38). The lumen of each thylakoid is thought to be connected with the lumen of other thylakoids, thereby defining a third internal compartment called the *thylakoid space,* which is separated from the stroma by the *thylakoid membrane.*

The structural similarities and differences between mitochondria and chloroplasts are illustrated in Figure 7–40. In a general way, one might view the chloroplast as a greatly enlarged mitochondrion in which the cristae are converted into a series of interconnected submitochondrial particles in the matrix space. The knobbed end of the chloroplast ATP synthetase, where ATP is made, protrudes from the thylakoid membrane into the stroma, just as it protrudes into the matrix from the membrane of each mitochondrial crista (see Figure 7–51).

7-21 Two Unique Reactions in Chloroplasts: The Light-driven Production of ATP and NADPH and the Conversion of CO_2 to Carbohydrate[26]

The many reactions that occur during photosynthesis can be grouped into two broad categories. (1) In the **photosynthetic electron-transfer reactions** (sometimes called the "light reactions"), energy derived from sunlight energizes an electron in *chlorophyll,* enabling the electron to move along an oxidation chain in the thylakoid membrane in much the same way that an electron moves along the respiratory chain in mitochondria. This electron-transport process pumps protons across the thylakoid membrane, and the resulting proton-motive force drives the synthesis of ATP in the stroma. At the same time, the electron-transfer reactions generate high-energy electrons that convert $NADP^+$ to NADPH; in this process, water is oxidized to provide the electrons donated to NADPH, and O_2 is liberated. All of these reactions are confined to the chloroplast. (2) In the **carbon-fixation reactions** (sometimes called the "dark reactions"), the ATP and NADPH produced by the photosynthetic electron-transfer reactions serve as the source of energy and reducing power, respectively, to drive the conversion of CO_2 to carbohydrate. These reactions, which begin in the chloroplast stroma and continue in the cytosol, produce sucrose in the leaves of the plant, from where it is exported to other tissues as a source of both organic molecules and energy for growth.

Thus the formation of oxygen (which directly requires light energy) and the conversion of carbon dioxide to carbohydrate (which requires light energy only indirectly) are separate processes (Figure 7–41). As we shall see, however, elaborate feedback mechanisms interconnect the two in order to balance biosynthesis. Changes in the cell's ATP and NADPH requirements, for example, regulate the production of these molecules in the thylakoid membrane, and several of the chloroplast enzymes required for carbon fixation are inactivated in the dark and reactivated by light-stimulated electron-transport processes.

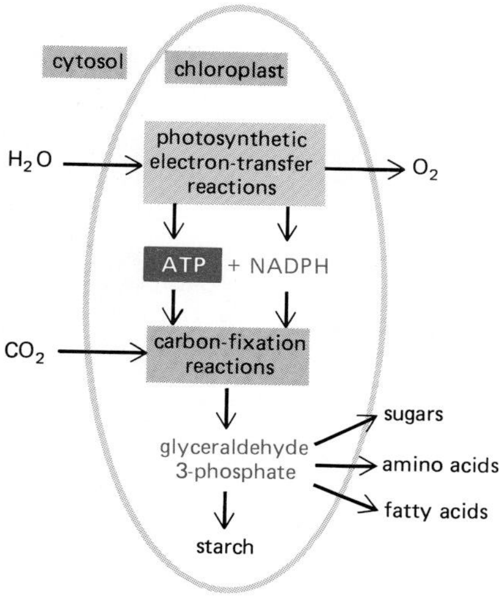

Figure 7–41 Photosynthesis in a chloroplast can be broadly separated into photosynthetic electron-transfer reactions and carbon-fixation reactions. Water is oxidized and oxygen is released in the first set of reactions, while carbon dioxide is assimilated (fixed) to produce organic molecules in the second set of reactions.

Carbon Fixation Is Catalyzed by Ribulose Bisphosphate Carboxylase[28]

We have seen earlier in this chapter how cells produce ATP by using the large amount of free energy released when carbohydrates are oxidized to CO_2 and H_2O. Clearly, therefore, the reverse reaction, in which CO_2 and H_2O combine to make carbohydrate, must be a very unfavorable one. Consequently, this synthesis must be coupled to other, very favorable reactions to drive it.

The central reaction, in which an atom of inorganic carbon is converted to organic carbon, is illustrated in Figure 7–42: CO_2 from the atmosphere combines with the five-carbon compound ribulose 1,5-bisphosphate plus water to give two molecules of the three-carbon compound 3-phosphoglycerate. This "carbon-fixing" reaction, which was discovered in 1948, is catalyzed in the chloroplast stroma by a large enzyme called *ribulose bisphosphate carboxylase* (~500,000 daltons). Since each molecule of the enzyme works sluggishly (processing about 3 molecules of substrate per second compared to 1000 molecules per second for a typical enzyme), many copies are needed. Ribulose bisphosphate carboxylase often represents more than 50% of the total chloroplast protein and is widely claimed to be the most abundant protein on earth.

Three Molecules of ATP and Two Molecules of NADPH Are Consumed for Each CO_2 Molecule That Is Fixed in the Carbon-Fixation Cycle[29]

The actual reaction in which CO_2 is fixed is energetically favorable, but it requires a continuous supply of the energy-rich compound *ribulose 1,5-bisphosphate,* to which each molecule of CO_2 is added (see Figure 7–42). The elaborate pathway by which this compound is regenerated was worked out in one of the most successful early applications of radioisotopes. As outlined in Figure 7–43, three molecules of CO_2 are fixed by ribulose bisphosphate carboxylase to produce six molecules of 3-phosphoglycerate (containing $6 \times 3 = 18$ carbon atoms in all: 3 from the CO_2 and 15 from ribulose 1,5-bisphosphate). The 18 carbon atoms then undergo a cycle of reactions that regenerate the three molecules of ribulose 1,5-bisphosphate used in the initial carbon-fixation step (containing $3 \times 5 = 15$ carbon atoms). This leaves one molecule of *glyceraldehyde 3-phosphate* (3 carbon atoms) as the net gain. In this **carbon-fixation cycle** (or Calvin-Benson cycle), three

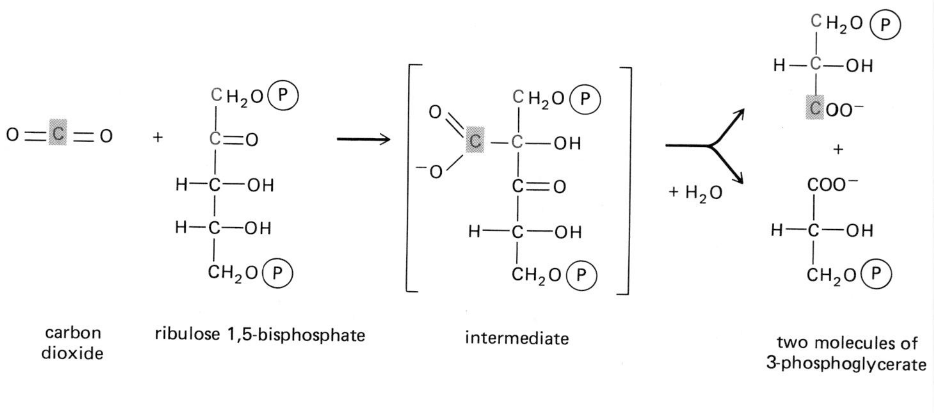

Figure 7–42 The initial reaction in the process by which carbon dioxide is converted into organic carbon. This reaction is catalyzed in the chloroplast stroma by the abundant enzyme *ribulose bisphosphate carboxylase,* and it forms 3-phosphoglycerate, which is also an important intermediate in glycolysis (see Figure 2–20, p. 66). The two colored carbon atoms are used to produce *phosphoglycolate* when the enzyme adds oxygen instead of CO_2 (see p. 371, below).

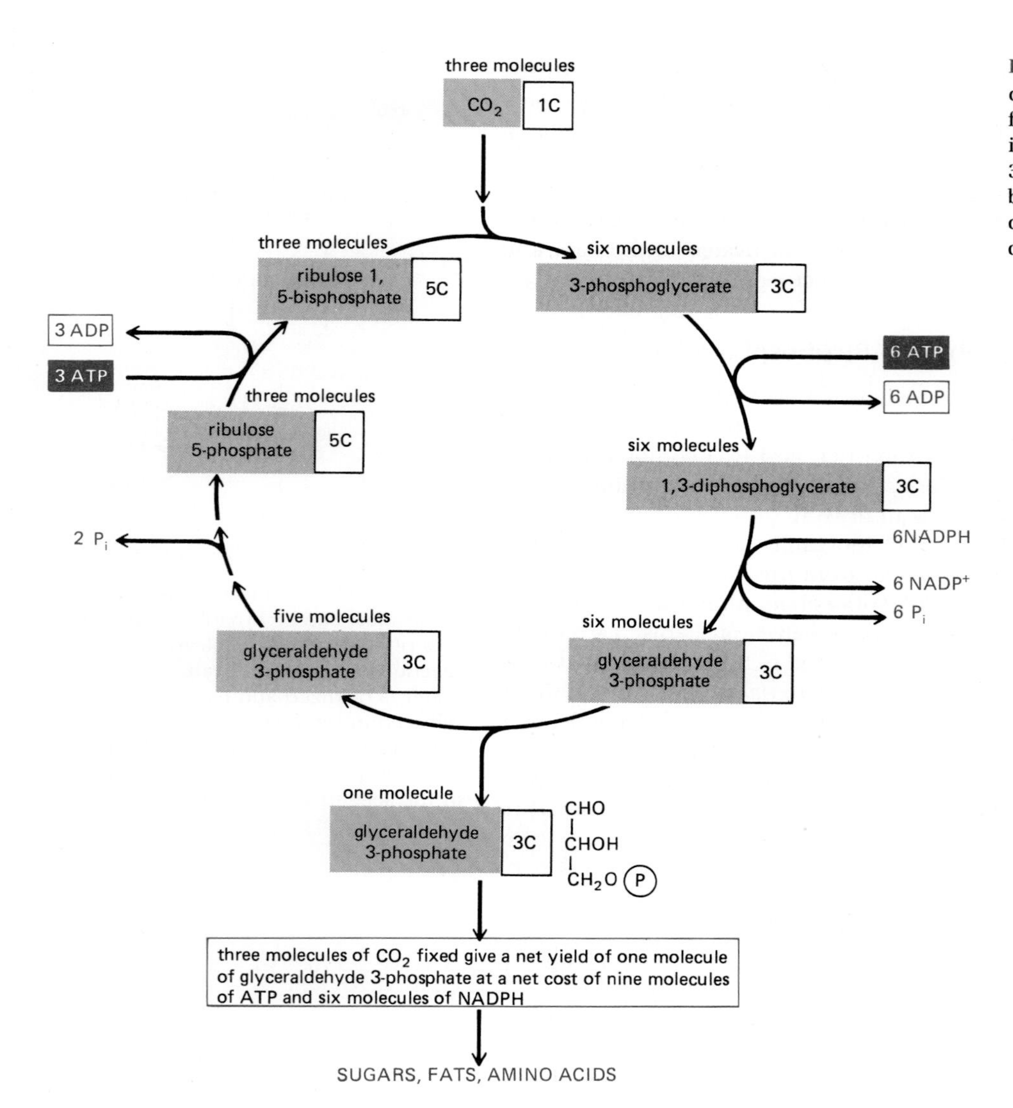

Figure 7–43 The carbon-fixation cycle, which forms organic molecules from CO_2 and H_2O. There are many intermediates between glyceraldehyde 3-phosphate and ribulose 5-phosphate, but they have been omitted here for clarity. The entry of water into the cycle is also not shown.

molecules of ATP and two molecules of NADPH are consumed for each CO_2 molecule converted into carbohydrate. The net equation is

$$3CO_2 + 9ATP + 6NADPH + \text{water} \rightarrow \text{glyceraldehyde 3-phosphate} + 8P_i + 9ADP + 6NADP^+$$

Thus both *phosphate-bond energy* (as ATP) and *reducing power* (as NADPH) are required for the formation of organic molecules from CO_2 and H_2O. We shall return to this important point later.

The glyceraldehyde 3-phosphate produced in chloroplasts by the carbon-fixation cycle is a three-carbon sugar that serves as a central intermediate in glycolysis (p. 65). Much of it is exported to the cytoplasm, where it can be rapidly converted into fructose 6-phosphate and glucose 1-phosphate by reversal of several reactions in glycolysis (see p. 83). Glucose 1-phosphate is then converted to the sugar nucleotide UDP-glucose, and this combines with fructose 6-phosphate to form sucrose phosphate, the immediate precursor of the disaccharide **sucrose.** Sucrose is the major form in which sugar is transported between plant cells, acting in plant cells as glucose acts in animal cells: just as glucose is transported in the blood of animals, sucrose is exported from the leaves via vascular bundles (see Figure 7–45, below), providing the carbohydrate required by the rest of the plant.

Most of the glyceraldehyde 3-phosphate that remains in the chloroplast is converted to *starch* in the stroma. Like glycogen in animal cells, **starch** is a large polymer of glucose that serves as a carbohydrate reserve. The production of starch is regulated so that it is produced and stored as large grains in the chloroplast stroma (see Figure 7–39B) during periods of excess photosynthetic capacity. This occurs through reactions in the stroma that are the reverse of those in glycolysis: they convert glyceraldehyde 3-phosphate to glucose 1-phosphate, which is then used to produce the sugar nucleotide ADP-glucose, the immediate precursor of starch. At night the starch is broken down to help support the metabolic needs of the plant.

7-20 7-22 Carbon Fixation in Some Tropical Plants Is Compartmentalized to Facilitate Growth at Low CO_2 Concentrations[30]

Although ribulose bisphosphate carboxylase preferentially adds CO_2 to ribulose 1,5-bisphosphate, if the concentration of CO_2 is low, it will add O_2 instead. This is an apparently wasteful pathway, which produces one molecule of 3-phosphoglycerate and one molecule of the two-carbon compound phosphoglycolate rather than two molecules of 3-phosphoglycerate (see Figure 7–42). The phosphoglycolate is converted to glycolate and shuttled into peroxisomes, which begin the process of converting two molecules of glycolate into one molecule of 3-phosphoglycerate (three carbons) plus one molecule of CO_2. Because the entire process uses up O_2 and liberates CO_2, it is termed *photorespiration*. In many plants about one-third of the CO_2 fixed is lost again as CO_2 because of photorespiration. It is not known whether photorespiration has some function in the plant or is simply a means of returning to the carbon-fixation pathway some of the carbon diverted into phosphoglycolate by the undesirable reactivity of oxygen with ribulose 1,5-bisphosphate.

Photorespiration can be a serious liability for plants in hot, dry conditions, where they close their stomata—the gas exchange pores in their leaves—to avoid excessive water loss. This causes the CO_2 levels in the leaf to fall precipitously and favors photorespiration. However, a special adaptation occurs in the leaves of many plants, such as corn and sugar cane, that live in hot, dry environments. In these plants the carbon-fixation cycle shown in Figure 7–43 occurs only in the chloroplasts of specialized *bundle-sheath cells,* which contain all of the plant's ribulose bisphosphate carboxylase. These cells are protected from the air and are surrounded by a specialized layer of mesophyll cells that "pump" CO_2 into the bundle-sheath cells, supplying the ribulose bisphosphate carboxylase with a high concentration of CO_2, which greatly reduces photorespiration.

The CO_2 pump is a reaction cycle that begins with a CO_2-fixation step catalyzed in the cytosol of the mesophyll cells by an enzyme that binds carbon dioxide (as bicarbonate) with high affinity. The product is a four-carbon compound that diffuses into the bundle-sheath cells, where it is broken down to give one molecule of CO_2 and one molecule of a three-carbon compound. The latter is then returned to the mesophyll cells where, in a reaction requiring ATP hydrolysis, it is converted to an activated form that can pick up another CO_2 molecule to start the CO_2 pumping cycle again (Figure 7–44).

When a pulse of radioactive $^{14}CO_2$ is given to a plant that pumps CO_2, it is assimilated in the CO_2-fixation step in the mesophyll cells so that the first organic compound that is labeled contains four carbons. In all other plants the three-carbon compound 3-phosphoglycerate is labeled first (see Figure 7–43). Consequently, CO_2-pumping plants are called *C_4 plants* and all other plants are called *C_3 plants* (Figure 7–45).

As for any vectorial transport process, pumping CO_2 into the bundle-sheath cells in C_4 plants costs energy. However, in hot, dry environments this cost is often much less than the energy lost by photorespiration in C_3 plants, and so C_4 plants are at an advantage. Moreover, because C_4 plants can carry out photosynthesis at a lower concentration of CO_2 inside the leaf, they need to open their stomata less and can therefore fix about twice as much net carbon as C_3 plants per unit of water lost.

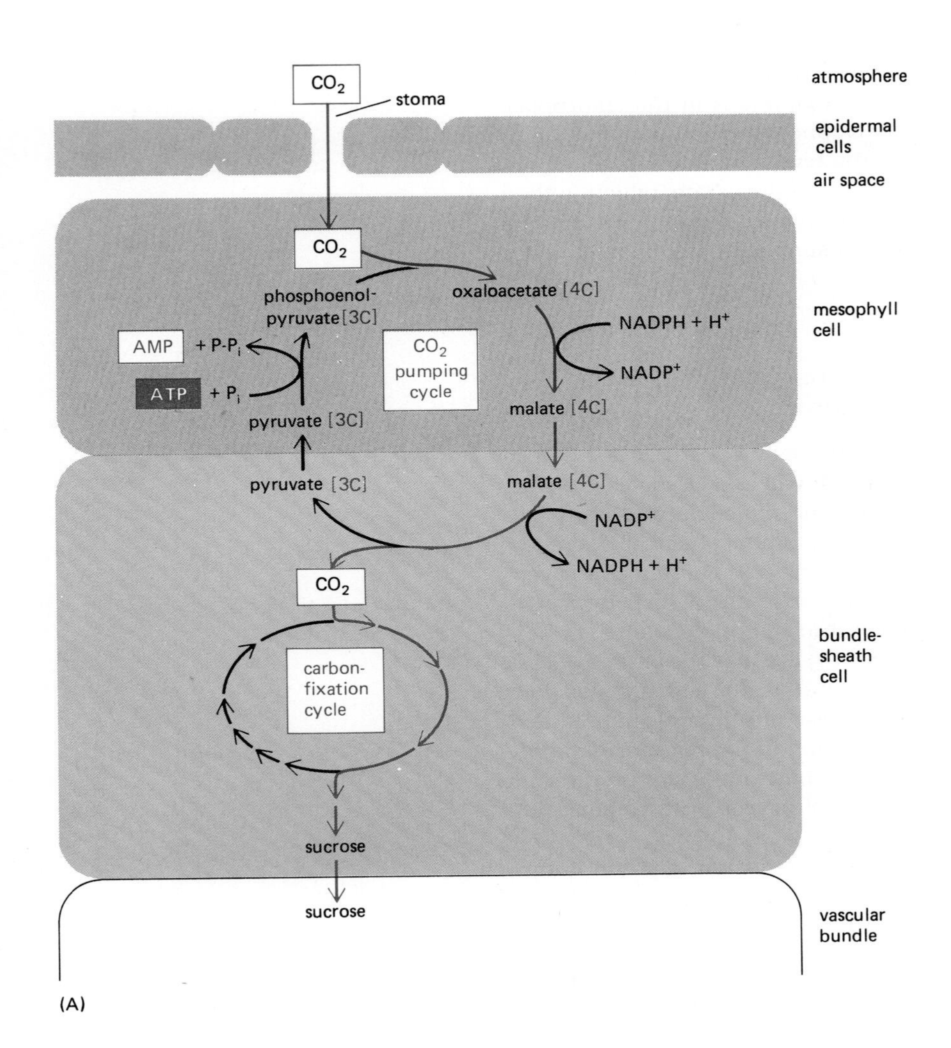

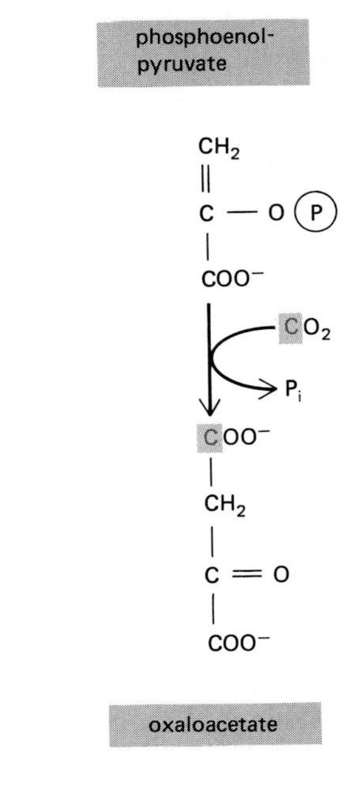

Figure 7–44 The CO_2 pumping cycle in a plant such as corn. (A) The net production of carbohydrate is confined to bundle-sheath cells, which contain all of the ribulose bisphosphate carboxylase. The mesophyll cells initiate the CO_2 pumping cycle, which involves the four-carbon and three-carbon compounds indicated. Variations of this pumping cycle are used in other CO_2-pumping plants. (B) The reaction of phosphoenolpyruvate with CO_2 in mesophyll cells.

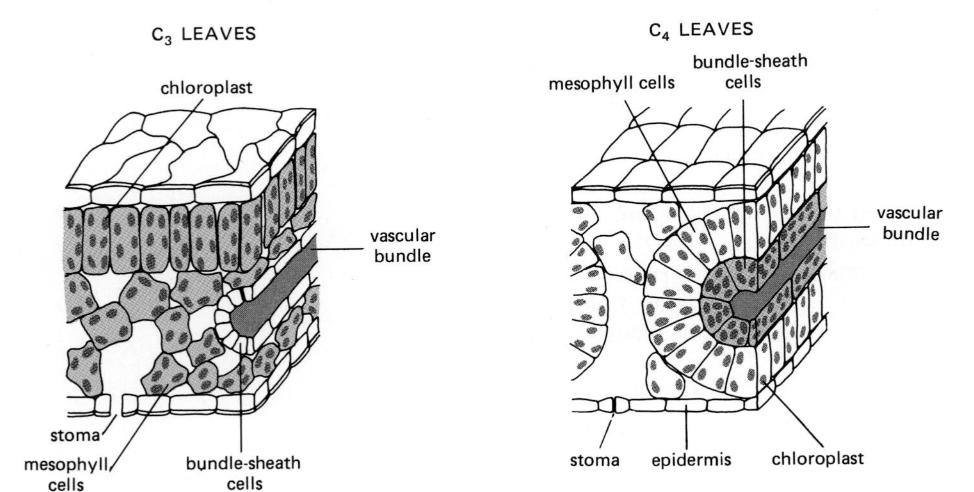

Figure 7–45 A comparison of the anatomy of the leaf in a C_3 plant and a C_4 plant. The colored cells contain chloroplasts that carry out the normal carbon-fixation cycle. In C_4 plants, the mesophyll cells are specialized for CO_2 pumping rather than for carbon fixation, and they create a high CO_2:O_2 ratio in the bundle-sheath cells, which are the only cells in these plants where the carbon-fixation cycle occurs (see Figure 7–44). The vascular bundles carry the sucrose made in the leaf to other tissues.

Figure 7–46 The structure of chlorophyll. A magnesium atom is held in a porphyrin ring, which is related to the porphyrin ring that binds iron in heme (compare with Figure 7–27). Electrons are delocalized over the bonds shown in color.

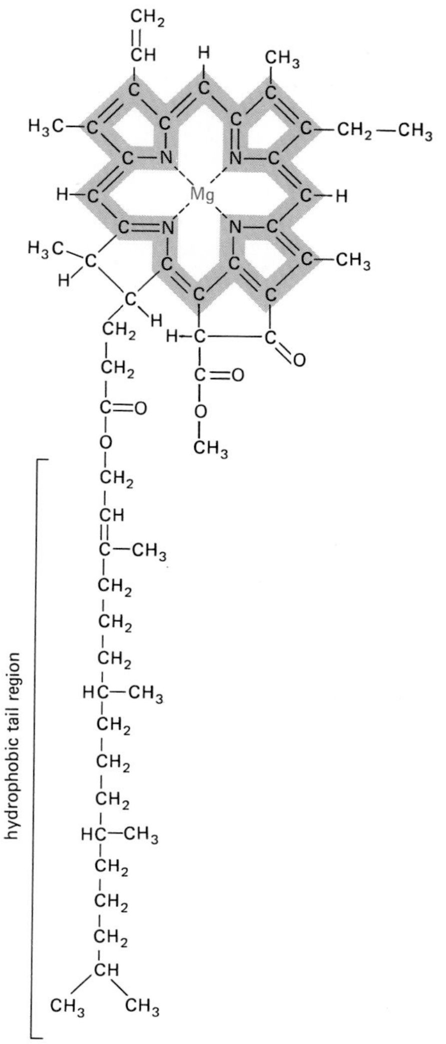

7-23 Photosynthesis Depends on the Photochemistry of Chlorophyll Molecules[31]

Having discussed the carbon-fixation reactions, we now return to the question of how the photosynthetic electron-transfer reactions in the chloroplast generate the ATP and the NADH needed to drive the production of carbohydrates from CO_2 and H_2O (see Figure 7–41). The required energy is derived from sunlight absorbed by **chlorophyll** molecules (Figure 7–46). The process of energy conversion begins when a chlorophyll molecule is excited by a quantum of light (a photon) and an electron is moved from one molecular orbital to another of higher energy. Such an excited molecule is unstable and will tend to return to its original, unexcited state in one of three ways: (1) by converting the extra energy into heat (molecular motions) or to some combination of heat and light of a longer wavelength (fluorescence), as when light energy is absorbed by an isolated chlorophyll molecule in solution; (2) by transferring the energy—but not the electron—directly to a neighboring chlorophyll molecule by a process called *resonance energy transfer;* or (3) by transferring the high-energy electron to another nearby molecule (an *electron acceptor*) and then returning to its original state by taking up a low-energy electron from some other molecule (an *electron donor,* Figure 7–47). The last two mechanisms play an essential part in photosynthesis.

A Photosystem Contains a Reaction Center Plus an Antenna Complex[32]

Multiprotein complexes called **photosystems** catalyze the conversion of the light energy captured in excited chlorophyll molecules to useful forms. A photosystem consists of two closely linked components: a *photochemical reaction center* and an *antenna complex* (Figure 7–48).

The **antenna complex** is important for light harvesting. In chloroplasts it consists of a cluster of several hundred chlorophyll molecules linked together by proteins that hold them tightly on the thylakoid membrane. Depending on the

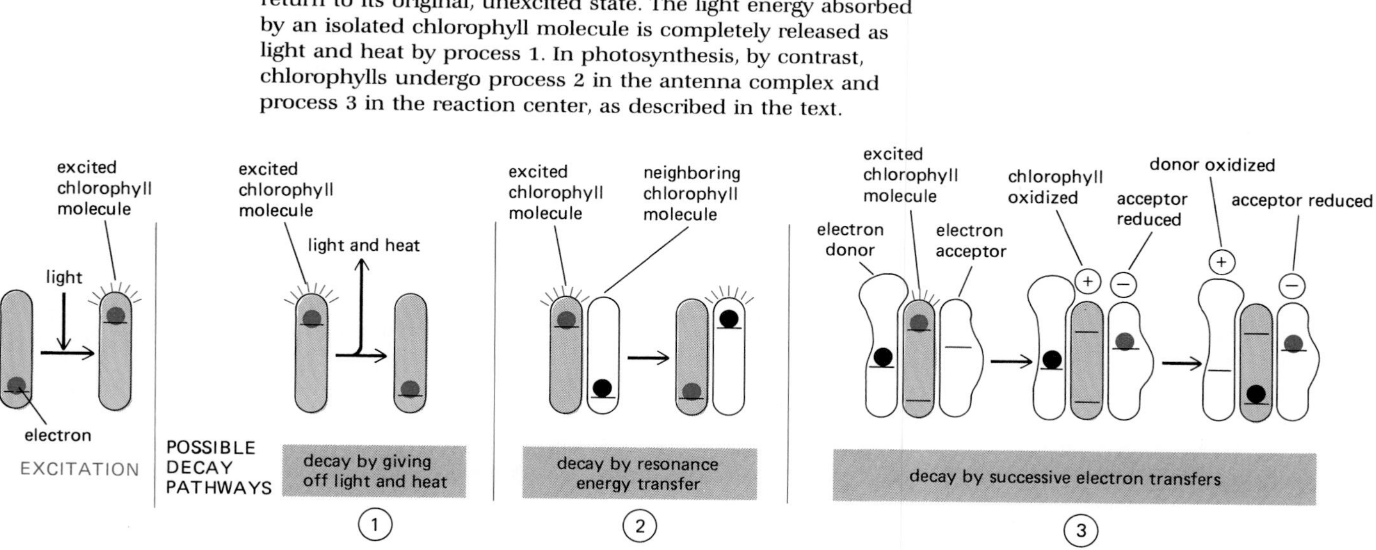

Figure 7–47 Three possible ways for an excited chlorophyll molecule (a molecule containing a high-energy electron) to return to its original, unexcited state. The light energy absorbed by an isolated chlorophyll molecule is completely released as light and heat by process 1. In photosynthesis, by contrast, chlorophylls undergo process 2 in the antenna complex and process 3 in the reaction center, as described in the text.

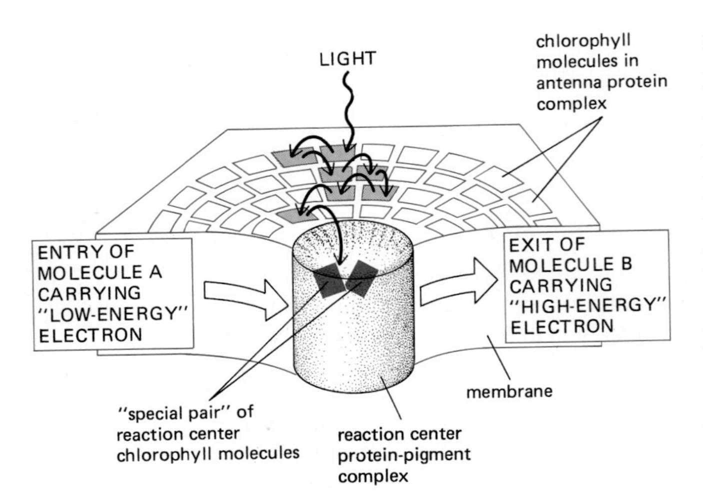

Figure 7–48 A photosystem consists of a reaction center and an antenna. The reaction center is a transmembrane protein complex that holds a "special pair" of chlorophyll molecules in a fixed spatial relation to other electron carriers (see Figure 7–49). It catalyzes process 3 in Figure 7–47. If viewed as an enzyme, the substrates of a reaction center would be a weak election donor (molecule A) and a weak electron acceptor (molecule B), and its products would be a strong electron acceptor (oxidized molecule A) and a strong electron donor (reduced molecule B). The antenna complex contains most of the chlorophylls in the thylakoid membrane, and it serves as a funnel for transferring the energy of an excited electron to the reaction center. Many of these energy transfers are between identical chlorophyll molecules, through which the excitation wanders randomly (process 2 in Figure 7–47). Nevertheless, the average time needed for the excitation quantum to encounter and "fall into" the reaction center is only 10^{-10} to 10^{-9} seconds, so very few of the quanta absorbed are lost by the random decay pathway shown as process 1 in Figure 7–47.

plant, varying amounts of accessory pigments called *carotenoids,* which can help collect light of other wavelengths, are also located in each complex. When a chlorophyll molecule in the antenna complex is excited, the energy is rapidly transferred from one molecule to another by resonance energy transfer until it reaches a special pair of chlorophyll molecules in the photochemical reaction center. Each antenna complex thereby acts as a "funnel," collecting light energy and directing it to a specific site where it can be used effectively (see Figure 7–48).

The **photochemical reaction center** is a transmembrane protein-pigment complex that lies at the heart of photosynthesis. It is thought to have evolved more than 3 billion years ago in primitive photosynthetic bacteria. The special pair of chlorophyll molecules in the reaction center acts as an irreversible trap for excitation quanta because its excited electron is immediately passed to a chain of electron acceptors that are precisely positioned as neighbors in the same protein complex (Figure 7–49). By moving the high-energy electron rapidly away from the chlorophylls, the reaction center transfers it to an environment where it is much more stable. The electron is thereby suitably positioned for subsequent photochemical reactions, which require more time to complete. As we shall see, the net result of these slower reactions is the conversion of a "low-energy" electron on a weak electron donor (such as water) to a "high-energy" electron on a strong electron donor (such as a *quinone*).

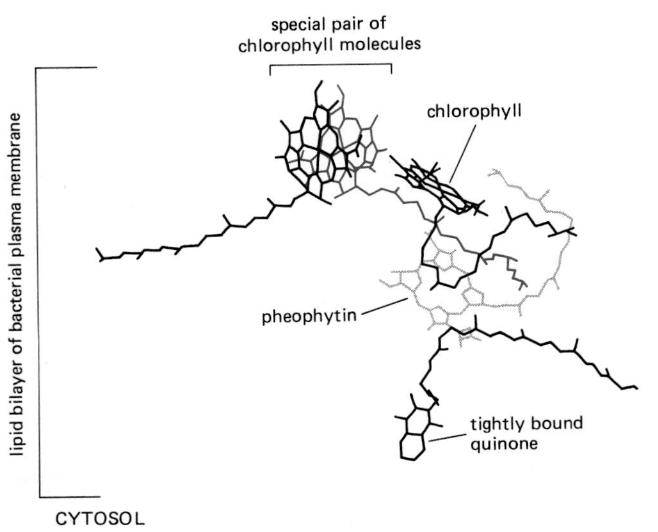

Figure 7–49 The arrangement of the electron carriers in a bacterial photochemical reaction center as determined by x-ray crytallography. The pigment molecules shown are held in the interior of a transmembrane protein and are surrounded by the lipid bilayer, as indicated. An electron in the special pair is excited by resonance from an antenna complex chlorophyll (process 2 in Figure 7–47), and the excited electron is then transferred stepwise from the special pair to the quinone (see Figure 7–50).

In a Reaction Center, Light Energy Captured by Chlorophyll Creates a Strong Electron Donor from a Weak One[33]

The electron transfers involved in the photochemical reactions just outlined have been analyzed extensively by rapid spectroscopic methods, especially in the photosystem of purple bacteria, which is simpler than the evolutionarily related photosystem in chloroplasts. The bacterial reaction center is a large protein-pigment complex that can be solubilized with detergent and purified in active form. In 1985 its complete three-dimensional structure was determined by x-ray crystallography (see Figure 6–32, p. 294, and Figure 7–49). This structure, combined with kinetic data, provides the best picture we have of the initial electron-transfer reactions that underlie photosynthesis.

The sequence of transfers that take place in the reaction center of purple bacteria is shown diagrammatically in Figure 7–50. An electron excited by the absorption of light is passed rapidly from the special pair of reaction-center chlorophylls through the other pigments in Figure 7–49 to a tightly bound quinone electron acceptor, designated as Q_A. This electron transfer, which occurs in less than 10^{-9} seconds and is virtually irreversible, leaves a positively charged "hole" with a very high affinity for electrons in the chlorophyll, which is refilled by the capture of an electron from a nearby cytochrome (normally a weak electron donor). The high-energy electron held by Q_A is subsequently passed to a second quinone, Q_B, from which it leaves the reaction center and passes to a mobile quinone molecule (Q) in the photosynthetic membrane. This reduced quinone is a strong electron donor whose reducing power can be harnessed to pump protons.

The essential principle illustrated here is that a photosystem enables light to cause a net electron transfer from a weak electron donor—that is, a molecule with a strong affinity for electrons (in this case a cytochrome)—to a molecule such as a quinone, which is a strong electron donor in its reduced form. In this way the excitation energy that would otherwise be released as fluorescence and/or heat is used instead to raise the energy of an electron and create a strong electron donor where none had been before. In the chloroplasts of higher plants, as we shall see, water, rather than cytochrome, serves as the initial electron donor, which is why oxygen is released by photosynthesis in plants. But before returning to consider the processes by which the more complex photosystems of chloroplasts ultimately drive the production of ATP and NADPH, we shall consider how these end products are generated in a less sophisticated way, although by essentially similar machinery, in purple bacteria.

Bacterial Photosynthesis Establishes an Electrochemical Proton Gradient Across the Plasma Membrane That Drives the Production of Both ATP and NADPH[34]

The energy stored in the electrons carried by reduced quinones in the plasma membranes of purple photosynthetic bacteria is used in two different ways to generate ATP and NADPH, respectively. ATP is produced by a proton-pumping mechanism very similar to the one we have already encountered in mitochondria (see p. 352): protons are pumped across the bacterial plasma membrane when the quinone transfers its high-energy electrons through a *b-c complex* embedded in this membrane. The b-c complex then transfers electrons to a soluble cytochrome

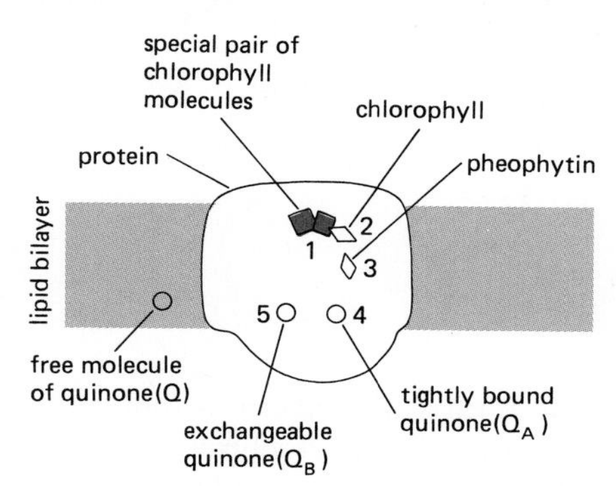

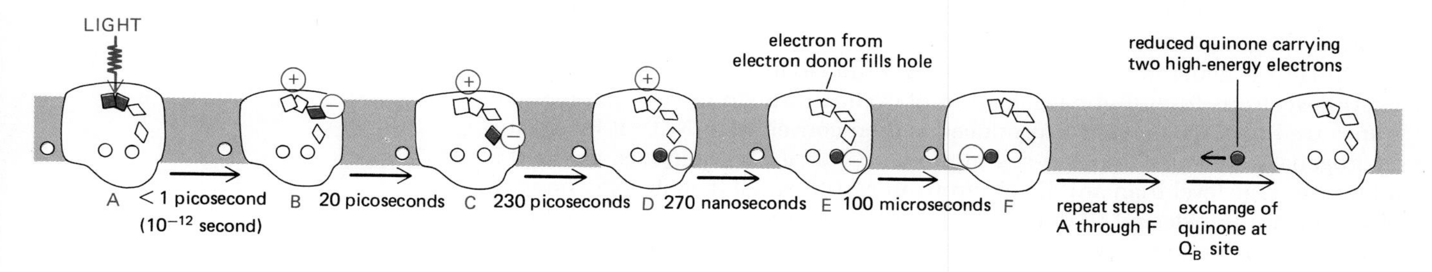

Figure 7–50 The electron transfers that occur in the photochemical reaction center of a purple bacterium. A similar set of reactions is believed to occur in the evolutionarily related photosystem II in plants. At the top right is a schematic diagram showing the molecules that carry electrons, which are those in Figure 7–49, plus an exchangeable quinone (Q_B) and a freely mobile quinone dissolved in the lipid bilayer (Q). Electron carriers 1 through 5 are bound in a specific position on a 596-amino-acid transmembrane protein formed from two separate subunits (see Figure 6–32, p. 294). Following excitation by a photon of light, a high-energy electron passes from pigment molecule to pigment molecule, creating a charge separation as shown in the sequence in steps B through D below, where the pigment molecule carrying high-energy electrons is indicated in color. Once released into the bilayer, the quinone with two electrons picks up two protons and loses its charge (see Figure 7–30).

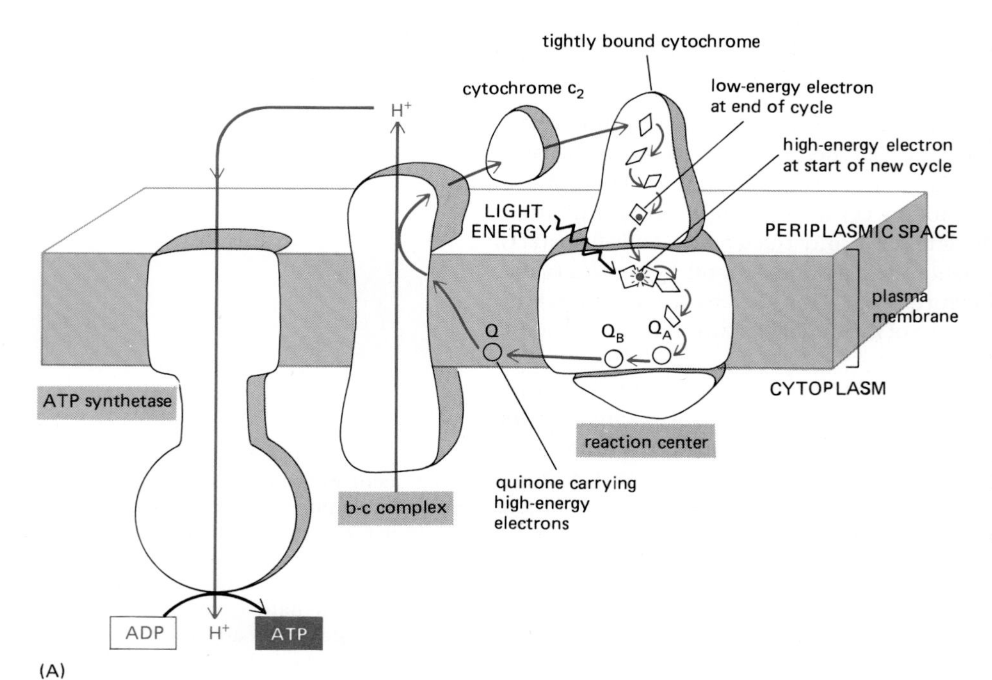

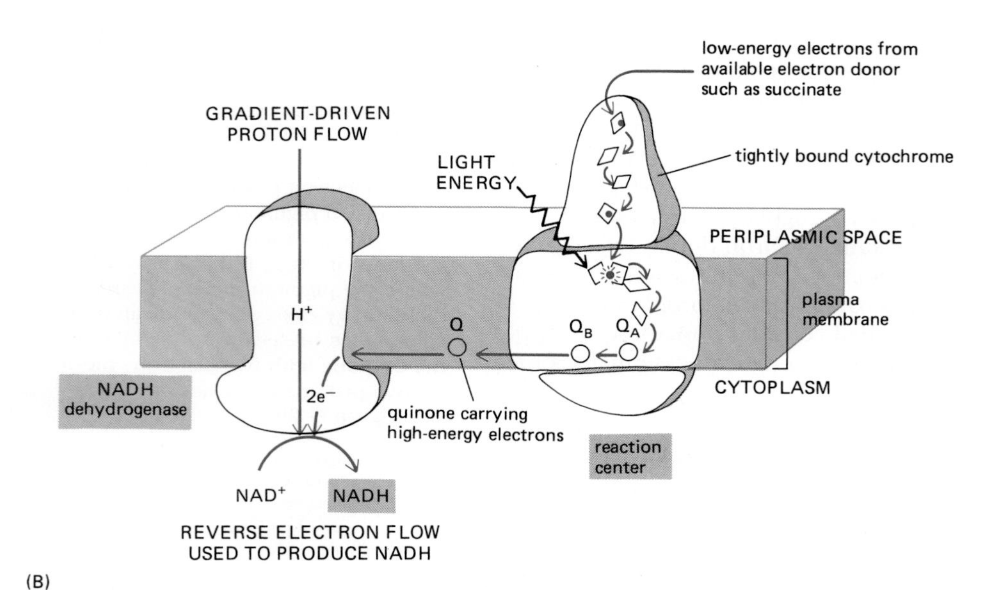

Figure 7–51 Two photosynthetic electron-transfer reactions in purple bacteria. These reactions occur in the inner (plasma) membrane of the bacterium, and cytochrome c_2 exists in soluble form in the periplasmic space beneath the outer membrane (see Figure 6–54, p. 312). (A) Cyclic electron flow produces an electrochemical proton gradient across the plasma membrane. This gradient is used here to drive the production of ATP by an ATP synthetase in the bacterial plasma membrane. (B) Reverse electron flow through NADH dehydrogenase, which is also driven by the large electrochemical proton gradient produced in (A), produces NADH. The reversibility of the reactions linked to the electrochemical gradient is discussed on p. 364.

that returns the electrons (now at low energy) to the reaction center complex via a tightly bound cytochrome, completing a cyclic process (Figure 7–51A). The electrochemical proton gradient created by this cyclic electron transport is then used by an ATP synthetase in the plasma membrane to produce ATP, just as in mitochondria (Figure 7–51A).

NADPH is generated by a second electron-transport reaction, in which high-energy electrons from quinones, instead of cycling through the b-c complex, are transferred to NAD; the NADH produced is then converted to NADPH by a transhydrogenase. Because the high-energy electrons carried by quinones are held at a lower energy level than are the electrons in NADH (recall that, in mitochondria, electrons are transferred to quinone from NADH and not vice versa—see Figure 7–34), the production of NADH from NAD requires a source of energy. In purple photosynthetic bacteria the energy is provided by the electrochemical proton gradient across the plasma membrane, which drives H^+ back into the cell through an NADH-dehydrogenase complex, enabling this complex to catalyze the energetically unfavorable **reverse electron flow** from quinone to NAD (Figure 7–51B).

In summary, the purple photosynthetic bacterium uses its reaction center to produce a large pool of reduced quinone molecules in its plasma membrane. Some of this quinone is used to generate a large electrochemical proton gradient across the bacterial plasma membrane. This gradient is then used in two ways: (1) to drive an ATP synthetase to make ATP and (2) to drive a reverse electron flow from the remaining reduced quinone to NAD, thereby generating the reducing power needed for the synthesis of the bacterium's organic molecules.

7-24 7-25 7-26

In Plants and Cyanobacteria, Noncyclic Photophosphorylation Produces Both NADPH and ATP[31,35]

Photosynthesis in plants and cyanobacteria is more complex. It produces both ATP and NADPH directly by a two-step process called **noncyclic photophosphorylation.** Because two photosystems in series are used to energize an electron, the electron can be transferred all the way from water to NADPH. As the high-energy electrons pass through the coupled photosystems to generate NADPH, some of their energy is siphoned off for ATP synthesis.

In the first of the two photosystems—called *photosystem II* for historical reasons—the oxygens of two water molecules bind to a cluster of manganese atoms in a poorly understood water-splitting enzyme, and electrons are removed one at a time to fill the holes created by light in reaction-center chlorophyll molecules. As soon as four electrons have been removed (requiring four quanta of light), O_2 is released by the enzyme; photosystem II thus catalyzes the reaction $2H_2O \rightarrow 4H^+ + 4e^- + O_2$.

The core of the reaction center in photosystem II is homologous to the bacterial reaction center just described, and it likewise produces strong electron donors in the form of reduced quinone molecules in the membrane. The quinones pass their electrons to a *b_6-f complex*, which closely resembles the b-c complex of bacteria and the b-c_1 complex in the respiratory chain of mitochondria. As in mitochondria, the complex pumps protons into the thylakoid space across the thylakoid membrane (in chloroplasts) or out of the cytosol across invaginations of the plasma membrane (in cyanobacteria), and the resulting electrochemical gradient drives the synthesis of ATP by an ATP synthetase (Figures 7–52 and 7–53). The final electron acceptor in this electron-transport chain is the second photosystem in the scheme (*photosystem I*), which accepts the electron into the hole left by light excitation of its reaction-center chlorophyll molecule. Whereas the electrons energized by photosystem II are at too low an energy to be passed to $NADP^+$, each electron that leaves photosystem I has been boosted to a very high energy level by the two quanta of light that have sequentially activated it. Consequently, these electrons can be passed to the iron-sulfur center in ferredoxin to drive the reduction of $NADP^+$ to NADPH (Figure 7–53), which also involves the uptake of a proton from the medium.

The zigzag scheme for photosynthesis shown in Figure 7–53 is known as the **Z scheme.** By means of its two electron-energizing steps, one catalyzed by each photosystem, an electron is passed from water, which normally holds onto its electrons very tightly (redox potential = +820 mV), to NADPH, which normally holds onto its electrons rather loosely (redox potential = −320 mV). There is not enough energy in a single quantum of visible light to energize an electron all the way from the bottom of photosystem II to the top of photosystem I, which is probably the energy change required to pass an electron efficiently from water to $NADP^+$. The use of two separate photosystems in series also means that there is enough energy left over to enable the electron-transport chain that links the two photosystems to pump H^+ across the thylakoid membrane (or the plasma membrane of cyanobacteria), which allows ATP synthetase to harness some of the light-derived electron energy for ATP production.

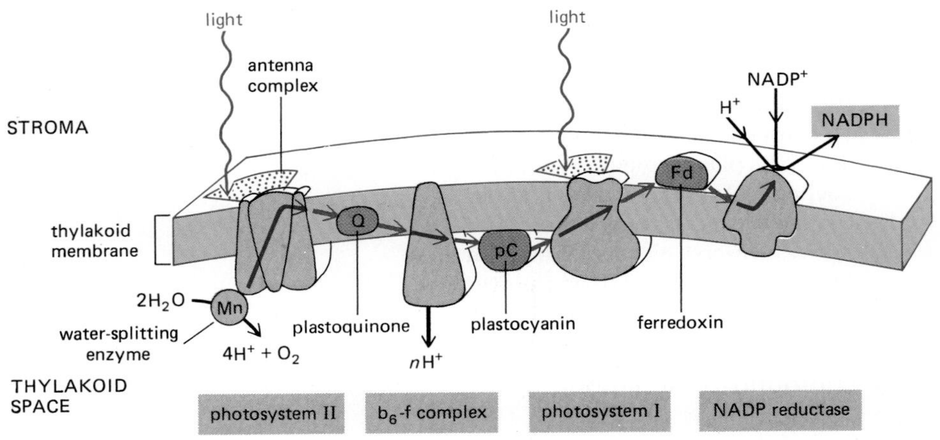

Figure 7–52 Electron flow during photosynthesis in the thylakoid membrane. The mobile electron carriers in the chain are plastoquinone (which closely resembles the ubiquinone of mitochondria), plastocyanin (a small copper-containing protein), and ferredoxin (a small protein containing an iron-sulfur center). The *b_6-f complex* closely resembles the b-c_1 complex of mitochondria and the b-c complex of bacteria (see Figure 7–63): all three complexes accept electrons from quinones and pump protons. Note that the H^+ released by water oxidation and the H^+ taken up during NADPH formation also contribute to the generation of the electrochemical proton gradient that drives ATP synthesis.

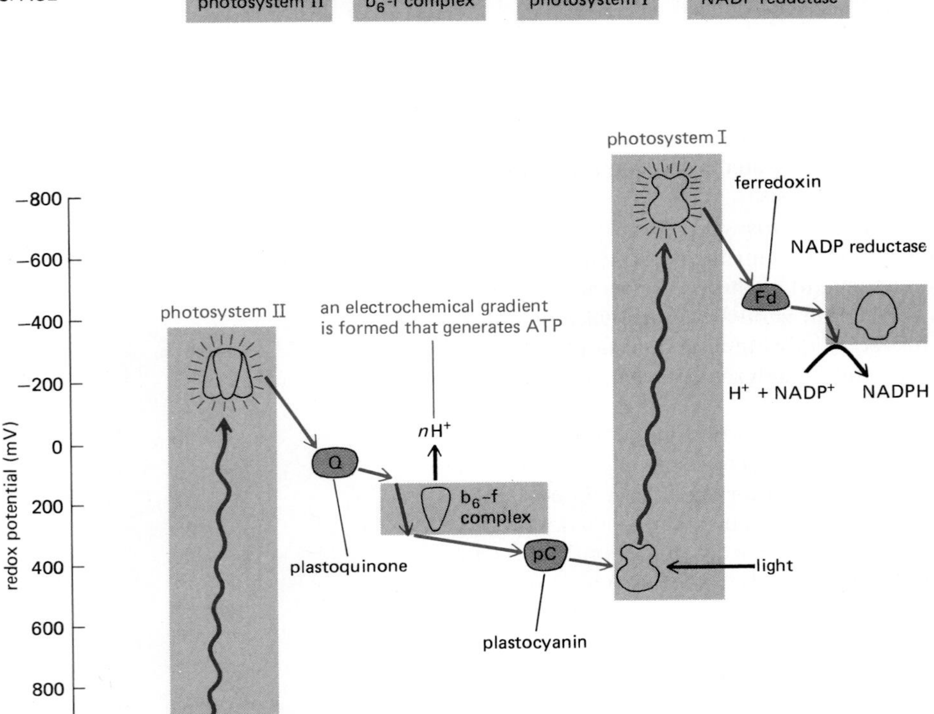

Figure 7–53 Redox potential changes for the passage of electrons during the photosynthetic production of NADPH and ATP in plants and cyanobacteria. Photosystem II closely resembles the reaction center in purple bacteria (see Figure 7–50), to which it is evolutionarily related. Photosystem I differs and is thought to be evolutionarily related to the photosystems of a different class of bacteria, the green bacteria; it passes electrons from its excited chlorophyll through a series of tightly bound iron-sulfur centers. The net electron flow through the two photosystems joined in series is from water to $NADP^+$, and it produces NADPH. In addition, ATP is synthesized by an ATP synthetase (not shown) that harnesses the electrochemical proton gradient produced by the electron-transport chain linking photosystem II and photosystem I. This *Z scheme* for ATP production is called *noncyclic photophosphorylation* to distinguish it from the cyclic scheme shown in Figure 7–54 (see also Figure 7–52).

Chloroplasts Can Make ATP by Cyclic Photophosphorylation Without Making NADPH[31,36]

In the noncyclic photophosphorylation scheme just discussed, high-energy electrons leaving photosystem II are harnessed to generate ATP, while those leaving photosystem I drive the production of NADPH. This produces slightly more than one molecule of ATP for every pair of electrons that passes from H_2O to $NADP^+$ to generate a molecule of NADPH. But considerably more ATP than NADPH is needed for carbon fixation (see Figure 7–43). To produce the extra ATP, chloroplasts can switch photosystem I into a cyclic mode in which its energy is directed into the synthesis of ATP instead of NADPH. This process, called **cyclic photophosphorylation,** involves an electron flow much like that used by photosynthetic bacteria to make ATP (see Figure 7–51A). Here the high-energy electrons from photosystem I are transferred back to the b_6-f complex rather than being passed

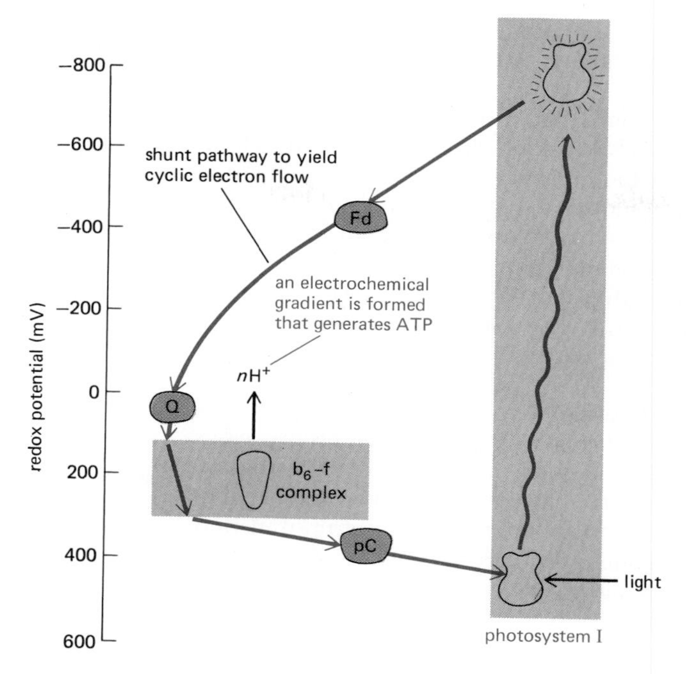

Figure 7–54 The path of electron flow in cyclic photophosphorylation. This pathway allows ATP to be made without producing either NADPH or O_2. Whether noncyclic or cyclic electron flow occurs depends on whether ferredoxin (Fd) donates its reactive electron to $NADP^+$, as in Figure 7–53, or to components leading back to the b_6-f complex. Whenever NADPH accumulates, $NADP^+$ levels will be low, tending to favor the cyclic scheme; other, less direct controls also ensure that appropriate proportions of ATP and NADPH are produced by photosynthesis.

on to $NADP^+$, so that protons are pumped across the thylakoid membrane, and the resulting electrochemical gradient drives the synthesis of ATP (Figure 7–54).

To review briefly, noncyclic photophosphorylation involves the photoreduction of $NADP^+$ by water, is mediated by the combined action of photosystem I and photosystem II, and produces NADPH, ATP, and O_2. Cyclic photophosphorylation, in contrast, involves only photosystem I and produces ATP without the formation of either NADPH or O_2. Thus the relative activities of cyclic and noncyclic electron flows determine how much light energy is converted into reducing power (NADPH) and how much into high-energy phosphate bonds (ATP). The balance is regulated according to the need for NADPH. Whether the electron flow is noncyclic or cyclic depends on whether ferredoxin donates its reactive electron to $NADP^+$ or to components that lead back to the *b_6-f complex* (compare Figures 7–53 and 7–54). At low $NADP^+$ concentrations, caused by an accumulation of NADPH, the cyclic scheme that produces only ATP is favored.

The effect of NADPH levels on cyclic photophosphorylation is part of an extensive regulatory network that balances the activities of photosystem I and photosystem II. An excess activity of photosystem II, for example, will increase the ratio of reduced to oxidized quinones in the thylakoid membrane, while an excess activity of photosystem I will have the opposite effect (see Figure 7–53). Whenever the ratio of reduced to oxidized quinones increases above a certain threshold, however, a protein kinase is activated that phosphorylates the major light-harvesting pigment protein in the antenna complex. This tends to dissociate the antenna complex from photosystem II and may even cause it to move in the thylakoid membrane from the stacked regions (the grana), where photosystem II is concentrated, to the unstacked regions, where photosystem I is concentrated. An increased proportion of light energy is thereby transferred to photosystem I until the quinone pool returns to normal.

7-27 The Geometry of Proton Translocation Is Similar in Mitochondria and Chloroplasts[37]

The presence of the thylakoid space separates a chloroplast into three rather than two internal compartments, making it seem quite different from a mitochondrion. However, the geometry of H^+ translocation in the two organelles is similar. As illustrated in Figure 7–55, in chloroplasts protons are pumped out of the stroma (pH 8) into the thylakoid space (pH about 5), creating a gradient of 3 to 3.5 pH units. This represents a proton-motive force of about 200 mV across the thylakoid membrane (nearly all of which is contributed by the pH gradient rather than by a membrane potential), which drives ATP synthesis by the ATP synthetase embedded in this membrane.

Like the stroma, the mitochondrial matrix has a pH of about 8, but this is created by pumping protons out of the mitochondrion into the cytosol (pH about 7), rather than into an interior space in the organelle. Thus the pH gradient is relatively small, and most of the proton-motive force across the mitochondrial inner membrane, which is about the same as that across the chloroplast thylakoid membrane, is caused by the resulting membrane potential (see p. 352). For both mitochondria and chloroplasts, however, the catalytic site of the ATP synthetase is at a pH of about 8 and is located in a large organelle compartment (matrix or stroma) packed full of soluble enzymes. Consequently, it is here that all of the organelle's ATP is made (Figure 7–55).

Although there are many similarities between mitochondria and chloroplasts, the structure of chloroplasts makes their electron- and proton-transport processes easier to study: by breaking both the inner and outer membranes of a chloroplast, isolated thylakoid discs can be obtained intact. These thylakoids resemble submitochondrial particles in that they have a membrane whose electron-transport chain has its $NADP^+$-, ADP-, and phosphate-utilization sites all freely accessible to the outside. But isolated thylakoids retain their undisturbed native structure and are much more active than isolated submitochondrial particles. For this reason, several of the experiments that first demonstrated the central role of chemiosmotic mechanisms were carried out with chloroplasts rather than with mitochondria.

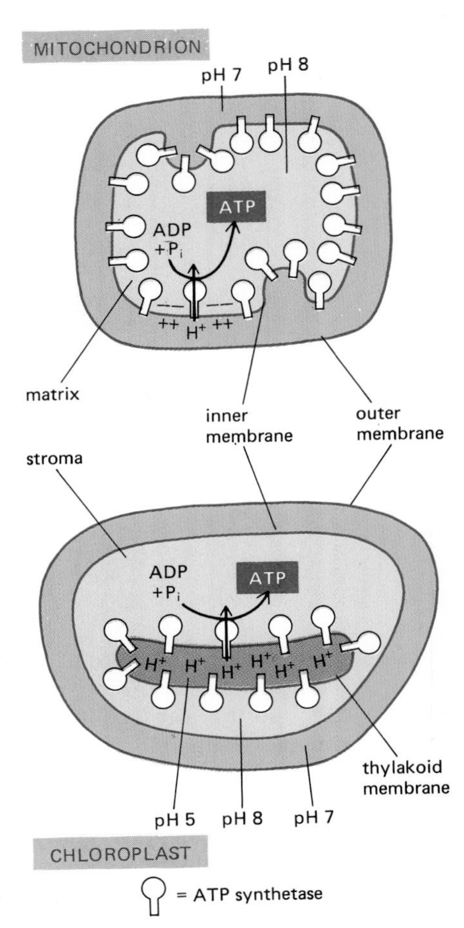

Figure 7–55 Comparison of proton flows and ATP synthetase orientations in mitochondria and chloroplasts. Those compartments with a similar pH have been colored similarly. The proton-motive force across the thylakoid membrane consists almost entirely of the pH gradient; the high permeability of this membrane to Mg^{2+} and Cl^- ions allows the flow of these ions to dissipate most of the membrane potential. Presumably, mitochondria could not tolerate having their matrix at pH 10, as would be required to generate their proton-motive force without a membrane potential.

Like the Mitochondrial Inner Membrane, the Chloroplast Inner Membrane Contains Carrier Proteins That Facilitate Metabolite Exchange with the Cytosol[38]

Although the photosynthetic electron- and proton-transfer reactions of photosynthesis are most readily studied in chloroplast preparations in which the inner and outer membranes have been broken or removed, such chloroplasts fail to carry out photosynthetic CO_2 fixation because of the absence of important substances that are normally present in the stroma. Chloroplasts can also be isolated in a way that leaves their inner membrane intact. In such chloroplasts the inner membrane can be shown to have a selective permeability, reflecting the presence of specific carrier proteins. Most notably, much of the glyceraldehyde 3-phosphate produced by CO_2 fixation in the chloroplast stroma is transported out of the chloroplast by an efficient antiport system that exchanges three-carbon sugar-phosphates for inorganic phosphate.

Glyceraldehyde 3-phosphate normally provides the cytosol with an abundant source of carbohydrate, which is used by the cell as the starting point for many other biosyntheses—including the production of sucrose for export. But this is not all it provides. Once the glyceraldehyde 3-phosphate reaches the cytosol, it is readily converted (by part of the glycolytic pathway) to 3-phosphoglycerate, generating one molecule of ATP and one of NADH. (The same two-step reaction working in reverse forms glyceraldehyde 3-phosphate in the carbon-fixation cycle—see Figure 7–43.) As a result, the export of glyceraldehyde 3-phosphate from the chloroplast provides not only the main source of fixed carbon to the rest of the cell, but also the reducing power and ATP needed for metabolism outside the chloroplast.

Chloroplasts Carry out Other Biosyntheses[39]

The chloroplast carries out many biosyntheses in addition to photosynthesis. All of the cell's fatty acids and a number of amino acids, for example, are made by enzymes in the chloroplast stroma. Similarly, the reducing power of light-activated electrons drives the reduction of nitrite (NO_2^-) to ammonia (NH_3) in the chloroplast; this ammonia provides the plant with nitrogen for the synthesis of amino acids and nucleotides. The metabolic importance of the chloroplast for plants and algae therefore extends far beyond its role in photosynthesis.

Summary

Chloroplasts and photosynthetic bacteria obtain high-energy electrons by means of photosystems that capture the electrons excited when sunlight is absorbed by chlorophyll molecules. Photosystems are composed of an antenna complex attached to a photochemical reaction center, which is a precisely ordered complex of proteins and pigments in which the photochemistry of photosynthesis occurs. By far the best-understood photochemical reaction center is that of the purple photosynthetic bacteria, for which the complete three-dimensional structure is known. In these bacteria a single photosystem produces an electrochemical proton gradient that is used to drive both ATP and NADPH synthesis. There are two photosystems in chloroplasts and cyanobacteria. Depending on the cell's needs, two types of electron flow occur in different ratios: (1) a noncyclic flow, mediated by the two photosystems linked in series, transfers electrons from water to $NADP^+$ to produce NADPH, with the concomitant production of ATP; and (2) a cyclic flow, mediated by a single photosystem through which electrons circulate in a closed loop, produces only ATP. In chloroplasts, all electron-transport processes occur in the thylakoid membrane: to make ATP, protons are pumped into the thylakoid space, and a backflow of protons through an ATP synthetase then produces the ATP in the stroma.

The ATP and NADPH made by photosynthesis drive many biosynthetic reactions in the chloroplast stroma, including the all-important carbon-fixation cycle, which creates carbohydrate from CO_2. This carbohydrate is exported to the cell cytosol where—as glyceraldehyde 3-phosphate—it provides organic carbon, ATP, and reducing power to the rest of the cell.

The Evolution of Electron-Transport Chains[40]

Much of the structure, function, and evolution of cells and organisms can be related to their need for energy. We have seen that the fundamental mechanisms for harnessing energy from such disparate sources as light and the oxidation of glucose are the same. Apparently, an effective method for synthesizing ATP arose early in evolution and has since been conserved with only small variations. How did the crucial individual components—ATP synthetase, redox-driven proton pumps, and photosystems—first arise? Hypotheses about events occurring on an evolutionary time scale are difficult to test. But clues abound, both in the many different primitive electron-transport chains that survive in some present-day bacteria and in geological evidence concerning the environment of the earth billions of years ago.

The Earliest Cells Probably Produced ATP by Fermentation[41]

As explained in Chapter 1, the first living cells are thought to have arisen roughly 3.5×10^9 years ago, when the earth was about 10^9 years old. Because the environment lacked oxygen but was rich in geochemically produced organic molecules, the earliest metabolic pathways for producing ATP presumably resembled present-day forms of fermentation.

In the process of **fermentation,** ATP is made by a substrate-level phosphorylation event (see p. 66) that harnesses the energy released by a reaction pathway in which a hydrogen-rich organic molecule, such as glucose, is partly oxidized. Without oxygen to serve as a hydrogen acceptor, the hydrogens lost from the oxidized molecules must be transferred (via NADH or NADPH) to a different organic molecule (or to a different part of the same molecule), which thereby becomes more reduced. At the end of the fermentation process, one (or more) of the organic molecules produced is excreted into the medium as a metabolic waste product; others, such as pyruvate, are retained by the cell for biosynthesis.

The excreted end products are different in different organisms, but they tend to be organic acids (carbon compounds that carry a COOH group). The most important of such products in bacterial cells include lactic acid (which also ac-

(A) FERMENTATION LEADING TO EXCRETION OF LACTIC ACID

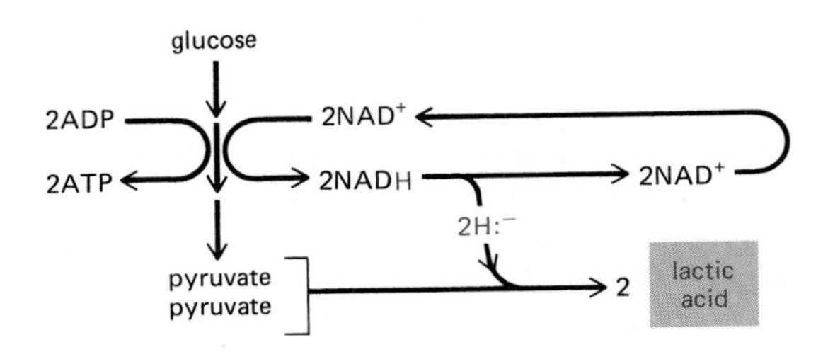

(B) FERMENTATION LEADING TO EXCRETION OF SUCCINIC ACID

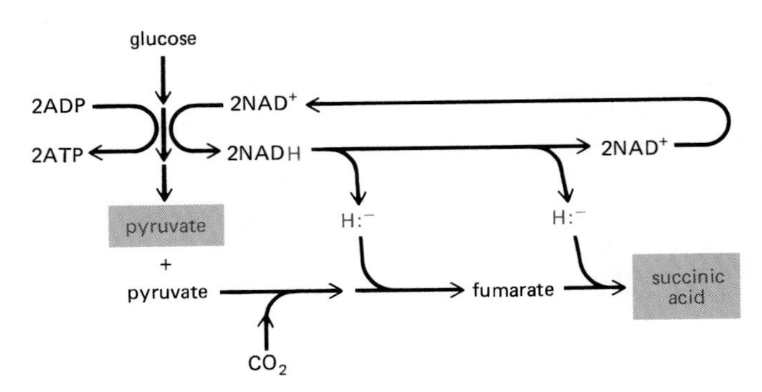

Figure 7–56 Two types of fermentation processes, with the end products highlighted by colored boxes. (A) The two molecules of NAD^+ used for each molecule of glucose that undergoes glycolysis are regenerated by the transfer of hydride ions from NADH to pyruvate to produce two molecules of lactic acid (the reverse of the type of reaction shown previously in Figure 7–18). The lactic acid is excreted. (B) The two molecules of NAD^+ used for each molecule of glucose that undergoes glycolysis are regenerated by successive transfers of hydride ions from two NADH molecules to compounds derived from pyruvate that produce succinic acid. For each molecule of succinic acid excreted, a molecule of pyruvate (in colored box) is saved for biosyntheses inside the cell. In both (A) and (B) an organic acid must be excreted to re-form NAD^+ and thereby enable glycolysis to continue in the absence of oxygen.

cumulates in anaerobic mammalian glycolysis; see p. 67) and formic, acetic, propionic, butyric, and succinic acids. Two fermentation pathways of present-day bacteria are illustrated in Figure 7–56.

7-28 The Evolution of Energy-conserving Electron-Transport Chains Enabled Anaerobic Bacteria to Use Nonfermentable Organic Compounds as a Source of Energy[42]

The early fermentation processes would have provided not only the ATP but also the reducing power (as NADH or NADPH) required for essential biosyntheses, and many of the major metabolic pathways probably evolved while fermentation was the only mode of energy production. With time, however, the metabolic activities of these procaryotic organisms must have changed the local environment, so that they were forced to evolve new biochemical pathways. The accumulation of waste products of fermentation might have resulted in the following series of changes:

Stage 1. Because of the continuous excretion of organic acids, the pH of the environment was lowered, and proteins that function as transmembrane proton pumps evolved to pump H^+ out of the cell to prevent death from intracellular acidification. One of these pumps may have used the energy available from ATP hydrolysis and could have been the ancestor of the present-day ATP synthetase.

Stage 2. At the same time that nonfermentable organic acids were accumulating and favoring the evolution of an ATP-consuming proton pump, the supply of geochemically generated fermentable nutrients, which provided the energy both for the pumps and for all other cellular processes, was dwindling. The resulting selective pressures strongly favored bacteria that could excrete H^+ without hydrolyzing ATP, allowing the ATP to be conserved for other cellular activities. Such pressures might have led to the first membrane-bound proteins that could use electron transport between molecules of different redox potential as the energy source for transporting H^+ across the plasma membrane. Some of these proteins must have found their electron donors and electron acceptors among the nonfermentable organic acids that had accumulated. Many such electron-transport proteins can, in fact, be found in present-day bacteria: for example, some present-

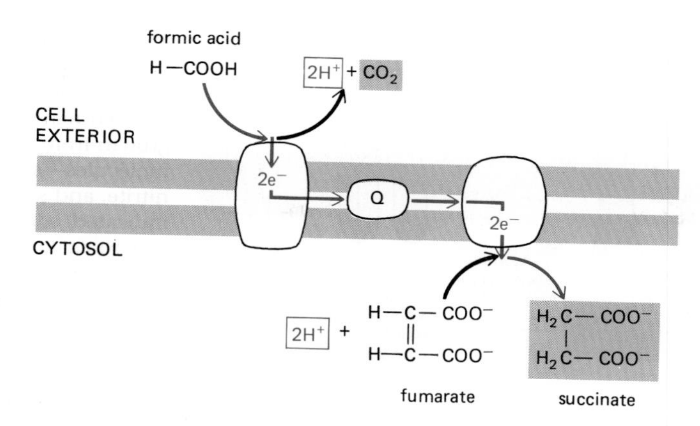

Figure 7–57 The oxidation of formic acid by fumarate is mediated by an energy-conserving electron-transport chain in the plasma membrane of some present-day bacteria that grow anaerobically, including *E. coli.* As indicated, the products are succinate and CO_2. Note that protons are used up inside the cell and generated outside the cell, which is equivalent to pumping protons to the cell exterior. Thus this membrane-bound electron-transport system can generate an electrochemical proton gradient across the plasma membrane. The redox potential of the formic-acid–CO_2 pair is −420 mV, while that of the fumarate-succinate pair is +30 mV.

day bacteria that grow on formic acid pump protons by using the relatively small amount of redox energy derived from the transfer of electrons from formic acid to fumarate (Figure 7–57). Other bacteria have developed similar electron-transport components devoted solely to the oxidation and reduction of inorganic substrates (for example, see Figure 7–59, below).

Stage 3. Eventually some bacteria evolved H^+-pumping electron-transport systems that were efficient enough to harness more redox energy than they needed just to maintain their internal pH. A large electrochemical proton gradient generated by excessive H^+ pumping allowed protons to leak back into the cell through the ATP-driven proton pumps, thereby running them in reverse so that they functioned as ATP synthetases to make ATP. Because such bacteria required much less of the increasingly scarce supply of fermentable nutrients, they proliferated at the expense of their neighbors.

These three hypothetical stages in the evolution of oxidative phosphorylation mechanisms are summarized in Figure 7–58.

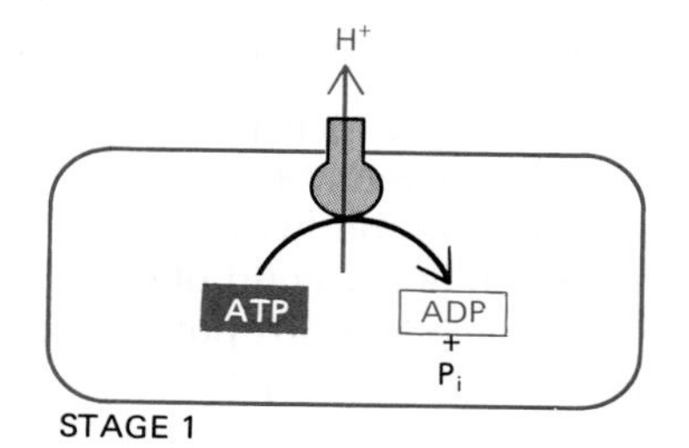

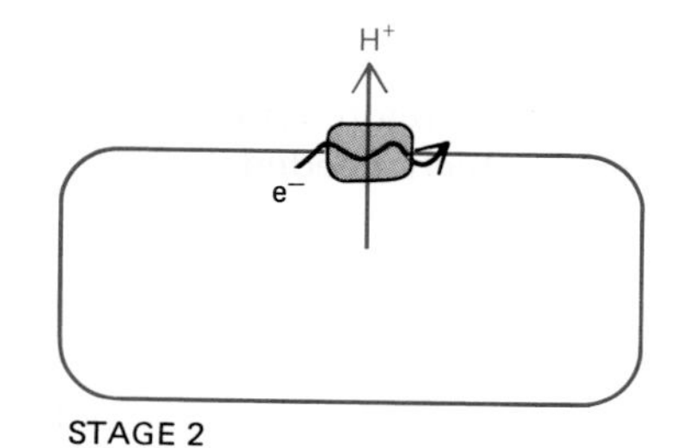

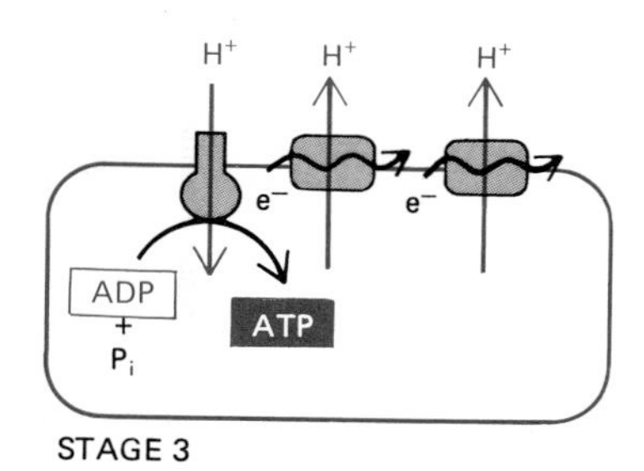

Figure 7–58 One possible sequence for the evolution of oxidative phosphorylation mechanisms (see text).

Photosynthetic Bacteria, by Providing an Inexhaustible Source of Reducing Power, Overcame a Major Obstacle in the Evolution of Cells[43]

The evolutionary steps just outlined would have solved the problem of maintaining both a neutral intracellular pH and an abundant store of energy, but they would have left unsolved another equally serious problem. The depletion of fermentable organic nutrients meant that some alternative source of carbon had to be found to make the sugars that served as the precursors of so many other cellular molecules. The carbon dioxide in the atmosphere provided an abundant potential carbon source. But to convert carbon dioxide into an organic molecule such as a carbohydrate requires that the fixed carbon dioxide be reduced by a strong electron donor, such as NADH or NADPH, which can provide the high-energy electrons needed to generate each (CH_2O) unit from CO_2 (see Figure 7–43). Early in cellular evolution, such strong reducing agents would have been plentiful as products of fermentation. But as the supply of fermentable nutrients dwindled and a membrane-bound ATP synthetase began to produce most of the ATP, the plentiful supply of NADH and other reducing agents would also have disappeared. It thus became imperative for cells to evolve a new way of generating a source of strong reducing power.

After most of the fermentable molecules in the environment had been used up, the main electron donors still available were the organic acids produced by the anaerobic metabolism of carbohydrates, inorganic molecules such as hydrogen sulfide (H_2S) generated geochemically, and water. But the reducing power of all of these molecules is far too weak to be useful for carbon dioxide fixation. An early supply of strong electron donors was probably generated by using the electrochemical proton gradient across the plasma membrane to drive a reverse electron flow, requiring the evolution of membrane-bound enzyme complexes resembling

in animals as diverse as *Drosophila* and sea urchins (Figure 7–68). Plants, however, contain a circular mitochondrial genome that is 10 to 150 times larger, depending on the plant. The largest of these are about half the size of typical bacterial genomes, which are also circular DNA molecules.

All mitochondria and chloroplasts contain multiple copies of the organelle DNA molecule (Table 7–3). These DNA molecules are usually distributed in several clusters in the matrix of the mitochondrion and in the stroma of the chloroplast, where they are thought to be attached to the inner membrane. Although it is not known how the DNA is packaged, the genome structure is likely to resemble that in bacteria rather than eucaryotic chromatin. As in bacteria, for example, there are no histones.

In mammalian cells, mitochondrial DNA makes up less than 1% of the total cellular DNA. However, in other cells—such as the leaves of higher plants or the very large egg cells of amphibia—a much larger fraction of the cellular DNA may be present in the energy-converting organelles (see Table 7–3), and a larger fraction of RNA and protein synthesis takes place there.

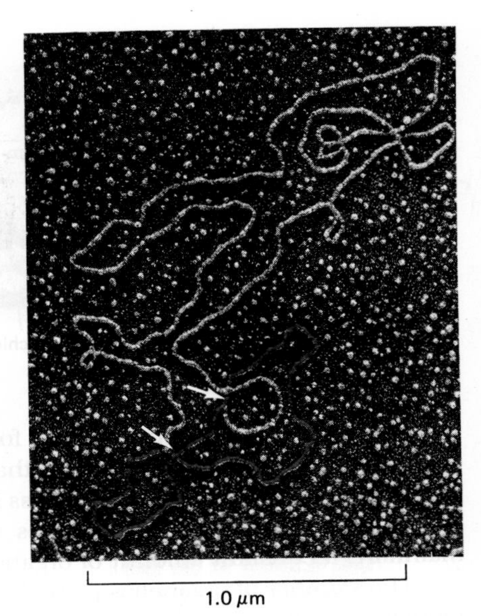

Figure 7–68 Electron micrograph of an animal mitochondrial DNA molecule caught during the process of DNA replication. The circular DNA genome has replicated only between the two points marked by arrows (*colored strands*). (Courtesy of David Clayton.)

Mitochondria and Chloroplasts Contain Complete Genetic Systems[48]

Despite the small number of proteins encoded in their genomes, energy-converting organelles carry out their own DNA replication, DNA transcription, and protein synthesis. These processes take place in the matrix in mitochondria and in the stroma in chloroplasts. The proteins that mediate these genetic processes are unique to the organelle, but most of them are encoded in the nuclear genome rather than in the organelle DNA (see p. 400). This is all the more surprising because the protein-synthesis machinery of the organelles resembles that of bacteria rather than that of eucaryotes. The resemblance is particularly close in the case of chloroplasts:

1. Chloroplast ribosomes are very similar to *E. coli* ribosomes, both in their sensitivity to various antibiotics (such as chloramphenicol, streptomycin, erythromycin, and tetracycline) and in their structure. Not only are the nucleotide sequences of the ribosomal RNAs of chloroplasts and *E. coli* strikingly similar, but chloroplast ribosomes are able to use bacterial tRNAs in protein synthesis. In all these respects, chloroplast ribosomes differ from those found in the cytosol of the same plant cell.
2. Protein synthesis in chloroplasts starts with *N*-formylmethionine, as in bacteria, and not with methionine, as in the cytosol of eucaryotic cells.
3. Unlike nuclear DNA, chloroplast DNA can be transcribed by the RNA polymerase enzyme from *E. coli* to produce chloroplast mRNAs, which are efficiently translated by an *E. coli* protein-synthesizing system.

Table 7–3 Relative Amounts of Organelle DNA in Some Cells and Tissues

Organism	Tissue or Cell Type	DNA Molecules per Organelle	Organelles per Cell	Organelle DNA as Percent of Total Cellular DNA
Mitochondrial DNA				
Rat	liver	5–10	1000	1
Mouse	L-cell line	5–10	100	<1
Yeast*	vegetative	2–50	1–50	15
Frog	egg	5–10	10^7	99
Chloroplast DNA				
Chlamydomonas	vegetative	80	1	7
Maize	leaves	20–40	20–40	15

*The large variation in the number and size of mitochondria per cell in yeast is due to mitochondrial fusion and fragmentation.

Although mitochondrial genetic systems are much less similar to those of present-day bacteria than are the genetic systems of chloroplasts, their ribosomes are also sensitive to antibacterial antibiotics, and protein synthesis in mitochondria also starts with *N*-formylmethionine.

The Chloroplast Genome of Higher Plants Contains About 120 Genes[49]

The best-studied chloroplast genomes are those of plants and green algae, whose chloroplasts are very similar. They are circular DNA molecules, and the complete nucleotide sequence has been determined for the chloroplasts of tobacco and liverwort. The results indicate that these two distantly related higher plants contain nearly identical chloroplast genes. In addition to four ribosomal RNAs, these genomes encode about 20 chloroplast ribosomal proteins, selected subunits of the chloroplast RNA polymerase, several proteins that are part of photosystems I and II, subunits of the ATP synthetase, portions of enzyme complexes in the electron-transport chain, one of the two subunits of ribulose bisphosphate carboxylase, and 30 tRNAs (Figure 7–69). In addition, the DNA sequences present seem to encode at least 40 proteins whose functions are unknown. Paradoxically, all of the known proteins encoded in the chloroplast are part of larger protein complexes that also contain one or more subunits encoded in the nucleus. Possible reasons will be discussed later (see p. 400).

The similarities between the genomes of chloroplasts and bacteria are striking. The basic regulatory sequences, such as transcription promoters and terminators, are virtually identical in the two cases. Protein sequences encoded in chloroplasts are clearly recognizable as bacterial, and several clusters of genes with related functions (for example, those encoding ribosomal proteins) are organized in the same way in the genomes of chloroplasts, *E. coli*, and cyanobacteria.

Detailed comparisons of homologous nucleotide sequences will be required to trace the evolutionary pathway from bacteria to chloroplasts, but several conclusions can already be drawn: (1) Chloroplasts in higher plants arose from photosynthetic bacteria. (2) The chloroplast genome has been stably maintained for at least several hundred million years, the estimated time of divergence of liverwort and tobacco. (3) Many of the genes of the original bacterium can be identified in the nuclear genome, where they have been transferred and stably maintained. In higher plants, for example, two-thirds of the 60 or so chloroplast ribosomal proteins are encoded in the cell nucleus, although the genes have a clear bacterial ancestry, and the chloroplast ribosomes retain their original bacterial properties.

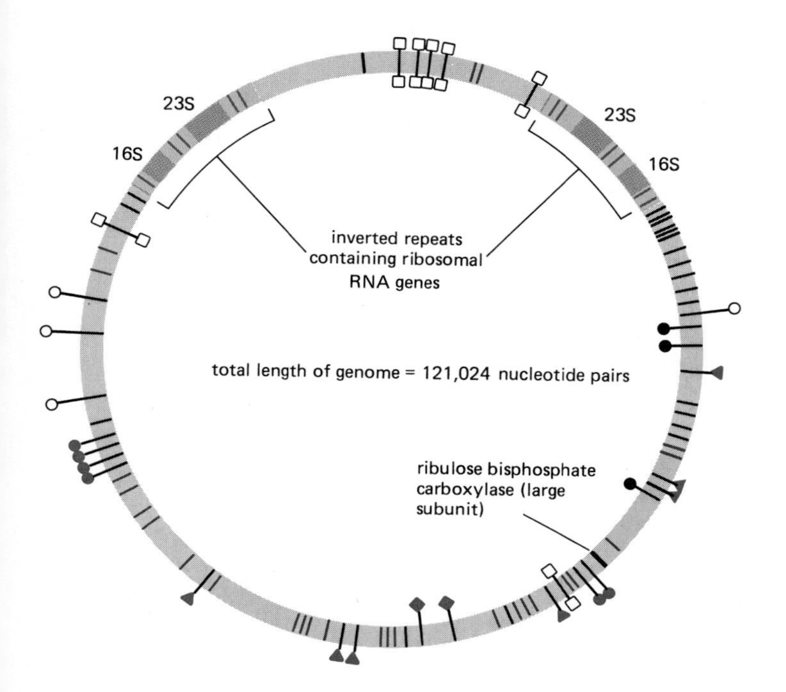

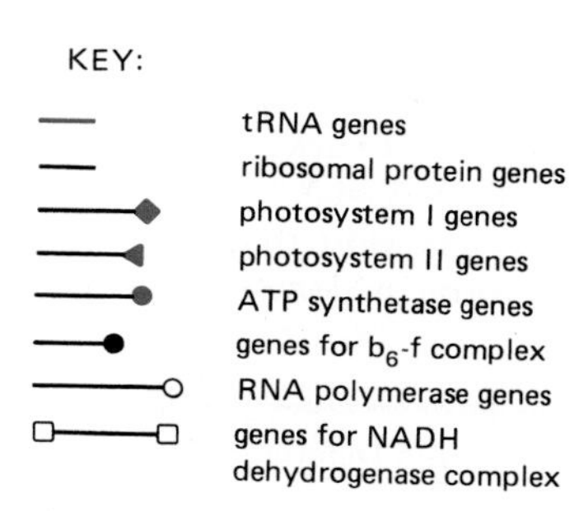

Figure 7–69 The organization of the liverwort chloroplast genome, whose complete nucleotide sequence has been determined. The organization of the chloroplast genome is very similar in all higher plants. However, the size of these circular DNA molecules varies from species to species depending on how much of the DNA surrounding the genes encoding the chloroplast's 16S and 23S ribosomal RNAs is present in two copies.

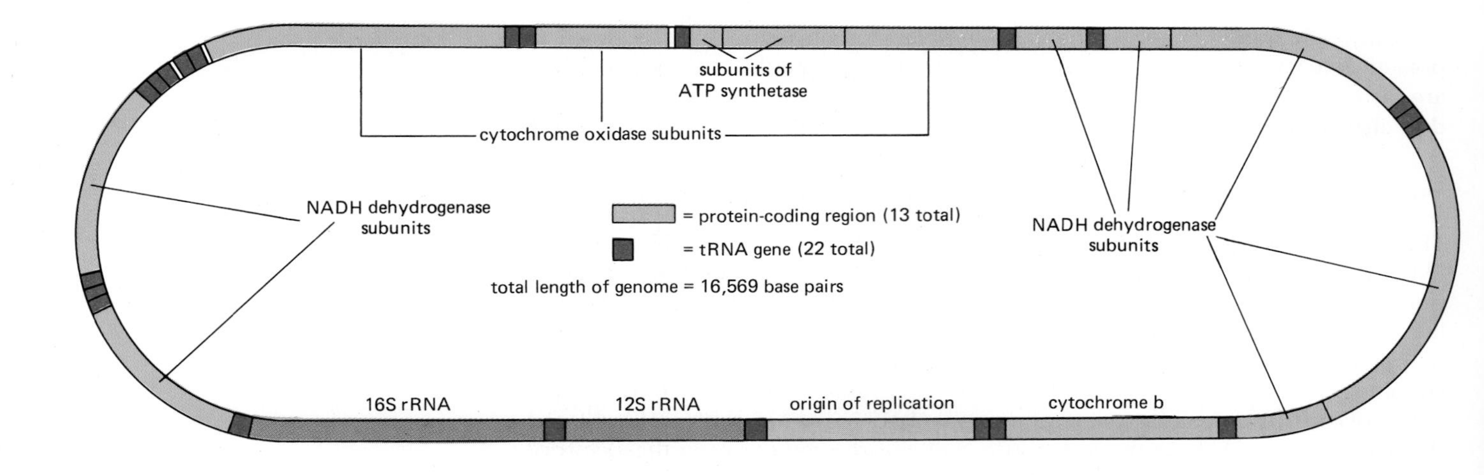

Figure 7–70 The organization of the human mitochondrial genome, based on determination of its complete nucleotide sequence. The genome contains 2 rRNA genes, 22 tRNA genes, and 13 protein-coding sequences. The DNAs of the bovine and mouse mitochondrial genomes have also been completely sequenced and have the same genes and gene organization.

7-34 Mitochondrial Genomes Have Several Surprising Features[50]

The chloroplast genome was not the first organelle genome to be completely sequenced. The relatively small size of the human mitochondrial genome made it a particularly attractive target for molecular geneticists equipped with newly devised DNA-sequencing techniques (see p. 185), and in 1981 the complete sequence of its 16,569 nucleotides was published. By comparing this sequence with known mitochondrial tRNA sequences and with the partial amino acid sequences available for proteins encoded by the mitochondrial DNA, it has been possible to locate all of the human mitochondrial genes on the circular DNA molecule (Figure 7–70).

Compared to nuclear, chloroplast, and bacterial genomes, the human mitochondrial genome has several surprising features: (1) Unlike other genomes, nearly every nucleotide appears to be part of a coding sequence, either for a protein or for one of the rRNAs or tRNAs. Since these coding sequences run directly into each other, there is very little room left for regulatory DNA sequences. (2) Whereas at least 31 tRNAs specify amino acids in the cytosol, and 30 in chloroplasts, only 22 tRNAs are required for mitochondrial protein synthesis. The normal codon-anticodon pairing rules are relaxed in mitochondria, so that many tRNA molecules recognize any one of the four nucleotides in the third (wobble) position (see p. 209). Such "2 out of 3" pairing allows one tRNA to pair with any one of four codons and permits protein synthesis with fewer tRNA molecules. (3) Perhaps most surprising, comparison of mitochondrial gene sequences and the amino acid sequences of the corresponding proteins indicates that the genetic code is altered, so that 4 of the 64 codons have "meanings" different from those they have in other genomes (Table 7–4).

The observation that the genetic code is nearly the same in all organisms provides strong evidence that all cells have evolved from a common ancestor. How,

Table 7–4 Some Differences Between the "Universal" Code and Mitochondrial Genetic Codes*

		Mitochondrial Codes			
Codon	**"Universal" Code**	Mammals	*Drosophila*	Yeasts	Plants
UGA	STOP	*Trp*	*Trp*	*Trp*	STOP
AUA	Ile	*Met*	*Met*	*Met*	Ile
CUA	Leu	Leu	Leu	*Thr*	Leu
AGA, AGG	Arg	*STOP*	*Ser*	Arg	Arg

*Italics and color shading indicate that the code differs from the "universal" code.

then, does one explain the few differences in the genetic code in mitochondria? A hint comes from the recent finding that the mitochondrial genetic code is different in different organisms. Thus UGA, which is a stop codon elsewhere, is read as tryptophan in mitochondria of mammals, fungi, and protozoans but as *stop* in plant mitochondria. Similarly, the codon AGG normally codes for arginine, but it codes for *stop* in the mitochondria of mammals and for serine in *Drosophila* (see Table 7–4). Such variation suggests that a random drift can occur in the genetic code in mitochondria. Presumably the unusually small number of proteins encoded by the mitochondrial genome makes an occasional change in the meaning of a rare codon tolerable, whereas such a change in a large genome would alter the function of many proteins and thereby destroy the cell.

Animal Mitochondria Contain the Simplest Genetic Systems Known[51]

Comparisons of DNA sequences in different organisms reveal that the rate of nucleotide substitution during evolution has been 10 times greater in mitochondrial genomes than in nuclear genomes (see p. 220), which is presumably due to a reduced fidelity of mitochondrial DNA replication, DNA repair, or both. Because only about 16,500 DNA nucleotides need to be replicated and expressed as RNAs and proteins in animal cell mitochondria, the error rate per nucleotide copied by DNA replication, maintained by DNA repair, transcribed by RNA polymerase, or translated into protein by mitochondrial ribosomes can be relatively high without adversely affecting the organelle. This is thought to explain why the mechanisms that carry out these processes are relatively simple compared to those used for the same purpose elsewhere in cells. For example, although this has not yet been tested adequately, one would expect that the presence of only 22 tRNAs and the unusually small size of the rRNAs (less than two-thirds the size of the *E. coli* rRNAs) would reduce the fidelity of protein synthesis in mitochondria.

The relatively high rate of evolution of mitochondrial genes makes mitochondrial DNA sequence comparisons especially useful for estimating the dates of relatively recent evolutionary events, such as the steps in primate development (see p. 13).

7-35 Why Are Plant Mitochondrial Genomes So Large?[52]

Mitochondrial genomes are much larger in plant than in animal cells and vary remarkably in their DNA content, ranging from about 150,000 to about 2.5×10^6 nucleotide pairs. Yet these genomes seem to encode only a few more proteins than do animal mitochondrial genomes. The paradox is compounded by the observation that in one family of plants, the cucurbits, mitochondrial genomes vary in size by as much as sevenfold. The green alga *Chlamydomonas* has a linear mitochondrial genome of only 16,000 nucleotide pairs, the same size as in animals.

Although very little sequence information is available for higher plant mitochondrial DNA molecules, almost all of the 78,000 nucleotide pairs in the large mitochondrial genome of the yeast *Saccharomyces cerevisiae* have been sequenced, and only about one-third of them code for protein. This finding raises the possibility that much of the extra DNA in yeast mitochondria, and possibly in plant mitochondria as well, is "junk DNA" of little consequence to the organism.

Some Organelle Genes Contain Introns[53]

The processing of precursor RNAs plays an important role in the two mitochondrial systems studied in most detail—human and yeast. In human cells both strands of the mitochondrial DNA are transcribed at the same rate from a single promoter region on each strand, producing two different giant RNA molecules, each containing a full-length copy of one DNA strand. Transcription is, therefore, completely symmetric. The transcripts made on one strand—called the *heavy strand (H strand)* because of its density in CsCl—are extensively processed by nuclease cleavage to yield the two rRNAs, most of the tRNAs, and about 10 poly-A-containing RNAs. In contrast, the *light strand (L strand)* transcript is processed to produce only eight tRNAs and one small poly-A-containing RNA; the remaining 90% of this transcript apparently contains no useful information (being complementary to coding sequences synthesized on the other strand) and is degraded. The poly-A-containing RNAs are the mitochondrial mRNAs: although they lack a cap structure at their 5′ end, they carry a poly-A tail at their 3′ end that is added post-transcriptionally by a mitochondrial poly-A polymerase.

Unlike human mitochondrial genes, some plant and fungal (including yeast) mitochondrial genes contain *introns*, which must be removed by RNA splicing (see p. 102). Introns have also been found in about 20 plant chloroplast genes. Many of the introns in organelle genes consist of related nucleotide sequences that are capable of splicing themselves out of the RNA transcripts by RNA-mediated catalysis (see p. 538), although these self-splicing reactions are generally aided by proteins. The presence of introns in organelle genes is surprising in view of the endosymbiont theory of the origin of the energy-converting organelles, since similar introns have not been found in the genes of the bacteria whose ancestors are thought to have given rise to mitochondria and plant chloroplasts.

In yeasts the same mitochondrial gene may have an intron in one strain but not in another. Such "optional introns" seem to be able to move in and out of genomes like transposable elements. On the other hand, introns in other yeast mitochondrial genes have been found in a corresponding position in the mitochondria of *Aspergillus* and *Neurospora*, implying that they were inherited from a common ancestor of these three fungi. It seems likely that the intron sequences themselves are of ancient origin and that, while they have been lost from many bacteria, they have been preferentially retained in those organelle genomes where RNA splicing is regulated to help control gene expression (see p. 397 and p. 603).

7-36 Mitochondrial Genes Can Be Distinguished from Nuclear Genes by Their Non-Mendelian (Cytoplasmic) Inheritance[54]

Most experiments on the mechanisms of mitochondrial biogenesis are performed with *Saccharomyces carlsbergensis* (brewer's yeast) and *Saccharomyces cerevisiae* (baker's yeast). There are several reasons for this. First, when grown on glucose, these yeasts have a unique ability to live by glycolysis alone and can therefore survive without functional mitochondria, which are required for oxidative phosphorylation. This makes it possible to grow cells with mutations in mitochondrial or nuclear DNA that drastically interfere with mitochondrial biogenesis; such mutations are lethal in nearly all other organisms. Second, yeasts are simple unicellular eucaryotes that are easy to grow and characterize biochemically. Finally, these yeast cells normally reproduce asexually by budding (asymmetrical mitosis), but they can also reproduce sexually. During sexual reproduction, two haploid cells mate and fuse to form a diploid zygote, which can either grow mitotically or divide by meiosis to produce new haploid cells. The ability to control the alternation between asexual and sexual reproduction in the laboratory (see p. 739) greatly facilitates genetic analyses. Because mutations in mitochondrial genes are not inherited according to the Mendelian rules that govern the inheritance of nuclear genes, genetic studies reveal which of the genes involved in mitochondrial function are located in the nucleus and which in the mitochondria.

An example of **non-Mendelian (cytoplasmic) inheritance** of mitochondrial genes in a haploid yeast cell is illustrated in Figure 7–71. The mutant gene makes mitochondrial protein synthesis resistant to chloramphenicol, so yeast cells that contain the mutant gene can be detected by their ability to grow in the presence of chloramphenicol on a substrate, such as glycerol, that cannot be used for glycolysis. With glycolysis blocked, ATP must be provided by functional mitochondria, and therefore only cells that carry chloramphenicol-resistant mitochondria will grow. When a chloramphenicol-resistant haploid cell mates with a chloramphenicol-sensitive wild-type haploid cell, the resulting diploid zygote will contain a mixture of mutant and wild-type mitochondria. But when the zygote undergoes mitosis to produce a diploid daughter by budding, only a limited number of mitochondria enter the bud. With continuing mitotic division, an occasional bud will receive all mutant or all wild-type mitochondria. Thereafter, all of the progeny

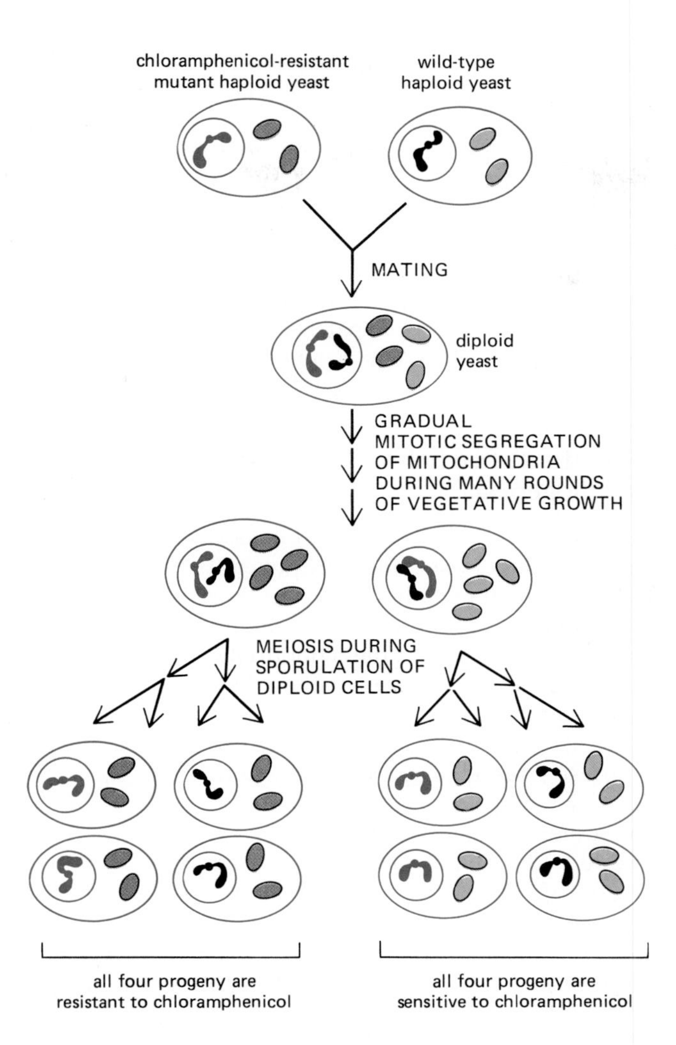

Figure 7–71 The difference in the pattern of inheritance between mitochondrial and nuclear genes of yeast. For each nuclear gene, two of the four cells that result from meiosis inherit the gene from one of the original haploid parent cells and the remaining two cells inherit the gene from the other (*Mendelian inheritance*). In contrast, because of the gradual mitotic segregation of mitochondria during vegetative growth (see text), it is possible for all four of the cells that result from meiosis to inherit their mitochondrial genes from only one of the two original haploid cells (*non-Mendelian* or *cytoplasmic inheritance*). In this example the mitochondrial gene is one that can mutate to make protein synthesis in the mitochondrion resistant to chloramphenicol, a protein synthesis inhibitor that acts specifically on energy-converting organelles and bacteria (see p. 218).

from that bud will have mitochondria that are genetically identical. Eventually this random process, called *mitotic segregation*, produces diploid yeast progeny with only a single type of mitochondrial DNA. When such diploid cells undergo meiosis to form four haploid daughter cells, each of the four daughters receives the same mitochondrial genes. This type of inheritance is called *non-Mendelian* or *cytoplasmic* to contrast it with the Mendelian inheritance of nuclear genes (see Figure 7–71). When it occurs, it demonstrates that the gene in question is located outside the nuclear chromosomes and, therefore, probably in the yeast mitochondria.

Organelle Genes Are Maternally Inherited in Many Organisms[55]

The consequences of cytoplasmic inheritance are more profound for some organisms, including ourselves, than they are for yeasts. In yeasts, when two haploid cells mate, they are equal in size and contribute equal amounts of mitochondrial DNA to the zygote (see Figure 7–71). Mitochondrial inheritance in yeasts is therefore *biparental:* both parents contribute equally to the mitochondrial gene pool of the progeny (although, as we have just seen, after several generations of vegetative growth the *individual* progeny often contain mitochondria from only one parent). In higher animals, by contrast, the egg cell always contributes much more cytoplasm to the zygote than does the sperm, and in some animals the sperm may contribute no cytoplasm at all. One would expect mitochondrial inheritance in higher animals, therefore, to be *uniparental* (or more precisely, *maternal*). Such *maternal inheritance* has been demonstrated in laboratory animals using two strains that differ in their mitochondrial DNA. When animals carrying type A mitochondrial DNA are crossed with animals carrying type B, the progeny contain only the maternal type of mitochondrial DNA. Similarly, by following the distribution of variant mitochondrial DNA sequences in large families, human mitochondrial DNA has been shown to be maternally inherited.

In about two-thirds of higher plants, the chloroplasts from the male parent (contained in pollen grains) do not enter the zygote, so that chloroplast as well as mitochondrial DNA is maternally inherited. In other plants the pollen chloroplasts enter the zygote, making chloroplast inheritance biparental. In such plants, defective chloroplasts are a cause of *variegation:* a mixture of normal and defective chloroplasts in a zygote may sort out by mitotic segregation during plant growth and development, thereby producing alternating green and white patches in leaves; the green patches contain normal chloroplasts, while the white patches contain defective chloroplasts.

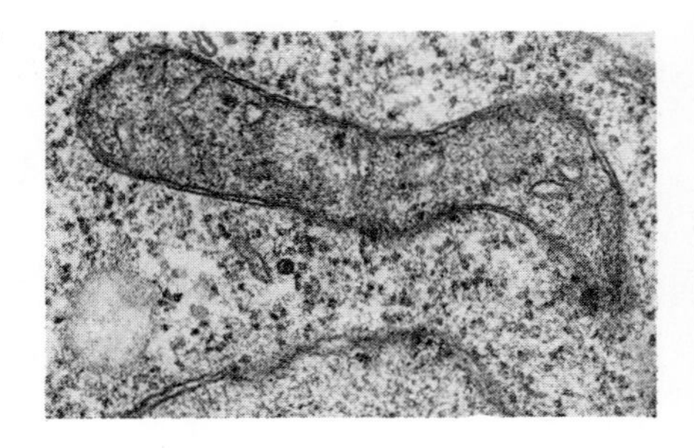

(A)

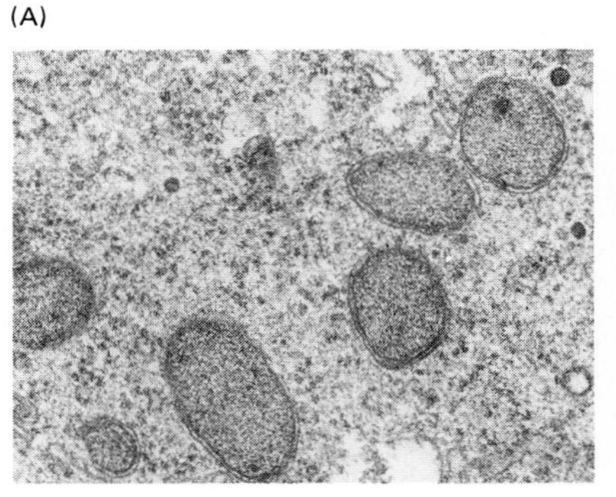

(B)

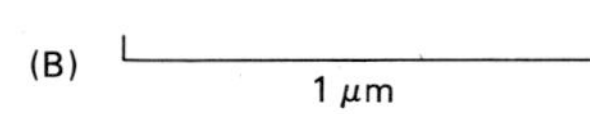

Figure 7–72 Electron micrographs of yeast cells showing the structure of normal mitochondria (A) and mitochondria in a petite mutant in which all of the mitochondrion-encoded gene products are missing (B). In the latter case the organelle is constructed entirely from nucleus-encoded proteins. (Courtesy of Barbara Stevens.)

Petite Mutants in Yeasts Demonstrate the Overwhelming Importance of the Cell Nucleus in Mitochondrial Biogenesis[56]

Genetic studies of yeasts have played a crucial part in the analysis of mitochondrial biogenesis. A striking example is provided by studies of yeast mutants that contain large deletions in their mitochondrial DNA so that all mitochondrial protein synthesis is abolished. Not surprisingly, these mutants cannot make respiring mitochondria. A rare but important subclass of such mutants lacks mitochondrial DNA altogether. Because they form unusually small colonies when grown in media with low glucose, all mutants with such defective mitochondria are called *cytoplasmic petite mutants.*

Although petite mutants cannot synthesize proteins in their mitochondria and therefore cannot make mitochondria that produce ATP, they nevertheless contain mitochondria that have a normal outer membrane and an inner membrane with poorly developed cristae (Figure 7–72). These mitochondria contain virtually all of the mitochondrial proteins that are specified by nuclear genes and imported from the cytosol, including DNA and RNA polymerases, all of the citric acid cycle enzymes, and most inner membrane proteins. Such mutants dramatically demonstrate the overwhelming importance of the nucleus in mitochondrial biogenesis. They also show that an organelle that divides by fission can replicate indefinitely in the cytoplasm of proliferating eucaryotic cells even in the complete absence of its own genome. Many biologists believe that peroxisomes normally replicate in this way (see p. 433).

For chloroplasts the nearest equivalent to yeast mitochondrial petite mutants are mutants of unicellular algae such as *Euglena.* Cells in which no chloroplast protein synthesis occurs still contain chloroplasts and are perfectly viable if oxidizable substrates are provided. However, if the development of mature chloroplasts is blocked in plants, either by raising the plants in the dark (see p. 1162) or because chloroplast DNA is defective or absent, the plants die as soon as their food stores run out.

7-37 Nuclear Gene Products Regulate the Synthesis of Mitochondria and Chloroplasts[57]

Nuclear and organelle genetic systems must communicate in order to coordinate their contributions to the formation of the energy-converting organelles. Overall control clearly resides in the nucleus, inasmuch as mitochondria and chloroplasts are made in normal amounts, although not with normal functions, in mutants in which organelle protein synthesis is blocked. In some of these functionally defective organelles, DNA synthesis and some RNA synthesis can also continue normally, showing that the proteins required for these processes are all encoded by nuclear genes.

The nucleus must regulate the number of mitochondria and chloroplasts in a cell according to need; it must also control the amount of protein made on organelle ribosomes so that a proper balance is maintained between nuclear and organelle contributions. Although these regulatory aspects are crucial to our understanding of eucaryotic cells, we know relatively little about them.

The nuclear regulation of mitochondrial protein synthesis has been analyzed extensively in yeast mutants. In *Saccharomyces cerevisiae*, large numbers of nuclear (as well as mitochondrial—see p. 394) gene mutants have been isolated that produce nonrespiring mitochondria. Each of these *nuclear petite mutants* has a defect in one nucleus-encoded protein that is required for mitochondrial function. The effect of each of these nuclear mutations on mitochondrial gene expression can be determined by labeling yeast cells with radioactive amino acids in the presence of cycloheximide so that only organelle-encoded proteins are made (see p. 388). Most of the mutations are found to have no effect on the synthesis of proteins in the mitochondrion, as expected for a mutant nuclear gene encoding a mitochondrial protein with a direct function in respiration, such as an ATP synthetase subunit or one of the citric acid cycle enzymes. Other mutations block the synthesis of all proteins in the mitochondrion, as expected for a mutant nuclear gene encoding a mitochondrial ribosomal protein or one of the subunits of the mitochondrial RNA polymerase.

It is a third class of yeast nuclear petite mutants, in which one or a few products of mitochondrial genes are absent or altered, that is most relevant to the regulatory process. More than 50 of these nuclear genes have been discovered, and several of them that are required for the expression of a single mitochondrial gene have been cloned and characterized. Some of these genes encode proteins that appear to act directly on a specific mitochondrial mRNA molecule to increase its stability or the efficiency with which it is used for protein synthesis in the mitochondrion. Others help catalyze mitochondrial RNA splicing and are therefore required for the expression of those yeast mitochondrial genes that contain introns. Both types of nuclear genes have been postulated to help regulate the function of mitochondrion-encoded proteins according to the cell's metabolic needs, although how the regulatory network operates is unknown.

Despite the dominance of the nucleus, there is evidence that the interactions between the nuclear and organelle genetic systems occur in both directions. When mitochondrial protein synthesis is blocked in intact cells, for example, some imported enzymes involved in mitochondrial DNA, RNA, and protein synthesis are overproduced, as though the cell were trying to overcome the block. The nature of the signal from the mitochondria to the nucleus remains to be determined.

The Energy-converting Organelles Contain Tissue-specific Proteins[58]

Mitochondrial function is also regulated by the cell in more conventional ways. The *urea cycle*, for example, is the central metabolic pathway in mammals for disposing of cellular breakdown products that contain nitrogen. These products are excreted in the urine as urea. Nuclear-encoded enzymes in the mitochondrial matrix carry out several steps in the cycle. Urea synthesis occurs in only a few tissues, such as the liver, and these enzymes are synthesized and imported into mitochondria only in these tissues. In addition, the respiratory enzyme complexes in the mitochondrial inner membrane of mammals contain several tissue-specific, nuclear-encoded subunits, which are thought to act as regulators of electron transport. Thus some humans with a genetic muscle disease have a defective subunit of cytochrome oxidase; since the subunit is specific to skeletal muscle cells, their heart muscle cells function normally, allowing the individuals to survive. As would be expected, tissue-specific differences are also found among the nuclear-encoded proteins in chloroplasts.

We shall now consider the general question of how specific cytosolic proteins are imported into mitochondria and chloroplasts, a subject that will be discussed in more detail in Chapter 8.

Proteins Are Imported into Mitochondria and Chloroplasts by an Energy-requiring Process[59]

The finding that most chloroplasts and mitochondrial proteins are imported from the cell cytosol (see p. 426) raises two related questions: how does the cell direct proteins to the appropriate organelle, and how do they enter the organelle?

The general answer was first provided by studies on the import of the small subunit (S) of the abundant enzyme *ribulose bisphosphate carboxylase* into the chloroplast stroma. When mRNA isolated from the cytoplasm of either the unicellular alga *Chlamydomonas* or pea leaves is translated into protein *in vitro,* one of the many proteins produced is the precursor of the S protein, called pro-S, which is larger than mature S by about 50 amino acids. When the completed pro-S protein is incubated with intact chloroplasts, it is taken up into the organelle and converted into mature S by an endopeptidase in the chloroplast. Mature S then associates with the large subunit of ribulose 1,5-bisphosphate carboxylase, which is made on chloroplast ribosomes, to form the active enzyme in the chloroplast stroma. As expected for a process of this type, the translocation of the pro-S protein into chloroplasts requires energy, and, as discussed in Chapter 8, the energy is provided by ATP hydrolysis (see p. 430).

The import of proteins into mitochondria is generally similar. If purified yeast mitochondria are incubated with cell extracts containing newly synthesized yeast proteins in a radioactive form, the nuclear-encoded mitochondrial proteins are specifically incorporated into the mitochondria in a manner that faithfully mimics their selective uptake within the cell. The outer membrane proteins, inner membrane proteins, matrix proteins, and proteins of the intermembrane space each find their way to their own special compartments within the mitochondrion (see Figure 8–30, p. 429).

The transport of proteins through the mitochondrial and chloroplast membranes seems to occur at special *contact sites* (also called contact zones) where the inner and outer membranes are joined (Figure 7–73), and it involves precursor proteins that contain a special *signal peptide.* The transported proteins must be unfolded in order to move into the organelle at these sites (see p. 428).

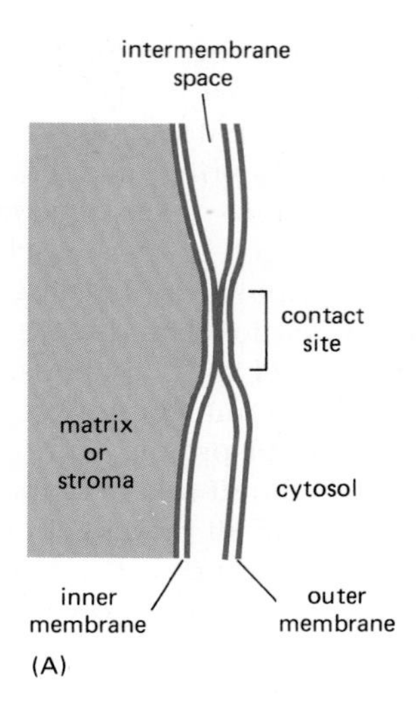

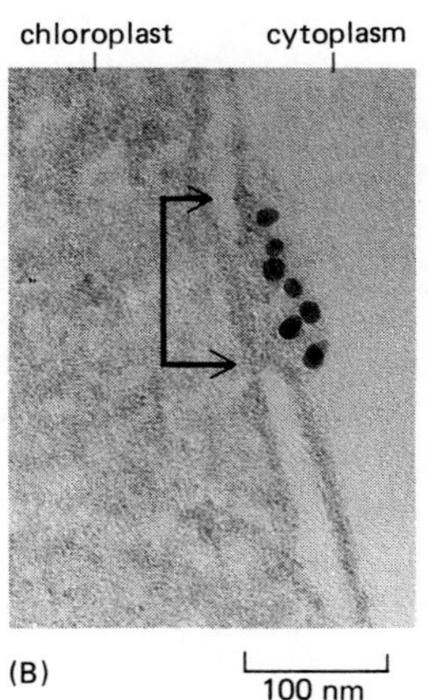

Figure 7–73 Contact sites. (A) Drawing of a small region of a mitochondrion or chloroplast containing a membrane contact site. These sites appear to be involved in selective protein import into these organelles. Contact sites (also called contact zones) have recently been isolated, and their special protein components are being identified. The proteins imported through these sites are those encoded by the cell nucleus and synthesized in the cytosol. (B) Electron micrograph of a small portion of a pea chloroplast, in which a contact site (*arrow*) has been labeled with a gold-conjugated antibody that is thought to localize an integral membrane protein involved in protein import. (B, from D. Pain, Y.S. Kanwar, and G. Blobel, *Nature* 331:232–237, 1988.)

Mitochondria Import Most of Their Lipids; Chloroplasts Make Most of Theirs[60]

The biosynthesis of new mitochondria and chloroplasts requires lipids in addition to nucleic acids and proteins. Chloroplasts tend to make the lipids they require. In spinach leaves, for example, all cellular fatty acid synthesis takes place in the chloroplast, although desaturation of the fatty acids occurs elsewhere. The major glycolipids of the chloroplast are also synthesized locally.

Mitochondria, on the other hand, import most of their lipids. In animal cells the phospholipids phosphatidylcholine and phosphatidylserine are synthesized in the endoplasmic reticulum and then transferred to the outer membrane of mitochondria. Although proof is lacking, the transfer reactions are thought to be mediated by phospholipid exchange proteins (see p. 450); the imported lipids then move into the inner membrane, presumably at contact sites. In addition to decarboxylating imported phosphatidylserine to phosphatidylethanolamine, the main reaction of lipid biosynthesis catalyzed by the mitochondria themselves is the conversion of imported lipids to cardiolipin (diphosphatidylglycerol). Cardiolipin is a "double" phospholipid that contains four fatty-acid tails; it is found mainly in the mitochondrial inner membrane, where it constitutes about 20% of the total lipid.

7-38 Both Mitochondria and Chloroplasts Probably Evolved from Endosymbiotic Bacteria[61]

As discussed in Chapter 1, the procaryotic character of the organelle genetic systems, especially striking in chloroplasts, suggests that mitochondria and chloroplasts evolved from bacteria that were at one time endocytosed. According to the **endosymbiont hypothesis,** eucaryotic cells started out as anaerobic organisms without mitochondria or chloroplasts and then established a stable endosymbiotic relation with a bacterium, whose oxidative phosphorylation system they subverted for their own use (Figure 7–74). The endocytic event that led to the development of mitochondria is presumed to have occurred when oxygen entered the atmosphere in substantial amounts, about 1.5×10^9 years ago, before animals and

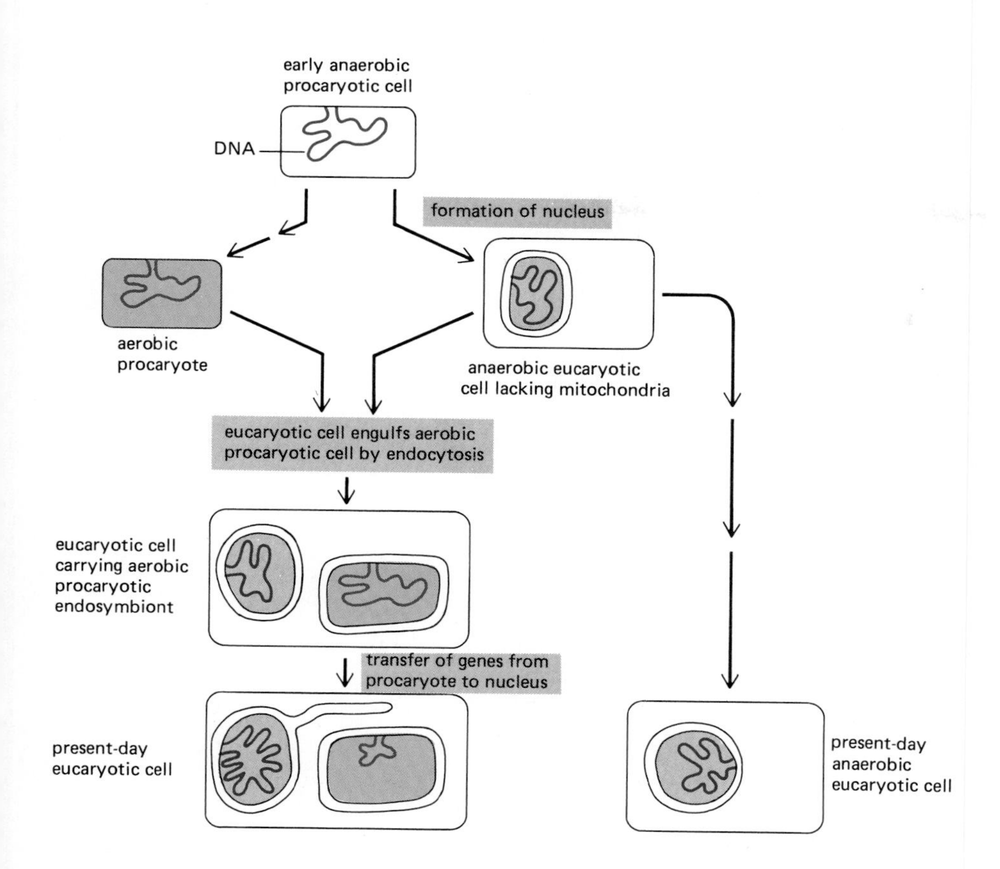

Figure 7–74 A suggested evolutionary pathway for the origin of mitochondria (shown in color). Although a single origin for all mitochondria is sometimes postulated, separate endosymbiotic events may have led to the mitochondria found in such distantly related eucaryotes as trypanosomes and euglenoids (see Figure 1–16, p. 13). *Microsporidia* are present-day anaerobic single-celled eucaryotes (protozoa) without mitochondria that live in the gut of many animals. Because they have an rRNA sequence that suggests a great deal of evolutionary distance from all other known eucaryotes, it has been postulated that their ancestors were also anaerobic and resembled the eucaryote that first engulfed the precursors of mitochondria (see p. 18).

plants separated (see Figure 7–61). Plant and algal chloroplasts seem to have been derived later from an endocytic event involving an oxygen-evolving photosynthetic bacterium. In order to explain the different pigments and properties of the chloroplasts found in present-day higher plants and green algae, red algae, and brown algae (see Figure 7–62), it is usually assumed that at least three separate events of this kind occurred.

Since most of the genes encoding present-day mitochondrial and chloroplast proteins are in the cell nucleus, it seems that an extensive transfer of genes from organelle to nuclear DNA has occurred during eucaryote evolution. This would explain why some of the nuclear genes encoding mitochondrial proteins resemble bacterial genes: the amino acid sequence of the amino terminus of the chicken mitochondrial enzyme *superoxide dismutase*, for example, resembles the corresponding segment of the bacterial enzyme much more than it resembles the superoxide dismutase found in the cytosol of the same eucaryotic cells. Further evidence that such DNA transfers have occurred during evolution comes from the discovery of some noncoding DNA sequences in nuclear DNA that seem to be of recent mitochondrial origin; they have apparently integrated into the nuclear genome as "junk DNA."

What type of bacterium gave rise to the mitochondrion? Complete amino acid sequence and three-dimensional x-ray crystallographic analyses of the c-type cytochromes isolated from many types of bacteria provided an early clue by showing that these proteins are all closely related to one another and to the cytochrome c of animal and plant mitochondrial respiratory chains. These findings, together with more recent nucleotide-sequence analyses, provide the main evidence for the evolutionary tree shown previously in Figure 7–62. It appears that mitochondria are descendants of a particular type of purple photosynthetic bacterium, which had previously lost its ability to carry out photosynthesis and was left with only a respiratory chain. However, as for chloroplasts, it is not clear that all mitochondria have originated from a single endosymbiotic event. While the mitochondria from protozoans have distinctly procaryotic features, for example, some of them are sufficiently different from plant and animal mitochondria to suggest a separate origin.

Why Do Mitochondria and Chloroplasts Have Their Own Genetic Systems?[62]

Why do mitochondria and chloroplasts require their own separate genetic systems, when other organelles—such as peroxisomes and lysosomes—do not? The question is not trivial, because maintaining a separate genetic system is costly: more than 90 proteins—including many ribosomal proteins, aminoacyl-tRNA synthetases, DNA and RNA polymerases, and RNA processing and modifying enzymes—must be encoded by nuclear genes specifically for this purpose (Figure 7–75). The amino acid sequences of most of these proteins in mitochondria and chloroplasts differ from those of their counterparts in the nucleus and cytosol, and there is reason to think that these organelles have relatively few proteins in common with the rest of the cell. This means that the nucleus must provide at least 90 genes just to maintain each organelle genetic system. The reason for such a costly arrangement is not clear, and the hope that the nucleotide sequences of mitochondrial and chloroplast genomes would provide the answer has proved unfounded. We cannot think of compelling reasons why the proteins made in mitochondria and chloroplasts should be made there rather than in the cytosol.

At one time it was suggested that some proteins have to be made in the organelle because they are too hydrophobic to get to their site in the membrane from the cytosol. More recent studies, however, make this explanation implausible. In many cases even highly hydrophobic subunits are synthesized in the cytosol. Moreover, although the individual protein subunits in the various mitochondrial

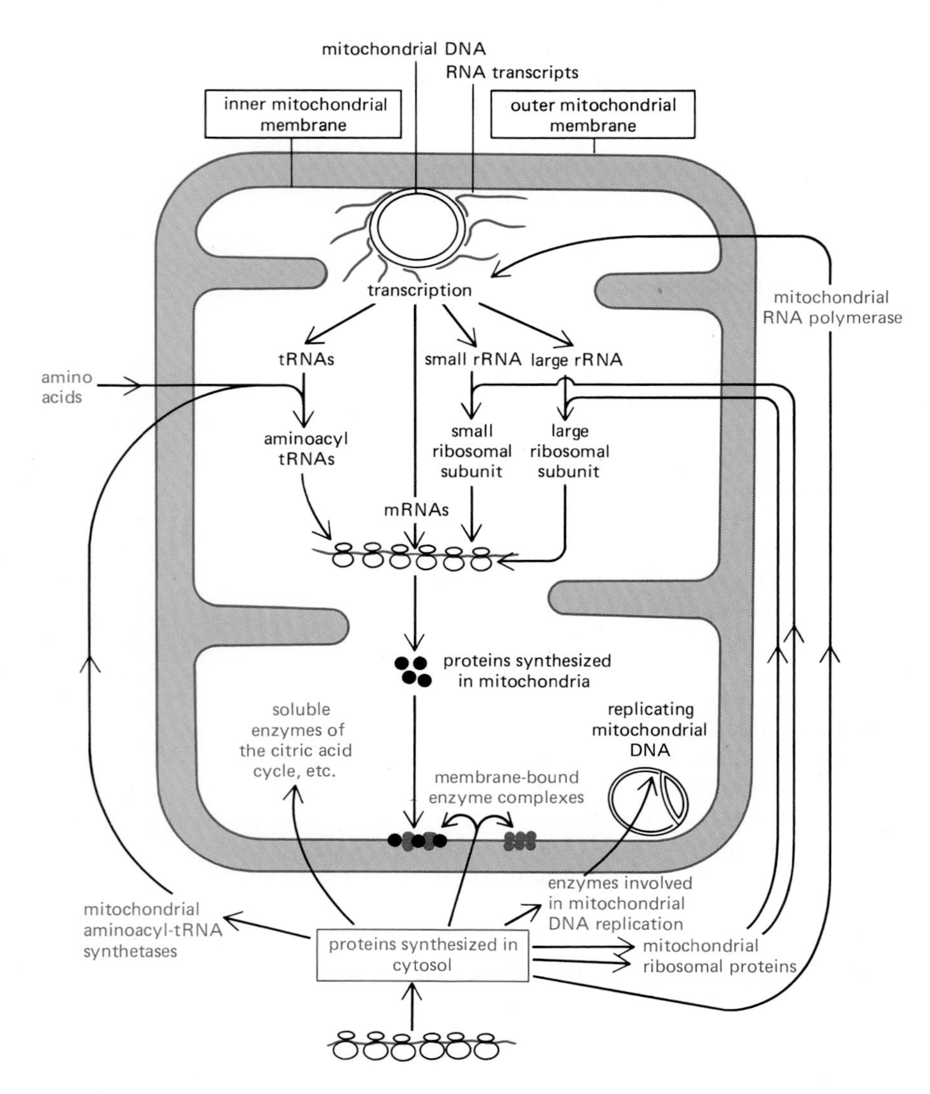

Figure 7–75 The proteins synthesized in the cytosol and then imported into the mitochondrion play a major part in creating the genetic system of the mitochondrion in addition to contributing most of the organelle protein. The mitochondrion itself contributes only mRNAs, rRNAs, and tRNAs to its genetic system.

enzyme complexes are highly conserved in evolution, their site of synthesis is not. The diversity in the location of the genes coding for the subunits of functionally equivalent proteins in different organisms is difficult to explain by any hypothesis that postulates a specific evolutionary advantage of present-day mitochondrial or chloroplast genetic systems.

Perhaps the organelle genetic systems are an evolutionary dead end. In terms of the endosymbiont hypothesis, this would mean that the process whereby the endosymbionts transferred most of their genes to the nucleus stopped before it was complete, perhaps becoming frozen—in the case of mitochondria—by recent alterations in the mitochondrial genetic code. Such alterations would probably make the remaining mitochondrial genes nonfunctional if they were transferred to the nucleus.

Summary

Mitochondria and chloroplasts grow and divide in two in a coordinated process that requires the contribution of two separate genetic systems—that of the organelle and that of the cell nucleus. Most of the proteins in these organelles are encoded by nuclear DNA, synthesized in the cytosol, and then imported individually into the organelle. However, some organelle proteins—along with their RNAs—are encoded by the organelle DNA and are synthesized in the organelle itself. The human mitochondrial genome contains about 16,500 nucleotides and encodes 2 ribosomal RNAs, 22 transfer RNAs, and 13 different polypeptide chains. Chloroplast genomes are about 10 times larger and contain about 120 genes. But partially functional organelles will form in normal numbers even in mutants that lack a functional organelle genome, demonstrating the overwhelming importance of the nucleus for the biogenesis of both organelles.

The ribosomes of chloroplasts closely resemble bacterial ribosomes, while mitochondrial ribosomes show both similarities and differences that make their origin more difficult to trace. Protein similarities, however, suggest that both organelles originated when a primitive eucaryotic cell entered into a stable endosymbiotic relationship with a bacterium: a purple bacterium is thought to have given rise to the mitochondrion, and (later) a relative of a cyanobacterium is thought to have given rise to the plant chloroplast. Although many of the genes of these ancient bacteria still function to make organelle proteins, most of them have become integrated into the nuclear genome, where they encode bacterial-like enzymes that are synthesized on cytosolic ribosomes and then imported into the organelle.

References

General

Becker, W.M. The World of the Cell, pp. 117–284, Menlo Park, CA: Benjamin-Cummings, 1986.

Ernster, L., ed. Bioenergetics. New York: Elsevier, 1984.

Harold, F.M. The Vital Force: A Study of Bioenergetics. New York: W.H. Freeman, 1986.

Lehninger, A.L. Principles of Biochemistry, Chapters 16, 17, 23. New York: Worth, 1982.

Nicholls, D.G. Bioenergetics: An Introduction to the Chemiosmotic Theory. New York: Academic Press, 1982.

Stryer, L. Biochemistry, 3rd ed., Chapters 16, 17, and 22. New York: W.H. Freeman, 1988.

Cited

1. Ernster, L.; Schatz, G. Mitochondria: a historical review. *J. Cell Biol.* 91:227s–255s, 1981.

   Fawcett, D.W. The Cell, 2nd ed., pp. 410–485. Philadelphia: Saunders, 1981.

   Harold, F.M. The Vital Force: A Study of Bioenergetics, Chapter 7. New York: W.H. Freeman, 1986.

   Tzagoloff, A. Mitochondria. New York: Plenum, 1982.

2. DePierre, J.W.; Ernster, L. Enzyme topology of intracellular membranes. *Annu. Rev. Biochem.* 46:201–262, 1977.

   Srere, P.A. The structure of the mitochondrial inner membrane-matrix compartment. *Trends Biochem. Sci.* 7:375–378, 1982.

3. Krstic, R.V. Ultrastructure of the Mammalian Cell, pp. 28–57. New York: Springer-Verlag, 1979.

   Pollak, J.K.; Sutton, R. The differentiation of animal mitochondria during development. *Trends Biochem. Sci.* 5:23–27, 1980.

4. Geddes, R. Glycogen: a metabolic viewpoint. *Biosci. Rep.* 6:415–428, 1986.

   McGilvery, R.W. Biochemistry: A Functional Approach, 3rd ed. Philadelphia: Saunders, 1983.

   Newsholme, E.A.; Start, C. Regulation in Metabolism. New York: Wiley, 1973.

5. Baldwin, J.E.; Krebs, H. The evolution of metabolic cycles. *Nature* 291:381–382, 1981.

Krebs, H.A. The history of the tricarboxylic acid cycle. *Perspect. Biol. Med.* 14:154–170, 1970.

Reed, I.J.; Damuni, Z.; Merryfield, M.L. Regulation of mammalian pyruvate and branched-chain α-keto acid dehydrogenase complexes by phosphorylation-dephosphorylation. *Curr. Top. Cell Regul.* 27:41–49, 1985.

Willamson, J.R.; H. Cooper, R.H. Regulation of the citric acid cycle in mammalian systems. *FEBS Lett.*, Suppl. 117:K73–K85, 1980.

6. Mitchell, P. Coupling of phosphorylation to electron and hydrogen transfer by a chemi-osmotic type of mechanism. *Nature* 191:144–148, 1961.

Racker, E. From Pasteur to Mitchell: a hundred years of bioenergetics. *Fed. Proc.* 39:210–215, 1980.

7. Hatefi, Y. The mitochondrial electron transport and oxidative phosphorylation system. *Annu. Rev. Biochem.* 54:1015–1070, 1985.

8. Nicholls, D.G. Bioenergetics: An Introduction to the Chemiosmotic Theory, Chapter 3. New York: Academic Press, 1982.

Wood, W.B.; Wilson, J.H.; Benbow, R.M.; Hood, L.E. Biochemistry: A Problems Approach, 2nd ed. Menlo Park, CA: Benjamin-Cummings, 1981. (See problems, Chapters 9, 12, and 14).

9. Al-Awqati, Q. Proton-translocating ATPases. *Annu. Rev. Cell Biol.* 2:179–199, 1986.

Hinkle, P.C.; McCarty, R.E. How cells make ATP. *Sci. Am.* 238(3):104–123, 1978.

10. Durand, R.; Briand, Y.; Touraille, S.; Alziari, S. Molecular approaches to phosphate transport in mitochondria. *Trends Biochem. Sci.* 6:211–214, 1981.

Klingenberg, M. The ADP, ATP shuttle of the mitochondrion. *Trends Biochem. Sci.* 4:249–252, 1979.

LaNoue, K.F.; Schoolwerth, A.C. Metabolite transport in mitochondria. *Annu. Rev. Biochem.* 48:871–922, 1979.

11. Eisenberg, D.; Crothers, D. Physical Chemistry with Applications to the Life Sciences, Chapters 4 and 5. Menlo Park, CA: Benjamin-Cummings, 1979.

12. Hatefi, Y. The mitochondrial electron transport and oxidative phosphorylation system. *Annu. Rev. Biochem.* 54:1015–1069, 1985.

Wikstrom, M.; Saraste, M. The mitochondrial respiratory chain. In Bioenergetics (L. Ernster, ed.), pp. 49–94. New York: Elsevier, 1984.

13. Racker, E. A New Look at Mechanisms in Bioenergetics. New York: Academic Press, 1976. (A personal account of the concepts and history.)

14. Amzel, L.M.; McKinney, M.; Narayanan, P.; Pedersen, P.L. Structure of the mitochondrial F_1 ATPase at 9-Å resolution. *Proc. Natl. Acad. Sci. USA* 79:5852–5856, 1982.

Futai, M.; Kanazawa, H. Structure and function of proton-translocating adenosine triphosphatase (F_oF_1): biochemical and molecular biological approaches. *Microbiol. Rev.* 47:285–312, 1983.

Racker, E.; Stoeckenius, W. Reconstitution of purple membrane vesicles catalyzing light-driven proton uptake and adenosine triphosphate formation. *J. Biol. Chem.* 249:662–663, 1974.

Schneider, E.; Altendorf, K. The proton-translocating portion (F_o) of the *E. coli.* ATP synthase. *Trends Biochem. Sci.* 9:51–53, 1984.

15. Hammes, G.G. Mechanism of ATP synthesis and coupled proton transport: studies with purified chloroplast coupling factor. *Trends Biochem. Sci.* 8:131–134, 1983.

Ogawa, S.; Lee, T.M. The relation between the internal phosphorylation potential and the proton motive force in mitochondria during ATP synthesis and hydrolysis. *J. Biol. Chem.* 259:10004–10011, 1984.

Pederson, P.L.; Carafoli, E. Ion motive ATPases. II. Energy coupling and work output. *Trends Biochem. Sci* 12:186–189, 1987.

Senior, A.E. ATP synthesis by oxidative phosphorylation. *Physiol. Rev.* 68:177–231, 1988.

16. Fillingame, R.H. The proton-translocating pumps of oxidative phosphorylation. *Annu. Rev. Biochem.* 49:1079–1113, 1980.

17. Chance, B.; Williams, G.R. A method for the localization of sites for oxidative phosphorylation, *Nature* 176:250–254, 1955.

Dickerson, R.E. The structure and history of an ancient protein. *Sci. Am.* 226(4):58–72, 1972. (The conformation and evolution of cytochrome *c*.)

Keilin, D. The History of Cell Respiration and Cytochromes. Cambridge, U.K.: Cambridge University Press, 1966.

Spiro, T.G., ed. Iron-Sulfur Proteins. New York: Wiley-Interscience, 1982.

18. Capaldi, R.A.; Darley-Usmar, V.; Fuller, S.; Millet, F. Structural and functional features of the interaction of cytochrome *c* with complex III and cytochrome *c* oxidase. *FEBS Lett.* 138:1–7, 1982.

Casey, R.P. Membrane reconstitution of the energy-conserving enzymes of oxidative phosphorylation. *Biochim. Biophys. Acta* 768:319–347, 1984.

Leonard, K.; Haiker, H.; Weiss, H. Three-dimensional structure of NADH: ubiquinone reductase (complex I) from *Neurospora* mitochondria determined by electron microscopy of membrane crystals. *J. Mol. Biol.* 194:277–286, 1987.

Weiss, H.; Linke, P.; Haiker, H.; Leonard, K. Structure and function of the mitochondrial ubiquinol: cytochrome *c* reductase and NADH: ubiquinone reductase. *Biochem. Soc. Trans.* 15:100–102, 1987.

19. Hackenbrock, C.R. Lateral diffusion and electron transfer in the mitochondrial inner membrane. *Trends Biochem. Sci.* 6:151–154, 1981.

20. Dutton, P.L; Wilson, D.F. Redox potentiometry in mitochondrial and photosynthetic bioenergetics. *Biochim. Biophys. Acta* 346:165–212, 1974.

Hamamoto, T.; Carrasco, N.; Matsushita, K.; Kabak, H.R.; Montal, M. Direct measurement of the electrogenic activity of O-type cytochrome oxidase from *E. coli* reconstituted into planar lipid bilayers. *Proc. Natl. Acad. Sci. USA* 82:2570–2573, 1985.

Lehninger, A.L. Bioenergetics: The Molecular Basis of Biological Energy Transformations, 2nd ed. Menlo Park, CA: Benjamin-Cummings, 1971.

21. Prince, R.C. The proton pump of cytochrome oxidose. *Trends Biochem. Sci.* 13:159–160, 1988.

Slater, E.C. The Q Cycle, an ubiquitous mechanism of electron transfer. *Trends Biochem. Sci.* 8:239–242, 1983.

22. Hanstein, W.G. Uncoupling of oxidative phosphorylation. *Trends Biochem. Sci.* 1:65–67, 1976.

23. Brand, M.D.; Murphy M.P. Control of electron flux through the respiratory chain in michondria and cells. *Biol. Rev. Cambridge Philsophic Soc.* 62:141–193, 1987.

Erecinska, A.; Wilson, D.F. Regulation of cellular energy metabolism. *J. Membr. Biol.* 70:1–14, 1982.

Racker, E. A New Look at Mechanisms in Bioenergetics. New York: Academic Press, 1976.

24. Klingenberg, M. Principles of carrier catalysis elucidated by comparing two similar membrane translocators from mitochondria, the ADP/ATP carrier and the uncoupling protein. *Ann. N.Y. Acad. Sci.* 456:279–288, 1985.

Nicholls, D.G.; Rial, E. Brown fat mitochondria. *Trends Biochem. Sci.* 9:489–491, 1984.

25. Gottschalk, G. Bacterial Metabolism, 2nd ed. New York: Springer-Verlag, 1986.

MacNab, R.M. The bacterial flagellar motor. *Trends Biochem. Sci.* 9:185–189, 1984.

Neidhardt, F.C.; et al, eds. *Escherichia coli* and *Salmonella typhimurium*: Cellular and Molecular Biology. Washington, DC: American Society for Microbiology, 1987.

Skulachev, V.P. Sodium bioenergetics. *Trends Biochem. Sci.* 9:483–485, 1984.

Thauer, R.; Jungermann, K.; Decker, K. Energy conservation in chemotrophic anaerobic bacteria. *Bacteriol. Rev.* 41:100–180, 1977.

26. Bogorad, L. Chloroplasts. *J. Cell Biol.* 91:256s–270s, 1981. (A historical review.)

Clayton, R.K. Photosynthesis: Physical Mechanisms and Chemical Patterns. Cambridge, U.K.: Cambridge University Press, 1980 (Excellent general treatment.)

Haliwell, B. Chloroplast Metabolism—The Structure and Function of Chloroplasts in Green Leaf Cells. Oxford, U.K.: Clarendon, 1981.

Hoober, J.K. Chloroplasts. New York: Plenum, 1984.

27. Cramer, W.A.; Widger, W.R.; Herrmann, R.G.; Trebst, A. Topography and function of thylakoid membrane proteins. *Trends Biochem. Sci.* 10:125–129, 1985.

Miller, K.R. The photosynthetic membrane. *Sci. Am. 241(4)102–113, 1979.*

28. Akazawa, T.; Takabe, T.; Kobayashi, H. Molecular evolution of ribulose-1,5-bisphosphate carboxylase/oxygenase (RuBisCO). *Trends Biochem. Sci.* 9:380–383, 1984.

Barber, J. Structure of key enzyme refined. *Nature* 325:663–664, 1987.

Lorimer, G.H. The carboxylation and oxygenation of ribulose-1, 5-bisphosphate: the primary events in photosynthesis and photorespiration. *Annu. Rev. Plant Physiol.* 32:349–383, 1981.

29. Bassham, J.A. The path of carbon in photosynthesis. *Sci. Am.* 206(6):88–100, 1962.

Preiss, J. Starch, sucrose biosynthesis and the partition of carbon in plants are regulated by orthophosphate and triose-phosphates. *Trends Biochem. Sci.* 9:24–27, 1984.

30. Bjorkman, O.; Berry J. High-efficiency photosynthesis. *Sci. Am.* 229(4):80–93, 1973. (C^4 plants.)

Chollet, R. The biochemistry of photorespiration. *Trends Biochem. Sci.* 2:155–159, 1977.

Edwards, G.; Walker, D. C^3, C^4 Mechanisms, and Cellular and Environmental Regulation of Photosynthesis. Berkeley: University of California Press, 1983.

Heber, U.; Krause, G.H. What is the physiological role of photorespiration? *Trends Biochem. Sci.* 5:32–34, 1980.

31. Clayton, R.K. Photosynthesis: Physical Mechanisms and Chemical Patterns. Cambridge, U.K.: Cambridge University Press, 1980.

Parson, W.W. Photosynthesis and other reactions involving light. In Biochemistry (G. Zubay, ed.), 2nd ed., pp. 564–597. New York: Macmillan, 1988.

32. Barber, J. Photosynthethic reaction centres: a common link. *Trends Biochem. Sci.* 12:321–326, 1987.

Govindjee; Govindjee, R. The absorption of light in photosynthesis. *Sci. Am.* 231(6):68–82, 1974.

Li, J. Light-harvesting chlorophyll *a/b*-protein: three-dimensional structure of a reconstituted membrane lattice in negative stain. *Proc. Natl. Acad. Sci. USA* 82:386–390, 1985.

Zuber, H. Structure of light-harvesting antenna complexes of photosynthetic bacteria, cyanobacteria, and red algae. *Trends Biochem. Sci.* 11:414–419, 1986.

33. Deisenhofer, J. Epp, O.; Miki, K.; Huber, R.; Michel, H. Structure of the protein subunits in the photosynthetic reaction centre of *Rhodopseudomonas virdis* at 3Å resolution *Nature* 318:618–624, 1985.

Deisenhofer, J.; Michel, H.; Huber, R. The structural basis of photosynthetic light reactions in bacteria. *Trends Biochem. Sci.* 10:243–248, 1985.

Knaff, D.B. Reaction centers of photsynthetic bacteria. *Trends Biochem. Sci.* 13:157–158, 1988.

Michel, H.; Epp, O.; Deisenhofer, J. Pigment-protein interactions in the photosynthetic reaction center from *Rhodopseudomonas viridis*.

EMBO J. 5:2445–2451, 1986. (Three-dimensional structure by x-ray diffraction.)

34. Govindjee, ed. Photosynthesis: Energy Conversion by Plants and Bacteria. New York: Academic Press, 1982. (Two volumes of review articles at an advanced level.)

Nugent, J.H.A. Photosynthetic electron transport in plants and bacteria. *Trends Biochem Sci.* 9:354–357, 1984.

Scolnik, P.A.; Marrs, B.L. Genetic research with photosynthetic bacteria. *Annu. Rev. Microbiol.* 41:703–726, 1987.

35. Blankenship, R.E.; Prince, R.C. Excited-state redox potentials and the Z scheme of photosynthesis. *Trends Biochem. Sci.* 10:382–383, 1985.

Prince, R.C. Manganese at the active site of the chloroplast oxygen-evolving complex. *Trends Biochem. Sci.* 11:491–492, 1986.

36. Anderson, J.M. Photoregulation of the composition, function, and structure of thylakoid membranes. *Annu. Rev. Plant Physiol.* 37:93–136, 1986.

Carrillo, N., Vallejos, R.H. The light-dependent modulation of photosynthetic electron transport. *Trends Biochem. Sci.* 8:52–56, 1983.

Miller, K.R.; Lyon, M.K. Do we really know why chloroplast membranes stack? *Trends Biochem. Sci.* 10:219–222, 1985.

37. Hinkle, P.C.; McCarty, R.E. How cells make ATP. *Sci. Am.* 238(3):104–123, 1978.

Jagendorf, A.T. Acid-base transitions and phosphorylation by chloroplasts. *Fed. Proc.* 26:1361–1369, 1967.

38. Flügge, U.I.; Heldt, H.W. The phosphate-triose phosphate-phosphoglycerate translocator of the chloroplast. *Trends Biochem. Sci.* 9:530–533, 1984.

Heber, U.; Heldt, H.W. The chloroplast envelope: structure, function and role in leaf metabolism. *Annu. Rev. Plant Physiol.* 32:139–168, 1981.

39. Raven, J.A. Division of labor between chloroplast and cytoplasm. In The Intact Chloroplast (J. Barber, ed.), pp. 403–443. Amsterdam: Elsevier, 1976.

40. Wilson, T.H.; Lin. E.C.C. Evolution of membrane bioenergetics. *J. Supramol. Struct.* 13:421–446, 1980.

Woese, C.R. Bacterial Evolution *Microbiol. Rev.* 51:221–271, 1987.

41. Gest, H. The evolution of biological energy-transducing systems. *FEMS Microbiol. Lett.* 7:73–77, 1980.

Gottschalk, G. Bacterial Metabolism, 2nd ed. New York: Springer-Verlag, 1986. (Chapter 8 covers fermentations.)

Miller, S.M.; Orgel, L.E. The Origins of Life on the Earth. Englewood Cliffs, NJ: Prentice-Hall, 1974.

42. Danson, M.J. Archaebacteria: the comparative enzymology of their central metabolic pathways. *Adv. Microb. Physiol.* 29:165–231, 1988.

Knowles, C.J., ed. Diversity of Bacterial Respiratory Systems, Vol. 1. Boca Raton, FL: CRC Press, 1980.

43. Clayton, R.K.; Sistrom, W.R., eds. The Photosynthetic Bacteria. New York: Plenum, 1978.

Deamer, D.W., ed. Light Transducing Membranes: Structure, Function and Evolution. New York: Academic Press, 1978.

Gromet-Elhanan, Z. Electrochemical gradients and energy coupling in photosynthetic bacteria. *Trends Biochem. Sci.* 2:274–277, 1977.

Olson, J.M.; Pierson, B.K. Evolution of reaction centers in photosynthetic prokaryotes. *Int. Rev. Cytol.* 108:209–248, 1987.

44. Dickerson, R.E. Cytochrome *c* and the evolution of energy metabolism. *Sci. Am.* 242(3):136–153, 1980.

Gabellini, N. Organization and structure of the genes for the cytochrome *b/c* complex in purple photosynthetic bacteria. A phylogenetic study describing the homology of the b/c1 subunits between prokaryotes, miotochondria, and chloroplasts. *J. Bioenerg. Biomembr.* 20:59–83, 1988.

Schopf, J.W.; Hayes, J.M.; Walter, M.R. Evolution of earth's earliest ecosystems: recent progress and unsolved problems. In Earth's Earliest Biosphere: Its Origin and Evolution (J.W. Schopf, ed.), pp. 361–384. Princeton, NJ: Princeton University Press, 1983.

45. Attardi, G.; Schatz, G. Biogenesis of mitochondria. *Annu. Rev. Cell Biol.* 4:289–333, 1988.

Ellis, R.J., ed. Chloroplast biogenesis. Cambridge, U.K.: Cambridge University Press, 1984.

46. Clayton, D.A. Replication of animal mitochondrial DNA. *Cell* 28:693–705, 1982.

Posakony, J.W.; England, J.M.; Attardi, G. Mitochondrial growth and division during the cell cycle in HeLa cells. *J. Cell Biol.* 74:468–491, 1977.

47. Attardi, G.; Borst, P.; Slonimski, P.P. Mitochondrial Genes. Cold Spring Harbor, NY: Cold Spring Harbor Laboratory, 1982.

Borst, P.; Grivell, L.A.; Groot, G.S.P. Organelle DNA. *Trends Biochem. Sci.* 9:128–130, 1984.

Palmer, J.D. Comparative organization of chloroplast genomes. *Annu. Rev. Genet.* 19:325–354, 1985.

48. Grivell, L.A. Mitochondrial DNA. *Sci. Am.* 248(3):60–73, 1983.

Hoober, J.K. Chloroplasts, New York: Plenum, 1984.

49. Ohyama, K.; et al. Chloroplast gene organization deduced from complete sequence of liverwort. *Marchantia polymorpha* chloroplast DNA. *Nature* 322:572–574, 1986.

Rochaix, J.D. Molecular genetics of chloroplasts and mitochondria in the unicellular green alga. *Chlamydomonas. FEMS Microbiol. Rev.* 46:13–34, 1987.

Shinozaki, K.; et al. The complete nucleotide sequence of the tobacco chloroplast genome: its gene organization and expression. *EMBO J.* 5:2034–2049, 1986.

Umesono, K.; Ozeki, H. Chloroplast gene organization in plants. *Trends Genet.* 3:281–287, 1987.

50. Anderson, S.; et al. Sequence and organization of the human mitochondrial genome. *Nature* 290:457–465, 1981.

Bibb, M.J.; Van Etten, R.A.; Wright, C.T.; Walberg, M.W.; Clayton, D.A. Sequence and gene organization of mouse mitochondrial DNA. *Cell* 26:167–180, 1981.

Breitenberger, C.A.; RajBhandary, U.L. Some highlights of mitchondrial research based on analysis of *Neurospora crassa* mitchondrial DNA. *Trends Biochem. Sci.* 10:478–482, 1985.

Fox, T.D. Natural variation in the genetic code. *Annu. Rev. Genet.* 21:67–91, 1987.

51. Attardi, G. Animal mitochondrial DNA: an extreme example of genetic economy. *Int. Rev. Cytol.* 93:93–145, 1985.

Wilson, A. The molecular basis of evolution. *Sci. Am.* 253(4):164–173, 1985.

52. Levings, C.S. The plant mitochondrial genome and its mutants. *Cell* 32:659–661, 1983.

Mulligan, R.M.; Walbot, V. Gene expression and recombination in plant mitochondrial genomes. *Trends Genet.* 2:263–266, 1986.

Newton, K.J. Plant mitochondrial genomes: organization, expression and variation. *Annu. Rev. Plant Physiol. Plant Mol. Biol.* 39:503–532, 1988.

53. Clayton, D.A. Transcription of the mammalian mitochondrial genome. *Annu. Rev. Biochem.* 53:573–594, 1984.

Gruissem, W.; Barken, A.; Deng, S.; Stern, D. Transcriptional and post-transcriptional control of plastid mRNA in higher plants. *Trends Genet.* 4:258–262, 1988.

Mullet, J.E. Chloroplast development and gene expression. *Annu. Rev. Plant Physiol. Plant Mol. Biol.* 39:475–502, 1988.

Tabak, H.F.; Grivell, L.A. RNA catalysis in the excision of yeast mitochondrial introns. *Trends Genet.* 2:51–55, 1986.

54. Birky, C.W., Jr. Transmission genetics of mitochondria and chloroplasts. *Annu. Rev. Genet.* 12:471–512, 1978.

55. Giles, R.E.; Blanc, H.; Cann, H.M.; Wallace, D.C. Maternal inheritance of human mitochondrial DNA. *Proc. Natl. Acad. Sci. USA* 77:6715–6719, 1980.

56. Bernardi, G. The petite mutation in yeast. *Trends Biochem. Sci.* 4:197–201, 1979.

Locker, J.; Lewin, A.; Rabinowitz, M. The structure and organization of mitochondrial DNA from petite yeast. *Plasmid* 2:155–181, 1979.

Montisano, D.F.; James, T.W. Mitochondrial morphology in yeast with and without mitochondrial DNA. *J. Ultrastruct. Res.* 67:288–296, 1979.

57. Attardi, G.; Schatz, G. Biogenesis of mitochondria. *Annu. Rev. Cell Biol.* 4:289–333, 1988.

Fox, T.D. Nuclear gene products required for translation of specific mitochondrially coded mRNAs in yeast. *Trends Genet.* 2:97–99, 1986.

Tzagoloff, A.; Myers, A.M. Genetics of mitochondrial biogenesis. *Annu. Rev. Biochem.* 55:249–285, 1986.

58. Capaldi, R.A. Mitochondrial myopathies and respiratory chain proteins. *Trends Biochem. Sci.* 13:144–148, 1988.

DiMauro, S.; et al. Mitochondrial myopathies. *J. Inherited Meta. Dis.* 10(Suppl. 1):113–28, 1987.

59. Chua, N.-H.; Schmidt, G.W. Post-translational transport into intact chloroplasts of a precursor to the small subunit of ribulose-1,5-bisphosphate carboxylase. *Proc. Natl. Acad. Sci. USA* 75:6110–6114, 1978.

Douglas, M.G.; McCammon, M.T.; Vassarotti A. Targeting proteins into mitochondria. *Microbiol. Rev.* 50:166–178, 1986.

Keegstra, K.; Bauerle, C. Targeting of proteins into chloroplasts. *Bioessays* 9:15–19, 1988.

Schmidt, G.W.; Mishkind, M.L. The transport of proteins into chloroplasts. *Annu. Rev. Biochem.* 55:879–912, 1986.

60. Bishop, W.R.; Bell, R.M. Assembly of phospholipids into cellular membranes: biosynthesis, transmembrane movement, and intracellular translocation. *Annu. Rev. Cell Biol.* 4:579–611, 1988.

61. Cavalier-Smith, T. The origin of eukaryotic and archaebacterial cells. *Ann. N.Y. Acad. Sci.* 503:17–54, 1987.

Butow, B.A.; Doeherty, R.; Parikh, V.S. A path from mitochondria to the yeast nucleus. *Philos. Trans. R. Soc. Lond. (Biol.)* 319:127–133, 1988.

Gellissen, G.; Michaelis, G. Gene transfer: Mitochondria to nucleus. *Ann. N.Y. Acad. Sci.* 503:391–401, 1987.

Margulis, I. Symbiosis in Cell Evolution. New York: W.H. Freeman, 1981.

Schwartz, R.M.; Dayhoff, M.O. Origins of prokaryotes, eukaryotes, mitochondria, and chloroplasts. *Science* 199:395–403, 1978.

Whatley, J.M.; John, P.; Whatley, F.R. From extracellular to intracellular: the establishment of mitochondria and chloroplasts. *Proc. R. Soc. Lond. (Biol.)* 204:165–187, 1979.

62. von Heijne, G. Why mitochondria need a genome. *FEBS Lett.* 198:1–4, 1986

Intracellular Sorting and the Maintenance of Cellular Compartments

8

Unlike a bacterium, which generally consists of a single compartment surrounded by a plasma membrane, a eucaryotic cell is elaborately subdivided into functionally distinct, membrane-bounded compartments. Each compartment, or *organelle,* contains its own distinct set of enzymes and other specialized molecules, and complex distribution systems convey specific products from one compartment to another. To understand the eucaryotic cell, it is essential to know what occurs in each of these compartments, how molecules move between them, and how the compartments themselves are created and maintained.

Proteins play a central part in the compartmentalization of a eucaryotic cell. They catalyze the reactions that occur in each organelle and selectively transport small molecules into and out of its interior, or *lumen.* They also serve as organelle-specific surface markers that direct new deliveries of proteins and lipids to the appropriate organelle. A mammalian cell contains about 10 billion (10^{10}) protein molecules of perhaps 10,000 kinds, and the synthesis of almost all of these begins in the cytosol, the common space that surrounds the organelles. Each newly synthesized protein is then delivered specifically to the cellular compartment that requires it. By tracing the protein traffic from one compartment to another, one can begin to make sense of the otherwise bewildering maze of intracellular membranes. We shall therefore make the intracellular traffic of proteins the central theme of this chapter. Although almost every compartment of the cell will be discussed, most attention will be devoted to the endoplasmic reticulum (ER) and Golgi apparatus, which together play a crucial role in modifying, sorting, and dispatching many of the newly synthesized proteins.

The Compartmentalization of Higher Cells

In this introductory section we give a brief overview of the compartments of the cell and of the relationships between them. In doing so, we consider how the traffic of proteins between compartments can be traced experimentally and review the general strategies by which protein molecules pass from one compartment to another.

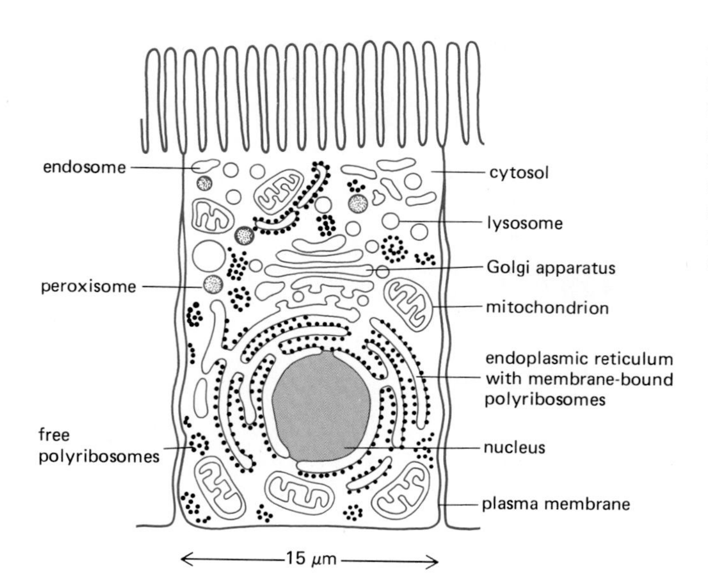

Figure 8–1 A drawing of an animal cell emphasizing the major intracellular compartments. The cytosol, endoplasmic reticulum, Golgi apparatus, nucleus, mitochondrion, endosome, lysosome, and peroxisome are distinct compartments isolated from the rest of the cell by at least one selectively permeable membrane.

All Eucaryotic Cells Have a Basic Set of Membrane-bounded Organelles[1]

Many vital biochemical processes take place in or on membrane surfaces. Oxidative phosphorylation and photosynthesis, for example, both require a semipermeable membrane in order to couple the transport of H^+ to the synthesis of ATP. Moreover, membranes themselves provide the framework for the synthesis of new membrane components. The internal membranes of eucaryotic cells not only provide extra membrane area but also allow functional specialization of different membranes, which, as we shall see, is crucial in separating the many different processes that occur in the cell.

The major intracellular compartments common to eucaryotic cells are illustrated in Figure 8–1. The *nucleus* contains the main genome and is the principal site of DNA and RNA synthesis. The surrounding *cytoplasm* consists of the *cytosol* and the cytoplasmic organelles suspended in it. The cytosol constitutes a little more than half the total volume of the cell, and it is the site not only of protein synthesis but also of most of the cell's intermediary metabolism—that is, the many reactions by which some small molecules are degraded and others are synthesized to provide the building blocks of macromolecules (see p. 64). About half the total area of membrane in a cell is utilized to enclose the labyrinthine spaces of the *endoplasmic reticulum (ER)*. The ER has many ribosomes bound to its cytosolic surface; these are engaged in the synthesis of integral membrane proteins and soluble proteins destined for secretion or for various other organelles. The ER also produces the lipid for the rest of the cell. The *Golgi apparatus* consists of organized stacks of disclike compartments called Golgi *cisternae;* it receives lipids and proteins from the ER and dispatches these molecules to a variety of intracellular destinations, usually covalently modifying them en route. *Mitochondria* and (in plants) *chloroplasts* generate most of the ATP utilized to drive biosynthetic reactions that require an input of free energy. *Lysosomes* contain digestive enzymes that degrade defunct intracellular organelles, as well as macromolecules and particles taken up from outside the cell by endocytosis. On their way to lysosomes, endocytosed macromolecules and particles must first pass through a series of compartments called *endosomes.* Finally, *peroxisomes* (also known as *microbodies*) are small vesicular compartments that contain enzymes utilized in a variety of oxidative reactions.

In addition to the major membrane-bounded organelles just described, cells contain many small vesicles that act as carriers between organelles, as well as other vesicles that communicate with the plasma membrane in the secretory and the endocytic pathways.

In general, each membrane-bounded organelle has a set of properties that are the same in all cell types, as well as properties that vary from cell type to cell type

Table 8–1 The Relative Volumes Occupied by the Major Intracellular Compartments in a Typical Liver Cell (Hepatocyte)

Intracellular Compartment	Percent of Total Cell Volume	Approximate Number per Cell*
Cytosol	54	1
Mitochondria	22	1700
Rough ER cisternae	9	1
Smooth ER cisternae plus Golgi cisternae	6	
Nucleus	6	1
Peroxisomes	1	400
Lysosomes	1	300
Endosomes	1	200

*All the cisternae of the rough and smooth endoplasmic reticulum are thought to be joined to form a single large compartment. The Golgi apparatus, in contrast, is organized into a number of discrete sets of stacked cisternae in each cell, and the extent of interconnection between these sets has not been clearly established.

and contribute to the specialized functions of differentiated cells. Together these organelles occupy nearly half the volume of the cell (Table 8–1).

As might be expected, a large amount of intracellular membrane is required to make all of these organelles. In the two mammalian cells analyzed in Table 8–2, for example, the endoplasmic reticulum has a total membrane surface that is, respectively, 25 times and 12 times that of the plasma membrane. In terms of its area and mass, the plasma membrane is only a minor membrane in most eucaryotic cells (Figure 8–2).

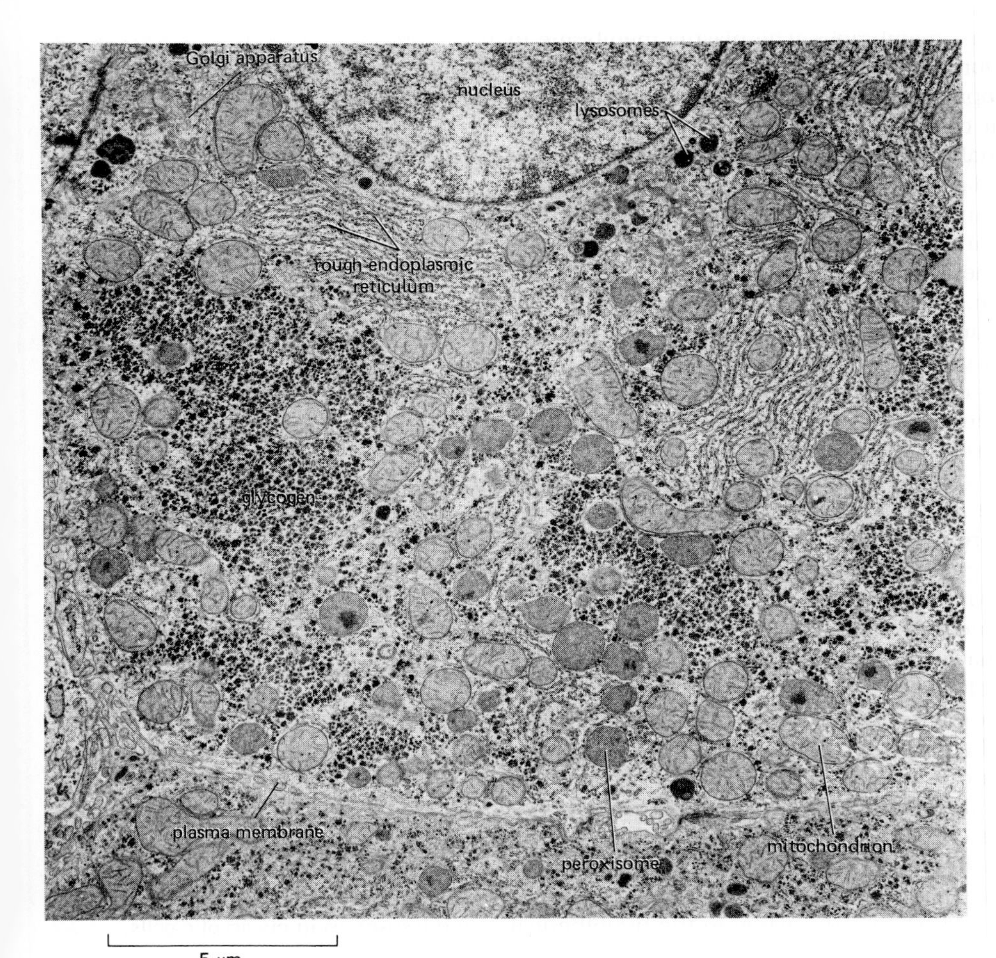

Figure 8–2 Electron micrograph of part of a liver cell seen in cross-section. Examples of most of the major intracellular compartments are indicated. (Courtesy of Daniel S. Friend.)

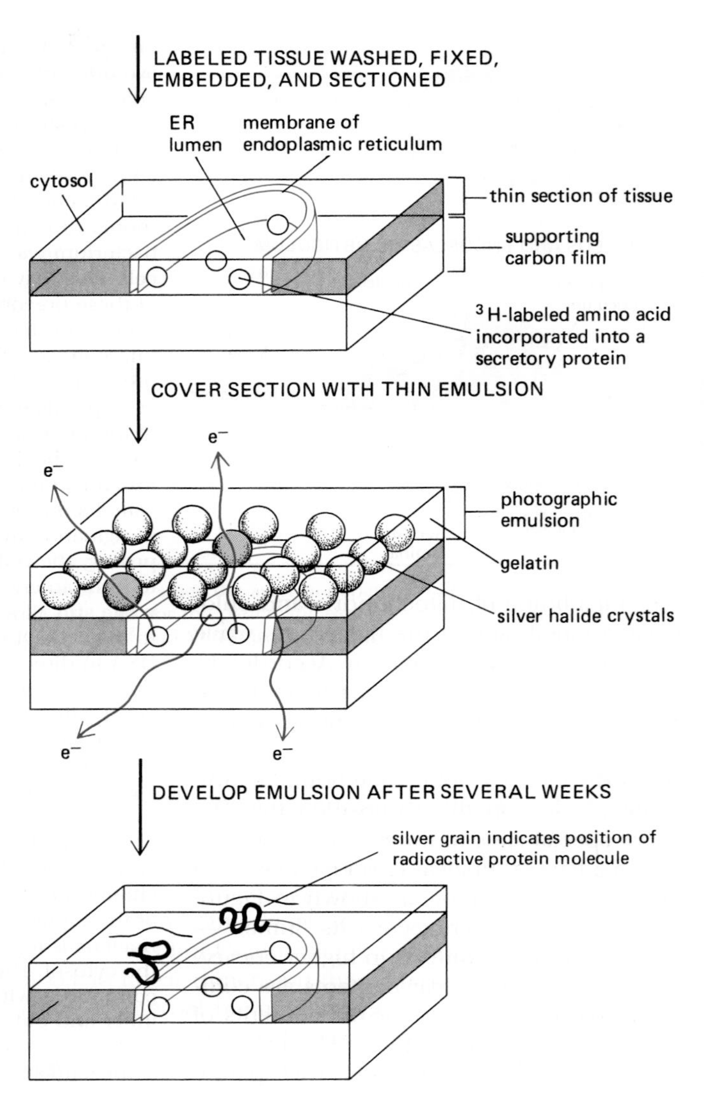

Figure 8–5 The principles of electron microscopic autoradiography, used here to trace secretory proteins by locating the ^{3}H-labeled amino acids they have incorporated. It is important that all free radioactive amino acids are washed away prior to embedding, so that the only radioactivity remaining in the tissue is that incorporated into proteins. The tissue is then covered with a thin layer of photographic emulsion. As indicated, electrons emitted by ^{3}H decay activate silver grains in a range of positions slightly displaced from the point of emission. (The developed silver grains appear as dark squiggles in the electron microscope.) Therefore the localization of a radioactive molecule is imprecise in comparison with the resolution of the electron microscope.

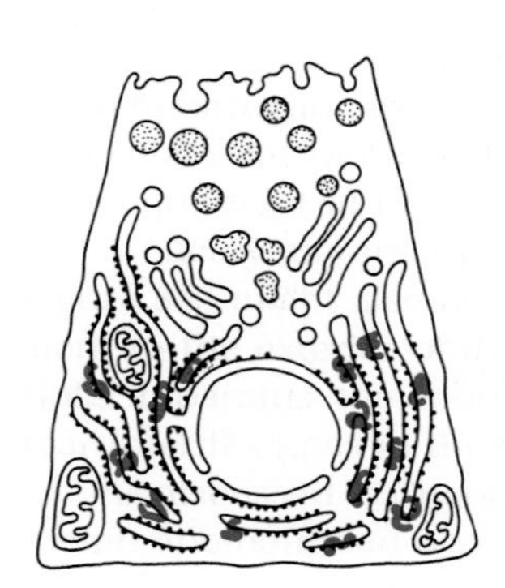

3 minutes:
silver grains over the ER

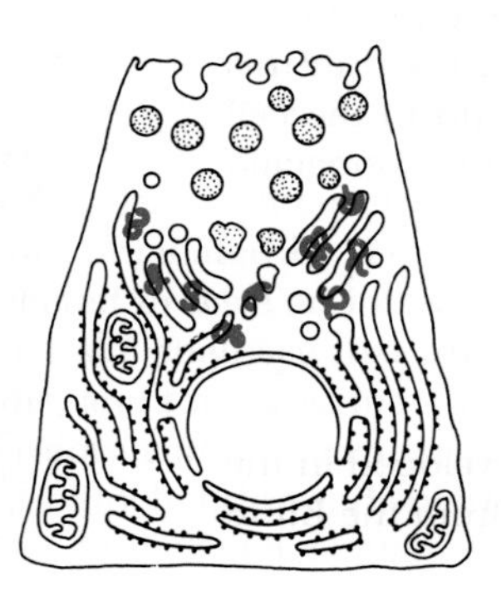

20 minutes:
silver grains over the Golgi apparatus

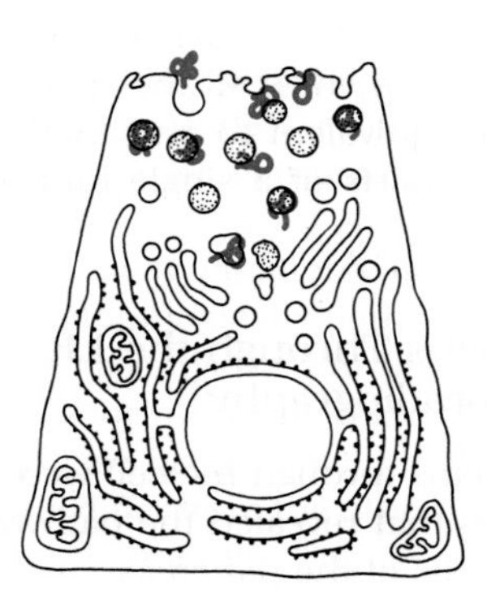

90 minutes:
silver grains over secretory vesicles

Figure 8–6 Results obtained when electron microscopic autoradiography (see Figure 8–5) is used to examine a pancreatic acinar cell that has been briefly pulse-labeled with ^{3}H-amino acids and then incubated in unlabeled media ("chased") for various lengths of time. With time, the activated silver grains (*colored*) are found ever closer to the cell exterior, indicating the path followed by newly synthesized secretory protein molecules. This cell is unusual in that about 85% of the protein it synthesizes is secreted; the proteins are stored in secretory vesicles until the cell is stimulated to secrete. A stimulated cell is shown.

Figure 8–7 Electron micrograph of the Golgi apparatus in a pancreatic acinar cell, showing secretory vesicles in various stages of maturation. Immature secretory vesicles (called *condensing vacuoles* in these cells) form by budding from the Golgi apparatus. The secretory proteins become progressively concentrated in these vesicles (hence the name "condensing" vacuole) to form mature, dense-cored secretory vesicles (*top left*). These secretory vesicles are called *zymogen granules* in pancreatic acinar cells. (Courtesy of George Palade.)

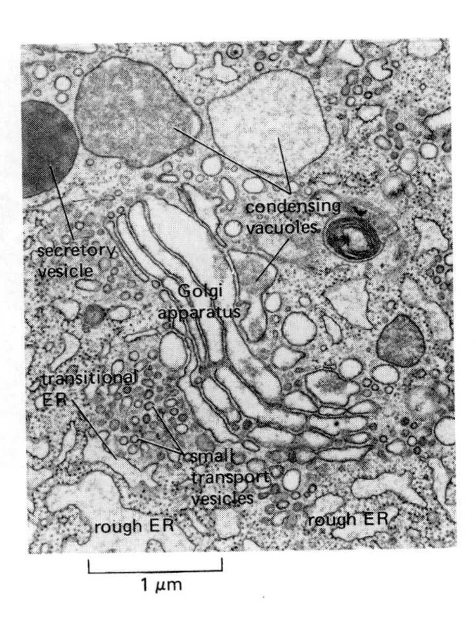

guish it from *constitutive secretion,* which is another form of exocytosis that occurs continuously in the absence of a stimulatory signal (see p. 465). The acinar cell of the pancreas secretes a variety of digestive enzymes (such as amylase, lipase, deoxyribonuclease, and ribonuclease) as well as enzyme precursors called *zymogens* (such as trypsinogen and chymotrypsinogen), which are activated by specific proteolytic cleavages after they are secreted into ducts that lead to the small intestine.

Because the bulk of the protein synthesized in the pancreatic acinar cell is destined for secretion, the problem of tracing the flow of traffic is simplified, and the path followed by secreted proteins from their site of synthesis to their discharge from the cell can be revealed by using autoradiography in conjunction with electron microscopy. The autoradiographic technique developed for this purpose is described in Figure 8–5. When cells are pulse-labeled with ^{3}H-amino acids, followed by a "chase" of varying length in nonradioactive medium, the newly synthesized radioactive proteins are seen first in the ER and then in the Golgi apparatus (Figure 8–6). The ^{3}H-proteins are next found in large, immature secretory vesicles close to the Golgi stack (Figure 8–7), where they become progressively concentrated to form mature secretory vesicles, which are easily identified in electron micrographs by the high density of their contents (Figures 8–7 and 8–8).

The secretory vesicles are stored in the apical region of the acinar cell (the side facing the lumen of the duct system) between the Golgi apparatus and the plasma membrane. They release their contents to the outside by exocytosis when the vesicles fuse with the plasma membrane (see p. 325). The vesicles fuse only with the apical portions of the plasma membrane, thereby avoiding fruitless and dangerous discharges into the spaces between cells or into other organelles in the cell. Furthermore, exocytosis occurs only in response to an appropriate extracellular chemical signal, which is released by nerves or by intestinal cells when pancreatic enzymes are needed for digestion.

Many secreted proteins have been studied in a variety of cell types, and all of them follow a similar pathway: ribosome → ER → Golgi → cell exterior. Moreover, those proteins destined to reside in the plasma membrane, in lysosomes, or in various cisternae of the Golgi apparatus are likewise initially imported into the ER—passing from there via the Golgi apparatus to their final locations.

More complicated techniques are needed to reveal the finer details of the pathways that begin in the ER. While some of the imported proteins will remain in the ER to catalyze ER functions, most are packaged into *transport vesicles* (typically 50–100 nm in diameter) that pinch off from specialized regions of the ER called the *transitional elements* (Figure 8–9) and fuse specifically with nearby cisternal elements of the Golgi apparatus. After fusion the constituents of the vesicle membrane become part of the Golgi membrane, while the soluble proteins in the vesicle lumen are delivered to the lumen of a Golgi cisterna (Figure 8–10). In this way soluble proteins are selectively carried from one membrane-bounded compartment to another without actually passing across a membrane. By similar cycles of vesicle budding and fusion, proteins are thought to be transported from one Golgi cisterna to another and then from the Golgi apparatus to the various final destinations appropriate to their different functions, passing through a whole series of spaces that are topologically equivalent to one another and to the cell exterior (Figure 8–11).

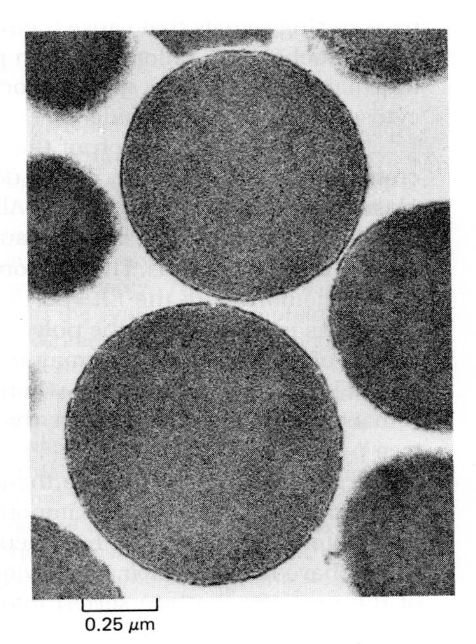

Figure 8–8 Electron micrograph of a purified preparation of large secretory vesicles. Such vesicles are found only in cells specialized for secretion. (Courtesy of Daniel S. Friend.)

perform its function, each vesicle must take up only the appropriate proteins and must fuse only with the appropriate target membrane: a vesicle carrying cargo from the ER to the Golgi apparatus, for example, must exclude proteins that are to remain in the ER and fuse only with the Golgi apparatus and not with other organelles. We shall consider how transport vesicles might achieve this selectivity later in the chapter.

Signal Peptides and Signal Patches Specify a Protein's Fate[6]

There are thought to be two types of sorting signals on proteins that direct them step by step through the branching pathways outlined in Figure 8–12. For some steps the sorting signal resides in a continuous stretch of amino acid sequence, typically 15–60 residues long. This **signal peptide** is often (but not always) removed from the finished protein once the sorting decision has been executed. The signal for other sorting steps is thought to consist of a particular three-dimensional arrangement of atoms on the protein's surface that forms when the protein folds up. The amino acid residues that comprise this **signal patch** may be quite distant from one another in the linear amino acid sequence, and they generally remain in the finished protein (Figure 8–13). *Signal peptides* are used to direct proteins from the cytosol into the ER, mitochondria, chloroplasts, and nucleus; they are also used to retain certain proteins in the ER. *Signal patches* are thought to be used for some other sorting steps, including the recognition of certain lysosomal proteins by a special sorting enzyme in the Golgi apparatus (see p. 463).

Different types of signal peptides are used to specify different destinations in the cell (Table 8–3). Proteins destined for initial transfer to the ER usually have amino-terminal signal peptides with a central part of the sequence composed of 5–10 hydrophobic amino acid residues. Most of these proteins will pass from the ER to the Golgi apparatus; those with a specific sequence of four amino acids at their carboxyl terminus, however, are retained as permanent ER residents. Proteins destined for mitochondria have signal peptides in which positively charged amino acid residues alternate with hydrophobic ones. Many proteins destined for the nucleus carry signal peptides formed from a cluster of positively charged amino acid residues. Finally, some cytosolic proteins have signal peptides that cause a fatty acid to be covalently attached to them, which directs the proteins to membranes without insertion into the ER (see p. 417).

The importance of each of these signal peptides for protein targeting has been shown by experiments in which the peptide is transferred from one protein to another by genetic engineering techniques: placing the amino-terminal ER signal peptide at the beginning of a cytosolic protein, for example, redirects the protein to the ER. The signal peptides attached to all proteins having the same destination are functionally interchangeable, even though their amino acid sequences may vary greatly. Physical properties, such as hydrophobicity, often appear to be more important in the signal-recognition process than the exact amino acid sequence.

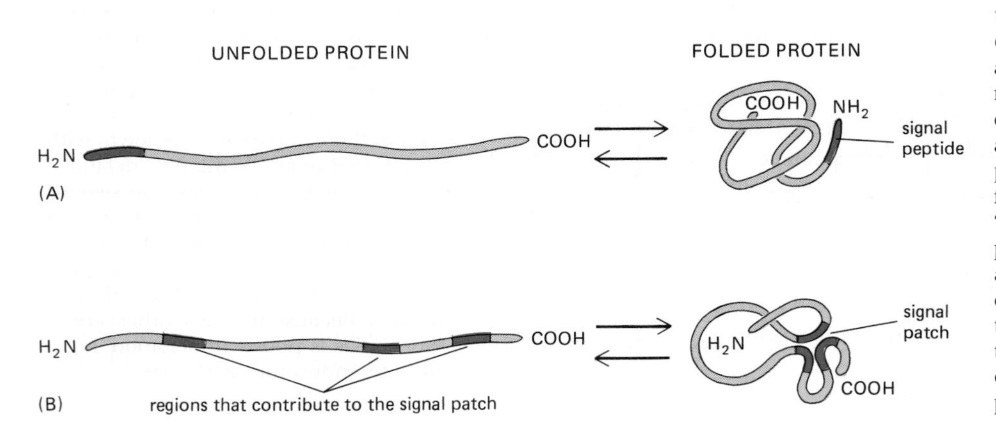

Figure 8–13 Two ways that a transport signal can be built into a protein. (A) The signal is in a single discrete stretch of amino acid sequence, called a *signal peptide,* that is exposed in the folded protein. Signal peptides often occur at the end of the polypeptide chain (as shown), but they can also be located elsewhere. They are normally detected experimentally by their effect on the intracellular sorting of other proteins when they are attached to them by recombinant DNA methods (see p. 266). (B) A *signal patch* can be formed by the juxtaposition of amino acids from regions that are physically separated before the protein folds (as shown); alternatively, separate "patches" on the surface of the folded protein that are spaced a fixed distance apart could form the signal. In either case the transport signal depends on the three-dimensional conformation of the protein. For this reason it is very difficult to locate this type of signal precisely.

Table 8–3 Typical Signal Peptide Sequences

Function of Signal Peptide	Example of Signal Peptide
Import into ER	H_3N-Met-Met-Ser-Phe-Val-Ser- Leu-Leu-Leu-Val ⊕ Gly-Ile-Leu-Phe-Trp-Ala -Thr-Glu-Ala-Glu- ⊖ ⊖ Gln-Leu-Thr-Lys-Cys-Glu-Val-Phe-Gln- ⊕ ⊖
Retain in lumen of ER	-Lys-Asp-Glu-Leu-COO^- ⊕ ⊖ ⊖ ⊖
Import into mitochondria	H_3N-Met-Leu-Ser-Leu-Arg-Gln-Ser-Ile-Arg-Phe- ⊕ ⊕ ⊕ Phe-Lys-Pro-Ala-Thr-Arg-Thr-Leu-Cys-Ser- ⊕ ⊕ Ser-Arg-Tyr-Leu-Leu- ⊕
Import into nucleus	-Pro-Pro-Lys-Lys-Lys-Arg-Lys-Val- ⊕ ⊕ ⊕ ⊕ ⊕
Attach to membranes via the covalent linkage of a myristic acid to the amino terminus	H_3N-Gly-Ser-Ser-Lys-Ser-Lys-Pro-Lys- ⊕ ⊕ ⊕ ⊕

Charged residues are indicated by a ⊕ or a ⊖. An extended block of hydrophobic residues is enclosed in a box. H_3N- indicates the amino terminus of a protein; —COO^- indicates the carboxyl terminus.

Signal patches are far more difficult to analyze than signal peptides; consequently, much less is known about their structure. Because they result from a complex three-dimensional protein-folding pattern, they cannot simply be transferred from one protein to another. Moreover, experimental alteration of a signal patch will often disturb the conformation of the protein as a whole.

Cells Cannot Construct Their Membrane-bounded Organelles de Novo: They Require Information in the Organelle Itself[7]

When a cell reproduces and divides, it has to duplicate its membrane-bounded organelles. In general, cells do this by enlarging these organelles by incorporating new molecules into them; the enlarged organelles then divide and are distributed to the two daughter cells. It is unlikely that cells could make all of these organelles de novo. If the ER were completely removed from a cell, for example, how could the cell reconstruct it? The membrane proteins that define the ER and carry out many of its key functions are themselves products of the ER: without an existing ER—at the very least, without a membrane that contains the translocators required to import proteins into the ER (and that lacks the translocators required to import proteins into other organelles)—a new ER could not be made.

Thus it seems that the information required to construct a membrane-bounded organelle does not reside exclusively in the DNA that specifies the organelle proteins. "Epigenetic" information in the form of at least one distinct protein in the organelle membrane is also required, and this information is passed from parent cell to progeny cell in the form of the organelle itself. Presumably, such information is essential for the propagation of the cell's compartmental organization, just as the information in DNA is essential for the propagation of nucleotide and amino acid sequences.

Summary

Eucaryotic cells contain intracellular membranes that enclose nearly half their total volume in separate intracellular compartments. The main types of membrane-bounded organelles in all eucaryotic cells are the endoplasmic reticulum, Golgi apparatus, nucleus, mitochondria, lysosomes, endosomes, and peroxisomes; plant cells also contain chloroplasts. Each organelle contains distinct proteins that mediate its unique functions.

Each newly synthesized organelle protein finds its way from the ribosome to the organelle by following a specific pathway, guided by a signal in its amino acid sequence that functions either as a signal peptide or as a signal patch. Protein sorting begins with a primary segregation event in which the protein either remains in the cytosol or is transferred to another compartment (such as the nucleus, a mitochondrion, or the endoplasmic reticulum). Proteins that enter the ER undergo secondary sorting processes as they are transported to the Golgi apparatus and from the Golgi apparatus to lysosomes, to secretory vesicles, or to the plasma membrane. Some resident proteins are retained specifically in the ER and in the various cisternae of the Golgi apparatus. Those destined for other compartments are thought to enter transport vesicles, which bud from one compartment and fuse with another.

The Cytosolic Compartment

The **cytosol**—that is, the part of the cytoplasm that occupies the space between the membrane-bounded organelles—generally accounts for about half of the total cell volume (see Table 8–1). It teems with thousands of enzymes involved in intermediary metabolism and is packed with ribosomes making proteins. About half of the proteins made on these ribosomes are destined to remain in the cytosol as permanent residents. In this section we discuss the fate of these cytosolic proteins as well as some of the devices that are used to control their lifetime and to direct them to specific locations within the cytosol.

The Cytosol Is Organized by Protein Filaments[8]

As we shall discuss in Chapter 11, the cytosol contains a variety of protein filaments arranged in a fibrous *cytoskeleton.* The cytoskeleton imparts shape to the cell, mediates coherent cytoplasmic movements, and provides a general framework that could help organize enzymatic reactions. Moreover, inasmuch as about 20% of its weight is protein, it is more appropriate to think of the cytosol as a highly organized gelatinous mass rather than as a simple solution of enzymes. Studies of diffusion rates, however, have shown that small molecules and some small proteins diffuse nearly as quickly in the cytosol as they do in pure water. This suggests that, from the point of view of intermediary metabolism (in which the substrates and products are all small molecules), we can treat the cytosol as a simple solution.

Large particles such as transport vesicles and organelles, on the other hand, diffuse very slowly, partly because they collide frequently with components of the cytoskeleton. To move at useful rates, therefore, such particles usually are actively transported by protein "motors" that hydrolyze ATP and use the energy released to propel the particles along microtubules or actin filaments (see p. 660 and p. 632). In principle, one could imagine a highly organized cytosol (see p. 668), in which specific filaments act as "highways" to direct each type of transport vesicle to its appropriate target membrane for fusion. Most cell biologists, however, believe that the cytoskeleton generally plays a less specific role and that most of the specificity of vesicular traffic resides in receptor systems located on the cytosolic surface of the vesicles themselves (see p. 463).

Many Proteins Are Covalently Modified in the Cytosol[9]

More than 100 different *post-translational modifications* of amino acid side chains have been described in proteins. The functions of most of these modifications are not known; some are chemical accidents and have no function at all, but many must be important for the operation of the cell, since they are tightly controlled by specific enzymes. We shall see later that some of these modifications occur in the ER and the Golgi apparatus. Glycosylating enzymes in these organelles, for example, add a complex series of sugar residues to proteins to produce glycoproteins (see p. 446). The only type of protein glycosylation known to occur in the cytosol of mammalian cells is the much simpler attachment of a single *N*-acetylglucosamine to an occasional protein (Figure 8–14). Many other covalent modifications, however, occur primarily in the cytosol. Some are permanent and required for activity, such as the covalent attachment of coenzymes (for example, biotin, lipoic acid, or pyridoxal phosphate) to some enzymes (see p. 75). Other covalent modifications that occur in the cytosol, such as phosphorylation, are reversible and serve to regulate the activity of many proteins (Figure 8–15).

From the perspective of protein targeting, one type of covalent modification that takes place in the cytosol is especially important. The attachment of a fatty acid to a protein can direct the protein to a specific membrane surface that is exposed to the cytosol.

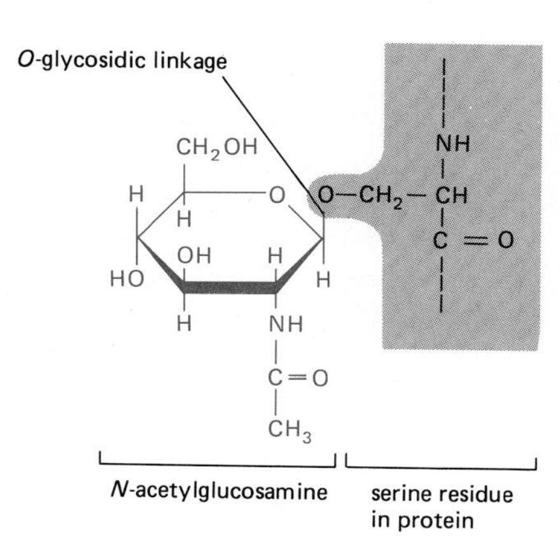

Figure 8–14 The attachment of *N*-acetylglucosamine to serine (or threonine) residues of a protein is the only form of protein glycosylation known to occur in the cytosol of mammalian cells. Many gene regulatory proteins and several nuclear pore proteins are modified in this way; the function of this modification is not known. Much more complex glycosylations occur in the ER and Golgi apparatus (see Figure 8–52).

Some Cytosolic Proteins Are Attached to the Cytoplasmic Face of Membranes by a Fatty Acid Chain[10]

Cells have a special mechanism for directing selected water-soluble proteins from the cytosol to membranes. The protein is covalently attached to a fatty acid chain, which inserts into the cytoplasmic leaflet of the lipid bilayer, anchoring the protein to the membrane. Attaching a protein to a membrane by a fatty acid can have important functional consequences. The *src* oncogene of Rous sarcoma virus, for example, encodes a tyrosine-specific protein kinase that is normally bound to membranes by a covalently attached myristic acid chain (an unsaturated fatty acid with 14 carbons). In this configuration the kinase can transform a cell into a cancer cell. If the attachment of this fatty acid is prevented by converting the amino-terminal glycine of the protein to an alanine, the src protein is still perfectly active as a protein kinase, but it remains in the cytosol and does not transform the cell. Evidently the kinase needs to be attached to a membrane in order to find its substrates efficiently (see p. 757). Similar experiments suggest that a different oncogene product, the ras protein (see p. 705), must be bound to membranes by a covalently attached palmitic acid chain (an unsaturated fatty acid with 16 carbons) in order to transform cells.

What determines whether a given cytosolic protein receives a fatty acid chain and whether it is myristic acid or palmitic acid? The enzymes that catalyze these modifications recognize distinct signal peptides in the protein: a myristic acid

Figure 8–15 Three types of reversible covalent modifications that occur in proteins to regulate their activity. Each modification shown changes the charge on an amino acid side chain. The most common modification is the phosphorylation of an OH group on a serine, threonine, or tyrosine side chain of a protein; it has been estimated that about 10% of the cytosolic proteins in animal cells are modified in this way (see p. 129).

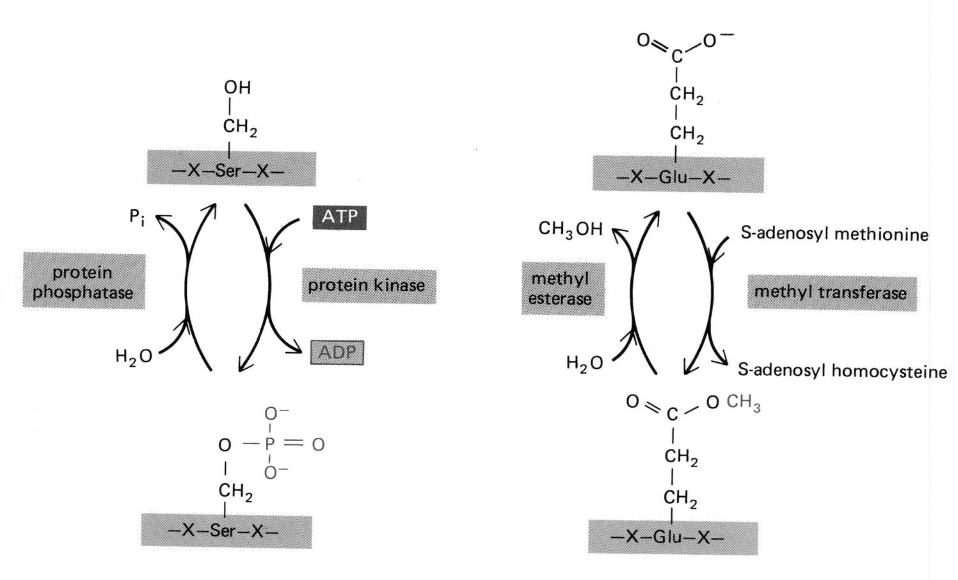

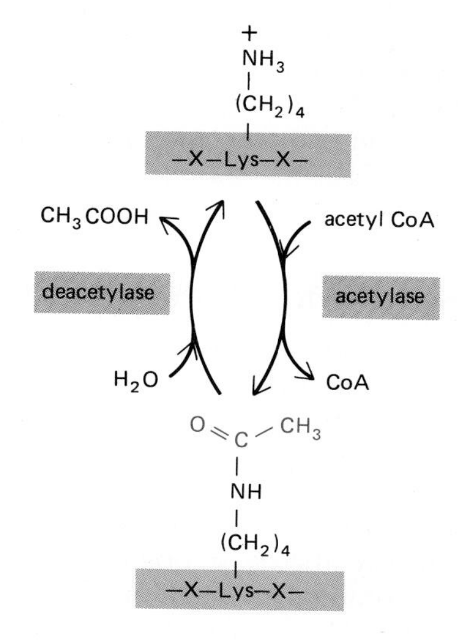

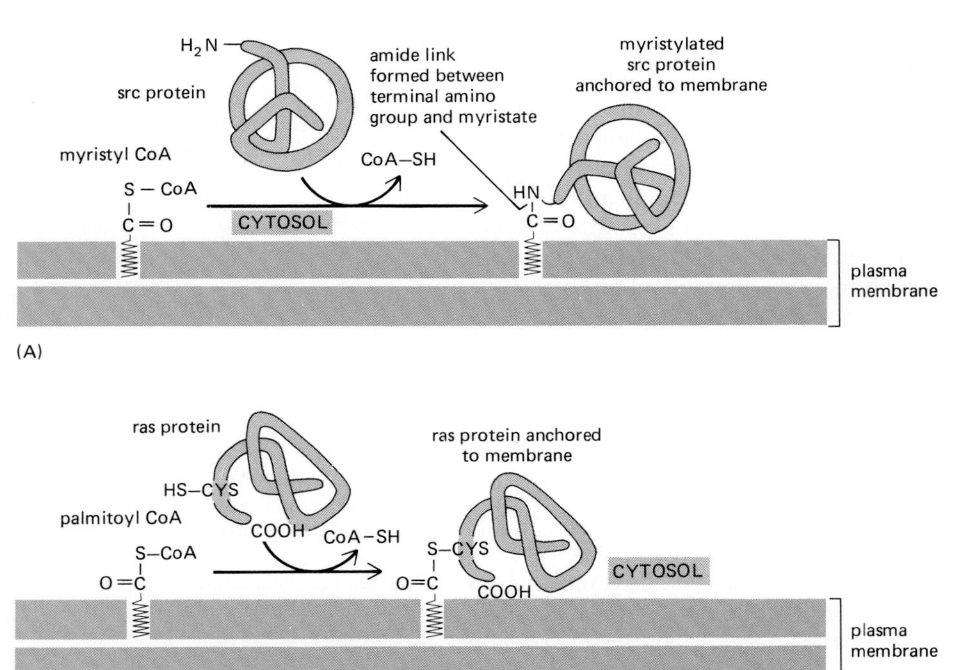

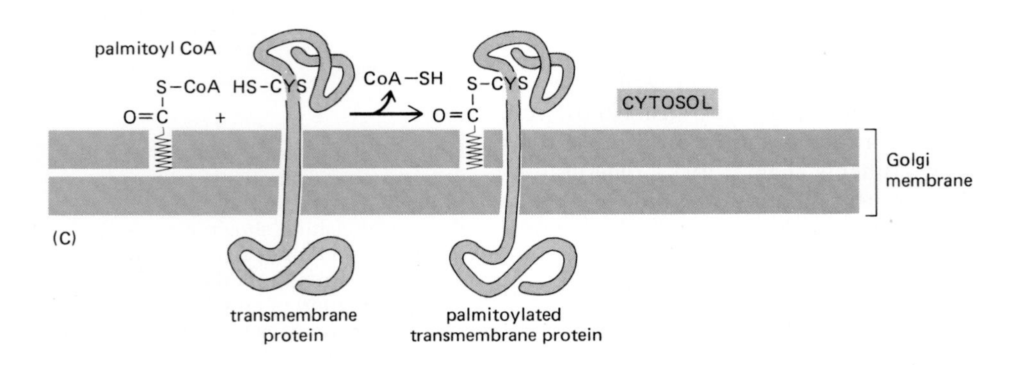

Figure 8–16 The covalent attachment of a fatty acid to a protein can localize a water-soluble protein to a membrane. (A) An amide linkage between an amino-terminal glycine and myristic acid anchors the src protein to the cytoplasmic side of a membrane after its synthesis in the cytosol. (B) A thiolester linkage between palmitic acid and a cysteine near the carboxyl terminus anchors the ras protein to the plasma membrane after its synthesis in the cytosol. (C) Palmitic and other fatty acids are frequently attached by a thiolester bond to a selected cysteine residue located in the cytoplasmic domains of a transmembrane protein. This occurs during the transport of the protein from the ER to the Golgi apparatus. The acylated cysteine is usually preceded by three hydrophobic amino acids (X) in the sequence NH_2-. . .-X-X-X-Cys-. . .-COOH. The protein in (C), unlike those in (A) and (B), will remain membrane-bound even without the fatty acid attachment.

chain (Figure 8–16A) is added to an amino-terminal glycine residue that is located in a particular context (see Table 8–3), while a palmitic acid chain is added to a cysteine side chain four residues from the carboxyl terminus in another signal peptide (Figure 8–16B). In addition, a reaction in the cytosol, which is catalyzed by a different enzyme, adds a palmitic acid chain to the exposed cytosolic tails of many transmembrane proteins as they pass from the ER to the Golgi apparatus on their way to the plasma membrane and elsewhere (Figure 8–16C).

8-8 Some Cytosolic Proteins Are Programmed for Rapid Destruction[11]

In addition to carrying signals that determine their location, proteins in cells carry signals that determine their lifetime. Cellular proteins are subject to continuous turnover: some molecules of each protein are randomly selected for degradation and are replaced by new copies. Most of the resident proteins of the cytosol are relatively long-lived, functioning for several days before they are degraded. Others, however, are degraded much more rapidly, sometimes within a few minutes of their synthesis. Among the short-lived proteins are enzymes that catalyze a rate-determining step in a metabolic pathway; the rates of synthesis of these enzymes are usually regulated according to environmental conditions to promote efficient use of the metabolic pathway. Other short-lived proteins are the products of cellular oncogenes such as *fos* and *myc,* which are thought to play important parts in the control of cell growth and division (see p. 757). Because these types of proteins are continuously and rapidly degraded, their concentrations can be quickly changed by changing their rates of synthesis (see p. 714). In most cases such regulation also requires an unusually rapid turnover of the mRNAs that encode these proteins (see p. 595).

Most misfolded, denatured, or otherwise abnormal proteins are also rapidly degraded in the cytosol. They are generally destroyed within minutes, while intact copies of the same proteins are spared. Abnormal proteins arise by accidents of protein synthesis in which incorrect amino acids are incorporated, and they are also produced by chemical damage such as the oxidation of certain amino acid side chains. Various mutant forms of normal proteins can also be recognized as abnormal. There is increasing evidence that both abnormal proteins and the proteins that are genetically programmed for rapid turnover are ultimately destroyed by the same proteolytic machinery in the cytosol.

8-6 A Ubiquitin-dependent Proteolytic Pathway Is Responsible for Selective Protein Turnover in Eucaryotes[12]

The cytosolic proteins that are programmed for rapid destruction carry signals that are recognized by the proteolytic machinery responsible for their degradation. One of these signals is remarkably simple, consisting only of the first amino acid in the polypeptide chain. When present as the amino-terminal residue, the amino acids Met, Ser, Thr, Ala, Val, Cys, Gly, and Pro are stabilizing, while the remaining 12 amino acids attract a proteolytic attack. These *destabilizing* amino acids are virtually never found at the amino terminus of stable cytosolic proteins. They are often present, however, at the amino terminus of proteins that are transported into other compartments, such as the ER; since the cytosolic degradation machinery is not present in the lumen of the ER or the Golgi apparatus, such proteins are normally long-lived in their respective compartments. A destabilizing amino-terminal amino acid on these noncytosolic proteins may provide the cell with a simple way of disposing of copies that have been misrouted: those molecules that fail to be rapidly translocated out of the cytosol are rapidly degraded. A similar single-residue code is apparently used in bacteria to signal the rapid destruction of specific proteins.

The proteolytic machinery responsible for selective protein degradation is complex and guarantees the complete destruction of a protein once the machinery is engaged. In eucaryotes a **ubiquitin-dependent pathway** is used. In this pathway numerous copies of a small protein called *ubiquitin* (Figure 8–17) are covalently linked to the target protein to be degraded. The ubiquitination of target proteins is catalyzed by a multienzyme complex that is thought to bind to the amino terminus of proteins that have a destabilizing amino-terminal amino acid. The enzyme complex attaches a ubiquitin molecule to a nearby exposed lysine residue in the polypeptide chain and then adds a series of additional ubiquitin molecules to this first ubiquitin, producing a branched multiubiquitin chain (Figure 8–18). Subsequently a large ATP-dependent protease rapidly degrades these proteins. Because only proteins that contain branched multiubiquitin chains appear to be substrates for the protease, proteins that contain a single ubiquitin molecule attached to a lysine, such as certain histones are spared.

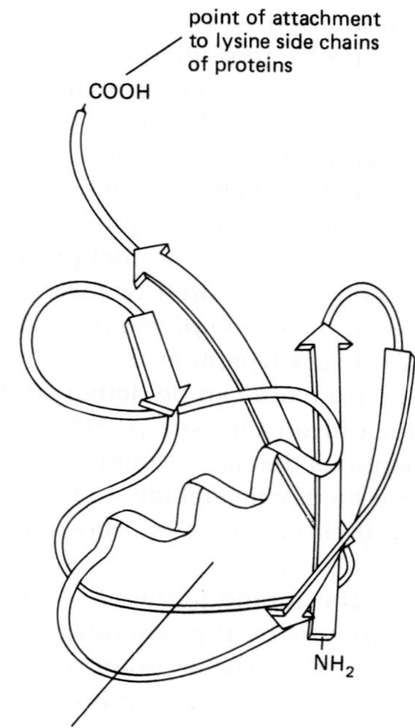

Figure 8–17 The three-dimensional structure of ubiquitin, a heat-stable protein of 76 amino acid residues. The attachment of a single ubiquitin molecule to a protein is a reversible modification with a regulatory role (see also Figure 8–15). The addition of a branched ubiquitin chain to a protein, however, condemns that protein to immediate and complete degradation (see Figure 8–18). (Based on S. Vijay-Kumar, C.E. Bugg, K.D. Wilkinson and W.J. Cook, *Proc. Natl. Acad. Sci. USA* 82:3582–3585, 1985.)

8-7 A Protein's Stability Can Be Determined by Enzymes That Alter Its Amino Terminus[13]

Since the amino-terminal amino acid of a cytosolic protein determines whether the protein will be degraded by the ATP-dependent protease, it is important to know how this crucial amino acid residue is acquired. Proteins that are genetically programmed for rapid destruction seem to acquire a destabilizing amino acid at their amino terminus immediately after their synthesis. As discussed in Chapter 5 (see p. 213), all proteins are initially synthesized with methionine as their amino-terminal amino acid (formyl methionine in bacteria). This methionine, which is a

vesicles are sealed, enclosing a lumen equivalent to that of the rough ER, while their cytoplasmic surface is easily accessible to components that can be added *in vitro.* To the biochemist, rough microsomes represent small authentic versions of the rough endoplasmic reticulum, still capable of protein synthesis, protein glycosylation, and lipid synthesis.

Rough Regions of ER Contain Proteins Responsible for Binding Ribosomes[35]

Because the ER membrane, like all membranes, is a two-dimensional fluid, most proteins and lipids will equilibrate freely between rough and smooth regions in the absence of special restraints. When isolated from liver, however, rough microsomes contain more than 20 proteins that are not present in smooth microsomes, showing that some restraining mechanism must in fact exist. Some of the nonequilibrating proteins in the rough ER membrane help bind ribosomes to it, while others presumably produce its flattened shape (see Figure 8–37). It is not clear whether these membrane proteins are retained by forming large two-dimensional aggregates in the lipid bilayer or whether they are held in place by interactions with a network of structural proteins on one or the other face of the ER membrane (see p. 297).

The ribosomes of the rough ER are held on the membrane in part by their growing polypeptide chains, which are threaded across the ER membrane as they are synthesized (see below). However, when the synthesis of polypeptide chains is terminated with a drug (such as puromycin) that releases the nascent chains from the ribosomes, the ribosomes still retain some affinity for the membrane of rough microsomes. This affinity is artificially increased at low salt concentrations, and when purified ribosomes are mixed under these conditions with rough microsomal membranes from which ribosomes have been removed, the "stripped" membrane regains the same number of ribosomes it had when originally isolated. The binding site on the ribosome is located on the large ribosomal subunit, but it is not yet clear to which of the many proteins in the rough ER membrane the ribosome binds. We shall see, however, that an additional, more specific attachment is required to bind ribosomes to the ER membrane under physiological conditions, where a nascent protein that contains a signal peptide is required.

Signal Peptides Were First Discovered in Proteins Imported into the ER[36]

Signal peptides (and the signal peptide strategy of protein import) were first discovered in the early 1970s in secreted proteins that are translocated across the ER membrane prior to their transport to the Golgi apparatus and eventual discharge from the cell. The observations that led to this discovery came from an experiment in which the mRNA encoding the secreted protein was translated by free ribosomes *in vitro.* When microsomes were omitted from this cell-free system, the protein synthesized was slightly larger than the normal secreted protein, the extra length being due to the presence of an amino-terminal *leader peptide.* In the presence of microsomes derived from the rough endoplasmic reticulum, however, a protein of the correct size was produced. These results were explained by the **signal hypothesis,** which postulates that the leader serves as a signal peptide that directs the secreted protein to the ER membrane and is then cleaved off by a special protease in the ER membrane before the polypeptide chain is completed (Figure 8–41).

According to the signal hypothesis, the secreted protein should be extruded into the lumen of the microsome during its *in vitro* synthesis. This can be demonstrated with protease treatment: a newly synthesized protein made in the absence of microsomes is degraded by the addition of a protease, whereas the same protein made in the presence of microsomes remains intact because of the protection afforded by the microsomal membrane. When proteins without ER signal peptides are similarly translated *in vitro,* they are not imported into microsomes and therefore remain susceptible to protease treatment.

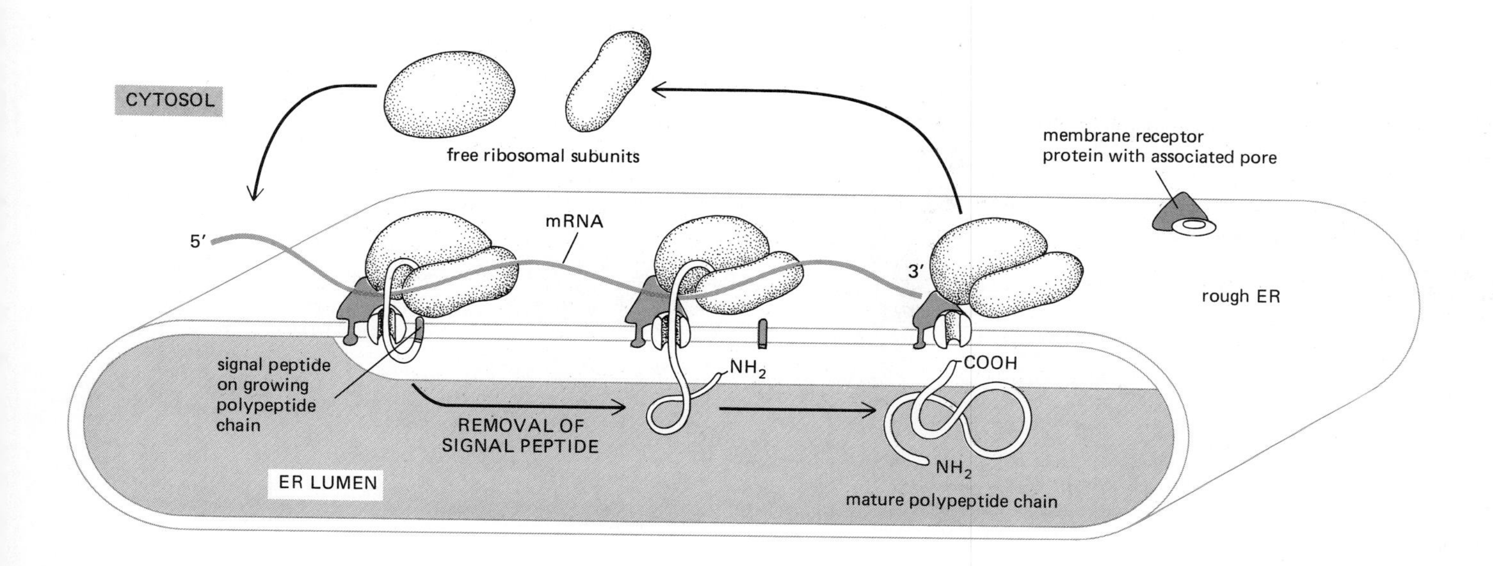

Figure 8–41 A simplified view of protein translocation across the ER membrane, as proposed in the original "signal hypothesis." When the signal peptide emerges from the ribosome, it directs the ribosome to a receptor protein on the ER membrane. As it is synthesized, the polypeptide is postulated to be translocated across the ER membrane through a protein pore associated with the receptor. The signal peptide is clipped off during translation, and the mature protein is released into the lumen of the ER immediately after being synthesized.

The signal hypothesis has been thoroughly tested by genetic and biochemical experiments and found to apply to both plant and animal cells, as well as to protein translocations across procaryotic plasma membranes. Moreover, amino-terminal leader peptides have been found not only in secreted proteins but also in precursors of plasma membrane and lysosomal proteins, which are also imported by the ER. As mentioned earlier, the signaling function of these leader peptides has been demonstrated directly by using recombinant DNA techniques to attach signal sequences to proteins that do not normally have them; the resulting fusion proteins are directed to the ER.

Cell-free systems in which protein import occurs have provided powerful assay procedures for identifying, purifying, and studying the various components of the molecular machinery responsible for the ER import process.

8-25 A Signal-Recognition Particle Directs ER Signal Peptides to a Specific Receptor in the ER Membrane[37]

The signal peptide is guided to the ER membrane by at least two components: a **signal-recognition particle (SRP),** which cycles between the ER membrane and the cytosol and binds to the signal peptide, and an *SRP receptor,* also known as a *docking protein,* in the ER membrane. The SRP was discovered when it was found that washing microsomes with salt eliminated their ability to import secreted proteins. Import could be restored by adding back the supernatant containing the salt extract. The "translocation factor" in the salt extract was then purified and found to be a complex particle consisting of six different polypeptide chains bound to a single molecule of 7SL RNA (Figure 8–42).

The signal-recognition particle binds to the signal peptide as soon as the peptide emerges from the ribosome. This causes a pause in protein synthesis and sometimes stops it completely. The translational pause presumably gives the ribo-

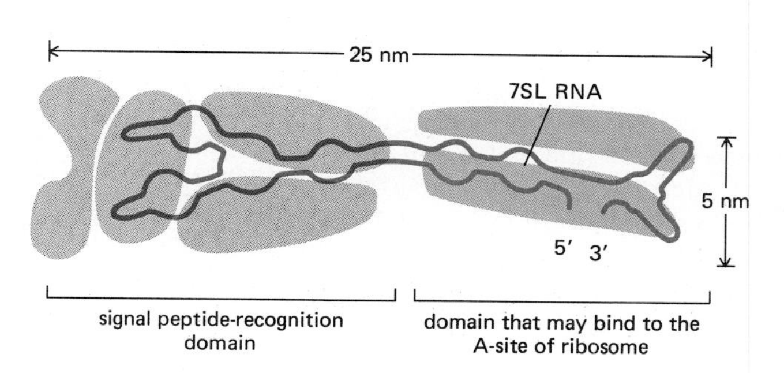

Figure 8–42 A highly schematic drawing of a signal-recognition particle (SRP). It is an elongated complex containing six polypeptide chains (*shaded areas*) and one molecule of 7SL RNA. One end of the particle binds to a ribosome; the other binds to an ER signal peptide on a nascent polypeptide chain. It has been suggested that a portion of the 7SL RNA may fold into a tRNA-like structure that competes at the A-site of the ribosome with incoming aminoacyl tRNAs, thereby causing a pause in translation. (Based on V. Siegel and P. Walter, *Nature* 320:82–84, 1986.)

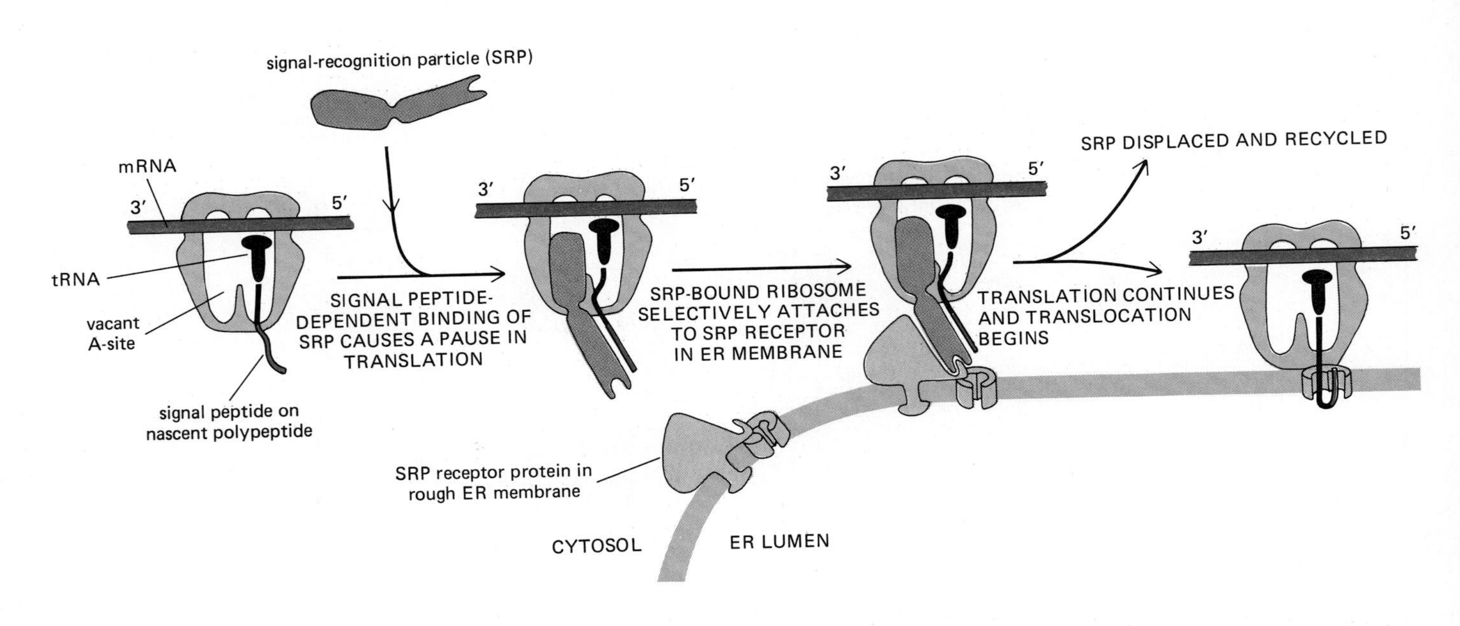

Figure 8–43 The signal-recognition particle and the SRP receptor protein are thought to act in concert to direct a ribosome synthesizing a protein with an ER signal peptide to the ER. The SRP binds to the exposed signal peptide and to the ribosome, probably covering the A-site. Because entry of the next aminoacyl-tRNA is blocked, translation pauses. The SRP receptor in the ER membrane binds the SRP-ribosome complex, and, in a complex reaction that is poorly understood, SRP is displaced and translation continues, with the ribosome now located on the ER membrane. The mechanism that initially inserts the polypeptide chain into the membrane involves a separate transmembrane protein that binds to the signal peptide (a signal-peptide receptor) as well as other protein components involved in translocation that have not been well characterized.

some a chance to engage the ER membrane before the polypeptide is completed, thereby avoiding the inappropriate release of the protein into the cytosol.

The SRP is an elongated molecule consisting of two groups of proteins bound to an RNA scaffold (see Figure 8–42). In one possible model the SRP straddles the ribosome, binding both to the signal peptide as it emerges from the large ribosomal subunit and to the ribosomal site for aminoacyl tRNAs (see p. 212), thus halting translation by preventing the next aminoacyl tRNA from entering the ribosome (Figure 8–43).

Translational arrest is lifted when the ribosome-carrying SRP binds to the **SRP receptor,** which is exposed on the cytosolic surface of the rough ER membrane. The SRP receptor, like the SRP, was initially identified as a component needed to reconstitute *in vitro* protein translocation into the ER. It is a two-chain integral membrane protein that interacts with SRP-bound ribosomes in such a way that the SRP is displaced and translation resumes. Simultaneously the ribosome becomes bound to the ER membrane, and its growing polypeptide chain is transferred to a poorly understood translocation apparatus in the membrane that includes a second signal-peptide-receptor protein distinct from the SRP (see Figure 8–43). This mechanism ensures that a ribosome that begins synthesizing a protein with an ER signal peptide will bind to the ER membrane and begin protein translocation across it.

8-26 Translocation Across the ER Does Not Always Require Ongoing Polypeptide Chain Elongation[38]

As we have seen, translocation of proteins into mitochondria, chloroplasts, and peroxisomes occurs *post-translationally,* after the protein is completed and released into the cytosol, whereas translocation across the ER membrane usually occurs during translation (*co-translationally*). This explains why ribosomes are bound to the ER membrane but not to the cytoplasmic surface of the other organelles. For many years it was thought that the ribosomes of the rough ER might use the energy released during protein synthesis to "eject" their growing polypeptide chains through the ER membrane. Recent studies of ER translocation *in vitro,* however, have shown that selected protein precursors can be imported into the ER after their synthesis has been completed. The import requires ATP hydrolysis but not ongoing protein synthesis (Figure 8–44). As for mitochondrial protein import (see p. 428), the ATP hydrolysis is thought to be required to unfold the protein as it passes through the membrane, and both genetic and biochemical experiments in yeasts indicate that a subclass of hsp70 stress-response proteins is required (see p. 420).

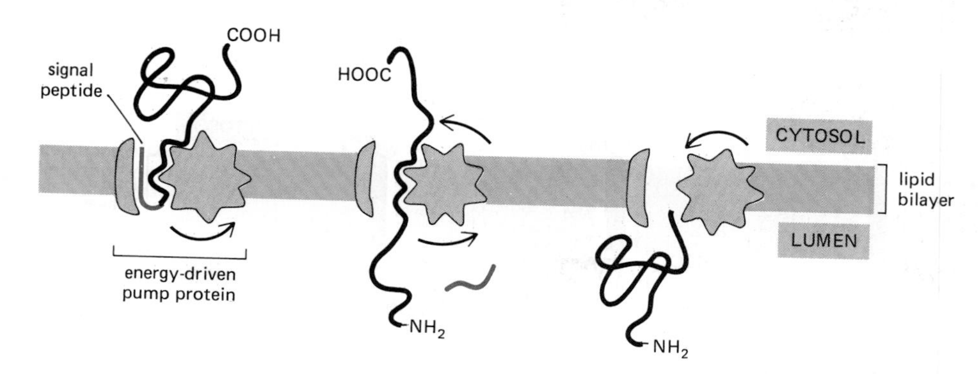

Figure 8–44 One view of the translocation of a protein across a membrane. After a receptor recognizes some special feature of an amino-terminal peptide, an energy-driven protein pump is activated that forces the entire protein through the membrane; in the process the polypeptide chain is transiently unfolded. An alternative possibility is that the unfolding of the protein is catalyzed by an ATP-dependent process on the cytosolic side of the membrane and that only the free energy of refolding on the luminal side drives the protein through the membrane.

Most ER precursor proteins in mammalian cells, however, cannot be imported into the ER once their synthesis has progressed beyond a certain point. It seems that these proteins fold up in a way that either masks the signal peptide or makes it impossible for the ER translocation machinery to unfold the protein. Co-translational import may have allowed these proteins to evolve without the folding constraints that presumably exist for proteins imported into other organelles.

8-27 Combinations of Start- and Stop-Transfer Peptides Can Determine the Many Different Topologies Observed for Transmembrane Proteins[39]

Although most amino-terminal signal peptides are removed by a specific **signal peptidase** bound to the ER membrane, the signal peptide is not in itself a sufficient cue for the peptidase to act: removal of the signal peptide requires an adjacent cleavage site that is not necessary for translocation; indeed, some proteins have signal peptides well within the polypeptide chain, where they are never cleaved.

Uncleaved signal peptides are thought to be essential to achieve the various modes of membrane insertion found in transmembrane proteins (see p. 284). All modes of insertion can be considered as variants of the sequence of events by which a soluble protein is transferred into the lumen of the ER. According to the current view of this process, the hydrophobic ER signal peptide of a soluble protein, in addition to its other functions, serves as a **start-transfer signal** that remains anchored to the membrane throughout translocation, while the rest of the protein is threaded continuously through the membrane in the form of a large loop (Figure 8–45A). Once the carboxyl terminus of the protein has passed through the membrane, only the signal peptide keeps the protein membrane-bound. Therefore, if this peptide is cleaved off, the protein is released into the ER lumen.

Membrane proteins have more complex requirements because some parts of the polypeptide chain are translocated, whereas others are not. In the simplest case the protein is translocated by exactly the same means as the soluble proteins just described—except that the signal peptide lacks an adjacent cleavage site and so is not removed by the signal peptidase. As a result, the translocated protein remains as a single-pass transmembrane protein anchored to the ER membrane at its amino terminus, where the signal peptide forms a membrane-spanning segment of 20–30 hydrophobic amino acid residues in the form of an α helix (see Figure 8–48A).

A more elaborate mechanism is needed in the case of single-pass transmembrane proteins that have their amino terminus, rather than their carboxyl terminus, on the luminal side of the ER. For these proteins, too, an amino-terminal signal peptide initiates translocation; but now an additional hydrophobic segment in the protein stops the process before the entire polypeptide chain is translocated. In these proteins it is this **stop-transfer peptide** that anchors the protein in the membrane, and the signal (start-transfer) peptide is cleaved off (Figure 8–45B).

Many "multipass" transmembrane proteins are known, in which the polypeptide chain passes repeatedly back and forth across the lipid bilayer (see

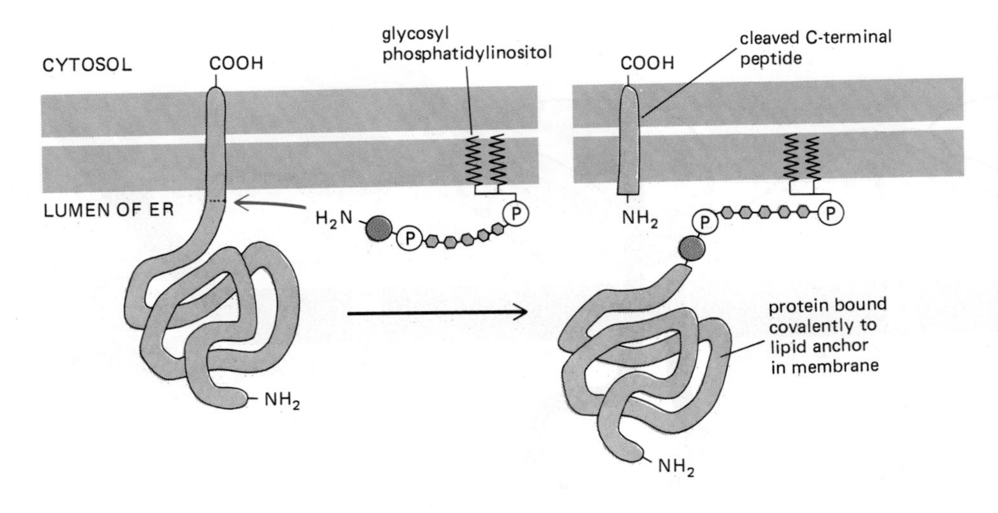

Figure 8–55 The synthesis of proteins linked to membranes by phosphatidylinositol anchors. Immediately after the completion of protein synthesis, the precursor protein remains anchored to the ER membrane only by a hydrophobic carboxyl-terminal sequence of 15 to 20 amino acids, with the rest of the protein in the ER lumen. Within less than a minute, an enzyme in the ER cuts the protein free from its membrane-bound carboxyl terminus while simultaneously attaching the new carboxyl terminus to an amino group on a preassembled glycosyl-phosphatidylinositol intermediate. Because of this covalently linked lipid anchor, the protein remains membrane bound; all of its amino acids are exposed on the luminal side of the ER and will therefore protrude on the cell exterior if the protein is transported to the plasma membrane. The exact structure of the glycolipid head group is not known.

its 22 five-carbon units can span the thickness of a lipid bilayer more than three times, so that the attached oligosaccharide is firmly anchored in the membrane.

All of the diversity of the *N*-linked oligosaccharide structures on mature glycoproteins results from extensive modification of the original precursor structure. While still in the ER, three glucose residues and one mannose residue are quickly removed from the oligosaccharides of most glycoproteins (see Figure 8–63). This oligosaccharide "trimming" or "processing" continues in the Golgi apparatus and is discussed on page 452.

The *N*-linked oligosaccharides are by far the most common ones found in glycoproteins. Less frequently, oligosaccharides are linked to the hydroxyl group on the side chain of a serine, threonine, or hydroxylysine residue. These *O-linked oligosaccharides* are formed in the Golgi apparatus by pathways that are not yet fully understood (see p. 456).

Some Membrane Proteins Exchange a Carboxyl-Terminal Transmembrane Tail for a Covalently Attached Inositol Phospholipid Shortly After They Enter the ER[44]

As discussed previously, several cytosolic enzymes catalyze the covalent addition of a single fatty acid to selected proteins (see p. 417). It has recently been discovered that a related process is catalyzed by enzymes in the ER: the carboxyl terminus of some plasma membrane proteins is covalently attached to a sugar residue of a glycolipid. This linkage forms in the lumen of the ER by the mechanism illustrated in Figure 8–55, and it adds a glycosylated phosphatidylinositol molecule, which contains two fatty acids, to the protein. An increasing number of plasma membrane proteins have been shown to be modified in this way, including one form of the neural cell adhesion molecule N-CAM and the major coat protein of a trypanosome (see p. 830). Since these proteins are attached to the exterior of the plasma membrane only by this means, in principle they could be released from cells in soluble form in response to signals that activate a specific phospholipase in the plasma membrane (see p. 702); so far such release has not been demonstrated.

Most Membrane Lipid Bilayers Are Assembled in the ER[45]

The ER membrane produces nearly all of the lipids required for the elaboration of new cellular membranes, including both phospholipids and cholesterol. The major phospholipid made is *phosphatidylcholine* (also called lecithin), which can be formed in three steps from two fatty acids, glycerol phosphate, and choline (Figure 8–56). Each step is catalyzed by enzymes in the ER membrane that have their active sites facing the cytosol, where all of the required metabolites are found.

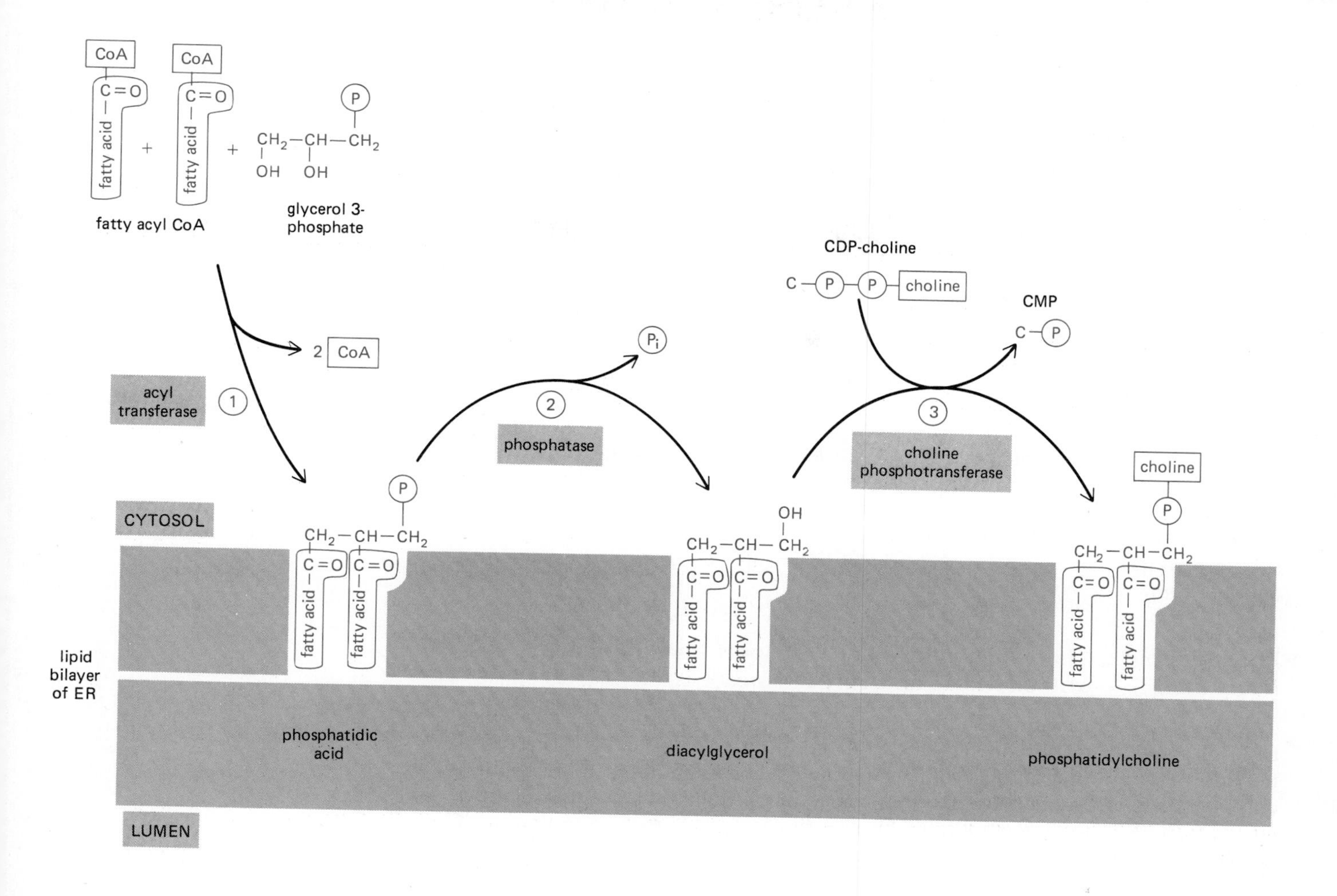

Figure 8–56 The synthesis of phospholipids occurs on the cytosolic side of the ER membrane. Each enzyme in the pathway is an integral membrane protein of the ER with its active site facing the cytosol, where all the intermediates required for phospholipid assembly are located. In the pathway illustrated here, *phosphatidylcholine* is synthesized from fatty acyl-coenzyme A, glycerol 3-phosphate, and cytidine-diphosphocholine (CDP-choline).

In the first step, acyl transferases successively add two fatty acids to glycerol phosphate to produce phosphatidic acid, a compound sufficiently water-insoluble to remain in the lipid bilayer after it has been synthesized. It is this step that enlarges the lipid bilayer. The later steps determine the head group of a newly formed lipid molecule, and therefore the chemical nature of the bilayer, but do not result in net membrane growth (see Figure 8–56). The major membrane phospholipids—phosphatidylcholine (PC), phosphatidylethanolamine (PE), phosphatidylserine (PS), and phosphatidylinositol (PI)—are all synthesized in this way.

The initial formation of phosphatidic acid and its subsequent modifications to form the different types of phospholipid molecules all take place in the cytosolic half of the ER lipid bilayer. This process might eventually turn the lipid bilayer into a monolayer if it were not for a mechanism that transfers some of the newly formed phospholipid molecules to the other half of the ER bilayer. In synthetic lipid bilayers, lipids do not "flip-flop" in this way (see p. 278). In the ER, however, phospholipids equilibrate across the membrane within minutes, which is almost 100,000 times faster than can be accounted for by spontaneous "flip-flop." This rapid transbilayer movement is thought to be mediated by *phospholipid translocators* that are head-group-specific. In particular, the ER membrane seems to contain a translocator (a "*flippase*") that transfers choline-containing phospholipids—but not ethanolamine-, serine-, or inositol-containing phospholipids—between cytoplasmic and luminal faces. This means that PC reaches the luminal face much more readily than PE, PS, or PI. The translocator maintains the bilayer and is responsible for the asymmetric distribution of the lipids in it (Figure 8–57).

The ER also produces cholesterol and ceramide. The ceramide is exported to the Golgi apparatus, where it serves as the precursor for the synthesis of two types of lipids: oligosaccharide chains are added to form *glycosphingolipids* (see p. 282),

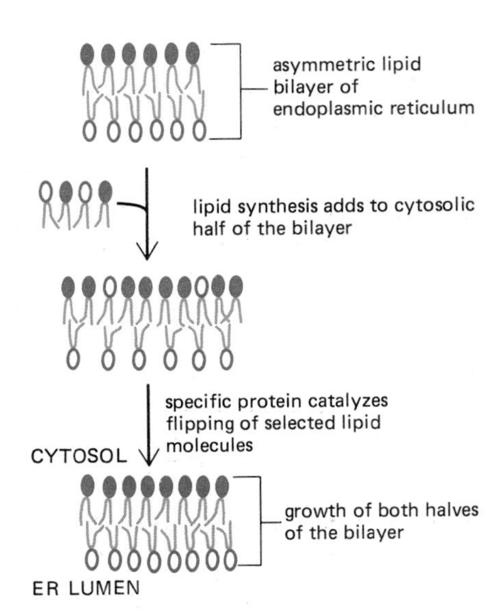

Figure 8–57 The growth of both halves of the ER lipid bilayer requires the protein-catalyzed "flipping" of phospholipid molecules from one half of the ER bilayer to the other. Since new lipid molecules are added only to the cytosolic half and lipid molecules do not flip spontaneously from one monolayer to the other, membrane-bound phospholipid translocator proteins ("flippases") are required to transfer selected lipid molecules to the luminal half so that the membrane grows as a bilayer. Because these proteins preferentially recognize and transfer only certain species of lipid, an asymmetric bilayer is generated in the ER. In particular, the luminal monolayer (which produces the outer half of the plasma membrane bilayer) becomes highly enriched for phosphatidylcholine.

and phosphocholine head groups are transferred from phosphatidylcholine to other ceramide molecules to form *sphingomyelin* (see p. 280). Thus both glycolipids and sphingomyelin are produced relatively late in the process of membrane synthesis. Because they are produced by enzymes exposed to the Golgi lumen, they are found exclusively in the noncytosolic half of the lipid bilayers that contain them.

8-29 Phospholipid Exchange Proteins May Transport Phospholipids from the ER to Mitochondria and Peroxisomes[46]

The plasma membrane and the membranes of the Golgi apparatus and lysosomes all form part of a membrane system that communicates with the ER by means of transport vesicles that supply both their proteins and their lipids. Mitochondria and peroxisomes do not belong to this system and require different mechanisms for the import of proteins and lipids for growth. We have already seen that most of the proteins in these organelles are imported post-translationally from the cytosol. Although mitochondria modify some of the lipids they import (see p. 398), their lipids have to be transferred from the ER, where they are synthesized, or obtained less directly from the ER by means of other cellular membranes.

Water-soluble carrier proteins—called *phospholipid exchange proteins* (or *phospholipid transfer proteins*)—have been shown in *in vitro* experiments to have the ability to transfer individual phospholipid molecules between membranes. Each exchange protein recognizes only specific types of phospholipids. Transfer between membrane bilayers is achieved when the protein "extracts" a molecule of phospholipid from a membrane and diffuses away with the lipid buried within its binding site. When it encounters another membrane, the exchange protein tends to discharge the bound phospholipid molecule into the new lipid bilayer (Figure 8–58). It has been proposed that phosphatidylserine is imported into mitochondria in this way and then decarboxylated to yield phosphatidylethanolamine, while phosphatidylcholine is imported intact.

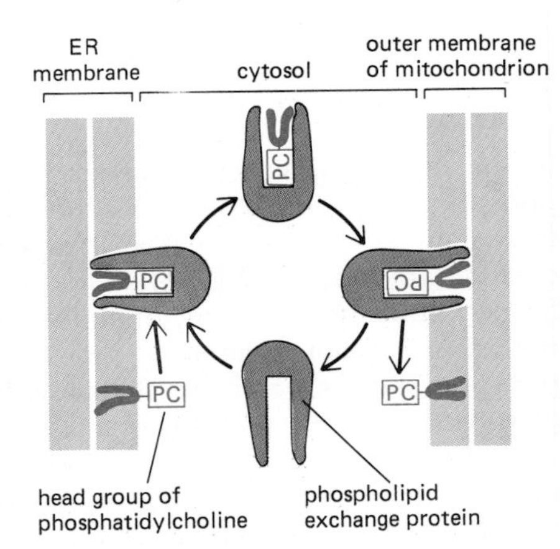

Figure 8–58 Soluble *phospholipid exchange proteins* can redistribute phospholipids between membrane-bounded compartments. Phospholipids are insoluble in water, so their passage between membranes requires a carrier protein. The exchange proteins carry a single molecule of phospholipid at a time and can pick up a lipid molecule from one membrane and release it at another. The transfer of phosphatidylcholine (PC) from ER to mitochondria can in principle occur spontaneously because the concentration of PC is high in the ER membrane (where it is made) and low in the mitochondrial outer membrane.

Exchange proteins act to distribute phospholipids at random among all membranes present. In principle, such a random exchange process can result in a net transport of lipids from a lipid-rich to a lipid-poor membrane, allowing phosphatidylcholine and phosphatidylserine molecules to be transferred from the ER, where they are synthesized, to a mitochondrial or peroxisomal membrane. It may be that mitochondria and peroxisomes are the only "lipid-poor" organelles in the cytoplasm and that such an exchange process is sufficient, although other more specific mechanisms may exist for transporting phospholipids to these organelles.

Summary

The ER serves as a factory for the production of the protein and lipid components of many organelles. Its extensive membrane contains numerous biosynthetic enzymes, including those responsible for almost all of the cell's lipid synthesis and for the addition of an N-linked oligosaccharide to many proteins on the luminal side of the ER. Newly synthesized proteins destined for secretion, as well as those destined for the ER itself and for the Golgi apparatus, the lysosomes, and the plasma membrane, must first be imported into the ER from the cytosol. Only proteins that carry a special hydrophobic signal peptide are imported into the ER. The signal peptide is recognized by a signal-recognition particle (SRP), which binds the nascent polypeptide chain and the ribosome and directs them to a receptor protein on the surface of the ER membrane. This binding to the membrane initiates an ATP-dependent translocation process that threads a loop of polypeptide chain across the ER membrane.

Soluble proteins destined for the ER lumen, for secretion, or for transfer to the lumen of other organelles pass completely into the ER lumen. Transmembrane proteins destined for the ER or for other cell membranes are translocated across the ER membrane but are not released into the lumen; instead, they remain anchored in the bilayer by one or more membrane-spanning α-helical regions in their polypeptide chain. These hydrophobic portions of the protein can act either as start-transfer or stop-transfer peptides during the translocation process. When a polypeptide contains multiple alternating start-transfer and stop-transfer peptides, it will pass back and forth across the bilayer many times.

The asymmetry of lipid synthesis, protein insertion, and glycosylation in the ER establishes the polarity of the membranes of all of the other organelles that the ER supplies with lipids and membrane proteins.

8-32 The Golgi Apparatus[47]

The **Golgi apparatus** (also called the *Golgi complex*) is usually located near the cell nucleus, and in animal cells it is often close to the centrosome, or cell center. It contains a collection of flattened membrane-bounded *cisternae* resembling a stack of plates. These **Golgi stacks** (called *dictyosomes* in plants) usually consist of four to six cisternae, each typically about 1 μm in diameter (Figure 8–59). The number of Golgi stacks per cell varies greatly depending on the cell type: some cells contain one large one, while others contain hundreds of very small ones.

Swarms of small vesicles (~50 nm in diameter) are associated with the Golgi stacks, clustered on the side abutting the ER and along the dilated rims of each cisterna (see Figure 8–59). These *Golgi vesicles* are thought to transport proteins and lipids to and from the Golgi and between the Golgi cisternae. Many vesicles

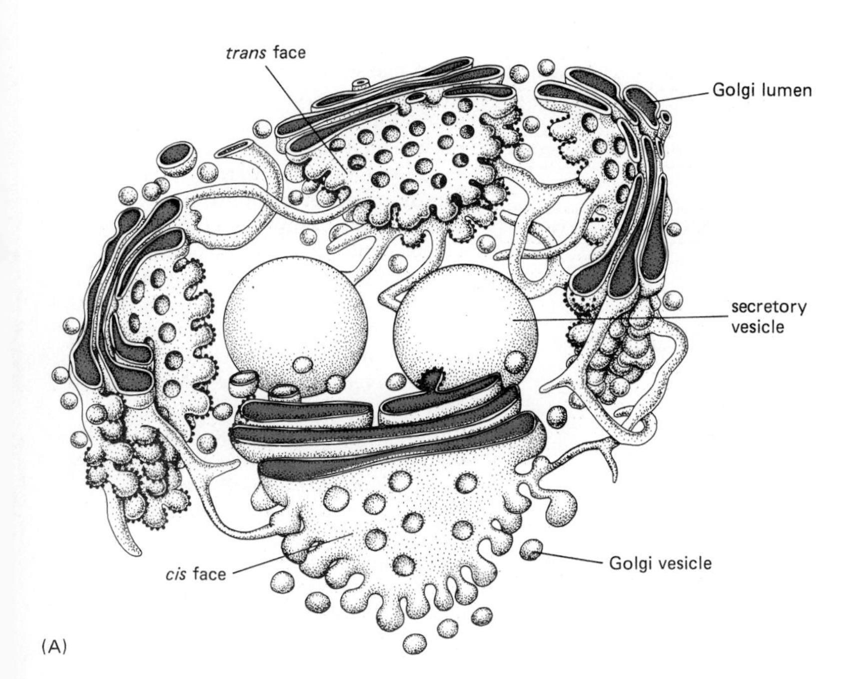

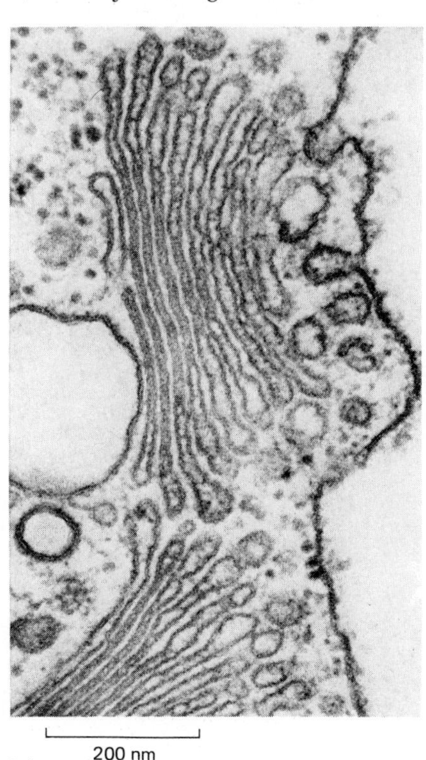

Figure 8–59 (A) Three-dimensional drawing of a Golgi apparatus, derived from electron micrographs of a secretory animal cell. The stacks of flattened Golgi cisternae have dilated edges from which small vesicles appear to be budding. The large secretory vesicles form by budding from the *trans* Golgi. (B) Electron micrograph of a Golgi apparatus in a plant cell seen in cross-section (the green alga *Chlamydomonas*). In plant cells the Golgi apparatus is generally more distinct and more clearly separated from other intracellular membranes than in animal cells. (A, after R.V. Krstić, Ultrastructure of the Mammalian Cell. New York: Springer-Verlag, 1979; B, courtesy of George Palade.)

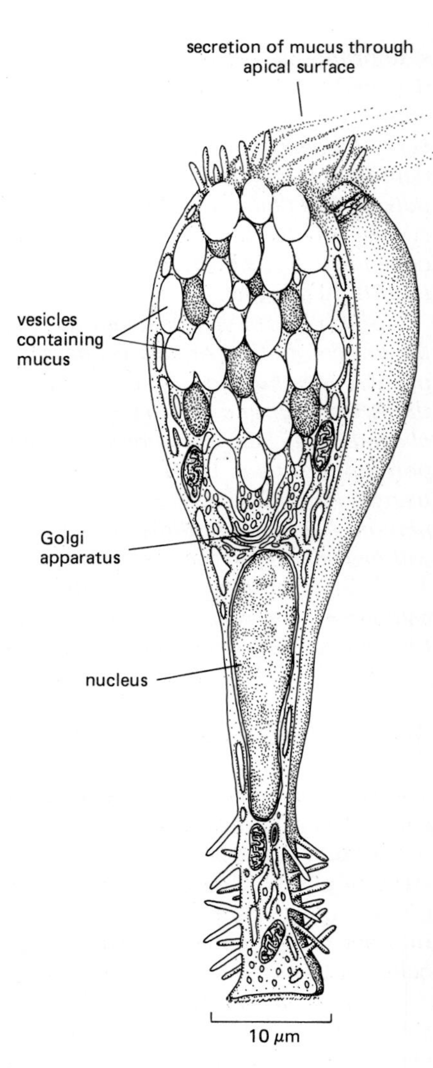

Figure 8–60 A goblet cell of the small intestine. This cell is specialized for secreting mucus, which is a mixture of glycoproteins and proteoglycans synthesized in the ER and Golgi apparatus. The Golgi apparatus is highly polarized, which facilitates the discharge of mucus by exocytosis at the apical surface. (After R.V. Krstić, Illustrated Encyclopedia of Human Histology. New York: Springer-Verlag, 1984.)

are *coated,* some by clathrin and others by other types of coat protein (see p. 472). Such **coated vesicles** are often seen budding from the Golgi cisternae.

The Golgi stack has two distinct faces: a ***cis* face** (or entry face) and a ***trans* face** (or exit face). The *cis* face is closely associated with the *transitional elements* of the ER (see p. 411); the *trans* face is distended into a tubular reticulum called the *trans Golgi network* (TGN). Proteins and lipids enter a Golgi stack in small vesicles from the ER on the *cis* side and exit for various destinations in vesicles that form on the *trans* side; these transported molecules undergo an ordered series of modifications as they pass from one Golgi cisterna to another.

The Golgi apparatus is prominent in cells that are specialized for secretion, such as the goblet cells of the intestinal epithelium, which secrete large amounts of mucus into the gut. In such cells, unusually large vesicles form from the *trans* side of the Golgi apparatus, which faces the plasma membrane domain where secretion occurs (Figure 8–60).

8-33 Oligosaccharide Chains Are Processed in the Golgi Apparatus[48]

As described previously, a single species of ***N*-linked oligosaccharide** is attached to many proteins in the ER, and this oligosaccharide is extensively trimmed while the protein is still in the ER (see p. 446). Further modifications occur in the Golgi apparatus.

Two broad classes of *N*-linked oligosaccharides, the *complex oligosaccharides* and the *high-mannose oligosaccharides,* are found in mature glycoproteins (Figure

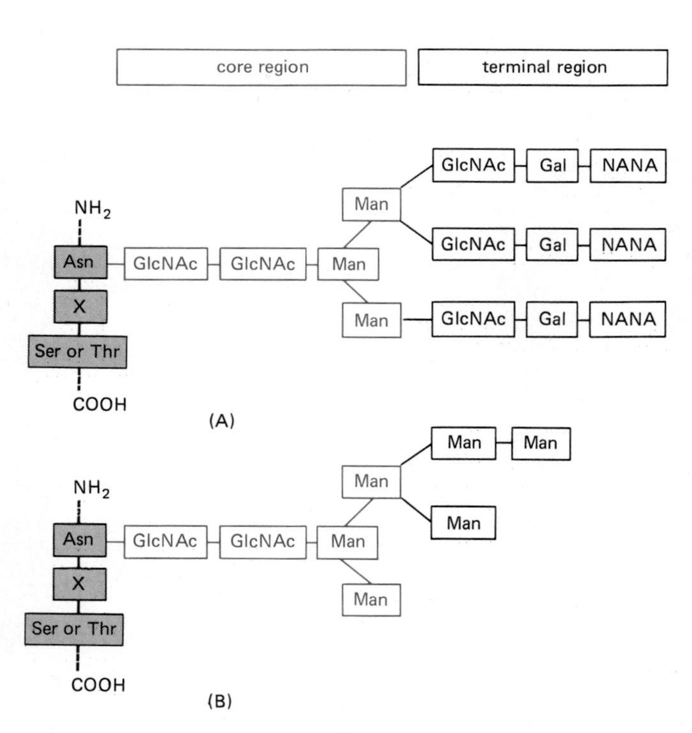

Figure 8–61 Examples of the two main classes of asparagine-linked (*N*-linked) oligosaccharides found in mature glycoproteins: a "complex" oligosaccharide (A) and a "high-mannose" oligosaccharide (B). Many variations occur on these themes. For example, while the complex oligosaccharide shown has three terminal branches, two and four branches are also common, depending on the glycoprotein and the cell in which it is made. "Hybrid" oligosaccharides with one Man branch and one GlcNAc-Gal branch are also found. The three colored amino acids constitute the sequence recognized by the enzyme that adds the initial oligosaccharide to the protein. Abbreviations: Asn, asparagine; Man, mannose; GlcNAc, *N*-acetylglucosamine; NANA, *N*-acetylneuraminic acid (sialic acid); Gal, galactose; X, any amino acid.

8–61). Sometimes both types are attached (in different places) to the same polypeptide chain. **High-mannose oligosaccharides** have no new sugars added to them in the Golgi apparatus. They contain just two *N*-acetylglucosamines and many mannose residues, often approaching the number originally present in the lipid-linked oligosaccharide precursor added in the ER. **Complex oligosaccharides,** by contrast, can contain more than the original two *N*-acetylglucosamines as well as a variable number of galactose and sialic acid residues and, in some cases, fucose. Sialic acid is of special importance because it is the only sugar residue of glycoproteins that bears a net negative charge (see p. 283).

The complex oligosaccharides are generated by a combination of further trimming of the original oligosaccharide added in the ER and the addition of further sugars. Thus each complex oligosaccharide consists of a *core region,* derived from the original *N*-linked oligosaccharide and typically containing two *N*-acetylglucosamine and three mannose residues (shown in color in Figure 8–61), plus a *terminal region* consisting of a variable number of *N*-acetylglucosamine–galactose–sialic acid trisaccharide units linked to the core mannose residues. Frequently the terminal region is truncated, containing only *N*-acetylglucosamine and galactose, or even just *N*-acetylglucosamine. In addition, a fucose residue may or may not be added, usually to the core *N*-acetylglucosamine residue attached to the asparagine. All of the terminal-region sugars are added in the *trans* Golgi by a series of glycosyl transferases that act in a rigidly determined sequence (Figure 8–62). The substrates are specific sugars activated by linkage to a nucleotide. These are transported into the lumen of the Golgi from the cytosol by a set of membrane-bound carriers that also export the nucleotide by-products of the glycosylation. One of the glycosyl transferases (*galactosyl transferase*) is commonly used as a marker to identify membrane vesicles derived from the Golgi after they have been purified by differential centrifugation of tissue homogenates (see p. 163).

The processing that generates complex oligosaccharide chains follows the highly ordered pathway shown in Figure 8–63. A number of drugs and antibiotics that inhibit specific steps have helped in defining the details of this pathway (Table 8–4). Whether a given oligosaccharide remains high-mannose or is processed is determined largely by its configuration on the protein to which it is attached: if the oligosaccharide is sterically accessible to the processing enzymes in the Golgi, it is likely to be converted to a complex form; if it is inaccessible, it is likely to remain in a high-mannose form.

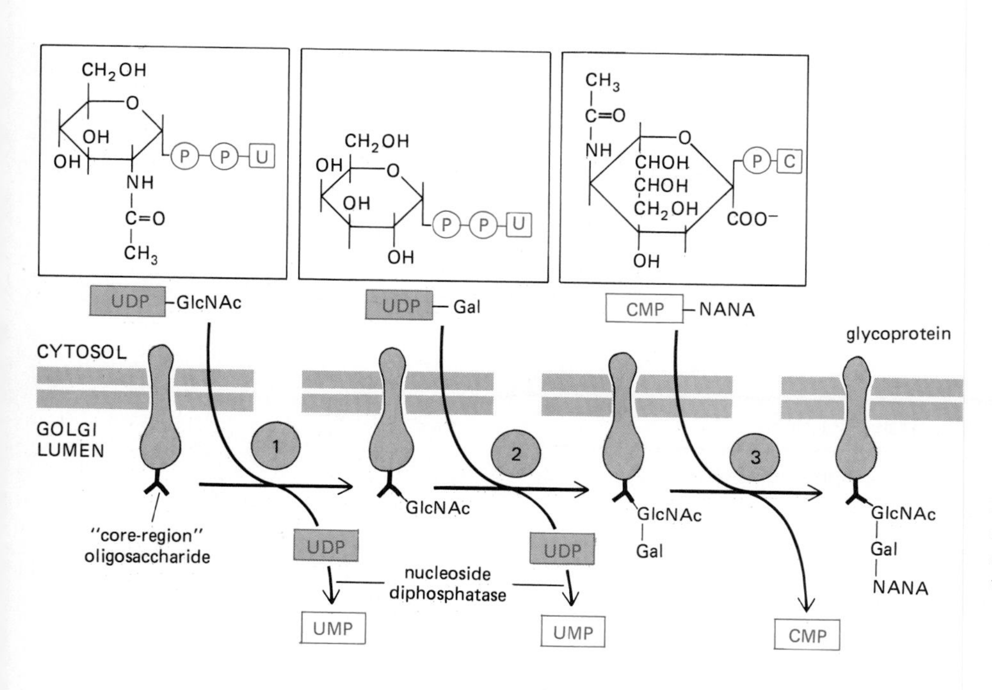

Figure 8–62 The stepwise addition of sugar residues that occurs to finish the synthesis of complex oligosaccharides in the cisternal compartments of the Golgi apparatus. Three types of glycosyl transferase enzymes act sequentially, using sugar substrates that have been activated by linkage to the indicated nucleotide. The membranes of the Golgi cisternae contain specific transmembrane carrier proteins (see p. 302) that allow each sugar nucleotide to enter in exchange for its nucleoside monophosphate product, thus permitting glycosylations to occur on the luminal face. The structures of the sugar nucleotides UDP-*N*-acetylglucosamine, UDP-galactose, and CMP-*N*-acetylneuraminic acid are shown at the top. For abbreviations used, see legend for Figure 8–61.

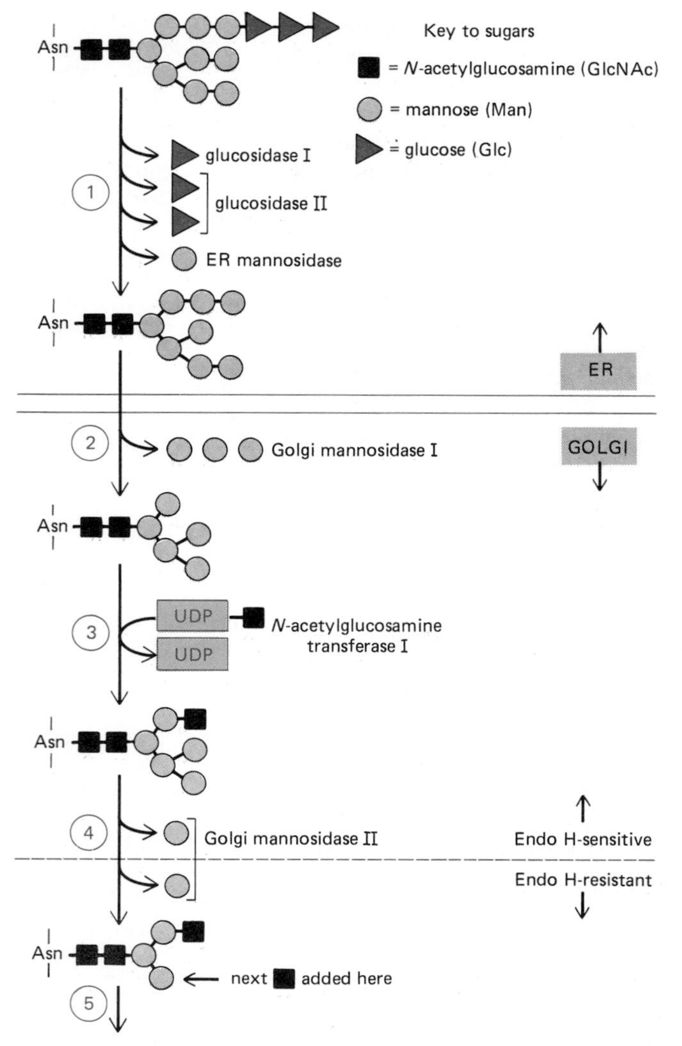

Figure 8–63 The oligosaccharide processing that takes place in the ER and the Golgi apparatus. The pathway is highly ordered so that each step shown is dependent on the previous reaction in the series. Processing begins in the ER with the removal of the glucose residues from the oligosaccharide initially transferred to the protein. All three glucose residues can be removed even before protein synthesis is completed. Then a mannosidase in the ER membrane removes a particular mannose residue. In the Golgi stack, mannosidase I removes three more mannose residues and *N*-acetylglucosamine transferase I adds a residue of GlcNAc, which enables mannosidase II to remove two additional mannose residues. This yields the final core of three mannose residues that is present in a "complex" type of oligosaccharide. At this stage the bond between the two GlcNAc residues in the core becomes resistant to attack by a highly specific endoglycosidase (*Endo H*). Since all later structures in the pathway are also Endo H-resistant, treatment with this enzyme is widely used to distinguish complex from high-mannose oligosaccharides. Finally, additional GlcNAc, galactose, and sialic acid residues are added. The extent of processing depends on the protein and on the location of the asparagine residue within the protein to which the oligosaccharide is attached. Some oligosaccharides will escape processing in the Golgi apparatus, whereas others will follow the pathway shown to varying extents. For abbreviations used, see legend for Figure 8–61.

The Carbohydrate in Cell Membranes Faces the Side of the Membrane That Is Topologically Equivalent to the Outside of the Cell

Because all oligosaccharide chains are added on the luminal side of the ER and Golgi apparatus (see p. 447), the distribution of carbohydrate on membrane proteins and lipids is asymmetrical. As with the asymmetry of the lipid bilayer itself, the asymmetric orientation of these glycosylated molecules is maintained during transport to the plasma membrane, secretory vesicles, or lysosomes. As a result, the oligosaccharides of all the glycoproteins and glycolipids in the corresponding intracellular membranes face the luminal side, while those in the plasma membrane face the outside of the cell (Figure 8–64).

Table 8–4 Drugs That Inhibit Steps in *N*-linked Glycosylation

Drug(s)	**Step(s) Inhibited**
Tunicamycin	dolichol-P → dol-P-P-GlcNAc
Castanospermine and N-Methyl deoxynojirimycin	glucose$_3$-Man$_9$-GlcNAc$_2$-Asn → glucose$_2$-Man$_9$-GlcNAc$_2$-Asn
Bromoconduritol	glucose$_2$-Man$_9$-GlcNAc$_2$-Asn → Man$_9$-GlcNAc$_2$-Asn
Deoxymannonojirimycin	Man$_8$-GlcNAc$_2$-Asn → Man$_5$-GlcNAc$_2$-Asn
Swainsonine	GlcNAc-Man$_5$-GlcNAc$_2$-Asn → GlcNAc-Man$_3$-GlcNAc$_2$-Asn

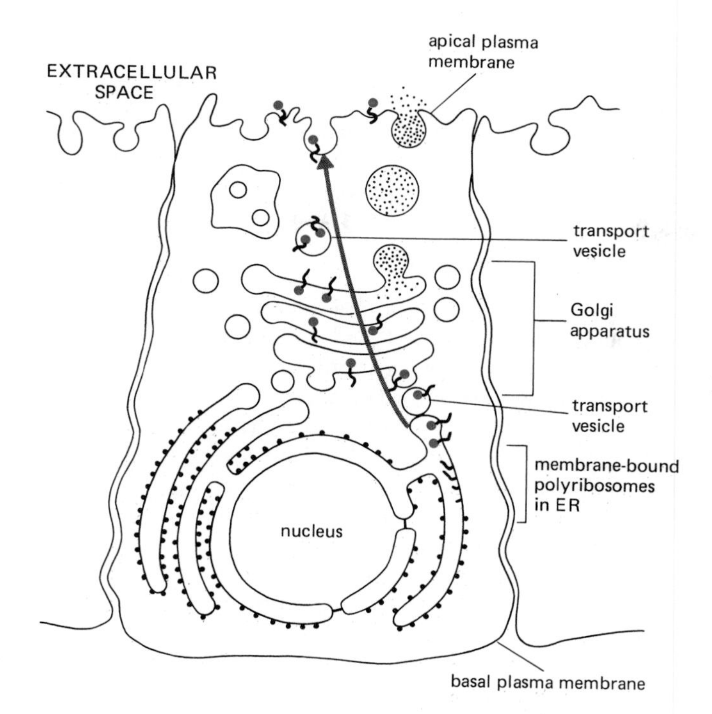

Figure 8–64 The orientation of a transmembrane protein in the ER membrane is preserved when that protein is transported to other membranes. The colored ball on the end of each glycoprotein molecule represents the *N*-linked oligosaccharide that is added to proteins in the ER lumen. Note that these sugar residues are confined to the lumen of each of the internal organelles and become exposed to the extracellular space after a transport vesicle fuses with the plasma membrane.

What Is the Purpose of *N*-linked Glycosylation?[49]

There is an important difference between the construction of an oligosaccharide and the synthesis of other macromolecules such as DNA, RNA, and protein. Whereas nucleic acids and proteins are copied from a template in a repeated series of identical steps using the same enzyme(s), complex carbohydrates require a different enzyme at each step, each product being recognized as the exclusive substrate for the next enzyme in the series. Given the complicated pathways that have evolved to synthesize them, it seems likely that the oligosaccharides on glycolipids and glycoproteins have important functions, but for the most part these functions are not known.

N-linked glycosylation, for example, is prevalent in all eucaryotes, including yeasts, but is absent from eubacteria. Because one or more *N*-linked oligosaccharides are present on most proteins transported through the ER and Golgi apparatus—a process that is unique to eucaryotic cells—it was once thought that their function was to aid this transport process. Drugs that block steps in glycosylation (Table 8–4), however, do not generally interfere with transport (with the important exception of transport to lysosomes, which will be discussed below—see p. 459), and mutant cells in culture that are blocked in various glycosylation steps in the Golgi are nevertheless viable and transport proteins normally. Although some proteins do not fold correctly without their normal oligosaccharide and therefore precipitate in the ER and fail to be transported, most proteins retain their normal activities in the absence of glycosylation.

Because chains of sugars have limited flexibility, even a small *N*-linked oligosaccharide protrudes from the surface of a glycoprotein (Figure 8–65) and can

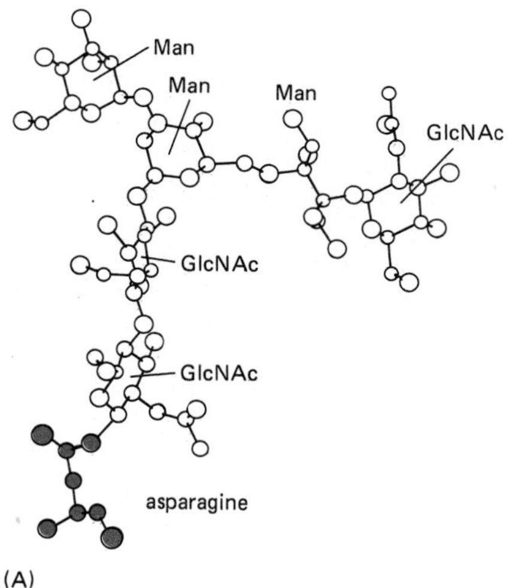

Figure 8–65 The three-dimensional structure of a small *N*-linked oligosaccharide, as determined by x-ray crystallographic analysis of a glycoprotein. This oligosaccharide contains only 6 sugar residues, whereas there are 14 sugar residues in the *N*-linked oligosaccharide that is initially transferred to proteins in the ER (see Figure 8–52). (A) Backbone model showing all atoms except hydrogens; (B) space-filling model, with the asparagine shown as dark atoms. (Courtesy of Richard Feldmann.)

thus limit the approach of other macromolecules to the surface of the glycoprotein. In this way, for example, the presence of oligosaccharide tends to make a glycoprotein relatively resistant to protease digestion. It may be that the oligosaccharides originally provided an ancestral eucaryotic cell with a protective coat that, unlike the rigid bacterial cell wall, allowed the cell freedom to change shape and move. They may have since become modified to serve other purposes as well.

Proteoglycans Are Assembled in the Golgi Apparatus[50]

It is not only the *N*-linked oligosaccharide chains on proteins that are altered as the proteins pass through the Golgi cisternae en route from the ER to their final destinations; many proteins are also modified in other ways. As mentioned earlier, for example, some proteins have sugars added to selected serine or threonine side chains. This ***O*-linked glycosylation,** like the extension of *N*-linked oligosaccharide chains, is catalyzed by a series of glycosyl transferase enzymes that use the sugar nucleotides in the Golgi lumen to add one sugar residue at a time to a protein. Usually *N*-acetylgalactosamine is added first, followed by a variable number of additional sugar residues, ranging from just a few to 10 or more.

The most heavily glycosylated proteins of all are some *proteoglycan core proteins,* which are modified in the Golgi apparatus to produce *proteoglycans.* As discussed in Chapter 14 (see p. 806), this involves the polymerization of one or more glycosaminoglycan chains (long unbranched polymers composed of repeating disaccharide units) to serines on the core protein, with xylose rather than *N*-acetylgalactosamine added first. Many proteoglycans are secreted as components of the extracellular matrix, while others remain anchored to the plasma membrane as integral membrane proteoglycans (see p. 808). In addition, the mucus that is secreted to form a protective coating over many epithelia consists of a concentrated mixture of proteoglycans and heavily glycosylated glycoproteins (see, for example, Figure 8–60).

The sugars incorporated into glycosaminoglycans are heavily sulfated immediately after these polymers are made in the Golgi apparatus, which helps to give proteoglycans their high negative charge. The sulfate is added from the activated sulfate donor 3′-phosphoadenosine–5′-phosphosulfate (PAPS), which is transported from the cytosol to the lumen of a late Golgi compartment. A more subtle protein modification carried out in the Golgi is the addition of sulfate from PAPS to the hydroxyl group of selected tyrosine residues in proteins. Sulfated tyrosines are frequently found in secreted proteins and occasionally in the extracellular domains of plasma membrane proteins.

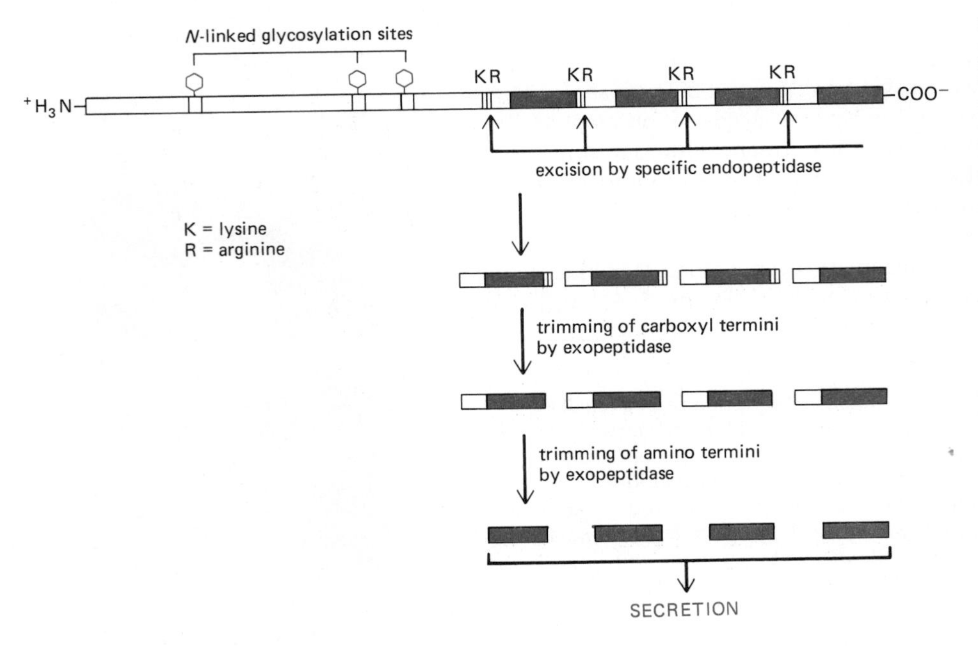

Figure 8–66 An example of a *polyprotein* that is cleaved to produce multiple copies of the same peptide signaling molecule. As indicated, the processing generally begins with cleavages at pairs of basic amino acids (here Lys-Arg pairs), which are catalyzed by a specific membrane-bound protease located in secretory vesicles or possibly in the *trans* Golgi network. Shown here is the processing pathway that produces the 13-amino acid α-factor in the yeast *Saccharomyces cerevisiae,* a secreted peptide that controls the mating behavior of this single-celled eucaryote. (After R. Fuller, A. Brake, and J. Thorner, in Microbiology 1986 [L. Lieve, ed.], pp. 273–278. Washington, D.C.: American Society for Microbiology, 1986.)

Proteins Are Often Proteolytically Processed During the Formation of Secretory Vesicles[51]

The most drastic of the many modifications that can occur to proteins before they are secreted are made last. Many polypeptide hormones and neuropeptides are synthesized as inactive protein precursors from which the active molecules are liberated by proteolysis. These cleavages are thought to begin in the *trans Golgi network,* and they continue in the secretory vesicles that form from this network to store these proteins. The initial cleavages are made by membrane-bound proteases that cut next to pairs of basic amino acid residues (Lys-Arg, Lys-Lys, Arg-Lys, or Arg-Arg pairs), and trimming reactions then produce the final secreted product (Figure 8–66). In the simplest case a polypeptide often will have a single amino-terminal *pro-piece* that is cleaved off shortly before secretion to yield the mature protein. These proteins are thus synthesized as *pre-pro-proteins,* the *pre-piece* consisting of the ER signal peptide that is cleaved off in the rough ER (see p. 438). Somewhat more complex are peptide signaling molecules that are made as *polyproteins* containing multiple copies of the same amino acid sequence (see Figure 8–66). Finally, a variety of peptide signaling molecules are synthesized as parts of a polyprotein that acts as a precursor for multiple end products, which are individually cleaved from the initial polypeptide chain. In these cases the same polyprotein can be processed in various ways to produce different peptides in different cell types, thereby increasing the diversity of molecules that can be used for chemical signaling between cells.

Why is this type of delayed proteolytic processing observed for so many polypeptides? Some of the peptides produced in this way, such as the enkephalins (five-amino acid neuropeptides), are undoubtedly too short in their mature forms to be synthesized efficiently by ribosomes, while even much longer peptides might be expected to lack the necessary signals for packaging into secretory vesicles (see p. 465). Moreover, delaying the production of an active protein until it reaches a secretory vesicle has the potential advantage of preventing it from acting inside the cell that synthesizes it.

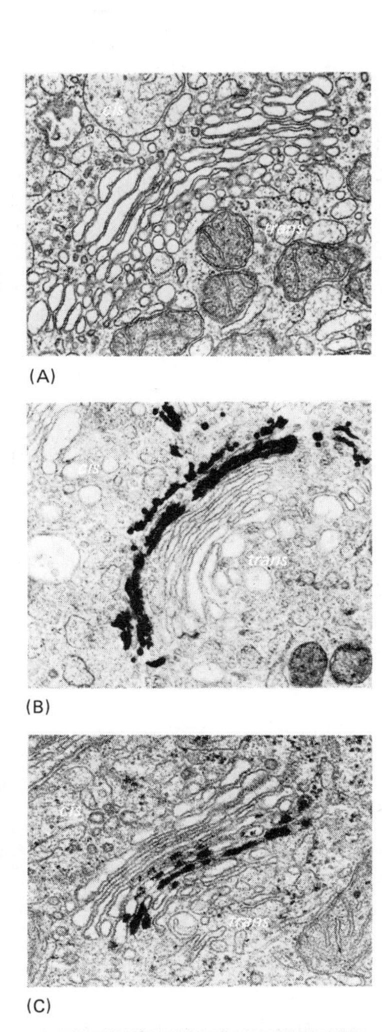

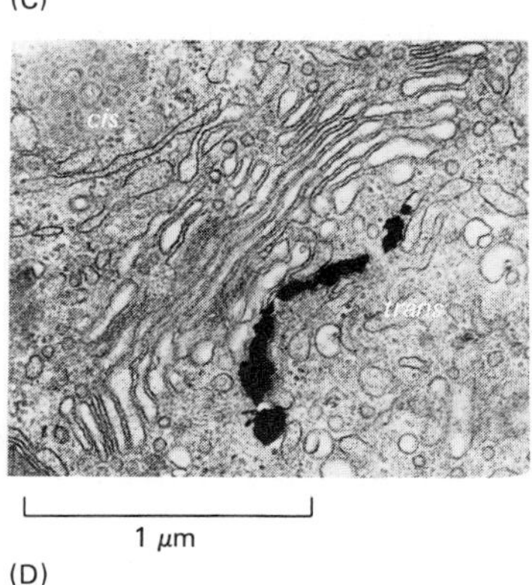

Figure 8–67 Histochemical stains demonstrate that the Golgi apparatus is biochemically polarized. (A) Unstained. (B) Osmium is preferentially reduced by the cisternae of the *cis* compartment. (C) The enzyme nucleoside diphosphatase (see Figure 8–62) is found in the *trans* Golgi cisternae; this enzyme was formerly called "thiamine pyrophosphatase." (D) The enzyme acid phosphatase marks the *trans* Golgi network (formerly referred to as the "Golgi endoplasmic reticulum lysosomes," or "GERL"). (Courtesy of Daniel S. Friend.)

The Golgi Cisternae Are Organized as a Sequential Series of Processing Compartments[52]

The processing pathways just outlined are highly organized in the Golgi stack. Each cisterna is a distinct compartment, with its own set of processing enzymes, so that the stack forms a multistage processing unit. Proteins are modified in successive stages as they move from compartment to compartment across the stack.

Proteins exported from the ER enter the first of the Golgi compartments (the ***cis* compartment**), then move to the next compartment (the ***medial* compartment,** consisting of the central cisternae of the stack), and finally move to the ***trans* compartment** (consisting of the last cisternae), where glycosylation is completed. From the *trans* compartment the proteins move to the ***trans* Golgi network (TGN);** in this tubular reticulum they are segregated into different transport vesicles and dispatched to their final destination—the plasma membrane, lysosomes, or secretory vesicles.

The functional differences between the *cis, medial,* and *trans* subdivisions of the Golgi stack were first discovered by localizing the enzymes involved in processing *N*-linked oligosaccharides in distinct regions of the stack, both by physical fractionation of the organelle and by immunoelectron microscopy. The removal of mannose residues and the addition of *N*-acetylglucosamine, for example, were shown to occur in the *medial* compartment, while the addition of galactose and sialic acid was found to occur in the *trans* compartment (Figures 8–67 and 8–68).

The proteins that enter the Golgi from the ER (except those destined to remain in one of the Golgi compartments) flow together across the stack from *cis* to *medial* to *trans* compartments, permitting their stepwise processing. Although the mechanism of protein and lipid transfer from one cisterna to another is uncertain, coated vesicles are thought to bud from the cisternal rims to carry this traffic from

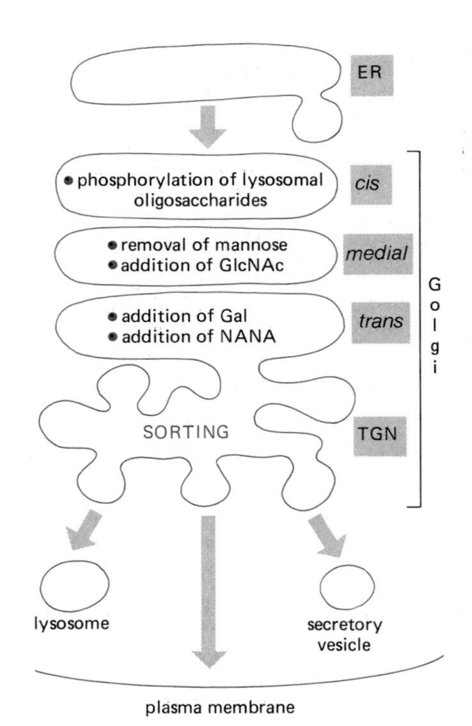

Figure 8–68 The compartmentalization of the Golgi apparatus. The covalent modification of proteins occurs sequentially during their passage through the clustered compartments of the Golgi stack; the *trans* Golgi network (TGN) is a tubular reticulum that acts primarily as a sorting station. The localization of each processing step shown was determined by a combination of techniques including subfractionation of the Golgi apparatus membranes and electron microscopy after staining with antibodies specific to some of the processing enzymes. The locations of many other processing reactions have not yet been determined.

cisterna to cisterna across the stack. Since experiments demonstrate that exported proteins move unidirectionally from *cis* to *medial* to *trans* compartments and never skip an intervening compartment, this scheme requires that the vesicles be programmed to fuse only with the next compartment in the sequence. Although only three functionally distinct cisternal compartments have so far been demonstrated, each of these sometimes consists of a block of two or more cisternae in sequence, and it is possible that there are finer subdivisions still to be discovered. Alternatively, it may be that there are only three fundamental compartments and that the several cisternae within the *cis*, *medial*, and *trans* divisions are simply multiple copies of the same functional unit.

Summary

The Golgi apparatus receives newly synthesized proteins and lipids from the ER and distributes them to the plasma membrane, lysosomes, and secretory vesicles. It is a polarized structure made up of one or more stacks of disc-shaped cisternae surrounded by a swarm of small vesicles. The cisternae are organized as a series of at least three distinct and sequential processing compartments, termed cis, *medial, and* trans *Golgi. Proteins are transferred from the lumen and membrane of the ER to the* cis *face of the Golgi stack by means of transport vesicles. Proteins targeted for secretory vesicles, the plasma membrane, and lysosomes move across the stack in the* cis-*to*-trans *direction, passing from one cisterna to the next in series. Finally, they reach the* trans *Golgi network, from which each type of protein departs for its final destination in a particular type of vesicle.*

The Golgi, unlike the ER, contains many sugar nucleotides; a variety of glycosyl transferase enzymes use these substrates to carry out glycosylation reactions on both lipid and protein molecules as they pass through the Golgi apparatus. N-*linked oligosaccharides on proteins, for example, are often trimmed by removal of mannose residues, and additional sugars—including* N-*acetylglucosamine, galactose, and sialic acid residues—are added. In addition, the Golgi is the site of both* O-*linked glycosylation and the conversion of proteoglycan core proteins to proteoglycans. Sulfation of the sugars in proteoglycans and of selected tyrosines on proteins also occurs in a late Golgi compartment.*

Transport of Proteins from the Golgi Apparatus to Lysosomes

All of the proteins that pass through the Golgi apparatus, except those that are retained there as permanent residents, are thought to be sorted in the *trans* Golgi network according to their final destination. The mechanism of sorting is particularly well understood for those proteins destined for the lumen of lysosomes, and in this section we shall consider this selective transport process. We begin with a brief account of lysosomal structure and function.

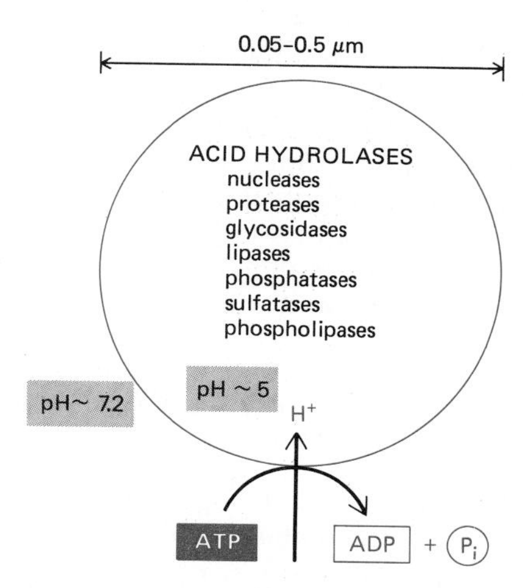

Figure 8–69 Lysosomes are defined as membrane-bounded vesicles involved in intracellular digestion that contain a variety of hydrolytic enzymes that are active under acidic conditions. The lumen is maintained at an acidic pH (around 5) by a H^+ pump in the membrane that uses the energy of ATP hydrolysis to pump H^+ into the vesicle.

Lysosomes Are the Principal Sites of Intracellular Digestion[53]

In the course of studies of enzymes in liver homogenates in 1949, certain irregularities were noticed in the assay of acid phosphatase. The activity of the enzyme, for example, was higher in extracts prepared with distilled water than in extracts prepared with an osmotically balanced sucrose solution. It was also higher in aged than in fresh preparations, and in aged preparations it was no longer associated with sedimentable particles. Similar findings were soon reported for several other hydrolytic enzymes and led to the discovery of a new organelle, named the **lysosome,** a membranous bag of hydrolytic enzymes used for the controlled intracellular digestion of macromolecules. Damage to lysosomal membranes in cell extracts, induced by osmotic lysis or aging, releases the enzymes in a nonsedimentable form.

About 40 hydrolytic enzymes are now known to be contained in lysosomes. They include proteases, nucleases, glycosidases, lipases, phospholipases, phosphatases, and sulfatases. All are **acid hydrolases,** optimally active near the pH 5 maintained within lysosomes. The membrane of the lysosome normally keeps the enzymes out of the cytosol (whose pH is about ~7.2), but the acid dependence of the enzymes protects the contents of the cytosol against damage even if leakage should occur.

Like all other intracellular organelles, the lysosome not only contains a unique collection of enzymes but also has a unique surrounding membrane. The lysosomal membrane, for example, contains transport proteins that allow the final products of the digestion of macromolecules to escape so that they can be either excreted or reutilized by the cell. It also contains a H^+ pump that utilizes the energy of ATP hydrolysis to pump H^+ into the lysosome, thereby maintaining the lumen at its pH of about 5 (Figure 8–69). Most of the lysosomal membrane proteins are unusually highly glycosylated, which may help protect them from the lysosomal proteases in the lumen.

Lysosomes Are Heterogeneous Organelles[54]

Lysosomes were seen clearly in the electron microscope about a decade after they had first been described. They are extraordinarily diverse in shape and size but can be identified as a single family of organelles by histochemistry, using the precipitate formed by the reaction product of a hydrolase with its substrate to show which organelles contain the enzyme (Figure 8–70). By this criterion, lysosomes are found in all eucaryotic cells.

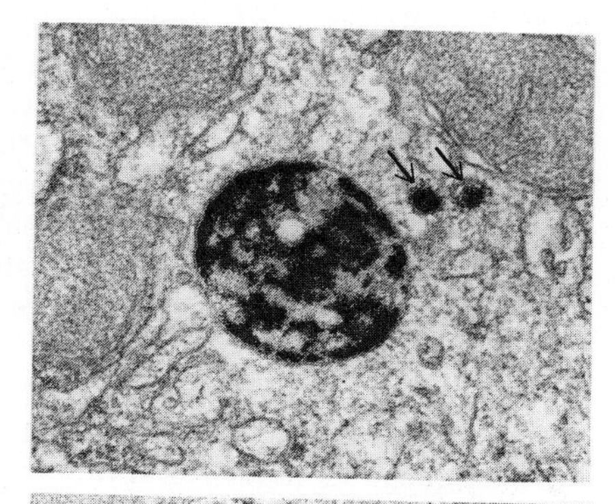

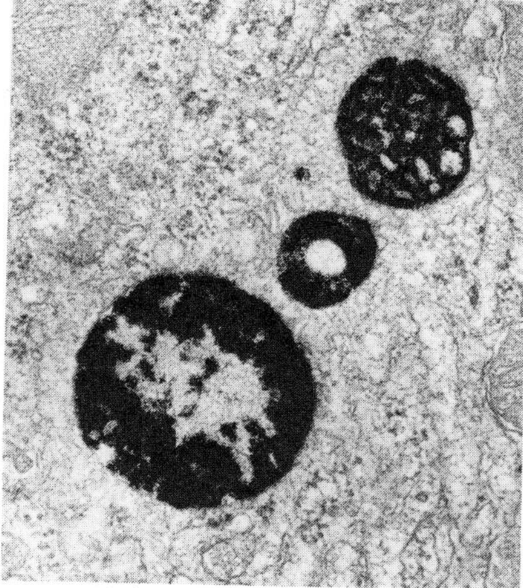

Figure 8–70 Electron micrographs of two sections of a cell stained to reveal the location of acid phosphatase, a marker enzyme for lysosomes. The larger membrane-bounded organelles, containing dense precipitates of lead phosphate, are lysosomes, whose diverse morphology reflects variations in the amount and nature of the material they are digesting. Two small vesicles thought to be carrying hydrolases from the Golgi apparatus are indicated by arrows in the top panel. The precipitates are produced when tissue fixed with glutaraldehyde (to fix the enzyme in place) is incubated with a phosphatase substrate in the presence of lead ions. (Courtesy of Daniel S. Friend.)

The heterogeneity of lysosomal morphology contrasts with the relatively uniform structures of most other cellular organelles. The diversity reflects the wide variety of digestive functions mediated by acid hydrolases, including the digestion of intra- and extracellular debris, the digestion of phagocytosed microorganisms, and even cell nutrition (since lysosomes are the principal site of cholesterol assimilation from endocytosed serum lipoprotein—see p. 328). For this reason, lysosomes are sometimes viewed as a heterogeneous collection of distinct organelles whose common feature is a high content of hydrolytic enzymes.

Lysosomes Can Obtain the Materials They Degrade by at Least Three Pathways[55]

The traffic of materials to lysosomes follows different paths depending on the source of the materials to be digested. The best-studied pathway is used to digest materials taken up by *endocytosis* in the pathway that leads from *coated pits* to *endosomes* to lysosomes. As discussed in Chapter 6 (see p. 323), materials endocytosed by this pathway pass sequentially from a *peripheral* to a *perinuclear* endosomal compartment. Those materials that have not been specifically retrieved from these endosomes for recycling to the plasma membrane then enter a third distinct "intermediate compartment," which receives newly synthesized lysosomal hydrolases and lysosomal membrane proteins from the Golgi apparatus. Because this *endolysosomal* compartment is mildly acidic, it is thought to be the site where the hydrolytic digestion of endocytosed materials begins. Conversion of the endolysosome into a mature lysosome requires that it lose its distinct endosomal membrane components and further decrease its internal pH. It is not known how the conversion occurs (see p. 331).

All cells have a second pathway for supplying materials to lysosomes for degradation, in which obsolete parts of the cell itself can be destroyed—a process called *autophagy*. In a liver cell, for example, an average mitochondrion has a lifetime of about 10 days, and electron microscopic images of normal cells reveal lysosomes containing (and presumably digesting) recognizable organelles such as mitochondria and secretory vesicles. The digestion process appears to involve the enclosure of an organelle by membranes derived from the ER, creating an *autophagosome*. The autophagosome is thought to fuse then with a lysosome (or endolysosome), initiating the digestion of its contents in an *autophagolysosome*. The process is highly regulated, and selected cell components can be targeted for destruction during cell remodeling: the smooth ER that has proliferated in a liver cell in response to a drug, for example, is selectively removed by autophagy when the drug is withdrawn (see p. 436).

The third pathway that provides materials to lysosomes occurs only in cells that are specialized for the *phagocytosis* of large particles and microorganisms. Cells such as macrophages and neutrophils can engulf such large objects to form a *phagosome* (see p. 334). The phagosome is thought to be converted to a *phagolysosome* in the manner discussed for the autophagosome. These three pathways are summarized in Figure 8–71; the three types of lysosomes that result may not be different from one another except for the materials they digest. All are normally referred to simply as *lysosomes*.

8-37 Lysosomal Enzymes Are Sorted in the Golgi Apparatus by a Membrane-bound Receptor Protein That Recognizes Mannose 6-Phosphate[56]

The biosynthesis of lysosomes requires the synthesis of specialized lysosomal hydrolases and membrane proteins. Both classes of proteins are synthesized in the ER and transported through the Golgi apparatus. Transport vesicles that deliver these proteins to the endolysosome—and then to lysosomes—bud from the *trans* Golgi network. These vesicles must incorporate lysosomal proteins while excluding the many other proteins being packaged into different transport vesicles for delivery elsewhere.

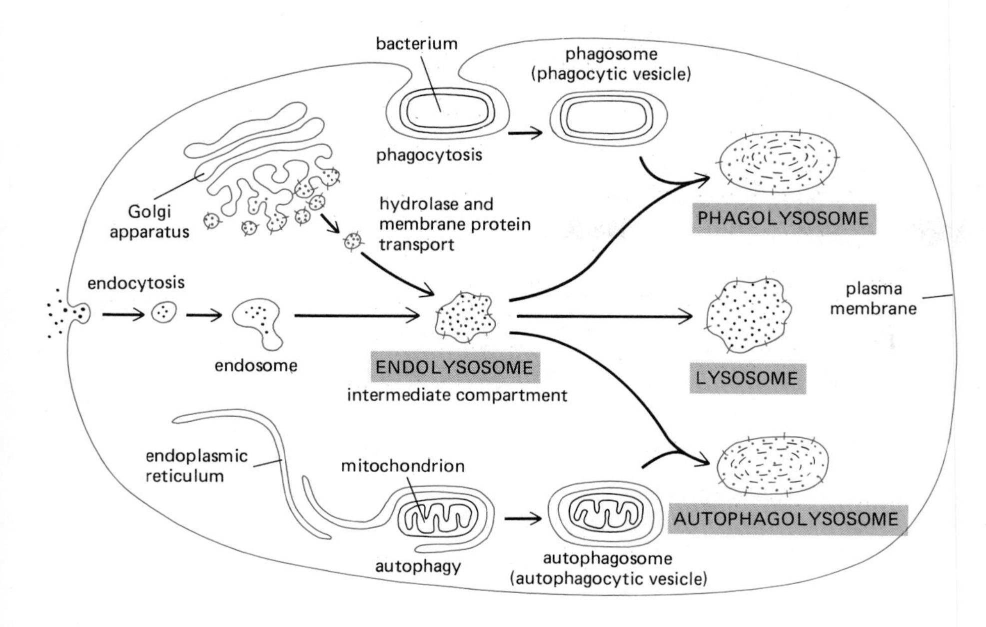

Figure 8–71 Three ways in which lysosomes are thought to form. Each pathway leads to the intracellular digestion of materials derived from a different source and produces a morphologically distinct lysosome. At the center of these pathways is an "intermediate compartment," designated here as an *endolysosome* (see p. 331).

Until recently it was traditional to make a distinction between *primary lysosomes* and *secondary lysosomes,* with the former term being used to describe newly formed lysosomal vesicles that have not yet encountered materials to be digested. Recent evidence, however, shows that lysosomal hydrolases and lysosomal membrane proteins are sorted by different receptors, raising the possibility that they leave the Golgi apparatus in separate transport vesicles and first meet in the endolysosome, which already contains endocytosed materials for digestion.

How are lysosomal proteins recognized and selected with the required accuracy? The same question can be asked for many other vesicle-mediated sorting processes in the cell. At the molecular level the answer is known in only one case—that of the lysosomal hydrolases. These carry a unique marker in the form of *mannose 6-phosphate (M6P)* groups, which are added exclusively to the *N*-linked oligosaccharides of these soluble lysosomal enzymes. This reaction occurs in the lumen of the *cis* Golgi. Complementary **M6P receptor proteins,** which cluster in the membrane and become concentrated in clathrin-coated vesicles budding from the *trans* Golgi network, have also been isolated and characterized. These receptors are transmembrane proteins that bind the lysosomal enzymes, thereby separating them from all of the other proteins present and concentrating them in coated transport vesicles. These vesicles rapidly lose their coats and fuse with an endolysosome, delivering their contents to this organelle.

In some cells a small fraction of the mannose 6-phosphate receptors are present in the plasma membrane, where they function in receptor-mediated endocytosis of lysosomal enzymes that have been released into the extracellular medium. These receptors transport the enzymes via coated pits into endosomes, from where the enzymes are eventually delivered to lysosomes. This *scavenger pathway* recaptures lysosomal hydrolases that have escaped the normal packaging process in the *trans* Golgi network and have therefore been transported to the cell surface and secreted.

The Mannose 6-Phosphate Receptor Shuttles Back and Forth Between Specific Membranes[57]

The mannose 6-phosphate receptor protein has been purified and characterized in *in vitro* studies. It binds its specific oligosaccharide at pH 7 and releases it at pH 6, which is the pH in the interior of the endolysosome. Thus the lysosomal enzymes dissociate from the M6P receptor proteins in the endolysosome and begin to digest the endocytosed material delivered from endosomes. Having released their enzymes, the receptors are retrieved and returned to the membrane of the *trans* Golgi network, probably by coated-vesicle-mediated transport (Figure 8–72). This process of *membrane recycling* from endolysosome back to the Golgi apparatus is similar to the recycling that occurs between endosomes and the plasma membrane during receptor-mediated endocytosis (see p. 331). Indeed, the receptor-mediated transport of lysosomal hydrolases from the Golgi apparatus to endolysosomes is analogous to the receptor-mediated endocytosis of extracellular molecules from the plasma membrane to endosomes. In both processes, receptors

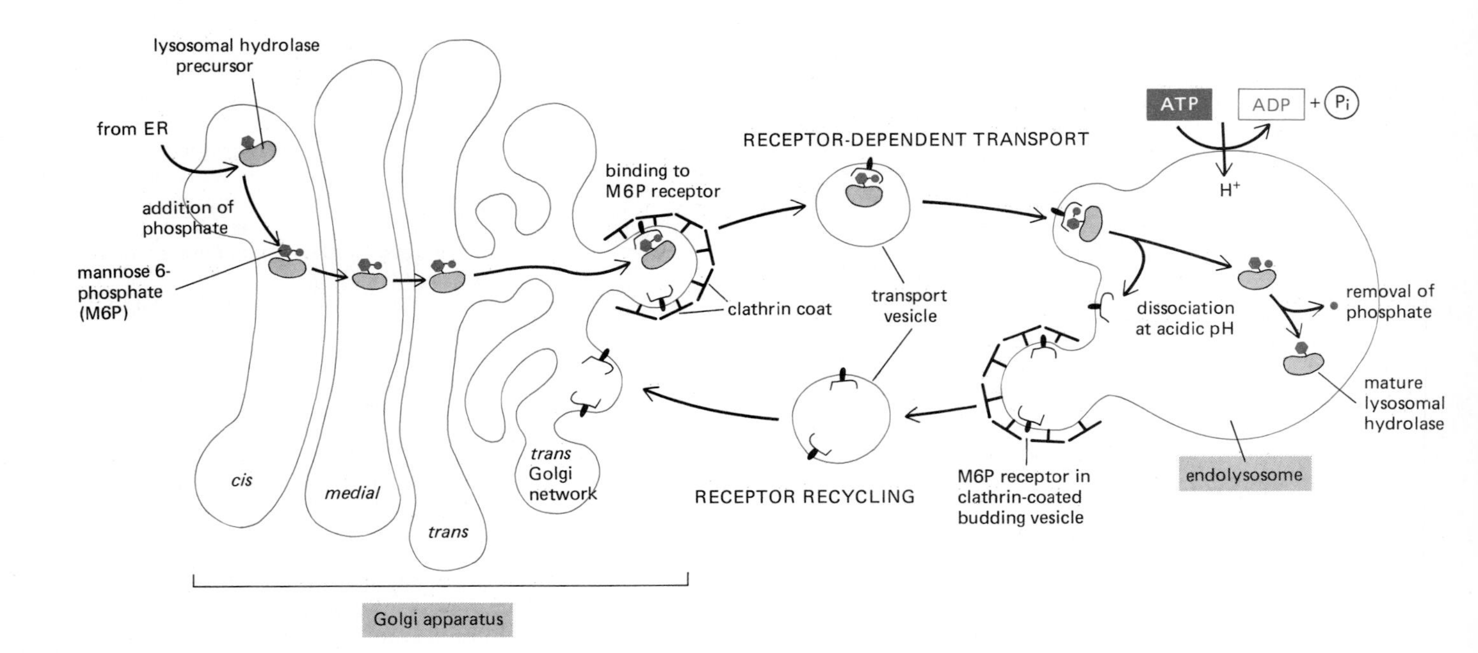

Figure 8–72 The transport of newly synthesized lysosomal hydrolases to lysosomes. The precursors of lysosomal hydrolases are tagged with mannose 6-phosphate groups in the *cis* Golgi and segregated from all other types of proteins in the *trans* Golgi network. The segregation occurs because clathrin-coated vesicles budding from the *trans* Golgi network concentrate mannose 6-phosphate-specific receptors, which bind the lysosomal hydrolases. These coated vesicles lose their coats and fuse with endolysosomes (see Figure 8–71). At the low pH of the endolysosome, the hydrolases dissociate from the receptors, which are recycled to the Golgi apparatus for further rounds of transport. The removal of phosphate from the mannose on the hydrolases further decreases the chance that the hydrolases will return to the Golgi apparatus with the receptor. Although there are two structurally distinct M6P receptor glycoproteins of very different size, they have a related amino acid sequence and appear to have similar functions.

cluster in clathrin-coated regions of membrane (called coated pits); these regions bud off to form clathrin-coated vesicles that transport ligands to a second compartment, which is acidic, from where the receptors are recycled back to their membrane of origin.

The recycling of the mannose 6-phosphate receptor has been followed using specific antibodies to locate the protein in the cell. The receptors are normally found in membranes of the Golgi apparatus and endolysosomes but not in mature lysosomes. When some cultured cells are treated with a weak base such as *ammonia* or *chloroquine,* which specifically accumulates in the interior of acidified organelles and raises their pH toward neutrality, the receptors disappear from the Golgi apparatus and accumulate in endolysosomes. The return of the receptors to the Golgi can be triggered in such cells either by removing the weak base or by adding high concentrations of mannose 6-phosphate to the culture medium. Both treatments cause the receptors to release their bound enzymes in the endolysosome, in one case by reacidifying the organelle and in the other by competitive binding of endocytosed mannose 6-phosphate to the receptors. These experiments suggest that transport back to the Golgi apparatus is facilitated by a conformational change in the receptor that occurs when it releases its bound hydrolase.

The shuttle system for the mannose 6-phosphate receptor shown in Figure 8–72 is specific—the vesicles carrying the receptor fuse with their specific target organelles but not, for example, with the ER membrane. The clathrin coat on the forming vesicles is thought to act as a "molecular filter," sequestering the receptor and its ligand into vesicles (see p. 328), but it cannot be responsible for the specificity of vesicle targeting because the coat is quickly removed after the vesicle is formed. *In vitro* experiments suggest that the removal of clathrin is catalyzed by an hsp70-like protein in a reaction that requires ATP hydrolysis. One or more of the proteins left exposed on the outer surface of the vesicle membrane then presumably serve as a specific "docking marker" that is recognized by a complementary "acceptor" on the target organelle membrane. A highly schematic view of such a process is illustrated in Figure 8–73.

The sorting of lysosomal hydrolases is the best model available for understanding the many transport-vesicle-mediated sorting events that occur in a eucaryotic cell. Although an oligosaccharide marker is not likely to be used elsewhere, "cargo" recognition by a membrane-bound receptor during vesicle budding, fusion of the vesicle with a specific target membrane, cargo release in the target compartment, and recycling of the empty receptor back to the original compartment are all likely to be common themes.

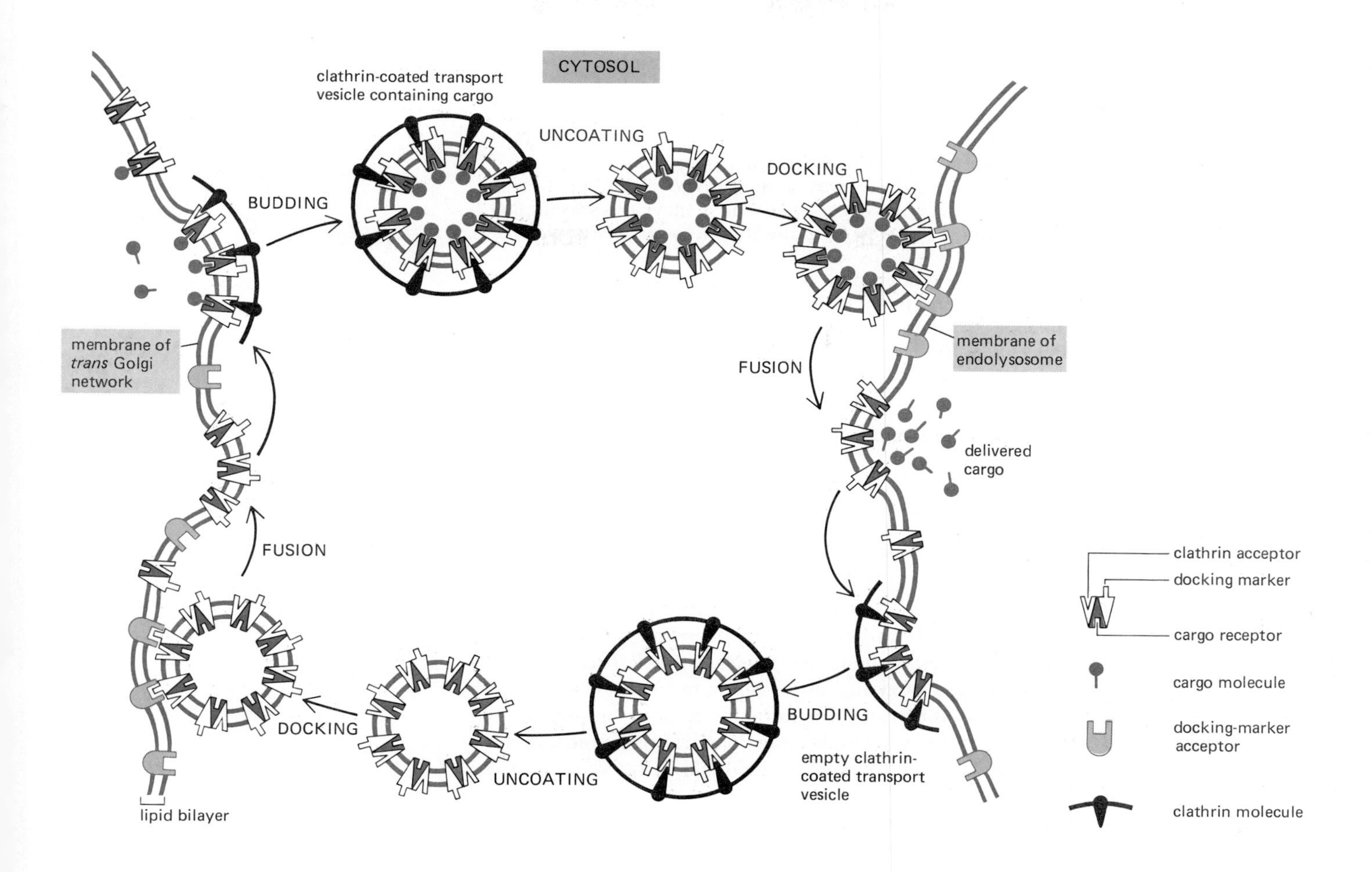

Figure 8–73 A possible mechanism for directing clathrin-coated transport vesicles back and forth between two specific membranes. In this hypothetical model the "cargo molecules" are lysosomal hydrolases and the "cargo receptor" is the mannose 6-phosphate receptor protein. The postulated "docking marker" and "docking-marker acceptor" molecules have not been well characterized, although the recent discovery of clathrin-binding proteins that also bind to the cytoplasmic tails of selected membrane proteins fits the description given here of the docking marker acceptor. (See Pearse, B.M., *EMBO J.* 7:3331–3336, 1988.)

The Attachment of Multiple Mannose 6-Phosphate Groups to a Lysosomal Enzyme Amplifies the Sorting Signal[58]

The sorting system that segregates lysosomal hydrolases and dispatches them to endolysosomes works because M6P groups are added only to the appropriate glycoproteins in the Golgi apparatus. This requires specific recognition of the hydrolases by the Golgi enzyme responsible for adding M6P. Since all glycoproteins arrive in the *cis* Golgi with identical *N*-linked oligosaccharide chains, the signal for adding the M6P units to oligosaccharides must reside somewhere in the polypeptide chain of each hydrolase.

Two enzymes act sequentially to catalyze the addition of M6P groups to lysosomal hydrolases: *N*-acetylglucosamine phosphotransferase (*GlcNAc-P-transferase*) transfers the GlcNAc-P portion of the sugar nucleotide UDP-GlcNAc to a mannose unit on the oligosaccharide while the hydrolase is in the *cis* Golgi; a second enzyme—a phosphoglycosidase—then removes the terminal GlcNAc, exposing the phosphate to form the final M6P marker (Figure 8–74). The phosphotransferase specifically binds the hydrolase by means of a *recognition site* that is separate from the *catalytic site* for the reaction (Figure 8–75). The signal recognized by the recognition site is believed to be a conformation-dependent *signal patch* rather than a signal peptide (see p. 414), since recognition is virtually eliminated when the hydrolase is partially unfolded.

Once the phosphotransferase has recognized the signal on the hydrolase, it adds GlcNAc-P to one or two of the mannoses on each oligosaccharide chain. Since most lysosomal hydrolases have multiple oligosaccharides, they acquire many M6P residues; this greatly amplifies the signal. Thus, while a lysosomal hydrolase typically binds to the recognition site of the phosphotransferase with an affinity constant (K_a) of about 10^5 liters/mole, the multiply phosphorylated hydrolase binds to the M6P receptor with a K_a of about 10^9 liters/mole, a 10,000-fold amplification.

Figure 8–74 Synthesis of the mannose 6-phosphate marker on a lysosomal hydrolase occurs in two steps. First, GlcNAc phosphotransferase transfers P-GlcNAc residues to the 6 position of several mannose residues on the *N*-linked oligosaccharides of the lysosomal precursor glycoprotein. Second, a phosphoglycosidase cleaves off the GlcNAc residue, creating the mannose 6-phosphate marker. The first enzyme is specifically activated by a signal patch present on lysosomal hydrolases (see Figure 8–75), while the phosphoglycosidase is a nonspecific enzyme. This modification of selected mannose residues in the *cis* Golgi compartment protects these mannoses from removal by the mannosidases that will be encountered later in the *medial* Golgi compartment.

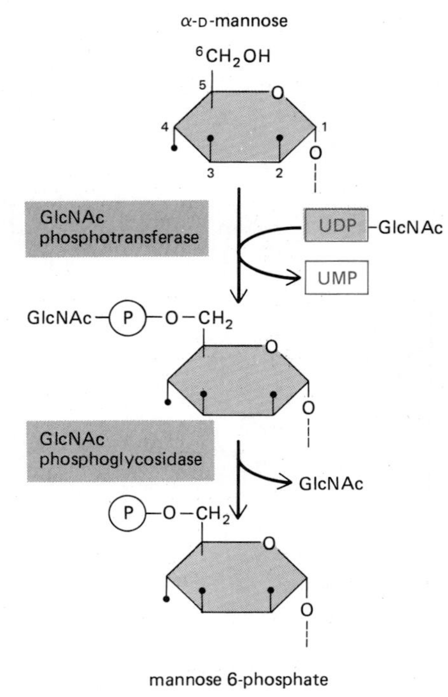

8-38 Defects in the GlcNAc Phosphotransferase Cause a Lysosomal Storage Disease in Humans[59]

Lysosomal storage diseases played a crucial part in the discovery of the lysosomal hydrolase-sorting mechanism. These diseases are caused by genetic defects that affect one or more of the lysosomal hydrolases and result in accumulation of their undigested substrates in lysosomes, with profound pathological consequences. They usually result from a mutation in a structural gene that codes for an individual lysosomal hydrolase. The most dramatic form of lysosomal storage disease, however, is a very rare disorder called inclusion cell disease (*I-cell disease*). In this disease almost all of the hydrolytic enzymes are missing from the lysosomes of fibroblasts, and their undigested substrates accumulate as large "inclusions" in the patients' cells. I-cell disease is due to a single gene defect, and like most genetic enzyme deficiencies, it is recessive—that is, only individuals with two bad copies of the gene have the disease.

In I-cell disease all the hydrolases missing from lysosomes are found in the blood, indicating that the structural genes encoding them are unaffected. The abnormality results from a missorting in the Golgi apparatus that causes the hydrolases to be secreted rather than transported to lysosomes. The missorting has been traced to a defective or missing GlcNAc-P-transferase. Because lysosomal enzymes are not phosphorylated in the *cis* Golgi, they are not segregated by M6P receptors into coated vesicles in the *trans* Golgi network. Instead they are carried to the cell surface and secreted. The oligosaccharides that would contain M6P in normal lysosomal enzymes are converted to the "complex" type, containing GlcNAc, Gal, and sialic acid. This indicates that the phosphorylation of mannose in the *cis* Golgi normally prevents the subsequent processing of the oligosaccharides on the hydrolases to complex forms in the *medial* and *trans* Golgi.

Experiments on I-cell disease in the late 1960s provided the first clue that all lysosomal enzymes have a common recognition marker. The marker was identified as mannose 6-phosphate in the late 1970s, when the same hydrolase was com-

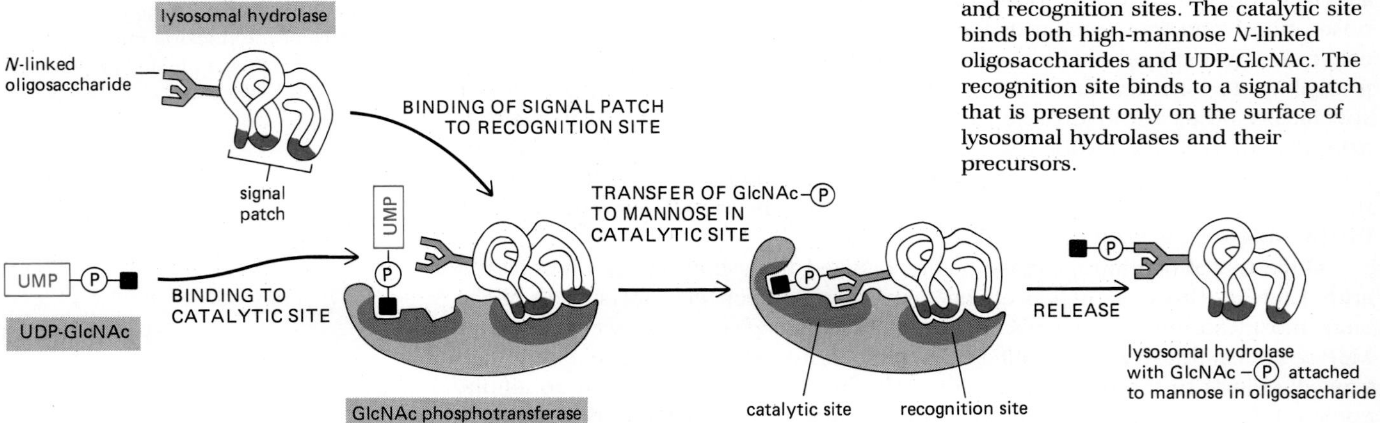

Figure 8–75 The GlcNAc phosphotransferase enzyme that recognizes lysosomal hydrolases in the Golgi apparatus has separate catalytic and recognition sites. The catalytic site binds both high-mannose *N*-linked oligosaccharides and UDP-GlcNAc. The recognition site binds to a signal patch that is present only on the surface of lysosomal hydrolases and their precursors.

pared in normal and mutant individuals. This led quickly to the purification of mannose 6-phosphate receptors and of GlcNAc-P-transferase and to an understanding of the role of the Golgi apparatus in the lysosomal hydrolase-sorting pathway.

In I-cell disease the lysosomes in some cell types, such as hepatocytes, contain a normal complement of lysosomal enzymes. This implies that there is another pathway for directing hydrolases to lysosomes that is used by some cell types but not others. The nature of this M6P-independent pathway is unknown. Perhaps in this case the hydrolases are sorted by direct recognition of their signal patch. Similarly, the lysosomal membrane proteins are sorted from the *trans* Golgi network to endolysosomes by an M6P-independent pathway in all cells. It is unclear why cells should need more than one sorting pathway to construct a lysosome.

Summary

Lysosomes are specialized for intracellular digestion. They contain unique membrane proteins and a wide variety of hydrolytic enzymes that operate best at pH 5, the internal pH of lysosomes. The acidic pH in lysosomes is maintained by an ATP-driven proton pump in their membranes. Newly synthesized lysosomal proteins are transferred into the lumen of the ER, transported through the Golgi apparatus, and then transported from the trans *Golgi network to an intermediate compartment (an endolysosome) by means of transport vesicles.*

The lysosomal hydrolases contain N*-linked oligosaccharides that are processed in a unique way in the* cis *Golgi so that their mannose residues are phosphorylated. These mannose 6-phosphate (M6P) groups are recognized by an M6P receptor protein in the* trans *Golgi network that segregates the hydrolases and helps to package them into budding clathrin-coated vesicles, which quickly lose their coats. These transport vesicles containing the mannose 6-phosphate receptor act as shuttles that move the receptor back and forth between the* trans *Golgi network and endolysosomes. The low pH in the endolysosome dissociates the lysosomal hydrolases from this receptor, making the transport of the hydrolases unidirectional.*

Transport from the Golgi Apparatus to Secretory Vesicles and to the Cell Surface[60]

Transport vesicles designed for immediate fusion with the plasma membrane normally leave the Golgi apparatus in a steady stream. The transmembrane proteins and lipids in these vesicles provide the new proteins and lipids for the cell's plasma membrane, while the soluble proteins in the vesicles are secreted to the cell exterior. In this way, for example, cells produce the proteoglycans and proteins of the extracellular matrix (see p. 802).

Whereas all cells require this **constitutive secretory pathway,** specialized secretory cells have a second secretory pathway in which soluble proteins and other substances are stored in secretory vesicles for later release—the so-called *triggered,* or **regulated secretory pathway** (Figure 8–76).

In this section we shall consider the role of the Golgi apparatus in these two secretory pathways and compare the mechanisms involved. We shall also consider how viruses exploit the sorting apparatus of their host cells and can be used to reveal the diversity of intracellular transport pathways.

8-41 Secretory Vesicles Bud from the *Trans* Golgi Network[61]

In cells in which secretion occurs in response to an extracellular signal, secreted proteins are concentrated and stored in **secretory vesicles** (frequently called *secretory granules* because of their dense cores), from which they are released by exocytosis in response to the signal. Secretory vesicles form by budding from the *trans* Golgi network. Their formation is thought to involve clathrin and its associated coat proteins, since a portion of the surface of forming secretory vesicles is

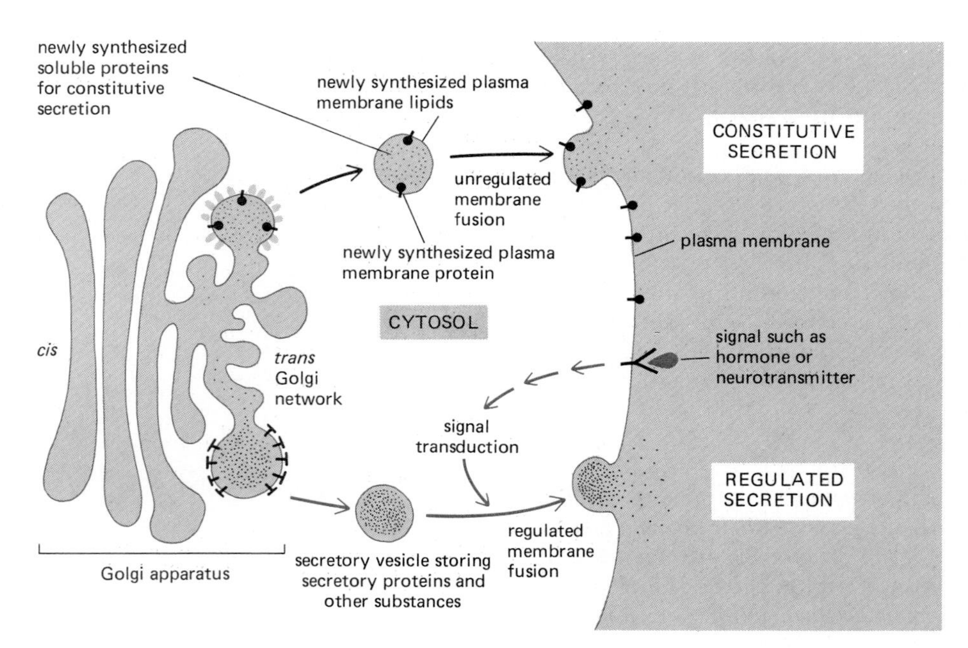

Figure 8–76 The regulated and constitutive pathways of secretion diverge in the *trans* Golgi network. Many soluble proteins are continually secreted from the cell by the *constitutive secretory pathway,* which operates in all cells. This pathway also supplies the plasma membrane with newly synthesized lipids and transmembrane proteins. Specialized secretory cells also have a *regulated secretory pathway,* by which selected proteins in the *trans* Golgi network are diverted into secretory vesicles, where the proteins are concentrated and stored until an extracellular signal stimulates their secretion.

usually clathrin coated. The coat is lost by the time the vesicle is fully formed (Figure 8–77).

Like the lysosomal hydrolases discussed in the preceding section, proteins destined for secretory vesicles (often called *secretory proteins*) must be sorted and packaged into appropriate vesicles in the *trans* Golgi network. In this case the mechanism is believed to involve the selective aggregation of secretory proteins, which can be detected in the electron microscope as electron-dense material in the lumen of the *trans* Golgi network. The "sorting signal" that directs proteins into such an aggregate is not known, but it is thought to be a signal patch that is shared by many secretory proteins: when a gene encoding a secretory protein is transferred to a different type of secretory cell that normally does not make the protein, the foreign protein is appropriately packaged into secretory vesicles.

How the aggregates containing secretory proteins are segregated into forming secretory vesicles is also unclear. Secretory vesicles contain unique membrane proteins, some of which might serve as receptors (in the *trans* Golgi network) to bind the aggregated material that will be packaged. Secretory vesicles are larger than the transport vesicles that carry lysosomal hydrolases, and the aggregates they contain are much too large for each molecule of the secreted protein to be bound by a receptor in the vesicle membrane—as proposed for transport of the lysosomal enzymes (see Figure 8–73); the uptake of the aggregates into secretory vesicles may more closely resemble the uptake of particles by phagocytosis at the cell surface (see p. 335), which can also be mediated by clathrin-coated membranes.

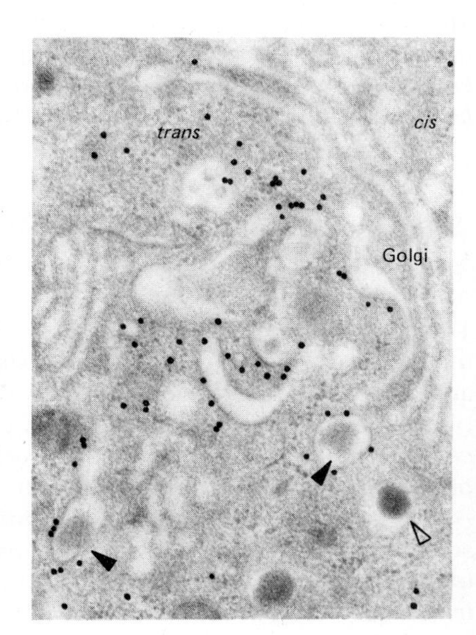

Figure 8–77 Electron micrograph of secretory vesicles forming from the *trans* Golgi network in an insulin-secreting cell of the pancreas. An antibody conjugated to gold spheres (*black dots*) has been used to locate clathrin molecules. The immature secretory vesicles (*black arrowheads*), which contain proinsulin molecules, are coated with clathrin. The clathrin coat is rapidly shed once the vesicle has formed and is absent from the mature secretory vesicles (*open arrowhead*). (Micrograph courtesy of Lelio Orci.)

After the immature secretory vesicles bud from the *trans* Golgi network, the clathrin coat is removed and their contents become greatly condensed. The condensation occurs suddenly and is believed to be caused by an acidification of the vesicle lumen induced by an ATP-driven H^+ pump in the vesicle membrane. The aggregation of secreted proteins (or other compounds, see p. 325) and their subsequent condensation in secretory vesicles can concentrate these substances by as much as 200-fold from the Golgi lumen, enabling the secretory cell to release large amounts of material on demand.

Secretory-Vesicle Membrane Components Are Recycled[62]

Many secretory cells, such as the pancreatic acinar cell, are polarized, and exocytosis occurs only at the apical surface, which typically faces the lumen of a duct system that collects the secretions. When a secretory vesicle fuses with the plasma membrane, its contents are discharged from the cell by exocytosis and its membrane becomes part of the plasma membrane (see p. 324). Although this should greatly increase the surface area of the plasma membrane, it does so only transiently because membrane components are removed from the surface (or *recycled*) by endocytosis almost as fast as they are added by exocytosis (Figure 8–78). There is evidence that this removal returns the proteins of the secretory vesicle membrane to the Golgi apparatus, where they can be used again. Such recycling maintains a steady-state distribution of membrane components among the various cellular compartments.

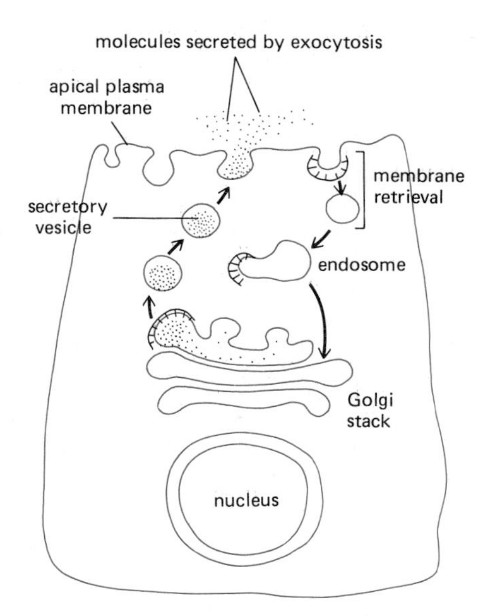

Figure 8–78 The membrane of a secretory vesicle is recycled after it fuses with the apical plasma membrane. The amount of membrane added during regulated secretion can be massive, but because membrane is retrieved in clathrin-coated vesicles and returned to the *trans* Golgi, probably via endosomes, the apical surface maintains an essentially constant area.

Proteins and Lipids Seem to Be Carried Automatically from the ER and Golgi Apparatus to the Cell Surface in Unpolarized Cells[63]

In a cell capable of regulated secretion, at least three types of proteins must be separated before they leave the *trans* Golgi network: those destined for endolysosomes, those destined for secretory vesicles, and those destined for immediate delivery to the cell surface. Proteins destined for endolysosomes are selected for packaging into departing vesicles on the basis of positive signals (M6P for lysosomal hydrolases), and a special signal must also be present on each of the proteins packaged into secretory vesicles. The proteins transferred to the cell surface could therefore in principle be transported by a nonselective "default pathway," provided that there is no need to target a particular plasma membrane protein to a selected region of the cell (Figure 8–79). Thus, in an unpolarized cell such as a white blood cell or most cells in culture, it has been proposed that any protein in the ER will automatically be carried through the Golgi apparatus to the cell surface by the constitutive secretory pathway unless it is either specifically retained as a resident of the ER or Golgi or selected for transport elsewhere. An attraction of this model lies in its simplicity and in the opportunity that such a default pathway offers for discarding damaged or misdirected proteins to the outside of the cell.

In an attempt to test for such an unselected "bulk-flow" pathway to the plasma membrane, cultured cells were incubated with a simple tripeptide (Asn-Tyr-Thr) containing the glycosylation code Asn-X-Thr. This small peptide was able to diffuse into cells and across intracellular membranes, and it became glycosylated on its asparagine residue in the lumen of the ER. The addition of the *N*-linked oligosaccharide prevented the tripeptide from diffusing back into the cytosol. Instead it was transported unidirectionally from the ER through the Golgi and to the cell surface in about 10 minutes, which is the rate of transport of the fastest normal plasma membrane proteins. This result is consistent with the hypothesis that there is an unselected flow of luminal fluid that automatically carries any soluble molecule in the lumen of the ER promptly to the cell surface unless it is specifically retained or directed elsewhere.

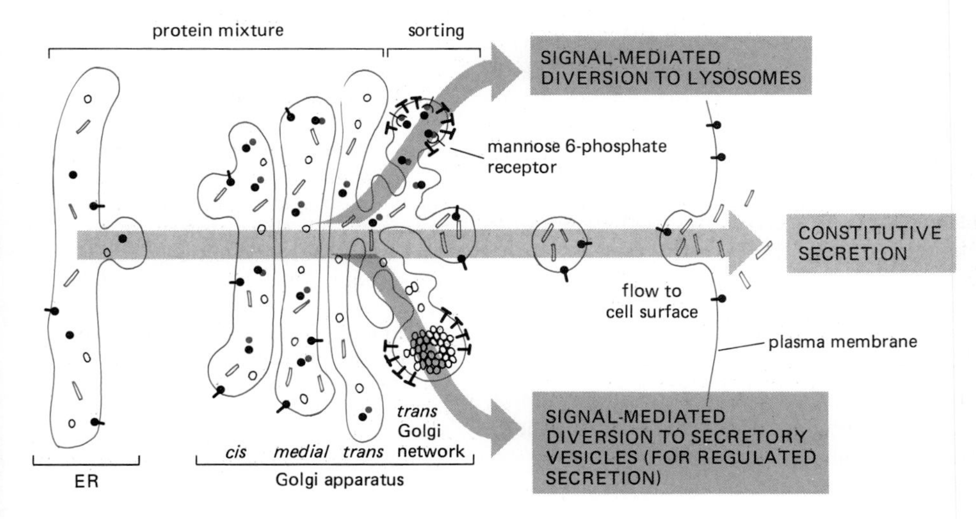

Figure 8–79 The best-understood pathways of protein sorting in the *trans* Golgi network. Proteins with the mannose 6-phosphate marker are diverted to lysosomes (via endolysosomes) in clathrin-coated vesicles (see Figure 8–72). Proteins with signals directing them to secretory vesicles are concentrated in large clathrin-coated vesicles that lose their coats to become secretory vesicles—a pathway that occurs only in specialized secretory cells. In unpolarized cells, proteins with no special features are thought to be delivered to the cell surface by default via the constitutive secretory pathway. In polarized cells, however, secreted and plasma membrane proteins are selectively directed to either the apical or the basolateral plasma membrane domain, so that at least one of these two pathways must be signal mediated.

In principle, the same type of unselected flow could carry transmembrane proteins and lipids that lack sorting signals automatically from the ER to the cell surface. It could also transport proteins destined for secretory vesicles and lysosomes from the ER to the end of the Golgi apparatus; only in the *trans* Golgi network would specific signals be required to separate these proteins from those going to the plasma membrane. A special sorting mechanism would also be needed to retain all proteins intended for permanent residence in the ER or Golgi apparatus (see p. 413). The observation that several ER resident proteins (including BiP and protein disulfide isomerase) contain a signal peptide that is responsible for their retention in the ER (see p. 415) supports this general view of protein traffic through the cell.

Some constitutively secreted proteins take a long time to leave the ER and require many hours to be secreted. To reconcile this observation with the unselected-flow hypothesis, it has been suggested that these proteins take a long time to fold correctly in the ER and are therefore delayed there, either because they stick to the ER membrane or because they are bound by special proteins such as BiP (see p. 445); once folded correctly, these retained proteins would also be carried forward in the unselected flow.

8-42 Polarized Cells Must Direct Proteins from the Golgi Apparatus to the Appropriate Domain of the Plasma Membrane[64]

Most cells in tissues are *polarized* and have two (and sometimes more) distinct plasma membrane domains (see p. 310). A typical epithelial cell, for example, has two physically continuous but compositionally distinct plasma membrane domains (see Figure 6–36, p. 297): the *apical domain* faces the lumen and often has specialized features such as cilia or a brush border of microvilli; the *basolateral domain* covers the rest of the cell. The two domains are joined at their border by a ring of *tight junctions* (see p. 793), which prevent proteins (and lipids in the outer leaflet of the lipid bilayer) from diffusing between the two membrane domains. Therefore, although the two domains appear as one continuous membrane when viewed in the electron microscope, the tight junctions effectively separate the domains so that the apical membrane contains one set of proteins while the basolateral membrane contains another. The lipid composition of the two bilayers is also different; in particular, glycolipids are found only in the apical membrane domain.

Epithelial cells can also secrete one set of proteins at their apical surface and a different set from their basolateral surface. Thus polarized cells must have ways of directing both membrane-bound and secreted molecules specifically to each plasma membrane domain. It has been shown in polarized cells in culture that proteins destined for different domains pass together from the ER to the *trans* Golgi network, where they are separated and dispatched in secretory or transport vesicles to the appropriate plasma membrane domain. It is possible that both the basolateral and apical proteins have distinct sorting signals that direct them to the appropriate domain; alternatively, only one of these pathways may require a sorting signal, with the other operating by default. The nature of the signals is unknown.

Viruses Exploit the Sorting Mechanisms of Their Host Cells[65]

Many animal viruses have only a small amount of nucleic acid in their genome and contain no more than four or five genes. Most of these genes code for structural proteins of the mature viral particle (or virion), so these viruses must parasitize host-cell pathways for most of the steps in their replication (see p. 248). Because viral products are usually synthesized in large amounts during infection, and because during its life cycle the virus follows a sequential route through the compartments of the host cell, virus-infected cells have been very useful for tracing the pathways of intracellular transport and for studying how essential biosynthetic reactions are compartmentalized in eucaryotic cells.

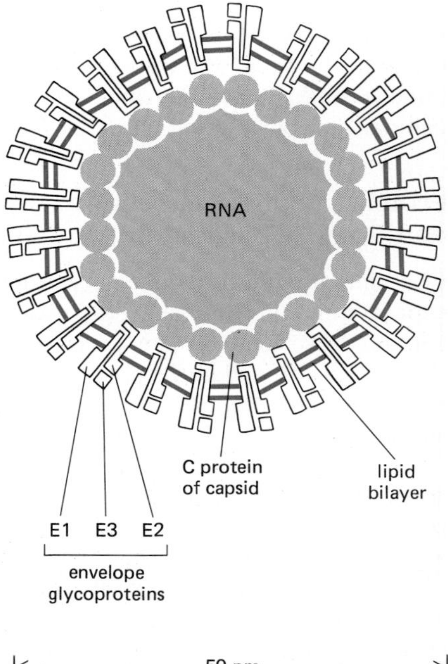

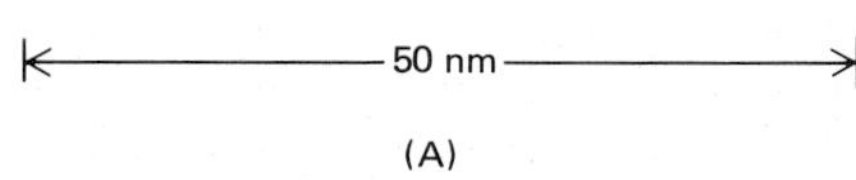

(A)

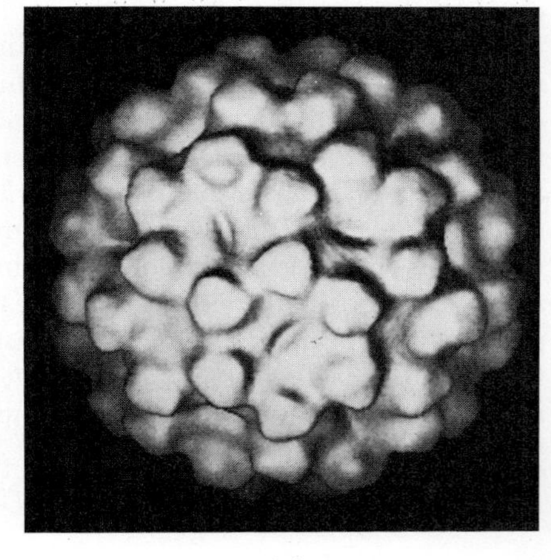

(B)

Figure 8–80 The structure of Semliki forest virus seen in a schematic drawing of a cross-section of the virus (A) and in a three-dimensional reconstruction of the surface of the virus derived from cryoelectron micrographs of unstained specimens (B). There are 180 copies of the capsid C protein (arranged as 60 trimers) and 240 copies of each of the three envelope proteins (arranged as 80 trimers) in each virus particle. The outer envelope of the virus consists of the envelope proteins embedded in a lipid bilayer membrane. The virus has a total mass of 46 million daltons. (B, from S.D. Fuller, *Cell*, 48:923–934, 1987.)

Enveloped animal viruses, in which the genome is enclosed in a lipid bilayer membrane (see p. 249), have exploited the compartmentalization of the cell to an especially fine degree. To follow the life cycle of an enveloped virus is to take a tour through the cell. A well-studied example is *Semliki forest virus,* which consists of an RNA genome surrounded by a **capsid** formed by a regularly arranged icosahedral (20-faced) shell of a protein (called C protein). The *nucleocapsid* (genome + capsid) is surrounded by a closely apposed lipid bilayer that contains only three proteins (called E1, E2, and E3). These **envelope proteins** are glycoproteins that span the lipid bilayer and interact with the C protein of the nucleocapsid, linking the membrane and nucleocapsid together (Figure 8–80A). The glycosylated portions of the envelope proteins are always on the outside of the lipid bilayer, and complexes of these proteins form "spikes" that can be seen in electron micrographs projecting outward from the surface of the virus (Figure 8–80B).

Infection is initiated when the virus binds to receptor proteins on the host-cell plasma membrane. The virus uses the cell's normal endocytic pathway to enter the cell by receptor-mediated endocytosis and is delivered to endosomes (see p. 329). But instead of being delivered to lysosomes, the virus escapes from the endosome by virtue of the special properties of one of its envelope proteins. At the acidic pH of the endosome, this protein causes the viral envelope to fuse with the endosome membrane, releasing the bare nucleocapsid into the cytosol (Figure 8–81). The nucleocapsid is "uncoated" in the cytosol, releasing the viral RNA, which is then translated by host-cell ribosomes to produce a virus-coded RNA polymerase. This in turn makes many copies of the RNA, some of which serve as mRNA molecules to direct the synthesis of the four structural proteins of the virus—the capsid C protein and the three envelope proteins E1, E2, and E3.

The capsid and envelope proteins follow separate pathways through the cytoplasm. The envelope proteins, like normal cellular membrane glycoproteins, are

Figure 8–81 The life cycle of Semliki forest virus. The virus parasitizes the host cell for most of its biosyntheses.

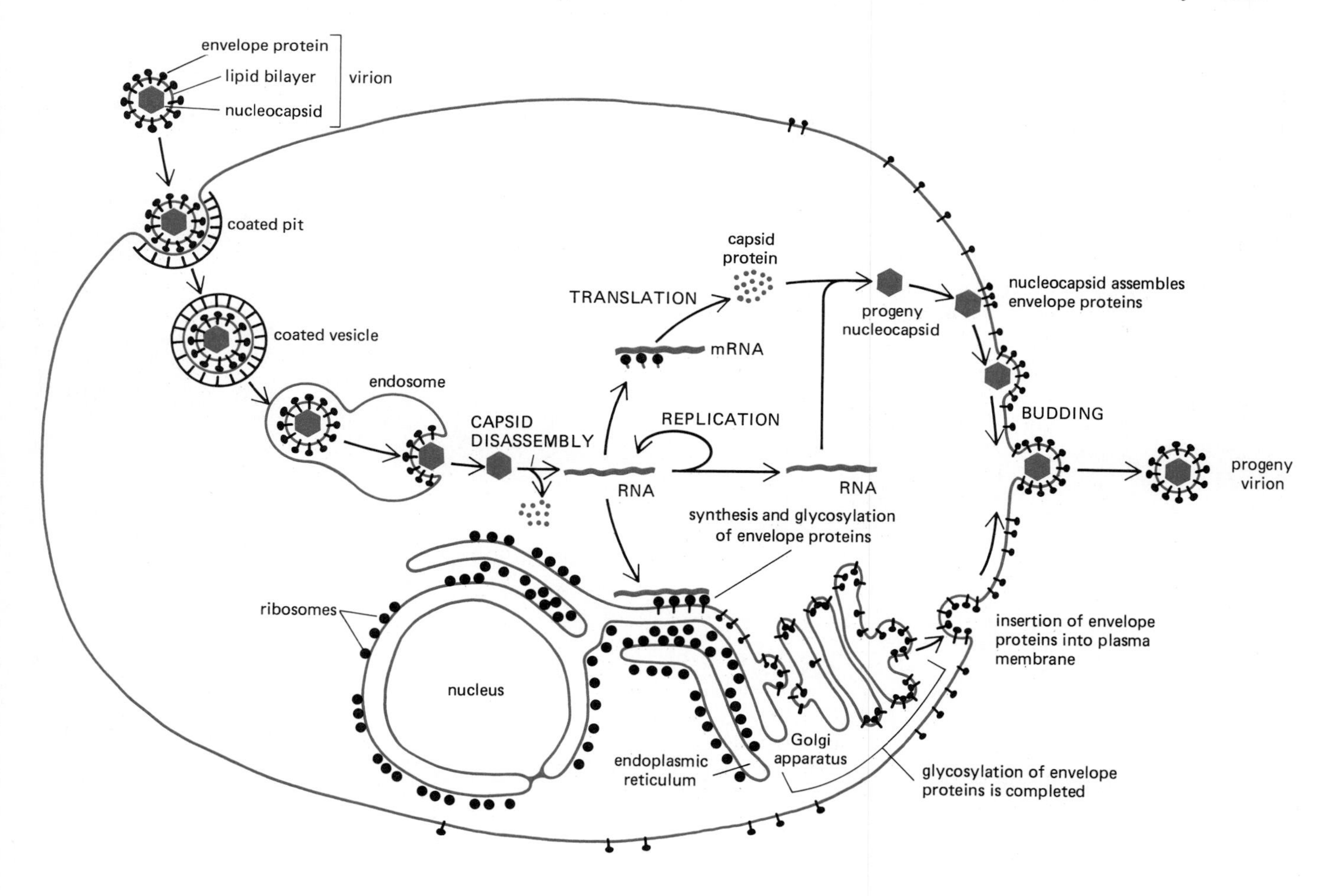

synthesized by ribosomes that are bound to the rough ER; the capsid protein, like a normal cytosolic protein, is synthesized by ribosomes that are not membrane bound. The newly synthesized capsid protein binds to the recently replicated viral RNA to form new nucleocapsids. The envelope proteins, in contrast, are inserted into the membrane of the ER, where they are glycosylated, transported to the Golgi apparatus (where their oligosaccharides are modified), and then delivered to the plasma membrane.

The viral nucleocapsids and envelope proteins finally meet at the plasma membrane (see Figure 8–81). As the result of a specific interaction with a cluster of envelope proteins, the nucleocapsid becomes wrapped in a portion of the plasma membrane, forming a bud that is highly enriched in the envelope proteins but contains host-cell lipids. Finally, the bud pinches off and a free virus is released on the outside of the cell. The clustering of envelope proteins in the lipid bilayer as they assemble around the nucleocapsid during viral budding has been suggested as a model for the segregation of specific membrane proteins into coated vesicles.

Viral Envelope Proteins Carry Signals That Direct Them to Specific Intracellular Membranes[66]

Different enveloped viruses bud from different host-cell membranes. Their envelope proteins, which are transmembrane proteins synthesized in the ER, must therefore carry signals directing them from the ER to the appropriate cell membrane. Some epithelial cell lines form polarized cell sheets when they are cultured on an appropriate surface such as a collagen-coated porous filter. When viruses infect such polarized cells, which maintain distinct apical and basolateral plasma membrane domains, some of them (such as influenza virus) bud exclusively from the apical plasma membrane, whereas others (such as Semliki forest virus and vesicular stomatitis virus) bud only from the basolateral plasma membrane (Figure 8–82). This polarity of budding reflects the presence on the envelope proteins of distinct apical or basolateral sorting signals, which direct the proteins to only one cell-surface domain; the proteins in turn cause the virus to assemble in that domain.

Other viruses have envelope proteins with different kinds of sorting signals. *Herpes virus*, for example, is a DNA virus that replicates in the nucleus, where its nucleocapsid assembles, and then acquires an envelope by budding through the inner nuclear membrane into the ER lumen; the envelope proteins therefore must

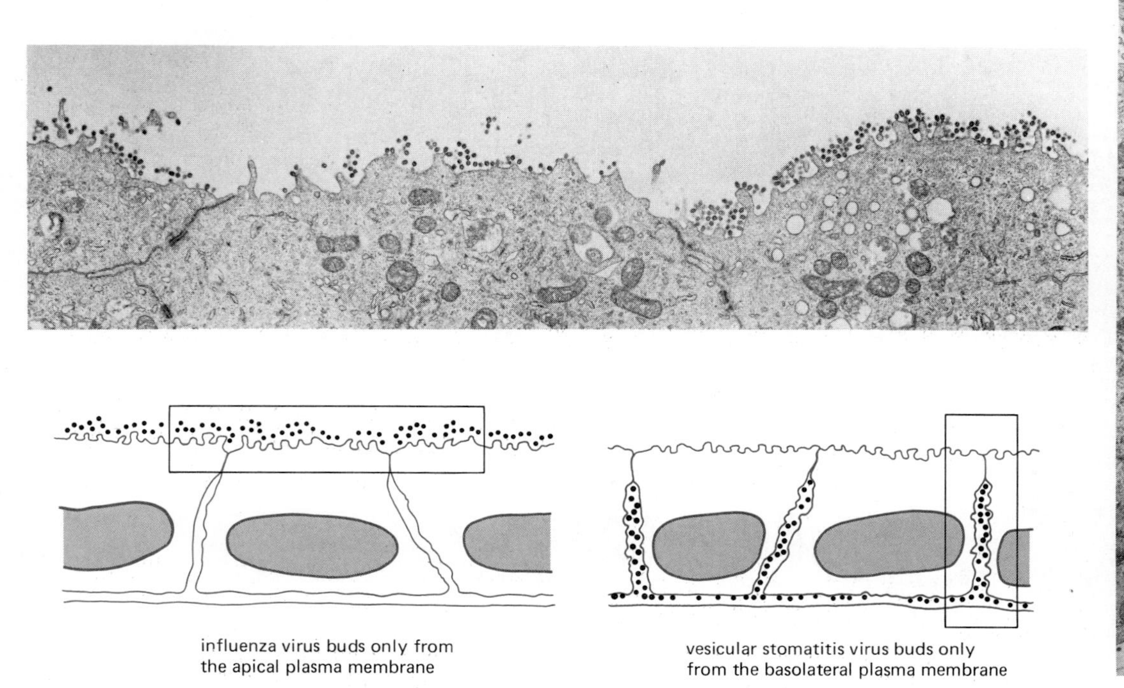

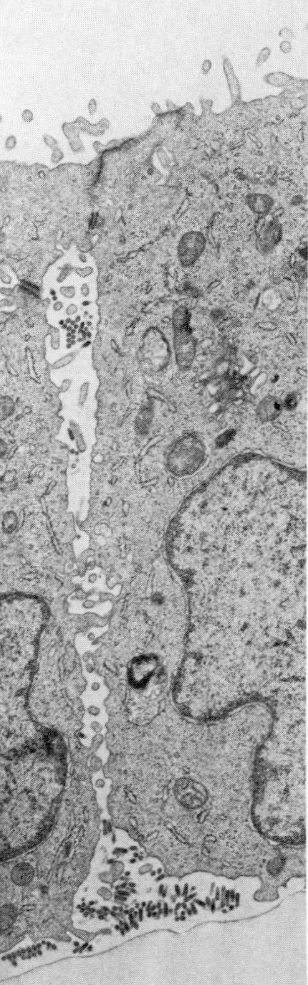

Figure 8–82 Electron micrographs showing that one type of enveloped virus buds from the apical plasma membrane, while another type buds from the basolateral plasma membrane, of the same epithelial cell line grown in culture. These cells grow with their basal surface attached to the culture dish. (Courtesy of E. Rodriguez-Boulan and D.D. Sabatini.)

be specifically transported from the ER membrane to the inner nuclear membrane. *Flavivirus,* in contrast, buds directly into the ER lumen, and *bunyavirus* buds into the Golgi apparatus, indicating that their envelope proteins carry signals for retention in the ER and Golgi membranes, respectively. The herpes virus, flavivirus, and bunyavirus particles are soluble in the ER and Golgi lumen, and they move outward toward the cell surface exactly as if they were secreted proteins; in the *trans* Golgi network they are incorporated into transport vesicles and secreted from the cell by the constitutive secretory pathway.

Summary

Proteins can be secreted from cells by exocytosis in either a constitutive or a regulated fashion. In the regulated pathway, molecules are stored in secretory vesicles, which do not fuse with the plasma membrane to release their contents until an extracellular signal is received. A selective condensation of the proteins targeted to secretory vesicles accompanies their packaging into these vesicles in the trans *Golgi network. The regulated pathway operates only in specialized secretory cells, but a constitutive secretory pathway operates in all cells, mediated by a continual process of vesicular transport from the* trans *Golgi network to the plasma membrane. For unpolarized cells there is evidence that proteins made in the ER are automatically delivered to the* trans *Golgi network and then to the plasma membrane by this constitutive pathway unless they are otherwise diverted or retained by specific sorting signals. In polarized cells, however, the transport pathways from the* trans *Golgi network to the plasma membrane must operate selectively to ensure that different sets of membrane proteins, secreted proteins, and lipids are delivered to the apical and basolateral domains.*

Vesicular Transport and the Maintenance of Compartmental Identity

Intracellular sorting seems to require at least 10 distinct types of transport vesicles, each with a unique set of "molecular address labels" on its surface that allows it to deliver its contents only to specific cell membranes. Thus the transport vesicles leaving the ER must fuse only with the *cis* Golgi compartment, those leaving the *cis* Golgi compartment must fuse only with the *medial* Golgi compartment, and so on. In each step, vesicle budding, vesicle docking, and vesicle fusion are involved, each requiring highly specific recognition events. At present we know little about the molecular mechanisms involved (see Figure 8–73). In this final section we examine some speculative views of the mechanisms that maintain compartmental identity and describe some new experimental approaches to these problems.

The ER and Golgi Compartments Retain Selected Proteins as Permanent Residents[67]

In ferrying cargo from one compartment to another, transport vesicles necessarily transfer membrane as well as their luminal content. Yet in the face of this homogenizing influence, the vital differences of membrane composition are maintained between the different compartments: the SRP receptor protein (see p. 440) is found only in the ER membrane, whereas glycosyltransferases and oligosaccharide-processing enzymes are located only in the membranes of the correct Golgi cisterna, and so on. The membranes of the ER and each type of Golgi cisterna must therefore have special mechanisms for maintaining their unique compositions. One possible mechanism would be to have each step in the forward movement through the ER and Golgi compartments selected by a signal, in the way that plasma membrane proteins that enter the cell by receptor-mediated endocytosis are selected by coated pits. As discussed previously, however, biosynthetic transport through the ER and Golgi apparatus is thought to work in the opposite way, with forward movement being automatic and retention requiring signals. In this view, each permanent resident of the ER or a Golgi compartment must carry a sorting signal that is responsible for its selective retention. As described previously, a signal peptide that causes the selective retention of proteins in the ER has been identified (see p. 415), although the mechanism of retention is unknown. A strategy of automatic forward movement and selective retention is attractive in part because the number of proteins passing through the ER and Golgi en route to other destinations greatly exceeds the number retained. Moreover, this strategy allows proteins that have either lost their sorting signals or been misdirected in an earlier step to be secreted to the cell exterior, as observed (see p. 464). Finally, if forward transport required specific signals, a protein presumably would need a different signal for each forward transfer step it undergoes—from ER to Golgi and from each Golgi cisterna to the next. The structure of many proteins might then be overly constrained by the large number of sorting signals required on their surface.

8-45 There Are at Least Two Kinds of Coated Vesicles[68]

Clathrin is found on the cytoplasmic surface of both the plasma membrane and the *trans* Golgi network and seems to be associated with transport decisions that are signaled. From the plasma membrane, clathrin-coated vesicles carry receptor-mediated endocytic traffic to endosomes; from the *trans* Golgi network, they are known to carry receptor-mediated traffic to endolysosomes. In the former case the vesicles are known to transport a highly selected set of cell-surface receptors (see p. 328), and the same is presumably true for each of the different types of clathrin-coated vesicles that bud from the Golgi apparatus.

Clathrin-coated vesicles do not bud from the ER or from *cis* or *medial* cisternae of the Golgi apparatus. Instead, another type of coated vesicle buds from these locations (as well as from the *trans* Golgi network). The coats of these vesicles do not stain with anti-clathrin antibodies, and they appear different from clathrin coats in the electron microscope (Figure 8–83). The coat proteins of these vesicles have not yet been identified.

A possible model for protein transport in unpolarized cells is presented in Figure 8–84. In this model, proteins move by default from ER to Golgi, from Golgi cisterna to Golgi cisterna, and from the *trans* Golgi network to the cell surface in an unselected process that is postulated to be mediated by non-clathrin-coated vesicles. In the *trans* Golgi network the signal-mediated diversion to lysosomes is mediated by clathrin-coated vesicles, while the signal-mediated diversion to secretory vesicles is mediated by a membrane that is partly clathrin coated. How can one test such ideas?

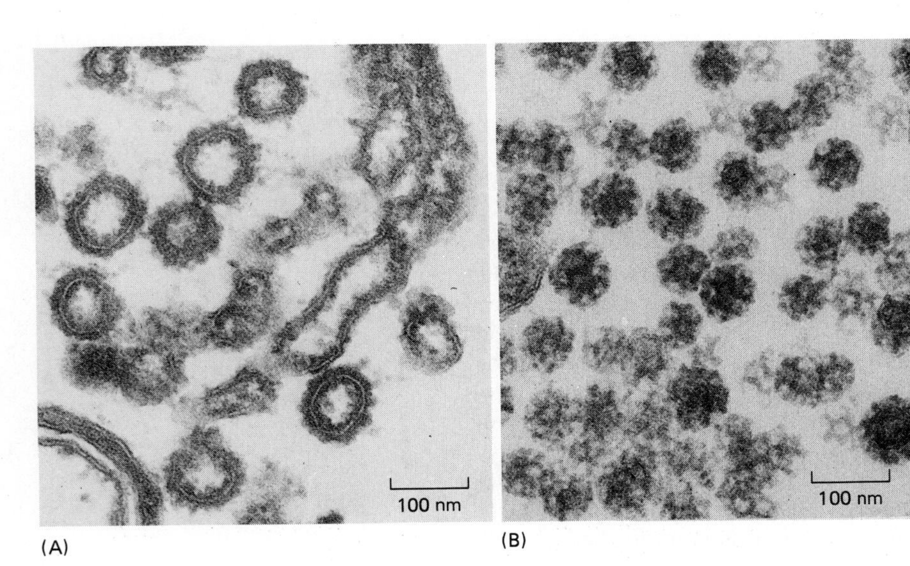

Figure 8–83 Comparison of clathrin- and non-clathrin-coated vesicles. (A) Electron micrograph of Golgi cisternae from a cell-free system in which non-clathrin-coated vesicles bud in the test tube. The clathrin-coated vesicles shown in (B) have a much more regular structure. (Electron micrographs courtesy of Lelio Orci, from L. Orci, B. Glick, and J. Rothman, *Cell* 46:171–184, 1986.)

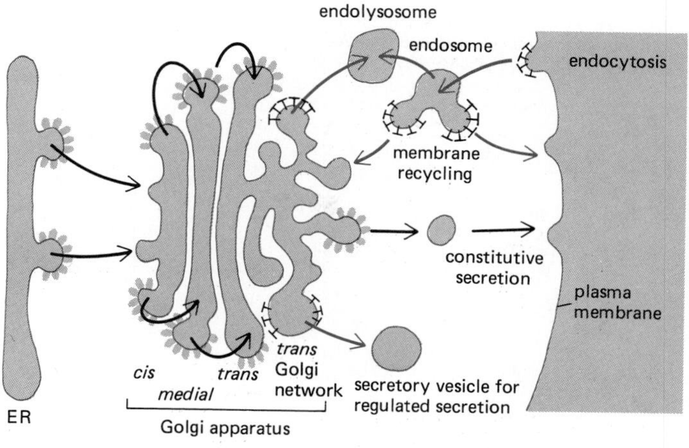

Figure 8–84 A hypothetical model for protein transport in nonpolarized cells. All unselected (constitutive) transport is postulated to be mediated by non-clathrin-coated vesicles. The various forms of signal-mediated transport are postulated to be carried out by clathrin-coated vesicles budding from the *trans* Golgi network, the endosomes, and the plasma membrane. In polarized cells an additional signaled pathway from the *trans* Golgi network is required.

Mutants in the Secretory Pathway Help to Identify the Transport Machinery[69]

Since vesicular transport is crucial to a eucaryotic cell, a mutant cell defective in a central component of this process—such as a docking marker or an acceptor protein (see Figure 8–73)—would probably die. If, however, the mutant protein is defective only at high temperatures, the mutant cell would survive at normal temperatures. More than 25 such *temperature-sensitive* mutations in genes involved in the secretory pathway have been identified in yeasts. When grown at high temperatures, some of these mutants fail to transport proteins from the ER to the Golgi apparatus, while others fail to transport proteins from one Golgi cisterna to another, or from the Golgi apparatus to the vacuole (the yeast lysosome) or to the plasma membrane.

The wild-type versions of these essential yeast genes can be readily cloned by transfecting wild-type DNA into the temperature-sensitive secretory mutants and then screening the transfected cells for survival at high temperatures. This approach is extremely powerful because one is led directly to the central proteins in the transport machinery without needing to know how secretion works. One gene required in the yeast secretory pathway that was isolated in this way is the *sec 4* gene, whose nucleotide sequence suggests that it encodes a GTP-binding protein of the *ras* family (see p. 705 and p. 757). Biochemical experiments in mammalian cells suggest that a similar GTP-binding protein functions in mammalian vesicular transport; it seems to regulate the uncoating of non-clathrin-coated vesicles prior to their fusion with membranes. These studies have depended on the ability to reconstitute vesicular transport in a cell-free system.

Cell-free Systems Provide Another Powerful Way to Analyze the Molecular Mechanisms of Vesicular Transport[70]

The key to understanding the molecular mechanisms that underlie the flow of traffic between membrane compartments is to elucidate the working parts of transport vesicles. How do transport vesicles bud off from membranes? What guides them to their targets? How do they fuse? In addition to the genetic approaches just described, a crucial step in trying to answer these questions is to reconstitute vesicular transport in a cell-free system. This was first achieved for the Golgi stack. When Golgi stacks are isolated from cells and incubated with cytosol and with ATP as a source of energy, non-clathrin-coated vesicles bud from their rims and apparently transport proteins between the cisternae (see Figure 8–83A). By following the progressive processing of the oligosaccharides on a glycoprotein as it moves from one Golgi compartment to the next, it was possible to reconstruct the process of vesicular transport *in vitro* (Figure 8–85).

The budding of the non-clathrin-coated vesicles in this cell-free system is found to require both ATP and a mixture of proteins from the cytosol, establishing

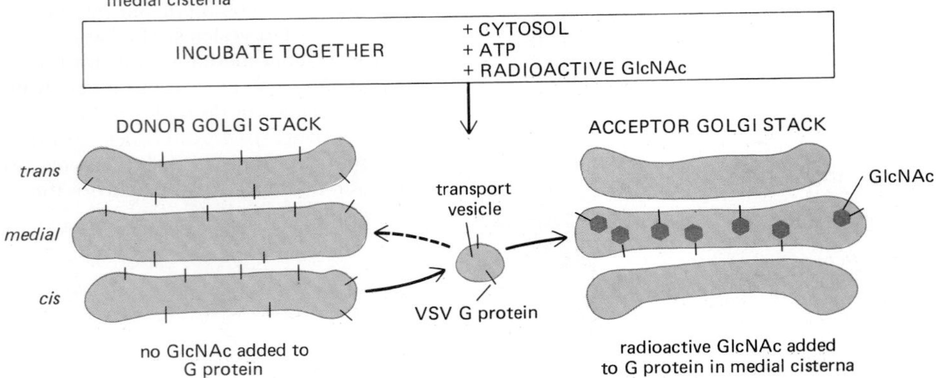

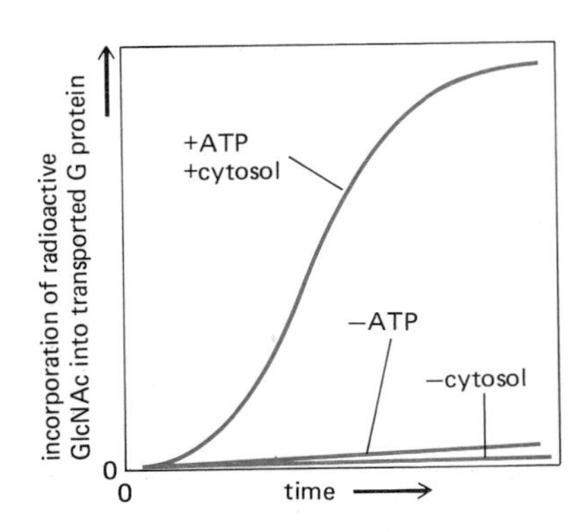

Figure 8–85 A cell-free system that reconstitutes the vesicular transport of a viral protein between the cisternal compartments of the Golgi apparatus. When isolated Golgi stacks are incubated with a cytosolic extract and ATP, non-clathrin-coated vesicles bud from the cisternae. Vesicular stomatitis virus (VSV) encodes an envelope glycoprotein called *G protein*. To measure the transport of this transmembrane protein between the cisternae of the Golgi apparatus, two populations of Golgi stacks are incubated together. The "donor" population is from VSV-infected cells of a mutant type lacking GlcNAc transferase I. As a result, the donor Golgi cannot incorporate GlcNAc residues into the *N*-linked oligosaccharides on the G protein (see Figure 8–63). The "acceptor" Golgi stacks are from uninfected wild-type cells and thus contain a good copy of GlcNAc transferase I but lack G protein. The addition of GlcNAc to G protein therefore requires the transport of G protein from the *cis* compartment of the donor Golgi stack to the *medial* compartment of the acceptor Golgi stack. This transport-coupled glycosylation is monitored by measuring the incorporation into G protein of ^{3}H-GlcNAc from UDP-^{3}H-GlcNAc added to the incubation medium. As shown on the right, transport occurs only when both ATP and cytosol are added, which are the same conditions needed for the non-clathrin-coated vesicles to bud. This type of scheme, first used to measure transport from *cis* to *medial* Golgi, has also allowed *in vitro* reconstitutions of transport from ER to *cis* Golgi, *medial* to *trans* Golgi, *trans* Golgi to plasma membrane, endosomes to lysosomes, and *trans* Golgi to endolysosomes.

that vesicle budding is an active, rather than a simple, self-assembly process. The necessary components have been highly conserved during evolution, since the cytosol from yeasts and from plants can be substituted for animal cell cytosol in promoting transport in the Golgi stacks isolated from animal cells. This suggests that it will be possible to combine the genetic approach in yeasts and mammalian cell-free systems to identify the molecules that constitute the transport machinery.

Summary

The ER and each of the compartments in the Golgi apparatus contain their own unique sets of proteins. It seems likely that these proteins are selectively retained in these organelles by a mechanism that depends on "retention signals" in the proteins, while much of the flow of materials from ER to the Golgi apparatus, through the Golgi stack, and from the trans *Golgi network to the cell surface occurs by default. It has been postulated that transport along the unselected default pathway is mediated by non-clathrin-coated transport vesicles that are nonselective with regard to their content, while signal-directed transport (sorting) is mediated by clathrin-coated vesicles. To test these ideas, the molecular mechanisms involved in transport vesicle budding, targeting, and fusion will have to be deciphered. Genetic studies in yeast cells have identified more than 25 genes whose products are required for particular steps in the pathway. In addition, reconstituted cell-free mammalian systems have been developed in which selective vesicular budding and fusion occur. The combination of genetics and biochemistry should allow the many proteins that mediate these processes to be identified and obtained in pure form so that their structures and activities can be ascertained.*

References

General

Burgess, T.L.; Kelly, R.B. Constitutive and regulated secretion of proteins. *Annu. Rev. Cell Biol.* 3:243–293, 1987.

Dingwall, C.; Laskey, R.A. Protein import into the cell nucleus. *Annu. Rev. Cell Biol.* 2:367–390, 1986.

Kornfeld, S. Trafficking of lysosomal enzymes. *FASEB J.* 1:462–468, 1987.

Pfeffer, S.R.; Rothman, J.E. Biosynthetic protein transport and sorting by the endoplasmic reticulum and Golgi. *Annu. Rev. Biochem.* 56:829–852, 1987.

Verner, K.; Schatz, G. Protein translocation across membranes. *Science* 241:1307–1313, 1988.

Cited

1. Bolender, R.P. Stereological analysis of the guinea pig pancreas. *J. Cell Biol.* 61:269–287, 1974.

   Palade, G.E.; Farquhar, M.G. Cell biology. In Pathophysiology: The Biological Principles of Disease (L.H. Smith, S.D. Thier, eds.), pp. 1–56. Philadelphia: Saunders, 1981.

   Weibel, E.R.; Staubli, W.; Gnagi, H.R.; Hess, F.A. Correlated morphometric and biochemical studies on the liver cell. *J. Cell Biol.* 42:68–91, 1969.
2. Blobel, G. Intracellular protein topogenesis. *Proc. Natl. Acad. Sci. USA* 77:1496–1500, 1980.

   Gray, M.W.; Doolittle, W.F. Has the endosymbiont hypothesis been proven? *Microbiol. Rev.* 46:1–42, 1982.

   Schwarz, R.M.; Dayhoff, M.O. Origins of prokaryotes, eukaryotes, mitochondria, and chloroplasts. *Science* 199:395–403, 1978.
3. Palade, G. Intracellular aspects of the process of protein synthesis. *Science* 189:347–358, 1975.
4. Kelly, R.B. Pathways of protein secretion in eukaryotes. *Science* 230:25–31, 1985.

   Sabatini, D.D.; Kreibich, G.; Morimoto, T.; Adesnik, M. Mechanisms for the incorporation of proteins in membranes and organelles. *J. Cell Biol.* 92:1–22, 1982.
5. Pfeffer, S.R.; Rothman, J.E. Biosynthetic protein transport and sorting by the endoplasmic reticulum and Golgi. *Annu. Rev. Biochem.* 56:829–852, 1987.

   Wickner, W.T.; Lodish, H.E. Multiple mechanisms of protein insertion into and across membranes. *Science*, 230:400–406, 1985.
6. Blobel, G. Intracellular protein topogenesis. *Proc. Natl. Acad. Sci. USA* 77:1496–1500, 1980.

   Garoff, H. Using recombinant DNA techniques to study protein targeting in the eucaryotic cell. *Annu. Rev. Cell Biol.* 1:403–445, 1985.
7. Warren, G. Membrane traffic and organelle division. *Trends Biochem. Sci.* 10:439–443, 1985.
8. Allen, R.D. The microtubule as an intracellular engine. *Sci. Am.* 238(2):42–49, 1987.

   Fulton, A.B. How crowded is the cytoplasm? *Cell* 30:345–347, 1982.

   Luby-Phelps, K.; Taylor, D.L.; Lanni, F. Probing the structure of cytoplasm. *J. Cell Biol.* 102:2015–2022, 1986.

   Vale, R.D. Intracellular transport using microtubule-based motors. *Annu. Rev. Cell Biol.* 3:347–378, 1987.
9. Chock, P.B.; Rhee, S.G.; Stadtman, E.R. Interconvertible enzyme cascades in cellular regulation. *Annu. Rev. Biochem.* 49:813–843, 1980.

   Holt, G.D.; et al. Nuclear pore complex glycoproteins contain cytoplasmically disposed O-linked *N*-acetylglucosamine. *J. Cell Biol.* 104:1157–1164, 1987.

   Wold, F. *In vivo* chemical modification of proteins (post-translational modification). *Annu. Rev. Biochem.* 50:783–814, 1981.
10. Kamps, M.P.; Buss, J.E.; Sefton, B.M. Mutation of NH_2-terminal glycine of $p60^{src}$ prevents both myristoylation and morphological transformation. *Proc. Natl. Acad. Sci. USA* 82:4625–4628, 1985.

    Schultz, A.M.; Henderson, L.E.; Orozlan, S. Fatty acylation of proteins. *Annu. Rev. Cell Biol.* 4:611–648, 1988.

    Willumsen, B.M.; Norris, K.; Papageorge, A.G.; Hubbert, N.L.; Lowy, D.R. Harvey murine sarcoma virus p21 *ras* protein: biological and biochemical significance of the cysteine nearest the carboxy terminus. *EMBO J.* 3:2582–2585, 1984.
11. Dice, J.F. Molecular determinants of protein half-lives in eucaryotic cells. *FASEB J.* 1:349–357, 1987.

    Goldberg, A.L.; Goff, S.A. The selective degradation of abnormal proteins in bacteria. In Maximizing Gene Expression (W. Reznikoff, L. Gold, eds.), pp. 287–314. Stoneham, MA: Butterworth, 1986.
12. Bachmair, A.; Finley, D.; Varshavsky, A. *In vivo* half-life of a protein is a function of its amino-terminal residue. *Science* 234:179–186, 1986.

    Hershko, A.; Ciechanover, A. The ubiquitin pathway for the degradation of intracellular proteins. *Prog. Nucleic Acid Res. Mol. Biol.* 33:19–56, 1986.

    Rechsteiner, M. Ubiquitin-mediated pathways for intracellular proteolysis. *Annu. Rev. Cell Biol.* 3:1–30, 1987.
13. Augen, J.; Wold, F. How much sequence information is needed for the regulation of amino-terminal acetylation of eukaryotic proteins? *Trends Biochem. Sci.* 11:494–497, 1986.

    Ferber, S.; Ciechanover, A. Role of arginine-tRNA in protein degradation by the ubiquitin pathway. *Nature* 326:808–811, 1987.

    Varshavsky, A.; Bachmair, A.; Finley, D.; Gonda, D.; Wünning, I. The N-end rule of selective protein turnover: mechanistic aspects and functional implications. In Ubiquitin (M. Rechsteiner, ed.), pp. 284–324. New York: Plenum, 1988.
14. Craig, E.A. The heat-shock response. *CRC Crit. Rev. Biochem.* 18:239–280, 1985.

    Lindquist, S. The heat-shock response. *Annu. Rev. Biochem.* 55:1151–1191, 1986.

Pelham, H.R.B. Speculations on the functions of the major heat shock and glucose-regulated proteins. *Cell* 46:959–961, 1986.

15. Franke, W.W.; Scheer, U.; Krohne, G.; Jarasch, E.D. The nuclear envelope and the architecture of the nuclear periphery. *J. Cell Biol.* 91:39s–50s, 1981.

Newport, J.W.; Forbes, D.J. The nucleus: structure, function, and dynamics. *Annu. Rev. Biochem.* 56:535–565, 1987.

16. Bonner, W.M. Protein migration and accumulation in nuclei. In The Cell Nucleus (H. Busch, ed.), Vol. 6, Part C, pp. 97–148. New York: Academic, 1978.

Lang, I.; Scholz, M.; Peters, R. Molecular mobility and nucleocytoplasmic flux in hepatoma cells. *J. Cell Biol.* 102:1183–1190, 1986.

17. Dingwall, C.; Laskey, R.A. Protein import into the cell nucleus. *Annu. Rev. Cell Biol.* 2:367–390, 1986.

Feldherr, C.M.; Kallenbach, E.; Schultz, N. Movement of a karyophilic protein through the nuclear pores of oocytes, *J. Cell Biol.* 99:2216–2222, 1984.

Newmeyer, D.D.; Forbes, D.J. Nuclear import can be separated into distinct steps *in vitro:* nuclear pore binding and translocation. *Cell* 52:641–653, 1988.

18. Goldfarb, D.S.; Gariépy, J.; Schoolnik, G.; Kornberg, R.D. Synthetic peptides as nuclear localization signals. *Nature* 322:641–644, 1986.

Kalderon, D.; Roberts, B.L.; Richardson, W.D.; Smith, A.E. A short amino acid sequence able to specify nuclear location. *Cell* 39:499–509, 1984.

Lanford, R.E.; Butel, J.S. Construction and characterization of an SV40 mutant defective in nuclear transport of T antigen. *Cell* 37: 801–813, 1984.

19. Clawson, G.A.; Feldherr, C.M.; Smuckler, E.A. Nucleocytoplasmic RNA transport. *Mol. Cell. Biochem.* 67:87–100, 1985.

Dworetzky, S.I.; Feldherr, C.M. Translocation of RNA-coated gold particles through the nuclear pores of oocytes. *J. Cell Biol.* 106:575–584, 1988.

20. Attardi, G.; Schatz, G. Biogenesis of mitochondria. *Annu. Rev. Cell Biol.* 4:289–333, 1988.

Tzagoloff, A. Mitochondria. New York: Plenum, 1982.

21. Hawlitschek, G.; et al. Mitochondrial protein import: identification of processing petidase and of PEP, a processing enhancing protein. *Cell* 53:795–806, 1988.

Hurt, E.C.; van Loon, A.P.G.M. How proteins find mitochondria and intramitochondrial compartments. *Trends Biochem. Sci.* 11:204–207, 1986.

Pfanner, N.; Neupert, W. Biogenesis of mitochondrial energy transducing complexes. *Curr. Top. Bioenerg.* 15:177–219, 1987.

Roise, D.; Schatz, G. Mitochondrial presequences. *J. Biol. Chem.* 263:4509–4511, 1988.

22. Eilers, M.; Schatz, G. Protein unfolding and the energetics of protein translocation across biological membranes. *Cell* 52:481–483, 1988.

Pfanner, N.; Neupert, W. Transport of proteins into mitochondria: a potassium diffusion potential is able to drive the import of ADP/ATP carrier. *EMBO J.* 4:2819–2825, 1985.

Roise, D.; Horvath, S.J.; Tomich, J.M.; Richards, J.H.; Schatz, G. A chemically synthesized pre-sequence of an imported mitochondrial protein can form an amphiphilic helix and perturb natural and artificial phospholipid bilayers. *EMBO J.* 5:1327–1334, 1986.

23. Schleyer, M.; Neupert, W. Transport of proteins into mitochondria: translocational intermediates spanning contact sites between outer and inner membranes. *Cell* 43: 339–350, 1985.

Schwaiger, M.; Herzog, V.; Neupert, W. Characterization of tranlocation contact sites involved in the import of mitochondrial proteins. *J. Cell Biol.* 105:235–246, 1987.

24. Deshaies, R.J.; Koch, B.D.; Werner-Washburne, M.; Craig, E.A.; Schekman, R. A subfamily of stress proteins facilitates translocation of secretory and mitochondrial precursor polypeptides. *Nature* 332:800–805, 1988.

Eilers, M.; Schatz, G. Binding of a specific ligand inhibits import of a purified precursor protein into mitochondria. *Nature* 322:228–232, 1986.

Pfanner, N.; Tropschug, M.; Neupert, W. Mitochondrial protein import: nucleoside triphosphates are involved in conferring import competence to precursors. *Cell* 49:815–823, 1987.

25. Hartl, F.U.; Ostermann, J.; Guiard, B.; Neupert, W. Successive translocation into and out of the mitochondrial matrix: targeting of proteins to the intermembrane space by a bipartite signal peptide. *Cell* 51:1027–1037, 1987.

van Loon, A.P.G.M.; Brandli, A.W.; Schatz, G. The presequences of two imported mitochondrial proteins contain information for intracellular and intramitochondrial sorting. *Cell* 44:801–812, 1986.

26. Pfaller, R.; Neupert, W. High-affinity binding sites involved in the import of porin into mitochondria. *EMBO J.* 6:2635–2642, 1987.

Pfanner, N.; et al. Role of ATP in mitochondrial protein import. *J. Biol. Chem.* 263:4049–4051, 1988.

27. Boutry, M.; Nagy, F.; Poulsen, G.; Aoyagi, K.; Chua, N.H. Targeting of bacterial chloramphenicol acetyltransferase to mitochondria in transgenic plants. *Nature* 328:340–342, 1987.

Pain, D.; Kanwar, Y.S.; Blobel, G. Identification of a receptor for protein import into chloroplasts and its localization to envelope contact zones. *Nature* 331:232–237, 1988.

Schmidt, G.W.; Mishkind, M.L. The transport of proteins into chloroplasts. *Annu. Rev. Biochem.* 55:879–912, 1986.

Smeekens, S.; Bauerie, C.; Hageman, J.; Keegstra, K.; Weisbeek, P. The role of the transit peptide in the routing of precursors toward different chloroplast compartments. *Cell* 46:365–375, 1986.

28. de Duve, C. Microbodies in the living cell. *Sci. Am.* 248(5):74–84, 1983.

de Duve, C.; Baudhuin, P. Peroxisomes (microbodies and related particles). *Physiol. Rev.* 46:323–357, 1966.

Fahimi, H.D.; Sies, H., eds. Peroxisomes in Biology and Medicine. Heidelberg: Springer, 1987.

29. Tolbert, N.E.; Essner, E. Microbodies: peroxisomes and glyoxysomes. *J. Cell Biol.* 91:271s–283s, 1981.

Veenbuis, M.; Van Dijken, J.P.; Harder, W. The significance of peroxisomes in the metabolism of one-carbon compounds in yeasts. *Adv. Microb. Physiol.* 24:1–82, 1983.

30. Gould, S.J.; Keller, G.A.; Subramani, S. Identification of a peroxisomal targeting signal at the carboxy terminus of four peroxisomal proteins. *J. Cell Biol.* 107:897–905, 1988.

Imanaka, T.; Small, G.M.; Lazarow, P.B. Translocation of acyl-CoA oxidase into peroxisomes requires ATP hydrolysis but not a membrane potential. *J. Cell Biol.* 105:2915–2922, 1987.

Lazarow, P.B.; Fujiki, Y. Biogenesis of peroxisomes. *Annu. Rev. Cell Biol.* 1:489–530, 1985.

31. DePierre, J.W.; Dallner, G. Structural aspects of the membrane of the endoplasmic reticulum. *Biochim. Biophys. Acta* 415: 411–472, 1975.

Fawcett, D. The Cell, 2nd ed., pp. 303–352. Philadelphia: Saunders, 1981.

Lee, C.; BoChen, L. Dynamic behavior of endoplasmic reticulum in living cells. *Cell* 54:37–46, 1988.

32. Adelman, M.R.; Sabatini, D.D.; Blobel, G. Ribosome-membrane interaction: nondestructive disassembly of rat liver rough microsomes into ribosomal and membranous components. *J. Cell Biol.* 56:206–229, 1973.

Blobel, G.; Dobberstein, B. Transfer of proteins across membranes. *J. Cell Biol.* 67:852–862, 1975.

33. Jones, A.L.; Fawcett, D.W. Hypertrophy of the agranular endoplasmic reticulum in hamster liver induced by phenobarbital. *J. Histochem. Cytochem.* 14:215–232, 1966.

Mori, H.; Christensen, A.K. Morphometric analysis of Leydig cells in the normal rat testis. *J. Cell Biol.* 84:340–354, 1980.

34. Dallner, G. Isolation of rough and smooth microsomes—general. *Methods Enzymol.* 31:191–201, 1974.

de Duve, C. Tissue fractionation past and present. *J. Cell Biol.* 50:20d–55d, 1971.

35. Hortsch, M.; Avossa, D.; Meyer, D.I. Characterization of secretory protein translocation: ribosome-membrane interaction in endoplasmic reticulum. *J. Cell Biol.* 103:241–253, 1986.

Kreibich, G.; Ulrich, B.L.; Sabatini, D.D. Proteins of rough microsomal membranes related to ribosome binding. *J. Cell Biol.* 77:464–487, 1978.

36. Blobel, G.; Dobberstein, B. Transfer of proteins across membranes. *J. Cell Biol.* 67:835–851, 1975.

Garoff, H. Using recombinant DNA techniques to study protein targeting in the eucaryotic cell. *Annu. Rev. Cell Biol.* 1:403–445, 1985.

Milstein, C.; Brownlee, G.; Harrison, T.; Mathews, M.B. A possible precursor of immunoglobulin light chains. *Nature New Biol.* 239:117–120, 1972.

von Heijne, G. Signal sequences: the limits of variation. *J. Mol. Biol.* 184:99–105, 1985.

37. Meyer, D.I.; Krause, E.; Dobberstein, B. Secretory protein translocation across membranes—the role of the "docking protein." *Nature* 297:647–650, 1982.

Tajima, S.; Lauffer, L.; Rath, V.L.; Walter, P. The signal recognition particle receptor is a complex that contains two distinct polypeptide chains. *J. Cell Biol.* 103:1167–1178, 1986.

Walter, P.; Blobel, G. Signal recognition particle contains a 7S RNA essential for protein translocation across the endoplasmic reticulum. *Nature* 299:691–698, 1982.

Walter, P.; Lingappa, V.R. Mechanism of protein translocation across the endoplasmic reticulum membrane. *Annu. Rev. Cell Biol.* 2:499–516, 1986.

Wiedmann, M.; Kurzchalia, T.V.; Hartmann, E.; Rapoport, T.A. A signal sequence receptor in the endoplasmic reticulum membrane. *Nature* 328:830–833, 1987.

38. Chirico, W.J.; Waters, M.G.; Blobel, G. 70K heat shock related proteins stimulate protein translocation into microsomes. *Nature* 332:805–810, 1988.

Perara, E.; Rothman, R.E.; Lingappa, V.R. Uncoupling translocation from translation: implications for transport of proteins across membranes. *Science* 232:348–352, 1986.

Zimmermann, R.; Meyer, D.I. 1986: A year of new insights into how proteins cross membranes. *Trends Biochem. Sci.* 11: 512–515, 1986.

39. Rapoport, T.A. Extensions of the signal hypothesis—sequential insertion model versus amphipathic tunnel hypothesis. *FEBS Lett.* 187:1–10, 1985.

Wickner, W.T.; Lodish, H.F. Multiple mechanisms of protein insertion into and across membranes. *Science* 230:400–406, 1985.

40. Engelman, D.M.; Steitz, T.A.; Goldman, A. Identifying nonpolar transbilayer helices in amino acid sequences of membrane proteins. *Annu. Rev. Biophys. Biophys. Chem.* 15:321–353, 1986.

Kaiser, C.A.; Preuss, D.; Grisafi, P.; Botstein, D. Many random sequences functionally replace the secretion signal sequence of yeast invertase. *Science* 235:312–317, 1987.

Kyte, J.; Doolittle, R.F. A simple method for displaying the hydropathic character of a protein. *J. Mol. Biol.* 157:105–132, 1982.

Zerial, M.; Huylebroeck, D.; Garoff, H. Foreign transmembrane peptides replacing the internal signal sequence of transferrin receptor allow its translocation and membrane binding. *Cell* 48:147–155, 1987.

41. Bole, D.G.; Hendershot, L.M.; Kearney, J.F. Posttranslational association of immunoglobulin heavy chain binding protein with nascent heavy chains in nonsecreting and secreting hydridomas. *J. Cell Biol.* 102:1558–1566, 1986.

Lodish, H.F. Transport of secretory and membrane glycoproteins from the rough endoplasmic reticulum to the Golgi. *J. Biol. Chem.* 263:2107–2110, 1988.

Munro, S.; Pelham, H.R.B. A C-terminal signal prevents secretion of luminal ER proteins. *Cell* 48:899–907, 1987.

42. Freedman, R. Native disulphide bond formation in protein biosynthesis: evidence for the role of protein disulphide isomerase. *Trends Biochem. Sci.* 9:438–441, 1984.

Holmgren, A. Thioredoxin. *Annu. Rev. Biochem.* 54:237–272, 1985.

43. Hirschberg, C.B.; Snider, M.D. Topography of glycosylation in the rough endoplasmic reticulum and Golgi apparatus. *Annu. Rev. Biochem.* 56:63–87, 1987.

Kornfeld, R.; Kornfeld, S. Assembly of asparagine-linked oligosaccharides. *Annu. Rev. Biochem.* 54:631–664, 1985.

Torres, C.; Hart, G. Topography and polypeptide distribution of terminal *N*-acetylglucosamine residues on the surface of intact lymphocytes. *J. Biol. Chem.* 259:3308–3317, 1984.

44. Cross, G.A.M. Eukaryotic protein modification and membrane attachment via phosphatidylinositol. *Cell* 48:179–181, 1987.

Ferguson, M.A.J.; Williams, A.F. Cell-surface anchoring of proteins via glycosyl-phosphatidylinositol structures. *Annu. Rev. Biochem.* 57:285–320, 1988.

Low, M.G.; Saltiel, A.R. Structural and functional roles of glycosyl-phosphatidylinositol in membranes. *Science* 239: 268–275, 1988.

45. Bishop, W.R.; Bell, R.M. Assembly of phospholipids into cellular membranes: biosynthesis, transmembrane movement, and intracellular translocation. *Annu. Rev. Cell Biol.* 4:579–610, 1988.

Bishop, W.R.; Bell, R.M. Assembly of the endoplasmic reticulum phospholipid bilayer: the phosphatidylcholine transporter. *Cell* 42:51–60, 1985.

Dawidowicz, E.A. Dynamics of membrane lipid metabolism and turnover. *Annu. Rev. Biochem.* 56:43–61, 1987.

Pagano, R.E.; Sleight, R.G. Defining lipid transport pathways in animal cells. *Science* 229:1051–1057, 1985.

Rothman, J.E.; Lenard, J. Membrane asymmetry. *Science* 195: 743–753, 1977.

46. Dawidowicz, E.A. Lipid exchange: transmembrane movement, spontaneous movement, and protein-mediated transfer of lipids and cholesterol. *Curr. Top. Memb. Transp.* 29:175–202, 1987.

Yaffe, M.P.; Kennedy, E.P. Intracellular phospholipid movement and the role of phospholipid transfer proteins in animal cells. *Biochemistry* 22:1497–1507, 1983.

47. Farquhar, M.G.; Palade, G.E. The Golgi apparatus (complex)—(1954–1981)—from artifact to center stage. *J. Cell Biol.* 91: 77s–103s, 1981.

Pavelka, M. Functional morphology of the Golgi apparatus. *Adv. Anat. Embryol. Cell Biol.* 106:1–94, 1987.

Rothman, J.E. The compartmental organization of the Golgi apparatus. *Sci. Am.* 253(3):74–89, 1985.

48. Hubbard, S.C.; Ivatt, R.J. Synthesis and processing of asparagine-linked oligosaccharides. *Annu. Rev. Biochem.* 50:555–583, 1981.

Kornfeld, R.; Kornfeld, S. Assembly of asparagine-linked oligosaccharides. *Annu. Rev. Biochem.* 54:631–664, 1985.

Schachter, H.; Roseman, S. Mammalian glycosyltransferases: their role in the synthesis and function of complex carbohydrates and glycolipids. In The Biochemistry of Glycoproteins and Proteoglycans (W.J. Lennarz, ed.), Chapter 3. New York: Plenum, 1980.

49. Elbein, A.D. Inhibitors of the biosynthesis and processing of N-linked oligosaccharide chains. *Annu. Rev. Biochem.* 56:497–534, 1987.

 Stanley, P. Glycosylation mutants and the functions of mammalian carbohydrates. *Trends Genet.* 3:77–81, 1987.

 West, C.M. Current ideas on the significance of protein glycosylation. *Mol. Cell. Biochem.* 72:3–20, 1986.

50. Hassell, J.R.; Kimura, J.H.; Hascall, V.C. Proteoglycan core protein families. *Annu. Rev. Biochem.* 55:539–567, 1986.

 Huttner, W.B. Tyrosine sulfation and the secretory pathway. *Annu. Rev. Physiol.* 50:363–376, 1988.

 Ruoslahti, F. Structure and biology of proteoglycans. *Annu. Rev. Cell Biol.* 4:229–255, 1988.

 Wagh, P.V.; Bahl, O.P. Sugar residues on proteins. *CRC Crit. Rev. Biochem.* 10:307–377, 1981.

51. Douglass, J.; Civelli, O.; Herbert, E. Polyprotein gene expression: generation of diversity of neuroendocrine peptides. *Annu. Rev. Biochem.* 53:665–715, 1984.

 Orci, L.; et al. Conversion of proinsulin to insulin occurs coordinately with acidification of maturing secretory vesicles. *J. Cell Biol.* 103:2273–2281, 1986.

52. Dunphy, W.G.; Rothman, J.E. Compartmental organization of the Golgi stack. *Cell* 42:13–21, 1985.

53. Bainton, D. The discovery of lysosomes. *J. Cell Biol.* 91:66s–76s, 1981.

 de Duve, C. Exploring cells with a centrifuge. *Science* 189:186–194, 1975.

54. Holzman, E. Lysosomes: A Survey. New York: Springer-Verlag, 1976.

55. Griffiths, G.; Hoflack, B.; Simons, K.; Mellman, I.; Kornfeld, S. The mannose 6-phosphate receptor and the biogenesis of lysosomes. *Cell* 52:329–341, 1988.

 Helenius, A.; Mellman, I.; Wall, D.; Hubbard, A. Endosomes. *Trends Biochem. Sci.* 8:245–250, 1983.

 Mayer, R.J.; Doherty, F. Intracellular protein catabolism: state of the art. *FEBS Lett.* 198:181–193, 1986.

 Mellman, I.; Fuchs, R.; Helenius, A. Acidification of the endocytic and exocytic pathways. *Annu. Rev. Biochem.* 55: 663–700, 1986,

 Silverstein, S.C.; Steinman, R.M.; Cohn, Z.A. Endocytosis. *Annu. Rev. Biochem.* 46:669–722, 1977.

56. Dahms, N.M.; Lobel, P.; Breitmeyer, J.; Chirgwin, J.M.; Kornfeld, S. 46 kd mannose 6-phosphate receptor: cloning, expression, and homology to the 215 kd mannose 6-phosphate receptor. *Cell* 50:181–192, 1987.

 Kornfeld, S. Trafficking of lysosomal enzymes. *FASEB J.* 1: 462–468, 1987.

 Pfeffer, S.R. Mannose 6-phosphate receptors and their role in targeting of proteins to lysosomes. *J. Membr. Biol.* 103:7–16, 1988.

 von Figura, K.; Hasilik, A. Lysosomal enzymes and their receptors. *Annu. Rev. Biochem.* 55:167–193, 1986.

57. Brown, W.J.; Goodhouse, J.; Farquhar, M.G. Mannose 6-phosphate receptors for lysosomal enzymes cycle between the Golgi complex and endosomes. *J. Cell Biol.* 103:1235–1247, 1986.

 Duncan, J.R.; Kornfeld, S. Intracellular movement of two mannose 6-phosphate receptors: return to the Golgi apparatus. *J. Cell Biol.* 106:617–628, 1988.

 Geuze, H.J.; Slot, J.W.; Strous, G.J.A.M.; Hasilik, A.; von Figura, K. Possible pathways for lysosomal enzyme delivery. *J. Cell Biol.* 101:2253–2262, 1985.

 Rothman, J.E.; Schmid, S.L. Enzymatic recycling of clathrin from coated vesicles. *Cell* 46:5–9, 1986.

58. Lang, L.; Reitman, M.; Tang, J.; Roberts, R.M.; Kornfeld, S. Lysosomal enzyme phosphorylation. *J. Biol. Chem.* 259: 14663–14671, 1984.

 Reitman, M.L.; Kornfeld, S. Lysosomal enzyme targeting. N-acetylglucosaminylphosphotransferase selectively phosphorylates native lysosomal enzymes. *J. Biol. Chem.* 256:11977–11980, 1981.

59. Kornfeld, S. Trafficking of lysosomal enzymes in normal and disease states. *J. Clin. Invest.* 77:1–6, 1986.

 Neufeld, E.F.; Lim, T.W.; Shapiro, L.J. Inherited disorders of lysosomal metabolism. *Annu. Rev. Biochem.* 44:357–376, 1975.

60. Burgess, T.L.; Kelly, R.B. Constitutive and regulated secretion of proteins. *Annu. Rev. Cell Biol.* 3:243–293, 1987.

61. Griffiths, G.; Simons, K. The *trans*-Golgi network: sorting at the exit site of the Golgi complex. *Science* 234:438–443, 1986.

 Orci, L.; et al. The trans-most cisternae of the Golgi complex: a compartment for sorting of secretory and plasma membrane proteins. *Cell* 51:1039–1051, 1987.

62. Herzog, V.; Farquhar, M.G. Luminal membrane retrieved after exocytosis reaches most Golgi cisternae in secretory cells. *Proc. Natl. Acad. Sci. USA* 74:5073–5077, 1977.

 Snider, M.D.; Rogers, O.C. Membrane traffic in animal cells: cellular glycoproteins return to the site of Golgi mannosidase I. *J. Cell Biol.* 103:265–275, 1986.

63. Lodish, H.F. Transport of secretory and membrane glycoproteins from the rough endoplasmic reticulum to the Golgi. *J. Biol. Chem.* 263:2107–2110, 1988.

 Rothman, J.E. Protein sorting by selective retention in the endoplasmic reticulum and Golgi stack. *Cell* 50:521–522, 1987.

 Wieland, F.T.; Gleason, M.L.; Serafini, T.A.; Rothman, J.E. The rate of bulk flow from the endoplasmic reticulum to the cell surface. *Cell* 50:289–300, 1987.

64. Bartles, J.R.; Hubbard, A.L. Plasma membrane protein sorting in epithelial cells: do secretory pathways hold the key? *Trends Biochem. Sci.* 13:181–184, 1988.

 Matlin, K.S. The sorting of proteins to the plasma membrane in epithelial cells. *J. Cell Biol.* 103:2565–2568, 1986.

 Mostov, K.E.; Breitfeld, P.; Harris, J.M. An anchor-minus form of the polymeric immunoglobulin receptor is secreted predominantly apically in Madin-Darby canine kidney cells. *J. Cell Biol.* 105:2031–2036, 1987.

 Simons, K.; Fuller, S.D. Cell surface polarity in epithelia. *Annu. Rev. Cell Biol.* 1:243–288, 1985.

65. Simons, K.; Garoff, H.; Helenius, A. How an animal virus gets into and out of its host cell. *Sci. Am.* 246(2):58–66, 1982.

 Simons, K.; Warren, G. Semliki forest virus: a probe for membrane traffic in the animal cell. *Adv. Protein Chem.* 36:79–132, 1984.

66. Rodriguez-Boulan, E.J. Membrane biogenesis, enveloped RNA viruses, and epithelial polarity. In Modern Cell Biology. Vol. 1 (J.R. McIntosh, B.H. Satir, eds.), pp.119–170, 1983.

 Rodriguez-Boulan, E.J.; Sabatini, D.D. Asymmetric budding of viruses in epithelial monolayers: a model system for the study of epithelial polarity. *Proc. Natl. Acad. Sci. USA* 75:5071–5075, 1978.

 Roth, M.G.; Sirnivas, R.V.; Compans, R.W. Basolateral maturation of retroviruses in polarized epithelial cells. *J. Virol.* 45:1065–1073, 1983.

 Strauss, E.G.; Strauss, J.H. Assembly of enveloped animal viruses. In Virus Structure and Assembly (S. Casjens, ed.), Chapter 6. Boston: Jones and Bartlett, 1985.

67. Rothman, J.E. Protein sorting by selective retention in the endoplasmic reticulum and Golgi stack. *Cell* 50:521–522, 1987.

68. Griffiths, G.; Pfeiffer, S.; Simons, K.; Matlin, K. Exit of newly synthesized membrane proteins from the *trans* cisterna of the Golgi complex to the plasma membrane. *J. Cell Biol.* 101:949–964, 1985.

Orci, L.; Glick, B.S.; Rothman, J.E. A new type of coated vesicular carrier that appears not to contain clathrin: its possible role in protein transport within the Golgi stack. *Cell* 46:171–184, 1986.

69. Bourne, H. Do GTPases direct membrane traffic in secretion? *Cell* 53:669–671, 1988.

Novick, P.; Field, C.; Schekman, R. Identification of 23 complementation groups required for post-translational events in the yeast secretory pathway. *Cell* 21:205–215, 1980.

Schekman, R. Protein localization and membrane traffic in yeast. *Annu. Rev. Cell Biol.* 1:115–143, 1985.

70. Balch, W.E.; Dunphy, W.G.; Braell, W.A.; Rothman, J.E. Reconstitution of the transport of protein between successive compartments of the Golgi measured by the coupled incorporation of *N*-acetylglucosamine. *Cell* 39:405–416, 1984.

Dunphy, W.G.; et al. Yeast and mammals utilize similar cytosolic components to drive protein transport through the Golgi complex. *Proc. Natl. Acad. Sci. USA* 83:1622–1626, 1986.

Fries, E.; Rothman, J.E. Transport of vesicular stomatitis virus glycoprotein in a cell-free extract. *Proc. Natl. Acad. Sci. USA* 77:3870–3874, 1980.

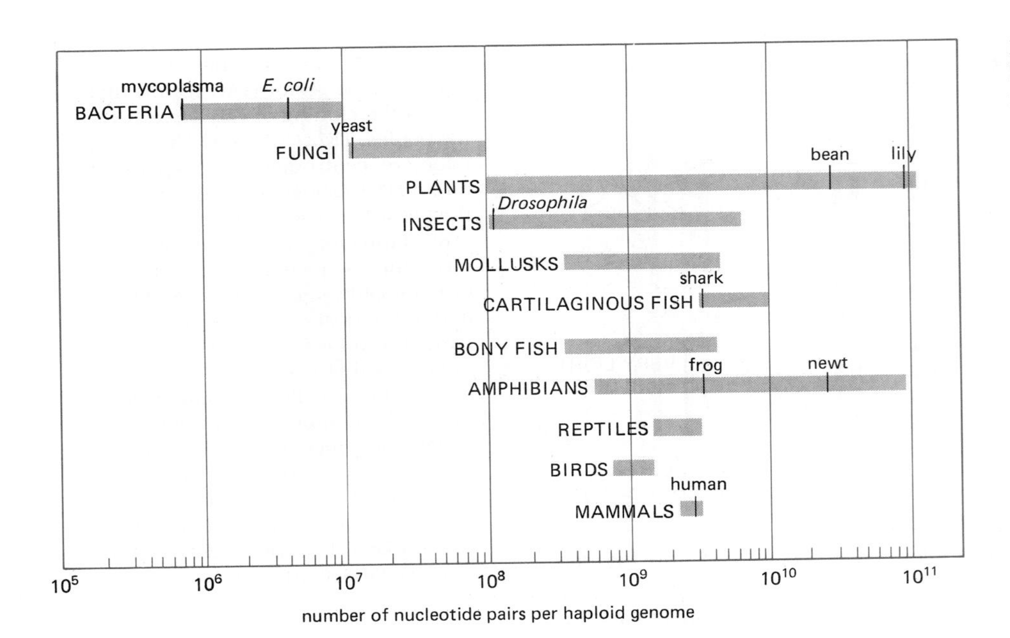

Figure 9–6 The amount of DNA in a haploid genome varies over a 100,000-fold range from the smallest procaryotic cell, the mycoplasma, to the large cells of some plants and amphibia. Note that the genome size of humans (3×10^9 nucleotide pairs) is much smaller than that of some other organisms.

about 10 times more complex than the fruit fly *Drosophila*, which is estimated to have about 5000 essential genes (see p. 510).

Whatever the nonessential DNA in higher eucaryotic chromosomes may do (see Chapter 10, p. 607), the data shown in Figure 9–6 make it clear that it is not a great handicap for a higher eucaryotic cell to carry a large amount of extra DNA. Indeed, even the essential coding regions are often interrupted by long stretches of noncoding DNA.

Each Gene Is a Complex Functional Unit for the Regulated Production of an RNA Molecule

The primary function of the genome is to produce RNA molecules. Selected portions of the DNA nucleotide sequence are copied into a corresponding RNA nucleotide sequence, which either (as mRNA) encodes a protein or forms a "structural" RNA, such as a tRNA or rRNA molecule. Each region of the DNA helix that produces a functional RNA molecule constitutes a **gene.**

Genes in a chromosome of a higher eucaryote can contain as many as 2 million DNA nucleotide pairs, and genes more than 100,000 nucleotide pairs in length are common (Table 9–1); yet only about 1000 nucleotide pairs are required to encode

Table 9–1 The Size of Some Human Genes in Thousands of Nucleotides

	Gene Size	mRNA Size	Number of Introns
β-Globin	1.5	0.6	2
Insulin	1.7	0.4	2
Protein kinase C	11	1.4	7
Albumin	25	2.1	14
Catalase	34	1.6	12
LDL receptor	45	5.5	17
Factor VIII	186	9	25
Thyroglobulin	300	8.7	36
Dystrophin*	more than 2000	17	more than 50

The size specified here for a gene includes both its transcribed portion and nearby regulatory DNA sequences. (Compiled from data supplied by Victor McKusick.)

*An altered form of this gene causes Duchenne muscular dystrophy.

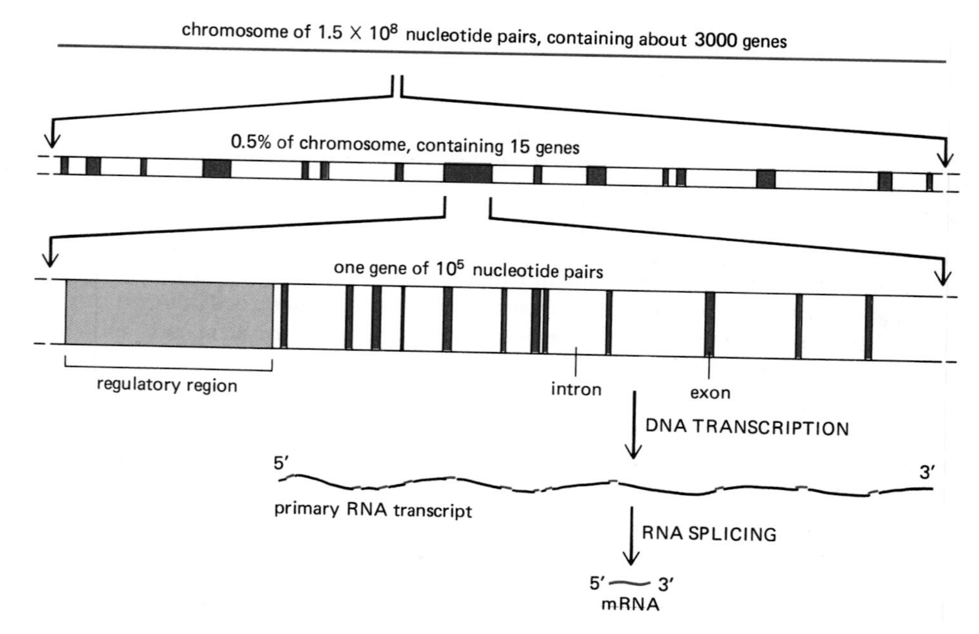

Figure 9–7 The organization of genes on a typical vertebrate chromosome. Proteins that bind to the DNA in regulatory regions determine whether a gene is transcribed; although often located on the 5′ side of a gene, as shown here, regulatory regions can also be located in introns, in exons, or on the 3′ side of a gene. The intron sequences are removed from the primary RNA transcripts that encode protein molecules to produce a messenger RNA (mRNA) molecule. The figure given here for the number of genes per chromosome is only a minimal estimate.

a protein of average size (one containing 300 to 400 amino acid residues). Most of the extra length consists of long stretches of noncoding DNA that interrupt the relatively short segments of coding DNA. The coding sequences are called **exons,** the intervening (noncoding) sequences are called **introns.** The RNA molecule (called a *primary RNA transcript*) synthesized from such a gene is altered to remove the intron sequences during its conversion to an mRNA molecule (see Figure 9–2) in the process of *RNA splicing* (see p. 531).

Large genes consist of a long string of alternating exons and introns, with most of the gene consisting of introns. In addition, each gene contains *regulatory DNA sequences,* which bind *gene regulatory proteins* that control transcription of the gene. Many regulatory sequences are located "upstream" (on the 5′ side) of the site where the RNA transcript begins, but they can also be located in introns, "downstream" (on the 3′ side) of the site where the RNA transcript ends, or even in exons. A typical vertebrate chromosome is illustrated schematically in Figure 9–7, along with one of its many genes.

9-6 Comparisons Between the DNAs of Related Organisms Distinguish Conserved and Nonconserved Regions of DNA Sequence[5]

Technical improvements in DNA sequencing are expected to allow the routine sequencing of stretches of chromosomal DNA that are millions of nucleotide pairs long, so that one can foresee the eventual determination of the sequence of all 3×10^9 nucleotides of the human genome. If more than 90% of this sequence is unimportant, however, it will be crucial to have some way of identifying the small proportion of sequence that is important. One way to achieve this is by the simultaneous sequencing of the corresponding regions of a related genome, such as that of the mouse. Human beings and mice are thought to have diverged from a common mammalian ancestor about 80×10^6 years ago, which is long enough for roughly two out of every three nucleotides to have been changed by random mutational events (see p. 220). Consequently, the only regions that will have remained closely similar (*conserved* regions) in the two genomes are those where mutations would impair function. (The organisms with these deleterious mutations would have been eliminated from the population by natural selection—see p. 221.) Thus, in general, *nonconserved* regions represent noncoding DNA—both between genes and in introns—whose DNA sequence is not critical for function. Conserved regions, in contrast, represent functionally important exons and regulatory regions. By revealing in this way the results of a very long natural "experiment," comparative DNA sequencing studies highlight the most interesting re-

Highly Condensed Chromatin Replicates Late in S Phase[37]

In higher eucaryotic cells, some regions of the DNA are more condensed than others. We have seen, for example, that *heterochromatin* remains in a highly condensed conformation (similar to that at mitosis) during interphase, while *active chromatin* assumes an especially decondensed conformation, which is apparently required to allow RNA synthesis (see p. 580).

One important clue to the mechanism that determines the timing of DNA replication is the observation that the blocks of heterochromatin, including the regions near the centromere that remain condensed throughout interphase, are replicated very late in the S phase. Late replication could thus be related to the packing of the DNA in chromatin. This conclusion is supported by the timing of replication of the two X chromosomes in a female mammalian cell. While these two chromosomes contain essentially the same DNA sequences, one is active and the other is not (see p. 577). Nearly all the inactive X chromosome is condensed into heterochromatin and its DNA replicates late in the S phase, whereas its active homologue is less condensed and replicates throughout S phase. It therefore seems that those regions of the genome whose chromatin is least condensed during interphase, and therefore most accessible to the replication machinery, are replicated first.

Autoradiography shows that replication forks move at comparable rates throughout S phase, so that the extent of chromosome condensation does not appear to influence replication forks once they have formed. It seems, however, that the order in which replication origins are activated depends, at least in part, on the chromatin structure in which the origins reside.

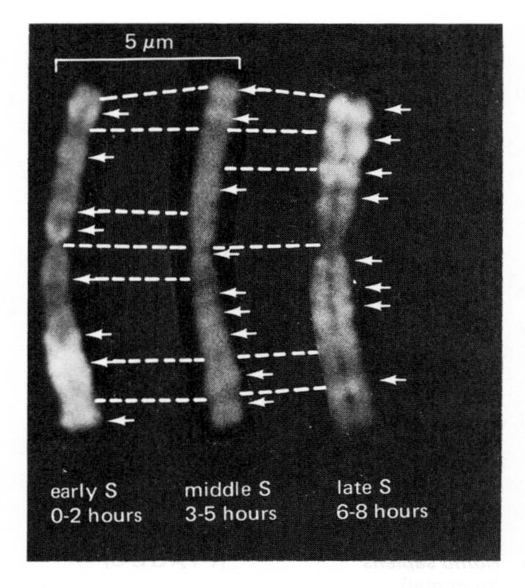

Figure 9–60 Light micrographs of stained mitotic chromosomes in which the replicating DNA has been differentially labeled during different defined intervals of the preceding S phase. In these experiments, cultured cells grown in the presence of the synthetic nucleoside 5-bromodeoxyuridine (BrdU) were briefly pulsed with thymidine during early, middle, or late S phase. Because the DNA made during the thymidine pulse is a double helix with thymidine on one strand and BrdU on the other, it stains more darkly than the remaining DNA (which has BrdU on both strands) and shows up as a bright band (*arrows*) on these negatives. Dashed lines connect corresponding positions on the three copies of the chromosome shown. (Courtesy of Elton Stubblefield.)

Genes in Active Chromatin Replicate Early in S Phase[38]

The suggested relationship between chromatin structure and the time of DNA replication is supported by studies in which the replication times of specific genes are measured. In these studies a growing cell population is briefly labeled with a pulse of BrdU, and the cells are immediately separated by centrifugation according to size. Because cells grow as they proceed through the cell cycle, the larger cells will be "older" and their DNA will therefore have been labeled later in S phase. For each size class of cells, the BrdU-labeled DNA is isolated and analyzed by hybridization with a series of specific DNA probes for the genes it contains. (Because DNA containing BrdU is denser, it can be easily separated from normal DNA by sedimentation to equilibrium in a cesium chloride density gradient—see p. 165.)

This method gives the replication time for any gene for which a DNA probe is available. The results show that so-called "housekeeping" genes, which are those active in all cells (see p. 585), replicate very early in S phase in all cells tested. Genes that are active in only a few cell types, in contrast, generally replicate early in the cells in which they are active and later in other types of cells. When a continuous stretch of 300,000 nucleotide pairs of an immunoglobulin gene was studied in this way, for example, all regions of its chromatin completed their replication near the beginning of S phase in cells in which the gene was active, suggesting that there are several replication origins within the gene and that these are all activated at about the same time. When replication times were measured with the same DNA probes in cells where no immunoglobulin is made, a single replication fork appeared to enter from one end of this chromosomal region about an hour after the start of S phase and then move steadily across the DNA at the expected rate of about 3000 nucleotides per minute.

A simple model that can explain these results is shown in Figure 9–61. In this view, all replication origins located in active chromatin are utilized very early in S phase. Because the replication forks formed at these origins will eventually move into adjacent chromosomal regions that have a more condensed chromatin structure, any gene located less than a million nucleotide pairs away from a replication origin in active chromatin will replicate by mid S phase. To explain how inactive chromosomal regions located far away from a patch of active chromatin eventually

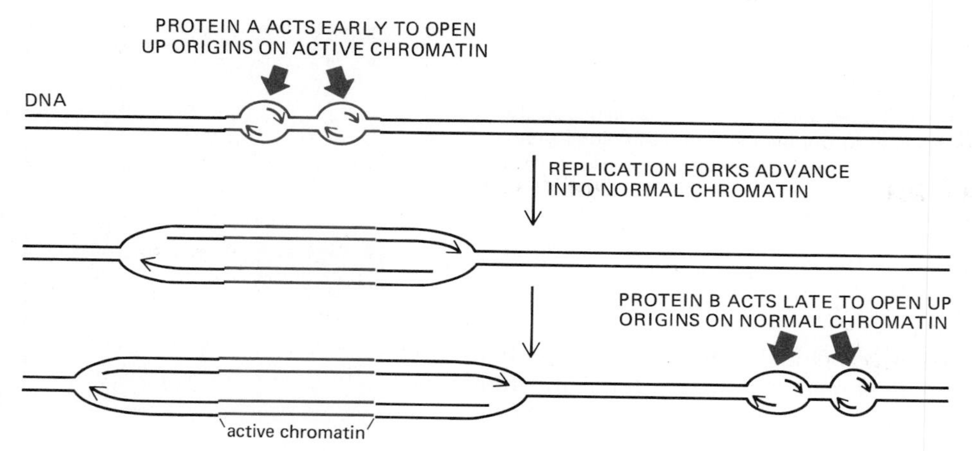

Figure 9–61 One model to explain why active chromatin (*colored*) is replicated early in S phase, whereas normal ("inactive") chromatin is replicated late in S phase. Different initiator proteins are postulated to act on the replication origins in active chromatin and in normal chromatin. Alternatively, the two sets of origins might be utilized by the same molecular machinery but at different times simply because the condensed structure of the normal chromatin delays access of the requisite replication proteins.

replicate (as with the DNA in the inactive X chromosome in females), it is necessary to postulate that a second group of replication origins is activated in mid to late S phase that can initiate replication forks in any form of chromatin.

The Late-replicating Replication Units Coincide with the A-T-rich Bands on Metaphase Chromosomes[23]

Many of the replication units seem to correspond to distinct chromosome bands made visible by the various fixation and staining procedures used for karyotyping. As discussed previously (see p. 503), as many as 2000 dark-staining *A-T-rich bands (G bands)* can be detected early in mitosis in the haploid set of mammalian chromosomes, and these are separated by an equal number of light-staining *G-C-rich bands (R bands)*. It is intriguing that the A-T-rich DNA and the G-C-rich DNA differ in the time of their replication during S phase. Experiments like the one shown in Figure 9–60 suggest that most G-C-rich bands replicate during the first half of S phase, while most A-T-rich bands replicate during the second half of S phase. It has therefore been suggested that housekeeping genes are located mostly in G-C-rich bands, while many cell-type-specific genes—the vast majority of which will be inactive in most cells—are located in A-T-rich bands. As stated earlier, it is a complete mystery why the mammalian genome should be segregated into such large alternating blocks of chromatin—many nearly equal in size to an entire bacterial genome. It is also not known how the many replication origins present in each replication unit are activated all at once. Perhaps the chromatin in a late-replicating unit remains condensed even after the end of M phase and decondenses only in mid S phase, making all the replication origins in the unit simultaneously accessible. In this case the all-or-none coordinated replication of the DNA in a single replication unit could reflect the cooperative nature of the chromatin decondensation process (see p. 499).

Does the Controlled Timing of Utilization of Replication Origins Have a Function?[39]

The S phase is completed extremely rapidly in the cleaving eggs of many species, where large stores of chromatin components (such as histones) are present, as required for the rapid manufacture of new nuclei (see p. 881). As illustrated in Figure 9–62, a short S phase also requires the use of an exceptionally large number of replication origins spaced at intervals of only a few thousand nucleotide pairs (rather than the tens or hundreds of thousands of nucleotide pairs found between the replication origins later in development). Since any foreign DNA injected into a fertilized frog egg is replicated (including small circular fragments of bacterial DNA), if a specific DNA sequence is required to form a replication origin in this cell, it must be a very short one that is present in any DNA molecule.

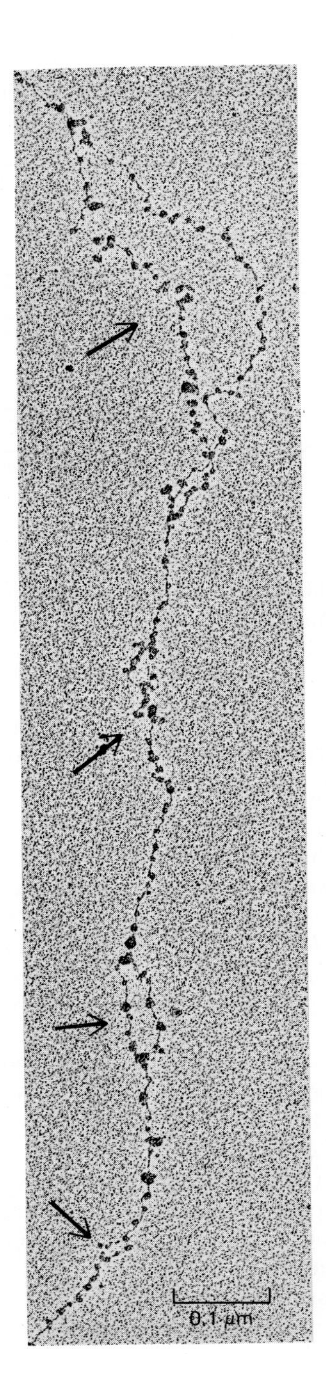

Figure 9–62 Electron micrograph of spread chromatin from an early *Drosophila* embryo showing that replication bubbles (*arrows*) are extremely closely spaced. Only about 10 minutes elapse between some of the successive nuclear divisions in this embryo. Since the replication origins used here are so closely spaced (separated by a few thousand nucleotide pairs), it should require only about a minute to replicate all the DNA between them. (Courtesy of Victoria Foe.)

Thus DNA replication can be viewed as a potentially very rapid process that in most cells is subject to a complex system of regulation that restrains the initiation of replication forks in a way that causes different portions of the genome to replicate at very different times. One possible advantage of such a prolonged S phase may be that it allows replication time to influence chromatin structure. It has been suggested, for example, that the chromatin that replicates early is assembled in part from a special store of chromosomal proteins produced during the G1 phase, which could help to maintain active chromatin and thereby facilitate the transcription of expressed genes.

9-21
9-22

Chromatin-bound Factors Ensure That Each Region of the DNA Is Replicated Only Once During Each S Phase, Providing a DNA Re-replication Block[40]

In a normal S phase the whole genome must be replicated exactly once and no more. As we have just seen, DNA replication in most eucaryotic cells is an asynchronous process that takes a relatively long time to complete. Because the replication origins are used at different times in different chromosomal regions, in the middle of S phase some parts of a chromosome will not yet have begun replication, while other parts will have replicated completely. An enormous "bookkeeping" problem therefore arises during the middle and late stages of the S phase. Those replication origins already used have been duplicated and, at least with respect to their DNA sequences, are presumably identical to other replication origins not yet used. But each replication origin must be used only once in each S phase. How is this accomplished?

Cell fusion experiments have provided an important clue. When an S-phase cell is fused with a G_1-phase cell, DNA synthesis is induced in the G_1-phase nucleus, suggesting that the transition from G_1 to S phase is mediated by a diffusible activator of DNA synthesis (see p. 732). In contrast, when the S phase cell is fused with a G_2-phase cell (that is, a cell that has just completed S phase—see p. 728), the G_2 nucleus is not stimulated to synthesize DNA. Since DNA synthesis continues undisturbed in the S-phase nucleus, the G_2 nucleus—having replicated all its DNA once—acts as though it is prevented from entering further rounds of replication by some nondiffusable inhibitor that is tightly bound to its DNA. Such an inhibitor, if applied locally during S phase in the wake of each replication fork, would neatly solve the bookkeeping problem: by modifying the chromatin of freshly replicated DNA, it would ensure that once-replicated DNA is not replicated again in the same S period (Figure 9–63A). As an equally plausible alternative, a mechanism based on tightly bound initiator proteins that are inactivated by the passage of a replication fork has been proposed (Figure 9–63B). Whatever its nature, the **DNA re-replication block** must be removed at or near the time of mitosis, since after cell division the DNA in the G_1 nuclei that emerge in the daughter cells is no longer protected.

The fragments of bacterial DNA that replicate when injected into a fertilized frog egg can be shown to be affected by the re-replication block. Therefore the mechanism responsible for the block cannot require a highly specific replication origin. The block does not affect the SV40 virus, presumably because its T-antigen supplies both initiator and DNA helicase functions to substitute for analogous host cell components that are as yet uncharacterized (see p. 517).

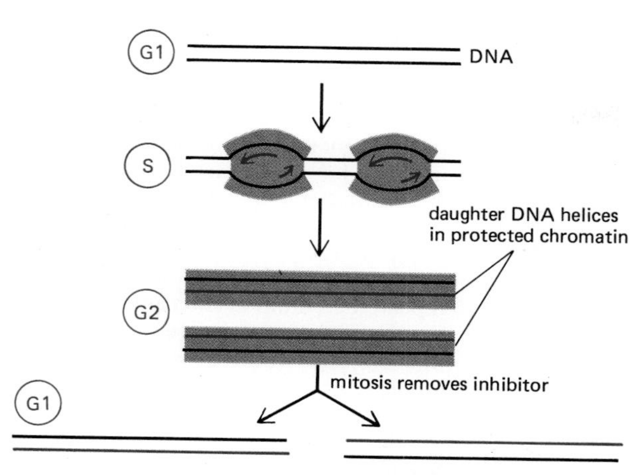

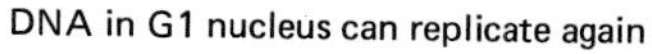

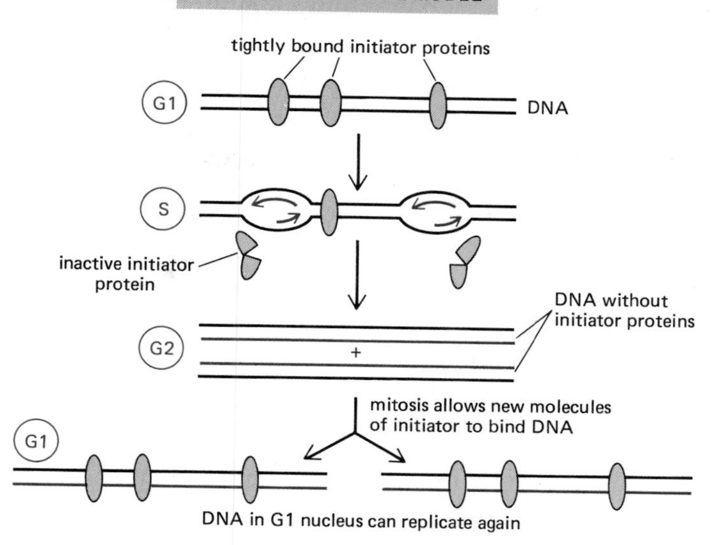

Figure 9–63 Two possible mechanisms that have been proposed to explain the "re-replication block," which protects replicated DNA from further replication in the same cell cycle. This block is crucial to replication bookkeeping, but its molecular nature is not known. Normally the block is removed at mitosis, but in a few specialized types of cells (the salivary gland cells of *Drosophila* larvae, for example), it is removed without mitosis, leading to the formation of giant *polytene* chromosomes (see p. 507). (A) A model based on the addition of an inhibitor to all newly replicated chromatin; (B) a model based on tightly bound initiator proteins that act only once and that can be added to the DNA only during mitosis.

Summary

In vitro *studies using the monkey virus SV40 as a model system suggest that, in eucaryotes as in procaryotes, DNA replication begins with the loading of a DNA helicase onto the DNA by an initiator protein bound to a replication origin. A replication bubble forms at such an origin as two replication forks move away from each other. During S phase in higher eucaryotes, neighboring replication origins appear to be activated in clusters known as replication units, with the origins spaced about one looped domain of chromatin apart. Since the replication fork moves at about 50 nucleotides per second, only about an hour should be required to complete the DNA synthesis in a replication unit. Throughout a typical 8-hour S phase, different replication units are activated in a sequence determined in part by their chromatin structure, the most condensed regions of chromatin being replicated last. The correspondence between replication units and the bands containing millions of nucleotide pairs seen on mitotic eucaryotic chromosomes suggests that replication units may correspond to structurally distinct domains in interphase chromatin.*

After the replication fork passes, chromatin structure is re-formed by the addition of new histones and other chromosomal proteins to the old histones inherited on the daughter DNA molecules. A DNA re-replication block of unknown nature acts locally to prevent a second round of replication from occurring until a chromosome has passed through mitosis; this block is needed to ensure that each region of the DNA is replicated only once in each S phase.

RNA Synthesis and RNA Processing[41]

We have thus far considered how chromosomes are organized as very large DNA-protein complexes and how they are duplicated before a cell divides. But the main function of a chromosome is to act as a template for the synthesis of RNA molecules, since only in this way does the genetic information stored in chromosomes become directly useful to the cell. RNA synthesis is extensive: the total rate at which nucleotides are incorporated into RNA during interphase is about 20 times the rate at which nucleotides are incorporated into DNA during S phase.

RNA synthesis (**DNA transcription**) is a highly selective process. In most mammalian cells, for example, only about 1% of the DNA nucleotide sequence is copied into functional RNA sequences (mature messenger RNA or structural RNA). The selectivity occurs at two levels, which we discuss in turn in this section: (1) only part of the DNA sequence is transcribed to produce nuclear RNAs, and

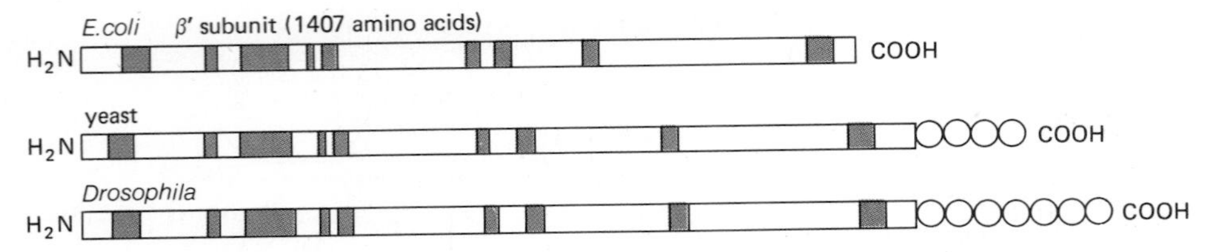

Figure 9–64 Amino acid sequence similarities between RNA polymerase subunits suggest a common evolutionary origin for the bacterial and eucaryotic enzymes. The largest subunit of the *E. coli*, yeast, and *Drosophila* polymerases (compared here) is thought to bind to DNA. The shaded regions of the sequence are more than 70% identical between yeast and *Drosophila* and more than 40% identical between *Drosophila* and *E. coli*. A uniquely eucaryotic sequence of unknown function (Ser-Pro-Ser-Tyr-Ser-Pro-Thr) is repeated 26 times at the carboxyl terminus of the yeast subunit and more than 40 times in the *Drosophila* subunit (indicated here by circles). (After A.L. Greenleaf et al., in RNA Polymerase and the Regulation of Transcription [W.S. Reznikoff et al., eds.], pp. 459–464. New York: Elsevier, 1987.)

(2) only a minor proportion of the nucleotide sequences in nuclear RNAs survives the RNA processing steps that precede the export of RNA molecules to the cytoplasm. We begin by describing RNA polymerases, the enzymes that catalyze all DNA transcription.

RNA Polymerase Exchanges Subunits as It Begins Each RNA Chain[42]

A general outline of DNA transcription was given in Chapter 5. Transcription begins when an *RNA polymerase* molecule binds to a *promoter* DNA sequence. First, in a complicated initiation step, the two strands of the DNA are locally separated to form an *open complex*, in which the template strand is exposed. After this complex has formed, the polymerase begins to move along the DNA, extending its growing RNA chain in the 5′-to-3′ direction by the stepwise addition of ribonucleoside triphosphates until it reaches a stop (termination) signal, at which point the newly synthesized RNA chain and the polymerase are released from the DNA. Each RNA molecule thus represents a single-strand copy of the nucleotide sequence of one DNA strand in a relatively short region of the genome (see Figure 5–1, p. 202).

RNA polymerases are generally formed from multiple polypeptide chains and have masses of 500,000 daltons or more. The enzymes in bacteria and eucaryotes are evolutionarily related (Figure 9–64). Since the bacterial enzyme has been far easier to study, its properties provide a basis for understanding its eucaryotic relatives. The *E. coli* enzyme contains five subunits, α, β, β′, σ, and ω, there being two copies of α and one each of the others. The complete amino acid sequence of each subunit has been determined from the nucleotide sequence of its gene, and DNA-footprinting studies indicate that the enzyme covers 60 nucleotide pairs when it binds to DNA.

The sigma (σ) subunit of the *E. coli* polymerase has a specific role as an **initiation factor** for transcription: it enables the enzyme to find the consensus promoter sequences described on page 204. A subset of RNA polymerases that recognize different promoters contain variant forms of the sigma subunit (see p. 562). Once bound to a promoter, the enzyme undergoes a series of reactions to start an RNA chain (Figure 9–65). After about eight nucleotides of an RNA molecule

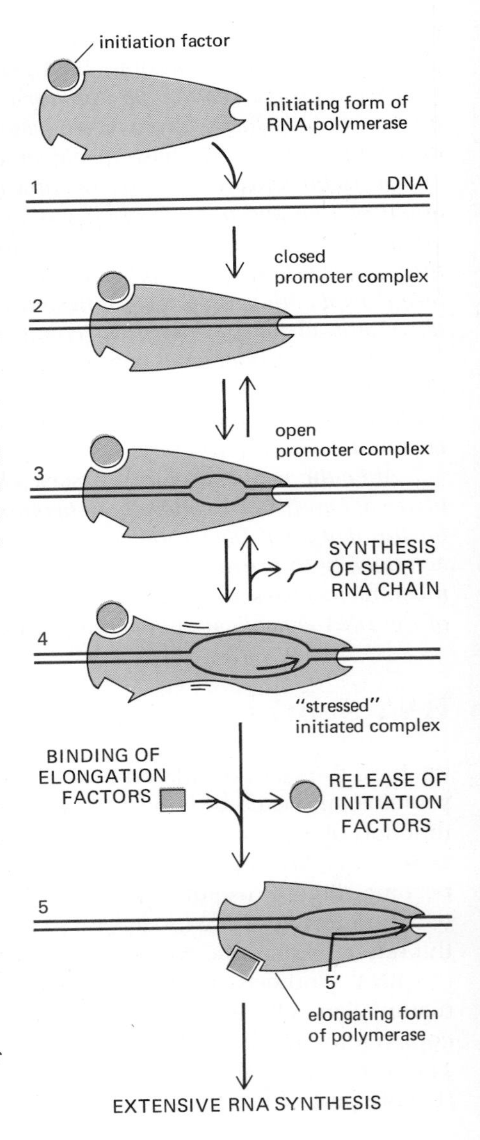

Figure 9–65 Highly schematic diagram of the steps in the initiation of RNA synthesis catalyzed by RNA polymerase. The steps indicated have been discovered in studies of the *E. coli* enzyme. A DNA molecule containing a promoter sequence for the *E. coli* polymerase is shown (see Figure 5–6, p. 205). The enzyme first forms a *closed complex* in which the two DNA strands remain fully base-paired. In the next step the enzyme catalyzes the opening of a little more than one turn of the DNA helix to form an *open complex*, in which a template is exposed for the initiation of an RNA chain. The polymerase containing the bound sigma subunit, however, behaves as though it is tethered to the promoter site: it seems unable to proceed with the elongation of the RNA chain and on its own frequently reverts to the closed complex with the release of a short RNA chain. As indicated, the conversion to an actively elongating polymerase requires the release of initiation factors (the sigma subunit in the case of the *E. coli* enzyme) and generally involves the binding of other proteins that serve as elongation factors (for example, the *E. coli* nusA protein). (Modified from D.C. Straney and D.M. Crothers, *J. Mol. Biol.* 193:267–278, 1987.)

have been synthesized (step 4 in Figure 9–65), the sigma subunit dissociates and a number of **elongation factors**—important for chain elongation and termination—become associated with the enzyme instead. The elongation factors include several proteins that, although well characterized, have functions that are incompletely understood.

In all organisms the various active genes are transcribed at very different rates. At some promoters a new RNA chain begins once every 1 or 2 seconds, while at others more than an hour is required to start an RNA chain. As described in detail in Chapter 10, *gene regulatory proteins* that affect transcriptional initiation usually determine the level of transcription of each gene. In general, these proteins act by accelerating or retarding one or more of the steps illustrated in Figure 9–65 (see p. 563).

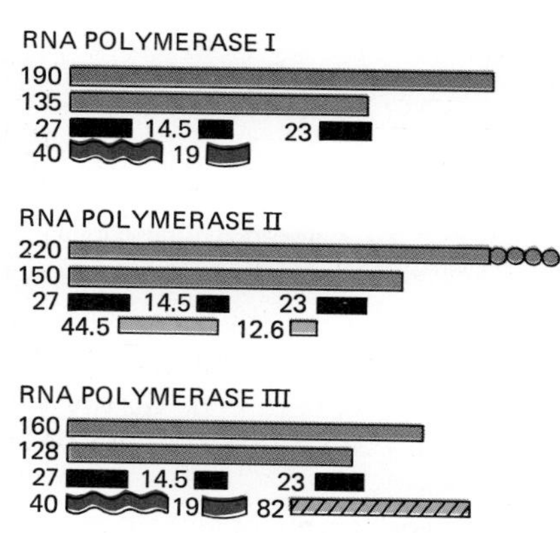

Figure 9–66 Some of the protein subunits that form the three eucaryotic RNA polymerases in yeasts. Only the best-characterized subunits are shown. Related subunits are colored the same, and their molecular mass is given in kilodaltons. The three subunits with identical mass (shown in black) are shared by all three enzymes.

9-25 Three Different RNA Polymerases Make RNA in Eucaryotes[43]

Although the mechanism of DNA transcription is similar in eucaryotes and procaryotes such as *E. coli,* the machinery is considerably more complex in eucaryotes. In eucaryotes as diverse as yeasts and humans, for example, there are three types of RNA polymerase, each responsible for transcribing different sets of genes. These enzymes—denoted as RNA polymerases I, II, and III—are structurally similar to one another and have some common subunits, although other subunits are unique (Figure 9–66). Each is more complex than *E. coli* RNA polymerase and is thought to contain 10 or more polypeptide chains. The most important distinction, however, is that, whereas the purified bacterial enzyme binds directly to the promoter, the eucaryotic enzymes can bind to their promoters only in the presence of additional protein factors already on the DNA. Partly for this reason, it was not until 1979 that systems became available in which eucaryotic initiation mechanisms could be analyzed *in vitro.*

The three eucaryotic RNA polymerases were initially distinguished by their chemical differences during purification and by their sensitivity to *α-amanitin,* a poison isolated from mushrooms. RNA polymerase I is unaffected by α-amanitin; RNA polymerase II is very sensitive to this poison; and RNA polymerase III is moderately sensitive to it. The sensitivity of RNA synthesis to α-amanitin is still used to determine which polymerase transcribes a gene. Such studies indicate that only **RNA polymerase II** transcribes the genes whose RNAs will be translated into proteins. The other two polymerases synthesize only RNAs that have structural or catalytic roles, chiefly as part of the protein synthetic machinery: **polymerase I** makes the large ribosomal RNAs and **polymerase III** makes a variety of very small, stable RNAs—including the small 5S ribosomal RNA and the transfer RNAs. Most of the small RNAs that form snRNPs (see p. 532), however, are made by polymerase II.

Mammalian cells typically contain about 40,000 molecules of RNA polymerase II, about the same number of RNA polymerase I molecules, and about 20,000 molecules of RNA polymerase III—although studies with cultured cells indicate that the concentrations of the RNA polymerases are regulated individually according to the rate of cell growth.

9-26 9-27 Transcription Factors Form Stable Complexes at Eucaryotic Promoters[44]

Eucaryotic RNA polymerases, as mentioned previously, do not recognize their promoters on purified DNA molecules. Instead, one or more sequence-specific DNA-binding proteins must be bound to the DNA to form a functional promoter. These are called **transcription factors (TF)** and are necessary for the initiation of RNA synthesis. They are distinct from procaryotic initiation factors (the sigma factors) in that they bind to DNA independently of the RNA polymerase. Polymerases I, II, and III recognize different promoters and for the most part require different transcription factors—designated by the prefixes **TFI, TFII,** and **TFIII,** respectively, followed by an alphabetical suffix assigned according to their order of discovery. TFIIIA, for example, was discussed previously as the prototype for a

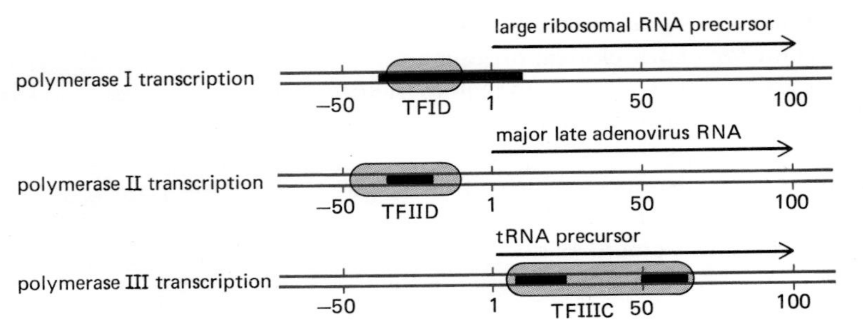

Figure 9–67 Some stable complexes formed by transcription factors on eucaryotic genes. The black boxes indicate the location of the specific DNA sequences thought to be essential for the function of each promoter. The approximate region of each gene covered by the bound factors (as determined by DNA-footprinting studies—see Figure 4–69, p. 188) is shown in color. Each of the transcription factors shown is a protein molecule that may contain multiple subunits, although the precise structures of these factors are not known. The numbers indicate positions on the DNA molecule (in nucleotide pairs) relative to the start site for transcription, which is set at +1.

"zinc finger" family of sequence-specific DNA-binding proteins (see p. 490); it was the first transcription factor to be characterized that acts on a polymerase-III-transcribed gene (the 5S rRNA gene).

Transcription factors *in vitro* appear to form relatively stable transcription complexes that selectively attract RNA polymerase molecules to their promoter. The different factors bind at different positions in relation to the transcription start site. Both polymerase I and polymerase II form a complex with a transcription factor that binds just upstream from the transcription start site. The major transcription factor for polymerase III genes, however, binds just *downstream* from the start site, so RNA polymerase III must transcribe over the protein without displacing it from the DNA (Figure 9–67). This factor (TFIIIC) is thought to fold the DNA around itself to form a large nucleoprotein particle.

Although many of the transcription factors have been difficult to purify to homogeneity, it has been possible to compare the transcriptional activity of DNA molecules that have been preincubated with transcription factors with that of DNA molecules that have not been preincubated. This has shown that a DNA molecule complexed with the appropriate factors is used repeatedly for RNA synthesis by RNA polymerase molecules, whereas a second DNA molecule that is otherwise identical is ignored (Figure 9–68). The observed pattern of transcription from DNA molecules injected into *Xenopus* eggs makes it very likely that the same types of complexes form *in vivo* as well.

A crucial transcription factor for many polymerase II promoters is TFIID. This is a large protein complex that is more commonly called the **TATA factor** because it can bind to a conserved A-T-rich sequence called the **TATA box,** centered about

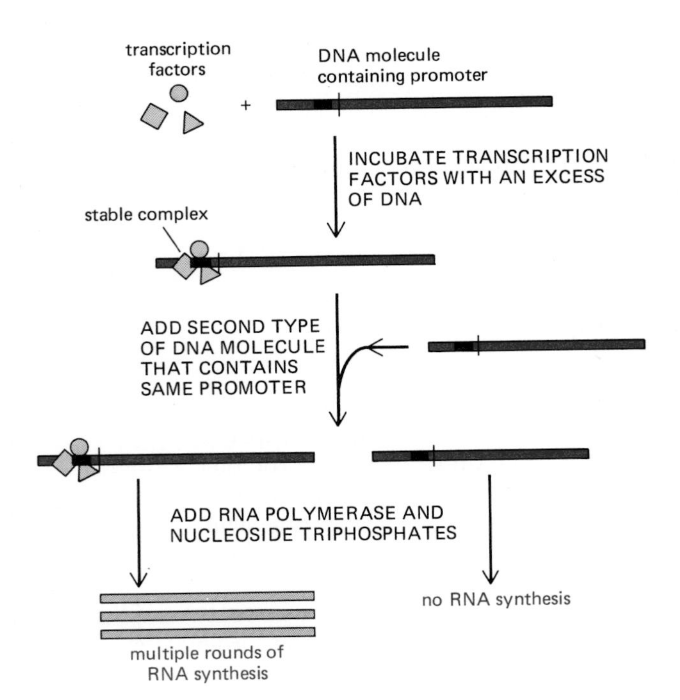

Figure 9–68 An *in vitro* experiment that demonstrates the importance of the formation of a stable transcription complex at a eucaryotic promoter. This type of experiment has been performed with transcription factors and promoters that are specific for each of the three eucaryotic RNA polymerases. Examples of the cloned genes used for such studies are shown in Figure 9–67.

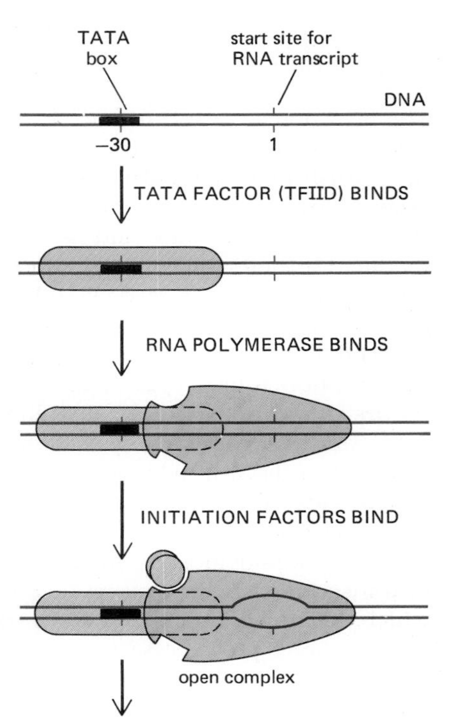

Figure 9–69 Minimum requirements for promoter recognition by a eucaryotic RNA polymerase II molecule. A TATA box-binding factor (TFIID) must form a stable transcription complex before the promoter can be recognized by the polymerase. The consensus sequence of the TATA box is T_{82} A_{97} T_{93} A_{85} (A or T)$_{100}$ A_{83} (A or T)$_{83}$, where the subscript indicates the percent occurrence of the indicated nucleotide. See also Figure 10–27, page 569. (After data in J.L. Workman and R.G. Roeder, *Cell* 51:613–622, 1987.)

25 nucleotides upstream from the start site for transcription. The activity of the TATA factor in stimulating polymerase II transcription is outlined in Figure 9–69.

Because RNA polymerase II makes all of the mRNA precursors and thus determines which proteins a cell will make, we shall focus most of our discussion on the synthesis and fate of the RNA transcripts synthesized by this enzyme.

9-28 RNA Polymerase II Transcribes Some DNA Sequences Much More Often Than Others[45]

Although experiments with purified polymerases and transcription factors *in vitro* are essential for establishing the mechanism of transcription, much can also be learned about the pattern of DNA transcription in a cell by using the electron microscope to examine genes in action, with their bound RNA polymerases caught in the act of transcription.

Ordinary thin-section electron micrographs of interphase nuclei show granular clumps of chromatin (see Figure 9–100), but they reveal very little about how genes are transcribed. A much more detailed picture emerges if the nucleus is ruptured and its contents spilled out onto an electron microscope grid (Figures 9–70 and 9–71). At the farthest point from the center of the lysed nucleus, the chromatin is diluted sufficiently to make individual chromatin strands visible in the expanded, beads-on-a-string form shown previously in Figure 9–22B.

RNA polymerase molecules actively engaged in transcription appear as globular particles with a single RNA molecule trailing behind. Particles representing active RNA polymerase II molecules are usually seen as single units, without nearby neighbors. This indicates that most genes are transcribed into mRNA precursors only infrequently, so that one polymerase finishes transcription before another one begins. Occasionally, however, many polymerase molecules (and their associated RNA transcripts) are seen clustered together. These clusters occur on the relatively few genes that are transcribed at high frequency (Figure 9–72). The length of the attached RNA molecules in such a cluster increases in the direction of transcription, producing a characteristic pattern. This pattern defines the RNA polymerase II start site and stop site for a specific **transcription unit** (Figure 9–73).

Biochemical studies have confirmed and extended the results obtained by electron microscopy, leading to three major conclusions:

1. Eucaryotic RNA polymerase molecules, like those in procaryotes, begin and end transcription at specific sites on the chromosome.
2. The average length of the finished RNA molecule produced by RNA polymerase II in a transcription unit is about 8000 nucleotides, and RNA molecules 10,000 to 20,000 nucleotides long are quite common. These lengths, which are much longer than the 1200 nucleotides of RNA needed to code for an average protein of 400 amino acid residues, reflect the complex structure of eucaryotic genes, and in particular the presence of long introns, as will be discussed later.
3. Although chain elongation rates of about 30 nucleotides per second are observed for all RNAs, different RNA polymerase II start sites function with very different efficiencies, so that some genes are transcribed at much higher rates than others. The pattern of transcription observed in electron micrographs agrees well with the results of biochemical studies showing that while many different messenger RNA molecules accumulate in a cell, most of them are present at relatively low frequency (Table 9–2).

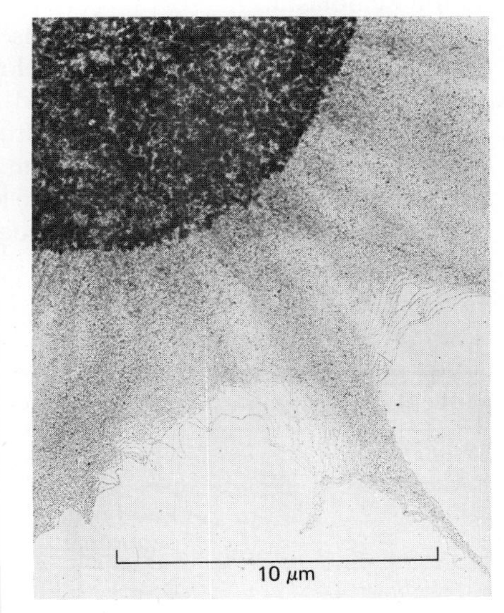

Figure 9–70 A typical cell nucleus visualized by electron microscopy using the procedure shown in Figure 9–71. An enormous tangle of chromatin can be seen spilling out of the lysed nucleus; only the chromatin at the outermost edge of this tangle will be sufficiently dilute for meaningful examination at higher power. (Courtesy of Victoria Foe.)

The Precursors of Messenger RNA Are Covalently Modified at Both Ends[46]

The RNA polymerase II transcripts in the nucleus are known as **heterogeneous nuclear RNA (hnRNA)** molecules because one of the first characteristics used to distinguish them from other RNAs in the nucleus was the heterogeneity of their sizes. Many of these transcripts are destined to leave the nucleus as **messenger RNA (mRNA)** molecules, and as they are being synthesized, they are covalently modified at both their 5′ end and their 3′ end in ways that clearly distinguish them from transcripts made by other RNA polymerases. These modifications will be useful later for their function as mRNA molecules in the cytoplasm.

The 5′ end of the RNA molecule (which is the end synthesized first during transcription) is first *capped* by the addition of a methylated G nucleotide. Capping occurs almost immediately, after about 30 nucleotides of RNA have been synthesized, and it involves condensation of the triphosphate group of a molecule of GTP with a diphosphate left at the 5′ end of the initial transcript (Figure 9–74). This **5′ cap** will later play an important part in the initiation of protein synthesis (see p. 214); it also seems to protect the growing RNA transcript from degradation.

The 3′ end of most polymerase II transcripts is defined not by the termination of transcription (which overshoots this point) but by a second modification, in which the growing transcript is cleaved at a specific site and a **poly-A tail** is added by a separate polymerase to the cut 3′ end. The signal for the cleavage is the appearance in the RNA chain of the sequence AAUAAA located 10 to 30 nucleotides upstream from the site of cleavage, plus a less well-defined downstream sequence. Immediately after cleavage, a *poly-A polymerase* enzyme adds 100 to 200 residues of adenylic acid (as *poly A*) to the 3′ end of the RNA chain to complete the **primary RNA transcript.** Meanwhile the polymerase fruitlessly continues transcribing for hundreds or thousands of nucleotides, until termination occurs at one of several later sites; the extra piece of RNA transcript thus generated presumably lacks a 5′ cap and is rapidly degraded (Figure 9–75).

The function of the poly-A tail is not known for certain, but it may play a role in the export of mature mRNA from the nucleus; there is also evidence that it helps to stabilize at least some mRNA molecules by retarding their degradation in the cytoplasm.

Even though polymerase II transcripts comprise more than half of the RNA synthesized by a cell, we shall see below that most of the RNA in these transcripts is unstable and therefore short-lived. Consequently, hnRNA in the cell nucleus and the cytoplasmic mRNA derived from it constitute only a minor fraction of the total RNA in a cell (Table 9–3). Despite their relative scarcity, these RNA molecules can be readily purified because of the long stretch of poly A at their 3′ ends. When the total cellular RNA is passed through a column containing poly dT linked to a

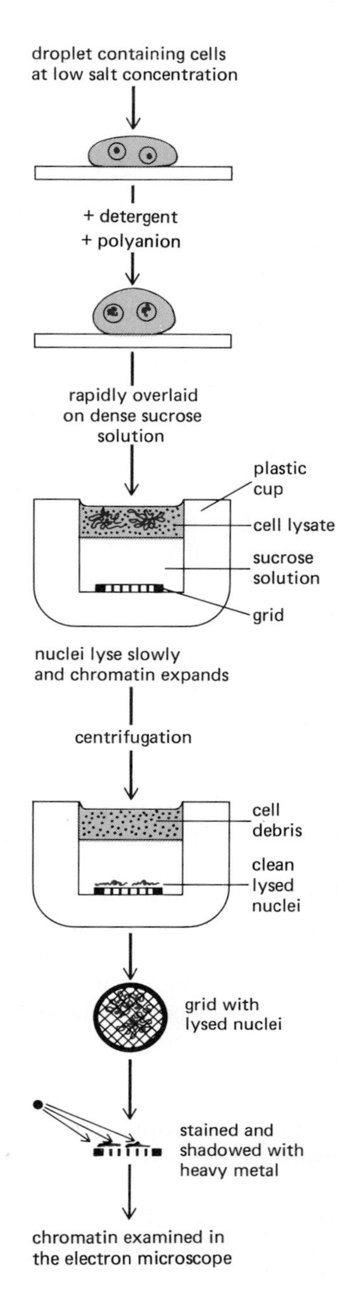

Figure 9–71 A method for examining the chromatin of a cell nucleus by electron microscopy after it has been gently spread out and freed from cellular debris.

Table 9–2 The Population of mRNA Molecules in a Typical Mammalian Cell

	Copies per Cell of Each mRNA Sequence		Number of Different mRNA Sequences in Each Class		Total Number of mRNA Molecules in Each Class
Abundant class	12,000	×	4	=	48,000
Intermediate class	300	×	500	=	150,000
Scarce class	15	×	11,000	=	165,000

This division of mRNAs into just three discrete classes is somewhat arbitrary, and in many cells a more continuous spread in abundances is seen. However, a total of 10,000 to 20,000 different mRNA species is normally observed in each cell, most species being present at a low level (5 to 15 molecules per cell). Most of the total cytoplasmic RNA is rRNA and only 3% to 5% is mRNA, a ratio consistent with the presence of about 10 ribosomes per mRNA molecule. This particular cell type contains a total of about 360,000 mRNA molecules in its cytoplasm.

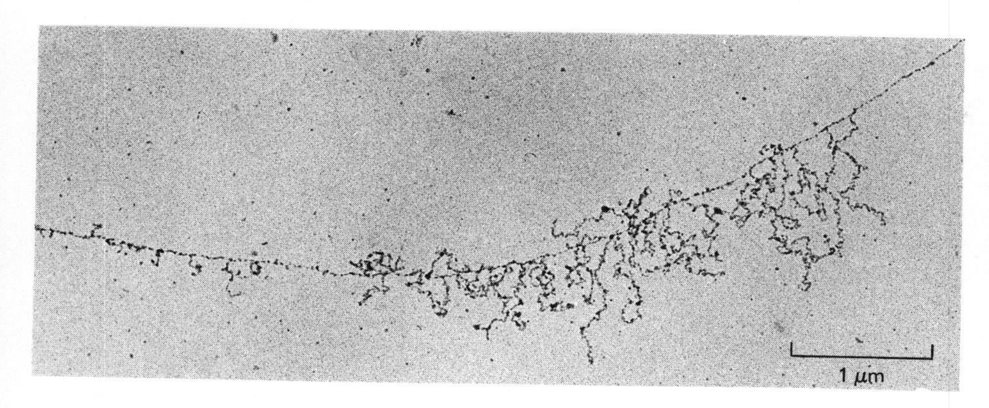

Figure 9–72 A region of chromatin containing a gene being transcribed at unusually high frequency, so that many RNA polymerase II molecules with their growing RNA transcripts are visible at the same time. The direction of transcription is from left to right (see Figure 9–73). (From V.E. Foe, L.E. Wilkinson, and C.D. Laird, *Cell* 9:131–146, 1976. © 1976 Cell Press.)

solid support, the complementary base-pairing between T and A residues selectively binds the molecules with poly-A tails to the column; the bound molecules can then be released for further analysis. This procedure is widely used to separate the hnRNA and mRNA molecules from the ribosomal and transfer RNA molecules that predominate in cells.

9-30 Capping and Poly-A Addition Require RNA Polymerase II[47]

Only RNA polymerase II transcripts have 5′ caps and 3′ poly-A tails. A gene normally transcribed by polymerase II can be separated from its promoter by recombinant DNA methods and fused to a promoter recognized by polymerase I or by polymerase III. If the modified gene is inserted back into a cell, it is transcribed by the polymerase that recognizes the promoter, but the RNA molecule produced is neither capped nor polyadenylated. It therefore seems that both the capping and the cleavage plus poly-A addition reactions are mediated by enzymes that bind selectively to polymerase II and function only when associated with it. These enzymes can thus be viewed as elongation factors for polymerase II. The requirement for capping and polyadenylation of mRNA precursors may explain why these RNAs are synthesized by a separate type of RNA polymerase molecule in eucaryotes.

Early hints that the cleavage plus poly-A addition reactions might require an elongation factor bound to the polymerase came from studies implying that each transcribing polymerase II molecule is able to generate only one polyadenylated 3′ end—that is, all AAUAAA signals encountered by the polymerase downstream from the initial cleavage and poly-A addition site are ignored. This suggests that the polymerase carries a factor that is lost upon completion of the cleavage and polyadenylation reaction (see Figure 9–75).

Most of the RNA Synthesized by RNA Polymerase II Is Rapidly Degraded in the Nucleus[48]

The first evidence that RNA polymerase II transcripts in the nucleus are unstable came from studies of cultured cells exposed to ^{3}H-uridine for a short period. This exposure introduced radioactivity into the hnRNA molecules, which could then be followed over a longer period of time. These and later experiments resulted in two remarkable discoveries:

1. The length of the newly made hnRNA molecules decreases rapidly, reaching the size of cytoplasmic mRNA molecules after only about 30 minutes. The primary RNA transcripts contain about 6000 nucleotides on average, while the mRNA molecules contain about 1500 nucleotides.
2. After about 30 minutes, radioactively labeled RNA molecules begin to leave the nucleus as mRNA molecules. Only about 5% of the mass of the labeled hnRNA, however, ever reaches the cell cytoplasm. The remainder is degraded into small fragments in the cell nucleus over a period of about an hour.

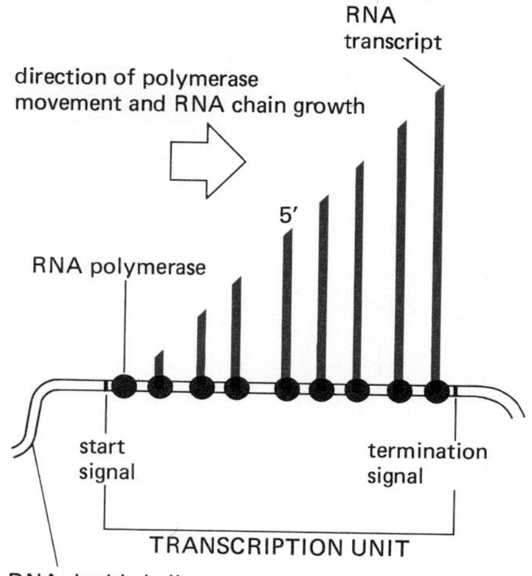

Figure 9–73 An idealized transcription unit, showing how the electron microscope appearance demonstrates the direction of transcription, as well as the start and stop sites of the unit.

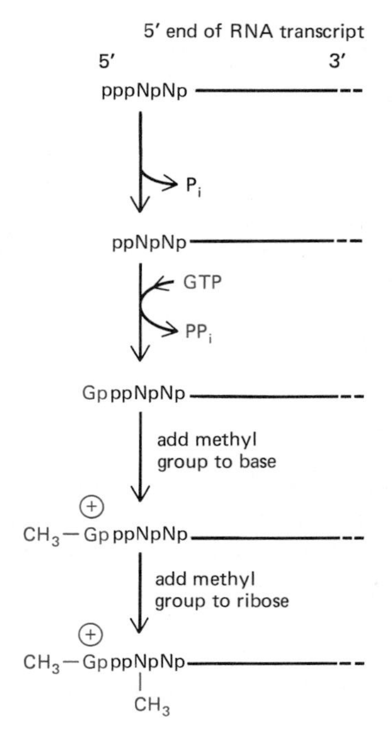

Figure 9–74 The reactions that cap the 5′ end of each RNA molecule synthesized by RNA polymerase II. The final cap contains a novel 5′-to-5′ linkage between the positively charged 7-methyl G residue and the original 5′ end of the RNA transcript (see Figure 5–24, p. 215). At least some of the enzymes required for this process are thought to be bound to polymerase II, since polymerase I and III transcripts are not capped and the indicated reaction occurs almost immediately following initiation of each RNA chain. The letter N is used here to represent any one of the four ribonucleotides, although the nucleotide that starts an RNA chain is usually a purine (an A or a G). (After A.J. Shatkin, *Bioessays* 7:275–277, 1987.)

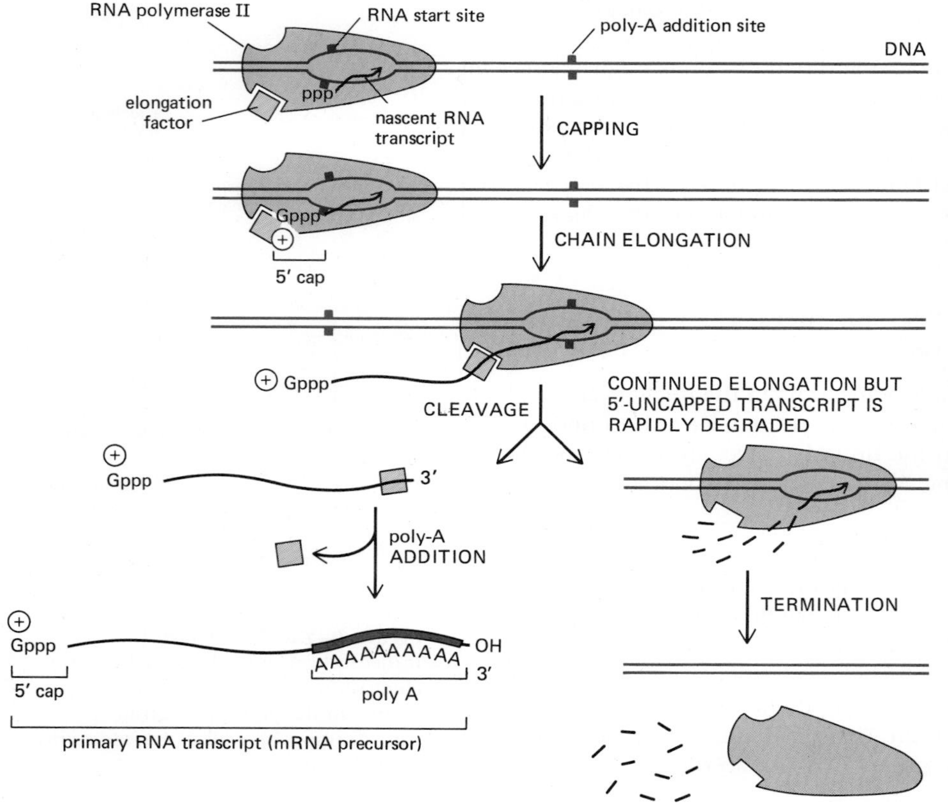

Figure 9–75 Synthesis of an hnRNA molecule (an mRNA precursor) by RNA polymerase II. This diagram starts with a polymerase that has just begun synthesizing an RNA chain (step 5 of Figure 9–65). Recognition of a poly-A addition signal in the growing RNA transcript causes the chain to be cleaved and then polyadenylated as shown. In yeasts the polymerase terminates its RNA synthesis almost immediately thereafter, but in higher eucaryotes it often continues transcription for thousands of nucleotides. It seems likely that the polymerase changes its properties once one RNA chain cleavage has occurred; thus it cannot cause poly-A addition to the downstream RNA transcript, and it seems to have a greater probability of responding to the sequences that cause chain termination and polymerase release. The simplest hypothesis, represented here, is that an elongation factor (or factors) is released from the polymerase after cleavage of the transcript.

Table 9–3 Selected Data on Amounts of RNA in a Typical Mammalian Cell

	Steady-State Amount (percent of total cell RNA)	Percent of Total RNA Synthesis
Nuclear rRNA precursors ↓	4	39
Cytoplasmic rRNA	71	—
Nuclear hnRNA ↓	7	58
Cytoplasmic mRNA	3	—
Small stable RNAs (mostly tRNAs)	15	3

The figures shown here were derived from the analysis of a mouse fibroblast cell line (L cells) in culture. Each cell contained 26 pg of RNA (5×10^{10} nucleotides of RNA), of which about 14% was located in the cell nucleus. (The cell nucleus thus contains about twice as much DNA as RNA.) An average of about 200×10^6 nucleotides is polymerized into RNA every minute during interphase. This is about 20 times the average rate at which DNA is synthesized during S phase. Note that although most of the RNA synthesized is hnRNA, most of this RNA is rapidly degraded in the nucleus. As a result, the mRNA produced from the hnRNA is only a minor fraction of the total RNA in the cell. (Modified from B. P. Brandhorst and E. H. McConkey, *J. Mol. Biol.* 85:451–563, 1974.)

When, in the early 1970s, both mRNAs and hnRNAs were discovered to have poly-A tails at their 3′ ends, it was natural to assume that mRNAs are derived from hnRNAs by extensive degradation at 5′ ends—in other words, that most of an hnRNA molecule consists of a very long "5′ leader sequence" upstream of the coding sequence. But the hypothesis had to be abandoned when the 5′ caps were discovered and shown to be largely preserved during the conversion of hnRNA molecules to mRNA. With hindsight, the logical conclusion would have been that the *middle* of an hnRNA molecule is removed, leaving both the 3′ and the 5′ ends intact. At the time, however, such a postulate seemed absurd. Moreover, researchers were at a loss to explain why a cell should discard most of the RNA that it synthesizes. The solution to this puzzle came only when it became possible to compare the nucleotide sequence of a particular mRNA molecule with the sequence of the genomic DNA that encodes it.

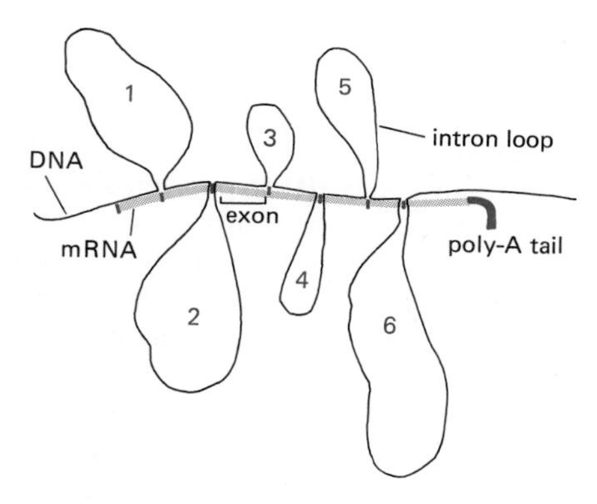

Figure 9–76 Early evidence for the existence of introns in eucaryotic genes was provided by the "R-loop technique," in which a base-paired complex between mRNA and DNA molecules is visualized in the electron microscope. An unusually abundant mRNA molecule, such as β-globin mRNA or ovalbumin mRNA, is readily purified from the specialized cells that produce it. When this single-stranded mRNA preparation is annealed in a suitable solvent to a cloned double-stranded DNA molecule containing the gene that encodes the mRNA, the RNA can displace a DNA stand wherever the two sequences match and form regions of RNA-DNA helix. Regions of DNA where no match to the mRNA sequence is possible are clearly visible as large loops of double-stranded DNA. Each of these loops (numbered 1 to 6) represents an intron in the gene sequence.

RNA Processing Removes Long Nucleotide Sequences from the Middle of RNA Molecules[48]

The discovery of interrupted genes in 1977 was entirely unexpected. Previous studies in bacteria had shown that their genes are composed of a continuous string of the nucleotides needed to encode the amino acids of a protein, and there seemed to be no obvious reason why a gene should be organized in any other way. The first indication that eucaryotic genes are not continuous like bacterial genes came when the new methods allowing an accurate comparison of mRNA and DNA sequences were applied to mRNAs produced by a human *adenovirus* (a large DNA virus). The region of the viral DNA producing these RNAs turned out to contain sequences that are not present in the mature RNAs. The possibility that this situation was unique to viruses was quickly eliminated by the finding of similar interruptions in the ovalbumin and β-globin genes of vertebrates. As discussed earlier, the sequences present in the DNA but omitted from the mRNA are called *intron* sequences, while those present in the mRNA are called *exon* sequences (Figures 9–76 and 9–77).

It then remained to determine how exon and intron sequences are sorted out by the cell: only then did the significance of the hnRNA become clear. We now know that the primary RNA transcript is a faithful copy of the gene, containing both exon and intron sequences, and that the latter sequences are cut out of the middle of the RNA transcript to produce an mRNA molecule that codes directly for a protein (see Figure 3–13, p. 101). Because the coding RNA sequences on either

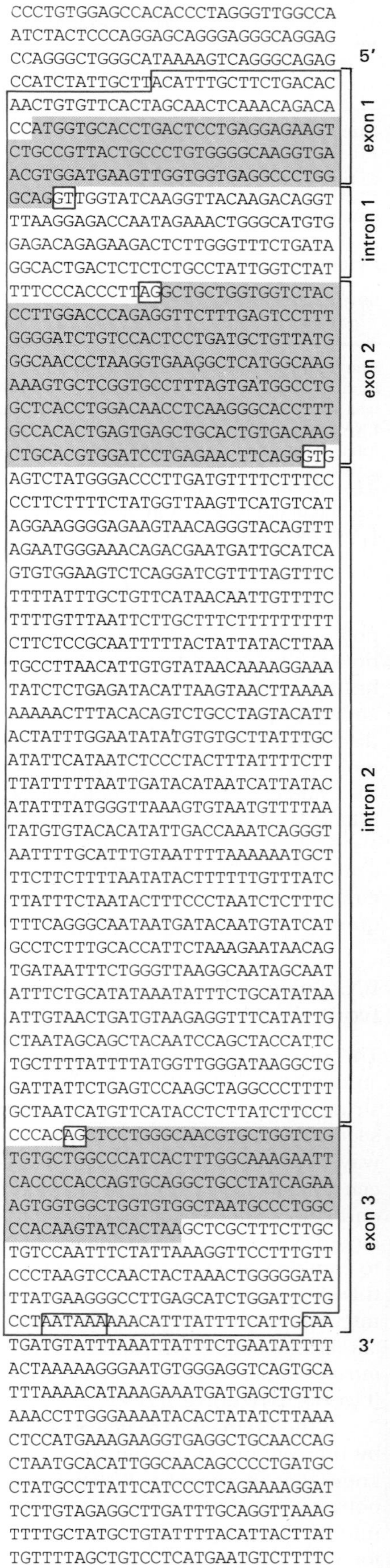

Figure 9–77 The transcribed portion of the human β-globin gene. The sequence of the DNA strand corresponding to the mRNA sequence is given, with the primary RNA transcript surrounded by a colored line and the nucleotides in the three coding regions (exons) shaded. Note that exon 1 includes a *5′-leader sequence* and that exon 3 includes a *3′-trailer sequence;* although these sequences are included in the mRNA, they do not code for amino acids. The highly conserved GT and AG nucleotides at the ends of each intron are boxed (see Figure 9–79), along with the cleavage and polyadenylation signal near the 3′ end of the gene (AATAAA, see Figure 9–75).

side of an intron sequence are joined to each other after the intron sequence has been cut out, this reaction is known as **RNA splicing.** RNA splicing occurs in the cell nucleus, out of reach of the ribosomes, and RNA is exported to the cytoplasm only when processing is complete (see p. 538).

Because most mammalian genes contain much more intron than exon sequence (see Table 9–1, p. 486), RNA splicing can account for the conversion of the very long nuclear hnRNA molecules (up to more than 50,000 nucleotides) to the much shorter cytoplasmic mRNA molecules (usually 500 to 3000 nucleotides).

Before discussing the distribution of introns in eucaryotic genes and some of their consequences for cell function, it is necessary to explain how intron sequences are recognized and removed by the splicing machinery.

9-31 hnRNA Transcripts Are Immediately Coated with Proteins and snRNPs[49]

Newly made RNA in eucaryotes, unlike that in bacteria, appears to become immediately condensed into a series of closely spaced protein-containing particles. These particles consist of about 500 nucleotides of RNA wrapped around an abundant protein complex that serves to condense and package each growing RNA transcript in a manner reminiscent of the DNA-protein complexes of nucleosomes. The resulting **hnRNP particles** (*heterogeneous nuclear ribonucleoprotein particles*) can be purified after nuclei have been treated with ribonucleases at levels just sufficient to destroy the linker RNA between them. The particles sediment at about 30S and have a diameter about twice that of nucleosomes (20 nm). Their protein core is correspondingly larger and more complex, being composed of a set of at least 8 different proteins of mass 34,000 to 120,000 daltons. Except for histones, the proteins in this core are the most abundant proteins in the cell nucleus. Those characterized thus far contain one or more copies of a short sequence of amino acids that is shared by many RNA-binding proteins (Figure 9–78).

The hnRNP particles are generally distorted by the standard spreading techniques used to view transcribing genes in the electron microscope (see Figure 9–70). These micrographs, however, reveal especially stable particles of a less common type on many RNA transcripts, whose location strongly implicates a role for them in RNA splicing. The stable particles form very quickly at the junctions between intron and exon sequences, and, as the RNA transcript elongates, these particles coalesce in pairs to form a larger assembly that is thought to be the *spliceosome* that catalyzes RNA splicing (Figure 9–79).

Biochemical analysis has revealed that the cell nucleus contains many complexes of proteins with small RNAs (generally RNAs of 250 nucleotides or less), which have arbitrarily been designated U1, U2,. . .,U12 RNAs. These complexes, called **small nuclear ribonucleoproteins (snRNPs),** resemble ribosomes in that each contains a set of proteins complexed to a stable RNA molecule. They are much smaller than ribosomes, however—about 250,000 daltons compared to 4.5 million daltons for a ribosome—and have higher protein-to-RNA ratios. Some proteins are present in several types of snRNPs, whereas others are unique to one type. This was first demonstrated using serum from patients with the disease *systemic lupus erythematosus,* who make antibodies directed against one or more snRNP proteins: a single antibody was found that binds the U1, U2, U5, and U4/U6 snRNPs, for example, suggesting that they all contain a common protein.

Figure 9–78 An amino acid sequence found in many eucaryotic RNA-binding proteins. This consensus—found in proteins from organisms as diverse as yeasts, *Drosophila,* and humans—is present in the proteins of hnRNP particles, in the protein bound to the poly-A tail of hnRNAs, in several snRNP proteins, and in the abundant nucleolar protein, nucleolin. When this sequence is found in a protein of unknown function, it suggests that the protein binds to RNA.

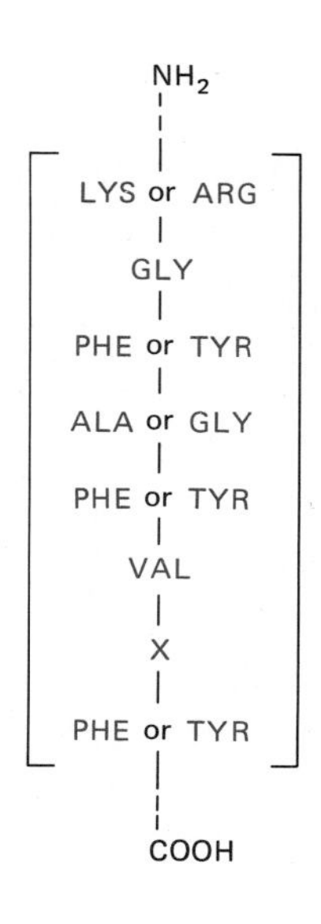

Individual snRNPs are believed to recognize specific nucleic acid sequences through RNA-RNA base-pair complementarity. Some mediate RNA splicing, some are involved in the cleavage reactions that generate the 3′ ends of some newly formed RNAs (see p. 596), while the function of others is unknown. The evidence for the role of snRNPs in RNA splicing comes from experiments on RNA processing *in vitro.*

Intron Sequences Are Removed as Lariat-shaped RNA Molecules[50]

Introns range in size from about 80 nucleotides to 10,000 nucleotides or more. They differ dramatically from exons in that their exact nucleotide sequences seem to be unimportant. Thus introns have accumulated mutations rapidly during evolution, and it is often possible to alter most of the nucleotide sequence of an intron without greatly affecting gene function. This has led to the suggestion that intron sequences have no function at all and are largely genetic "junk," a proposition we shall examine later (see p. 602). The only highly conserved sequences in introns are those required for intron removal. Thus there are consensus sequences at each end of an intron that are nearly the same in all known intron sequences, and these cannot be altered without affecting the splicing process that normally removes the intron sequence from the primary RNA transcript. These conserved boundary sequences at the **5′ splice site (donor site)** and the **3′ splice site (acceptor site)** are shown in Figure 9–80. The RNA breaking and rejoining reactions must be carried out precisely because an error of even one nucleotide would shift the reading frame in the resulting mRNA molecule and make nonsense of its message.

The pathway by which the intron sequences are removed from primary RNA transcripts has been elucidated by *in vitro* studies in which a pure RNA species containing a single intron is prepared by incubating an appropriately designed DNA fragment with a highly active, purified RNA polymerase (Figure 9–81). When these RNA molecules are added to a cell extract, they become spliced in a two-step enzymatic reaction that requires prolonged incubation with ATP, selected proteins in the extract, and the U1, U2, U5, and U4/U6 snRNPs; these components assemble into a large multicomponent ribonucleoprotein complex, or **spliceosome.** Characterization of the RNA species that appear as intermediates during the reaction, as well as the snRNPs required to produce them, led to the discovery

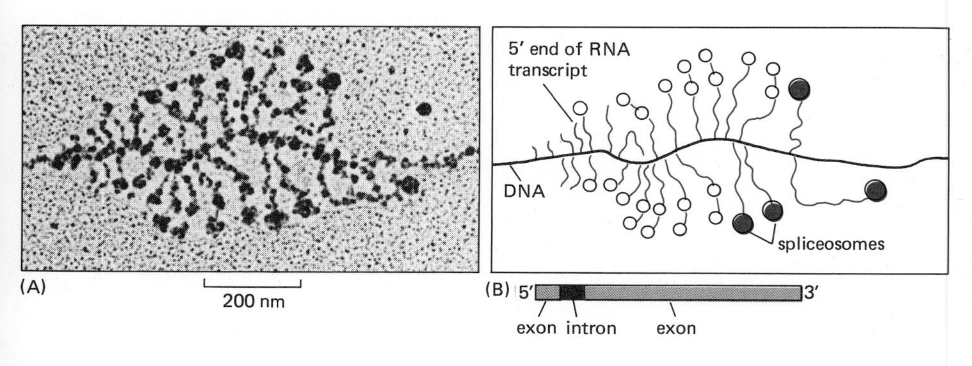

Figure 9–79 Electron micrograph of a chromatin spread showing large ribonucleoprotein particles assembling at the 5′ and 3′ splice sites to form a *spliceosome.* In the micrograph (A) a gene encoding a *Drosophila* chorion protein has been identified, so that the positions of the splice sites on the primary RNA transcript are known. (B) Most of the RNA transcripts have either one or two large RNP particles near their 5′ ends. When there are two particles on a transcript [open circles in (B)], they average 25 nm in diameter and occur at or very near the positions of the 5′ and 3′ splice sites for the single small intron sequence (228 nucleotides long) near the 5′ end of the transcripts. The more mature, longer transcripts on the two genes frequently display a single larger particle [colored circles in (B)] in the region of the intron, which probably results from the stable association of the two smaller particles and represents the assembled spliceosome. Since splicing occurs in some cases while the 3′ end of the RNA chain is still being transcribed, the poly A at the 3′ end of hnRNA molecules cannot be required for splicing. Most of the major hnRNP proteins have been removed from these transcripts by the spreading conditions used. (Adapted from Y.N. Osheim, O.L. Miller, and A.L. Beyer, *Cell* 43:143–151, 1985.)

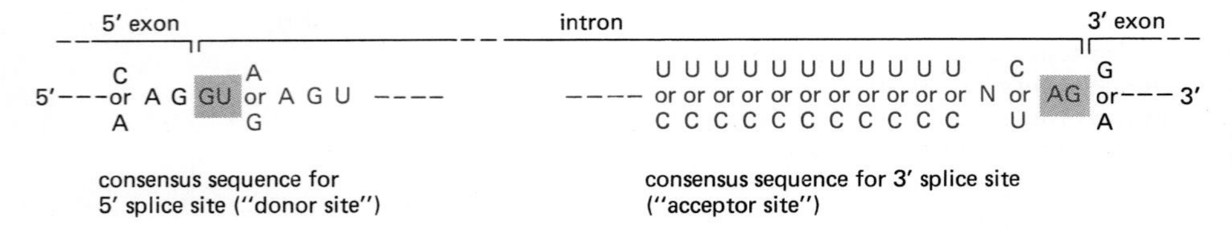

Figure 9–80 Consensus sequences for the 5′ and 3′ splice sites used in RNA splicing. The sequence given is that for the RNA chain; the nearly invariant GU and AG dinucleotides at either end of the intron are shaded in color (see also Figure 9–77).

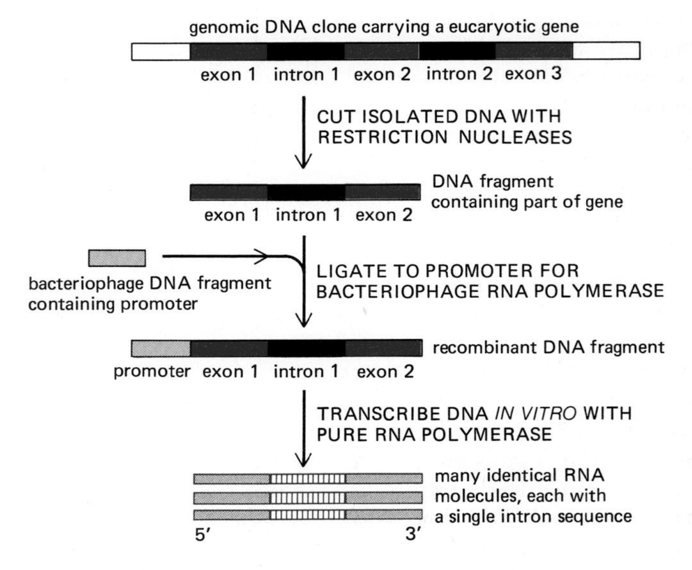

Figure 9–81 Outline of the procedure used to produce abundant amounts of pure RNA molecules for the analysis of RNA splicing *in vitro*. The method depends on the ability to produce large amounts of any desired DNA sequence by genetic engineering and DNA cloning (see p. 258) as well as on the availability of relatively simple RNA polymerases from bacteriophages T7 or SP6, which transcribe DNA with high efficiency *in vitro*. By coupling a eucaryotic DNA fragment to a bacteriophage promoter, a bacteriophage RNA polymerase can be used to generate *in vitro* large amounts of the RNA encoded by the eucaryotic DNA fragments. The 5′ cap present on hnRNAs can be incorporated into such RNAs by using a chemically synthesized, capped nucleotide to initiate the transcription process (not shown).

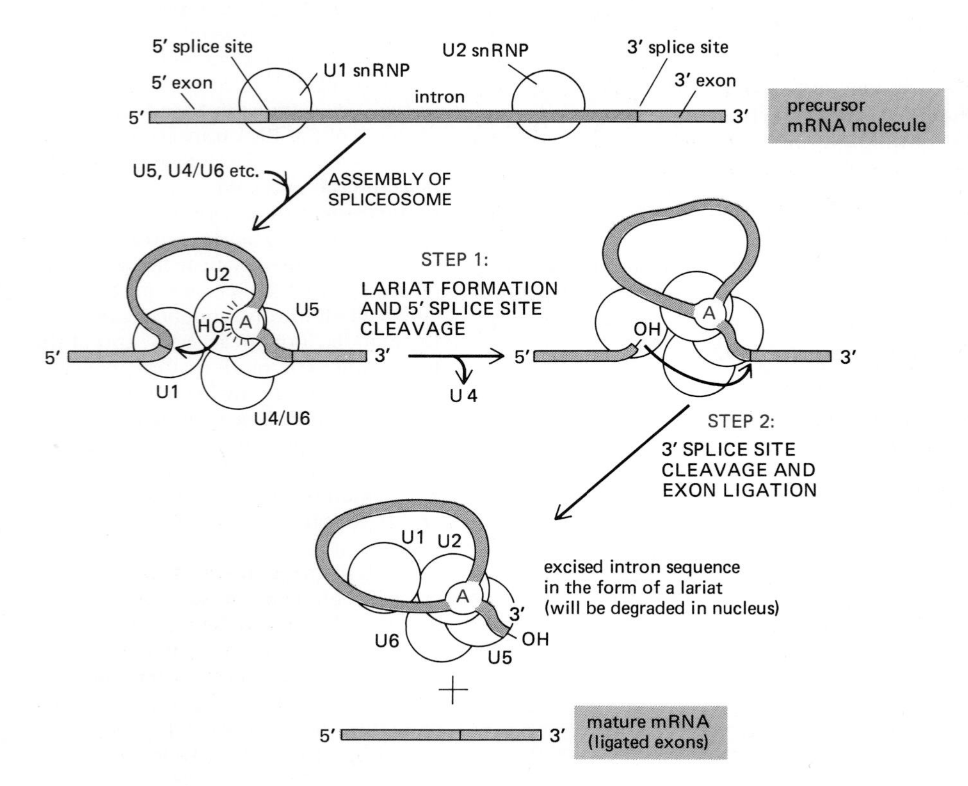

Figure 9–82 Catalysis of RNA splicing by a spliceosome formed from the assembly of U1, U2, U5, and U4/U6 snRNPs (shown as circles), plus other components (not shown). After assembly of the spliceosome, the reaction occurs in two steps: in step 1 a special A nucleotide in the intron sequence located close to the 3′ splice site attacks the 5′ splice site, which is cleaved; the cut 5′ end of the intron sequence joins covalently to this A nucleotide, forming the branched nucleotide shown in Figure 9–83. In step 2 the 3′-OH end of the first exon, which was exposed in the first step, adds to the beginning of the second exon, cleaving the RNA molecule at the 3′ splice site; the two exon sequences are thereby joined to each other and the intron sequence is released as a lariat. The complete spliceosome complex sediments at 60S, indicating that it is nearly as large as a ribosome. These splicing reactions occur in the nucleus and generate mRNA molecules from primary RNA transcripts (mRNA precursor molecules).

that the intron is excised in the form of a *lariat*, according to the splicing pathway shown in Figures 9–82 and 9–83.

Individual roles have been defined for several of the snRNPs. The U1 snRNP, for example, binds to the 5′ splice site, guided by a nucleotide sequence in the U1 RNA that is complementary to the nine-nucleotide splice-site consensus sequence (see Figure 9–80). Since RNA is capable of acting like an enzyme (see p. 105), either the RNA or the protein components of the spliceosome could be responsible for catalyzing the breakage and formation of covalent bonds required for RNA splicing.

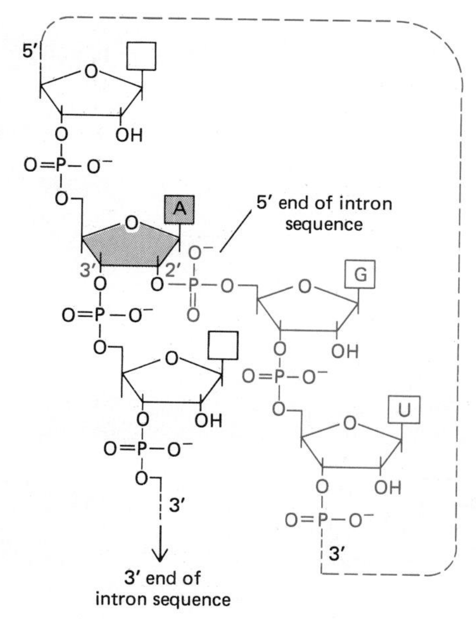

Figure 9–83 Structure of the branched RNA chain that forms during nuclear RNA splicing. The shaded A is the nucleotide highlighted in Figure 9–82, and the branch is formed in step 1 of the splicing reaction illustrated there. In this step the 5′ end of the intron sequence is cleaved and its phosphate group couples covalently to the 2′-OH ribose group of the A nucleotide, which is located about 30 nucleotides from the 3′ end of the intron sequence. The branched chain remains in the final excised intron sequence and is responsible for its lariat form (see Figure 9–82).

9-32 Multiple Intron Sequences Are Usually Removed from Each RNA Transcript[51]

Because the spliceosome seems mainly to recognize a consensus sequence at each intron boundary, the *5′ splice site* (*donor site*) at the end of any one intron can in principle be joined to the *3′ splice site* (*acceptor site*) of any other intron in the splicing process. Thus, when the 5′ and 3′ halves of two different introns are experimentally combined, the resulting hybrid intron sequence is often recognized by the RNA-splicing enzymes and removed.

In view of this result, it is surprising that vertebrate genes can contain as many as 50 introns (see Table 9–1, p. 486). If any two 5′ and 3′ splice sites were mispaired for splicing, some functional mRNA sequences would be lost, with disastrous consequences. Somehow such mistakes are avoided: the RNA processing machinery normally guarantees that each 5′ splice site pairs only with the 3′ splice site that is closest to it in the downstream (5′-to-3′) direction of the linear RNA sequence (Figure 9–84). How this sequential pairing of splice sites is accomplished is not known, although the assembly of the spliceosome while the RNA transcript is still growing (see Figure 9–79) is presumed to play a major part in ensuring an orderly pairing of the appropriate splice sites. There is also evidence that the exact three-dimensional conformations adopted by the intron and exon sequences in the RNA transcript are important. We shall see, however—both below and in Chapter 10—that splicing can be controlled, and in selected cases the simple pattern of 5′-to-3′ splicing does not hold.

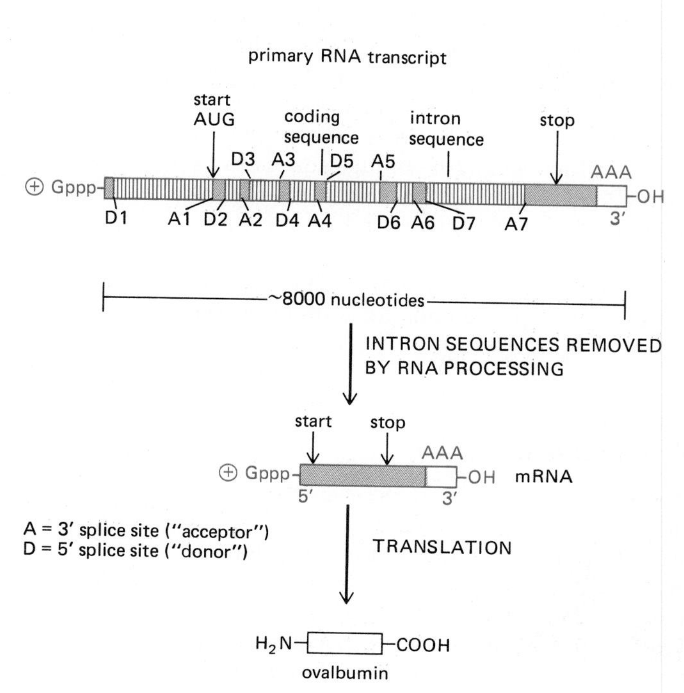

Figure 9–84 The primary RNA transcript for the chicken ovalbumin gene, showing the organized removal of seven introns required to obtain a functional mRNA molecule. The 5′ splice sites (donor sites) are denoted by D, and 3′ splice sites (acceptor sites) are denoted by A.

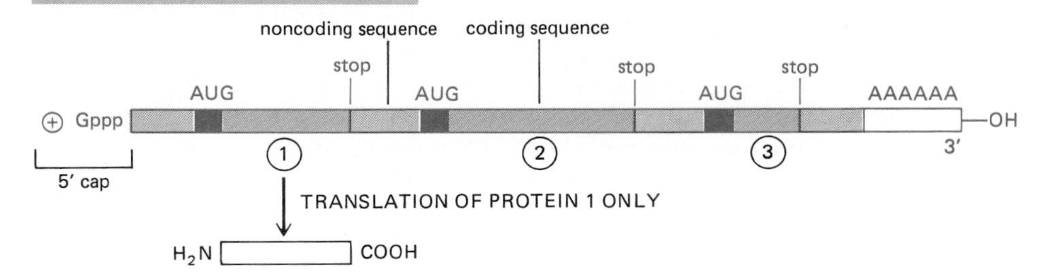

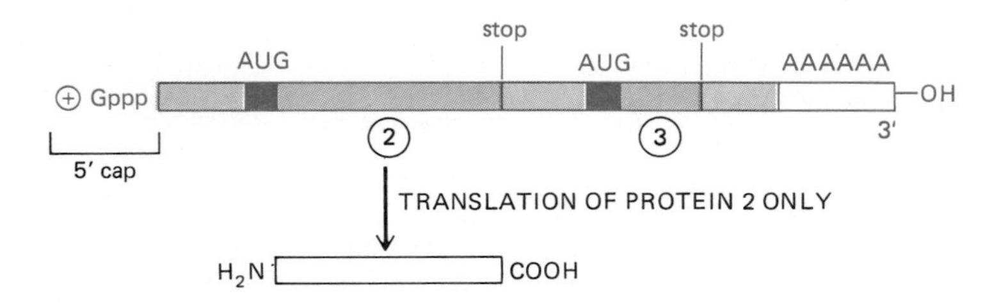

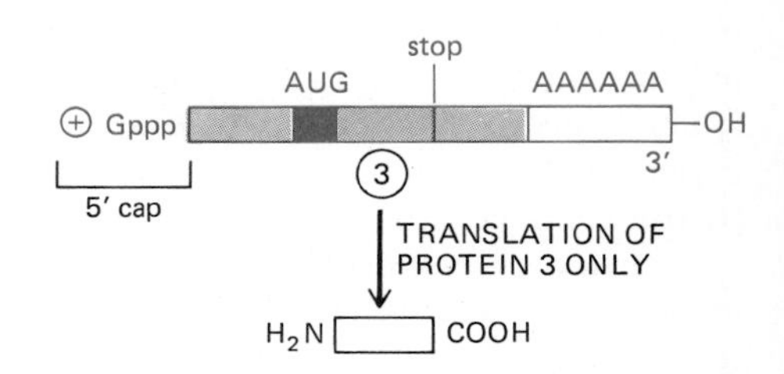

Figure 9–85 For some viruses the same primary RNA transcript is spliced in several ways to produce three (or more) different mRNA molecules, each coding for a different protein. In each case, only the coding sequence closest to the 5′ cap is translated from the mRNA molecule.

9-33 10-23 The Same RNA Transcript Can Be Processed in Different Ways to Produce mRNAs Coding for Several Proteins[52]

Although most intron sequences themselves appear to have no specific function, the existence of RNA splicing makes it possible to generate several different mRNAs, and thereby several different proteins, from the same RNA transcript, thus conferring extra genetic flexibility on the cell. We shall see in Chapter 10 that changes in the pattern of splicing of many RNA transcripts occur in the course of cell differentiation, so that the same DNA coding sequences are put to different uses as cells develop (see p. 589).

The versatility conferred by RNA splicing was first discovered in the adenovirus in which splicing itself was discovered. The adenovirus genome directs the synthesis of some very long RNA transcripts that contain the coding sequences for several proteins. This does not normally occur in a eucaryotic cell, where an individual mRNA molecule generally codes for only one protein (see p. 214); translation is initiated only near the 5′ cap site (see p. 528), and it usually will stop as soon as the first stop codon is encountered. In adenovirus, however, this limitation is overcome by the RNA-splicing machinery, which can treat coding sequences as introns and remove them, so that the same 5′ cap is spliced to each of the downstream coding sequences in a proportion of the mRNAs produced by the virus. This *alternative RNA splicing* allows the same 5′ cap to serve as the initiation signal for the synthesis of different proteins (Figure 9–85). This device is widely used by viruses to enable only a few different RNA transcripts to code for a much larger number of proteins.

9-34 The mRNA Changes in Thalassemia Reveal How RNA Splicing Can Allow New Proteins to Evolve[53]

Recombinant DNA techniques have made human mutants an increasingly important source of material for genetic studies of cellular mechanisms (see p. 181). In a group of human genetic diseases called the **thalassemia syndromes,** for example, patients have an abnormally low level of hemoglobin—the oxygen-carrying protein in red blood cells. The change in the DNA sequence has been determined for more than 50 such mutants, and a large proportion of these cause alterations in the pattern of RNA splicing. Thus single nucleotide changes have been detected that either inactivate a splice site or create a new splice site by changing a se-

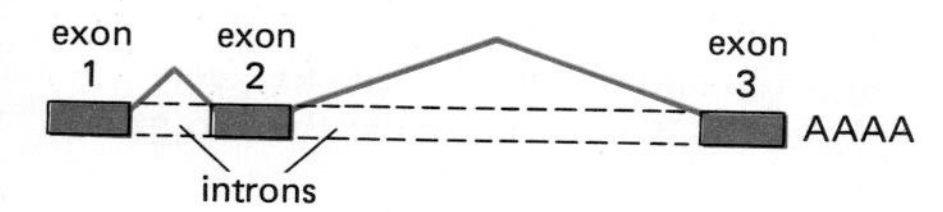

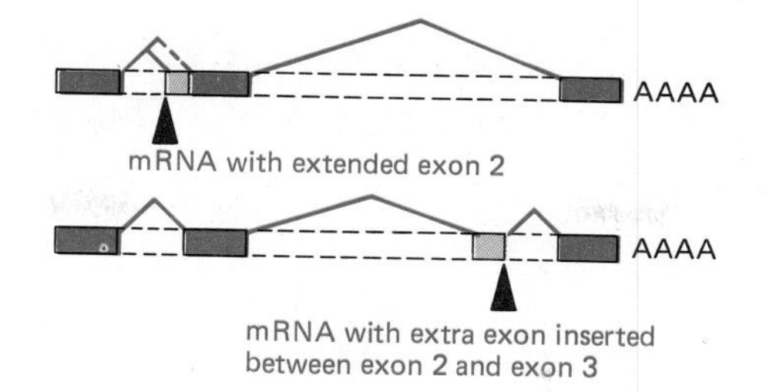

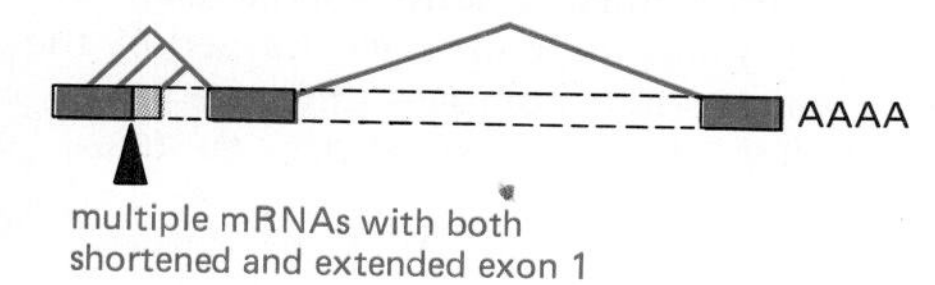

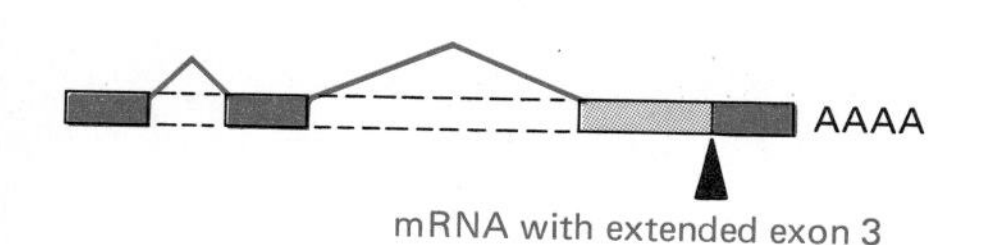

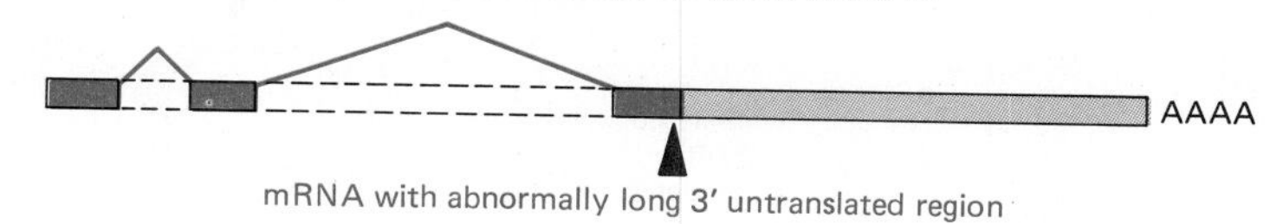

Figure 9–86 Examples of the abnormal processing of the β-globin primary RNA transcript that have been observed in mutant humans with β thalassemia. The site of each mutation is denoted by the black arrowhead. The colored boxes represent the three normal exons illustrated previously in Figure 9–77, and the colored lines join the 5′ and 3′ splice sites utilized in splicing the primary RNA transcript produced by the gene. The open boxes depict new nucleotide sequences included in the final mRNA molecule as a result of a mutation. Note that when a mutation leaves a normal splice site without a partner, one or more abnormal "cryptic" splice sites nearby are used as the partner site, as in (C). (After S.H. Orkin, in The Molecular Basis of Blood Diseases [G. Stanatoyannopoulos et al., eds.], pp. 106–126. Philadelphia: Saunders, 1987.)

quence in an intron or an exon into a consensus splice site. Surprisingly, analysis of the mRNAs produced in these mutant individuals reveals that the loss of a splice site does not prevent splicing but instead causes its normal partner site to seek out and become joined to a new "cryptic" site nearby; often a number of alternative splices are made in these mutants, causing the mutant gene to produce a set of altered proteins rather than just one (Figure 9–86). These results demonstrate that RNA splicing is a very flexible process in higher eucaryotic cells.

Because a simple mutation will often cause a gene to produce a variety of new proteins, the cell can test possible improvements to its genetic dowry in a very efficient manner. For this reason the flexibility of RNA splicing might have played a crucial part in the evolution of higher eucaryotes. Lower eucaryotes, such as yeasts, have more highly specified splicing rules, which should greatly constrain the evolution of new mRNAs through changes in patterns of RNA splicing; perhaps as a result, their rate of divergence in form and function seems to have been very much slower than that of the higher eucaryotes.

Spliceosome-catalyzed RNA Splicing Probably Evolved from Self-splicing Mechanisms[54]

When the lariat intermediate in nuclear RNA splicing was first discovered, it puzzled molecular biologists. Why was this bizarre pathway used, rather than the apparently simpler alternative of bringing the 5′ and 3′ splice sites together in an initial step, followed by their direct cleavage and rejoining? The answer seems to lie in the way the spliceosome evolved.

As explained in Chapter 1, it is thought that early cells used RNA molecules rather than proteins as their major catalysts and stored their genetic information in RNA rather than DNA sequences (see p. 7). RNA-catalyzed splicing reactions presumably played important roles in these early cells, and some self-splicing RNA introns remain today—for example, in the nuclear rRNA genes of *Tetrahymena* (see p. 105), in bacteriophage T4, and in some mitochondrial and chloroplast genes (see p. 393). In these cases, large parts of the intron sequence have been highly conserved because of their need to fold to create a catalytic surface in the RNA molecule. Two major classes of self-splicing introns can be readily distinguished: *group I introns* begin the splicing reaction by binding a G nucleotide to

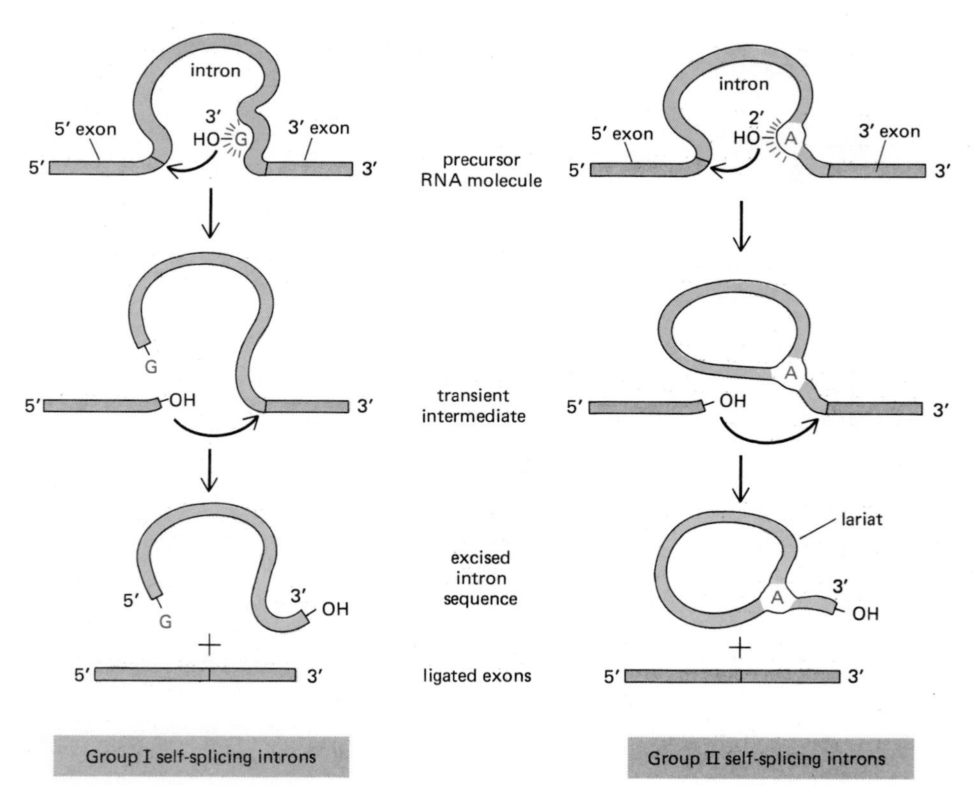

Figure 9–87 The two known classes of self-splicing introns. The group I introns bind a free G nucleotide to a specific site to initiate splicing (see Figure 3–19, p. 105), while the group II introns use a specially reactive A nucleotide in the intron sequence itself for the same purpose. The two mechanisms have been drawn in a way that emphasizes their similarities. Both are normally aided by proteins that speed up the reaction, but the catalysis is nevertheless mediated by the RNA in the intron. The mechanism used by group II introns forms a lariat and resembles the pathway catalyzed by the spliceosome (compare to Figure 9–82.) (After T.R. Cech, *Cell* 44:207–210, 1986.)

the intron sequence; the G is thereby activated to form the attacking group that will break the first of the phosphodiester bonds cleaved during splicing (the bond at the 5′ splice site); in *group II introns* a specially reactive A residue in the intron sequence has this role, and a lariat intermediate is generated. Otherwise the reaction pathways are the same, and both are presumed to represent vestiges of very ancient mechanisms (Figure 9–87).

In the evolution of nuclear RNA splicing, the reaction pathway used by the group II self-splicing introns seems to have been retained, but the catalytic role of the intron sequences has been replaced by separate spliceosome components. Thus the small RNAs U1 and U2, for example, may well be remnants of catalytic sequences that were originally present in introns. Shifting the catalysis from intron to spliceosome presumably lifted most of the constraints on the evolution of introns, allowing many new intron sequences to evolve.

The Transport of mRNAs to the Cytoplasm Is Delayed Until Splicing Is Complete[55]

As discussed in Chapter 8, finished mRNA molecules are thought to be recognized by receptor proteins in the nuclear pore complex that help move the mRNAs to the cytoplasm by an active transport process (see p. 425). The major proteins of the hnRNP particles and various processing molecules bound to the RNA in the nucleus, however, never seem to leave the nucleus, which suggests that they are stripped off the RNA as it passes through the nuclear pore (Figure 9–88). Studies of mutant yeasts suggest that, for RNAs that have splice sites, this transport process can occur only after the splicing reaction has been completed. In conditionally lethal yeast mutants that fail to splice their RNAs at high temperature because of a defect in their splicing machinery, all unspliced mRNA precursors remain in the nucleus, while those mRNAs that do not require splicing (which includes most of the mRNAs in this single-celled eucaryote) are transported normally to the cytoplasm. This observation is consistent with the idea that RNAs are retained by their bound spliceosome components, which seem to form numerous large aggregates throughout the interior of the nucleus of higher eucaryotes. These aggregates may serve as "splicing islands," although it is not known how they form or function (Figure 9–89). They could, however, be analogous to the *nucleolus*, a much larger

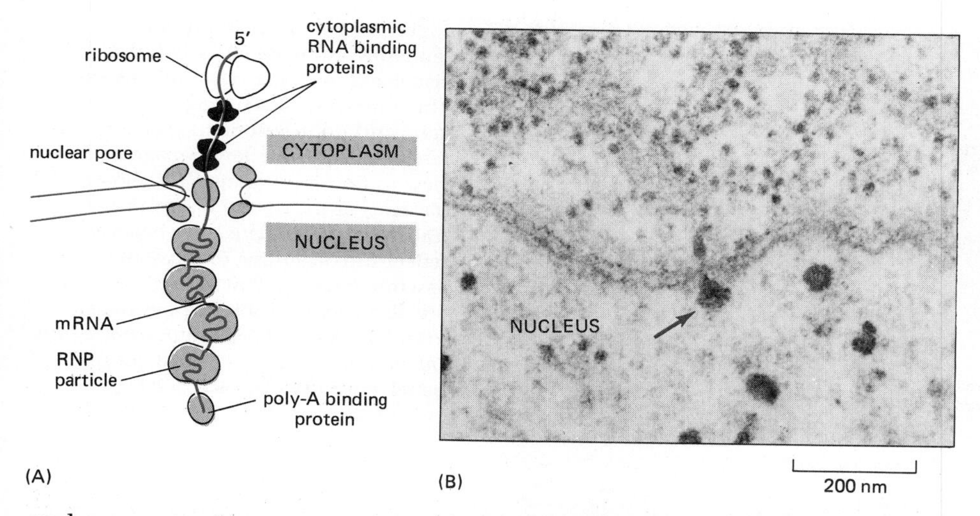

Figure 9–88 The movement of mRNA molecules through nuclear pores. (A) Schematic illustration of the change thought to occur in the proteins bound to the RNA molecule as it moves out of the nucleus. (B) Electron micrograph of a large mRNA molecule produced in an insect salivary gland cell; this molecule has apparently been caught in the process of moving to the cytoplasm (*arrow*). (B, from B.J. Stevens and H. Swift, *J. Cell Biol.* 31:55–77, 1966.)

and more prominent structure in the nucleus, whose organization and function are better understood.

The nucleolus is the site where ribosomal RNA (rRNA) molecules are processed from a larger precursor RNA and assembled into ribosomes by the binding of ribosomal proteins. Before discussing nucleolar structure, however, we need to consider how the precursor rRNA molecules are synthesized from rRNA genes.

10-23 Ribosomal RNAs Are Made on Tandemly Arranged Sets of Identical Genes[56]

Many of the most abundant proteins of a differentiated cell, such as hemoglobin in the red blood cell and myoglobin in a muscle cell, are synthesized from genes that are present in only a single copy per haploid genome. These proteins are abundant because each of the many mRNA molecules transcribed from the gene can be translated into as many as 10 protein molecules per minute. This will normally produce more than 10,000 protein molecules per mRNA molecule in each cell generation. Such an amplification step is not available for the synthesis of the RNA components of ribosomes, however, since these RNA molecules are the final gene products. Yet a growing higher eucaryotic cell must synthesize 10 million copies of each type of ribosomal RNA molecule in each cell generation in order to construct its 10 million ribosomes. Adequate quantities of ribosomal RNAs can, in fact, be produced only because the cell contains multiple copies of the genes that code for ribosomal RNAs (**rRNA genes**).

Even *E. coli* needs seven copies of its rRNA genes to keep up with the the cell's need for ribosomes. Human cells contain about 200 rRNA gene copies per haploid genome, spread out in small clusters on five different chromosomes; while cells of the frog *Xenopus* contain about 600 rRNA gene copies per haploid genome in a single cluster on one chromosome. In eucaryotes the multiple copies of the highly conserved rRNA genes on a given chromosome are located in a tandemly arranged series in which each gene (8,000 to 13,000 nucleotide pairs long, depending on the organism) is separated from the next by a nontranscribed region known as *spacer DNA*, which can vary greatly in length and sequence. We shall see in Chapter 10 that such multiple copies of tandemly arranged genes tend to co-evolve (see p. 600).

Because of their repeating arrangement, and because they are transcribed at a very high rate, the tandem arrays of rRNA genes can easily be seen in spread chromatin preparations. The RNA polymerase molecules and their associated transcripts are so densely packed (typically about 100 per gene) that the transcripts fan out perpendicularly from the DNA to give each transcription unit a "Christmas tree" appearance (Figure 9–90). As noted earlier (see Figure 9–73), the tip of each of these "trees" represents the point on the DNA at which transcription begins and where the transcripts are thus shortest, while the other end of the rRNA transcription unit is sharply demarcated by the sudden disappearance of RNA polymerase molecules and their transcripts.

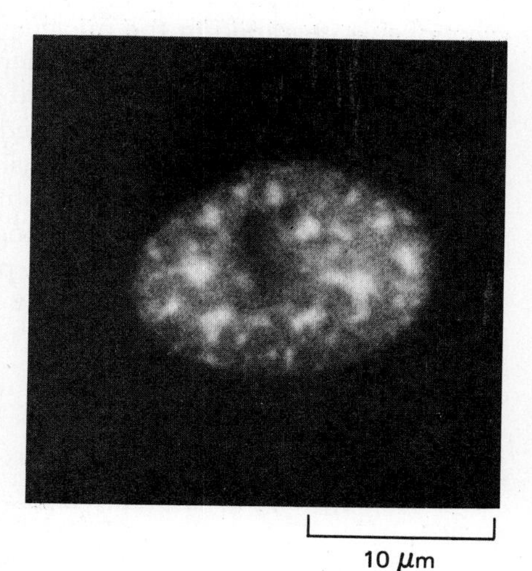

Figure 9–89 Immunofluorescence staining of a human fibroblast nucleus with a monoclonal antibody that detects the snRNP particles involved in nuclear splicing of mRNA precursor molecules. The snRNP particles are present in large aggregates, which could function as "splicing islands." The antibody detects specific proteins that are present in several of the snRNPs that function in the spliceosome. (Courtesy of N. Ringertz.)

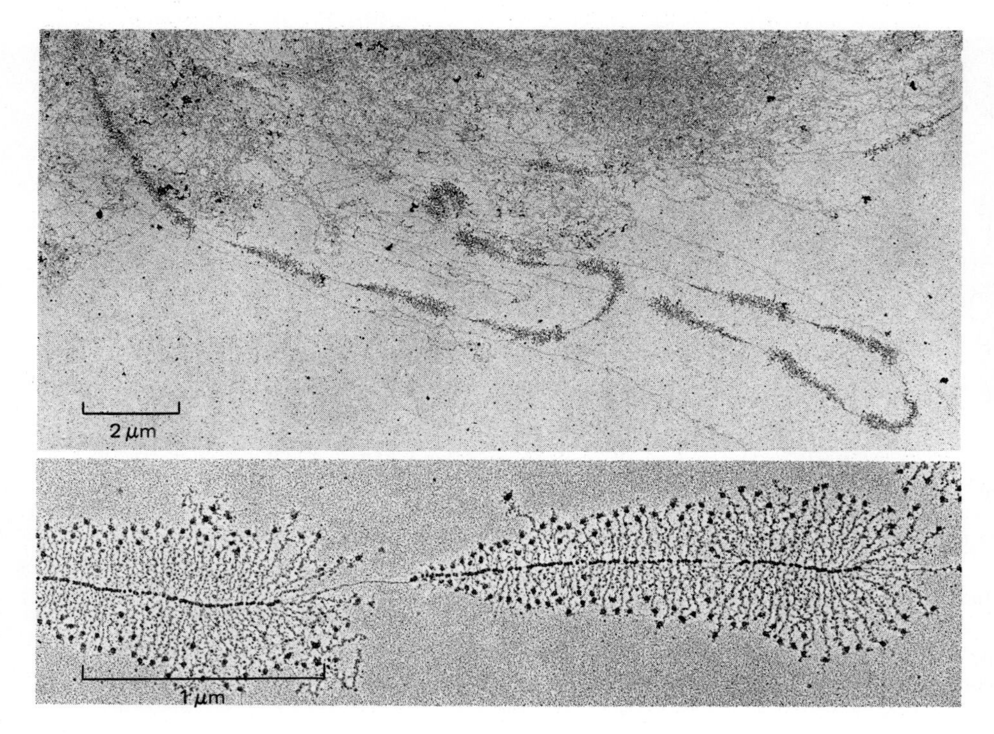

Figure 9–90 Transcription from tandemly arranged rRNA genes, as visualized in the electron microscope. The pattern of alternating transcribed gene and nontranscribed spacer is readily seen in the lower-magnification view in the upper panel. The large particles at the 5′ end of each rRNA transcript (*lower panel*) are believed to reflect the beginning of ribosome assembly; RNA polymerase molecules are also clearly visible. (Upper panel, from V.E. Foe, *Cold Spring Harbor Symp. Quant. Biol.* 42:723–740, 1978; lower panel, courtesy of Ulrich Scheer.)

The rRNA genes are transcribed by RNA polymerase I, and each gene produces the same primary RNA transcript. In humans this RNA transcript, known as *45S rRNA,* is about 13,000 nucleotides long. Before it leaves the nucleus in assembled ribosomal particles, the 45S rRNA is cleaved to give one copy each of the 28S rRNA (about 5000 nucleotides), the 18S rRNA (about 2000 nucleotides), and the 5.8S rRNA (about 160 nucleotides) of the final ribosome (see p. 221). The derivation of these three rRNAs from the same primary transcript ensures that they will be made in equal quantities. The remaining part of each primary transcript (about 6000 nucleotides) is degraded in the nucleus (Figure 9–91). Some of these extra RNA sequences are thought to play a transient part in ribosome assembly, which begins immediately as specific proteins bind to the growing 45S rRNA transcripts.

Another set of tandemly arranged genes with nontranscribed spacers codes for the 5S rRNA of the large ribosomal subunit (the only rRNA that is transcribed separately). The 5S rRNA genes are only about 120 nucleotide pairs in length, and like a number of other genes encoding small stable RNAs (most notably the tRNA genes), they are transcribed by RNA polymerase III. Humans have about 2000 5S rRNA genes tandemly arranged in a single cluster far from all the other rRNA genes. It is not known why this one type of rRNA is transcribed separately.

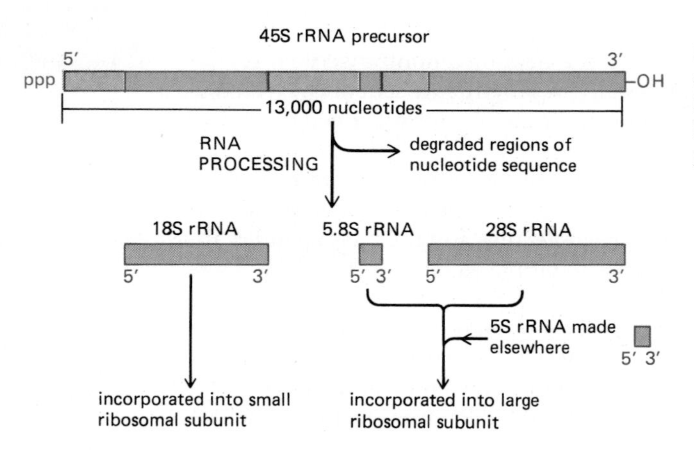

Figure 9–91 The pattern of processing of a 45S rRNA precursor molecule into three separate ribosomal RNAs. Nearly half of the nucleotide sequences in this precursor are degraded in the nucleus.

The Nucleolus Is a Ribosome-producing Machine[57]

The continuous transcription of multiple gene copies ensures an adequate supply of the rRNAs, which are immediately packaged with ribosomal proteins to form ribosomes. The packaging takes place in the nucleus, in a large, distinct structure called the **nucleolus.** The nucleolus contains large loops of DNA emanating from several chromosomes, each of which contains a cluster of rRNA genes. Each such gene cluster is known as a **nucleolar organizer** region. Here the rRNA genes are transcribed at a rapid rate by RNA polymerase I. The beginning of the rRNA packaging process can be seen in electron micrographs of these genes: the 5′ tail of each transcript is encased by a protein-rich granule (see Figure 9–90). These granules, which do not appear on other types of RNA transcripts, presumably reflect the first of the protein-RNA interactions that take place in the nucleolus.

The biosynthetic functions of the nucleolus can be traced by briefly labeling newly made RNA with ^{3}H-uridine. After various intervals of further incubation in unlabeled media, a cell fractionation procedure can be used to break the rRNA genes free of their chromosomes, thereby allowing the radioactive nucleoli to be isolated in relatively pure form (Figure 9–92). Such experiments show that the intact 45S transcript is first packaged into a large complex containing many different proteins imported from the cytoplasm, where all proteins are synthesized. Most of the 70 different polypeptide chains that will make up the ribosome, as well as the 5S rRNAs, are incorporated at this stage. Other molecules are needed to guide the assembly process. Thus the nucleolus also contains other RNA-binding proteins and certain small ribonucleoprotein particles (including U3 snRNP) that are believed to help catalyze the construction of ribosomes. These components remain in the nucleolus when the ribosomal subunits are exported to the cytoplasm in finished form. An especially notable component is *nucleolin,* an abundant, well-characterized RNA-binding protein that seems to coat only ribosomal transcripts; this protein stains with silver in the characteristic manner of the nucleolus itself.

As the 45S rRNA molecule is processed, it gradually loses some of its RNA and protein and then splits to form separate precursors of the large and small ribosomal subunits (Figure 9–93). Within 30 minutes of radioactive pulse labeling, the first mature small ribosomal subunits, containing their 18S rRNA, emerge from the nucleolus and appear in the cytoplasm. Assembly of the mature large ribosomal subunit, with its 28S, 5.8S, and 5S rRNAs, takes about an hour to complete. The nucleolus therefore contains many more incomplete large ribosomal subunits than small ones.

The last steps in ribosome maturation occur only as these subunits are transferred to the cytoplasm. This delay prevents functional ribosomes from gaining access to the incompletely processed hnRNA molecules in the nucleus.

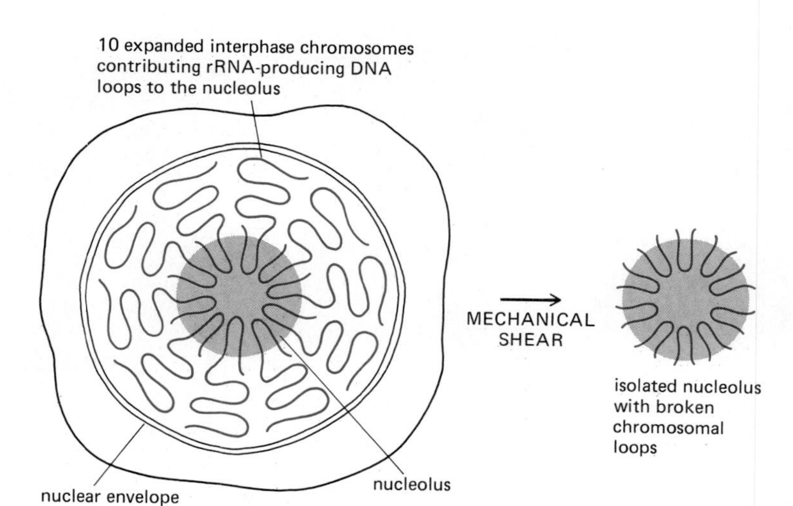

Figure 9–92 Highly schematic view of a human cell, showing the contributions to a single large nucleolus of loops of chromatin containing rRNA genes from 10 separate chromosomes. Purified nucleoli are very useful for biochemical studies of nucleolar function; to obtain such nucleoli, the loops of chromatin are mechanically sheared from their chromosomes, as shown.

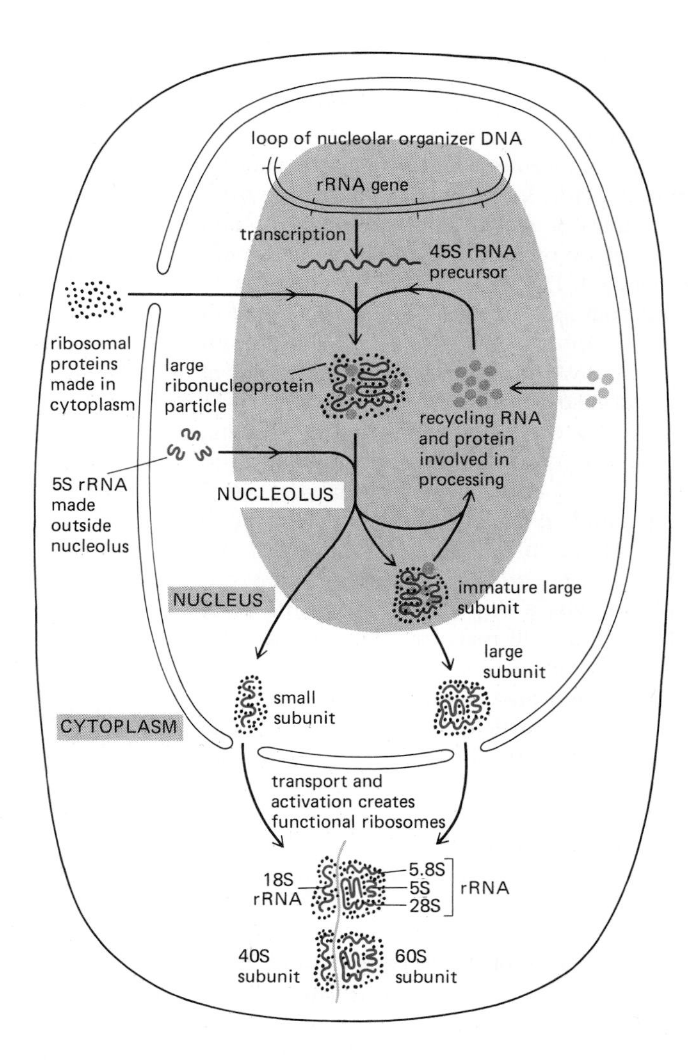

Figure 9–93 The function of the nucleolus in ribosome synthesis. The 45S rRNA transcript is packaged in a large ribonucleoprotein particle containing many ribosomal proteins imported from the cytoplasm. While this particle remains in the nucleolus, selected pieces are discarded as it is processed into immature large and small ribosomal subunits. These two subunits are thought to attain their final functional form only as they are individually transported through the nuclear pores into the cytoplasm.

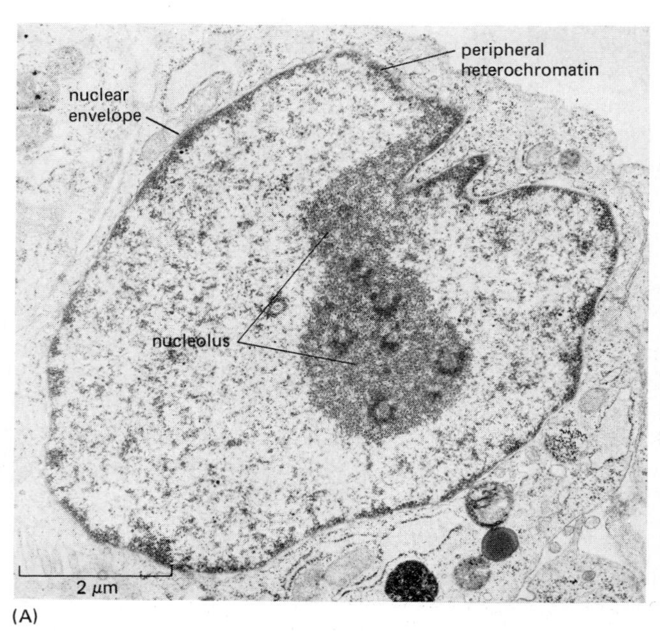

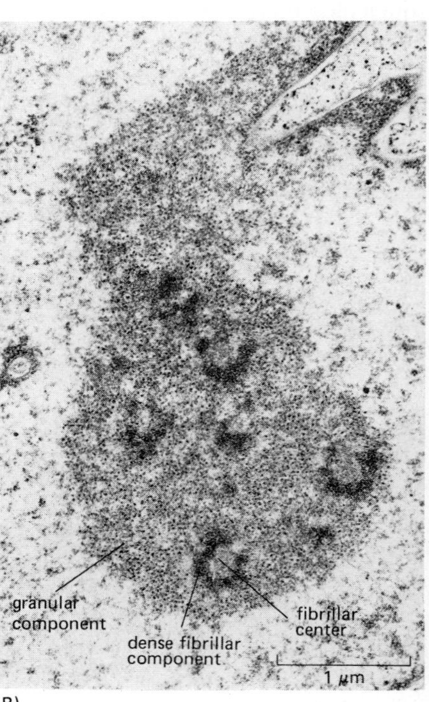

Figure 9–94 Electron micrograph of a thin section of a nucleolus in a human fibroblast, showing its three distinct zones. (A) View of entire nucleus. (B) High-power view of the nucleolus. (Courtesy of E.G. Jordan and J. McGovern.)

The Nucleolus Is a Highly Organized Subcompartment of the Nucleus[57]

As seen in the light microscope, the large spheroidal nucleolus is the most obvious structure in the nucleus of a nonmitotic cell. Consequently, it was so closely scrutinized by early cytologists that an 1898 review could list some 700 references. By the 1940s cytologists had demonstrated that the nucleolus contains high concentrations of RNA and proteins; but its major function in ribosomal RNA synthesis and ribosome assembly was not discovered until the 1960s.

Some of the details of nucleolar organization can be seen in the electron microscope. Unlike the cytoplasmic organelles, the nucleolus is not bounded by a membrane; instead, it seems to be constructed by the specific binding of unfinished ribosome precursors to each other to form a large network. In a typical electron micrograph, three partially segregated regions can be distinguished (Figure 9–94): (1) a pale-staining *fibrillar center,* which contains DNA that is not being actively transcribed; (2) a *dense fibrillar component,* which contains RNA molecules in the process of transcription; and (3) a *granular component,* which contains maturing ribosomal precursor particles.

The size of the nucleolus reflects its activity and therefore varies greatly in different cells and can change in a single cell. It is very small in some dormant plant cells, for example, but can occupy up to 25% of the total nuclear volume in cells that are making unusually large amounts of protein. The differences in size are due largely to differences in the amount of the granular component, which is probably controlled at the level of ribosomal gene transcription: electron microscopy of spread chromatin shows that both the fraction of activated ribosomal genes and the rate at which each gene is transcribed can vary according to circumstances.

The Nucleolus Is Reassembled on Specific Chromosomes After Each Mitosis[58]

The appearance of the nucleolus changes dramatically during the cell cycle. As the cell approaches mitosis, the nucleolus first decreases in size and then disappears as the chromosomes condense and all RNA synthesis stops, so that generally there is no nucleolus in a metaphase cell. When ribosomal RNA synthesis restarts at the end of mitosis (in telophase), tiny nucleoli reappear at the chromosomal locations of the ribosomal RNA genes (Figure 9–95).

In humans the ribosomal RNA genes are located near the tips of each of 5 different chromosomes, as shown previously in Figure 9–40 (that is, on 10 of the 46 chromosomes in a diploid cell). Correspondingly, 10 small nucleoli form after mitosis in a human cell, although they are rarely seen as separate entities because they quickly grow and fuse to form the single large nucleolus typical of many interphase cells (Figure 9–96).

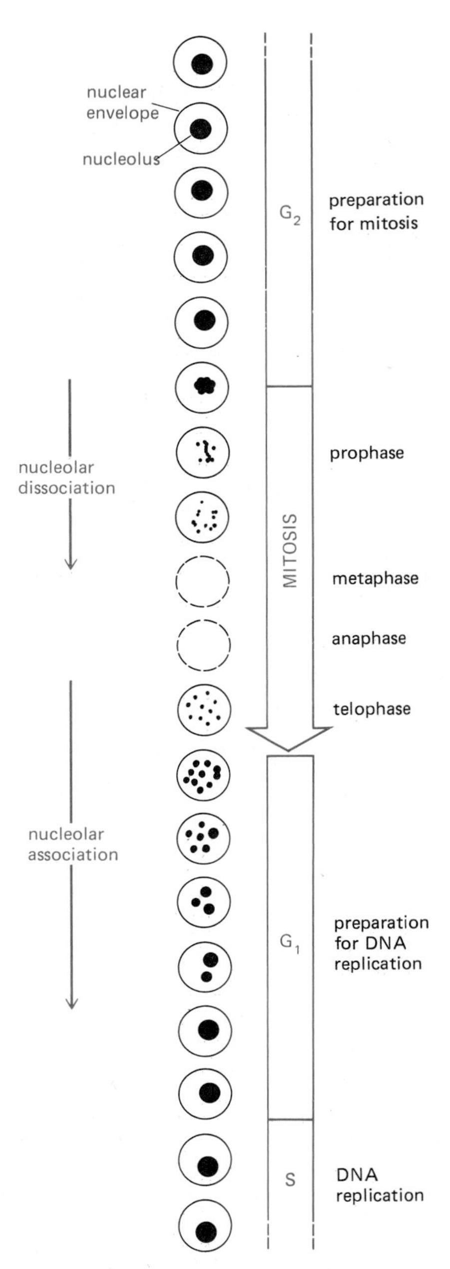

Figure 9–95 Changes in the appearance of the nucleolus in a human cell during the cell cycle. Only the cell nucleus is represented in this diagram.

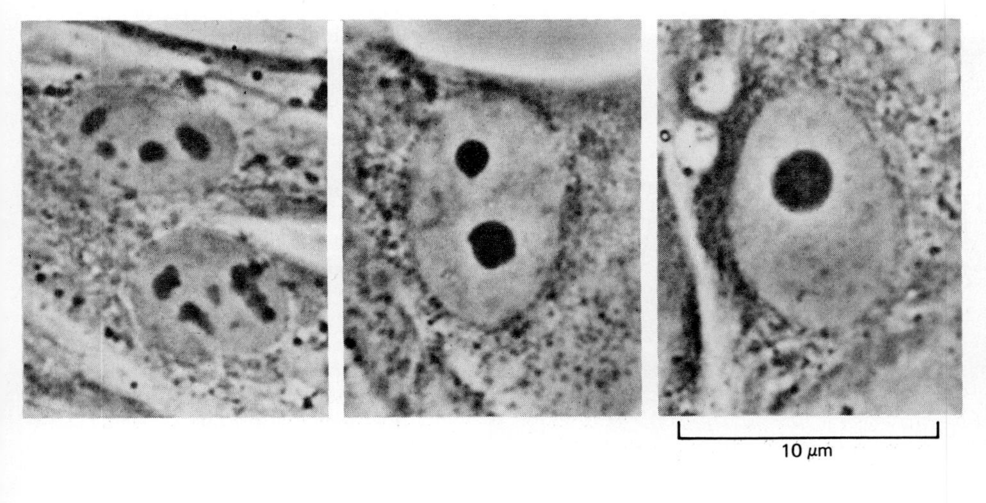

Figure 9–96 Light micrographs of human fibroblasts grown in culture, showing various stages of nucleolar fusion. (Courtesy of E.G. Jordan and J. McGovern.)

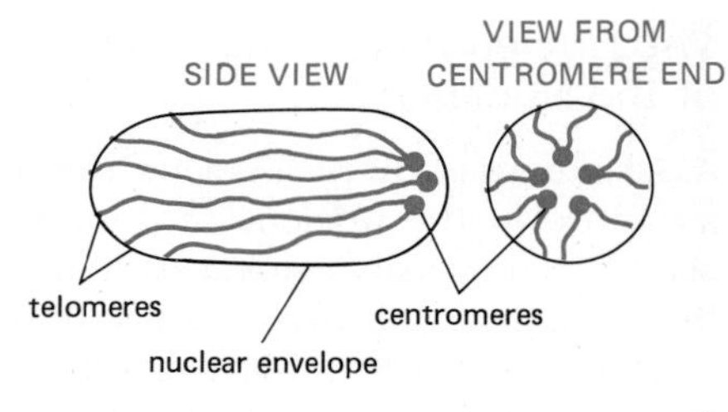

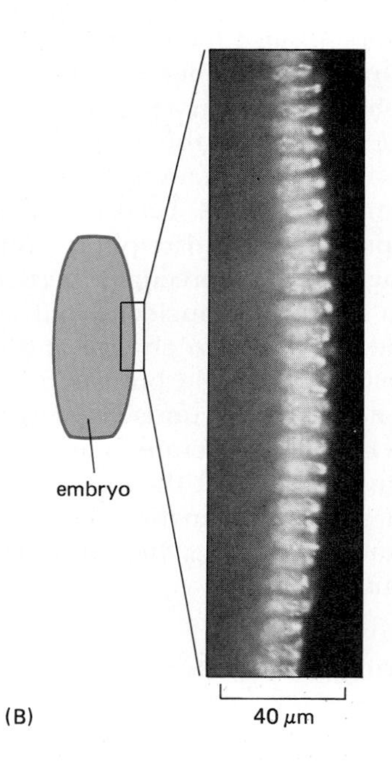

Figure 9–97 The polarized orientation of chromosomes in interphase cells of the early *Drosophila* embryo. (A) Diagrams of the *Rabl orientation*, with all centromeres facing one nuclear pole and all telomeres pointing toward the opposite pole. In the embryo each nucleus is elongated as shown. (B) Low-magnification light micrograph of a *Drosophila* embryo at the cellular blastoderm stage, in which the chromosomes in each interphase nucleus have been stained with a fluorescent dye. Note that the most brightly staining region (the chromocenter), which is known to contain the centromeric regions of each of the four chromosomes (see Figure 9–43), is oriented toward the outer surface of the embryo and thus faces the apical plasma membrane of every cell. (Courtesy of John Sedat.)

What happens to the RNA and protein components of the disassembled nucleolus during mitosis? It seems that at least some of them become distributed over the surface of all of the metaphase chromosomes and are carried as cargo to each of the two daughter cell nuclei. As the chromosomes decondense at telophase, these "old" nucleolar components help reestablish the newly emerging nucleoli.

Individual Chromosomes Occupy Discrete Territories in the Nucleus During Interphase[59]

As we have just seen, specific genes from separate interphase chromosomes are brought together at a single site in the nucleus when the nucleolus forms. This raises the question of whether other parts of chromosomes are also nonrandomly ordered in the nucleus. First raised by biologists in the late nineteenth century, this fundamental question has still not been answered satisfactorily.

A certain degree of chromosomal order results from the configuration that the chromosomes always have at the end of mitosis. Just before a cell divides, the condensed chromosomes are pulled to each spindle pole by microtubules attached to the centromeres; thus the centromeres lead the way and the distal arms of the chromosomes (terminating in telomeres) lag behind (see p. 773). The chromosomes in many nuclei tend to retain this so-called *Rabl orientation* throughout interphase, with their centromeres facing one pole of the nucleus and their telomeres pointing toward the opposite pole (Figure 9–97A). In some cases the nuclear poles are specifically oriented in the cell: in the early *Drosophila* embryo, for example, all the centromeres face apically (Figure 9–97B). Such fixed nuclear orientations might have important effects on cell polarity, but it is difficult to design experiments to test this possibility.

In most cells the various chromosomes are indistinguishable from one another during interphase, so that it is difficult to assess their arrangement in more detail than just described. The giant interphase chromosomes of the polytene

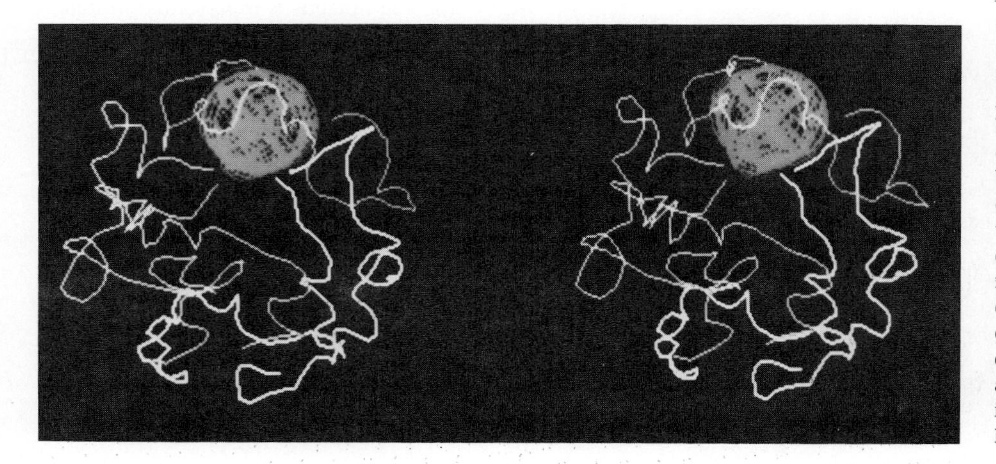

Figure 9–98 A stereo pair that displays the three-dimensional arrangement of the polytene chromosomes in a single nucleus of a *Drosophila* larval gland cell. The large ball is the nucleolus, and the course of each chromosome arm is represented by a line running along the chromosome axis. The telomeres tend to be on the surface of the nuclear envelope opposite the surface that is nearest the nucleolus, where all the centromeres are located. The chromosomes in such nuclei are never entangled, but their detailed foldings and neighbors are different in otherwise identical nuclei. (Courtesy of Mark Hochstrasser and John W. Sedat.)

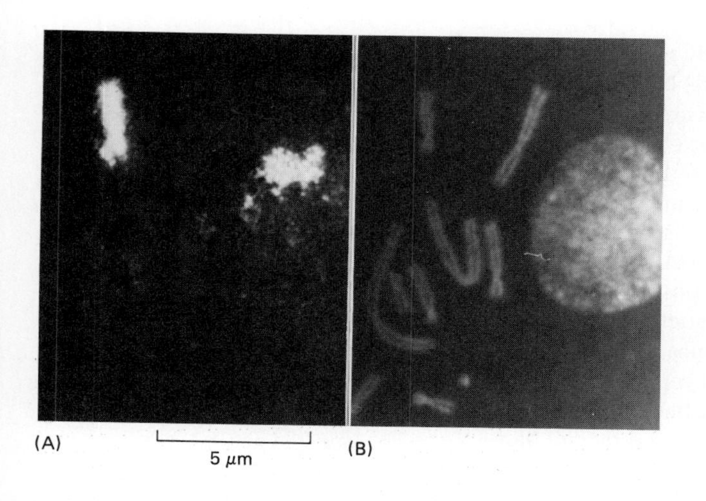

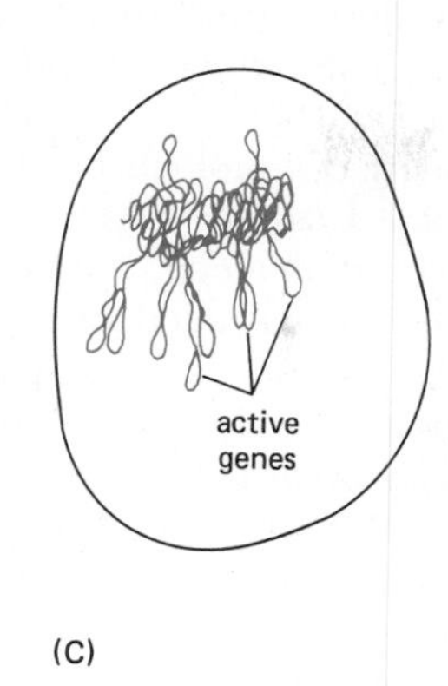

Figure 9–99 Selective labeling of a single chromosome in a cultured mammalian cell nucleus during interphase. (A) The results of *in situ* hydridization (using a fluorescent probe) to outline the single human chromosome in a human-hamster hybrid cell line. The same preparation is shown with all of the DNA fluorescently labeled in (B). In both (A) and (B), an interphase nucleus freed from its cytoplasm is shown at the right with scattered mitotic chromosomes released from a second cell on the left. (C) Schematic drawing of the human chromosome in the interphase nucleus shown in (A). (A and B, courtesy of Joyce A. Kobori and David R. Cox.)

cells of *Drosophila* larvae, however, are an exception. Here the individual chromosome bands can be resolved clearly enough to determine the precise positions of specific genes in intact nuclei by optical-sectioning and reconstruction techniques. The results of such analyses suggest that the interphase chromosome set is not highly ordered: although the Rabl orientation tends to be maintained, two apparently identical cells often have different chromosomes as nearest neighbors.

These analyses of polytene chromosomes have also indicated that each chromosome occupies its own territory in the interphase nucleus—that is, the individual chromosomes are not extensively intertwined (Figure 9–98). Other experiments have shown that nonpolytene chromosomes also tend to occupy discrete domains in interphase nuclei. *In situ* hybridization experiments with an appropriate DNA probe, for example, can outline a single chromosome in hybrid mammalian cells grown in culture (Figure 9–99). Most of the DNA of such a chromosome is seen to occupy only a small portion of the interphase nucleus, suggesting that each individual chromosome remains compact and organized while allowing portions of its DNA to be active in RNA synthesis.

How Well Ordered Is the Nucleus?[60]

The interior of the nucleus is not a random jumble of its many RNA, DNA, and protein components. We have seen that the nucleolus is organized as an efficient ribosome-construction machine, and clusters of spliceosome components are apparently organized as discrete RNA-splicing islands (see Figure 9–89). Order is also seen in the electron microscope when one focuses on the regions around nuclear pores: the chromatin that lines the inner nuclear membrane (which is unusually condensed chromatin and therefore clearly visible in electron micrographs) is excluded from a considerable region beneath and around each nuclear pore, clearing a path between the cytoplasm and the nucleoplasm (Figure 9–100). In some

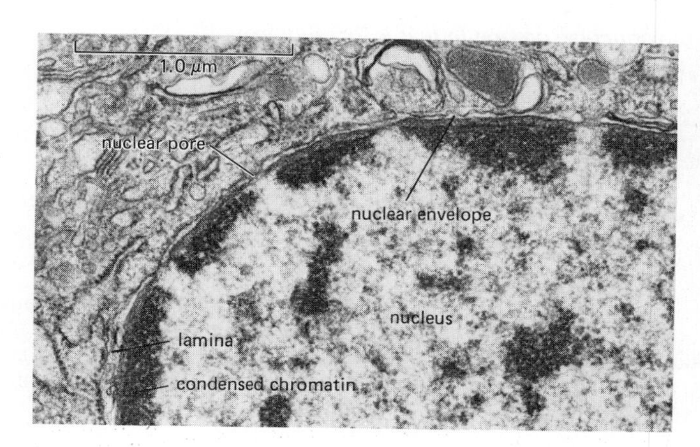

Figure 9–100 Electron micrograph of a mammalian cell nucleus, showing that the condensed chromatin underlying the nuclear envelope is excluded from regions around the nuclear pores. (Courtesy of Larry Gerace.)

special cases, moreover, the nuclear pores are found to be highly organized in the nuclear envelope (Figure 9–101). Such ordering presumably reflects a corresponding organization within the nuclear lamina to which the pores are attached.

Is there an intranuclear framework, analogous to the cytoskeleton, on which nuclear components are organized? Many cell biologists believe there is. The *nuclear matrix* or *scaffold* has been defined as the insoluble material left in the nucleus after a series of biochemical extraction steps. The proteins that constitute it can be shown to bind specific DNA sequences called *SARs* or *MARs* (for scaffold- or matrix-associated regions). Such DNA sequences have been postulated to form the base of chromosomal loops (see Figure 9–34). By means of such chromosomal attachment sites, the matrix might help organize chromosomes, localize genes (see, for example, p. 583), and regulate DNA transcription and replication within the nucleus. Because the structural components of the matrix have not yet been identified, however, it remains uncertain whether the matrix isolated by cell biologists represents a structure that is present in intact cells.

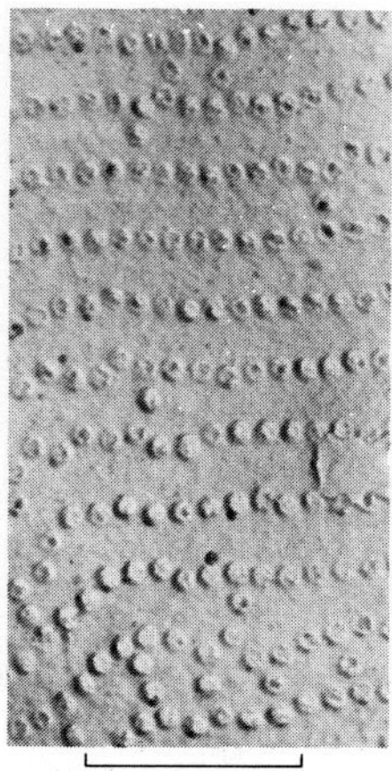

Figure 9–101 Freeze-fracture electron micrograph of the elongated nuclear envelope of a fern spore, illustrating the ordered arrangement of the nuclear pore complexes in parallel rows. In other cells, either concentrated clusters of nuclear pores or unusual areas free of nuclear pores have been detected in the nuclear envelope, and these are specifically oriented with respect to other structures in the cell. (Courtesy of Don H. Northcote; from K. Roberts and D.H. Northcote, *Microsc. Acta* 71:102–120, 1971.)

Summary

RNA polymerase, the enzyme that catalyzes DNA transcription, is a complex molecule containing many polypeptide chains. In eucaryotic cells there are three RNA polymerases, designated polymerases I, II, and III; they are evolutionarily related to one another and to bacterial RNA polymerase, and they have some subunits in common. After initiating transcription, each enzyme is thought to release one or more subunits (called initiation factors) and to bind other subunits (called elongation factors) that are required for RNA chain elongation, termination, and modification. The elongation factors are presumably different for each type of polymerase, which would explain why the transcripts that each enzyme synthesizes are differently modified.

Most of the cell's mRNA is produced by a complex process beginning with the synthesis of heterogeneous nuclear RNA (hnRNA). The primary hnRNA transcript is made by RNA polymerase II. It is then capped by the addition of a special nucleotide to its 5′ end and is cleaved and then polyadenylated at its 3′ end. The modified RNA molecules are usually then subjected to one or more RNA splicing events, in which intron sequences are removed from the middle of the hnRNA by a reaction catalyzed by a large ribonucleoprotein complex known as a spliceosome. In this process most of the mass of the primary RNA transcript is removed and degraded in the nucleus. As a result, although the rate of production of hnRNA typically accounts for about half of a cell's RNA synthesis, the mRNA produced represents only about 3% of the steady-state quantity of RNA in a cell.

Unlike genes that code for proteins, which are transcribed by polymerase II, the genes that code for most structural RNAs are transcribed by polymerase I and III. These genes are usually repeated many times in the genome and are often clustered in tandem arrays. RNA polymerase III makes a variety of stable small RNAs, including the tRNAs and the small 5S rRNA of the ribosome. RNA polymerase I makes the large rRNA precursor molecule (45S rRNA) containing the major rRNAs. Except for those in mitochondria and chloroplasts, all the cell's ribosomes are assembled in the nucleolus—a distinct intranuclear organelle that is formed around the tandemly arranged rRNA genes, which are brought together from several chromosomes.

References

General

Lewin, B. Gene Expression, Vol. 2: Eucaryotic Chromosomes, 2nd ed. New York: Wiley, 1980.

Lewin, B. Genes, 3rd ed. New York: Wiley, 1987.

Newport, J.W.; Forbes, D.J. The nucleus: structure, function, and dynamics. *Annu. Rev. Biochem.* 56:535–566, 1987.

Watson, J.D.; Hopkins, N.H.; Roberts, J.W.; Steitz, J.A.; Weiner, A.M. Molecular Biology of the Gene, 4th ed. Menlo Park, CA: Benjamin-Cummings, 1987.

Cited

1. Adolph, K.W., ed. Chromosomes and Chromatin, Vols. 1–3. Boca Raton, FL: CRC Press, 1988.
   Felsenfeld, G. DNA. *Sci. Am.* 253(4):58–67, 1985.
   Hsu, T.C. Human and Mammalian Cytogenetics: A Historical Perspective. New York: Springer-Verlag, 1979.
2. Kavenoff, R.; Klotz, L.C.; Zimm, B.H. On the nature of chromosome-sized DNA molecules. *Cold Spring Harbor Symp. Quant. Biol.* 38:1–8, 1974.
3. Burke, D.T.; Carle, G.F.; Olson, M.V. Cloning of large segments of exogenous DNA into yeast by means of artificial chromosome vectors. *Science* 236:806–812, 1987.
   Murray, A.W. Chromosome structure and behavior. *Trends Biochem. Sci.* 10:112–115, 1985.
4. Gall, J.G. Chromosome structure and the C-value paradox. *J. Cell Biol.* 91:3s–14s, 1981.
   Ohta, T.; Kimura, M. Functional organization of genetic material as a product of molecular evolution. *Nature* 233:118–119, 1971.
5. Mapping and Sequencing the Human Genome. Washington, DC: National Academy Press, 1988.
   Wilson, A.C.; Ochman, H.; Prager, E.M. Molecular time scale for evolution. *Trends Genet.* 3:241–247, 1987.
6. Fried, M.; Crothers, D.M. Equilibria and kinetics of *lac* repressor-operator interactions by polyacrylamide gel electrophoresis. *Nucleic Acids Res.* 9:6505–6525, 1981.
7. Kadonaga, J.T.; Tjian, R. Affinity purification of sequence-specific DNA binding proteins. *Proc. Natl. Acad. Sci. USA* 83:5889–5893, 1986.
   Rosenfeld, P.J.; Kelly, T.J. Purification of nuclear factor I by DNA recognition site affinity chromatography. *J. Biol. Chem.* 261:1398–1408, 1986.
   Staudt, L.M.; et al. Cloning of a lymphoid-specific cDNA encoding a protein binding the regulatory octamer DNA motif. *Science* 241:577–580, 1988.
8. Pabo, C.T.; Sauer, R.T. Protein-DNA interactions. *Annu. Rev. Biochem.* 53:293–321, 1984.
   Schleif, R. DNA binding by proteins. *Science* 241:1182–1187, 1988.
9. Ptashne, M. A Genetic Switch: Gene Control and Phage Lambda. Palo Alto, CA: Blackwell, 1986.
   Takeda, Y.; Ohlendorf, D.H.; Anderson, W.F.; Matthews, B.W. DNA-binding proteins. *Science* 221:1020–1026, 1983.
10. Berg, O.G.; von Hippel, P.H. Selection of DNA binding sites by regulatory proteins. *Trends Biochem. Sci.* 13:207–211, 1988.
    von Hippel, P.H.; Bear, D.G.; Morgan, W.D.; McSwiggen, J.A. Protein-nucleic acid interaction in transcription. *Annu. Rev. Biochem.* 53:389–446, 1984.
11. Dickerson, R.E. The DNA helix and how it is read. *Sci. Am.* 249(6):94–111, 1983.
    Drew, H.R.; McCall, M.J.; Callandine, C.R. Recent studies of DNA in the crystal. *Annu. Rev. Cell Biol.* 4:1–20, 1988.
12. Koo, H.S.; Crothers, D.M. Calibration of DNA curvature and a unified description of sequence-directed bending. *Proc. Natl. Acad. Sci. USA* 85:1763–1767, 1988.
    Lilley, D. Bent molecules—how and why? *Nature* 320:487, 1986.
    Marini, J.C.; Levene, S.D.; Crothers, D.M.; Englund, P.T. Bent helical structure in kinetoplast DNA. *Proc. Natl. Acad. Sci. USA* 80:7678–7682, 1982.
13. Echols, H. Multiple DNA-protein interactions governing high-precision DNA transactions. *Science* 233:1050–1056, 1986.
    Thompson, J.F.; de Vargas, L.M.; Koch, C.; Kahmann, R.; Landy, A. Cellular factors couple recombination with growth phase: characterization of a new component in the lambda site-specific recombination pathway. *Cell* 50:901–908, 1987.
14. Isenberg, I. Histones. *Annu. Rev. Biochem.* 48:159–191, 1979.
    von Holt, C. Histones in perspective. *Bioessays* 3:120–124, 1985.
    Wells, D.E. Compilation analysis of histones and histone genes. *Nucleic Acids Res.* 14:r119–r149, 1986.
    Wu, R.S.; Panusz, H.T.; Hatch, C.L.; Bonner, W.M. Histones and their modifications. *CRC Crit. Rev. Biochem.* 20:201–263, 1986.
15. Chromatin. *Cold Spring Harbor Symp. Quant. Biol.*, Vol. 42, 1978.
    Kornberg, R.D.; Klug A. The nucleosome. *Sci. Am.* 244(2):52–64, 1981.
    McGhee, J.D.; Felsenfeld, G. Nucleosome structure. *Annu. Rev. Biochem.* 49:1115–1156, 1980.
    Richmond, T.J.; Finch, J.T.; Rushton, B.; Rhodes, D.; Klug, A. Structure of the nucleosome core particle at 7 Å resolution. *Nature* 311:532–537, 1984.
16. Simpson, R.T. Nucleosome positioning *in vivo* and *in vitro*. *Bioessays* 4:172–176, 1986.
    Travers, A.A. DNA bending and nucleosome positioning. *Trends Biochem. Sci.* 12:108–112, 1987.
17. Eissenberg, J.C.; Cartwright, I.L.; Thomas, G.H.; Elgin, S.C. Selected topics in chromatin structure. *Annu. Rev. Genet.* 19:485–536, 1985.
    Emerson, B.M.; Lewis, C.D.; Felsenfeld, G. Interaction of specific nuclear factors with the nuclease-hypersensitive region of the chicken adult β-globin gene: nature of the binding domain. *Cell* 41:21–30, 1985.
    Gross, D.S.; Garrard, W.T. Nuclease hypersensitive sites in chromatin. *Annu. Rev. Biochem.* 57:159–198, 1988.
18. Belmont, A.S.; Sedat, J.W.; Agard, D.A. A three-dimensional approach to mitotic chromosome structure: evidence for a complex hierarchical organization. *J. Cell Biol.* 105:77–92, 1987.
    Pederson, D.S.; Thoma, F.; Simpson, R. Core particle, fiber, and transcriptionally active chromatin structure. *Annu. Rev. Cell Biol.* 2:117–147, 1986.
19. Allan, J.; Hartman, P.G.; Crane-Robinson, C.; Aviles, F.X. The structure of histone H1 and its location in chromatin. *Nature* 288:675–679, 1980.
    Clark, D.J.; Thomas, J.O. Salt-dependent cooperative interaction of histone H1 with linear DNA. *J. Mol. Biol.* 187:569–580, 1986.
    Coles, L.S.; Robins, A.J.; Madley, L.K.; Wells, J.R. Characterization of the chicken histone H1 gene complement: generation of a complete set of vertebrate H1 protein sequences. *J. Biol. Chem.* 262:9656–9663, 1987.
    Thoma, F.; Koller, T.; Klug, A. Involvement of histone H1 in the organization of the nucleosome and of the salt-dependent superstructures of chromatin. *J. Cell Biol.* 83:403–427, 1979.
20. De Bernardin, W.; Koller, T.; Sogo, J.M. Structure of *in vivo* transcribing chromatin as studied in simian virus 40 minichromosomes. *J. Mol. Biol.* 191:469–482, 1986.
    Losa, R.; Brown, D.D. A bacteriophage RNA polymerase transcribes *in vitro* through a nucleosome core without displacing it. *Cell* 50:801–808, 1987.
21. Benyajati, C.; Worcel, A. Isolation, characterization, and structure of the folded interphase genome of *Drosophila melanogaster*. *Cell* 9:393–408, 1976.
    Gasser, S.M.; Laemmli, U.K. A glimpse at chromosomal order. *Trends Genet.* 3:16–22, 1987.
    Schmid, M.B. Structure and function of the bacterial chromosome. *Trends Biochem. Sci.* 13:131–135, 1988.
22. Marsden, M.; Laemmli, U.K. Metaphase chromosome structure: evidence for a radial loop model. *Cell* 17:849–858, 1979.
    Georgiev, G.P.; Nedospasov, S.A.; Bakayev, V.V. Supranucleosomal levels of chromatin organization. In The Cell Nucleus (H. Busch, ed.), Vol. 6, pp. 3–34. New York: Academic Press, 1978.

23. Holmquist, G. DNA sequences in G-bands and R-bands. In Chromosomes and Chromatin Structure (K.W. Adolph, ed.), Vol. 2, pp. 75–122. Boca Raton, FL: CRC Press, 1988.
Lewin, B. Gene Expression, Vol. 2: Eucaryotic Chromosomes, 2nd ed., pp. 428–440. New York: Wiley, 1980.
24. Bostock, C.J.; Sumner, A.T. The Eucaryotic Chromosome, pp. 347–374. Amsterdam: North-Holland, 1978.
Callan, H.G. Lampbrush chromosomes. *Proc. R. Soc. Lond. (Biol.)* 214:417–448, 1982.
Roth, M.B.; Gall, J.G. Monoclonal antibodies that recognize transcription unit proteins on newt lampbrush chromosomes. *J. Cell Biol.* 105:1047–1054, 1987.
25. Agard, D.A.; Sedat, J.W. Three-dimensional architecture of a polytene nucleus. *Nature* 302:676–681, 1983.
Beermann, W. Chromosomes and genes. In Developmental Studies on Giant Chromosomes (W. Beermann, ed.), pp. 1–33. New York: Springer-Verlag, 1972.
26. Ashburner, M.; Chihara, C.; Meltzer, P.; Richards, G. Temporal control of puffing activity in polytene chromosomes. *Cold Spring Harbor Symp. Quant. Biol.* 38:655–662, 1974.
Lamb, M.M.; Daneholt, B. Characterization of active transcription units in Balbiani rings of *Chironomus tentans. Cell* 17:835–848, 1979.
27. Bossy, B.; Hall, L.M.; Spierer, P. Genetic activity along 315 kb of the *Drosophila* chromosome. *EMBO J.* 3:2537–2541, 1984.
Hill, R.J.; Rudkin, G. Polytene chromosomes: the status of the band-interband question. *Bioessays* 7:35–40, 1987.
Judd, B.H.; Young, M.W. An examination of the one cistron: one chromomere concept. *Cold Spring Harbor Symp. Quant. Biol.* 38:573–579, 1974.
28. Garel, A.; Zolan, M.; Axel, R. Genes transcribed at diverse rates have a similar conformation in chromatin. *Proc. Natl. Acad. Sci. USA* 74:4867–4871, 1977.
Weintraub, H.; Groudine, M. Chromosomal subunits in active genes have an altered conformation. *Science* 193:848–856, 1976.
Yaniv, M.; Cereghini, S. Structure of transcriptionally active chromatin. *CRC Crit. Rev. Biochem.* 21:1–26, 1986.
29. Allis, C.D.; et al. hv1 is an evolutionarily conserved H2A variant that is preferentially associated with active genes. *J. Biol. Chem.* 261:1941–1948, 1986.
Dorbic, T.; Wittig, B. Chromatin from transcribed genes contains HMG17 only downstream from the starting point of transcription. *EMBO J.* 6:2393–2399, 1987.
Hebbes, T.R.; Thorne, A.W.; Crane-Robinson, C. A direct link between core histone acetylation and transcriptionally active chromatin. *EMBO J.* 7:1395–1402, 1988.
Rose, S.M.; Garrard, W.T. Differentiation-dependent chromatin alterations precede and accompany transcription of immunoglobulin light chain genes. *J. Biol. Chem.* 259:8534–8544, 1984.
30. Brown, S.W. Heterochromatin. *Science* 151:417–425, 1966.
James, T.C.; Elgin, S.C.R. Identification of a nonhistone chromosomal protein associated with heterochromatin in *Drosophila melanogaster* and its gene. *Mol. Cell. Biol.* 6:3862–3872, 1986.
Pimpinelli, S.; Bonaccorsi, S.; Gatti, M.; Sandler, L. The peculiar genetic organization of *Drosophila* heterochromatin. *Trends Genet.* 2:17–20, 1986.
31. Hand, R. Eucaryotic DNA: organization of the genome for replication. *Cell* 15:317–325, 1978.
Huberman, J.A.; Riggs, A.D. On the mechanism of DNA replication in mammalian chromosomes. *J. Mol. Biol.* 32:327–341, 1968.
32. Campbell, J. Eucaryotic DNA replication: yeast bares its ARSs. *Trends Biochem. Sci.* 13:212–217, 1988.
Palzkill, T.G.; Newlon, C.S. A yeast replication origin consists of multiple copies of a small conserved sequence. *Cell* 53:441–450, 1988.
Struhl, K.; Stinchcomb, D.T.; Sherer, S.; Davis, R.W. High-frequency transformation of yeast: autonomous replication of hybrid DNA molecules. *Proc. Natl. Acad. Sci. USA* 76:1035–1039, 1979.
33. Dodson, M.; Dean, F.B., Bullock, P.; Echols, H.; Hurwitz, J. Unwinding of duplex DNA from the SV40 origin of replication by T antigen. *Science* 238:964–967, 1987.
Li, J.J.; Kelly, T.J. Simian virus 40 DNA replication *in vitro*. *Proc. Natl. Acad. Sci. USA* 81:6973–6977, 1984.
So, A.G.; Downey, K.M. Mammalian DNA polymerases alpha and delta: current status in DNA replication. *Biochemistry* 27:4591–4595, 1988.
Stahl, H.; Dröge, P.; Knippers, R. DNA helicase activity of SV40 large T antigen. *EMBO J.* 5:1939–1944, 1986.
34. Russev, G.; Hancock, R. Assembly of new histones into nucleosomes and their distribution in replicating chromatin. *Proc. Natl. Acad. Sci. USA* 79:3143–3147, 1982.
Sariban, E.R.; Wu, R.S.; Erickson, L.C.; Bonner, W.M. Interrelationships of protein and DNA synthesis during replication in mammalian cells. *Mol. Cell. Biol.* 5:1279–1286, 1985.
Weintraub, H.; Worcel, A.; Alberts, B.M. A model for chromatin based upon two symmetrically paired half-nucleosomes. *Cell* 9:409–417, 1976.
Worcel, A.; Han, S.; Wong, M.L. Assembly of newly replicated chromatin. *Cell* 15:969–977, 1978.
35. Blackburn, E.H.; Szostak, J.W. The molecular structure of centromeres and telomeres. *Annu. Rev. Biochem.* 53:163–194, 1984.
Greider, C.W.; Blackburn, E.H. The telomere terminal transferase of *Tetrahymena* is a ribonucleoprotein enzyme with two kinds of primer specificity. *Cell* 51:887–898, 1987.
36. Stubblefield, E. Analysis of the replication pattern of Chinese hamster chromosomes using 5-bromodeoxyuridine suppression of 33258 Hoechst fluorescence. *Chromosoma* 53:209–221, 1975.
37. Lima-de-Faria, A.; Jaworska, H. Late DNA synthesis in heterochromatin. *Nature* 217:138–142, 1968.
38. Brown, E.H.; et al. Rate of replication of the murine immunoglobulin heavy-chain locus: evidence that the region is part of a single replicon. *Mol. Cell. Biol.* 7:450–457, 1987.
39. Callan, H.G. DNA Replication in the chromosomes of eukaryotes. *Cold Spring Harbor Symp. Quant. Biol.* 38:195–203, 1974.
Kriegstein, H.J.; Hogness, D.S. Mechanism of DNA replication in *Drosophila* chromosomes: structure of replication forks and evidence for bidirectionality. *Proc. Natl. Acad. Sci. USA* 71:135–139, 1974.
Mechali, M.; Kearsey, S. Lack of specific sequence requirement for DNA replication in *Xenopus* eggs compared with high sequence specificity in yeast. *Cell* 38:55–64, 1984.
40. Blow, J.J.; Laskey, R.A. A role for the nuclear envelope in controlling DNA replication within the cell cycle. *Nature* 332:546–548, 1988.
Harland, R. Initiation of DNA replication in eukaryotic chromosomes. *Trends Biochem. Sci.* 6:71–74, 1981.
Rao, P.N.; Johnson, R.T. Mammalian cell fusion: studies on the regulation of DNA synthesis and mitosis. *Nature* 225:159–164, 1970.
41. Watson, J.D.; Hopkins, N.H.; Roberts, J.W.; Steitz, J.A.; Weiner, A.M. Molecular Biology of the Gene, 4th ed. Menlo Park, CA: Benjamin-Cummings, 1987. (Chapters 13, 20, and 21.)
42. Chamberlin, M. Bacterial DNA-dependent RNA polymerases. In The Enzymes, 3rd ed. (P. Boyer, ed.), Vol. 15B, pp. 61–108. New York: Academic Press, 1982.
Greenblatt, J.; Li, J. Interaction of the sigma factor and the *nusA* gene protein of *E. coli* with RNA polymerase in the initiation-termination cycle of transcription. *Cell* 24:421–428, 1981.

McClure, W. Mechanism and control of transcription initiation in prokaryotes. *Annu. Rev. Biochem.* 54:171–204, 1985.
Yager, T.D.; von Hippel, P.H. Transcript elongtation and termination in *E. coli.* In *Escherichia coli* and *Salmonella typhimurium:* Cellular and Molecular Biology (F.C. Neidhardt, ed.), pp. 1241–1275. Washington, DC: American Society for Microbiology, 1987.

43. Chambon, P. Eucaryotic nuclear RNA polymerases. *Annu. Rev. Biochem.* 44:613–638, 1975.
Geiduschek, E.P.; Tocchini-Valentini, G.P. Transcription by RNA polymerase III. *Annu. Rev. Biochem.* 57:873–914, 1988.
Sentenac, A. Eucaryotic RNA polymerases. *CRC Crit. Rev. Biochem.* 18:31–91, 1985.
Sollner-Webb, B.; Tower, J. Transcription of cloned eucaryotic ribosomal RNA genes. *Annu. Rev. Biochem.* 55:801–830, 1986.

44. Brown, D.D. The role of stable complexes that repress and activate eucaryotic genes. *Cell* 37:359–365, 1984.
Workman, J.L.; Roeder, R.G. Binding of transcription factor TFIID to the major late promoter during *in vitro* nucleosome assembly potentiates subsequent initiation by RNA polymerase II. *Cell* 51:613–622, 1987.

45. Foe, V.E.; Wilkinson, L.E.; Laird, C.D. Comparative organization of active transcription units in *Oncopeltus fasciatus. Cell* 9:131–146, 1976.
Hastie, N.D.; Bishop, J.O. The expression of three abundance classes of mRNA in mouse tissues. *Cell* 9:761–774, 1976.
Lewin, B. Gene Expression, Vol. 2: Eucaryotic Chromosomes, 2nd ed., pp. 708–719. New York: Wiley, 1980.
Miller, O.L. The nucleolus, chromosomes, and visualization of genetic activity. *J. Cell Biol.* 91:15s–27s, 1981. (A review.)

46. Birnstiel, M.L.; Busslinger, M.; Strub, K. Transcription termination and 3′ processing: the end is in site! *Cell* 41:349–359, 1985.
Friedman, D.I.; Imperiale, M.J.; Adhya, S.L. RNA 3′ end formation in the control of gene expression. *Annu. Rev. Genet.* 21:453–488, 1987.
Nevins, J.R. The pathway of eukaryotic mRNA formation. *Annu. Rev. Biochem.* 52:441–466, 1983.
Takagaki, Y.; Ryner, L.C.; Manley, J.L. Separation and characterization of a poly(A) polymerase and a cleavage/specificity factor required for pre-mRNA polyadenylation. *Cell* 52:731–742, 1988.

47. Sisodia, S.S.; Sollner-Webb, B.; Cleveland, D.W. Specificity of RNA maturation pathways: RNAs transcribed by RNA polymerase III are not substrates for splicing or polyadenylation. *Mol. Cell Biol.* 7:3602–3612, 1987.
Smale, S.T.; Tjian, R. Transcription of herpes simplex virus tk sequences under the control of wild-type and mutant human RNA polymerase I promoters. *Mol. Cell. Biol.* 5:352–362, 1985.

48. Chambon, P. Split genes. *Sci. Am.* 244(5):60–71, 1981.
Crick, F. Split genes and RNA splicing. *Science* 204:264–271, 1979.
Darnell, J.E., Jr. Variety in the level of gene control in eucaryotic cells. *Nature* 297:365–371, 1982.
Perry, R.P. RNA processing comes of age. *J. Cell Biol.* 91:28s–38s, 1981. (Includes a historical review.)

49. Guthrie, C.; Patterson, B. Spliceosomal snRNAs. *Annu. Rev. Genet.* 22:387–419, 1988.
Dreyfuss, G.; Swanson, M.S.; Piñol-Roma, S. Heterogeneous nuclear ribonucleoprotein particles and the pathway of mRNA formation. *Trends Biochem. Sci.* 13:86–91, 1988.
Osheim, Y.N.; Miller, O.L.; Beyer, A.L. RNP particles at splice junction sequences on *Drosophila* chorion transcripts. *Cell* 43:143–151, 1985.
Samarina, O.P.; Krichevskaya, A.A.; Georgiev, G.P. Nuclear ribonucleoprotein particles containing messenger ribonucleic acid. *Nature* 210:1319–1322, 1966.
Steitz, J.A. "Snurps." *Sci. Am.* 258(6):56–63, 1988.

50. Edmonds, M. Branched RNA. *Bioessays* 6:212–216, 1987.
Maniatis, T.; Reed, R. The role of small nuclear ribonucleoprotein particles in pre-mRNA splicing. *Nature* 325:673–678, 1987.
Padgett, R.A.; Grabowski, P.J.; Konarska, M.M.; Seiler, S.; Sharp, P.A. Splicing of messenger RNA precursors. *Annu. Rev. Biochem.* 55:1119–1150, 1986.

51. Aebi, M.; Weissman, C. Precision and orderliness in splicing. *Trends Genet.* 3:102–107, 1987.

52. Andreadis, A.; Gallego, M.E.; Nadal-Ginard, B. Generation of protein isoform diversity by alternative splicing: mechanistic and biological implications. *Annu. Rev. Cell Biol.* 3:207–242, 1987.

53. Orkin, S.H.; Kazazian, H.H. The mutation and polymorphism of the human β-globin gene and its surrounding DNA. *Annu. Rev. Genet.* 18:131–171, 1984.

54. Cech, T.R. The generality of self-splicing RNA: relationship to nuclear mRNA splicing. *Cell* 44:207–210, 1986.

55. Ringertz, N.; et al. Computer analysis of the distribution of nuclear antigens: studies on the spatial and functional organization of the interphase nucleus. *J. Cell Sci.* Suppl. 4:11–28, 1986.
Warner, J. Applying genetics to the splicing problem. *Genes Dev.* 1:1–3, 1987.

56. Long, E.O.; Dawid, I.B. Repeated genes in eucaryotes. *Annu. Rev. Biochem.* 49:727–764, 1980.
Miller, O.L. The nucleolus, chromosomes, and visualization of genetic activity. *J. Cell Biol.* 91:15s–27s, 1981.

57. Hadjiolov, A.A. The Nucleolus and Ribosome Biogenesis. New York: Springer-Verlag, 1985.
Jordan, E.G.; Cullis, C.A., eds. The Nucleolus. Cambridge, U.K.: Cambridge University Press, 1982.
Sommerville, J. Nucleolar structure and ribosome biogenesis. *Trends Biochem. Sci.* 11:438–442, 1986.

58. Anastassova-Kristeva, M. The nucleolar cycle in man. *J. Cell Sci.* 25:103–110, 1977.
McClintock, B. The relation of a particular chromosomal element to the development of the nucleoli in Zea Mays. *Z. Zellforsch. Mikrosk. Anat.* 21:294–323, 1934.

59. Comings, D.E. Arrangement of chromatin in the nucleus. *Hum. Genet.* 53:131–143, 1980.
Cremer, T.; et al. Rabl's model of the interphase chromosome arrangement tested in Chinese hamster cells by premature chromosome condensation and laser-UV-microbeam experiments. *Hum. Genet.* 60:46–56, 1980.
Hochstrasser, M.; Sedat, J.W. Three-dimensional organization of *Drosophila melanogaster* interphase nuclei. II. Chromosome spatial organization and gene regulation. *J. Cell Biol.* 104:1471–1483, 1987.
Manuelidis, L. Individual interphase chromosome domains revealed by *in situ* hybridization. *Hum. Genet.* 71:288–293, 1985.

60. Gasser, S.M.; Laemmli, U.K. A glimpse at chromosome order. *Trends Genet.* 3:16–22, 1987.
Gerace, L.; Burke, B. Functional organization of the nuclear envelope. *Annu. Rev. Cell Biol.* 4:335–374, 1988.
Newport, J.W.; Forbes, D.J. The nucleus: structure, function, and dynamics. *Annu. Rev. Biochem.* 56:535–566, 1987.

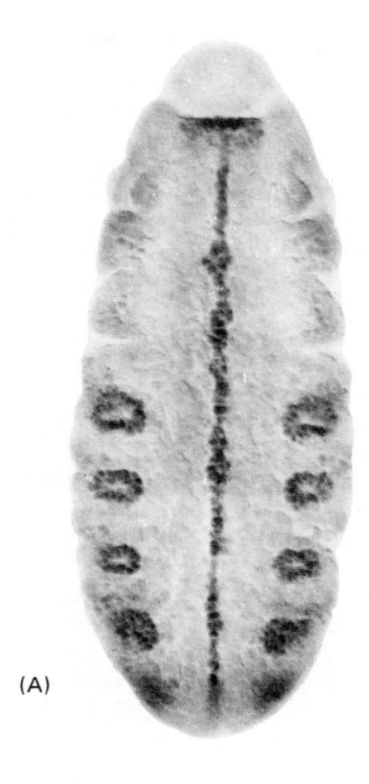
(A)

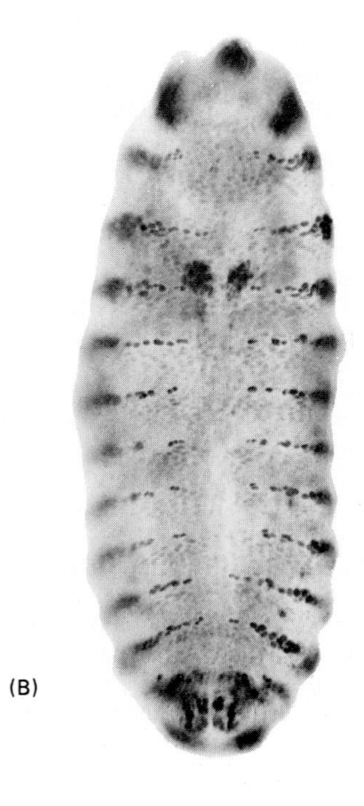
(B)

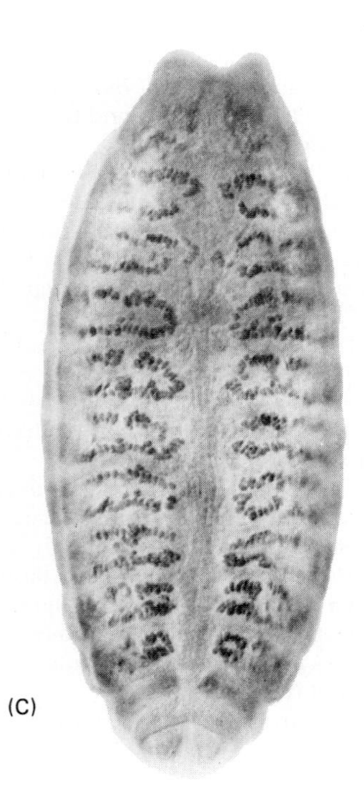
(C)

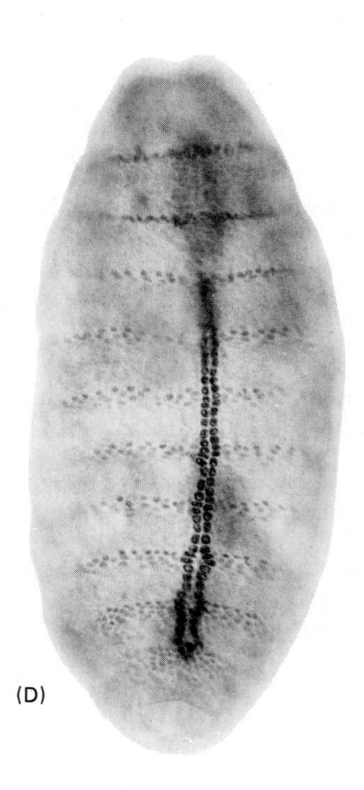
(D)

Position-dependent patterns of gene expression in four different transgenic *Drosophila* embryos. Each embryo has a single copy of a bacterial B-galactosidase gene inserted into its genome. Depending on the chromosomal insertion site, nearby *Drosophila* enhancers cause the gene to be expressed in a pattern that corresponds either to (A) trachea and mesectoderm, (B) sensory organs, (C) muscle, or (D) dorsal epidermal stripe plus a segmentally repeated pattern. (Courtesy of Yuh-Nung Jan and Larry Ackerman.)

Control of Gene Expression

10

An organism's DNA sequences encode all of the RNA and protein molecules that are available to construct its cells. Yet a complete description of the DNA sequence of a genome—be it the few million nucleotides of a bacterium or the 3 billion nucleotides of a human—would provide relatively little understanding of the organism itself. It has been said that the genome represents a complete "dictionary" for an organism, containing all of the "words" available for its construction. But we can no more reconstruct the logic of an organism from such a dictionary than we can reconstruct a play by Shakespeare from a dictionary of English words. In both cases the problem is to know how the elements in the dictionary are used; the number of possible combinations of elements is so vast that obtaining the dictionary itself is the relatively easy part and only a start toward solving the problem.

Of course, we are still very far from being able to "write" an organism from the sequence of its genome. This will require a much more complete understanding of all of cell biology, including knowledge of how the thousands of large and small molecules in a cell behave once they have been synthesized. In this chapter we discuss a more limited aspect of the problem of how the genome determines the form and behavior of an organism: we consider the rules by which a subset of the genes are selectively activated in each cell. We shall see that the mechanisms that control the *expression* of genes operate at a variety of levels, and in the central sections of this chapter we discuss the different levels in turn. At the end of the chapter we shall consider how genes and their elaborate regulatory networks might have evolved. We begin, however, with an overview of some basic principles of gene control in higher organisms.

Strategies of Gene Control

The different cell types in a higher organism often differ dramatically in both morphology and function. If we compare a mammalian neuron with a lymphocyte, for example (see Figure 13–29, p. 750), the differences are so extreme that it is difficult to imagine that the two cells contain the same genome. For this reason, and because cell differentiation is usually irreversible, biologists originally suspected that genes might be selectively lost when a cell differentiates. We now know, however, that cell differentiation generally depends on changes in gene expression rather than on gene loss.

nucleosomes, or 10^4 copies per mammalian cell) and recognizes a particular DNA sequence that is usually about 8 to 15 nucleotides long (see p. 566). The binding of these proteins to the DNA can either facilitate (**positive regulation**) or inhibit (**negative regulation**) transcription of an adjacent gene (Figure 10–6). We shall describe some of the mechanisms involved later (see p. 564). The different cell types in a multicellular organism have different mixtures of gene regulatory proteins, which causes each cell type to transcribe different sets of genes.

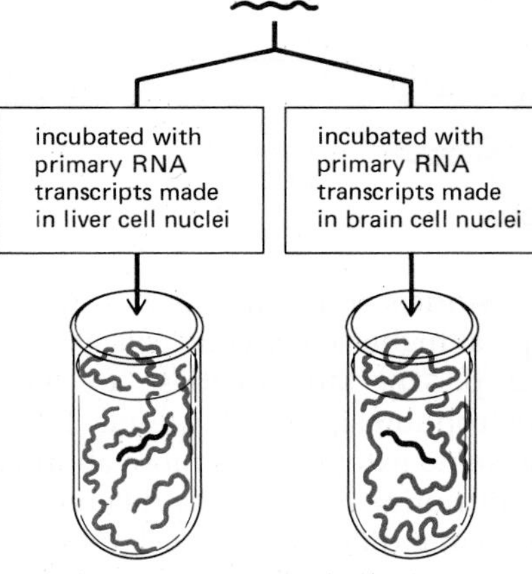

Figure 10–5 An experiment showing that gene expression in vertebrate cells is controlled largely at the level of gene transcription (see also Figure 10–4).

Combinations of a Few Gene Regulatory Proteins Could Specify a Large Number of Cell Types[4]

The essence of **combinatorial gene regulation** is illustrated in Figure 10–7, in which each numbered element represents a different gene regulatory protein. In this purely hypothetical scheme, one initial cell type gives rise to two types of cells, A and B, which differ only in the presence of gene regulatory protein ① in one but not in the other. The subsequent development of each of these cells leads to the additional production in some cells first of gene regulatory proteins ② and ③ and later of gene regulatory proteins ④ and ⑤. In the end, 8 cell types (G through cell N) have been created with 5 different gene regulatory proteins. With the addition of 2 more gene regulatory proteins to the scheme shown in Figure 10–7 (⑥ and ⑦), 16 cell types could be generated at the next step. By the time 10 more such steps occurred, slightly more than 10,000 cell types could have been specified, in principle, through the action of only 25 different gene regulatory proteins.

Combinatorial regulation of this sort is thus a very efficient way to generate biological complexity with relatively few regulatory elements. In Chapter 16 we shall see how it is used during the development of the *Drosophila* embryo to determine specific cell fates, as the interactions of a set of regulatory genes define a progressive subdivision of the early embryo into distinct regions (see p. 925).

Input from Several Different Regulatory Proteins Usually Determines the Activity of a Gene[5]

The scheme for combinatorial gene regulation shown in Figure 10–7, at first glance, seems to predict simple additive differences between the cells of successive generations. One might imagine, for example, that the addition of regulatory protein ② to cell C and to cell E would add to each of these cells the same set of additional proteins—namely, those encoded by genes activated by the regulatory protein ②. Such a view is incorrect for a simple reason. Combinatorial gene regulation is more complicated than that because different gene regulatory proteins interact with one another. Even in bacteria the interaction of two different regulatory proteins is often needed to turn on a single gene (see p. 560). In higher eucaryotes whole clusters of gene activator proteins generally act in concert to determine whether a gene is to be transcribed (see p. 567). By interacting with gene regulatory protein ①, for example, protein ② can turn on a different set of genes in cell E from those it turns on in cell C. This is presumably why a single steroid-hormone-receptor protein (an example of a gene regulatory protein) modulates the synthesis of different sets of proteins in different types of mammalian cells (see p. 692). The particular change in gene expression caused by the synthesis of a given gene

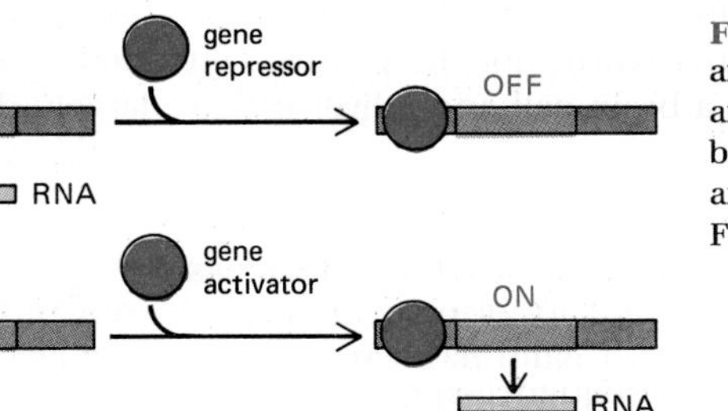

Figure 10–6 A comparison of positive and negative regulation. In this example an effect on transcription is shown, but the same two categories apply to any of the levels of control illustrated in Figure 10–2.

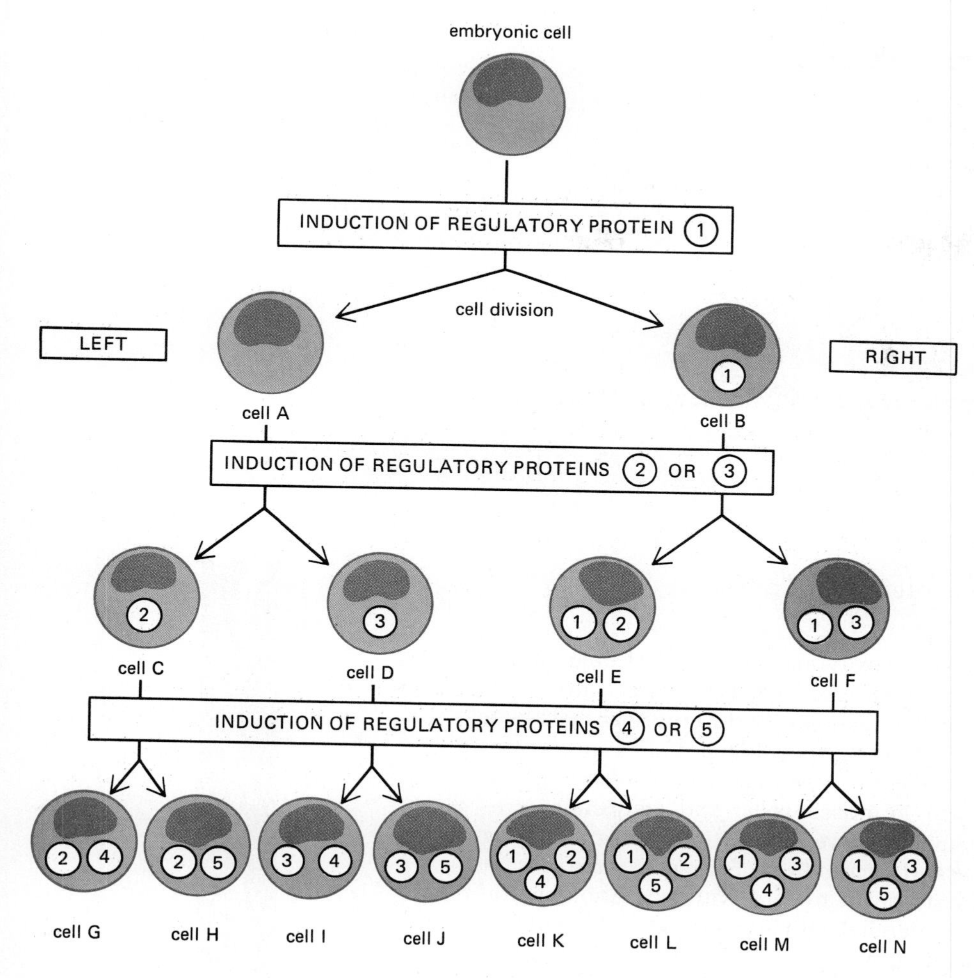

Figure 10–7 A highly schematic scheme for cell development illustrating how combinations of a few gene regulatory proteins can generate many cell types in embryos.

In this simple scheme a "decision" to make one of a pair of different gene regulatory proteins (shown as numbered circles) is made after each cell division. Sensing its relative position in an embryonic field, the daughter cell toward the left side of the embryo (the left side of the page) is always induced to synthesize the even-numbered protein of each pair, while the daughter cell toward the right side of the embryo is induced to synthesize the odd-numbered protein. The production of each gene regulatory protein is assumed to be self-perpetuating (see Figures 10–33 and 10–35). Therefore, the cells in the enlarging clone contain an increasing number of regulatory proteins, each of which is assumed to control a whole battery of genes.

regulatory protein in a developing cell will depend, in general, on the cell's past history, since this history will determine which gene regulatory proteins are already present in the cell (Figure 10–8).

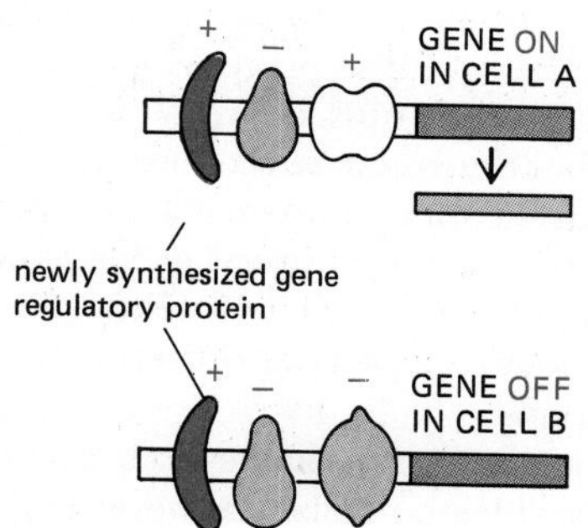

Figure 10–8 The effect of a newly synthesized gene regulatory protein on a cell. The effect depends on the regulatory proteins already present and, therefore, on the cell's past history. In this diagram the same gene is illustrated in cells A and B. This gene is initially off in both cells; the production of the leftmost protein, however, turns on the gene in cell A but fails to do so in cell B. For simplicity, each gene regulatory protein individually is assumed to have either a positive or negative effect on transcription, and the effects combine to determine the transcription of the gene. In reality, the net effect need not simply be additive; in some cases, for example, two gene regulatory proteins interact with each other when they bind to the DNA to change their individual activities.

Master Gene Regulatory Proteins Activate Multiple Genes[6]

As we have just seen, a cell contains many gene regulatory proteins, each of which acts in combination with others to control numerous other genes. Thus a branching network of interactions is possible in which each gene regulatory protein controls genes that produce other gene regulators, and so on.

Not all gene regulatory proteins are equal, however. The regulatory network contains **master gene regulatory proteins,** each of which has a decisive coordinating effect in controlling many other genes (Figure 10–9). We shall see in Chapter 16, for example, that a number of single gene mutations in *Drosophila* convert one part of the fly's body to another. Mutations that bring about such transformations are known as *homeotic mutations.* One such mutation in *Drosophila,* called *Antennapedia,* produces a single master gene regulatory protein aberrantly in the group of cells that would normally make an antenna and thereby causes these cells to switch to making a leg instead, so that the adult fly has a leg growing out of its head (see p. 920). Early evidence for a similar role for master gene regulatory proteins in vertebrates came from the observation that the absence of a single gene regulatory protein (the receptor protein for the steroid hormone testosterone) causes a human with a male (XY) genotype to develop as an almost perfect female (see p. 693). The most direct evidence for the use of master gene regulators in vertebrate development, however, comes from recent studies on the development of skeletal muscle cells.

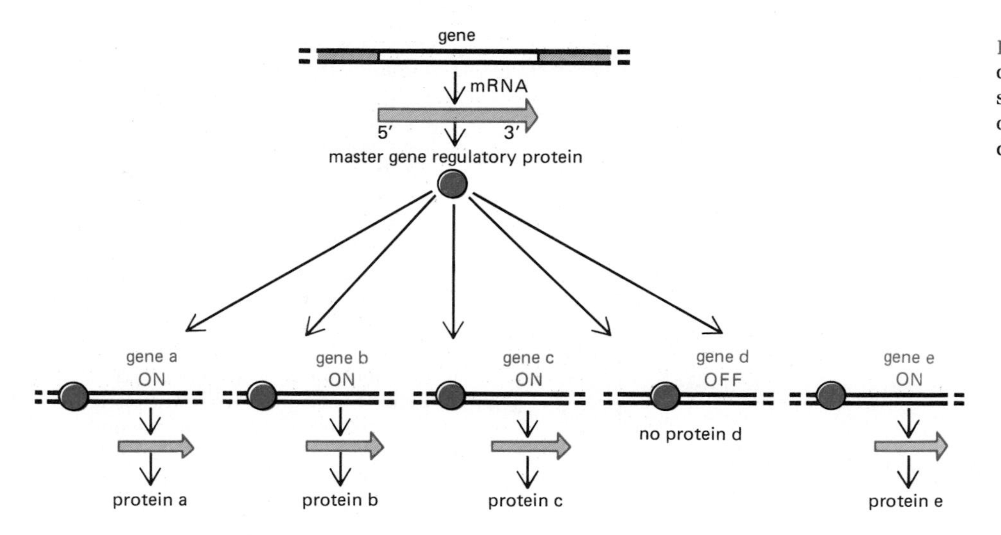

Figure 10–9 A schematic illustration of how the "decision" to produce a single master gene regulatory protein can affect the production of many different proteins in a cell.

10-5 A Single Master Gene Regulatory Protein Can Convert a Fibroblast to a Myoblast[7]

A mammalian skeletal muscle cell is typically extremely large and contains many nuclei. It is formed by the fusion of many muscle cell precursor cells called *myoblasts* (see p. 983). The mature muscle cell is distinguished from other cells by a large number of characteristic proteins, including specific types of actin, myosin, tropomyosin, and troponin (all part of the contractile apparatus), creatine phosphokinase (for the specialized metabolism of muscle cells), and acetylcholine receptors (to make the membrane sensitive to nerve stimulation). In proliferating myoblasts these muscle-specific proteins and their mRNAs are absent, or are present in very low concentrations. As myoblasts begin to fuse with one another, the production of these proteins increases in parallel with an increase in the concentration of their mRNAs, which indicates that the expression of the corresponding genes is controlled at the level of transcription. The rate of synthesis of many of the muscle-specific proteins increases by a factor of at least 500. Studies using two-dimensional polyacrylamide-gel electrophoresis indicate that the rates of production of many other proteins change at the same time: some are switched off, others rise to a peak and then fall, others shift from one steady level to another, and so on.

This entire series of events, including cell fusion, can be triggered in cultured skin fibroblasts—cells that never express these muscle-specific genes otherwise—by transfecting them with DNA that contains a gene that has been engineered to produce a gene regulatory protein called **myoD1**, which normally is expressed only in myoblasts and skeletal muscle cells (Figure 10–10). Apparently, myoD1 is a master gene regulatory protein that normally specifies "myoblast"; when expressed at a high enough concentration in a fibroblast, it subverts the normal gene controls of the fibroblast and converts it to a muscle cell.

The myoD1 protein is concentrated in the cell nucleus, and its amino acid sequence has been deduced from DNA-sequencing studies. Moreover this protein has been shown to bind to the regulatory regions of several muscle-specific genes. Discovery of this remarkable protein raises the possibility that a small number of master gene regulatory proteins also specify other cell types.

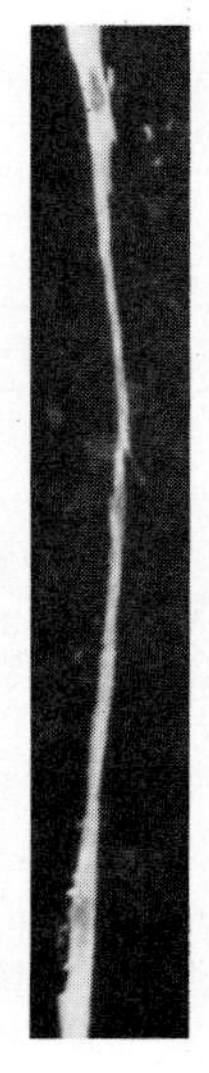

Figure 10–10 Immunofluorescence micrograph of skin fibroblasts from a chick embryo that have been converted to muscle cells by the experimentally induced expression of the *myoD1* gene product. These fibroblasts were grown in culture and transfected 3 days earlier with a recombinant DNA plasmid containing the *myoD* coding sequence linked to a viral promoter/enhancer that is active in chick fibroblasts. A few percent of the fibroblasts take up the DNA and produce the myoD protein; these cells have fused to form elongated myotubes, which are stained here with an antibody that detects a muscle-specific protein. The stained cells are intermixed with a confluent layer of fibroblasts, which are not visible in this micrograph. Control cultures transfected with another plasmid contain no muscle cells. (Courtesy of Stephen Tapscott, Andrew Lassar, Robert Davis, and Harold Weintraub.)

Summary

The cell types in multicellular organisms become different by expressing different genes from the same genome, although surprisingly few differences in protein content distinguish one cell type from another. The expression of most genes is controlled predominantly at the transcriptional level, although post-transcriptional controls are also important. Transcriptional controls depend on gene regulatory proteins that bind to specific DNA sequences. These proteins can help to turn a gene either on (positive control) or off (negative control). Genes in higher eucaryotes usually are regulated by the combinatorial effects of several such positive and negative gene regulatory proteins. Master gene regulatory proteins play a special part in this gene control network by regulating large sets of genes: the experimentally induced expression of the myoD1 protein, for example, can convert a fibroblast to a myoblast.

Controlling the Start of Transcription[8]

Only 40 years ago the idea that a gene could be specifically turned on or off was revolutionary. This concept, which was a major advance in our understanding of cells, originated from studies of *E. coli* growing in a mixture of glucose and lactose (a disaccharide). Given this choice of carbon source, the bacteria first used up all the glucose and only then began to metabolize the lactose. The switch to lactose utilization was accompanied by a pause in bacterial growth, during which the enzyme *β-galactosidase,* which hydrolyzes lactose to glucose and galactose, was synthesized. The isolation and characterization of mutant bacteria with specific defects in the regulation of this switch led to biochemical studies that, in 1966, resulted in the identification and isolation of the *lactose repressor protein.*

Biochemical and genetic studies of the lactose repressor, the bacteriophage lambda repressor (see p. 492), and several other bacterial gene regulatory proteins quickly led to a general model for transcriptional regulation in procaryotes. Sequence-specific DNA-binding proteins were presumed either to inhibit or to stimulate the initiation of RNA synthesis from a gene by binding next to the **promoter,** where RNA polymerase binds to start transcription (see p. 524). Changes in the binding of these gene regulatory proteins to the DNA were presumed to turn genes on and off.

For many years it remained unclear how well this bacterial model of gene control might apply to eucaryotic cells. The DNA of eucaryotes, unlike that of procaryotes, is packaged into nucleosomes by histone proteins, whose mass is equal to that of the DNA (see p. 496). The presence of nucleosomes suggested new possibilities for gene regulation; besides, some novel mechanism seemed to be needed to explain why eucaryotic gene regulatory proteins often bind thousands of nucleotide pairs away from the promoters they affect. Progress in elucidating the mechanisms of eucaryotic gene control was slow, largely because most eucaryotic gene regulatory proteins are present in small quantities (about one part in 50,000 of total cell protein). As late as 1983 only two unusually abundant (and perhaps atypical) examples—the SV40 virus *large T-antigen* and the 5S rRNA gene transcription factor *TFIIIA*—had been well characterized.

Soon thereafter the situation changed dramatically. New methods based on recombinant DNA technology made large numbers of eucaryotic gene regulatory proteins available for detailed biochemical and genetic studies. In addition, studies in bacteria provided clear examples of gene regulatory proteins affecting genes from a distance and revealed their mechanism of action. In this section, therefore, we emphasize the similarities rather than the differences between the mechanisms that regulate gene transcription in procaryotes and eucaryotes. Additional mechanisms that may be unique to eucaryotes are discussed later when we consider some of the mechanisms responsible for creating different cell types (see p. 577).

Bacterial Gene Repressor Proteins Bind Near Promoters and Thereby Inhibit Transcription of Specific Genes[9]

The chromosome of *E. coli* consists of a single circular DNA molecule of about 4.7 $\times$ 10^6 nucleotide pairs. This is enough DNA to code for about 4000 different proteins, though only a fraction of these proteins are made at any one time. *E. coli* regulates the expression of many of its genes according to the intracellular

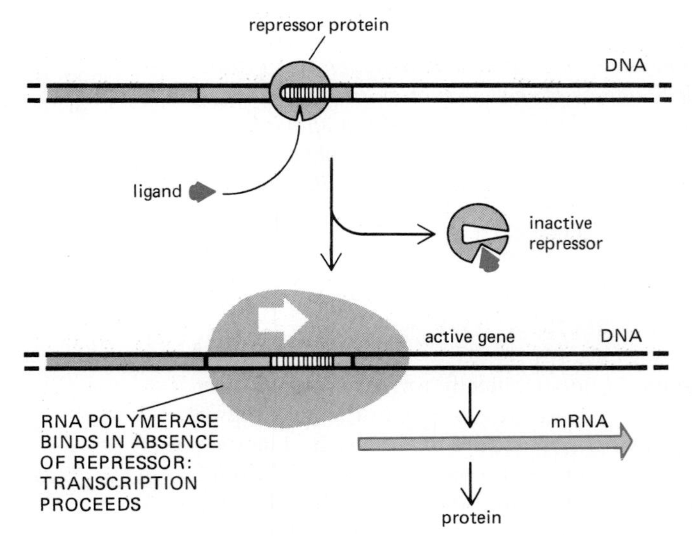

Figure 10–11 Gene derepression in bacteria. A specific small molecule binds to a repressor protein, thereby altering the conformation of the protein, which dissociates from the DNA, allowing the adjacent gene(s) to be transcribed. In the example in the text, allolactose binds to the lactose repressor protein to allow the synthesis of a single RNA transcript that codes for three proteins involved in lactose metabolism (β-galactosidase, galactoside permease, and galactoside acetylase). Because the three genes that encode these proteins are adjacent and coordinately controlled as part of the same transcription unit, the gene cluster is called the lactose *operon.*

levels of specific metabolites, which often vary depending on the food sources in the cell's environment.

By 1959, genetic studies of *E. coli* lactose utilization had provided strong indirect evidence for the existence of a repressor protein that binds to a transcription unit called the *lac* operon to turn β-galactosidase production off when lactose is absent. The subsequent purification of the **lactose repressor protein** allowed the mechanism of this regulation to be deciphered. In experiments *in vitro* with purified components, the repressor was shown to inhibit transcription of the *lac* operon by binding to a specific DNA sequence of 21 nucleotide pairs (called the *operator*) that overlaps an adjacent RNA polymerase binding site at which RNA synthesis begins (the *promoter*). When the repressor binds to the operator sequence, it prevents RNA polymerase from starting RNA synthesis at the promoter, thereby blocking transcription of the adjacent region of DNA (see Figure 10–14).

The lactose repressor protein acts to adjust the production of β-galactosidase to the needs of the cell. In the presence of lactose a small sugar molecule called *allolactose* is formed in the cell. Allolactose binds to the repressor protein, and when it reaches a high enough concentration, it induces an allosteric conformational change (see p. 127) that causes the repressor to loosen its hold on the DNA so that transcription can proceed: the gene is then said to be *derepressed.* As a result, an *E. coli* cell is able to make the enzymes it needs for the breakdown of lactose only when lactose is present (Figure 10–11).

Figure 10–12 The binding of tryptophan to the tryptophan repressor protein changes the conformation of the repressor. The conformational change enables this gene regulatory protein to bind tightly to a specific DNA sequence, thereby blocking transcription of the genes encoding the enzymes required to produce tryptophan (the *trp* operon). The three-dimensional structure of this bacterial helix-turn-helix protein, as determined by x-ray diffraction with and without tryptophan bound, is illustrated. Tryptophan binding increases the distance between the two recognition helices (*colored cylinders*) in the dimer (see p. 492), allowing the formation of symmetrically arranged hydrogen-bond interactions, indicated schematically here as colored rays. (Adapted from R. Zhang et al., *Nature* 327:591–597, 1987.)

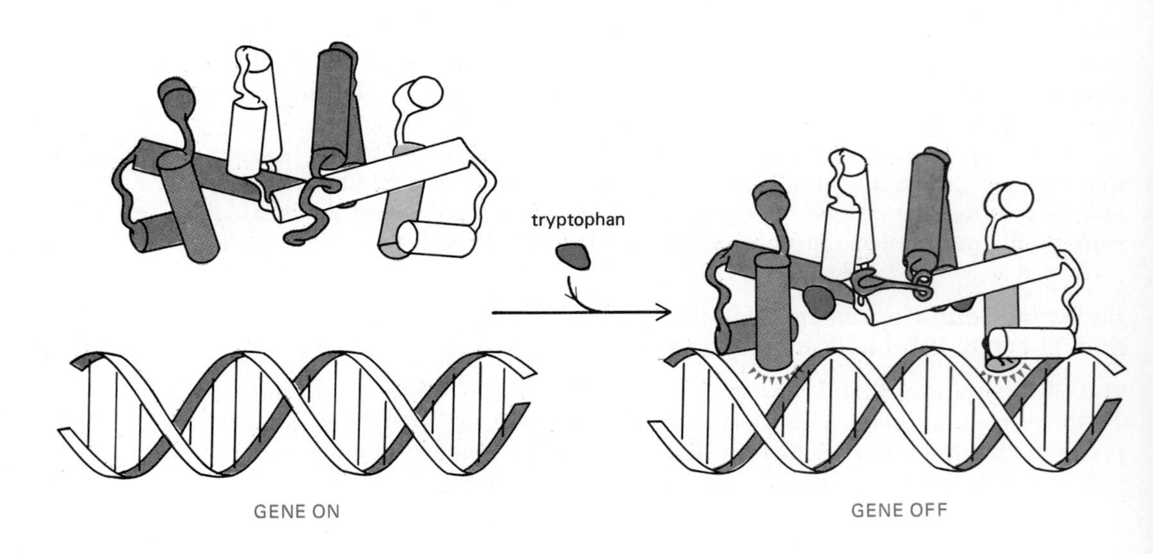

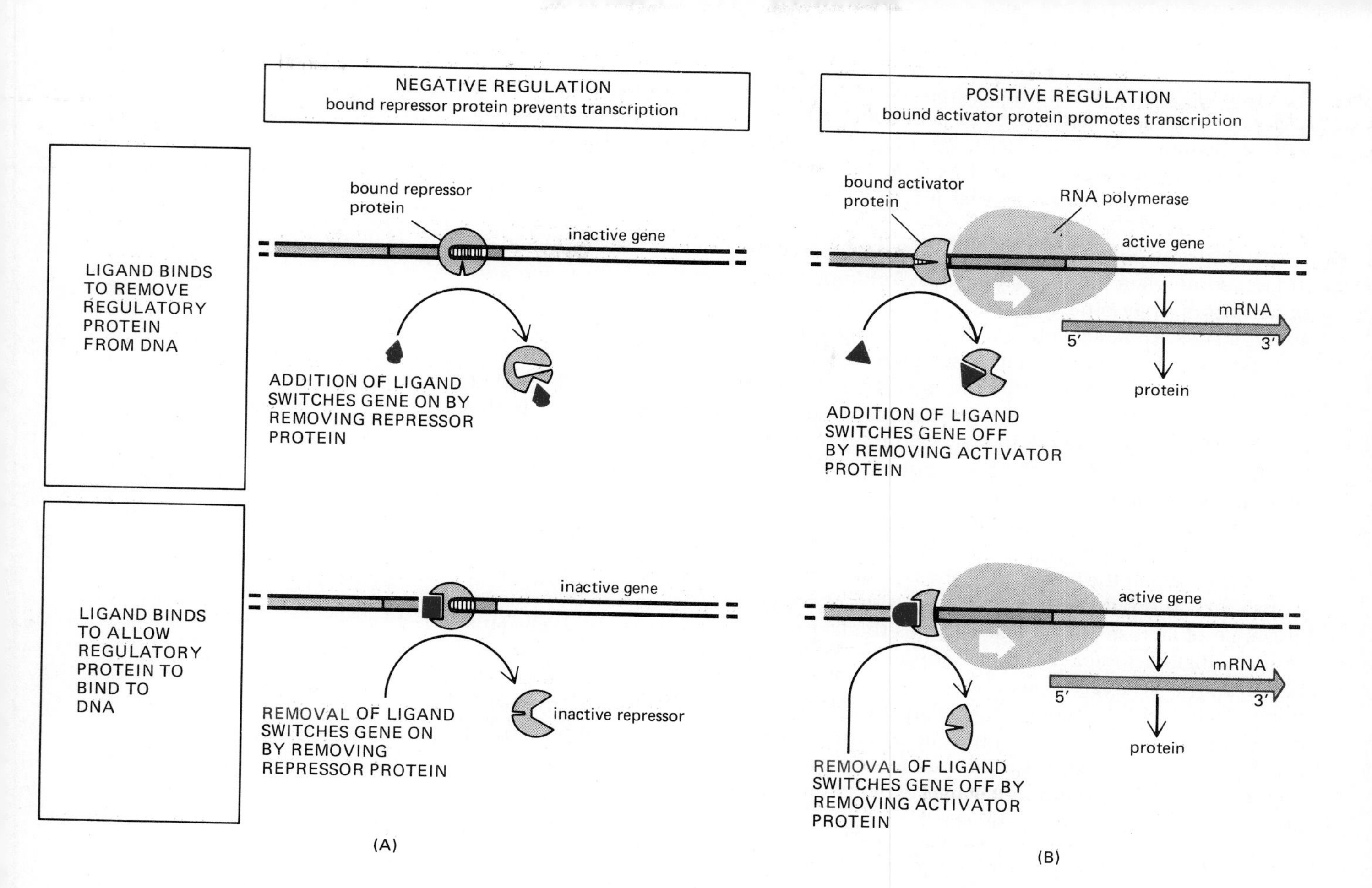

Figure 10–13 Summary of the mechanisms by which specific gene regulatory proteins control gene transcription in procaryotes. (A) Negative regulation; (B) positive regulation. Note that the addition of an inducing ligand can turn on a gene either by removing a gene repressor protein from the DNA (*upper left panel*) or by causing a gene activator protein to bind (*lower right panel*). Likewise, the addition of an inhibitory ligand can turn off a gene either by removing a gene activator protein from the DNA (*upper right panel*) or by causing a gene repressor protein to bind (*lower left panel*).

Many other examples of this type of specific gene repression in bacteria are now known. In each case the binding of a **gene repressor protein** to a specific DNA sequence turns a gene (or transcription unit) off. The binding is always regulated by specific signaling molecules, such as allolactose. Sometimes, as for the lactose repressor, the presence of the signaling molecule in the cell turns on a gene or transcription unit by decreasing the affinity of the repressor protein for its specific DNA sequence. But a signaling molecule can equally be used to turn *off* a gene or transcription unit by means of a repressor protein. An allosteric change caused by binding of the signaling molecule to a repressor, for example, can *increase,* instead of decrease, the affinity with which the repressor binds to its specific DNA sequence. This mechanism operates in the control of a set of five adjacent genes (the *trp* operon) encoding enzymes producing the amino acid tryptophan in *E. coli.* The synthesis of the single large mRNA molecule that encodes these five proteins is controlled by the **tryptophan repressor protein,** whose binding to DNA requires that tryptophan (the signaling molecule that turns off this operon) be bound to the repressor (Figure 10–12).

For the *lac* operon an increase in the concentration of a signaling molecule releases the repressor from the DNA and activates transcription, whereas for the *trp* operon it induces DNA binding and suppresses transcription. Because in both cases the binding of the regulatory protein suppresses transcription, this type of gene control is called **negative regulation** (Figure 10–13A).

Bacterial Gene Activator Proteins Contact the RNA Polymerase and Help It Start Transcription[10]

For the *negative regulation* discussed so far, a gene repressor protein binds near the promoter and interferes with the activity of the RNA polymerase. In **positive regulation,** by contrast, a **gene activator protein** facilitates the local action of

RNA polymerase. Positive regulation is important for some RNA transcription units in *E. coli* that have relatively weak promoters, which on their own do not readily bind polymerase. Such transcription units are activated by binding an activator protein to an adjacent specific DNA sequence, from which the protein can touch the RNA polymerase in a way that increases its likelihood of initiating transcription.

In other respects gene activator proteins closely resemble repressor proteins. In fact, some bacterial gene regulatory proteins function both as repressors and as activators: they bind at several sites in the genome and repress transcription at some sites while activating it at others. Like repressors, activators often bind to specific signaling ligands that either increase or decrease the affinity of the activator protein for DNA and thereby turn genes on or off, respectively. This type of gene control is called *positive regulation* because more transcription occurs in the presence of the gene regulatory protein than in its absence (Figure 10–13B).

A particularly well-studied example of a gene activator protein is the **catabolite activator protein (CAP)** of *E. coli*, whose structure was described in Chapter 3 (see p. 114). This protein enables the bacterium to use alternative carbon sources only if glucose, a preferred carbon source, is not available.

In the case of the *lac* operon, the lactose repressor protein and CAP cooperate to produce the pattern of gene expression observed. We have seen that, in the presence of lactose, allolactose releases the lactose repressor from the DNA. This is not sufficient to activate transcription of the *lac* operon, however, because the *lac* promoter is a very poor match to the *E. coli* promoter consensus sequence (see p. 204) and thus binds RNA polymerase only weakly. Transcription from this promoter requires that RNA polymerase binding be enhanced by the binding of CAP just upstream from the promoter, as illustrated in Figure 10–14. The same applies to the promoters for the genes encoding maltose, galactose, and several other sugar-metabolizing enzymes.

The DNA binding of CAP is regulated by glucose to ensure that alternative carbon sources are used only in the absence of glucose. Glucose starvation induces an increase in the intracellular levels of cyclic AMP, which acts as an intracellular signaling molecule in bacteria as well as in eucaryotic cells. Cyclic AMP binds to CAP, inducing a conformational change in the protein that enables it to bind to its specific DNA sequence and thereby activate transcription of adjacent genes. When glucose is plentiful, cyclic AMP levels drop; cyclic AMP therefore dissociates from CAP, which reverts to an inactive form that can no longer bind DNA, so that the cell switches to metabolize only glucose (see Figure 10–14).

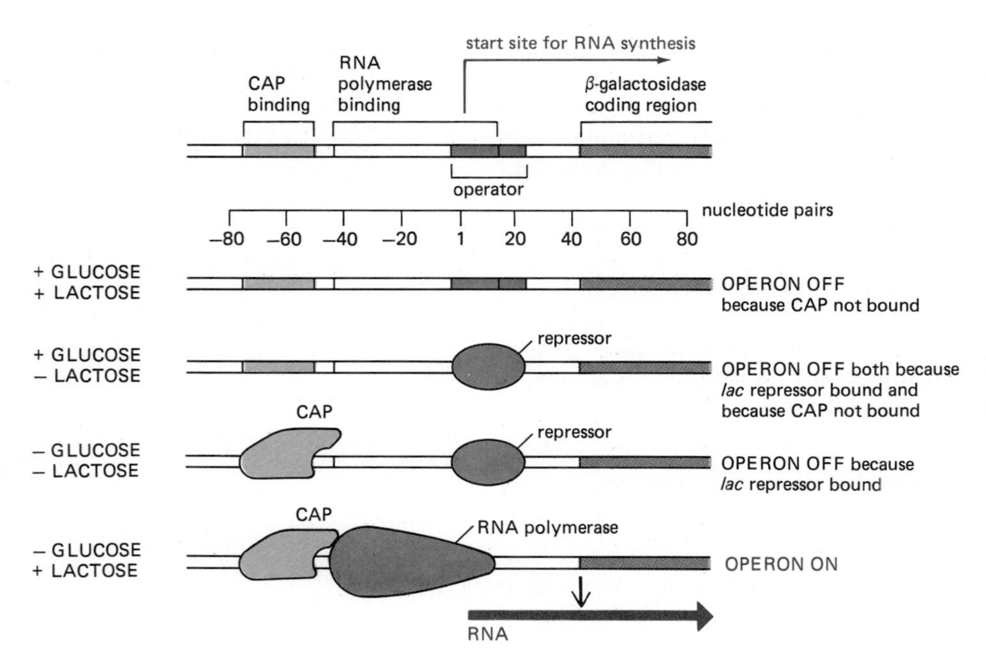

Figure 10–14 Glucose and lactose levels control the initiation of transcription of the *lac* operon through their effects on the lactose repressor protein and CAP. Lactose addition increases the concentration of allolactose, which removes the repressor protein from the DNA (see Figure 10–11). Glucose addition decreases the concentration of cyclic AMP; because cyclic AMP no longer binds to CAP, this gene activator protein dissociates from the DNA, turning off the operon. Binding sites and proteins are drawn approximately to scale; as indicated, CAP is thought to touch the polymerase to help it to begin RNA synthesis.

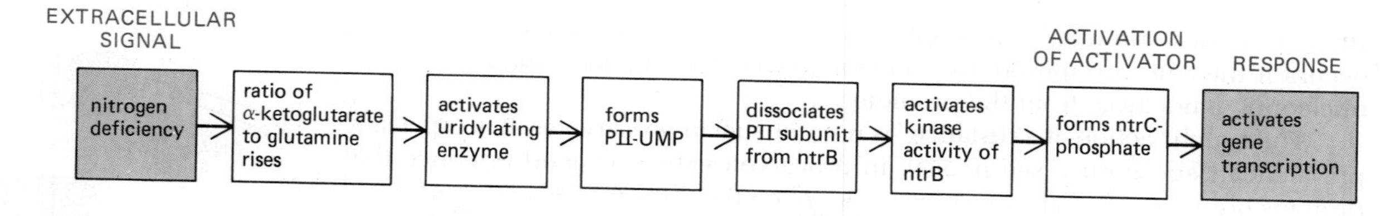

Figure 10–15 Part of the regulatory cascade that allows nitrogen deficiency to activate the genes needed for nitrogen metabolism in bacteria. The protein PII serves as a regulatory subunit of the ntrB protein. This multistep regulatory pathway has the advantage that it can produce a large change in metabolism in response to a relatively small change in nitrogen availability. Side branches of the pathway also reversibly modify other enzymes to control their catalytic activities in response to nitrogen.

10-9 Changes in Protein Phosphorylation Can Regulate Genes[11]

The first gene regulatory proteins identified were all bacterial. Like the lactose repressor, the tryptophan repressor, and CAP, these proteins are controlled by the reversible binding of specific small molecules. While small intracellular signaling molecules, such as cyclic AMP, also control the activity of gene regulatory proteins in eucaryotic cells, they usually do so indirectly, often by affecting protein phosphorylation and dephosphorylation (see p. 695). Although phosphorylation is used much less extensively for regulation in bacteria, there is one bacterial regulatory system dependent on protein phosphorylation that has been especially well studied. We shall use it to introduce several aspects of gene regulation that may help in understanding the more complicated regulatory systems in higher eucaryotes.

Evolutionarily related proteins control aspects of nitrogen metabolism, phosphate metabolism, membrane protein synthesis, chemotaxis and sporulation in various bacteria. We shall focus on nitrogen metabolism in *E. coli,* where—when nitrogen becomes limiting—the synthesis of a number of proteins is increased. These proteins include the enzyme *glutamine synthetase,* which is the most important enzyme of nitrogen assimilation, catalyzing the reaction glutamic acid + ammonia → glutamine.

Gene activation in nitrogen metabolism involves two central regulatory components: the *ntrC protein,* a gene activator protein that turns on genes only in its phosphorylated form (ntrC-phosphate), and an enzyme, the *ntrB protein,* which can either phosphorylate (through a kinase activity) or dephosphorylate (through a phosphatase activity) the ntrC protein. The degree of ntrC phosphorylation is determined by an enzymatic cascade that is triggered by a drop in nitrogen levels to add uridine monophosphate (UMP) residues to a regulatory subunit of ntrB. This modification increases the relative kinase activity of the ntrB enzyme.

A need for nitrogen is measured as an increase in the ratio of α-ketoglutarate to glutamine, which stimulates the phosphorylation of ntrC and thereby induces transcription of the glutamine synthetase gene (Figure 10–15). Similar cascades of protein modification reactions occur in response to changes in the environment of higher eucaryotic cells and lead to the phosphorylation or dephosphorylation of gene regulatory proteins, although the details are less well understood.

10-10 DNA Flexibility Allows Gene Regulatory Proteins That Are Bound to a Remote Site to Affect Gene Transcription[11,12]

A second aspect of the two-component bacterial regulatory system just described is even more similar to another important feature of eucaryotic gene control. The ntrC protein has two strong binding sites in DNA that are located 100 nucleotide pairs or more upstream from the glutamine synthetase gene promoter (Figure 10–16). When the phosphorylated form of ntrC (ntrC-phosphate) is bound to them,

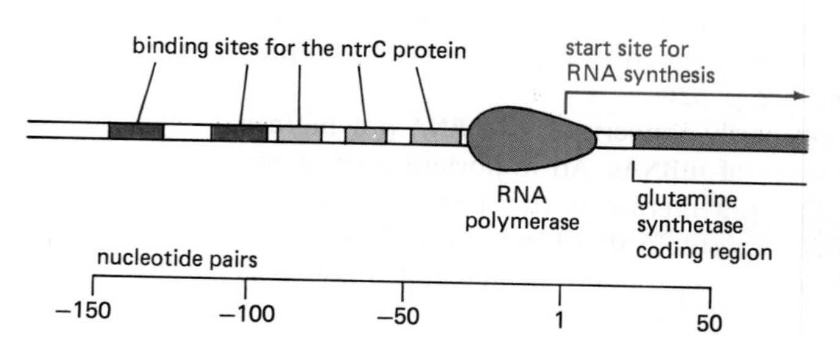

Figure 10–16 The regulatory region of the bacterial glutamine synthetase gene. Glutamine synthetase catalyzes the reaction glutamic acid + ammonia → glutamine. The two darkly colored sites bind the ntrC protein with particularly high affinity and are essential for transcriptional activation.

generate it (see p. 236). Consequently, most of the DNA in eucaryotic cells is not under tension. But helix unwinding is associated with the initiation step of DNA transcription (see Figure 9–65, p. 524). Moreover, a moving RNA polymerase molecule (as well as other proteins tracking along DNA) will tend to generate positive superhelical tension in the DNA in front of it and negative superhelical tension behind it (Figure 10–42). Through such topological effects, an event that occurs at a single DNA site can produce forces that are felt throughout an entire looped domain of chromatin. It is not yet known, however, whether an effect of this kind triggers further changes, as would be required for it to have any causative role in the control of gene expression in eucaryotes.

The Mechanisms That Form Active Chromatin Are Not Understood

The model for gene activation outlined in Figure 10–40 implies that some of the gene regulatory proteins in higher eucaryotes have functions that differ from those of their bacterial analogues. Instead of loading RNA polymerase (or its transcription factors) directly onto a nearby promoter sequence (see Figure 10–27), some sequence-specific DNA-binding proteins may function solely to decondense the chromatin in a local chromosomal domain or to remove a nucleosome from an adjacent enhancer or promoter in order to provide access for gene regulatory proteins of a more familiar kind. We are not certain, however, that this is so; it remains possible that the observed differences in the chromatin structure of active genes are an automatic consequence of the assembly of transcription factors and/or RNA polymerase onto a promoter sequence rather than being a required prerequisite for any of these events.

The identification of specific DNA sequences that act as a domain control region for the human β-globin gene should allow the proteins that bind to these sequences to be isolated and their genes cloned (see p. 489). At present we can only guess how they might act. It is possible that these regions simply serve as unusually potent enhancer elements of the conventional type (see p. 569). Three other possibilities are outlined in Figure 10–43.

The very different types of models proposed in Figure 10–43 indicate how far we are from understanding the transition from inactive to active chromatin. It is not known how many forms of chromatin there are or exactly what structural feature makes some regions more condensed than others. Simple views of chromatin function fail to explain why the amino acid sequences of the histones (especially H3 and H4) have been so highly conserved (see p. 221). Some unique chemical features of active chromatin have been discovered (see p. 511), however, and monoclonal antibodies that recognize either acetylated histones or the HMG 17 protein can now be used to separate the nucleosomes (with their associated DNA sequences) present in active chromatin from all other nucleosomes. If applied to the chromatin prepared from transgenic animals, such tools should in principle make it possible to distinguish those regulatory regions that create active chromatin from those that do not, which seems an essential step in deciphering exactly how higher eucaryotic genes are controlled.

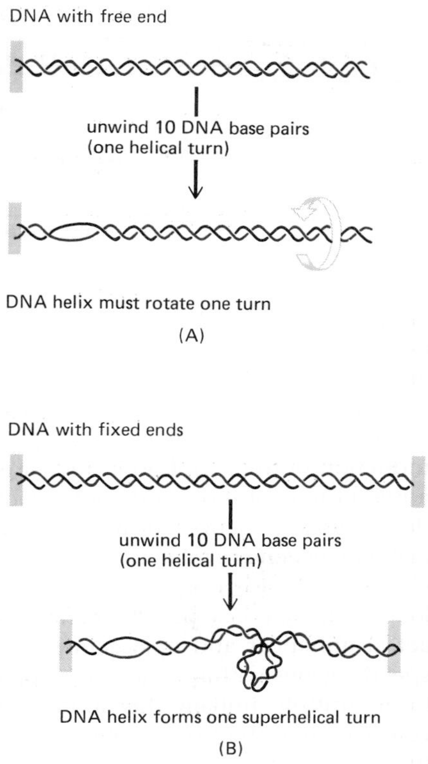

Figure 10–41 Superhelical tension in DNA causes DNA supercoiling. (A) For a DNA molecule with one free end (or a nick in one strand that serves as a swivel—see p. 236), the DNA double helix rotates by one turn for every 10 base pairs opened. (B) If rotation is prevented, superhelical tension is introduced into the DNA by helix opening. As a result, one DNA supercoil forms in the DNA double helix for every 10 base pairs opened. The supercoil formed here is a *positive supercoil* (see Figure 10–42).

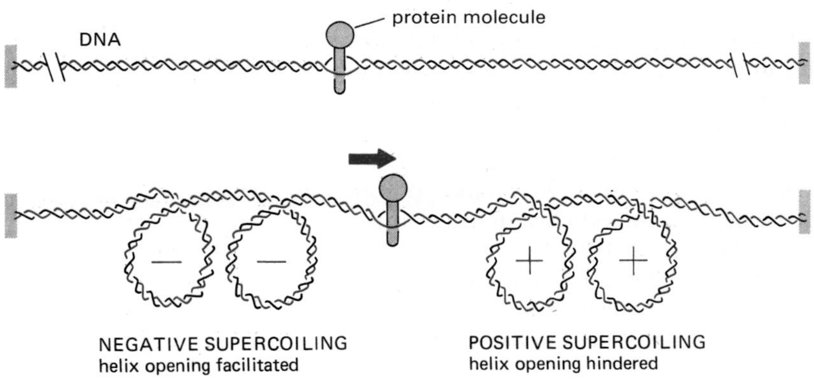

Figure 10–42 Supercoiling of a DNA segment by a protein tracking along the DNA double helix. The two ends of the DNA are fixed as in Figure 10–41B, and the protein molecule is assumed to be anchored relative to these ends or prevented by frictional forces from rotating freely as it moves. The movement thereby causes an excess of helical turns to accumulate in the DNA helix ahead of the protein and a deficit of helical turns to arise in the DNA behind the protein. Experimental evidence suggests that a moving RNA polymerase molecule causes supercoiling in this way; the positive superhelical tension ahead of it makes the DNA helix somewhat more difficult to open, but this tension should facilitate the unwrapping of the DNA from around nucleosomes, as required for the polymerase to transcribe chromatin (see Figure 9–32, p. 501).

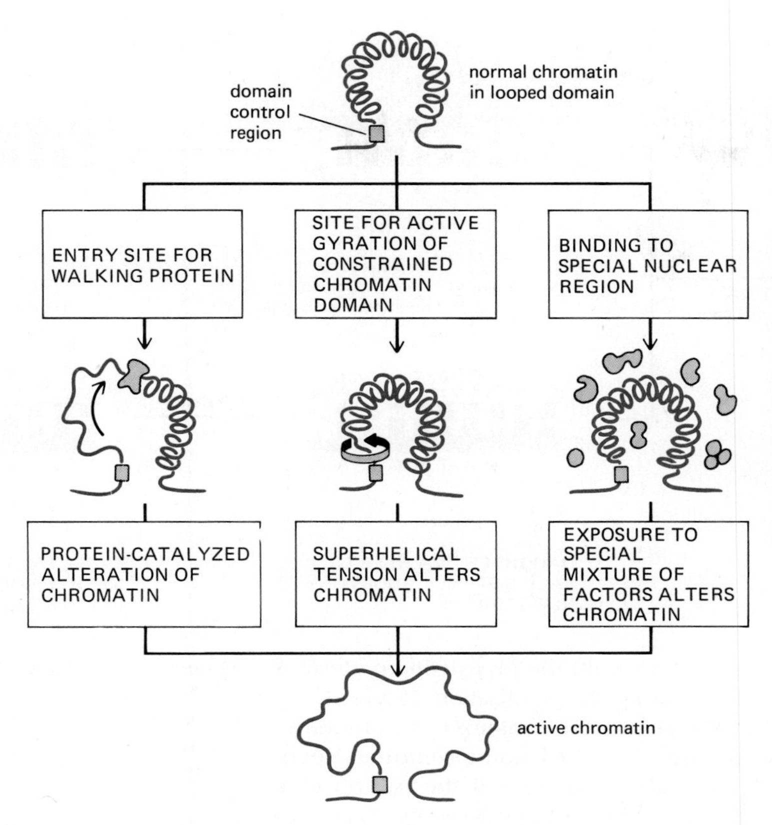

Figure 10–43 Three models proposed to explain the long-range influence of a domain control region on gene activity. The loop of chromatin shown usually is postulated to represent an entire chromosomal looped domain, which can contain 100,000 or more nucleotide pairs of DNA. The actual mechanism that causes long-range effects is unknown, and other mechanisms are also conceivable. Evidence for a chromatin decondensation step that is independent of DNA transcription has come from observations of the polytene chromosome band that contains the *Drosophila Sgs-3* gene. Several mutants that fail to produce RNA from this region nevertheless form a chromosomal puff from the band as the larva develops (see Figure 9–48, p. 510).

New Layers of Gene Control Have Been Added During the Evolution of Multicellular Eucaryotes[33]

Yeast cells provide important models for understanding higher eucaryotic cells. In fact, the high degree of functional relatedness between yeast and human proteins has been one of the great surprises of recent years. Not all aspects of gene control can be studied in yeasts, however. Yeast cells do not seem to contain H1 histone, and nearly all of their chromatin appears to be in an active form. Likewise, control regions of DNA that influence genes from great distances have not been observed. Nor do yeast cells make use of the alternative RNA splicing to be discussed shortly (see p. 589).

Invertebrates such as *Drosophila* have larger genomes than yeasts and contain H1 histone, heterochromatin, and at least some form of active chromatin (see Figure 9–54, p. 514). They also utilize alternative RNA splicing. But invertebrates lack at least one level of gene control that seems to be unique to vertebrates: they do not possess a general repression system based on DNA methylation.

The Pattern of DNA Methylation Is Inherited When Vertebrate Cells Divide[34]

The bases in the DNA helix can be covalently modified. We have discussed previously how the methylation of A in the sequence GATC makes possible a mismatch proofreading system in bacterial DNA replication (see p. 233), while the methylation of either an A or a C at a specific site protects a bacterium from the action of its own restriction nuclease (see p. 182). Vertebrate DNAs contain **5-methylcytosine (5-methyl C),** which has the same relation to cytosine that thymine has to uracil and likewise has no effect on base-pairing (Figure 10–44A). The methylation is restricted to C bases in the sequence CG; since this sequence is base-paired to exactly the same sequence (in opposite orientation) on the other strand of the DNA helix, a simple mechanism permits a preexisting pattern of DNA methylation to be inherited directly by a templating process. An enzyme called a *maintenance methylase* acts only on those CG sequences that are base-paired with

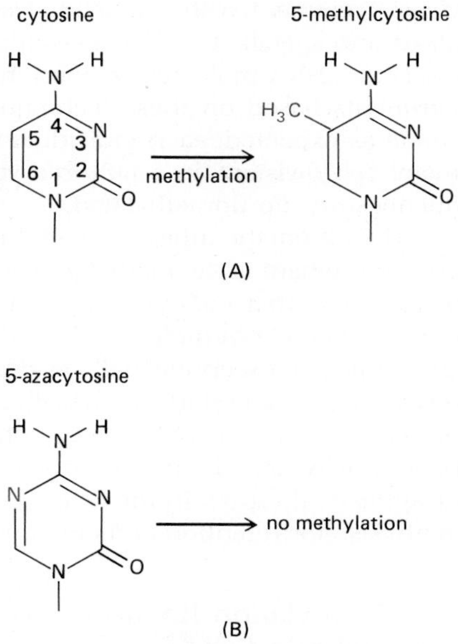

cytosine → methylation → 5-methylcytosine

(A)

5-azacytosine → no methylation

(B)

Figure 10–44 (A) Formation of 5-methylcytosine occurs by methylation of a cytosine base in the DNA double helix. In vertebrates this event is confined to selected cytosine (C) residues located in the sequence CG. (B) A synthetic nucleotide carrying the base 5-azacytosine (5-aza C) cannot be methylated. Furthermore, when small amounts of 5-aza C are incorporated into DNA, they inhibit the methylation of normal C residues.

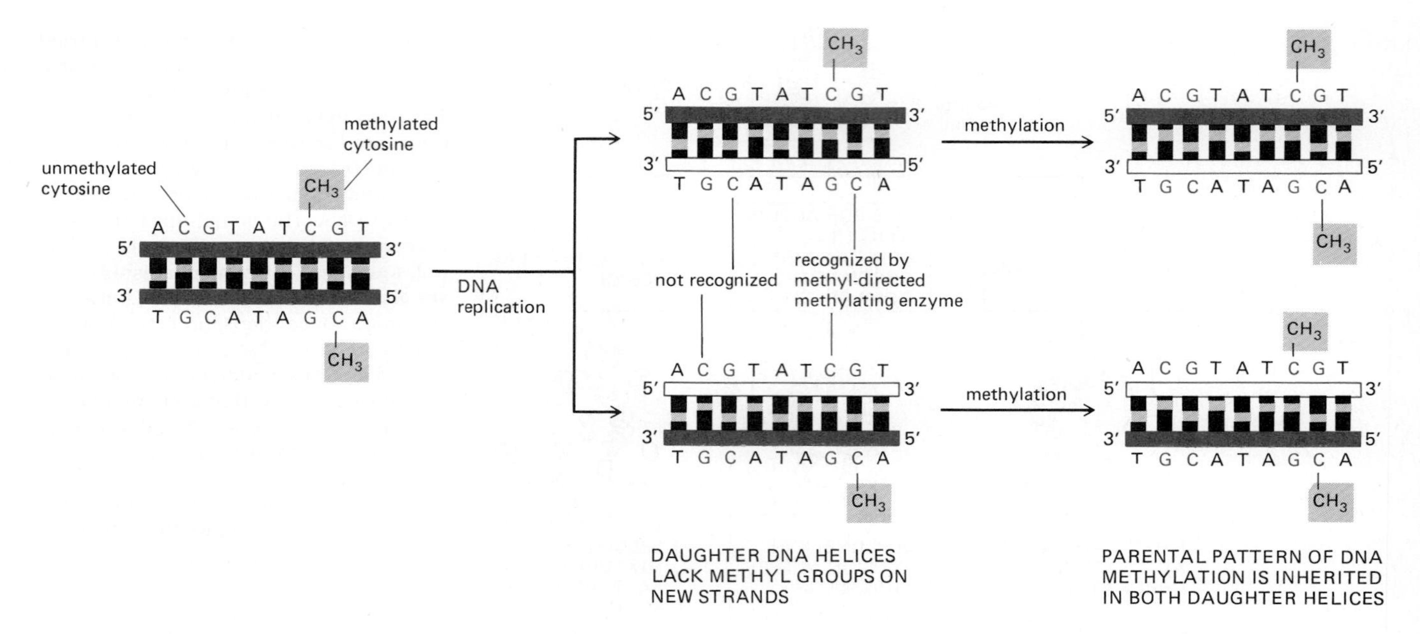

Figure 10–45 How DNA methylation patterns are faithfully inherited. In vertebrate DNAs a large fraction of the cytosine bases in the sequence CG are methylated (see Figure 10–44). Because of the existence of a methyl-directed methylating enzyme (the maintenance methylase), once a pattern of DNA methylation is established, each site of methylation is inherited in the progeny DNA, as shown. This means that changes in DNA methylation patterns will be perpetuated in a clonally inherited manner.

a CG sequence that is *already* methylated. As a result, the preexisting pattern of DNA methylation will be inherited directly following DNA replication (Figure 10–45).

Certain restriction nucleases cut DNA at sequences containing an unmethylated CG dinucleotide and require the absence of methylation for cutting. HpaII, for example, cuts the sequence CCGG but fails to cleave it if the central C is methylated. Thus the susceptibility of a DNA molecule to cleavage by HpaII can be used to detect whether specific DNA sites are methylated. The enzyme that normally protects the bacterium against its own HpaII enzyme is the HpaII-methylase. Treatment with this methylase thus can be used to introduce 5-methyl C bases into specific CG sequences (those in the sequence CCGG) on a cloned DNA molecule (DNA molecules cloned in *E. coli* lose all of their CG methylation). Experiments based on these techniques confirm that the maintenance methylase works as expected: each individual methylated CG is generally retained through many cell divisions in a cultured vertebrate cell, whereas unmethylated CG sequences remain unmethylated.

The automatic inheritance of 5-methyl C residues raises a "chicken and egg" problem: where is the methyl group first added in a vertebrate organism? Experiments show that methyl groups will be added to nearly every CG site in a fully unmethylated DNA molecule that is injected into a fertilized mouse egg (although an important exception will be described below). Thus the vast majority of a vertebrate genome starts out heavily methylated. Since the maintenance methylase normally cannot methylate fully unmethylated DNA, a novel *establishment methylase* activity must be present in the egg. Since the establishment methylase soon disappears, the DNA in the cells of developing tissues relies on the maintenance methylase for retention of its methylated nucleotides.

10-5 10-20 DNA Methylation Reinforces Developmental Decisions in Vertebrate Cells[35]

What is the effect of CG methylation? Tests with the HpaII enzyme indicate that, in general, the DNA of inactive genes is more heavily methylated than that of active genes. Moreover, an inactive gene that contains methylated DNA usually will lose many of its methyl groups after the gene has been activated. Evidence that the change in methylation affects gene expression comes from experiments in which a nucleoside containing the base analogue *5-azacytosine* (*5-aza C*, see Figure 10–44B) is added for a brief period to cells in culture. The 5-aza C, which cannot be methylated, is incorporated into DNA, where it acts as an inhibitor of the maintenance methylase, thereby reducing the general level of DNA methylation. In cells treated in this way, selected genes that were previously inactive become active and, at the same time, acquire unmethylated C residues. Once activated, the active state of these genes usually is maintained for many cell generations in the absence of 5-aza C, implying that the initial methylation of the genes helped to maintain their inactivity.

When the effect of 5-aza C on cultured cells was discovered, it was proposed that DNA methylation might play a dominant part in generating different cell types. This would require that cell specialization occur by different mechanisms in vertebrates and invertebrates. Subsequent studies have suggested that DNA methylation plays a more subsidiary part in cell diversification. The important developmental decisions apparently are made by gene regulatory proteins that can turn genes on or off regardless of their methylation status. The female X chromosome, for example, is first condensed and inactivated and only later acquires an increased level of methylation on some of its genes. Conversely, several liver-specific genes are turned on during development while they are fully methylated; only later does their level of methylation decrease.

DNA transfection experiments have helped to reconcile these contrasting observations on the role of DNA methylation in gene expression. A tissue-specific gene coding for muscle actin, for example, can be prepared in both its fully methylated and fully unmethylated form. When these two versions of the gene are introduced into cultured muscle cells, both are transcribed at the same high rate. When they are introduced into fibroblasts, which normally do not transcribe the gene, the unmethylated gene is transcribed at a low level but still one much higher than either the exogenously added methylated gene or the endogenous gene of the fibroblast (which is also methylated). These experiments suggest that DNA methylation is used in vertebrates to reinforce developmental decisions made in other ways.

In a few cases it has been possible to test with high sensitivity whether a DNA sequence that is transcribed at high levels in one vertebrate cell type is transcribed at all in another cell type. Such experiments have demonstrated rates of gene transcription differing between two cell types by a factor of more than 10^6. Unexpressed vertebrate genes are much less "leaky" in terms of transcription than are unexpressed genes in bacteria, in which the largest known differences in transcription rates between expressed and unexpressed gene states are about 1000-fold. By further reducing the transcription of genes that are turned off in other ways, DNA methylation appears to account for at least part of this difference.

CG-rich Islands Reveal About 30,000 "Housekeeping Genes" in Mammals[36]

Because of the way DNA repair enzymes work, methylated C residues in the genome tend to be eliminated in the course of evolution. Accidental deamination of an unmethylated C gives rise to U, which is not normally present in DNA and thus is recognized easily by the DNA repair enzyme uracil DNA glycosylase, excised, and then replaced with a C (see p. 225). But accidental deamination of a 5-methyl C cannot be repaired in this way, for the deamination product is a T and so indistinguishable from the other, nonmutant T residues in the DNA. Thus those C residues in the genome that are methylated tend to mutate to T over evolutionary time.

Since the divergence of vertebrates from invertebrates about 400 million years ago, more than three out of every four CGs have been lost in this way, leaving vertebrates with a remarkable deficiency of this dinucleotide. The CG sequences that remain are very unevenly distributed in the genome; they are present at 10 to 20 times their average density in selected regions that are 1000 to 2000 nucleotide pairs long, called **CG islands.** These islands surround the promoters of so-called "**housekeeping genes**"—those genes that encode the many proteins that are essential for cell viability and are therefore expressed in most cells (Figure 10–46). These genes are to be contrasted with **tissue-specific genes,** which encode proteins needed only in selected types of cells.

The distribution of CG islands can be explained readily as a secondary effect of the introduction of CG methylation as a way of reducing the expression of

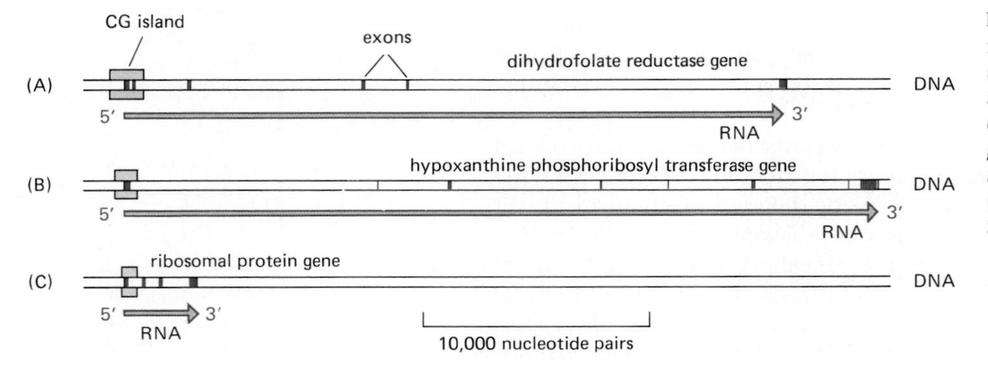

Figure 10–46 The CG islands in three mammalian housekeeping genes. Boxes show the extent of each island, which are seen to surround the promoter of each gene. Note also that, as for most genes in mammals, the exons (*color*) are short relative to the introns. (Adapted from A.P. Bird, *Trends Genet.* 3:342–347, 1987.)

inactive genes in vertebrates (Figure 10–47). In germ cells, all tissue-specific genes (except those specific to eggs and sperm) are inactive and methylated; over long periods of evolutionary time, their methylated CG sequences are lost through accidental deamination events that are not correctly repaired. The CG sequences in the regions surrounding the promoters of all active genes in germ cells, however, including all housekeeping genes, are kept demethylated, and so they can be readily repaired after spontaneous deamination events in the germ line. These genes are thought to be recognized by sequence-specific DNA-binding proteins present in the germ cells that remove any methylation near their promoters. Experiments with cloned genes show that only the CGs in CG islands remain unmethylated when fully unmethylated DNAs are injected into a fertilized mouse egg.

The mammalian genome (about 3×10^9 nucleotide pairs) contains an estimated 30,000 CG islands, each about a thousand or so nucleotide pairs in length. Most of the islands mark the 5′ ends of a transcription unit and thus, presumably, a gene. Since it is possible to clone specifically the DNA surrounding the CG islands, it is relatively easy to identify and characterize housekeeping genes. Presumably, some tens of thousands of other genes are cell-type-specific and not expressed in germ cells. Because they have lost most of their CG sequences, these tissue-specific genes are more difficult to find in the genome.

Complex Patterns of Gene Regulation Are Required to Produce a Multicellular Organism[37]

As we shall discuss in Chapter 16, the cells in an embryo resemble tiny computers in that they constantly receive information about their present location and integrate it with remembered information from their past in order to act appropriately at each stage of development (see p. 914). Genetic studies in *Drosophila* indicate that a relatively small number (perhaps 100 or so) of interacting master regulatory genes play major roles in establishing and maintaining the basic body plan (see p. 923). In any multicellular organism the vast majority of genes, including both housekeeping genes and tissue-specific genes, are presumably regulated by complex control pathways that lead from such master regulatory genes. If mechanisms very different from those known in bacteria play a central role in eucaryotic

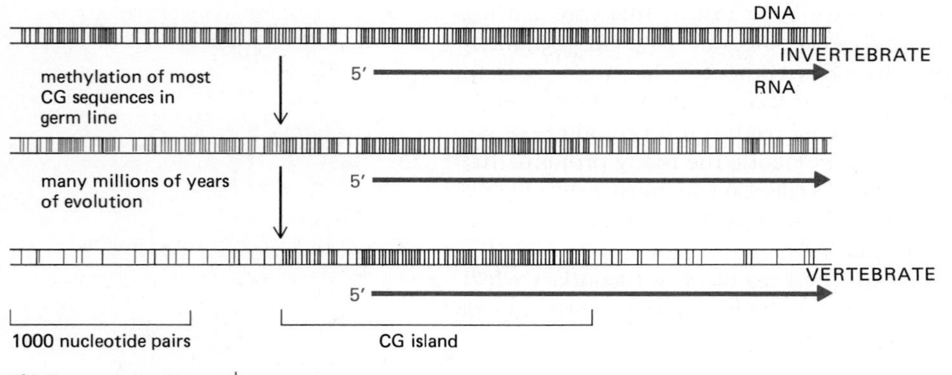

Figure 10–47 A mechanism to explain both the marked deficiency of CG sequences and the presence of CG islands in vertebrate genomes. A black line marks the location of an unmethylated CG dinucleotide in the DNA sequence, while a red line marks the location of a methylated CG dinucleotide. No examples of methylated invertebrate genes are known.

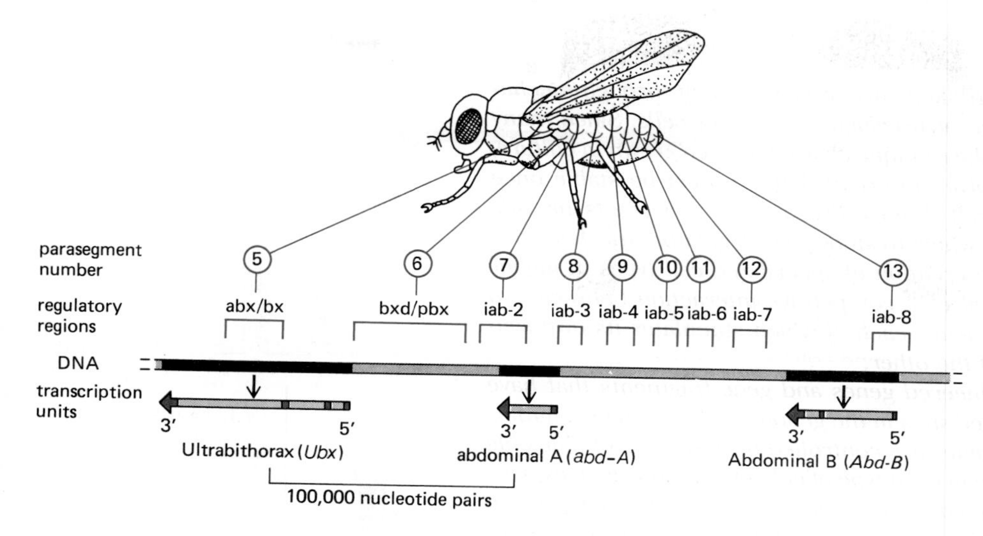

Figure 10–48 Organization of the *Drosophila* bithorax complex. This important chromosomal region of 300,000 nucleotide pairs contains three genes—*Ubx, abd-A,* and *Abd-B*—encoding master gene regulatory proteins that control the development of the thoracic and abdominal regions of the fly. Homeotic mutations (see p. 930) have further defined nine groups of regulatory DNA sequences. Each is required for the development of the indicated parasegment as well as more posterior parasegments. These regulatory DNA sequences are believed to act as enhancers to control the expression of one of the nearby genes, and their order on the DNA corresponds to the order of the body segments that they affect (see Figure 10–49). (Adapted from M. Peifer, F. Karch, and W. Bender, *Genes Devel.* 1:891–898, 1987.)

gene regulation—for example, mechanisms that depend on a directly inherited chromatin structure (see p. 576)—we might expect to find these mechanisms controlling some of the master regulatory genes.

Although nothing is known about how master regulatory genes are controlled in vertebrates, we are beginning to learn how these genes are controlled in *Drosophila.* The homeotic genes that determine the different identities of the fly's body segments, for example, are present in two complex loci, known as the *Antennapedia complex* and the *bithorax complex* (see p. 929). The bithorax complex, which is responsible for the differentiation of two thoracic and eight abdominal segments, contains three transcription units—called *Ubx, abd-A,* and *Abd-B.* Each transcription unit appears to encode a family of gene regulatory proteins produced by alternative RNA splicing. Whereas the protein-coding regions of the complex are thought to contain fewer than 20,000 nucleotide pairs, the regulatory regions span about 300,000 nucleotide pairs. Remarkably, the regulatory regions seem to be formed from sets of enhancers that are arranged on the chromosome in the same order as the body segments that the enhancers affect (Figure 10–48). This finding, taken together with the properties of a mutant fly in which a single enhancer has been translocated from one part of the bithorax complex to another, has suggested a model for gene regulation within the bithorax complex based on changes in chromatin structure. In this model, successive chromatin domains are exposed along the bithorax complex in the cells of progressively more posterior segments, allowing the enhancers in these domains to be successively activated (Figure 10–49). It seems likely that the mechanisms controlling master regulatory genes in vertebrates will be at least as complex.

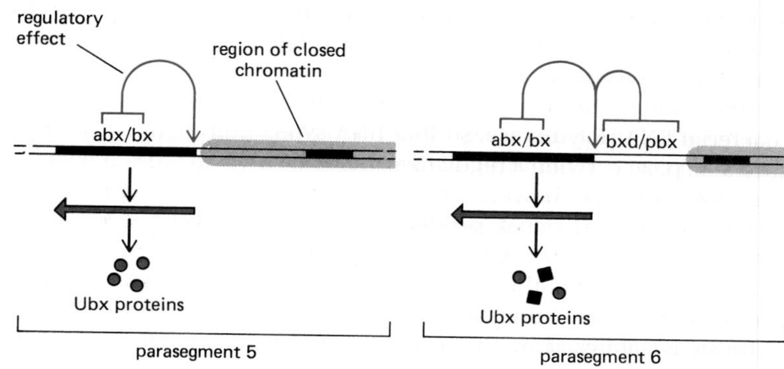

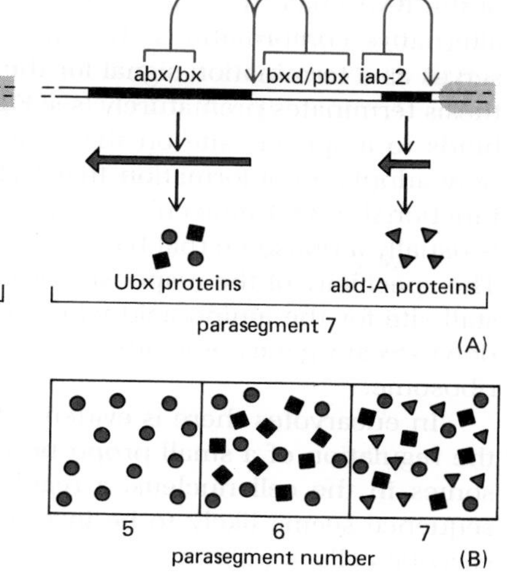

Figure 10–49 A model explaining the exact correspondence between the chromosomal position of each regulatory region in the bithorax complex and the location in the fly of the most anterior parasegment that is affected by a mutation in that region. (A) Control by effects on chromatin structure. The chromatin is postulated to become decondensed or otherwise activated progressively in each of the more posterior parasegments, so that only the abx/bx regulatory region is exposed in parasegment 5, while all of the regulatory regions are exposed in the most posterior parasegment affected by the complex (parasegment 13, see Figure 10–48). Only three of the parasegments are illustrated here. The *Ubx* gene can produce several different transcripts (symbolized here by squares and circles), whose selection is governed by the regulatory regions; the bithorax complex therefore produces a different mixture of gene regulatory proteins in each parasegment (B).

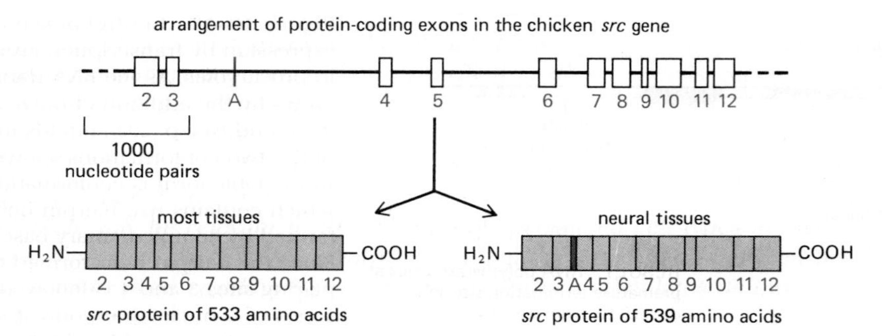

Figure 10–51 Regulated alternative RNA splicing produces two slightly different tyrosine protein kinases from the *src* gene (see p. 754); exon A is included only in neural tissues. Since this tissue-specific splicing difference has been conserved evolutionarily (being found in both birds and mammals), the resulting difference in the src protein is assumed to be important for the biological function of this regulatory protein. Only the protein-coding exons are shown (exon 1 forms the 5′ leader on the mRNA). (After J.B. Levy et al., *Mol. Cell Biol.* 7:4142–4145, 1987.)

Alternative RNA Splicing Can Be Used to Turn Genes On and Off[40]

Some genes are constantly transcribed in all cells, but because the constitutive RNA-splicing mechanism produces an mRNA that codes for a nonfunctional protein, the gene is expressed only in selected cells in which a specialized splicing reaction occurs.

This type of gene regulation has been especially well characterized in *Drosophila.* The ability of the P element (see p. 268) to transpose only in germ cells, for example, is due to its failure to produce a functional transposase in somatic cells; this failure has in turn been traced to the presence of an intron in the transposase mRNA that seems to be removed only in germ cells. As another example, genetic analyses indicate that the sex of the fly is determined by a cascade of gene activations, each leading to the production of a protein that makes possible the correct splicing of the RNA that is synthesized by the next gene in the series (Figure 10–52). The DNA sequences encoding several of these sex-determining proteins have been cloned and sequenced, which should greatly facilitate studies of the mechanisms regulating the splice-site choice.

10-23 The Mechanisms Responsible for Splice-Site Choice in Regulated RNA Splicing Are Not Understood

Regulated changes in the choice of RNA splice sites are presumed to be mediated by the binding of tissue- and gene-specific proteins or RNA molecules to the growing RNA transcripts. Since the splice sites selected in both the constitutive

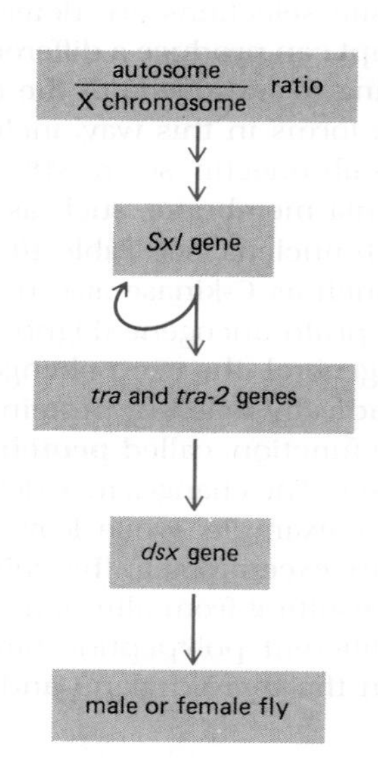

Figure 10–52 The cascade of changes in gene expression that determine the sex of a fly depends on alternative RNA splicing. In *Drosophila,* individuals with an autosome to X-chromosome ratio of one (normally, two sets of autosomes and two X chromosomes) develop as females, while those with a ratio of two (normally, two sets of autosomes and one X chromosome) develop as males. The ratio is assessed early in development and is remembered in each cell thereafter. The function of the genes shown is to transmit the information about this ratio to the many other genes that are involved in creating the sex-related phenotypes. These other genes function as two alternative sets: those that specify female features and those that specify male features. The *dsx* gene derives its name (*doublesex*) from the observation that, in mutants that do not express the gene, both the genes specifying female features and those specifying male features are expressed.

Events are shown as they occur in females, with arrows indicating the effects of each functional gene. In males the *Sxl, tra,* and *tra2* genes are transcribed but produce only nonfunctional mRNAs, and the *dsx* gene transcript is spliced to produce a protein that turns off the genes that specify female features. In females the *Sxl* transcript is spliced in a different way, so that it now produces a splicing-control protein that both maintains its own synthesis and turns on the two *tra* genes (*arrows*). The products of the *tra* genes in turn cooperate to change the pattern of RNA splicing for the *dsx* gene transcript; the resulting *dsx* mRNA produces an altered form of the dsx protein that turns off the genes that specify male features.

and regulated pathways of RNA splicing all seem to share the standard consensus sequences described previously (see Figure 9–80, p. 534), the binding of the specific component must change the conformation of the RNA transcript so as preferentially to mask or expose preexisting splice sites. The mechanism is likely to be complex, since no simple scheme, such as the masking of a splice site by the binding of a protein to it, seems able to account for the variety of splicing alternatives observed (Figure 10–53). Deciphering the molecular details of alternative RNA splicing will require that regulated splicing be reconstructed in a cell-free system so that all of the required components can be isolated and the effect of each of them on the spliceosome can be analyzed (see p. 533).

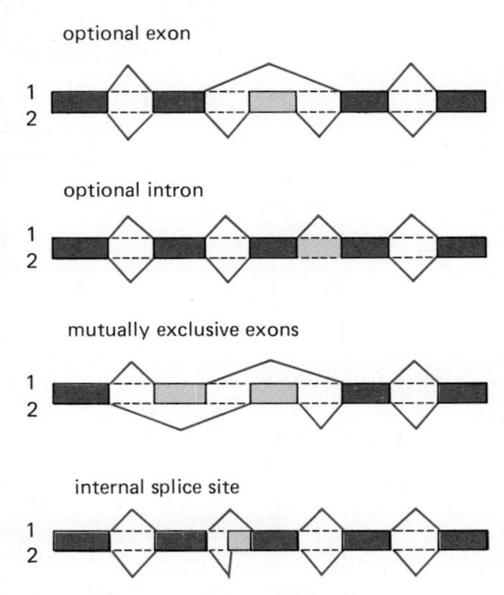

Figure 10–53 Four patterns of alternative RNA splicing that have been observed. In each case a single type of RNA transcript is spliced in two alternative ways to produce two distinct mRNAs (1 and 2). The darkly colored boxes mark RNA sequences that are retained in both mRNAs; the lightly colored boxes mark sequences that are included in only one of the mRNAs. Adjacent boxes are joined by colored lines denoting intron sequences. There seems to be no simple mechanism or rule that can explain all of these choices. (Adapted from A. Andreadis, M.E. Gallego, and B. Nadal-Ginard, *Annu. Rev. Cell Biol.* 3:207–242, 1987.)

A Change in the Site of RNA Transcript Cleavage and Poly-A Addition Can Change the Carboxyl Terminus of a Protein[41]

In eucaryotes the 3′ end of an mRNA molecule is not determined by the termination of RNA synthesis by the RNA polymerase; instead, it is determined by an RNA cleavage reaction that is catalyzed by additional factors while the transcript is elongating (see p. 528). The site of this cleavage can be controlled so as to change the carboxyl terminus of the resultant protein (which is encoded by the 3′ end of the mRNA). In procaryotes, producing a longer RNA transcript can only add more amino acids onto a protein chain. In eucaryotes, however, RNA splicing can create mRNAs that cause the original carboxyl terminus of a protein to be removed entirely and to be replaced with a new one after a longer transcript is made.

A well-studied change of this type mediates the switch from the synthesis of membrane-bound to secreted antibody molecules during the development of B lymphocytes. Early in the life history of a B cell, the antibody it produces is anchored in the plasma membrane, where it serves as a receptor for antigen. Antigen stimulation causes these cells both to multiply and to start secreting their antibody (see p. 1028). The secreted form of the antibody is identical to the membrane-bound form except at the extreme carboxyl terminus, where the membrane-bound form has a long string of hydrophobic amino acids that traverses the lipid bilayer of the membrane, and the secreted form has a much shorter string of water-soluble amino acids. The switch from membrane-bound to secreted antibody therefore requires a different nucleotide sequence at the 3′ end of the mRNA.

The membrane-bound form of the protein is generated by the transcription of all of the coding DNA sequences into RNA to make a long primary transcript. The nucleotides coding for the hydrophobic carboxyl terminus of the membrane-bound protein are located in the last exon of this long transcript (left side of Figure 10–54). The intron that precedes this exon contains the nucleotides coding for the water-soluble tail of the secreted molecule, and they are therefore removed by RNA splicing during production of the mRNA.

The secreted form of the molecule is generated from a shorter primary transcript that ends before the beginning of the last exon. Thus no acceptor splice site is present in this transcript to combine with the donor splice site just upstream from the nucleotides coding for the water-soluble tail; these nucleotides therefore remain in the final mRNA molecule (right side of Figure 10–54).

It is not known how the cleavage reaction is controlled to produce the kind of programmed switch in RNA transcript length just described.

The Definition of a Gene Has Had to Be Modified Since the Discovery of Alternative RNA Splicing[42]

The discovery that eucaryotic genes usually contain introns and that their coding sequences can be put together in more than one way has raised new questions about the definition of a gene. A gene was first clearly defined in molecular terms in the early 1940s from work on the biochemical genetics of the fungus *Neurospora*. Until then, a **gene** had been defined operationally as a region of the genome that segregates as a single unit during meiosis and gives rise to a definable phenotypic trait, such as a red or a white eye in *Drosophila* or a round or wrinkled seed in peas. After the work on *Neurospora* it became clear that most genes correspond to a region of the genome that directs the synthesis of a single enzyme.

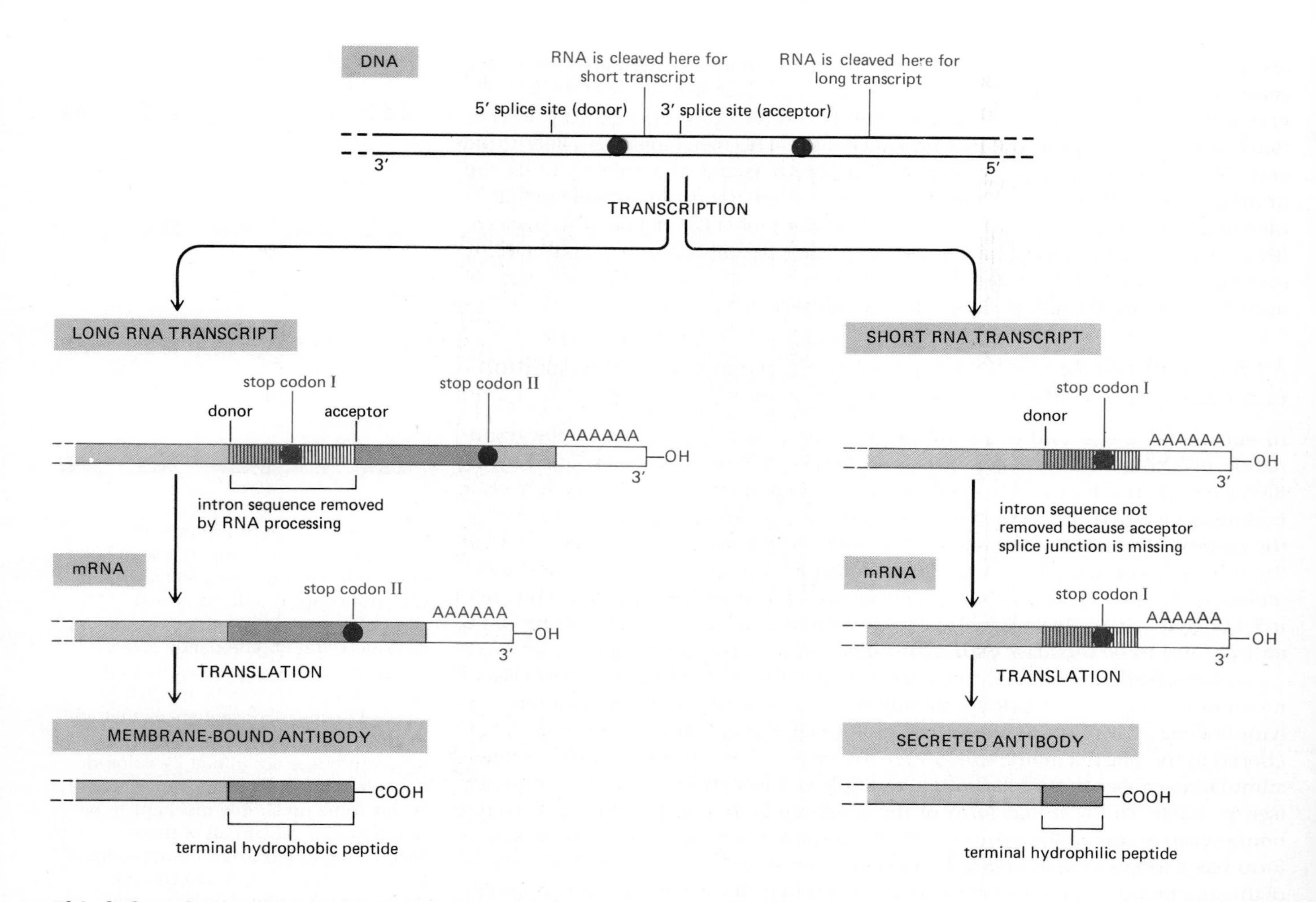

Figure 10–54 Regulation of the site of RNA cleavage and poly-A addition has an important effect on antibody synthesis. In unstimulated B cells (*left side*) a long RNA transcript is produced, and the intron sequence near its 3′ end is removed by RNA splicing to give rise to an mRNA molecule that codes for a membrane-bound antibody molecule. In contrast, after antigen stimulation (*right side*) the primary RNA transcript is cleaved upstream from the acceptor splice site of the last exon. As a result, some of the intron sequence that is removed from the long transcript remains as coding sequence in the short transcript. These are the sequences that encode the hydrophilic carboxyl-terminal portion of the secreted antibody molecule.

This led to the hypothesis that one gene encodes one polypeptide chain. The hypothesis proved to be extremely fruitful for subsequent research; and, as more was learned about the mechanism of gene expression in the 1960s, a gene became identified as that stretch of DNA that was transcribed into the RNA coding for a single polypeptide chain (or a single structural RNA such as a tRNA or rRNA molecule). The discovery of split genes in the late 1970s could be readily accommodated by the original definition of a gene, provided that a single polypeptide chain was specified by the RNA transcribed from any one DNA sequence. But it is now clear that many DNA sequences in higher eucaryotic cells produce two or more distinct proteins by means of alternative RNA splicing. How then is a gene to be defined?

In those relatively rare cases in which two very different eucaryotic proteins are produced from a single transcription unit, the two proteins are considered to be produced by distinct genes that overlap on the chromosome. It seems unnecessarily complex, however, to consider most of the modified proteins produced by alternative RNA splicing as being derived from overlapping genes. A more sensible alternative is to modify the original definition to include as a gene any DNA sequence that is transcribed as a single unit and encodes one set of closely related polypeptide chains (*protein isoforms*).

RNA Transport from the Nucleus Can Be Regulated[43]

An average primary RNA transcript seems to be at most 10 times longer than the mature mRNA molecule generated from it by RNA splicing. Yet it has been estimated that only about one-twentieth of the total mass of the hnRNA made ever leaves the cell nucleus (see p. 531). It seems, therefore, that a substantial fraction of the primary transcripts (perhaps half) may be completely degraded in the nucleus without ever generating an mRNA molecule for export. The discarded RNAs may be those whose sequence cannot be made into an mRNA molecule; on the other hand, some may represent potential mRNA molecules that are appropriately processed into mRNA only in other cell types.

RNA export through the nuclear pores is an active process (see p. 425). If this export depends on the specific recognition of the transported RNA molecule (or of a protein or RNA molecule bound to it) by a receptor protein in the nuclear pore complex, RNAs that lack this recognition signal would be selectively retained in the nucleus. Alternatively, RNA export may not require recognition signals; all RNAs might be automatically transported unless they are specifically retained. A third possibility is that a combination of selective export and selective retention operates. Since an RNA molecule seems to be selectively retained in the nucleus until all of the spliceosome components have dissociated from it (see p. 538), selective retention could be caused by a mechanism that prevents the completion of RNA splicing on particular RNA molecules. At present, however, these ideas remain largely speculative, and it seems unlikely that RNA export from the nucleus provides an important regulatory point for the expression of most eucaryotic genes.

Because a virus parasitizes normal intracellular pathways, studies of viral development often help to decipher these pathways (see p. 469). Adenovirus, for example, has a double-stranded DNA genome, which is replicated and transcribed in the host-cell nucleus. Late in infection the transport of host-cell RNAs from the nucleus is blocked, so that most of the RNAs that reach the cytoplasm are encoded by the adenovirus. Genetic analysis has shown that two adenovirus proteins that are produced early in infection are required for this change in the selectivity of transport of RNAs from the nucleus, providing a promising model system for analyzing how RNA transport is controlled.

10-24 Proteins That Bind to the 5′ Leader Region of mRNAs Mediate Negative Translational Control[44]

Not all mRNA molecules that reach the cytoplasm are translated into protein. The translation of some is blocked by specific *translation repressor proteins* that bind near their 5′ end, where translation would otherwise begin (Figure 10–55). This type of mechanism was first discovered in bacteria, where it enables excess ribosomal proteins to repress the translation of their own mRNAs.

In eucaryotic cells a particularly well-studied form of **negative translational control** allows the synthesis of the intracellular iron storage protein *ferritin* to be adjusted rapidly to the level of soluble iron atoms present. The ferritin mRNA in the cytoplasm can be shown to shift from an inactive ribonucleoprotein complex to a translationally active polyribosome complex after exposure of a cell to iron. Recombinant DNA experiments indicate that the iron regulation depends on a sequence of about 30 nucleotides in the 5′ leader of the ferritin mRNA molecule. This *iron-response element* folds into a stem-loop structure (see Figure 10–60B) that binds a regulatory iron-binding protein when that protein is not bound to iron. When the protein is bound to the iron-response element, the translation of any RNA sequence downstream is repressed (see Figure 10–55). The addition of iron dissociates the protein from the mRNA, increasing the rate of translation of the mRNA by as much as 100-fold.

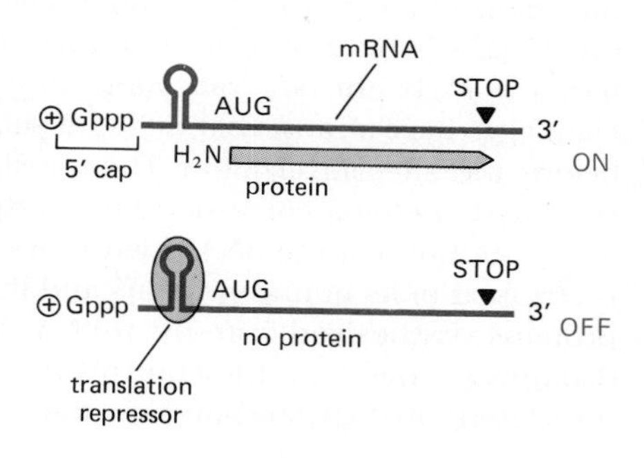

Figure 10–55 Negative translational control mediated by a sequence-specific RNA-binding protein (a *translation repressor*). Binding of the protein to an mRNA molecule decreases the translation of the mRNA. Several cases of this type of translational control are known; the illustration is modeled on the mechanism that causes more ferritin protein to be synthesized when the free iron concentration rises in the cell (see also Figure 10–60).

The Translation-Enhancer Sequences in Some Viral mRNAs Suggest a Mechanism for Positive Translational Control[45]

In principle, **positive translational control** could be mediated by a special "translation-enhancer" region in an mRNA molecule that preferentially attracts ribosomes. Certain RNA viruses—the *picornaviruses*—can be shown to contain such a region, whose presence can cause translation to begin at internal AUG sites that would not otherwise be used to start protein synthesis in a eucaryotic cell (Figure 10–56).

Positive translational control has also been demonstrated in yeast cells. Genetic studies have identified specific proteins required to activate the translation of the mRNA produced from a yeast gene called *GCN4;* without them the mRNA remains untranslated. The *GCN4* mRNA resembles a class of poorly translated mRNAs from higher eucaryotes, whose translation is suspected to be similarly

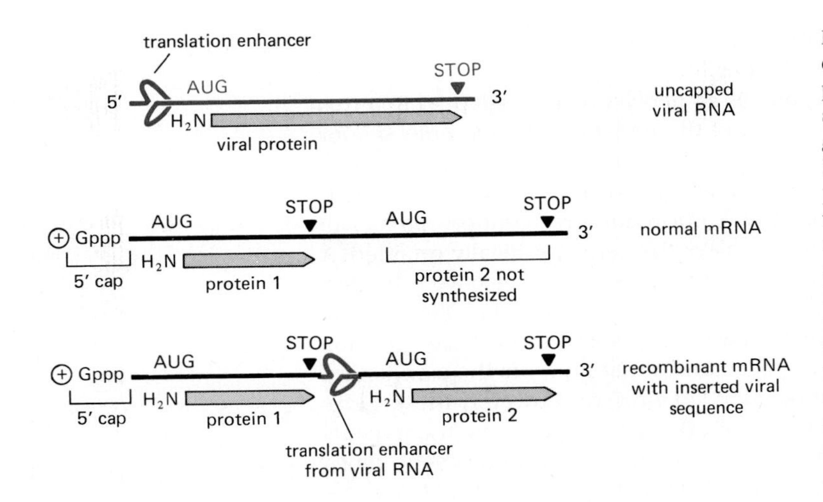

Figure 10–56 Diagram of an experiment that demonstrates the presence of an RNA sequence that serves as a translation enhancer in the genomes of certain RNA viruses. Picornaviruses (which includes poliovirus) are positive-strand RNA viruses—that is, their genomes can serve directly as mRNAs for the synthesis of viral proteins (see p. 251). Because of their mode of RNA synthesis, they lack the 5′ caps that are required for the initiation of protein synthesis on most mRNAs (see p. 215). By measuring the protein synthesis catalyzed by various recombinant RNAs, a translation-enhancer sequence several hundred nucleotides long can be identified in the viral RNA molecule. As illustrated, moving this sequence to an internal position in an mRNA chain will cause a ribosome to start translation at a neighboring internal AUG, so that the rules that normally cause only the first AUG in the mRNA chain to be used to start protein synthesis (see p. 214) are bypassed.

controlled. (This class contains about 5% of all mRNAs characterized thus far.) These RNAs have an unusually long 5′ leader sequence that contains a series of AUGs that interfere with the translation of the major coding sequence downstream by initiating the synthesis of short peptides; a stop codon occurring before the major coding sequence prevents readthrough. By analogy with the picornaviruses, the translation of the major coding sequence in these mRNAs might depend on the binding of translation-activator molecules to a translation-enhancer sequence next to the appropriate AUG, so that translation reinitiates (Figure 10–57). The mechanism responsible for these types of translational activations, however, is not known.

10-25 Many mRNAs Are Subject to Translational Control[46]

How widespread is translational control in higher eucaryotes? According to one estimate, the expression of about one gene in ten is strongly regulated in this way. Translational control enables a cell to change the concentration of a protein rapidly and reversibly without rapid turnover of its mRNA (see p. 714). The expression of some proto-oncogenes seems to be regulated at this level.

Translational controls are especially important in many fertilized eggs, which need to switch from making proteins required to maintain a quiescent oocyte to making proteins required for rapid cell division. These eggs have a large store of mRNA that is built up during the development of the oocyte. Many of these *maternal mRNAs* are not translated until the egg has been fertilized. Studies with clam eggs have shown that different sets of mRNAs are associated with ribosomes before and after fertilization. The translation of these mRNAs in a cell-free system produces proteins corresponding to the appropriate stage of clam egg development—but only if the RNA is left in its native, ribonucleoprotein form. If the RNA is stripped of its bound proteins and then translated, the differences between the proteins synthesized from the mRNAs taken from different stages of development disappear. Therefore, the controls that determine whether a particular mRNA is translated must depend on how the RNA is packaged with regulatory molecules.

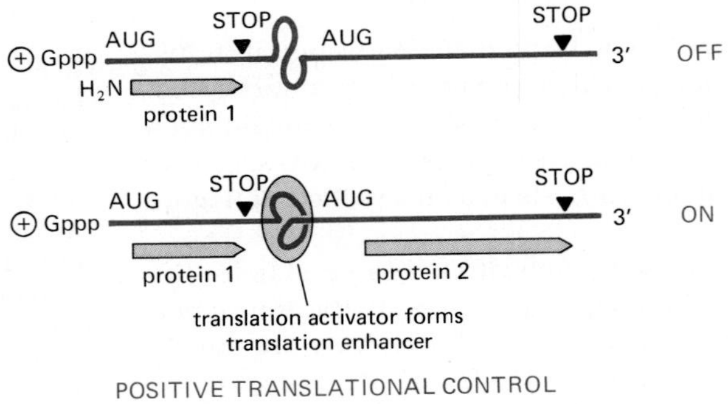

Figure 10–57 A model for positive translational control in which the binding of a protein (a *translation activator*) is needed for extensive translation of an mRNA. Although positive controls are known to cause the translation of specific mRNAs, the mechanisms involved are unclear. However, positive control is associated with the synthesis of short peptides whose translation is initiated upstream from the correct AUG. This suggests an analogy with the picornavirus mechanism in Figure 10–56, on which this illustration is modeled.

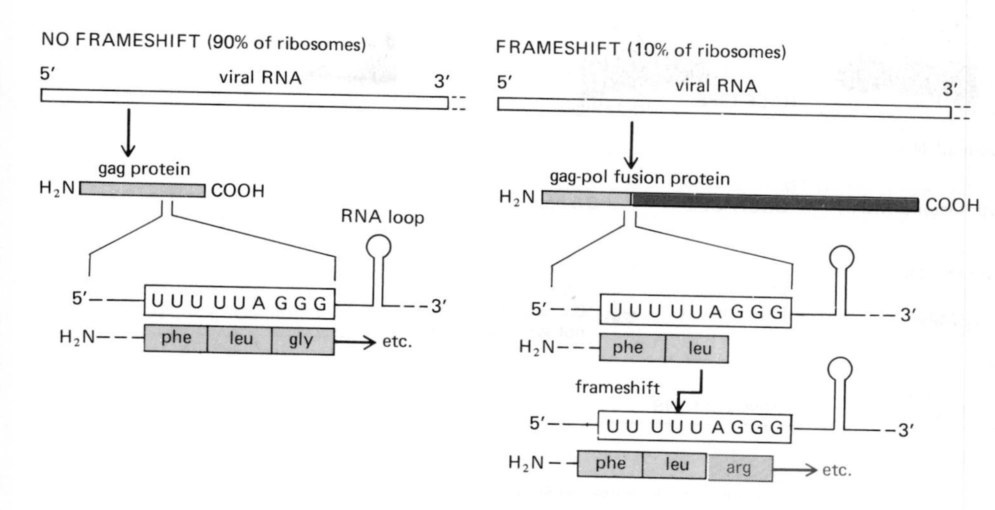

Figure 10–58 Translational frameshifting is necessary to produce the reverse transcriptase of a retrovirus. The viral reverse transcriptase and integrase are produced by cleavage of the large gag-pol fusion protein, whereas the viral capsid proteins are produced by cleavage of the more abundant gag protein. Both proteins start identically, but the gag protein terminates at an in-frame stop codon; the indicated shift to the −1 frame upstream from this codon allows the synthesis of the fusion protein. The frameshift occurs because features in the local RNA structure (including the RNA loop shown) cause the $tRNA^{Leu}$ attached to the carboxyl terminus of the growing polypeptide chain (see p. 208) occasionally to slip backward by one nucleotide on the ribosome, so that it pairs with a UUU codon instead of the UUA codon that had specified its incorporation. The sequence shown is from the human immunodeficiency virus (HIV-1), which causes AIDS. (Adapted from T. Jacks et al., *Nature* 331:280–283, 1988.)

5-8

Translational Frameshifting Allows Two Proteins to Be Made from One mRNA Molecule[47]

The translational controls thus far discussed affect the rate at which new protein chains are initiated on an mRNA molecule. Usually, the completion of the synthesis of a protein is automatic once this synthesis has begun. In special cases, however, a process called **translational frameshifting** can alter the final protein that is made.

Translational frameshifting is commonly used by retroviruses; it allows different amounts of two or more proteins to be synthesized from a single mRNA. These viruses commonly make a large *polyprotein* that is cleaved by a viral protease to produce a group of capsid proteins (*gag proteins*) and the viral reverse transcriptase and integrase (*pol proteins*). In many cases the *gag* and *pol* genes are in different reading frames, so that a translational frameshift is required to produce the much less abundant *pol* proteins. The frameshift occurs at a particular codon in the mRNA and requires specific sequences, some of which are upstream and some downstream from this site (Figure 10–58).

10-26
10-27

Gene Expression Can Be Controlled by a Change in mRNA Stability[48]

Most mRNAs in a bacterial cell are very unstable, having a half-life of about 3 minutes. Because bacterial mRNAs are both rapidly synthesized and rapidly degraded, a bacterium can adjust its gene expression quickly in response to environmental changes.

The mRNAs in eucaryotic cells are more stable. Some, such as that encoding β globin, have a half-life of more than 10 hours. Others have a half-life of only 30 minutes or less. The unstable mRNAs often code for regulatory proteins whose levels change rapidly in cells, such as growth factors and the products of the proto-oncogenes *fos* and *myc* (see p. 1207). The 3′ untranslated region of many of these unstable mRNAs contains a long sequence rich in A and U nucleotides, which seems to be responsible for their instability (Figure 10–59).

The stability of an mRNA can be changed in response to extracellular signals. Thus, for example, steroid hormones affect a cell not only by increasing the transcription of specific genes (see p. 690), but also by increasing the stability of several of their mRNAs. Conversely, the addition of iron to cells decreases the stability of the mRNA that encodes the *transferrin receptor,* causing less of this iron-scavenging protein to be made. Interestingly, the destabilization of the transferrin receptor mRNA seems to be mediated by the same iron-sensitive RNA-binding protein that controls ferritin mRNA translation; in the case of the transferrin receptor, the protein binds to the opposite end of the mRNA (the 3′ untranslated region) and causes an increase rather than a decrease in protein production (Figure 10–60).

is achieved by a large number of sequence-specific RNA-binding molecules, most of which are still uncharacterized. The sites on the RNA to which they bind usually include a group of nucleotides whose bases are exposed in the single-stranded RNA chain (see Figure 10–60); this sequence-specific binding therefore differs from that which occurs on DNA, where the nucleotide sequence is generally recognized with the bases paired in a double helix (see p. 488). Moreover, whereas the known molecules that bind to specific DNA sequences are all proteins, the binding to specific RNA sequences can be mediated either by proteins or by other RNA molecules, which use complementary RNA-RNA base-pairing as part of their recognition mechanism. Thus, in attempting to dissect posttranscriptional mechanisms, we have largely entered an RNA world.

The reactions catalyzed by RNA molecules have been more difficult to study than the reactions catalyzed by proteins. Large RNA molecules are degraded readily during isolation by traces of ribonucleases, and they are difficult to purify to homogeneity in functional form. Recombinant DNA technology, however, now allows large amounts of pure RNAs of any sequence to be produced *in vitro* with purified RNA polymerases (see Figure 9–81, p. 534). This has made it possible to study the detailed chemistry of RNA-catalyzed self-splicing reactions (see p. 105) and to define the minimal sequences required for the RNA-mediated self-cleavage of a plant viroid (Figure 10–61). Another reaction catalyzed directly by RNA in both procaryotes and eucaryotes is the cleavage of tRNA precursors by a protein-RNA complex known as *RNAse P.* The RNA components of some of the snRNP particles in the spliceosome may likewise make and break covalent bonds (although this has not yet been demonstrated), and the active center of the *peptidyl transferase* enzyme complex that polymerizes amino acids on the ribosome is suspected to reside in the rRNA (see p. 212).

RNA molecules also have regulatory roles in cells. The *antisense RNA* strategy for experimentally manipulating cells so that they fail to express a particular gene (see p. 195) mimics a normal mechanism that is known to regulate the expression of a few selected genes in bacteria and may be used much more widely than is now realized. A particularly well-understood example of this kind of mechanism provides a feedback control on the initiation of DNA replication for a large family of bacterial DNA plasmids. The control system limits the copy number of the plasmid, thereby preventing the plasmid from killing its host cell (Figure 10–62).

Studies of RNA-catalyzed reactions are of special interest from an evolutionary perspective. As discussed in Chapter 1 (see p. 7), the first cells are thought to have contained no DNA and may have contained very few, if any, proteins. Many of the RNA-catalyzed reactions in present-day cells may represent molecular fossils—descendants of the complex network of RNA-mediated reactions that are presumed to have dominated cellular metabolism more than 3.5 billion years ago. By understanding these reactions, biologists may be able to trace the paths by which a living cell first evolved.

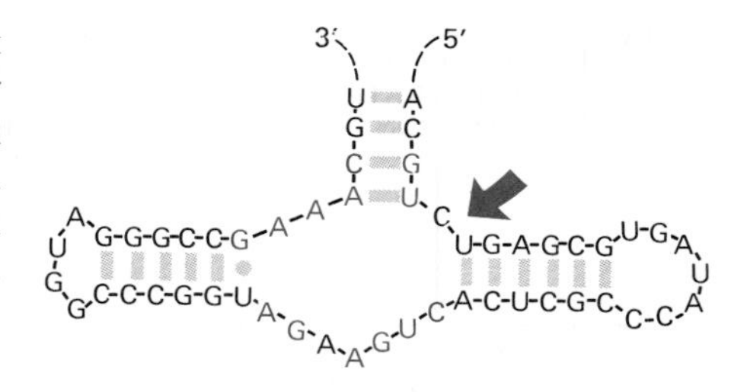

Figure 10–61 Structure of an active center of a plant viroid RNA. This short RNA molecule cleaves itself at the colored arrow. The colored nucleotides are identical in seven self-cleaving RNAs, six of which are found in plants (viroids and virusoids) and one of which is present in an animal (newts). This RNA-catalyzed reaction converts a tandemly repeated single-stranded RNA sequence that forms as a replication intermediate in viroidlike RNA molecules into a single-copy linear RNA molecule, which then circularizes. (After A.C. Forster et al., *Nature* 334:265–267, 1988).

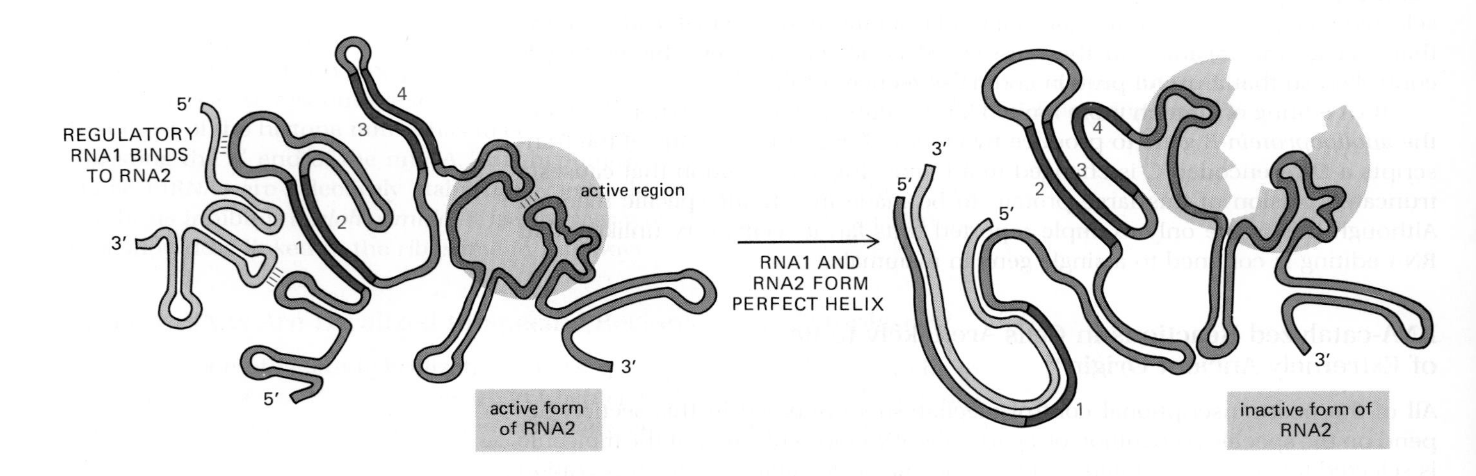

Figure 10–62 A regulatory interaction between two RNA molecules maintains a constant plasmid copy number in the ColE1 family of bacterial DNA plasmids. RNA 1 (about 100 nucleotides long) is a regulatory RNA that inhibits the activity of RNA 2 (about 500 nucleotides long) in the initiation of plasmid DNA replication. RNA 1 is complementary in sequence to the 5′ end of RNA 2, and its concentration increases in proportion to the number of plasmid DNA molecules in a cell. In RNA 2, sequence 2 is complementary to both sequence 1 and sequence 3 (compare with Figure 10–50), and it is displaced from one to the other by the binding of RNA 1; RNA 1 thereby alters the conformation of sequence 4, inactivating RNA 2. (After H. Masukata and J. Tomizawa, *Cell* 44:125–136, 1986.)

Summary

Many steps in the pathway from RNA to protein are regulated by cells to control gene expression. Most genes are thought to be regulated at multiple levels, although the control of the initiation of transcription usually predominates. Some genes, however, are transcribed at a constant level and turned on and off solely by processes that subsequently affect the RNA. These posttranscriptional regulatory processes include (1) attenuation of the transcript by its premature termination, (2) alternative splice-site selection, (3) control of 3′-end formation by cleavage and poly-A addition, (4) control of translational initiation, and (5) regulated mRNA degradation. Most of these control processes require the recognition of specific sequences or structures in the RNA molecule being regulated. This recognition can be accomplished by either a regulatory protein or a regulatory RNA molecule.

The Organization and Evolution of the Nuclear Genome[53]

Much of evolutionary history is recorded in the genomes of present-day organisms and can be deciphered from a careful analysis of their DNA sequences. Tens of millions of DNA nucleotides have been sequenced thus far, and we can now see in outline how the genes coding for certain proteins have evolved over hundreds of millions of years. Studies of the occasional changes that occur in present-day chromosomes provide additional clues to the mechanisms that have brought about evolutionary change in the past. In this section we present some of the general principles that have emerged from such molecular genetic studies, with emphasis on the organization and evolution of the nuclear genome in higher eucaryotes.

Genomes Are Fine-Tuned by Point Mutation and Radically Remodeled or Enlarged by Genetic Recombination[54]

DNA nucleotide sequences must be accurately replicated and conserved. In Chapter 5 we discussed the elaborate DNA-replication and DNA-repair mechanisms that enable DNA sequences to be inherited with extraordinary fidelity: only about one nucleotide pair in a thousand is randomly changed every 200,000 years (see p. 220). Even so, in a population of 10,000 individuals, every possible nucleotide substitution will have been "tried out" on about fifty occasions in the course of a million years, which is a short span of time in relation to the evolution of species. If a variant sequence is advantageous, it will be rapidly propagated by natural selection. Consequently, it can be expected that in any given species the function of most genes will be optimized with respect to variation by point mutation.

While point mutation is an efficient mechanism for fine-tuning the genome, evolutionary progress in the long term must depend on more radical types of genetic change. Genetic recombination causes major rearrangements of the genome with surprising frequency: the genome can expand or contract by duplication or deletion, and its parts can be transposed from one region to another to create new combinations. Component parts of genes—their individual exons and regulatory elements—can be shuffled as separate modules to create proteins that serve entirely new roles. In addition, duplicated copies of genes tend to diverge by further mutation and become specialized and individually optimized for subtly different functions. By these means the genome as a whole can evolve to become increasingly complex and sophisticated. In a mammal, for example, multiple variant forms of almost every gene exist—different actin genes for the different types of contractile cells, different opsin genes for the perception of lights of different colors, different collagen genes for the different types of connective tissues, and so on. The expression of each gene is regulated according to its own precise and specific rules. Moreover, DNA sequencing reveals that many genes share related modular segments but are otherwise very different. Thus certain portions of the rhodopsin genes share a common ancestry with portions of the genes for

The end result of the gene duplication processes that have given rise to the diversity of globin chains is seen clearly in the genes that arose from the original β gene, which are arranged as a series of homologous DNA sequences located within 50,000 nucleotide pairs of one another (see Figure 10–39A). A similar cluster of α-globin genes is located on a separate human chromosome. Because the α- and β-globin gene clusters are on separate chromosomes in birds and mammals but are together in the frog *Xenopus*, it is believed that a translocation event separated the two genes about 300 million years ago (see Figure 10–66). Such translocations probably help stabilize duplicated genes with distinct functions by protecting them from the homogenizing processes that act on closely linked genes of similar DNA sequence (see Figure 10–64).

There are several duplicated globin DNA sequences in the α- and β-globin gene clusters that are not functional genes. They are examples of *pseudogenes*, which have a close homology to the functional genes but have been disabled by mutations that prevent their expression. The existence of such pseudogenes should not be surprising since not every DNA duplication would be expected to lead to a new functional gene and nonfunctional DNA sequences are not rapidly discarded, as indicated by the large excess of noncoding DNA in mammalian genomes (see p. 485).

A great deal of our evolutionary history will be discernible in our chromosomes once the DNA sequences of many gene families have been compared in animals of increasing complexity (see also Figure 4–62, p. 183).

Genes Encoding New Proteins Can Be Created by the Recombination of Exons[54]

The role of DNA duplication in evolution is not confined to the generation of large gene families. It can also be important in generating new single genes. The proteins encoded by genes so generated can be recognized by the presence of repeating, similar protein domains, which are covalently linked to one another in series. The immunoglobulins (Figure 10–67) and albumins, for example, as well as most fibrous proteins (such as spectrins and collagens) are encoded by genes that have evolved by repeated duplications of a primordial DNA sequence.

In genes that have evolved in this way, as well as in many other genes, each separate exon often encodes an individual protein folding unit, or domain (see p. 112). It is believed that the organization of DNA coding sequences as a series of such exons separated by long introns has greatly facilitated the evolution of new proteins. The duplications necessary to form a single gene coding for a protein with repeating domains, for example, can occur by breaking and rejoining the DNA anywhere in the long introns on either side of an exon encoding a useful protein domain; without introns there would be only a few sites in the original gene at which a recombinational exchange between sister DNA molecules could duplicate the domain. By enabling the duplication to occur at many potential recombination sites rather than at just a few, introns greatly increase the probability of a favorable duplication event.

For the same reason the presence of introns greatly increases the probability that a chance recombination event will join two initially separated DNA sequences that code for different protein domains (for example, see Figure 10–71). The presumed results of such recombinations are seen in many present-day proteins (see Figure 3–38, p. 118). Thus the large separation between the exons encoding individual domains in higher eucaryotes is thought to accelerate the process by which random genetic-recombination events generate useful new proteins. This could help to explain the successful evolution of these very complex organisms.

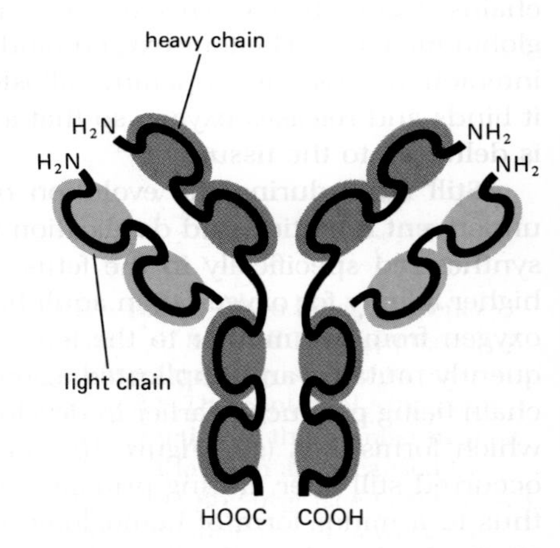

Figure 10–67 Schematic view of an antibody (immunoglobulin) molecule. This molecule is a complex of two identical heavy chains and two identical light chains (*colored*). Each heavy chain contains four similar, covalently linked domains; each light chain contains two such domains. Each domain is encoded by a separate exon, and all of the exons are thought to have evolved by the serial duplication of a single ancestral exon.

Most Proteins Probably Originated from Highly Split Genes Containing a Series of Small Exons[57]

The discovery of split genes in 1977 was unexpected. Previously all genes analyzed in detail were bacterial genes, which lack introns. Bacteria also lack nuclei and internal membranes and have smaller genomes than eucaryotic cells, and traditionally they were considered to resemble the simpler cells from which eucaryotic

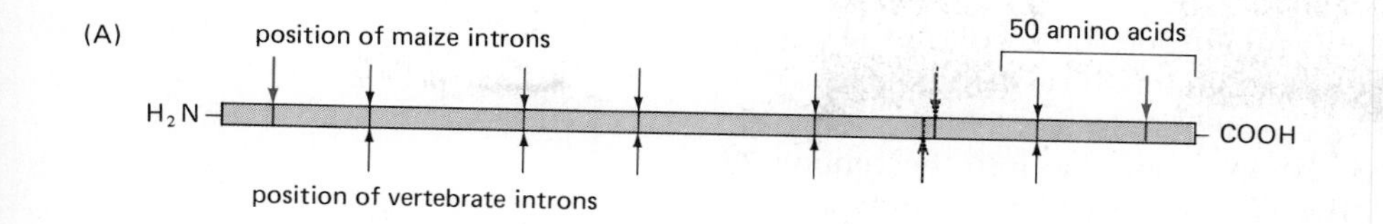

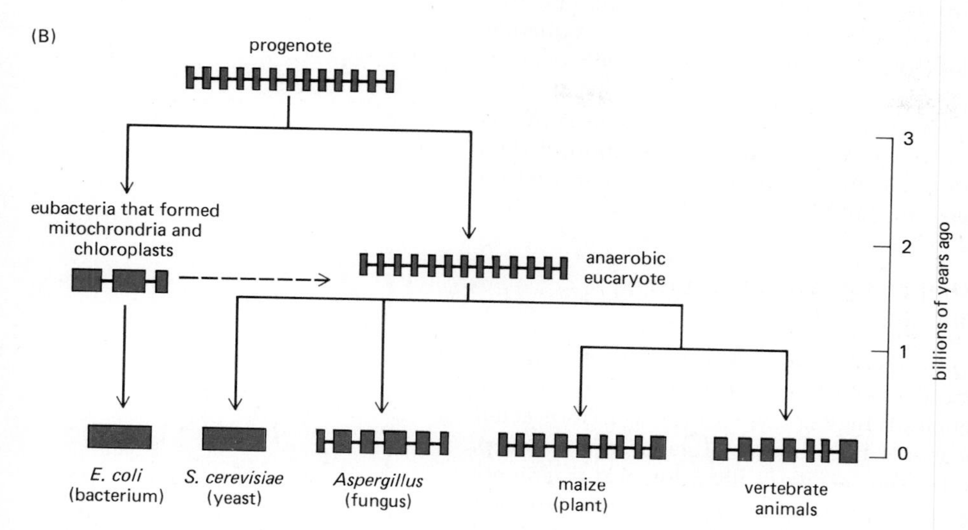

Figure 10–68 The ancient evolution of split genes. (A) A comparison of the exon structure of the *triosephosphate isomerase* gene in plants and animals. The intron positions that are identical in maize (corn) and vertebrates are marked with black arrows, while the intron positions that differ are marked with red arrows. Since plants and animals are thought to have diverged from a common ancestor about one billion years ago, the introns that they share must be of very ancient origin. (B) An outline of how a particular gene may have evolved. The exon sequences are shown in color and the intron sequences in black. The gene illustrated here codes for a protein that is required in all cells. Like *triosephosphate isomerase,* this protein must have evolved to its final three-dimensional structure before the eubacterial, archaebacterial, and eucaryotic lineages split off from a common ancestor cell—designated here as a "progenote." The dotted line marks the approximate time of the endosymbiotic events that gave rise to mitochondria and chloroplasts (see p. 398). (A, after W. Gilbert, M. Marchionni, and G. McKnight, *Cell* 46:151–154, 1987.)

cells must have been derived. Not surprisingly, most biologists initially assumed that introns were a bizarre and late evolutionary addition to the eucaryotic line. It now seems likely, however, that split genes are the ancient condition and that bacteria lost their introns only after most of their proteins had evolved.

The idea that introns are very old is consistent with current concepts of protein evolution by the trial-and-error recombination of separate exons that encode distinct protein domains. Moreover, evidence for the ancient origin of introns has been obtained by examination of the gene that encodes the ubiquitous enzyme *triosephosphate isomerase.* Triosephosphate isomerase has an essential role in the metabolism of all cells, catalyzing the interconversion of glyceraldehyde-3 phosphate and dihydroxyacetone phosphate—a central step in glycolysis and gluconeogenesis (see Figure 2–38, p. 83). By comparing the amino acid sequence of this enzyme in various organisms, it is possible to deduce that the enzyme evolved before the divergence of procaryotes and eucaryotes from a common ancestor; the human and bacterial amino acid sequences are 46% identical. The gene encoding the enzyme contains six introns in vertebrates (chickens and humans), and five of these are in precisely the same positions in maize. This implies that these five introns were present in the gene before plants and animals diverged in the eucaryotic lineage, an estimated 10^9 years ago (Figure 10–68).

In general, small unicellular organisms are under a strong selection pressure to reproduce by cell division at the maximum rate permitted by the levels of nutrients in the environment. For this, they must minimize the amount of unnecessary DNA that they have to synthesize in each cell division cycle. For larger organisms that live by predation, where size is an advantage, and for multicellular organisms in general, where rates of cell division are constrained by other requirements, there will not be such strong selection pressure to eliminate superfluous DNA from the genome. This argument may help to explain why bacteria should have lost their introns while eucaryotes retain them. It also tallies with another finding from the study of triosephosphate isomerase: whereas the multicellular fungus *Aspergillus* has five introns in its gene for this enzyme, its unicellular relative, the yeast *Saccharomyces,* has none.

What is the mechanism for loss of introns? While it is possible to lose introns by piecemeal random deletions of short segments of DNA, eucaryotic cells (and perhaps the ancestors of bacteria also) appear to have ways of precisely and selectively deleting entire introns from their genomes. Whereas most vertebrates contain only a single insulin gene with two introns, for example, rats contain a second, neighboring insulin gene with only one intron. The second gene apparently arose by gene duplication relatively recently and subsequently lost one of its introns. Because intron loss requires the exact rejoining of DNA coding sequences, it is presumed to arise from the rare incorporation into the genome of a DNA copy of the mRNA of the appropriate gene, from which the introns have been precisely removed. Such intronless copies presumably arise on occasion through the activity of reverse transcriptases (see p. 260), and it is thought that recombination enzymes allow them to become paired with the original sequence, which is then "corrected" to an intronless form by a gene-conversion type of event (see p. 244).

Reverse transcriptases are produced in cells by specific transposable elements (see Table 10–3, p. 605) as well as by all retroviruses, and the generation of DNA copies of segments of the genome by reverse transcription has also contributed in other ways to the evolution of the genomes of higher organisms (see p. 608).

A Major Fraction of the DNA of Higher Eucaryotes Consists of Repeated, Noncoding Nucleotide Sequences[58]

Eucaryotic genomes contain not only introns but also large numbers of copies of other seemingly nonessential DNA sequences that do not code for protein. The presence of such repeated DNA sequences in higher eucaryotes was first revealed by a hybridization technique that measures the number of gene copies (see p. 188). In this procedure the genome is broken mechanically into short fragments of DNA double helix about 1000 nucleotide pairs long, and the fragments are then denatured to produce DNA single strands. The speed with which the single-stranded fragments in the mixture reanneal under conditions in which the double-helical conformation is stable depends on how many complementary strands each fragment finds. For the most part, the reaction is very slow. The haploid genome of a mammalian cell, for example, is represented by about 6 million different 1000-nucleotide-long DNA fragments, and any fragment whose sequence is present in only one copy must randomly collide with 6 million noncomplementary strands for every complementary partner strand that it happens to find.

When the DNA from a human cell is analyzed in this way under conditions that require near perfect matching (high stringency conditions, see p. 191), about 70% of the DNA strands reanneal as slowly as one would expect for a large collection of unique (nonrepeated) DNA sequences, requiring days for complete annealing. But most of the remaining 30% of the DNA strands anneal much more quickly. These strands contain sequences that are repeated many times in the genome, and they thus collide with a complementary partner relatively rapidly. Most of these highly repeated DNA sequences do not encode proteins, and they are of two types: about one-third are the tandemly repeated *satellite DNAs*, to be discussed next; the rest are *interspersed repeated DNAs*. Most of the latter DNAs derive from a few transposable DNA sequences that have multiplied to especially high copy numbers in our genome (see p. 608).

Satellite DNA Sequences Have No Known Function[59]

The most rapidly annealing DNA strands in an experiment of the type just described usually consist of very long tandem repetitions of a short nucleotide sequence (Figure 10–69). The repeat unit in a sequence of this type may be composed of only one or two nucleotides, but most repeats are longer, and in mammals they are typically composed of variants of a short sequence organized into a repeat of a few hundred nucleotides. These tandem repeats of a simple sequence are called **satellite DNAs** because the first DNAs of this type to be discovered had an unusual ratio of nucleotides that made it possible to separate them from the bulk of the cell's DNA as a minor component (or "satellite"). Satellite DNA sequences generally

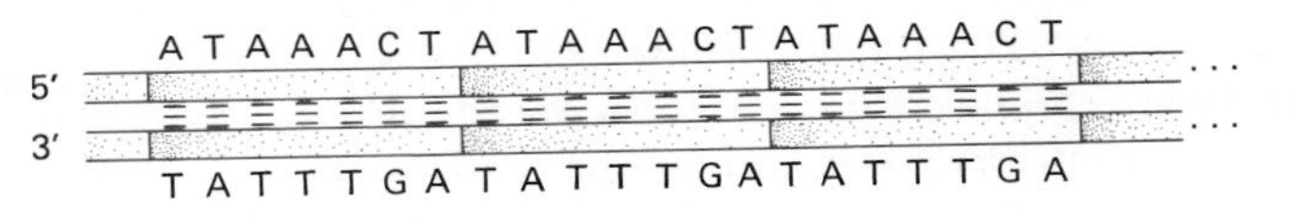

Figure 10–69 A simple satellite DNA sequence consisting of many serially arranged repetitions of a sequence seven nucleotide pairs long. This particular DNA sequence is found in *Drosophila*.

are not transcribed and are located most often in the heterochromatin associated with the centromeric regions of chromosomes (see p. 577). In some mammals a single type of satellite DNA sequence constitutes 10% or more of the DNA and may even occupy a whole chromosome arm, so that the cell contains millions of copies of the basic repeated sequence.

Satellite DNA sequences seem to have changed unusually rapidly and even to have shifted their positions on chromosomes in the course of evolution. When two homologous mitotic chromosomes of any human are compared, for example, some of the satellite DNA sequences usually are found arranged in a strikingly different manner on the two chromosomes. Moreover, in contrast to the high degree of conservation of DNA sequences elsewhere in the genome, generally there are marked differences in the satellite DNA sequences of two closely related species. No function has yet been found for satellite DNA sequences: tests designed to demonstrate a role in chromosome pairing or nuclear organization have failed thus far to reveal any evidence for such a role. It has therefore been suggested that they are an extreme form of "selfish DNA" sequences, whose properties ensure their own retention in the genome but which do nothing to help the survival of the cells containing them. Other sequences that are commonly viewed as selfish are the *transposable elements,* which we discuss next.

10-36 The Evolution of Genomes Has Been Accelerated by Transposable Elements of at Least Three Types[60]

Genomes generally contain many varieties of **transposable elements.** These elements were first discovered in maize, where several have been sequenced and characterized. Transposable elements have been studied most extensively in *Drosophila,* where more than 30 varieties are known, varying in length between 2000 and 10,000 nucleotide pairs; most are present in 5 to 10 copies per diploid cell.

At least three broad classes of transposable elements can be distinguished by the peculiarities of their sequence organization (Table 10–3). Some elements move from place to place within chromosomes directly as DNA, while many others move via an RNA intermediate, as described in Chapter 5 (see p. 255). In either case they can multiply and spread from one site in a genome to a multitude of other sites, sometimes behaving as disruptive parasites.

Transposable elements seem to make up at least 10% of higher eucaryotic genomes. Although most of these elements move only very rarely, so many elements are present that their movement has a major effect on the variability of a species. More than half of the spontaneous mutations examined in *Drosophila,*

Table 10–3 Three Major Families of Transposable Elements

Structure	Genes in Complete Element	Mode of Movement	Examples
① short inverted repeats at each end	encodes transposase	moves as DNA, either excising or following a replicative pathway	P element (*Drosophila*) Ac-Ds (maize) tn3 and IS1 (*E. coli*) Tam3 (*Antirrhinum*)
② directly repeated long terminal repeats (LTRs) at ends	encodes reverse transcriptase and resembles retrovirus	moves via an RNA intermediate produced by promoter in LTR	Copia Ty THE-1 bs1
③ AAAA / TTTT Poly A at 3′ end of RNA transcript; 5′ end is often truncated	encodes reverse transcriptase	moves via an RNA intermediate that is presumably produced from a neighboring promotor	F element (*Drosophila*) L1 (human) cin-4 (maize)

These elements range in length from 2000 to about 12,000 nucleotide pairs; each family contains many members, only a few of which are listed here.

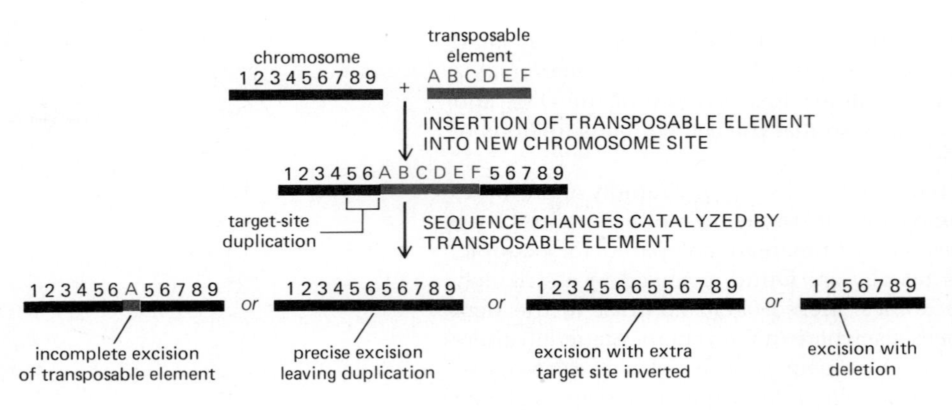

Figure 10–70 Some changes in chromosomal DNA sequences caused by transposable elements. The insertion of a transposable element always produces a short target-site duplication of the chromosomal sequence, which is generally 3 to 12 nucleotide pairs in length. The site-specific recombination enzymes associated with the element can also cause its subsequent excision. This excision often fails to restore the original chromosomal DNA sequence, as in the four examples shown.

for example, are due to the insertion of a transposable element in or near the mutant gene.

Mutations can occur either when an element inserts into a gene or when it exits to move elsewhere. All known transposable elements cause a short "target-site duplication" because of their mechanism of insertion (see Figure 5–67B, p. 248); when they exit, they generally leave behind part of this duplication—often with other local sequence changes as well (Figure 10–70). Thus, as transposable elements move in and out of chromosomes, they cause a variety of short additions and deletions of nucleotide sequences.

Transposable elements have also contributed to genome diversity in another way. When two transposable elements that are recognized by the same site-specific recombination enzyme (*transposase*) integrate into neighboring chromosomal sites, the DNA between them can become a substrate for transposition by the transposase. Because this provides a particularly effective pathway for the duplication and movement of exons, these elements may help to create new genes (Figure 10–71).

10-36 Transposable Elements Can Affect Gene Regulation[61]

The DNA sequence rearrangements caused by transposable elements often alter the pattern of expression of nearby genes, thereby affecting various aspects of animal or plant development, such as pigmentation (Figure 10–72) or morphogenesis (the shape of an eye or a flower, for example). While most of these changes in gene regulation would be expected to be detrimental to an organism, some of them will bring benefits.

Several aspects of the mutations caused by transposable elements are unusual and seem to distinguish them from the mutations caused by errors in DNA rep-

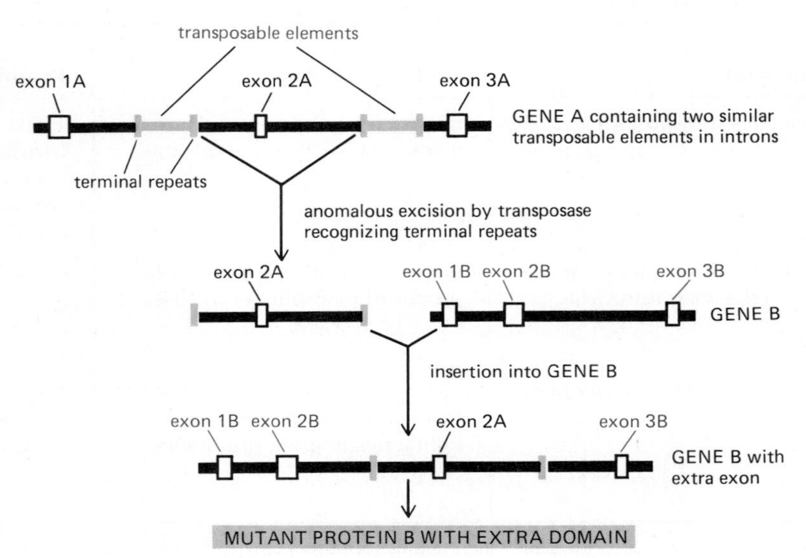

Figure 10–71 An example of the exon shuffling that can be caused by transposable elements. When two elements of the same type (*colored* DNA) happen to insert near each other in a chromosome, the transposition mechanism may occasionally use the ends of two different elements (instead of the two ends of the same element) and thereby move the chromosomal DNA between them to a new chromosomal site. Since introns are very large relative to exons (see Figure 9–7, p. 487), the illustrated insertion of a new exon into a preexisting intron is not an improbable outcome.

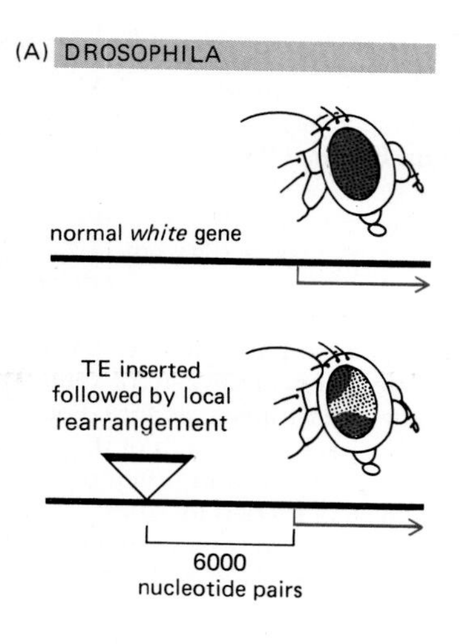

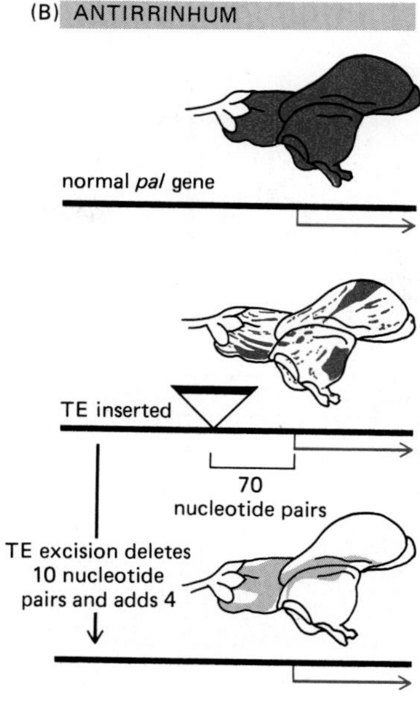

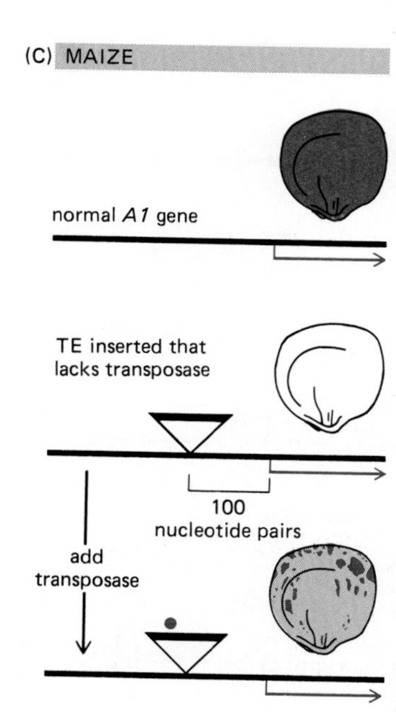

Figure 10–72 Striking changes in gene regulation can be caused by transposable elements. Examples of heritable changes in the pigmentation pattern caused by transposable-element (TE) insertion into the regulatory regions of genes are shown for each of three organisms; similar types of events can cause morphological changes in the organism by affecting cell growth and differentiation. (A) An insertion into an upstream regulatory region of the *white* gene causes red eye pigmentation to appear only near the dorsal and the ventral edges of the fly eye. (B) An insertion into the upstream promoter of a gene needed for pigment production produces a snapdragon flower that lacks red pigment everywhere except in those patches of cells where the element has been excised by a transposition event. A subsequent excision of the element from the genome of the entire plant creates the spatially restricted pattern of pale pigmentation shown in the bottom panel. (C) An example of a regulated change in the pigmentation of the maize kernel that is caused by a transposable element. In this case the transposase serves as a gene regulatory protein that restores some pigmentation to all cells in an otherwise unpigmented kernel. In addition, the transposase catalyzes the occasional excision of the element, thereby producing isolated patches of more darkly colored cells. (A, after G.M. Rubin et al., *Cold Spring Harbor Symp. Quant. Biol.* 50:329–335, 1985; B, after E.S. Coen, R. Carpenter, and C. Martin, *Cell* 47:285–296, 1986; C, after Zs. Schwarz-Sommer et al., *EMBO J.* 6:287–294, 1987.)

lication or DNA repair. One major difference is that the movement of a transposable element often will bring to the vicinity of a gene new sequences that act as binding sites for sequence-specific DNA-binding proteins, including a transposase and the proteins that regulate the transcription of the transposable element DNA. These sequences can thereby act as enhancers and affect the transcription of genes located thousands of nucleotide pairs away. An example of this type of effect on the expression of a pigment gene in maize is illustrated in Figure 10–72. Similar effects commonly contribute to the evolution of cancer cells, where oncogenes can be created by the transposition of such regulatory sequences into the neighborhood of a proto-oncogene (see p. 1208).

The organization of higher eucaryotic genomes, with long noncoding DNA sequences interspersed with comparatively short coding sequences, provides an accommodating "playground" for the integration and excision of mobile DNA sequences. Because gene transcription can be regulated from distances that are tens of thousands of nucleotide pairs away from a promoter (see p. 580), many of the resulting changes in the genome would be expected to affect gene expression; by contrast, relatively few would be expected to disrupt the short exons that contain the coding sequences.

Might the vast excess of noncoding DNA in higher eucaryotes have been favored by selection during evolution because of the regulatory flexibility that it has provided to organisms with a large variety of transposable elements? What is known about the regulatory systems that control higher eucaryotic genes is consistent with this possibility. Enhancers, like exons, seem to function as separate modules, and the activity of a gene depends on a summation of the influences received at its promoter from a set of enhancers (Figure 10–73). Transposable elements, by moving such enhancer modules around in a genome, may allow gene regulation to be optimized for the long-term survival of the organism.

Transposition Bursts Cause Cataclysmic Changes in Genomes and Increase Biological Diversity[62]

Another unique feature that distinguishes transposable elements as mutagens is their tendency to undergo long quiescent periods, during which they remain fixed in their chromosomal positions, followed by a period of intense movement. Their transposition, and therefore their mutagenic action, is activated from time to time in a few individuals in a population of organisms. Such cataclysmic changes in genomes, called **transposition bursts,** can involve near simultaneous transpositions of several types of transposable elements. Transposition bursts were first observed in developing maize plants that were subjected to repeated chromosome

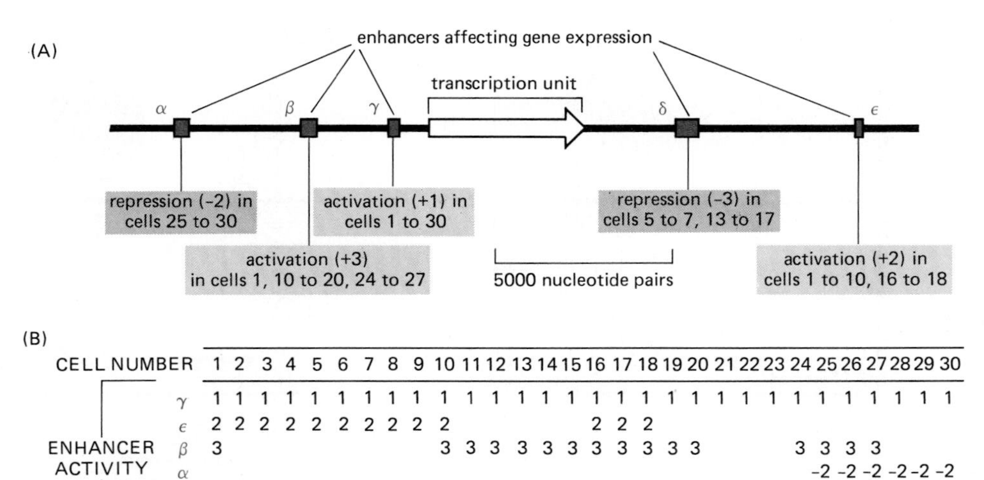

Figure 10–73 How the combined action of separate enhancer modules creates cell-specific patterns of gene expression. Because the mixture of gene regulatory proteins that binds to each enhancer varies from cell to cell, the effect of an enhancer is different in different cells. This example is modeled after results obtained in *Drosophila,* where the many enhancers that control a single gene can be analyzed for their separate effects in transgenic flies. For simplicity we have given each of the stimulatory (+ values) and inhibitory (− values) effects caused by each enhancer (α, β, γ, δ, or ε) a number between +3 and −3 and assumed that these numbers can be arithmetically summed to obtain a total enhancer activity, which determines the level of gene expression.

breakage. They also are observed in crosses between certain strains of flies—a phenomenon known as *hybrid dysgenesis.* When they occur in the germ line, they induce multiple changes in the genome of an individual progeny fly or plant.

By simultaneously changing several properties of an organism, transposition bursts increase the probability that two new traits that are useful together but of no selective value by themselves will appear in a single individual in a population. In several types of plants there is evidence that transposition bursts can be activated by a severe environmental stress, creating a variety of randomly modified progeny organisms, some of which may be better suited than the parent to survive in the new conditions. It seems that, at least in these plants, a mechanism has evolved to activate transposable elements to serve as mutagens that create an enhanced range of variant organisms when this variation is most needed. Thus transposable elements are not necessarily just disruptive parasites; rather, they may on occasion act as useful symbionts that aid the long-term survival of the species whose genomes they inhabit.

10-37 About 10% of the Human Genome Consists of Two Families of Transposable Elements That Appear to Have Multiplied Relatively Recently[63]

Primate DNA is unusual in at least one respect: it contains a remarkably large number of copies of two transposable DNA sequences that seem to have overrun our chromosomes. Both of these sequences move by an RNA-mediated process that requires a reverse transcriptase. One is the **L1 transposable element,** which resembles the F element in *Drosophila* and the cin4 element in maize and is thought to encode a reverse transcriptase (see Table 10–3, p. 605). Transposable elements have generally evolved with feedback control systems that severely limit their numbers in each cell (thereby saving the cell from potential disaster); the L1 element in humans, however, constitutes about 4% of the mass of the genome.

Even more unusual is the ***Alu* sequence,** which is very short (about 300 nucleotide pairs) and moves like a transposable element, creating target-site duplications when it inserts. It was derived, however, from an internally deleted host cell 7SL RNA gene, which encodes the RNA component of the signal-recognition particle (SRP) that functions in protein synthesis (see p. 439); it is therefore not clear whether the *Alu* sequence should be considered as a transposable element or as an unusually mobile pseudogene. It is present in about 500,000 copies in the haploid genome and constitutes about 5% of human DNA; thus it is present on average about once every 5000 nucleotide pairs. The *Alu* DNA is transcribed from the 7SL RNA promoter, a polymerase-III promoter that is internal to the transcript (see p. 526), so that it carries the information necessary for its own transcription wherever it moves. It needs to use a borrowed reverse transcriptase, however, to transpose.

Comparisons of the sequence and locations of the L1- and *Alu*-like sequences in different mammals suggest that these sequences have multiplied to high copy numbers relatively recently (Figure 10–74). It is hard to imagine that these highly abundant sequences scattered throughout our genome have not had major effects on the expression of many nearby genes. How many of our uniquely human qualities, for example, do we owe to these parasitic elements?

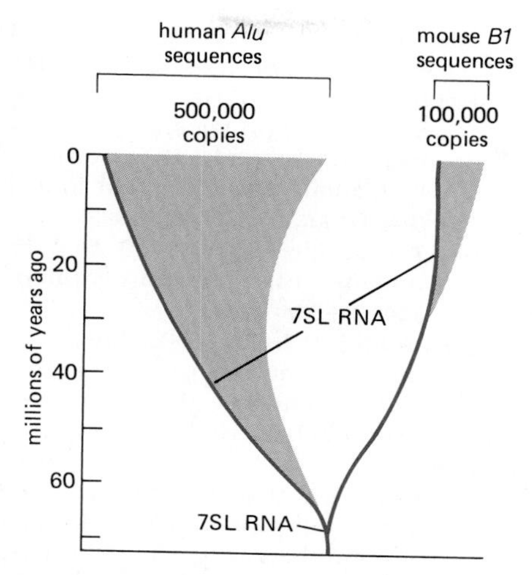

Figure 10–74 The proposed pattern of evolution of the abundant *Alu*-like sequences found in human and mouse genomes. Both of these transposable DNA sequences are thought to have evolved from the essential 7SL RNA gene. Based on the species distribution and sequence homology of these highly repeated elements, however, the major expansion in copy numbers seems to have occurred independently. (Adapted from P.L. Deininger and G.R. Daniels, *Trends in Genetics* 2:76–80, 1986.)

Summary

The functional DNA sequences in the genomes of higher eucaryotes appear to be constructed from small genetic modules of at least two kinds. Modules of coding sequence are combined in myriad ways to produce proteins, whereas modules of regulatory sequences are scattered throughout long stretches of noncoding sequences and regulate the expression of genes. Both the coding sequences (exons) and the regulatory sequences (enhancers) are typically less than a few hundred nucleotide pairs long. A variety of genetic-recombination processes occur in genomes, causing the random duplication and translocation of DNA sequences. Some of these changes create duplicates of entire genes, which can then evolve new functions. Others produce new proteins by shuffling exons or alter the expression of genes by exposing them to new combinations of enhancers. This type of sequence shuffling, which is of great importance for the evolution of organisms, is greatly facilitated by the split structure of higher eucaryotic genes and by the fact that these genes are subject to multiple activating and repressing influences from a combination of distant enhancers.

Many types of transposable elements are present in genomes. Collectively, they constitute more than 10% of the mass of both Drosophila *and vertebrate genomes. Occasionally, "transposition bursts" occur in germ cells and cause many heritable changes in gene expression in the same individual. Transposable elements are thought to have had a special evolutionary role in the generation of organismal diversity.*

References

General

Lewin, B. Genes, 3rd ed. New York: Wiley, 1987.

Schleif, R. Genetics and Molecular Biology. Reading, MA: Addison-Wesley, 1986.

Stent, G.S. Molecular Genetics: An Introductory Narrative. San Francisco: Freeman, 1971.

Watson, J.D.; Hopkins, N.H.; Roberts, J.W.; Steitz, J.A.; Weiner, A.M. Molecular Biology of the Gene, 4th ed. Menlo Park, CA: Benjamin-Cummings, 1987.

Cited

1. Gurdon, J.B. The developmental capacity of nuclei taken from intestinal epithelium cells of feeding tadpoles. *J. Embryol. Exp. Morphol.* 10:622–640, 1962.

   Steward, F.C.; Mapes, M.O.; Mears, K. Growth and organized development of cultured cells. *Am. J. Bot.* 45:705–713, 1958.

2. Garrels, J.I. Changes in protein synthesis during myogenesis in a clonal cell line. *Dev. Biol.* 73:134–152, 1979.

3. Darnell, J.E., Jr. Variety in the level of gene control in eucaryotic cells. *Nature* 297:365–371, 1982.

   Derman, E.; et al. Transcriptional control in the production of liver-specific mRNAs. *Cell* 23:731–739, 1981.

4. Gierer. A. Molecular models and combinatorial priniciples in cell differentiation and morphogenesis. *Cold Spring Harbor Symp. Quant. Biol.* 38:951–961, 1974.

   Scott, M.P.; O'Farrell, P.H. Spatial programming of gene expression in early *Drosophila* embryogenesis. *Annu. Rev. Cell Biol.* 2:49–80, 1986.

5. Maniatis, T.; Goodbourn, S.; Fischer, J.A. Regulation of inducible and tissue-specific gene expression. *Science* 236:1237–1244, 1987.

   Yamamoto, K. Steroid receptor regulated transcription of specific genes and gene networks. *Annu. Rev. Genet.* 19:209–252, 1985.

6. Gehring, W.J.; Hiromi, Y. Homeotic genes and the homeobox. *Annu. Rev. Genet.* 20:147–173, 1986.

7. Blau, H.M.; et al. Plasticity of the differentiated state. *Science* 230:758–766, 1985.

   Davis, R.L.; Weintraub, H.; Lassar, A.B. Expression of a single transfected cDNA converts fibroblasts to myoblasts. *Cell* 51:987–1000, 1987.

8. Miller, J.H.; Reznikoff, W.S., eds. The Operon. Cold Spring Harbor, NY: Cold Spring Harbor Laboratory, 1978.

   Neidhardt, F.C.; et al., eds. *Escherichia coli* and *Salmonella typhimurium*: Cellular and Molecular Biology, Vol. 2, pp. 1439–1526. Washington, DC: American Society for Microbiology, 1987. (Paradigms of operon regulation in bacteria.)

   Ptashne, M. A Genetic Switch. Palo Alto, CA: Blackwell, 1986.

9. Gilbert, W.; Müller-Hill, B. The *lac* operator is DNA. *Proc. Natl. Acad. Sci. USA* 58:2415–2421, 1967.

Gottesman, S. Bacterial regulation: global regulatory networks. *Annu. Rev. Genet.* 18:415–441, 1984.

Jacob, F.; Monod, J. Genetic regulatory mechanisms in the synthesis of proteins. *J. Mol. Biol.* 3:318–356, 1961.

Reznikoff, W.S.; Siegele, D.A.; Cowing, D.W.; Gross, C.A. The regulation of transcription initiation in bacteria. *Annu. Rev. Genet.* 19:355–388, 1985.

10. de Combrugghe, B.; Busby, S.; Buc, H. Cyclic AMP receptor protein: role in transcription activation. *Science* 224:831–838, 1984.

Hochschild, A.; Irwin, N.; Ptashne, M. Repressor structure and the mechanism of positive control. *Cell* 32:319–325, 1983.

Raibaud, O.; Schwartz, M. Positive control of transcription initiation in bacteria. *Annu. Rev. Genet.* 18:173–206, 1984.

11. Keener, J.; Wong, P.; Popham, D.; Wallis, J.; Kustu, S. A sigma factor and auxiliary proteins required for nitrogen-regulated transcription in enteric bacteria. In RNA Polymerase and the Regulation of Transcription. (W.S. Reznikoff, et al, eds.), pp. 159–175. New York: Elsevier, 1987.

Ninfa, A.J.; Reitzer, L.J.; Magasanik, B. Initiation of transcription at the bacterial *glnAp2* promoter by purified *E. coli* components is facilitated by enhancers. *Cell* 50:1039–1046, 1987.

12. Dunn, T.M.; Hahn, S.; Ogden, S.; Schleif, R.F. An operator at −280 base pairs that is required for the repression of araBAD operon promoter. *Proc. Natl. Acad. Sci. USA* 81:5017–5020, 1984.

Griffith, J.; Hochschild, A.; Ptashne, M. DNA loops induced by cooperative binding of lambda repressor. *Nature* 322:750–752, 1986.

Mossing, M.C.; Record, M.T. Upstream operators enhance repression of the *lac* promoter. *Science* 233:889–892, 1986.

13. Helmann, J.D.; Chamberlin, M.J. Structure and function of bacterial sigma factors. *Annu. Rev. Biochem.* 57:839–872, 1988.

14. Davison, B.L.; Egly, J.M.; Mulvihill, E.R.; Chambon, P. Formation of stable preinitiation complexes between eucaryotic class B transcription factors and promoter sequences. *Nature* 301:680–686, 1983.

Sawadogo, M.; Roeder, R.G. Interaction of a gene-specific transcription factor with the adenovirus major late promoter upstream of the TATA box region. *Cell* 43:165–175, 1985.

Workman, J.L.; Roeder, R.G. Binding of transcription factor TFIID to the major late promoter during *in vitro* nucleosome assembly potentiates subsequent initiation by RNA polymerase II. *Cell* 51:613–622, 1987.

15. Atchison, M.L. Enhancers: mechanisms of action and cell specificity. *Annu. Rev. Cell Biol.* 4:127–153, 1988.

Maniatis, T.; Goodbourn, S.; Fischer, J. Regulation of inducible and tissue-specific gene expression. *Science* 236:1237–1245, 1987.

McKnight, S.L.; Kingsbury, R. Transcriptional control signals of a eukaryotic protein-coding gene. *Science* 217:316–324, 1982.

Serfling, E.; Jasin, M.; Schaffner, W. Enhancers and eucaryotic gene transcription. *Trends Genet.* 1:224–230, 1985.

16. Emerson, B.M.; Nickol, J.M.; Jackson, P.D.; Felsenfeld, G. Analysis of the tissue-specific enhancer at the 3′ end of the chicken adult β-globin gene. *Proc. Natl. Acad. Sci. USA* 84:4786–4790, 1987.

Evans, T.; Reitman, M.; Felsenfeld, G. An erythrocyte-specific DNA-binding factor recognizes a regulatory sequence common to all chicken globin genes. *Proc. Natl. Acad. Sci. USA* 85:5976–5980, 1988.

Jones, N.C.; Rigby, P.W.J.; Ziff, E.B. *Trans*-acting protein factors and the regulation of eukaryotic transcription: lessons from studies on DNA tumor viruses. *Genes Dev.* 2:267–281, 1988.

Nomiyama, H.; Fromental, C.; Xiao, J.H.; Chambon, P. Cell-specific activity of the constituent elements of the Simian virus 40 enhancer. *Proc. Natl. Acad. Sci. USA* 84:7881–7885, 1987.

17. Brent. R.; Ptashne, M. A eukaryotic transcriptional activator bearing the DNA specificity of a prokaryotic repressor. *Cell* 43:729–736, 1985.

Evans, R.M. The steroid and thyroid hormone receptor superfamily. *Science* 240:889–895, 1988.

Godowski, P.J.; Picard, D.; Yamamoto, K. Signal transduction and transcriptional regulation by glucocorticoid receptor—lex A fusion proteins. *Science* 241:812–816, 1988.

Kumar, V.; et al. Functional domains of the human estrogen receptor. *Cell* 51:941–951, 1987.

18. Sen, R.; Baltimore, D. Inducibility of kappa immunoglobin enhancer-binding protein NF-kappa B by a posttranslational mechanism. *Cell* 47:921–928, 1986.

Yamamoto, K.K.; Gonzalez, G.A.; Biggs, W.H.; Montminy, M.R. Phosphorylation-induced binding and transcriptional efficacy of nuclear factor CREB. *Nature* 334:494–498, 1988.

Zimarino, V.; Wu, C. Induction of sequence-specific binding of *Drosophila* heat shock activator protein without protein synthesis *Nature* 327:727–730, 1987.

19. Metzger, D.; White, J.H.; Chambon, P. The human estrogen receptor functions in yeast. *Nature* 334:31–36, 1988.

Ptashne, M. Gene regulation by proteins acting nearby and at a distance. *Nature* 322:697–701, 1986.

Struhl, K. Promoters, activator proteins, and the mechanism of transcriptional initiation in yeast. *Cell* 49:295–297, 1987.

20. Borst, P.; Greaves, D.R. Programmed gene rearrangements altering gene expression. *Science* 235:658–667, 1987.

Meyer, T.F. Molecular basis of surface antigen variation in *Neisseria. Trends Genet.* 3:319–324, 1987.

Simon, M.; Zieg, J.; Silverman, M.; Mandel, G.; Doolittle, R. Phase variation: evolution of a controlling element. *Science* 209:1370–1374, 1980.

21. Cross, F.; Hartwell, L.H.; Jackson, C.; Konopka, J.B. Conjugation in *Saccharomyces cerevisiae. Annu. Rev. Cell Biol.* 4:429–457, 1988.

Herskowitz, I. Master regulatory loci in yeast and lambda. *Cold Spring Harbor Symp. Quant. Biol.* 50:565–574, 1985.

Kushner, P.J.; Blair, L.C.; Herskowitz, I. Control of yeast cell types by mobile genes: a test. *Proc. Natl. Acad. Sci. USA* 76:5264–5268, 1979.

22. Kostriken, R.; Strathern, J.N.; Klar, A.; Hicks, J.B.; Heffron, F. A site-specific endonuclease essential for mating-type switching in *Saccharomyces cerevisiae. Cell* 35:167–174, 1983.

23. Nasmyth, K.; Shore, D. Transcriptional regulation in the yeast life cycle. *Science* 237:1162–1170, 1987.

24. Brand, A.H.; Breeden, L.; Abraham, J.; Sternglanz, R.; Nasmyth, K. Characterization of a "silencer" in yeast: a DNA sequence with properties opposite to those of a transcriptional enhancer. *Cell* 41:41–48, 1985.

25. Friedman, D.I.; et al. Interactions of bacteriophage and host macromolecules in the growth of bacteriophage lambda. *Microbiol. Rev.* 48:299–325, 1984.

Ptashne, M.; et al. How the lambda repressor and cro work. *Cell* 19:1–11, 1980.

26. Brown, D.D. The role of stable complexes that repress and activate eucaryotic genes. *Cell* 37:359–365, 1984.

Weintraub, H. Assembly and propagation of repressed and derepressed chromsomal states. *Cell* 42:705–711, 1985.

27. Brown, S.W. Heterochromatin. *Science* 151:417–425, 1966.

Hsu, T. C.; Cooper, J.E.K., Mace, M.L., Brinkley, B.R. Arrangement of centromeres in mouse cells. *Chromosoma* 34:73–87, 1971.

28. Gartler, S.M.; Riggs, A.D. Mammalian X-chromosome inactivation. *Annu. Rev. Genet.* 17:155–190, 1983.

Lock, L.F.; Takagi, N.; Martin G.R. Methylation of the *Hprt* gene on the inactive X occurs after chromosome inactivation. *Cell* 48:39–46, 1987.

Lyon, M.F. X-chromosome inactivation and developmental patterns in mammals. *Biol. Rev.* 47:1–35, 1972.

29. Baker, W.K. Position-effect variegation. *Adv. Genet.* 14:133–169, 1968.

Spofford, J.B. Position-effect variegation in *Drosophila*. In The Genetics and Biology of *Drosophila* (M. Ashburner, E. Novitski, eds.), Vol. 1C, pp. 955–1018. New York: Academic Press, 1976.

30. Goldberg, D.A.; Posakony, J.W.; Maniatis, T. Correct developmental expression of a cloned alcohol dehydrogenase gene transduced into the *Drosophila* germ line. *Cell* 34:59–73, 1983.

Grosveld, F.; van Assendelft, G.B.; Greaves, D.R.; Kollias, G. Position-independent, high-level expression of the human β-globin gene in transgenic mice. *Cell* 51:975–985, 1987.

Meyerowitz, E.M.; Raghavan, K.V.; Mathers, P.H.; Roark, M. How *Drosophila* larvae make glue: control of *Sgs-3* gene expression. *Trends Genet.* 3:288–293, 1987.

Palmiter, R.D.; Brinster, R.L. Germ-line transformation of mice. *Annu. Rev. Genet.* 20:465–499, 1986.

31. Ephrussi, A.; Church, G.M.; Tonegawa, S.; Gilbert, W. B lineage-specific interactions of an immunolglobulin enhancer with celluar factors *in vivo*. *Science* 227:134–140, 1985.

Garel, A.; Zolan, M.; Axel, R. Genes transcribed at diverse rates have a similar conformation in chromatin. *Proc. Natl. Acad. Sci. USA* 74:4867–4871, 1977.

Karlsson, S.; Nienhuis, A.W. Developmental regulation of human globin genes. *Annu. Rev. Biochem.* 54:1071–1108, 1985.

Weintraub, H.; Groudine, M. Chromosomal subunits in active genes have an altered conformation. *Science* 193:848–856, 1976.

32. Brill, S.J.; Sternglanz, R. Transcription-dependent DNA supercoiling in yeast DNA topoisomerase mutants. *Cell* 54:403–411, 1988.

Wang, J.C. Superhelical DNA. *Trends Biochem. Sci.* 5:219–221, 1980.

Wang, J.C.; Giaever, G.N. Action at a distance along a DNA. *Science* 240:300–304, 1988.

33. Guarente, L. Regulatory proteins in yeast. *Annu. Rev. Genet.* 21:425–452, 1987.

34. Razin, A.; Cedar, H.; Riggs, A.D., eds. DNA Methylation: Biochemistry and Biological Significance. New York: Springer-Verlag, 1984.

35. Cedar, H. DNA methylation and gene activity. *Cell* 53:3–4, 1988.

Ivarie, R.D.; Schacter, B.S.; O'Farrell, P.H. The level of expression of the rat growth hormone gene in liver tumor cells is at least eight orders of magnitude less than that in anterior pituitary cells. *Mol. Cell. Biol.* 3:1460–1467, 1983.

Yisraeli, J.; et al. Muscle-specific activation of a methylated chimeric actin gene. *Cell* 46:409–416, 1986.

36. Bird, A.P. CpG islands as gene markers in the vertebrate nucleus. *Trends Genet.* 3:342–347, 1987.

37. Duncan, I. The bithorax complex. *Annu. Rev. Genet.* 21:285–319, 1987.

Peifer, M.; Karchi, F.; Bender, W. The bithorax complex: control of segmental identity. *Genes Dev.* 1:891–898, 1987.

38. Landrick, R.; Yanofsky, C. Transcription attenuation. In *Escherichia coli* and *Salmonella typhimurium*: Cellular and Molecular Biology (F.C. Neidhardt; et al, eds.), Vol. 2, pp. 1276–1301, Washington, DC: American Society for Microbiology, 1987.

Platt, T. Transcription termination and the regulation of gene expression. *Annu. Rev. Biochem.* 55:339–372, 1986.

Yanofsky, C. Operon-specific control by transcription attenuation. *Trends Genet.* 3:356–360, 1987.

39. Andreadis, A.; Gallego, M.E.; Nadal-Ginard, B. Generation of protein isoform diversity by alternative splicing: mechanistic and biological implications. *Annu. Rev. Cell Biol.* 3:207–242, 1987.

Leff, S.; Rosenfeld, M.; Evans, R. Complex transcriptional units: diversity in gene expression by alternative RNA processing. *Annu. Rev. Biochem.* 55:1091–1117, 1986.

Schwarz, T.L.; Tempel, B.L.; Papazian, D.M.; Jan, Y.N.; Jan, L.Y. Multiple potassium-channel components are produced by alternative splicing at the *Shaker* locus in *Drosophila*. *Nature* 331:137–142, 1988.

40. Baker, B.S.; Belote, J.M. Sex determination and dosage compensation in *Drosophila melanogaster*. *Annu. Rev. Genet.* 17:345–393, 1983.

Bingham, P.M.; Chou, T.; Mims, I.; Zachar, Z. On/off regulation of gene expression at the level of splicing. *Trends Genet.* 4:134–138, 1988.

Boggs, R.T.; Gregor, P.; Idriss, S.; Belote, J.M.; McKeown, M. Regulation of sexual differentiation in *D. melanogaster* via alternative splicing of RNA from the transformer gene. *Cell* 50:739–747, 1987.

Laski, F.A.; Rio, D.C.; Rubin, G.M. Tissue specificity of *Drosophila* P element transposition is regulated at the level of RNA splicing. *Cell 44:7–19, 1986.*

41. Early, P.; et al. Two mRNAs can be produced from a single immunoglobulin μ gene by alternative RNA processing pathways. *Cell* 20:313–319, 1980

Peterson, M.L.; Perry, R.P. Regulated production of μ_m and μ_s mRNA requires linkage of the poly(A) addition sites and is dependent on the length of the μ_m-μ_s intron. *Proc. Natl. Acad. Sci. USA* 83:8883–8887, 1986.

42. Beadle, G. Genes and the chemistry of the organism. *Am. Sci.* 34:31–53, 1946.

43. Newport, J.W.; Forbes, D.J. The nucleus: structure, function, and dynamics. *Annu. Rev. Biochem.* 56:535–565, 1987.

Schneider, R.J.; Shenk, T. Impact of virus infection on host cell protein synthesis. *Annu. Rev. Biochem.* 56:317–332, 1987.

44. Aziz, N.; Munro, H.N. Iron regulates ferritin mRNA translation through a segment of its 5' untranslated region. *Proc. Natl. Acad. Sci. USA* 84:8478–8482, 1987.

Gold, L. Posttranscriptional regulatory mechanisms in *Escherichia coli*. *Annu. Rev. Biochem.* 57:199–234, 1988.

Nomura, M.; Gourse, R.; Baughman, G. Regulation of the synthesis of ribosomes and ribosomal components. *Annu. Rev. Biochem.* 53:75–117, 1984.

Walden, W.E.; et al. Translational repression in eukaryotes: partial purification and characterization of a repressor of ferritin in RNA translation. *Proc. Natl. Acad. Sci. USA* 85:9503–9507, 1988.

45. Hunt, T. False starts in translational control of gene expression. *Nature* 316:580–581, 1985.

Kozak, M. Bifunctional messenger RNAs in eucaryotes. *Cell* 47:481–483, 1986.

Pelletier, J.; Sonenberg, N. Internal initiation of translation of eukaryotic mRNA directed by a sequence derived from poliovirus RNA. *Nature* 334:320–325, 1988.

46. Ilan, J., ed. Translational Regulation of Gene Expression. New York: Plenum Press, 1987.

Rosenthal, E.T.; Hunt, T.; Ruderman, J.V. Selective translation of mRNA controls the pattern of protein synthesis during early development of the surf clam, *Spisula solidissima*. *Cell* 20:487–494, 1980.

Walden, W.E.; Thach, R.E. Translational control of gene expression in a normal fibroblast: characterization of a subclass of mRNAs with unusual kinetic properties. *Biochemistry* 25:2033–2041, 1986.

47. Craigen, W.J.; Caskey, C.T. Translational frameshifting: where will it stop? *Cell* 50:1–2, 1987.

48. Casey, J.L.; et al. Iron-responsive elements: regulatory RNA

sequences that control mRNA levels and translation. *Science* 240:924–928, 1988.
Raghow, R. Regulation of messenger RNA turnover in eukaryotes. *Trends Biochem. Sci.* 12:358–360, 1987.
Shaw, G.; Kamen, R. A conserved AU sequence from the 3' untranslated region of GM-CSF mRNA mediaters selective mRNA degradation. *Cell* 46:659–667, 1986.
49. Graves, R.A.; Pandey, N.B.; Chodchoy, N.; Marzluff, W.F. Translation is required for regulation of histone mRNA degradation. *Cell* 48:615–626, 1987.
Marzluff, W.F.; Pandey, N.B. Multiple regulatory steps control histone mRNA concentrations. *Trends Biochem. Sci.* 13:49–52, 1988.
Mowry, K.L.; Steitz, J.A. Identification of the human U7 snRNP as one of several factors involved in the 3' end maturation of histone premessenger RNAs. *Science* 238:1682–1687, 1987.
50. Driever, W.; Nüsslein-Volhard, C. A gradient of bicoid protein in *Drosophila* embryos. *Cell* 54:83–93, 1988.
Lawrence, J.B.; Singer, R.H. Intracellular localization of messenger RNAs for cytoskeletal proteins. *Cell* 45:407–415, 1986.
Weeks, D.L.; Melton, D.A. A maternal mRNA localized to the vegetal hemisphere in *Xenopus* eggs codes for a growth factor related to TGF-beta. *Cell* 51:861–867, 1987.
51. Borst, P. Discontinous transcription and antigenic variation in trypanosomes. *Annu. Rev. Biochem.* 55:701–732, 1986.
Eisen, H. RNA editing: who's on first? *Cell* 53:331–332, 1988.
Powell, L.M.; et al. A novel form of tissue-specific RNA processing produces apolipoprotein-B48 in intestine. *Cell* 50:831–840, 1987.
Sharp, P.A. *Trans* splicing: variation on a familiar theme. *Cell* 50:147–148, 1987.
52. McClain, W.H.; Guerrier-Takada, C.; Altman, S. Model substrates for an RNA enzyme. *Science* 238:527–530, 1987.
Pines, O.; Inouye, M. Antisense RNA regulation in prokaryotes. *Trends Genet.* 2:284–287, 1986.
Tomizawa, J. Control of ColE1 plasmid replication: binding of RNA I to RNA II and inhibition of primer formation. *Cell* 47:89–97, 1986.
Wu, H.N.; Uhlenbeck, O.C. Role of a bulged A residue in a specific RNA-protein interaction. *Biochemistry* 26:8221–8227, 1987.
53. Clarke, B.C.; Robertson, A.; Jeffreys, A.J., eds. The Evolution of DNA Sequences. London: The Royal Society, 1986.
Nei, M.; Koehn, R.K., eds. Evolution of Genes and Proteins. Sunderland, MA: Sinauer, 1983.
54. Doolittle, R.F. Proteins. *Sci. Am.* 253(4):88–99, 1985.
Holland, S.K.; Blake, C.C. Proteins, exons, and molecular evolution. *Biosystems* 20:181–206, 1987.
Maeda, N.; Smithies; O. The evolution of multigene families: human haptoglobin genes. *Annu. Rev. Genet.* 20:81–108, 1986.
55. Kourilsky, P. Molecular mechanisms for gene conversion in higher cells. *Trends Genet.* 2:60–63, 1986.
Roth, D.B.; Porter, T.N.; Wilson, J.H. Mechanisms of nonhomologous recombination in mammalian cells. *Mol. Cell. Biol.* 5:2599–2607, 1985.
Smith, G.P. Evolution of repeated DNA sequences by unequal crossovers. *Science* 191:528–535, 1976.
Stark, G.R.; Wahl, G.M. Gene amplification. *Annu. Rev. Biochem.* 53: 447–491, 1984.
56. Dickerson, R.E.; Geis, I. Hemoglobin: Structure, Function, Evolution, and Pathology. Menlo Park, CA: Benjamin-Cummings, 1983.
Efstratiadis, A.; et al. The structure and evolution of the human β-globin gene family. *Cell* 21:653–668, 1980.
Vollrath, D.; Nathans, J.; Davis, R.W. Tandem array of human visual pigment genes at Xq28. *Science* 240:1669–1672, 1988.
57. Doolittle, W.F. RNA mediated gene conversion? *Trends Genet.* 1:64–65, 1985.
Gilbert, W.; Marchionni, M.; McKnight, G. On the antiquity of introns. *Cell* 46:151–153, 1986.
Sharp, P. On the origin of RNA splicing and introns. *Cell* 42:397–400, 1985.
58. Britten, R.J.; Kohne, D.E. Repeated sequences in DNA. *Science* 161:529–540, 1968.
Jelinek, W.R.; Schmid, C.W. Repetitive sequences in eukaryotic DNA and their expression. *Annu. Rev. Biochem.* 51:813–844, 1982.
59. Craig-Holmes, A.P.; Shaw, M.W. Polymorphism of human constitutive heterochromatin. *Science* 174:702–704, 1971.
Hsu, T.C. Human and Mammalian Cytogenetics: A Historical Perspective. New York: Springer-Verlag, 1979.
John B.; Miklos, G.L.G. Functional aspects of satellite DNA and heterochromatin. *Int. Rev. Cytol.* 58:1–114, 1979.
Orgel L.E.; Crick, F.H.C. Selfish DNA: the ultimate parasite. *Nature* 284:604–607, 1980.
60. Berg, D.E.; Howe, M.M.; eds. Mobile DNA. Washington, DC: American Society for Microbiology, 1989.
Döring, H.-P.; Starlinger, P. Molecular genetics of transposable elements in plants. *Annu. Rev. Genet.* 20:175–200, 1986.
Finnegan, D.J. Transposable elements in eukaryotes. *Int. Rev. Cytol.* 93:281–326, 1985.
McClintock, B. Controlling elements and the gene. *Cold Spring Harbor Symp. Quant. Biol.* 21:197–216, 1956.
61. Coen, E.S.; Carpenter, R. Transposable elements in *Antirrhinum majus*: generators of genetic diversity. *Trends Genet.* 2:292–296, 1986.
Georgiev, G.P. Mobile genetic elements in animal cells and their biological significance. *Eur. J. Biochem.* 145:203–220, 1984.
O'Kane, C.J.; Gehring, W. Detection *in situ* of genomic regulatory elements in *Drosophila. Proc. Natl. Acad. Sci. USA* 84:9123–9127, 1987.
Hiromi, Y.; Gehring, W.J. Regulation and function of the Drosophila segmentation gene *fushi tarazu*. Cell 50:963–974, 1987.
62. Gerasimova, T.I.; Mizrokhi, L.J.; Georgiev, G.P. Transposition bursts in genetically unstable. *Drosophila. Nature* 309:714–716, 1984.
McClintock, B. The significance of responses of the genome to challenge. *Science* 226:792–801, 1984.
Walbot, V.; Cullis, C.A. Rapid genomic change in higher plants. *Annu. Rev. Plant Physiol.* 36:367–396, 1985.
63. Deininger, P.L.; Daniels, G.R. The recent evolution of mammalian repetitive DNA elements. *Trends Genet.* 2:76–80, 1986.
Ruffner, D.E.; Sprung, C.N.; Minghetti, P.P.; Gibbs, P.E.; Dugaiczyk, A. Invasion of the human albumin-α-fetoprotein gene family by Alu, Kpn, and two novel repetitive DNA elements. *Mol. Biol. Evol.* 4:1–9, 1987.
Weiner, A.M.; Deininger, P.L.; Efstratiadis, A. Nonviral retroposons: genes, pseudogenes, and transposable elements generated by the reverse flow of genetic information. *Annu. Rev. Biochem.* 55:631–661, 1986.

The Cytoskeleton

11

The ability of eucaryotic cells to adopt a variety of shapes and to carry out coordinated and directed movements depends on the **cytoskeleton,** a complex network of protein filaments that extends throughout the cytoplasm. The cytoskeleton might equally well be called the "cytomusculature," because it is directly responsible for such movements as the crawling of cells on a substratum, muscle contraction, and the many changes in shape of a developing vertebrate embryo; it also provides the machinery for actively moving organelles from one place to another in the cytoplasm. Since the cytoskeleton is apparently absent from bacteria, it may have been a crucial factor in the evolution of eucaryotic cells.

The diverse activities of the cytoskeleton depend on just three principal types of protein filaments: *actin filaments, microtubules,* and *intermediate filaments.* Each type of filament is formed from a different protein monomer and can be built into a variety of structures according to its associated proteins. Some of the associated proteins link filaments to one another or to other cell components, such as the plasma membrane. Others control where and when actin filaments and microtubules are assembled in the cell by regulating the rate and extent of their polymerization. Yet other associated proteins interact with filaments to produce movements, the two best-understood examples being muscle contraction, which depends on actin filaments, and the beating of cilia, which depends on microtubules.

We begin this chapter by considering structures built from actin filaments, proceeding from the specialized *myofibril* in a muscle cell to the ubiquitous actin-rich *cortex* beneath the plasma membrane of all animal cells. We then examine microtubules, proceeding from those in organized bundles that are responsible for ciliary beating to those that extend throughout the cytoplasm of cells and control organelle movements and define cell polarity. After discussing the diverse family of intermediate filaments, which give the cell tensile strength and form the nuclear lamina, we finally consider how the cytoskeleton functions as an integrated network to control and coordinate the movements and shapes of single cells and tissues.

Muscle Contraction[1]

Many of the protein molecules that constitute the actin-based structures common to all cells were first discovered in muscle, and muscle contraction is the most familiar and the best understood of all the kinds of movement of which animals are capable. In vertebrates, for example, running, walking, swimming, and flying all depend on the ability of *skeletal muscle* to contract rapidly on its scaffolding

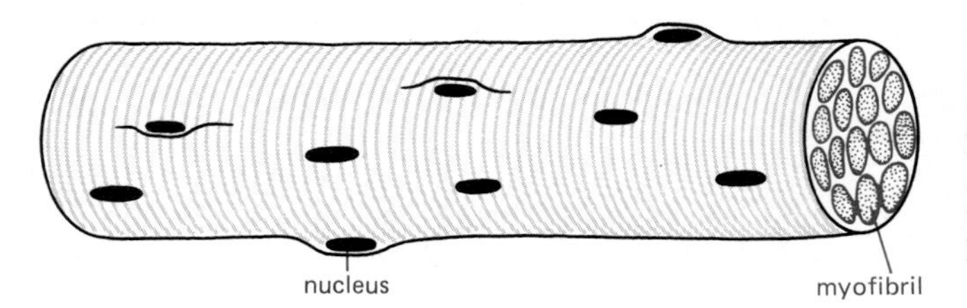

Figure 11–1 Schematic drawing of a short section of a skeletal muscle cell (also called a muscle fiber). In an adult human these huge multinucleated cells are typically 50 μm in diameter, and they can be up to 500,000 μm (500 mm) long.

of bone, while involuntary movements, such as heart pumping and gut peristalsis, depend on the contraction of *cardiac* and *smooth muscle*, respectively.

Muscle contraction is mediated by a sophisticated and powerful intracellular protein apparatus that is present in a more rudimentary form in almost all eucaryotic cells. During the evolution of muscle cells, parts of the cytoskeleton became greatly amplified and specialized to make the contractile machinery in muscle unusually stable and efficient. In *striated muscles* (including skeletal muscle, cardiac muscle, and comparable invertebrate tissues such as insect flight muscle) this machinery is so highly organized that its contraction can be visualized directly, revealing some important properties of its molecules.

Myofibrils Are the Contractile Elements of a Skeletal Muscle Cell

The long thin *muscle fibers* of skeletal muscle are huge single cells formed during development by the fusion of many separate cells (see p. 983). The nuclei of the contributing cells are retained in this large cell and lie just beneath the plasma membrane. But the bulk of the cytoplasm (about two-thirds of its dry mass) is made up of *myofibrils*—cylindrical elements 1 to 2 μm in diameter, which are often as long as the muscle cell itself (Figure 11–1). Isolated myofibrils have a series of prominent bands along their length that are responsible for the striated appearance of skeletal muscle cells. If ATP and Ca^{2+} are added to isolated myofibrils, the myofibrils instantly contract, indicating that they are the force generators in muscle cells. Each myofibril consists of a chain of tiny contractile units composed of miniature, precisely arranged assemblies of thick and thin filaments.

Myofibrils Are Composed of Repeating Assemblies of Thick and Thin Filaments

Each of the regular repeating units, or *sarcomeres*, that give the vertebrate myofibril its striated appearance is about 2.5 μm long. At high magnification a series of broad light and dark bands can be seen in each sarcomere; a dense line in the center of each light band separates one sarcomere from the next and is known as the Z line or *Z disc* (Figure 11–2).

Figure 11–2 (A) Low-magnification electron micrograph of a longitudinal section through a skeletal muscle cell of a rabbit, showing the regular pattern of cross-striations. The cell contains many myofibrils aligned in parallel (see Figure 11–1). (B) Detail of the skeletal muscle cell shown in (A), showing portions of two adjacent myofibrils and the definition of a sarcomere. (C) Schematic diagram of a single sarcomere, showing the origin of the dark and light bands seen in the electron micrographs. The dark bands are sometimes referred to as *A bands* because they appear anisotropic in polarized light (that is, their refractive index changes with the plane of polarization). The light bands are relatively *i*sotropic in polarized light and are sometimes called *I bands*. (A and B, courtesy of Roger Craig.)

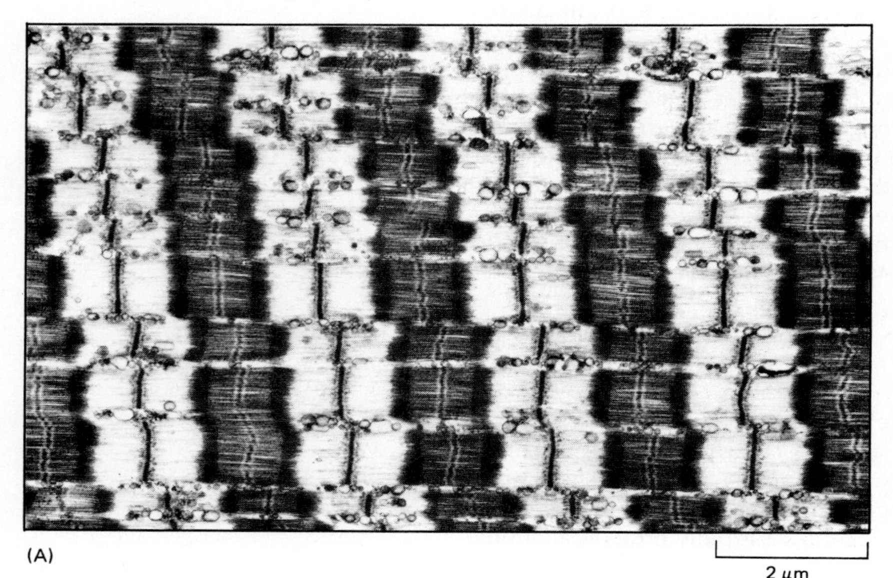

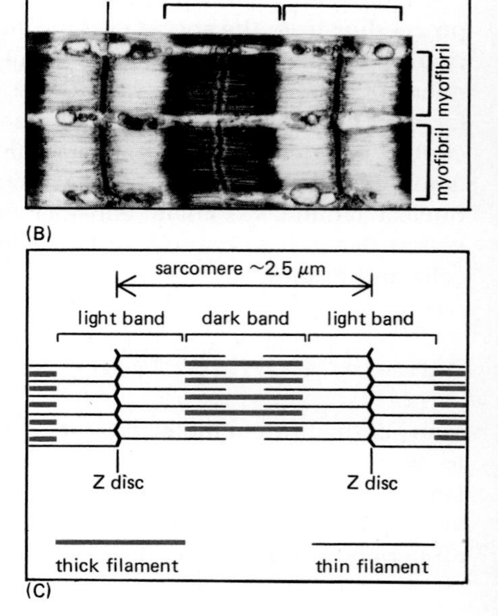

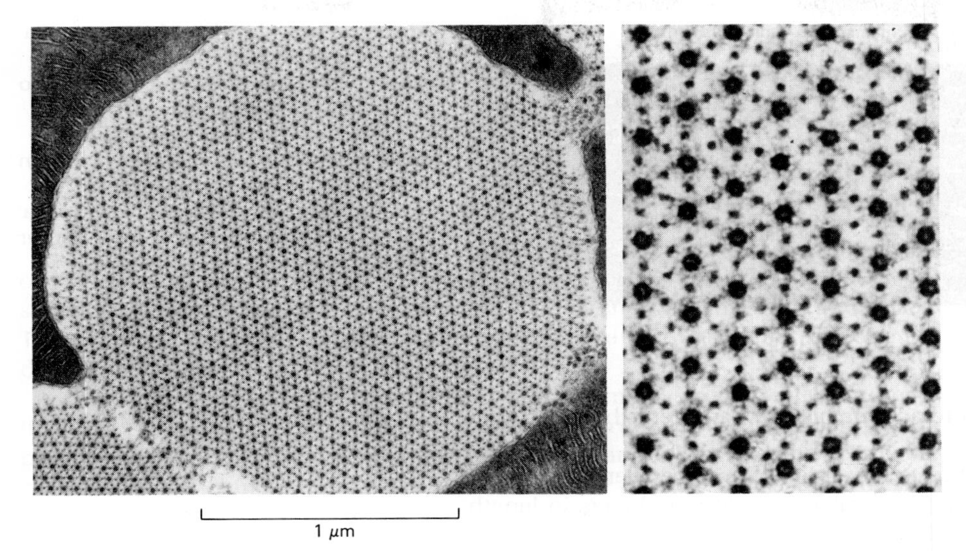

Figure 11–3 Electron micrographs of an insect flight muscle viewed in cross-section, showing how the thick and thin filaments are packed together with crystalline regularity. Unlike their vertebrate counterparts, these thick filaments have a hollow center, as seen in the enlargement on the right. The geometry of the hexagonal lattice is slightly different in vertebrate muscle. (From J. Auber, *J. de Microsc.* 8:197–232, 1969.)

The molecular basis of the cross-striations, and a strong clue to their functional significance, was revealed in 1953 in one of the first applications of the electron microscope to biological material. Each sarcomere was found to contain two sets of parallel and partly overlapping filaments: *thick filaments*, extending from one end of the dark band to the other, and *thin filaments*, extending across each light band and partway into the two neighboring dark bands (Figure 11–2C). When the region of the dark band where thick and thin filaments overlap was viewed in cross-section, the thick filaments were seen to be arranged in a regular hexagonal lattice, with the thin filaments placed regularly between them (Figure 11–3).

11-3 Contraction Occurs as the Thick and Thin Filaments Slide Past Each Other[2]

If a source of monochromatic light is directed through a living muscle cell, a series of interference fringes is produced that provides a sensitive measure of sarcomere spacing. Such measurements reveal that each sarcomere shortens proportionately as the muscle contracts: if a myofibril containing a chain of 20,000 sarcomeres contracts from 5 cm to 4 cm (that is, by 20%), the length of each sarcomere decreases correspondingly from 2.5 to 2.0 μm.

When a sarcomere shortens, only the light band decreases in length; the dark band remains unchanged. This is easily explained if the contraction is caused by thick filaments sliding past the thin filaments with no change in the length of either type of filament (Figure 11–4). This *sliding filament model*, first proposed in 1954, was crucial to understanding the contractile mechanism. In particular, it

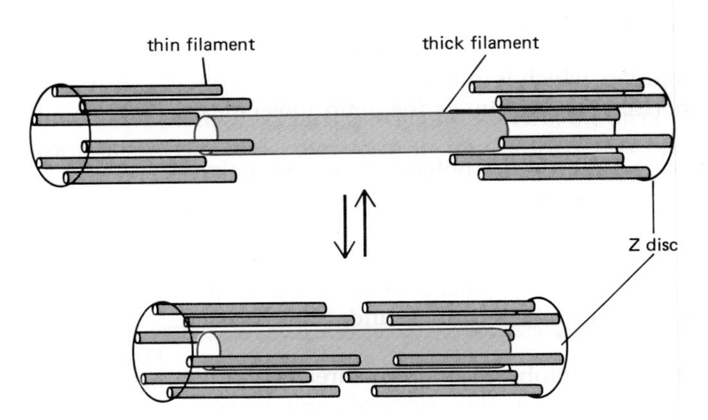

Figure 11–4 The sliding filament model of muscle contraction, in which the thin and thick filaments slide past one another without shortening.

ionic interactions between the tails of the individual molecules. This is why solutions of high salt concentration, which disrupt ionic interactions but do not affect hydrophobic interactions, release individual myosin molecules from muscle. As the salt concentration is reduced to physiological ionic strength, the tails of the myosin molecules associate to form large filaments that may closely resemble muscle thick filaments. In muscle cells these interactions are stabilized by various accessory proteins, and the thick filaments that form are composed of hundreds of myosin tails packed together in a regular staggered array from which the myosin heads project in a repeating pattern (Figure 11–12). The structure is bipolar, with a bare central region where two oppositely oriented sets of myosin tails come together. The globular heads of the myosin molecules interact with actin, forming the cross-bridges between the thick and thin filaments.

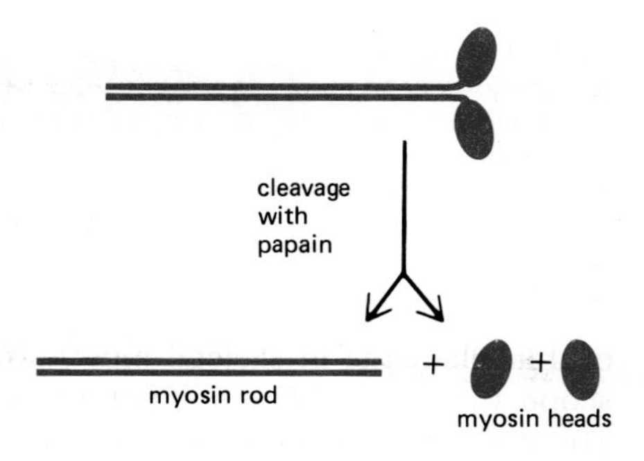

Figure 11–10 Limited digestion with the proteolytic enzyme papain cleaves the myosin molecule into a rod and two heads.

11-5 ATP Hydrolysis Drives Muscle Contraction[4]

Skeletal muscle converts chemical energy into mechanical work with very high efficiency—only 30% to 50% of the energy is wasted as heat. (An automobile engine, by contrast, typically wastes 80% to 90% of the energy available from gasoline.)

The energy for muscle contraction comes from ATP hydrolysis. Yet one detects no major difference in ATP levels between a resting muscle and one that is actively contracting, because a muscle cell has a very efficient backup system for regenerating ATP. The enzyme *phosphocreatine kinase* catalyses a reaction between an even more reactive phosphate compound, *phosphocreatine* (Figure 11–13), and ADP to form creatine and ATP. It is the intracellular level of phosphocreatine that drops after a short burst of muscle activity, even though the contractile machinery itself consumes ATP. The pool of phosphocreatine acts like a battery—storing ATP energy and recharging itself from the new ATP that is generated by cellular oxidations when the muscle is resting.

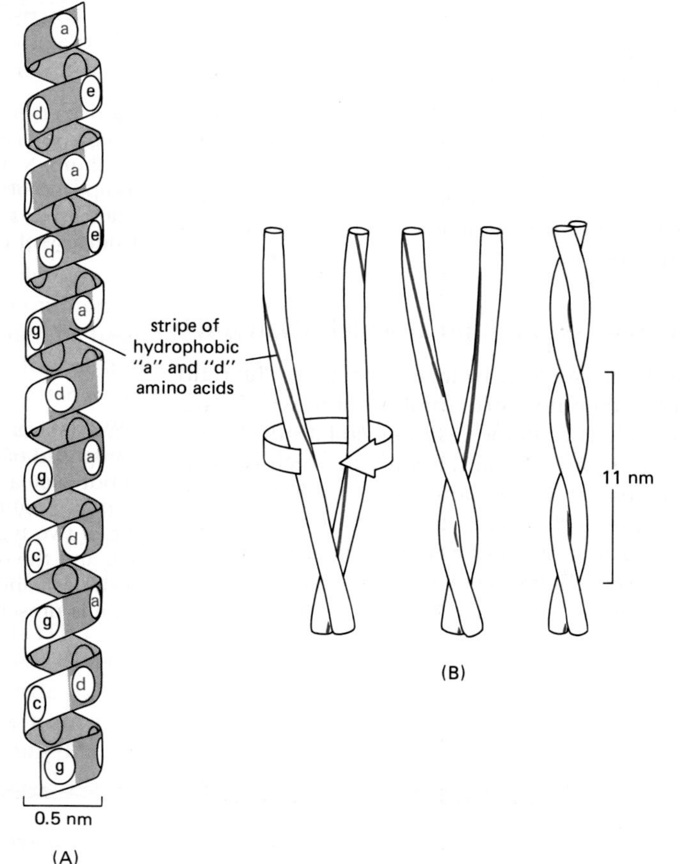

Figure 11–11 Topology of a coiled coil. In (A) a single α helix is represented as a cylinder with successive amino acid side chains labeled in a sevenfold sequence "abcdefg" (from bottom to top). Amino acids "a" and "d" in such a sequence lie close together on the cylinder surface, forming a "stripe" that winds slowly around the α helix (*shaded in color*). Proteins that form coiled coils typically have hydrophobic amino acids at positions "a" and "d." Consequently, as shown in (B), the two α helices can wrap around each other with the hydrophobic side chains of one α helix intercalated in the spaces between the hydrophobic side chains of the other, while the more hydrophilic amino acid side chains are left exposed to the aqueous environment.

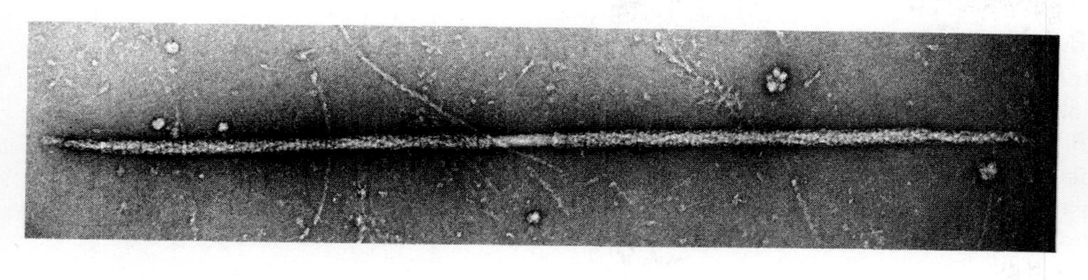

(A)

500 nm

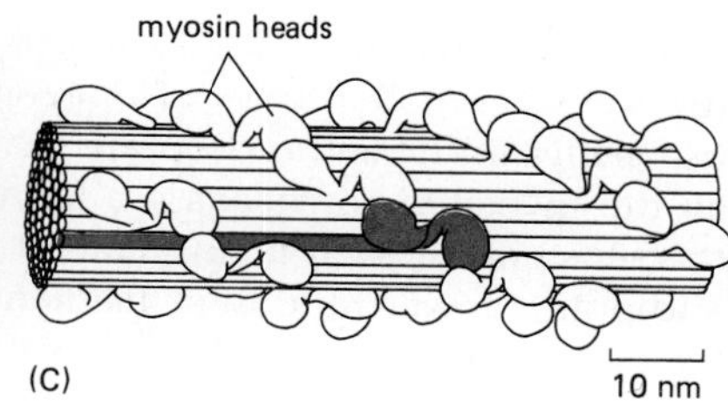

bare zone

myosin heads

(B)

Figure 11–12 The myosin thick filament. (A) Electron micrograph of a myosin thick filament isolated from scallop muscle. Note the central bare zone. (B) Schematic diagram, not drawn to scale. The myosin molecules aggregate together by means of their tail regions, with their heads projecting to the outside. The bare zone in the center of the filament consists entirely of myosin tails. (C) A small section of a thick filament as reconstructed from electron micrographs. An individual myosin molecule is highlighted in color. (A, courtesy of Roger Craig; C, based on R.A. Crowther, R. Padron, and R. Craig, *J. Mol. Biol.* 184:429–439, 1985.)

Myosin Is an Actin-activated ATPase[5]

ATP hydrolysis during muscle contraction is a direct consequence of the interaction between myosin and actin. Even on its own, myosin acts as an ATPase. But purified myosin by itself works relatively slowly. Each molecule takes about 30 seconds to complete a cycle in which an ATP molecule is hydrolyzed to completion. The rate-limiting step is not the initial binding of ATP to the myosin nor the hydrolysis of its terminal phosphate group—both of which occur rapidly—but the release of the products of ATP hydrolysis (ADP and inorganic phosphate, P_i), which remain noncovalently bound to the myosin molecule and prevent further ATP binding and hydrolysis.

In the presence of actin filaments, the ATPase activity of myosin is greatly stimulated. Each myosin molecule now hydrolyzes 5 to 10 molecules of ATP every second, which is comparable to the rates measured in contracting muscle. The stimulation of myosin ATPase by actin filaments reflects a physical association between the two that is at the heart of muscle contraction. The binding of myosin to an actin filament causes a rapid release of ADP and P_i from the myosin molecule, which is thus freed to bind another molecule of ATP and start the cycle again.

Myosin Heads Bind to Actin Filaments[6]

It is the globular head of the myosin molecule that both binds to actin filaments and hydrolyzes ATP. Isolated myosin heads, which can be obtained separately after papain digestion (see Figure 11–10), retain both the ATPase activity and the actin-filament-binding properties of the intact myosin molecule and therefore can be used to analyze the interaction between actin and myosin.

Each actin molecule in an actin filament is capable of binding one myosin head to form a complex that reveals the structural polarity of the actin filament. With negative staining, such complexes can be seen in the electron microscope to have a regular and distinctive form: each myosin head forms a lateral projection, and the superimposed image of many such projections gives the appearance of arrowheads along the actin filament (Figure 11–14). Because the heads bind with the same orientation to each actin subunit, all of the actin molecules face in the same direction along the filament. The actin filament therefore has two structurally distinct ends, which are called the **minus end** (also called the *pointed end* because it corresponds to the point of the arrow after decoration) and the **plus end** (also called the *barbed end* because it corresponds to the barb of the arrow). The terminology "plus" and "minus" derives from the observation that the two ends of an actin filament elongate at different rates *in vitro* (see p. 637).

As shown in Figure 11–12, myosin heads face in opposite directions on either side of the bare central region of a thick filament. Since the heads must interact

$$^{-}O-P(=O)(O^{-})-NH-C(=NH)-N(CH_3)-CH_2-COO^{-}$$

phosphocreatine

Figure 11–13 Phosphocreatine acts as a reserve source of high-energy phosphate groups in vertebrate muscle and other tissues. The high-energy phosphate group (*color*) is transferred to ADP by the enzyme creatine kinase to produce ATP when needed.

with thin filaments in the region of overlap, the thin filaments on either side of the sarcomere should be of opposite polarity. This has been demonstrated by using myosin heads to decorate the actin filaments attached to isolated Z discs: all the myosin arrowheads are found to point away from the Z disc. Therefore, the plus end of each actin filament is embedded in the Z disc, while the minus end points toward the thick filaments (Figure 11–15).

A Myosin Head "Walks" Toward the Plus End of an Actin Filament[7]

Muscle contraction is driven by the interaction between myosin heads and adjacent actin filaments. During this interaction, the myosin head hydrolyzes ATP. The ATP hydrolysis and subsequent dissociation of the tightly bound products (ADP and P_i) produce an ordered series of allosteric changes in the conformation of myosin. As a result, part of the energy released is coupled to the production of movement. For a discussion of the general principles involved in coupling ATP hydrolysis to the directed movement of protein molecules, see page 130.

Kinetic analyses of ATP hydrolysis during muscle contraction, together with electron microscopic and x-ray diffraction studies, suggest the sequence of events illustrated in Figure 11–16. A free myosin head binds ATP (state 1) and hydrolyzes it; the process is reversible because the energy of ATP hydrolysis is initially stored in a highly strained protein conformation with ADP and P_i bound (state 2). While alternating between these two states, a myosin head, as a result of random motions, can move to a neighboring actin subunit and bind to it weakly; this triggers the release of P_i, which causes the head to bind very tightly to the actin filament (state 3). Once bound in this way, the head undergoes a conformational change that generates a "power stroke" that pulls on the rest of the thick filament. At the end of the power stroke (state 4), ADP is released and a fresh molecule of ATP binds to the head, detaching it from the actin filament and returning the head to state 1. Hydrolysis of the bound ATP then prepares the myosin head for a second cycle.

Because each turn of the cycle illustrated in Figure 11–16 results in the hydrolysis and release of one ATP molecule, the series of conformational changes just described is driven by a large favorable change in free energy, making it unidirectional (see p. 130). Each individual myosin head, therefore, "walks" in a single direction along an adjacent actin filament, always moving toward the filament's plus end (see Figure 11–15). As it undergoes its cyclical change in conformation, the myosin head pulls against the actin filament, causing this filament to slide against the thick filament. Once an individual myosin head has detached from the actin filament, it is carried along by the action of other myosin heads in the same thick filament, so that a snapshot of an entire thick filament in a contracting muscle would show some of the myosin heads attached to actin filaments and others unattached. (A certain amount of springlike elasticity in the myosin molecule is essential to allow this to happen.) Each thick filament has about 500 myosin heads, and each head cycles about five times per second in the course of a rapid contraction—sliding the thick and thin filaments past each other at rates up to 15 μm/second.

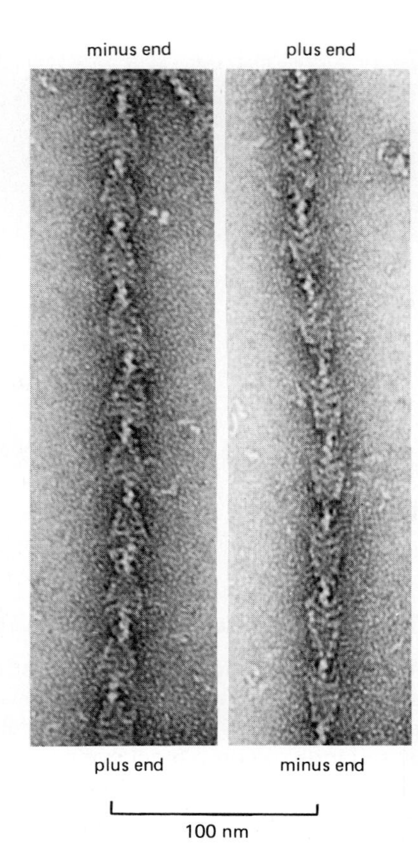

Figure 11–14 Electron micrograph of actin filaments decorated with isolated myosin heads. The helical arrangement of the bound myosin heads, which are tilted in one direction, gives the appearance of arrowheads and indicates the polarity of the actin filament. The pointed end is called the *minus end*, the barbed end the *plus end* because of the different rates of assembly of actin monomers at the two ends (see Figure 11–40). (Courtesy of Roger Craig.)

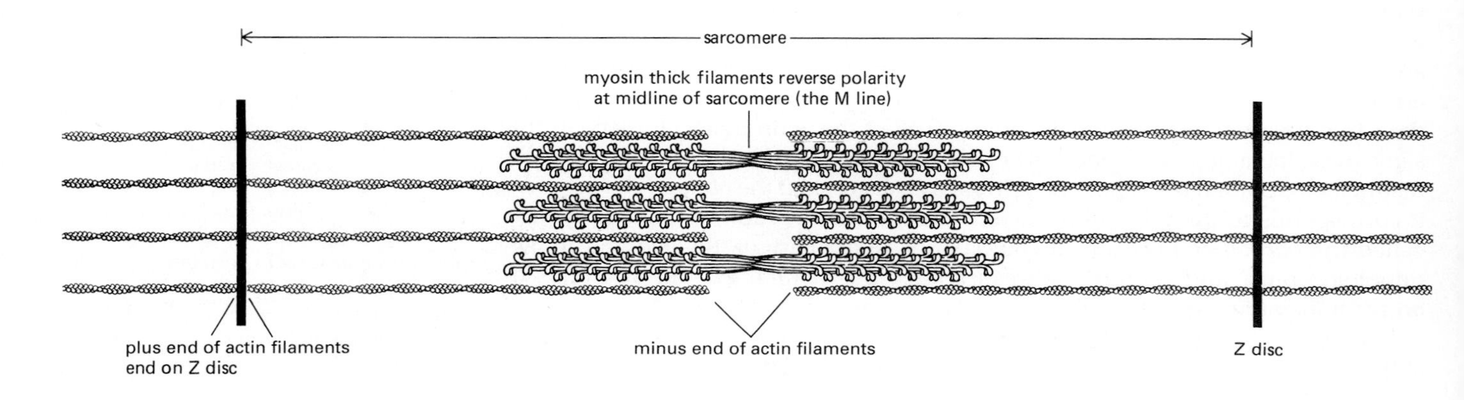

Figure 11–15 The thick and thin filaments of a sarcomere overlap with the same relative polarity on either side of the midline.

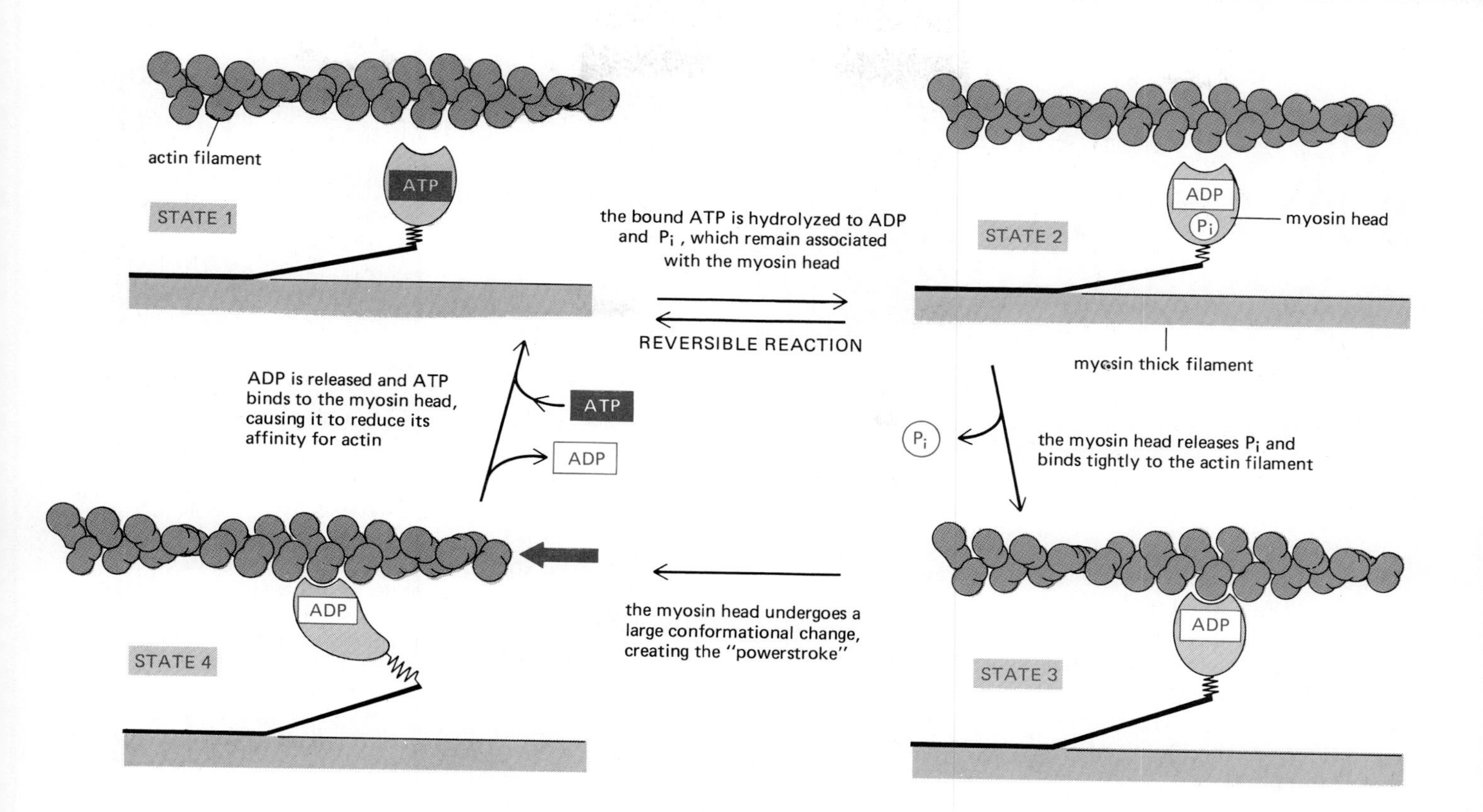

Figure 11–16 Diagram showing how a myosin molecule is thought to use the energy of ATP hydrolysis to move from the minus to the plus end of an actin filament. In the transition from state 2 to state 3, an initial binding to actin causes the myosin head to release its bound phosphate and bind more tightly to the actin filament. The myosin head then undergoes a poorly understood change in shape accompanied by the release of ADP, which makes the myosin head pull against the actin filament (the power stroke). Each of the two heads on a myosin molecule is thought to cycle independently of the other.

Muscle Contraction Is Initiated by a Sudden Rise in Cytosolic Ca^{2+}[8]

The force-generating molecular interaction just described takes place only when a signal passes to the skeletal muscle from its motor nerve. The signal from the nerve triggers an action potential in the muscle cell plasma membrane, and this electrical excitation spreads rapidly into a series of membranous folds, the *transverse tubules,* or *T tubules,* that extend inward from the plasma membrane around each myofibril. The signal is then somehow relayed to the *sarcoplasmic reticulum,* an adjacent sheath of anastomosing flattened vesicles that surrounds each myofibril like a net stocking (Figure 11–17).

The gap between the T tubule and the sarcoplasmic reticulum is only 10–20 nm, but it is unclear how the signal passes between them. When the T tubules are electrically excited, large *Ca^{2+} release channels* in the sarcoplasmic reticulum membrane (see Figure 11–17) are somehow opened, allowing Ca^{2+} to escape into the cytosol from the sarcoplasmic reticulum, where Ca^{2+} is stored in large quantities. The resulting sudden rise in free Ca^{2+} concentration in the cytosol initiates the contraction of each myofibril. Because the signal from the muscle-cell plasma membrane is passed within milliseconds (via the T tubules and sarcoplasmic reticulum) to every sarcomere in the cell, all of the myofibrils in the cell contract at the same time. The increase in Ca^{2+} concentration in the cytosol is transient because the Ca^{2+} is rapidly pumped back into the sarcoplasmic reticulum by an abundant Ca^{2+}-ATPase in its membrane (see p. 307). Typically, the cytosolic Ca^{2+} concentration is restored to resting levels within 30 milliseconds, causing the myofibrils to relax.

Troponin and Tropomyosin Mediate the Ca^{2+} Regulation of Skeletal Muscle Contraction[9]

The Ca^{2+} dependence of vertebrate skeletal muscle contraction, and hence its dependence on motor commands transmitted via nerves, is due entirely to a set of specialized accessory proteins closely associated with actin filaments. If myosin is mixed with pure actin filaments in a test tube, myosin ATPase is activated whether or not Ca^{2+} is present; in a normal myofibril, on the other hand, where

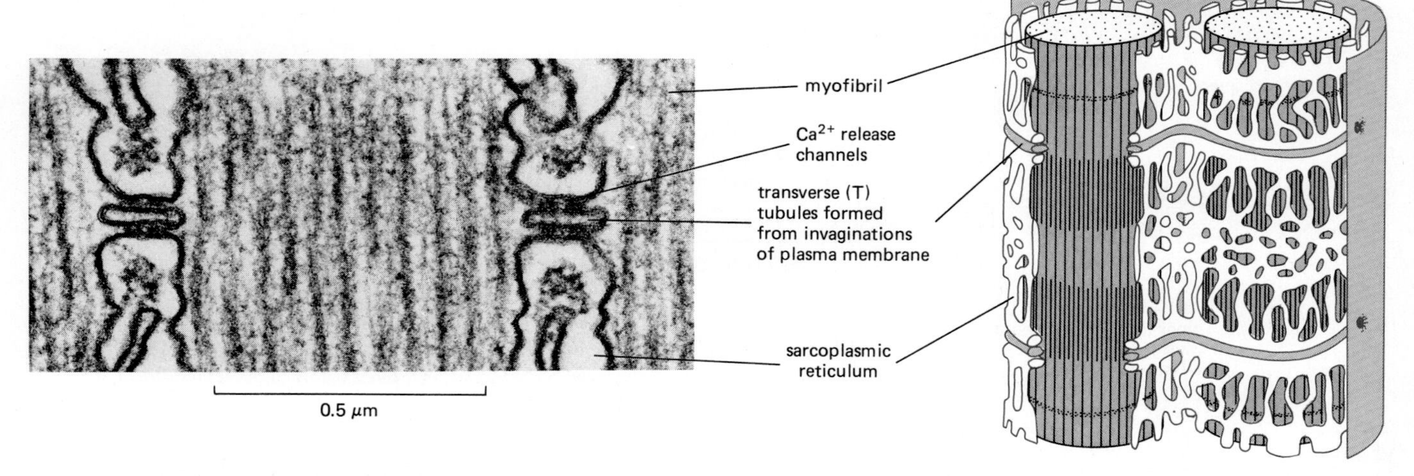

Figure 11–17 The system of membranes involved in relaying the signal to contract from the muscle cell plasma membrane to all of the myofibrils in the cell. The electron micrograph illustrates two T tubules and shows the large Ca^{2+} release channels in the sarcoplasmic reticulum membrane that look like square-shaped "feet" and seem to connect to the adjacent T-tubule membrane. (Micrograph courtesy of Clara Franzini-Armstrong.)

the actin filaments are associated with accessory proteins, the activation of the myosin ATPase depends on Ca^{2+}.

One of these accessory proteins is a rigid, 41-nm-long rod-shaped molecule, called *tropomyosin* because of similarities to myosin in its x-ray diffraction pattern. Like the myosin tail, tropomyosin is a dimer of two identical α-helical chains (284 amino acids each), which wind around each other in a coiled coil (see p. 618). By binding along the length of an actin filament, tropomyosin stabilizes and stiffens the filament (Figure 11–18).

The other major accessory protein involved in Ca^{2+} regulation in vertebrate skeletal muscle is *troponin,* a complex of three polypeptides—troponins T, I, and C (named for their *T*ropomyosin-binding, *I*nhibitory, and *C*alcium-binding activities). The troponin complex has an elongated shape, with subunits C and I forming a globular head region and T forming a long tail. The tail of *troponin T* binds to tropomyosin and is thought to be responsible for positioning the complex on the thin filament (see Figure 11–18). *Troponin I* binds to actin, and when it is added to troponin T and tropomyosin, the complex inhibits the interaction of actin and myosin, even in the presence of Ca^{2+}.

The further addition of *troponin C* completes the troponin complex and makes its effects sensitive to Ca^{2+}. Troponin C binds up to four molecules of Ca^{2+}, and with Ca^{2+} bound, it relieves the inhibition of myosin binding to actin produced by the other two troponin components. Troponin C is closely related to *calmodulin,* which mediates Ca^{2+}-signaled responses in all cells, including the activation of smooth muscle myosin (see p. 626). Troponin C may therefore be regarded as a specialized form of calmodulin that has evolved permanent binding sites for troponin I and troponin T, thereby ensuring that the myofibril responds extremely rapidly to an increase in Ca^{2+} concentration.

There is only one molecule of the troponin complex for every seven actin monomers in an actin filament (see Figure 11–18). Structural studies suggest that in a resting muscle the binding of troponin I to actin moves the tropomyosin molecules to a position on the actin filaments that in an actively contracting muscle is occupied by the myosin heads and thus inhibits the interaction of actin

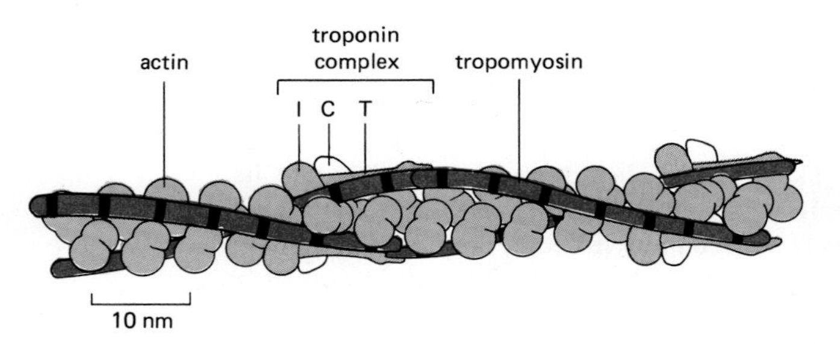

Figure 11–18 A muscle thin filament showing the position of tropomyosin and troponin along the actin filament. Each tropomyosin molecule has seven evenly spaced regions of homologous sequence, each of which is thought to bind to an actin monomer as shown. Note that the ends of adjacent tropomyosin molecules overlap slightly as they polymerize in a head-to-tail fashion along the actin filament. (Adapted from G.N. Phillips, J.P. Fillers, and C. Cohen, *J. Mol. Biol.* 192:111–131, 1986.)

and myosin. When the level of Ca^{2+} is raised, troponin C causes the troponin I to release its hold on actin, thereby allowing the tropomyosin molecules to shift their position slightly so that the myosin heads can bind to the actin filament (Figure 11–19).

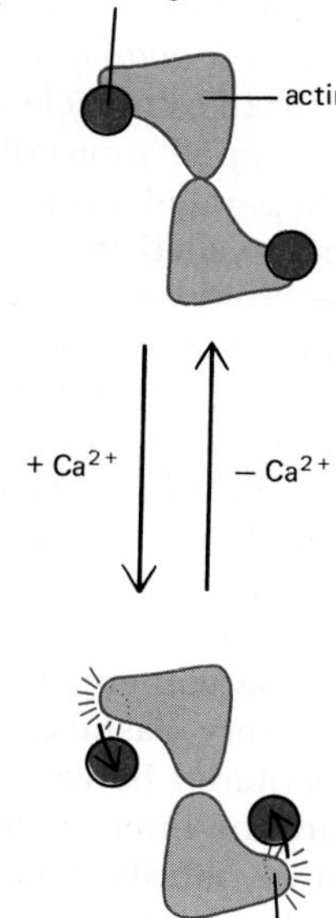

Other Accessory Proteins Maintain the Architecture of the Myofibril and Provide It with Elasticity[10]

The remarkable speed and power of muscle contraction depend on the filaments of actin and myosin in each myofibril being held at the optimal distance from each other and in correct alignment. The precise organization of the myofibril is maintained by a number of structural proteins, more than a dozen of which have so far been identified (Table 11–1); the positions of most of these proteins in the sarcomere have been determined by immunocytochemical methods (see p. 177).

Table 11–1 Major Protein Components of Vertebrate Skeletal Myofibrils

Protein	Percent Total Protein	MW (kDa)	Subunits (kDa)	Function
Myosin	44	510	2 × 223 (heavy chains) 22 + 18 (light chains)	Major component of thick filaments. Interacts with actin filaments with hydrolysis of ATP to develop mechanical force
Actin	22	42	—	Major component of muscle thin filaments, against which muscle thick filaments slide during muscle contraction
Tropomyosin	5	64	2 × 32	Rodlike protein that binds along the length of actin filaments
Troponin	5	78	30 (Tn-T) 30 (Tn-I) 18 (Tn-C)	Complex of three muscle proteins positioned at regular intervals along actin filaments and involved in the Ca^{2+} regulation of muscle contraction
Titin	9	about 2500	—	Very large flexible protein that forms an elastic network linking thick filaments to Z discs
Nebulin	3	600	—	Elongated, inextensible protein attached to Z disc, oriented parallel to actin filaments
α-Actinin	1	190	2 × 95	Actin-bundling protein that links actin filaments together in the region of the Z disc
Myomesin	1	185	—	Myosin-binding protein present at the central "M line" of the muscle thick filament
C protein	1	140	—	Myosin-binding protein found in distinct stripes on either side of the thick-filament M line

The vertebrate striated myofibril also contains at least 20 other proteins not included in this table.

Figure 11–19 A cross-sectional view of a thin filament showing how, in the absence of Ca^{2+}, tropomyosin is thought to block the interaction of the myosin head with actin.

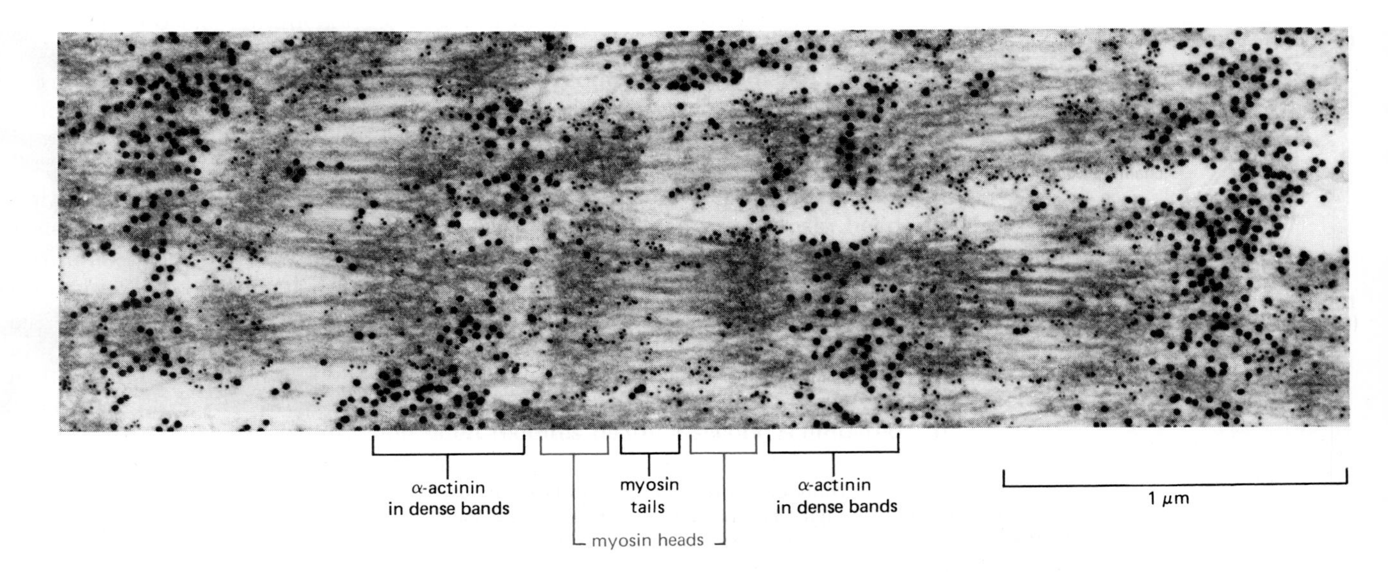

Figure 11–28 An electron micrograph of a cultured fibroblast, showing the musclelike arrangement of proteins in a stress fiber revealed by immunogold staining. The positions of two types of actin-binding protein molecules are shown: α-actinin (large gold particles) is seen in association with periodic dense bands in the stress fiber (α-actinin is associated with the Z disc in striated muscle), while myosin heads (small gold particles) are seen on either side of the α-actinin-containing bands. This pattern is reminiscent of a sarcomere (compare with Figure 11–2) and indicates that the myosin molecules are arranged in filaments. (Courtesy of M. de Brabander, J. de Mey, and G. Langanger.)

Muscle Proteins Are Encoded by Multigene Families[15]

We have seen that a very similar type of contractile apparatus is present in three kinds of muscle cells as well as in most nonmuscle cells. The important differences between the types of contractions that occur in these cell types depend in part on the tissue-specific expression of the genes that encode the proteins of the contractile apparatus. A mammal, for example, has at least six actin genes, six myosin heavy-chain genes, three tropomyosin genes, and three troponin T genes. In some cases the different genes encode proteins that are known to have somewhat different functions; in others, functional differences have not yet been detected.

Of the six actin proteins expressed in mammals, one is restricted to skeletal muscle and another to heart muscle; two are restricted to smooth muscle (one to vascular and the other to nonvascular smooth muscle); and two, known as nonmuscle or cytoplasmic actins, appear to be universal components of the cytoskeleton and are abundant in most nonmuscle cells. These closely related species of actin, also known as actin "isoforms," are very similar in sequence. Muscle and cytoplasmic actins, for example, differ in fewer than 7% of their amino acids. Apart from some differences in the amino-terminal end of the molecule that may have a subtle influence on actin polymerization, it is not certain that the amino acid differences have any functional consequence. If a heart-muscle actin gene is expressed in a cultured fibroblast, it makes a protein that integrates readily with the indigenous population of actin molecules without changing the shape or behavior of the cell. By contrast, as just discussed, the variation among the myosins affects the rate of contraction, the regulatory mechanisms that control contraction, and the extent of myosin self-assembly in the cell.

Muscle Proteins Are Further Diversified by Alternative RNA Splicing[16]

The variations of muscle type described so far give only a partial indication of the diversity that exists. An adult human skeletal muscle, for example, contains a mixture of three types of muscle cells: *white* muscle cells specialized for fast, anaerobic contraction, in which the ATP for contraction is generated mainly by glycolysis; *red* muscle cells specialized for slower, longer-lasting contraction, which use mainly aerobic metabolism and whose color is due to the high concentration of the oxygen-carrying protein myoglobin; and an *intermediate* type of muscle cell, which utilizes both aerobic and anaerobic metabolism (see p. 65). Within each category there are further subtypes that "fine-tune" the muscle to a particular metabolic and physiological function. The same muscle in the fetus is different again.

Each of these types of muscle has somewhat different proteins. A particularly rich source of protein variation is provided by tissue-specific regulation of RNA splicing, which can combine different sets of exons to make slightly different RNA molecules from the same gene (see p. 589). Hormonal, neuronal, and other influences can alter the pattern of RNA splicing and thereby change the amino acid sequences of particular muscle proteins according to the tissue and stage of development. The single skeletal muscle gene that encodes troponin T, for example, can produce at least 10 distinct forms of the protein by alternative RNA splicing. This variability is likely to modify interactions of troponin T with troponin C and tropomyosin, and hence to provide subtle variations in the regulation of muscle contraction.

Summary

Muscle contraction is produced by the sliding of actin filaments against myosin filaments. The head regions of myosin molecules, which project from myosin filaments, engage in an ATP-driven cycle in which they attach to adjacent actin filaments, undergo a conformational change that pulls the myosin filament against the actin filament, and then detach. This cycle is facilitated by special accessory muscle proteins that hold the actin and myosin filaments in parallel overlapping arrays with the correct orientation and spacing for sliding to occur. Two other accessory proteins—troponin and tropomyosin—allow the contraction of skeletal and cardiac muscle to be regulated by Ca^{2+}.

Actin and myosin are also found in smooth muscle cells and in most nonmuscle cells, where they produce contraction in fundamentally the same way as in skeletal and cardiac muscle. The contractile units, however, are smaller and less highly ordered in such cells, and their activity and state of assembly is controlled by the Ca^{2+}-regulated phosphorylation of one of the myosin light chains.

The contractile apparatus in muscle and nonmuscle cells is fine-tuned to fit each cell type by the tissue-specific expression of different genes encoding muscle proteins and by tissue-specific regulation of RNA splicing, by which the same gene can produce slightly different forms of a protein.

Actin Filaments and the Cell Cortex[17]

Actin is the most abundant protein in many eucaryotic cells, often constituting 5% or more of the total cell protein. While actin is distributed throughout the cytoplasm, most animal cells possess an especially dense network of actin filaments and associated proteins just beneath the plasma membrane. This network constitutes the **cell cortex,** which gives mechanical strength to the surface of the cell and enables the cell to change its shape and to move. The precise form of the cortex varies from cell to cell and in different regions of the same cell. In some cells it consists of a thick three-dimensional network of cross-linked actin filaments that excludes large particles and organelles in the underlying cytoplasm (Figure 11–29); in other cells the cortex is a thinner, more two-dimensional net-

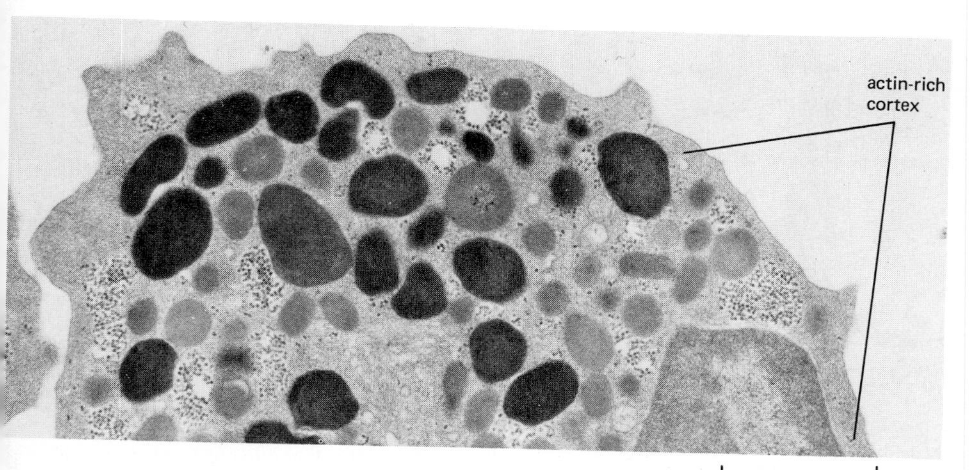

Figure 11–29 The actin cortex in a thin-section electron micrograph of a white blood cell. Although various kinds of granules fill the cytoplasm, they are excluded from the layer just beneath the plasma membrane (the cortex), which contains a network of actin filaments and associated proteins that determines the surface movements of the cell. (Courtesy of Dorothy Bainton.)

work. In selected areas of animal cells, small bundles of actin filaments project outward from the cortex to form the stiff core of cell-surface extensions; while in other areas, actin filaments pull inward on the membrane. Because the plasma membrane is so closely integrated with the cortical actin network, for some purposes these two entities are best considered as a single functional unit.

Approximately 50% of the actin molecules in most animal cells is unpolymerized, existing either as free monomers or as small complexes with other proteins. A dynamic equilibrium exists between this pool of unpolymerized actin molecules and actin filaments, which helps drive many of the surface movements of cells. In this section we shall discuss how actin-binding proteins control the assembly of actin filaments, link them together into bundles or networks, and determine their position, length, and other properties.

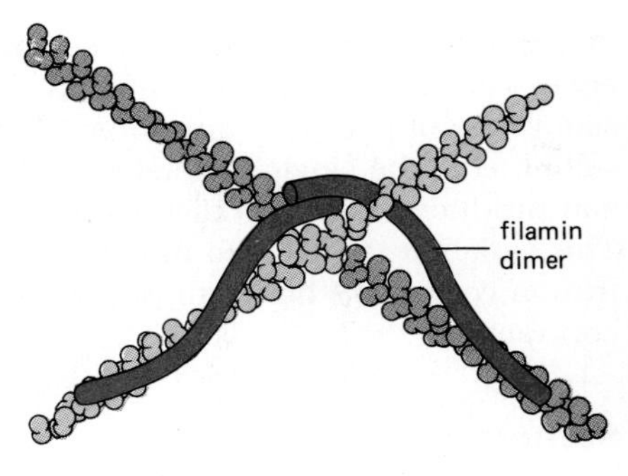

Figure 11–30 By forming a flexible link between two adjacent actin filaments, filamin produces a three-dimensional network of actin filaments with the physical properties of a gel. Each filamin dimer is about 160 nm long when fully extended.

Abundant Actin-binding Proteins Cross-link Actin Filaments into Large Networks[18]

Actin filaments are often linked together into a stiff three-dimensional network by *cross-linking proteins*. The most abundant of these is **filamin,** a long flexible molecule composed of two identical polypeptide chains joined head to head, with a binding site for actin filaments at each tail end (Figure 11–30). Proteins of this type

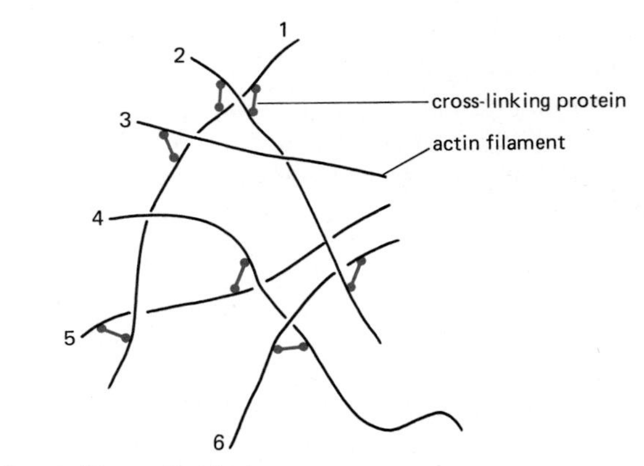

(A) actin-filament-based gel

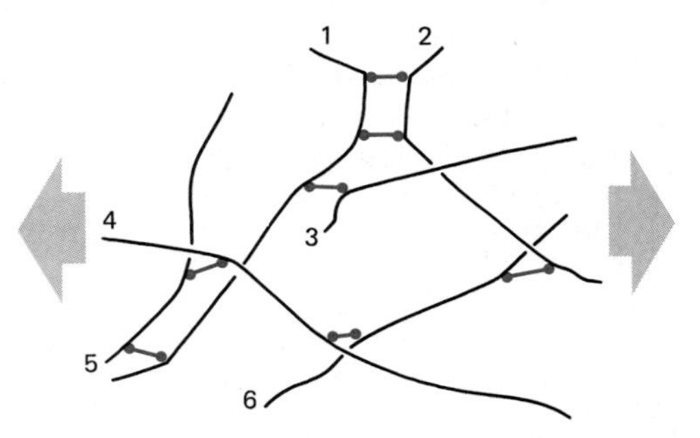

(B) gel resists sudden pull because of cross-linking proteins

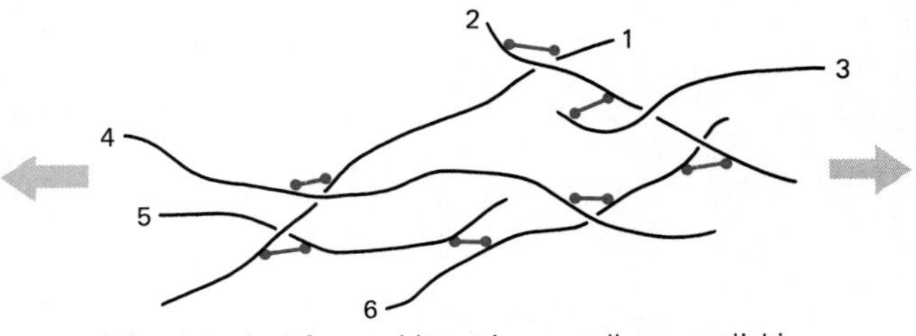

(C) gel slowly deforms with continuous pull, as cross-linking proteins rearrange their contacts with actin filaments

Figure 11–31 The mechanical properties of an actin filament gel formed by actin cross-linking proteins. A rapid deformation (B) is strongly resisted by the gel because cross-linking proteins do not have time to dissociate from their bound actin filaments. A slow deformation (C) encounters little resistance because the cross-linking protein molecules have time to dissociate and reassociate repeatedly, readjusting their positions. (After M. Sato, W.H. Schwartz, and T.D. Pollard, *Nature* 325:828–830, 1987.)

constitute almost 1% of the total protein of many cells (about 1 dimer for every 50 actin monomers).

Gels made *in vitro* from actin filaments and a cross-linking protein have the interesting mechanical property of preserving their form when subjected to a sudden force but being readily deformed by a slow steady pressure. Figure 11–31 suggests a molecular basis for this distinctive property, which is also displayed by the cortical cytoplasm. It is presumably because of such actin networks that the cell surface recoils elastically from small rapid insults, while being capable of large deformations if a slight force is steadily applied.

Gelsolin Fragments Actin Filaments When Activated by Ca^{2+} [19]

Extracts prepared from many types of animal cells form a gel in the presence of ATP when they are warmed to 37°C. Although this gelation depends on both actin filaments and a cross-linking protein such as filamin, the gels exhibit more complex behavior than simple mixtures of actin filaments and filamin. If the Ca^{2+} concentration is raised above 10^{-7} M, for example, the semisolid actin gel begins to liquefy—a process known as solation—and regions of the solating gel show vigorous local streaming when examined under a microscope. Clearly there must be components besides actin and filamin in the extracts to account for its Ca^{2+}-dependent solation and streaming. These components are likely to be involved in the *cytoplasmic streaming* observed in some large cells, where vigorous flowing movements are required to maintain an even distribution of metabolites and other cytoplasmic components. These movements seem to be associated with a sudden local change in the cytoplasm from a solid gel-like consistency to a more fluid state.

A number of proteins have been isolated from cell extracts that, when added to a gel of actin filaments and filamin, cause it to change to a more fluid state in the presence of Ca^{2+}. The best characterized of these is **gelsolin,** a compact protein of 90,000 daltons. When activated by the binding of Ca^{2+}, gelsolin severs an actin filament and forms a cap on the newly exposed plus end of the filament, thus breaking up the cross-linked network of actin filaments. Similar proteins are found in the cortex of many types of vertebrate cells. These *severing proteins* are activated by concentrations of Ca^{2+} (about 10^{-6} M) that occur only transiently in the cytosol, and they are believed to help mediate responses of the cell cortex to extracellular signals. When a phagocytic white blood cell contacts a microorganism, for example, the network of actin filaments in the white cell's cortex locally disassembles, enabling the surface of the cell to move and engulf the microorganism. We shall return later to the actin-based mechanism underlying such movements.

Myosin Can Mediate Cytoplasmic Movements[20]

While an artificial mixture of actin filaments, filamin, and gelsolin is capable of undergoing Ca^{2+}-dependent gel-to-sol transitions, it will not contract or show the streaming movements displayed by the cruder actin-rich gels obtained from cells. These activities seem to require *myosin:* if myosin is selectively removed from the crude actin-rich gels, contractions and streaming no longer occur, suggesting that an interaction between actin and myosin generates the force for cytoplasmic streaming.

How can actin and myosin produce coherent movements when the actin filaments in extracts are initially distributed in an apparently random three-dimensional network? As we have seen, an actin filament has a well-defined polarity, and myosin heads can bind and move along it only if they are oriented correctly with regard to its polarity. The small bipolar assemblies of nonmuscle myosin molecules (see Figure 11–26) may be able to generate some long-range order in the solution simply by pulling one set of actin filaments against another, even though the actin filaments and myosin assemblies are initially not well oriented (Figure 11–32).

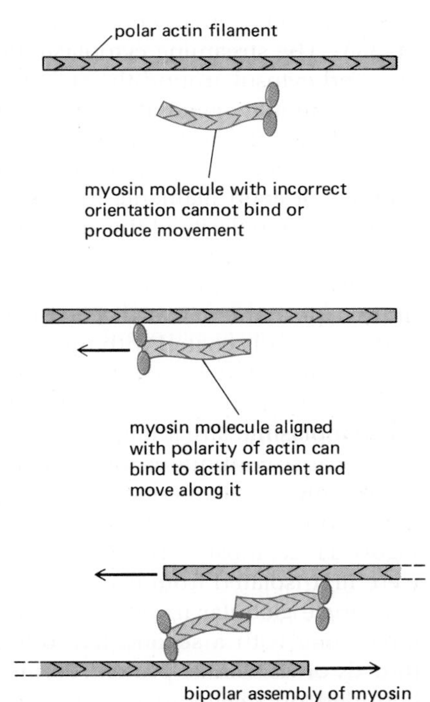

Figure 11–32 A bipolar assembly of nonmuscle myosin molecules (see Figure 11–26) produces a sliding of two actin filaments of opposite polarity, as in muscle. In this way myosin can cause contraction even in a randomly oriented network of actin filaments.

Figure 11–35 A bundle of parallel actin filaments, held together by actin-bundling proteins, forms the core of a microvillus. Lateral arms connect the sides of the actin filament bundle to the overlying plasma membrane. The plus ends of the actin filaments are all at the tip of the microvillus, where they are embedded in an amorphous, densely staining substance of unknown composition.

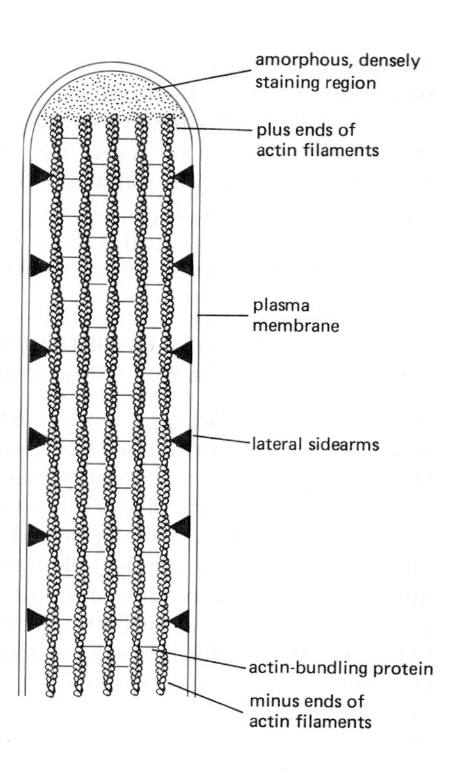

is about 0.08 μm wide and 1 μm long, making the cell's absorptive surface area 20 times greater than it would be without them. The plasma membrane that covers these microvilli is highly specialized, containing a thick extracellular coat of polysaccharide and digestive enzymes.

The core of each intestinal microvillus contains a rigid bundle of 20 to 30 parallel actin filaments that extends from the tip of the microvillus down into the cell cortex. The actin filaments in the bundle are all oriented with their plus ends pointing away from the cell body and are held together at regular intervals by several *actin-bundling proteins,* including *fimbrin* and *fascin* (Figure 11–35). In contrast to filamin and other flexible actin-cross-linking proteins that tie actin filaments into loose networks (see p. 630), these actin-bundling proteins are relatively small, compact molecules with two distinct actin-binding sites on a single polypeptide chain. Consequently, they tie actin filaments into tight bundles, with adjacent actin filaments held rigidly about 10 nm apart in parallel arrays.

The lower portion of the actin filament bundle in the microvillus core is anchored in the specialized cortex at the apex of the intestinal epithelial cell. This cortex, known as the *terminal web,* contains a dense network of spectrin molecules that overlies a layer of intermediate filaments (Figure 11–36); it is thought to stiffen

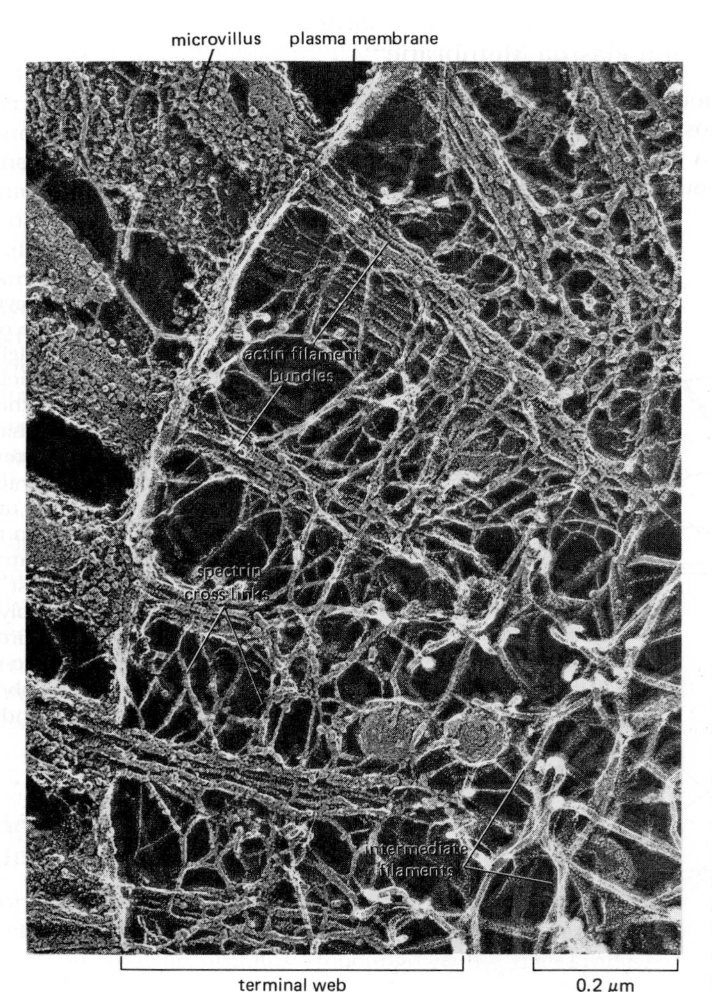

Figure 11–36 Freeze-etch electron micrograph of an intestinal epithelial cell, showing the terminal web beneath the apical plasma membrane. Bundles of actin filaments forming the core of microvilli extend into the terminal web, where they are linked together primarily by spectrin. Beneath the terminal web is a layer of intermediate filaments. (From N. Hirokawa and J.E. Heuser, *J. Cell Biol.* 91:399–409, 1981. Reproduced by permission of the Rockefeller University Press.)

the microvilli above it, keeping their actin bundles projecting outward at a right angle to the apical cell surface.

How is the bundle of actin filaments at the core of a microvillus attached to the overlying plasma membrane? Electron microscopy reveals two distinct regions of membrane association: a helical "staircase" of lateral arms extends from the sides of the actin filament bundle to make contact with the plasma membrane, and an amorphous "cap" of densely staining material connects the plus ends of the actin filaments to the plasma membrane at the tip of the microvillus (see Figure 11–35).

If the plasma membrane is removed from such an epithelial cell with a non-ionic detergent, the lateral arms remain attached to the denuded cytoskeleton. They can, however, be removed by subsequent exposure to ATP; each arm is then found to consist of a minimyosin molecule tightly bound to the calcium-binding protein calmodulin. The minimyosin binds with its ATP-dependent head region to actin filaments in the microvillus core and by its truncated tail region to the plasma membrane. Why a motility-generating protein is used to form this linkage is not known; but since myosin molecules move toward the plus ends of actin filaments, those in the microvillus would be expected to carry membrane components toward the microvillus tip. Perhaps this movement assists in the continual release ("sloughing") of plasma membrane from the microvillus into the lumen of the intestine, where the digestive enzymes associated with the plasma membrane continue their action.

Less is known about the amorphous, densely staining material at the tip of a microvillus than about the lateral arms. As we shall see, there are reasons to suspect that proteins in this region control the length and diameter of the actin filament bundle and thereby determine the size of the microvillus (see p. 676). If so, the mode of actin filament insertion into the membrane at the microvillus tip is likely to be complex.

Focal Contacts Allow Actin Filaments to Pull Against the Substratum[24]

Bundles of actin filaments often bind to the plasma membrane in a way that allows them to pull on the extracellular matrix or on another cell. We have already mentioned such attachment sites for the ends of actin filament bundles in fibroblasts and smooth muscle cells (see p. 627 and p. 625). Attachments of this type are mediated by transmembrane linker glycoproteins in the plasma membrane. Those formed by cultured fibroblasts with the extracellular matrix are the best characterized. When fibroblasts grow on a culture dish, most of their cell surface is separated from the substratum by a gap of more than 50 nm. In certain regions called **focal contacts,** or *adhesion plaques,* however, this gap is reduced to 10 to 15 nm. These regions appear as dark areas in an interference reflection microscope, in which only the light reflected from the under surface of the cell is collected. They can be shown by staining with anti-actin antibodies to be the sites where the ends of stress fibers attach to the plasma membrane (Figure 11–37).

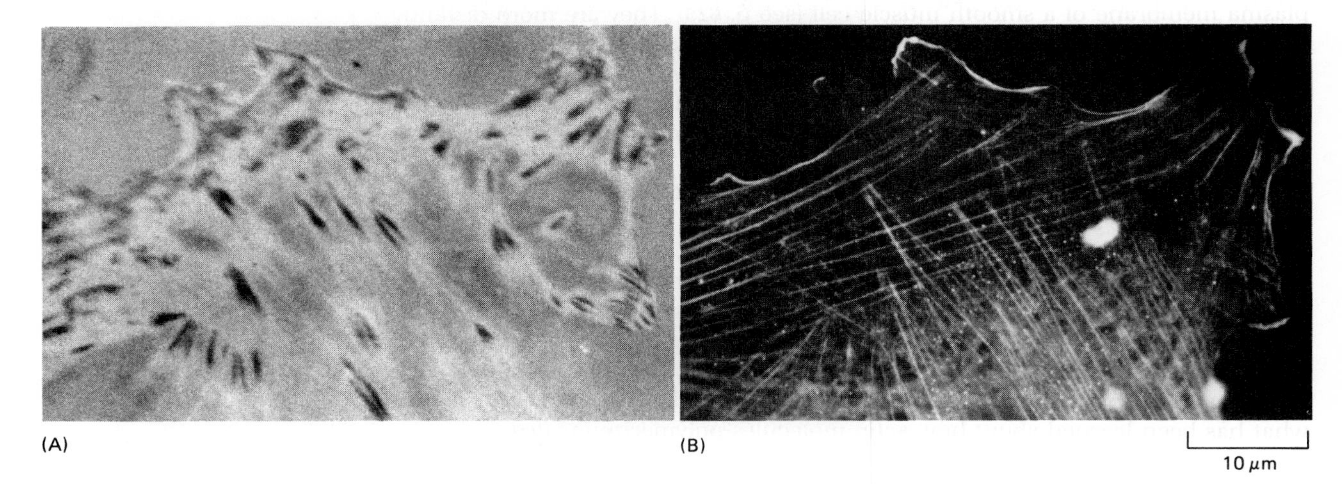

Figure 11–37 The relation between focal contacts and stress fibers in cultured fibroblasts. Focal contacts are best seen in living cells by reflection-interference microscopy (A). In this technique, light is reflected from the lower surface of a cell attached to a glass slide, and the focal contacts appear as dark patches. (B) Staining of the same cell (after fixation) with antibodies to actin shows that most of the cell's actin-filament bundles (or stress fibers) terminate at or close to a focal contact. (Courtesy of Grenham Ireland.)

11-12 Actin Filaments Can Undergo a "Treadmilling" of Subunits[25]

In vitro experiments like that shown in Figure 11–40 indicate that the actin concentration at which filament growth ceases—the critical concentration—is different at the plus and minus ends of the filament. Consequently, when actin polymerizes in a test tube, a steady state is reached at which actin monomers dissociate predominantly at the minus end and associate predominantly at the plus end. At this steady state the rate of subunit addition at the plus end equals the rate of loss from the minus end, so that a constant concentration of the actin monomer is maintained. Even though the net length of the polymer does not change, individual actin molecules are continually tranferred from one end of the filament to the other, a process called **treadmilling** (Figure 11–41).

The treadmilling of actin filaments requires energy, since otherwise it could be used to create a perpetual motion machine that does work without an input of energy, violating the laws of thermodynamics. The energy in this case is provided by ATP hydrolysis. Each actin monomer binds a molecule of ATP, which is hydrolyzed shortly after the monomer is added to a filament. As explained in Panel 11–1, this hydrolysis of ATP makes treadmilling possible. As we shall see shortly, treadmilling may be one mechanism by which actin filaments and components associated with them generate movements in cells (see p. 640).

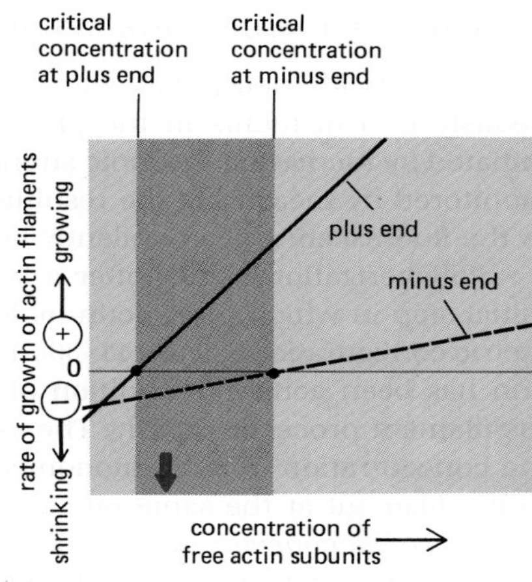

Figure 11–41 The rate of growth observed at the two ends of an actin filament at different concentrations of free actin molecules. This graph shows that the two ends have different growth rates and different critical concentrations. As a consequence, there is a range of free actin concentrations (*dark color*) in which the plus end of the actin filament is polymerizing while the minus end is depolymerizing. When free actin is in apparent equilibrium with actin filaments *in vitro*, no net growth occurs because the rate at which actin molecules are coming off the minus end is equal to the rate at which they are adding to the plus end. At this concentration of free actin subunits (intermediate between the critical concentrations of the two ends—*colored arrow*), actin filaments do not change their length, but individual actin molecules continually move from one end of the polymer to the other in a process called treadmilling (see Panel 11–1).

Many Cells Extend Dynamic Actin-containing Microspikes and Lamellipodia from Their Surface[26]

Dynamic surface extensions containing actin filaments are a common feature of animal cells, especially when the cells are migrating or changing shape. Cells in culture, for example, often extend many thin, stiff protrusions called **microspikes,** which are about 0.1 μm wide and 5 to 10 μm long and contain a loose bundle of about 20 actin filaments oriented with their plus ends pointing outward. The growing tip (growth cone) of a developing nerve cell axon extends even longer microspikes, called *filopodia,* which can be up to 50 μm long (see p. 1115). These extensions are motile structures that can form and retract with great speed. It is thought that they act as feelers by which cells explore their environment: those microspikes that adhere strongly to their surroundings act to steer the cell toward the adhesive region; those that fail to adhere are carried backward over the upper surface of the cell and are retracted.

In addition to microspikes, crawling cells and growth cones periodically extend thin sheetlike processes known as **lamellipodia** from their advancing edge (the *leading edge*). Like microspikes, some of these adhere to the substratum, while others fail to adhere, curl back over the top of the cell, and are swept backward as a "ruffle" (Figure 11–42).

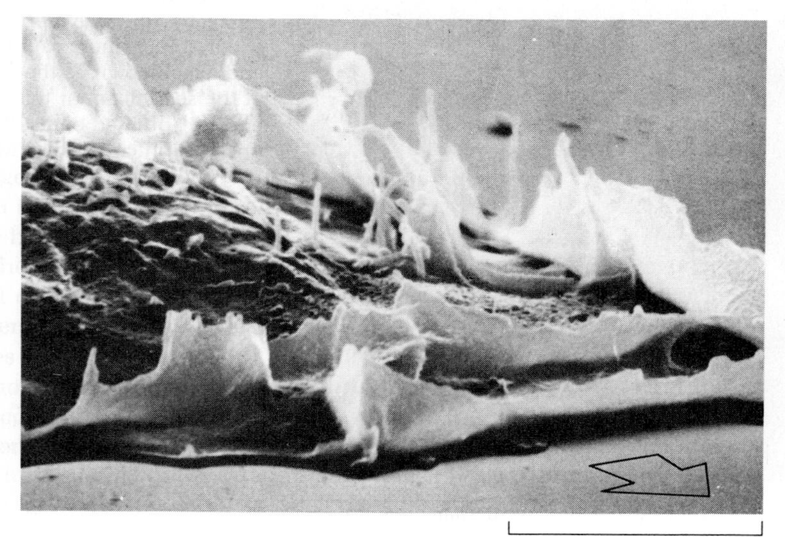

Figure 11–42 Scanning electron micrograph of lamellipodia and microspikes at the leading edge of a human fibroblast migrating in culture. The arrow shows the direction of cell movement. As the cell moves forward, lamellipodia and microspikes sweep backward over its dorsal surface—a movement known as ruffling. (Courtesy of Julian Heath.)

DIFFERENT GROWTH RATES AT THE PLUS AND MINUS ENDS

The assembly (polymerization) and disassembly (depolymerization) of actin filaments occurs by the addition and removal of actin subunits from the filament ends. During assembly, one end of the filament grows faster than the other. The fast-growing end is called the plus end, while the slow-growing end is called the minus end. The difference in the rates of growth at the two ends is made possible by changes in the conformation of each subunit as it enters the polymer (→).
free subunit subunit in polymer

This conformational change affects subunit addition at one end preferentially.

SLOW ADDITION TO MINUS END

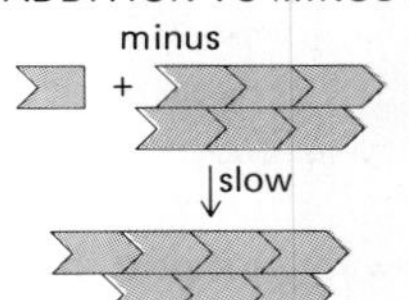

FAST ADDITION TO PLUS END

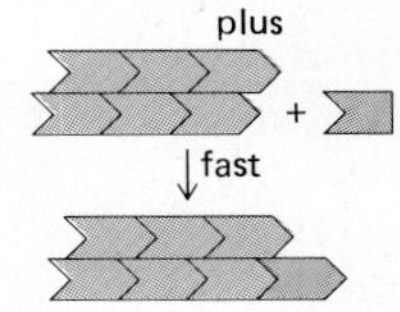

THE CRITICAL CONCENTRATION OF SUBUNITS

The number of subunits that add to the polymer per second will be proportional to the concentration of free subunit ($k_{on}[C]$), but the subunits will leave the polymer end at a constant rate (k_{off}) that does not depend on [C]. As the polymer grows, subunits are used up and [C] drops until it reaches a constant value, called the critical concentration (C_c), at which the rate of subunit addition equals the rate of subunit loss.

At this equilibrium point,

$$k_{on}[C_c] = k_{off}$$

so that

$$\text{critical conc.} = C_c = \frac{k_{off}}{k_{on}} = \frac{1}{\mathbf{K}}$$ (where **K** is the equilibrium constant for subunit addition)

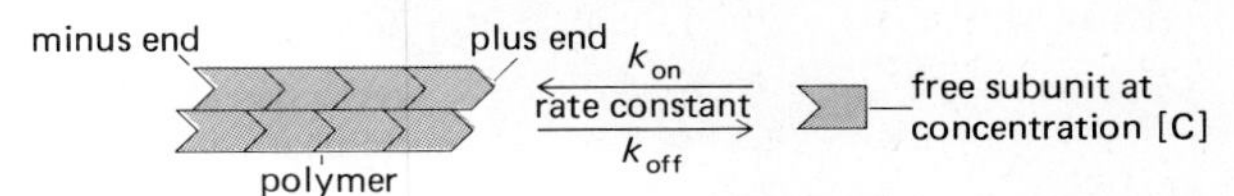

Even though k_{on} and k_{off} will have different values for the plus and minus ends of the polymer, their ratio, k_{off}/k_{on}—and hence C_c—must be the same at both ends. This is because exactly the same subunit interactions are broken when a subunit is lost at either end, and the final state of the subunit after dissociation is identical. Therefore the ΔG for subunit loss, which determines the equilibrium constant for its association with the end (see Table 3–3, p. 95), is identical at both ends: if the plus end grows four times faster than the minus end, it must also shrink four times faster.

Thus: for $[C] > C_c$ both ends grow
for $[C] < C_c$ both ends shrink (see Figure 11–41)

ATP HYDROLYSIS CHANGES THE EQUILIBRIUM CONSTANT AT EACH END

Each actin molecule carries a tightly bound ATP molecule that is hydrolyzed to a tightly bound ADP molecule soon after the conformational change occurs that accompanies subunit assembly into the polymer.

free subunit	subunit in polymer	subunit in polymer
actin (ATP) →	actin (ATP)	→ actin (ADP) + P_i

These changes will be abbreviated as
T → T → D

ATP hydrolysis reduces the binding affinity of the subunit for neighboring subunits and makes it more likely to dissociate from each end of the filament. It is usually the T form that adds to the filament and the D form that leaves (changing to D in solution), so that the reaction that takes place when the subunit leaves the polymer is no longer the exact reverse of the reaction that occurs on assembly.

Considering events at the plus end only:

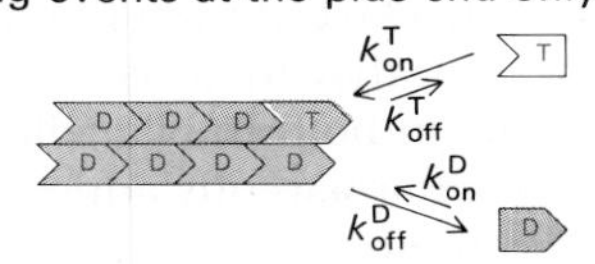

As before, the polymer will grow until $[C] = C_c$. For illustrative purposes, we can ignore k^D_{on} and k^T_{off} as insignificant, so that polymer growth ceases when

$$k^T_{on}[C_c] = k^D_{off} \quad \text{or} \quad C_c = \frac{k^D_{off}}{k^T_{on}} = \frac{1}{\mathbf{K}}$$

This is a steady state and not a true equilibrium because the ATP that is hydrolyzed must be replenished by a nucleotide-exchange reaction on the free subunit (D → T).

TREADMILLING

Since k^D_{off} and k^T_{on} refer to different reactions, the ratio k^D_{off}/k^T_{on} need not be the same at the two ends of the polymer. Thus the ATP hydrolysis is found to create a different critical concentration at each end, with

$$C_c\ (\textit{minus}\ \text{end}) > C_c\ (\textit{plus}\ \text{end})$$

Assuming that both ends of the polymer are exposed, polymerization will proceed until [C] reaches a value that is above C_c for the plus end but below C_c for the minus end (see Figure 11–41). At this steady state, subunits will assemble at the plus end and disassemble at the minus end at an identical rate, so that the polymer maintains a constant length even though there is a net flux of subunits through the polymer known as treadmilling:

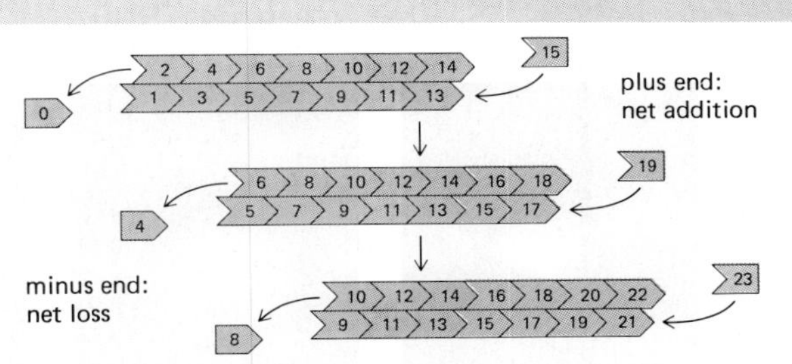

ATP hydrolysis is not required for filament assembly. Actin subunits that are bound to a nonhydrolyzable analogue of ATP will polymerize under appropriate conditions. The filaments formed in these cases, however, will not treadmill; treadmilling is made possible by the ATP hydrolysis that accompanies polymerization.

Specific Drugs Change the State of Actin Polymerization and Thereby Affect Cell Behavior[29]

One would expect that many of the movements produced by the cell cortex, such as phagocytosis and cell locomotion, would depend on the dynamic equilibrium between actin molecules and actin filaments. But compared to the explosive changes that take place in activated sperm, changes in actin polymerization during the course of such movements are usually small and transient and therefore difficult to detect. The importance of actin polymerization and depolymerization in these cell movements is indicated by the effects of drugs that prevent changes in the the state of actin polymerization and thereby disrupt such movements. For example, the **cytochalasins** (Figure 11–46), a family of metabolites excreted by various molds, paralyze many kinds of vertebrate cell movement—including cell locomotion, phagocytosis, cytokinesis, the production of microspikes and lamellipodia, and the folding of epithelial sheets into tubes. These drugs do not inhibit chromosome separation on the mitotic spindle, which depends principally on microtubules; or muscle contraction, which depends on stable actin filaments that do not undergo assembly and disassembly. The principal action of the cytochalasins is to bind specifically to the fast-growing plus ends of actin filaments, preventing the addition of actin molecules there.

Phalloidin is a highly poisonous alkaloid produced by the toadstool *Amanita phalloides.* By contrast with the cytochalasins, it stabilizes actin filaments and inhibits their depolymerization. Also unlike the cytochalasins, it does not readily cross the plasma membrane and therefore must be injected into a cell in order

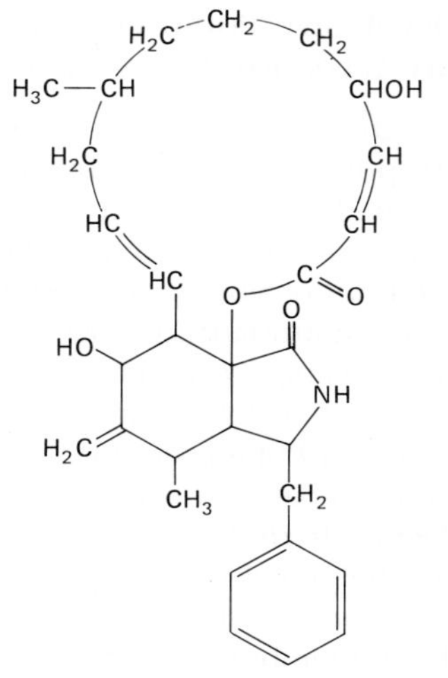

cytochalasin B

Figure 11–46 Chemical structure of cytochalasin B.

Figure 11–47 Some major classes of actin-binding proteins that are found in most vertebrate cells.

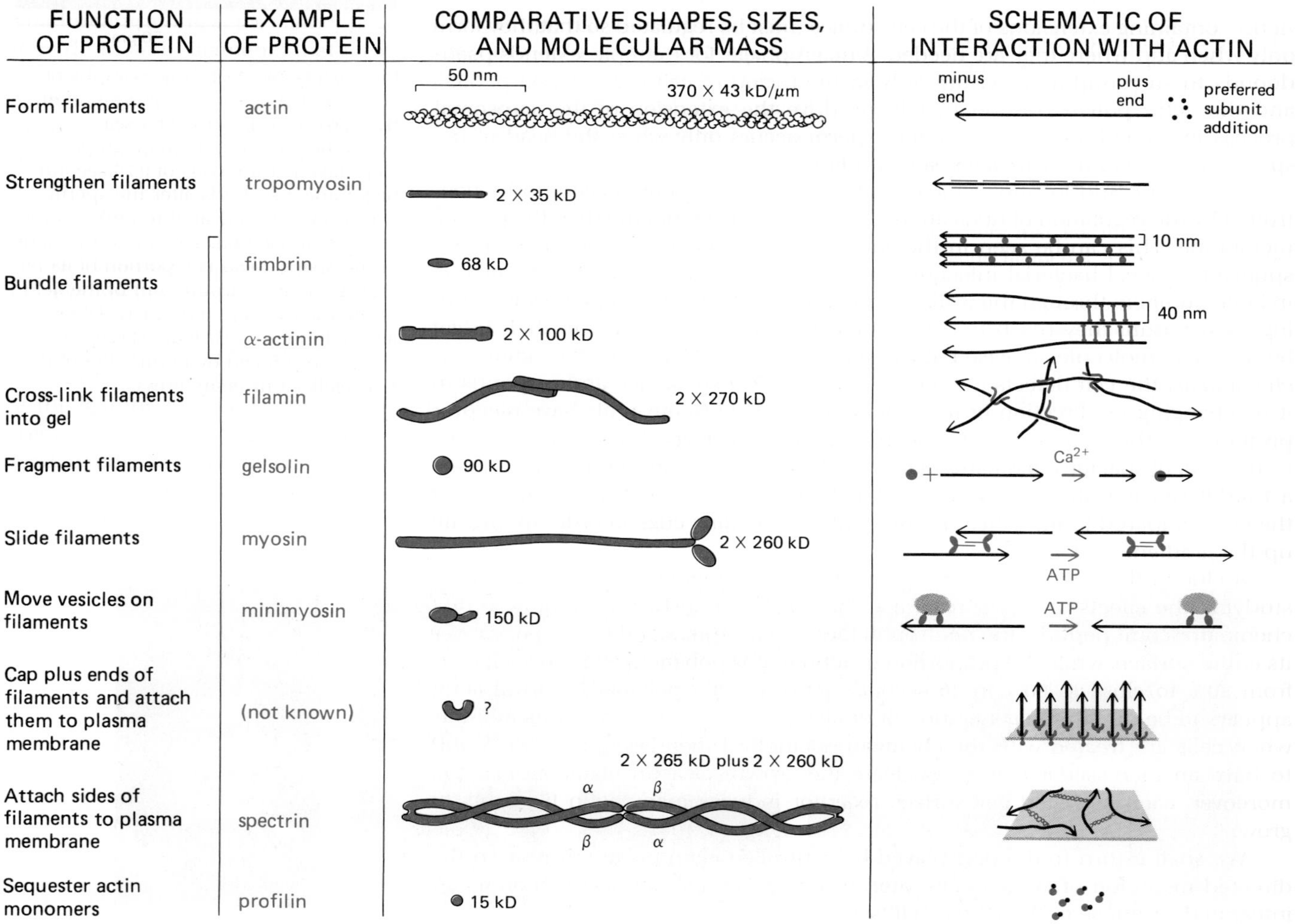

FUNCTION OF PROTEIN	EXAMPLE OF PROTEIN	COMPARATIVE SHAPES, SIZES, AND MOLECULAR MASS	SCHEMATIC OF INTERACTION WITH ACTIN
Form filaments	actin	50 nm; 370 × 43 kD/μm	minus end; plus end; preferred subunit addition
Strengthen filaments	tropomyosin	2 × 35 kD	
Bundle filaments	fimbrin	68 kD	10 nm
	α-actinin	2 × 100 kD	40 nm
Cross-link filaments into gel	filamin	2 × 270 kD	
Fragment filaments	gelsolin	90 kD	Ca^{2+}
Slide filaments	myosin	2 × 260 kD	ATP
Move vesicles on filaments	minimyosin	150 kD	ATP
Cap plus ends of filaments and attach them to plasma membrane	(not known)	?	
Attach sides of filaments to plasma membrane	spectrin	2 × 265 kD plus 2 × 260 kD; α β β α	
Sequester actin monomers	profilin	15 kD	

to act efficiently. When this is done, phalloidin blocks the migration of both amoeba and vertebrate cells in culture, suggesting that the dynamic assembly and disassembly of actin filaments is crucial for these movements. In addition, because phalloidin binds specifically to actin filaments, fluorescent derivatives of the drug are often used instead of anti-actin antibodies to stain actin filaments in cells.

The Behavior of the Cell Cortex Depends on a Balance of Cooperative and Competitive Interactions Among a Large Set of Actin-binding Proteins

Much more is known about actin-binding proteins than about the proteins that associate with microtubules or with intermediate filaments, the other two major classes of cytoskeletal filaments. A summary of the actin-binding proteins in vertebrate nonmuscle cells that we have discussed in this chapter is presented in Figure 11–47. The list is far from complete. Each of the functional categories indicated is thought to have several members, each with slightly different properties. Moreover, there are some actin-binding proteins with no obvious function, and there are undoubtedly other actin-binding proteins that have not yet been identified. In particular, cells must have mechanisms by which they can distinguish the two ends of an actin filament and control its orientation and location in the cytoplasm. These mechanisms presumably involve *"capping proteins"* that bind selectively to the plus end of an actin filament, hold it close to the plasma membrane, and regulate the addition of actin monomers. In contrast, since the minus ends of actin filaments are relatively inactive with respect to both polymerization and depolymerization, it is possible that these ends are often left free. Much remains to be learned about the nature of the membrane-bound organizing centers for actin: virtually nothing is known, for example, about the molecular nature of the densely staining material at the tip of a microvillus, which is presumably responsible for organizing the actin core and controlling the growth and regeneration of the microvillus (see p. 634).

Even after all the components of the actin filament network have been defined, there will remain the difficult task of deciphering the consequences of their many interactions. The proteins shown in Figure 11–47 do not bind simultaneously or randomly to actin filaments; they cooperate and compete to produce ordered patterns of interactions that have profound consequences for the cell (Figure 11–48). Moreover, the interactions among the components of the network are modulated by local changes in the concentration of ions in the cytosol and by physical forces that stretch or compress the network and hence move its constituent molecules in relation to one another. It is not suprising, therefore, that we are only beginning to understand this crucial part of the cytoskeleton.

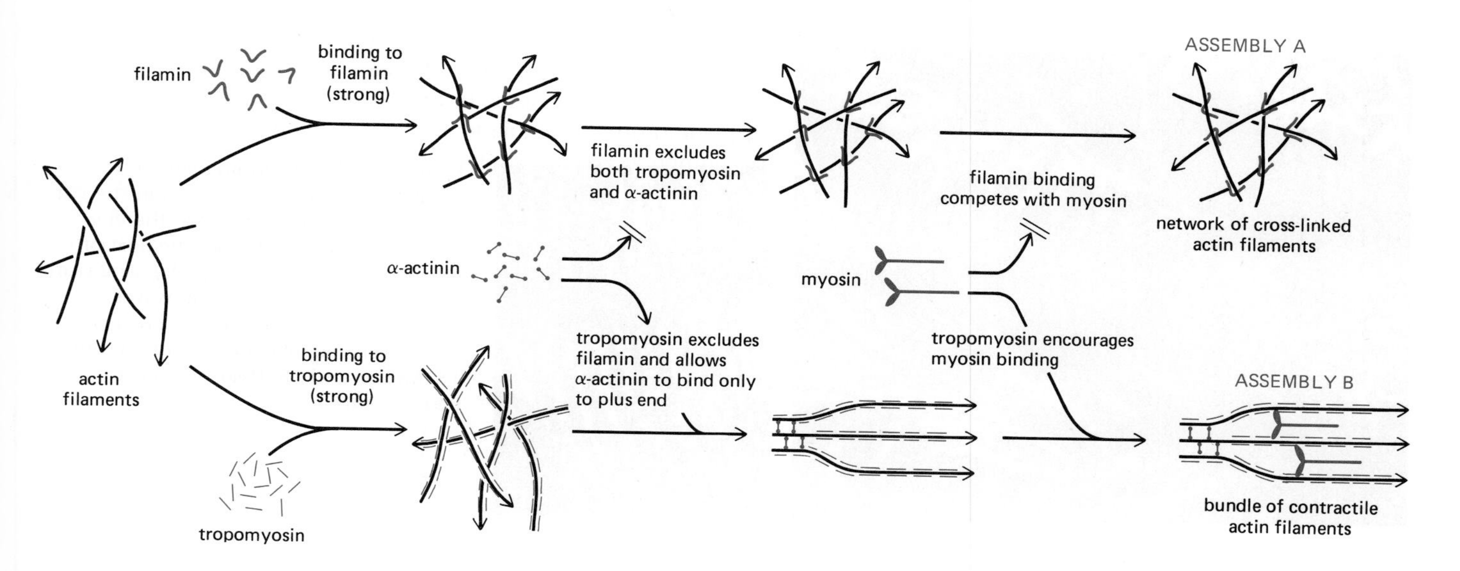

Figure 11–48 Some examples of competitive and cooperative interactions between actin-binding proteins. Tropomyosin and filamin both bind strongly to actin filaments, but their binding is competitive. Because tropomyosin binds cooperatively to actin filaments, either tropomyosin or filamin will predominate over large regions of the actin filament network. Other actin-binding proteins, such as α-actinin or myosin, will be excluded from specific sites by a competitive interaction; thus, for example, α-actinin binds all along pure actin filaments *in vitro*, but it binds relatively weakly to actin filaments in cells, where it is largely confined to sites near the plus ends because of competition with other proteins. Alternatively, binding can be enhanced through cooperative interaction; thus tropomyosin appears to enhance the binding of myosin to actin filaments. Multiple interactions of these types between the actin-binding proteins in Figure 11–47 (and others) are thought to be responsible for the complex variety of actin networks found in all eucaryotic cells.

Defects in axonemes also occur in humans, where various hereditary forms of male sterility have been shown to be due to nonmotile sperm. Depending on the particular mutation, the sperm flagella lack dynein arms, the radial-spoke heads, or the inner sheath together with one or both of the central pair of microtubules. Exactly the same defects occur in the respiratory cilia of such individuals, who commonly have long histories of respiratory tract disease—with recurrent bronchitis and chronic sinusitis—because their nonmotile respiratory cilia are unable to clear mucus from their lungs and sinuses. Remarkably, about half of the individuals with this *immotile cilia syndrome* also have an extremely rare condition known as situs inversus, in which the normal asymmetry of the body—as seen in the position of the heart, intestine, liver, and appendix—is reversed (the whole complex of abnormalities is known as *Kartagener's syndrome*). It has therefore been suggested that the unidirectional beating of cilia during early human development could play a critical part in determining the normal left-right asymmetry of the body.

Having discussed how flagella and cilia move, we shall now consider how they are formed.

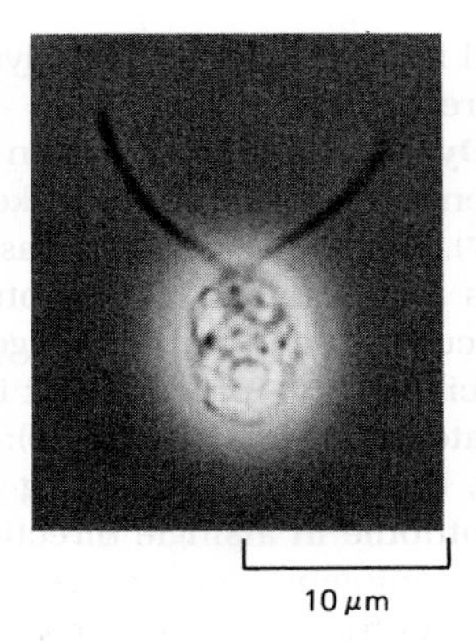

Figure 11–58 The unicellular green alga *Chlamydomonas reinhardtii.* This organism swims by means of its two flagella, which cooperate to produce a cycle of whiplike movements reminiscent of the breast stroke of human swimmers (unlike the flagella of sperm). (Courtesy of John Hopkins.)

Centrioles Play Two Distinct Roles in the Cell[38]

If the twin flagella of *Chlamydomonas* are sheared from the cell, the centrioles remain and the flagella rapidly re-form. Nearly all of the necessary protein components can be found in a soluble form in the cytoplasm of the cell, and these are utilized to construct the new flagella. A few of the individual steps of the assembly can take place in cell-free extracts: tubulin molecules will polymerize into microtubules, as we shall see in greater detail in the following section, and dynein arms can be added back to an axoneme stripped of its own arms by extraction in salt solutions of high ionic strength. By themselves, however, axonemal proteins cannot generate the distinctive 9 + 2 pattern of the axoneme; a nucleating structure is needed to serve as a template on which further growth can take place. In the cell the required template is provided by a *centriole.*

A **centriole** is a small cylindrical organelle about 0.2 μm wide and 0.4 μm long. Nine groups of three microtubules, fused into triplets, form the wall of the centriole, each triplet being tilted inward toward the central axis at an angle of about 45° to the circumference, like the blades of a turbine (Figure 11–59). Adjacent triplets are linked at intervals along their length, while faint protein spokes can often be seen in electron micrographs to radiate out to each triplet from a central core, forming a pattern like a cartwheel (see Figure 11–59A). Centrioles often exist in pairs, with the two centrioles at right angles to each other (Figure 11–60).

The centriole is a permanent feature of the ciliary axoneme, where it is traditionally called a *basal body.* Appendages known as *striated rootlets* are often attached to this centriole, linking it to other components of the cytoskeleton. During the formation or regeneration of the cilium, each doublet microtubule of the axoneme grows from two of the microtubules in the triplet microtubules of the centriole so that the ninefold symmetry of the centriolar microtubules is pre-

Figure 11–59 (A) Electron micrograph of a cross-section through three centrioles in the cortex of a protozoan. Each centriole (also called a basal body—see text) forms the lower portion of a ciliary axoneme. (B) Schematic drawing of a centriole. It is composed of nine sets of *triplet* microtubules, each triplet containing one complete microtubule (the A tubule) fused to two incomplete microtubules (the B and C tubules). Other proteins form links that hold the cylindrical array of microtubules together (*color*). (Micrograph courtesy of D.T. Woodrum and R.W. Linck.)

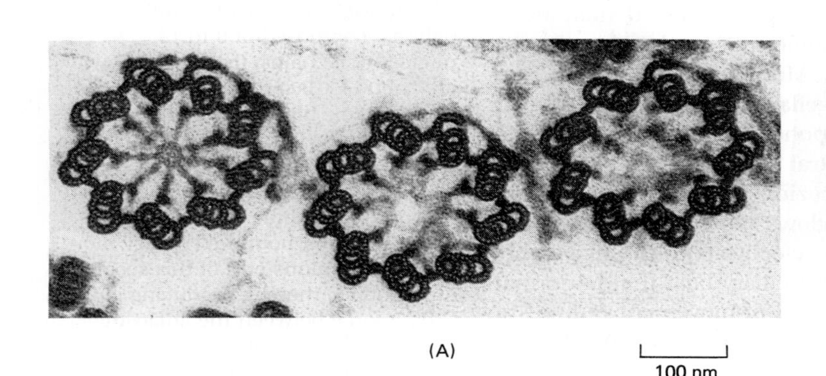

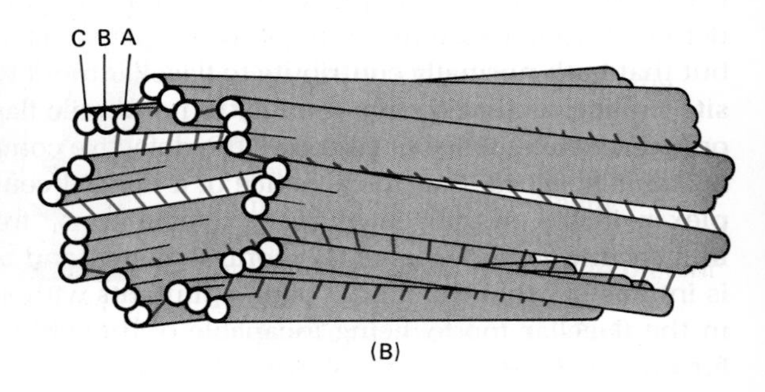

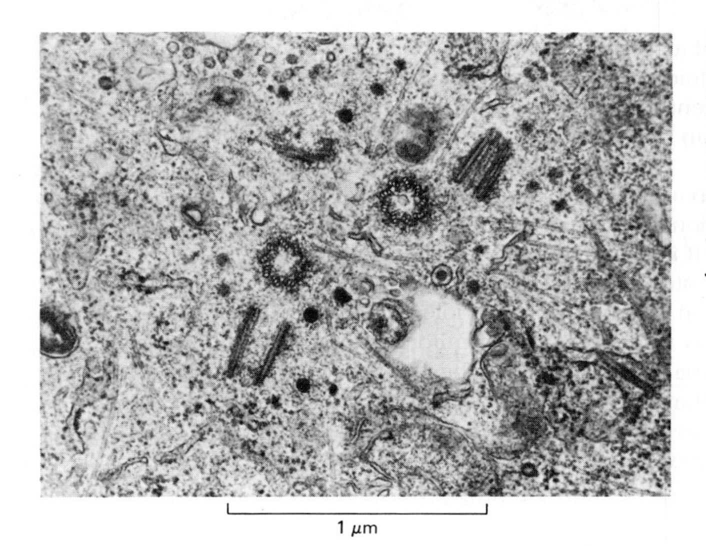

Figure 11–60 An electron micrograph showing a newly replicated pair of centrioles. One centriole of each pair has been cut in cross-section and the other in longitudinal section, indicating that the two members of each pair are aligned at right angles to each other. (From M. McGill, D.P. Highfield, T.M. Monahan, and B.R. Brinkley, *J. Ultrastruct. Res.* 57:43–53, 1976.)

served in the ciliary axoneme. Autoradiographic evidence suggests that the addition of tubulin and other proteins of the axoneme takes place at the distal tip of the structure, at the plus end of the microtubules. How the central pair of single microtubules forms in the axoneme is not known; there is no central pair in centrioles.

The centrioles that form the basal bodies of cilia are performing a specialized function in the cell, just as the cilia themselves are specialized structures. Nearly all animal cells, however, have a pair of centrioles that acts as a focal point for the *centrosome.* The centrosome (also called the *cell center*—see p. 763) organizes the array of cytoplasmic microtubules during interphase and duplicates at mitosis to nucleate the two poles of the mitotic spindle, as we shall see in the next section. Sometimes centrioles can serve first one function and then another in turn: prior to each division in *Chlamydomonas,* for example, the two flagella resorb and the basal bodies leave their position to act as mitotic spindle poles.

Centrioles Usually Arise by the Duplication of Preexisting Centrioles[39]

The otherwise continuous increase in cell mass throughout the animal cell cycle is punctuated by two discrete duplication events: the replication of DNA and the doubling of the centrioles. In cultured fibroblasts, centriole doubling commences at around the time that DNA synthesis begins. First the two members of a pair separate; then a daughter centriole is formed perpendicular to each original centriole (see Figure 11–60). An immature centriole contains a ninefold symmetric array of *single* microtubules; each microtubule then presumably acts as a template for the assembly of the triplet microtubule of mature centrioles.

In a ciliated vertebrate cell, which may contain hundreds of cilia, the centrioles of the precursor cell give rise to the many basal bodies required to nucleate the cilia in the mature cell. During the differentiation of the ciliated epithelial cells that line the oviduct and the trachea, for example, the centriole pair migrates from its normal location near the nucleus to the apical region of the cell where the cilia will form. There, instead of forming a single daughter centriole in the typical manner, each centriole in the pair forms numerous electron-dense "satellites." Many basal bodies then arise from these satellites and migrate to the membrane to initiate the formation of cilia.

There are also cases where centrioles seem to arise de novo. For example, although unfertilized eggs of many animals lack functional centrioles and use the sperm centriole for the first mitotic division (see p. 874), under certain experimental conditions—such as extreme ionic imbalance or electrical stimulation—the unfertilized egg can produce a variable number of centrioles. Each of these also seem to exist in these two states. The average lifetime of a microtubule in cultured fibroblasts in interphase, for example, is short—less than 10 minutes. Thus the array of microtubules radiating from the centrosome is continually changing, as new microtubules grow and replace others that have depolymerized.

The inherent instability of microtubules helps to explain how they can be induced to grow in specific directions in a cell—toward the leading edge of a crawling cell, for example, or toward a condensed chromosome in a dividing cell (see p. 770). A naked microtubule with both ends free in the cytoplasm rapidly disappears. But an organizing center continually produces new microtubules in random directions, with their minus ends anchored in the organizing center and thus protected from depolymerization. A microtubule that grows from such a center can be stabilized if its plus end is somehow capped so as to prevent its depolymerization. If capped by a structure in a particular region of the cell, it will establish a relatively stable link between that structure and the organizing center. Microtubules originating in the organizing center can thus be selectively stabilized by events elsewhere in the cell. Cell polarity is thought to be determined in this way by unknown structures or factors localized in specific regions of the cell cortex that "capture" the plus ends of microtubules. The plane of cell division may be determined in a similar way (see p. 779).

In many cells the initial stabilization of microtubules at their plus ends is consolidated to produce a more permanent polarization of the cell, as we shall now discuss.

Figure 11–68 A large and unusually highly ordered microtubule-organizing center found in the cytopharyngeal basket of the ciliate *Nassula*. Microtubules grow in a regular hexagonal array from the lower surface of a flat three-layered sheet, forming one element of the complex mouthpart of this cell. (From J.B. Tucker, *J. Cell Sci.* 6:385–429, 1970.)

Microtubules Undergo a Slow "Maturation" as a Result of Posttranslational Modifications of Their Tubulin Subunits[46]

The continual formation and loss of microtubules is characteristic of cells undergoing a major internal reorganization—such as cells that are dividing or crawling over a substratum. When cells have become part of an established tissue, the microtubules they contain become relatively permanent features, especially in those cells, such as nerve cells, that no longer divide after they differentiate. This microtubule "maturation" depends partly on the posttranslational modification of the tubulin molecules and partly on the interaction of microtubules with specific microtubule-associated proteins.

A number of enzymes modify selected amino acids in tubulin. One of these is *tubulin acetyl transferase,* which acetylates a specific lysine of the α-tubulin subunit. In *Chlamydomonas* this enzyme is located principally in the flagellar axoneme and appears to acetylate tubulin molecules after they assemble into the distal tip of the axoneme (see p. 645). A specific deacetylating enzyme is located in the cytoplasm and removes acetyl groups from unpolymerized tubulin. As a consequence of the localization of these two enzymes in *Chlamydomonas,* tubulin molecules in the axoneme microtubules are acetylated whereas those in cytoplasmic microtubules, most of which are turning over rapidly, are mainly nonacetylated. We shall discuss the possible consequence of tubulin acetylation below.

A second, more unusual form of modification is the removal of the carboxyl-terminal tyrosine residue from an α-tubulin molecule that has become incorporated into a microtubule. This modification is catalyzed by a detyrosinating enzyme in the cytoplasm of many vertebrate cells, and, as in the case of acetylation, there is an oppositely-acting enzyme that restores the tyrosine to unpolymerized tubulin. In cells with highly unstable microtubules, tubulin is generally not present in polymerized microtubules long enough to be detyrosinated, and so "tyrosine-tubulin" is the predominant form. By contrast, "detyrosinated tubulin" becomes enriched in any "older" microtubules that survive the normal rapid turnover of newly formed microtubules. Both detyrosination and acetylation thus mark the conversion of transiently stabilized microtubules into a much more permanent form (Figure 11–69). When cultured fibroblasts are treated with drugs that depolymerize microtubules (see p. 652), the small population of microtubules that is spared during the drug treatment can be selectively labeled with antibodies that recognize either acetylated or detyrosinated tubulin.

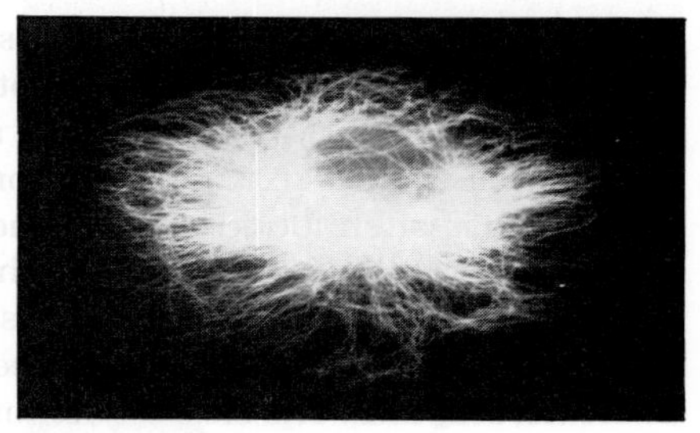
(A)

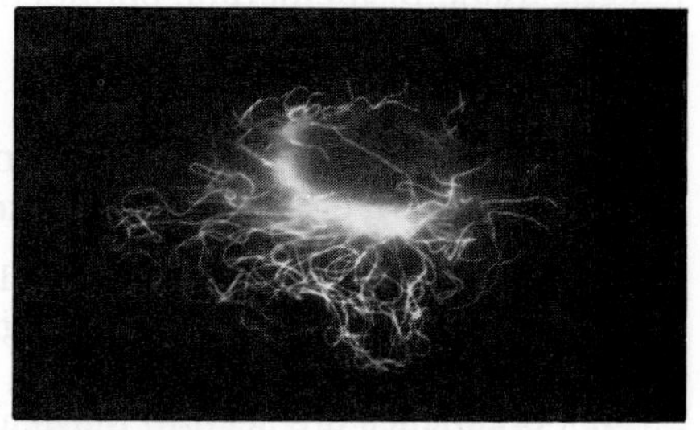
(B)

Figure 11–69 Immunofluorescence micrographs showing that while most of the microtubules in the cytoplasm of a cultured cell are highly dynamic (A), others are relatively stable (B). The single cell shown was injected with tubulin that was covalently coupled to the small molecule biotin; after 1 hour the cell was labeled first with anti-biotin antibodies coupled to a fluorescent molecule (fluorescein) and then with anti-tubulin antibodies coupled to another fluorescent molecule (rhodamine). Microtubules that are rapidly depolymerizing and repolymerizing incorporate the biotinylated tubulin and are therefore labeled by the anti-biotin antibodies (A), which cover the microtubules and prevent anti-tubulin antibodies from binding. Stable microtubules, on the other hand, do not incorporate biotinylated tubulin and therefore are not stained by the anti-biotin antibodies, but they are stained by the anti-tubulin antibodies (B). The stable microtubules are also found to stain with antibodies that recognize detyrosinated or acetylated tubulin. (Courtesy of Eric Schulze and Marc Kirschner.)

Microtubules formed *in vitro* from either acetylated or detyrosinated tubulin are not detectably more stable than microtubules formed from unmodified tubulin. These modifications are therefore thought to act as signals for the binding of specific proteins that stabilize microtubules and modify their properties inside the cell.

The Properties of Cytoplasmic Microtubules Are Modified by Microtubule-associated Proteins (MAPs)[47]

Posttranslational modification of tubulin appears to mark certain microtubules as "mature" and to add to their stability. But the most far-reaching and versatile modifications of microtubules are believed to be those conferred by other proteins. These **microtubule-associated proteins,** or **MAPs,** serve both to stabilize microtubules against disassembly and to mediate their interaction with other cell components. As one might expect from the diverse functions of microtubules, there are many kinds of MAPs.

Two major classes of MAPs can be isolated from brain in association with microtubules: *HMW proteins* (high-molecular-weight proteins), which have molecular weights of 200,000 to 300,000 or more; and *tau proteins,* with molecular weights of 40,000 to 60,000. Both classes of proteins have two domains, one of which binds to microtubules; because this domain binds to several unpolymerized tubulin molecules simultaneously, MAPs speed up the nucleation step of tubulin polymerization *in vitro.* The other domain is thought to be involved in linking the microtubule to other cell components (Figure 11–70). Antibodies to HMW and tau proteins show that both proteins bind along the entire length of cytoplasmic microtubules.

Many other MAPs have been isolated as proteins that bind selectively to microtubules. The functions of most of these are unknown. Some presumably act as structural components to stabilize microtubules and provide permanent links to other cell components (including other parts of the cytoskeleton and selected organelle membranes). Others are responsible for moving organelles along microtubules.

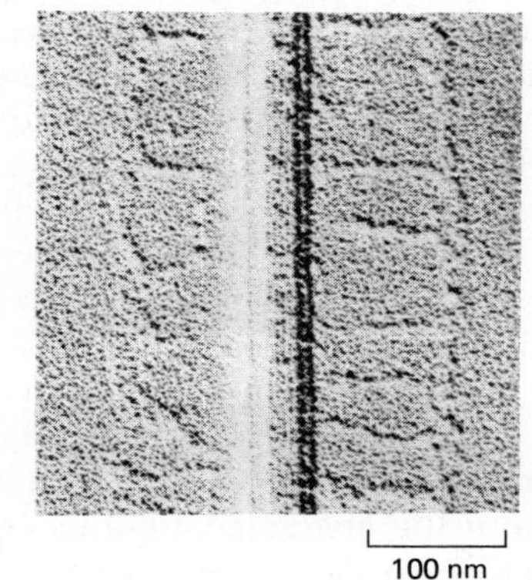

(A)

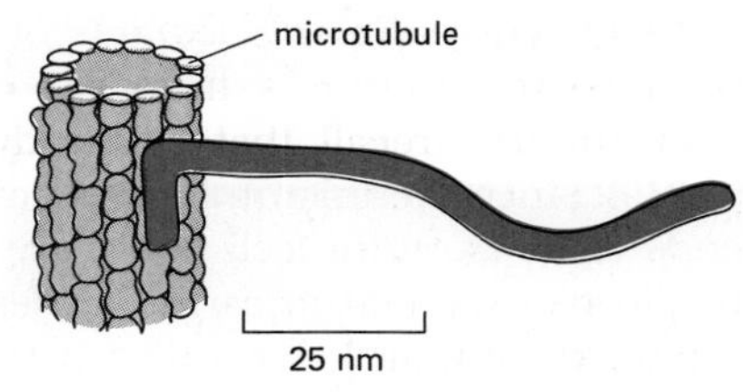

(B)

Figure 11–70 The regularly spaced side arms formed on a microtubule by a large microtubule-associated protein (known as MAP-2) isolated from vertebrate brain. The electron micrograph in (A) shows a portion of a microtubule to which many molecules of MAP-2 are bound. Portions of the protein project away from the microtubule, as shown schematically in (B). (Electron micrograph courtesy of William Voter and Harold Erickson.)

The Transport of Organelles in the Cytoplasm Is Often Guided by Microtubules[48]

If a living vertebrate cell is observed in a phase-contrast or a differential-interference-contrast microscope (see p. 141), its cytoplasm is seen to be in continual motion. Over the course of minutes, mitochondria and smaller membrane-bounded organelles change their positions by periodic *saltatory movements,* which are too

Keratin Filaments Are Remarkably Diverse[52]

Among the most stable and long-lived types of intermediate filaments are those formed from keratins. They are also the most diverse. The simplest epithelia, such as those found in developing embryos and in some adult tissues such as the liver, contain just two types of keratin, one acidic and one neutral. Epithelia in other locations (for example, tongue, bladder, and sweat glands) contain six or more keratins—the particular "blend" depending on their anatomical location. Because of the variety and stability of keratin filaments, they provide a distinctive "fingerprint," which is useful for tracing the origins of tumors to particular types of epithelial cells.

The diversity of keratins is even more pronounced in the epidermis of the skin, which consists of a tough, stratified epithelium (see p. 968). Distinct sets of keratin proteins are expressed by the cells in the different layers of the epidermis. The keratin filaments eventually become covalently cross-linked to one another and to associated proteins, and as cells in the outermost layers of the epidermis die, the cross-linked keratins persist as a major part of the protective outer layer of the animal. Specialized epithelial cells at particular locations in the skin provide regional variation by generating surface appendages such as hairs, nails, and feathers. Thus the intermediate filaments help provide the animal with its primary barrier against heat and water loss, as well as supplying it with camouflage, armament, and ornamentation.

What Is the Function of Intermediate Filaments?

Animal cells can survive without cytoplasmic intermediate filaments. The glial cells that make myelin in the central nervous system do not have any, and those in cultured fibroblasts can be disrupted by an intracellular injection of antibodies against IF proteins without apparent effects on cell organization or behavior. It seems likely that the principal function of most intermediate filaments is to provide mechanical support to the cell and its nucleus. Intermediate filaments in epithelia form a transcellular network that seems designed to resist external forces. The neurofilaments in the nerve cell axon probably resist stresses caused by the motion of the animal, which would otherwise break these long, thin cylinders of cytoplasm. Desmin filaments provide mechanical support for the sarcomeres in muscle cells, and vimentin filaments surround (and probably support) the large fat droplets in fat cells.

But if the function of intermediate filaments is simply to resist tension, why are there so many types of subunit proteins? And what is the function of the variable parts of the molecule, which do not appear to be involved in formation of the filament itself? A detailed answer to these questions cannot be given at present, but it is clear from our examples that the type of support provided by intermediate filaments, and the way it is harnessed to other components, varies greatly among cell types. The desmin filaments that appear to "tie" the edges of the Z discs together in striated muscle cells are likely to have binding sites for specific proteins in the Z disc. Neurofilaments are subjected to lesser forces but may have to be linked together side by side to provide a continuous "rope" a meter or more in length; it is perhaps for this reason that neurofilaments have large projections along their length while other intermediate filaments do not (Figure 11–77).

The different requirements for binding to other proteins must be provided by the variable regions of the IF proteins. By modulating the properties of the intermediate filament, the variable regions determine not only its ability to self-associate but also how it interacts with other cellular components, such as microtubules or the plasma membrane. This strategy contrasts with that used by the two other major elements of the cytoskeleton—actin filaments and microtubules. As we have seen, these polymers are largely invariant in structure, and their properties are adapted to different functions by diverse sets of actin-binding proteins and microtubule-associated proteins. In a sense, therefore, the variable regions of

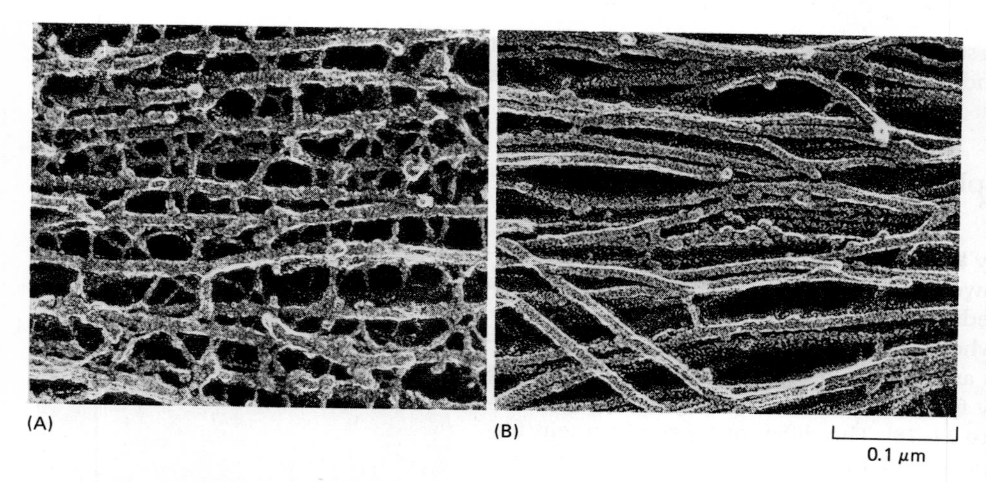

Figure 11–77 Electron micrographs of two types of intermediate filaments seen in neural tissues prepared by fast-freezing and deep-etching. (A) Neurofilaments in a nerve cell axon are extensively cross-linked through protein cross-bridges—an arrangement believed to provide great tensile strength in this long cell process. The cross-links are thought to be formed by the long nonhelical extensions at the carboxyl terminus of the largest neurofilament protein (see Figure 11–74). (B) The intermediate filaments (called glial filaments) in an astrocyte are subjected to less mechanical stress. They are smooth and have few cross-bridges. (Courtesy of N. Hirokawa.)

IF proteins may serve a function similar to some of the accessory proteins of actin filaments and microtubules, with the primary difference that they are covalently linked to the subunit of the filament itself rather than being a separate protein.

Summary

Intermediate filaments are ropelike polymers of fibrous polypeptides that are thought to play a structural or tension-bearing role in the cell. A variety of tissue-specific forms are known that differ in the type of polypeptide they contain: these include the keratin filaments of epithelial cells, the neurofilaments of nerve cells, the glial filaments of astrocytes and Schwann cells, the desmin filaments of muscle cells, and the vimentin filaments of fibroblasts and many other cell types. The nuclear lamins, which form the fibrous lamina that underlies the nuclear envelope, are a separate family of intermediate filament proteins that are present in all eucaryotic cells.

The polypeptides of the different types of intermediate filaments differ in amino acid sequence and have very different molecular weights. But they all contain a homologous central domain that forms a rigid coiled-coil structure when the protein dimerizes. These dimeric subunits associate with one another in large overlapping arrays to form the intermediate filament. The rodlike domains of the subunits form the structural core of the intermediate filaments, while the globular domains at either end project from the surface and allow the individual filaments to vary in character. This variation allows the mechanical properties of the intermediate filaments and their associations with other components of the cell to match the requirements of the particular cell type.

Organization of the Cytoskeleton[57]

Up to this point we have discussed microtubules, actin filaments, and intermediate filaments as though they were independent cytoskeletal components in the cell. But the different parts of the cytoskeleton must, of course, be linked together and their functions coordinated in order to mediate changes in cell shape and produce various types of cell movement. When a fibroblast in culture rounds up to divide, for example, the entire cytoskeleton is reorganized: stress fibers and cytoplasmic microtubules are disassembled, while a mitotic spindle and then a contractile ring are formed, all as part of a controlled sequence of events.

In this section we shall discuss the interactions among the major filament systems of the cytoskeleton in relation to three of its functions. First we shall examine how the cytoskeleton helps to organize the contents of the cytoplasm, including components that are usually thought to be freely soluble. Next we shall consider how the coordinated behavior of the cytoskeleton enables an animal cell to crawl in a directed fashion over a solid surface. Lastly we shall discuss how the cytoskeleton generates the many morphological changes involved in the de-

Saito, A.; Inui, M.; Radermacher, M.; Frank, J.; Fleischer, S. Ultrastructure of the calcium release channel of sarcoplasmic reticulum. *J. Cell Biol.* 107:211–219, 1988.

9. Murray, J.M.; Weber, A. The cooperative action of muscle proteins. *Sci. Am.* 230(2):59–71, 1974.

Phillips, G.N.; Fillers, J.P.; Cohen, C. Tropomyosin crystal structure and muscle regulation. *J. Mol. Biol.* 192:111–131, 1986.

Zot, A.S.; Potter, J.D. Structural aspects of troponin-tropomyosin regulation of skeletal muscle contraction. *Annu. Rev. Biophys. Biophys. Chem.* 16:535–559, 1987.

10. Wang, K. Sarcomere-associated cytoskeletal lattices in striated muscle. Review and hypothesis. In Cell and Muscle Motility (J.W. Shay, ed.), Vol. 6, pp. 315–369. New York: Plenum Press, 1985.

11. Fawcett, D.W. A Textbook of Histology, 11th ed. Philadelphia: Saunders, 1986.

12. Korn, E.D.; Hammer, J.A. Myosins of nonmuscle cells. *Annu. Rev. Biophys. Biophys. Chem.* 17:23–45, 1988.

Sellers, J.R.; Adelstein, R.S. Regulation of contractile activity. In The Enzymes (P. Boyer, E.G. Krebs, ed.), Vol. 18, pp. 381–418. San Diego, CA: Academic Press, 1987.

13. Citi, S.; Kendrick-Jones, J. Regulation of non-muscle myosin structure and function. *Bioessays* 7:155–159, 1987.

14. Byers, H.R.; Fujiwara, K. Stress fibers in cells*in situ*: immunofluorescence visualization with anti-actin, anti-myosin and anti-alpha-actinin. *J. Cell Biol.* 93:804–811, 1982.

Langanger, B.; et al. The molecular organization of myosin in stress fibers of cultured cells. *J. Cell Biol.* 102:200–209, 1986.

Schroeder, T.E. Actin in dividing cells: contractile ring filaments bind heavy meromyosin. *Proc. Natl. Acad. Sci. USA* 70:1688–1692, 1973.

15. Buckingham, M.E. Actin and myosin multigene families: their expression during the formation of skeletal muscle. *Essays Biochem.* 20:77–109, 1985.

Emerson, C.P.; Bernstein, S.I. Molecular genetics of myosin. *Annu. Rev. Biochem.* 56:695–726, 1987.

Otey, C.A.; Kalnoski, M.H.; Bulinski, J.C. Identification and quantification of actin isoforms in vertebrate cells and tissues. *J. Cell. Biochem.* 34:113–124, 1987.

16. Breitbart, R.E.; Andreadis, A.; Nadal-Ginard, B. Alternative splicing: a ubiquitous mechanism for the generation of multiple protein isoforms from single genes. *Annu. Rev. Biochem.* 56:467–95, 1987.

17. Bray, D.; Heath, J.; Moss, D. The membrane-associated "cortex" of animal cells: its structure and mechanical properties. *J. Cell Sci.*, Suppl. 4:71–88, 1986.

Korn, E.D. Actin polymerization and its regulation by proteins from nonmuscle cells. *Physiol. Rev.* 62:672–737, 1982.

Pollard, T.D.; Cooper, J.A. Actin and actin-binding proteins. A critical evaluation of mechanisms and functions. *Annu. Rev. Biochem.* 55:987–1035, 1986.

Tilney, L.G. Interactions between actin filaments and membranes give spatial organization to cells. In Modern Cell Biology. Vol. 2: Spatial Organization of Eukaryotic Cells (J.R. McIntosh, B.H. Satir ed.) pp. 163–199, New York: Liss, 1983.

18. Sato, M.; Schwartz, W.H.; Pollard, T.D. Dependence of the mechanical properties of actin/alpha-actinin gels on deformation rate. *Nature* 325:828–830, 1987.

Stossel, T.P.; et al. Non-muscle actin binding proteins. *Annu. Rev. Cell Biol.* 1:353–402, 1985.

19. Matsudaira, P.; Janmey, P. Pieces in the actin-severing protein puzzle. *Cell* 54:139–140, 1988.

Yin, H.L. Gelsolin: calcium and polyphosphoinositide-regulated actin modulating protein. *Bioessays* 7:176–179, 1987.

20. Korn, E.D.; Hammer, J.A. Myosins of nonmuscle cells. *Annu. Rev. Biophys. Biophys. Chem.* 17:23–45, 1988.

Warrick, H.M.; Spudich, J.A. Myosin structure and function in cell motility. *Annu. Rev. Cell Biol.* 3:379–421, 1987.

21. Adams, R.J.; Pollard. T.D. Propulsion of organelles isolated from *Acanthamoeba* along actin filaments by myosin-I. *Nature* 322:754–756, 1986.

Sheetz, M.P., Spudich, J.A. Movement of myosin-coated fluorescent beads on actin cables *in vitro*. *Nature* 303:31–35, 1983.

22. Bennett, V. The membrane skeleton of human erythrocytes and its implications for more complex cells. *Annu. Rev. Biochem.* 54:273–304, 1985.

23. Conzelman, K.A.; Mooseker, M.S. The 110-kD protein-calmodulin complex of the intestinal microvillus in an actin-activated MgATPase. *J. Cell Biol.* 105:313–324, 1987.

Mooseker, M.S. Organization, chemistry, and assembly of the cytoskeletal apparatus of the intestinal brush border. *Annu. Rev. Cell Biol.* 1:209–241, 1985.

24. Burridge, K.; et al. Focal adhesions: transmembrane junctions between the extracellular matrix and the cytoskeleton. *Annu. Rev. Cell Biol.* 4:487–525, 1988.

Horwitz, A.; Duggan, K.; Buck, C.; Beckerle, M.C.; Burridge, K. Interaction of plasma membrane fibronectin receptor with talin—a transmembrane linkage. *Nature* 320:531–533, 1986.

25. Bonder, E.M.; Fishkind, D.J.; Mooseker, M.S. Direct measurement of critical concentrations and assembly rate constants at the two ends of an actin filament. *Cell* 34:491–501, 1983.

Korn, E.D.; Carlier, M-F.; Pantaloni, D. Actin polymerization and ATP hydrolysis. *Science* 238:638–644, 1987.

26. Abercrombie, M. The crawling movement of metazoan cells. *Proc. R. Soc. Lond. (Biol.)* 207:129–147, 1980.

Small, J.V.; Rinnerthaler, G.; Hinssen, H. Organization of actin meshworks in cultured cells: the leading edge. *Cold Spring Harbor Symp. Quant. Biol.* 46:599–611, 1982.

Wang, Y. Exhange of actin subunits at the leading edge of living fibroblasts: possible role of treadmilling. *J. Cell Biol.* 101:597–602, 1985.

27. Tilney, L.G.; Inoué, S. Acrosomal reaction of *Thyone* sperm. II. The kinetics and possible mechanism of acrosomal process elongation. *J. Cell Biol.* 93:820–827, 1982.

28. Carson, M.; Weber, A.; Zigmond, S.H. An actin-nucleating activity in polymorphonuclear leukoyctes is modulated by chemotactic peptides. *J. Cell. Biol.* 103:2707–2714, 1986.

Devreotes, P.; Zigmond, S. Chemotaxis in eucaryotic cells. *Annu. Rev. Cell Biol.* 4:649–686, 1988.

Tilney, L.G.; Bonder, E.M.; DeRosier, D.J. Actin filaments elongate from their membrane-associated ends. *J. Cell Biol.* 90:485–494, 1981.

29. Cooper, J.A. Effects of cytochalasin and phalloidin on actin. *J. Cell Biol.* 105:1473–1478, 1987.

30. Dustin, P. Microtubules, 2nd ed., pp. 127–164. New York: Springer-Verlag, 1984.

Gibbons, I.R. Cilia and flagella of eukaryotes. *J. Cell Biol.* 91:107s–124s, 1981.

Roberts, K.; Hyams, J.S.; eds. Microtubules. New York: Academic Press, 1979.

Satir, P. How cilia move. *Sci. Am* 231(4):44–63, 1974.

31. Amos, L.A.; Baker, T.S. The three dimensional structure of tubulin protofilaments. *Nature* 279:607–612, 1979.

Mandelkow, E-M., Schultheiss, R., Rapp, R., Múller, M., Mandelkow, E. On the surface lattice of microtubules: helix starts, protofilament number, seam and handedness. *J. Cell Biol.* 102:1067–1073, 1986.

Raff, E.C. Genetics of microtubule systems. *J. Cell Biol.* 99:1–10, 1984.

Sullivan, K.F. Structure and utilization of tubulin isotypes. *Annu. Rev. Cell Biol.* 4:687–716, 1988.

32. Linck, R.W.; Amos, L.A.; Amos, W.B. Localization of tektin filaments in microtubules of sea urchin sperm flagella by immunoelectron microscopy. *J. Cell Biol.* 100:126–135, 1985.
33. Goodenough, U.W.; Heuser, J.E. Substructure of inner dynein arms, radial spokes, and the central pair/projection complex of cilia and flagella. *J. Cell Biol.* 100:2008–2018, 1985.
34. Summers, K.E.; Gibbons, I.R. ATP-induced sliding of tubules in trypsin-treated flagella of sea urchin sperm. *Proc. Natl. Acad. Sci, USA* 68:3092–3096, 1971.
Warner, F.D.; Satir, P. The structural basis of ciliary bend formation. *J. Cell Biol.* 63:35–63, 1974.
35. Johnson, K.A. Pathway of the microtubule-dynein ATPase and structure of dynein: a comparison with actomyosin. *Annu. Rev. Biophys. Biophys. Chem.* 14:161–188, 1985.
36. Brokaw, C.J. Future directions for studies of mechanisms for generating flagellar bending waves. *J. Cell Sci.*, Suppl. 4:103–113, 1986.
Brokaw, C.J.; Luck, D.J.L.; Huang, B. Analysis of the movement of *Chlamydomonas* flagella: the function of the radial-spoke system is revealed by comparison of wild-type and mutant flagella. *J. Cell Biol.* 92:722–732, 1982.
37. Afzelius, B.A. The immotile-cilia syndrome: a microtubule-associated defect. *CRC Crit. Rev. Biochem.* 19:63–87, 1985.
Huang, B. *Chlamydomonas reinhardtii*: a model system for genetic analysis of flagellar structure and motility. *Int. Rev. Cytol.* 99:181–215, 1986.
Luck, D.J.L. Genetic and biochemical dissection of the eucaryotic flagellum. *J. Cell Biol.* 98:789–794, 1984.
38. Lefebvre, P.A.; Rosenbaum, J.L. Regulation of the synthesis and assembly of ciliary and flagellar proteins during regeneration. *Annu. Rev. Cell Biol.* 2:517–546, 1986.
Wheatley, D.N. The Centriole: A Central Enigma of Cell Biology. New York: Elsevier, 1982.
39. Karsenti, E.; Maro, B. Centrosomes and the spatial distribution of microtubules in animal cells. *Trends Biochem. Sci.* 11:460–463, 1986.
Ramanis, Z.; Luck, D.J.L. Loci affecting flagellar assembly and function map to an unusual linkage group in *Chlamydomonas reinhardtii. Proc. Natl. Acad. Sci. USA* 83:423–436, 1986.
Vorobjev, I.A.; Chentsov, Y.S. Centrioles in the cell cycle. 1. Epithelial cells. *J. Cell. Biol.* 93:938–949, 1982.
40. Dustin, P. Microtubules, 2nd ed. Berlin: Springer-Verlag, 1984.
41. De Brabander, M.; et al. Microtubule dynamics during the cell cycle: the effects of taxol and nocodazole on the microtubule system of Ptk2 cells at different stages of the mitotic cycle. *Int. Rev. Cytol.* 101:215–274, 1986.
Inoué, S. Cell division and the mitotic spindle. *J. Cell Biol.* 91:131s–147s, 1981.
Salmon, E.D.; McKeel, M.; Hays, T. Rapid rate of tubulin dissociation from microtubules in the mitotic spindle *in vivo* measured by blocking polymerization with colchicine. *J. Cell Biol.* 99:1066–1075, 1984.
42. Farrell, K.W.; Jordan, M.A.; Miller, H.P.; Wilson, L. Phase dynamics at microtubule ends: the coexistence of microtubule length changes and treadmilling. *J. Cell Biol.* 104:1035–1046, 1987.
McIntosh, J.R.; Euteneuer, U. Tubulin hooks as probes for microtubule polarity: an analysis of the method and evaluation of data on microtubule polarity in the mitotic spindle. *J. Cell Biol.* 98:525–533, 1984.
43. Carlier, M-F. Role of nucleotide hydrolysis in the polymerization of actin and tubulin. *Cell Biophys.* 12:105–117, 1988.
Horio, H. Hotani, H. Visualization of the dynamic instability of individual microtubules by dark field microscopy. *Nature* 321:605–607, 1986.
Mitchison, T., Kirschner, M. Dynamic instability of microtubule growth. *Nature* 312:237–242, 1984.
44. Karsenti, E.; Maro, B. Centrosomes and the spatial distribution of microtubules in animal cells. *Trends Biochem. Sci.* 11:460–463, 1986.
Mitchison, T., Kirschner, M. Microtubule assemby nucleated by isolated centrosomes. *Nature* 312:232–237, 1984.
45. Kirschner, M.; Mitchison, T. Beyond self-assembly: from microtubules to morphogenesis. *Cell* 45:329–342, 1986.
Sammak, P.J. Borisy, G.G. Direct observation of microtubule dynamics in living cells. *Nature* 332:724–726, 1988.
46. Barra, H.S.; Arce, C.A.; Argarana, C.E. Posttranslational tyrosination detyrosination of tubulin. *Molec. Neurobiol.* 2:133–153, 1988.
Gundersen, G.G., Khawja, S., Bulinski, J.C. Postpolymerization detyrosination of [α]-tubulin: a mechanism for subcelluar differentiation of microtubules. *J. Cell Biol.* 105:251–264, 1987.
Maruta, H., Greer, K., Rosenbaum, J.L. The acetylation of α-tubulin and its relationship to the assembly and disassembly of microtubules. *J. Cell Biol.* 103:571–579, 1986.
Schulze, E.; Asai, D.J.; Bulinski, J.C.; Kirschner, M. Post-translational modification and microtubule stability. *J. Cell Biol.* 105:2167–2177, 1987.
47. Olmsted, J.B. Microtubule-associated proteins. *Annu. Rev. Cell Biol.* 2:421–457, 1986.
Vallee, R.B.; Bloom, G.S.; Theurkauf, W.E. Microtubule-associated proteins: subunits of the cytomatrix. *J. Cell Biol.* 99:38s–44s, 1984.
48. Allen, R.D. The microtubule as an intracellular engine. *Sci. Am.* 256(2):42–49, 1987.
Allen, R.D.; et al. Gliding movement of and bidirectional transport along single native microtubules from squid axoplasm: evidence for an active role of microtubules in cytoplasmic transport. *J. Cell Biol.* 100:1736–1752, 1985.
49. Vale, R. Intracelluar transport using microtubule-based motors. *Annu. Rev. Cell Biol.* 3:347–378, 1987.
Vale, R.D.; Reese, T.S.; Sheetz, M.P. Indentification of a novel force-generating protein, kinesin, involved in microtubule-based motility. *Cell* 42:39–50, 1985.
Vallee, R.B.; Wall, J.S.; Paschal, B.M.; Shpetner, H.S. Microtubule-associated protein 1C from brain is a two-headed cytosolic dynein. *Nature* 332:561–563, 1988.
50. Allan, V.J.; Kreis, T.E. A microtubule-binding protein associated with membranes of the Golgi apparatus. *J. Cell Biol.* 103:2229–2239, 1986.
Dabora, S.L.; Sheetz, M.P. The microtubule-dependent formation of a tubulovesicular network with characteristics of the ER from cultured cell extracts. *Cell* 54:27–35, 1988.
Lee, C.; Chen, L.B. Dynamic behavior of endoplasmic reticulum in living cells. *Cell* 54:37–46, 1988.
Lucocq, J.M.; Warren, G. Fragmentation and partitioning of the Golgi apparatus during mitosis in Hela cells. *EMBO J.* 6:3239–3246, 1987.
51. Geiger, B. Intermediate filaments: looking for a function. *Nature* 329:392–393, 1987.
Steinert, P.M.; Roop, D.R. Molecular and cellular biology of intermediate filaments. *Annu. Rev. Biochem.* 57:593–626, 1988.
Traub, P. Intermediate Filaments: A Review. New York: Springer-Verlag, 1985.
Wang, E.; Fischman, D.; Liem, R.K.H.; Sun, T.-T., eds. Intermediate Filaments. *Ann. N.Y. Acad. Sci.* 455, 1985.
52. Osborn, M.; Weber, K. Tumor diagnosis by intermediate filament typing: a novel tool for surgical pathology. *Lab. Invest.* 48:372–394, 1983.
53. Ip. W.; Hartzer, M.K.; Pang, S.Y.-Y.; Robson, R.M. Assembly of vimentin *in vitro* and its implications concerning the structure of intermediate filaments. *J. Mol. Biol.* 183:365–375, 1985.

Quinlan, R.A.; et al. Characterization of dimer subunits of intermediate filament proteins. *J. Mol. Biol.* 192:337–349, 1986.

54. Geuens, G.; De Brabander, M.; Nuydens, R.; De Mey, J. The interaction between microtubules and intermediate filaments in cultured cells treated with taxol and nocodazole. *Cell Biol. Int. Rep.* 7:35–47, 1983.

Goldman, R.; et al. Intermediate filaments: possible functions as cytoskeletal connecting links between the nucleus and the cell surface. *Ann. N.Y. Acad. Sci.* 455:1–17, 1985.

55. Geisler, N.; Weber, K. Phosphorylation of desmin *in vitro* inhibits formation of intermediate filaments: identification of three kinase A sites in the aminoterminal head domain. *EMBO J.* 7:15–20, 1988.

Inagaki, M,; Nishi, Y.; Nishizawa, K.; Matsuyama, M.; Sato, C. Site-specific phosphorylation induces disassembly of vimentin filaments *in vitro*. *Nature* 328:649–652, 1987.

56. Aebi, U.; Cohn, J.; Buhle, L.; Gerace, L. The nuclear lamina is a meshwork of intermediate-type filaments. *Nature* 323:560–564, 1986.

McKeon, F.D.; Kirschner, M.W.; Caput, D. Homologies in both primary and secondary structure between nuclear envelope and intermediate filament proteins. *Nature* 319:463–468, 1986.

57. Abercrombie, M. The crawling movment of metazoan cells. *Proc. R. Soc. Lond. (Biol.)* 207:129–147, 1980.

Bridgman, P. Structure of cytoplasm as revealed by modern electron microscopy techniques. *Trends Neurosci.* 10:321–325, 1987.

Singer, S.J.; Kupfer, A. The directed migration of eukaryotic cells. *Annu. Rev. Cell Bio.* 2:337–365, 1986.

Trinkaus, J.P. Cells into Organs: The Forces that Shape the Embryo. 2nd ed. Englewood Cliffs, NJ: Prentice-Hall, 1984.

58. Bridgman, P.C.; Reese, T.S. The structure of cytoplasm in directly frozen cultured cells. 1. Filamentous meshworks and the cytoplasmic ground substance. *J. Cell Biol.* 99:1655–1668, 1984.

Heuser, J.; Kirschner, M.W. Filament organization revealed in platinum replicas of freeze-dried cytoskeletons. *J. Cell Biol.* 86:212–234, 1980.

59. Fulton, A.B. How crowded is the cytoplasm? *Cell* 30:345–347, 1982.

Luby-Phelps, K.; Taylor, D.L.; Lanni, F. Probing the structure of the cytoplasm. *J. Cell Biol.* 102:2015–2022, 1986.

60. Kolega, J. Effects of mechanical tension on protrusive activity and microfilament and intermediate filament organization in an epidermal epithelium moving in culture. *J. Cell Biol.* 102:1400–1411, 1986.

Trinkaus, J.P. Cells into Organs: The Forces That Shape the Embryo, 2nd. ed., pp. 157–244. Englewood Cliffs, NJ: Prentice-Hall, 1984.

61. Bray, D.; Hollenbeck, P.J. Growth cone motility and guidance. *Annu. Rev. Cell Biol.* 4:43–62, 1988.

Euteneuer, U.; Schliwa, M. Persistent, directional motility of cells and cytoplasmic fragments in the absence of microtubles. *Nature* 310:58–61, 1984.

Malawista, S.E.; De Boisfleury Chevance, A. The cytokinetplast: purified, stable, and functional motile machinery from human blood polymorphonuclear leukocytes. *J. Cell Biol.* 95:960–973, 1982.

Marsh, L.; Letourneau, P.C. Growth of neurites without filopodial or lamellipodial activity in the presence of cytochalasin B. *J. Cell Biol.* 99:2041–2047, 1984.

Vasiliev, J.M.; et al. Effect of colcemid on the locomotion of fibroblasts. *J. Embryol. Exp. Morphol.* 24:625–640, 1970.

62. Bray, D.; White, J.G. Cortical flow in animal cells. *Science* 239:883–888, 1988.

De Lozanne, A.; Spudich, J.A. Disruption of the *Dictyostelium* myosin heavy chain gene by homologous recombination. *Science* 236:1086–1091, 1987.

Knecht, D.A.; Loomis, W.F. Antisense RNA inactivation of myosin heavy chain gene expression in *Dictyostelium discoideum*. *Science* 236:1081–1086, 1987.

63. Bergmann, J.E.; Kupfer, A.; Singer, S.J. Membrane insertion at the leading edge of motile fibroblasts. *Proc. Natl. Acad. Sci. USA* 80:1367–1371, 1983.

Bretscher, M.S. How animal cells move. *Sci. Am.* 257(6):72–90, 1987.

64. Lackie, J.M. Cell Movement and Cell Behaviour. pp. 253–275 London: Allen and Unwin, 1986.

Trinkaus, J.P. Cells into Organs: The Forces That Shape the Embryo. 2nd ed., pp. 157–244. Englewood Cliffs, NJ: Prentice-Hall, 1984.

65. Odell, G.M.; Oster, G.; Alberch, P.; Burnside, B. The mechanical basis of morphogenesis. 1. Epithelial folding and invagination. *Dev. Biol.* 85:446–462, 1981.

66. Tilney LG; Tilney MS; Cotanche DA. Actin filaments, stereocilia, and hair cellls of the bird cochlea. V. How the staircase pattern of stereociliary lengths is generated. *J Cell Biol.* 106:355–365, 1988.

Tilney, L.G.; De Rosier, D.J. Actin filaments, stereocilia, and hair cells of the bird cochlea. IV. How the actin filaments become organized in developing stereocilia and in the cuticular plate. *Dev. Biol.* 116:119–129, 1986.

Cell Signaling

12

Cells in a multicellular organism need to communicate with one another in order to regulate their development and organization into tissues, to control their growth and division, and to coordinate their functions. Animal cells communicate in three ways: (1) they secrete chemicals that signal to cells some distance away; (2) they display plasma-membrane-bound signaling molecules that influence other cells in direct physical contact; and (3) they form gap junctions that directly join the cytoplasms of the interacting cells, thereby allowing exchange of small molecules (Figure 12–1).

Communication that depends on cell-cell contact through gap junctions will be discussed in Chapter 14. In this chapter we shall be concerned primarily with communication at a distance that is mediated by secreted chemical signals. This

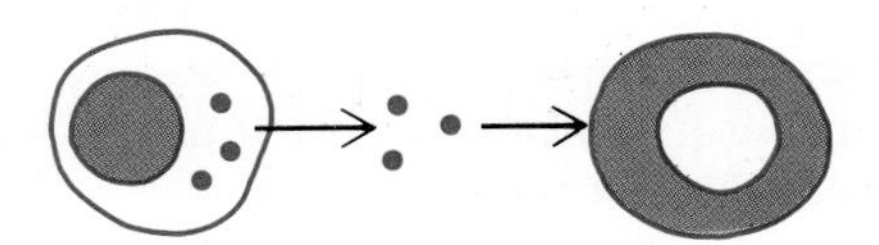

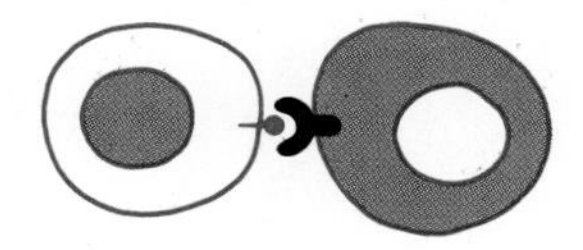

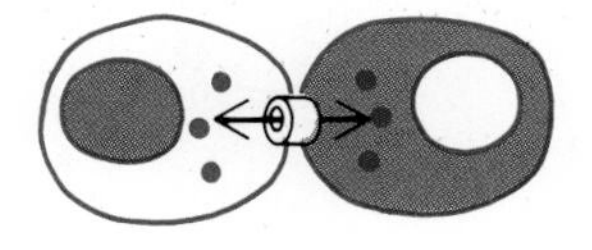

Figure 12–1 Three ways in which cells communicate with one another.

emphasis reflects the state of current knowledge. Secreted molecules are very much easier to study than those that are membrane-bound, and much is known about how they work. Contact-dependent signaling via membrane-bound molecules, although harder to demonstrate and less well studied, may nonetheless be important, especially during development and in immune responses (see p. 1037); its molecular basis is thought to be closely related to that of signaling at a distance. Specialized aspects of chemical signaling in the nervous system and the unique principles that apply to signaling in plants will be discussed separately, in Chapters 19 and 20, respectively.

Three Strategies of Chemical Signaling: Endocrine, Paracrine, and Synaptic

Chemical signaling mechanisms vary in the distances over which they operate: (1) In **endocrine signaling,** specialized endocrine cells secrete **hormones,** which travel through the bloodstream to influence target cells that are distributed widely throughout the body. (2) In **paracrine signaling,** cells secrete **local chemical mediators,** which are so rapidly taken up, destroyed, or immobilized that the mediators act only on cells in the immediate environment, perhaps within a millimeter or so. (3) In **synaptic signaling,** which is confined to the nervous system, cells secrete **neurotransmitters** at specialized junctions called *chemical synapses;* the neurotransmitter diffuses across the synaptic cleft, typically a distance of about 50 nm, and acts only on the adjacent postsynaptic target cell (Figure 12–2). In each case the target cell responds to a particular extracellular signal by means of specific proteins, called **receptors,** that bind the signaling molecule and initiate the response. Many of the same signaling molecules and receptors are used in endocrine, paracrine, and synaptic signaling. The crucial differences lie in the speed and selectivity with which the signals are delivered to their targets.

12-3 12-4 Endocrine Cells and Nerve Cells Are Specialized for Different Types of Chemical Signaling[1]

Endocrine cells and nerve cells work together to coordinate the diverse activities of the billions of cells in a higher animal. The endocrine cells are usually organized in discrete glands, and they secrete their hormone molecules into the extracellular (interstitial) fluid that surrounds all cells in tissues. From there the molecules diffuse into capillaries to enter the bloodstream, which carries them to tissues throughout the body. Within each tissue, hormone molecules escape from capillaries into the interstitial fluid, where they can bind to their target cells. Because endocrine signaling relies on diffusion and blood flow, it is relatively slow: it usually takes minutes for a hormone to reach its target cells after secretion. Moreover, the specificity of signaling in the endocrine system depends entirely on the chemistry of the signal and the receptors on the target cell: each type of endocrine cell secretes a different hormone into the blood, and each cell that has complementary receptors will respond in a way appropriate for the cell type (Figure 12–3A).

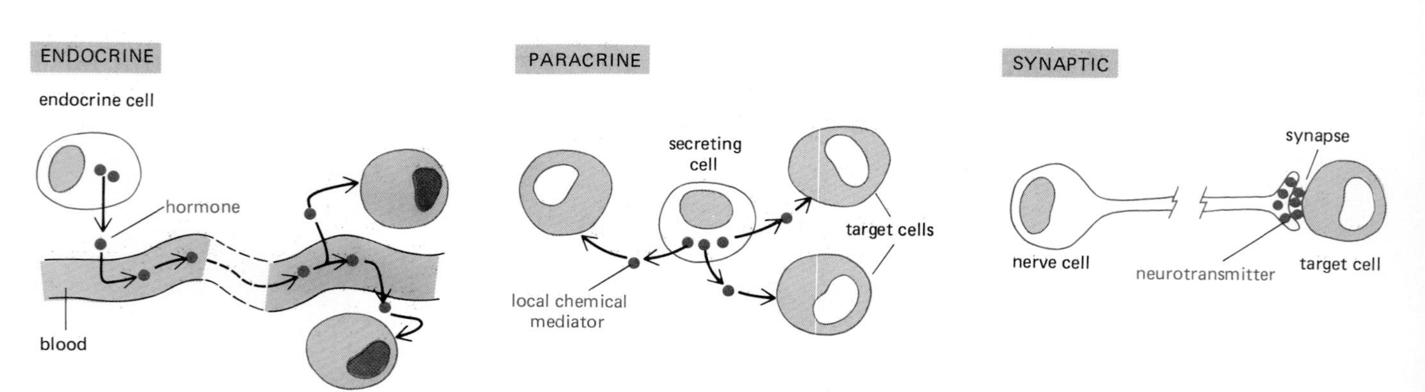

Figure 12–2 Three forms of signaling mediated by secreted molecules. Not all neurotransmitters act in the strictly synaptic mode shown; some act in a paracrine mode as local chemical mediators that influence multiple target cells in the area.

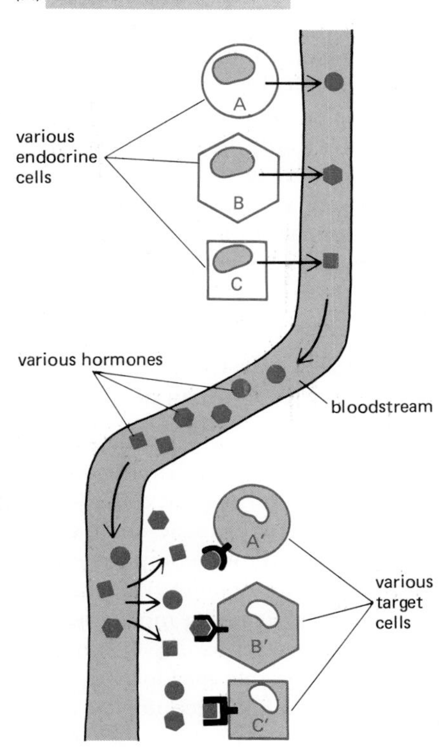

Figure 12–3 The contrast between endocrine (A) and synaptic signaling (B). Endocrine cells secrete many different hormones into the blood and signal specific target cells, which have receptors for binding specific hormones and thereby "pull" the appropriate hormones from the extracellular fluid. In synaptic signaling, by contrast, the specificity arises from the contacts between nerve processes and the specific target cells they signal: only a target cell that is in synaptic contact with a nerve cell is exposed to the neurotransmitter released from the nerve terminal. Different endocrine cells must use different hormones in order to communicate specifically with their target cells, but many nerve cells can use the same neurotransmitter and still communicate in a specific manner.

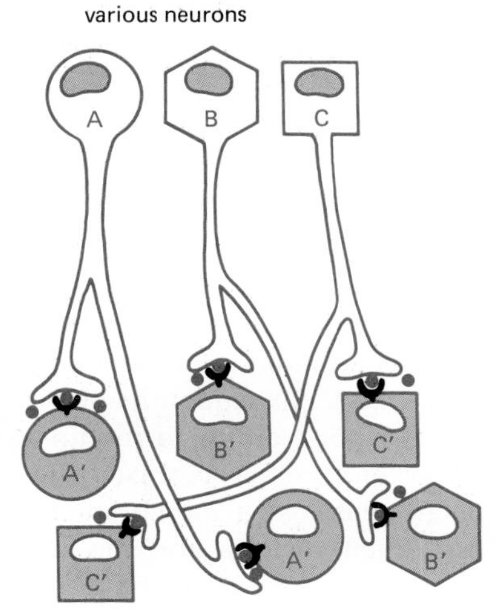

Nerve cells, by contrast, can achieve much greater speed and precision. They can transmit information over long distances by means of electrical impulses that carry signals along nerve processes at rates of up to 100 meters per second. Only when a neurotransmitter is released at the nerve terminal is the electrical impulse converted into a chemical signal. Chemical signals released by nerve cells can act in either the paracrine or the synaptic mode. In the paracrine mode the neurotransmitter functions as a local chemical mediator, diffusing outward to influence any target cells in the neighborhood that have receptors for the released molecule. Signaling in the synaptic mode is much more precise, and the effect of the neurotransmitter is confined to a single target cell even if adjacent cells have receptors for the same neurotransmitter (see Figure 12–3B); here the neurotransmitter has to diffuse less than 100 nm to the target cell, a process that takes less than a millisecond (see Figure 12–2).

Hormones are greatly diluted in the bloodstream and interstitial fluid and therefore must be able to act at very low concentrations (typically $<10^{-8}$ M), but neurotransmitters are diluted much less in their short journey to the target cell and can achieve high local concentrations. For example, the concentration of the neurotransmitter acetylcholine in the synaptic cleft of an active neuromuscular junction is about 5×10^{-4} M. Correspondingly, in synaptic signaling the neurotransmitter receptors have a relatively low affinity for their ligand, and, as a result, they do not respond significantly to the low concentrations of transmitter that reach them by diffusion from neighboring synapses. Moreover, after being secreted, the neurotransmitter is quickly removed from the synaptic cleft either by specific hydrolytic enzymes or by specific membrane transport proteins that pump the neurotransmitter back into the nerve terminal. Rapid removal ensures not only spatial precision of the signal but also temporal precision: a brief pulse of neurotransmitter release evokes a prompt and brief response, so that the timing of the signal can be faithfully relayed from cell to cell (see p. 1081).

Neuroendocrine Cells in the Hypothalamus Regulate the Endocrine System[1,2]

The endocrine system and nervous system in vertebrates are physically and functionally linked by a specific region of the brain called the **hypothalamus.** The hypothalamus communicates directly with the *pituitary gland* via a bridge called the *pituitary stalk.* The linking function of the hypothalamus is mediated by cells that have properties of both nerve cells and endocrine cells; for this reason they are called *neuroendocrine cells.* Most types of hypothalamic neuroendocrine cells respond to stimulation by other nerve cells in the brain by secreting a specific peptide hormone into the special blood vessels of the pituitary stalk; on reaching the pituitary gland, each hormone specifically stimulates or suppresses the secretion of a second hormone into the main bloodstream. (Other hypothalamic neuroendocrine cells send their axons along the pituitary stalk and discharge their secretions directly into the main bloodstream.) Because many of the pituitary hormones released under the control of the hypothalamus stimulate another endocrine gland to secrete a third hormone into the blood, the hypothalamus serves as the main regulator of the endocrine system in vertebrates. As an example, Figure 12–4 illustrates how the hypothalamus regulates the secretion of *thyroid hormone.*

Selected examples of local chemical mediators, neurotransmitters, and hormones are listed in Table 12–1, together with their sites of origin, structures, and principal actions. It can be seen that signaling molecules are as varied in structure as they are in function. They include small peptides, larger proteins and glycoproteins, amino acids and related compounds, steroids (molecules derived from cholesterol and closely related in structure), and fatty acid derivatives. Although each signaling molecule is listed in only one category in Table 12–1, many of them can act in more than one mode. Many peptide hormones, for example, also act as neurotransmitters (in the paracrine mode) in the vertebrate brain.

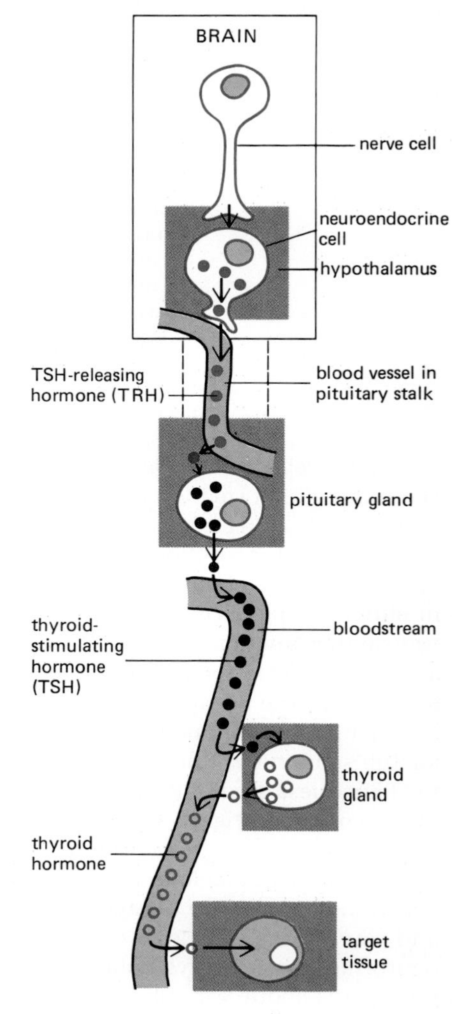

Figure 12–4 Thyroid hormone secretion is regulated indirectly by the nervous system. When stimulated by nerve cells in higher centers of the brain, specific neuroendocrine cells in the hypothalamus secrete TSH-releasing hormone (TRH) into blood vessels of the pituitary stalk, which transport the hormone to the pituitary gland. Here the TRH stimulates specific cells to release TSH (thyroid-stimulating hormone) into the main bloodstream, which transports the hormone to the thyroid gland. TSH in turn stimulates the cells in the thyroid gland to synthesize and secrete thyroid hormone, which is transported in the blood to most cells in the body. Thyroid hormone then stimulates a variety of metabolic processes in these cells. The secretion of both TRH and TSH is suppressed by increased concentrations of thyroid hormone in the blood (not shown). This *feedback inhibition* prevents the level of thyroid hormone in the blood from rising too high. Many hormones are regulated by a similar feedback mechanism.

Different Cells Respond in Different Ways to the Same Chemical Signal

Most cells in mature animals are specialized to perform one primary function, and they contain a characteristic array of receptor proteins that allows them to respond to each of the different chemical signals that initiate or modulate that function. Many of these chemical signals act at very low concentration (typically $\leq 10^{-8}$ M), and their complementary receptors usually bind the signaling molecule with high affinity (affinity constant $K_a \geq 10^8$ liters/mole; see p. 98).

A single signaling molecule often has different effects in different target cells. Acetylcholine, for example, stimulates the contraction of skeletal muscle cells but decreases the rate and force of contraction in heart muscle cells. This is because the acetylcholine receptor proteins on skeletal muscle cells are different from those on heart muscle cells. But receptor differences are not always the explanation for the different effects. In many cases the same signaling molecule binds to identical receptor proteins and yet produces very different responses in different types of target cells (Figure 12–5). This indicates that the responses to a signaling molecule are programmed in two ways—through the receptors the target cells carry, and through the internal machinery to which the receptors are coupled.

Some Cellular Responses to Chemical Signals Are Rapid and Transient, While Others Are Slow and Long-lasting[2,3]

In coordinating the responses of cells to changes in an animal's environment, chemical signals generally induce rapid and transient responses. An increase in blood glucose levels, for instance, stimulates endocrine cells in the pancreas to secrete the protein hormone *insulin* into the blood. Within minutes the resulting increase in insulin concentration stimulates fat and muscle cells to take up more glucose, and, consequently, blood glucose levels fall. There are three parts to the response, none of which requires new protein synthesis: (1) In the pancreas the elevated glucose levels trigger the exocytic release of stored insulin. (2) In fat and muscle cells, extra membrane-bound glucose transport proteins are stored in intracellular vesicles and the elevated insulin levels cause these proteins to be added by exocytosis to the plasma membrane, where they increase the rate of glucose uptake. This causes blood glucose levels to drop, and, as a result, the rate of insulin secretion decreases. (3) Since the extra glucose carriers are rapidly removed from the cell surface by endocytosis and returned to the intracellular pool, when insulin levels decrease, the rate of glucose uptake by fat and muscle cells returns to its previous level. In this way insulin helps to maintain a relatively constant blood glucose concentration. Neurotransmitters elicit even more rapid responses than hormones do: skeletal muscle cells contract and relax again within milliseconds in response to acetylcholine released from nerve terminals at neuromuscular junctions.

Chemical signals also play an important part in animal development, often influencing when and how certain cells differentiate. Some of these effects are slow in onset and long-lasting. For example, the female sex hormone *estradiol,* a steroid, is secreted in large amounts by cells in the ovary around the time of puberty. It induces changes in a wide variety of cells in different parts of the body,

Table 12–1 Some Examples of Extracellular Signaling Molecules

Local Chemical Mediators	Site of Origin	Structure	Major Effects
Proteins			
Nerve growth factor	skin; all tissues innervated by sympathetic nerves	2 identical chains of 118 amino acids	survival and growth of sensory and sympathetic neurons and some neurons in central nervous system
Small Peptides			
Eosinophil chemotactic factor	mast cells	4 amino acids	chemotactic signal for a particular class of white blood cells (eosinophils)
Amino Acid Derivatives			
Histamine	mast cells	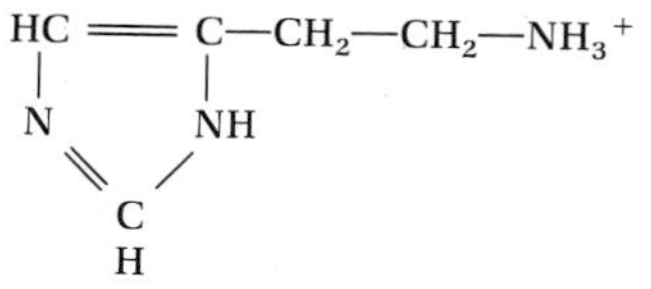 $HC{=}C{-}CH_2{-}CH_2{-}NH_3^+$; ring: HC, N, C–H, NH	causes blood vessels to dilate and become leaky
Fatty Acid Derivatives			
Prostaglandin E_2	many cell types	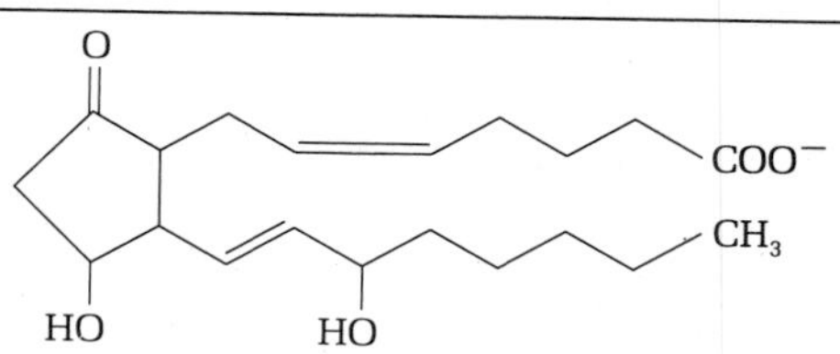 O; COO^-; CH_3; HO; HO	contraction of smooth muscle

Neurotransmitters*	Site of Origin	Structure	Major Effects
Amino Acids and Related Compounds			
Glycine	nerve terminals	$^+H_3N{-}CH_2{-}COO^-$	inhibitory transmitter in central nervous system
Norepinephrine (noradrenaline)	nerve terminals	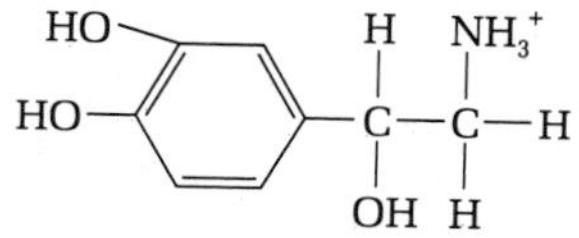 HO, HO (ring); $-\overset{H}{\underset{OH}{C}}-\overset{NH_3^+}{\underset{H}{C}}-H$	excitatory and inhibitory transmitter in central and peripheral nervous system
γ-Aminobutyric acid (GABA)	nerve terminals	$^+H_3N{-}CH_2{-}CH_2{-}CH_2{-}COO^-$	inhibitory transmitter in central nervous system
Acetylcholine	nerve terminals	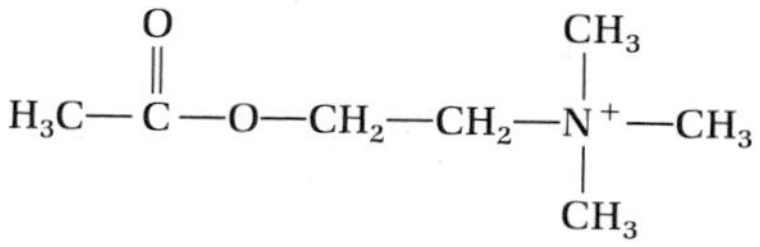 $H_3C{-}\overset{O}{\overset{\Vert}{C}}{-}O{-}CH_2{-}CH_2{-}\overset{CH_3}{\underset{CH_3}{N^+}}{-}CH_3$	excitatory transmitter at neuromuscular junction; excitatory and inhibitory transmitter in central and peripheral nervous system
Small Peptides			
Enkephalin	nerve terminals	5 amino acids	morphinelike action (inhibits pain pathways in central nervous system)

*Norepinephrine and enkephalin act in paracrine rather than synaptic signaling; acetylcholine can act in either mode.
Excitatory neurotransmitters stimulate target cell activity, whereas inhibitory neurotransmitters suppress target cell activity.

(Continued)

Table 12–1 *Continued*

Hormones**	Site of Origin	Structure	Major Effects
Proteins			
Insulin	beta cells of pancreas	protein α-chain = 21 amino acids β-chain = 30 amino acids	utilization of carbohydrate (including uptake of glucose into cells); stimulation of protein synthesis; stimulation of lipid synthesis in fat cells
Somatotropin (growth hormone)	anterior pituitary	protein 191 amino acids	stimulation of liver to produce somatomedin-1, which in turn causes growth of muscle and bone; stimulation of fat, muscle, and cartilage cell differentiation
Somatomedin-1 (insulinlike growth factor-1)	mainly liver	protein 70 amino acids	growth of bone and muscle; influences metabolism of Ca^{2+}, phosphate, carbohydrate, and lipid
Adrenocorticotropic hormone (ACTH)	anterior pituitary	protein 39 amino acids	stimulation of adrenal cortex to produce cortisol; fatty acid release from fat cells
Parathormone	parathyroid	protein 84 amino acids	increase in bone resorption, thereby increasing blood Ca^{2+} and phosphate; increase in resorption of Ca^{2+} and Mg^{2+} and decrease in resorption of phosphate in kidney tubules
Follicle-stimulating hormone (FSH)	anterior pituitary	glycoprotein α-chain = 92 amino acids β-chain = 118 amino acids	stimulation of ovarian follicles to grow and secrete estradiol; stimulation of spermatogenesis in testis
Luteinizing hormone (LH)	anterior pituitary	glycoprotein α-chain = 92 amino acids β-chain = 115 amino acids	stimulation of oocyte maturation and ovulation and progesterone secretion from ovary; stimulation of testis to produce testosterone
Epidermal growth factor	unknown	protein 53 amino acids	stimulation of epidermal and other cells to divide
Thyroid-stimulating hormone (TSH)	anterior pituitary	glycoprotein α-chain = 92 amino acids β-chain = 112 amino acids	stimulation of thyroid to produce thyroid hormone; fatty acid release from fat cells
Small Peptides			
TSH-releasing hormone (TRH)	hypothalamus	3 amino acids	stimulation of anterior pituitary to secrete thyroid-stimulating hormone (TSH)

**Many of the nonsteroid hormones are also made by some nerve cells in the brain.

(Continued)

Table 12–1 *Continued*

Hormones	Site of Origin	Structure	Major Effects
LH-releasing hormone	hypothalamus	10 amino acids	stimulation of anterior pituitary to secrete luteinizing hormone (LH)
Vasopressin (antidiuretic hormone, ADH)	posterior pituitary	9 amino acids	elevation of blood pressure by constriction of small blood vessels; increase in water resorption in kidney tubules
Somatostatin	hypothalamus	14 amino acids	inhibition of somatotropin release from anterior pituitary
Amino Acid Derivatives			
Epinephrine (adrenaline)	adrenal medulla	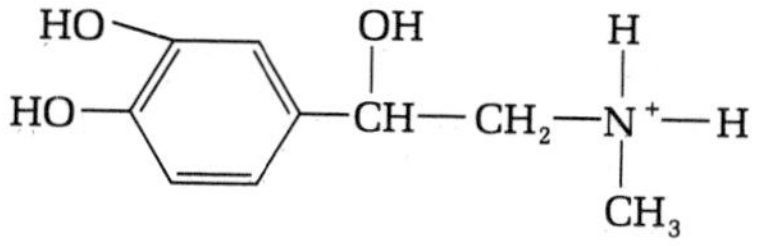 HO, HO, OH, —CH—CH_2—N^+—H, H, CH_3	increase in blood pressure and heart rate; increase in glycogenolysis in liver and muscle; fatty acid release from fat cells
Thyroid hormone (thyroxine)	thyroid	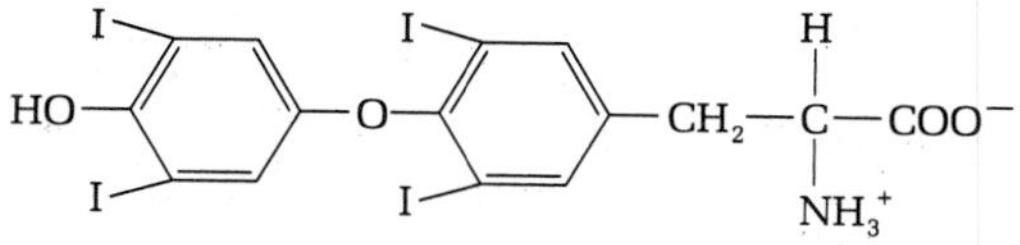 I, I, HO—, —O—, I, I, —CH_2—C—COO^-, H, NH_3^+	increase in metabolic activity in most cells
Steroids			
Cortisol	adrenal cortex	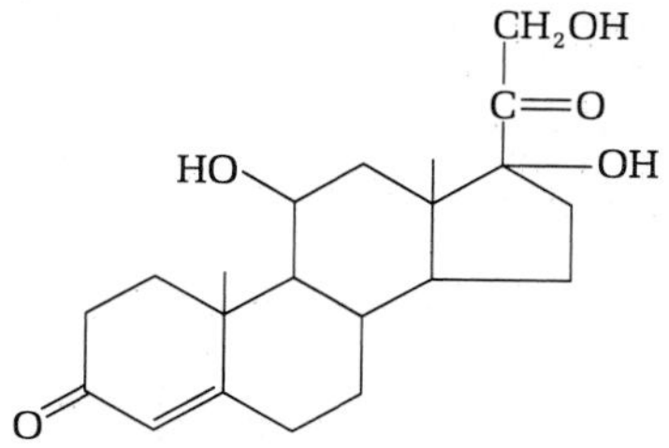 CH_2OH, C=O, OH, HO, O	effects on metabolism of proteins, carbohydrates and lipids in most tissues; suppression of inflammatory reactions
Estradiol	ovary, placenta	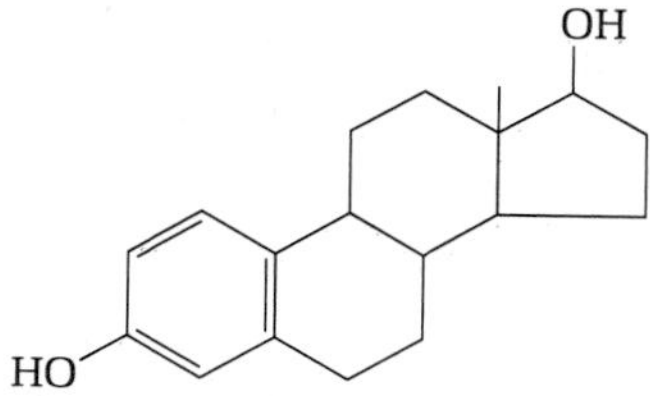 OH, HO	development and maintenance of secondary female sex characteristics; maturation and cyclic function of accessory sex organs; development of duct system in mammary glands
Testosterone	testis	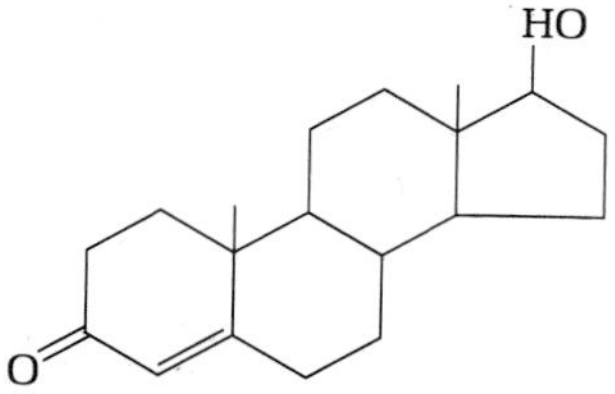 HO, O	development and maintenance of secondary male sex characteristics; maturation and normal function of accessory sex organs
Progesterone	ovary (corpus luteum), placenta	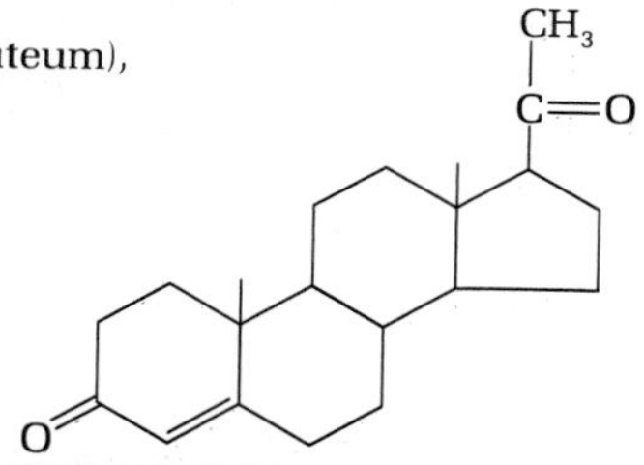 CH_3, C=O, O	preparation of uterus for pregnancy; maintenance of pregnancy; development of alveolar system in mammary glands

changes that eventually lead to the development of secondary female characteristics such as breast enlargement. Although this effect is slowly reversed if estradiol secretion stops, some of the responses to steroid sex hormones during very early mammalian development are irreversible. Similarly, a tenfold increase in thyroid hormone levels in the blood of a tadpole induces the dramatic and irreversible changes that result in its transformation into a frog (Figure 12–6).

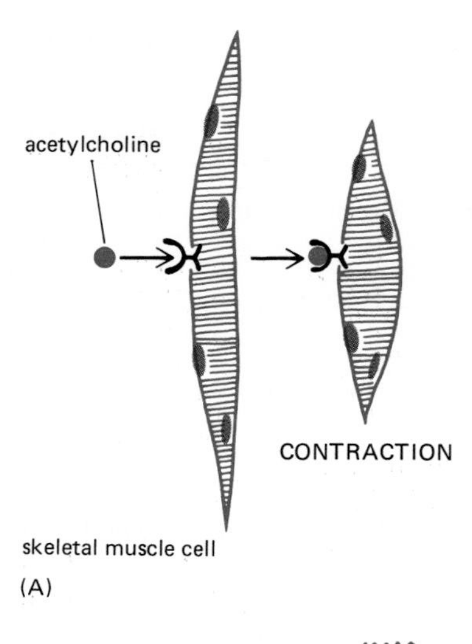

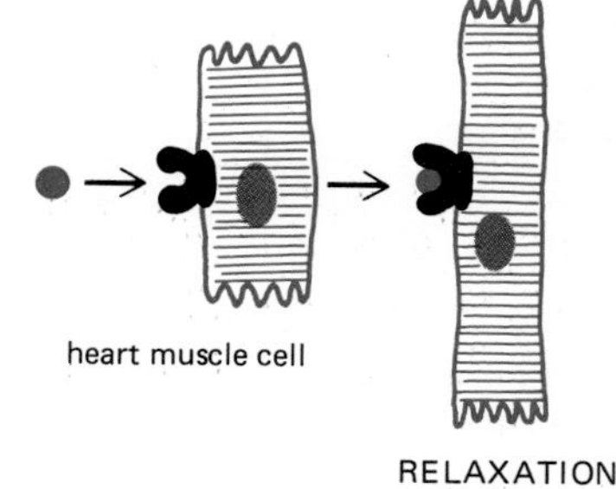

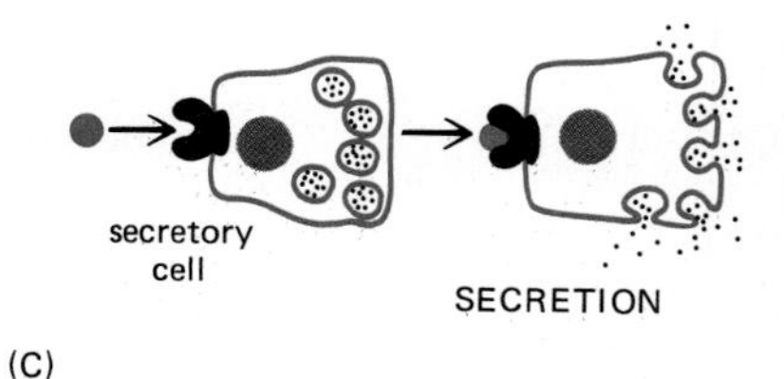

Figure 12–5 The same signaling molecule can induce different responses in different target cells. In some cases this is because the signaling molecule binds to different receptor proteins, as illustrated in (A) and (B). In other cases the signaling molecule binds to identical receptor proteins but these activate different response pathways in different cells, as illustrated in (B) and (C).

Only Lipid-soluble Signaling Molecules Can Enter Cells Directly

Most hormones and local chemical mediators, and all known neurotransmitters, are water-soluble. There are some exceptions, however, that form a distinctive class of signaling molecules. Important examples are the water-insoluble steroid and thyroid hormones, which are made soluble for transport in the bloodstream by binding to specific carrier proteins. This difference in solubility gives rise to a fundamental difference in the mechanism by which the two classes of molecules influence target cells. Water-soluble molecules are too hydrophilic to pass directly through the lipid bilayer of a target-cell plasma membrane; instead they bind to specific receptor proteins on the cell surface. The steroid and thyroid hormones, on the other hand, are lipid-soluble, and once released from their carrier proteins, they can pass easily through the plasma membrane of the target cells; these hormones then bind to specific receptor proteins *inside* the cell (Figure 12–7).

Another important difference between these two classes of signaling molecules is the length of time that they persist in the bloodstream or tissue fluids. Most water-soluble hormones are removed and/or broken down within minutes of entering the blood, and local chemical mediators and neurotransmitters are removed from the extracellular space even faster—within seconds or milliseconds. Steroid hormones, by contrast, persist in the blood for hours, and thyroid hormone for days. Consequently, water-soluble signaling molecules usually mediate responses of short duration, whereas the water-insoluble molecules tend to mediate longer lasting responses. As usual, there are exceptions to these general rules: *prostaglandins,* for example, are local chemical mediators that are hydrophobic, yet they bind to cell-surface receptors and mediate rapid, short-lasting responses, as we shall see below.

12-3 Local Chemical Mediators Are Rapidly Destroyed, Retrieved, or Immobilized After They Are Secreted[4]

The molecules that mediate paracrine signaling act only on cells in the immediate vicinity of the cells that secrete them. These *local chemical mediators* are so rapidly taken up by cells or destroyed by extracellular enzymes or immobilized in the extracellular matrix that they generally do not enter the blood in significant amounts.

Some cells are specialized for paracrine signaling. For example, *histamine* (a derivative of the amino acid histidine, see Table 12–1) is secreted mainly by *mast cells.* These cells, which are found in connective tissues throughout the body, store histamine in large secretory vesicles and release it rapidly by exocytosis when stimulated by injury, local infection, or certain immunological reactions (see p. 1016). Histamine causes local blood vessels to dilate and become leaky, which facilitates the access of both serum proteins (such as antibodies and components of the complement system, see Chapter 18) and phagocytic white blood cells to the site of injury. Mast cells also release two tetrapeptides that attract a class of white blood cells called *eosinophils* from the blood to the site of tetrapeptide release; eosinophils contain a variety of enzymes that help inactivate histamine and other chemical mediators released by mast cells and thereby help to terminate the response.

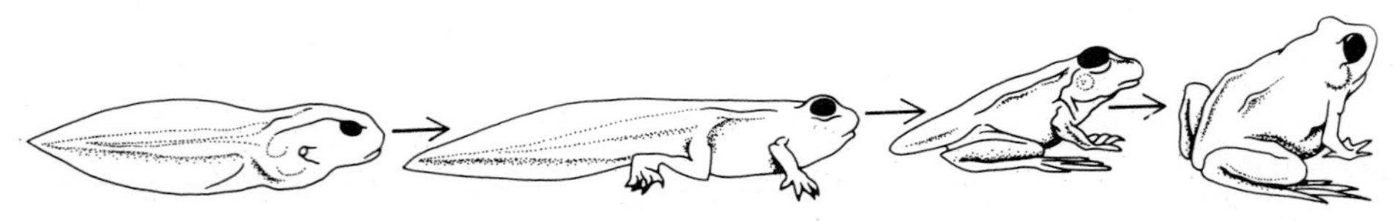

Figure 12–6 Various stages in the metamorphosis of a tadpole into a frog. All of the dramatic changes shown are signaled by thyroid hormone. If the presumptive thyroid gland is removed from a developing embryo, the animal fails to undergo metamorphosis and continues to grow as a tadpole. If thyroid hormone is injected into such a giant tadpole, the tadpole transforms into a frog.

Figure 12–7 Depending on their solubility, extracellular signaling molecules bind to either cell-surface receptors or intracellular receptors. Hydrophilic signaling molecules are unable to cross the plasma membrane directly and bind to receptors on the surface of the target cell. Many hydrophobic signaling molecules are able to diffuse across the plasma membrane and bind to receptors inside the target cell—either in the cytoplasm or in the nucleus (as shown). Because they are insoluble in aqueous solutions, hydrophobic hormones are transported in the bloodstream bound to specific carrier proteins from which they dissociate before entering the target cell.

HYDROPHILIC SIGNALS

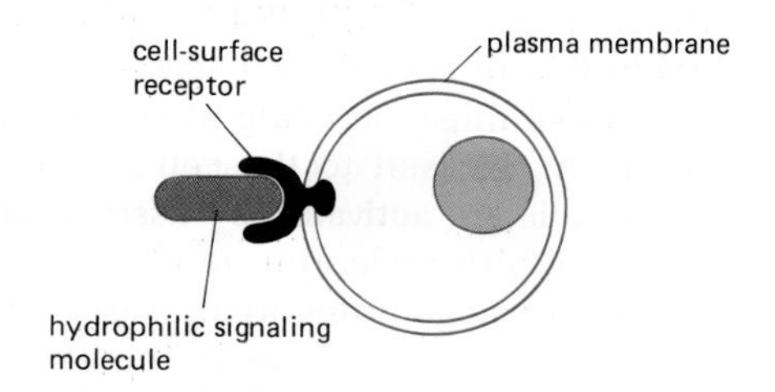

HYDROPHOBIC SIGNALS

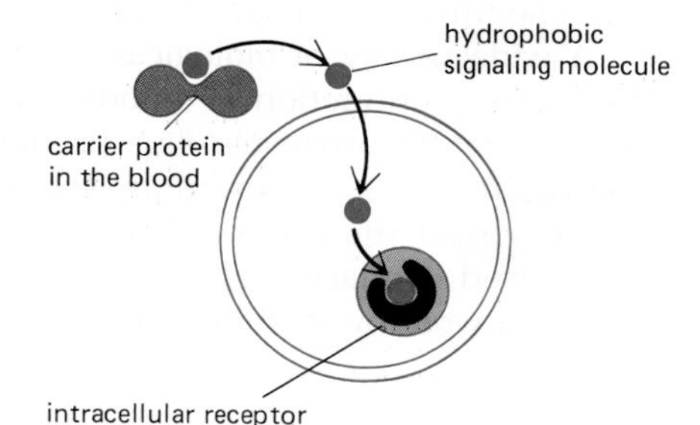

Some local chemical mediators are rapidly immobilized rather than eliminated after they have been secreted. Fibronectin, proteoglycans, and other macromolecules of the extracellular matrix are a case in point. These secreted macromolecules can be considered as special types of local mediators because they signal neighboring cells to alter their behavior (see p. 803). Unlike other local chemical mediators, they assemble into large insoluble networks in the extracellular space and thus become immobilized near the site where they are produced; thus their effects, although local, can be long-lasting. The extracellular matrix may also bind soluble signaling molecules, immobilizing them so that they act only in a particular location. *Fibroblast growth factor (FGF),* for example, which is a small protein that stimulates a wide variety of cells to divide in culture, binds strongly to a matrix proteoglycan in a test tube and might be immobilized in this way in tissues.

Cells in All Mammalian Tissues Continuously Release Prostaglandins[5]

Many local chemical mediators are secreted by cells that are specialized for this purpose, but others are of more widespread origin. The **prostaglandins,** a family of 20-carbon fatty acid derivatives, are an important example, being made by cells in all mammalian tissues. These local mediators are continuously synthesized in membranes from precursors cleaved from membrane phospholipids by phospholipases (Figure 12–8). They are also continuously degraded by enzymes in extra-

membrane phospholipid

phospholipase

arachidonic acid (20 carbons), extended conformation

arachidonic acid, folded conformation

OXIDATION STEPS

prostaglandin (PGE_2)

Figure 12–8 Prostaglandins are continuously synthesized in membranes from 20-carbon fatty acid chains that contain at least three double bonds, as shown for the synthesis of PGE_2. The subscript refers to the two carbon-carbon double bonds outside the ring of PGE_2. Prostaglandins, together with the chemically related signaling molecules *thromboxanes, leukotrienes,* and *lipoxins,* are all made mainly from arachidonic acid and are collectively called *eicosanoids.* This metabolic pathway is an important target for therapeutic drugs, since eicosanoids play an important part in inflammation. Corticosteroid hormones such as cortisone, for example, are widely used clinically as drugs to treat noninfectious inflammatory diseases such as some forms of arthritis. One way they might act is by inducing white blood cells to synthesize and/or secrete local chemical mediators called *lipocortins* (or calpactins), proteins that somehow inhibit the activity of the phospholipase in the first step of the eicosanoid synthesis pathway shown. Nonsteroid anti-inflammatory drugs, such as aspirin, block the oxidation steps of prostaglandin synthesis. Both corticosteroids and aspirin are used in the treatment of arthritis.

culties and have revolutionized our understanding of receptor structure and function.

Unlike intracellular receptors for steroid and thyroid hormones, cell-surface receptors do not regulate gene expression directly. Instead, they relay a signal across the plasma membrane, and the influence they exert on events in the cytosol or nucleus generally depends on the production of new intracellular signals. It might be imagined that cell-surface receptors could simply transfer the extracellular signaling molecule across the membrane into the cytosol to serve there as an intracellular signal, but this is not the case. Even though many protein signaling molecules, such as insulin, are ingested by receptor-mediated endocytosis (see p. 718), they do not escape from the endosomal or lysosomal compartments into the cytosol. The task of the extracellular ligand seems to be simply to force an appropriate conformational change in the cell-surface receptor protein. In fact, antibodies that bind to the receptor can often mimic the effects of the normal ligand—a phenomenon that underlies some disease states. The usual cause of hyperthyroid disease in humans, for example, is the abnormal production of antibodies that bind to thyroid-stimulating-hormone (TSH) receptors, activating the receptor and causing too much thyroid hormone to be produced.

In this section we consider how a conformational change induced in a cell-surface receptor protein by the binding of an extracellular ligand enables the receptor to act directly or indirectly as a transducer, converting the extracellular signal into a signal inside the cell.

There Are at Least Three Known Classes of Cell-Surface Receptor Proteins: Channel-linked, G-Protein-linked, and Catalytic[9]

Most cell-surface receptor proteins belong to one of three classes, which are defined by the transduction mechanism used. **Channel-linked receptors** are transmitter-gated ion channels (see p. 1074) involved mainly in rapid synaptic signaling between electrically excitable cells. This type of signaling is mediated by a small number of neurotransmitters that transiently open or close the ion channel to which they bind, briefly changing the ion permeability of the plasma membrane and thereby the excitability of the postsynaptic cell. DNA sequencing studies have shown that channel-linked receptors belong to a family of homologous, multipass

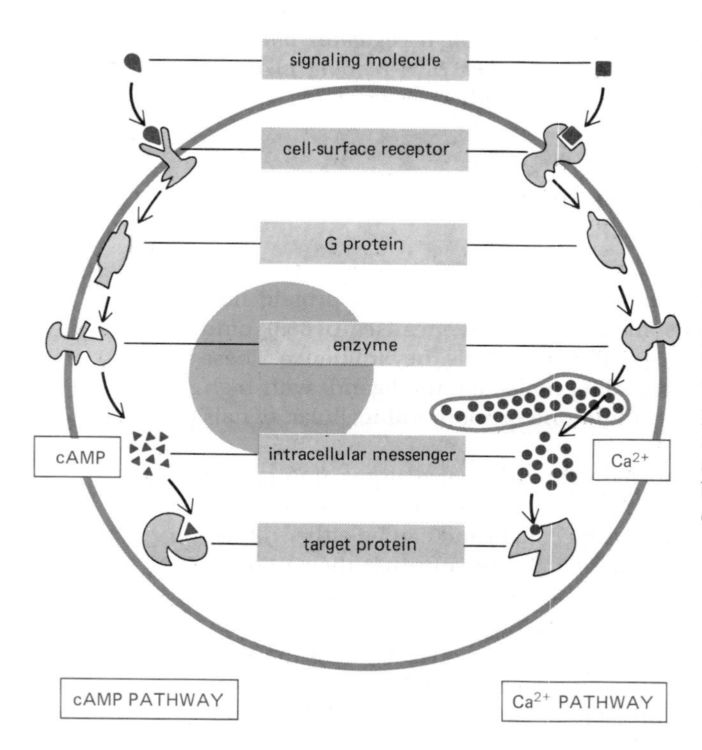

Figure 12–13 Two major pathways by which G-protein-linked cell-surface receptors generate intracellular messengers. In both cases the binding of an extracellular ligand alters the conformation of the cytoplasmic domain of the receptor so that it binds to a G protein, which in turn activates (or inactivates) a plasma membrane enzyme. In other cases the G protein binds to an ion channel rather than to an enzyme. In the cyclic AMP pathway, the enzyme produces cyclic AMP. In the Ca^{2+} pathway, the enzyme produces a soluble mediator that releases Ca^{2+} from an intracellular storage site. Both cyclic AMP and Ca^{2+} bind to other specific proteins in the cell, thereby altering their activity.

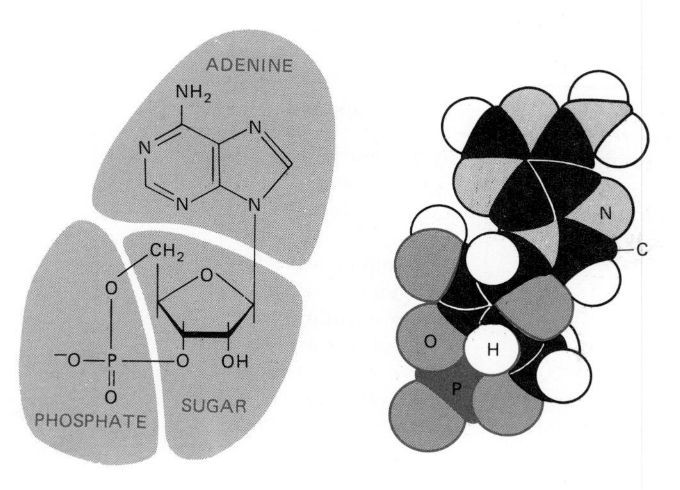

Figure 12–14 Cyclic AMP shown as a formula and as a space-filling model. (C, H, N, O, and P indicate carbon, hydrogen, nitrogen, oxygen, and phosphorus atoms, respectively.)

transmembrane proteins. These receptors are discussed in Chapter 19 (see p. 1074) and will not be considered further here.

Catalytic receptors, when activated by their ligand, operate directly as enzymes. Almost all of the known catalytic receptors are transmembrane proteins with a cytoplasmic domain that functions as a tyrosine-specific protein kinase.

G-protein-linked receptors indirectly activate or inactivate a separate plasma-membrane-bound enzyme or ion channel. The interaction between the receptor and the enzyme or ion channel is mediated by a third protein, called a *GTP-binding regulatory protein* (or *G protein*). The G-protein-linked receptors usually activate a chain of events that alters the concentration of one or more small intracellular signaling molecules, often referred to as **intracellular messengers** (or **intracellular mediators**). These intracellular messengers act in turn to alter the behavior of yet other target proteins in the cell. Two of the most important intracellular messengers are *cyclic AMP (cAMP)* and Ca^{2+}. Cyclic AMP and Ca^{2+} signals are generated by different pathways, both involving G proteins, and are used by almost all animal cells (Figure 12–13). We shall discuss these pathways before returning to consider catalytic receptors with tyrosine-specific protein kinase activity. We begin with the experiments that led to the discovery of cyclic AMP and paved the way to our present understanding of the coupling of its production to extracellular signals.

Cyclic AMP Is a Ubiquitous Intracellular Messenger in Animal Cells[10]

When muscle or liver cells are exposed to the hormone *epinephrine* (*adrenaline*), they are stimulated to break down their stores of glycogen. It was found that epinephrine causes activation of the enzyme *glycogen phosphorylase,* which catalyzes glycogen breakdown. It was then shown that treatment of the isolated membranes of liver cells with epinephrine (in the presence of ATP) induced the production of a small heat-stable mediator that could activate the phosphorylase present in a membrane-free extract of liver cells. The mediator was identified in 1959 as **cyclic AMP** (Figure 12–14), which has since been found to regulate intracellular reactions in all procaryotic and animal cells that have been studied.

The identification of cyclic AMP led to the study of the enzymes that make and degrade it. For cyclic AMP to function as an intracellular mediator, its intracellular concentration (normally $\leq 10^{-6}$ M) must be able to change rapidly up or down in response to extracellular signals: upon hormonal stimulation, cyclic AMP levels can change by fivefold in seconds. As explained on page 714, such responsiveness requires that rapid synthesis of the molecule be balanced by rapid breakdown or removal. Cyclic AMP is synthesized from ATP by the plasma-membrane-bound enzyme **adenylate cyclase,** and it is rapidly and continuously destroyed by one or more **cyclic AMP phosphodiesterases,** which hydrolyze cyclic AMP to adenosine 5′-monophosphate (5′-AMP) (Figure 12–15).

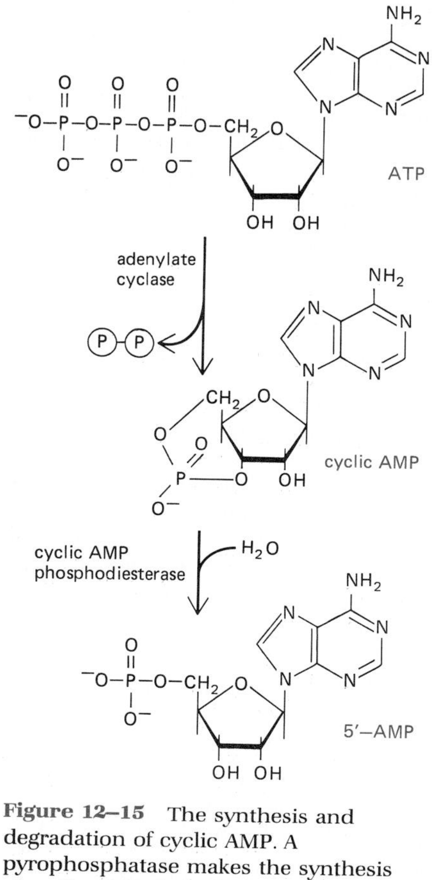

ATP

adenylate cyclase

Ⓟ—Ⓟ

cyclic AMP

cyclic AMP phosphodiesterase

H_2O

5′–AMP

Figure 12–15 The synthesis and degradation of cyclic AMP. A pyrophosphatase makes the synthesis of cyclic AMP an irreversible reaction by hydrolyzing the released pyrophosphate (Ⓟ—Ⓟ).

Table 12–2 Some Hormone-induced Cellular Responses Mediated by Cyclic AMP

Target Tissue	Hormone	Major Response
Thyroid	thyroid-stimulating hormone (TSH)	thyroid hormone synthesis and secretion
Adrenal cortex	adrenocorticotropic hormone (ACTH)	cortisol secretion
Ovary	luteinizing hormone (LH)	progesterone secretion
Muscle, liver	epinephrine	glycogen breakdown
Bone	parathormone	bone resorption
Heart	epinephrine	increase in heart rate and force of contraction
Kidney	vasopressin	water resorption
Fat	epinephrine, ACTH, glucagon, TSH	triglyceride breakdown

Receptor and Adenylate Cyclase Molecules Are Separate Proteins That Functionally Interact in the Plasma Membrane

12-16

Many hormones and local chemical mediators work by controlling cyclic AMP levels, and they do so by activating (or in some cases by inhibiting) adenylate cyclase rather than by altering phosphodiesterase activity. Just as the same steroid hormone produces different effects in different target cells, so different target cells respond very differently to external signals that change intracellular cyclic AMP levels (Table 12–2). All ligands that activate adenylate cyclase in a given type of target cell, however, usually produce the same effect. For example, at least four hormones activate adenylate cyclase in fat cells, and all of them stimulate the breakdown of triglyceride (the storage form of fat) to fatty acids (see Table 12–2). The different receptors for these hormones seem to activate a common pool of adenylate cyclase molecules. That the receptors and the adenylate cyclase are separate molecules can be shown by receptor "transplantation" experiments. Epinephrine receptors isolated from detergent-solubilized plasma membranes have no adenylate cyclase activity, but when they are transplanted to the plasma membrane of cells that do not have their own epinephrine receptors, the transplanted receptors are able to interact functionally with adenylate cyclase molecules of the recipient cell, causing their activation in the presence of the hormone (Figure 12–16).

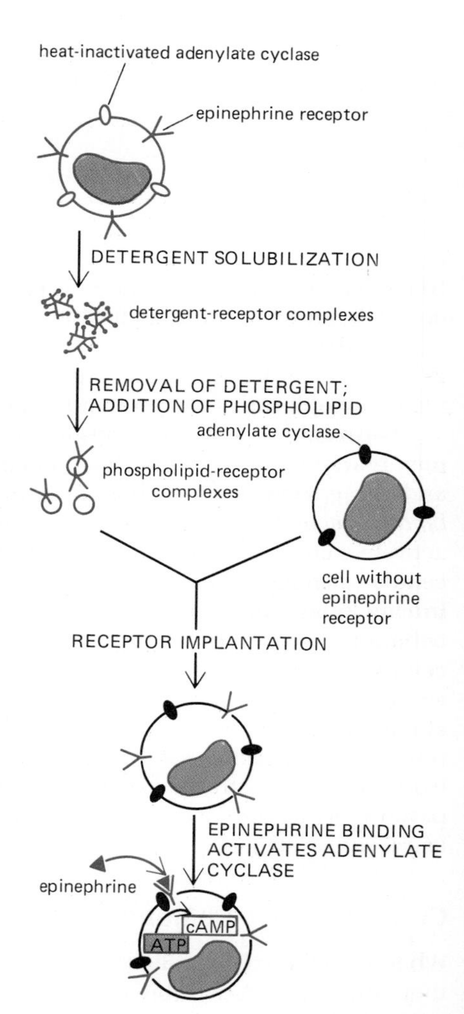

Figure 12–16 Functional epinephrine receptors can be extracted from cells (in which adenylate cyclase molecules have been inactivated by heat) and implanted into the plasma membrane of cells that lack such receptors. When activated by epinephrine, the transplanted receptors activate adenylate cyclase molecules in the recipient cell plasma membrane. The intermediary G protein is not shown. More recently, purified epinephrine receptors have been used in these experiments with the same results. There are at least three types of epinephrine receptors (also called adrenergic receptors)—α_1, α_2, and β; only β-adrenergic receptors activate adenylate cyclase.

Receptors Activate Adenylate Cyclase Molecules via a Stimulatory G Protein (G_s)[12]

In bacteria, receptors and adenylate cyclase molecules interact directly, but in animal cells another protein intervenes to couple the activated receptor to the enzyme. The first hint of this complication came from the observation that hormonal activation of adenylate cyclase in disrupted animal cells requires GTP. Later, mutant cell lines were isolated in which binding of epinephrine failed to activate adenylate cyclase in spite of normal levels of epinephrine receptors and of adenylate cyclase. By mixing plasma membrane preparations from such "uncoupled" cells with detergent extracts of plasma membranes from normal cells, a hormone-sensitive adenylate cyclase system requiring GTP could be reconstituted. The detergent extracts proved to contain a **GTP-binding regulatory protein,** or **G protein,** that was missing from the "uncoupled" mutant cells. Because this G protein is involved in enzyme *activation*, it is called **stimulatory G protein (G_s).** Individuals who are genetically deficient in G_s have decreased responses to many hormones and, consequently, fail to grow or mature sexually, are mentally retarded, and have many metabolic abnormalities. Recently an epinephrine-activated adenylate cyclase system has been reconstituted in synthetic phospholipid vesicles (see p. 286) from purified epinephrine receptor, G_s, and adenylate cyclase molecules, indicating that no other proteins are required for the activation process.

If the G_s protein is to relay a signal from the receptor to the cyclase, it must have some way of changing its structure when a signal has been received. This is the function of the GTP. When G_s is activated by the receptor-hormone complex, it simultaneously binds a molecule of GTP. Carrying the GTP, it now activates an adenylate cyclase molecule. The G_s keeps the cyclase active so long as the GTP is intact. Eventually, hydrolysis of the GTP to GDP by the G_s protein (which is a GTPase) terminates the activation of the cyclase.

12-17 The G_s Protein is a Heterotrimer That Is Thought to Disassemble When Activated[13]

The function of a G protein depends critically on its subunit structure. G_s is composed of three polypeptides: an *α chain* ($G_{s\alpha}$), which binds and hydrolyzes GTP and activates adenylate cyclase, and a tight complex of a *β chain* and a *γ chain* ($G_{\beta\gamma}$), which anchors G_s to the cytoplasmic face of the plasma membrane. A current model of how G_s couples receptor activation to adenylate cyclase activation is shown in Figure 12–17. In its inactive form, G_s exists as a trimer with GDP bound to $G_{s\alpha}$. When activated by binding to a receptor-hormone complex, the guanyl-nucleotide-binding site on $G_{s\alpha}$ is altered, allowing GTP to bind in place of GDP. The binding of GTP is thought to cause $G_{s\alpha}$ to dissociate from $G_{\beta\gamma}$, allowing $G_{s\alpha}$ to bind tightly to an adenylate cyclase molecule, which is thus activated to produce cyclic AMP. Within less than a minute, the $G_{s\alpha}$ hydrolyzes its bound GTP to GDP, causing $G_{s\alpha}$ to dissociate from the adenylate cyclase (which thereby becomes inactive) and reassociate with $G_{\beta\gamma}$ to reform an inactive G_s molecule.

The adenylate cyclase system in bacteria lacks a G_s intermediate. Why, then, have animal cells evolved such a complex multistep mechanism for signal transduction, with a G protein interposed between the receptor and the enzyme it is to activate? One reason may lie in a need for amplification of the signal (see p. 713), another in a need for additional levels of control.

G_s allows for two sorts of amplification. Most simply, a single activated receptor protein can, in principle, collide with and activate many molecules of G protein, thereby activating many molecules of adenylate cyclase. In some cases, however, the extracellular ligand may not remain bound to its receptor long enough for this amplification mechanism to operate: some ligands, for example, may dissociate from their receptors in less than a second. G_s itself, however, is thought to remain active for up to 10 or 15 seconds before hydrolyzing its bound GTP. In this way it can keep an adenylate cyclase molecule active long after such an extracellular ligand has dissociated. This amplification effect can be demonstrated in an exaggerated form if cells are broken open and exposed to an analogue of GTP in which the terminal phosphate cannot be hydrolyzed. Hormone treatment then has a greatly prolonged effect on cyclic AMP production.

In addition to amplification, G proteins provide an important step where the activation process can be regulated. In principle, the efficiency of coupling between receptors and enzyme can be altered by covalently modifying the G protein or by changing its concentration in the plasma membrane. This is illustrated most dramatically by the effects of the bacterial toxin that is responsible for the symptoms of cholera. **Cholera toxin** is an enzyme that catalyzes the transfer of ADP-ribose from intracellular NAD^+ to the α subunit of the G_s protein, altering it so that it can no longer hydrolyze its bound GTP. An adenylate cyclase molecule activated by such an altered G_s protein thus remains in the active state indefinitely. The resulting prolonged elevation in cyclic AMP levels within intestinal epithelial cells causes a large efflux of Na^+ and water into the gut, which is responsible for the severe diarrhea that is characteristic of cholera.

G_s is only one member of a large family of G proteins that couple receptors to a variety of enzymes and ion channels in eucaryotic cell membranes. As we shall now discuss, one of these G proteins inhibits rather than activates adenylate cyclase.

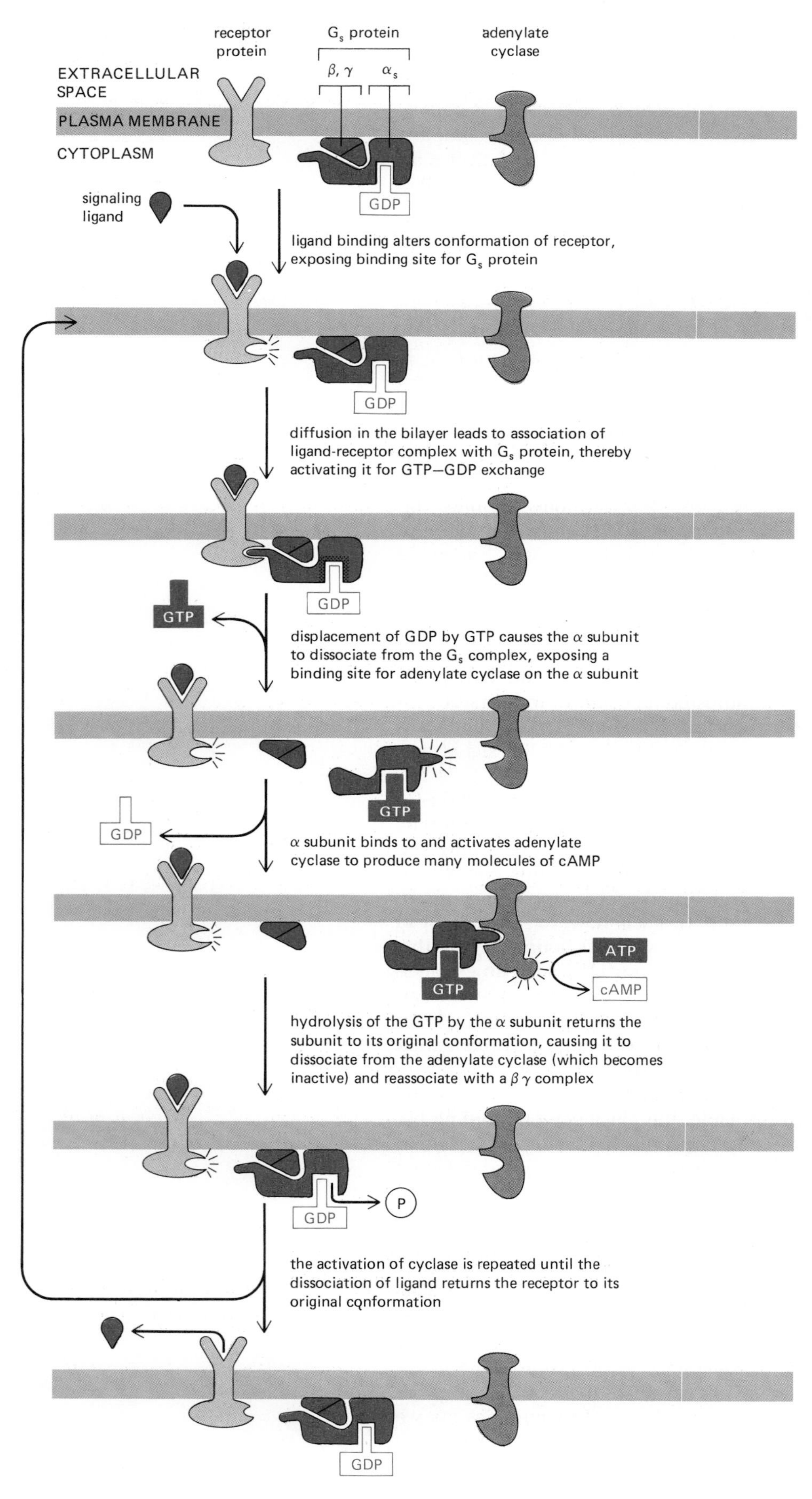

Figure 12–17 A current model illustrating how receptor proteins may be functionally coupled to adenylate cyclase via the stimulatory G protein G_s. As long as the signaling ligand remains bound, the receptor protein can continue to activate molecules of G_s protein, thereby amplifying the response. An additional amplification mechanism (that is more important in some signaling systems) is the persistence of bound GTP on the α subunit of the G_s protein for many seconds, during which adenylate cyclase remains activated. According to an alternative model, the G_s and adenylate cyclase molecules remain permanently associated during the indicated activation and deactivation processes.

Receptors Inactivate Adenylate Cyclase via an Inhibitory G Protein (G_i)[13]

The same signaling molecule can either increase or decrease the intracellular concentration of cyclic AMP depending on the type of receptor to which it binds. For example, there are several types of epinephrine (adrenergic) receptors: *β-adrenergic receptors* activate adenylate cyclase, whereas *α_2-adrenergic receptors* inhibit it. The difference is due to the G proteins that couple these receptors to the cyclase. The β receptors are functionally coupled to adenylate cyclase by G_s; the α_2 receptors are coupled to the same enzyme by an **inhibitory G protein (G_i)**, which contains the same $G_{\beta\gamma}$ complex as G_s but a different α subunit ($G_{i\alpha}$). When activated, α_2-adrenergic receptors bind to G_i, causing $G_{i\alpha}$ to bind GTP in place of GDP. This is thought to cause $G_{i\alpha}$ to dissociate from the $G_{\beta\gamma}$, and both the released $G_{i\alpha}$ and $G_{\beta\gamma}$ are believed to contribute to the inhibition of adenylate cyclase: $G_{i\alpha}$ inhibits the cyclase directly, whereas $G_{\beta\gamma}$ acts indirectly by binding to free α_s subunits, thereby preventing them from activating cyclase molecules.

Just as cholera toxin maintains high levels of cyclic AMP by ADP-ribosylating $G_{s\alpha}$ and inactivating its GTPase activity, so *pertussis toxin*, made by the bacterium that causes whooping cough, produces the same effect by ADP-ribosylating $G_{i\alpha}$. In this case, however, the G_i complex is prevented from interacting with receptors and therefore fails to inhibit adenylate cyclase in response to receptor activation.

Although G proteins were first discovered because of their effects on adenylate cyclase, they can also act in other ways, as summarized in Table 12–3. In particular, by activating phospholipase C (see p. 702), other G proteins can couple receptor activation to changes in the concentration of Ca^{2+} in the cytosol, and Ca^{2+} is even more widely used as an intracellular messenger than cyclic AMP.

Ca^{2+} Is Stored in a Special Intracellular Calcium-sequestering Compartment[14]

The concentration of free Ca^{2+} in the cytosol of any cell is extremely low ($\sim 10^{-7}$ M), whereas its concentration in the extracellular fluid ($>10^{-3}$ M) and in a specialized intracellular calcium-sequestering compartment is high. Thus there is a large gradient tending to drive Ca^{2+} into the cytosol across the plasma membrane and the membrane of the intracellular compartment. When a signal tran-

Table 12–3 Some GTP-binding Regulatory Proteins Involved in Cell Signaling

Type of G Protein	α Subunit*	Function	Modified by Bacterial Toxin
G_s	α_s	activates adenylate cyclase	cholera
G_i	α_i	inactivates adenylate cyclase	pertussis
G_p	?	activates phosphoinositide-specific phospholipase C	pertussis (only in some cells)
G_o	α_o	main G protein in brain; may regulate ion channels	pertussis
Transducin	T_α	activates cyclic GMP phosphodiesterase in vertebrate rod cells (see p. 705)	pertussis and cholera
ras proteins	—**	involved in unknown way in growth-factor stimulation of cell proliferation (see pp. 705 and 757)	no

*Except for *ras* proteins (and G_p, whose structure is unknown), G proteins are heterotrimers composed of an α chain loosely bound to a βγ dimer. All of the known α subunits (40,000–50,000 daltons) are homologous, and several of the G proteins share the same (or very similar) β (35,000 daltons) and γ (8000 daltons) subunits.
**The *ras* proteins are single polypeptide chains (21,000 daltons) with relatively little homology to the subunits of the other G proteins; it is not known if they act as intermediaries in receptor coupling the way the other listed G proteins do.

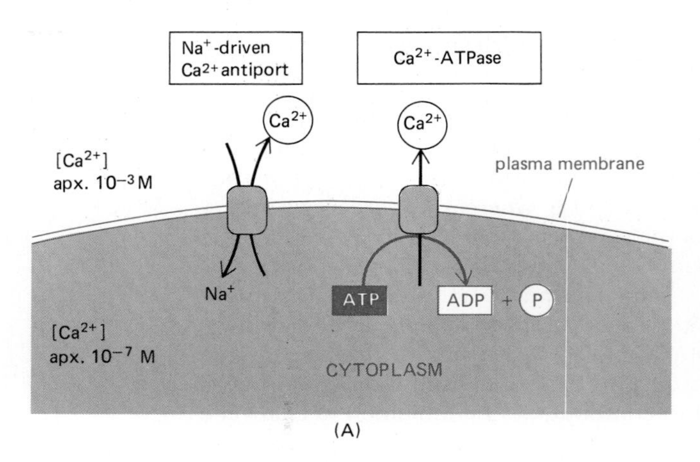

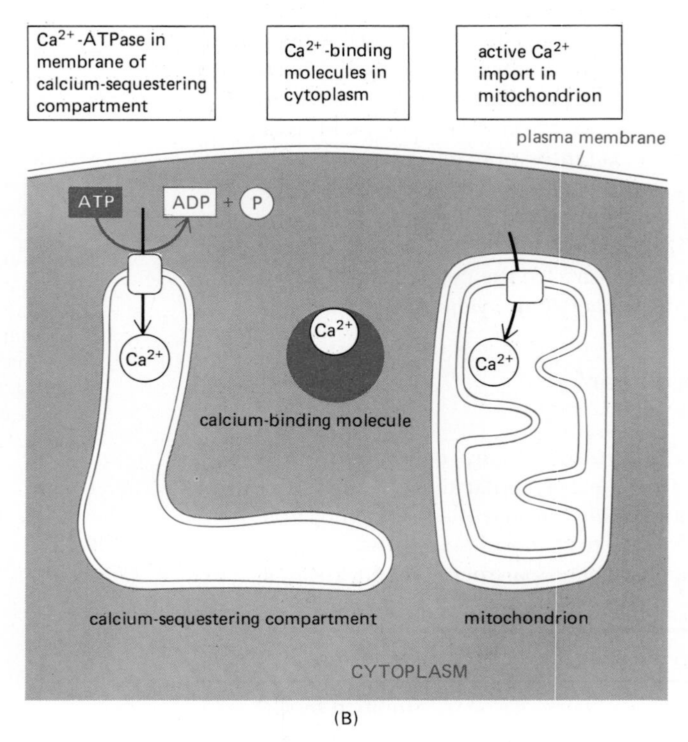

Figure 12–18 The main ways in which cells maintain a very low concentration of free Ca^{2+} in the cytosol in the face of high concentrations of Ca^{2+} in the extracellular fluid. Ca^{2+} is actively pumped out of the cytosol to the cell exterior (A) and into a special internal compartment, the calcium-sequestering compartment (B). In addition, various molecules in the cell bind free Ca^{2+} tightly. Mitochondria can also pump Ca^{2+} out of the cytosol, but they do so efficiently only when Ca^{2+} levels are extremely high—usually as a result of cell damage.

siently opens Ca^{2+} channels in either of these membranes, Ca^{2+} rushes into the cytosol, dramatically increasing the local Ca^{2+} concentration and activating Ca^{2+}-sensitive response mechanisms in the cell.

For this signaling mechanism to work, the concentration of Ca^{2+} in the cytosol must be kept low, and this is achieved in several ways (Figure 12–18). All eucaryotic cells have a Ca^{2+}-ATPase in their plasma membrane that uses the energy of ATP hydrolysis to pump Ca^{2+} out of the cytosol. Muscle and nerve cells, which make extensive use of Ca^{2+} signaling, have an additional Ca^{2+} pump in their plasma membrane that couples the efflux of Ca^{2+} to the influx of Na^+. This Na^+-Ca^{2+} exchanger has a relatively low affinity for Ca^{2+} and therefore begins to operate efficiently only when cytosolic Ca^{2+} levels rise to about 10 times their normal level, as occurs after repeated muscle or nerve cell stimulation.

A Ca^{2+} pump in the membrane of the specialized intracellular compartment also plays an important part in keeping the cytosolic Ca^{2+} concentration low: this Ca^{2+}-ATPase enables the intracellular compartment to take up large amounts of Ca^{2+} from the cytosol against a steep concentration gradient, even when Ca^{2+} levels in the cytosol are low. This Ca^{2+} is stored in the lumen of the compartment, loosely bound to a Ca^{2+}-binding protein called *calsequestrin,* which has a low affinity ($K_a \simeq 10^3$ liters/mole) but high capacity (~50 calcium ions/molecule) for Ca^{2+}. When antibodies against calsequestrin and the Ca^{2+}-ATPase are used to

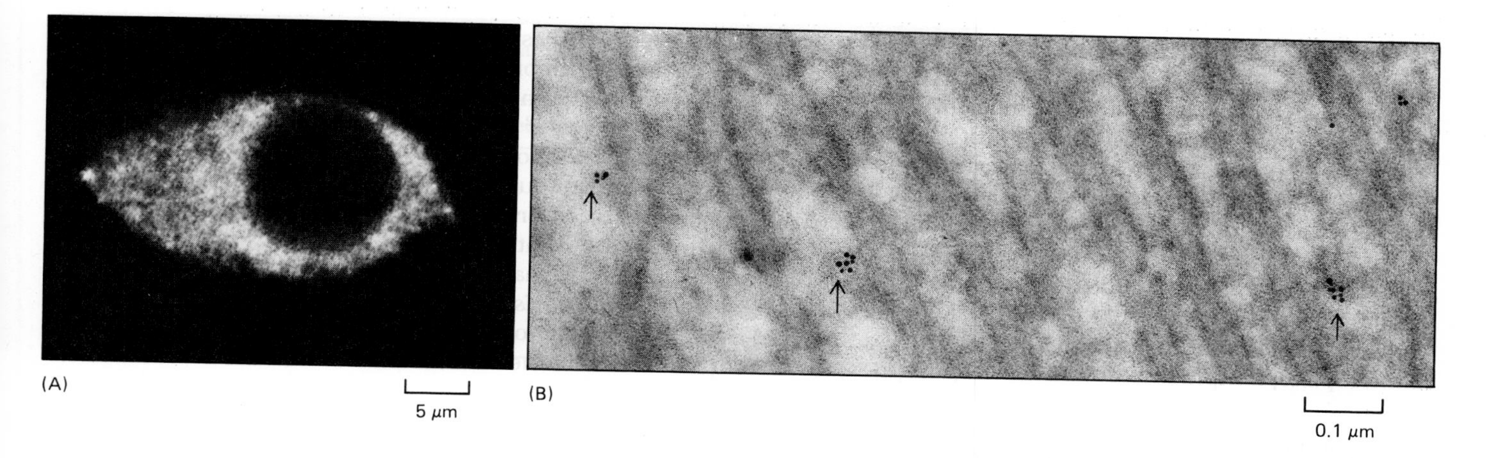

stain cells, the two antibodies label the same membrane-bounded compartment, which is distinct from the rough endoplasmic reticulum and much more limited in extent (Figure 12–19). This recently discovered compartment is homologous with the more extensive sarcoplasmic reticulum in muscle cells (see p. 621) and, like the sarcoplasmic reticulum, is specialized for calcium storage and release. We shall therefore refer to it as the **calcium-sequestering compartment.**

Normally the concentration of free Ca^{2+} in the cytosol varies from about 10^{-7} M when the cell is at rest to about 5×10^{-6} M when the cell is activated by an extracellular signal. But when a cell is damaged and cannot pump Ca^{2+} out of the cytosol efficiently, the Ca^{2+} concentration can rise to dangerously high levels ($>10^{-5}$ M). In these circumstances a low-affinity, high-capacity Ca^{2+} pump in the inner mitochondrial membrane comes into action and uses the electrochemical gradient generated across this membrane during the electron-transfer steps of oxidative phosphorylation to take up Ca^{2+} from the cytosol (see p. 353).

Figure 12–19 The intracellular calcium-sequestering compartment revealed by labeling with antibodies against the Ca^{2+}-binding protein calsequestrin. (A) Immunofluorescence micrograph of a rat neural cell in culture, showing that the compartment is distributed throughout the cytoplasm. (B) Immunogold electron micrograph of a frozen thin section of rat liver, showing that the calcium-sequestering compartment is not the rough endoplasmic reticulum (ER), although it might be part of the smooth ER; the gold-coupled antibodies are indicated by arrows. Quantitative analyses of such electron micrographs indicate that the compartment occupies less than 1% of the cell volume. (A, from P. Volpe et al. *Proc. Natl. Acad. Sci. USA*, 85:1091–1095, 1988. B, courtesy of J. Meldolesi.)

Ca^{2+} Functions as a Ubiquitous Intracellular Messenger[15]

The first direct evidence that Ca^{2+} functions as an intracellular mediator came from an experiment done in 1947 showing that the intracellular injection of a small amount of Ca^{2+} causes a skeletal muscle cell to contract. In recent years it has become clear that Ca^{2+} acts as an intracellular messenger in a wide variety of cellular responses, including secretion and cell proliferation. Two pathways of Ca^{2+} signaling have been defined (Figure 12–20), one used mainly by electrically

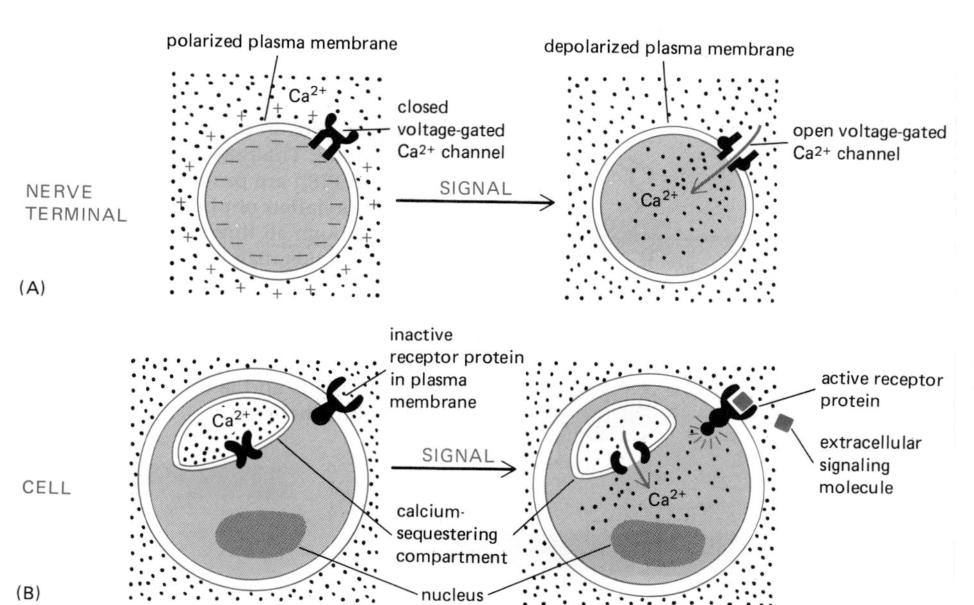

Figure 12–20 Two common pathways by which Ca^{2+} can enter the cytosol to act as an intracellular mediator of extracellular signals. In (A), Ca^{2+} enters a nerve terminal from the extracellular fluid through voltage-gated Ca^{2+} channels when the nerve terminal membrane is depolarized by an action potential. In (B), the binding of an extracellular signaling molecule to a cell-surface receptor stimulates the release of Ca^{2+} from the calcium-sequestering compartment inside the cell.

(see p. 713); the resulting fall in the concentration of this intracellular mediator causes an electrical change in the photoreceptor cell (see p. 1006). Whereas all of these G proteins can modify ion channels indirectly, some G proteins can interact with ion channels directly. The binding of acetylcholine to receptors on heart muscle cells, for example, activates a G protein (related to G_i) that directly activates a K^+ channel in the plasma membrane. (These receptors, which are sensitive to the fungal alkaloid muscarine, are called *muscarinic acetylcholine receptors* to distinguish them from the very different *nicotinic acetylcholine receptors,* which are channel-linked receptors on skeletal muscle cells—see p. 319.)

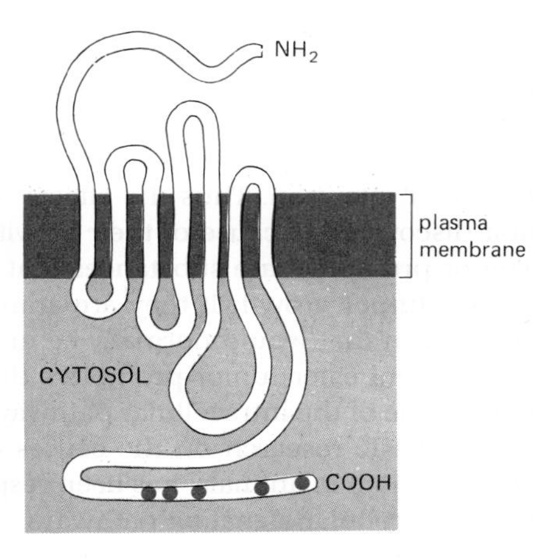

Figure 12–24 The β-adrenergic receptor as it is thought to be oriented in the plasma membrane. The colored regions of the cytoplasmic tail indicate the location of serine residues, which are potential sites for the phosphorylations that mediate receptor desensitization (see p. 718). Other G-protein-linked receptor proteins have been shown to have a similar seven-pass transmembrane structure. (Based on data from R.A.F. Dixon et al., *Nature* 321:75–79, 1986.)

The G proteins involved in these disparate systems that have been well characterized are evolutionarily related, with similar subunit structure and amino acid sequence; the α subunits of transducin and G_i, for example, are about 65% identical in amino acid sequence. Perhaps not surprisingly, many, if not all, of the receptors that interact with these G proteins are themselves homologous, as has become clear from DNA sequencing experiments. The deduced amino acid sequences of an increasing number of these receptors reveal a common structure consisting of a single polypeptide chain that threads back and forth across the lipid bilayer seven times. This family of seven-pass transmembrane receptor proteins includes β-adrenergic receptors (Figure 12–24), muscarinic acetylcholine receptors, several neuropeptide receptors, and even rhodopsin. It seems likely that these glycoproteins are part of a very large family of evolutionarily related receptors. The structural motif probably arose early in evolution, as it is shared by bacteriorhodopsin, a bacterial light-activated proton pump (which, however, does not act via a G protein—see p. 293) and by receptor proteins used by yeasts during mating (see p. 571). Not all cell-surface receptors are multipass transmembrane proteins, however, and we turn now to another class: the tyrosine-specific protein kinase family of receptors.

Many Catalytic Receptors Are Single-Pass Transmembrane Glycoproteins with Tyrosine-specific Protein Kinase Activity[20]

Many cell-surface receptors seem to convert an extracellular signal into an intracellular one by regulating the activity of a G protein, but some seem to signal the cell more directly. These are catalytic receptor proteins, and the best-studied examples in animal cells are single-pass transmembrane **tyrosine-specific protein kinases** with their catalytic domain exposed on the cytoplasmic side of the plasma membrane. When activated by ligand binding, they transfer the terminal phosphate group from ATP to the hydroxyl group on a tyrosine residue of selected proteins in the target cell. Included in this family of receptors are those for *insulin* and for a number of growth factors, including *platelet-derived growth factor* (*PDGF,* see p. 746) and *epidermal growth factor* (*EGF,* which stimulates epidermal cells and a variety of other cell types to divide) (Figure 12–25). Most other protein kinases phosphorylate serine or (less often) threonine residues on proteins, so that less than 0.1% of phosphorylated proteins in cells contain phosphotyrosine. In all cases studied, receptor proteins with tyrosine kinase activity phosphorylate themselves when activated; in the case of the insulin receptor, this autophosphorylation enhances the activity of the kinase—an example of *positive feedback regulation.*

How does binding of a ligand to the extracellular domain of these receptors activate the catalytic domain on the other side of the plasma membrane? It is difficult to imagine how a conformational change could propagate across the lipid bilayer through the single transmembrane α helix. In the case of the EGF receptor, ligand binding induces a conformational change in the extracellular domain of the receptor protein that causes the receptor to assemble into dimers. It is possible that the resulting interaction between the two adjacent cytoplasmic domains in such a dimer activates the catalytic activity.

There is good evidence that the kinase activity of these receptors is important in the signaling process. For example, cells containing a mutant insulin receptor with a single amino acid change that selectively inactivates the kinase activity are unresponsive to insulin. However, it has been exceedingly difficult to identify the key substrates that become phosphorylated besides the receptor itself, so the exact

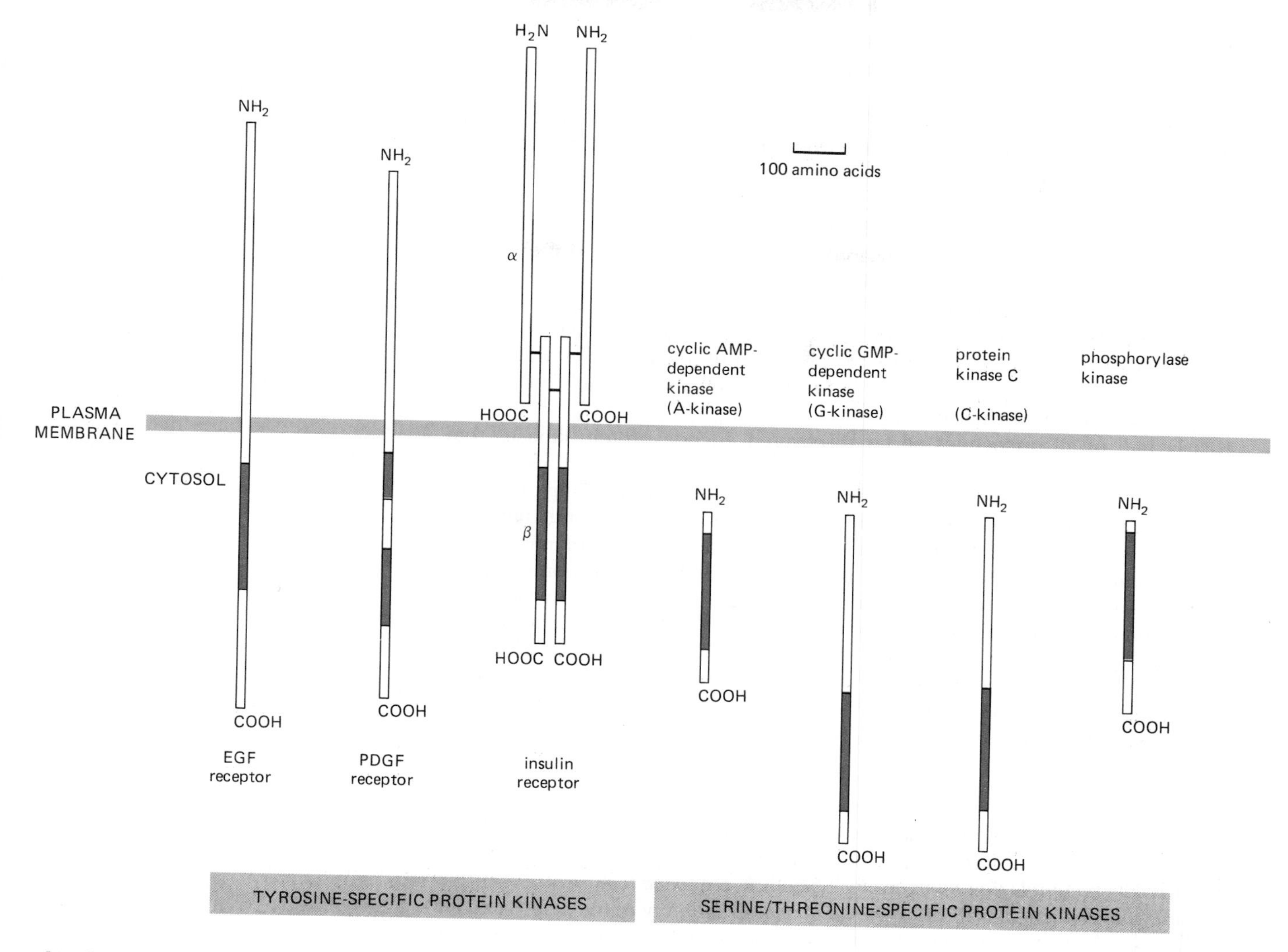

role of tyrosine phosphorylation in signal transduction remains uncertain. In the case of the PDGF receptor, however, one substrate seems to be the kinase that phosphorylates phosphatidylinositol (*PI-kinase*—see Figure 12–21). This may explain the paradoxical finding that PDGF causes a slow activation of the inositol phospholipid signaling pathway, even though it binds to a receptor that is not thought to be coupled to G_p.

Following ligand binding, many catalytic receptors are endocytosed via coated vesicles as receptor-ligand complexes (see p. 328). In some cases, this receptor-mediated endocytosis seems to depend on receptor autophosphorylation, and it can play an important role both in degrading the signaling molecule and in regulating the concentration of the receptor on the target-cell surface (see p. 718). It also translocates the tyrosine kinase domain to new locations in the cell, which could be important in the signaling process, but this has not been demonstrated.

Figure 12–25 Some of the protein kinases discussed in this chapter, showing the size and location of their catalytic domain. In each case the catalytic domain (*colored*) is about 250 amino acid residues long and is similar in amino acid sequence, suggesting that they have all evolved from a common primordial kinase. The three tyrosine-specific kinases shown are transmembrane receptor proteins that, when activated by the binding of specific extracellular ligands, phosphorylate proteins (including themselves) on tyrosine residues inside the cell. Both chains of the insulin receptor are encoded by a single gene, which produces a precursor protein that is cleaved into the two disulfide-linked chains. The extracellular domain of the PDGF receptor is thought to be folded into five immunoglobulin (Ig)-like domains, suggesting that this protein belongs to the Ig superfamily (see p. 1053). The regulatory subunits normally associated with A-kinase (see Figure 12–27) and with phosphorylase kinase (see Figure 12–31) are not shown.

Some Oncogenes Encode Abnormal Catalytic Receptors with Constitutive Kinase Activity[21]

The first tyrosine-specific protein kinase was discovered in 1979. It was not a cell-surface receptor but an intracellular product of a viral oncogene, a protein called $pp60^{v\text{-}src}$ (see p. 754). The EGF receptor was the first receptor protein to be shown to be a tyrosine-specific kinase (in 1982), and several years later the viral *erbB* oncogene was shown to encode a truncated version of an EGF receptor. This truncated protein lacks the extracellular EGF-binding domain and has an intracellular tyrosine kinase domain that is constitutively active: cells carrying this faulty receptor behave as though they are constantly being signaled to proliferate.

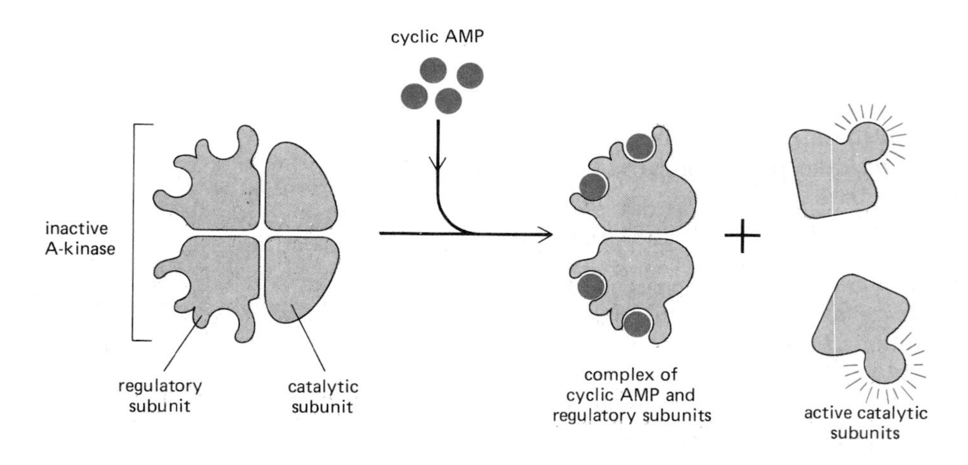

Figure 12–27 The activation of cyclic AMP-dependent protein kinase (A-kinase). The binding of cyclic AMP to the regulatory subunits induces a conformational change, causing these subunits to dissociate from the complex and thereby activating the catalytic subunits. Each regulatory subunit has two cyclic AMP-binding sites, and the release of the catalytic subunits is a cooperative process requiring the binding of more than two cyclic AMP molecules to the tetramer. This greatly sharpens the response of the kinase to changes in cyclic AMP concentration, as discussed on page 716. Many cells have two types of A-kinase, with identical catalytic subunits but different regulatory subunits.

tostatin. The promoter region of the somatostatin gene contains a short DNA sequence (about 30 nucleotides long) that is also found in the promoter region of several other genes that are activated by cyclic AMP. This sequence is recognized by a specific gene regulatory protein that activates transcription from these genes when it is phosphorylated by A-kinase.

In the inactive state the A-kinase consists of a complex of two regulatory subunits that bind cyclic AMP and two catalytic subunits. The binding of cyclic AMP alters the conformation of the regulatory subunits, causing them to dissociate from the complex. The released catalytic subunits are thereby activated to phosphorylate substrate protein molecules (Figure 12–27).

A-kinase is found in all animal cells and is thought to account for almost all of the effects of cyclic AMP in these cells. While most of the substrates for the kinase have not yet been characterized, it is clear that many of them differ in different cell types, explaining why the effects of cyclic AMP vary depending on the target cell.

Cyclic AMP Inhibits an Intracellular Protein Phosphatase[24]

Since the effects of cyclic AMP are usually transient, it is clear that cells must dephosphorylate the proteins that have been phosphorylated by A-kinase. The dephosphorylation is catalyzed by two main *protein phosphatases,* one of which is itself regulated by cyclic AMP. The level of phosphorylation at any instant will depend on the balance between the kinase and phosphatase activities.

In skeletal muscle cells, the cyclic AMP-regulated protein phosphatase is most active in the absence of cyclic AMP, and it dephosphorylates each of the three key enzymes in the glycogen pathway that were mentioned earlier—phosphorylase kinase, glycogen phosphorylase, and glycogen synthase. These dephosphorylation reactions tend to counteract the protein phosphorylations stimulated by cyclic AMP. However, when A-kinase is activated by cyclic AMP, it also phosphorylates a specific *phosphatase inhibitor protein,* which is thereby activated. This activated inhibitor protein binds to the protein phosphatase and inactivates it (Figure 12–28). By both activating phosphorylase kinase and inhibiting the opposing action of the protein phosphatase, the A-kinase causes a rise in cyclic AMP levels to have a much larger and sharper effect on glycogen metabolism than could be obtained if the A-kinase acted on one of these enzymes alone.

12-22 Calmodulin Is a Ubiquitous Intracellular Receptor for Ca^{2+}[25]

Since the free Ca^{2+} concentration in the cytosol is usually about 10^{-7} M and generally does not rise above 5×10^{-6} M even when the cell is activated by an influx of Ca^{2+}, any structure in the cell that is to serve as a direct target for Ca^{2+}-dependent regulation must have an affinity constant (K_a) for Ca^{2+} of around 10^6 liters/mole. Moreover, since the concentration of free Mg^{2+} in the cytosol is relatively constant at about 10^{-3} M, these Ca^{2+}-binding sites must have a selectivity

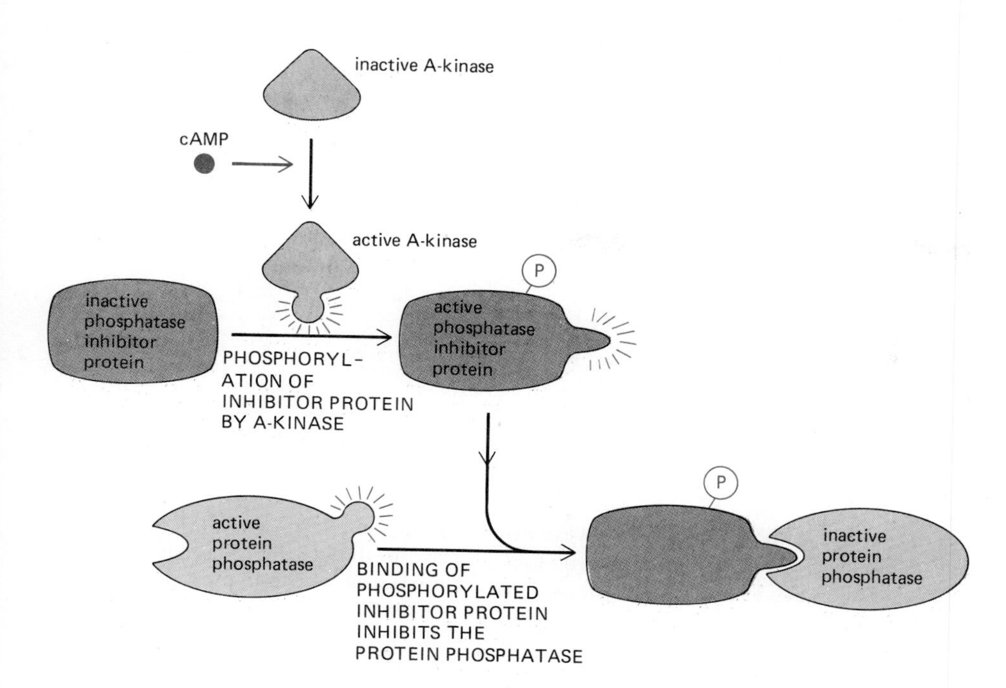

Figure 12–28 Cyclic AMP inhibits the protein phosphatase that would otherwise oppose the phosphorylation reactions stimulated by cyclic AMP. It does so by activating A-kinase to phosphorylate a phosphatase inhibitor protein, which can then bind to and inhibit the protein phosphatase.

for Ca^{2+} over Mg^{2+} of at least 1000-fold. Several specific Ca^{2+}-binding proteins fulfill these criteria.

The first such protein to be discovered was *troponin C* in skeletal muscle cells; its role in muscle contraction has been discussed in Chapter 11 (see p. 621). A closely related Ca^{2+}-binding protein, known as **calmodulin,** is found in all animal and plant cells that have been examined. A typical animal cell contains more than 10^7 molecules of calmodulin, which can constitute as much as 1% of the total protein mass of the cell. Calmodulin functions as a multipurpose intracellular Ca^{2+} receptor, mediating most Ca^{2+}-regulated processes. It is a highly conserved, single polypeptide chain of about 150 amino acid residues, with four high-affinity Ca^{2+}-binding sites; and it undergoes a large conformational change when it binds Ca^{2+} (Figure 12–29).

The allosteric activation of calmodulin by Ca^{2+} is analogous to the allosteric activation of A-kinase by cyclic AMP, except that Ca^{2+}-calmodulin complexes have no enzyme activity themselves but act by binding to other proteins. In some cases calmodulin serves as a permanent regulatory subunit of an enzyme complex (as in the case of phosphorylase kinase—see below), but in most cases the binding of Ca^{2+} induces calmodulin to bind to various target proteins in the cell and thereby alter their activity (Figure 12–30).

Among the targets regulated by Ca^{2+}-calmodulin complexes are many enzymes and membrane transport proteins. Prominent among these are **Ca^{2+}/calmodulin-dependent protein kinases (Ca-kinases),** which phosphorylate serine and threonine residues on proteins. The first Ca-kinases to be discovered, including *myosin light-chain kinase* and *phosphorylase kinase,* had narrow substrate specificities. More recently a broad-specificity Ca-kinase has been identified; called *Ca^{2+}/calmodulin-regulated "multifunctional" kinase* (or *Ca-kinase II*), it may mediate many of the actions of Ca^{2+} in mammalian cells. As in the case of cyclic AMP, the response of a target cell to an increase in free Ca^{2+} concentration in the cytosol depends on which Ca^{2+}-calmodulin regulated target proteins are present in the cell.

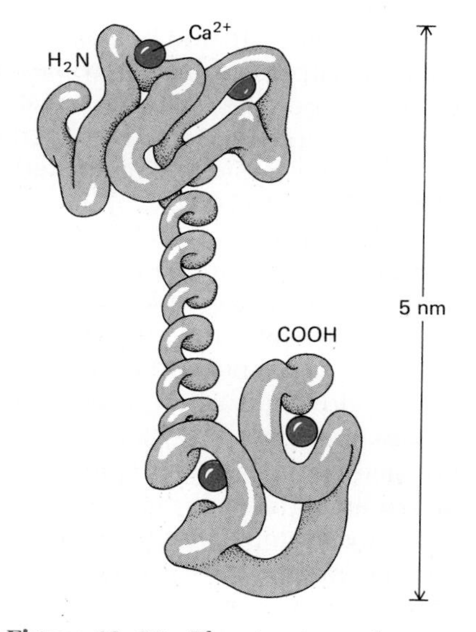

Figure 12–29 The structure of calmodulin based on x-ray diffraction studies. The molecule has a "dumbbell" shape, with two globular ends connected by a long, exposed α helix. Each end has two Ca^{2+}-binding domains, each with a loop of 12 amino acid residues in which aspartic acid and glutamic acid side chains form ionic bonds with Ca^{2+}. The two Ca^{2+}-binding sites in the carboxyl-terminal part of the molecule have a tenfold higher affinity for Ca^{2+} than those in the amino-terminal part. (Based on data from Y.S. Babu et al., *Nature* 315:37–40, 1985.)

12-23 The Cyclic AMP and Ca^{2+} Pathways Interact[26]

The cyclic AMP and Ca^{2+} second messenger pathways interact in at least three ways. First, intracellular Ca^{2+} and cyclic AMP levels can influence each other. In some cells, for example, Ca^{2+}-calmodulin complexes bind to and regulate en-

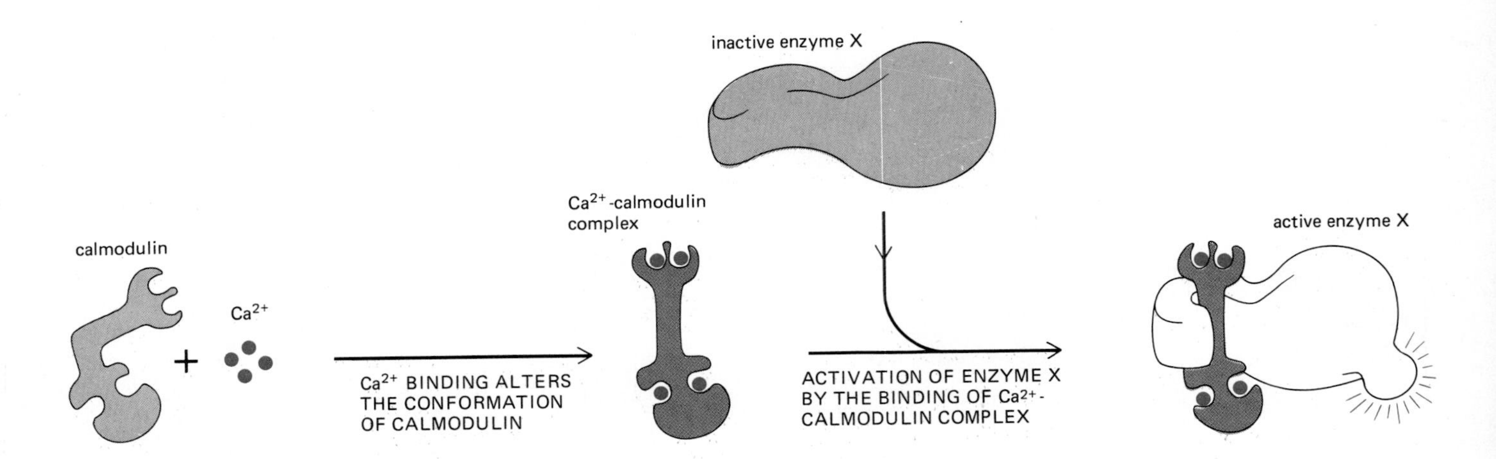

Figure 12–30 How an increase in free Ca^{2+} in the cytosol indirectly activates an enzyme by altering the conformation of calmodulin molecules. As is the case for cyclic AMP binding to A-kinase, the binding of Ca^{2+} to calmodulin is cooperative, requiring more than one Ca^{2+} to activate the protein. This sharpens the response of calmodulin to changes in Ca^{2+} concentration (see p. 716).

zymes that break down and make cyclic AMP—cyclic AMP phosphodiesterase and adenylate cyclase, respectively. Conversely, A-kinase can phosphorylate some Ca^{2+} channels and pumps and alter their activity. Second, some Ca-kinases are phosphorylated by A-kinase. Third, A-kinase and Ca-kinases frequently phosphorylate different sites on the same proteins, which are thereby regulated by both cyclic AMP and Ca^{2+}.

As an example of how Ca^{2+} and cyclic AMP pathways can interact, consider the *phosphorylase kinase* of skeletal muscle, whose role in glycogen degradation we have already discussed. This kinase phosphorylates glycogen phosphorylase, which breaks down glycogen (see Figure 12–26). It is a multisubunit enzyme, but only one of its four subunits actually catalyzes the phosphorylation reaction: the other three subunits are regulatory and enable the enzyme complex to be activated both by cyclic AMP and by Ca^{2+}. The four subunits are designated α, β, γ, and δ, and each is present in four copies in the phosphorylase kinase complex. The γ subunit carries the catalytic activity; the δ subunit is calmodulin and is largely responsible for the Ca^{2+} dependence of the enzyme. The α and β subunits are the targets for cyclic AMP-mediated regulation, both being phosphorylated by the A-kinase (Figure 12–31).

The same Ca^{2+} signal that initiates muscle contraction ensures that there is adequate glucose to power the contraction. The large influx of Ca^{2+} into the cytosol from the sarcoplasmic reticulum that initiates myofibril contraction (see p. 621) also alters the conformation of the δ subunit, increasing the activity of phosphorylase kinase and thereby increasing the rate of glycogen breakdown several hundredfold within seconds. In addition, the Ca^{2+} influx activates two Ca-kinases that phosphorylate and inhibit glycogen synthase, thereby shutting off glycogen synthesis. The epinephrine-induced phosphorylations previously discussed adjust muscle cell metabolism in anticipation of an increased energy demand; for example, the phosphorylation of phosphorylase kinase by A-kinase allows the enzyme to be activated when fewer calcium ions are bound to calmodulin, thereby making the enzyme more sensitive to Ca^{2+}.

Hundreds of different phosphorylated proteins are revealed by two-dimensional gel electrophoresis in a eucaryotic cell. Only a small fraction of these are phosphorylated by known protein kinases, which suggests that most kinases remain to be discovered. Indeed, it has been estimated that a single mammalian cell may contain more than 100 distinct protein kinases, which regulate the myriad reaction pathways in the cell. There are also many other reversible covalent modifications that regulate the activity of proteins: methylation-demethylation, acetylation-deacetylation, uridylation-deuridylation, and adenylation-deadenylation, among others. It seems likely that most cellular proteins that catalyze the rate-limiting steps in biological processes are regulated in one or another of these ways. In view of the complexities of the feedback loops known to exist in the glycogen pathways alone, it is clear that the unraveling of the mechanisms and kinetics of these regulatory processes poses a formidable challenge.

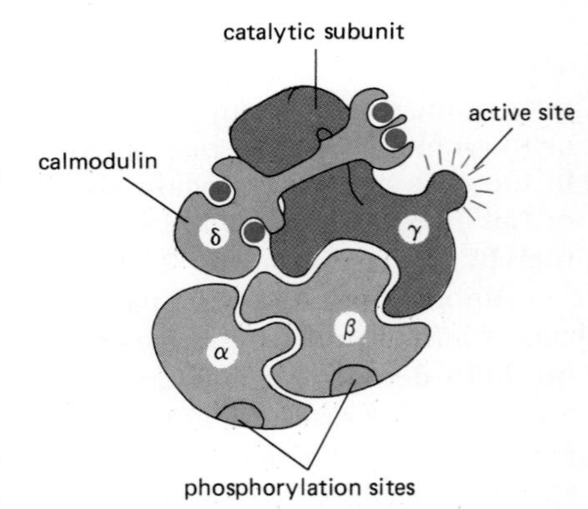

Figure 12–31 Highly schematized drawing of the four subunits of the enzyme *phosphorylase kinase* from mammalian muscle. The γ subunit has the catalytic activity of the active enzyme; the α and β subunits and the δ subunit (calmodulin) mediate the regulation of the enzyme by cyclic AMP and Ca^{2+}, respectively. The actual enzyme complex contains four copies of each subunit.

12-16 Cyclic GMP Is Also an Intracellular Messenger[27]

Cyclic AMP is not the only cyclic nucleotide to participate in intracellular signaling. Most animal cells also contain **cyclic guanosine monophosphate (cyclic GMP)** (Figure 12–32), although at a concentration one-tenth or less that of cyclic AMP. Cyclic GMP is known to activate a specific protein kinase (*G-kinase*) that phosphorylates target proteins in the cell (see Figure 12–25), but the role of cyclic GMP in signaling by cell-surface receptors is, with a few exceptions, unclear. Unlike adenylate cyclase, **guanylate cyclase,** which catalyzes the production of cyclic GMP from GTP, is usually a soluble enzyme that is not obviously coupled to cell-surface receptors, although cyclic GMP levels often increase when the inositol-phospholipid pathway is activated. However, recently plasma-membrane-bound forms of guanylate cyclase have been demonstrated in some cells that act as, or are coupled to, cell-surface receptors.

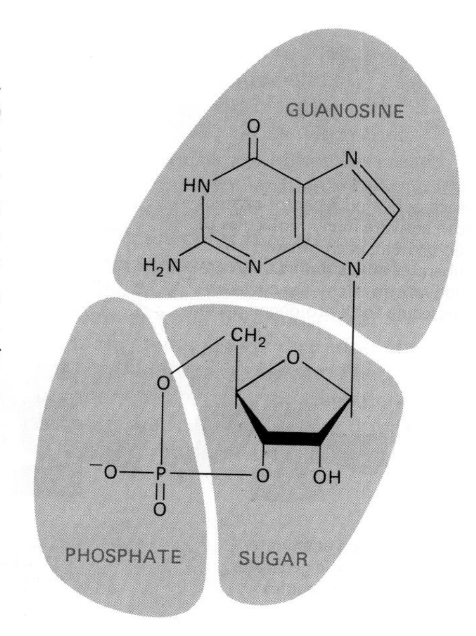

Figure 12–32 Cyclic GMP.

The signaling role of cyclic GMP is especially well understood in the response to light of rod cells in the vertebrate retina, where cyclic GMP acts directly on Na^+ channels in the plasma membrane of the rod cells. In the absence of a light signal, cyclic GMP is bound to the Na^+ channels, keeping them open. Light activates rhodopsin in the disc membrane of the rod cell, and activated rhodopsin then binds to and activates the G protein *transducin.* The α subunit of transducin (T_α) in turn activates a *cyclic GMP phosphodiesterase* that hydrolyzes cyclic GMP, so that cyclic GMP levels in the cytosol drop and the cyclic GMP bound to the Na^+ channels dissociates—allowing the channels to close. In this way the light signal is converted into an electrical signal, as explained in detail in Chapter 19 (see p. 1006). This direct effect of cyclic GMP on an ion channel is one of the few known examples where a cyclic nucleotide acts independently of a protein kinase in a vertebrate cell. But it is not the only example: olfactory receptor cells in the nose provide another. The binding of some odorants to specific olfactory receptor proteins on these cells activates adenylate cyclase via a G_s protein, increasing the concentration of cyclic AMP in the cytosol. The cyclic AMP then directly opens Na^+ channels in the plasma membrane, causing an excitatory depolarization.

Despite the differences in molecular details, all the signaling systems that are triggered by G-protein-dependent receptors share certain features and are governed by similar general principles. All of them, for example, depend on complex cascades or relay chains of intracellular messengers. What are the advantages of such seemingly complex systems that cause them to be used by so many types of cells for such a wide range of purposes?

Extracellular Signals Are Greatly Amplified by the Use of Intracellular Messengers and Enzymatic Cascades[28]

By contrast with more direct signaling systems, such as those involving steroid hormones, catalytic cascades of intracellular mediators provide numerous opportunities for amplifying and regulating the responses to extracellular signals. As illustrated in Figure 12–33, for example, when a ligand activates adenylate cyclase indirectly by binding to a receptor, each receptor protein can activate many molecules of G_s protein, each of which can activate a cyclase molecule. Each cyclase molecule, in turn, catalyzes the conversion of a large number of ATP molecules to cyclic AMP molecules. The same type of amplification operates in the inositol-phospholipid pathway. As a result, a nanomolar (10^{-9} M) concentration of an extracellular signal often induces micromolar (10^{-6} M) concentrations of an intracellular second messenger such as cyclic AMP or Ca^{2+}. Since these molecules themselves function as allosteric effector molecules to activate specific enzymes, a single extracellular signaling molecule can cause many thousands of molecules to be altered within the target cell. Moreover, each regulatory protein in the relay chain of signals can be a separate target for metabolic control, as, for example, in the glycogen breakdown cascade in skeletal muscle cells (see p. 709).

Such metabolically explosive cascades require tight regulation. Therefore, it is not surprising that cells have efficient mechanisms for rapidly degrading cyclic

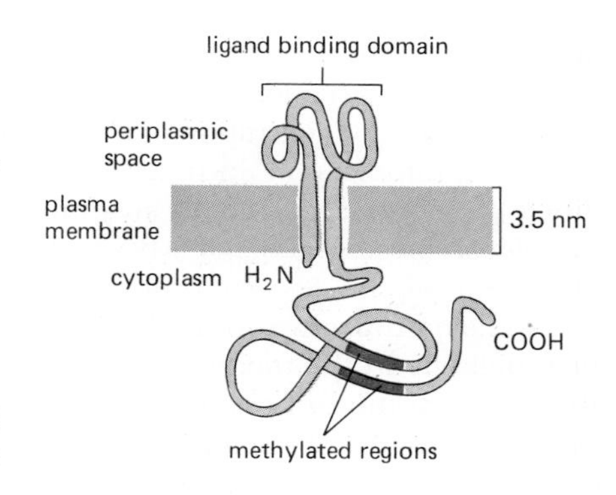

Figure 12–45 Structure of a chemotaxis receptor protein. The two regions of the polypeptide chain that become methylated consist of highly conserved 13-amino acid sequences. Although this drawing is based on the structure of the aspartate receptor, all four types of chemotaxis receptors have similar structures. (Based on A.F. Russo and D.E. Koshland, Jr., *Science* 220:1016–1020, 1983. Copyright 1983 by the AAAS.)

to serine and aspartate, respectively, by binding these amino acids directly and transducing the binding event into a signal in the cytosol. The other two receptors mediate responses to sugars and dipeptides, respectively, but are activated indirectly through *periplasmic substrate-binding proteins,* which also serve to mediate the transport of sugars and dipeptides across the plasma membrane (see p. 312). These proteins, which are dissolved in the periplasmic space (the space between the outer membrane and the plasma membrane), specifically bind the sugars and dipeptides and then form a complex with the appropriate chemotaxis receptors in the plasma membrane to activate them (Figure 12–46). Although the transport and chemotaxis systems for these sugars and dipeptides use common periplasmic substrate-binding proteins, the other parts of the machinery are different, as indicated by mutations that inactivate transport without affecting chemotaxis, and vice versa.

12-32 Receptor Methylation Is Responsible for Adaptation[35]

There is strong evidence that adaptation in bacterial chemotaxis results from the covalent methylation of the chemotaxis receptor proteins. When methylation is blocked by mutation, adaptation is markedly inhibited and exposure of the mutant bacteria to an attractant results in the suppression of tumbling for days instead of for a minute or so. Therefore, the activation of chemotaxis receptors by a chemoattractant has two separable consequences: (1) a rapid excitation occurs because the activated receptor generates an intracellular signal that causes the flagellar motor to continue to rotate counterclockwise, resulting in the suppression of tumbling and continuous smooth swimming; (2) a slower adaptation occurs because, while activated, the receptor is methylated by enzymes in the cytoplasm, reversing its activation over a period of a few minutes (Figure 12–47).

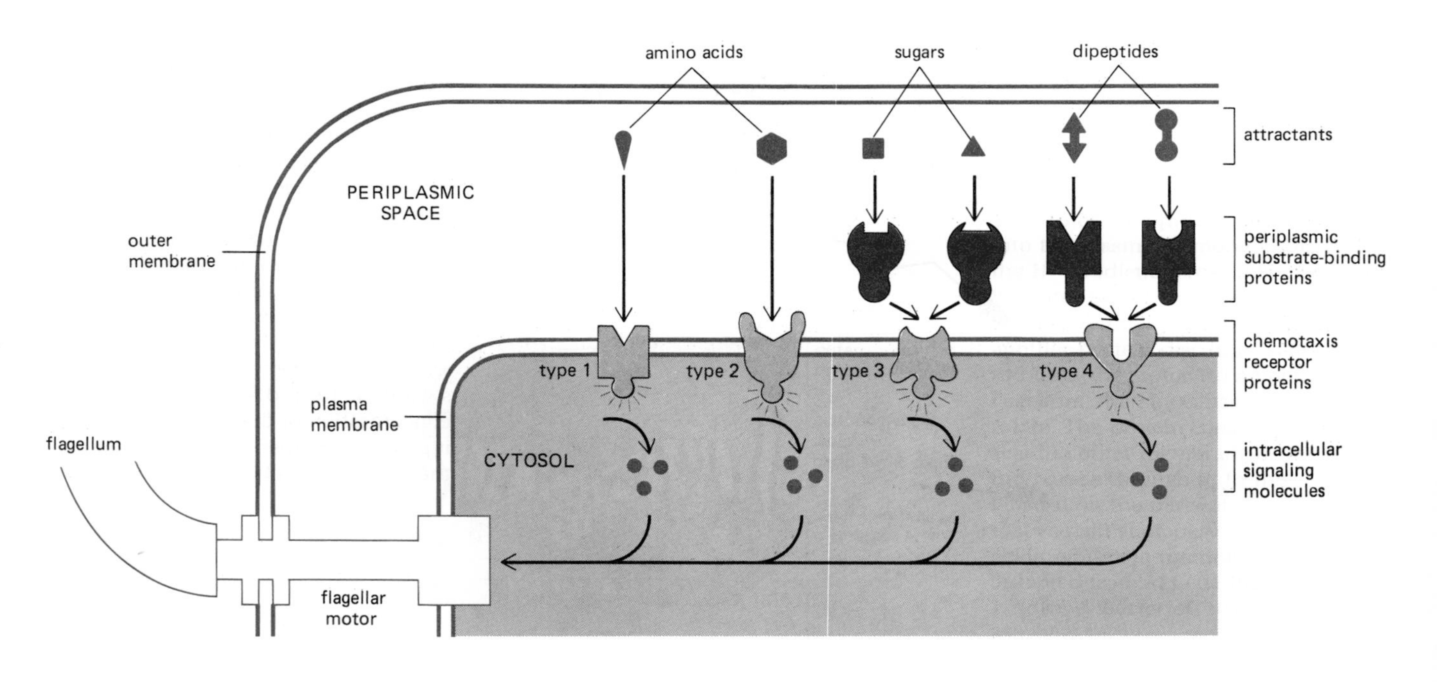

Figure 12–46 The steps in signal transduction during bacterial chemotaxis. Chemical attractants bind to type 1 or type 2 chemotaxis receptors in the plasma membrane or to periplasmic substrate-binding proteins that then bind to type 3 or type 4 chemotaxis receptors. This binding activates the chemotaxis receptors to produce an intracellular signal that causes the flagellar motor to continue to rotate counterclockwise, thereby suppressing tumbling and causing continuous smooth swimming. The attractants diffuse into the periplasmic space from outside the cell through large channels in the outer membrane (not shown).

Figure 12–47 The sequential activation and adaptation (via methylation) of a chemotaxis receptor. Note that the state of the receptor, and therefore the tumbling frequency of the bacterium, is the same in the resting and adapted states. The receptor is shown with two methylation sites for simplicity; in fact, there are four methylation sites on each receptor. As the concentration of ligand increases, the fraction of time that the receptor is occupied by the ligand increases. A higher level of ligand will thereby initially cause a greater change in the conformation of the receptor than a low level, pushing the receptor more toward its fully altered state. However, a slower increase in methylation ensues, so that within minutes the conformational strain on the receptor is exactly reversed—with more methyl groups being present at higher attractant concentrations. The receptor has now adapted. Although the ligand is shown here binding directly to the receptor, in some cases it binds first to a periplasmic substrate-binding protein, which then binds to the receptor.

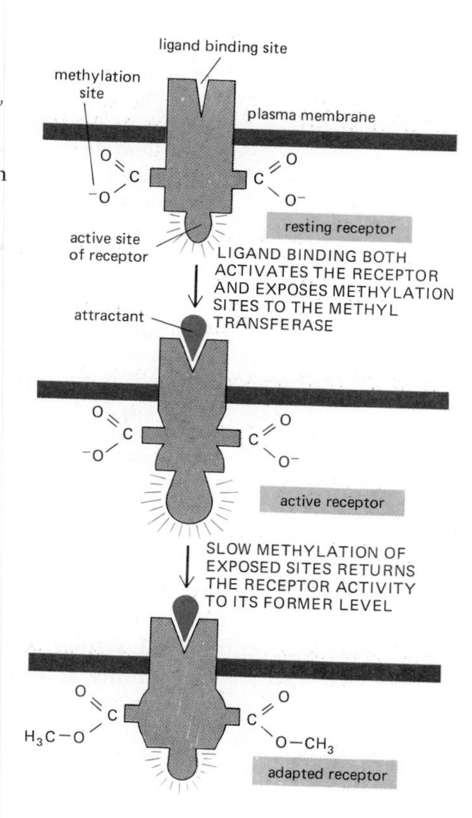

Receptor methylation is catalyzed by a soluble enzyme (*methyl transferase*) that transfers a methyl group to a free carboxyl group on a glutamic acid residue of the activated receptor protein (Figure 12–48). As many as four methyl groups can be transferred to a single receptor, the extent of methylation increasing at higher concentrations of attractant (where each receptor spends a larger proportion of its time with ligand bound). When the attractant is removed, the receptor is demethylated by a soluble demethylating enzyme (see Figure 12–48). Although the level of methylation changes during chemotactic responses, it remains constant once a bacterium is adapted because an exact balance is reached between the rates of methylation and demethylation.

A Cascade of Protein Phosphorylation Couples Receptor Activation to Changes in Flagellar Rotation[36]

The activation of chemotaxis receptors by attractants or repellents must lead to the generation of an intracellular signal that affects the direction of rotation of the flagellar motor. Genetic studies indicate that four cytoplasmic proteins—CheA, CheW, CheY, and CheZ—are involved in this intracellular signaling process. CheY and CheZ act at the effector end of the pathway to control the direction of flagellar rotation, apparently by binding to the flagellar motor. CheY signals the motor to rotate clockwise, resulting in tumbling; mutants that lack this protein swim continuously without tumbling. CheZ antagonizes the action of CheY, causing the motor to rotate counterclockwise, resulting in smooth swimming. CheA, and possibly CheW, are thought to relay the signal from the chemotaxis receptors to CheY and CheZ by a mechanism that involves protein phsophorylation and dephosphorylation.

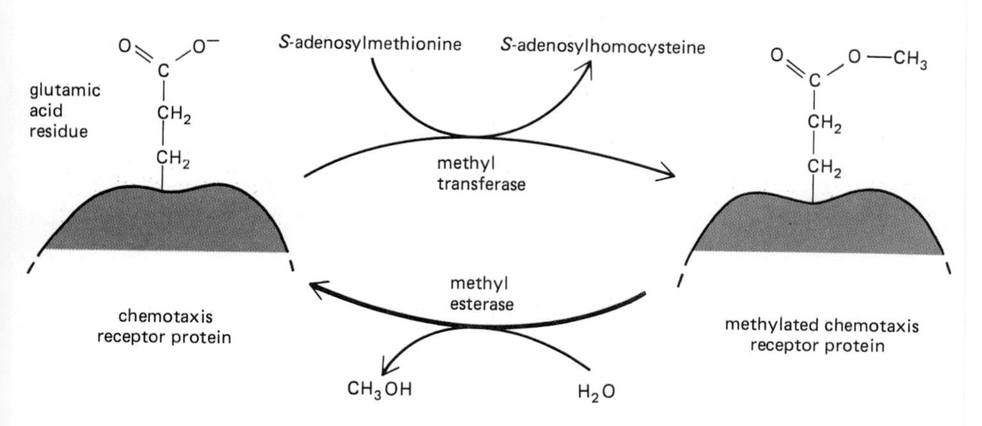

Figure 12–48 Methylation and demethylation reactions involving chemotaxis receptor proteins. Up to four methyl groups can be added to each receptor by ester linkages.

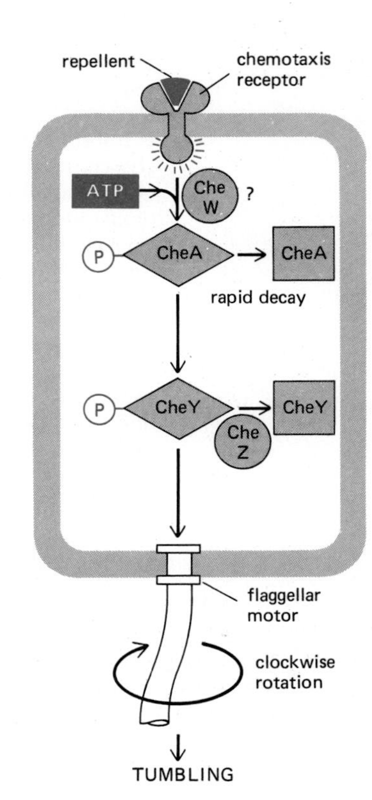

Figure 12–49 The phosphorylation relay system that is thought to enable the chemotaxis receptors to control the flagellar motor. The binding of a repellent activates the receptor, leading to the transient phosphorylation of CheA. CheA quickly transfers its covalently-bound, high-energy phosphate directly to CheY to generate CheY-phosphate, which binds to the flagellar motor and causes it to rotate clockwise, resulting in tumbling. The binding of an attractant has the opposite effect, leading to a decrease in the phosphorylation of CheA and CheY, counterclockwise flagellar rotation, and smooth swimming. CheZ accelerates the dephosphorylation of CheY-phosphate, thereby antagonizing the action of CheY. Each of these phosphorylated intermediates decays in about 10 seconds, enabling the bacterium to respond very quickly to changes in its environment (see Figure 12–34). It is not known how the chemotaxis receptors communicate with CheA or what role CheW plays in the process.

In vitro studies with purified proteins show that CheA is a protein kinase that phosphorylates itself in the presence of ATP and then quickly transfers the phosphate to CheY. The phosphorylated CheY is then dephosphorylated by a reaction that is accelerated by CheZ, which, as mentioned above, antagonizes CheY function *in vivo*. CheY is apparently activated (to cause tumbling) by CheA-mediated phosphorylation and inactivated (to cause smooth swimming) by CheZ-mediated dephosphorylation. It is thought that the phosphorylation of CheA and CheY decreases when chemotaxis receptors bind an attractant; the resulting inactivation of CheY leads to a decrease in tumbling and prolonged smooth swimming. By contrast, repellents are thought to activate CheA-mediated phosphorylation of CheY, thereby activating the protein to induce tumbling (Figure 12–49).

The same network of proteins also regulates the adaptation process. CheA phosphorylates the enzyme that demethylates the chemotaxis receptors (see Figure 12–48), increasing its activity and thereby providing feedback regulation to the chemotaxis receptors.

All of the genes and proteins involved in bacterial chemotaxis may now have been identified, and in most cases the proteins have been sequenced and are available in large quantities. It seems that we are rapidly approaching an almost complete molecular understanding of this highly adaptive behavior.

Summary

By adapting to high concentrations of a signaling ligand in a time-dependent, reversible manner, cells can adjust their sensitivity to the level of the stimulus and thereby respond to changes in a ligand's concentration instead of to its absolute level. Adaptation occurs in various ways: (1) ligand binding can induce the internalization of receptors, which are then transiently sequestered inside the cell or are degraded in lysosomes; (2) activated receptors can be reversibly inactivated by being phosphorylated or methylated; (3) nonreceptor proteins in the signal transduction pathway (such as G proteins) can be reversibly inactivated by mechanisms that are still uncertain. At a molecular level the best-understood example of adaptation occurs in bacterial chemotaxis, in which the reversible methylation of key signal-transducing proteins in the plasma membrane helps the cell to swim toward an optimal environment.

References

Cited

1. Smith, E.L.; et al. Principles of Biochemistry: Mammalian Biochemistry, 7th ed., pp. 355–619. New York: McGraw-Hill, 1983.
   Snyder, S.H. The molecular basis of communication between cells. *Sci. Am.* 253(4):132–140, 1985.
2. Norman, A.W.; Litwack, G. Hormones. San Diego, CA: Academic, 1987.
   Wilson, J.D.; Foster, D.W. Williams' Textbook of Endocrinology, 7th ed. Philadelphia: Saunders, 1985.
3. Simpson, I.A.; Cushman, S.W. Hormonal regulation of mammalian glucose transport. *Annu. Rev. Biochem.* 55:1059–1089, 1986.
4. Beer, D.J.; Matloff, S.M.; Rocklin, R.E. The influence of histamine in immune and inflammatory responses. *Adv. Immunol.* 35:209–268, 1984.
   Gospodarowicz, D.; Cheng, J.; Lui, G.M.; Baird, A.; Bohlen, P. Isolation of brain fibroblast growth factor by heparin sepharose affinity chromatography: identity with pituitary fibroblast growth factor. *Proc. Natl. Acad. Sci. USA* 81:6963–6967, 1984.
5. Smith, W.L.; Borgeat, P. The eicosanoids: prostaglandins, thromboxanes, leukotrienes, and hydroxyeicosaenoic acids. In Biochemistry of Lipids and Membranes (D.E. Vance, J.E. Vance, eds.), pp. 325–360. Menlo Park, CA: Benjamin-Cummings, 1985.
6. Evans, R.M. The steroid and thyroid hormone receptor superfamily. *Science* 240:889–895, 1988.
   Gehring, U. Steroid hormone receptors: biochemistry, genetics, and molecular biology. *Trends Biochem. Sci.* 12:399–402, 1987.
   Ivarie, R.D.; O'Farrell, P.H. The glucocorticoid domain: steroid-mediated changes in the rate of synthesis of rat hepatoma proteins. *Cell* 13:41–55, 1978.
   Yamamoto, K.R. Steroid receptor regulated transcription of specific genes and gene networks. *Annu. Rev. Genet.* 19:209–252, 1985.
7. Ashburner, M.; Chihara, C.; Meltzer, P.; Richards, G. Temporal control of puffing activity in polytene chromosomes. *Cold Spring Harbor Symp. Quant. Biol.* 38:655–662, 1974.
8. Attardi, B.; Ohno, S. Physical properties of androgen receptors in brain cytosol from normal and testicular feminized (Tfm/y♂) mice. *Endocrinology* 103:760–770, 1978.
9. Berridge, M. The molecular basis of communication within the cell. *Sci. Am.* 253(4):142–152, 1985.
   Kahn, C.R. Membrane receptors for hormones and neurotransmitters. *J. Cell Biol.* 70:261–286, 1976.
   Levitski, A. Receptors: A Quantitative Approach. Menlo Park, CA: Benjamin-Cummings, 1984.
   Rees Smith, B.; Buckland, P.R. Structure-function relations of the thyrotropin receptor. In Receptors, Antibodies and Disease, Ciba Foundation Symposium 90 (D. Evered, J. Whelan, eds.), pp. 114–132. London: Pitman, 1982.
   Snyder, S.H. The molecular basis of communication between cells. *Sci. Am.* 253(4):132–140, 1985.
10. Pastan, I. Cyclic AMP. *Sci. Am.* 227(2):97–105, 1972.
    Sutherland, E.W. Studies on the mechanism of hormone action. *Science* 177:401–408, 1972.
11. Schramm, M.; Selinger, Z. Message transmission: receptor controlled adenylate cyclase system. *Science* 225:1350–1356, 1984.
12. Casperson, G.F.; Bourne, H.R. Biochemical and molecular genetic analysis of hormone-sensitive adenylate cyclase. *Annu. Rev. Pharmacol. Toxicol.* 27:371–384, 1987.
    Feder, D.; et al. Reconstitution of beta$_1$-adrenoceptor-dependent adenylate cyclase from purified components. *EMBO J.* 5:1509–1514, 1986.
    Rodbell, M. The role of hormone receptors and GTP-regulatory proteins in membrane transduction. *Nature* 284:17–22, 1980.
13. Gilman, A.G. G proteins and dual control of adenylate cyclase. *Cell* 36:577–579, 1984.
    Gilman, A.G. G proteins: transducers of receptor-generated signals. *Annu. Rev. Biochem.* 56:615–649, 1987.
    Lai, C.-Y. The chemistry and biology of cholera toxin. *CRC Crit. Rev. Biochem.* 9:171–206, 1980.
    Levitzki, A. From epinephrine to cyclic AMP. *Science* 241:800–806, 1988.
    Stryer, L; Bourne, H.R. G proteins: a family of signal transducers. *Annu. Rev. Cell Biol.* 2:391–419, 1986.
14. Carafoli, E. Intracellular calcium homeostasis. *Annu. Rev. Biochem.* 56:395–433, 1987.
    Carafoli, E.; Penninston, J.T. The calcium signal. *Sci. Am.* 253(5):70–78, 1985.
    Evered, D.; Whelan, J., eds. Calcium and the Cell, Ciba Foundation Symposium 122. Chichester, U.K.: Wiley, 1986.
    Volpe, P.; et al. "Calciosome," a cytoplasmic organelle: the inositol 1,4,5-trisphosphate-sensitive Ca^{2+} store of nonmuscle cells? *Proc. Natl. Acad. Sci. USA* 85:1091–1095, 1988.
15. Augustine, G.J.; Charlton, M.P.; Smith, S.J. Calcium action in synaptic transmitter release. *Annu. Rev. Neurosci.* 10:633–693, 1987.
    Heilbrunn, L.V.; Wiercenski, F.J. The action of various cations on muscle protoplasm. *J. Cell. Comp. Physiol.* 29:15–32, 1947.
16. Berridge, M.J. Inositol lipids and calcium signalling. *Pro. R. Soc. Lond. (Biol.)* 234:359–378, 1988.
    Cockcroft, S. Polyphosphoinositide phosphodiesterase: regulation by a novel guanine nucleotide binding protein, Gp. *Trends Biochem. Sci.* 12:75–78, 1987.
    Majerus, P.W.; et al. The metabolism of phosphoinositide-derived messenger molecules. *Science* 234:1519–1526, 1986.
    Michell, R.H.; Putney, J.W., eds. Inositol Lipids in Cellular Signaling. Current Communications in Molecular Biology. Cold Spring Harbor, NY: Cold Spring Harbor Laboratory, 1987.
    Sekar, M.C.; Hokin, L.E. The role of phosphoinositides in signal transduction. *J. Memb. Biol.* 89:193–210, 1986.
    Woods, N.M.; Cuthbertson, K.S.R.; Cobbold, P.H. Repetitive transient rises in cytoplasmic free calcium in hormone-stimulated hepatocytes. *Nature* 319:600–602, 1986.
17. Angel, P.; et al. Phorbol ester-inducible genes contain a common *cis* element recognized by a TPA-modulated *trans*-acting factor. *Cell* 49:729–739, 1987.
    Bell, R.M. Protein kinase C activation by diacylglycerol second messengers. *Cell* 45:631–632, 1986.
    Lee, W.; Mitchell, P.; Tijan, R. Purified transcription factor AP-1 interacts with TPA-inducible enhancer elements. *Cell* 49:741–752, 1987.
    Nishizuka, Y. Studies and perspectives of protein kinase C. *Science* 233:305–312, 1986.
    Parker, P.J.; et al. The complete primary structure of protein kinase C—the major phorbol ester receptor. *Science* 233:853–859, 1986.
18. Barbacid, M. *ras* genes. *Annu. Rev. Biochem.* 56:779–827, 1987.
19. Dohlman, H.G.; Caron, M.G.; Lefkowitz, R.J. A family of receptors coupled to guanine nucleotide regulatory proteins. *Biochemistry* 26: 2657–2664, 1987.

Dunlap, K.; Holz, G.G.; Rane, S.G. G proteins as regulators of ion channel function. *Trends Neurosci.* 10:241–244, 1987.

Kubo, T.; et al. Cloning, sequencing and expression of complementary DNA encoding the muscarinic acetylcholine receptor. *Nature* 323:411–416, 1986.

Masu, Y.; et al. cDNA cloning of bovine substance-K receptor through oocyte expression system. *Nature* 329:836–838, 1987.

Stryer, L. The molecules of visual excitation. *Sci. Am.* 257(1):42–50, 1987.

20. Carpenter, G. Receptors for epidermal growth factor and other polypeptide mitogens. *Annu. Rev. Biochem.* 56:881–914, 1987.

Kaplan, D.R.; et al. Common elements in growth factor stimulation and oncogenic transformation: 85 kd phosphoprotein and phosphatidylinositol kinase activity. *Cell* 50:1021–1029, 1987.

Rosen, O.M. After insulin binds. *Science* 237:1452–1458, 1987.

Schlessinger, J. Allosteric regulation of the epidermal growth factor receptor kinase. *J. Cell Biol.* 103:2067–2072, 1986.

Yarden, Y.; Ullrich, A. Growth factor receptor tyrosine kinases. *Annu. Rev. Biochem.* 57:443–478, 1988.

21. Deuel, T.F. Polypeptide growth factors: roles in normal and abnormal cell growth. *Annu. Rev. Cell Biol.* 3:443–492, 1987.

Hanks, S.K.; Quinn, A.M.; Hunter, T. The protein kinase family: conserved features and deduced phylogeny of the catalytic domains. *Science* 241:42–52, 1988.

Hunter, T. A thousand and one protein kinases. *Cell* 50:823–829, 1987.

Ullrich, A.; et al. Human insulin receptor and its relationship to the tyrosine kinase family of oncogenes. *Nature* 313:756–761, 1985.

22. Cohen, P. Control of Enzyme Activity, 2nd ed. London: Chapman & Hall, 1983.

Edelman, A.M.; Blumenthal, D.K.; Krebs, E.G. Protein serine/threonine kinases. *Annu. Rev. Biochem.* 56:567–613, 1987.

23. Cohen, P. Protein phosphorylation and the control of glycogen metabolism in skeletal muscle. *Philos. Trans. R. Soc. Lond. (Biol.)* 302:13–25, 1983.

Montminy, M.R.; Bilezikjian, L.M. Binding of a nuclear protein to the cyclic-AMP response element of the somatostatin gene. *Nature* 328:175–178, 1987.

Pilkis, S.J.; El-Maghrabi, M.R; Claus, T.H. Hormonal regulation of hepatic gluconeogenesis and glycolysis. *Annu. Rev. Biochem.* 57:755–784, 1988.

Smith, S.B.; White, H.D.; Siegel, J.B.; Krebs, E.G. Cyclic AMP-dependent protein kinase I: cyclic nucleotide binding, structural changes, and release of the catalytic subunits. *Proc. Natl. Acad. Sci. USA* 78:1591–1595, 1981.

24. Alemany, S.; Pelech, S.; Brierley, C.H.; Cohen, P. The protein phosphatases involved in cellular regulation. Evidence that dephosphorylation of glycogen phosphorylase and glycogen synthase in the glycogen and microsomal fractions of rat liver are catalysed by the same enzyme: protein phosphatase-1. *Eur. J. Biochem.* 156:101–110, 1986.

Ingebritsen, T.S.; Cohen P. Protein phosphatases: properties and role in cellular regulation. *Science* 221:331–338, 1983.

25. Babu, Y.S.; et al. Three-dimensional structure of calmodulin. *Nature* 315:37–40, 1985.

Cheung, W.Y. Calmodulin. *Sci. Am.* 246(6):48–56, 1982.

Gerday, C.; Gilles, R.; Bolis, L., eds. Calcium and Calcium Binding Proteins. Berlin: Springer-Verlag, 1988.

Klee, C.B.; Crouch, T.H.; Richman, P.G. Calmodulin. *Annu. Rev. Biochem.* 49:489–515, 1980.

26. Cohen, P. Protein phosphorylation and hormone action. *Proc. R. Soc. Lond. (Biol.)* 234:115–144, 1988.

27. Goldberg, N.D.; Haddox, M.K. Cyclic GMP metabolism and involvement in biological regulation. *Annu. Rev. Biochem.* 46:823–896, 1977.

Nakamura, T.; Gold, G.H. A cyclic nucleotide-gated conductance in olfactory receptor cilia. *Nature* 325:442–444, 1987.

Schnapf, J.L.; Baylor, D.A. How photoreceptor cells respond to light. *Sci. Am.* 256(4):40–47, 1987.

Stryer, L. Cyclic GMP cascade of vision. *Annu. Rev. Neurosci.* 9:87–119, 1986.

28. Cohen, P. Protein phosphorylation and hormone action. *Proc. R. Soc. Lond. (Biol.)* 234:115–144, 1988.

29. Schimke, R.T. On the roles of synthesis and degradation in regulation of enzyme levels in mammalian tissues. *Curr. Top. Cell. Regul.* 1:77–124, 1969.

30. Lewis, J.; Slack, J.; Wolpert, L. Thresholds in development. *J. Theor. Biol.* 65:579–590, 1977.

Miller, S.G.; Kennedy, M.B. Regulation of brain type II Ca^{2+}/calmodulin-dependent protein kinase by autophosphorylation: a Ca^{2+}-triggered molecular switch. *Cell* 44:861–870, 1986.

Mulvihill, E.R.; Palmiter, R.D. Relationship of nuclear estrogen receptor levels to induction of ovalbumin and conalbumin mRNA in chick oviduct. *J. Biol. Chem.* 252:2060–2068, 1977.

31. Lefkowitz, R.J., ed. Receptor regulation. Receptors and Recognition, Series B, Vol. 13. London: Chapman & Hall, 1981.

Soderquist, A.M.; Carpenter, G. Biosynthesis and metabolic degradation of receptors for epidermal growth factor. *J. Memb. Biol.* 90:97–105, 1986.

32. Sibley, D.R.; Benovic, J.L.; Caron, M.G.; Lefkowitz, R.J. Regulation of transmembrane signaling by receptor phosphorylation. *Cell* 48:913–922, 1987.

33. Kassis, S.; Fishman, P.H. Different mechanisms of desensitization of adenylate cyclase by isoproterenol and prostaglandin E_1 in human fibroblasts: role of regulatory components in desensitization. *J. Biol. Chem.* 257:5312–5318, 1982.

Klee, W.A.; Sharma, S.K.; Nirenberg, M. Opiate receptors as regulators of adenylate cyclase. *Life Sci.* 16:1869–1874, 1975.

Snyder, S.H. Opiate receptors and internal opiates. *Sci. Am.* 236(3):44–56, 1977.

34. Adler, J. The sensing of chemicals by bacteria. *Sci. Am.* 234(4):40–47, 1976.

Berg, H. How bacteria swim. *Sci. Am.* 233(2):36–44, 1975.

35. Koshland, D.E., Jr. Biochemistry of sensing and adaptation in a simple bacterial system. *Annu. Rev. Biochem.* 50:765–782, 1981.

Russo, A.F.; Koshland, D.E. Receptor modification and absolute adaptation in bacterial sensing. In Sensing and Response in Microorganisms (M. Eisenbach, M. Balaban, eds.), pp. 27–41. Amsterdam: Elsevier, 1985.

Springer, M.S.; Goy, M.F.; Adler, J. Protein methylation in behavioral control mechanisms and in signal transduction. *Nature* 280:279–284, 1979.

36. Hess, J.F.; Oosawa, K.; Kaplan, N.; Simon, M.I. Phosphorylation of three proteins in the signaling pathway of bacterial chemotaxis. *Cell* 53:79–87, 1988.

Oosawa, K.; Hess, J.F.; Simon, M.I. Mutants defective in bacterial chemotaxis show modified protein phosphorylation. *Cell* 53:89–96, 1988.

Cell Growth and Division

13

Cells reproduce by duplicating their contents and then dividing in two. Complex sequences of cell divisions, punctuated periodically by sexual cell fusion, generate multicellular organisms. Even after a higher animal or plant has reached maturity, cell division is usually required in order to make up for losses due to wear and tear. Thus an adult human being must manufacture many millions of new cells each second, simply to maintain the status quo; and if all cell division is halted—for example, by a large dose of ionizing radiation—he or she will die within a few days.

For most of the constituents of a cell, duplication need not be controlled exactly. If there are many copies of a particular type of molecule or organelle, it is sufficient that the number of copies be approximately doubled in one cycle and that the dividing parent cell allocate approximately equal shares to each daughter. But there is at least one obvious exception: the DNA must always be duplicated exactly and divided precisely between the two daughter cells, and this requires special machinery. In discussing the cell cycle, therefore, it is sometimes convenient to distinguish between the *chromosome cycle* and the parallel *cytoplasmic cycle*. In the **chromosome cycle,** *DNA synthesis,* in which the nuclear DNA is duplicated, alternates with *mitosis,* in which the duplicate copies of the genome are separated. In the **cytoplasmic cycle,** *cell growth,* in which the many other components of the cell double in quantity, alternates with *cytokinesis,* in which the cell as a whole divides in two.

We begin this chapter by discussing the coordination and control of these interdependent cycles. We examine the mechanisms that ensure that all the nuclear DNA gets replicated once and only once between one cell division and the next, and we consider how the events of the chromosome cycle are coordinated with those of the cytoplasmic cycle. We then explore the regulation of cell division in multicellular animals by factors in the cell's environment—a topic that has been greatly illuminated by recent advances in cancer research. Finally, we discuss the molecular machinery responsible for mitosis and cytokinesis. These two processes require that the *centrosome* (see p. 763) be inherited reliably and duplicated precisely in order to form the two poles of the mitotic spindle; this *centrosome cycle* can be considered a third component of the cell cycle.

The Steps of the Cell Cycle and Their Causal Connections

The division of a eucaryotic cell presents a striking spectacle under the microscope. In **mitosis** the contents of the nucleus condense to form visible chromosomes, which through an elaborately orchestrated series of movements are pulled apart into two equal sets; then, in **cytokinesis,** the cell itself splits into two daughter cells, each receiving one of the two sets of chromosomes. Because they are so easily seen, mitosis and cytokinesis were the chief focus of interest to early investigators. These two events, however, together occupy only a brief period, known as the **M phase** (M = mitosis), in the cell's reproductive cycle. The much longer time that elapses between one M phase and the next is known as **interphase.** Under the microscope interphase appears, deceptively, as an uneventful interlude in which the cell simply grows slowly in size. More sophisticated techniques reveal that interphase is actually a period in which elaborate preparations for division are occurring in a carefully ordered sequence. In this section we discuss how the sequence of events in interphase can be investigated and how the steps of the cell cycle are causally connected.

13-3 13-4 Replication of the Nuclear DNA Occurs During a Specific Part of Interphase[1]

In most cells the DNA in the nucleus is replicated during only a limited portion of interphase; this period of DNA synthesis is called the **S phase** of the cell cycle. Between the end of the M phase and the beginning of DNA synthesis, there is usually an interval, known as the G_1 *phase* (G = gap); a second interval, known as the G_2 *phase,* separates the end of DNA synthesis from the beginning of the next M phase. Interphase is thus composed of successive G_1, S, and G_2 phases, and it normally comprises 90% or more of the total cell-cycle time. For example, in rapidly proliferating cells of higher eucaryotes, M phases generally occur only once every 16 to 24 hours, and each M phase itself lasts only 1 to 2 hours. A typical cell cycle with its four successive phases is illustrated in Figure 13–1, and some of the major events are outlined in the legend.

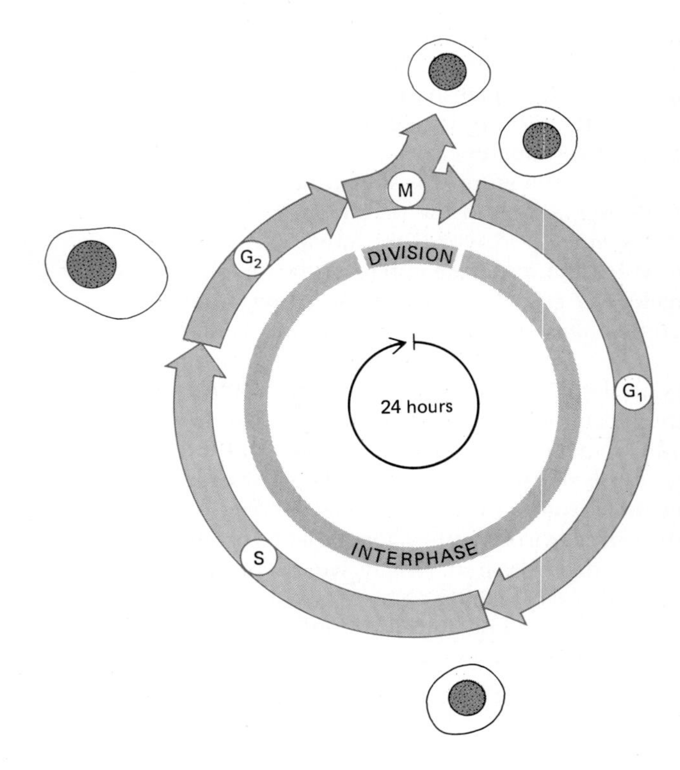

Figure 13–1 The four successive phases of a typical eucaryotic cell cycle. After the *M phase,* which consists of nuclear division (*mitosis*) and cytoplasmic division (*cytokinesis*), the daughter cells begin interphase of a new cycle. Interphase starts with the G_1 *phase,* in which the biosynthetic activities of the cells, which proceed very slowly during mitosis, resume at a high rate. The *S phase* begins when DNA synthesis starts, and ends when the DNA content of the nucleus has doubled and the chromosomes have replicated (each chromosome now consists of two identical "sister chromatids"). The cell then enters the G_2 *phase,* which continues until mitosis starts, initiating the M phase. During the M phase, the replicated chromosomes condense and are easily seen in the light microscope. The nuclear envelope breaks down (except in some unicellular eucaryotes such as yeasts, where it remains intact), the sister chromatids separate, two new nuclei form, and the cytoplasm divides to generate two daughter cells, each with a single nucleus. Cytokinesis terminates the M phase and marks the beginning of the interphase of the next cell cycle. A typical 24-hour cycle is illustrated here, although cell-cycle times in eucaryotic cells vary widely, from less than 8 hours to more than a year in adult animals, with most of the variability being in the length of the G_1 phase.

The time of DNA synthesis in the cell cycle was first demonstrated in the early 1950s by exploiting the technique of autoradiography to mark specifically those cells that are synthesizing DNA. The standard method employs ^{3}H-thymidine, a radioactive precursor of a compound that every cell uses exclusively for the synthesis of DNA. The ^{3}H-thymidine can either be injected into an animal to study the division cycles of cells in tissues or added to the culture medium of cells *in vitro* (Figure 13–2). In the former case, tissue is removed from the animal at a measured time after the injection of ^{3}H-thymidine, and autoradiographs are prepared from it. Those cells that have synthesized DNA at any time during the labeling period (and thus have been in S phase) can be identified by the silver grains over their nuclei. From the fraction of cells labeled in this way after different periods of exposure of an animal to ^{3}H-thymidine, and by scoring those cells that are in M phase, it is possible to show that the cell cycle has the four distinct phases described above and to measure the duration of each.

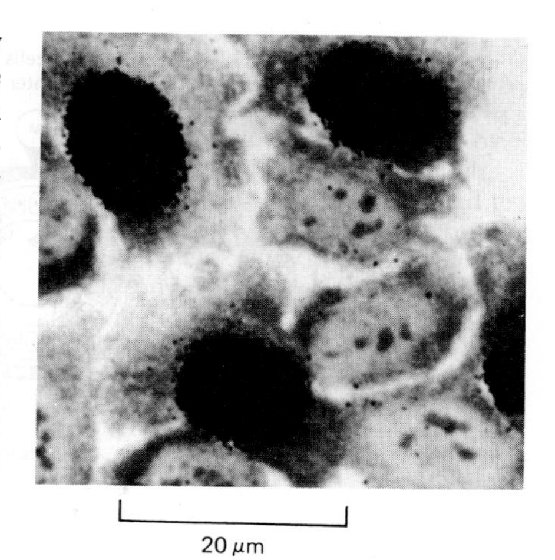

Figure 13–2 An autoradiograph of cells that have been exposed for a short period to ^{3}H-thymidine. The technique is explained on page 176. The presence of silver grains in the photographic emulsion over a cell nucleus (*blackened area*) indicates that the cell incorporated ^{3}H-thymidine into its DNA, and thus was in S phase, sometime during the labeling period. (Courtesy of James Cleaver.)

Suppose that a single injection of ^{3}H-thymidine is given and that the cells are fixed for autoradiography after a short time interval—say, half an hour. In a typical population of cells that are all proliferating rapidly but asynchronously, about 30% of the cells will be radioactively labeled. These are the cells that were synthesizing DNA during the brief exposure to ^{3}H-thymidine, and their frequency in the population reflects the fraction of the cell cycle that is occupied by S phase (Figure 13–3). Only around 5% of cells will be caught in mitosis at the moment of fixation (the small value of this *mitotic index* indicates that mitosis occupies only a small fraction of the cell cycle), and none of these will be radiolabeled, indicating that M phase and S phase are separate parts of the cycle. If, on the other hand, samples are fixed several hours after the injection of ^{3}H-thymidine, some of the cells in mitosis will be radiolabeled. These cells must have been synthesizing DNA at the time of injection. The minimum delay between injection and the time when radiolabeled mitotic cells appear will be equal to the duration of the G_2 phase. Proceeding along these lines, one can discover the durations of all four phases of the cycle. An example of the use of this method is explained in Figure 13–4.

Cells in different tissues, in different species, and at different stages of embryonic development have division cycles that vary enormously in duration, from less than an hour (for example, in the early frog embryo) to more than a year (for example, in the adult human liver). Although all phases of the cell cycle vary to some extent, by far the greatest variation occurs in the duration of G_1, which may be practically zero (as in the early frog embryo) or so long that the cell appears to have altogether ceased progressing through the division cycle and to have withdrawn into a quiescent state (as in the adult liver). Cells in such a quiescent G_1 state are often said to be in the *G_0 state*, as described below (see p. 749).

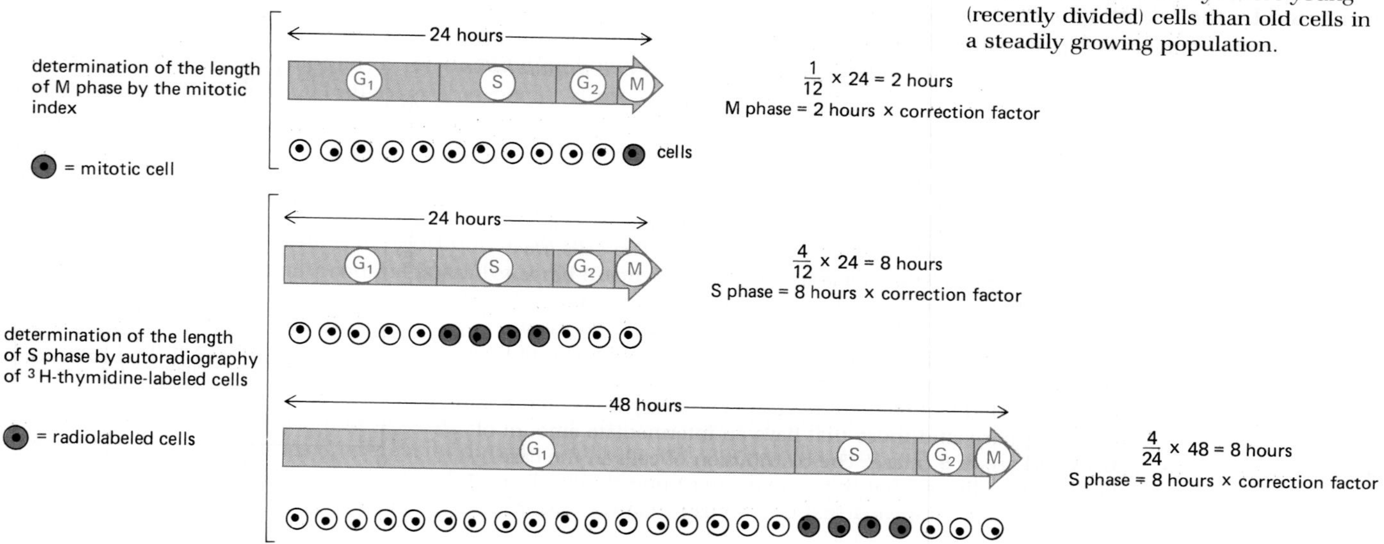

Figure 13–3 The length of each phase of the cell cycle is approximately equal to the fraction of cells in that phase at any instant multiplied by the total cell-cycle time, assuming that the population of cells is growing steadily and that all the cells are proliferating at the same rate. A precise calculation of the length of each phase, however, involves a "correction factor" (ranging from approximately 0.7 for early G_1 cells to 1.4 for mitotic cells), which is needed because there are always more young (recently divided) cells than old cells in a steadily growing population.

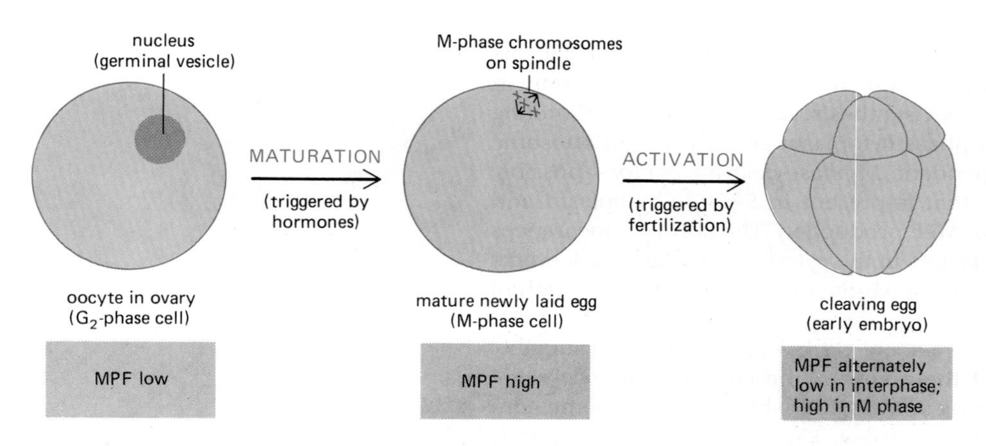

Figure 13–12 The levels of M-phase-promoting factor (MPF) in the *Xenopus* oocyte, egg, and early embryo. The oocyte is arrested in a meiotic G_2 phase, with a low level of MPF; the mature newly laid egg is arrested in a meiotic M phase, with a high level of MPF; following fertilization, the early embryo passes through alternate S and M phases, oscillating between low and high levels of MPF activity.

of DNA replication and division are abridged by speeding up both S phase and M phase and making the G_1 and G_2 phases so short as to be imperceptible.

MPF Induces Mitosis in a Wide Variety of Cells[10]

Because the *Xenopus* oocyte and egg are so big, it is easy to inject substances into their cytoplasm. Moreover, the oocyte, egg, and early embryo provide abundant sources of cytoplasm from defined stages of the cell cycle. This has been particularly important in the study of the **M-phase-promoting factor (MPF)** mentioned previously. MPF was first discovered in mature unfertilized *Xenopus* eggs (which are arrested in M phase). When cytoplasm from such an egg is injected into an oocyte, it releases the oocyte from its G_2 arrest and drives it into M phase. This initiates oocyte maturation, and the initials MPF originally stood for "maturation-promoting factor" (see p. 860). Active MPF also appears in the cleaving egg (embryo) during each M phase (Figure 13–12). Thus the *Xenopus* egg and oocyte provide both a good source of material for attempts to purify MPF and a convenient means to assay for it (Figure 13–13).

MPF is of universal importance to eucaryotic cells and has been highly conserved during evolution: extracts prepared from mitotic cells of the most diverse species, including mammals, sea urchins, clams, and yeasts, can be injected into *Xenopus* oocytes and will drive them into M phase. Material with MPF activity has been purified from mature *Xenopus* eggs. It behaves as a large protein that includes two types of subunits, one of which is a protein kinase and apparently can phosphorylate the other. Correspondingly, MPF seems to be able to activate itself: when a small amount of material with MPF activity is injected into a *Xenopus* oocyte, the cell responds by generating a very much larger amount of MPF from its own inactive reserves (see p. 860). This and other evidence suggest that the appearance and disappearance of MPF activity during the normal cell cycle depend on modification of the protein by phosphorylation and dephosphorylation, rather than on *de novo* synthesis and degradation. The normal triggering of MPF activation does, however, require the synthesis of another protein, identified as *cyclin* (see below); thus cells of all types are unable to progress from interphase to M phase when protein synthesis is blocked.

Many of the molecular changes that occur in mitosis seem to be brought about by phosphorylation. The MPF kinase directly phosphorylates several substrates, including, in particular, histone H1, thereby probably promoting chromosome condensation (see p. 503); and it may be through a cascade of phosphorylations that MPF triggers all the complex events of mitosis.

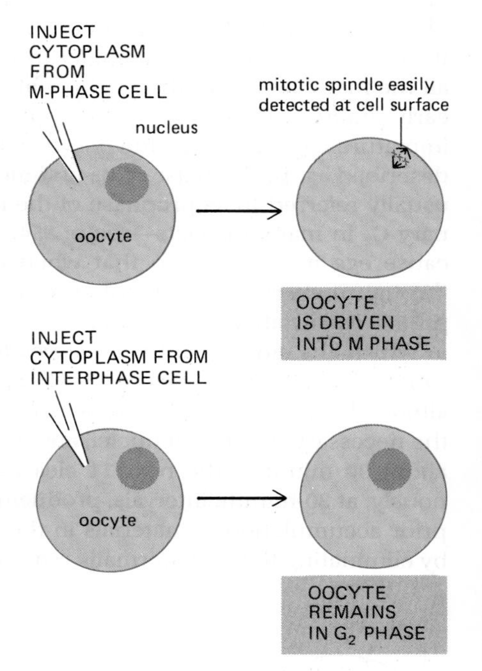

Figure 13–13 Assaying for MPF by injection into a *Xenopus* oocyte. MPF can be detected because it drives the oocyte into M phase. The large nucleus (or "germinal vesicle") of the oocyte breaks down as the mitotic spindle forms.

MPF Is Generated by a Cytoplasmic Oscillator[8,10,11]

The surge of MPF that occurs every 30 minutes in the cleaving *Xenopus* embryo is generated by a cytoplasmic oscillator that operates even in the absence of a nucleus. By constricting the activated egg before it has completed its first division, one can split it into two completely separate parts, one containing a nucleus, the

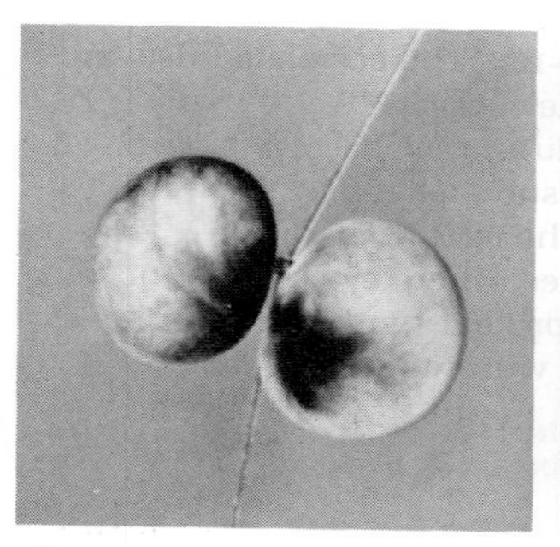

Figure 13–14 Technique for demonstrating a cytoplasmic oscillation associated with the cell-division cycle in the cleaving *Xenopus* egg. A newly fertilized egg is split in two by constricting it with a loop of fine human hair; one half contains the nucleus and proceeds to divide, while the other half lacks a nucleus and does not divide. Time-lapse photography then shows that the nonnucleated half periodically changes its height through changes in the stiffness of the cell cortex, oscillating in close synchrony with the divisions of the nucleated half. (From K. Hara, P. Tydeman, and M. Kirschner, *Proc. Natl. Acad. Sci. USA* 77:462–466, 1980.)

other not (Figure 13–14). The nucleated part continues with the normal program of rapid cleavages. Remarkably, the nonnucleated part also goes through a series of oscillations, manifest in repeated cycles of slight periodic contraction and stiffening of its cortical cytoplasm. These recurrent spasms occur in almost perfect synchrony with the cleavage divisions of the nucleated half-egg. By taking samples of cytoplasm at intervals from the oscillating nonnucleated cell and assaying them by injection into oocytes, it can be shown that the visible oscillations are accompanied by, and perhaps caused by, oscillations in the concentration of active MPF.

These and other experiments suggest that the cleavage divisions in the early *Xenopus* embryo involve two parallel cyclic processes—a chromosome-replication cycle and a cytoplasmic MPF cycle—that normally stay in step with each other because each new chromosome cycle can be initiated only when the DNA rereplication block is lifted following the pulse of MPF during M phase. This interaction between the two cycles prevents the chromosome cycle from running faster than the MPF cycle and will keep them in step as long as there is no risk of the chromosome cycle running too slowly to complete DNA replication before MPF levels rise. In the *Xenopus* egg, with its exceptionally rapid S phases and regular division cycles, this risk is presumably slight, and the single interaction between the cycles seems to be sufficient. However, in the mammalian cells discussed earlier (and probably in most eucaryotic cells other than cleaving eggs) there is also another, complementary effect: as we have seen, unreplicated DNA generates an M-phase-delaying signal that prevents the cytoplasmic MPF cycle from running faster than the chromosome cycle. Experiments in which DNA replication is blocked with inhibitors show that in the early *Xenopus* cell cycle this additional control does not operate. Moreover, as suggested by the absence of a G_1 phase, an S-phase activator seems to be present at all times. Thus the early *Xenopus* cycle appears to be simplified and stripped down for speed.

The phenomena just described imply that a cytoplasmic oscillator may exist in all cells, but they do not tell us its mechanism. A possible clue is provided by another protein, called **cyclin,** that has been identified in cleaving eggs of *Xenopus*, sea urchins, and surf clams. Cyclin, like MPF, belongs to the small group of proteins whose activity depends markedly on the phase of the division cycle. Although cyclin is synthesized at a rapid, constant rate throughout the cycle, it is abruptly destroyed halfway through M phase. Thus in each cycle its concentration rises steadily from zero and then suddenly drops back to zero again. Cyclin genes have been cloned, allowing the preparation of pure cyclin mRNA. When this mRNA is injected into a *Xenopus* oocyte, it has the same effect as an injection of MPF, driving the oocyte out of G_2 and into M phase. From these and other observations, it has been suggested that the surge of MPF in M phase may be triggered by the increase of cyclin concentration above a certain threshold; cyclin destruction is caused by an event in M phase; and the subsequent disappearance of MPF could be a consequence of cyclin destruction (Figure 13–15). In this view the time from one mitosis to the next would be determined chiefly by the time required for the cyclin concentration to build up from zero to the threshold value; and, as is indeed observed, the cell cycle should be halted in interphase by inhibitors of protein synthesis.

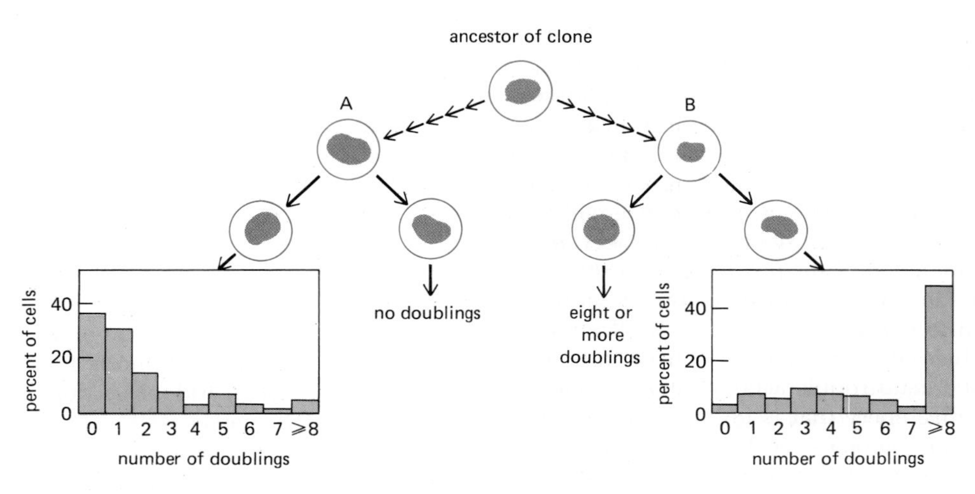

Figure 13–30 Evidence for cell variation in a heritable ability to divide. Individual cells in a clone, even though genetically identical, vary in the number of division cycles they will undergo. Here, different pairs of sister cells from the same clone have been studied, and histograms have been drawn to show the numbers of cells that divide a given number of times. If one sister fails to divide at all, the other sister usually does likewise or divides only a few times (*left side*); and if one sister undergoes eight or more doublings, the other also usually undergoes eight or more doublings (*right side*). This shows that there are heritable differences between genetically identical cells in the numbers of division cycles of which they are capable. The different heritable states are not perfectly stable, however, so sister cells sometimes behave differently. Further studies show that, as the cell population ages, the cells undergo random transitions toward states of reduced division capability. (Data from J.R. Smith and R.G. Whitney, *Science* 207:82–84, 1980.)

Summary

Cell division in multicellular organisms depends on complex social controls, and the proliferation of different cell types is governed by different combinations of protein growth factors. These act at very low concentrations, and most serve as local chemical mediators to help regulate cell population densities. In addition, most normal cells are unable to divide unless they are anchored to the extracellular matrix. Cells that are starved of growth factors or deprived of matrix adhesion come to a halt after mitosis, entering a special quiescent state, G_0, from which they take several hours to recover when reexposed to growth factors. Once a cell has left G_0 and passed the restriction point in G_1, it will complete S, G_2, and M promptly, irrespective of growth factors or adhesion. In a proliferating cell population, passage through the restriction point is an all-or-none event that can be characterized by a transition probability, like a radioactive decay. In addition to immediate controls of cell proliferation, there are long-term controls that cause the normal somatic cells of mammals to become senescent and cease dividing after a limited number of division cycles in culture.

Genes for the Social Control of Cell Division[26]

As we have seen for yeasts, genetics provides a powerful approach to the problem of determining the molecular basis of the controls on cell division, provided that there are ways of selecting for mutations of the relevant genes. In multicellular animals, mutations in the genes that are involved specifically in the social controls on cell division, which we shall call **social control genes,** are selected all too easily. A cell that undergoes a mutation or set of mutations that disrupts the social restraints on division will divide without regard to the needs of the organism as a whole, and its progeny will become apparent as a tumor.

Cancers, by definition, are *malignant* tumors; that is, the tumor cells not only divide in an ill-controlled way, they also invade and colonize other tissues of the body to create widespread secondary tumors, or *metastases.* To generate a cancer, a cell must first undergo a number of mutations to escape the multiple controls on cell division and then accumulate further changes to become endowed with the capacity for invasion and metastasis. These aspects of cancer are discussed in Chapter 21. Here we shall not attempt to explain cancer but rather to see what can be learned from cancer cells about the genes that normally control cell division.

Cell Transformation in Culture Provides an Assay for Genes Involved in the Social Control of Cell Division[27]

Given a clone of tumor cells that are presumed to have originated by mutation, how does one identify the mutant gene or genes? Classical genetic mapping is not feasible because the cells do not reproduce sexually. A more direct approach

Table 13–2 Some Changes Commonly Observed When a Normal Tissue-Culture Cell Is Transformed by a Tumor Virus

1. Plasma-membrane-related abnormalities
- A. Enhanced transport of metabolites
- B. Excessive blebbing of plasma membrane
- C. Increased mobility of plasma membrane proteins

2. Adherence abnormalities
- A. Diminished adhesion to surfaces; therefore able to maintain a rounded morphology
- B. Failure of actin filaments to organize into stress fibers
- C. Reduced external coat of fibronectin
- D. High production of plasminogen activator, causing increased extracellular proteolysis

3. Growth and division abnormalities
- A. Growth to an unusually high cell density
- B. Lowered requirement for growth factors
- C. Less "anchorage dependence" (can grow even without attachment to solid surface)
- D. "Immortal" (can continue proliferating indefinitely)
- E. Can cause tumors when injected into susceptible animals

is to take genetic material from the tumor cells and search for fragments of it that will, when introduced into normal cells, cause these cells to behave like tumor cells. Techniques for achieving this feat were first devised in the late 1970s, but their development depended on earlier studies of a very similar process that occurs naturally.

Certain types of tumors are caused by viruses. The viruses shed from these tumors can infect normal cells and, as a result of the introduction of the RNA or DNA carried by the virus, *transform* them into tumor cells. The first **tumor virus** to be discovered causes connective tissue tumors, or *sarcomas*, in birds; the infectious agent—the *Rous sarcoma virus*—has been an important object of study ever since, along with a variety of more recently discovered tumor viruses.

The nature of tumorigenic **cell transformation** is most easily observed in culture. A few days after tumor viruses have been added to a culture of normal cells, small colonies of abnormally proliferating cells appear. Each such colony is a clone derived from a single cell that has been infected with the virus and has stably incorporated the viral genetic material. Released from the social controls on cell division, the transformed cells outgrow normal ones in the culture dish just as in the body and are therefore usually easy to select. The transformed cells commonly show a complex syndrome of abnormalities (summarized in Table 13–2): they tend not to be constrained by density-dependent inhibition of cell division (see p. 748) but pile up in layer upon layer as they proliferate (Figure 13–31); they often do not depend on anchorage for growth and are capable of dividing even when held in suspension; they have an altered shape and adhere poorly to the substratum and to other cells, maintaining a rounded appearance reminiscent of a normal cell in mitosis; they may be able to proliferate even in the absence of growth factors; they are immortal and do not undergo senescence in culture; and when they are injected back into a suitable host animal, they can give rise to tumors.

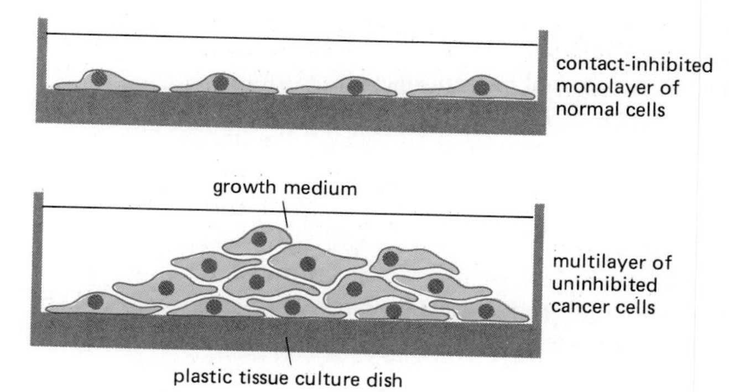

Figure 13–31 Cancer cells, unlike most normal cells, usually continue to grow and pile up on top of one another after they have formed a confluent monolayer.

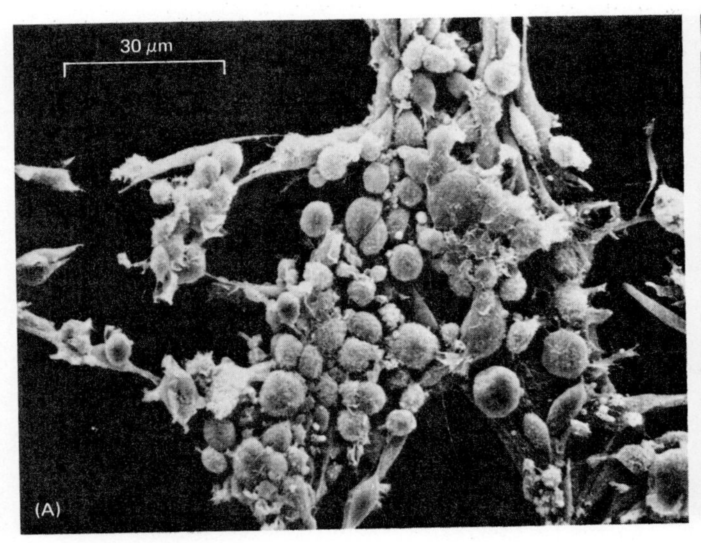

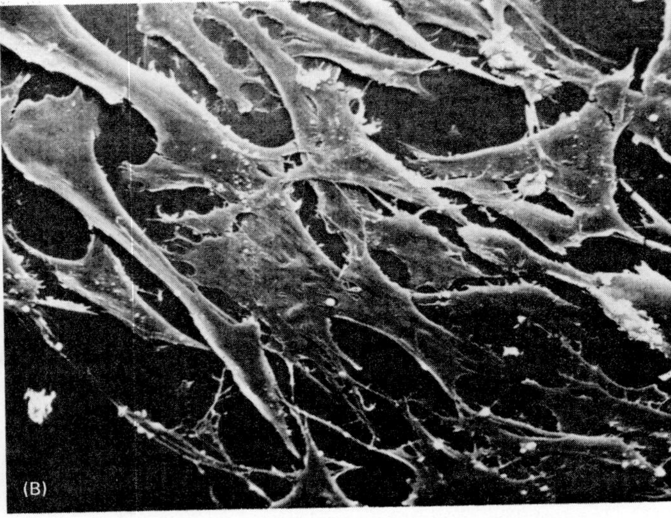

Figure 13–32 Scanning electron micrographs of cells in culture infected with a form of the Rous sarcoma virus that carries a temperature-sensitive mutation in the gene responsible for transformation (the v-*src* oncogene). (A) The cells are transformed and have an abnormal rounded shape at low temperature (34°C), where the oncogene product is functional. (B) The same cells adhere strongly to the culture dish and thereby regain their normal flattened appearance when the oncogene product is inactivated by a shift to higher temperature (39°C). (Courtesy of G. Steven Martin.)

13-20 Tumor Viruses Provide a Source of Ready-cloned Oncogenes[28]

A tumor virus subverts the normal controls on cell division by causing a permanent change in the genetic constitution of its host cell such that the cell begins to make a protein that overrides the normal controls. Such viruses therefore provide ways to identify the mechanisms that are normally responsible for cell division controls. Thus far the most important revelations have come from studies of RNA tumor viruses, also called **retroviruses.** After a retrovirus infects a cell, its RNA is copied into DNA by reverse transcription and the DNA is then inserted into the host genome. Figure 5–75 (p. 255) outlines the life cycle of a retrovirus and shows how its genome undergoes reverse transcription, integration into host DNA, and exit from and entry into host cells.

When a retrovirus transforms a normal cell into a tumor cell, the misbehavior is often brought about by a gene that is carried by the virus but is not necessary for the virus's own survival or reproduction. This was first demonstrated by the discovery of mutant Rous sarcoma viruses that multiply normally but no longer transform their host cells. Some of these nontransforming mutants were found to have deleted all or part of a gene coding for a protein of molecular weight 60,000. Other mutations in this gene can make the transforming effect of the virus temperature-sensitive: infected cells show a transformed phenotype at 34°C, but when the temperature is raised to 39°C, they return promptly (within a matter of hours) to the normal phenotype (Figure 13–32). Apparently this specific gene in the tumor virus is responsible for cell transformation (thereby bringing the virus that contains it to our attention) but is superfluous baggage from the point of view of the virus's own propagation.

The transforming gene of the Rous sarcoma virus identified by these experiments is called the **v-*src* gene.** It is classified as an **oncogene** (from the Greek *onkos,* a mass or tumor) because when it is introduced into a normal cell, it can transform it into a tumor cell. What is the origin of this gene, and what is its normal function? When a radioactive DNA copy of the viral *src* gene sequence was used as a probe to search for related sequences by DNA-DNA hybridization (see p. 188), it was found that the genomes of normal vertebrate cells contain a sequence that is closely similar, but not identical, to the *src* gene of the Rous sarcoma virus. This normal cellular counterpart of the viral *src* gene is called **c-*src*,** and it is classified as a **proto-oncogene.** Evidently the viral oncogene has been picked up from the genome of a previous host cell but has undergone mutation in the process. One suspects that the proto-oncogene is a normal social control gene that the retrovirus has, in effect, cloned for us. A large number of other oncogenes have now been identified and analyzed in similar ways, and each has led to the discovery of a corresponding proto-oncogene.

13-21 Tumors Caused in Different Ways Contain Mutations in the Same Proto-oncogenes[26,29]

Tumors often result from mutations occurring spontaneously or in response to chemical carcinogens or radiation rather than from viral infection. DNA can be purified from these tumor cells and assayed for the presence of oncogenes by introducing it into nontransformed cultured cells. 3T3 cells are often used for this assay because they divide indefinitely in culture and contain mutations that make them easy to transform with a single added oncogene. Using recombinant DNA techniques, the tumor-derived oncogene responsible for such a transformation can be identified, cloned, and sequenced (Figure 13–33). Remarkably, in most cases where this has been done, the oncogene has turned out to be a mutant form of one of the same proto-oncogenes identified by the retrovirus approach, although some new oncogenes have been discovered as well.

Related techniques reveal that transformation can also be caused by overproduction of certain normal gene products. Tumors often contain an unchanged proto-oncogene that is overexpressed, either because it is present in an abnormally large number of copies or because a chromosomal rearrangement has brought it under the control of an inappropriate promoter, as discussed in Chapter 21.

More than 50 proto-oncogenes have thus far been identified (see pp. 1207 and 1209). These may represent a substantial sample of the proto-oncogenes in the normal cell. It is likely, however, that many social control genes have yet to be identified. The fibroblastlike 3T3 cells commonly used in the transformation assay may fail to be transformed by an oncogene that helped to transform some other differentiated cell type that gave rise to a tumor. Moreover, the cell-transformation assay is capable of detecting only *dominant* mutations of social control genes—those mutations that deregulate cell proliferation even when normal gene copies are present in the cell. Recessive mutations in social control genes—resulting from loss of function—may be the most common type in cancer cells but will not be detected by the assay. Genes whose products normally help stimulate cell division can therefore be identified more readily with current techniques than those whose products normally help inhibit it. Nevertheless, there is evidence that genes with inhibitory effects on cell division exist and that recessive mutations in them are a common cause of cell transformation and cancer. For example, when transformed cells are fused with nontransformed cells, the resulting hybrid cells very often appear nontransformed, implying that normal cell division control has been restored by restoring a protein that was missing from the transformed cell. Bearing in mind, therefore, that many important social control genes remain to be discovered, we shall now consider what the known ones do.

Figure 13–33 A procedure by which human oncogenes can be identified and cloned. Oncogenes present in a sample of DNA taken from a human tumor are detected by their ability to transform mouse 3T3 cells. Transformed 3T3 cells proliferate in an unrestrained way and can be recognized by the colonies they form on a culture dish. Repetitive DNA sequences in the *Alu* family (see p. 608) are scattered throughout the human genome. They therefore provide a convenient marker that can be used to identify the human DNA in a nonhuman cell. Used as a DNA probe, the *Alu* sequences allow the human oncogene that has transformed the 3T3 cell to be cloned. The cloned DNA that contains the oncogene will transform 3T3 cells with very high efficiency when retested in the same assay.

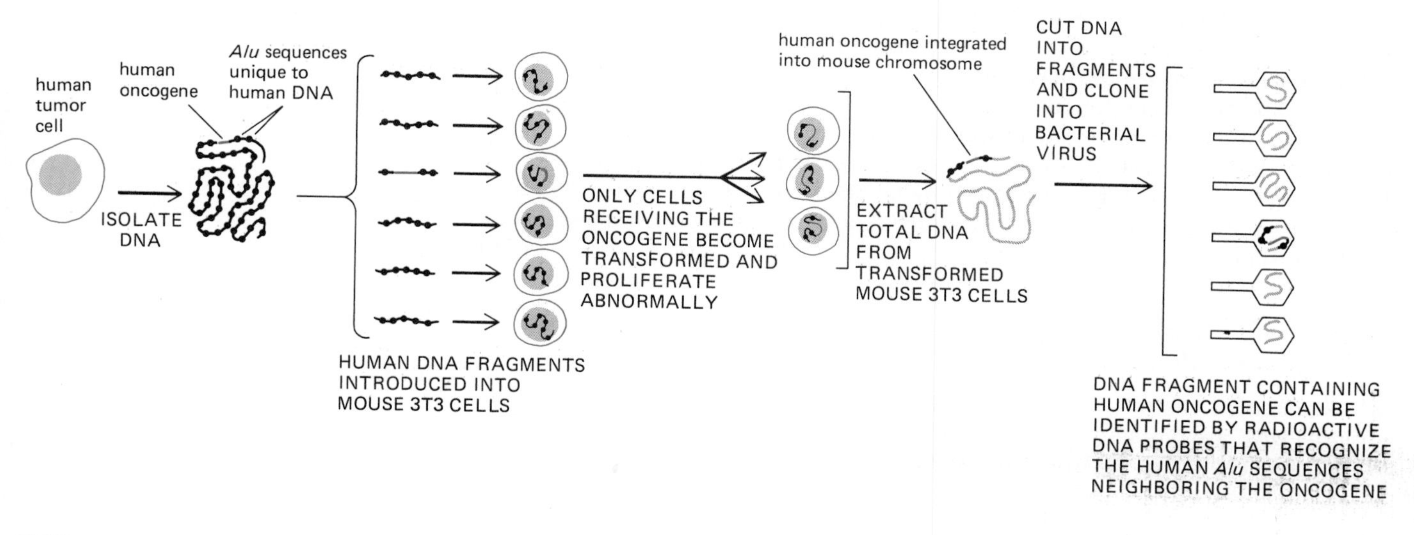

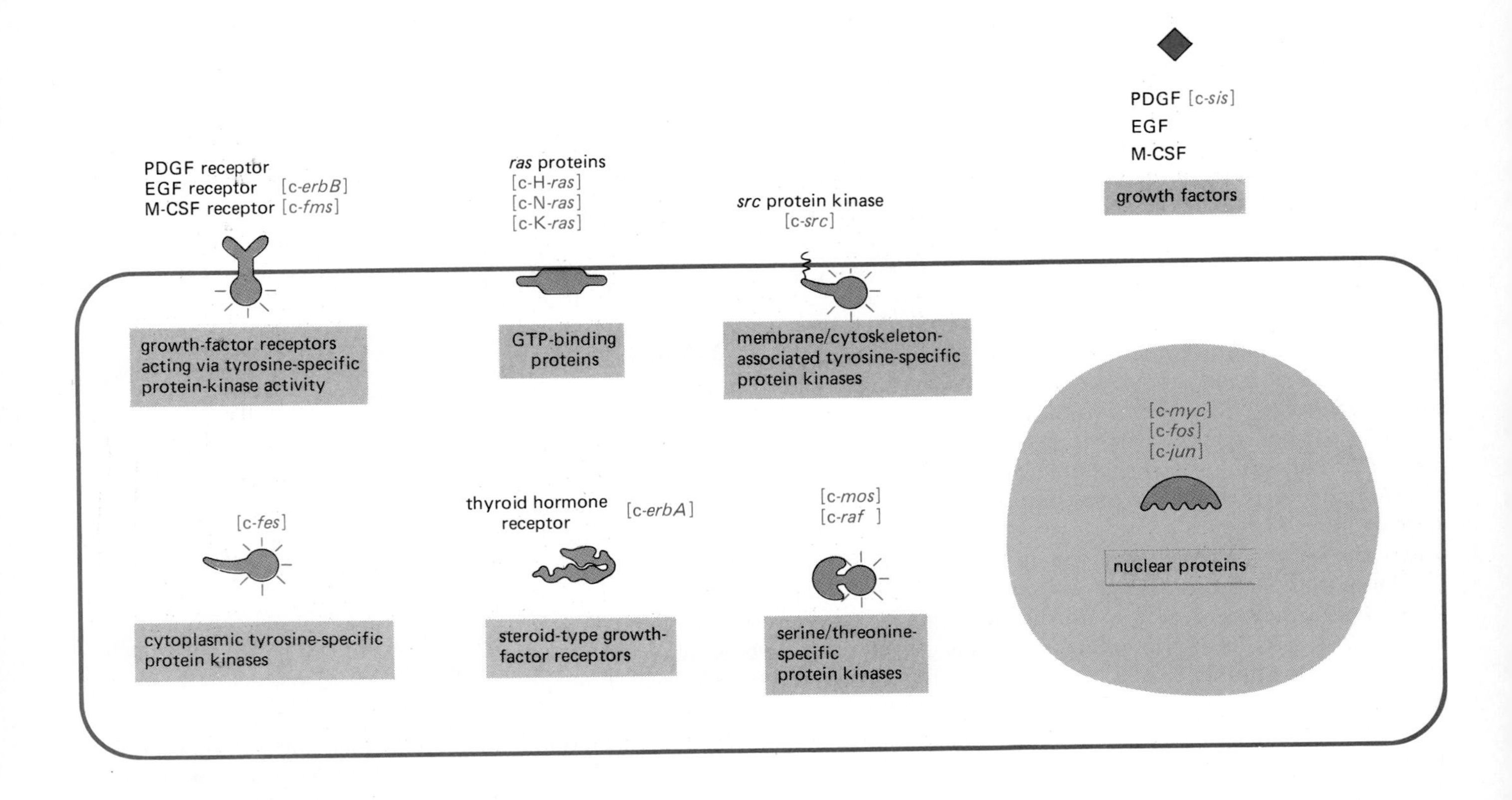

Figure 13–34 The activities and cellular locations of the main classes of known proto-oncogenes. The names of some representative proto-oncogenes in each class are shown in color. See also Figure 21–27, p. 1211.

13-16 13-22 Some Proto-oncogenes Code for Growth Factors or for Growth-Factor Receptors[26,30]

Once an oncogene has been cloned and sequenced, it is often possible to gain clues to the function of the corresponding social control gene (proto-oncogene) by looking for homologies between its sequence and the sequences of genes already known. In this way one proto-oncogene, known as c-*sis*, was discovered to be the gene that encodes a functionally active subunit of the growth factor PDGF. A cell that contains the corresponding v-*sis* oncogene manufactures the PDGF subunit constantly and inappropriately, and this protein, by binding to the cell's receptors for PDGF, constantly stimulates the cell to proliferate. At least three other proto-oncogenes, called c-*erbB*, c-*fms*, and c-*erbA*, code for receptors for growth factors or hormones: c-*erbB* encodes the receptor for epidermal growth factor (EGF, see p. 706); c-*fms* encodes the receptor for macrophage colony-stimulating factor (M-CSF)—a growth factor that stimulates proliferation of macrophage precursors (see p. 981); and c-*erbA* encodes the receptor for thyroid hormone (see p. 690). When mutated to oncogenes, these genes code for faulty receptors that behave as though a ligand were bound even when it is not, thereby stimulating the cell inappropriately (see p. 707).

Although the exact functions of most other proto-oncogenes are still unknown, one might expect many of them to code for proteins that are part of the signaling network inside the cell that enables growth factors to stimulate cell proliferation. We must now examine how far the known classes of proto-oncogenes, summarized in Figure 13–34, can account for the functions they might be expected to perform.

Some Proto-oncogenes Encode Intracellular Mediators Involved in Signaling Cells to Divide[26,31]

As discussed in Chapter 12 (p. 706), the receptors for many growth factors, including PDGF, are tyrosine-specific protein kinases that, when activated, phosphorylate themselves and various other proteins. One class of oncogenes encodes abnormal forms of such receptors; this class includes the altered EGF and M-CSF receptors just described (see Figure 13–34). Another small family of related proto-oncogenes, the c-*ras* genes (see Figure 13–34), codes for proteins that bind and hydrolyze GTP and may be distantly related to the G proteins involved in the transduction of many types of signals (see p. 705). Mutant *ras* genes that cause cell transformation are associated with raised concentrations or raised efficacy of the intracellular mediators inositol trisphosphate and diacylglycerol and make the cell hypersensitive to several growth factors that are believed to exert some of their effects by inducing production of inositol trisphosphate and diacylglycerol. Genes homologous to *ras* are present in budding yeasts, where they are involved in controlling the cell-division cycle according to the supply of nutrients in the medium.

In the normal response of an animal cell to growth-factor stimulation, many other intracellular changes occur—including changes in Ca^{2+}, pH, cyclic AMP, protein phosphorylation, gene transcription, mRNA processing and degradation, protein synthesis, and the cytoskeleton. While many of these occur within seconds, others take hours. Most of the proto-oncogene proteins that participate in this complex web of control systems can be classified only vaguely as yet. Some of them, like the growth-factor receptors mentioned above, are tyrosine-specific protein kinases that are associated with cell membranes. Others, like the *cdc2/28* gene product in yeasts, are serine/threonine-specific protein kinases found in the cytoplasm. A third category comprises proteins that are located chiefly in the nucleus (see Figure 13–34); one of these, the c-*jun* protein, has been identified as the transcriptional regulator AP-1 (see p. 566) and combines with another member of the family, the c-*fos* protein, to form a DNA-binding complex.

Another protein in the nuclear category, corresponding to a proto-oncogene called c-*myc*, seems to be a marker of whether the cell is in a proliferative mode: in a cell that is rapidly cycling, the c-*myc* protein is present at a constant low level throughout the cycle, but it disappears when the cell enters a G_0 state and becomes quiescent. When growth factors are added to the medium bathing a quiescent cell, the concentration of c-*myc* protein rises steeply, reaching a peak within a few hours, and then falls to a lower nonzero level. In contrast to c-*myc*, the concentrations of the vast majority of other proteins in the cell change little between the proliferative and the quiescent state.

The Effects of Oncogenes on Cell-Division Control Are Closely Coupled with Effects on Cell Adhesion[32,33]

One of the most extensively studied proto-oncogenes is c-*src*, corresponding to the oncogene v-*src* of the Rous sarcoma virus. It belongs to a small family of homologous proto-oncogenes. It encodes the *src* protein, a tyrosine-specific protein kinase of molecular weight 60,000 (hence the alternative name $p60^{src}$) that contains a covalently linked fatty acid that attaches it to the cytoplasmic side of the plasma membrane (see p. 417). The kinase is hyperactive in its oncogenic form (v-*src*), and the membrane attachment (Figure 13–35) is required for it to be able to cause cell transformation. Studies with antibodies indicate that the *src* protein becomes concentrated at *focal contacts*, where the cell is bound tightly to the substratum by a cell-matrix junction involving actin filaments on its intracellular side (see Figure 13–36). The *src* protein seems to be functionally involved with actin-membrane attachments, since activating a temperature-sensitive version of the v-*src* protein (by lowering the temperature) causes an immediate increase in membrane ruffling (an actin-mediated lamellipodial movement, see p. 638), as well as a general weakening of cell adhesions, including the disruption of focal contacts causing the cell to round up (see Figure 13–32). At least two separate actions of the *src* protein have been implicated in the observed change in cell adhesion. First, the active v-*src* kinase phosphorylates a tyrosine residue on the cytoplasmic tail of the fibronectin receptor molecules in the cell (see p. 636). *In vitro* studies suggest that this phosphorylation reduces the affinity of the receptor for both talin (on the inside of the cell) and fibronectin (on the outside). In addition, cells trans-

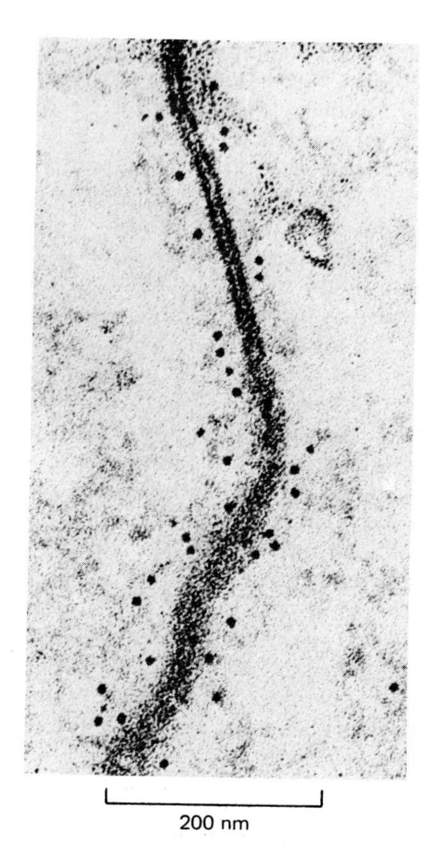

Figure 13–35 Electron micrograph demonstrating that the protein kinase synthesized by the v-*src* oncogene of Rous sarcoma virus is attached to the inner surface of the plasma membrane; the *src* protein synthesized by the c-*src* protein is thought to have a similar location but is harder to detect because it is normally present only in small quantities. The *src* protein has been localized in this preparation by reacting it with specific antibodies to which electron-dense ferritin particles are attached. (Courtesy of Ira Pastan; from M.C. Willingham, G. Jay, and I. Pastan, *Cell* 18:125–134, 1979. © Cell Press.)

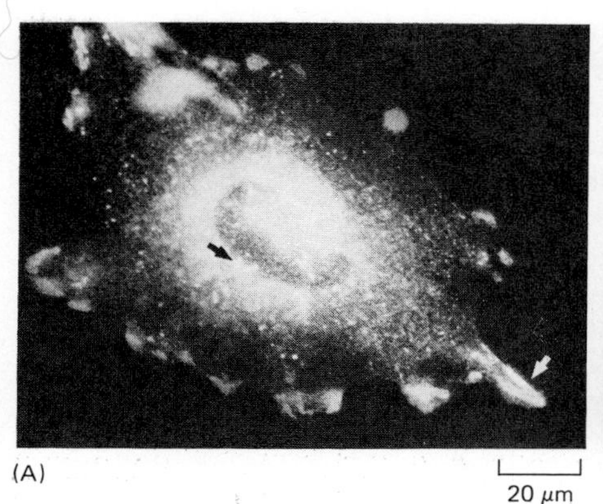

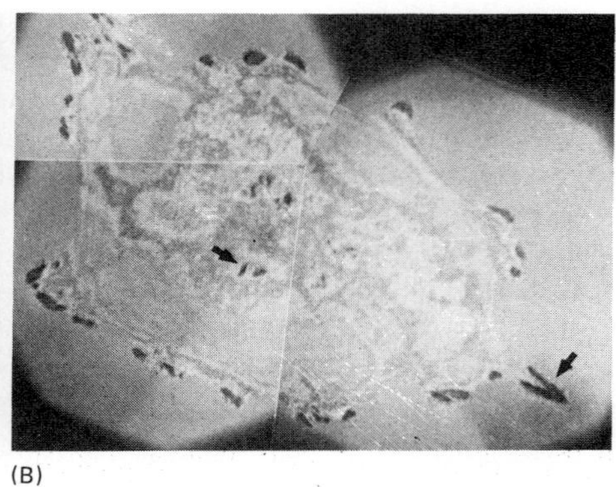

Figure 13–36 The *src* protein is present in many regions of the cell, but it appears especially concentrated at focal contacts and other sites of cell attachment to the extracellular matrix. (A) Immunofluorescence photograph showing the distribution of the *src* protein as revealed by the binding of *src*-specific antibodies. (B) A view of the same cell using an optical interference technique that makes sites of close adhesion of the cell to the substratum appear dark. The pattern of bright *src* protein spots in (A) matches the pattern of dark adhesion spots in (B) (see arrows). These pictures show the distribution of viral *src* protein in a cell transformed by Rous sarcoma virus; the *src* protein produced in a normal cell probably has a similar distribution but is harder to detect because it is present in smaller quantities. (From L.R. Rohrschneider, *Proc. Natl. Acad. Sci. USA* 77:3514–3518, 1980.)

formed by v-*src* secrete large amounts of a proteolytic enzyme called *plasminogen activator*. This enzyme is named for its ability to activate a second proteolytic enzyme, *plasmin*, by cleaving the precursor plasminogen; but plasminogen activator can also degrade other proteins directly, and, directly or indirectly, it evidently helps cells to dissolve their attachments (and to migrate through the extracellular matrix). When a monoclonal antibody against plasminogen activator is added to the culture medium, the cells become more adherent and tend to flatten down on the substratum. Thus the hyperactive v-*src* tyrosine-kinase activity appears to loosen cell adhesions in two separate ways: by phosphorylating the fibronectin receptor (and other transmembrane cell-matrix adhesion molecules in the integrin family—see p. 797), and by causing a protease to be secreted that destroys fibronectin (and other matrix molecules).

What might these findings mean for normal cell growth regulation? When a normal quiescent fibroblast is treated with PDGF, membrane ruffling is immediately induced and the cell's focal contacts change their structure within minutes: vinculin transiently disappears from focal contacts and the bundles of actin filaments that were anchored there are temporarily disrupted. Thus, in stimulating a quiescent cell to divide, PDGF induces many of the same types of changes as v-*src*. In fact, included among the immediate changes induced by PDGF is an increased phosphorylation of the c-*src* protein, which, by activating the *src* kinase activity, could explain the similarities directly (Figure 13–37). In this view the transformation of cells by v-*src* (and by many other oncogenes with similar effects) reflects an exaggeration of a normal mechanism of growth stimulation that involves loosening cell adhesions. The effects of oncogenes are dangerous for the organism because, unlike the PDGF stimulation, which is transitory and is soon turned off, proteins like v-*src* tend to drive cells out of G_0 permanently and thereby keep them in a proliferative state.

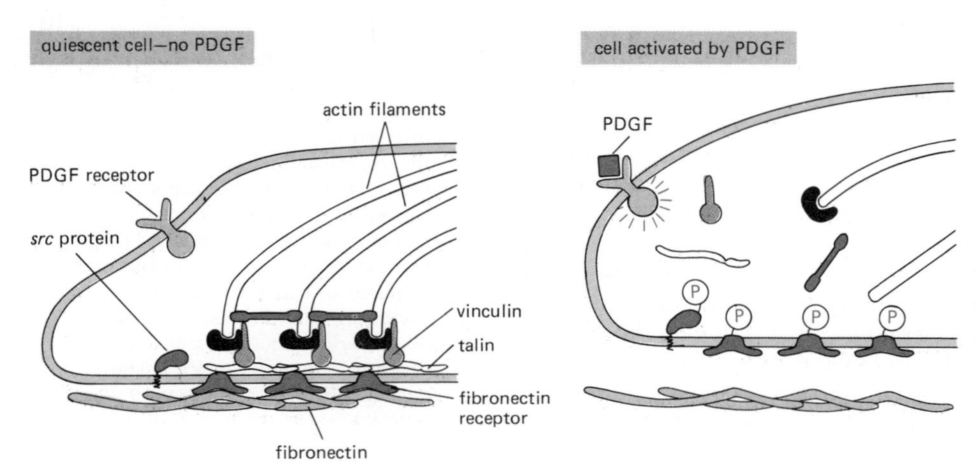

Figure 13–37 A speculative model of how rapid changes in cell adhesion might be mediated when a cell is stimulated to proliferate by PDGF. Binding of PDGF to its receptor leads (by an unspecified pathway) to phosphorylation of the c-*src* protein. This plasma-membrane-associated protein kinase is thereby activated and in turn phosphorylates tyrosine residues on neighboring transmembrane cell-adhesion proteins, including the fibronectin receptor. As a result, focal contacts and other sites of cell adhesion are partially disrupted and the associated actin filaments detach from the plasma membrane. The model is based partly on observations made on cells transformed by the Rous sarcoma virus, which carries a modified *src* protein (v-*src*) that is permanently active. The tyrosine protein kinases encoded by two other proto-oncogenes in the *src* family, c-*abl* and c-*yes*, may act in the same way as the c-*src* protein in the above pathway. However, many proteins are generally phosphorylated by such enzymes, and it is not certain which ones are crucial for the control of cell division. Some important targets may be present in only a few copies per cell—too few for detection by ordinary biochemical methods—and in different cell types the targets may be different. Above all, causal relationships are difficult to establish in a complex network of interacting components, where multiple factors may act in parallel and similar effects may be produced by different means.

The Connection Between Cell Proliferation and Cell Adhesion Is Not Understood[33]

Our discussion of the control of normal vertebrate cell proliferation leads here to a paradox. On the one hand, it is clear that normal cells must form adhesions to the substratum (cell-matrix adhesions) in order to exit from G_0 and proliferate. This suggests that the transmembrane proteins that bind cells to the extracellular matrix (including the fibronectin receptor and other members of the integrin family) generate an intracellular signal that facilitates cell division when they are properly engaged. On the other hand, adherence by itself is not enough to trigger cell division: growth factors are also required. The paradox is that the growth factors seem to act, in part, by transiently weakening the adhesions on which normal cell proliferation depends (see Figure 13–37). This effect is reminiscent of a second observation: in many cases a brief exposure of normal growth-arrested cells to a protease such as trypsin, which causes the attached cells to lose their adherence to the culture dish and round up, has the side effect of triggering a single round of cell division. It seems that cell proliferation is transiently stimulated both by extracellular proteases that loosen cell-matrix adhesions directly, by digesting the external proteins responsible for adhesion, and by growth factors that loosen such adhesions indirectly, by acting on the focal contacts via intracellular mediators.

Studies on tumor cells reinforce the paradox. Most tumor cells, including those transformed by the well-studied oncogenes shown in Figure 13–34, differ from their normal counterparts in that they proliferate without requiring adhesion to a substratum. Since this *anchorage independence* enables the transformed cells to grow in new environments where normal cell-cell and cell-matrix attachments cannot be made (see p. 748), one might expect it to be an outcome of natural selection for cells that form tumors. But why do many tumor cells not merely divide without regard to anchorage but also fail to make firm attachments to the extracellular matrix even where the opportunity exists? A hint comes from observations of transformed cells that are forced to attach to the culture dish by artificial means. As noted above when fibroblasts from a chick embryo are transformed with v-*src,* they secrete large amounts of plasminogen activator, which loosens their attachments to the culture dish. If these cells are grown in the presence of an antibody that blocks the activity of this protease, they attach more firmly to the dish and, at the same time, become more obedient to the normal social controls on cell division: instead of piling up in multiple layers, they tend to stop dividing on reaching confluence. Thus, for these transformed cells, the formation of firm attachments to the extracellular matrix seems to inhibit growth.

The different growth requirements for normal and transformed cells and the seemingly contradictory effects of cell-matrix adhesion (Figure 13–38) must have a logical explanation. In some way the complex structure that forms at a focal contact between a cell and its substratum must play a central part in generating the intracellular signals that regulate cell division. The observations might be explained by supposing that normal triggering of cell division requires three steps: (1) anchoring of the cell to the matrix through cell-matrix attachments at which

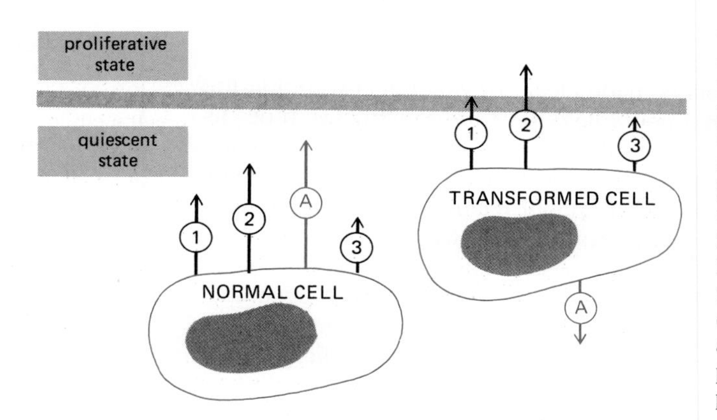

Figure 13–38 The general nature of the social control signals that act on normal and transformed cells. For both types of cells, different growth factors (here denoted by 1, 2, and 3) act in combination to "lift" the cell from G_0 to a proliferative state. Because the transformed cell is maintained at a position closer to the transition boundary (*gray line*), it can often be stimulated to proliferate by a single growth factor (or by an unusually low concentration of a mixture of growth factors). As discussed in the text, however, there is a marked difference in the effect of anchorage (denoted by A) on these two types of cells: whereas anchorage is required for normal cell proliferation, it tends to inhibit the proliferation of transformed cells.

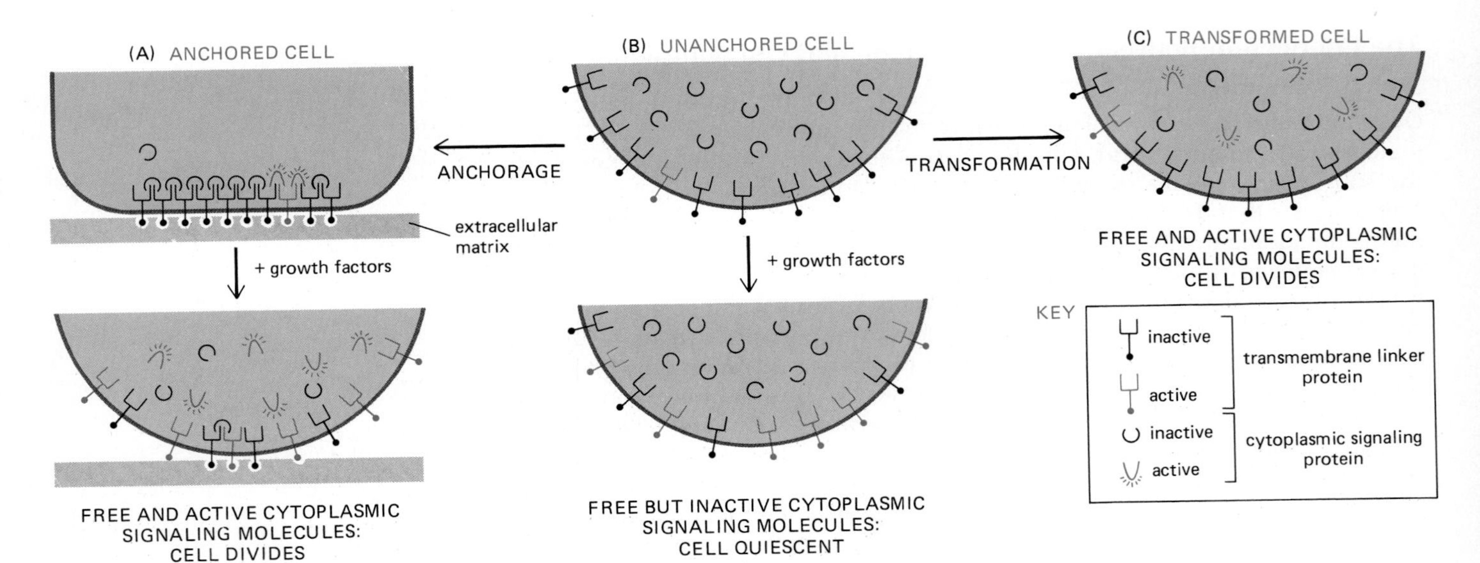

an ordered assembly of cytoskeletal proteins is created inside the cell (see p. 635); (2) activation of this assembly, typically by one or more growth factors, to release an intracellular signal for cell division; and (3) as a necessary part of the process of releasing the signal, partial disassembly of the cell-matrix attachments. In a transformed cell the requirement for step 1 would be bypassed, and disassembled intracellular components of the cell-matrix attachment structure would be necessary and sufficient to generate the signal to proliferate. A hypothetical model for such a growth control mechanism is presented in Figure 13–39.

Figure 13–39 A speculative scheme to account for the observed relationships between cell-matrix adhesion and the proliferation of normal and transformed cells. The model focuses on two types of molecule: (1) a cytoplasmic protein that provides the intracellular signal for cell division and (2) a transmembrane linker protein that can bind both to the cytoplasmic signaling molecule on one side of the plasma membrane and to the extracellular matrix on the other side. The binding is cooperative, so that signaling molecules bound to the linker protein intracellularly stabilize the transmembrane array and promote its binding to the extracellular matrix, and, reciprocally, binding to the extracellular matrix promotes binding of signal molecules to the linker proteins inside the cell. For the cell to be signaled to divide, the signaling molecules must be released into the cytoplasm and must be in an activated conformation, in which they bind less strongly to the linker proteins. It is presumed that the signaling molecules are activated by phosphorylation catalyzed by a kinase activity associated with the linker protein.

When an anchored cell is stimulated by growth factors (A), the kinase activity is switched on: the signaling molecules become phosphorylated and dissociate from the linker proteins, signaling the cell to divide and weakening the extracellular adhesion. Normal cells in suspension (B) fail to divide in response to growth factors because very few of the intracellular signaling molecules are bound to the linker proteins that would enable them to become phosphorylated. Transformed cells (C) have reduced adherence and can divide even in suspension because there is a "short circuit" in the control system such that the signaling molecules are phosphorylated constitutively.

Positional Signals and Cell-autonomous Programs Control Cell Division in the Growing Body[20,34]

Experiments conducted under the simplified artificial conditions of cell culture have told us most of what we know about the molecular mechanisms controlling cell growth and division in multicellular animals. Yet this work has so far revealed only some of the basic nuts and bolts of the much more complex social control systems that must operate in the intact body to regulate the proliferation of each group of cells according to its position and its developmental history.

As suggested by our discussion of cell senescence, cells often appear to be governed by long-term intracellular control programs, whereby the present proliferative behavior of a cell depends on its history of exposure to factors acting many cell generations before (see p. 751). While the relationship between the long-term and short-term controls is still mysterious, both seem to involve many of the same molecules—in particular, growth factors and products of proto-oncogenes. Programs of cell division can be remarkably complex and strictly defined during embryonic development. This is strikingly illustrated in the nematode worm *Caenorhabditis elegans,* whose fertilized egg divides so as to generate precisely 959 somatic cell nuclei in the adult body; some of the gene products that implement these programs are beginning to be characterized (see p. 905).

It should not be supposed, however, that the growth of embryos is regulated simply by counting cell divisions. This is made clear, for example, by comparing newts of differing ploidy. The cells of a pentaploid newt are roughly five times as big as those of a haploid, and yet, because there are roughly one-fifth as many cells in each tissue in the pentaploid, the dimensions of the body and of its organs are practically the same in the two types of animal (Figures 13–40 and 13–41). Evidently in vertebrates the mechanisms that control cell division to regulate bodily dimensions depend on measurements of distances rather than on simple counts of cell numbers or division cycles. Such mechanisms require complex positional controls, in which diffusible growth factors may well play an important part.

Positional control of cell division can operate with remarkable specificity. When a piece of epithelium from one cockroach leg is transplanted to a homologous site in another, it "heals in" without significant cell division. However, if it is

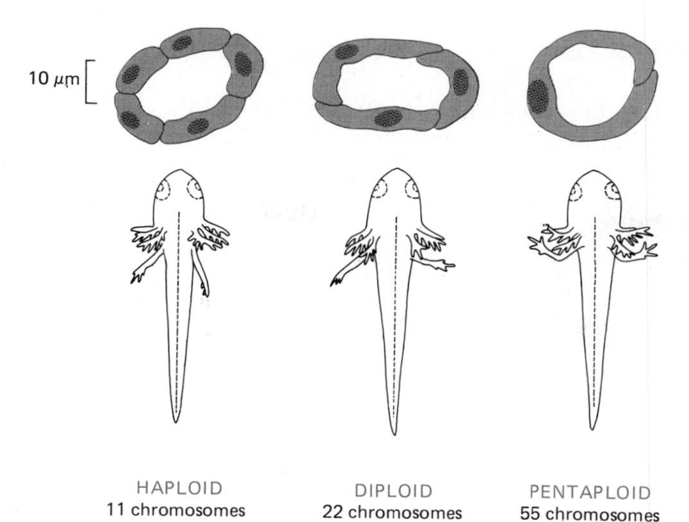

Figure 13–40 Drawings of representative sections of kidney tubules from salamander larvae of different ploidy. Pentaploid salamanders have cells that are bigger than those of haploid salamanders, but the animals and their individual organs are the same size because each tissue in the pentaploid animal contains fewer cells. This indicates that the number of cells is regulated by some mechanism based on size and distance rather than on the counting of cell divisions or of cell numbers. (After G. Fankhauser, in Analysis of Development [B.H. Willier, P.A. Weiss, and V. Hamburger, eds.], pp. 126–150. Philadelphia: Saunders, 1955.)

transplanted to a nonhomologous site, both the graft and the adjacent host cells proliferate and then differentiate to generate the cells that would normally lie between the region from which the graft was taken and the region to which it was transplanted (see p. 918). The molecular basis of this behavior is a complete mystery.

In general, cell division during embryonic development is governed by an interplay of cell-autonomous programs and cell-cell interactions, with the importance of each varying from species to species and from one part of the body to another. In adult tissues, too, cell division is regulated by a complex network of controls: when a deep skin wound heals in a vertebrate, about a dozen different cell types, ranging from fibroblasts to Schwann cells, must be regenerated in appropriate numbers to reconstruct the lost tissue. Moreover, there is redundancy in the social control system, with multiple restraints acting in parallel, so that the failure of a single control component in one cell (a common result of somatic cell mutation) will not endanger the whole organism by allowing the genesis of an enormous clone of cells that proliferates wildly. Studies of cancer incidence suggest that about four to six mutations must typically occur in a given cell lineage before it will give rise to a malignant tumor (see p. 1192).

Dissecting in molecular detail the elaborate social controls that enable an organ such as the kidney to develop and persist in an adult animal may well be an enterprise to occupy generations of cell biologists. But powerful tools are now available, such as antibodies to block specific growth factors or receptors, and transgenic animals engineered to produce such signaling molecules inappropriately in selected types of cells (see p. 268). With the help of these new approaches, the task, although formidable, does not seem impossible.

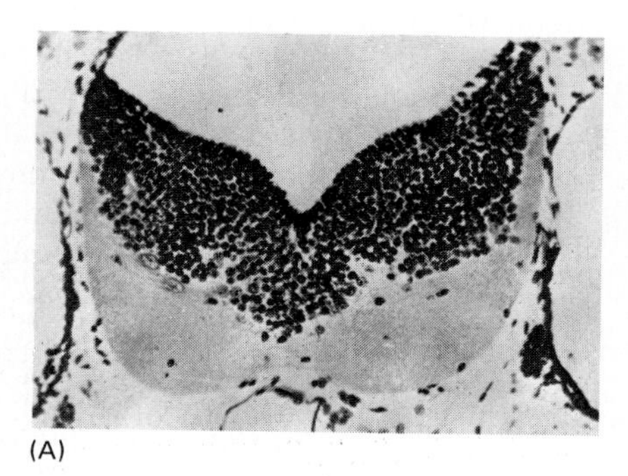
(A)

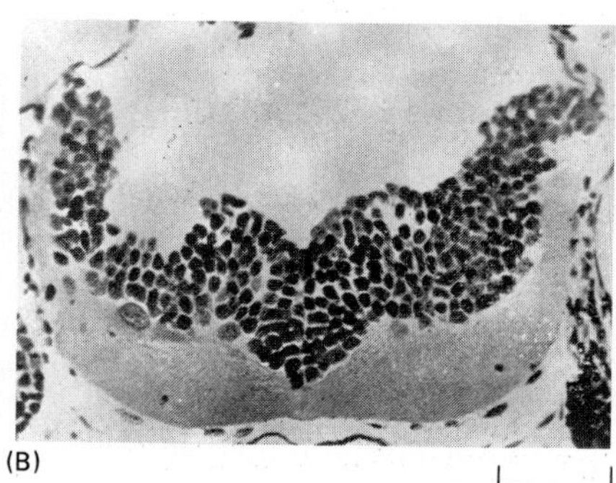
(B)

Figure 13–41 Micrographs comparing cells in the brains of haploid and tetraploid salamanders (see also Figure 13–40). (A) Cross-section of the hindbrain of a haploid salamander. (B) Corresponding cross-section through the hindbrain of a tetraploid salamander showing how reduced cell numbers compensate for increased cell size. (From G. Fankhauser, *Int. Rev. Cytol.* 1:165–193, 1952.)

Summary

Abnormal cells that disobey the social constraints on cell division proliferate to form tumors in the body, and they also appear transformed in cell culture. Although often lethal to the organism as a whole, as individual cells they are favored by natural selection and therefore are easy to isolate. Cell transformation is often accompanied by mutation or overexpression of specific oncogenes, identified in many cases because they are carried by RNA tumor viruses (retroviruses). The normal counterparts of these viral oncogenes in the healthy cell are known as proto-oncogenes and are thought to encode key components of the normal system of social controls of cell division. Some proto-oncogenes code for growth factors, some for growth-factor receptors, some for intracellular regulatory proteins that are involved in cell adhesion, and some for proteins that help relay signals for cell division to the cell nucleus. Cells must contain alterations in multiple social control genes in order to become cancerous, reflecting a redundancy in the complex control systems that affect cell proliferation in tissues.

13-29 The Mechanics of Cell Division[35]

In this last section we discuss the events of M phase—the culmination of the cell cycle. In a comparatively brief period the chromosomes condense and the contents of the parental cell, which were doubled by the biosynthetic activities of the preceding interphase, are segregated into two daughter cells (Figure 13–42).

At the molecular level, M phase is thought to be initiated by a cascade of protein phosphorylations triggered by the appearance of MPF and terminated by dephosphorylations that restore the proteins to their interphase state (see p. 742). The protein phosphorylations present during M phase are in turn likely to be responsible for the many morphological changes that accompany mitosis, including chromosome condensation, nuclear envelope breakdown, and the cytoskeletal changes to be described below. The first readily visible manifestation of an impending M phase is a progressive compaction of the dispersed interphase chromatin into threadlike chromosomes. This condensation of the chromosomes is required for their subsequent organized segregation into daughter cells, and it is accompanied by phosphorylation of the many histone H1 molecules present in the cell (up to six phosphates per H1 molecule). Since histone H1 is present in a concentration of about one molecule per nucleosome and is known to be involved in packing nucleosomes together (see p. 499), its phosphorylation by the MPF kinase (see p. 742) at the onset of the M phase could be a major cause of chromosome condensation. While still quite tentative, this type of molecular explanation illustrates the level at which one must ultimately explain the entire cell cycle.

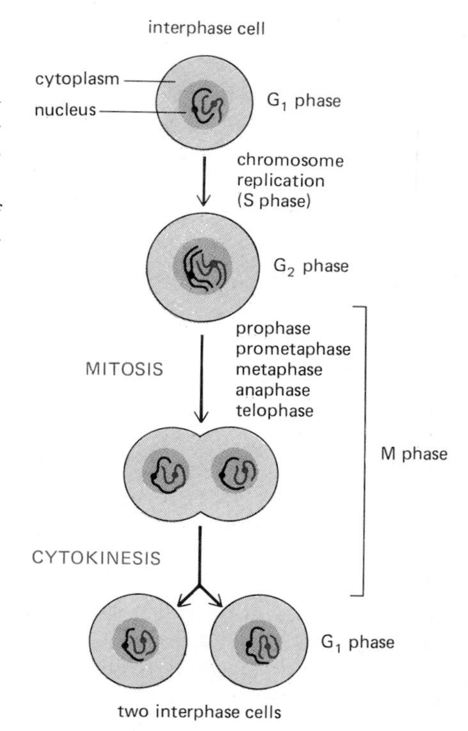

Figure 13–42 The M phase (or cell-division phase) of the cell cycle starts at the end of G_2 phase and ends at the start of the next G_1 phase. It includes the five stages of nuclear division (mitosis) and cytoplasmic division (cytokinesis).

It has been said that the chromosomes in mitosis are like the corpse at a funeral: they provide the reason for the proceedings but play no active part in them. The active role is played by two distinct cytoskeletal structures that appear transiently in M phase. The first to form is a bipolar **mitotic spindle,** composed of microtubules and their associated proteins. The mitotic spindle first aligns the replicated chromosomes in a plane that bisects the cell; each chromosome then separates into two daughter chromosomes, which are moved by the spindle to opposite ends of the cell. The second cytoskeletal structure required in M phase in animal cells is a **contractile ring** of actin filaments and myosin that forms slightly later just beneath the plasma membrane. This ring pulls the membrane inward so as to divide the cell in two, thereby ensuring that each daughter cell receives not only one complete set of chromosomes but also half of the cytoplasmic constituents and organelles in the parental cell. The two cytoskeletal structures contain different sets of proteins and can be formed separately in some specialized cells. Their formation is usually closely coordinated, however, so that cytoplasmic division (*cytokinesis*) occurs immediately after the end of nuclear division (*mitosis*). The same is true for plant cells, even though, as we shall see, their rigid walls necessitate a different mechanism for cytokinesis.

The description of M phase just given applies only to eucaryotic cells. Bacterial cells do not contain either actin filaments or microtubules; they generally have only one chromosome, whose replicated copies are segregated to daughter cells by a mechanism that involves chromosome attachment to the bacterial plasma membrane (see p. 785). The need for complex mitotic machinery probably arose only with the evolution of cells that contained greatly increased amounts of DNA packaged in a number of discrete chromosomes. The primary function of this machinery is to divide the replicated chromosomes precisely between the two daughter cells. Its accuracy has been determined in yeast cells, where an error in chromosome segregation is made only about once every 10^5 cell divisions.

M Phase Is Traditionally Divided into Six Stages[35]

The basic strategy of cell division is remarkably constant among eucaryotic organisms. The first five stages of the M phase constitute mitosis; the sixth is cytokinesis. These six stages form a dynamic sequence, the complexity and beauty of which are hard to appreciate from written descriptions or from a set of static pictures. The description of cell division is based on observations from two sources:

light microscopy of living cells (often combined with microcinematography) and light and electron microscopy of fixed and stained cells. A brief summary of the various stages of cell division is given in Panel 13–1. As defined, the five stages of mitosis—*prophase, prometaphase, metaphase, anaphase,* and *telophase*—occur in strict sequential order, while cytokinesis begins during anaphase and continues through the end of the mitotic cycle (Figure 13–43). Light micrographs of cell division in a typical animal and a typical plant cell are shown in Figures 13–44 and 13–45, respectively.

Innumerable variations in all of the stages of cell division shown schematically in Panel 13–1 occur in the animal and plant kingdoms. We shall mention some of these when we take a closer look at the mechanisms of cell division, since they can help us understand the different parts of the mammalian mitotic apparatus.

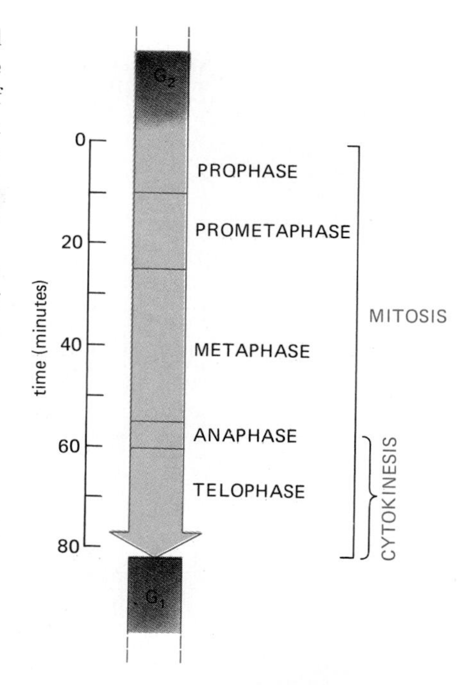

Figure 13–43 A typical time course for mitosis and cytokinesis (M phase) in a mammalian cell. The exact times vary for different cells. Note that cytokinesis begins before mitosis ends. The beginning of prophase (and therefore of M phase as a whole) is defined as the point in the cell cycle at which condensed chromosomes first become visible—a somewhat arbitrary criterion, since the extent of chromosome condensation appears to increase continuously during late G_2.

Formation of the Mitotic Spindle in an M-Phase Cell Is Accompanied by Striking Changes in the Dynamic Properties of Microtubules[36]

We saw in Chapter 11 that the principal *microtubule organizing center (MTOC)* in most animal cells is the **centrosome,** a cloud of amorphous material surrounding a pair of centrioles (see p. 654). During interphase the centrosomal material nucleates the growth of microtubules, which project outward toward the cell perimeter with their minus ends attached to the centrosome. This *interphase microtubule array* radiating from the centrosome is a dynamic, constantly changing structure in which microtubules continually form and disassemble. New microtubules grow by adding tubulin molecules to their plus ends; sporadically, and apparently at random, individual microtubules become unstable and undergo rapid, "catastrophic" disassembly, yielding their subunit molecules to the pool of unpolymerized tubulin already present in the cytoplasm (see Panel 11–2, p. 655).

Changes occur in the centrosome throughout the cell cycle, as illustrated in Figure 13–46. Sometime during S phase, the centriole pair replicates while remaining embedded in a single cloud of centrosomal material. In **prophase** the centrosome splits and each daughter centrosome becomes the focal point of a separate starlike aster of microtubules, which have their ends embedded in the cloud of centrosomal material. The microtubules in each aster elongate where they contact each other, and the two centrosomes move apart. Then, at prometaphase, the nuclear envelope breaks down, allowing the microtubules from each centrosome to enter the nucleus and interact with the chromosomes. The two daughter centrosomes are now said to constitute the two **spindle poles.**

The events just described are thought to result from major changes that occur at prophase in both the inherent stability of microtubules and the properties of the centrosome. As mentioned previously, there is evidence that an M-phase-promoting factor (MPF) triggers entry into M phase by initiating a cascade of multiple protein phosphorylations (see p. 742). Certain molecules that interact with microtubules are presumably affected, since the half-life of an average microtubule decreases about twentyfold (from about 5 minutes to 15 seconds) as a cell enters prophase (see Figure 13–48). This change is thought to reflect a greatly increased probability that a typical growing microtubule will convert to a shrinking one by a change at its plus end (see p. 655), combined with an alteration in the centrosome whereby it acquires, at prophase, a greatly increased capacity for nucleating the growth of microtubules (this can be observed *in vitro*). These two changes are sufficient to explain why the onset of M phase is characterized by a rapid transition from relatively few long microtubules that extend from the centrosome to the cell periphery (the interphase microtubule array) to large numbers of short microtubules that surround each centrosome (see Prophase in Figure 13–46).

As mitosis proceeds, the elongating ends of the microtubules emanating from the spindle poles are believed to encounter structures that bind to them and stabilize them against catastrophic disassembly. Because each pole nucleates microtubules in random directions, it is thought that selective stabilization gives rise to the characteristic bipolar form of the mitotic spindle in which the majority of

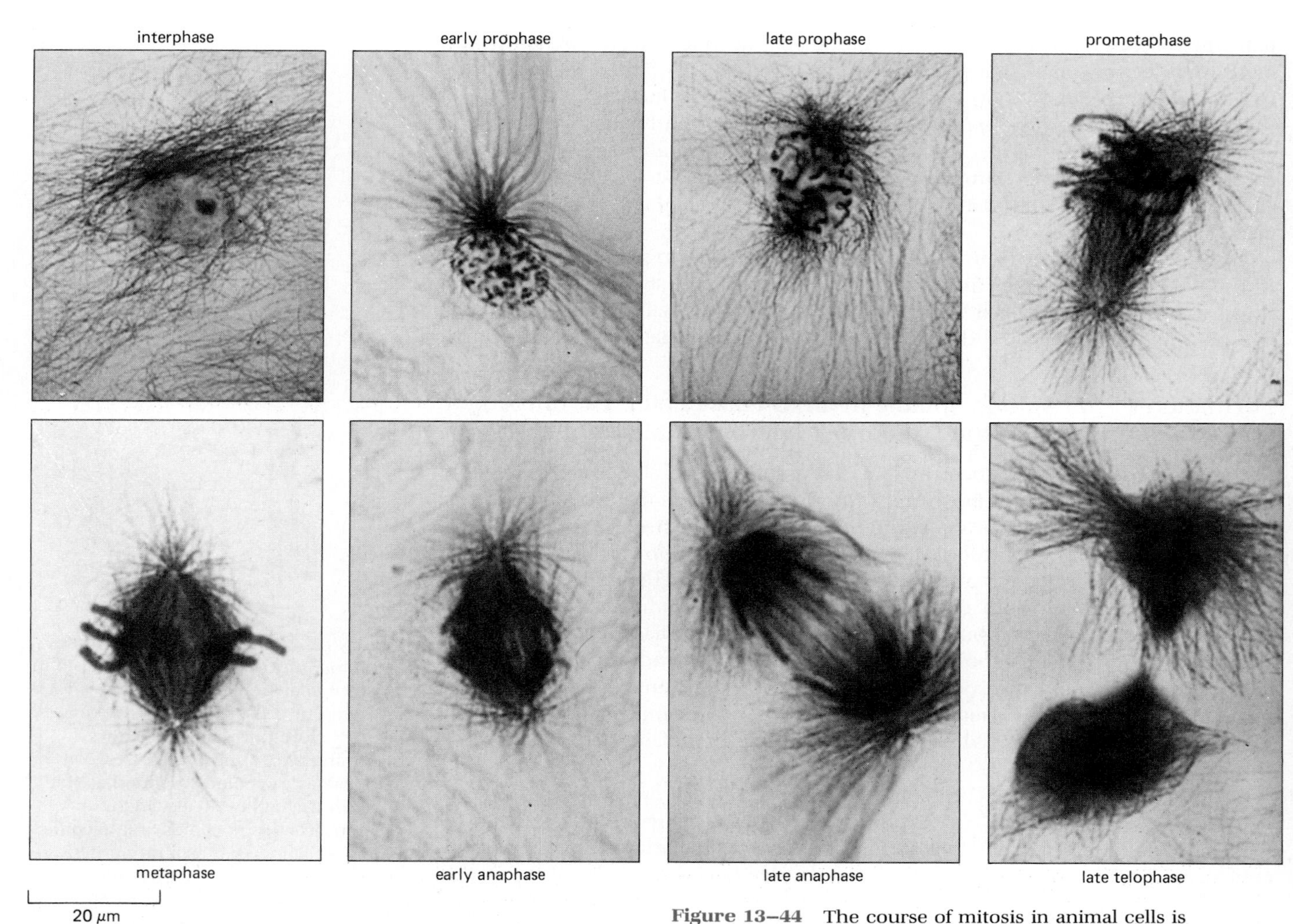

Figure 13–44 The course of mitosis in animal cells is illustrated here by selected micrographs of cultured marsupial (PtK) cells. The microtubules have been visualized using immunogold staining, while chromatin is stained with toluidine blue. At the light microscope level shown here, the major events of cell division have been known for more than 100 years. During *interphase* the centrosome, containing a centriole pair, forms the focus for the interphase microtubule array. By *early prophase* the single centrosome contains two centriole pairs (not visible); at *late prophase* the centrosome divides, and the resulting two asters move apart. The nuclear envelope breaks down at *prometaphase,* allowing the spindle microtubules to interact with the chromosomes. At *metaphase* the bipolar spindle structure is clear, and all the chromosomes are aligned across the middle of the spindle. The chromatids all separate synchronously at *early anaphase* and, under the influence of the spindle fibers, begin to move toward the poles. By *late anaphase* the spindle poles have moved farther apart, increasing the separation of the two groups of chromatids. At telophase the daughter nuclei re-form, and by *late telophase* cytokinesis is almost complete, with the midbody persisting between the daughter cells. (Photographs courtesy of M. deBrabander.)

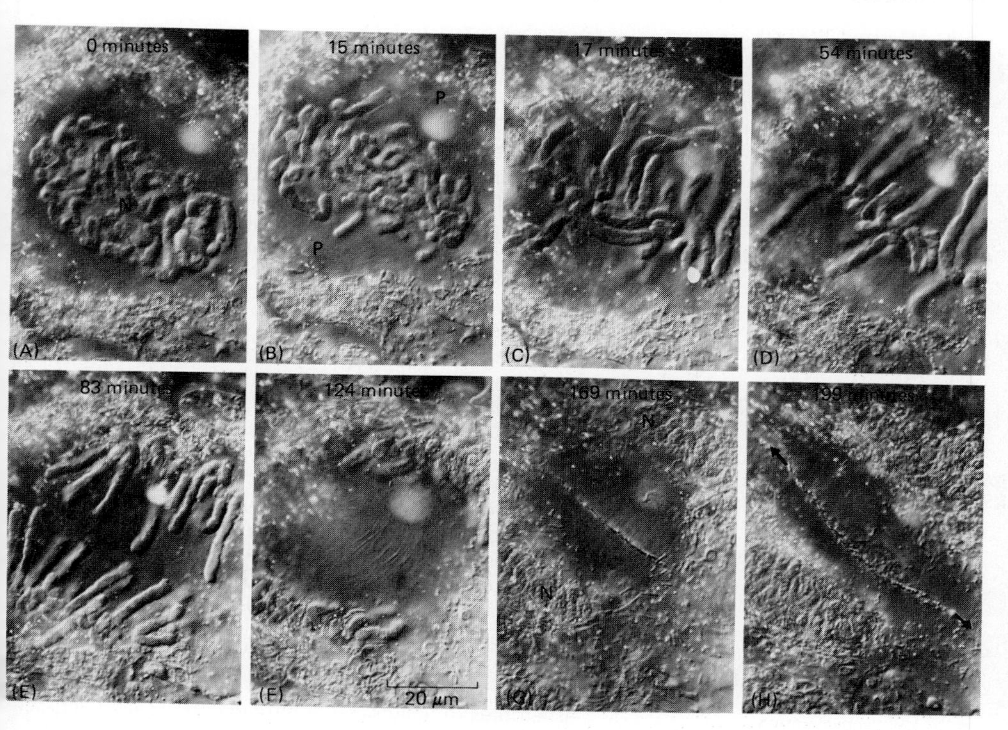

Figure 13–45 The course of mitosis in a typical plant cell. These micrographs were taken of a living *Haemanthus* (lily) cell at the times indicated using differential-interference-contrast microscopy (see p. 141). The cell has unusually large chromosomes, so they are easy to see. (A) *Prophase:* the chromosomes have condensed and are clearly visible in the cell nucleus (marked N). (B) and (C) *Prometaphase:* the nuclear envelope has broken down and the chromosomes are interacting with microtubules that emanate from the two spindle poles (marked P). Note that only 2 minutes have elapsed between the stages shown in (B) and (C). (D) *Metaphase:* the chromosomes have lined up at the metaphase plate with their kinetochores located halfway between the two spindle poles. (E) *Anaphase:* the chromosomes have separated into their two sister chromatids, which are moving to opposite poles. (F) *Telophase:* the chromosomes are decondensing to form the two nuclei that are seen later [marked N in (G)]. (G) and (H) *Cytokinesis:* two successive stages in the formation of the cell plate are shown; the cell plate appears as a line whose direction of outgrowth is indicated by arrows in (H). (Courtesy of Andrew Bajer.)

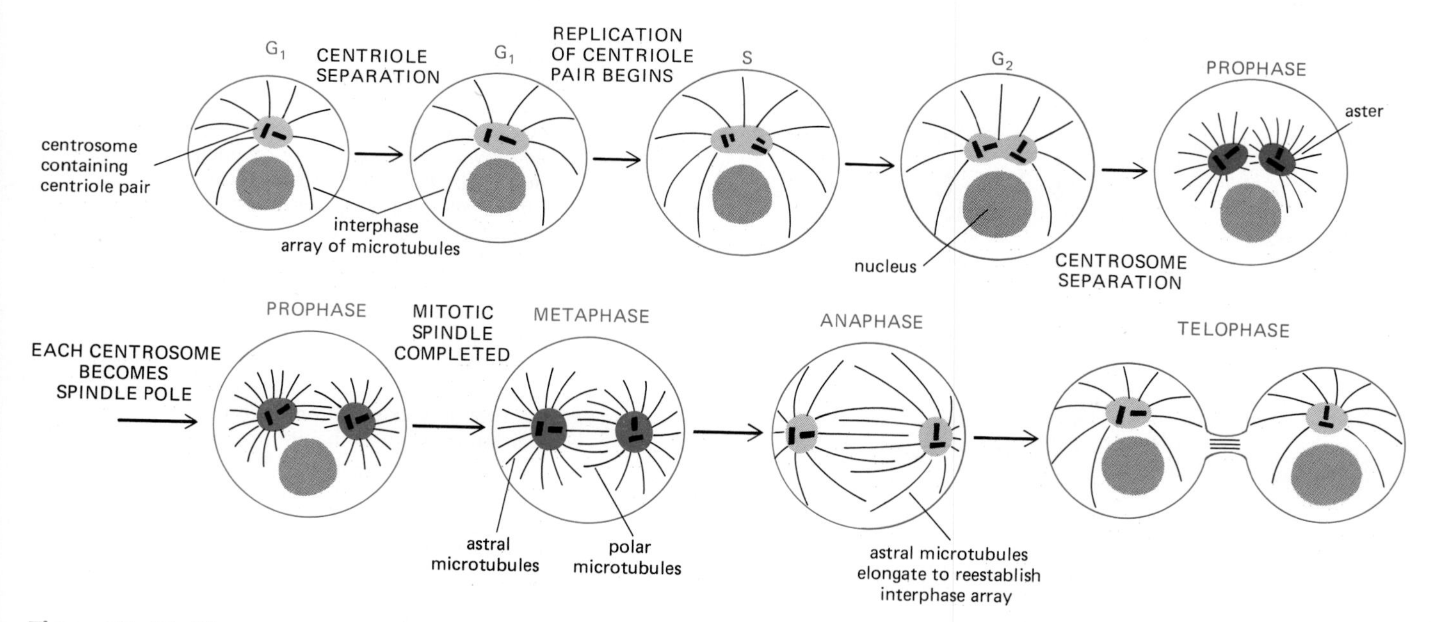

Figure 13–46 The centrosome cycle. The centrosome in an interphase cell duplicates to form the two poles of a mitotic spindle. In most animal cells (but not in plant cells) a centriole pair (shown here as a pair of black bars) is embedded in the centrosomal material (*color*) that nucleates microtubule outgrowth. At a discrete point in G_1 phase, the two centrioles separate by a few micrometers. During S phase a daughter centriole begins to grow near the base of each old centriole and at a right angle to it. The elongation of the daughter centriole is usually completed by G_2 phase. Initially the two centriole pairs remain embedded in a single mass of centrosomal material, forming a single centrosome. In early M phase each centriole pair becomes part of a separate microtubule-organizing center that nucleates a radial array of microtubules, the *aster* (aster = star). The two asters, which initially lie side by side and close to the nuclear envelope, move apart. By late prophase the bundles of *polar microtubules* that interact between the two asters preferentially elongate as the two centers move apart along the outside of the nucleus. In this way a mitotic spindle is rapidly formed.

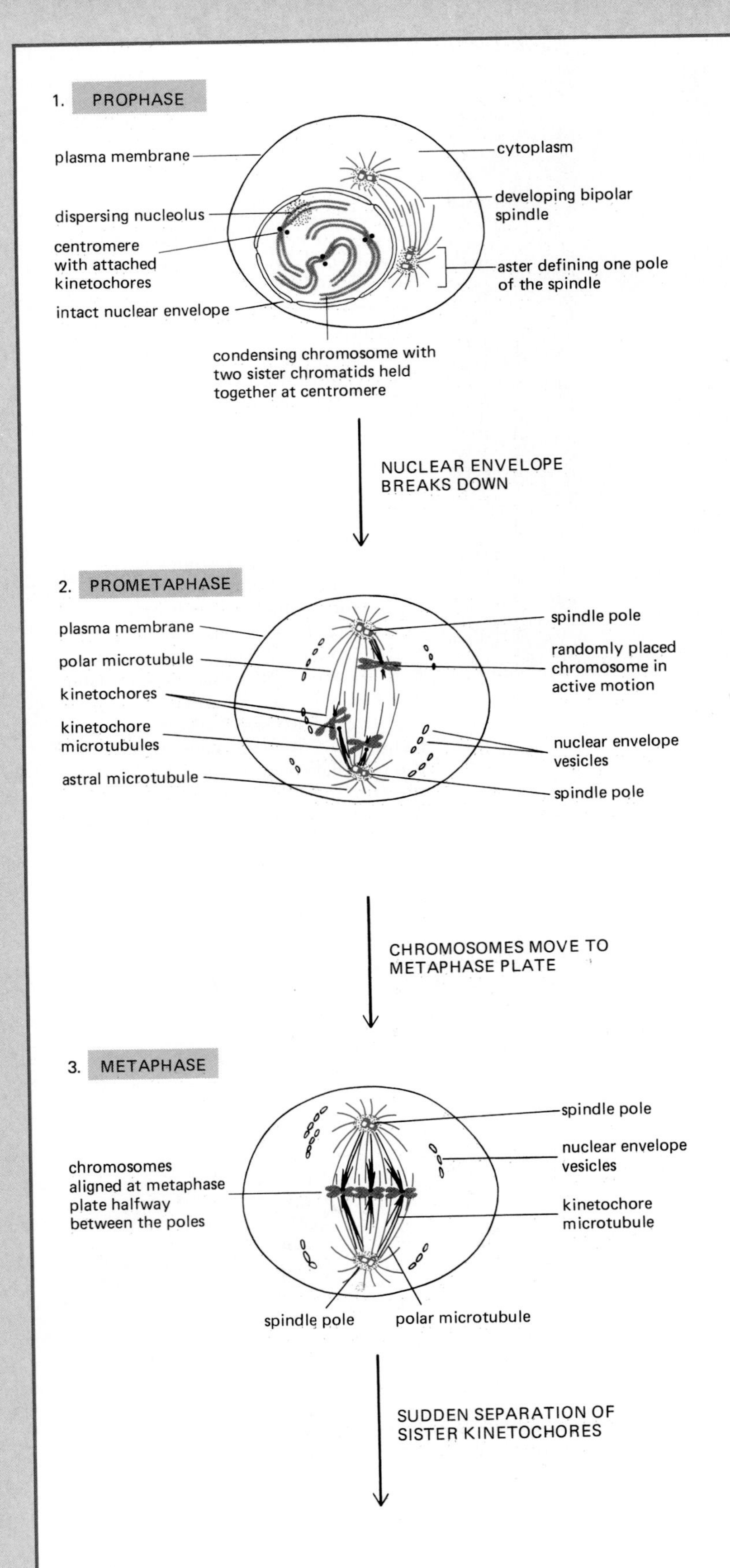

1 PROPHASE

As viewed in the microscope, the transition from the G_2 phase to the M phase of the cell cycle is not a sharply defined event. The chromatin, which is diffuse in interphase, slowly condenses into well-defined chromosomes, the exact number of which is characteristic of the particular species. Each chromosome has duplicated during the preceding S phase and consists of two sister *chromatids;* each of these contains a specific DNA sequence known as a *centromere,* which is required for proper segregation. Toward the end of prophase, the cytoplasmic microtubules that are part of the interphase cytoskeleton disassemble and the main component of the mitotic apparatus, the *mitotic spindle,* begins to form. This is a bipolar structure composed of microtubules and associated proteins. The spindle assembles initially outside the nucleus.

2 PROMETAPHASE

Prometaphase starts abruptly with disruption of the nuclear envelope, which breaks into membrane vesicles that are indistinguishable from bits of endoplasmic reticulum. These vesicles remain visible around the spindle during mitosis. The spindle microtubules, which have been lying outside the nucleus, can now enter the nuclear region. Specialized protein complexes called *kinetochores* mature on each centromere and attach to some of the spindle microtubules, which are then called *kinetochore microtubules.* The remaining microtubules in the spindle are called *polar microtubules,* while those outside the spindle are called *astral microtubules.* The kinetochore microtubules extend in opposite directions from the two sister chromatids in each chromosome, exerting tension on the chromosomes, which are thereby thrown into agitated motion.

3 METAPHASE

The kinetochore microtubules eventually align the chromosomes in one plane halfway between the spindle poles. Each chromosome is held in tension at this *metaphase plate* by the paired kinetochores and their associated microtubules, which are attached to opposite poles of the spindle.

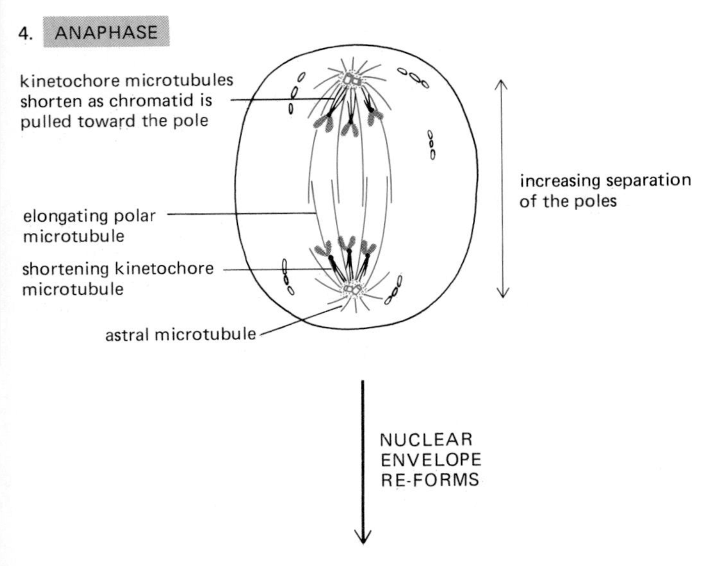

NUCLEAR ENVELOPE RE-FORMS

4 ANAPHASE

Triggered by a specific signal, anaphase begins abruptly as the paired kinetochores on each chromosome separate, allowing each chromatid to be pulled slowly toward the spindle pole it faces. All chromatids move at the same speed, typically about 1 μm per minute. Two categories of movement can be distinguished. During *anaphase A,* kinetochore microtubules shorten as the chromosomes approach the poles. During *anaphase B,* the polar microtubules elongate and the two poles of the spindle move farther apart. Anaphase typically lasts only a few minutes.

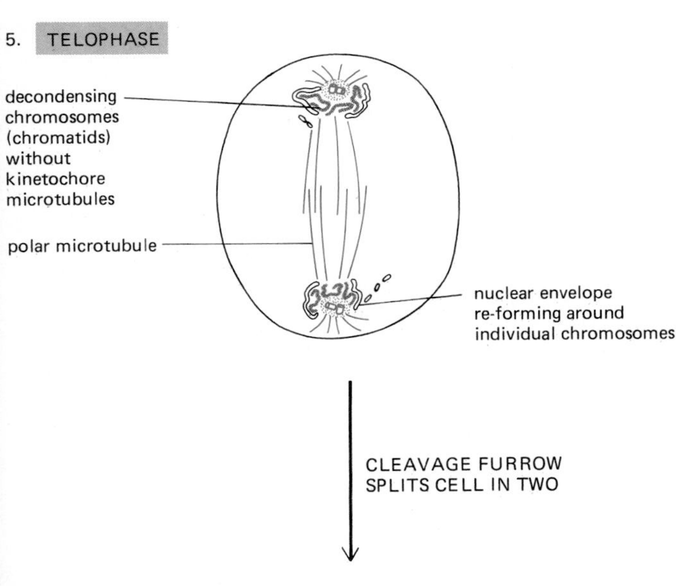

CLEAVAGE FURROW SPLITS CELL IN TWO

5 TELOPHASE

In telophase (*telos,* end) the separated daughter chromatids arrive at the poles and the kinetochore microtubules disappear. The polar microtubules elongate still more, and a new nuclear envelope re-forms around each group of daughter chromosomes. The condensed chromatin expands once more, the nucleoli—which had disappeared at prophase—begin to reappear, and mitosis is at an end.

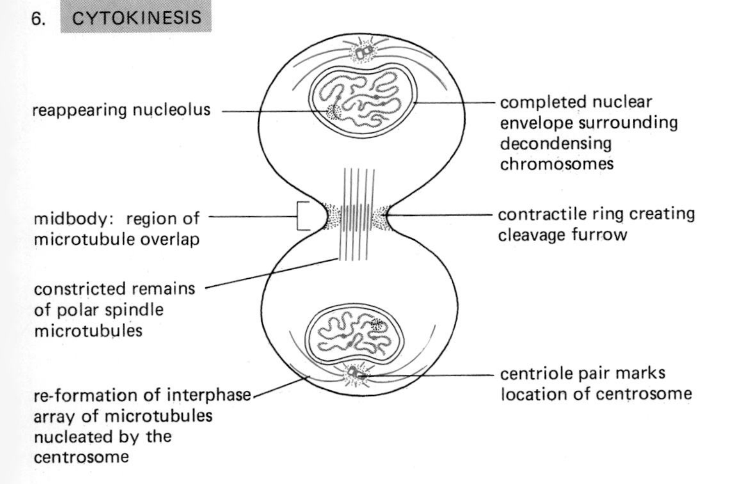

6 CYTOKINESIS

The cytoplasm divides by a process known as *cleavage,* which usually starts sometime during anaphase. The process is illustrated here as it occurs in animal cells. The membrane around the middle of the cell, perpendicular to the spindle axis and between the daughter nuclei, is drawn inward to form a *cleavage furrow,* which gradually deepens until it encounters the narrow remains of the mitotic spindle between the two nuclei. This thin bridge, or *midbody,* may persist for some time before it narrows and finally breaks at each end, leaving two separated daughter cells.

microtubules extend from the two poles of the spindle toward the equatorial plane halfway between them. Those microtubules that cross the equator may become selectively stabilized by microtubule-binding proteins that cross-link neighboring parallel microtubules of opposite polarity (Figure 13–47). At metaphase the spindles of higher animal and plant cells may contain up to several thousand microtubules, while the spindles of some fungi contain as few as 40.

Even though some of the microtubules in the spindle are partially stabilized against spontaneous disassembly, the majority continue to exchange their subunits with the pool of soluble tubulin molecules in the cytosol. The exchange can be measured directly by the method described in Figure 13–48. It can also be seen by subjecting mitotic cells to conditions that reversibly shift the equilibrium between tubulin polymerization and depolymerization and by observing the behavior of the birefringent spindle microtubules with polarized light (Figure 13–49). If mitotic cells are placed in heavy water (D_2O) or treated with taxol, either of which inhibits microtubule disassembly, the spindle fibers lengthen. Such stabilized spindles cannot move chromosomes, and mitosis is arrested. At the other extreme, mitosis is blocked when the spindle microtubules are reversibly disrupted by treatment with the drug colcemid, by low temperature, or by high hydrostatic pressure, all of which interfere with the assembly of tubulin molecules into microtubules. The observation that neither stabilized nor disassembled spindle microtubules can move chromosomes suggests that the spindle must be delicately poised at equilibrium between assembly and disassembly in order to perform mitotic movements. Before discussing the basis for these movements, we must describe in more detail how the spindle is organized and how the chromosomes are positioned.

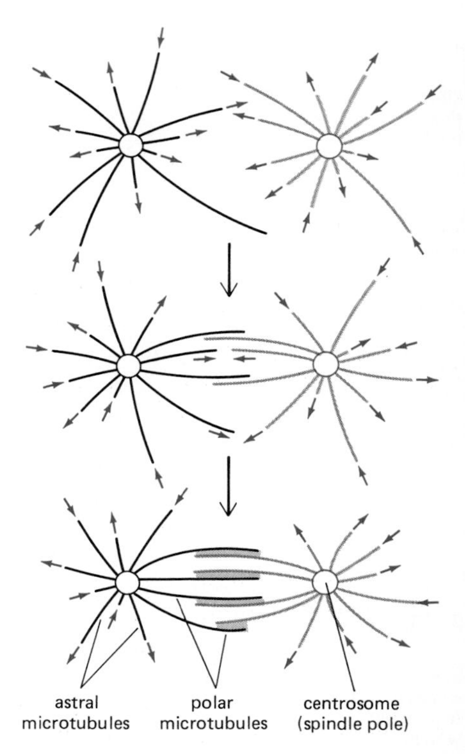

Figure 13–47 A model for how a bipolar mitotic spindle is thought to form by the selective stabilization of interacting microtubules. New microtubules grow out in random directions from two nearby centrosomes, to which they are anchored by their minus ends. Their plus ends are "dynamically unstable" and will switch suddenly from uniform growth to rapid shrinkage, during which the entire microtubule often depolymerizes (see p. 655). When two microtubules from opposite centrosomes interact in an overlap zone, microtubule-associated proteins are believed to cross-link the microtubules together (*gray shading*) in a way that caps their plus ends, stabilizing them by decreasing their probability of depolymerizing.

13-25 Chromosomes Attach to Microtubules by Their Kinetochores During Mitosis[37]

Replicated chromosomes bind to the mitotic spindle via structures called **kinetochores.** At the start of M phase, each chromosome consists of two sister chromatids paired along their length but primarily joined near their **centromeres,** which consist of a specific DNA sequence required for chromosome segregation. During late prophase a mature kinetochore develops on each centromere, with the two kinetochores (one on each sister chromatid) facing in opposite directions. By metaphase, microtubules have become attached to each kinetochore (Figure 13–50). In most organisms the kinetochore is a large multiprotein complex that can be seen in the electron microscope as a platelike trilaminar structure (Figure 13–51). The number of microtubules linked to a kinetochore varies widely depending on the species. Human kinetochores, for example, have 20 to 40 microtubules, while yeasts and some other microorganisms have only one; a single microtubule must therefore be sufficient to move a chromosome.

The information that specifies the construction of a kinetochore at a specific site on a chromosome must come directly from the DNA sequence at the centromere. In yeasts, centromeric DNA can be identified genetically by its ability to

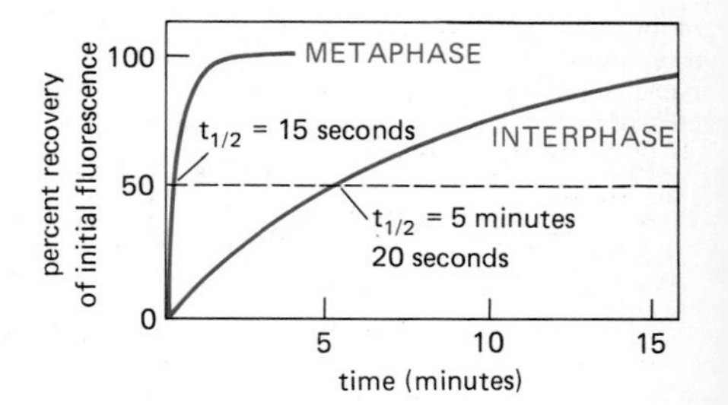

Figure 13–48 A study showing that the microtubules in an M-phase cell are much more dynamic, on average, than the microtubules at interphase. Mammalian cells in culture were injected with tubulin that had been covalently linked to a fluorescent dye. After the injected fluorescent tubulin had had time to become incorporated into a cell's microtubules, all of the fluorescence in a small region was bleached by an intense laser beam. The recovery of fluorescence in the bleached region of microtubules, caused by the exchange of their bleached tubulin subunits for unbleached fluorescent tubulin from the soluble pool, was then monitored as a function of time. The time $t_{1/2}$ for 50% recovery of fluorescence is thought to be equal to the time required for half of the microtubules in the region to depolymerize and re-form. (Data from W.M. Saxton *et al.*, *J. Cell Biol.* 99:2175–2187, 1984, by copyright permission of the Rockefeller University Press.)

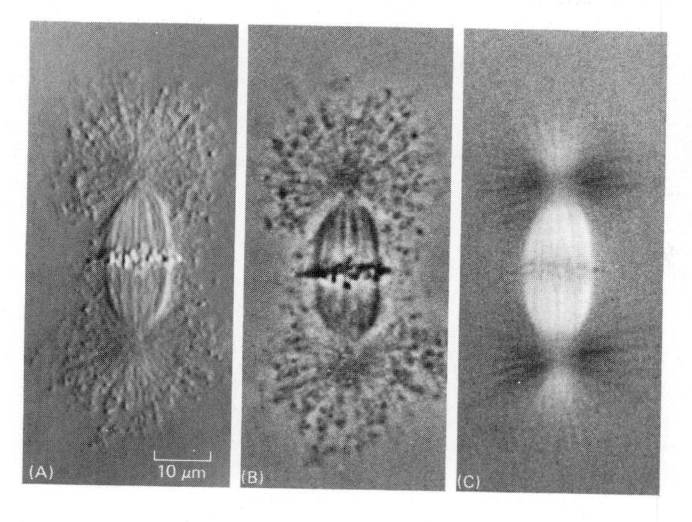

Figure 13–49 An isolated metaphase spindle viewed by three techniques of light microscopy: (A) differential-interference-contrast microscopy, (B) phase-contrast microscopy, and (C) polarized-light microscopy. (Courtesy of E.D. Salmon and R.R. Segall, from *J. Cell Biol.* 86:355–365, 1980. Reproduced by copyright permission of the Rockefeller University Press.)

allow plasmids to be stably inherited that otherwise segregate irregularly to daughter cells and are lost. Molecular genetic experiments show that each of the 17 chromosomes of the yeast *Saccharomyces cerevisiae* contains a different centromeric sequence about 110 base pairs long (Figure 13–52), although all the sequences contain substantial regions of homology and can be inverted or swapped from chromosome to chromosome without loss of function. The yeast centromere sequences bind specific proteins, which in turn are thought to initiate the formation of a multiprotein complex (the kinetochore) that binds the end of a single microtubule. Mammalian centromeres are thought to consist of different and much longer DNA sequences, and they form much larger kinetochores that bind multiple microtubules.

An unexpected opportunity to study the proteins of mammalian kinetochores came from the finding that human patients suffering from certain types of *scleroderma* (a disease of unknown cause that is associated with a progressive fibrosis of connective tissue in skin and other organs) produce autoantibodies that react specifically with kinetochores. When these antibodies are used to stain dividing cells by immunofluorescence, a pattern of fluorescent spots is obtained, each spot marking the position of a kinetochore. A similar speckled pattern is also obtained if nondividing cells are stained, and the number of spots per cell corresponds to the number of its chromosomes, suggesting that a kinetochore precursor is attached to each centromere even in interphase nuclei (Figure 13–53). Scleroderma antibodies have also made it possible to clone the genes that encode several of the many proteins associated with kinetochores, so that these normally rare proteins can now be produced in large quantities by recombinant DNA technologies; thus their interactions with one another, with DNA, and with microtubules can be characterized.

How are microtubules and kinetochores connected to each other? The linkage has some unique properties. If chemically marked tubulin is injected into mitotic

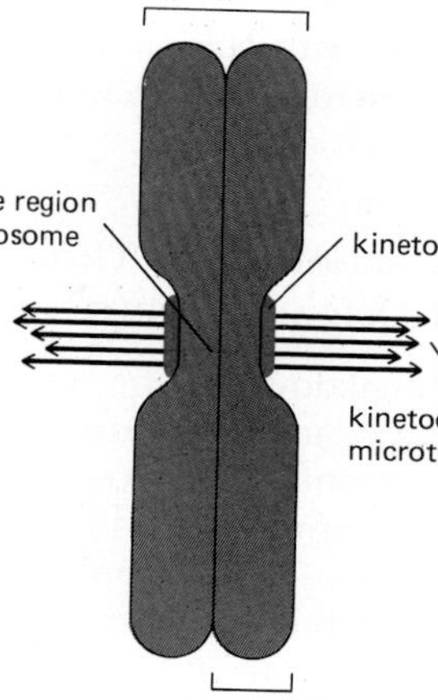

metaphase chromosome
centromere region of chromosome
kinetochore
kinetochore microtubules
chromatid

Figure 13–50 Schematic drawing of a metaphase chromosome showing its two sister chromatids attached to kinetochore microtubules.

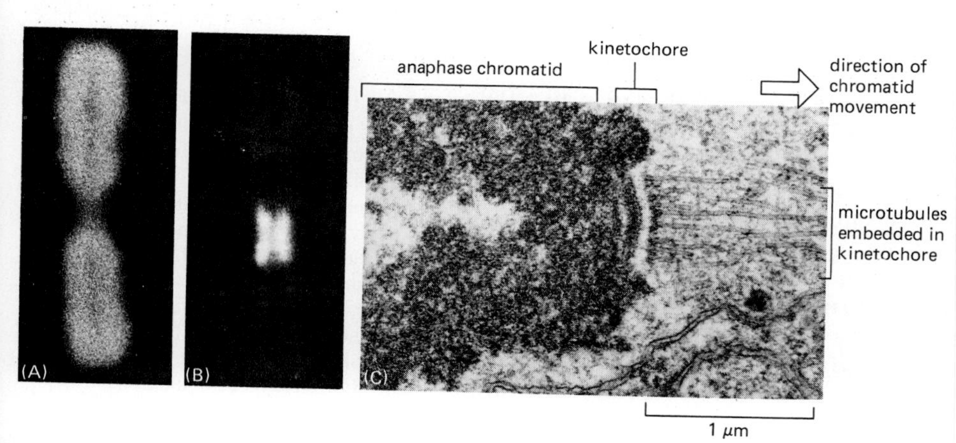

Figure 13–51 The kinetochore. A metaphase chromosome (A), when stained with human autoantibodies that react with specific kinetochore proteins, reveals two kinetochores, one associated with each chromatid (B). (C) Electron micrograph of an anaphase chromatid with microtubules attached to its kinetochore. While most kinetochores have a trilaminar structure, the one shown (from a green alga) has an unusually complex structure with additional layers. (A and B, courtesy of Bill Brinkley; C, from J.D. Pickett-Heaps and L.C. Fowke, *Aust. J. Biol. Sci.* 23:71–92, 1970. Reproduced by permission of CSIRO.)

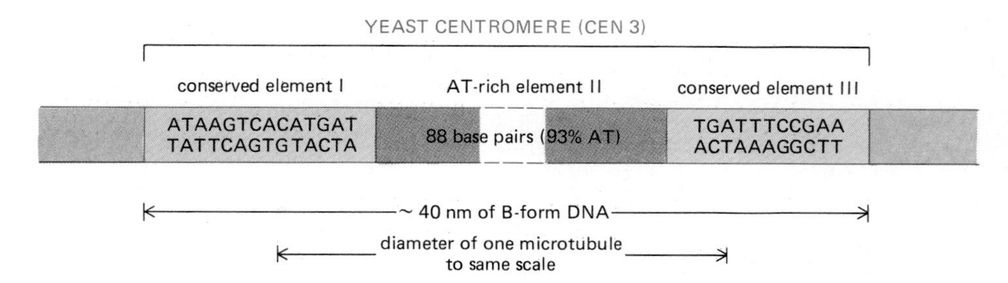

Figure 13–52 The DNA sequence of a typical centromere in the yeast *Saccharomyces cerevisiae*. The DNA sequence shown is sufficient to cause faithful chromosome segregation; it serves to assemble the kinetochore proteins that attract a single microtubule to the kinetochore.

cells during metaphase, it is found to be continually incorporated into microtubules near their point of attachment to the kinetochore (Figure 13–54). Conversely, as we shall see below, the reverse reaction takes place during anaphase, when tubulin molecules are lost from these microtubules in the region close to the kinetochore as this structure moves toward a spindle pole. The puzzle is that both the addition and loss of tubulin molecules must take place while the kinetochore maintains a firm mechanical attachment to the microtubules, since it is through this point of attachment that chromosomes are pulled through the cytoplasm. The kinetochore thus seems to act like a sliding collar, maintaining a lateral association with polymerized tubulin subunits near the end of the microtubule while also allowing addition or loss of tubulin molecules to occur at that end (see Figure 13–61, below).

13-28 Kinetochores Are Thought to Capture the Plus Ends of Microtubules That Originate from a Spindle Pole[38]

The breakdown of the nuclear envelope, which signals the end of prophase and the beginning of **prometaphase,** enables the mitotic spindle to interact with the chromosomes. The eventual outcome of this interaction is that one chromatid of each chromosome will be faithfully segregated to each daughter nucleus. The kinetochore microtubules play a major role in this segregation process: they (1) orient each chromosome with respect to the spindle axis so that one kinetochore faces each pole and (2) move each chromosome into a plane (the *metaphase plate*) that bisects the mitotic spindle, a process that takes between 10 and 20 minutes in mammalian cells and reaches completion at the end of prometaphase.

Prometaphase is characterized by a period of frantic activity during which the spindle appears to be trying to contain and align the chromosomes at the metaphase plate. In actuality, the chromosomes are violently rotating and oscillating back and forth between the spindle poles because their kinetochores are capturing microtubules growing from one or the other spindle pole and are being pulled by the captured microtubules. The initial attachment of a chromosome usually takes place when it is close to a spindle pole, when microtubules attach to only one kinetochore; eventually the other kinetochore captures microtubules growing from the other pole. These random prometaphase movements, and the final chance orientation of the chromosome that results, guarantee the random segregation of chromatids to the daughter cells, which is important for gene mixing during the analogous nuclear division in meiosis (see p. 854).

Only the plus ends of microtubules extend from the poles, and it is these ends that attach to kinetochores. The kinetochore thereby acts as a "cap" that tends to protect the plus end from depolymerizing, just as the centrosome at the spindle pole tends to protect the minus end from depolymerizing. It is perhaps not surprising, therefore, that kinetochore microtubules, which are capped at both ends, are unusually stable. Other microtubules in the spindle (called *polar microtubules*) are less stable.

While the kinetochore microtubules tend to *pull* the chromosome toward their respective pole (see below), a separate force apparently repels any chromosomes that come too close to the pole. If the arms of a chromosome are cut free from the kinetochore by laser microsurgery, they tend to move away from the closest spindle pole, even though they seem not to be attached to microtubules or to any

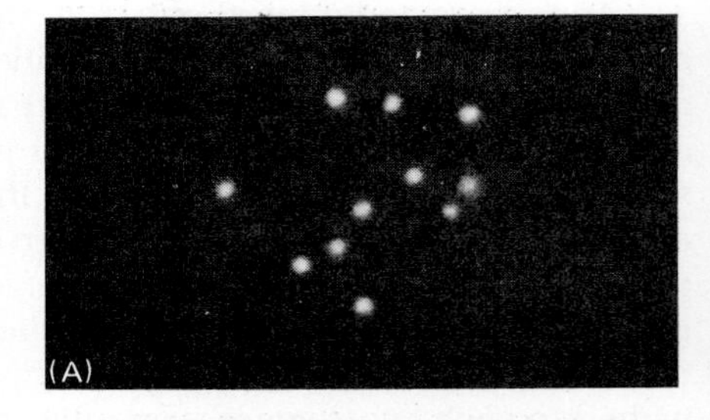

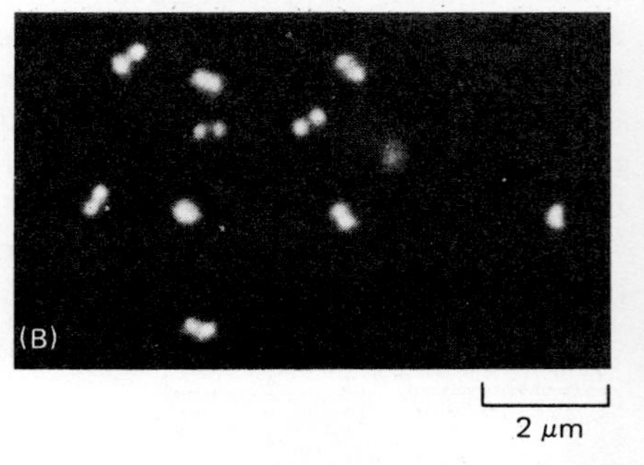

Figure 13–53 Immunofluorescent staining of kinetochores in interphase cells using an antibody that binds specifically to a kinetochore protein. These marsupial cells (PtK cells) contain relatively few chromosomes. (A) In cells in G_1 phase, one kinetochore is stained per chromosome. (B) In cells in G_2 phase, two kinetochores are stained per chromosome. (From S.L. Brenner and B.R. Brinkley, *Cold Spring Harbor Symp. Quant. Biol.* 46:241–254, 1982.)

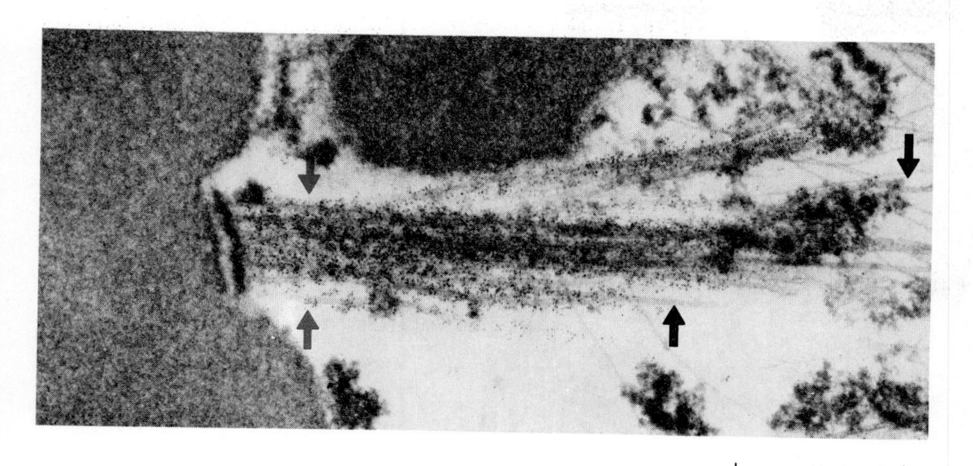

Figure 13–54 Experiment demonstrating that metaphase kinetochore microtubules grow at their kinetochore-attached (plus) end. Tubulin was covalently coupled to a small organic molecule (biotin) and then microinjected into a mammalian metaphase cell in culture. After 1 minute the cell was permeabilized, fixed, and stained with antibiotin antibodies bound to gold spheres, and then sections were prepared for electron microscopy. Regions of microtubules that incorporated the biotinylated tubulin during the minute following the injection are finely speckled with dark dots of gold (*colored arrows*), while regions of preexisting microtubule are unlabeled (*black arrows*). (Photograph courtesy of Louise Evans.)

other obvious structure in the cell. One possibility is that the rapid polymerization of spindle microtubules outward from each pole might produce a general "wind" that carries any large unattached structures, such as the chromosome arms, away from the poles.

Sister Chromatids Attach by Their Kinetochores to Opposite Spindle Poles[39]

In early prometaphase the two kinetochores on a chromosome can become attached to the same spindle pole. However, this and other incorrect configurations, which would cause a failure of the chromosome to segregate properly if they persisted, are almost always corrected. It seems that a balanced arrangement, in which each sister kinetochore is attached to a different spindle pole and only to that spindle pole, has the greatest stability. Why this might be is suggested by experiments designed to test how chromosomes are attached to the mitotic spindle.

Delicate micromanipulation experiments, in which extremely fine glass needles are used to poke and pull at chromosomes inside a living mitotic cell, demonstrate that kinetochores are not committed to face one particular spindle pole as if polarized like a magnet, since a kinetochore of a manipulated chromosome will reengage to either pole if it is turned around. Moreover, by micromanipulation during prometaphase, it is possible to force the two kinetochores on a single chromosome to engage with the same spindle pole. If this configuration persists, the entire chromosome (with joined sister chromatids) is drawn toward the pole to which it is attached. Such an arrangement is generally unstable, however, and new kinetochore microtubules are usually captured from the other pole to produce the correct balanced configuration. On the other hand, if the movement of the incorrectly associated chromosome is restrained by a glass needle, the association of the chromosome with only one spindle pole becomes stabilized, suggesting that microtubules strengthen their linkage to the kinetochore by pulling against tension. Thus only chromosomes attached to both poles will normally retain their kinetochore microtubules and thereby interact stably with the spindle.

The tension generated by opposing kinetochore microtubules not only stabilizes the kinetochore-microtubule interaction, it also eventually aligns each chromosome on the metaphase plate, as will now be described.

Balanced Bipolar Forces Hold Chromosomes on the Metaphase Plate[40]

Why do all chromosomes line up at an equal distance from the two spindle poles at **metaphase,** thereby defining the metaphase plate? Experiments in which chromosomes are displaced with a glass needle suggest that the force exerted on a kinetochore is proportional to the length of the kinetochore fibers: that is, the pull

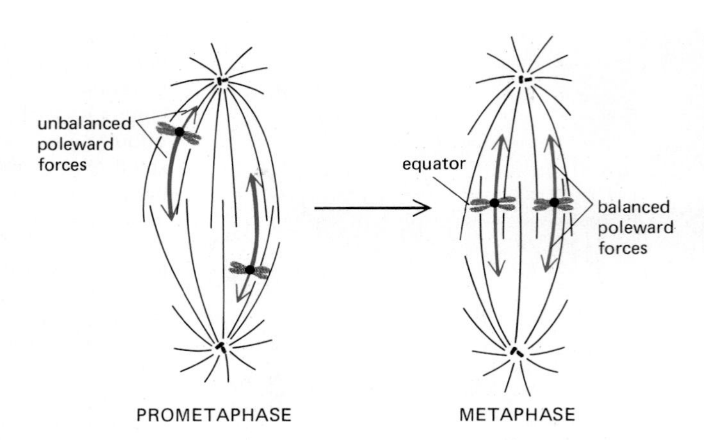

Figure 13–55 Chromosomes enter the spindle randomly during prometaphase and eventually line up at the equator (the metaphase plate) because the force on each kinetochore decreases as it gets closer to a pole. Thus the chromosomes are held under tension at the equator by balanced poleward forces. It is still unclear how the forces are generated.

on each kinetochore decreases the closer the kinetochore approaches to the attached pole (Figure 13–55). It is as though each chromosome is attached by a spring to each of the two spindle poles, so that any displacement toward one pole produces a restoring force in the opposite direction. The spindle that results at metaphase is illustrated in Figure 13–56.

The force on the chromosomes continues even after they have become aligned on the metaphase plate. Thus the chromosomes oscillate back and forth on the metaphase plate, continually adjusting their positions. Moreover, if one of a pair of metaphase kinetochore fibers is severed with a laser beam, the entire chromosome immediately moves toward the pole to which it is still attached. Similarly, if the attachment between the two chromatids is severed at metaphase, the two chromatids separate and move toward opposite poles, just as they do in anaphase. These experiments suggest that the same forces that move chromosomes to the metaphase plate also carry the chromatids to opposite poles as soon as the two kinetochores on each chromosome have separated.

Metaphase occupies a substantial portion of the mitotic period (see Figure 13–43), as if cells pause until all their chromosomes are lined up appropriately on the metaphase plate. Several experimental observations support this notion. Many cells arrest in mitosis for hours or days if treated with a drug, such as

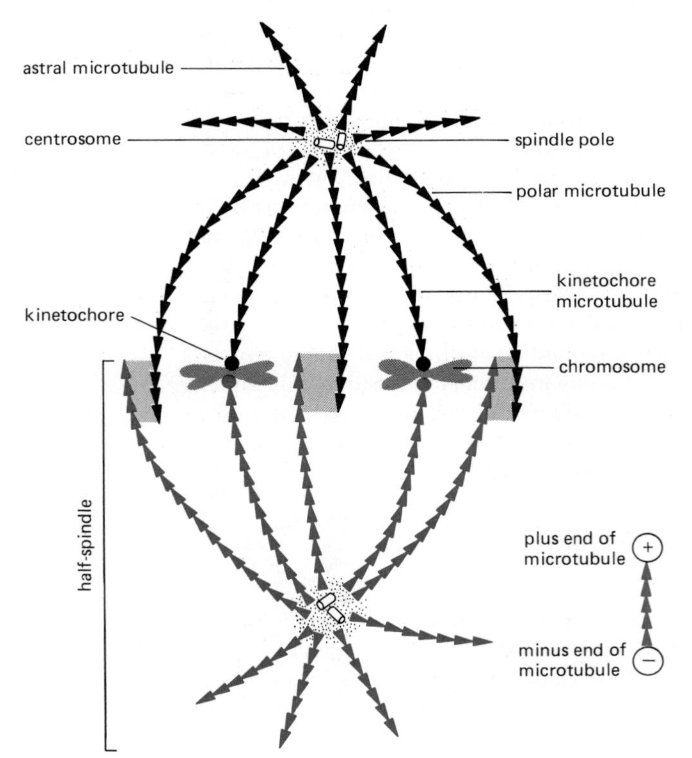

Figure 13–56 Simplified diagram of the mitotic spindle at metaphase. The spindle is constructed from two half-spindles (*black* and *color*), each composed of kinetochore, polar, and astral microtubules. The polarity of the microtubules is indicated by the arrowheads. The polar microtubules emanating from opposite spindle poles have a region of overlap (*shaded gray*), where microtubule-associated proteins may cross-link them; note that the microtubules are antiparallel in this overlap zone.

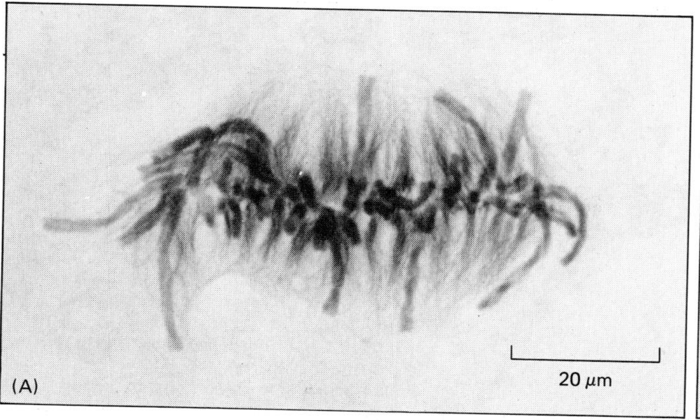

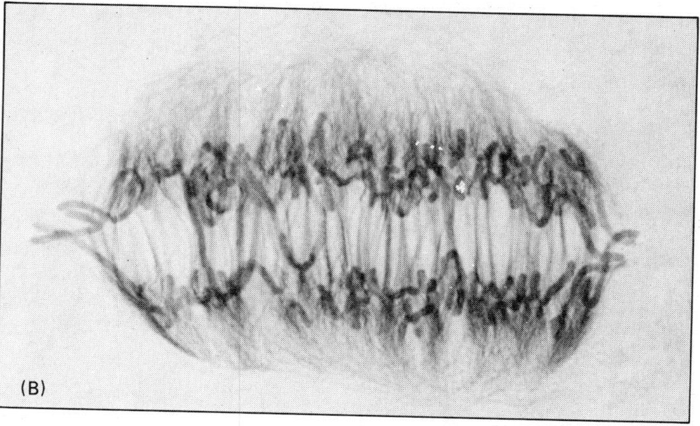

Figure 13–57 Chromosomal movement at anaphase. In the transition from metaphase (A) to anaphase (B), chromosomes are pulled apart by spindle microtubules—as seen in these *Haemanthus* (lily) endosperm cells stained with gold-labeled antibodies to tubulin. (Courtesy of Andrew Bajer.)

colchicine or vinblastine, that depolymerizes their microtubules; in fact, this method of cell-cycle arrest is commonly used to collect large numbers of cells in mitosis so that their condensed chromosomes can be subjected to cytological analysis (see p. 505). The mitotic spindle rapidly regenerates if the drug is removed, and a normal mitosis often resumes once the chromosomes have been correctly positioned on the metaphase plate. It has been suggested that a chromosome with an unattached kinetochore may generate a diffusible signal that normally delays entry into anaphase, providing extra time for it to be attached correctly. If such a signal exists, disrupting the spindle with drugs would be expected to generate a strong signal and thereby prolong metaphase.

Sister Chromatids Separate Suddenly at Anaphase[41]

Metaphase, as we have just seen, is a relatively stable state, and in normal circumstances many cells remain for an hour or more with their chromosomes aligned and oscillating back and forth on the metaphase plate. **Anaphase** then begins abruptly with the synchronous splitting of each chromosome into its sister chromatids, each with one kinetochore (Figure 13–57). The signal to initiate anaphase is not a force exerted by the spindle itself, since chromosomes that are detached from the spindle will separate into chromatids at the same time as those that are still attached. Some experiments suggest that the signal could involve an increase in cytosolic Ca^{2+}: (1) continuous monitoring of living cells containing a fluorescent Ca^{2+} indicator dye (see p. 156) reveals a rapid, transient, tenfold rise in intracellular Ca^{2+} at anaphase in some cells; (2) microinjection of low levels of Ca^{2+} into cultured cells at metaphase can cause anaphase to begin prematurely; and (3) accumulations of membrane vesicles are usually seen at the spindle poles, and specialized electron microscopic techniques indicate that the vesicles are rich in Ca^{2+}. It is therefore possible that the spindle-associated vesicles release Ca^{2+} to initiate anaphase (Figure 13–58), much as the sarcoplasmic reticulum releases Ca^{2+} to initiate skeletal muscle contraction (see p. 624).

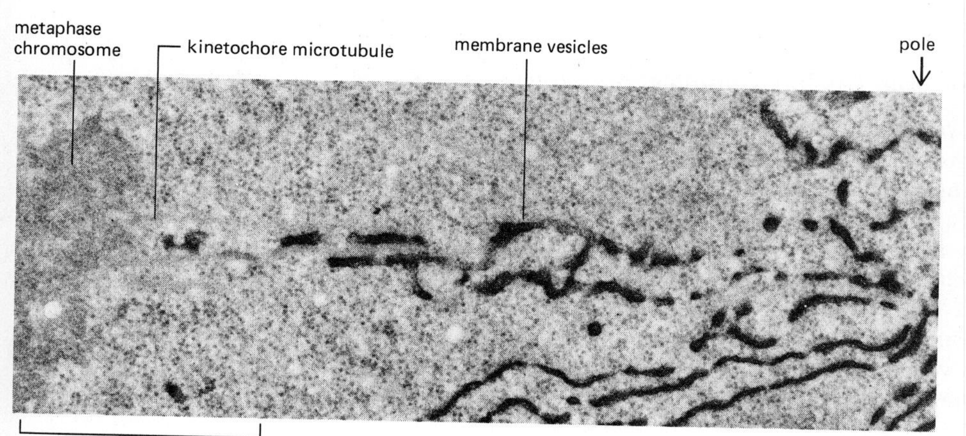

Figure 13–58 Electron micrograph showing the collection of membrane vesicles (resembling cytoplasmic reticulum), specially stained for clarity, that are present at the spindle poles and seem to stretch out along the microtubules in the mitotic spindle. The cell, at metaphase, is from a barley leaf. (Courtesy of Peter Hepler, reproduced from *J. Cell Biol.* 86:490–499, 1980, by copyright permission of the Rockefeller University Press.)

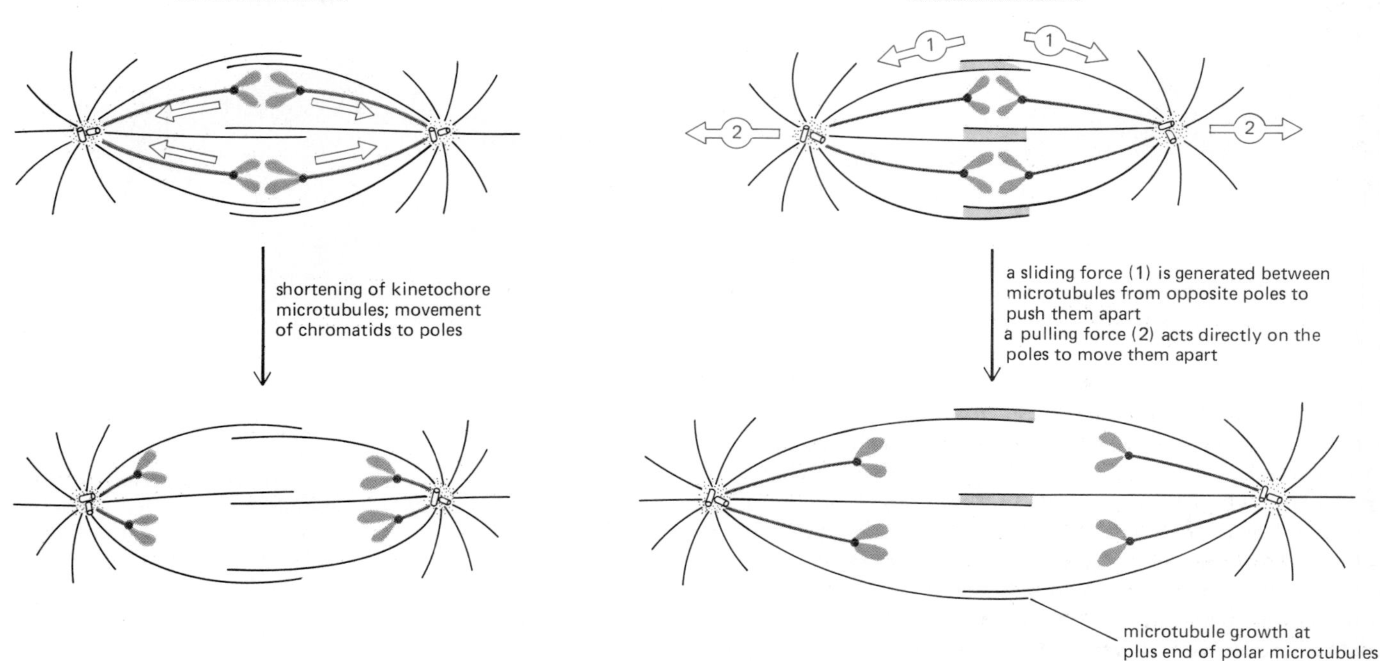

Figure 13–59 The several forces that act at anaphase to separate sister chromatids. (A) Chromatids are *pulled* toward opposite poles by forces associated with shortening of their kinetochore microtubules, a movement known as *anaphase A.* (B) Simultaneously the two spindle poles move apart, a movement known as *anaphase B.* It is likely that the forces that cause anaphase B are similar to those that cause the centrosome to split and separate into two spindle poles at prophase (see Figure 13–46). There is evidence that two separate forces are responsible for anaphase B: (1) the elongation and sliding of the polar microtubules *pushes* the two poles apart, while (2) outward forces act on the asters at each spindle pole to *pull* the poles away from each other.

Two Distinct Processes Separate Chromosomes at Anaphase[42]

Once each chromosome has split in response to the anaphase trigger, its two chromatids move to opposite spindle poles, where they will assemble into the nucleus of a new cell. Their movement appears to be the consequence of two independent processes within the spindle (Figure 13–59). The first is the poleward movement of chromatids accompanied by the shortening of the kinetochore microtubules, usually referred to as **anaphase A.** The second is the separation of the poles themselves accompanied by the elongation of the polar microtubules and is known as **anaphase B.** It is possible to distinguish the two processes by their differential sensitivities to certain drugs. A low concentration of chloral hydrate, for example, prevents the poles from separating and the polar microtubules from lengthening (anaphase B) but has no effect on the kinetochore microtubules or the poleward movement of the chromatids (anaphase A). The relative contribution of each of the two processes to the final separation of the chromosomes varies considerably depending on the organism. In mammalian cells, anaphase B begins shortly after the chromatids have begun their voyage to the poles and stops when the spindle is about 1.5–2 times its metaphase length. In some other cells, such as yeasts, anaphase B begins only after the chromatids reach their final destination; and in certain protozoa, anaphase B predominates and the spindles elongate to 15 times their metaphase length.

Kinetochore Microtubules Disassemble During Anaphase A[43]

A surprisingly large force acts on a chromosome as it moves from the metaphase plate to the spindle pole. Measurements using the deflection of fine glass needles give an estimate of 10^{-5} dynes per chromosome, which is more than 10,000 times greater than the force required simply to move chromosomes at their observed rate through the cytoplasm. Evidently there must be a powerful motor to move the chromosomes, but the speed of their movement must be limited by something other than viscous drag. As mentioned above, the same motor may be responsible for generating the tension on chromosomes on the metaphase plate.

As each chromosome moves poleward, its kinetochore microtubules disassemble, so that they have nearly disappeared at telophase. The site of subunit loss can be determined by injecting labeled tubulin into cells during metaphase. The

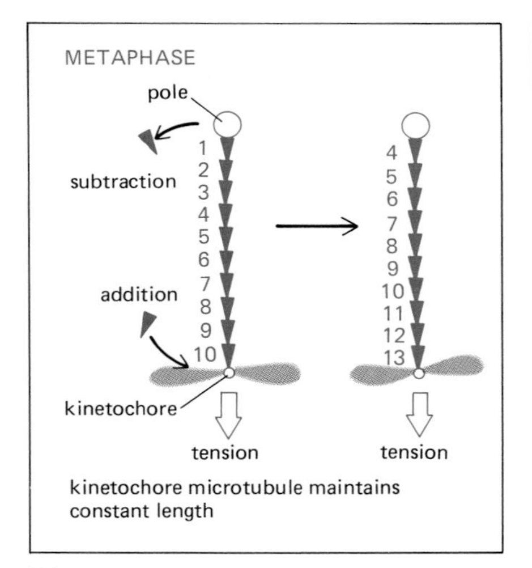

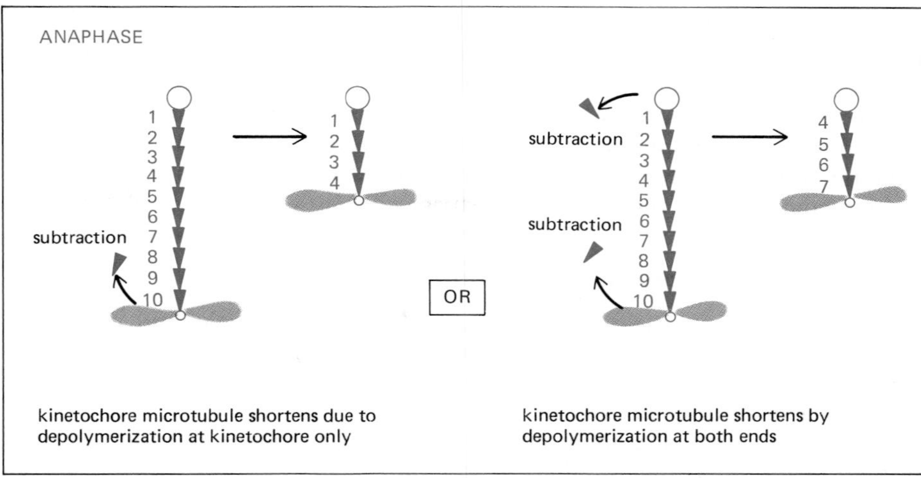

labeled subunits are found to be added to the kinetochore end of the kinetochore microtubules and then lost as anaphase A proceeds, indicating that the kinetochore "eats" its way poleward along its microtubules at anaphase. This conclusion is supported by the observation that anaphase kinetochores move toward a stationary mark generated on kinetochore microtubules by photobleaching. Microtubule disassembly at kinetochores, poles or both sites is probably necessary for chromosome-to-pole movement (Figure 13–60), since this movement is inhibited if microtubule depolymerization is blocked by the addition of taxol or D_2O.

The mechanism by which the kinetochore, and thus the chromosome, moves up the spindle during anaphase A is still unknown. Two possible models are presented schematically in Figure 13–61. In one the kinetochore hydrolyzes ATP to move along its attached microtubule, with the plus end of the microtubule depolymerizing as it becomes exposed. In the other, depolymerization of the microtubule itself causes the kinetochore to move passively to optimize its binding energy on the microtubule. A third possibility, not illustrated in Figure 13–61, is that microtubules are not directly responsible for the poleward force on kinetochores but serve merely to regulate movements generated by some other structure. It has been suggested, for example, that a system of elastic protein filaments—perhaps similar to the very long elastic filaments in striated muscle (see p. 624)—might connect the kinetochore to the pole and pull the kinetochore steadily poleward.

Figure 13–60 The behavior of kinetochore microtubules changes during the transition from metaphase to anaphase. (A) At metaphase, subunits are added to the plus end of a microtubule at the kinetochore and are removed from the minus end at the spindle pole. Thus a constant poleward flux of tubulin subunits occurs, with the microtubules remaining stationary and under tension. (B) At anaphase the tension is released, and the kinetochore moves rapidly up the microtubule, removing subunits from its plus end as it goes (*left side*). Its attached chromatid is thereby carried to a spindle pole. In at least some organisms, part of the chromatid movement is due to the simultaneous shortening of the microtubules at the pole (*right side*).

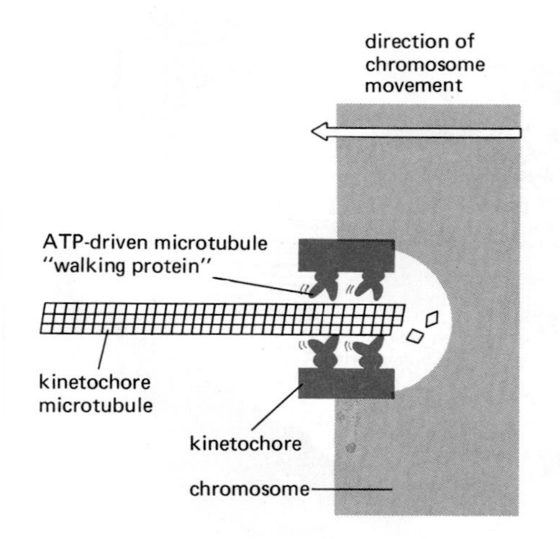

(A) ATP-driven chromosome movement drives microtubule disassembly

direction of chromosome movement
protein with high affinity for polymerised tubulin
kinetochore microtubule
kinetochore
chromosome

(B) Microtubule disassembly drives chromosome movement

Figure 13–61 Two alternative models of how the kinetochore might be able to generate a poleward force on its chromosome during anaphase. (A) Microtubule-walking proteins that resemble dynein or kinesin are part of the kinetochore, and they use the energy of ATP hydrolysis to pull the chromosome along its bound microtubules (see p. 660). (B) Chromosome movement is driven by microtubule disassembly: as tubulin subunits dissociate, the kinetochore tends to slide poleward in order to restore its binding to the walls of the microtubule. Similar mechanisms may be used at the spindle pole, which likewise seems to be able to hold onto microtubules while permitting their controlled depolymerization (see Figure 13–60).

Regardless of the force-generating mechanism, the dramatic change in microtubule polymerization at the kinetochore caused by the metaphase-to-anaphase

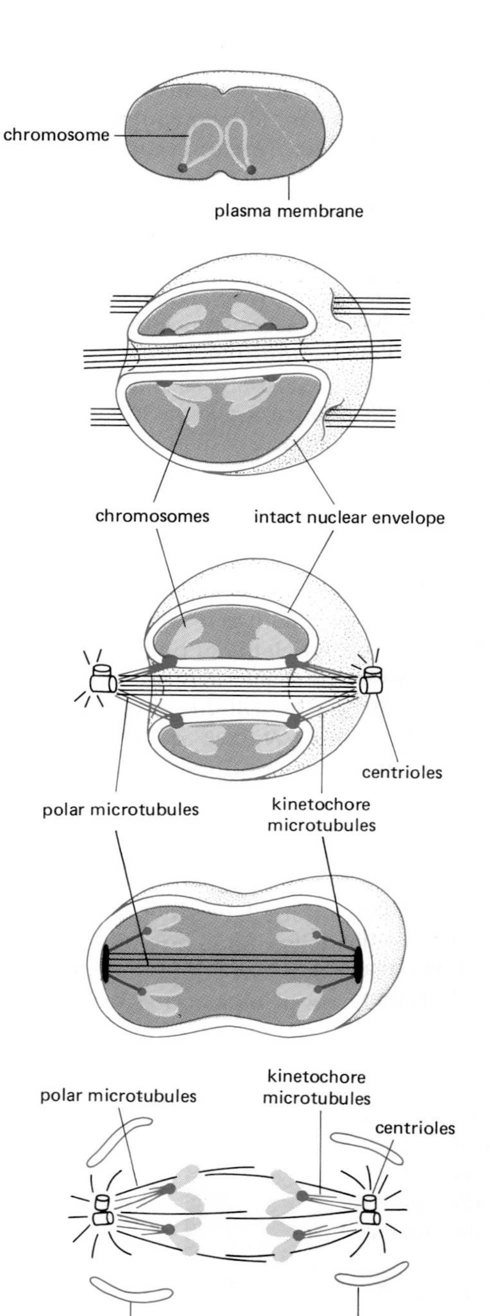

BACTERIA
daughter chromosomes attached to the plasma membrane are separated by the growth of membrane between them

TYPICAL DINOFLAGELLATES
several bundles of microtubules pass through tunnels in the intact nuclear envelope to establish the polarity of division; chromosomes move apart in association with the inner nuclear membrane without being attached to the microtubule bundles

HYPERMASTIGOTES AND SOME UNUSUAL DINOFLAGELLATES
a single central spindle between centrioles is formed in a tunnel through the intact nuclear envelope; chromosomes are attached by their kinetochores to the nuclear membrane and interact with the spindle poles via kinetochore microtubules

YEASTS AND DIATOMS
nuclear envelope remains intact; continuous polar spindle microtubules form inside the nucleus and run between the two spindle pole bodies associated with the nuclear envelope; a single kinetochore microtubule attaches each chromosome to a pole

HIGHER ORGANISMS
the spindle begins to form outside the nucleus; at prometaphase the nuclear envelope breaks down to allow chromosomes to capture spindle microtubules, which now become kinetochore microtubules

Figure 13–75 Different chromosome separation mechanisms are used by different organisms. Some of these may have been intermediary stages in the evolution of the mitotic spindle of higher organisms. For all the examples except bacteria, only the central nuclear region of the cell is shown.

digitate to form a zone of overlap. In both yeasts and diatoms the spindle is attached to chromosomes by their kinetochores, and the chromosomes are segregated in a way closely similar to that described for mammalian cells—except that the entire process generally occurs within the confines of the nuclear envelope. At present there is no convincing explanation for why higher plants and animals have instead evolved a mitotic apparatus that requires the controlled and reversible dissolution of the nuclear envelope.

Summary

The process of cell division consists of nuclear division (mitosis) followed by cytoplasmic division (cytokinesis). Mitosis begins with prophase, a transitional period during which the centrosome splits apart to form the two spindle poles that will organize the subsequent intracellular movements. At the same time, the beginning of M phase is accompanied by a marked increase in the phosphorylation of specific proteins. Perhaps as a result, a mitotic cell contains an unusually dynamic microtubule array. After the nuclear envelope breaks down in prometaphase, the kinetochores on condensed chromosomes can capture and stabilize subsets of microtubules from the large numbers that continually grow out from each spindle pole. Opposing pole-directed forces pull on these kinetochore microtubules, creating a tension that causes the chromosomes to line up at the spindle equator during metaphase. At anaphase this tension is suddenly released as sister chromatids detach from each other and are pulled to opposite poles. In addition, the two mitotic poles often move apart. In the final, telophase stage of mitosis, the nuclear envelope re-forms on the surface of each group of separated chromosomes as the proteins phosphorylated at the onset of M phase are dephosphorylated.

Cell division ends as the cytoplasmic contents are divided by the process of cytokinesis and the chromosomes decondense and resume RNA synthesis. Cytokinesis appears to be guided by organized bundles of actin filaments in eucaryotic cells as diverse as animals, plants, and fungi. Large membrane-bounded organelles such as the Golgi apparatus and the endoplasmic reticulum break up into smaller fragments and vesicles during M phase, which presumably ensures their even distribution into daughter cells. However, cytokinesis can also involve a programmed asymmetric distribution of materials. A particular cell can divide into one small cell and one large one, for example, or a specific cytoplasmic component can be moved to one side of a cell prior to cytokinesis so that it is inherited by only one of the two otherwise equal daughter cells.

References

General

Baserga, R. The Biology of Cell Reproduction. Cambridge, MA: Harvard University Press, 1985.

Beach, D.; Basilico, C.; Newport, J., eds. Cell Cycle Control in Eukaryotes. Cold Spring Harbor, NY: Cold Spring Harbor Laboratory, 1988.

John, P.C.L., ed. The Cell Cycle. Cambridge, U.K.: Cambridge University Press, 1981.

Mitchison, J.M. The Biology of the Cell Cycle. Cambridge, U.K.: Cambridge University Press, 1971.

Pollack, R. Readings in Mammalian Cell Culture, 2nd ed. Cold Spring Harbor, NY: Cold Spring Harbor Laboratory, 1981. (An anthology, including many important papers on cell growth and division.)

Prescott, D.M. Reproduction of Eucaryotic Cells. New York: Academic Press, 1976.

Cited

1. Baserga, R. The Biology of Cell Reproduction. Cambridge, MA: Harvard University Press, 1985.

   Howard, A.; Pelc, S.R. Nuclear incorporation of P^{32} as demonstrated by autoradiographs. *Exp. Cell Res.* 2:178–187, 1951.

   Lloyd, D.; Poole, R.K.; Edwards, S.W. The Cell Division Cycle. New York: Academic Press, 1982.

   Mitchison, J.M. The Biology of the Cell Cycle. Cambridge, U.K.: Cambridge University Press, 1971.

2. Van Dilla, M.A.; Trujillo, T.T.; Mullaney, P.F.; Coulter, J.R. Cell microfluorometry: a method for rapid fluorescence measurement. *Science* 163:1213–1214, 1969.

3. Creanor, J.; Mitchison, J.M. Patterns of protein synthesis during the cell cycle of the fission yeast *Schizosaccharomyces pombe*. *J. Cell Sci.* 58:263–285, 1982.

   Pardee, A.B.; Coppock, D.L.; Yang, H.C. Regulation of cell proliferation at the onset of DNA synthesis. *J. Cell Sci.*, Suppl. 4:171–180, 1986.

4. Rao, P.N.; Johnson, R.T. Mammalian cell fusion: studies on the regulation of DNA synthesis and mitosis. *Nature* 225:159–164, 1970.

5. Xeros, N. Deoxyriboside control and synchronization of mitosis. *Nature* 194:682–683, 1962.

6. Mitchison, J.M. The Biology of the Cell Cycle, pp. 234–244. Cambridge, U.K.: Cambridge University Press, 1971.

   Schlegel, R.; Pardee, A.B. Caffeine-induced uncoupling of mitosis from the completion of DNA replication in mammalian cells. *Science* 232:1264–1266, 1986.

7. Johnson, R.T.; Rao, P.N. Mammalian cell fusion: induction of premature chromosome condensation in interphase nuclei. *Nature* 226:717–722, 1970.

8. Blow, J.J.; Laskey, R.A. A role for the nuclear envelope in controlling DNA replication within the cell cycle. *Nature* 332:546–548, 1988.

9. Kirschner, M.; Newport, J.; Gerhart, J. The timing of early developmental events in *Xenopus*. *Trends Genet.* 1:41–47, 1985.

10. Gerhart, J.; Wu, M.; Kirschner, M. Cell cycle dynamics of an M-phase-specific cytoplasmic factor in *Xenopus laevis* oocytes and eggs. *J. Cell Biol.* 98:1247–1255, 1984.

    Lohka, M.J.; Hayes, M.K.; Maller, J.L. Purification of maturation-promoting factor, an intracellular regulator of early mitotic events. *Proc. Natl. Acad. Sci. USA* 85:3009–3013, 1988.

    Newport, J.; Kirschner, M. Regulation of the cell cycle during early *Xenopus* development. *Cell* 37:731–742, 1984.

11. Hara, K.; Tydeman, P.; Kirschner, M. A cytoplasmic clock with the same period as the division cycle in *Xenopus* eggs. *Proc. Natl. Acad. Sci. USA* 77:462–466, 1980.

    Kimelman, D.; Kirschner, M.; Scherson, T. The events of the midblastula transition in *Xenopus* are regulated by changes in the cell cycle. *Cell* 48:399–407, 1987.

    Pines, J.; Hunt, T. Molecular cloning and characterization of the mRNA for cyclin from sea urchin eggs. *EMBO J.* 6:2987–2995, 1987.

    Swenson, K.I.; Farrell, K.M.; Ruderman, J.V. The clam embryo protein cyclin A induces entry into M phase and the resumption of meiosis in *Xenopus* oocytes. *Cell* 47:861–870, 1986.

12. Nurse, P. Cell cycle control genes in yeast. *Trends Genet.* 1:51–55, 1985.

    Pringle, J.R.; Hartwell, L.H. The *Saccharomyces cerevisiae* cell cycle. In The Molecular Biology of the Yeast *Saccharomyces*, Life Cycle and Inheritance (J.N. Strathern, E.W. Jones, J.R. Broach, eds.), pp. 97–142. Cold Spring Harbor, NY: Cold Spring Harbor Laboratory, 1981.

    Watson, J.D.; Hopkins, N.H.; Roberts, J.W.; Weiner, A.M. Molecular Biology of the Gene, 4th ed., pp. 550–592. Menlo Park, CA: Benjamin-Cummings, 1987.

13. Hartwell, L.H. Cell division from a genetic perspective. *J. Cell Biol.* 77:627–637, 1978.

    Nurse, P. Genetic analysis of the cell cycle. *Symp. Soc. Gen. Microbiol.* 31:291–315, 1981.

14. Hayles, J.; Nurse, P. Cell cycle regulation in yeast. *J. Cell Sci.*, Suppl. 4:155–170, 1986.

    Johnston, G.C.; Pringle, J.R.; Hartwell, L.H. Coordination of growth with cell division in the yeast *Saccharomyces cerevisiae*. *Exp. Cell Res.* 105:79–98, 1977.

    Prescott, D.M. Changes in nuclear volume and growth rate and prevention of cell division in Amoeba proteus resulting from cytoplasmic amputations. *Exp. Cell Res.* 11:94–98, 1956.

15. Dunphy, W.G.; Brizuela, L.; Beach, D.; Newport, J. The *Xenopus cdc2* protein is a component of MPF, a cytoplasmic regulator of mitosis. *Cell* 54:423–431, 1988.

    Gautier, J.; Norbury, C.; Lohka, M.; Nurse, P.; Maller, J. Purified maturation-promoting factor contains the product of a *Xenopus* homolog of the fission yeast cell cycle control gene *cdc2*$^{+}$. *Cell* 54:433–439, 1988.

    Lee, M.G.; Nurse, P. Complementation used to clone a human homologue of the fission yeast cell cycle control gene *cdc2*. *Nature* 327:31–35, 1987.

    Murray, A.W. A mitotic inducer matures. *Nature* 335:207–208, 1988.

    Solomon, M.; et al. Cyclin in fission yeast. *Cell* 54:738–740, 1988.

16. Cheng, H.; LeBlond, C.P. Origin, differentiation and renewal of the four main epithelial cell types in the mouse small intestine. *Am. J. Anat.* 141:461–480, 1974.

    Goss, R.J. The Physiology of Growth. New York: Academic Press, 1978.

17. Pardee, A.B. A restriction point for control of normal animal cell proliferation. *Proc. Natl. Acad. Sci. USA* 71:1286–1290, 1974.

18. Brooks, R.F. The transition probability model: successes, limitations and deficiencies. In Temporal Order (L. Rensing, N.I. Jaeger, eds.). Berlin: Springer, 1985.

    Shields, R. Transition probability and the origin of variation in the cell cycle. *Nature* 267:704–707, 1977.

    Smith, J.A.; Martin, L. Do cells cycle? *Proc. Natl. Acad. Sci. USA* 70:1263–1267, 1973.

19. Deuel, T.F. Polypeptide growth factors: roles in normal and abnormal growth. *Annu. Rev. Cell Biol.* 3:443–492, 1987.

    Evered, D.; Nugent, J.; Whelan, J., eds. Growth Factors in Biology and Medicine. Ciba Foundation Symposium 116. London: Pitman, 1985.

    Sato, G., ed. Hormones and Cell Culture. Cold Spring Harbor, NY: Cold Spring Harbor Laboratory, 1979.

    Wang, J.L.; Hsu, Y.-M. Negative regulators of cell growth. *Trends Biochem. Sci.* 11:24–27, 1986.

20. Ross, R.; Raines, E.W.; Bowen-Pope, D.F. The biology of platelet- derived growth factor. *Cell* 46:155–169, 1986.

    Stiles, C.D. The molecular biology of platelet-derived growth factor. *Cell* 33:653–655, 1983.

21. Dunn, G.A.; Ireland, G.W. New evidence that growth in 3T3 cell cultures is a diffusion-limited process. *Nature* 312:63–65, 1984.

    Holley, R.W.; Kiernan, J.A. "Contact inhibition" of cell division in 3T3 cells. *Proc. Natl. Acad. Sci. USA* 60:300–304, 1968.

    Stoker, M.G.P. Role of diffusion boundary layer in contact inhibition of growth. *Nature* 246:200–203, 1973.

22. Folkman, J.; Moscona, A. Role of cell shape in growth control. *Nature* 273:345–349, 1978.

    O'Neill, C.; Jordan, P.; Ireland, G. Evidence for two distinct mechanisms of anchorage stimulation in freshly explanted and 3T3 mouse fibroblasts. *Cell* 44:489–496, 1986.

23. Brooks, R.F. Regulation of the fibroblast cell cycle by serum. *Nature* 260:248–250, 1976.

    Larsson, O.; Zetterberg, A.; Engstrom, W. Consequences of parental exposure to serum-free medium for progeny cell division. *J. Cell Sci.* 75:259–268, 1985.

    Zetterberg, A.; Larsson, O. Kinetic analysis of regulatory events in G1 leading to proliferation or quiescence of Swiss 3T3 cells. *Proc. Natl. Acad. Sci. USA* 82:5365–5369, 1985.

24. Baserga, R. The Biology of Cell Reproduction, pp. 103–113. Cambridge, MA: Harvard University Press, 1985.

    Larsson, O.; Dafgard, E.; Engstrom, W.; Zetterberg, A. Immediate effects of serum depletion on dissociation between growth in size and cell division in proliferating 3T3 cells. *J. Cell. Physiol.* 127:267–273, 1986.

25. Bauer, E.; et al. Diminished response of Werner's syndrome fibroblasts to growth factors PDGF and FGF. *Science* 234:1240–1243, 1986.

    Hayflick, L. The limited *in vitro* lifetime of human diploid cell strains. *Exp. Cell Res.* 37:614–636, 1965.

    Holliday, R., ed. Genes, Proteins, and Cellular Aging. New York: Van Nostrand, 1986. (An anthology of papers on cell senescence.)

    Loo, D.T.; Fuquay, J.I.; Rawson, C.L.; Barnes, D.W. Extended culture of mouse embryo cells without senescence: inhibition by serum. *Science* 236:200–202, 1987.

    Rheinwald, J.G.; Green, H. Epidermal growth factor and the multiplication of cultured human epidermal keratinocytes. *Nature* 265:421–424, 1977.

    Smith, J.R.; Whitney, R.G. Intraclonal variation in proliferative potential of human diploid fibroblasts: stochastic mechanism for cellular aging. *Science* 207:82–84, 1980.

26. Kahn, P.; Graf, T., eds. Oncogenes and Growth Control. Berlin: Springer, 1986.

    Watson, J.D.; Hopkins, N.H.; Roberts, J.W.; Weiner, A.M. Molecular Biology of the Gene, 4th ed., pp. 961–1096. Menlo Park, CA: Benjamin-Cummings, 1987.

27. Feramisco, J.; Ozanne, B.; Stiles, C., eds. Cancer Cells 3: Growth Factors and Transformation. Cold Spring Harbor, NY: Cold Spring Harbor Laboratory, 1985.

    Temin, H.M.; Rubin, H. Characteristics of an assay for Rous sarcoma virus and Rous sarcoma cells in tissue culture. *Virology* 6:669–688, 1958.

28. Bishop, J.M. Viral oncogenes. *Cell* 42:23–38, 1985.

    Martin, G.S. Rous sarcoma virus: a function required for the maintenance of the transformed state. *Nature* 227:1021–1023, 1970.

    Swanstrom, R.; Parker, R.C.; Varmus, H.E.; Bishop, J.M. Transduction of a cellular oncogene—the genesis of Rous sarcoma virus. *Proc. Natl. Acad. Sci. USA* 80:2519–2523, 1983.

    Varmus, H. Cellular and viral oncogenes. In The Molecular Basis of Blood Diseases (G. Stamatoyannopoulos, A.W. Nienhuis, P. Leder, P.W. Majerus, eds.), pp. 271–345. Philadelphia: Saunders, 1987.

29. Harris, H. The genetic analysis of malignancy. *J. Cell Sci.*, Suppl. 4:431–444, 1986.

    Klein, G. The approaching era of the tumor suppressor genes. *Science* 238:1539–1545, 1987.

Weinberg, R.A. A molecular basis of cancer. *Sci. Am.* 249(5):126–142, 1983.

30. Doolittle, R.F.; et al. Simian sarcoma virus oncogene, *v-sis*, is derived from the gene (or genes) encoding a platelet-derived growth factor. *Science*, 221:275–277, 1983.

Marshall, C.J. Oncogenes. *J. Cell Sci.*, Suppl. 4:417–430, 1986.

Waterfield, M.D.; et al. Platelet-derived growth factor is structurally related to the putative transforming protein $p28^{sis}$ of simian sarcoma virus. *Nature* 304:35–39, 1983.

31. Almendral, J.M.; et al. Complexity of the early genetic response to growth factors in mouse fibroblasts. *Mol. Cell. Biol.* 8:2140–2148, 1988.

Hunter, T. The proteins of oncogenes. *Sci. Am.* 251(2):70–79, 1984.

Robertson, M. Molecular associations and conceptual connections. *Nature* 334:100–102, 1988.

Yu, C.L.; Tsai, M.H.; Stacey, D.W. Cellular *ras* activity and phospholipid metabolism. *Cell* 52:63–71, 1988.

32. Herman, B.; Pledger, W.J. Platelet-derived growth factor-induced alterations in vinculin and actin distributions in BALB/c-3T3 cells. *J. Cell Biol.* 100:1031–1040, 1985.

Hirst, R.; Horwitz, A.; Buck, C.; Rohrschneider, L. Phosphorylation of the fibronectin receptor complex in cells transformed by oncogenes that encode tyrosine kinases. *Proc. Natl. Acad. Sci. USA* 83:6470–6474, 1986.

Jove, R.; Hanafusa, H. Cell transformation by the viral *src* oncogene. *Annu. Rev. Cell Biol.* 3:31–56, 1987.

33. Burger, M.M. Proteolytic enzymes initiating cell division and escape from contact inhibition of growth. *Nature* 227:170–171, 1970.

Sullivan, L.M.; Quigley, J.P. An anticatalytic monoclonal antibody to avian plasminogen activator: its effect on behavior of RSV-transformed chick fibroblasts. *Cell* 45:905–915, 1986.

34. Adamson, E.D. Oncogenes in development. *Development* 99:449–471, 1987.

Bryant, P.J.; Bryant, S.V.; French, V. Biological regeneration and pattern formation. *Sci. Am.* 237(1):67–81, 1977.

Fankhauser, G. Nucleo-cytoplasmic relations in amphibian development. *Int. Rev. Cytol.* 1:165–193, 1952.

35. Bajer, A.S.; Mole-Bajer, J. Spindle Dynamics and Chromosome Movements. New York: Academic Press, 1972.

McIntosh, J.R. Mechanisms of mitosis. *Trends Biochem. Sci.* 9:195–198, 1984.

Mazia, C. Mitosis and the physiology of cell division. In The Cell (J. Brachet, A.E. Mirsky, eds.), Vol. 3, pp. 77–412. London: Academic Press, 1961.

Wilson, E.B. The Cell in Development and Heredity, 3rd ed. with corrections. New York: Macmillan Company, 1928. (Reprinted, New York: Garland, 1987.)

36. Inoué, S.; Sato, H. Cell motility by labile association of molecules: the nature of mitotic spindle fibers and their role in chromosome movement. *J. Gen. Physiol.* 50:259–292, 1967.

Karsenti, E.; Maro, B. Centrosomes and the spatial distribution of microtubules in animal cells. *Trends Biochem. Sci.* 11:460–463, 1986.

Kirschner, M.; Mitchison, T. Beyond self-assembly: from microtubules to morphogenesis. *Cell* 45:329–342, 1986.

Kuriyama, R.; Borisy, G.G. Microtubule-nucleating activity of centrosomes in Chinese hamster ovary cells is independent of the centriole cycle but coupled to the mitotic cycle. *J. Cell Biol.* 91:822–826, 1981.

Saxton, W.M.; et al. Tubulin dynamics in cultured mammalian cells. *J. Cell Biol.* 99:2175–2186, 1984.

37. Clarke, L.; Carbon, J. The structure and function of yeast centromeres. *Annu. Rev. Genet.* 19:29–56, 1985.

Earnshaw, W.C.; et al. Molecular cloning of cDNA for CENP-B, the major human centromere autoantigen. *J. Cell Biol.* 104:817–829, 1987.

Mitchison, T.J.; Evans, L.; Schulze, E.; Kirschner, M. Sites of microtubule assembly and disassembly in the mitotic spindle. *Cell* 45:515–527, 1986.

Peterson, J.B.; Ris, H. Electron microscopic study of the spindle and chromosome movement in the yeast *Saccharomyces cerevisiae*. *J. Cell Sci.* 22:219–242, 1976.

Rieder, C.L. The formation, structure and composition of the mammalian kinetochore fiber. *Int. Rev. Cytol.* 79:1–58, 1982.

38. Euteneuer, U.; McIntosh, J.R. Structural polarity of kinetochore microtubules in PtK_1 cells. *J. Cell Biol.* 89:338–345, 1981.

Mitchison, T.J.; Kirschner, M.W. Properties of the kinetochore *in vitro*. II: Microtubule capture and ATP-dependent translocation. *J. Cell Biol.* 101:766–777, 1985.

Rieder, C.L.; Davison, E.A.; Jensen, C.W.; Cassimeris, L.; Salmon, E.D. Oscillatory movements monooriented of chromosomes and their position relative to the spindle pole result from the ejection properties of the aster and half-spindle. *J. Cell Biol.* 103:581–591, 1986.

Roos, U.-P. Light and electron microscopy of rat kangaroo cells in mitosis. III. Patterns of chromosome behavior during prometaphase. *Chromosoma* 54:363–385, 1976.

39. Begg, D.A.; Ellis, G.W. Micromanipulation studies of chromosome movement. *J. Cell Biol.* 82:528–541, 1979.

Nicklas, R.B. The forces that move chromosomes in mitosis. *Annu. Rev. Biophys. Biophys. Chem.* 17:431–450, 1988.

Nicklas, R.B.; Kubai, D.F. Microtubules, chromosome movement, and reorientation after chromosomes are detached from the spindle by micromanipulation. *Chromosoma* 92:313–324, 1985.

40. Hays, T.S.; Wise, D.; Salmon, E.D. Traction force on a kinetochore at metaphase acts as a linear function of kinetochore fiber length. *J. Cell Biol.* 93:374–382, 1982.

McNeill, P.A.; Berns, M.W. Chromosome behavior after laser microirradiation of a single kinetochore in mitotic PtK2 cells. *J. Cell Biol.* 88:543–553, 1981.

Ostergren, G. The mechanism of coordination in bivalents and multivalents. The theory of orientation by pulling. *Hereditas* 37:85–156, 1951.

41. Hepler, P.K.; Callaham, D.A. Free calcium increases during anaphase in stamen hair cells of *Tradescantia*. *J. Cell Biol.* 105:2137–2143, 1987.

Murray, A.W.; Szostak, J.W. Chromosome segregation in mitosis and meiosis. *Annu. Rev. Cell Biol.* 1:289–315, 1985.

Wolniak, S.M. The regulation of mitotic spindle function. *Biochem. Cell Biol.* 66:490–514, 1988.

42. Ris, H. The anaphase movement of chromosomes in the spermatocytes of grasshoppers. *Biol. Bull. (Woods Hole)* 96:90–106, 1949.

43. Gorbsky, G.J.; Sammak, P.J.; Borisy, G.G. Microtubule dynamics and chromosome motion visualized in living anaphase cells. *J. Cell Biol.* 106:1185–1192, 1988.

Mitchison, T.J. Microtubule dynamics and kinetochore function in mitosis. *Annu. Rev. Cell Biol.* 4:527–549, 1988.

Nicklas, R.B. Measurements of the force produced by the mitotic spindle in anaphase. *J. Cell Biol.* 97:542–548, 1983.

Spurck, T.P.; Pickett-Heaps, J.D. On the mechanism of anaphase A: Evidence that ATP is needed for microtubule disassembly and not generation of polewards force. *J. Cell Biol.* 105:1691–1705, 1987.

44. Aist, J.R.; Berns, M.W. Mechanics of chromosome separation during mitosis in *Fusarium* (Fungi imperfecti): new evidence from ultrastructural and laser microbeam experiments. *J. Cell Biol.* 91:446–458, 1981.

Masuda, H.; Cande, W.Z. The role of tubulin polymerization during spindle elongation *in vitro*. *Cell* 49:193–202, 1987.

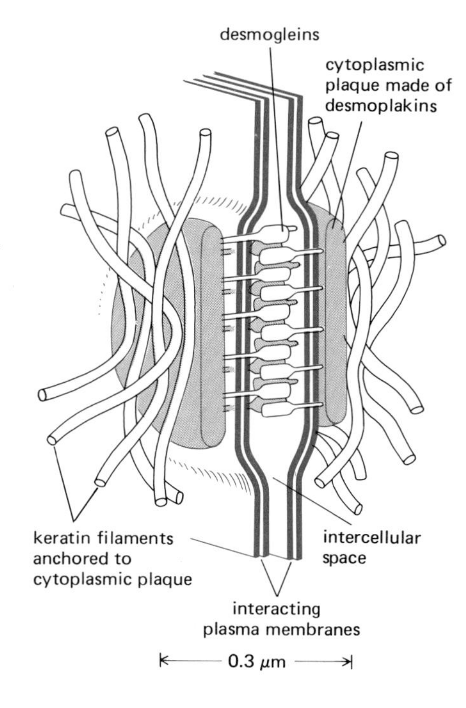

Figure 14–11 A highly schematized drawing of a desmosome. On the cytoplasmic surface of each interacting plasma membrane is a dense plaque composed of a mixture of intracellular attachment proteins called *desmoplakins*. Each plaque is associated with a thick network of keratin filaments, which pass along the surface of the plaque. Transmembrane linker glycoproteins called *desmogleins* bind to the plaques and interact through their extracellular domains to hold the adjacent membranes together by a Ca^{2+}-dependent mechanism. Although desmosomes and adhesion belts are morphologically and chemically distinct, they share at least one intracellular attachment protein (called plakoglobin).

fluids into the loosened epithelium. The antibodies disrupt desmosomes only in skin, suggesting that desmosomes in other tissues may be biochemically different.

Hemidesmosomes, or half-desmosomes, resemble desmosomes morphologically but are both functionally and chemically distinct. Instead of joining adjacent epithelial cell membranes, they connect the basal surface of epithelial cells to the underlying *basal lamina*—a specialized mat of extracellular matrix at the interface between the epithelium and connective tissue (see p. 818). Moreover, whereas the keratin filaments associated with desmosomes make lateral attachments to the desmosomal plaques (see Figure 14–11), many of those associated with hemidesmosomes terminate in the plaques (Figure 14–12).

Together desmosomes and hemidesmosomes act as rivets to distribute tensile or shearing forces through an epithelium and its underlying connective tissue.

14-6 Gap Junctions Allow Small Molecules to Pass Directly from Cell to Cell[5]

Perhaps the most intriguing cell junction is the **gap junction.** It is one of the most widespread, being found in large numbers in most tissues and in practically all animal species. It appears in conventional electron micrographs as a patch where the membranes of two adjacent cells are separated by a uniform narrow gap about 3 nm wide. Gap junctions mediate communication between cells by allowing inorganic ions and other small water-soluble molecules to pass directly from the cytoplasm of one cell to the cytoplasm of the other, thereby coupling the cells both electrically and metabolically. Such *cell coupling* has important functional implications, many of which are only beginning to be understood.

Cell-cell communication of this type was first demonstrated physiologically in 1958, but it took more than 10 years to show that this physiological coupling correlates with the presence of gap junctions seen in the electron microscope. The initial evidence for cell coupling came from electrophysiological studies of specific pairs of interacting nerve cells in the nerve cord of a crayfish. When a voltage gradient was applied across the junctional membrane after inserting an electrode into each of the two interacting cells, an unexpectedly large current flowed, indicating that inorganic ions (which carry current in living tissues) could pass freely from one cell interior to the other. Later experiments showed that small fluorescent dye molecules injected into one cell can likewise pass readily into adjacent cells without leaking into the extracellular space, provided that the molecules are no bigger than 1000 to 1500 daltons. This suggests a functional pore size for the connecting channels of about 1.5 nm (Figure 14–13), implying that coupled cells share their small molecules (such as inorganic ions, sugars, amino acids, nucleotides, and vitamins) but not their macromolecules (proteins, nucleic acids, and polysaccharides).

This sharing of small intracellular metabolites between cells is the basis of *metabolic cooperation,* which can be demonstrated in cells in culture. Mutant cell lines that lack the enzyme thymidine kinase, for example, can be cultured together with normal (wild-type) cells, which have thymidine kinase. The mutant cells on their own are unable to incorporate thymidine into their DNA because they cannot

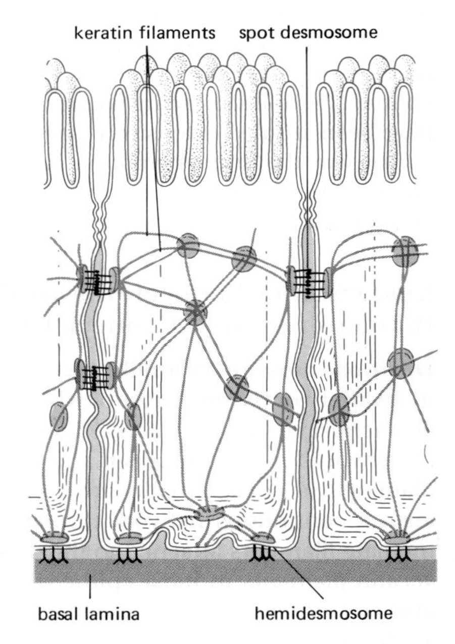

Figure 14–12 The distribution of desmosomes and hemidesmosomes in epithelial cells of the small intestine. The keratin filament networks of adjacent cells are indirectly connected to one another through desmosomes and to the basal lamina through hemidesmosomes. Whereas the keratin filaments make lateral attachments to the surface of the dense plaques associated with desmosomes, they tend to terminate in hemidesmosomes.

perform the initial step of converting the thymidine to thymidine triphosphate. But when these cells are co-cultured with wild-type cells and then exposed to radioactive thymidine, radioactivity is incorporated into the DNA of those mutant cells that are in direct contact with wild-type cells. This observation implies that a DNA precursor containing the radioactive thymidine—thymidine triphosphate, in fact—is passed directly from the wild-type cells into mutant cells in contact with them (Figure 14–14). Such metabolic cooperation does not occur when this type of experiment is performed with cells that cannot form gap junctions.

The evidence that gap junctions mediate electrical and chemical coupling between cells in contact with each other comes from several sources. Gap-junction structures can almost always be found where coupling can be demonstrated by electrical or chemical criteria. Conversely, coupling has not been demonstrated between vertebrate cells where there are no gap junctions. Moreover, dye and electrical coupling can be blocked if antibodies directed against the major gap-junction protein (see below) are microinjected into cells connected by gap junctions. Finally, when this gap-junction protein is reconstituted into synthetic lipid bilayers, or when mRNA encoding the protein is injected into frog oocytes, channels with many of the properties expected of gap-junction channels can be demonstrated electrophysiologically.

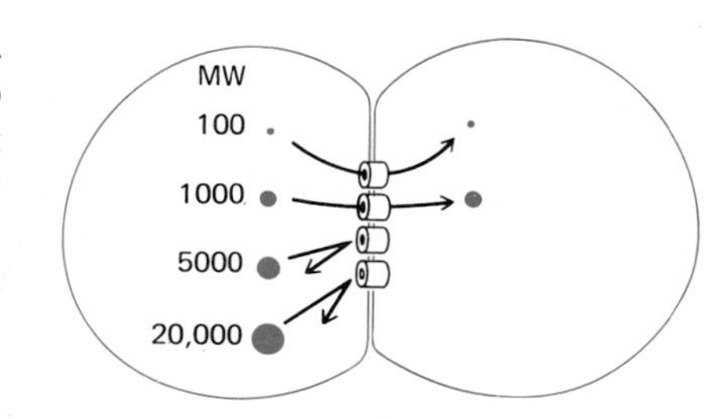

Figure 14–13 When fluorescent molecules of various sizes are injected into one of two cells coupled by gap junctions, molecules smaller than about 1000 to 1500 daltons (depending on the species and cell type) can pass into the other cell, but larger molecules cannot. This suggests that the functional diameter of the channel connecting the two cells is about 1.5 nm.

Gap-Junction Connexons Are Oligomers of a Multipass Transmembrane Protein[6]

Gap junctions are constructed from transmembrane proteins that form structures called *connexons.* When the connexons in the plasma membranes of two cells in contact are aligned, they form a continuous aqueous channel, which connects the two cell interiors (Figure 14–15). The connexons join in such a way that the interacting plasma membranes are separated by an interrupted gap—hence the term "gap junction"—emphasizing the contrast with a tight junction, where the membranes are more closely juxtaposed (compare Figures 14–5 and 14–15). Each connexon is seen as an intramembrane particle in freeze-fracture electron micrographs, and each gap junction can contain up to several hundred clustered connexons (Figure 14–16).

An unusual resistance to proteolytic enzymes and detergents has made it possible to isolate gap junctions from rodent liver (Figure 14–17). The junctions contain a single major protein of about 30,000 daltons. DNA sequencing studies suggest that the polypeptide chain (about 280 amino acid residues) crosses the lipid bilayer as four α helices. Six such protein molecules are thought to associate to form each connexon in a way similar to that postulated for the acetylcholine receptor channel, creating an aqueous pore lined by one α helix from each protein subunit (see Figure 6–64, p. 321).

Antibodies against the 30,000 dalton protein react with gap junctions in many tissues and species, suggesting that the connexon proteins in these tissues and organisms are similar (although biochemical and physiological analyses indicate that they are not identical). This accords with the finding that different cell types in culture will usually form gap junctions with each other, even across species.

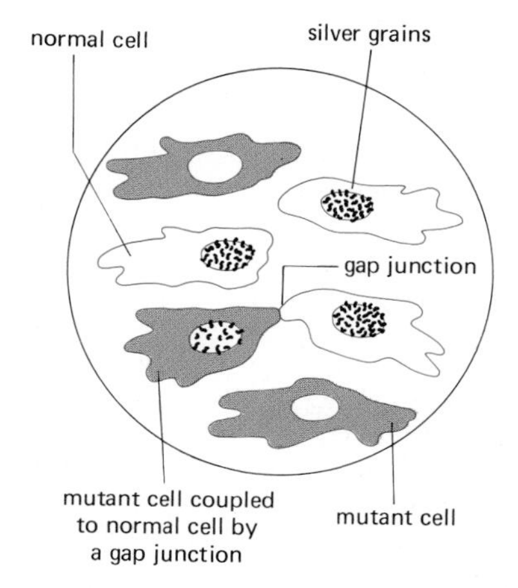

Figure 14–14 Schematic drawing of an autoradiograph demonstrating metabolic cooperation between cells in culture connected by gap junctions. The mutant cells lack the enzyme thymidine kinase and therefore cannot incorporate radioactive thymidine into DNA when thymidine is added to the medium. Normal cells can incorporate the thymidine into their DNA, and their nuclei are therefore stippled with black dots, representing developed silver grains in the autoradiograph. In mixed cultures of normal and mutant cells, if a mutant cell makes contact and forms gap junctions with a normal cell, its nucleus is also radiolabeled, as shown. This labeling occurs because the radioactive thymidine is phosphorylated by thymidine kinase to form thymidine triphosphate in the normal cell; the radioactive thymidine triphosphate then passes through the gap junctions into the mutant cell, where it is incorporated into DNA.

14-7 Most Cells in Early Embryos Are Coupled via Gap Junctions[7]

In some tissues, cell coupling via gap junctions serves an obvious function. For example, electrical coupling synchronizes the contractions of heart muscle cells and of smooth muscle cells responsible for the peristaltic movements of the intestine. Similarly, electrical coupling between nerve cells allows action potentials to spread rapidly from cell to cell without the delay that occurs at chemical synapses; this is advantageous where speed and reliability are crucial, as in certain escape responses in fish and insects. It is less obvious why gap junctions occur in tissues that are not electrically active. In principle, the sharing of small metabolites and ions provides a mechanism for coordinating the activities of individual cells in such tissues. For example, the activities of cells in an epithelial cell sheet, such as the beating of cilia, might be coordinated via gap junctions; and since intracellular mediators such as cyclic AMP can pass through gap junctions, responses of coupled cells to extracellular signaling molecules may be propagated and coordinated in this way.

Cell coupling via gap junctions appears to be important in embryogenesis. In early vertebrate embryos (beginning with the late eight-cell stage in mouse embryos), most cells are electrically coupled to one another. As specific groups of cells in the embryo develop their distinct identities and begin to differentiate, however, they commonly uncouple from surrounding tissue. As the neural tube closes, for instance, its cells uncouple from the overlying ectoderm (see Figure 14–9). Meanwhile the cells within each group remain coupled with one another and so tend to behave as a cooperative assembly, all following a similar developmental pathway in a coordinated fashion.

An attractive hypothesis is that the coupling of cells in embryos might provide a pathway for long-range cell signaling within a developing epithelium. For example, a small molecule could pass through gap junctions from a region of the tissue where its intracellular concentration is kept high to a region where it is kept low, thereby setting up a smooth concentration gradient. The local concentration could provide cells with "positional information" to control their differentiation according to their location in the embryo. Whether gap junctions actually function in this way is not known.

The importance of gap-junction-mediated communication in development is suggested by an experiment in which antibodies against the major gap-junction protein were microinjected into one cell of an eight-cell amphibian embryo; the injected antibodies not only selectively interrupted electrical coupling and dye transfer between progeny of the injected cell (assayed two cell cycles later in 32-cell embryos), they also grossly disrupted the development of the embryo (Figure 14–18). It is not clear how the inhibition of cell coupling at an early stage caused the later developmental defects seen in the injected embryos, but experiments of this type offer a promising beginning to an analysis of the role of gap junctions in embryonic development.

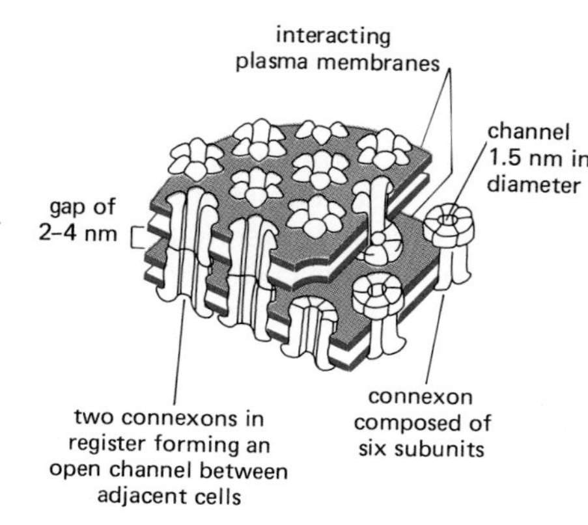

Figure 14–15 A model of a gap junction based on biochemical, electron microscopic, and x-ray diffraction observations. The drawing shows the interacting plasma membranes of two adjacent cells. The apposed lipid bilayers are penetrated by protein assemblies called *connexons*, each of which is thought to be formed by six identical protein subunits. Two connexons join across the intercellular gap to form a continuous aqueous channel connecting the two cells.

The Permeability of Gap Junctions Is Regulated[8]

The permeability of gap junctions is rapidly (within seconds) and reversibly decreased by experimental manipulations that decrease cytosolic pH or increase the cytosolic concentration of free Ca^{2+}, and in some tissues the permeability can be regulated by the voltage gradient across the junction or by extracellular chemical

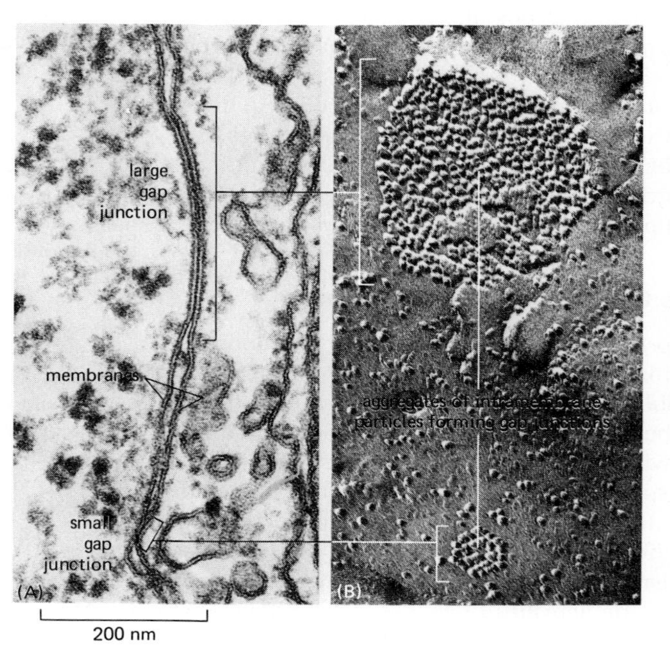

Figure 14–16 Thin-section (A) and freeze-fracture (B) electron micrographs of a large and a small gap junction between fibroblasts in culture. In (B) each gap junction is seen as a cluster of homogeneous intramembrane particles associated exclusively with the cytoplasmic fracture face (P face) of the plasma membrane. Each intramembrane particle corresponds to a connexon, illustrated in Figure 14–15. (From N.B. Gilula, in Cell Communication [R.P. Cox, ed.], pp. 1–29. New York: Wiley, 1974. Reprinted by permission of John Wiley & Sons, Inc.)

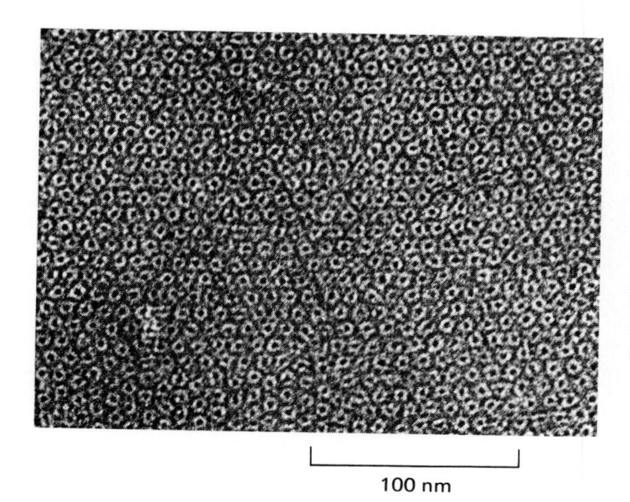

Figure 14–17 Electron micrograph of an area of an isolated gap junction from rat liver. The preparation has been negatively stained to show the connexons, which are organized as a hexagonal lattice. The densely stained central hole in each connexon has a diameter of about 2 nm. (From N.B. Gilula, in Intercellular Junctions and Synapses [Receptors and Recognition Series B, Vol. 2; J. Feldman, N.B. Gilula, and J.D. Pitts, eds.], pp. 3–22. London: Chapman & Hall, 1978.)

signals. These observations indicate that gap junctions are dynamic structures that can open and close in response to changes in the cell. Thus, like conventional ion channels, they are *gated* (see p. 313), although transitions between open and closed states occur much less frequently in a gap-junction channel than in most conventional ion channels.

The importance of voltage and pH regulation of gap-junction permeability to the normal function of cell assemblies is unknown. There is one case, however, where the reason for the Ca^{2+} control seems clear. When a cell dies or is damaged, its membrane becomes leaky. Ions such as Ca^{2+} and Na^{+} move into the cell, and valuable metabolites leak out. If the cell were to remain coupled to its healthy neighbors, these too would suffer a dangerous disturbance of their internal chemistry. But the influx of Ca^{2+} into the sick cell, by closing the gap junction channels, effectively isolates it and prevents damage from spreading in this way.

Where gap-junction permeability is increased by extracellular chemical signals, the effect is to spread the response to neighboring cells that are not in direct contact with the signal. The hormone *glucagon*, for example, which stimulates liver cells to break down glycogen and release glucose into the blood, can also be shown to increase gap-junction permeability in rat liver cells. It does so by increasing the concentration of intracellular cyclic AMP, which activates cyclic AMP-dependent protein kinase (see p. 709), which in turn is thought to phosphorylate the major gap-junction protein. The breakdown of glycogen by liver cells is also mediated by an increase in cyclic AMP, so the simultaneous increase in gap-junction permeability, by facilitating the diffusion of cyclic AMP from cell to cell, tends to spread the glycogen-breakdown response through neighboring groups of liver cells.

Figure 14–19 summarizes the various types of junctions formed between cells in an epithelium. In the most apical portion of the cell, the relative positions of the junctions are the same in nearly all epithelia: the tight junction occupies the most apical portion of the cell, followed by the adhesion belt and then by a special parallel row of desmosomes; together these form a "junctional complex." Gap junctions and additional desmosomes are less regularly organized.

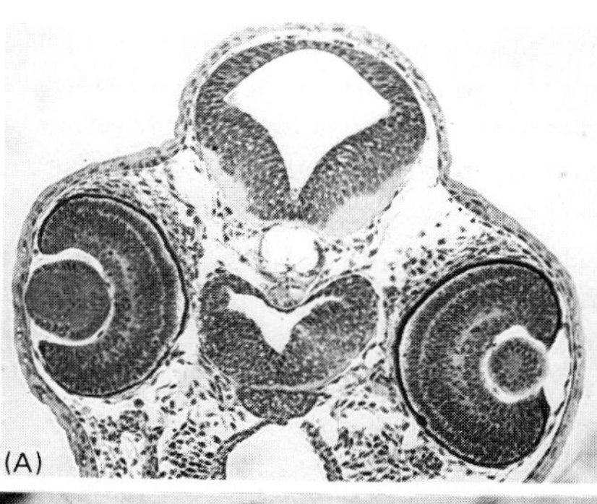

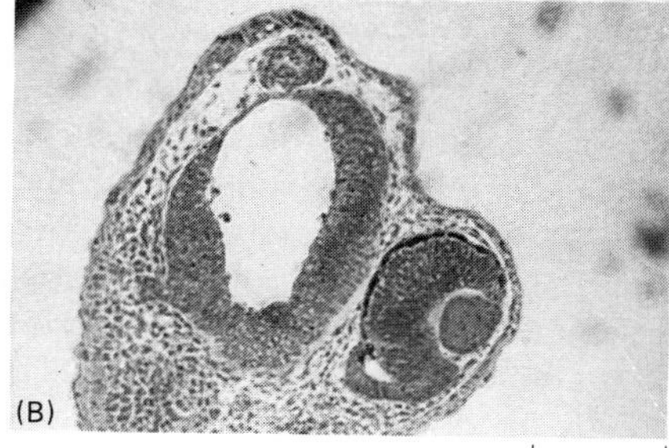

Figure 14–18 Effect of antibodies against the major gap-junction protein microinjected into one cell of an early *Xenopus* embryo. The micrographs are of transverse sections through a normal embryo (A) and through an embryo that was injected at the eight-cell stage (B). Note that the eye is missing and the brain is underdeveloped on the injected side of the embryo shown in (B). (From A. Warner, S. Guthrie, and N.B. Gilula, *Nature* 331:126–131, 1985. Copyright © 1985 Macmillan Journals Limited.)

Summary

Many cells in tissues are linked to each other and to the extracellular matrix at specialized contact sites called cell junctions. Cell junctions fall into three functional classes: occluding junctions, anchoring junctions, and communicating junctions. Tight junctions *are the main occluding junctions, and they play a critical part in maintaining the concentration differences of small hydrophilic molecules across epithelial cell sheets by (1) sealing the plasma membranes of adjacent cells together to create a continuous permeability barrier across the cell sheet and (2) acting as barriers in the lipid bilayer to restrict the diffusion of membrane transport proteins*

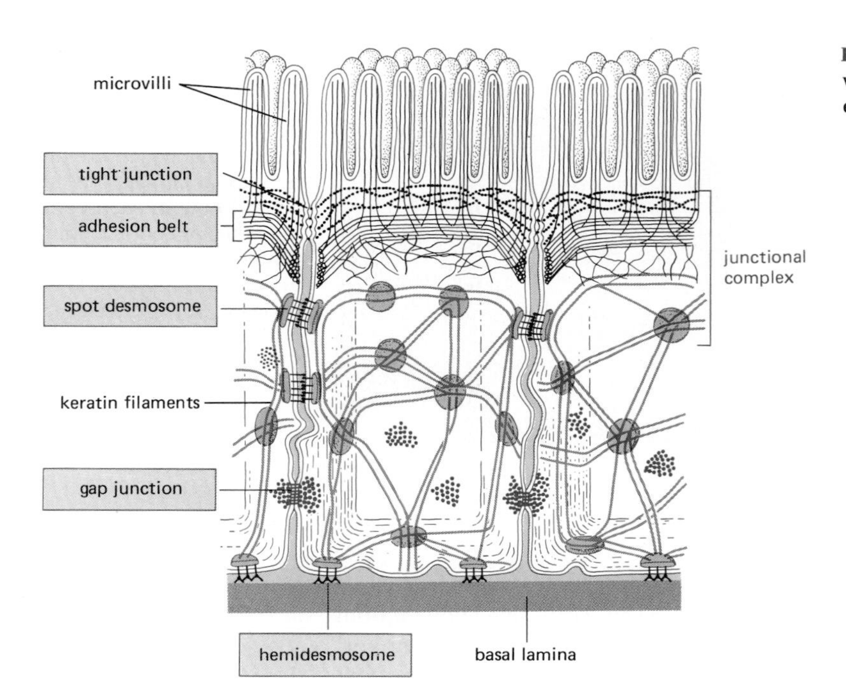

Figure 14–19 Distribution of the various cell junctions formed by epithelial cells of the small intestine.

between separate apical and basolateral domains of the plasma membrane in each epithelial cell.

There are two main types of anchoring junctions: adherens junctions *and* desmosomes. *Both join groups of cells together into strong structural units by connecting elements of their cytoskeletons; adherens junctions connect bundles of actin filaments, whereas desmosomes connect intermediate filaments.* Gap junctions *are communicating junctions composed of clusters of channel proteins that allow molecules of less than 1500 daltons to pass directly from the inside of one cell to the inside of the other. Cells connected by such junctions share many of their inorganic ions and other small molecules and are said to be chemically and electrically coupled. Gap junctions are important in coordinating the activities of electrically active cells, and they are thought to play a similar role in other groups of cells as well.*

The Extracellular Matrix[9]

Tissues are not composed solely of cells. A substantial part of their volume is *extracellular space,* which is largely filled by an intricate network of macromolecules constituting the **extracellular matrix** (Figure 14–20). This matrix comprises a variety of versatile polysaccharides and proteins that are secreted locally and assemble into an organized meshwork. Whereas we discussed cell junctions chiefly in the context of epithelial tissues, our account of extracellular matrix will focus chiefly on **connective tissues** (Figure 14–21). In these tissues the matrix is generally more plentiful than the cells and surrounds the cells on all sides, determining the tissue's physical properties. Connective tissues form the architectural framework of the vertebrate body, and the amounts found in different organs vary greatly: from skin and bone, in which they are the major component, to brain and spinal cord, in which they are only minor constituents.

Variations in the relative amounts of the different types of matrix macromolecules and the way they are organized in the extracellular matrix give rise to an amazing diversity of forms, each highly adapted to the functional requirements of the particular tissue. The matrix can become calcified to form the rock-hard structures of bone or teeth, or it can form the transparent matrix of the cornea, or it can adopt the ropelike organization that gives tendons their enormous tensile strength. At the interface between an epithelium and connective tissue, the matrix forms a *basal lamina,* an extremely thin but tough mat that plays an important

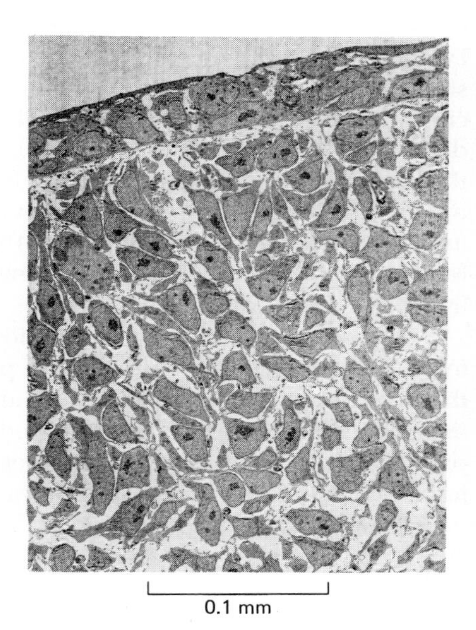

Figure 14–20 Low-power electron micrograph showing cells surrounded by spaces filled with extracelluiar matrix. The particular cells shown are those in an early chick limb. The cells have not yet acquired their specialized characteristics. (Courtesy of Cheryll Tickle.)

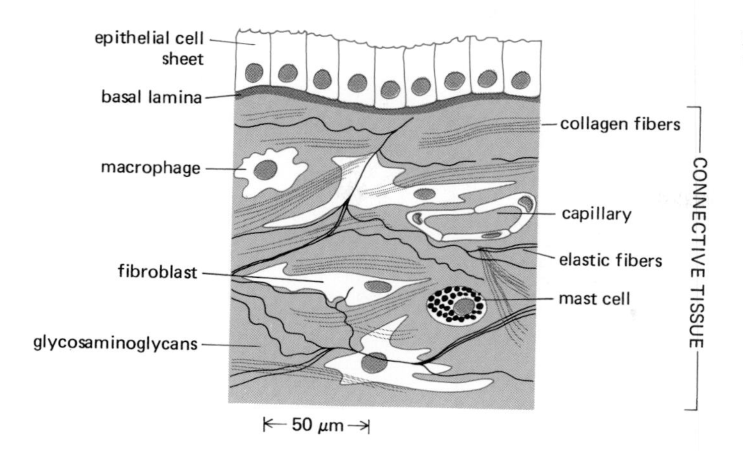

Figure 14–21 The connective tissue underlying an epithelial cell sheet.

part in controlling cell behavior. We shall confine our account to the extracellular matrix of vertebrates, but unique and interesting related structures are seen in many other organisms, such as the cell walls of bacteria and plants, the cuticles of worms and insects, and the shells of mollusks. Plant cell walls are considered in detail in Chapter 20.

Until recently the vertebrate extracellular matrix was thought to serve mainly as a relatively inert scaffolding to stabilize the physical structure of tissues. But now it is clear that the matrix plays a far more active and complex role in regulating the behavior of the cells that contact it—influencing their development, migration, proliferation, shape, and metabolic functions. The extracellular matrix has a correspondingly complex molecular composition; although our understanding of its organization is still fragmentary, there has been rapid progress in characterizing some of its major components.

The Extracellular Matrix Consists Primarily of Fibrous Proteins Embedded in a Hydrated Polysaccharide Gel

The macromolecules that constitute the extracellular matrix are mainly secreted locally by cells in the matrix. In most connective tissues these macromolecules are secreted largely by *fibroblasts* (Figure 14–22). In some specialized connective tissues, however, such as cartilage and bone, they are secreted by cells of the fibroblast family that have more specific names: chondroblasts, for example, form cartilage, and osteoblasts form bone. The two main classes of extracellular macromolecules that make up the matrix are (1) polysaccharide *glycosaminoglycans (GAGs)*, which are usually found covalently linked to protein in the form of *proteoglycans*, and (2) fibrous proteins of two functional types: mainly structural (for example, *collagen* and *elastin*) and mainly adhesive (for example, *fibronectin* and *laminin*). The glycosaminoglycan and proteoglycan molecules form a highly hydrated, gel-like "ground substance" in which the fibrous proteins are embedded. The aqueous phase of the polysaccharide gel permits the diffusion of nutrients, metabolites, and hormones between the blood and the tissue cells; the collagen fibers strengthen and help to organize the matrix, and rubberlike elastin fibers give it resilience. The adhesive proteins help cells attach to the extracellular matrix: fibronectin promotes the attachment of fibroblasts and related cells to the matrix in connective tissues, while laminin promotes the attachment of epithelial cells to the basal lamina.

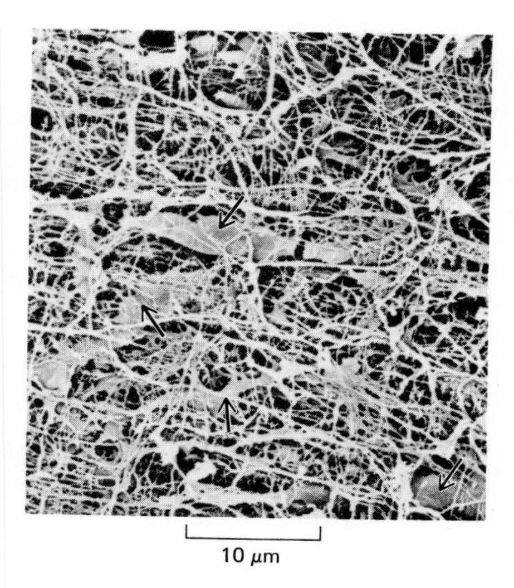

Figure 14–22 Scanning electron micrograph of fibroblasts (*arrows*) in the connective tissue of the cornea in a chick embryo. The extracellular matrix surrounding the fibroblasts is composed largely of collagen fibers (there are no elastic fibers in the cornea). The glycosaminoglycans, which normally form a hydrated gel filling the interstices of the fibrous network, have collapsed onto the surface of the collagen fibers during the dehydration process necessary for specimen preparation. (Courtesy of Robert Trelstad.)

Glycosaminoglycan Chains Occupy Large Amounts of Space and Form Hydrated Gels[10]

Glycosaminoglycans (GAGs) are long, unbranched polysaccharide chains composed of repeating disaccharide units. They are called glycosaminoglycans because one of the two sugar residues in the repeating disaccharide is always an amino sugar (*N*-acetylglucosamine or *N*-acetylgalactosamine). In most cases this

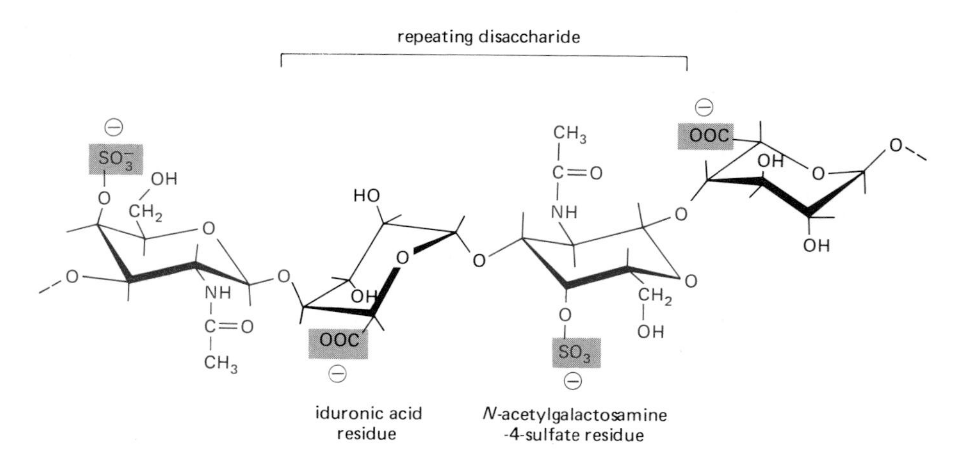

Figure 14–23 Glycosaminoglycans are long linear polymers composed of a repeating disaccharide sequence. A small part of a dermatan sulfate chain is shown here; these chains are typically 70 to 200 sugar residues long. There is a high density of negative charges along the chain resulting from the presence of both carboxyl and sulfate groups.

amino sugar is sulfated and the second sugar is a uronic acid. Because of the sulfate or carboxyl groups on most of their sugar residues, glycosaminoglycans are highly negatively charged (Figure 14–23). Four main groups of glycosaminoglycans have been distinguished by their sugar residues, the type of linkage between these residues, and the number and location of sulfate groups: (1) *hyaluronic acid,* (2) *chondroitin sulfate* and *dermatan sulfate,* (3) *heparan sulfate* and *heparin,* and (4) *keratan sulfate* (Table 14–2).

Polysaccharide chains are too inflexible to fold up into the compact globular structures that polypeptide chains typically form. Moreover, they are strongly hydrophilic. Thus glycosaminoglycans tend to adopt highly extended, so-called random-coil conformations, which occupy a huge volume relative to their mass (Figure 14–24), and they form gels even at very low concentrations. Their high density of negative charges attracts a cloud of cations, such as Na^+, that are osmotically active, causing large amounts of water to be sucked into the matrix. This creates a swelling pressure, or turgor, that enables the matrix to withstand compressive forces (in contrast to collagen fibrils, which resist stretching forces). Cartilage matrix, for example, resists compression by this mechanism.

The amount of glycosaminoglycan in connective tissue is usually less than 10% by weight of the amount of the fibrous proteins. Because they form porous hydrated gels, however, the glycosaminoglycan chains fill most of the extracellular space, providing mechanical support to tissues while still allowing the rapid diffusion of water-soluble molecules and the migration of cells.

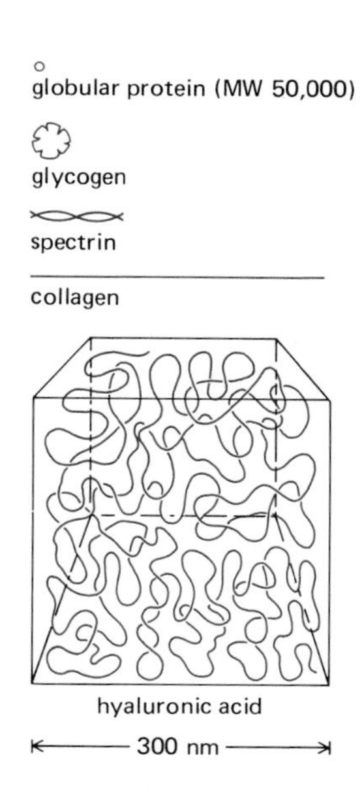

Figure 14–24 The relative volumes occupied by various proteins, a glycogen granule, and a single hydrated molecule of hyaluronic acid of about 8×10^6 daltons.

Hyaluronic Acid Is Thought to Facilitate Cell Migration During Tissue Morphogenesis and Repair[11]

Hyaluronic acid (also called hyaluronate or hyaluronan), which can contain up to several thousand sugar residues, is a relatively simple molecule consisting of a regular repeating sequence of nonsulfated disaccharide units (Figure 14–25). It is

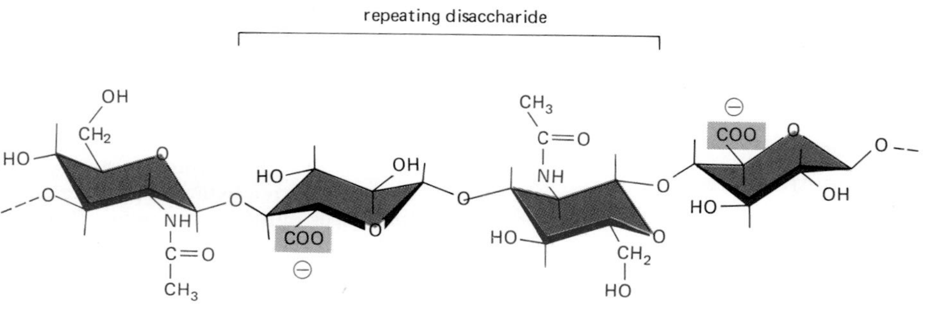

Figure 14–25 The repeating disaccharide sequence in hyaluronic acid, a relatively simple glycosaminoglycan that consists of a single long chain of up to several thousand sugar residues. Note the absence of sulfate groups.

Table 14–2 The Glycosaminoglycans

Group	Glycosaminoglycan	Molecular Weight	Repeating Disaccharide (A-B)$_n$ Monosaccharide A	Monosaccharide B	Sulfates per Disaccharide Unit	Linked to Protein	Other Sugar Components	Tissue Distribution
1	**Hyaluronic acid**	4,000 to 8×10^6	D-glucuronic acid	*N*-acetyl-D-glucosamine	0	–	0	various connective tissues, skin, vitreous body, cartilage, synovial fluid
2	**Chondroitin sulfate**	5,000–50,000	D-glucuronic acid	*N*-acetyl-D-galactosamine	0.2–2.3	+	D-galactose D-xylose	cartilage, cornea, bone, skin, arteries
	Dermatan sulfate	15,000–40,000	D-glucuronic acid or *L-iduronic acid	*N*-acetyl-D-galactosamine	1.0–2.0	+	D-galactose D-xylose	skin, blood vessels, heart, heart valves
3	**Heparan sulfate**	5,000–12,000	D-glucuronic acid or *L-iduronic acid	*N*-acetyl-D-glucosamine	0.2–2.0	+	D-galactose D-xylose	lung, arteries, cell surfaces, basal laminae
	Heparin	6,000–25,000	D-glucuronic acid or *L-iduronic acid	*N*-acetyl-D-glucosamine	2.0–3.0	+	D-galactose D-xylose	lung, liver, skin, mast cells
4	**Keratan sulfate**	4,000–19,000	D-galactose	*N*-acetyl-D-glucosamine	0.9–1.8	+	D-galactosamine D-mannose L-fucose, sialic acid	cartilage, cornea, intervertebral disc

*L-iduronic acid is produced by the epimerization of D-glucuronic acid at the position where the carboxyl group is located. Thus dermatan sulfate is a modified form of chondroitin sulfate, and the two types of repeating disaccharide sequences usually occur as alternating segments in the same glycosaminoglycan chain.

found in variable amounts in all tissues and fluids in adult animals and is especially abundant in early embryos. Because of its simplicity, hyaluronic acid is thought to represent the earliest evolutionary form of glycosaminoglycan, but it is not typical of the majority of glycosaminoglycans. All of the others (1) contain sulfated sugars, (2) tend to contain a number of different disaccharide units arranged in more complex sequences, (3) have much shorter chains, consisting of fewer than 300 sugar residues, and (4) are covalently linked to protein.

There is increasing evidence that hyaluronic acid has a special function in tissues where cells are migrating—such as during development or wound repair. It is produced in large amounts during periods of cell migration; and when cell migration ends, the excess hyaluronic acid is degraded by the enzyme *hyaluronidase.* This sequence of events has been demonstrated in a wide variety of tissues, suggesting that increased local production of hyaluronic acid, which attracts water and thereby swells the matrix, may be a general strategy for facilitating cell migration during morphogenesis and repair. Hyaluronic acid is also an important constituent of joint fluid, where it serves as a lubricant.

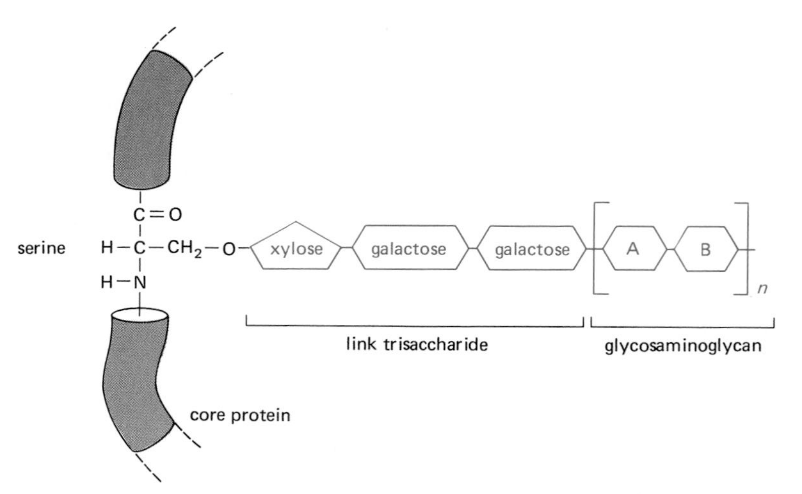

Figure 14–26 The linkage between a glycosaminoglycan chain and a serine residue of a core protein in a proteoglycan molecule. A specific "link trisaccharide" is first added to the serine, which is often located within the sequence Asp(or Glu)-Asp(or Glu)-X-*Ser*-Gly-X-Gly, where X is any amino acid. The rest of the glycosaminoglycan chain, consisting mainly of a repeating disaccharide unit (composed of the two monosaccharides A and B in Table 14–2), is then synthesized, one sugar residue being added at a time.

14-11 Proteoglycans Are Composed of Long Glycosaminoglycan Chains Covalently Linked to a Core Protein[12]

Except for hyaluronic acid, all glycosaminoglycans are found covalently attached to protein in the form of **proteoglycans.** As is the case for a glycoprotein (see p. 440), the polypeptide chain, or *core protein,* of a proteoglycan is made on membrane-bound ribosomes and threaded into the lumen of the endoplasmic reticulum. The polysaccharide chains are assembled on the core protein mainly in the Golgi apparatus: first a special *link trisaccharide* is attached to a serine residue on the core protein to serve as a primer for polysaccharide growth; then one sugar residue is added at a time by specific glycosyl transferases (Figure 14–26). While chain elongation is proceeding in the Golgi apparatus, many of the polymerized sugar residues are covalently modified by a sequential and coordinated series of sulfation reactions (see p. 456) and epimerization reactions that alter the configuration of the substituents around individual carbon atoms in the sugar molecule. The sulfation greatly increases the negative charge of proteoglycans.

Proteoglycans are usually easily distinguished from glycoproteins by the nature, quantity, and arrangement of their sugar side chains. Glycoproteins usually contain from 1% to 60% carbohydrate by weight in the form of numerous, relatively short, branched, *O*- and *N*-linked oligosaccharide chains, generally of fewer than 15 sugar residues and variable composition, which often terminate with sialic acid (see p. 453). Although the core protein in a proteoglycan can itself be a glycoprotein, proteoglycans can contain as much as 95% carbohydrate by weight, most of which takes the form of one to several hundred unbranched glycosaminoglycan chains, each typically about 80 sugar residues long and usually without sialic acid. Moreover, whereas glycoproteins are rarely larger than 3×10^5 daltons, proteoglycans can be much larger. For example, one of the best-characterized proteoglycan molecules is a major component of cartilage; it typically consists of about 100 chondroitin sulfate chains and about 50 keratan sulfate chains linked to a serine-rich core protein of more than 2000 amino acids. Thus its total mass is about 3×10^6 daltons, with approximately 1 glycosaminoglycan chain for every 20 amino acid residues (Figure 14–27). On the other hand, many proteoglycans are much smaller and have only 1 to 10 glycosaminoglycan chains.

In principle, proteoglycans have the potential for almost limitless heterogeneity. They can differ markedly in protein content, molecular size, and the number and types of glycosaminoglycan chains per molecule. Moreover, although there is always an underlying repeating pattern of disaccharides, the length and composition of the glycosaminoglycan chains can vary greatly, as can the spatial arrangement of hydroxyl, sulfate, and carboxyl side groups along the chains. It is thus an extremely complex problem to identify and classify proteoglycans in terms of their sugars. Many core proteins are now being sequenced with the aid of recombinant DNA techniques, and in the future it is likely that the classification of proteoglycans can be made more meaningful by characterizing them according to their core proteins rather than their glycosaminoglycan chains.

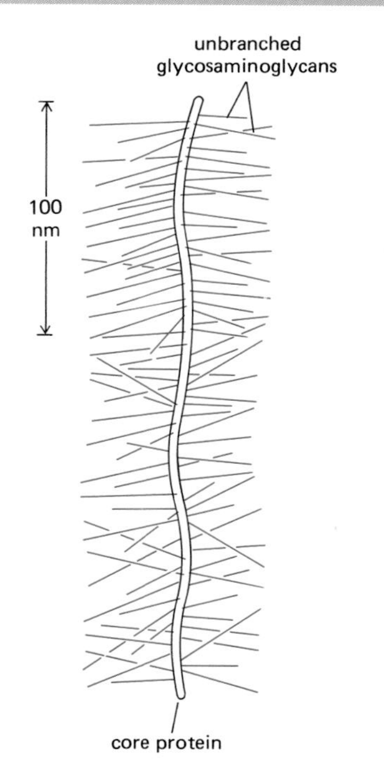

Figure 14–27 The main proteoglycan molecule in cartilage. It consists of many glycosaminoglycan chains covalently linked to a protein core. The core protein contains a number of *N*-linked and *O*-linked oligosaccharide chains (not shown) in addition to the glycosaminoglycan chains. Most proteoglycans are smaller than the one shown, and their glycosaminoglycan chains are often restricted to particular regions of the core polypeptide chain. The lower drawing shows a typical glycoprotein molecule (pancreatic ribonuclease B), drawn to scale for comparison.

Glycosaminoglycan Chains May Be Highly Organized in the Extracellular Matrix[13]

Given the structural heterogeneity of proteoglycan molecules, it seems highly unlikely that their function is limited to providing hydrated space around and between cells. Proteoglycans have been shown to bind various secreted signaling molecules in a test tube, and it seems likely that they do so in tissues, thereby localizing the action of the signaling ligand: fibroblast growth factor (FGF—see p. 747), for example, binds to heparan sulfate proteoglycans both *in vitro* and in tissues. Proteoglycans may form gels of varying pore size and charge density, thus functioning as sieves to regulate the traffic of molecules and cells according to their size and/or charge. There is evidence that proteoglycans function in this capacity in the basal lamina of the kidney glomerulus, which filters molecules passing into the urine from the bloodstream (see p. 820).

The organization of glycosaminoglycans and proteoglycans in the extracellular matrix is poorly understood. Biochemical studies indicate that these molecules bind to each other in specific ways, as well as to the fibrous proteins in the matrix. It would be surprising if such interactions are not important in organizing the matrix. The major keratan sulfate/chondroitin sulfate proteoglycan in cartilage discussed above has been shown to assemble in the extracellular space into large aggregates that are noncovalently bound through their core proteins to a large hyaluronic acid molecule. As many as 100 proteoglycan monomers are bound to a single hyaluronic acid chain, producing a giant complex with a molecular weight of 100 million or more and occupying a volume equivalent to that of a bacterium. When isolated from tissues, these complexes are readily seen in the electron microscope (Figure 14–28).

By contrast, attempts to determine the arrangement of proteoglycan molecules by electron microscopy while they are still in the tissues have been frustrating. Since they are highly water-soluble, they are readily washed out of the extracellular matrix when tissue sections are exposed to aqueous solutions during fixation. Recently proteoglycans have been seen in near-native state in cartilage that has been rapidly frozen at very low temperature (−196°C) under high pressure and then fixed and stained while still frozen (Figure 14–29). An alternative approach is to use a cationic dye with a relatively low charge density together with

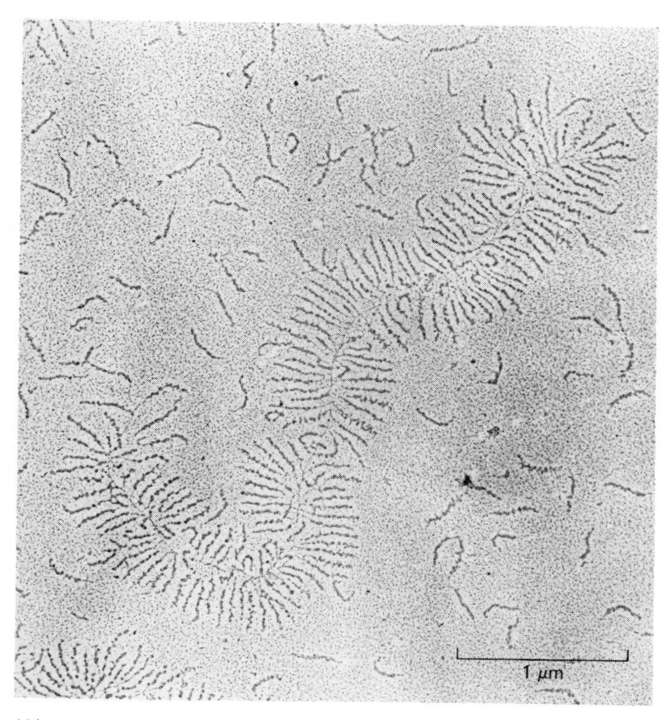

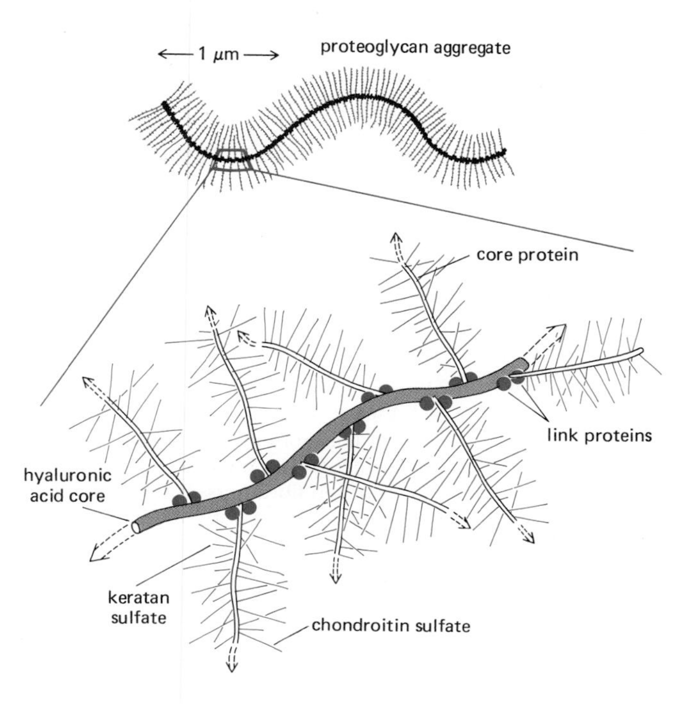

Figure 14–28 (A) Electron micrograph of a proteoglycan aggregate from fetal bovine cartilage shadowed with platinum. Many free proteoglycan molecules are also seen. (B) Schematic drawing of the giant proteoglycan aggregate shown in (A). It consists of about 100 proteoglycan monomers (each like that shown in Figure 14–27) noncovalently bound to a single hyaluronic acid chain through two link proteins that bind to both the core protein of the proteoglycan and to the hyaluronic acid chain, thereby stabilizing the aggregate. The molecular weight of such a complex can be 10^8 or more, and it occupies a volume equivalent to that of a bacterium, which is about 2×10^{-12} cm^3. (A, courtesy of Lawrence Rosenberg.)

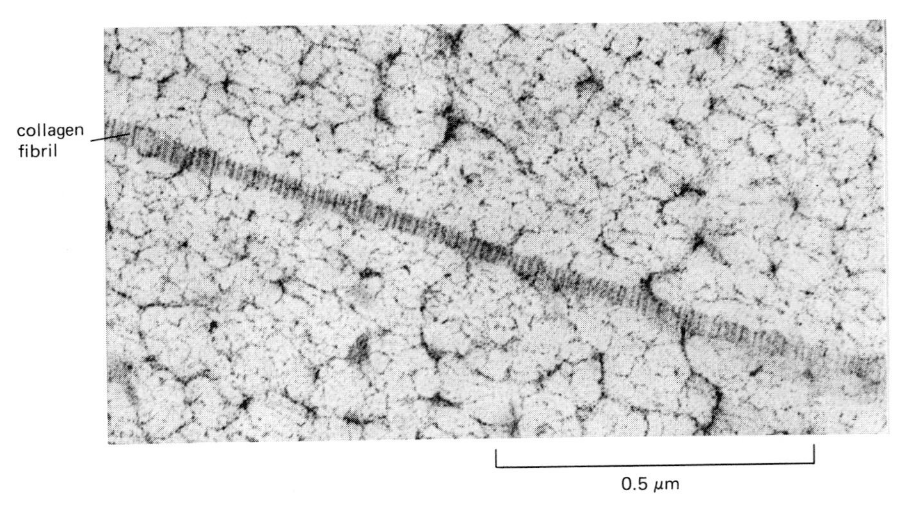

Figure 14–29 Electron micrograph of proteoglycans in the extracellular matrix of rat cartilage. The tissue was rapidly frozen at −196°C and fixed and stained while still frozen (a process called *freeze substitution*) to prevent the glycosaminoglycan chains from collapsing. The proteoglycan molecules are seen to form a fine filamentous network in which a single striated collagen fibril is embedded. The more darkly stained parts of the proteoglycan molecules are the core proteins; the faintly stained threads are the glycosaminoglycan chains. (Reproduced from E.B. Hunziker and R.K. Schenk, *J. Cell Biol.* 98:277–282, 1985 by copyright permission of the Rockefeller University Press.)

more conventional fixation. When the proteoglycans in rat tail tendon are stained with such a dye, they are seen as threadlike structures encircling collagen fibrils at regular intervals (Figure 14–30). They cross the collagen fibrils at intervals of about 65 nm, reflecting the staggered arrangement of collagen molecules in the collagen fibrils (see p. 812). Such ordered patterns of molecules are likely to be widespread in the extracellular matrix, and, given the diversity of both collagen molecules and proteoglycans, the patterns may be complex and varied.

Some polysaccharide chains are known to assemble into highly ordered helical or ribbonlike structures. In higher plants, for example, cellulose (polyglucose) chains are packed tightly together in ribbonlike crystalline arrays to form the microfibrillar component of the cell wall (see Figure 20–5, p. 1141). *In vitro,* two *different* polysaccharide chains can associate specifically with each other, producing regions with a regular helical structure (Figure 14–31); such polysaccharide-polysaccharide interactions may also occur in the extracellular matrix. If proteoglycan molecules can assume structural conformations as diverse as their chemistry, we have hardly begun to understand them.

Not all proteoglycans are secreted components of the extracellular matrix. Some are integral components of plasma membranes, and some of these have their core protein oriented across the lipid bilayer. The integral membrane proteoglycans usually contain only a small number of glycosaminoglycan chains, and they are thought to play a part in binding cells to the extracellular matrix and in organizing the matrix macromolecules that cells secrete.

14-12 Collagen Is the Major Protein of the Extracellular Matrix[14]

The **collagens** are a family of highly characteristic fibrous proteins found in all multicellular animals. They are secreted mainly by connective tissue cells and are the most abundant proteins in mammals, constituting 25% of their total protein. The characteristic feature of collagen molecules is their stiff, triple-stranded helical structure. Three collagen polypeptide chains, called *α chains* (each about 1000

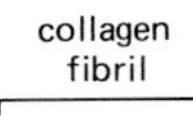

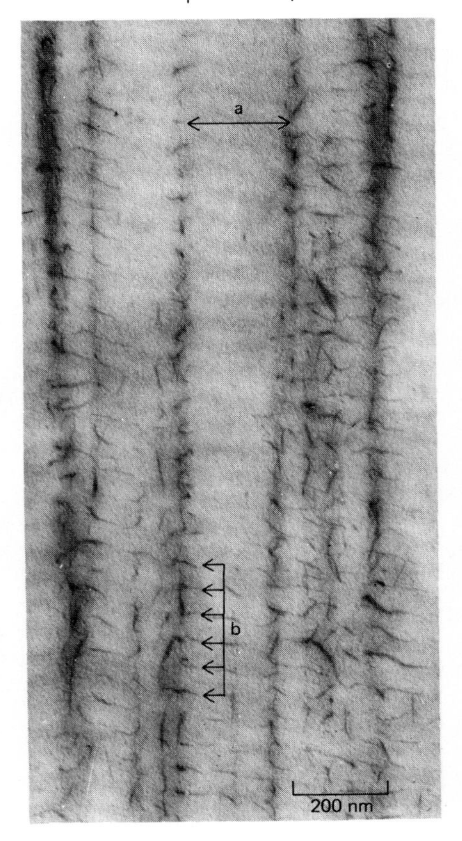

Figure 14–30 Electron micrograph of a longitudinal section of rat tail tendon stained with a copper-containing dye to visualize proteoglycan molecules. A tendon consists of closely packed collagen fibrils, several of which are shown here. Proteoglycan molecules are seen as fine filaments encircling each of the collagen fibrils at regular intervals of about 65 nm (for example, those indicated by arrows b), revealing a specific interaction between proteoglycan and collagen molecules. In those regions where proteoglycan threads are not seen crossing a collagen fibril (such as the region of fibril indicated by the double arrow a), the plane of section presumably cuts through the interior of the fibril. (Reproduced with permission from J.E. Scott, *Biochem. J.* 187:887–891, 1980. Copyright 1980 American Chemical Society.)

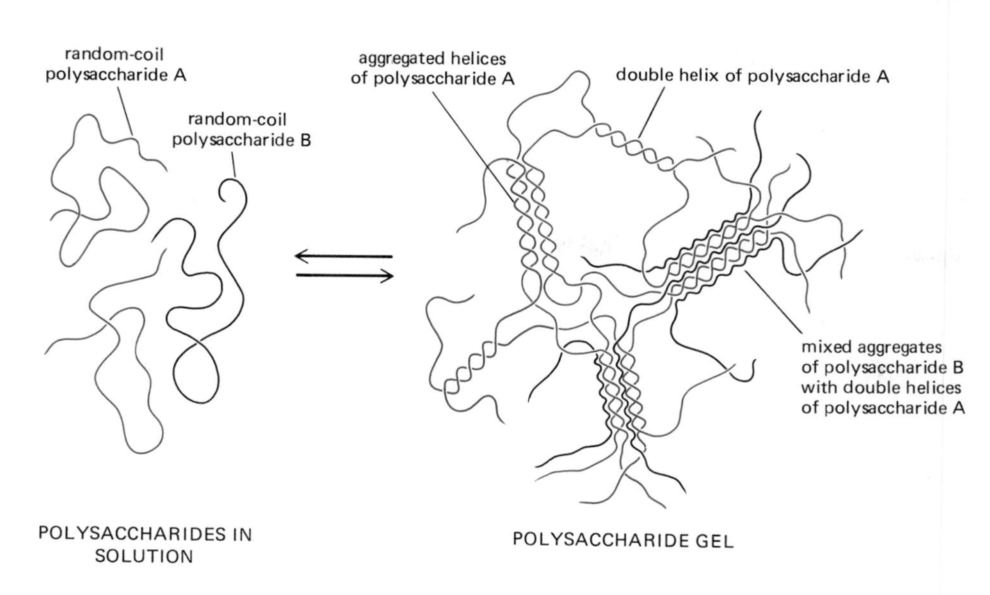

Figure 14–31 Some of the ordered conformations that two different polysaccharide chains, A and B, can assume in forming a gel *in vitro.* Since these interactions between molecules are confined to certain regions of the chains (so-called junctional regions) and are not propagated along the entire molecule, each chain can combine with more than one partner and thereby form a gel network. Examples of gel-forming polysaccharides are the agars (of algae) and the pectins (of higher plants).

amino acids long), are wound around one another in a regular superhelix to generate a ropelike collagen molecule about 300 nm long and 1.5 nm in diameter. Collagens are extremely rich in proline and glycine, both of which are important in the formation of the triple-stranded helix. Proline, because of its ring structure, stabilizes a left-handed helical conformation in each α chain, with three amino acid residues per turn. Glycine is the smallest amino acid (because it has only a hydrogen atom as a side chain); regularly spaced at every third residue throughout the central region of the α chain, it allows the three helical α chains to pack tightly together to form the final collagen superhelix (Figure 14–32).

So far, about 20 distinct collagen α chains have been identified, each encoded by a separate gene. Different combinations of these genes are expressed in different tissues. Although in principle more than 1000 types of triple-stranded collagen molecules could be assembled from various combinations of the 20 or so α chains, only about 10 types of collagen molecules have been found. The best defined are types I, II, III, and IV (Table 14–3). Types I, II, and III are the **fibrillar collagens.** They are the main types of collagen found in connective tissues, type I being by far the most common. After being secreted into the extracellular space, these three types of collagen molecules assemble into ordered polymers called **collagen fibrils,** which are thin (10–300 nm in diameter) cablelike structures, many micrometers long and clearly visible in electron micrographs (Figure 14–33). The collagen fibrils often aggregate into larger bundles, which can be seen in the light microscope as *collagen fibers* several micrometers in diameter. Type IV collagen molecules are found exclusively in basal laminae; instead of forming fibrils, they assemble into a sheetlike meshwork that constitutes a major part of all basal laminae (see p. 815). The arrangement of most of the other half-dozen or so types of collagen molecules in tissues is uncertain.

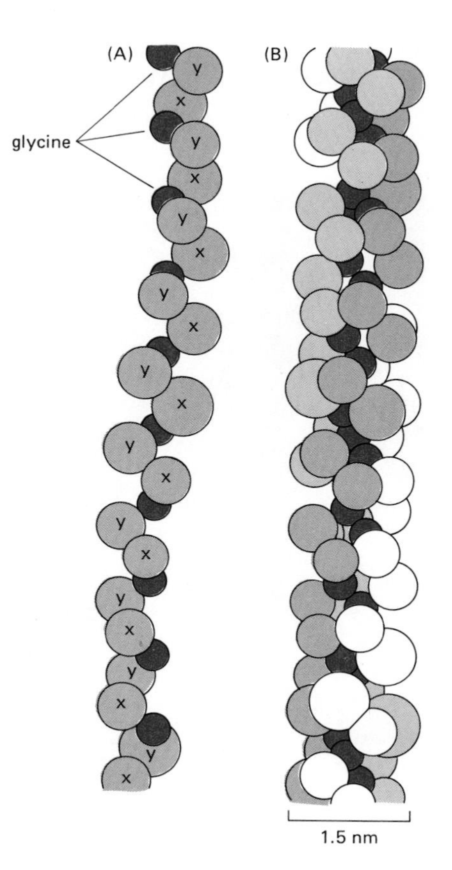

Figure 14–32 (A) A model of a single collagen α chain in which each amino acid is represented by a sphere. The chain is arranged as a left-handed helix with three amino acid residues per turn and with glycine (*dark color*) as every third residue. Therefore an α chain is composed of a series of triplet Gly-X-Y sequences in which X and Y can be any amino acid (although one is commonly proline). (B) A model of a part of a collagen molecule in which three α chains are wrapped around one another to form a triple-stranded helical rod. One α chain is shown in light color, one in gray, and one in white. Glycine is the only amino acid small enough to occupy the crowded interior of the triple helix. Only a short length of the molecule is shown; the entire molecule is 300 nm long, with each α chain containing about 1000 amino acid residues. (Drawn from model by B.L. Trus.)

Table 14–3. Four Major Types of Collagen and Their Properties

Type	Molecular Formula	Polymerized Form	Distinctive Features	Tissue Distribution
I	$[\alpha1(I)]_2\alpha2(I)$	fibril	low hydroxylysine, low carbohydrate, broad fibrils	skin, tendon, bone, ligaments, cornea, internal organs (accounts for 90% of body collagen)
II	$[\alpha1(II)]_3$	fibril	high hydroxylysine, high carbohydrate, usually thinner fibrils than type I	cartilage, intervertebral disc, notochord, vitreous body of eye
III	$[\alpha1(III)]_3$	fibril	high hydroxyproline, low hydroxylysine, low carbohydrate	skin, blood vessels, internal organs
IV	$[\alpha1(IV)]_2\alpha2(IV)$	basal lamina	very high hydroxylysine, high carbohydrate, retains procollagen extension peptides	basal laminae

Note that types I and IV are each composed of two types of α chain, whereas types II and III are composed of only one type of α chain each. Only the four major types of collagen are shown, but, more than 10 types of collagen and about 20 types of α chain have been defined so far.

Many proteins that contain a repeated pattern of amino acids have evolved by duplications of DNA sequences (see p. 602). The fibrillar collagens apparently arose in this way. Thus the genes that encode the α chains of these collagens are very large (30–40 kilobases in length) and contain about 50 exons. Most of the exons are 54, or multiples of 54, nucleotides long, suggesting that these collagens arose by multiple duplications of a primordial gene containing 54 nucleotides; this is not the case for type IV collagen, which may therefore have evolved differently.

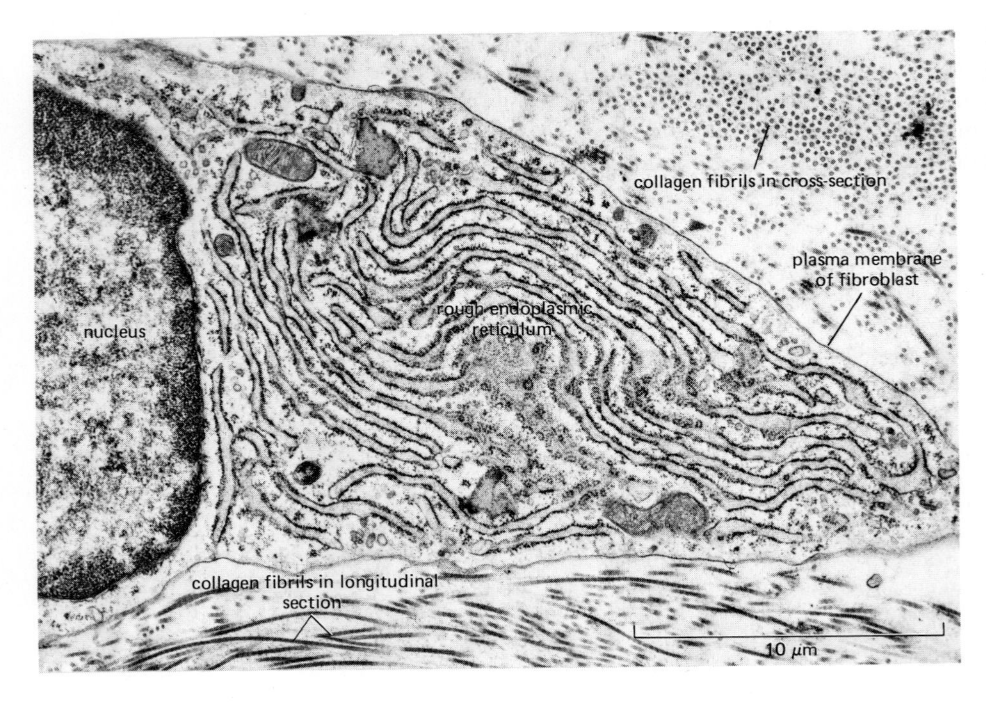

Figure 14–33 Electron micrograph showing part of a fibroblast surrounded by collagen fibrils in connective tissue. The extensive rough endoplasmic reticulum in the fibroblast cytoplasm reflects the cell's active synthesis and secretion of collagen and other extracellular matrix macromolecules. (Courtesy of Russell Ross.)

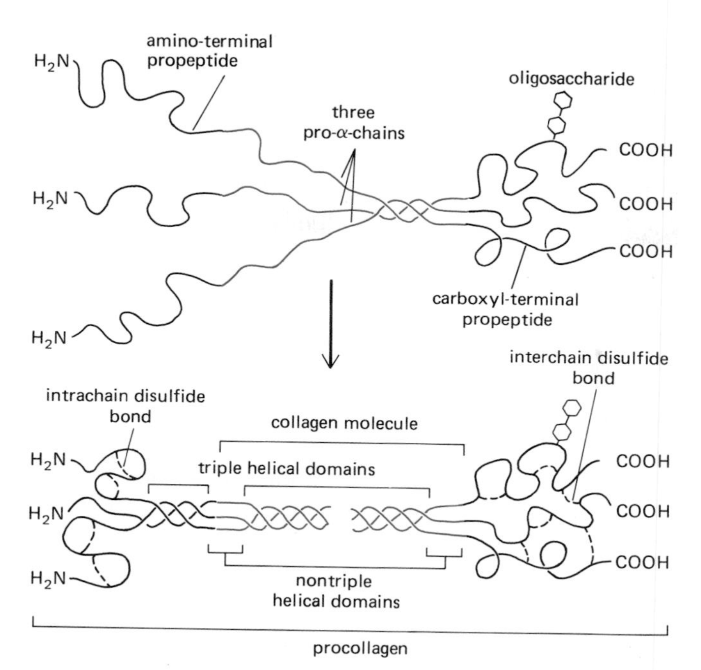

Figure 14–34 In fibrillar collagen molecules the α chains are initially synthesized in the form of pro-α chains, which contain extra propeptides at both ends (shown here in black) that will later be removed. The carboxyl-terminal propeptide is thought to help guide triple-helix formation during the assembly of the *procollagen* molecule. Note that the carboxyl-terminal propeptides in the procollagen molecule are covalently linked together by disulfide bonds and often contain an oligosaccharide chain. The amino-terminal propeptides form a short, triple-stranded "minicollagen" region. The final collagen molecule contains only the portion of the procollagen molecule that is shown in red; the rest is cleaved off.

14-13 Collagens Are Secreted with a Nonhelical Extension at Each End[14,15]

The individual collagen polypeptide chains are synthesized on membrane-bound ribosomes and injected into the lumen of the endoplasmic reticulum (ER) as larger precursors, called *pro-α chains.* These precursors not only have the short amino-terminal "signal peptide" required to thread secreted proteins through the membrane of the ER (see p. 438), they also have other extra amino acids, called *propeptides,* at both their amino- and carboxyl-terminal ends. In the lumen of the ER, selected proline and lysine residues are hydroxylated to form hydroxyproline and hydroxylysine, respectively. Each pro-α chain then combines with two others to form a hydrogen-bonded, triple-stranded helical molecule known as *procollagen* (Figure 14–34). The secreted forms of fibrillar collagens (but not of type IV collagen) are converted to *collagen molecules* in the extracellular space by the removal of the propeptides (see below).

Hydroxyproline and hydroxylysine residues (Figure 14–35) are rarely found in other proteins. Why are they present in collagen? There is indirect evidence that the hydroxyl groups of hydroxyproline residues form interchain hydrogen bonds that help stabilize the triple-stranded helix. For example, conditions that prevent proline hydroxylation (such as a deficiency of ascorbic acid [vitamin C]) inhibit procollagen helix formation. Normal collagens are continuously (albeit slowly) degraded by specific extracellular enzymes called *collagenases.* In scurvy, a human disease caused by a dietary deficiency of vitamin C, the defective pro-α chains that are synthesized fail to form a triple helix and are immediately degraded. Consequently, with the gradual loss of the preexisting normal collagen in the matrix, blood vessels become extremely fragile and teeth become loose in their sockets. This implies that in these particular tissues degradation and replacement of collagen is relatively rapid. In many other adult tissues, however, the "turnover" of collagen (and other extracellular matrix macromolecules) is thought to be normally very slow: in bone, to take an extreme example, collagen molecules persist for about 10 years before they are degraded and replaced. By contrast, most cellular proteins have half-lives of the order of hours or days.

The hydroxylation of lysine residues has a function different from that of the hydroxylation of proline residues. It is required for an unusual form of lysine-linked glycosylation found in collagen (the function of which is unknown) and is crucial for the extensive cross-linking of collagen molecules that occurs during collagen assembly in the extracellular space (see p. 813).

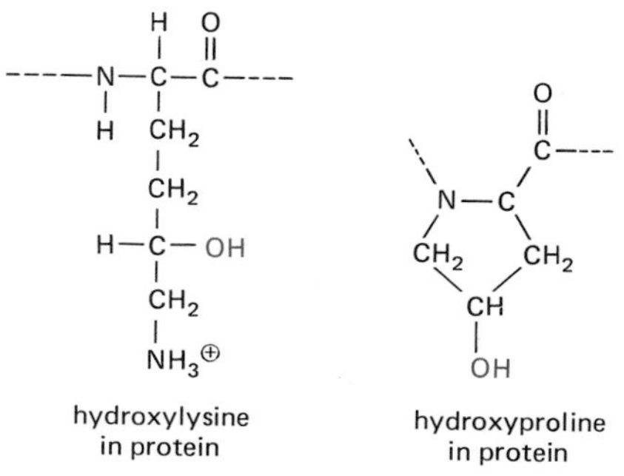

hydroxylysine in protein

hydroxyproline in protein

Figure 14–35 The structures of hydroxyproline and hydroxylysine residues, two modified amino acids that are common in collagen.

After Secretion, Types I, II, and III Procollagen Molecules Are Cleaved to Collagen Molecules, Which Assemble into Fibrils[16]

After secretion, the propeptides of types I, II, and III procollagen molecules are removed by specific proteolytic enzymes outside the cell. This converts the procollagen molecules to collagen (also called *tropocollagen*) molecules (1.5 nm in diameter), which then associate in the extracellular space to form the much larger collagen fibrils (10–300 nm in diameter). The process of fibril formation is driven, in part, by the tendency of the collagen molecules to self-assemble. However, the fibrils form close to the cell surface, often in deep recesses formed by the infolding of the plasma membrane, and the underlying cortical cytoskeleton can therefore influence the sites, rates, and orientation of fibril assembly (see p. 822).

The propeptides have at least two functions: (1) they guide the intracellular formation of the triple-stranded collagen molecules; and (2) because they are removed only after secretion, they prevent the intracellular formation of large collagen fibrils, which could be catastrophic for the cell. It is equally important, however, that the propeptides be removed from the fibrillar collagens once they have performed their functions. In some genetic diseases, such as Ehlers-Danlos syndrome, this process is defective and collagen fibril formation is impaired, resulting in fragile skin and hypermobile joints in affected individuals.

When isolated collagen fibrils are fixed, stained, and viewed in an electron microscope, they exhibit cross-striations every 67 nm. This pattern reflects the packing arrangement of the individual collagen molecules in the fibril: they are staggered, as shown in Figure 14–36, so that adjacent molecules are displaced longitudinally by almost one-quarter of their length (a distance of 67 nm). This arrangement presumably maximizes the tensile strength of the aggregate, and it gives rise to the striations seen in negatively stained fibrils (Figure 14–37). However, it is still not certain how these staggered molecules are packed in the three dimensions of a cylindrical fibril.

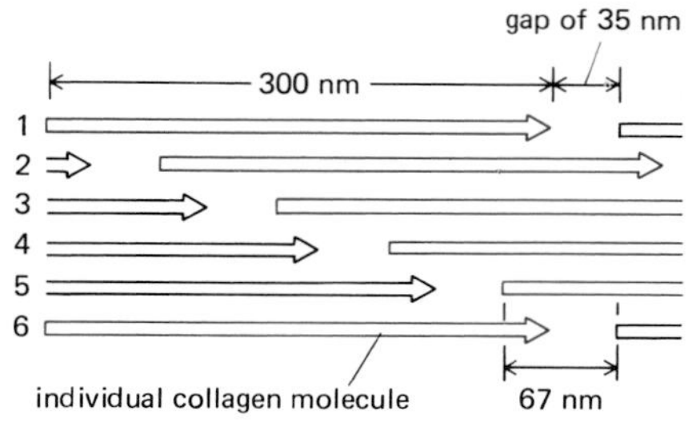

Figure 14–36 The staggered arrangement of collagen molecules in a collagen fibril. Adjacent molecules (shown as arrows) are displaced by 67 nm, with a 35-nm gap between successive molecules in a row. The gap size is such that the pattern repeats after five molecules have been lined up in this staggered fashion; thus the molecules in rows 1 and 6 are in register.

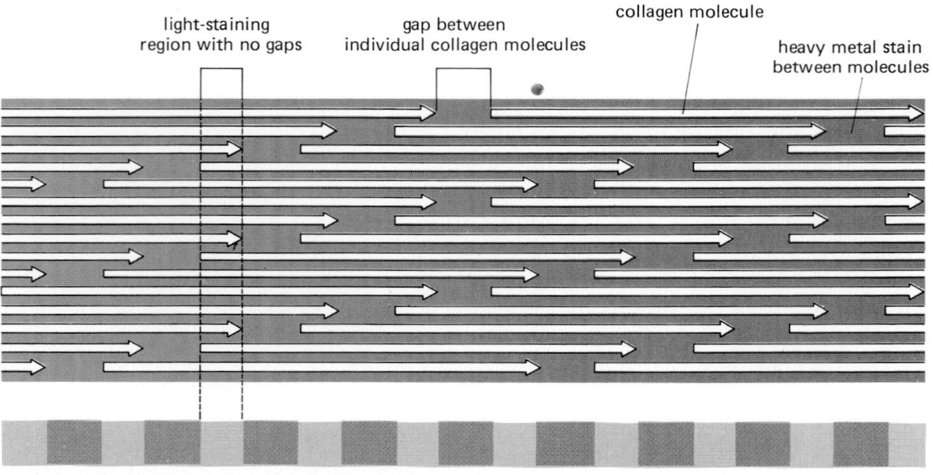

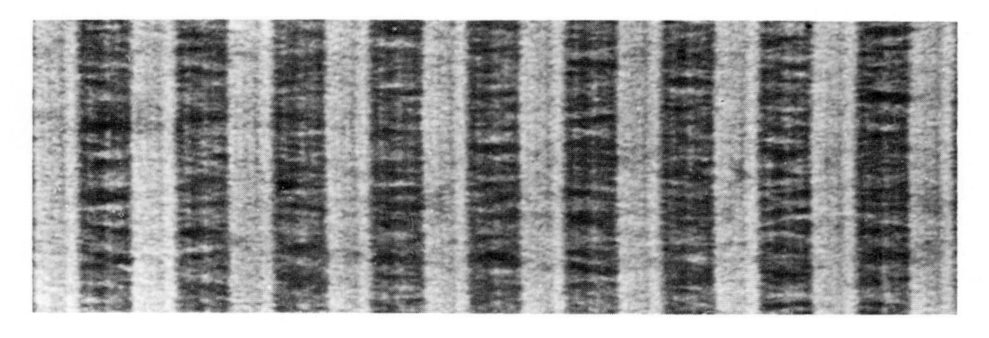

Figure 14–37 How the staggered arrangement of collagen molecules gives rise to the striated appearance of a negatively stained fibril. Since the negative stain fills only the space between the molecules, the stain in the gaps between the individual molecules in each row accounts for the dark staining bands. An electron micrograph of a negatively stained fibril is shown at the bottom of the figure. (Electron micrograph courtesy of Robert Horne.)

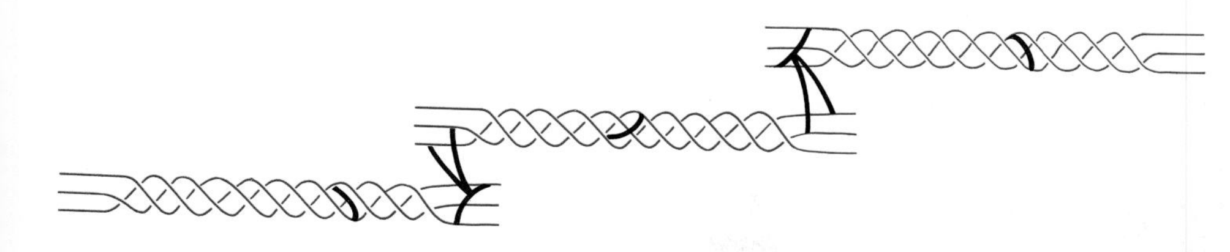

Figure 14–38 The covalent intramolecular and intermolecular cross-links formed between modified lysine side chains within a collagen fibril. The cross-links are formed in several steps. First, certain lysine and hydroxylysine residues are deaminated by the extracellular enzyme lysyl oxidase to yield highly reactive aldehyde groups. The aldehydes then react spontaneously to form covalent bonds with each other or with other lysine or hydroxylysine residues in which more than two amino acid side chains can be involved. Some of these bonds are relatively unstable and are ultimately modified to form a variety of more stable cross-links. Note that most of the cross-links form between the short nonhelical segments at each end of the collagen molecules (see Figure 14–35).

After collagen fibrils have formed in the extracellular space, they are greatly strengthened by the formation of covalent cross-links within and between lysine residues of the constituent collagen molecules (Figure 14–38). The types of covalent bonds involved are found only in collagen and elastin. If cross-linking is inhibited, collagenous tissues become fragile and structures such as skin, tendons, and blood vessels tend to tear. The extent and type of cross-linking varies from tissue to tissue. Collagen is especially highly cross-linked in the Achilles tendon, for example, where tensile strength is crucial.

The Organization of Collagen Fibrils in the Extracellular Matrix Is Adapted to the Needs of the Tissue[17]

Collagen fibrils come in a variety of diameters and are organized in different ways in different tissues. In mammalian skin, for example, they are woven in a wickerwork pattern so that they resist stress in multiple directions. In tendons they are organized in parallel bundles aligned along the major axis of stress on the tendon. And in mature bone and in the cornea, they are arranged like plywood in orderly layers, with the fibrils in each layer lying parallel to each other but nearly at right angles to the fibrils in the layers on either side. The same arrangement occurs in tadpole skin, which serves to illustrate this organization (Figure 14–39).

The connective tissue cells themselves determine the size and arrangement of the collagen fibrils. The cells can express one or more of the genes for the different types of fibrillar procollagen molecules (including minor types not listed in Table 14–3) and can regulate the disposition of the molecules after secretion. By controlling the order in which the amino- and carboxyl-terminal propeptides are sequentially cleaved, by secreting different kinds and amounts of noncollagen matrix macromolecules along with the collagen, and by guiding collagen fibril formation in close association with the plasma membrane, cells can determine the geometry and properties of the fibrils in their environment. Finally, the collagen is cross-linked to a greater or lesser degree depending on the tensile strength required. Figure 14–40 summarizes the steps in fibrillar collagen synthesis and assembly.

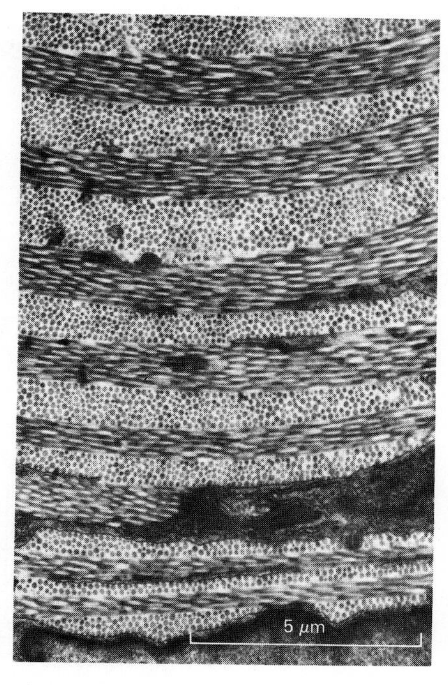

Figure 14–39 Electron micrograph of a cross section of tadpole skin showing the plywoodlike arrangement of collagen fibrils, in which successive layers of fibrils are laid down at right angles to each other. This arrangement is also found in mature bone and in the cornea. (Courtesy of Jerome Gross.)

Cells Can Help Organize the Collagen Fibrils They Secrete by Exerting Tension on the Matrix[18]

There is yet another way that collagen-secreting cells determine the spatial organization of the matrix they produce. Fibroblasts work on the collagen they have secreted, crawling over it and tugging on it—helping to compact it into sheets and draw it out into cables. This mechanical role of fibroblasts in shaping collagen matrices has been demonstrated dramatically in culture. When fibroblasts are mixed with a meshwork of randomly oriented collagen fibrils that form a gel in a culture dish, the fibroblasts tug on the meshwork, drawing in collagen from the surroundings and causing the gel to contract to a small fraction of its initial volume; by similar activities, a cluster of fibroblasts will surround itself with a capsule of densely packed and circumferentially oriented collagen fibers.

If two small pieces of embryonic tissue containing fibroblasts are placed far apart on a collagen gel, the collagen becomes organized into a compact band of aligned fibers that connect the two explants (Figure 14–41). The fibroblasts subsequently migrate out from the explants along the aligned collagen fibers. Thus

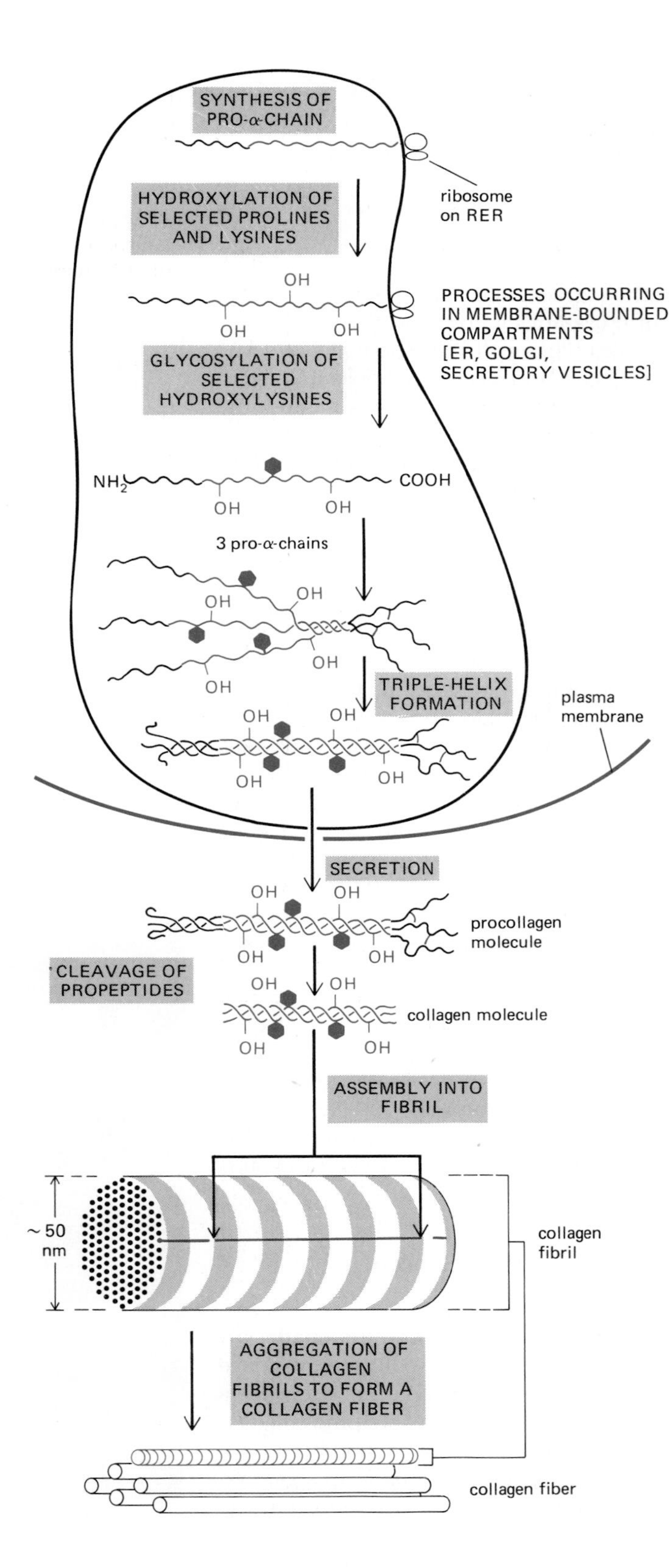

Figure 14–40 The intracellular and extracellular events involved in the formation of a collagen fibril. As one example of how the collagen fibrils can form ordered arrays in the extracellular space, they are shown further assembling into large collagen fibers, which are visible in the light microscope. The covalent cross-links that stabilize the extracellular assemblies are not shown. There are many human genetic diseases that affect the formation of collagen fibrils, which is not surprising given the large number of enzymatic steps involved.

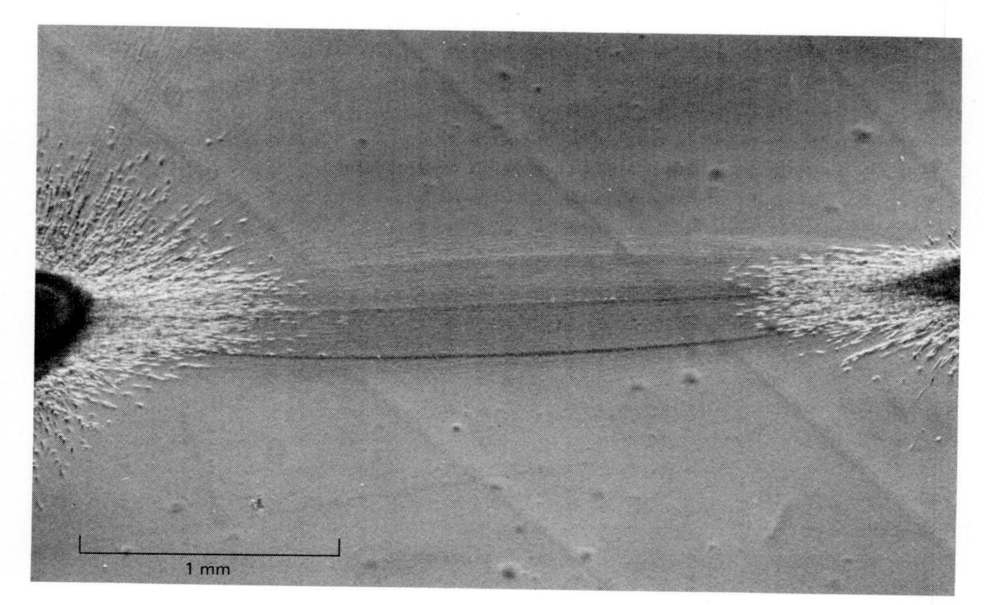

Figure 14–41 Photomicrograph of the region between two pieces of embryonic chick heart (rich in fibroblasts as well as muscle cells) growing in culture on a collagen gel for 4 days. Note that a dense tract of aligned collagen fibers has formed between the explants. (From D. Stopak and A.K. Harris, *Dev. Biol.* 90:383–398, 1982.)

the fibroblasts influence the alignment of the collagen fibers, and the collagen fibers in turn affect the distribution of the fibroblasts. Fibroblasts presumably play a similar role in generating long-range order in the extracellular matrix inside the body—in helping to create tendons and ligaments, for example, and the tough, dense layers of connective tissue that ensheathe and bind together most organs.

Type IV Collagen Molecules Assemble into a Laminar Meshwork[19]

Type IV collagen molecules differ from the fibrillar collagen molecules in several ways. First, the regular Gly-X-Y repeating amino acid sequence of the type IV α chain is interrupted in a number of regions, locally disrupting the triple-stranded helical structure of the collagen molecule. Second, type IV "procollagen" molecules are not cleaved after secretion and so retain their propeptides; the secreted molecules interact via their uncleaved propeptide domains to assemble into a sheetlike multilayered network rather than into fibrils. Electron microscopic studies of preparations of assembling type IV collagen molecules suggest that these molecules associate by their carboxyl-terminal propeptides to form head-to-head dimers, which then form an extended lattice by the further associations shown in Figure 14–42. Disulfide and other covalent cross-links between the collagen molecules stabilize these associations. Type IV collagen sheets are thought to form the core of all basal laminae; additional components of the basal lamina will be discussed below (see p. 818).

Elastin Is a Cross-linked, Random-Coil Protein That Gives Tissues Their Elasticity[20]

Tissues such as skin, blood vessels, and lungs require elasticity in addition to tensile strength in order to function. A network of **elastic fibers** in the extracellular matrix of these tissues gives them the required ability to recoil after transient stretch. The main component of elastic fibers is **elastin,** a highly hydrophobic, nonglycosylated protein (about 830 amino acid residues long), which, like collagen, is unusually rich in proline and glycine but, unlike collagen, contains little hydroxyproline and no hydroxylysine. Elastin molecules are secreted into the extracellular space, where they form filaments and sheets in which the elastin molecules are highly cross-linked to one another to generate an extensive network (Figure 14–43). The cross-links are formed between lysine residues by the same mechanism that operates in cross-linking collagen molecules (see Figure 14–38). Elastin molecules are unlike most other proteins in that their function requires

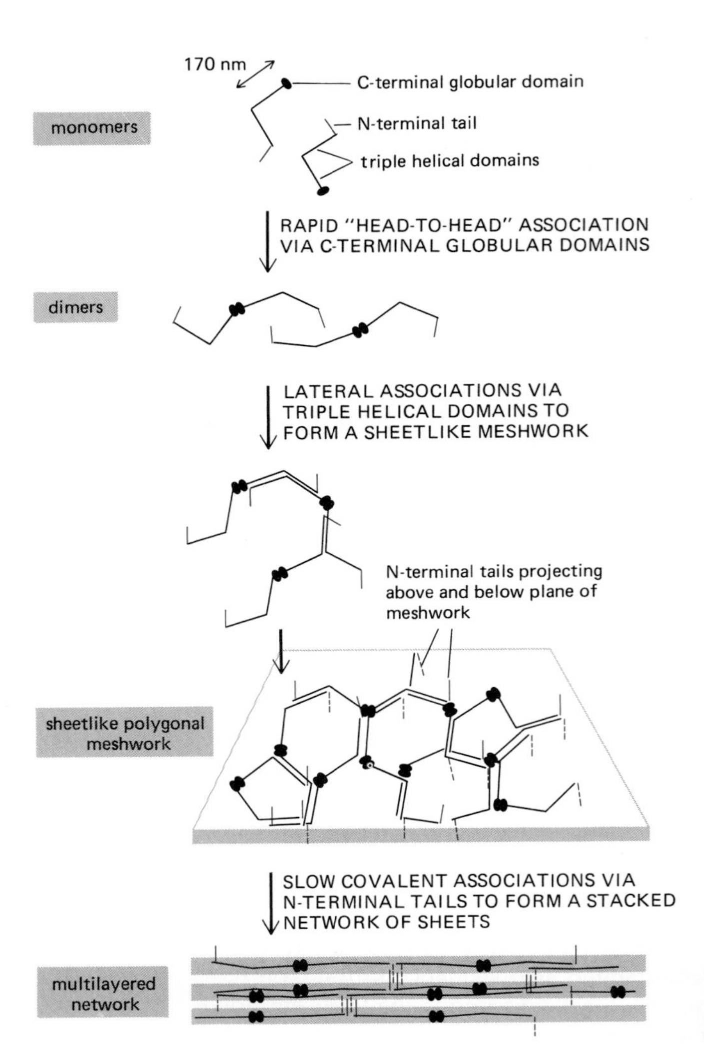

Figure 14–42 How type IV collagen molecules are thought to assemble into a multilayered network, which forms the core of all basal laminae. The model is based on electron micrographs of rotary-shadowed preparations of these molecules assembling *in vitro*. (Based on P.D. Yurchenco, E.C. Tsilibary, A.S. Charonis, and H. Furthmayr, *J. Histochem. Cytochem.* 34:93–102, 1986.)

their polypeptide backbones to remain unfolded as "random coils" (Figure 14–44). It is the cross-linked, random-coil structure of the elastic fiber network that allows the network to stretch and recoil like a rubber band (Figure 14–45). Elastic fibers are at least five times more extensible than a rubber band of the same cross-sectional area. Long, inelastic collagen fibrils are interwoven with the elastic fibers to limit the extent of stretching and thereby prevent the tissue from tearing.

Elastic fibers are not composed solely of elastin. They also contain a glycoprotein that is distributed mainly as microfibrils on the elastic fiber surface. Elastic fibers are assembled in close association with the plasma membrane of the cells that secrete elastin and the microfibrillar glycoprotein. The microfibrils appear before elastin and may help the cell organize the secreted elastin molecules into the elastic fibers and sheets that form in the extracellular matrix.

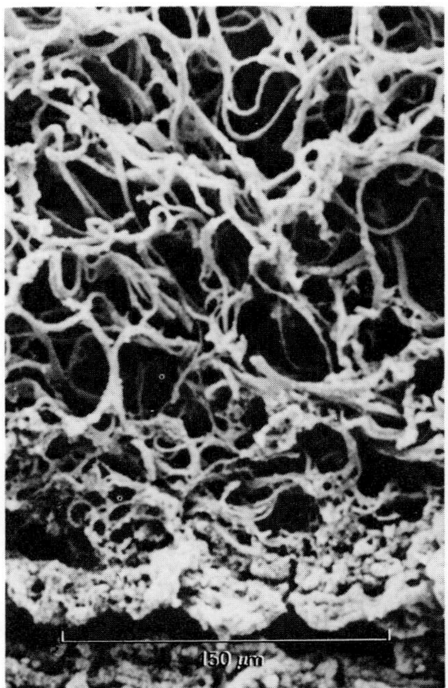

Figure 14–43 Scanning electron micrograph showing the extensive network of elastic fibers in a section of human skin connective tissue (dermis). The tissue has been heated under pressure to remove the collagen and glycosaminoglycans. (From T. Tsuji, R.M. Lavker, and A.M. Kligman, *J. Microscop.* 115:165–173, 1978.)

14-14 Fibronectin Is an Extracellular Adhesive Glycoprotein That Helps Mediate Cell-Matrix Adhesion[21]

The extracellular matrix contains a number of **adhesive glycoproteins** that bind to both cells and other matrix macromolecules and thereby help cells attach to the extracellular matrix. The best characterized of these is **fibronectin,** a large fibril-forming glycoprotein found throughout the animal kingdom. Fibronectin is a dimer composed of two similar subunits (each almost 2500 amino acid residues long); these subunits are joined by a pair of disulfide bonds near their carboxyl termini and are folded into a series of globular domains separated by regions of flexible polypeptide chain (Figure 14–46). Sequencing studies indicate that a fibronectin molecule is composed mainly of three types of short amino acid sequences repeated many times, suggesting that the fibronectin gene evolved by multiple duplications of three small genes.

Fibronectin exists in three forms: (1) a soluble dimeric form, called *plasma fibronectin*, circulates in the blood and other body fluids, where it is thought to enhance blood clotting, wound healing, and phagocytosis; (2) oligomers of fibronectin can be found transiently attached to the surface of cells (*cell-surface fibronectin*); and (3) highly insoluble fibronectin fibrils form in the extracellular matrix (*matrix fibronectin*). In the cell-surface and matrix aggregates, fibronectin dimers are cross-linked to one another by additional disulfide bonds.

Fibronectin is a multifunctional molecule in which the various globular domains play different roles. For example, one domain binds to collagen, another to heparin, another to specific receptors on the surface of various types of cells, and so on (see Figure 14–46). In this way fibronectin contributes to the organization of the matrix and helps cells attach to it.

The parts played by the different domains, and in particular by the cell-binding domains, have been analyzed by cleaving the molecule into its separate domains with proteolytic enzymes or by synthesizing specific protein fragments either chemically or by recombinant DNA techniques. Thus a domain responsible for cell-binding activity has been isolated from proteolytic fragments and its amino acid sequence determined. Synthetic peptides corresponding to different segments of this domain were prepared and used to localize the cell-binding activity to a specific tripeptide sequence (Arg-Gly-Asp, or R-G-D). Peptides containing this **RGD sequence** compete for the binding site on cells and so inhibit the attachment of cells to fibronectin; and when these peptides are coupled to a solid surface, they cause cells to adhere to that surface. The RGD sequence is not confined to fibronectin. It is a common motif in a variety of extracellular adhesive proteins, and it is recognized by a family of homologous cell-surface receptors that bind these proteins (see p. 821). Despite the common tripeptide sequence found at the sites recognized by these receptors, each receptor specifically recognizes its own small set of adhesive molecules. Thus receptor binding must also depend on other parts of the adhesive protein sequence.

Fibronectin is important not only for cell adhesion but also for cell migration. In both invertebrate and vertebrate embryos, it seems to guide cell migration in many cases. For example, large amounts of fibronectin are found along the pathway followed by migrating prospective mesodermal cells during amphibian gastrulation (see p. 882). The migration of these cells can be inhibited either by injecting antibodies against fibronectin into the blastocoel cavity or by injecting peptides containing the cell-binding tripeptide but lacking the matrix-binding domains of fibronectin. Fibronectin presumably promotes cell migration by helping cells attach to the matrix. The effect must be delicately balanced so that the

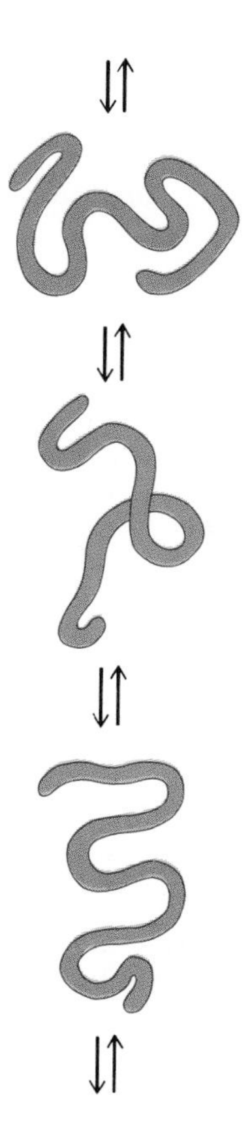

Figure 14–44 An elastin molecule in various "random-coil" conformations. Unlike most proteins, the elastin molecule does not adopt a unique structure but oscillates among a variety of partially extended, random conformations, as illustrated.

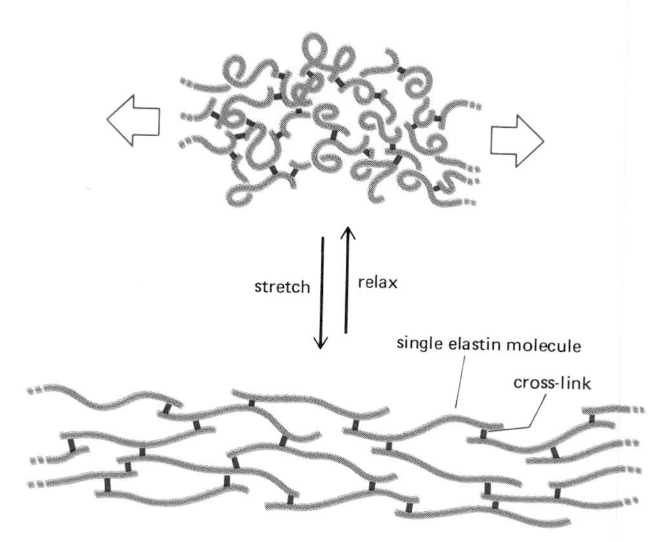

Figure 14–45 Elastin molecules are joined together by covalent bonds (indicated in color) to generate an extensive cross-linked network. Because each elastin molecule in the network can expand and contract as a random coil, the entire network can stretch and recoil like a rubber band.

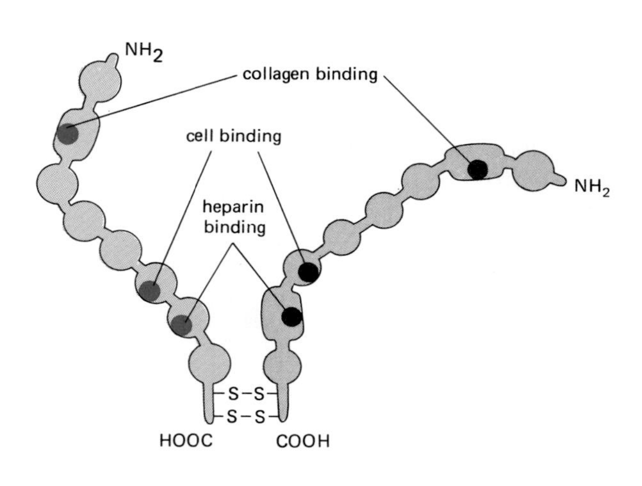

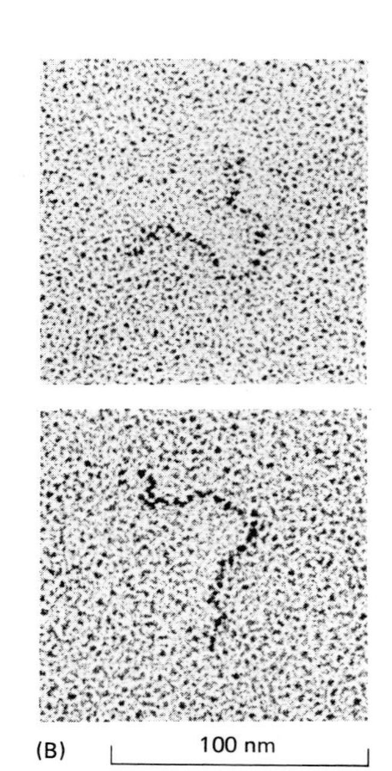

Figure 14–46 The structure of a fibronectin dimer shown schematically in (A) and in electron micrographs of individual molecules shadowed with platinum in (B). The two polypeptide chains are similar but not identical and are joined by two disulfide bonds near the carboxyl terminus. Each chain is folded into a series of globular domains connected by flexible polypeptide segments. Individual domains are specialized for binding to a particular molecule or to a cell, as indicated for three of the domains; for simplicity, not all of the known binding sites are shown. (B, from J. Engel et al., *J. Mol. Biol.* 150:97–120, 1981. © Academic Press Inc. [London] Ltd.)

migrating cells get a grip on the matrix without becoming immobilized on it. We shall return later to the question of how this balance may be achieved by the many adhesive molecules believed to play a part in guiding morphogenetic movements.

Multiple Forms of Fibronectin Are Produced by Alternative RNA Splicing[22]

Aside from being a member of a large family of RGD-containing adhesion molecules, fibronectin itself, as mentioned earlier, can have a variety of forms; even the polypeptide chains in a single dimer may have minor differences. Yet all the different fibronectin polypeptide chains are encoded by a single large gene that, in the rat, is more than 70 kilobases long and contains about 50 exons, making it one of the largest genes characterized so far. Transcription produces a single large RNA molecule that is spliced in different ways to produce one or more of about 20 different messenger RNAs, depending on the cell type. It is not clear how these patterns of RNA splicing are determined or how the various polypeptide chains that result differ in function. There is some evidence that one function of alternative splicing of the human fibronectin RNA transcript is to add to selected fibronectin molecules an additional cell-binding domain that is distinct from the RGD-containing cell-binding site.

Fibronectin is not the only secreted glycoprotein involved in cell-matrix adhesion. For example, **tenascin** is also an extracellular adhesive glycoprotein, but it has a much more restricted distribution than fibronectin and is most abundant in embryonic tissues. In the nervous system it is secreted by glial cells, and some neurons are thought to adhere to it by means of a specific cell-surface proteoglycan. Tenascin is a large complex of six disulfide-linked polypeptide chains, which radiate from a center like the spokes of a wheel (see Figure 14–51).

Some cells, especially epithelial cells, secrete another type of extracellular adhesion glycoprotein, called *laminin*, which is a major protein in all basal laminae. It binds both to epithelial cells (as well as to some other cell types) and to type IV collagen, the main collagen type in the basal lamina.

The Basal Lamina Is a Specialized Extracellular Matrix Composed Mainly of Type IV Collagen, Proteoglycans, and Laminin[23]

Basal laminae are continuous thin mats of specialized extracellular matrix that underlie all epithelial cell sheets and tubes; they also surround individual muscle cells, fat cells, and Schwann cells (which wrap around peripheral nerve cell axons to form myelin). The basal lamina thus separates these cells and cell sheets from the underlying or surrounding connective tissue. In other locations, such as the kidney glomerulus and lung alveolus, a basal lamina lies between two different cell sheets, where it functions as a highly selective filter (Figure 14–47). However, basal laminae serve more than simple structural and filtering roles. They are able to determine cell polarity, influence cell metabolism, organize the proteins in adjacent plasma membranes, induce cell differentiation, and, like fibronectin, serve as specific "highways" for cell migration.

The basal lamina is largely synthesized by the cells that rest on it (Figure 14–48). In essence it is a tough mat of type IV collagen (see Figure 14–42) with specific additional molecules on each face that help bind it to the adjacent cells or matrix. Although the precise composition of basal laminae varies from tissue to tissue and even from region to region in the same lamina (see p. 821), all basal laminae contain type IV collagen together with proteoglycans (primarily heparan sulfates) and the glycoproteins *laminin* and *entactin*. **Laminin** is a large (~850,000 daltons) complex of three very long polypeptide chains arranged in the shape of a cross and held together by disulfide bonds (Figure 14–49). Like fibronectin, it consists of a number of functional domains: one binds to type IV collagen, one to heparan sulfate, and one or more to laminin receptor proteins on the surface of cells. A single dumbell-shaped entactin molecule is thought to be tightly bound to each laminin molecule where the short arms meet the long one.

As seen in the electron microscope after conventional fixation and staining, most basal laminae consist of two distinct layers: an electron-lucent layer (*lamina lucida* or *rara*) adjacent to the basal plasma membrane of the cells that rest on the lamina—typically epithelial cells—and an electron-dense layer (*lamina densa*) just below. In some cases a third layer containing collagen fibrils (*lamina reticularis*) connects the basal lamina to the underlying connective tissue. Some cell biologists use the term *basement membrane* to describe the composite of all three layers (Figure 14–50), which is usually thick enough to be seen in the light microscope. The detailed molecular organization of the basal lamina is still uncertain, although electron microscopic studies using antibody labeling suggest that the lamina densa is composed primarily of type IV collagen, with proteoglycan molecules located on either side; laminin is thought to be present mainly on the plasma-membrane side of the lamina densa, where it helps to bind epithelial cells to the lamina, while fibronectin helps to bind the matrix macromolecules and connective tissue cells on the opposite side.

The shapes and sizes of some of the major components of the basal lamina and other forms of extracellular matrix are compared in Figure 14–51.

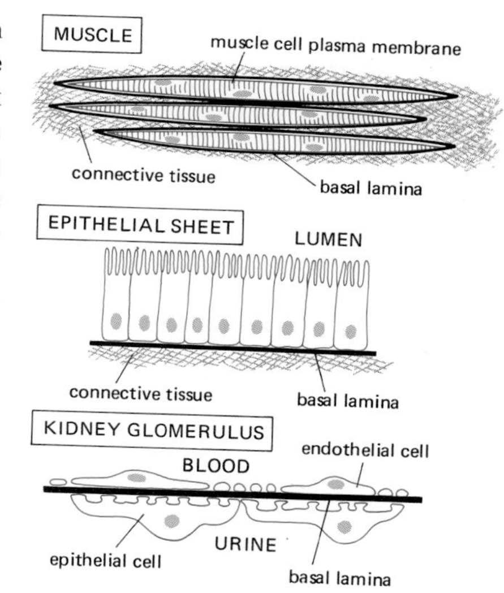

Figure 14–47 Three ways in which basal laminae (*black lines*) are organized: surrounding cells (such as muscle cells), underlying epithelial cell sheets, and interposed between two cell sheets (as in the kidney glomerulus). Note that in the kidney glomerulus, both cell sheets have gaps in them, so that the basal lamina serves as the permeability barrier determining which molecules will pass into the urine from the blood. Since the glomerular basal lamina develops as a result of the fusion of two basal laminae, one produced by the endothelial cells and the other by the epithelial cells, it is twice as thick as most basal laminae.

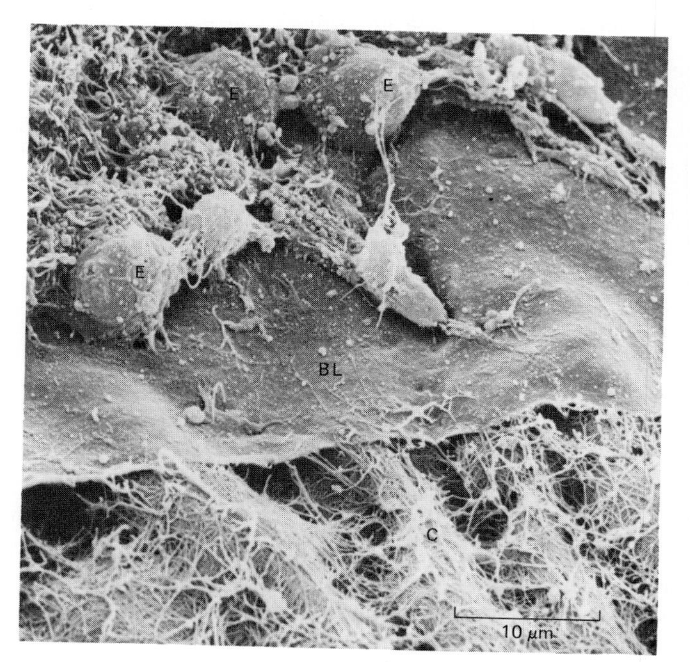

Figure 14–48 Scanning electron micrograph of a basal lamina in the cornea of a chick embryo. Some of the epithelial cells (E) have been removed to expose the upper surface of the matlike basal lamina (BL). Note the network of collagen fibrils (C) in the underlying connective tissue interacting with the lower face of the lamina. The macromolecules that comprise the basal lamina are synthesized by the epithelial cells that sit on it. (Courtesy of Robert Trelstad.)

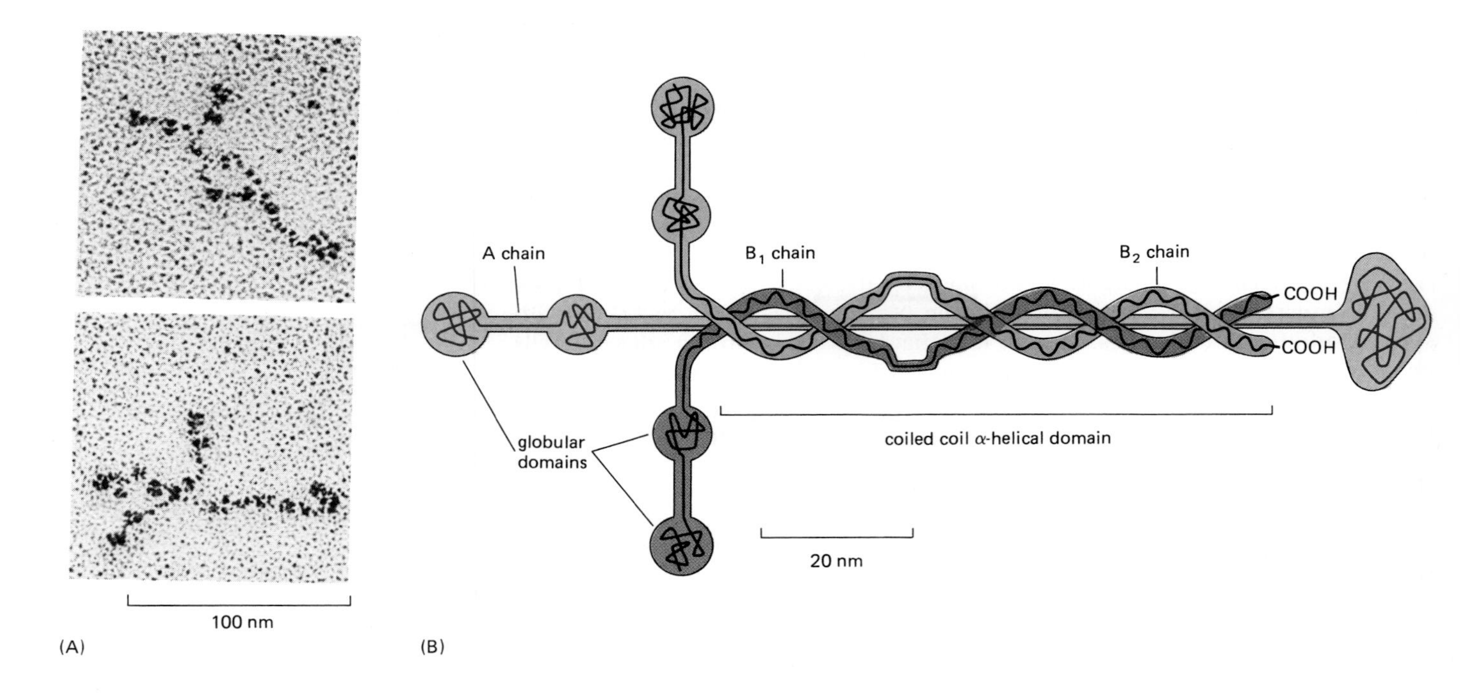

Figure 14–49 Electron micrographs of laminin molecules shadowed with platinum (A) and a schematic drawing of a model for the structure of laminin (B). The multidomain glycoprotein is composed of three polypeptides (A, B_1, and B_2) that are disulfide bonded into an asymmetric crosslike structure. Each of the polypeptide chains is more than 1500 amino acid residues long. (A, from J. Engel et al., *J. Mol. Biol.* 150:97–120, 1981. © Academic Press Inc. [London] Ltd.; B, based on B.L. Hogan *et al.,* in Basement Membranes [S. Shibata, ed.], pp. 147–154. Amsterdam: Elsevier, 1985.)

Basal Laminae Perform Diverse and Complex Functions[24]

The functions of basal laminae are surprisingly diverse. In the kidney glomerulus, an unusually thick basal lamina acts as a molecular filter, regulating the passage of macromolecules from the blood into the urine as urine is formed (see Figure 14–47). Proteoglycans seem to be important for this function: when they are removed by specific enzymes, the filtering properties of the lamina are destroyed. The basal lamina can also act as a selective cellular barrier. The lamina beneath epithelial cells, for example, usually prevents fibroblasts in the underlying connective tissue from making contact with the epithelial cells. It does not, however, stop macrophages, lymphocytes, or nerve processes from passing through it.

The basal lamina plays an important part in tissue regeneration after injury. When tissues such as muscles, nerves, and epithelia are damaged, the basal lamina survives and provides a scaffolding along which regenerating cells can migrate. In this way the original tissue architecture is readily reconstructed. A dramatic example of the importance of the basal lamina in regeneration comes from studies on the *neuromuscular junction,* where a nerve cell transmits its stimulus to a skeletal muscle cell.

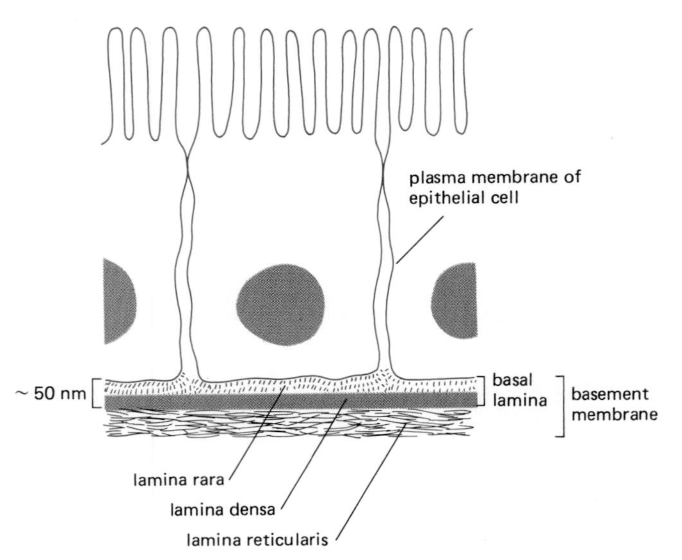

Figure 14–50 Schematic drawing of a basal lamina underlying an epithelial cell sheet as seen in cross section in an electron microscope.

At the site of neuromuscular contact (the synapse), the basal lamina has a chemically distinctive character, recognized, for example, by antibodies that bind to the lamina exclusively in this region. One of the functions of the junctional basal lamina, apparently, is to coordinate the spatial organization of the components on either side of the synapse. Evidence for the central role of the junctional basal lamina in reconstructing a synapse after nerve or muscle injury will be discussed in Chapter 19 (see p. 1124). These studies make it clear that we still have much to learn about the chemical and functional specializations of basal laminae. They also suggest that minor (but as yet undefined) components in the extracellular matrix may play a critical part in directing morphogenesis during embryonic development.

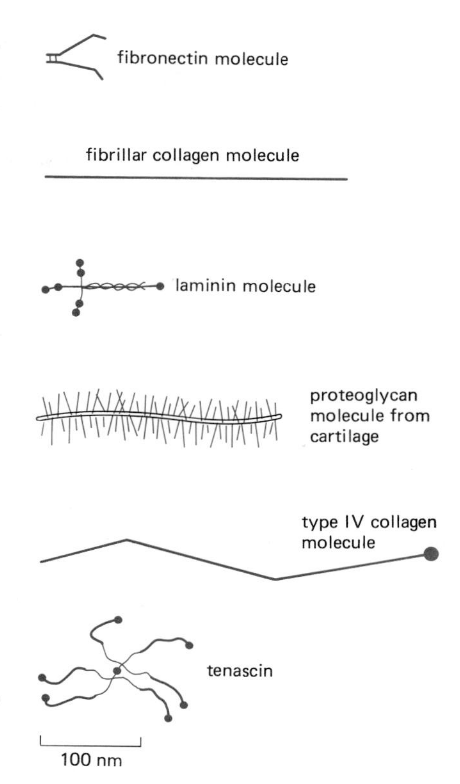

Figure 14–51 The comparative shapes and sizes of some of the major extracellular matrix macromolecules.

Integrins Help Bind Cells to the Extracellular Matrix[25]

To understand how the extracellular matrix interacts with cells, one has to define the cell-surface molecules that bind the matrix components as well as the extracellular matrix components themselves. As mentioned previously, some proteoglycans are integral components of the plasma membrane; their core protein may be either inserted across the lipid bilayer or covalently linked to it. By binding to most types of extracellular matrix components, these proteoglycans help link cells to the matrix. However, extracellular matrix components also bind to the cell surface via specific receptor glycoproteins. Because of the multiple interactions among matrix macromolecules in the extracellular space, it is largely a matter of semantics where the plasma membrane components end and the extracellular matrix begins. The glycocalyx of a cell, for example, often includes components of both (see p. 299).

The matrix receptors differ from cell-surface receptors for hormones and for other soluble signaling molecules in that they bind their ligand with relatively low affinity ($K_a = 10^6$–10^8 liters/mole) and are usually present at about 10- to 100-fold higher concentration on the cell surface. This suggests that the receptors might function cooperatively and that cells may respond to an organized group of ligands in the matrix rather than to individual molecules. In support of this suggestion, soluble cell-binding fragments of matrix components usually fail to elicit the cellular responses induced by the same components immobilized in a matrix.

A **fibronectin receptor** on mammalian fibroblasts is one of the best-characterized matrix receptors. It was initially identified as a plasma membrane glycoprotein that bound to a fibronectin affinity column and could be eluted with a small peptide containing the RGD cell-binding sequence (see p. 817). The receptor is a noncovalently associated complex of two distinct, high-molecular-weight polypeptide chains, called α and β. It functions as a transmembrane linker to mediate interactions between the actin cytoskeleton inside the cell and fibronectin in the extracellular matrix (Figure 14–52). We shall see later that these interactions across

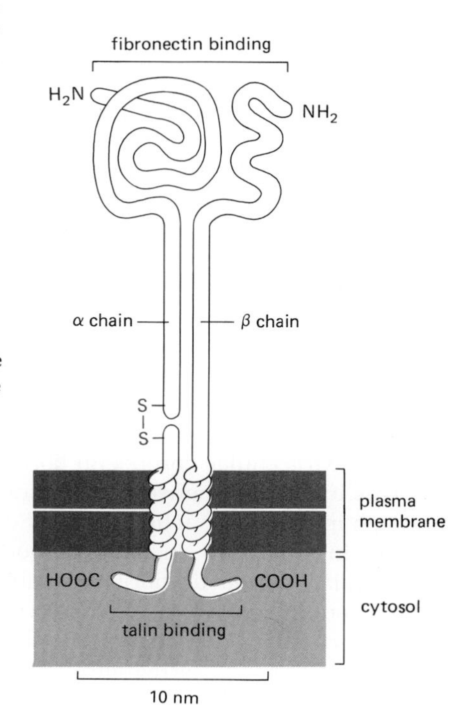

Figure 14–52 The subunit structure of a cell-surface fibronectin receptor. Electron micrographs of isolated receptors suggest that the molecule has approximately the shape shown, with the globular head projecting more than 20 nm from the lipid bilayer. By binding to fibronectin outside the cell and to the cytoskeleton (via the attachment protein talin) inside the cell, the protein serves as a transmembrane linker. The α and β chains are both glycosylated (not shown) and are held together by noncovalent bonds. The α chain is usually made initially as a single 140,000-dalton polypeptide chain, which is then cleaved into one small transmembrane chain and one large extracellular chain that remain held together by a disulfide bond. The extracellular part of the β chain contains a repeating cysteine-rich region, indicative of extensive intrachain disulfide bonding (not shown). The fibronectin receptor belongs to a large superfamily of homologous matrix receptors called *integrins*, most of which recognize RGD sequences in the extracellular proteins they bind.

the plasma membrane can orient both cells and matrix. Many other matrix receptors, including some that bind collagen and laminin, have been characterized and shown to be related to the fibroblast fibronectin receptor. Collectively called **integrins,** they are all heterodimers with α and β chains homologous to those of the fibronectin receptor. Most seem to recognize RGD sequences in the matrix components they bind.

There are at least three families within the large superfamily of integrins; the members of a family share a common β chain but differ in their α chains. One family includes a fibroblast fibronectin receptor and at least five other members. Another family includes a receptor found on blood platelets that binds several matrix components, including fibronectin and *fibrinogen,* which is a protein that interacts with platelets during blood clotting; humans with *Glanzmann's disease* are genetically deficient in these receptors and bleed excessively. A third family of integrins consists of receptors found mainly on the surface of white blood cells: one is called *LFA-1* (for lymphocyte function associated); another is called *Mac-1* because it is found mainly on macrophages. These receptors are involved in both cell-cell and cell-matrix interactions, and they are critically important in enabling these cells to fight infection. Humans with the disease called *leucocyte adhesion deficiency* are genetically unable to synthesize the β subunit. As a consequence, their white blood cells lack the entire family of receptors, and they suffer repeated bacterial infections. A number of cell-surface glycoproteins involved in position-specific cell adhesion in *Drosophila* larvae also belong to the integrin superfamily, but their relationship to the three families that are found in mammals is not certain.

Not all matrix receptors, however, belong to this superfamily. Some cells, for example, utilize an apparently unrelated transmembrane glycoprotein in binding to collagen; and many cells, as mentioned previously, have integral membrane proteoglycans that link cells to the extracellular matrix.

The Cytoskeleton and Extracellular Matrix Communicate Across the Plasma Membrane[26]

Extracellular matrix macromolecules have striking effects on the behavior of cells in culture, influencing not only their movement but also their shape, polarity, metabolism, and differentiation. Corneal epithelial cells, for example, make very little collagen when they are cultured on synthetic surfaces; but when they are cultured on laminin, collagen, or fibronectin, they accumulate and secrete large amounts of collagen. Other examples of extracellular matrix influences on cell metabolism and differentiation are discussed in Chapter 17 (see p. 987).

The matrix can also influence the organization of a cell's cytoskeleton. In general, the basal surfaces of epithelial cells cultured on plastic or glass are irregular, and the overlying cytoskeletons within the cells are disorganized. When the same cells are cultured on appropriate extracellular matrix macromolecules, the basal surfaces are smooth and the overlying cytoskeletons are highly organized, as they are in the intact tissue. Similar results have been obtained with neoplastically transformed fibroblasts in culture. Transformed cells often make less fibronectin than normal cultured cells and behave differently: for example, they adhere poorly to the substratum and fail to flatten out or develop the organized intracellular actin filament bundles known as *stress fibers* (see p. 627). In some of these cells, the fibronectin deficiency seems to be at least partly responsible for this abnormal behavior: if the cells are grown on a matrix of organized fibronectin fibrils, they will flatten and assemble intracellular stress fibers that are aligned with the extracellular fibronectin fibrils.

This interaction between the extracellular matrix and the cytoskeleton is reciprocal: intracellular actin filaments can influence the arrangement of secreted fibronectin molecules. In the neighborhood of cultured fibroblasts, for example, extracellular fibronectin fibrils assemble in alignment with adjacent intracellular stress fibers (Figure 14–53). If these cells are treated with the drug cytochalasin, which disrupts actin filaments, the fibronectin fibrils dissociate from the cell surface (just as they do during mitosis when a cell rounds up). Clearly there must be

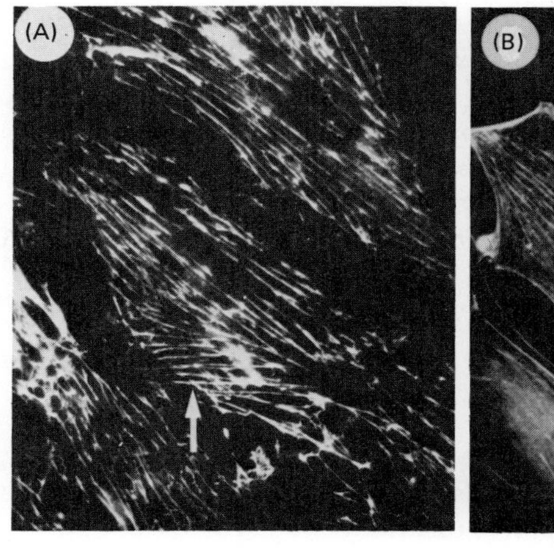

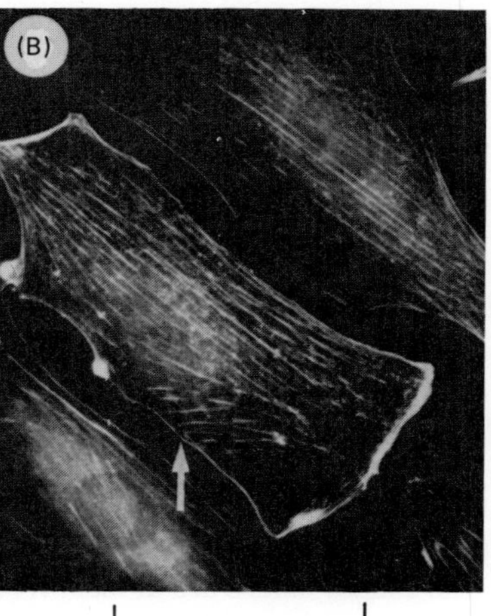

Figure 14–53 Immunofluorescence micrographs of extracellular fibronectin fibers (A) and intracellular actin filament bundles (B) in three rat fibroblasts in culture. The fibronectin is visualized by the binding of rhodamine-coupled anti-fibronectin antibodies and the actin by fluorescein-coupled anti-actin antibodies. Note that the orientation of the fibronectin fibers coincides with the orientation of the bundles of actin filaments. (From R.O. Hynes and A.T. Destree, *Cell* 15:875–886, 1978. © Cell Press.)

a connection between extracellular fibronectin and intracellular actin filaments across the fibroblast plasma membrane. The connection is mediated by the fibronectin receptors discussed previously, which serve as transmembrane linkers between fibronectin and intracellular actin filaments via a set of intracellular attachment proteins, including talin (see p. 636 and Figure 14–52). The part of the receptor that binds talin contains a tyrosine residue that, when phosphorylated by tyrosine-specific protein kinases, seems to inactivate the talin binding site, thereby breaking the link between fibronectin and cortical actin filaments. It is thought that the attachment of cells to the matrix may be regulated in this way by specific growth factors that activate tyrosine-specific kinases (see Figure 13–37, p. 758).

Since the cytoskeletons of cells can order the matrix macromolecules they secrete, and the matrix macromolecules can in turn organize the cytoskeletons of cells that contact them, the extracellular matrix can in principle propagate order from cell to cell (Figure 14–54). Thus the matrix is thought to play a central part in generating and maintaining the orientations of cells in tissues and organs during development: the parallel alignment of fibroblasts and collagen fibrils in tendons, for example, may in part reflect this type of interaction between cells and matrix. The transmembrane matrix receptors serve as "adaptors" in this ordering process, mediating the interactions between cells and the matrix around them.

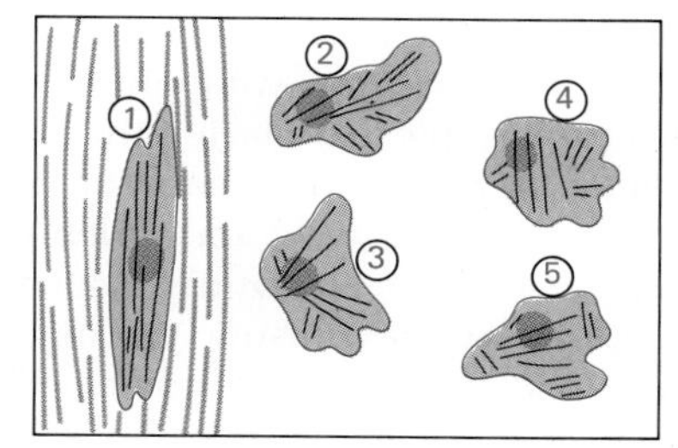

orientation of cytoskeleton in cell ① orients the assembly of secreted extracellular matrix molecules in the vicinity

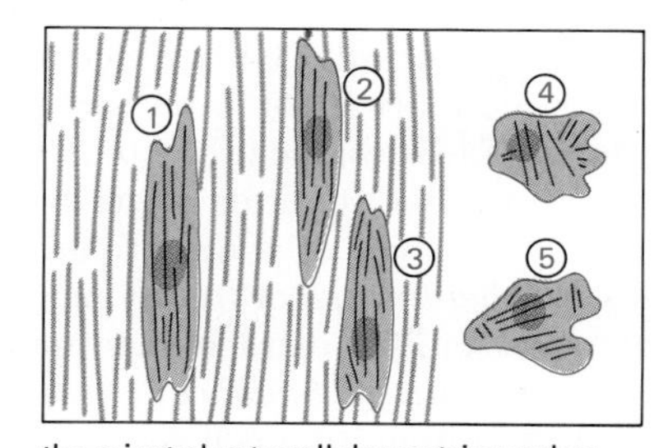

the oriented extracellular matrix reaches cells ② and ③ and orients the cytoskeleton of those cells

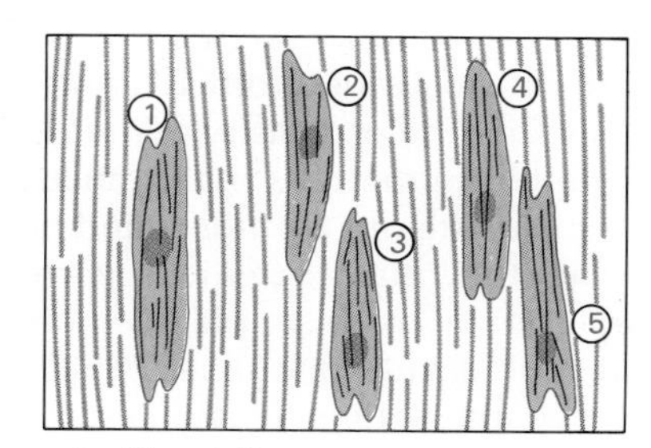

cells ② and ③ now secrete an oriented matrix in their vicinity; in this way the ordering of cytoskeletons is propagated to cells ④ and ⑤

Figure 14–54 A hypothetical scheme showing how the extracellular matrix could propagate order from cell to cell within a tissue. For simplicity, the figure shows one cell influencing the orientation of its neighboring cells, but by this scheme cells could mutually affect one another's orientation.

Summary

Cells in connective tissues are embedded in an intricate extracellular matrix that not only binds cells and tissues together but also influences the development, polarity, and behavior of the cells it contacts. The matrix contains various fiber-forming proteins interwoven in a hydrated gel composed of a network of glycosaminoglycan chains. The glycosaminoglycans are a heterogeneous group of long, negatively charged polysaccharide chains, which (except for hyaluronic acid) are covalently linked to protein to form proteoglycan molecules.

The fiber-forming proteins are of two functional types: mainly structural (collagens and elastin) and mainly adhesive (such as fibronectin and laminin). The fibrillar collagens (types I, II, and III) are ropelike, triple-stranded helical molecules that aggregate into long cablelike fibrils in the extracellular space; these in turn can assemble into a variety of highly ordered arrays. Type IV collagen molecules assemble into a sheetlike meshwork that forms the core of all basal laminae. Elastin molecules form an extensive cross-linked network of fibers and sheets that can stretch and recoil, imparting elasticity to the matrix. Fibronectin and laminin are examples of large adhesive glycoproteins in the matrix; fibronectin is widely distributed in connective tissues, whereas laminin is found mainly in basal laminae. By means of their multiple binding domains, such proteins help cells adhere to and become organized by the extracellular matrix. Many of these adhesive glycoproteins contain a common tripeptide sequence (RGD), which forms part of the structure recognized by a superfamily of homologous transmembrane matrix receptors, called integrins.

All of the matrix proteins and polysaccharides are secreted locally by cells in contact with the matrix, and they can be ordered by close associations with the exterior surface of the plasma membrane. Since the structure and orientation of the matrix in turn influence the orientation of the cells it contains, order is likely to be propagated from cell to cell through the matrix.

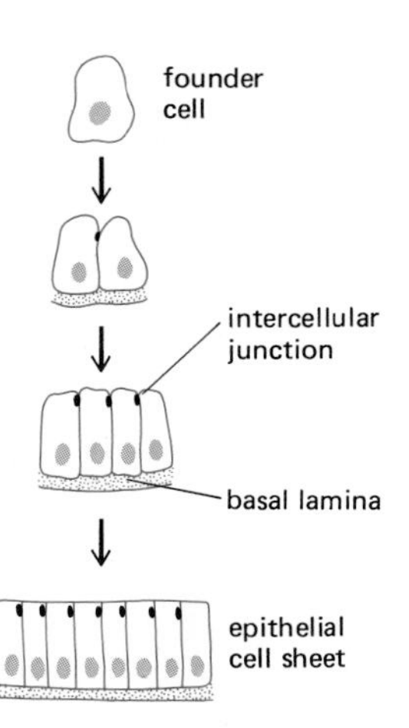

Figure 14–55 The simplest mechanism by which cells assemble to form a tissue. The progeny of the founder cells are retained in the epithelial sheet by the basal lamina and by cell-cell adhesion mechanisms, including the formation of intercellular junctions.

Cell-Cell Recognition and Adhesion[27]

We have so far considered how cell junctions and the extracellular matrix hold cells together in mature tissues and organs. But how do cells become associated with one another to form tissues in the first place? There are at least two distinctly different ways. Most commonly a tissue forms from "founder cells" whose progeny are prevented from wandering away by being attached to extracellular matrix macromolecules and/or to other cells (Figure 14–55). The precise pattern of adhesions determines the shape of the cell assembly. Epithelial cell sheets usually originate in this way, and much of animal development involves the formation, folding, and differentiation of such cell sheets to produce the tissues and organs of the mature organism. Typically, all of the cells in an early embryo are located in epithelia; only later do some cells change their adhesive properties and thereby escape to form other types of tissue (see pp. 882–888).

The other strategy for tissue formation seems more complex and involves cell migration: one population of cells invades another and assembles with them—and perhaps with other migrant cells—to form a tissue of mixed origin. In vertebrate embryos, for example, cells from the *neural crest* break away from the epithelial (neural) tube with which they are initially associated and migrate along specific paths to many other regions. There they assemble and differentiate into a variety of tissues, including those of the peripheral nervous system (Figure 14–56). Such a process requires some mechanism for directing the cells to their final destination, such as the secretion of a soluble chemical that attracts migrating cells (by *chemotaxis*) or the laying down of adhesive molecules, such as fibronectin (see p. 817), in the extracellular matrix to guide the migrating cells along the right paths (by *pathway guidance*).

Once a migrating cell reaches its destination, it must recognize other cells of the appropriate type in order to assemble into a tissue. Even in tissues that form without cell migration, there is evidence that the constituent cells specifically recognize one another: if the cells of such a developing tissue are dissociated into a suspension of single cells, they preferentially reassociate with one another rather than with cells of another tissue (see p. 827). Presumably this specific cell-cell recognition helps cells within a developing tissue stay together and remain segregated from the cells of neighboring tissues.

In an attempt to understand how cells in developing animal tissues recognize one another, many ingenious studies have been carried out on the social behavior displayed by some simple microorganisms that can alternate between unicellular and multicellular life-styles. Quite apart from their role as tentative models for cell-cell interactions in animals, these organisms are intriguing in their own right.

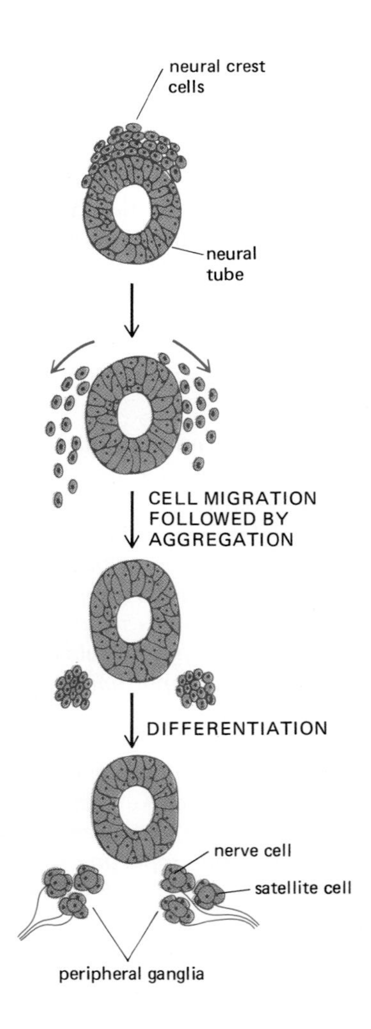

Figure 14–56 An example of a more complex mechanism by which cells assemble to form a tissue. Neural crest cells escape from the epithelium forming the upper surface of the neural tube and migrate away to form a variety of cell types and tissues throughout the embryo. Here they are shown assembling and differentiating to form two collections of nerve cells in the peripheral nervous system. Such a collection of nerve cells is called a *ganglion*. Other neural crest cells differentiate in the ganglion to become supporting (satellite) cells surrounding the neurons.

Slime Mold Amoebae Aggregate to Form Multicellular Fruiting Bodies When Starved[28]

The cellular slime mold, ***Dictyostelium discoideum,*** is a eucaryote whose genome is only 10 times larger than that of a bacterium and 100 times smaller than that of a human. These organisms live on the forest floor as independent motile cells called *amoebae,* which feed on bacteria and yeast and, under optimal conditions, divide every few hours. (In the laboratory they can be nourished with chemically defined liquid medium.) When their food supply is exhausted, the amoebae stop dividing and gather together to form tiny (1–2 mm), multicellular, wormlike structures, which crawl about as glistening slugs and leave trails of slime behind them (Figure 14–57).

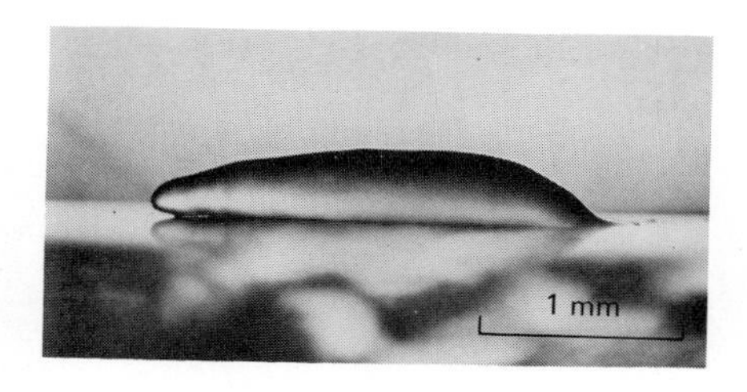

Figure 14–57 Light micrograph of a migrating slug of the cellular slime mold *Dictyostelium discoideum.* (Courtesy of David Francis.)

Each slug is formed by the aggregation of up to 100,000 cells and shows a variety of behaviors that are not displayed by the free-living amoebae. The slug is extremely sensitive to light and heat, for example, and will migrate toward a light source as feeble as a luminous watch; presumably this behavior helps guide the slug to an advantageous environment. As the slug migrates, the cells begin to differentiate, initiating a process that ends with the production of a minute plant-like structure consisting of a stalk and a *fruiting body* some 30 hours after the beginning of aggregation (Figure 14–58). The fruiting body contains large numbers of *spores,* which can survive for long periods of time even in extremely hostile environments. The complex cell migrations that occur in stalk and fruiting-body formation are diagrammed in Figure 14–59. The cells in the front of the slug become the stalk region, those behind differentiate into spores, and those right at the rear form the foot plate. Both the stalk cells and the spore cells become covered with extracellular matrix (in the form of cellulose walls), and, in the end, all except the spore cells die. Only when conditions are favorable do the spores germinate to produce the free-living amoebae that start the cycle again (Figure 14–60).

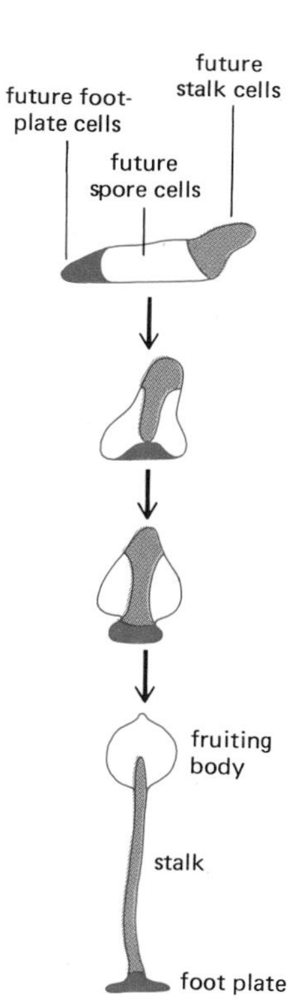

Figure 14–59 Cell migrations involved in the formation of a fruiting body in *Dictyostelium discoideum.* Cells in the front of the slug migrate down to become the stalk, while cells in the middle migrate up and differentiate into the collection of spores that forms the fruiting body.

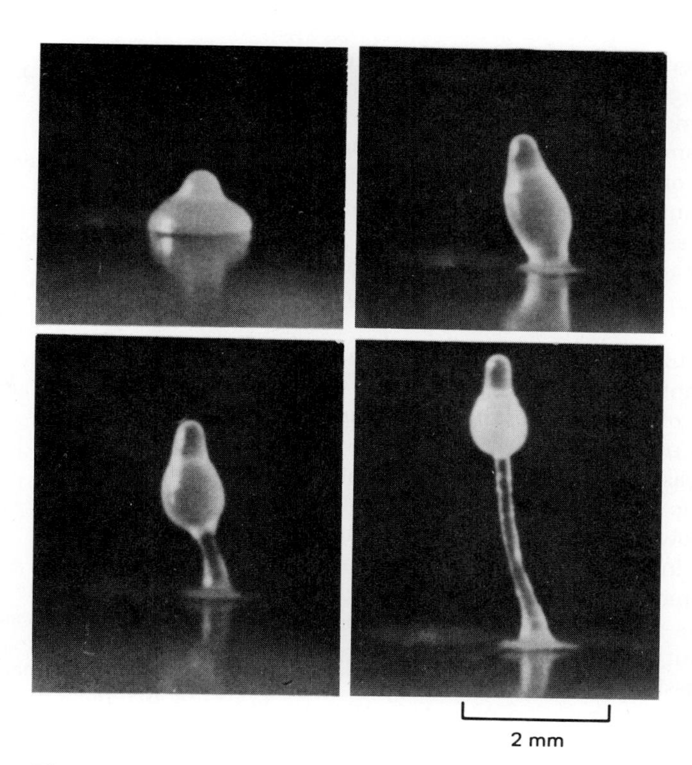

Figure 14–58 Light micrographs of *Dictyostelium discoideum* showing various stages in fruiting-body formation. (Courtesy of John Bonner.)

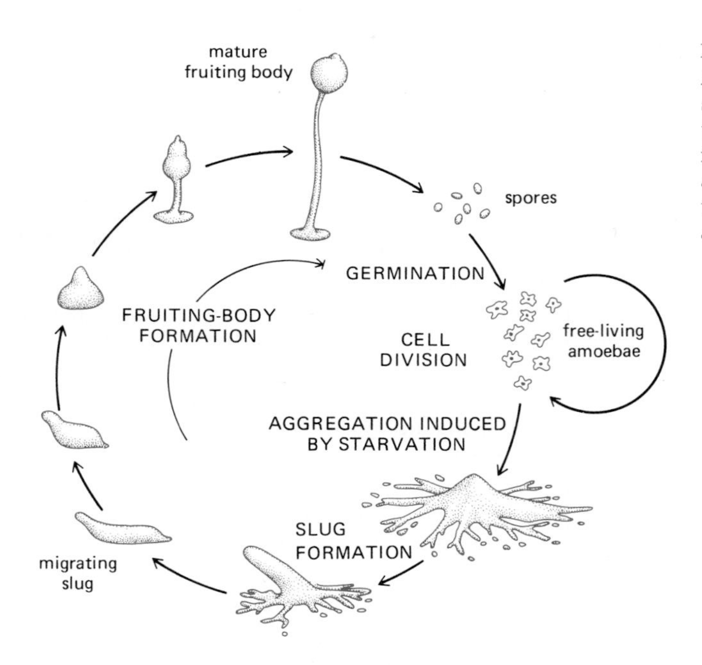

Figure 14–60 The life cycle of *Dictyostelium discoideum.* When starved, slime mold amoebae aggregate to form migrating slugs, which in turn form fruiting bodies. When conditions are favorable, the spores released from the fruiting bodies germinate to form amoebae again.

14-18 Slime Mold Amoebae Aggregate by Chemotaxis[29]

In forming the slug, individual slime mold amoebae aggregate by *chemotaxis,* which we shall digress briefly to discuss before returning to the part played by cell-cell adhesion. One response of the amoebae to starvation is to start making and secreting cyclic AMP, which serves as a chemotactic signal that attracts other amoebae. (Cyclic AMP, as we have seen in Chapter 12, serves as an intracellular signal in procaryotic and animal cells; *Dictyostelium* is the only known organism in which cyclic AMP also acts as an extracellular signaling molecule.) Aggregation is apparently initiated at random: whichever cells first begin secreting cyclic AMP attract other cells and thereby become *aggregation centers.* The cyclic AMP that these "initiator" cells make is secreted in pulses and binds to specific receptors on the surface of neighboring starved amoebae, thereby orienting their normal locomotion in the direction of the source of cyclic AMP. This chemotactic response can be demonstrated by applying a tiny amount of cyclic AMP with a micropipette to any point on the surface of a starved amoeba cell. The result is the immediate formation of a pseudopod, which grows toward the micropipette (Figure 14–61); the pseudopod adheres to the surface on which the cell is placed and pulls the cell along in the same direction.

Once an aggregation center starts to form, its area of influence is rapidly enlarged because the aggregating cells not only respond to the cyclic AMP signal but relay it from cell to cell. Each pulse of cyclic AMP induces surrounding cells both to move toward the source of the pulse and to secrete their own pulse of cyclic AMP. In turn, this new pulse, released after a slight delay, orients the cells just beyond and induces a pulse of cyclic AMP from them, and so on. In this way regular pulsating waves of cyclic AMP flow from each aggregation center, causing more distant amoebae to move inward in surging concentric or spiraling waves that can be seen in time-lapse motion pictures (Figure 14–62). The advantage of such a relay system is that the signal is repeatedly renewed as it spreads from the center, so that it propagates without diminution over a large area. A signal that merely diffuses, by contrast, progressively fades out as it spreads. The difference can be appreciated by comparing the aggregation process in *Dictyostelium discoideum* with that in *Dictyostelium minutum,* a strain that lacks the relay system. In *D. minutum* the range of the signal released from each aggregation center is greatly reduced, and the slug and fruiting bodies that form are very small.

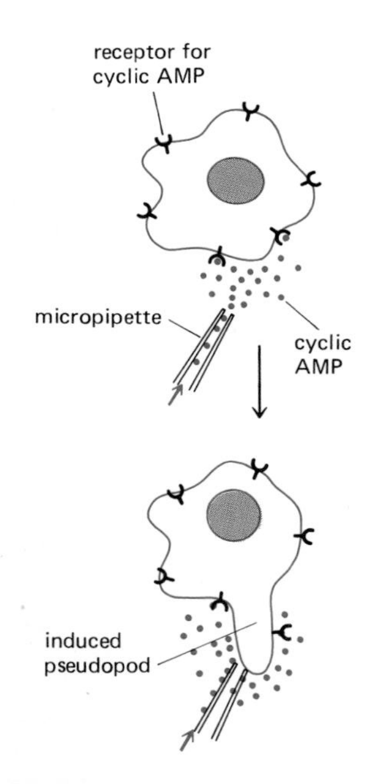

Figure 14–61 Application of a small amount of cyclic AMP to any point on the surface of a starved *Dictyostelium* amoeba (a single cell) results in the immediate formation of a pseudopod at that point. In this way the amoeba is induced to move toward a source of cyclic AMP. The cyclic AMP acts by binding to specific cell-surface receptors.

14-20 Cell-Cell Adhesion in Slime Molds Depends on Specific Cell-Surface Glycoproteins[30]

Besides activating the cyclic AMP signaling system, starvation of *Dictyostelium* amoebae induces the expression of hundreds of new genes, some of which encode the cell adhesion molecules involved in cell aggregation. For example, **discoidin-1,** which is a carbohydrate-binding protein (i.e., a lectin—see p. 299), is thought to be secreted by the starved cells to provide a primitive form of pathway guidance. By binding both to the surface of the amoebae and to the substratum on which they are migrating, it might help the amoebae move in streams toward aggregation centers, much as a trail of fibronectin guides cell migration during gastrulation in developing animals. In fact, cell binding to discoidin-1 depends on the same cell-binding tripeptide (RGD) sequence found in fibronectin and many other adhesive proteins (see p. 817).

Different newly synthesized proteins promote the cell-cell adhesion process that enables the migrating amoebae to adhere tightly to one another and assemble into a multicellular organism. During the first 8 hours of starvation, cells adhere by a Ca^{2+}-dependent mechanism involving a cell-cell adhesion molecule called **contact site B.** After 8 hours, a second adhesion system comes into play in which cells adhere by a Ca^{2+}-independent mechanism involving a cell-cell adhesion molecule called **contact site A.** Contact sites A and B were identified and characterized as integral plasma membrane glycoproteins by an ingenious immunological strategy, outlined in Figure 14–63. This method was later used to identify cell-cell adhesion molecules in vertebrates as well.

Figure 14–62 Light micrograph of waves of starved *Dictyostelium* amoebae moving toward an aggregation center. Individual amoebae cannot be distinguished at this low magnification. (Courtesy of Günter Gerisch.)

How do cell-surface glycoproteins such as contact sites A and B bind cells together? Three possibilities are illustrated in Figure 14–64: (1) molecules on one cell may bind to other molecules of the same kind on adjacent cells (so-called *homophilic* binding); (2) molecules on one cell may bind to molecules of a different kind on adjacent cells (so-called *heterophilic* binding); and (3) cell-surface receptors on adjacent cells may be linked to one another by secreted multivalent linker molecules. All of these mechanisms have been found to operate in animals.

Contact site A is thought to bind cells together by a homophilic mechanism: when the protein is coupled to synthetic beads, the beads bind only to cells that express contact site A, and the binding is blocked if the cells are pretreated with antibody against contact site A. DNA-sequencing studies show that contact site A is a single-pass transmembrane protein, apparently unrelated to any vertebrate cell-cell adhesion protein so far characterized (see below).

Dissociated Vertebrate Cells Can Reassemble into Organized Tissues Through Selective Cell-Cell Adhesion[31]

One of the attractions of using social microorganisms such as *Dictyostelium* to study cell aggregation is that the phenomenon proceeds normally in a culture dish, where it is accessible for investigation. Unfortunately, this is rarely the case with cell-cell recognition processes that occur during the development of multicellular animals. Usually the best one can do is to dissociate newly formed tissues into single cells, which can then be tested for their ability to reassemble *in vitro.* Unlike adult vertebrate tissues, which are difficult to dissociate, embryonic vertebrate tissues are easily dissociated by treatment with low concentrations of a proteolytic enzyme such as trypsin, sometimes combined with the removal of extracellular Ca^{2+} with a Ca^{2+} chelator (such as EDTA). These reagents disrupt the protein-protein interactions (many of which are Ca^{2+}-dependent—see p. 830) that hold cells together. Remarkably, such dissociated cells often reassemble *in vitro* into structures that resemble the original tissue. The tissue structure, therefore, is not just a product of history but is actively maintained and stabilized by the system of affinities that cells have for one another and for the extracellular matrix. Thus, by studying the reassembly of dissociated cells in culture, one can hope to illuminate the role of cell-cell and cell-matrix adhesion in creating and maintaining the organization of tissues in the body.

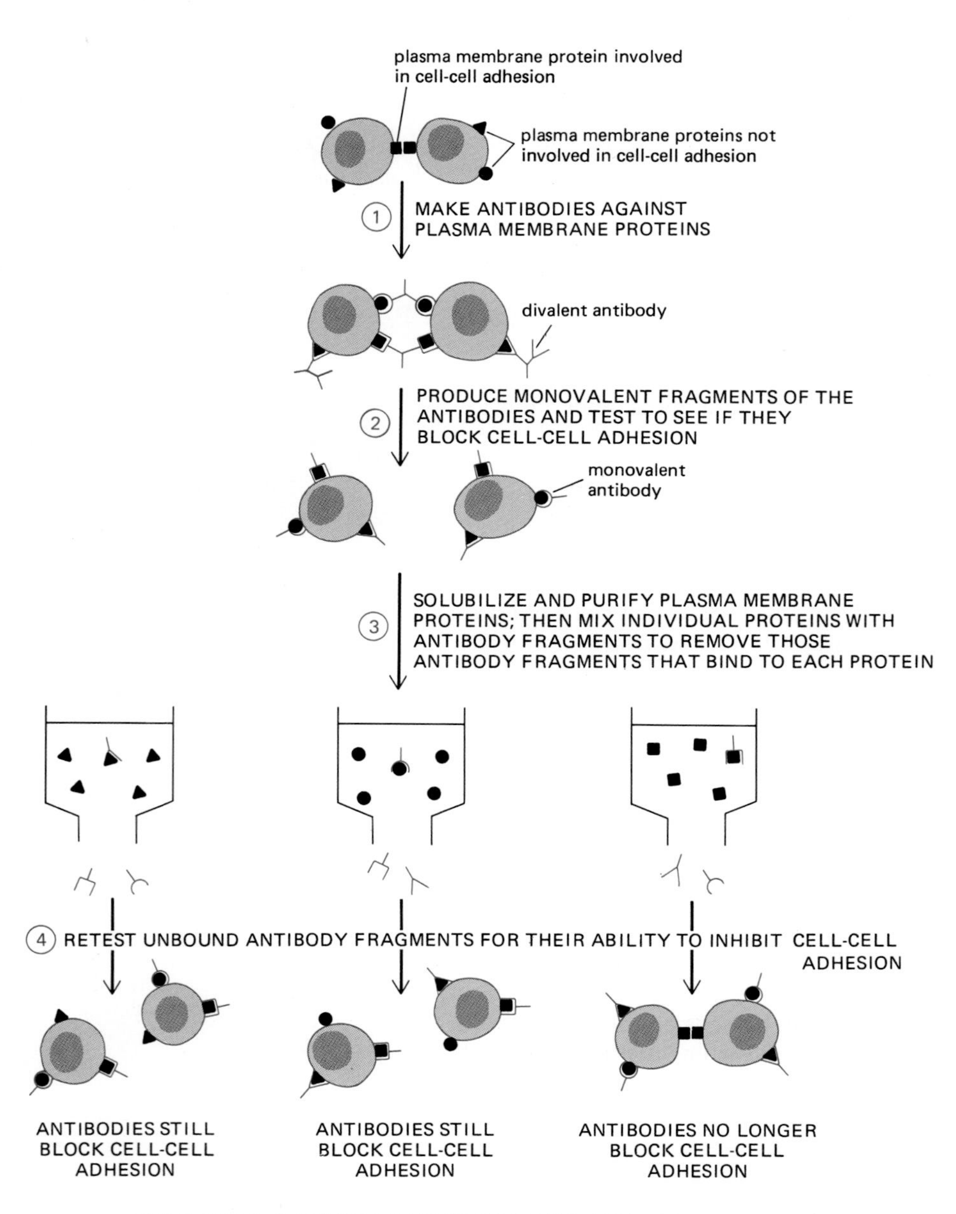

Figure 14–63 An immunological strategy for identifying plasma membrane proteins involved in cell-cell adhesion. In step 1, antibodies are made (usually in rabbits) against the cell of interest or against plasma membranes isolated from such cells. In step 2, monovalent fragments of the antibodies are produced and tested to find an antibody preparation that inhibits cell-cell adhesion. (Monovalent fragments, made by protease digestion [see p. 1013], are used because they cannot cross-link cells and cause irrelevant adhesions.) To find the cell-surface molecules involved in cell-cell adhesion, the plasma membrane proteins are solubilized from the cells of interest, separated from one another, and individual fractions are tested for their ability to neutralize the aggregation-blocking effect of the antibody fragments (steps 3 and 4). Fractions that contain blocking activity are then further purified and retested until a pure protein is obtained (not shown). An alternative immunological strategy involves making monoclonal antibodies (see p. 178) against cell-surface antigens and screening large numbers to find those few that block cell-cell adhesion. Both of these immunological strategies depend on an important general finding: simply coating the surface of cells with antibodies does not in itself interfere with normal cell adhesion; adhesion is inhibited only when specific cell-surface molecules involved in the adhesion process are targets for antibody binding.

Experiments on cultured cells from the *epidermis* (the epithelium of the skin) provide an instructive example. In this tissue, Ca^{2+}-dependent adhesion systems play a crucial part in holding the cells, known as *keratinocytes*, together in a multilayered sheet resting on a basal lamina. The keratinocytes in the basal layer of the skin are relatively undifferentiated and proliferate steadily, releasing progeny into the upper layers, where cell division halts and terminal differentiation occurs (see p. 968). Given a suitable substratum, dissociated keratinocytes will likewise proliferate and differentiate in culture. If the concentration of Ca^{2+} in the culture medium is kept abnormally low, however, the Ca^{2+}-dependent cell-cell adhesion systems cannot operate, and the keratinocytes grow as a monolayer in which proliferating and differentiating cells are intermingled. If the Ca^{2+} concentration is then raised, the spatial organization of the cells is soon transformed: the monolayer is converted into a multilayered epithelium in which the proliferating cells form the basal layer adherent to the substratum and the differentiating cells are segregated into the upper layers, just as in normal skin. This result suggests that the normal stratified arrangement of keratinocytes, ordered according to their state of differentiation, is maintained by Ca^{2+}-dependent cell-cell adhesion mechanisms (see Figure 14–68, p. 832).

The Reassembly of Dissociated Vertebrate Cells Depends on Tissue-specific Recognition Systems[32]

The normal development of most tissues does not involve sorting out of randomly mixed cell types (see p. 911). Nevertheless, when dissociated embryonic cells from two vertebrate tissues such as liver and retina are mixed together, the mixed aggregates initially formed gradually sort out according to their tissue of origin. This assay presumably detects tissue-specific cell-cell recognition systems that keep the cells in a developing tissue together. Such recognition systems can also be demonstrated in another way. As described in Figure 14–65, disaggregated cells are found to adhere more readily to aggregates of their own tissue than to aggregates of other tissues. Thus similar results are obtained by two different assays, one determining the selective affinities of cells in aggregates during prolonged incubation (the sorting-out experiments) and the other measuring the rate at which cells bind to preformed aggregates.

What is the molecular basis of this selective cell-cell adhesion in vertebrates? As with slime molds, two distinct classes of cell-cell adhesion mechanisms, one Ca^{2+}-independent and the other Ca^{2+}-dependent, seem to be responsible, and each class is mediated by its own family of homologous cell-surface glycoproteins.

Plasma-Membrane Glycoproteins of the Immunoglobulin Superfamily Mediate Ca^{2+}-independent Cell-Cell Adhesion in Vertebrates: The Neural Cell Adhesion Molecule (N-CAM)[33]

The immunological strategy outlined in Figure 14–63 has been used to identify a number of cell-surface glycoproteins involved in cell-cell adhesion in vertebrates. In one of the best-studied examples, monovalent antibody fragments were made

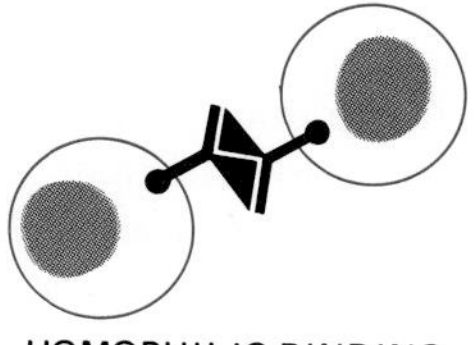

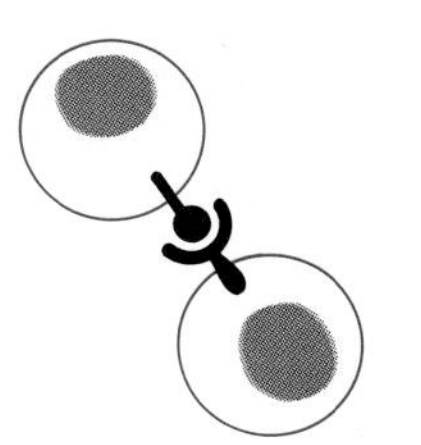

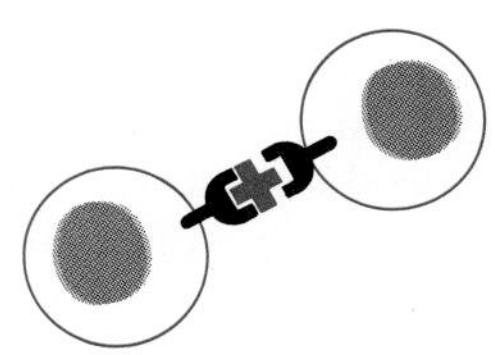

Figure 14–64 Three mechanisms by which cell-surface molecules can mediate cell-cell adhesion.

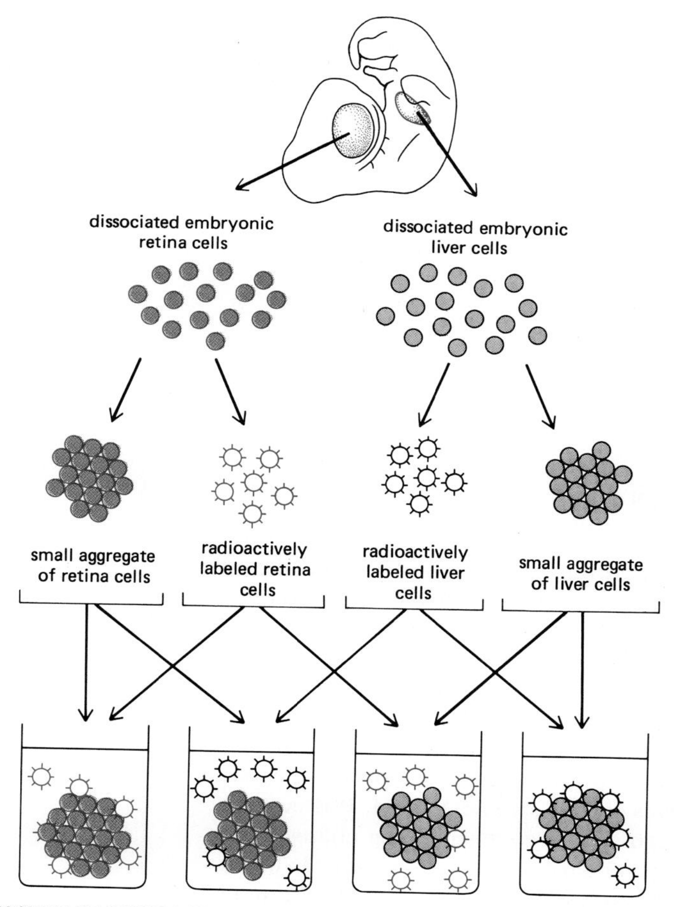

Figure 14–65 Tissue-specific adhesion of dissociated vertebrate embryo cells determined by a radioactive cell-binding assay. The rate of cell adhesion can be measured by determining the number of radioactively labeled cells bound to the cell aggregates after various periods of time. The rate of adhesion is greater between cells of the same kind. In a commonly used modification of this assay, cells labeled with a fluorescent or radioactive marker are allowed to bind to a monolayer of unlabeled cells in culture.

against cells from embryonic chick retina. Antibodies that inhibited the reaggregation of these cells *in vitro* were selected. Retinal cell membrane proteins were then fractionated and tested for their ability to neutralize the blocking activity of the antibodies. In this way a large, single-pass transmembrane glycoprotein (about 1000 amino acid residues long), called **neural cell adhesion molecule (N-CAM),** was identified. N-CAM is expressed on the surface of nerve cells and glial cells (see p. 1064) and causes them to stick together by a Ca^{2+}-independent mechanism. When these membrane proteins are purified and inserted into synthetic phospholipid vesicles, the vesicles bind to one another, as well as to cells that have N-CAM on their surface; the binding is blocked if the cells are pretreated with monovalent anti-N-CAM antibodies. These findings suggest that N-CAM binds cells together by a homophilic interaction that directly joins two N-CAM molecules (see Figure 14–64).

Anti-N-CAM antibodies disrupt the orderly pattern of retinal development in tissue culture and, when injected into the developing chick eye, disturb the normal growth pattern of retinal nerve cell axons. As explained in Chapter 19 (see p. 1116), these observations suggest that N-CAM plays an important part in the development of the central nervous system by promoting cell-cell adhesion. In addition, the neural crest cells that form the peripheral nervous system have large amounts of N-CAM on their surface when they are associated with the neural tube, lose it while they are migrating, and then reexpress it when they aggregate to form a ganglion (see Figure 14–56), suggesting that N-CAM may play a part in the assembly of the ganglion. N-CAM is also expressed transiently during critical stages in the development of many nonneural tissues, where it presumably helps specific cells stay together.

There are several forms of N-CAM, each encoded by a distinct mRNA. The different mRNAs are generated by alternative splicing of an RNA transcript produced from a single large gene. The large extracellular part of the polypeptide chain (~680 amino acid residues) is identical in most forms of N-CAM and is folded into five domains that are homologous to the immunoglobulin domains characteristic of antibody molecules (see p. 1021). Thus N-CAM belongs to the same ancient superfamily of recognition proteins to which antibodies belong (see p. 1053). The various forms of N-CAM differ mainly in their membrane-associated segments and cytoplasmic domains and therefore may interact with the cytoskeleton in different ways; in fact, one form does not cross the lipid bilayer and is attached to the plasma membrane only by a covalent linkage to phosphatidylinositol (see p. 448) (Figure 14–66), while another is secreted and incorporated into the extracellular matrix. It is not known how the various forms differ in function.

An increasing number of cell-surface glycoproteins that mediate Ca^{2+}-independent cell-cell adhesion in vertebrates are being discovered to belong to the immunoglobulin superfamily. But not all cell-surface proteins that mediate cell-cell adhesion are members of this superfamily: those that can function only in the presence of extracellular Ca^{2+} belong to another family.

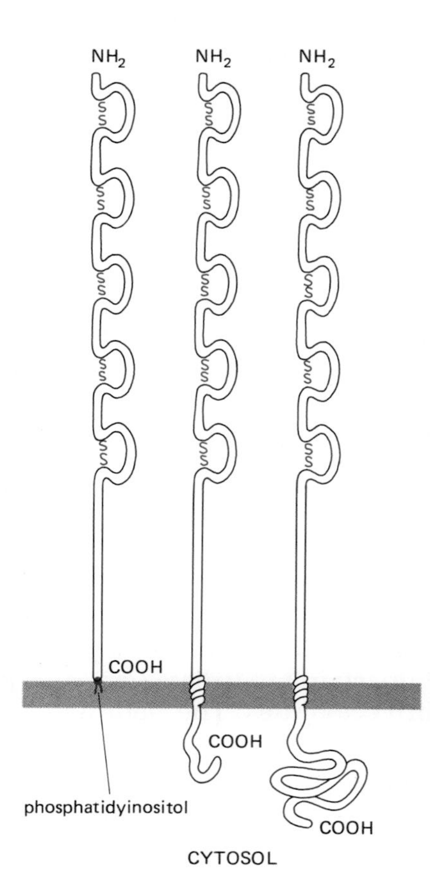

Figure 14–66 Schematic drawing of three forms of N-CAM. The extracellular part of the polypeptide chain is the same in each case and is folded into five immunoglobulinlike domains. Disulfide bonds connect the ends of each loop forming such a domain. (Based on data from B.A. Cunningham, J.J. Hemperly, B.A. Murray, E.A. Prediger, R. Brackenbury, and G.M: Edelman, *Science* 236:799–800, 1987. Copyright 1987 by the AAAS.)

The Cadherins, a Family of Homologous Cell-Surface Glycoproteins, Mediate Ca^{2+}-dependent Cell-Cell Adhesion in Vertebrates[34]

The immunological strategies explained in Figure 14–63 also played a crucial part in the discovery of three related cell-surface glycoproteins, collectively called **cadherins,** involved in Ca^{2+}-dependent cell-cell adhesion in vertebrate tissues. *E-cadherin* is found on many types of epithelial cells (and on cells in preimplantation mammalian embryos), *N-cadherin* on nerve, heart, and lens cells, and *P-cadherin* on cells in the placenta and epidermis; as for N-CAM, all are also found transiently on various other tissues during development. The three cadherins are homologous, single-pass transmembrane glycoproteins (each composed of about 700 amino acid residues). In these respects they resemble N-CAM. However, in the absence of Ca^{2+}, the cadherins undergo a large conformational change and, as a result, are rapidly degraded by proteolytic enzymes. Since some cells (such as endothelial cells) show Ca^{2+}-dependent adhesion but do not express any of the three known cadherins, it seems likely that additional members of the cadherin family remain to be discovered.

E-cadherin (also called liver cell adhesion molecule [*L-CAM*] or *uvomorulin*) is the best characterized. The large extracellular part of its polypeptide chain is folded into three homologous domains, which are apparently unrelated to immunoglobulin domains. It seems to play an important part in holding adjacent cells together in various epithelia. The Ca^{2+}-dependent reaggregation of dissociated epithelial cells from liver, for example, is blocked by antibodies to E-cadherin. In mature epithelial tissues, E-cadherin is usually concentrated in *adhesion belts*, where it is thought to function as a transmembrane linker, connecting the cortical actin cytoskeletons of the cells it holds together (see p. 796). It is also involved in the compaction of blastomeres in the early mouse embryo (see p. 894). During compaction, the blastomeres, at first loosely attached to one another, become tightly packed together and joined by intercellular junctions. Antibodies against E-cadherin block blastomere compaction, whereas antibodies that react with various other cell-surface molecules on these cells do not.

It seems likely that cadherins also play a crucial role in later stages of vertebrate development, since their appearance and disappearance correlate with major morphogenetic events in which tissues segregate from one another. For example, as the neural tube forms and pinches off from the overlying ectoderm (see p. 887), the cells of the developing neuroepithelium lose E-cadherin and acquire N-cadherin (as well as N-CAM) (Figure 14–67). As for N-CAM (see p. 830), when neural crest cells dissociate from the neural tube and migrate away, they lose N-cadherin, reexpressing it later when they aggregate to form a neural ganglion (see Figure 14–56).

The biological significance of the striking Ca^{2+}-dependence of the cadherin family of cell-cell adhesion proteins is unknown. There is no evidence as yet that the extracellular concentration of Ca^{2+} is regulated to control cell-cell adhesion during development, for example.

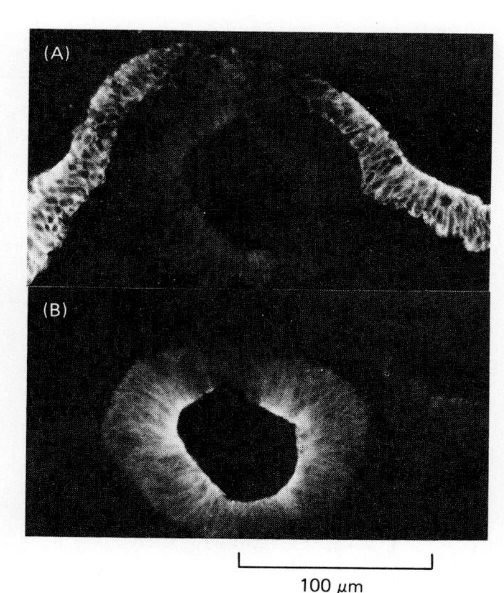

Figure 14–67 Immunofluorescence micrographs of a cross section of a chick embryo showing the developing neural tube labeled with antibodies against E-cadherin (A) and N-cadherin (B). Note that the overlying ectoderm cells express only E-cadherin, while the cells in the neural tube have lost E-cadherin and have acquired N-cadherin. (Courtesy of Kohei Hatta and Masatoshi Takeichi.)

Cell-Surface Molecules That Mediate Cell-Cell and Cell-Matrix Adhesion Can Be Viewed as Elements of a Morphogenetic Code[35]

Morphological, cell biological, and biochemical studies all indicate that even a single cell type utilizes multiple molecular mechanisms in adhering to other cells and to the extracellular matrix. Some of these mechanisms involve organized cell junctions; others do not (Figure 14–68). Because of the large number of adhesive systems used by an individual cell, nearly every cell type will share at least one cell-cell adhesion system with every other and therefore bind to it with some affinity. Usually cells from different tissues (and even from very different species) will also form desmosomes, gap junctions, and adherens junctions with each other, suggesting that the junctional proteins involved are highly conserved between tissues (and species). However, just as each cell in a multicellular animal contains an assortment of cell-surface receptors that enables the cell to respond specifically to a complementary assortment of soluble chemical signals (hormones and local chemical mediators), so each cell in a tissue is thought to have a particular combination (or concentration) of cell-surface receptors that enables it to bind in its own characteristic way to other cells and to the extracellular matrix.

Unlike receptors for soluble chemical signals, which bind their specific ligand with high affinity, the receptors that bind to molecules on cell surfaces or in the extracellular matrix usually do so with relatively low affinity. The latter receptors therefore rely on the enormous increase in binding strength gained through simultaneous binding of multiple receptors to multiple ligands on an opposing cell or in the adjacent matrix. The mixture of specific types of cell-cell adhesion molecules and matrix receptors present on any two cells, as well as their concentration and distribution on the cell surface, will therefore determine the total affinity with which the two cells bind to each other and to the matrix. It is this mixture that presumably constitutes a "morphogenetic code" to help determine how cells are organized in tissues. Since even closely related types of animal cells sort out correctly *in vitro*, cells must be able to detect relatively small differences in adhesion and to act on these differences so as to establish only the most adhesive of many possible cell-cell and cell-matrix contacts. Studies carried out on motile cells in culture suggest how this might occur.

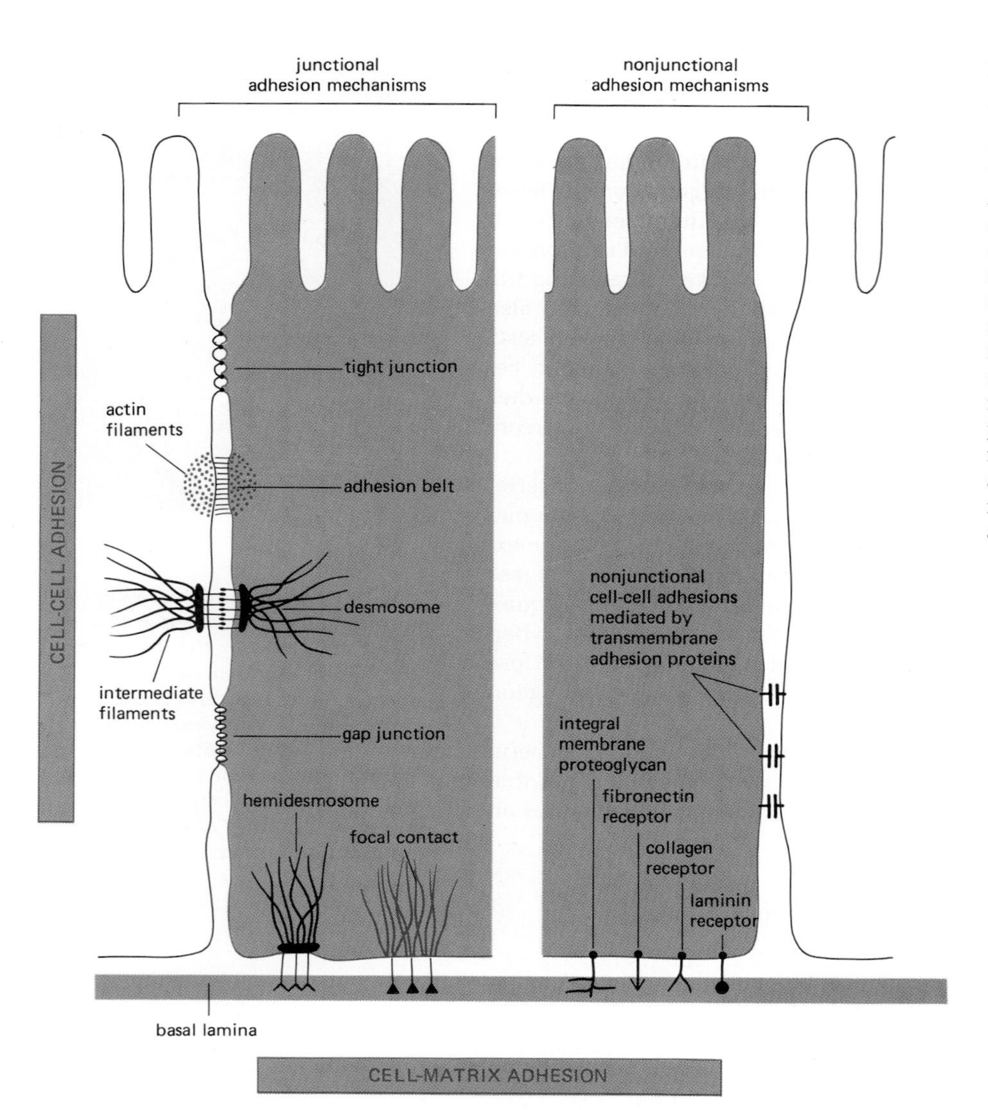

Figure 14–68 A summary of the junctional and nonjunctional adhesive mechanisms used by typical epithelial cells in binding to one another and to the extracellular matrix (basal lamina). A junctional interaction is operationally defined as one that can be seen as a specialized region of contact by conventional and/or freeze-fracture electron microscopy. In some cases the same cell-surface glycoproteins are involved in binding cells to one another (or to the matrix) at both junctional and nonjunctional contacts. Except for gap junctions, all of the junctional adhesion mechanisms are Ca^{2+}-dependent; as discussed in the text, some nonjunctional adhesion mechanisms are Ca^{2+}-dependent and others are not.

Highly Motile Cells Are Sensitive Detectors of Small Adhesive Differences[36]

The cells involved in morphogenetic processes in embryos are often highly motile. If such cells are dissociated from embryos and observed in a culture dish, they are seen initially to extend microspikes and lamellipodia in all directions and then to crawl actively along the surface of the dish. The onset of motile behavior often coincides with the beginning of cell diversification and, therefore, with the period when specific cell recognition would be expected to become important. The cells in a *Xenopus* embryo, for example, suddenly become highly motile at the mid-blastula transition stage, when gene transcription begins (see p. 881).

Extensive studies on the mechanism of cell motility have been carried out in culture on fibroblasts, neutrophils, and regenerating neurons. The results, summarized in Chapter 11, indicate that a motile cell is a highly sensitive detector of small adhesive differences. The microspikes and lamellipodia that it extends in all directions appear to engage in a "tug-of-war," causing the cell to become polarized and move steadily in the direction of the most adhesive part of the substratum, even when the differences in adhesiveness are small (see p. 669). Fibroblasts, for example, will move steadily up a small adhesive gradient created on the surface of a culture dish. From studies of chemotaxis in neutrophils, it seems that a motile cell can detect differences in adhesiveness on either side of it as small as 1%. By similar means, cells in tissues may be able to decipher the "morphogenetic code" on cell surfaces with great sensitivity, moving steadily to establish intimate contacts with those neighboring cells to which they are most adherent.

Nonjunctional Contacts May Initiate Tissue-specific Cell-Cell Adhesion That Junctional Contacts Then Stabilize[37]

Which, if any, of the many types of intercellular junctions discussed at the beginning of this chapter are involved as cells migrate and recognize one another during the formation of tissues and organs? One way to find out is to use an electron microscope to examine the contacts between adjacent cells when they are moving over each other in developing embryos or in adult tissues undergoing repair after injury. Such studies show that these contacts generally do not involve the formation of organized intercellular junctions. Nevertheless, the interacting plasma membranes often come close together and run parallel, separated by a space of 10–20 nm. This is about the distance (~13 nm) that the influenza virus hemagglutinin glycoprotein (the first plasma membrane glycoprotein whose three-dimensional structure was determined—see p. 446) projects from the plasma membrane. Two cell-surface glycoproteins of this size could interact with each other across the 10–20-nm gap to mediate the adhesion. This type of nonjunctional contact may be optimal for cell locomotion—close enough to give traction but not tight enough to immobilize the cell.

Since junctional contacts are generally not seen between moving embryonic cells (except perhaps for small gap junctions), the formation of intercellular junctions may be an important mechanism for immobilizing cells within an organized tissue once it has formed. A reasonable hypothesis is that nonjunctional cell-surface adhesion proteins initiate tissue-specific cell-cell adhesion, which is then stabilized by the assembly of intercellular cell junctions. Since many of the transmembrane glycoproteins involved can diffuse in the plane of the plasma membrane, they can accumulate at sites of cell-cell contact and therefore be used for junctional as well as nonjunctional adhesions. Thus some cell-cell adhesion proteins, such as E cadherins (see p. 831), may help initiate cell-cell adhesion and then later become integral parts of intercellular junctions.

Ideally, one would like to be able to inactivate the various types of cell-cell adhesion proteins and matrix receptors individually and in various combinations in order to decipher the rules of recognition and binding used in the morphogenesis of complex tissues. As an increasing number of monoclonal antibodies and peptide fragments are characterized, each of which blocks a single type of cell-cell adhesion molecule or matrix receptor—and as the genes that encode these cell-surface proteins become available for manipulation in cells in culture and in transgenic animals—this dream of developmental biologists is becoming a reality.

Summary

Cells dissociated from various tissues of vertebrate embryos preferentially reassociate with cells from the same tissue when they are mixed together. The initial difficulties in studying the molecular mechanisms underlying the normal assembly of cells into complex tissues in higher animals, however, encouraged the study of simpler systems. The free-living amoebae of the cellular slime mold Dictyostelium discoideum *aggregate to form multicellular fruiting bodies when they are starved. Their cell-cell adhesion is mediated by at least two cell-surface glycoproteins: one operates early in development and depends on extracellular* Ca^{2+}*, while the other operates later in development and does not require* Ca^{2+}*. The tissue-specific recognition process in vertebrates is likewise mediated by at least two families of cell-surface glycoproteins—one* Ca^{2+}*-dependent (the cadherins) and the other* Ca^{2+}*-independent (exemplified by N-CAM and other members of the immunoglobulin superfamily). Both families of cell-cell adhesion molecules seem to play important roles in guiding vertebrate morphogenesis. Since even a single cell type uses multiple molecular mechanisms in adhering to other cells (and to the extracellular* *matrix), the specificity of cell-cell adhesion seen in embryonic development must result from a mechanism that sums the affinity of a number of different adhesion systems. The demonstrated ability of motile cells to detect small adhesive differences suggests how the specific combination, concentration, and distribution of cell-cell adhesion molecules and matrix receptors present on each type of cell can function as the required "morphogenetic code."*

References

Cited

1. Bock, G.; Clark, S., eds. Junctional Complexes of Epithelial Cells. Ciba Symposium 125. New York: Wiley, 1987.
   Farquhar, M.G.; Palade, G.E. Junctional complexes in various epithelia. *J. Cell. Biol.* 17:375–412, 1963.
   Gilula, N.B. Junctions between cells. In Cell Communication (R.P. Cox, ed.), pp. 1–29. New York: Wiley, 1974.
   Goodenough, D.A.; Revel, J.P. A fine structural analysis of intercellular junctions in the mouse liver. *J. Cell Biol.* 45:272–290, 1970.
   Staehelin, L.A.; Hull, B.E. Junctions between living cells. *Sci. Am.* 238(5):141–152, 1978.
2. Diamond, J.M. The epithelial junction: bridge, gate and fence. *Physiologist* 20:10–18, 1977.
   Madara, J.L. Tight junction dynamics: is paracellular transport regulated? *Cell* 53:497–498, 1988.
   Madara, J.L.; Dharmsathaphorn, K. Occluding junction structure-function relationships in cultured epithelial monolayer. *J. Cell Biol.* 101:2124–2133, 1985.
   Simons, K.; Fuller, S.D. Cell surface polarity in epithelia. *Annu. Rev. Cell Biol.* 1:243–288, 1985.
   van Meer, G.; Gumbiner, B.; Simons, K. The tight junction does not allow lipid molecules to diffuse from one epithelial cell to the next. *Nature* 322:639–641, 1986.
3. Burridge, K.; Fath, K.; Kelly, T.; Nuckolls, G.; Turner, C. Focal adhesions: transmembrane junctions between the extracellular matrix and the cytoskeleton. *Annu. Rev. Cell Biol.* 4:487–526, 1988.
   Geiger, B.; Volk, T.; Volberg, T. Molecular heterogeneity of adherens junctions. *J. Cell Biol.* 101:1523–1531, 1985.
4. Franke, W.W.; Cowin, P.; Schmelz, M.; Kapprell, H.-P. The desmosomal plaque and the cytoskeleton. In Junctional Complexes of Epithelial Cells, Ciba Foundation Symposium 125 (G. Bock, S. Clark, eds.), pp. 26–48. New York: Wiley, 1987.
   Garrod, D.R. Desmosomes, cell adhesion molecules and the adhesive properties of cells in tissues. *J. Cell Sci.* Suppl. 4:221–237, 1986.
   Jones, J.C.R.; Yokoo, K.M.; Goldman, R.D. Further analysis of pemphigus autoantibodies and their use in studies on the heterogeneity, structure, and function of desmosomes. *J. Cell Biol.* 102:1109–1117, 1986.
   Steinberg, M.S.; et al. On the molecular organization, diversity and functions of desmosomal proteins. In Junctional Complexes of Epithelial Cells, Ciba Foundation Symposium 125 (G. Bock, S. Clark, eds.), pp. 3–25. New York: Wiley, 1987.
5. Bennett, M.; Spray, D., eds. Gap Junctions. Cold Spring Harbor, NY: Cold Spring Harbor Laboratory, 1985.
   Furshpan, E.J.; Potter, D.D. Low-resistance junctions between cells in embryos and tissue culture. *Curr. Top. Dev. Biol.* 3:95–127, 1968.
   Gilula, N.B.; Reeves, O.R.; Steinbach, A. Metabolic coupling, ionic coupling and cell contacts. *Nature* 235:262–265, 1972.
   Hooper, M.L.; Subak-Sharpe, J.H. Metabolic cooperation between cells. *Int. Rev. Cytol.* 69:45–104, 1981.
   Loewenstein, W.R. The cell-to-cell channel of gap junctions. *Cell* 48:725–726, 1987.
   Neyton, J.; Trautmann, A. Single-channel currents of an intercellular junction. *Nature* 317:331–335, 1985.
   Pitts, J.D.; Finbow, M.E. The gap junction. *J. Cell Sci.* Suppl. 4:239–266, 1986.
   Young, J.D.-E.; Cohn, Z.A.; Gilula, N.B. Functional assembly of gap junction conductance in lipid bilayers: demonstration that the major 27 kd protein forms the junctional channel. *Cell* 48:733–743, 1987.
6. Caspar, D.L.D.; Goodenough, D.; Makowski, L.; Phillips, W.C. Gap junction structures. I. Correlated electron microscopy and x-ray diffraction. *J. Cell Biol.* 74:605–628, 1977.
   Gilula, N.B. Topology of gap junction protein and channel function. In Junctional Complexes of Epithelial Cells, Ciba Foundation Symposium 125 (G. Bock, S. Clark, eds.), pp. 128–139. New York: Wiley, 1987.
   Paul, D.L. Molecular cloning of cDNA for rat liver gap junction protein. *J. Cell Biol.* 103:123–134, 1986.
   Unwin, P.N.T.; Zampighi, G. Structure of the junction between communicating cells. *Nature* 283:545–549, 1980.
7. Caveney, S. The role of gap junctions in development. *Annu. Rev. Physiol.* 47:319–335, 1985.
   Warner, A.E. The role of gap junctions in amphibian development. *J. Embryol. Exp. Morphol.* Suppl. 89:365–380, 1985.
   Warner, A.E.; Guthrie, S.C.; Gilula, N.B. Antibodies to gap-junctional protein selectively disrupt junctional communication in the early amphibian embryo. *Nature* 311:127–131, 1984.
8. Rose, B.; Loewenstein, W.R. Permeability of cell junction depends on local cytoplasmic calcium activity. *Nature* 254:250–252, 1975.
   Saez, S.C.; et al. Cyclic AMP increases junctional conductance and stimulates phosphorylation of the 27-kDa principal gap junction polypeptide. *Proc. Natl. Acad. Sci. USA* 83:2473–2477, 1986.
   Spray, D.C.; Bennett, M.V.L. Physiology and pharmacology of gap junctions. *Annu. Rev. Physiol.* 47:218–303, 1985.
   Turin, L.; Warner, A.E. Intracellular pH in early *Xenopus* embryo: its effect on current flow between blastomeres. *J. Physiol. (Lond.)* 300:489–504, 1980.
9. Hay, E.D., ed. Cell Biology of Extracellular Matrix. New York: Plenum, 1981.
   McDonald, J.A. Extracellular matrix assembly. *Annu. Rev. Cell Biol.* 4:183–208, 1988.
   Piez, K.A.; Reddi, A.H., eds. Extracellular Matrix Biochemistry. New York: Elsevier, 1984.

10. Evered, D.; Whelan, J., eds. Functions of the Proteoglycans. Ciba Foundation Symposium 124. New York: Wiley, 1986.
Hascall, V.C.; Hascall, G.K. Proteoglycans. In Cell Biology of Extracellular Matrix (E.D. Hay, ed.), pp. 39–63. New York: Plenum, 1981.
Wight, T.N.; Meeham, R.P., eds. Biology of Proteoglycans. San Diego, CA: Academic Press, 1987.
11. Laurent, T.C.; Fraser, J.R.E. The properties and turnover of hyaluronan. In Functions of the Proteoglycans, Ciba Foundation Symposium 124 (D. Evered, J. Whelan, eds.), pp. 9–29. New York: Wiley, 1986.
Toole, B.P. Glycosaminoglycans in Morphogenesis. In Cell Biology of Extracellular Matrix (E.D. Hay, ed.), pp. 259–294. New York: Plenum, 1981.
12. Dorfman, A. Proteoglycan biosynthesis. In Cell Biology of Extracellular Matrix (E.D. Hay, ed.), pp. 115–138. New York: Plenum, 1981.
Hassell, J.R.; Kimura, J.H.; Hascall, V.C. Proteoglycan core protein families. *Annu. Rev. Biochem.* 55:539–567, 1986.
Heinegård, D.; Paulsson, M. Structure and metabolism of proteoglycans. In Extracellular Matrix Biochemistry (K.A. Piez, A.H. Reddi, eds.), pp. 277–328. New York: Elsevier, 1984.
Ruoslahti, E. Structure and biology of proteoglycans. *Annu. Rev. Cell Biol.* 4:229–255, 1988.
13. Fransson, L.-Å. Structure and function of cell-associated proteoglycans. *Trends Biochem. Sci.* 12:406–411, 1987.
Höök, M.; Kjellén, L.; Johansson, S.; Robinson, J. Cell-surface glycosaminoglycans. *Annu. Rev. Biochem.* 53:847–869, 1984.
Rees, D.A. Polysaccharide Shapes, Outline Studies in Biology, pp. 62–73. London: Chapman & Hall, 1977.
Scott, J.E. Proteoglycan-collagen interactions. In Functions of the Proteoglycans, Ciba Foundation Symposium 124 (D. Evered, J. Whelan, eds.), pp. 104–124. New York: Wiley, 1986.
14. Burgeson, R.E. New collagens, new concepts. *Annu. Rev. Cell Biol.* 4:551–577, 1988.
Linsenmayer, T.F. Collagen. In Cell Biology of Extracellular Matrix (E.D. Hay, ed.), pp. 5–37. New York: Plenum, 1981.
Martin, G.R.; Timpl, R.; Muller, P.K.; Kühn, K. The genetically distinct collagens. *Trends Biochem. Sci.* 10:285–287, 1985.
15. Fleischmajer, R.; Olsen, B.R.; Kühn, K., eds. Biology, Chemistry, and Pathology of Collagen. *Ann. N.Y. Acad. Sci.*, Vol. 460, 1985.
Olsen, B.R. Collagen Biosynthesis. In Cell Biology of Extracellular Matrix (E.D. Hay, ed.), pp. 139–177. New York: Plenum, 1981.
Woolley, D.E. Mammalian collagenases. In Extracellular Matrix Biochemistry (K.A. Piez, A.H. Reddi, eds.), pp. 119–157. New York: Elsevier, 1984.
16. Eyre, D.R.; Paz, M.A.; Gallop, P.M. Cross-linking in collagen and elastin. *Annu. Rev. Biochem.* 53:717–748, 1984.
Piez, K.A. Molecular and aggregate structures of the collagens. In Extracellular Matrix Biochemistry (K.A. Piez, A.H. Reddi, eds.), pp. 1–39. New York: Elsevier, 1984.
17. Prockop, D.J.; Kivirikko, K.I. Heritable diseases of collagen. *New Engl. J. Med.* 311:376–386, 1984.
Trelstad, R.L.; Silver, F.H. Matrix assembly. In Cell Biology of Extracellular Matrix (E.D. Hey, ed.), pp. 179–215. New York: Plenum, 1981.
18. Stopak, D.; Harris, A.K. Connective tissue morphogenesis by fibroblast traction I. Tissue culture observations. *Dev. Biol.* 90:383–398, 1982.
19. Yurchenco, P.D.; Furthmayr, H. Self-assembly of basement membrane collagen. *Biochemistry* 23:1839–1850, 1984.
Yurchenco, P.D.; Ruben, G.C. Basement membrane structure *in situ:* evidence for lateral associations in the type IV collagen network. *J. Cell Biol.* 105:2559–2568, 1987.
20. Cleary, E.G.; Gibson, M.A. Elastin-associated microfibrils and microfibrillar proteins. *Int. Rev. Connect. Tissue Res.* 10:97–209, 1983.
Gosline, J.M.; Rosenbloom, J. Elastin. In Extracellulr Matrix Biochemistry (K.A. Piez, A.H. Reddi, eds.), pp. 191–227. New York: Elsevier, 1984.
Ross, R.; Bornstein, P. Elastic fibers in the body. *Sci. Am.* 224(6):44–52, 1971.
21. Dufour, S.; Duband, J.-L.; Kornblihtt, A.R.; Thiéry, J.P. The role of fibronectins in embryonic cell migrations. *Trends Genet.* 4:198–203, 1988.
Hynes, R.O. Molecular biology of fibronectin. *Annu. Rev. Cell Biol.* 1:67–90, 1985.
Hynes, R.O. Fibronectins. *Sci. Am.* 254(6):42–51, 1986.
Hynes, R.O.; Yamada, K.M. Fibronectins: multifunctional modular proteins. *J. Cell Biol.* 95:369–377, 1982.
Ruoslahti, E.; Pierschbacher, M.D. New perspectives in cell adhesion: RGD and integrins. *Science* 238:491–497, 1987.
22. Humphries, M.J.; Akiyama, S.K.; Komoriya, A.; Olden, K.; Yamada, K.M. Neurite extension of chicken peripheral nervous system neurons on fibronectin: relative importance of specific adhesion sites in the central cell-binding domain and the alternatively spliced type III connecting segment. *J. Cell Biol.* 106:1289–1297, 1988.
Tamkun, J.W.; Schwarzbauer, J.E.; Hynes, R.O. A single rat fibronectin gene generates three different mRNAs by alternative splicing of a complex exon. *Proc. Natl. Acad. Sci. USA* 81:5140–5144, 1984.
23. Farquhar, M.G. The glomerular basement membrane: a selective macromolecular filter. In Cell Biology of Extracellular Matrix (E.D. Hay, ed.), pp. 335–378. New York: Plenum, 1981.
Martin, G.R.; Timpl, R. Laminin and other basement membrane components. *Annu. Rev. Cell Biol.* 3:57–85, 1987.
Sasaki, M.; Kato, S.; Kohno, K.; Martin, G.R.; Yamada, Y. Sequence of the cDNA encoding the laminin B1 chain reveals a multidomain protein containing cysteine-rich repeats. *Proc. Natl. Acad. Sci. USA* 84:935–939, 1987.
24. Reist, N.E.; Magill, C.; McMahan, U.J. Agrin-like molecules at synaptic sites in normal, denervated, and damaged skeletal muscles. *J. Cell Biol.* 105:2457–2469, 1987.
25. Anderson, D.C.; Springer, T.A. Leukocyte adhesion deficiency: an inherited defect in the Mac-1, LFA-1 and P150,95 glycoprotein. *Annu. Rev. Med.* 38:175–194, 1987.
Buck, C.A.; Horwitz, A.F. Cell surface receptors for extracellular matrix molecules. *Annu. Rev. Cell Biol.* 3:179–205, 1987.
Hynes, R.O. Integrins: a family of cell surface receptors. *Cell* 48:549–554, 1987.
Ruoslahti, E. Fibronectin and its receptors. *Annu. Rev. Biochem.* 57:375–414, 1988.
26. Bornstein, P.; Duksin, D.; Balian, G.; Davidson, J.M.; Crouch, E. Organization of extracellular proteins on the connective tissue cell surface: relevance to cell-matrix interactions *in vitro* and *in vivo*. *Ann. N.Y. Acad. Sci.* 312:93–105, 1978.
Burridge, K.; Fath, K.; Kelly, T.; Nuckolls, G.; Turner, C. Focal adhesions: transmembrane junctions between the extracellular matrix and the cytoskeleton. *Annu. Rev. Cell Biol.* 4:487–526, 1988.
Horwitz, A.; Duggan, K; Buck, C.; Beckerle, M.C.; Burridge, K. Interaction of plasma membrane fibronectin receptor with talin—a transmembrane linkage. *Nature* 320:531–533, 1986.
Hynes, R. Structural relationships between fibronectin and cytoplasmic cytoskeletal networks. In Cytoskeletal Elements and Plasma Membrane Organization (G. Poste, G.L. Nicolson, eds.), Vol. 7, pp. 100–137. Amsterdam: Elsevier, 1981.
Watt, F.M. The extracellular matrix and cell shape. *Trends Biochem. Sci.* 11:482–485, 1986.
27. Le Douarin, N.; Smith, J. Development of the peripheral nervous system from the neural crest. *Annu. Rev. Cell Biol.* 4:375–404, 1988.

McClay, D.R.; Ettensohn, C.A. Cell adhesion and morphogenesis. *Annu. Rev. Cell Biol.* 3:319–346, 1987.

28. Loomis, W.F. *Dictyostelium discoideum.* A Developmental System. New York: Academic Press, 1975.
29. Bonner, J.T. Chemical signals of social amoebae. *Sci. Am.* 248(4):114–120, 1983.

    Gerisch, G. Cyclic AMP and other signals controlling cell development and differentiation in *Dictyostelium. Annu. Rev. Biochem.* 56:853–879, 1987.
30. Gerisch, G. Interrelation of cell adhesion and differentiation in *Dictyostelium discoideum. J. Cell Sci.* Suppl. 4:201–219, 1986.

    Gerisch, G. Univalent antibody fragments as tools for the analysis of cell interactions in *Dictyostelium. Curr. Top. Dev. Biol.* 14:243–270, 1980.
31. Hennings, H.; Holbrook, K.A. Calcium regulation of cell-cell contact and differentiation of epidermal cells in culture. An ultrastructural study. *Exp. Cell Res.* 143:127–142, 1983.
32. Moscona, A.A.; Hausman, R.E. Biological and biochemical studies on embryonic cell-cell recognition. In Cell and Tissue Interactions, Society of General Physiologists Series (J.W. Lash, M.M. Burger, eds.), Vol. 32, pp. 173–185. New York: Raven, 1977.

    Roth, S.; Weston, J. The measurement of intercellular adhesion. *Proc. Natl. Acad. Sci. USA* 58:974–980, 1967.
33. Cunningham, B.A.; et al. Neural cell adhesion molecule: structure, immunoglobulin-like domains, cell surface modulation, and alternative RNA splicing. *Science* 236:799–806, 1987.

    Edelman, G.M. Cell-adhesion molecules: a molecular basis for animal form. *Sci. Am.* 250(4):118–129, 1984.

    Edelman, G.M. Cell adhesion molecules in the regulation of animal form and tissue pattern. *Annu. Rev. Cell Biol.* 2:81–116, 1986.

    Rutishauser, U.; Goridis, C. N-CAM: the molecule and its genetics. *Trends Genet.* 2:72–76, 1986.

    Williams, A.F.; Barclay, A.N. The immunoglobulin superfamily—domains for cell surface recognition. *Annu. Rev. Immunol.* 6:381–406, 1988.
34. Takeichi, M. The cadherins: cell-cell adhesion molecules controlling animal morphogenesis. *Development* 102:639–655, 1988.
35. Ekblom, P.; Vestweber, D.; Kemler, R. Cell-matrix interactions and cell adhesion during development. *Annu. Rev. Cell Biol.* 2:27–48, 1986.

    Garrod, D.R. Desmosomes, cell adhesion molecules and the adhesive properties of cells in tissues. *J. Cell Sci.* Suppl. 4:221–237, 1986.

    Jessel, T.M. Adhesion molecules and the hierarchy of neural development. *Neuron* 1:3–13, 1988.

    Steinberg, M.S. The adhesive specification of tissue self-organization. In Morphogenesis and Pattern Formation (T.G. Connelly, et al, eds.), pp. 179–203. New York: Raven, 1981.
36. Devreotes, P.; Zigmond, S.H. Chemotaxis in eukaryotic cells. *Annu. Rev. Cell Biol.* 4:649–686, 1988.
37. Trinkaus, J.P. Cells into Organs, 2nd ed., pp. 69–178. Englewood Cliffs, NJ: Prentice-Hall, 1984.

A mouse embryo at 15 days of gestation.▶

From Cells to Multicellular Organisms

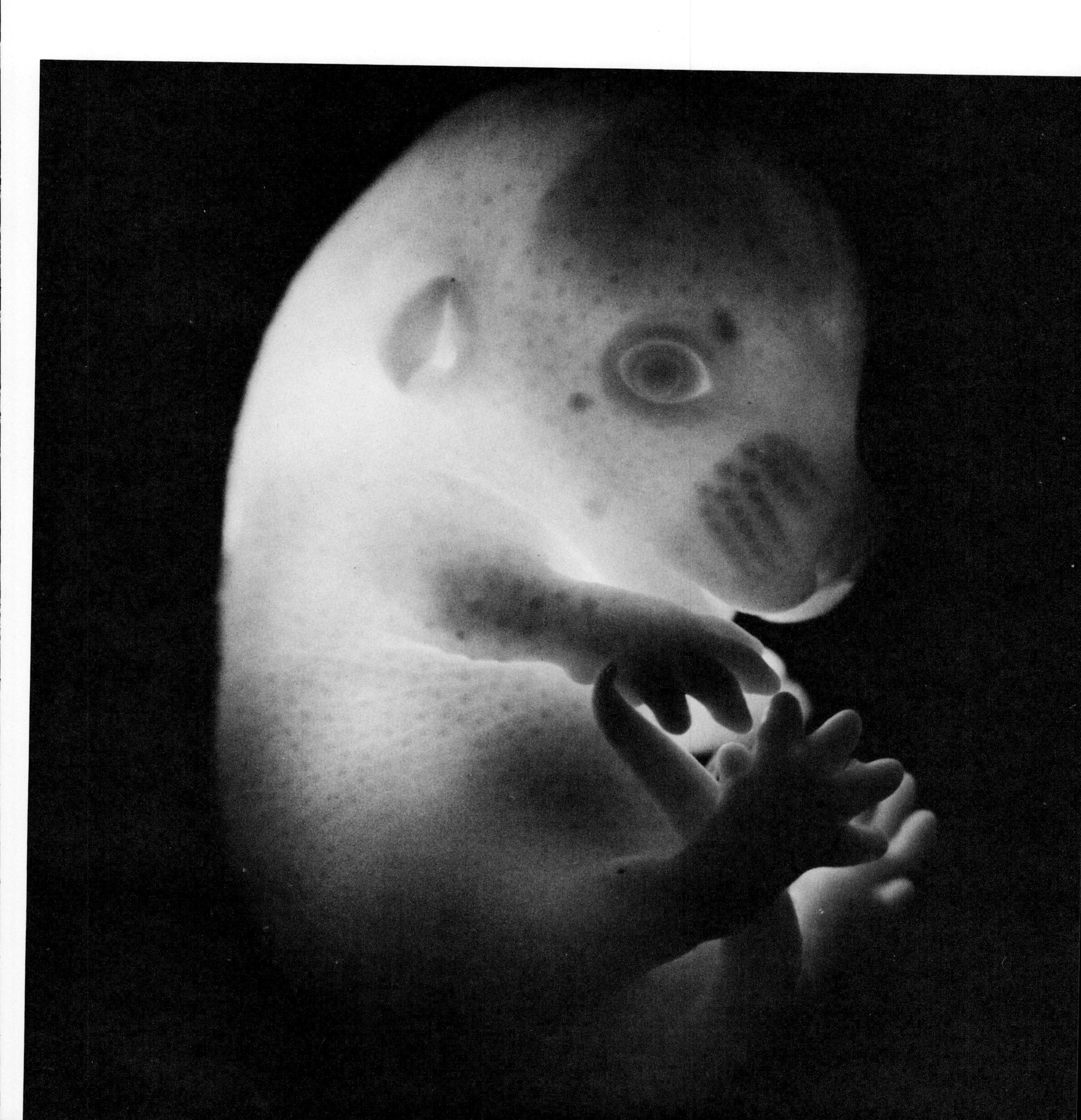

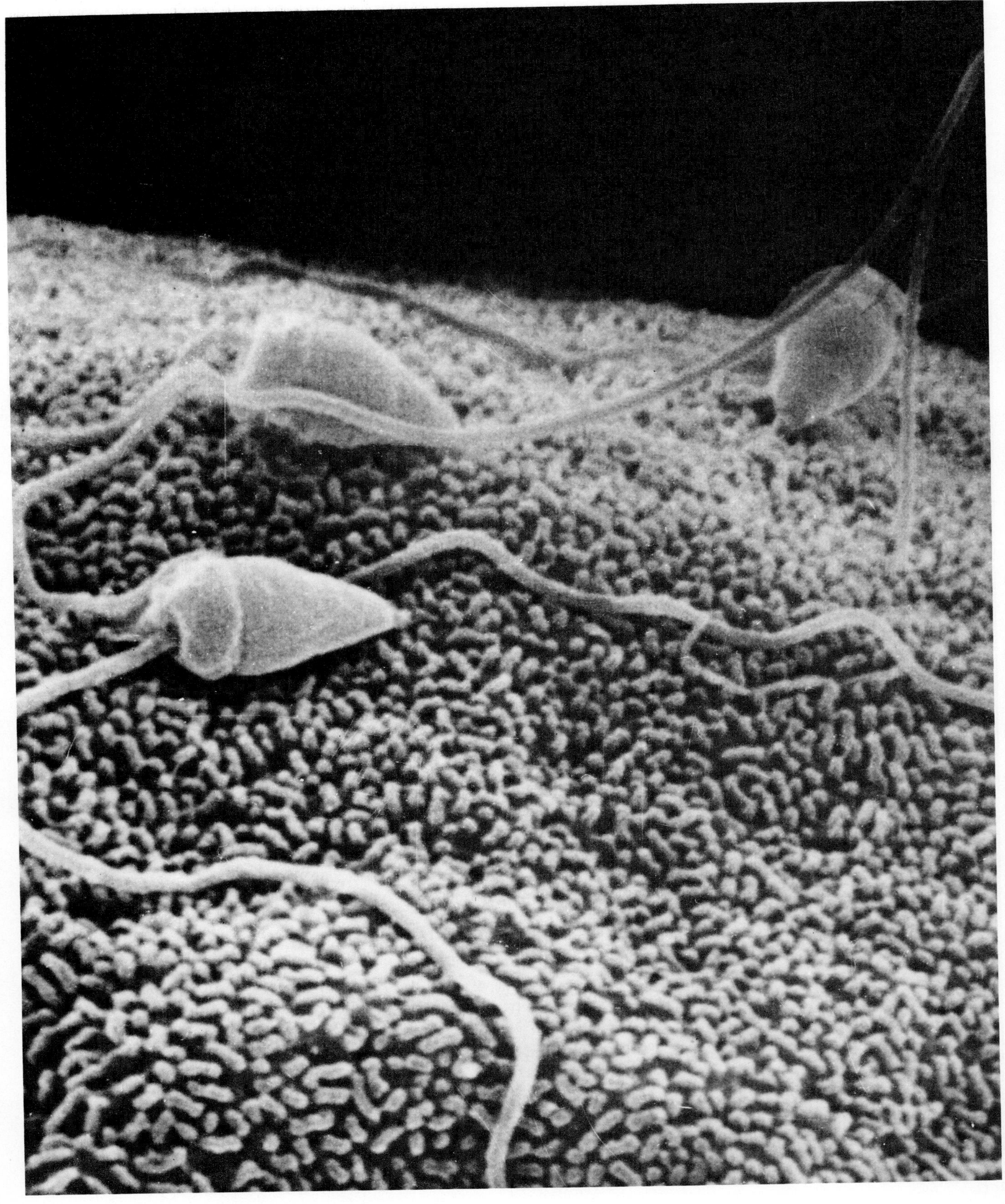

Scanning electron micrograph of sperm on the surface of a sea urchin egg. (Courtesy of Brian Dale.)

Germ Cells and Fertilization

15

Sex is not necessary for reproduction. Single-celled organisms can reproduce by simple mitotic division. Many plants propagate vegetatively by forming multicellular offshoots that later detach from the parent. Likewise, in the animal kingdom, a solitary multicellular *Hydra* can produce offspring by budding (Figure 15–1). Sea anemones and marine worms can split into two half-organisms, each of which then regenerates its missing half. There are even species of lizards that consist only of female individuals and reproduce without mating. While such **asexual reproduction** is simple and direct, it gives rise to offspring that are genetically identical to the parent organism. **Sexual reproduction,** on the other hand, involves the mixing of genomes from two different individuals to produce offspring that usually differ genetically from one another and from both their parents. This form of reproduction apparently has great advantages, since the vast majority of plants and animals have adopted it. Even many procaryotes and other organisms that normally reproduce asexually engage in occasional bouts of sexual reproduction, thereby creating new combinations of genes. This chapter is concerned with the cellular machinery of sexual reproduction. Before discussing in detail how the machinery works, we shall pause to consider why it exists and what benefits it brings.

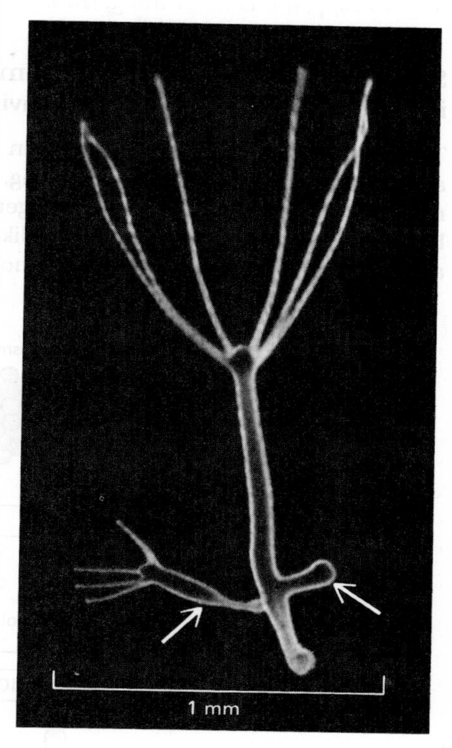

Figure 15–1 Photograph of a *Hydra* from which two new organisms are budding (*arrows*). The offspring, which are genetically identical to their parent, will eventually detach and live independently. (Courtesy of Amata Hornbruch.)

The Benefits of Sex

The sexual reproductive cycle involves an alternation of **haploid** generations of cells, each carrying a single set of chromosomes, with **diploid** generations of cells, each carrying a double set of chromosomes (Figure 15–2). The mixing of genomes is achieved by fusion of two haploid cells to form a diploid cell. Later, new haploid cells are created when a descendant of this diploid cell divides by the process of *meiosis*. During meiosis the chromosomes of the double chromosome set exchange DNA by genetic recombination before being shared out, in fresh combinations, into single chromosome sets (see p. 847). In this way each cell of the new haploid generation receives a novel assortment of genes, with some genes on each chromosome originating from one ancestral cell of the previous haploid generation and some from the other. Thus, through cycles of haploidy, fusion, diploidy, and meiosis, old combinations of genes are broken up and new combinations are created.

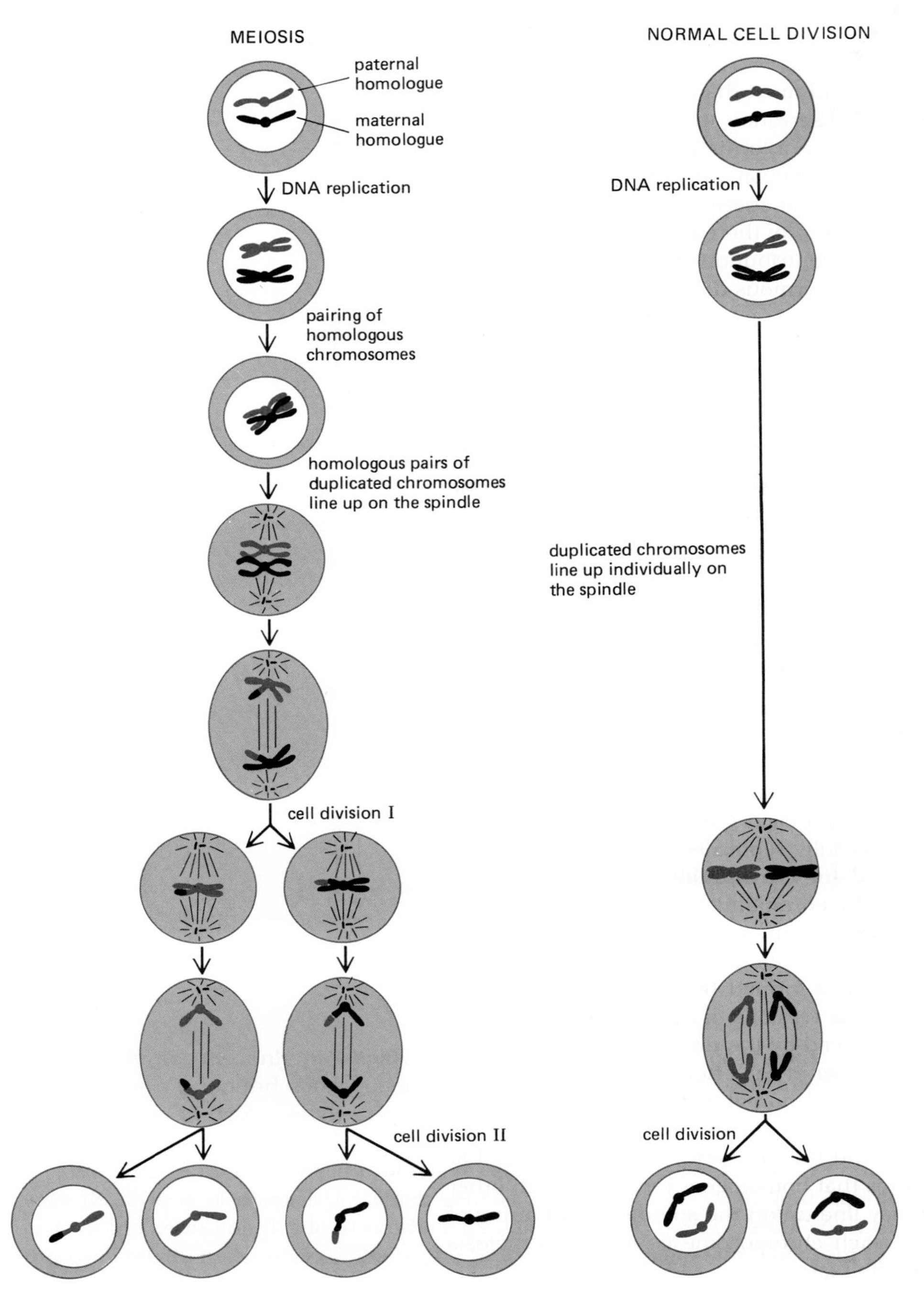

Figure 15–8 Comparison of meiosis and normal cell division. For clarity, only one set of homologous chromosomes is shown. The pairing of homologous chromosomes (homologues) is unique to meiosis. Each chromosome has been duplicated and exists as attached sister chromatids before the pairing occurs, so that two nuclear divisions are required to produce the haploid gametes. Each diploid cell that enters meiosis therefore produces four haploid cells. As shown, the chromosome pairing in meiosis involves crossing-over between homologous chromosomes, which will be explained below.

the meiotic cell divides. The two progeny of this division (**division I of meiosis**) therefore contain a diploid amount of DNA but differ from normal diploid cells in two ways: (1) both of the two DNA copies of each chromosome derive from only one of the two homologous chromosomes in the original cell (although, as we shall see, there has been some mixing of the maternal and paternal DNA caused by genetic recombination), and (2) these two copies are inherited as closely associated sister chromatids, as if they were a single chromosome (see Figure 15–8).

Formation of the actual gamete nuclei can now proceed simply through a second cell division, **division II of meiosis,** in which chromosomes align, without further replication, on a second spindle and the sister chromatids separate, as in normal mitosis, to produce cells with a haploid DNA content. Meiosis thus consists of two nuclear divisions following a single phase of DNA replication, so that four haploid cells are produced from each cell that enters meiosis (see Figure 15–8). Occasionally the meiotic process occurs abnormally and homologues fail to sepa-

Figure 15–9 Two major contributions to the reassortment of genetic material that occurs during meiosis. (A) The independent assortment of the maternal and paternal homologues during the first meiotic division produces 2^n different haploid gametes for an organism with n chromosomes. Here $n = 3$ and there are 8 different possible gametes, as indicated. (B) Crossing-over during meiotic prophase I exchanges segments of homologous chromosomes and thereby reassorts genes in individual chromosomes. Because of the many small differences in DNA sequence that always exist between any two homologues, both mechanisms increase the genetic variability of organisms that reproduce sexually.

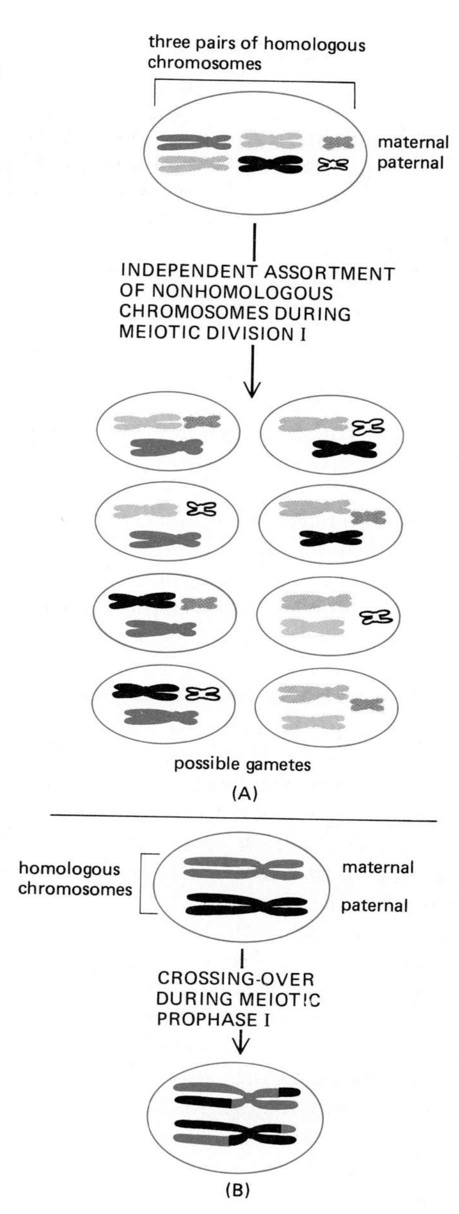

rate—a phenomenon known as **nondisjunction.** In this case some of the haploid cells that are produced lack a chromosome while others have more than one copy. Such gametes form abnormal embryos, most of which die.

Genetic Reassortment Is Enhanced by Crossing-over Between Homologous Nonsister Chromatids

We have seen that genes can be mixed by the fusion of gametes from two individuals. But genetic variation is not produced solely by this means. Unless they are identical twins, no two offspring of the same parents are genetically the same. This is because, long before the two gametes fuse, two kinds of genetic reassortment have already occurred during meiosis.

One kind of reassortment is a consequence of the random distribution of the maternal and paternal homologues between the daughter cells at meiotic division I, as a result of which each gamete acquires a different mixture of maternal and paternal chromosomes (Figure 15–9A). From this process alone, one individual could, in principle, produce 2^n genetically different gametes, where n is the haploid number of chromosomes. In humans, for example, each individual can produce at least $2^{23} = 8.4 \times 10^6$ genetically different gametes. But the actual number is very much greater than this because of **chromosomal crossing-over,** a process that takes place during the long prophase of meiotic division I, in which parts of homologous chromosomes are exchanged. On average, between two and three crossover events occur on each pair of human chromosomes. This process scrambles the genetic constitution of each of the chromosomes in gametes, as illustrated in Figure 15–9B.

Chromosomal crossing-over involves breaking the single maternal and paternal DNA double helices in each of the two chromatids and rejoining them to each other in a reciprocal fashion by a process known as **genetic recombination.** The molecular details of this process are outlined in Chapter 5 (see p. 239). Recombination takes place during prophase of meiotic division I at a time when the two sister chromatids are packed together so tightly that their individuality cannot be distinguished by light or electron microscopy (see below). Much later in this extended prophase, the two separate sister chromatids of each duplicated homologue become clearly visible, although they remain tightly apposed along their entire length and intimately connected to each other at their centromeres (Figure 15–10). The two duplicated homologues (maternal and paternal) remain attached to each other at those points where a crossover between a paternal and a maternal chromatid has occurred. At each attachment point, called a **chiasma** (plural **chiasmata**), two of the four chromatids are seen to have crossed over between

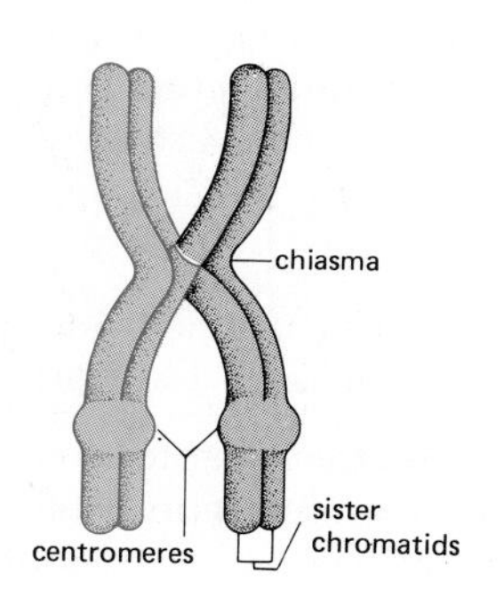

Figure 15–10 Paired homologous chromosomes during the transition to metaphase of meiotic division I. A single crossover event has occurred earlier in prophase to create one chiasma. Note that the four chromatids are arranged in two distinct pairs of sister chromatids and that the two chromatids in each pair are tightly aligned along their entire lengths as well as joined at their centromeres. The entire unit shown here is therefore frequently referred to as a *bivalent.*

the homologues (see Figure 15–10). Chiasmata are thus the morphological consequences of a prior, unobserved crossover event.

At this stage of meiosis, each pair of duplicated homologues, or *bivalent,* is held together by at least one chiasma. Many bivalents contain more than one chiasma, indicating that multiple crossovers can occur between homologues (Figures 15–11 and 15–12).

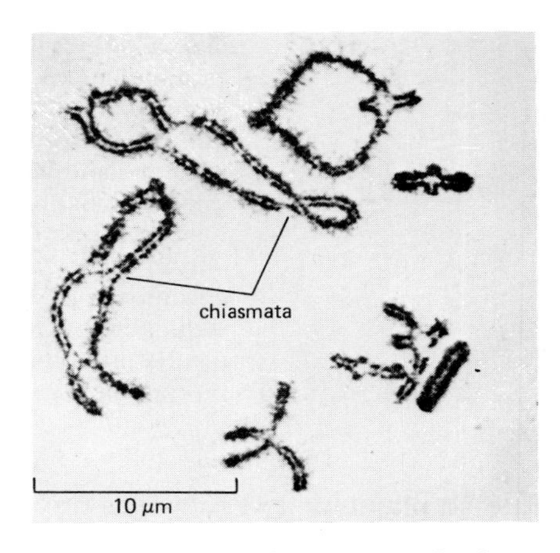

Figure 15–11 Light micrograph of several bivalents containing multiple chiasmata at diplotene, a late stage of meiotic prophase I (see Figure 15–18). These large grasshopper chromosomes provide particularly favorable material for cytological observations of this kind. (Courtesy of Bernard John.)

The Synaptonemal Complex Mediates Chromosome Pairing

Elaborate morphological changes occur in the chromosomes as they pair (*synapse*) and then begin to unpair (*desynapse*) during the first meiotic prophase. This prophase is traditionally divided into five sequential stages—*leptotene, zygotene, pachytene, diplotene,* and *diakinesis*—defined by these morphological changes (Figure 15–13). The most striking event is the initiation of intimate chromosome synapsis at **zygotene,** when a complex structure called the *synaptonemal complex* begins to develop between the two sets of sister chromatids in each bivalent. **Pachytene** is said to begin as soon as synapsis is complete, and it generally persists for days, until desynapsis begins the **diplotene** stage, in which the chiasmata are first seen.

Genetic recombination requires a close apposition between the recombining chromosomes. The **synaptonemal complex,** which forms just before pachytene and dissolves just afterward, keeps the homologous chromosomes in a bivalent together and closely aligned and is thought to be required for the crossover events to occur. It consists of a long ladderlike protein core, on opposite sides of which the two homologues are aligned to form a long linear chromosome pair (a bivalent, Figure 15–14). The sister chromatids in each homologue are kept tightly packed together, and their DNA extends from the same side of the protein ladder in a series of loops. Thus, while the homologous chromosomes are closely aligned along their length in the synaptonemal complex, the maternal and paternal chromatids that will recombine are kept separated by 100 nm on either side of the protein ladder.

From cytological studies, chromosome synapsis is seen to be preceded by the formation of a ropelike proteinaceous axis along each of the homologues. As pairing proceeds, the axes appear to become linked to each other to form the lateral elements of the synaptonemal complex—that is, the two sides of the protein ladder. Both the axes and the lateral elements contain a protein with unique silver-staining properties that make these structures visible by both light and electron microscopy (Figure 15–15).

It is not known what causes the homologous parts of chromosomes to become precisely aligned during zygotene. It is unlikely that homologous base-pairing all along the interacting chromosomes is required, since most of the chromatin of one homologue is positioned well away from the chromatin of its partner in the synaptonemal complex, and in special cases the synaptonemal complex is capable of joining regions of the two chromosomes that are not homologous. One possibility is that the initial pairing between chromosomes is mediated by complementary DNA base-pair interactions confined to specific regions on each chromosome; the synaptonemal complex then packs together the remainder of the prealigned chromosomes. An initial point-by-point matching mechanism of some type is required to explain the observation that the presence of an inverted section of chromosome in one of two pairing homologues usually (but not always) results in a temporary interruption of the normal zipperlike synapsis during zygotene, allowing homologous genes to synapse even within the inversion (Figures 15–16 and 15–17). The various stages of meiosis are outlined and described in detail in Figure 15–18.

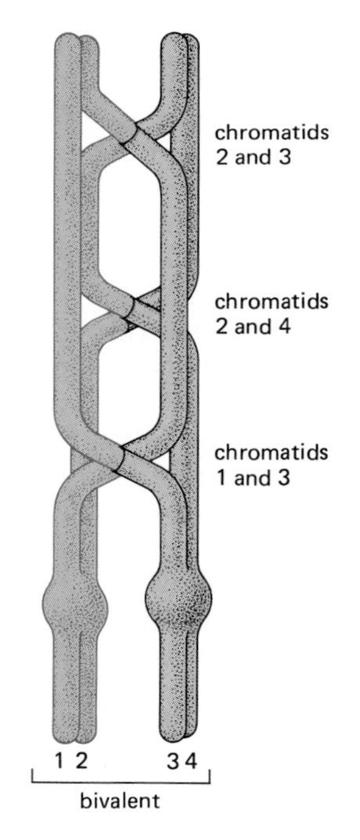

Figure 15–12 Three chiasmata resulting from three separate crossover events. Each of the two chromatids on each chromosome can cross over with either of the chromatids on the other chromosome in the bivalent. For example, here chromatid 3 has undergone an exchange with both chromatid 1 and chromatid 2.

Recombination Nodules Are Thought to Mediate the Chromatid Exchanges

Although the synaptonemal complex provides the structural framework for recombination events, it probably does not participate in them directly. The active recombination process is thought to be mediated instead by **recombination nodules,**

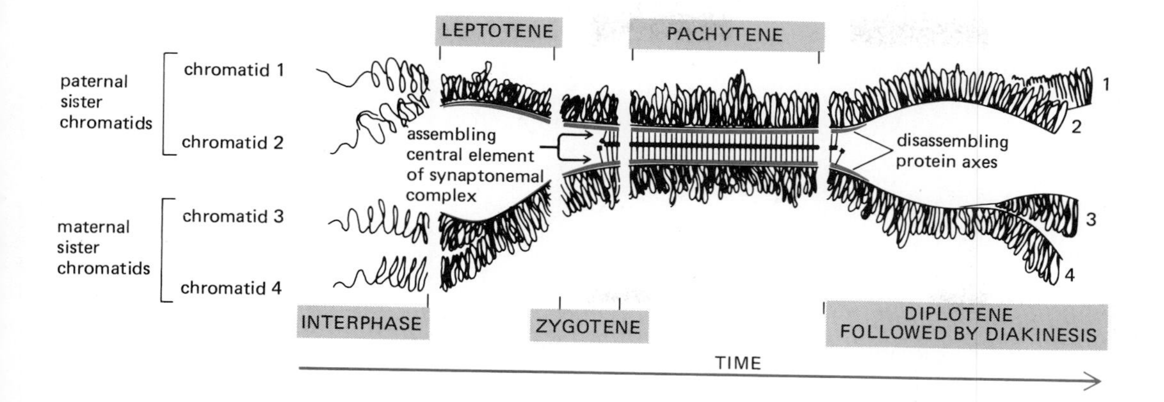

Figure 15–13 Time course of chromosome synapsis and desynapsis during meiotic prophase I. A single bivalent is shown. The pachytene stage is defined as the period during which a fully formed synaptonemal complex exists.

which are very large protein-containing assemblies with a diameter of about 90 nm. (For comparison, a large globular protein molecule of molecular weight 400,000 has a diameter of about 10 nm.) Recombination nodules sit at intervals on the synaptonemal complex, placed like basketballs on a ladder between the two homologous chromosomes (see Figure 15–14). They are thought to mark the site of a large multienzyme "recombination machine," which brings local regions of DNA on the maternal and paternal chromatids together across the 100-nm-wide synaptonemal complex.

The evidence that the recombination nodule serves this function is indirect: (1) The total number of nodules is about equal to the total number of chiasmata seen later in prophase. (2) The nodules are distributed along the synaptonemal complex in the same way that crossover events are distributed. For example, like the crossover events themselves, the nodules are absent from those regions of the synaptonemal complex that hold heterochromatin together. Moreover, both genetic and cytological measurements indicate that the occurrence of one crossover event prevents a second crossover event occurring at any nearby chromosomal site; similarly, the nodules tend not to occur very near one another. (3) Some *Drosophila* mutations cause an abnormal distribution of crossover events along the chromosomes, as well as a greatly diminished recombination frequency. In these mutants, correspondingly fewer recombination nodules are found, with a changed distribution that parallels the changed crossover distribution. This correlation strongly suggests that a recombination nodule determines the site of each crossover event. (4) Genetic recombination is thought to involve a limited amount of DNA synthesis at the site of each crossover event (see p. 244). Electron microscopic autoradiography shows that radioactive DNA precursors are preferentially incorporated into pachytene DNA at or near recombination nodules.

Because there are about as many recombination nodules as crossover events, their suggested role implies that recombination nodules are extremely efficient in causing the chromatids on opposite homologues to recombine. Nothing is yet known about their structure or mechanism of action.

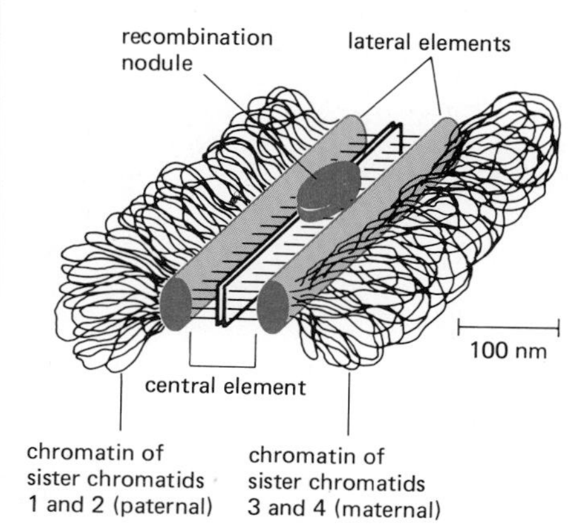

Figure 15–14 A typical synaptonemal complex, showing the lateral and central elements of the complex. A recombination nodule is also shown. Only a short section of the long ladderlike complex is shown. Although a similar synaptonemal complex is present in organisms as diverse as yeast and human, very little is known about the protein molecules that form it.

Chiasmata Play an Important Part in Chromosome Segregation in Meiosis

In addition to reassorting genes, chromosomal crossing-over is crucial for the segregation of the two homologues to separate daughter nuclei. This is because each crossover event creates a chiasma, which plays a role analogous to that of the centromere in an ordinary mitotic division, holding the maternal and paternal homologues together on the spindle until anaphase I. In mutant organisms that have a reduced frequency of meiotic chromosome crossing-over, some of the chromosome pairs lack chiasmata. These pairs fail to segregate normally, and a high proportion of the resulting gametes contain too many or too few chromosomes—an example of nondisjunction.

There are at least two major differences in the way chromosomes separate in meiotic division I and in normal mitosis. (1) During normal mitosis the kineto-

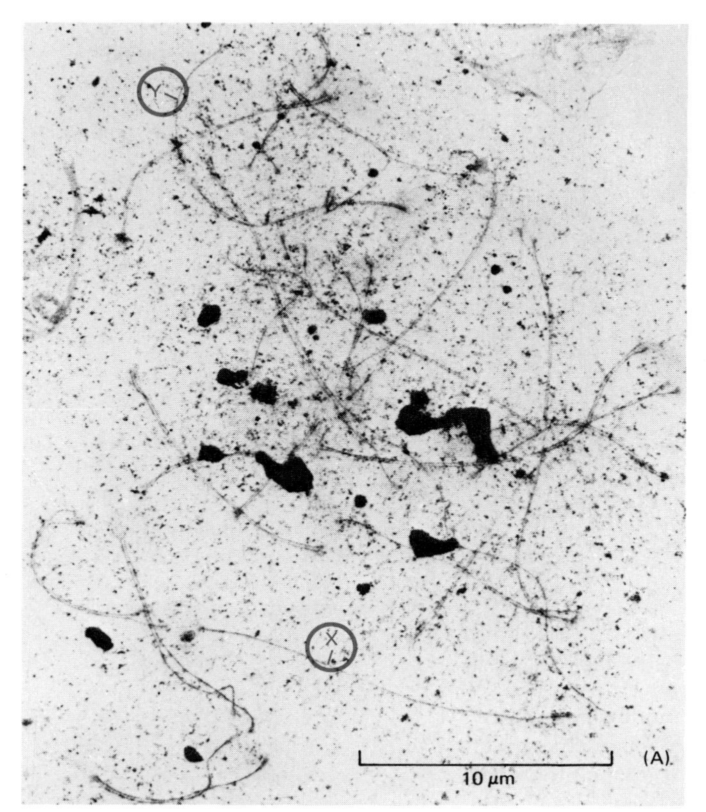

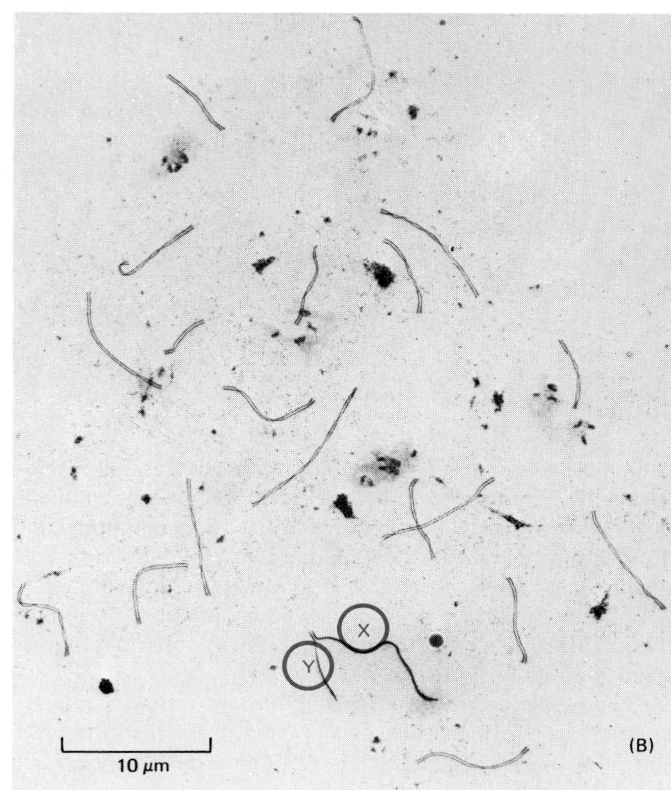

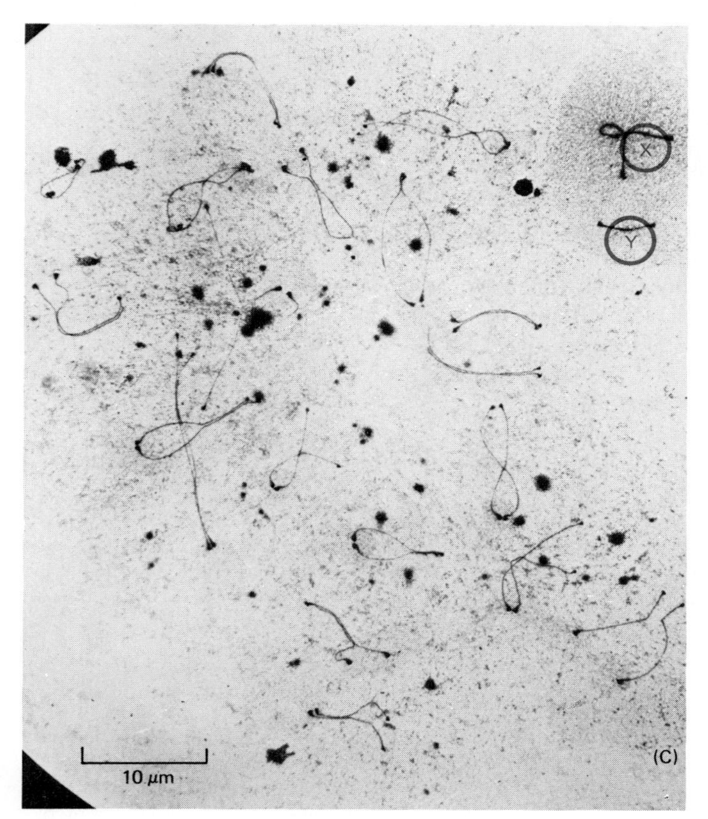

Figure 15–15 Electron micrographs of silver-stained, whole-mount spreads of chromosomes from mouse spermatocytes of (A) early (zygotene), (B) middle (pachytene), and (C) later (diplotene) stages of meiotic prophase I. (A) **Synapsing (zygotene):** The chromosome axes form before the chromosomes synapse, and they are initially seen as separate. The axes then move together and, when properly spaced at one or more synaptic initiation sites, the synaptonemal complex forms, often beginning at the end. The large separation between the sex chromosomes (X and Y) indicates that, in order to pair, chromosomes must often travel over substantial intranuclear distances, but it is not known how they do so. The dark bodies are nucleoli. (B) **Synapsed (pachytene):** Synapsis is complete when synaptonemal complexes have fully joined all homologous autosomes in pairs. The X and Y do not synapse fully. Crossing-over takes place between chromatids, which cannot be individually distinguished in these preparations.
(C) **Desynapsing (diplotene):** Just before the axes disassemble, they separate, marking the end of synapsis. In places they are held together by persistent segments of synaptonemal complex thought to represent sites where crossing-over has occurred. Later, when the chromatin condenses and chromatids are distinguishable, chiasmata will indicate the crossovers. (Original micrographs courtesy of Montrose J. Moses.)

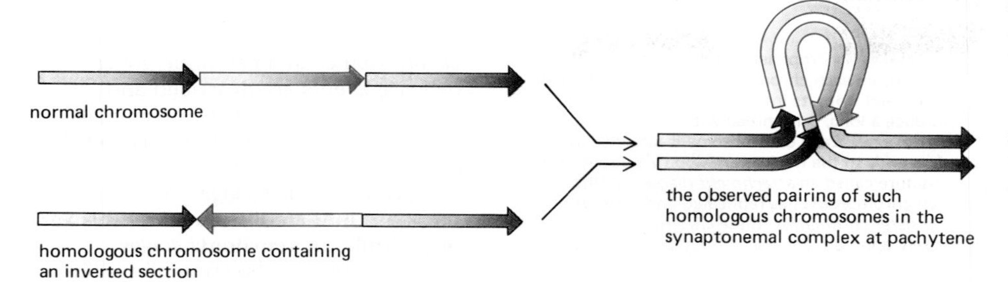

Figure 15–16 Formation of the synaptonemal complex between one normal chromosome and its homologue carrying an inverted section. Such structures demonstrate that a local like-with-like pairing process brings homologous chromosomes together. Compare with Figure 15–17.

chores on each sister chromatid have attached kinetochore fibers pointing in opposite directions, whereas at metaphase I of meiosis the kinetochores on both sister chromatids appear to have fused so that their attached kinetochore fibers all point in the same direction (Figure 15–19). (2) During normal mitosis the movement of chromatids to the poles is triggered by a mechanism that detaches the two sister kinetochores from each other (thus beginning anaphase, see p. 773); in anaphase I of meiosis, however, this movement is initiated by the disruption of the poorly understood forces that have previously kept the arms of sister chromatids closely apposed, and this in turn dissolves the chiasmata that were holding the homologous maternal and paternal chromosomes together (see Figure 15–19). This explains not only why the chiasmata are necessary for the normal alignment of the chromosomes at metaphase I in many organisms, but also why the chromosomes produced at anaphase I tend to have nonadherent sister chromatid arms, giving each pair of sister chromatids an unusual "splayed-out" appearance compared with the arrangement of the chromatids in normal metaphase chromosomes (see Figure 15–19).

Pairing of the Sex Chromosomes Ensures That They Also Segregate

We have explained how homologous chromosomes pair so that they segregate between the daughter cells. But what about the sex chromosomes, which in male mammals are not homologous? Females have two X chromosomes, which pair and segregate like other homologues. But males have one X and one Y chromo-

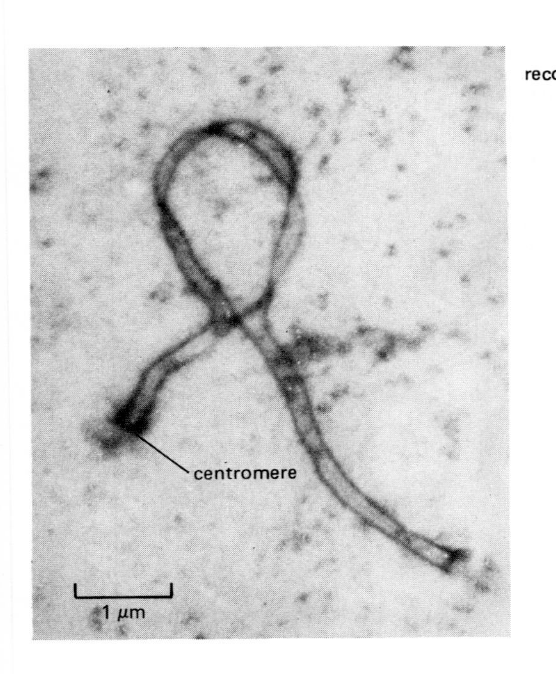

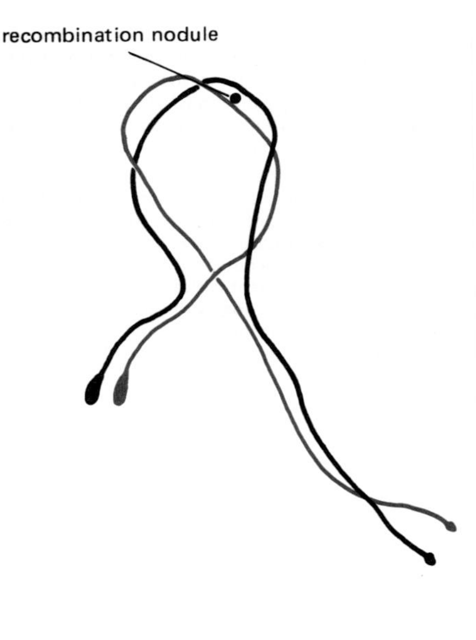

Figure 15–17 Electron micrograph and drawing of two tightly synapsed mouse homologues at pachytene, one of which contains an inversion. A recombination nodule (see p. 848) is seen in the loop. (From P.A. Poorman, M.J. Moses, T.H. Roderick, and M.T. Davisson, *Chromosoma* 83:419, 1981.)

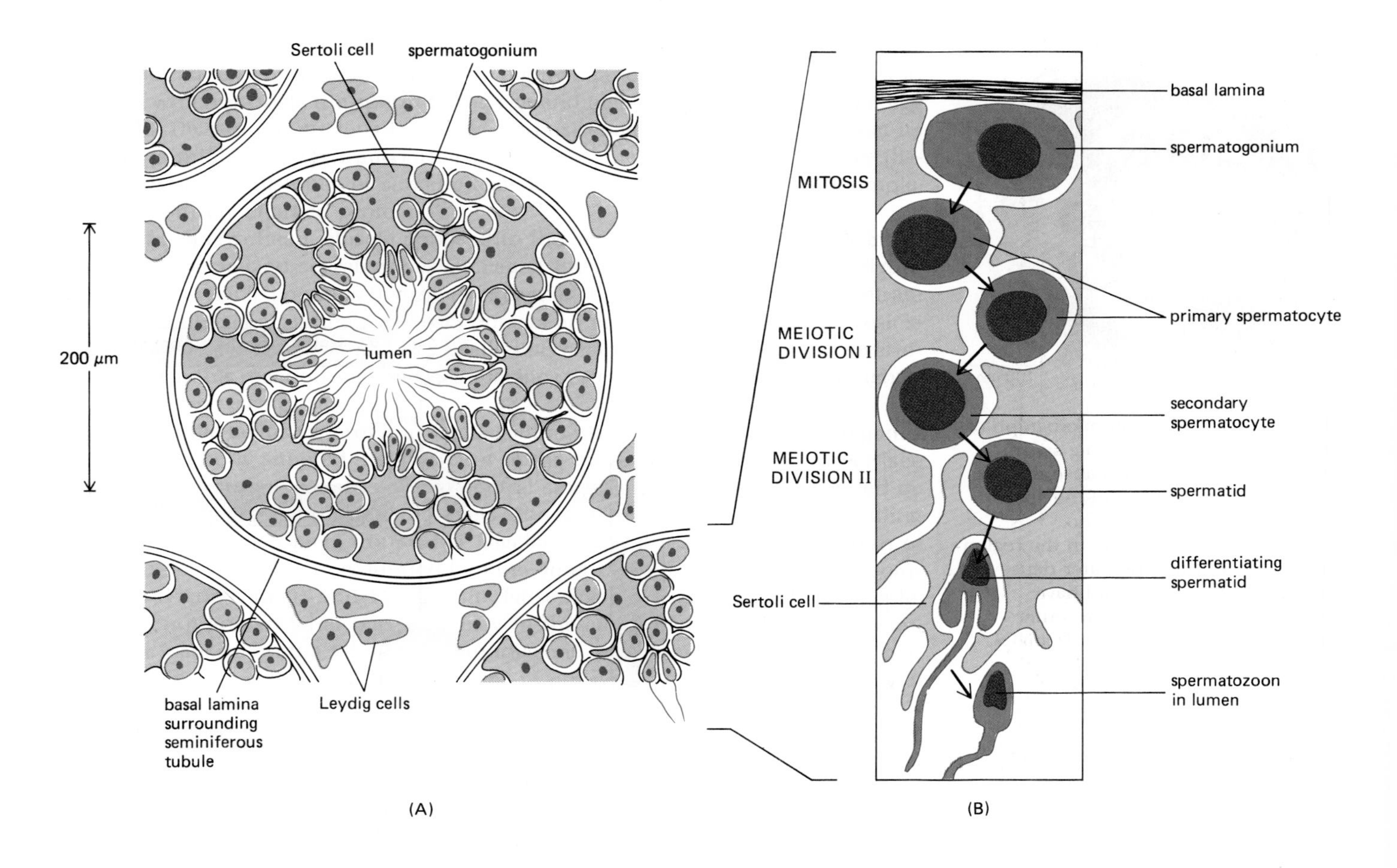

Figure 15–38 Highly simplified drawing of a cross-section of a seminiferous tubule in a mammalian testis. (A) All of the stages of spermatogenesis shown take place while the developing gametes are in intimate association with *Sertoli cells,* which are large cells that extend from the basal lamina to the lumen of the seminiferous tubule. Spermatogenesis depends on testosterone secreted by *Leydig cells,* located between the seminiferous tubules. (B) Dividing spermatogonia are found along the basal lamina. Some of these cells stop dividing and enter meiosis to become primary spermatocytes. Eventually spermatozoa are released into the lumen. In man it takes about 24 days for a spermatocyte to complete meiosis to become a spermatid and another 5 weeks for a spermatid to develop into a spermatozoon. Spermatozoa undergo further maturation and become motile in the epididymis and are only then fully mature sperm.

As in the case of oogenesis, spermatogenesis is under hormonal control. Beginning at puberty, the pituitary gland in human males secretes the same gonadotropin that we encountered earlier in discussing mammalian oogenesis—luteinizing hormone (LH). LH stimulates *Leydig cells* located between the seminiferous tubules in the testis to secrete large amounts of the male sex hormone, *testosterone;* testosterone in turn stimulates spermatogenesis, probably through its action on *Sertoli cells,* which completely envelop the developing sperm, protecting and nourishing them (see Figure 15–38).

An intriguing feature of spermatogenesis is that the developing male germ cells fail to complete cytoplasmic division (cytokinesis) during mitosis and meiosis, so that all the differentiating daughter cells descended from one maturing spermatogonium remain connected by cytoplasmic bridges (Figure 15–39). These cytoplasmic bridges persist until the very end of sperm differentiation, when individual sperm are released into the tubule lumen. A group of cells joined in this way is known as a *syncytium.* This accounts for the observation that mature sperm arise synchronously in any given area of a seminiferous tubule. But what is the function of the syncytial arrangement?

Sperm Nuclei Are Haploid, but Sperm Cell Differentiation Is Directed by the Diploid Genome[17]

Unlike oocytes, sperm undergo most of their differentiation after their nuclei have completed meiosis to become haploid. In principle, the cytoplasmic bridges between them could allow each developing haploid sperm, by sharing a common cytoplasm with its neighbors, to be supplied with all the products of a complete diploid genome. There are two reasons why it is important that the diploid genome direct sperm differentiation, just as it directs egg differentiation. First, the diploid genome from which the sperm derives generally will include some defective gene copies, corresponding to recessive lethal mutations (see p. 842); a haploid cell receiving one of these defective gene copies is likely to die unless it is provided

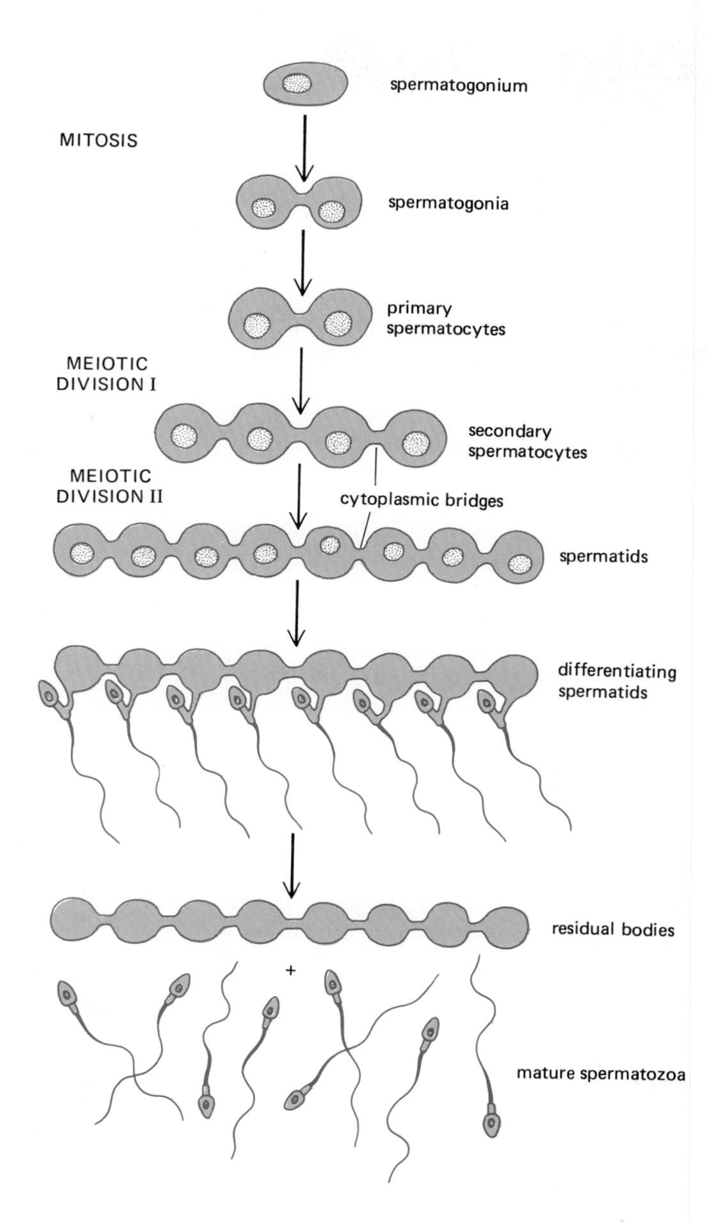

Figure 15–39 Illustration showing how the progeny of a single maturing spermatogonium remain connected to one another by cytoplasmic bridges throughout their differentiation into mature sperm. For the sake of simplicity, only two connected maturing spermatogonia are shown entering meiosis to eventually form eight connected haploid spermatids. In fact, the number of connected cells that go through two meiotic divisions and differentiate together is very much larger than shown here.

with the functional gene products encoded by other nuclei that have the good gene copy. Second, in organisms such as humans, some sperm inherit an X chromosome at meiosis, while others inherit a Y chromosome. Since the X chromosome carries many essential genes that are lacking on the Y chromosome, it is unlikely that the Y-bearing sperm would be able to survive and mature without products of the X chromosome.

There is direct experimental evidence that sperm differentiation is governed by products of the diploid genome. In *disjunction-defective mutants* of *Drosophila*, for example, chromosomes are divided unequally between daughter cells during meiosis. As a result, some sperm contain too few chromosomes, some too many, and some contain none at all. Yet the major features of sperm differentiation occur in all such cells, even in those without chromosomes. An obvious explanation is that the products of missing chromosomes are supplied by diffusion through the cytoplasmic bridges that connect adjacent germ cells. However, it is also possible that the diploid spermatogonia or primary spermatocytes produce stable instructions for sperm differentiation (presumably in the form of long-lived mRNA) before meiosis, so that there is no need for the haploid sperm genome to function during the period of differentiation. In either case it is evident that the differentiation of a sperm, even though it occurs when the sperm nucleus is haploid, uses products specified by both sets of parental chromosomes.

Summary

Eggs develop in stages from primordial germ cells that migrate into the ovary very early in development to become oogonia. After mitotic proliferation, oogonia become primary oocytes that begin meiotic division I and then arrest at prophase for days or years, depending on the species. During this prophase-I arrest period, primary oocytes grow and accumulate ribosomes, mRNAs, and proteins, often enlisting the help of other cells, including surrounding accessory cells. Further development (oocyte maturation) depends on polypeptide hormones (gonadotropins) that act on the surrounding accessory cells, causing them to induce a proportion of the primary oocytes to mature. These oocytes complete meiotic division I to form a small polar body and a large secondary oocyte and proceed into metaphase of meiotic division II, where, in many species, the oocyte is arrested until stimulated by fertilization to complete meiosis and begin embryonic development.

A sperm is usually a small, compact cell, highly specialized for the task of delivering its DNA to the egg. Whereas in many female organisms the total pool of oocytes is produced early in embryogenesis, in males new germ cells enter meiosis continually from the time of sexual maturation, each primary spermatocyte giving rise to four mature sperm. Sperm differentiation occurs after meiosis, when the nuclei are haploid. However, because the maturing spermatogonia and spermatocytes fail to complete cytokinesis, the progeny of a single spermatogonium develop as a large syncytium. This may be why sperm differentiation is directed by the products of both parental chromosomes.

Fertilization[18]

Once released, egg and sperm alike are destined to die within minutes or hours unless they find each other and fuse in the process of **fertilization.** Through fertilization the egg and sperm are saved: the egg is activated to begin its developmental program, and the nuclei of the two gametes fuse to form the genome of a new organism. Much of what we know about the mechanism of fertilization has been learned from studies of marine invertebrates, especially sea urchins (Figure 15–40). In these organisms, fertilization occurs in sea water, into which huge numbers of both sperm and eggs are released. Such *external fertilization* is much more accessible to study than the *internal fertilization* of mammals, which occurs in the female reproductive tract following mating. For this reason most of our discussion of fertilization will deal with sea urchins. However, despite the great evolutionary distance between sea urchins and mammals, many of the cellular and molecular mechanisms involved in fertilization are similar in both.

Figure 15–40 Photograph of two species of sea urchin that are commonly used in studies of fertilization. The one on top is *Strongylocentrotus purpuratus,* and the one below is *Strongylocentrotus franciscanus* (shown at approximate actual size). (Courtesy of Victor Vacquier.)

Contact with the Jelly Coat of an Egg Stimulates a Sea Urchin Sperm to Undergo an Acrosomal Reaction[19]

A typical female sea urchin has 10^7 eggs and a typical male 10^{12} sperm, so sea urchin gametes can be obtained as pure populations in very large numbers, all at the same stage of development. When the two types of gametes are mixed together, the events of sperm-egg interaction begin synchronously within seconds. Fertilization begins when the head of a sperm contacts the jelly coat (see Figure 15–24) of an egg. The contact triggers the sperm to undergo an **acrosomal reaction,** in which the contents of the acrosomal vesicle are released to the exterior surroundings. In sea urchins and many other marine invertebrates, the discharge of the contents of the acrosomal vesicle is accompanied by the formation of a long, actin-containing **acrosomal process,** which projects from the anterior end of the sperm. As shown in Figure 15–41, the tip of this process becomes covered by the components of the old acrosomal vesicle membrane. It also becomes coated with the secreted contents of the acrosomal vesicle, including (1) hydrolytic enzymes that help the sperm penetrate the jelly coat and thus gain access to the vitelline layer, (2) specific binding proteins that mediate the attachment of the acrosomal process to the vitelline layer (see below), and (3) hydrolytic enzymes that then permit the acrosomal process to bore through this layer to the egg's plasma membrane. When

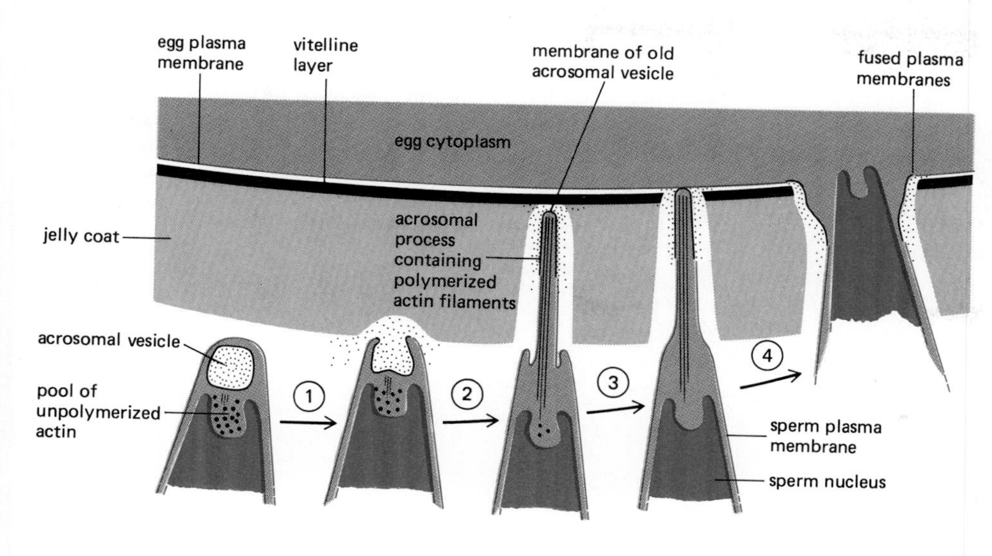

Figure 15–41 Details of the acrosomal reaction in sea urchins. When the sperm contacts the jelly coat, exocytosis of the acrosomal vesicle occurs (1), followed by the explosive polymerization of actin to form the long acrosomal process that penetrates the jelly coat (2). Proteins released from the acrosomal vesicle (*small black dots*) adhere to the surface of the acrosomal process and serve both to bind the sperm to the vitelline layer and to digest this layer (3). When the old acrosomal vesicle membrane (which forms the tip of the acrosomal process) contacts the egg plasma membrane (3), the two membranes fuse, the actin filaments disassemble, and the sperm enters the egg (4).

How do the sperm find an egg to fertilize after the gametes have been released into the seawater? Sea urchin eggs secrete a peptide called *resact,* which acts as a species-specific chemoattractant for sea urchin sperm. Resact binds to a transmembrane receptor on the sperm surface that has been shown to be an enzyme, guanylate cyclase, which catalyzes the production of cyclic GMP inside the sperm.

the membrane at the tip of the acrosomal process contacts the egg plasma membrane, the two membranes fuse, allowing the sperm nucleus to enter the egg (see Figure 15–41).

The trigger for the acrosomal reaction for sea urchin sperm is a large, fucose-sulfate-rich polysaccharide of the egg's jelly coat: when this polysaccharide is extracted from a sea urchin egg and added to sperm, it induces a normal acrosomal reaction within seconds. The jelly coat polysaccharide binds to a receptor glycoprotein in the sperm plasma membrane, causing the membrane to depolarize; the depolarization is thought to open voltage-gated Ca^{2+} channels in the membrane, allowing Ca^{2+} to enter the sperm. The jelly coat polysaccharide also activates a proton pump in the sperm plasma membrane that pumps H^+ out of the cell in exchange for Na^+. The resulting rise in pH inside the sperm head, together with the increase in cytosolic Ca^{2+}, initiates the acrosomal reaction. The rise in intracellular pH is thought to act, at least in part, by causing unpolymerized actin to dissociate from actin-binding proteins in the sperm cytoplasm that otherwise prevent actin polymerization (see p. 641); this initiates the explosive polymerization of actin, resulting in the formation of the acrosomal process.

Actin polymerization, however, is not the only mechanism driving the elongation of the acrosomal process. A net influx of ions (Ca^{2+}, Na^+, and Cl^-) increases the number of osmotically active molecules within the sperm head, thereby causing an influx of water. The sudden increase in hydrostatic pressure that results is thought to help extend the acrosomal process.

Sperm-Egg Adhesion Is Mediated by Species-specific Macromolecules[20]

The species-specificity of fertilization is especially important for aquatic animals that discharge their eggs and sperm into water, where they are liable to become mixed with eggs and sperm of other species. In sea urchins this specificity resides in the binding of the sperm to the vitelline layer beneath the jelly coat: sea urchin sperm will sometimes undergo an acrosomal reaction in response to eggs of other species, but they cannot bind to such eggs and therefore cannot fertilize them.

The molecule in sea urchin sperm that is thought to be responsible for the species-specific adhesion of the sperm to the egg's vitelline layer has been isolated. It is a protein called *bindin,* which is normally sequestered in the acrosomal vesicle. After its release in the acrosomal reaction, it coats the surface of the acrosomal process and mediates the attachment of the sperm to the egg. Each species of sea urchin makes a different type of bindin, which binds only to the vitelline layer of sea urchin eggs of the same species. The vitelline layer of one species of sea urchin egg has been found to contain a proteoglycan that acts as a *bindin receptor* in the adhesion process, and there is evidence that bindin acts as a lectin that rec-

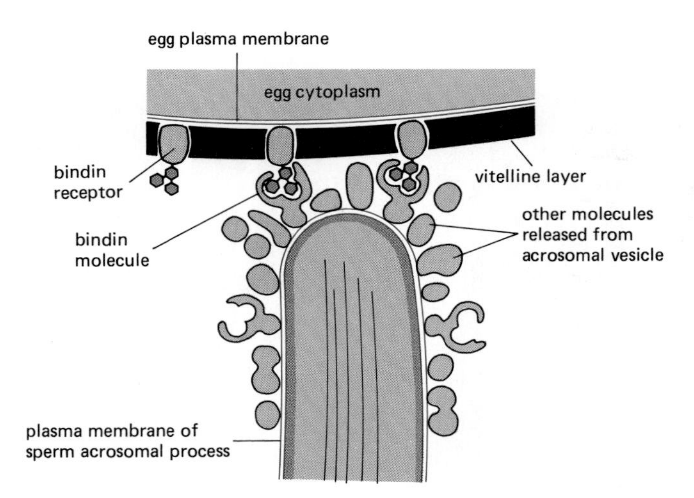

Figure 15–42 Schematic drawing of bindin molecules covering the surface of the acrosomal process of a sea urchin sperm. These proteins are thought to bind to a specific sugar sequence on a receptor molecule associated with the vitelline layer of the egg.

ognizes specific carbohydrate determinants on these proteoglycan molecules (Figure 15–42). Because bindin can induce the fusion of artificial lipid vesicles *in vitro*, it has been proposed that it may catalyze the fusion of the plasma membranes of the acrosomal process and the egg after bringing the two membranes together (see p. 336).

Egg Activation Involves Changes in Intracellular Ion Concentrations[21]

Once an activated sea urchin sperm attaches to an egg, the acrosomal process rapidly bores through the vitelline layer, and the membrane at the tip of the process fuses with the egg plasma membrane at the tip of a microvillus (Figure 15–43). Neighboring microvilli rapidly elongate and cluster around the sperm, which is then drawn head-first into the egg as the microvilli are resorbed.

The sperm activates the developmental program of the egg. Before fertilization an egg is metabolically dormant: it does not synthesize DNA, and it synthesizes RNA and protein at very low rates. Once released from the supportive environment of the ovary, an egg will die within hours unless rescued by fusion with a sperm. Metabolic activation by fusion with a sperm launches the dormant egg on a path leading to DNA synthesis and cleavage. The sperm, however, serves only to trigger a program that is already present in the egg. The sperm itself is not required. An egg can be activated artificially by a variety of nonspecific chemical or physical treatments; for example, a frog egg can be activated by pricking it with a needle. (The development of an egg that has been activated in the absence of a sperm is

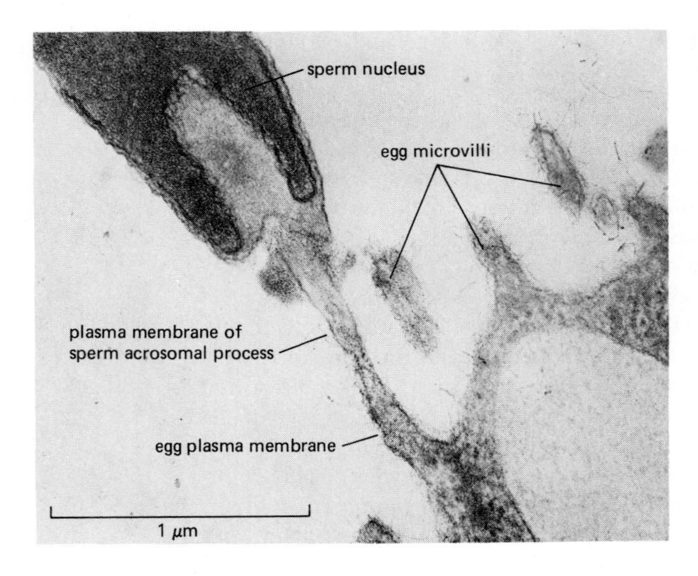

Figure 15–43 Electron micrograph showing a sea urchin sperm in the process of fertilizing an egg. The membrane at the tip of the acrosomal process of the sperm has fused with the egg plasma membrane at the tip of a microvillus on the egg surface. An unfertilized sea urchin egg is covered with more than 100,000 microvilli. (Courtesy of Frank Collins.)

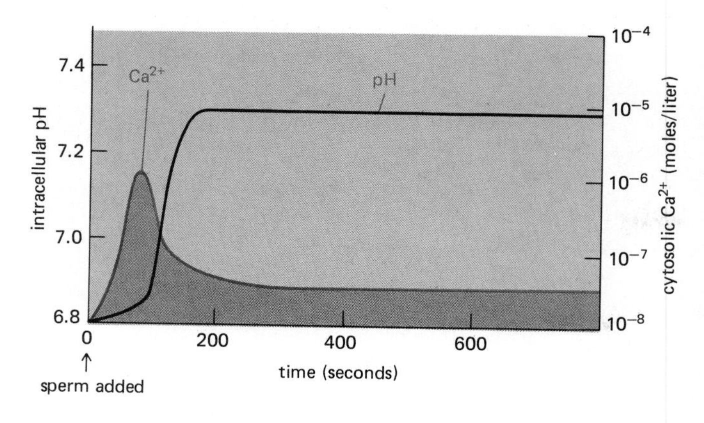

Figure 15–44 Two ionic changes involved in the activation of a sea urchin egg following fertilization. Beginning within 10 seconds after fertilization, Ca^{2+} is released into the cytosol from an intracellular Ca^{2+}-sequestering compartment, increasing the cytosolic concentration of free Ca^{2+} for about 2.5 minutes; afterwards the resting level of Ca^{2+} remains slightly higher than in the unfertilized egg. At about 60 seconds, a sustained efflux of H^+ coupled to an influx of Na^+ causes a permanent increase in intracellular pH.

called **parthenogenesis;** some organisms, including a few vertebrates, normally reproduce parthenogenically.) Moreover, the initial stages of egg activation cannot depend on the generation of any new proteins because they occur perfectly normally in the presence of drugs that inhibit protein synthesis.

In sea urchins many of the early steps in egg activation have been examined in detail and shown to be mediated by changes in ion concentrations within the egg. Three ionic changes occur within seconds or minutes of the addition of sperm to a suspension of eggs: (1) an increase in the permeability of the plasma membrane to Na^+ causes the membrane to depolarize within a few seconds; (2) a massive release of Ca^{2+} from an intracellular calcium-sequestering compartment (see p. 669) causes a marked increase in the concentration of Ca^{2+} in the cytosol within about 10 seconds; and (3) an efflux of H^+ coupled to an influx of Na^+ begins within 60 seconds and causes a large increase in intracellular pH (Figure 15–44). As we shall now describe, these three ionic changes have two consequences: first, they cause the egg to become impenetrable to further sperm; and second, they help trigger the initial steps in the developmental program of the egg.

The Rapid Depolarization of the Egg Plasma Membrane Prevents Further Sperm-Egg Fusions, Thereby Mediating the Fast Block to Polyspermy[22]

Although many sperm can attach to an egg, normally only one fuses with the egg plasma membrane and injects its nucleus into the cell. If more than one sperm fuses (a condition referred to as *polyspermy*), extra mitotic spindles are formed, resulting in the abnormal segregation of chromosomes during cleavage; nondiploid cells are produced, and development quickly stops. This means that eggs that are normally fertilized by the deposition of large numbers of sperm in their vicinity must somehow block the entry of extra sperm very soon after fertilization. Thus the eggs of many marine animals exhibit a **fast block to polyspermy,** brought about by different mechanisms, depending on the species.

Fish eggs have a small channel, called the *micropyle,* through which sperm must pass in single file. The passage of a single sperm through the channel stimulates the egg, causing the cortical granules to release their contents, which plug the hole so that no other sperm can enter. In most other organisms that reproduce using external fertilization, however, the eggs do not have a micropyle and can fuse with sperm over all or much of their surface. In some of these eggs (such as those of sea urchins and amphibians), the rapid depolarization of the plasma membrane caused by the fusion of the first sperm prevents further sperm from fusing. The membrane potential of a sea urchin egg is about −60 mV. Within a few seconds after sperm have been added, the membrane potential shifts precipitously to about +20 mV, where it remains for a minute or so before gradually returning to the original prefertilization level (Figure 15–45). If depolarization is prevented by fertilizing eggs in a solution containing an abnormally low concentration of Na^+, which reduces the sperm-triggered Na^+ influx that is largely responsible for depolarizing the membrane, there is an increased incidence of

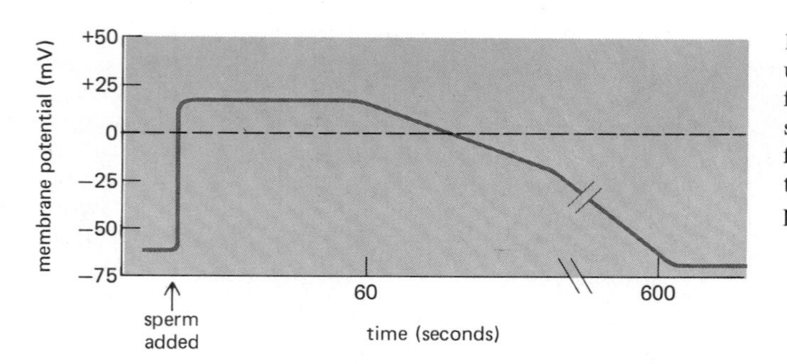

Figure 15–45 Changes in the sea urchin egg membrane potential after fertilization. The rapid depolarization somehow prevents further sperm from fusing with the egg plasma membrane, thereby creating the fast block to polyspermy.

polyspermy. Moreover, if an unfertilized egg is depolarized artificially by a current passed into it through a microelectrode, sperm can attach to the egg but cannot fuse; if the membrane is now repolarized with the microelectrode, the attached sperm fuse with and enter the egg. Although the molecular mechanism is unknown, it seems likely that the membrane depolarization that normally accompanies fertilization alters the conformation of a crucial protein in the egg plasma membrane so that a sperm membrane can no longer fuse with the egg membrane.

The egg membrane potential returns to normal within a few minutes after fertilization; therefore, a second mechanism must provide a longer-term barrier to polyspermy. In most eggs, including those of mammals, this barrier is provided by substances released from the cortical granules located just under the plasma membrane of the egg.

The Cortical Reaction Is Responsible for the Slow Block to Polyspermy[23]

The cortical granules in sea urchin eggs fuse with the plasma membrane and release their contents within 10 to 50 seconds of adding sperm. This **cortical reaction** is mediated by a large rise in the concentration of free Ca^{2+} in the cytosol. In an activated sea urchin egg, the Ca^{2+} concentration increases by about a hundredfold within less than a minute after the addition of sperm and then after a minute or so drops back toward normal (see Figure 15–44). The importance of Ca^{2+} in triggering the cortical reaction can be demonstrated directly in plasma membranes isolated from sea urchin eggs with cortical granules still attached to their cytoplasmic surfaces (Figure 15–46). When small amounts of Ca^{2+} are added to such preparations, exocytosis occurs within seconds.

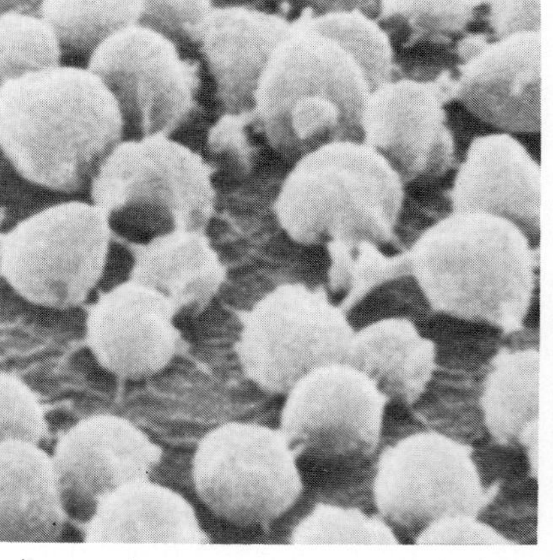

Figure 15–46 Scanning electron micrograph of cortical granules attached to the cytoplasmic surface of the isolated plasma membrane of an unfertilized sea urchin egg. When Ca^{2+} is added to this preparation, the cortical granules fuse with the plasma membrane and release their contents by exocytosis. Since there are about 15,000 cortical granules in each cell, the cortical reaction causes the surface area of the egg to more than double in less than a minute; some of the extra membrane is accommodated by a lengthening of each microvillus on the egg surface, while the rest is endocytosed via coated pits and vesicles (see p. 328). (From V.D. Vacquier, *Dev. Biol.* 43:62–74, 1975.)

In sea urchin eggs the cortical reaction has at least two separate effects: (1) proteolytic enzymes released from the cortical granules rapidly destroy the bindin receptors on the vitelline layer that are responsible for sperm attachment, and (2) components released from the cortical granules cause the vitelline layer to move away from the egg plasma membrane; at the same time, they enzymatically cross-link proteins in the vitelline layer, causing it to harden. In this way a *fertilization membrane* is formed that sperm cannot bind to or penetrate (Figure 15–47).

Egg Activation Is Mediated by the Inositol Phospholipid Cell-signaling Pathway[24]

Although membrane depolarization is the first detectable change following fertilization, it appears to serve only to prevent polyspermy. Artificially depolarizing the egg membrane does not activate the egg to begin biosynthesis; nor does blocking membrane depolarization at the time of fertilization inhibit this activation.

There is strong evidence that the transient increase in cytosolic Ca^{2+} concentration (which propagates as a wave across the egg from the site of sperm fusion—see Figure 4–35, p. 157) helps to initiate the program of egg development. If the cytosolic concentration of Ca^{2+} is increased artificially—either directly, by an injection of Ca^{2+}, or indirectly, by the use of Ca^{2+}-carrying ionophores, such as A23187 (see p. 322)—the eggs of all animals so far tested, including mammals, are activated. Moreover, preventing the increase in Ca^{2+} by injecting the Ca^{2+} chelator EGTA inhibits egg activation after fertilization. At least one way in which Ca^{2+} acts

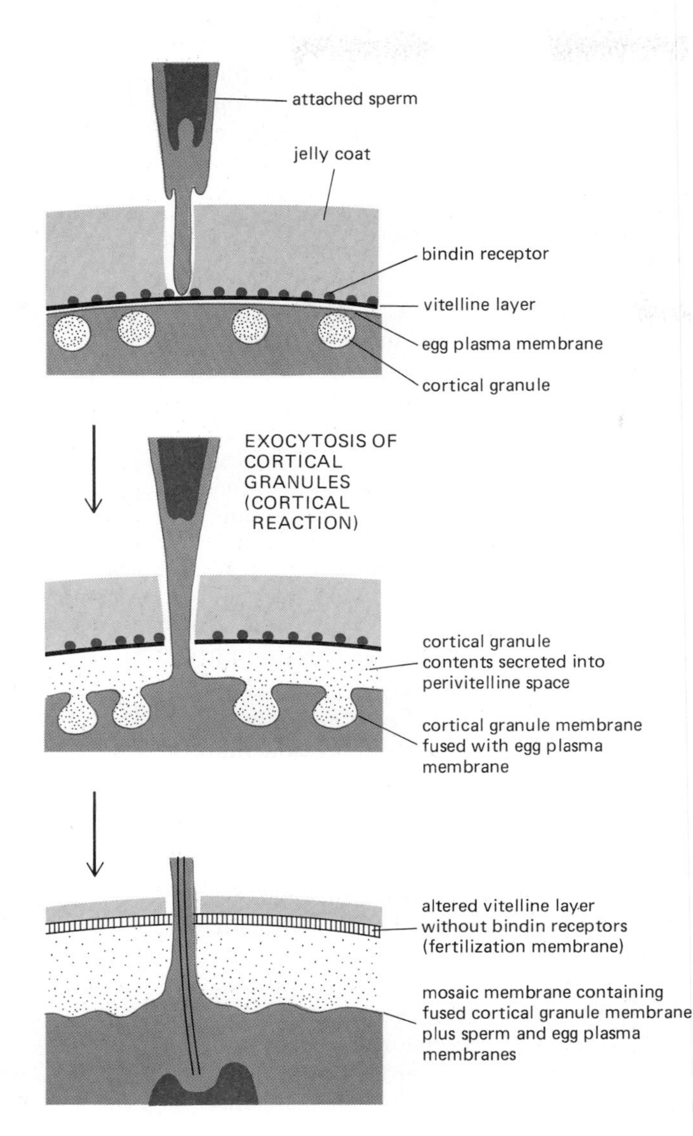

Figure 15–47 The cortical reaction in a sea urchin egg prevents additional sperm from entering the egg. The released contents of the cortical granules raise the vitelline layer and alter it so that it no longer contains bindin receptors and is converted into a *fertilization membrane* that sperm cannot penetrate. This "hardening" of the vitelline layer is due mainly to the formation of covalent cross-links between protein tyrosine residues, which generates an extensive, insoluble protein network.

in cells is by binding to the Ca^{2+}-binding protein *calmodulin,* which in turn activates a variety of cellular proteins (see p. 711). Calmodulin has been found in large amounts in all eggs that have been studied.

How does fertilization lead to an increase in the concentration of Ca^{2+} in the cytosol of the egg? We discussed in Chapter 12 how extracellular ligands binding to cell-surface receptor proteins can lead to the hydrolysis of *phosphatidylinositol bisphosphate* (PIP_2) in the plasma membrane to produce *inositol trisphosphate* ($InsP_3$) and *diacylglycerol:* the $InsP_3$ in turn releases Ca^{2+} from an intracellular calcium-sequestering compartment (see p. 702) into the cytosol, while the diacylglycerol activates *protein kinase C* (see p. 703). Evidence obtained in sea urchin eggs suggests that fertilization increases the concentration of cytosolic Ca^{2+} via this pathway. The concentration of $InsP_3$ increases within seconds after fertilization, just before the concentration of Ca^{2+} increases in the cytosol, and if $InsP_3$ is injected into an unfertilized egg, it increases the concentration of Ca^{2+} in the cytosol and thereby activates the egg. As expected, sperm activation of this pathway appears to be mediated by a G protein that activates a specific phospholipase C to hydrolyze PIP_2 (see p. 702). It is not known, however, whether the sperm binds to a receptor in the egg plasma membrane that is functionally coupled to the phospholipase C via a G protein, or whether the sperm injects a G protein activator at the time of sperm-egg fusion.

Since the increase in Ca^{2+} concentration in the cytosol following fertilization is transient, lasting only a minute or so, it is clear that it cannot *directly* mediate the events observed during the later stages of egg activation, which in sea urchins include a gradual increase in protein synthesis beginning at 8 minutes and the initiation of DNA synthesis beginning at about 30 minutes. There is increasing evidence that the activation of protein kinase C plays an important part in these later events, mainly by increasing intracellular pH.

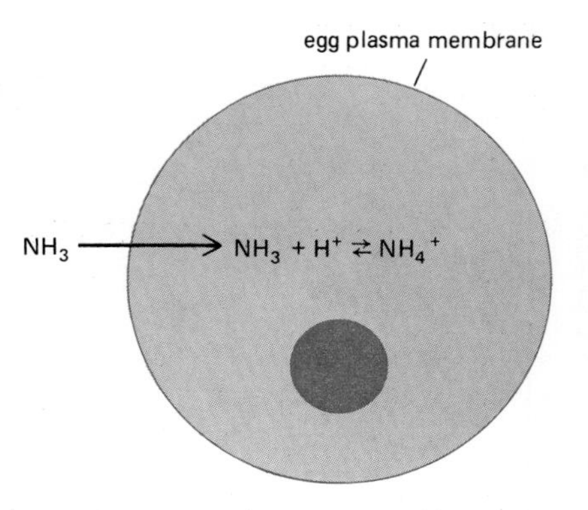

Figure 15–48 Increasing intracellular pH by incubating cells (such as eggs) in ammonia. The ammonia diffuses through the plasma membrane and combines with H^+ in the cytosol to form NH_4^+, thereby decreasing the intracellular concentration of H^+ and increasing the pH.

A Rise in the Intracellular pH in Some Organisms Induces the Late Synthetic Events of Egg Activation[25]

In sea urchins the activation of protein kinase C by diacylglycerol leads to the activation (presumably by phosphorylation) of a Na^+-H^+ exchanger in the egg plasma membrane. This membrane transport protein uses the energy stored in the Na^+ gradient across the membrane to pump H^+ out of the cell (see p. 309). The efflux of H^+ leads to an increase in the intracellular pH from 6.7 to 7.2, which is maintained throughout the rest of zygote development (see Figure 15–44). The unusually low intracellular pH in unfertilized sea urchin eggs is thought to be largely responsible for keeping the eggs metabolically inactive, and there is compelling evidence that the increase in pH following fertilization helps to induce the late synthetic events in these eggs: (1) When the intracellular pH is raised in unfertilized eggs by incubating them in a medium containing ammonia (Figure 15–48), protein synthesis is markedly increased and DNA replication is triggered, even in the absence of an increase in free intracellular Ca^{2+}. (2) When eggs are placed in Na^+-free sea water just after fertilization, so that there is no Na^+ gradient to drive the H^+ efflux, the intracellular pH does not rise and the late events do not occur. Such eggs can be rescued by adding ammonia to the medium: now the intracellular pH rises and the synthesis of protein and DNA is induced, even in the absence of extracellular Na^+.

The marked increase in protein synthesis in fertilized sea urchin eggs does not require RNA synthesis, since it is unaffected by the drug actinomycin D, which inhibits RNA synthesis. It is thought to result from at least two separate changes: (1) preexisting mRNA molecules stored in the egg are made available for protein synthesis, and (2) the egg ribosomes are activated so that they translate the available mRNA molecules more rapidly. By contrast, the increase in protein synthesis in unfertilized eggs treated with ammonia results solely from increased recruitment of preexisting mRNA molecules. This suggests that, while the increase in intracellular pH is responsible for mRNA recruitment, some other factor normally increases the rate of ribosome movement along mRNAs. The detailed mechanisms involved in these two types of activation are unknown.

Although fertilization is a very special event, it relies on the same signaling pathways that regulate intracellular processes in somatic cells (see Chapter 12). The sequence of some of the events in sea urchin egg activation after fertilization is summarized in Table 15–1.

The Fusion of Sea Urchin Sperm and Egg Pronuclei Depends on Centrioles Donated by the Sperm[26]

Once fertilized, the egg is called a **zygote.** In most species, including some sea urchins, fertilization is not complete until the two haploid nuclei (called *pronuclei*) have fused. Because the egg is large, the sperm and egg pronuclei have to migrate substantial distances to find each other. Not surprisingly, this migration depends on the cytoskeleton.

The sea urchin sperm contributes more than DNA to the zygote: it donates two centrioles. The sperm centrioles are crucial because the egg loses its own centrioles during the last meiotic division. The sperm centrioles become the center of a radiating array of microtubules called the *sperm aster,* which seems to guide the male pronucleus toward the female pronucleus: if the microtubules are depolymerized by treatment with colchicine, the migration of the two pronuclei toward the center of the egg does not occur. Eventually the two pronuclei make contact and their membranes fuse to form the diploid nucleus of the zygote. The pair of sperm centrioles, with its associated aster, then divides to form the two poles of the mitotic spindle for the first cleavage division.

Table 15–1 Sequence of Events Following Fertilization of Sea Urchin Eggs

Event	Time After Fertilization	Mediator
1. Plasma membrane depolarization	<5 seconds	sperm-induced increase in plasma membrane permeability to Na^+ (and to some extent to Ca^{2+})
2. Hydrolysis of phosphatidylinositol bisphosphate	<10 seconds	activation of phospholipase C
3. Increased concentration of free cytosolic Ca^{2+}	10–40 seconds	$InsP_3$-induced release of Ca^{2+} from intracellular Ca^{2+}-sequestering compartment
4. Cortical granule exocytosis	10–50 seconds	increased intracellular Ca^{2+}
5. Increased intracellular pH	60 seconds	activation of Na^+-H^+ exchanger by protein kinase C
6. Increased protein synthesis	8 minutes	increased intracellular pH
7. Fusion of sperm and egg nuclei	30 minutes	
8. Initiation of DNA replication	30–45 minutes	increased intracellular pH

Mammalian Eggs Can Be Fertilized *in Vitro*[27]

Compared with sea urchin eggs, mammalian eggs are difficult to study. Whereas sea urchin eggs are readily available by the millions, investigators must be content to work with tens or hundreds of mammalian eggs. Nonetheless, it is now possible to fertilize mammalian eggs *in vitro.* (Although we shall continue to use the term egg, it should be recalled that in mammals it is a secondary oocyte that is fertilized—see p. 862.) This brings a medical benefit: mammalian eggs that have been fertilized *in vitro* can develop into normal individuals when transplanted into the uterus; in this way many previously infertile women have been able to produce normal children. *In vitro* fertilization of mammalian eggs also allows one to study the events of fertilization and activation. Such studies indicate that, whereas the sequence of events just described in sea urchin fertilization is followed in broad outline during mammalian fertilization, there are important differences in many of the individual steps.

Some of the differences are related to the sperm. Mammalian sperm are unable to fertilize an egg until they have undergone a process referred to as **capacitation,** induced by secretions in the female genital tract. The mechanism of capacitation is unclear; it seems to involve both an alteration in the lipid and glycoprotein composition of the sperm plasma membrane and an increase in sperm metabolism and motility. Capacitated mouse sperm penetrate the shell of follicle cells and bind specifically to a major glycoprotein in the *zona pellucida*—the protective coat equivalent to the vitelline layer of sea urchin eggs (see p. 856). At least in some species, this same egg glycoprotein is thought to activate sperm to undergo the acrosomal reaction. The zona pellucida of the mouse egg, for example, is composed of only three glycoproteins, which are secreted by growing oocytes and self-assemble into a three-dimensional network of interconnected filaments (see Figure 15–23). One of these, called ZP3, serves both to bind the sperm and then to induce the acrosomal reaction. The sperm is thought to recognize specific carbohydrate determinants on the ZP3 glycoprotein in binding to the zona. Unlike the case in sea urchins, the sperm molecules involved in the recognition process are located on the plasma membrane rather than on the acrosomal membrane. The zona pellucida, like the vitelline layer in sea urchin eggs, serves as a barrier to fertilization across species, and removing the zona often removes this barrier.

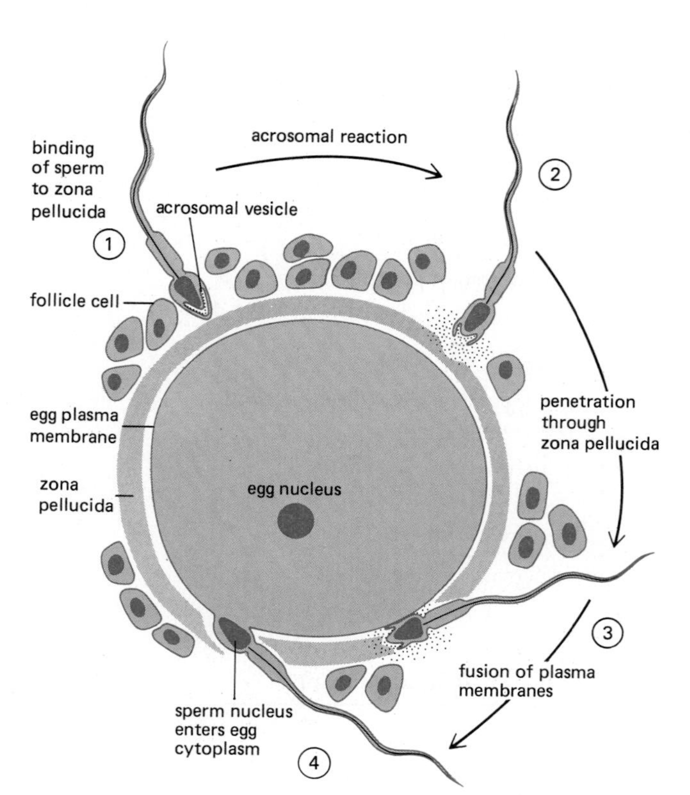

Figure 15–49 The acrosomal reaction that occurs when a mammalian sperm fertilizes an egg. In mice a single glycoprotein in the zona pellucida is thought to be responsible for both binding the sperm and inducing the acrosomal reaction. Note that a mammalian sperm interacts tangentially with the egg plasma membrane so that fusion occurs at the side rather than at the tip of the sperm head. In mice the zona pellucida is 7 μm in diameter and sperm cross it at a rate of about 1 μm/min.

For example, hamster eggs from which the zona pellucida has been removed with specific enzymes can be fertilized by human sperm. Not surprisingly, such hybrid "humsters" do not develop.

The acrosomal reaction of mammalian sperm releases proteases and hyaluronidase, which are essential for penetration of the zona pellucida. However, an acrosomal process similar to that formed by sea urchin sperm is not produced, and in most mammalian sperm it is the equatorial (postacrosomal) region of the plasma membrane that fuses with the egg, rather than the membrane of the acrosomal vesicle (Figure 15–49). Because relatively few sperm usually manage to reach an ovulated egg in mammals (fewer than 200 of the 3×10^8 human sperm released during coitus reach the site of fertilization), a fast block to polyspermy is apparently unnecessary: the rapid depolarization of the plasma membrane that accompanies fertilization and prevents polyspermy in sea urchin and amphibian eggs does not occur. On the other hand, the enzymes released by the cortical reaction of mammalian eggs change the structure of the egg coat to provide a slow block to polyspermy. They alter the ZP3 glycoprotein in the zona pellucida of the mouse egg, for example, so that it can no longer bind sperm or activate them to undergo an acrosomal reaction. In some other mammalian eggs, the cortical reaction alters the plasma membrane, rather than the zona, in a way that prevents further sperm from fusing with it.

Another distinctive feature of mammalian fertilization is that the egg contains centrioles while the sperm does not. Moreover, in fertilized mammalian eggs the two pronuclei do not fuse directly: they approach each other, but their chromosomes do not come together into one nucleus until after the membrane of each pronucleus breaks down in preparation for the first cleavage. Most of the other early events of fertilization in sea urchin eggs outlined in Table 15–1, such as the activation of the inositol phospholipid pathway and the resulting increase in cytosolic Ca^{2+} concentration, also occur in mammalian eggs. Subsequent events are part of the process of *embryogenesis,* in which the zygote develops into a new individual. It is perhaps the most remarkable phenomenon in all of biology and is the subject of the next chapter.

Summary

Fertilization begins when the head of a sperm makes contact with the protective coat surrounding the egg. This induces an acrosomal reaction in which the sperm releases the contents of its acrosomal vesicle, including proteins that help the sperm digest its way to the egg plasma membrane in order to fuse with it. Fertilization activates a cascade of changes in the egg that is initiated by the hydrolysis of phosphatidylinositol bisphosphate in the egg plasma membrane. Egg activation includes changes to the surface of the egg that prevent the fusion of additional sperm; one such block to polyspermy results from the cortical reaction in which cortical granules release their contents to the outside, altering the egg coat. Changes also take place in the interior of the egg in preparation for the subsequent development of the zygote after the sperm and egg pronuclei have come together.

References

General

Austin, C.R.; Short, R.V., eds. Reproduction in Mammals: I. Germ Cells and Fertilization, 2nd ed. Cambridge, U.K.: Cambridge University Press, 1982.

Browder, L. Developmental Biology, 2nd ed., Chapters 5, 6, and 8. Philadelphia: Saunders, 1980.

Epel, D. The program of fertilization. *Sci. Am.* 237(11):128–138, 1977.

Karp, G.; Berrill, N.J. Development, 2nd ed., Chapters 4 and 5. New York: McGraw-Hill, 1981.

Longo, F.J. Fertilization. London: Chapman & Hall, 1987.

Cited

1. Crow, J.F. The importance of recombination. In The Evolution of Sex: An Examination of Current Ideas (R.E. Michod, B.R. Levin, eds.), pp. 56–73. Sunderland, MA: Sinauer, 1988.

   Maynard Smith, J. Evolution of Sex. Cambridge, U.K.: Cambridge University Press, 1978.

   Williams, G.C. Sex and Evolution. Princeton, NJ: Princeton University Press, 1975.

2. Ayala, F.; Kiger, J. Modern Genetics, 2nd ed. Menlo Park, CA: Benjamin-Cummings, 1984.

   Ferris, S.D.; Whitt, G.S. Loss of duplicate gene expression after polyploidization. *Nature* 265:258–260, 1977.

   Fincham, J.R.S. Genetics. Boston: Jones and Bartlett, 1983.

3. Lewis, J.; Wolpert, L. Diploidy, evolution and sex. *J. Theor. Biol.* 78:425–438, 1979.

   Spofford, J.B. Heterosis and the evolution of duplications. *Am. Nat.* 103:407–432, 1969.

4. Evans, C.W.; Dickinson, H.G., eds. Controlling Events in Meiosis. *Symp. Soc. Exp. Biol.*, Vol. 38. Cambridge, U.K.: The Company of Biologists, 1984.

   Whitehouse, H.L. Towards an Understanding of the Mechanism of Heredity, 3rd ed. London: St. Martins. 1973. (Contains a lucid description of the development of our current understanding of chromosome behavior during meiosis.)

   Wolfe, S.L. Biology of the Cell, 2nd ed., pp. 432–470. Belmont, CA: Wadsworth, 1981.

5. John, B.; Lewis, K.R. The Meiotic Mechanism. Oxford Biology Readers (J.J. Head, ed.). Oxford, Eng.: Oxford University Press, 1976.

   Jones, G.H. The control of chiasma distribution. In Controlling Events in Meiosis (C.W. Evans; H.G. Dickinson, eds), *Symp. Soc. Exp. Biol.*, Vol. 38, pp. 293–320. Cambridge, U.K.: The Company of Biologists, 1984.

   Orr-Weaver, T.L.; Szostak, J.W. Fungal recombination. *Microbiol. Rev.* 49:33–58, 1985.

6. Heyting, C.; Dettmers, R.J.; Dietrich, A.J.; Redeker, E.J. Two major components of synaptonemal complexes are specific for meiotic prophase nuclei. *Chromosoma* 96:325–332, 1988.

   Moses, M.J. Synaptonemal complex. *Annu. Rev. Genet.* 2:363–412. 1968.

   Smithies, O.; Powers, P.A. Gene conversions and their relationship to homologous pairing. *Phil. Trans. R. Soc. Lond. (Biol.)* 312:291–302, 1986.

   von Wettstein, D.; Rasmussen, S.W.; Holm, P.B. The synaptonemal complex in genetic segregation. *Annu. Rev. Genet.* 18:331–413, 1984.

7. Carpenter, A.T.C. Gene conversion, recombination nodules, and the initiation of meiotic synapsis. *Bioessays* 6:232–236, 1987.

   Carpenter, A.T.C. Recombination nodules and synaptonemal complex in recombination-defective females of *Drosophila melanogaster*. *Chromosoma* 75:259–236, 1979.

8. Buckle, V.; Mondello, C.; Darling, S.; Craig, I.W.; Goodfellow, P.N. Homologous expressed genes in the human sex chromosome pairing region. *Nature* 317:739–741, 1985.

   Chandley, A. C. Meiosis in man. *Trends Genet.* 4:79–84, 1988.

   Solari, A.J. The behavior of the XY pair in mammals, *Int. Rev. Cytol.* 38:273–317, 1974.

9. Austin, C.R.; Short, R.V., eds. Reproduction in Mammals: I. Germ Cells and Fertilization. Cambridge, U.K.: Cambridge University Press, 1982.

10. Browder, L. Developmental Biology, pp. 173–231. Philadelphia: Saunders, 1980.

    Karp, G.; Berrill, N.J. Development, 2nd ed., pp. 116–138. New York: McGraw-Hill, 1981.

11. Browder, L.W., ed. Oogenesis. New York: Plenum, 1985.

    Davidson, E.H. Gene Activity in Early Development, 3rd ed., pp. 305–407. Orlando, FL: Academic, 1986.

    Metz, C.B.; Monroy, A., eds. Biology of Fertilization, Vol. 1: Model Systems and Oogenesis. Orlando, FL: Academic, 1985.

12. Bornslaeger, E.A.; Mattei, P.; Schultz, R.M. Involvement of cAMP-dependent protein kinase and protein phosphorylation in regulation of mouse oocyte maturation. *Dev. Biol.* 114:453–462, 1986.

    Maller, J.L. Regulation of amphibian oocyte maturation. *Cell Differ.* 16:211–221, 1985.

    Masui, Y.; Clarke, H.J. Oocyte maturation. *Int. Rev. Cytol.* 57:185–282, 1979.

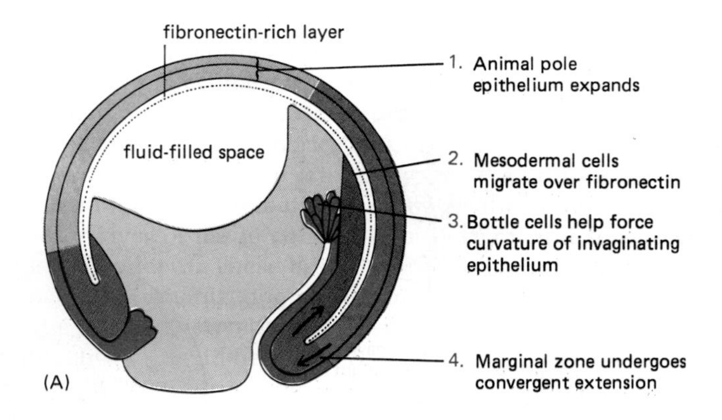

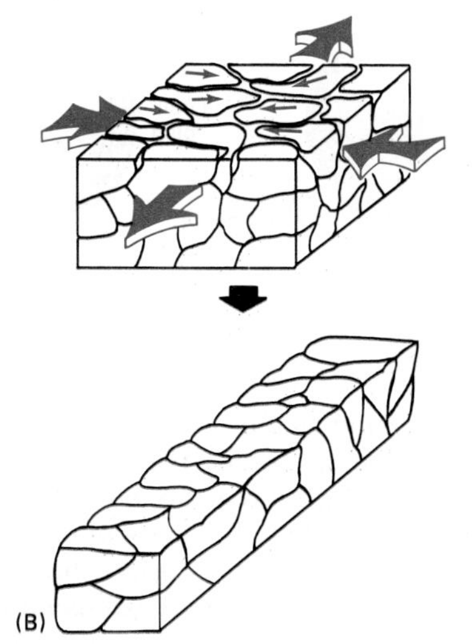

Figure 16–11 (A) A section through a gastrulating *Xenopus* embryo, cut in the same plane as in Figure 16–9, indicating the four main types of cell movement that gastrulation involves. (B) Cartoon of the cell repacking that brings about convergent extension, thought to be the chief driving force for gastrulation in Xenopus. (A, after R.E. Keller, *J. Exp. Zool.* 216:81–101, 1981; B, from J. Gerhart and R. Keller, *Annu. Rev. Cell Biol.* 2:201–229, 1986.)

Gastrulation Movements Are Organized Around the Blastopore[2,7,8]

The cell movements of gastrulation are complex but orderly, so that it is possible to plot on the surface of the pregastrulation embryo a *fate map* (Figure 16–9B) showing which cells of the very early stage will be carried into which parts of the mature animal. But how is the whole complex of gastrulation movements set in train and organized? In amphibians the rearrangement of the egg contents immediately after fertilization is an essential preliminary step. Invagination always begins at the site corresponding to the gray crescent (see p. 880): here, at the future dorsal lip of the blastopore, the fertilization-induced rotation of egg cortex relative to egg core (see Figure 16–2) evidently creates a combination of cell components with unique properties. If the dorsal lip of the blastopore is excised from a normal embryo at the beginning of gastrulation and grafted into another embryo but in a different position, the host embryo initiates gastrulation both at the site of its own dorsal lip and at the site of the graft (Figure 16–12). The movements of gastrulation at the second site entail the formation of a second whole set of body structures, and a double embryo (Siamese twins) results. By carrying out such grafts between species with differently pigmented cells, so that host tissue can be distinguished from implanted tissue, it has been shown that the grafted blastopore lip recruits host epithelium into its own system of invaginating endoderm and mesoderm. We shall see later that chemical, as well as physical, interactions between cells are crucial for the creation of the different germ layers at gastrulation. But first we must outline briefly the later development of the layers of endoderm, mesoderm, and ectoderm that constitute the vertebrate embryo just after gastrulation.

The Endoderm Gives Rise to the Gut and Associated Organs Such as the Lungs and Liver[9]

The endoderm forms a tube, the primordium of the digestive tract, from the mouth to the anus. It gives rise not only to the pharynx, esophagus, stomach, and intestines, but also to many associated glands. The salivary glands, the liver, the pancreas, the trachea, and the lungs, for example, all develop from extensions of the wall of the originally simple digestive tract and grow to become systems of branching tubes that open into the gut or pharynx. While the endoderm forms the epithelial components of these structures—the lining of the gut and the secretory cells of the pancreas, for example—the supporting muscular and fibrous elements arise from the mesoderm.

The Mesoderm Gives Rise to Connective Tissues, Muscles, and the Vascular and Urogenital Systems[9,10]

The mesodermal layer is divided in the post-gastrulation embryo into separate parts on the left and right of the body. Defining the central axis of the vertebrate body, and effecting this separation, is a very early specialization of the mesoderm known as the **notochord.** This is a slender rod of cells, about 80 μm in diameter, with ectoderm above it, endoderm below it, and mesoderm on either side (see Figure 16–15). The cells of the notochord become swollen with vacuoles, so that

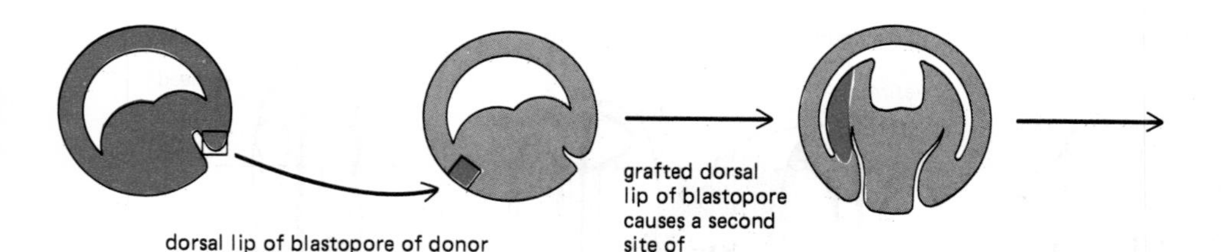

the rod elongates and stretches out the embryo. In the most primitive chordates, which have no vertebrae, the notochord is retained as a primitive substitute for a vertebral column. In vertebrates it serves as a core around which mesodermal cells gather to form the vertebrae. Thus the notochord is the precursor of the vertebral column, both in an evolutionary and in a developmental sense.

In general, the mesoderm gives rise to the connective tissues of the body—at first to the loose, space-filling, three-dimensional mesh of cells known as *mesenchyme* (see Figure 14–20, p. 802), and ultimately to cartilage, bone, and fibrous tissue, including the dermis (the inner layer of the skin). Muscle cells also derive from mesoderm. In addition, most of the tubules of the urogenital system form from it and so does the vascular system, including the heart and the cells of the blood.

Figure 16–12 (A) Diagram of an experiment showing that the dorsal lip of the blastopore initiates and controls the movements of gastrulation and thereby, if transplanted, organizes the formation of a second set of body structures. (B) Photograph of a two-headed, two-tailed axolotl tadpole resulting from such an operation; the results are similar for *Xenopus*, though the extent of duplication is somewhat variable. (B, courtesy of Jonathan Slack.)

The Ectoderm Gives Rise to the Epidermis and the Nervous System[9,11]

At the end of gastrulation the sheet of ectoderm covers the embryo and thus eventually forms the epidermis (the outer layer of the skin). But that is not all: the entire nervous system also derives from it. In a process known as **neurulation,** a broad central region of the ectoderm thickens, rolls up into a tube, and pinches off from the rest of the cell sheet. This transformation is induced by an interaction with the underlying notochord and the mesoderm adjacent to it (see p. 941). The tube thus created from the ectoderm is called the **neural tube;** it will form the brain and the spinal cord. Along the line where the neural tube pinches off from the future epidermis, a number of ectodermal cells break loose from the epithelium and migrate as individuals out through the mesoderm. These are the cells of the **neural crest;** they will form almost all of the peripheral nervous system (including the sensory and sympathetic ganglia and the Schwann cells that make the myelin sheaths of peripheral nerves—see p. 1060) as well as the epinephrine-secreting cells of the adrenal gland and the pigment cells of the skin. In the head, many of the neural crest cells will differentiate into cartilage, bone, and other connective tissues, which elsewhere in the body arise from the mesoderm. This is one of several instances that run counter to the general scheme in which the three germ layers give rise to cells in three corresponding concentric layers of the adult body.

The sense organs, by which light, sound, smell, and so forth impinge on the nervous system, also have ectodermal origins: some derive from the neural tube, some from the neural crest, and some from the exterior layer of ectoderm (see Figure 19–55, p. 1109). The retina, for example, originates as an outgrowth of the brain and so is derived from cells of the neural tube, while the olfactory cells of the nose differentiate directly from the ectodermal epithelium lining the nasal cavity.

The Neural Tube Is Formed by Coordinated Changes in Cell Shape[6,12]

The formation of the neural tube (Figure 16–13) is a dramatic event to watch. At first the surface of the gastrula appears more or less uniform. But subtle changes are occurring: the ectoderm close to the midline begins to thicken, forming the *neural plate.* Then the lateral edges of the neural plate start to rear up in folds; these *neural folds* gradually roll together, while the midline of the plate sinks deeper. Eventually the folds meet and fuse to form the hollow neural tube, roofed over by a continuous sheet of ectoderm. As in gastrulation, the whole process

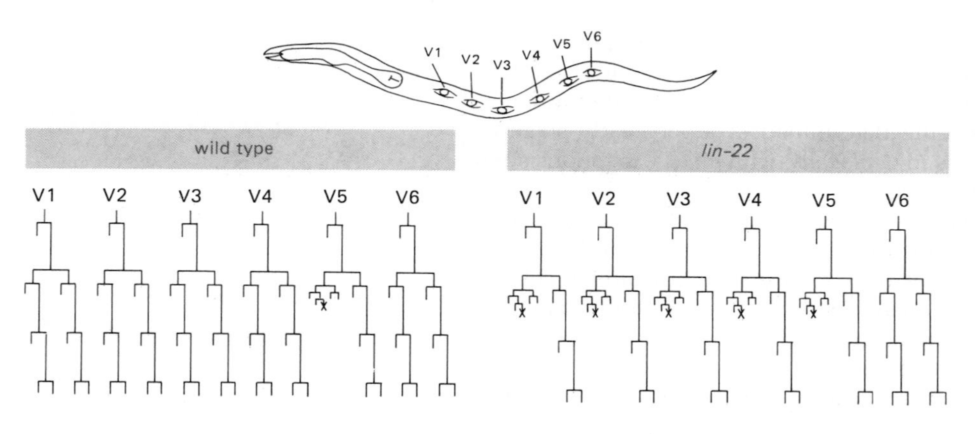

Figure 16–34 Lineage diagrams illustrating how a mutation in a developmental control gene (here a gene called *lin-22*) can cause the same substitution of one lineage motif for another in several separate branches of the lineage tree. Lines of descent are shown for the progeny of each of six precursor cells in the larval hypodermis of *C. elegans;* the locations of these cells are indicated above on a diagram of a young larva. The cross denotes a programmed cell death—a common feature of normal development, both in *C. elegans* and in other species. (After H.R. Horvitz et al., *Cold Spring Harbor Symp. Quant. Biol.* 48:453–463, 1983.)

As computer programmers know, small changes in a program can have drastic effects on the output produced when a program is executed. Likewise, a mutation in a single control gene can result in a grossly abnormal lineage tree. This is well illustrated by the *heterochronic mutations,* which cause certain sets of cells to behave in a way that would be appropriate for normal cells at a different stage in development. A daughter cell may behave like a parent or grandparent, for example, and the offspring of the daughter may behave again in the same way, and so on, with the result that a portion of the lineage pattern is reiterated several times and development is in effect retarded. Figure 16–35 shows the lineage diagrams for a set of mutations in a gene called *lin-14,* illustrating this phenomenon: instead of progressing through the normal series of cell divisions characteristic of the first, second, third, and fourth larval instars and then halting, many of the cells in certain *lin-14* mutants repeatedly go through the patterns of cell divisions characteristic of the first larval instar, continuing through as many as five or six molt cycles and persisting in the manufacture of an immature type of cuticle. Other mutations in the *lin-14* gene have the reverse effect, causing cells to adopt mature states precociously, skipping intermediate stages, so that the animal reaches its final state prematurely and with an abnormally small number of cells. This precocious development occurs in mutants deficient in normal *lin-14* activity, whereas retarded development occurs in mutants with abnormally high levels of *lin-14* activity. Thus the effect of the *lin-14* gene product is to keep the cells young, and normal development seems to depend on its progressive reduction as the animal matures.

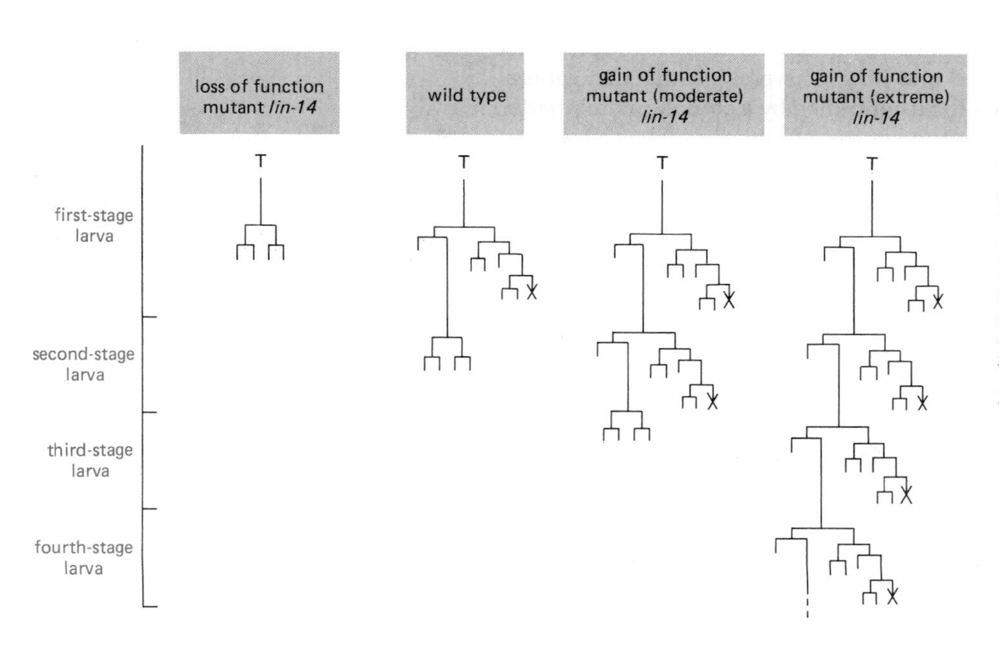

Figure 16–35 Heterochronic mutations in the gene *lin-14* of *C. elegans* and their effects on one of the many lineages they alter. The loss-of-function (recessive) mutation in this gene causes premature occurrence of the pattern of cell division and differentiation characteristic of a late larva; the gain-of-function (dominant) mutations in this gene have the opposite effect. The cross denotes a programmed cell death. (After V. Ambros and H.R. Horvitz, *Science* 226:409–416, 1984.)

The Program of Differentiation Is Coordinated with the Program of Cell Division[30]

Heterochronic mutations emphasize another important general point: the genome during development has to define a control program for cell division as well as cell differentiation, and the two aspects of cell behavior have to be synchronized. In the heterochronic mutant both division and differentiation are affected in a coordinated way, suggesting that both processes are regulated by one master timing mechanism that has been disturbed by the mutation.

One possibility is that the cell division cycle itself serves as the timer, but the evidence on the whole is against this: cells in developing embryos frequently differentiate on schedule even when cell division is artificially prevented by agents that block either cytokinesis or DNA synthesis. Thus for many species it seems that cell divisions usually are not in themselves the ticks of the clock that sets the tempo of biochemical development but rather are governed by this clock: the changing chemical state of the cell controls in parallel the decision to divide and the decision as to when and how to differentiate. The molecular mechanisms of cell-division control in embryos are almost completely unknown, and they pose one of the central unsolved problems of developmental biology. Cell-lineage mutants in the nematode may hold a key to the solution.

Development Depends on an Interplay Between Cell-autonomous Behavior and Cell-Cell Interactions[31]

One cannot begin to understand a cell's behavior during development until one knows whether it is autonomous or is governed by signals from the surroundings. The importance of signals from neighboring cells can be assessed in the nematode by observing the consequences when these cells are individually eliminated. This can be done using a focused laser beam with a diameter of about 0.5 μm (the average nucleus in *C. elegans* is about 2 μm in diameter). When a cell nucleus is exposed to repeated pulses of laser light, the cell dies with no apparent damage to other cells in the animal. Most often it is found that the surviving cells persist in their normal course of development even after their close neighbors have been destroyed. This suggests that most of the cells in the nematode, during much of their history, are following their developmental programs autonomously, without regard to signals from adjacent cells.

Nevertheless, intercellular signals play a crucial part in the development of *C. elegans*, as in that of other animals. A good example is provided by the development of the *vulva*, the egg-laying orifice in hermaphrodites. The vulva is a ventral opening in the hypodermis (skin) formed by 22 cells that arise by specific lineages from three precursor cells in the hypodermis. A single nondividing cell in the gonad, called the *anchor cell*, attaches or "anchors" the developing vulva to the overlying gonad (the uterus) to create a passageway through which the eggs can pass to the outside world. Laser destruction studies show that the anchor cell is responsible for inducing the three nearest hypodermal cells to form a vulva. If the anchor cell is killed, these cells, instead of following a vulval lineage, give rise to ordinary hypodermal cells. Thus the anchor cell induces vulval differentiation in *C. elegans* just as the vegetal blastomeres induce mesodermal differentiation in the early *Xenopus* embryo. Only the anchor cell is necessary for this induction: if all the gonadal cells except the anchor cell are killed, the vulva still develops normally (Figure 16–36A).

The inducing signal from the anchor cell ensures that the vulva develops in exactly the right place in relation to the gonad, and there is some potential for flexibility in the system: the three hypodermal cells that normally form the vulva are flanked by three others that are also capable of doing so if they are brought under the influence of the anchor cell. If the normal vulval precursor cells are destroyed with a laser beam, for example, these neighboring cells are diverted from their usual hypodermal fate (perhaps through a shift in their position) and form a vulva instead. Cells other than the three immediate neighbors, however, are unable to respond to the inducing signal and will not form a vulva under any

Threshold Reactions Can Convert a Smooth Morphogen Gradient into a Sharply Defined Spatial Pattern of Determination[37]

If the concentration gradient of a morphogen is smooth, one might expect that the consequent pattern of cell characters would also be smoothly graded. Smoothly graded patterns of cell character do indeed occur on a small scale in some tissues. But many of the differentiations of greatest interest in development are discrete. The ultimate cell types are sharply distinct; there is no graded series of mature kinds of cells intermediate between cartilage and muscle, for example. Sharp distinctions can arise in a population of initially uniform cells through a *threshold* in the response to a smoothly graded signal: positive feedback in each responding cell can amplify the effect of a small increment in the signal in such a way that cells exposed to only slightly different intensities of the signal are launched on radically different courses of development according to whether their exposure is above or below a certain threshold intensity. There may indeed be several thresholds of response to one signal, so that a single variable may control the pattern of several different choices. Once a cell is well launched on a given course, it will persist on that course even in the absence of the environmental influence that initially controlled the choice. In this way, transient, position-dependent influences can have effects that are "remembered" as discrete choices of cell state and thereby define the spatial pattern of determination. The choice of cell state represents the cell's *memory* of the positional information supplied. This record, registered as an intrinsic feature of the cell itself, may be called its **positional value.**

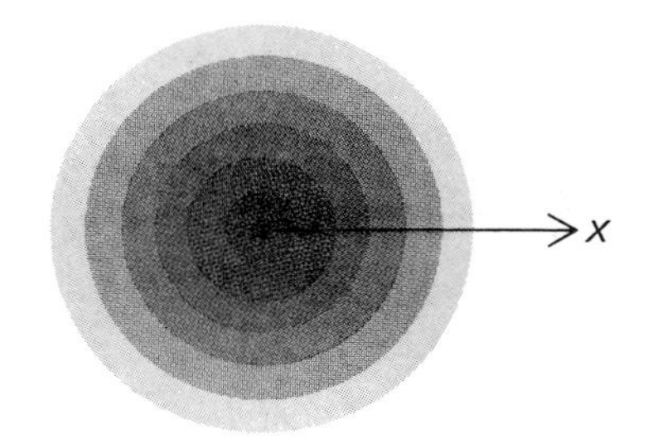

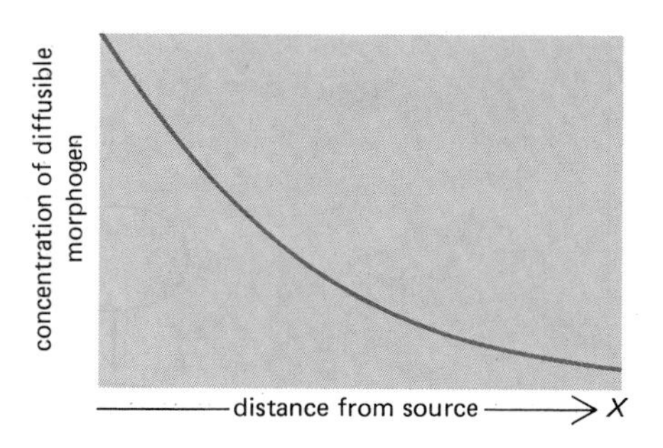

Figure 16–42 If a substance is produced at a point source and is degraded as it diffuses from that point, a concentration gradient results with a maximum at the source. The substance can serve as a morphogen, whose local concentration controls the behavior of cells according to their distance from the source.

Since Embryonic Fields Are Small, Gross Features of the Adult Must Be Determined Early Through Cell Memory[38]

Whatever their nature, the mechanisms for supplying positional information in an animal embryo generally act over only small regions, or *morphogenetic fields,* on the order of a millimeter long (or about 100 cell diameters) or less. There is clearly a limit to the amount of detail that can be defined in so small a space. For this fundamental reason, the final positional specification of a cell has to be built up as a composite of a sequence of items of positional information registered at different times. Cell memory, therefore, is crucial for the development of large complex animals. The distinction between head and tail has to be established when the rudiments of the head and the tail are no more than about a millimeter apart. The circumstances that gave rise to that distinction are ancient history by the time the animal is a centimeter or a meter long; if the distinction between head and tail is to be maintained, it must be through cell memory.

Thus the gross plan of the body is specified early, and successive levels of detail are filled in later as the rudiment of each part grows to a size at which

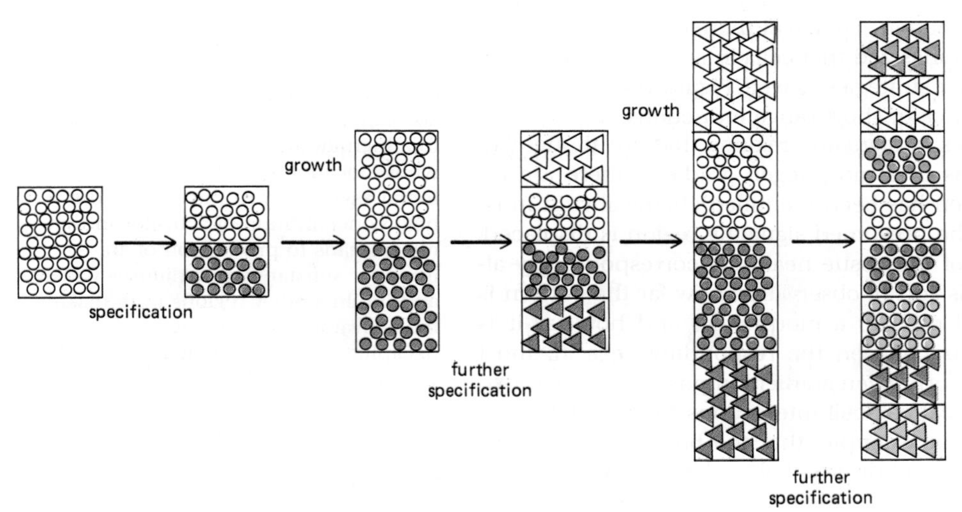

Figure 16–43 When an embryo is small, it becomes subdivided into a few distinct regions corresponding to the major subdivisions of the adult body. The cells in each of these coarse subdivisions are stamped with a crude positional value (represented here by the triangular or circular symbols). As the embryo grows, the subdivisions grow and themselves become subdivided, creating a progressively more fine-grained pattern of positional values.

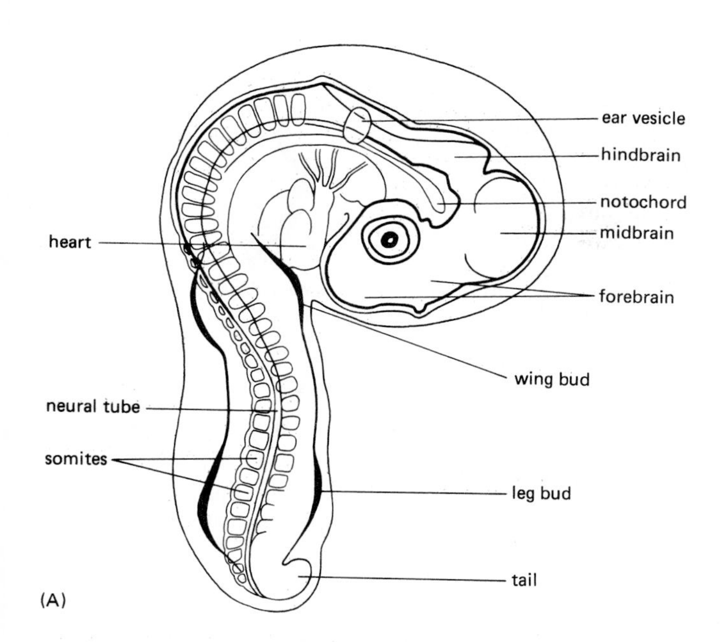

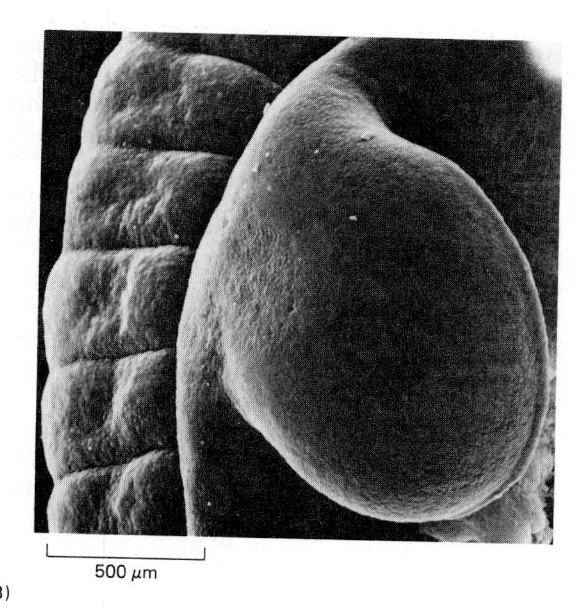

mechanisms for supplying additional positional information can conveniently act (Figure 16–43).

We shall now examine in more detail how this occurs, taking our examples from the development of limbs.

Figure 16–44 (A) A chick embryo after 3 days of incubation, illustrating the positions of the early limb buds. (B) Scanning electron micrograph showing a dorsal view of the wing bud and adjacent somites one day later; the bud has grown to become a tongue-shaped projection about 1 mm long, 1 mm broad, and 0.5 mm thick. (A, after W.H. Freeman and B. Bracegirdle, An Atlas of Embryology. London: Heinemann, 1967; B, courtesy of Paul Martin.)

Positional Information in Limb Development Is Refined by Installments[39]

The development of the limbs—and of many other organs such as teeth or vertebrae or skin—involves relatively few modes of differentiation. In the case of the limbs, for example, the chief cell types are those of muscle, cartilage, bone, and loose connective tissue. But these few differentiated cell types are arranged in a complex spatial pattern. The forelimb differs from the hindlimb not because it contains different types of tissues but because it contains a different spatial arrangement of tissues. Transplantation experiments with the developing limbs of a chick show that intrinsic differences between the limb cells with respect to the patterns they will generate are determined long before differentiation begins.

In the chick embryo, the leg and the wing originate at about the same time in the form of small tongue-shaped buds projecting from the flank (Figure 16–44). The cells in the two pairs of limb buds appear similar and are undifferentiated at first, showing no hint of the subsequent skeletal pattern (see Figure 14–20, p. 802). A small block of undifferentiated tissue at the base of the leg bud, from the region that would normally give rise to part of the thigh, can be cut out and grafted into the tip of the wing bud. Developing there, the graft forms not the appropriate part of the wing tip, nor a misplaced piece of thigh tissue, but a toe (Figure 16–45). This experiment shows, first, that early leg cells are already determined as leg and, second, that although they are determined as leg, they have not yet been assigned their detailed positional values along the limb axis and can respond to cues in the wing so that they form structures appropriate to the tip of the limb rather than the base. We can thus conclude that the full specification of position in vertebrates is not supplied all at once but is built up from a series of items of positional information registered in the cell memory at different times. The final cell state is arrived at by a sequence of decisions.

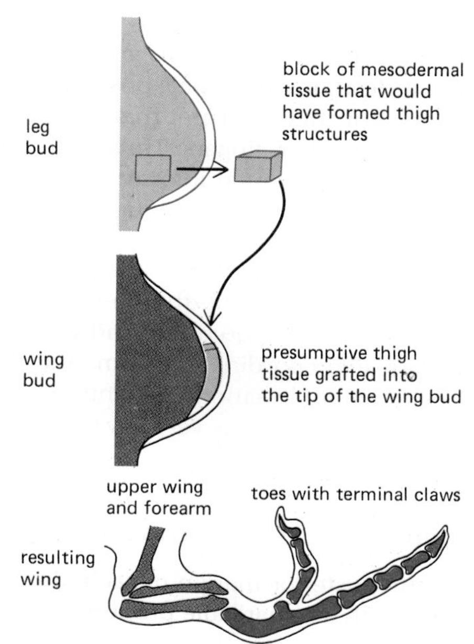

Figure 16–45 Prospective thigh tissue grafted into the tip of a chick wing bud forms toes. (After J.W. Saunders et al., *Dev. Biol.* 1:281–301, 1959.)

Positional Values Make Apparently Similar Cells Nonequivalent[40]

The cells of the forelimb bud and the hindlimb bud, although they give rise to the same range of differentiated types of cells, are evidently *nonequivalent:* they are in intrinsically different states, corresponding to different positional values. The cells of the limb buds may retain the positional values that distinguish leg from wing

The Pattern of Positional Values Controls Growth and Is Regulated by Intercalation[43]

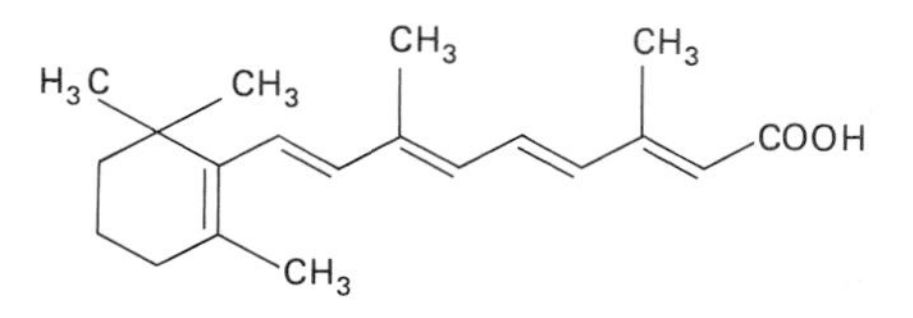

Figure 16–50 The structure of retinoic acid.

In the foregoing discussion of pattern formation and positional values, we have neglected one crucial aspect of the process: the regulation of growth, through which the parts of the pattern attain their appropriate sizes. In some cases this seems to depend on cell-autonomous programs initiated at an early stage in the creation of an organ rudiment. In many other cases, however, growth and the pattern of positional values both depend on continuing cell-cell interactions and depend on them in a closely coupled way. A simple and general rule has been deduced from studies of the regeneration that occurs in various organisms when fragments of tissue with different positional values are juxtaposed and allowed time to grow and adjust. The principles are perhaps most clearly illustrated by studies on the leg of the cockroach.

Cockroaches belong to the class of insects in which there is no radical metamorphosis from larva to adult but a gradual progression through a series of juvenile forms. The juvenile cockroach has well-differentiated limbs, but the differentiated cells—unlike those in human limbs—are still able to respond to the cues that governed the development of the limb pattern, and they can regenerate that pattern if it is disturbed. Thus the workings of the pattern-formation system can be tested by operations done long after the period of embryonic development.

The experiments to be described involve the epidermal sheet of cells and cuticle that covers the cockroach and forms the externally visible parts of the limbs. This outer covering grows by successive molts, in which the juvenile cockroach sheds its old cuticle and lays down a new and larger cuticle in its place. The cuticle is secreted by the epidermal cells, which are arranged underneath in a sheet one cell layer thick. Positional values in the epidermal sheet of cells are displayed in the pattern of the overlying cuticle that they lay down; the effect of experimental manipulation on the patterning of epidermal cells is detected in the cuticle after the animal has molted. Regeneration can be observed only in juveniles, since fully mature adults do not grow or molt.

The cockroach leg consists of several segments, called (in sequence from base to tip) coxa, trochanter, femur, tibia, and tarsus, the tarsus itself being a composite of several smaller segments and terminating in a pair of claws (Figure 16–51). If two legs are amputated through the tibia, say, but at different levels, the distal fragment of the one can be grafted onto the proximal stump of the other in such a way that the composite leg heals with the middle part of the tibia missing. Yet the leg that emerges after the animal has molted appears normal: the missing middle part has regenerated (Figure 16–52A). More surprising is the result of a variant of this operation. The tibia of one cockroach leg is cut through near the proximal end and that of another leg near the distal end. The large detached portion of the first leg is then stuck onto the large remaining stump of the second leg to give an excessively long leg with a middle part present in duplicate (Figure 16–52B). The animal is left to molt. The leg that results, far from being more nearly normal, is now even longer because a third middle part of a tibia has developed between the two already present. As shown in Figure 16–52B, the bristles on this freshly formed region point in the direction opposite to that of the bristles on the rest of the tibia.

Many different operations of this type can be performed. All of them point to the existence of a system of positional values that makes cells in different positions nonequivalent and that is intimately coupled to the control of cell proliferation. It is convenient to describe the positional value by a number that is graded from a maximum at one end of the limb segment to a minimum at the other. In the operations described above, epidermal cells with sharply different positional values are brought together. As a result, new cells are formed by proliferation of the epidermal cells in the neighborhood of the junction. These new cells acquire positional values smoothly interpolated between those of the two sets of cells that were brought into confrontation (Figure 16–52). This behavior is summed up in the **rule of intercalation:** *discontinuities of positional value provoke local cell proliferation, and the newly formed cells take on intermediate positional values so as to restore continuity in the pattern.* Cell proliferation ceases only when cells

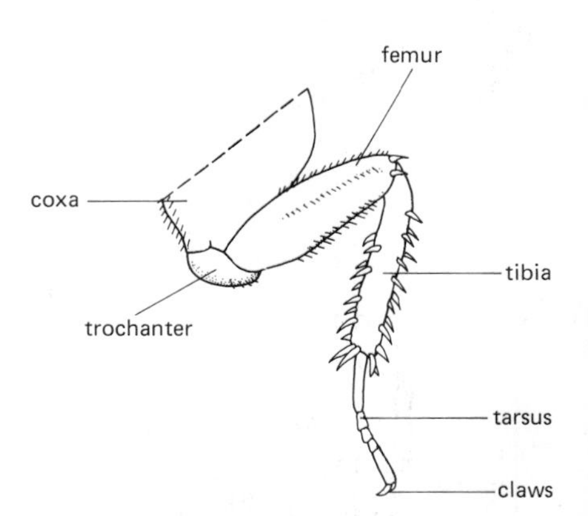

Figure 16–51 The cockroach leg. With each successive molt, the leg grows bigger but does not change its basic structure.

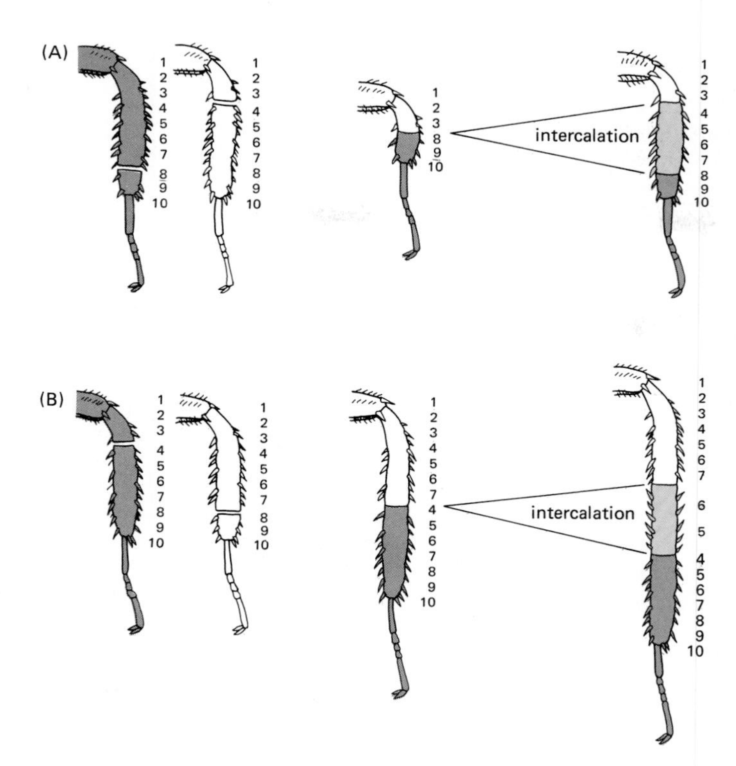

Figure 16–52 When mismatched portions of the cockroach tibia are grafted together, new tissue (*light color*) is intercalated to fill in the gap in the pattern of positional values (numbered from 1 to 10). In case (A), intercalation restores the missing part. In case (B), intercalation generates a third middle part of a tibia between the two middle parts already present. The bristles indicate the polarity of the intercalated tissue. In both cases, continuity is restored in the final pattern of positional values.

with all the missing positional values have been intercalated in the initial gap and have become spread out to the normal spatial separation from one another. This process as a whole is called **intercalary regeneration.**

The rule of intercalation, with the corollary that growth continues until a certain spacing of positional values has been attained, is a very powerful organizing principle in those systems to which it applies. Beginning with a pattern specified approximately and in miniature—for example, by a morphogen gradient—it can determine the construction of a complete accurate pattern of positional values and regulate the growth of each part of the pattern to a standard size: all that is necessary is that the initial pattern should be qualitatively—that is, topologically—correct. Applied to two- and three-dimensional patterns, the rule of intercalation accounts for a remarkably wide range of phenomena, including both normal regeneration of amputated parts and such bizarre effects as the genesis of supernumerary limbs after certain types of grafting procedures. It appears to govern many processes of organogenesis and regeneration not only in insects but also in crustaceans and amphibians. We shall see shortly that it can function in *Drosophila* to correct quite large early errors in pattern specification (see Figure 16–60). Even in creatures such as mammals, where lost structures generally do not regenerate in the adult, the rule of intercalation may help to regulate growth and pattern formation during embryonic development. The molecular mechanisms that underlie this crucial form of growth control are unknown.

Summary

The different kinds of cells in an embryo are produced in a regular spatial pattern. The formation of this pattern usually begins with asymmetries in the egg and continues by means of cell-cell interactions in the embryo. At each stage in the process, it is possible for a weak asymmetry to be amplified by positive feedback so as to create a well-defined pattern. The spatial signals that coordinate pattern formation may be said to supply cells with positional information. In the simplest case a graded concentration of a diffusible morphogen may serve to control the character of cells according to their distance from the source of the morphogen; retinoic acid appears to be a morphogen of this sort in the chick limb bud, controlling the pattern along the thumb-to-little-finger axis. Discrete differences of cell character would correspond to thresholds in the response to the morphogen.

The full positional specification of a cell may be built up from a combination of bits of positional information supplied at different times—a coarse specification first, when the embryo is small, and finer details later, as the parts grow. Cells in the early forelimb and hindlimb rudiments of a vertebrate embryo, for example, acquire different positional values, making forelimb and hindlimb cells nonequivalent in their intrinsic character, long before the detailed pattern of cell differentiation has been determined. To determine its fine-grained pattern, each organ rudiment then generates a more detailed internal framework of positional information. This system of detailed positional information seems to be the same in homologous organs, such as the forelimb and hindlimb. Because of cell memory, the cells in these separate fields interpret the same fine-grained positional information differently according to their different prior histories.

In many animals the pattern of positional values is closely coupled to the control of cell proliferation according to a simple rule of intercalation, derived mainly from studies of limb regeneration in insects and amphibians. According to the rule, discontinuities of positional value provoke local cell proliferation, and the newly formed cells take on intermediate positional values that restore continuity in the pattern. The same mechanism may also operate in normal embryonic development to correct inaccuracies in the initial specification of positional information.

Drosophila and the Molecular Genetics of Pattern Formation[44]

The structure of an organism is controlled by its genes: classical genetics is based on this proposition. Yet for almost a century, and even long after the role of DNA in inheritance had become clear, the mechanisms of the genetic control of body structure remained an intractable mystery. In recent years this chasm in our understanding has begun to be filled. The fly *Drosophila,* more than any other organism, has provided the new insights. Studies on *Drosophila* have revealed a class of developmental control genes whose specific function is to mark out the pattern of the body, and the combination of classical and molecular genetics has begun to show how these genes work. We shall see that not only the general strategies but also the specific genes controlling pattern in *Drosophila* may have close counterparts in vertebrates.

The first glimpses of this genetic system came with the discovery of mutations that cause bizarre disturbances of the adult *Drosophila* body plan. In the mutation *Antennapedia,* for example, legs sprout from the head in place of antennae (Figure 16–53); while in the mutation *bithorax,* portions of an extra pair of wings appear where normally there should be appendages called halteres. Such mutations, which transform parts of the body into structures appropriate to other positions, are called *homeotic,* and the normal genes that they disrupt are called *homeotic selector genes.* The discovery of homeotic mutants led to ingenious experiments showing that the body of the normal fly is formed as a patchwork of discrete regions, each expressing a different set of homeotic selector genes. The products of these genes act as molecular address labels, equipping the cells with a coarse-grained specification of their positional value. A homeotic mutation thus causes

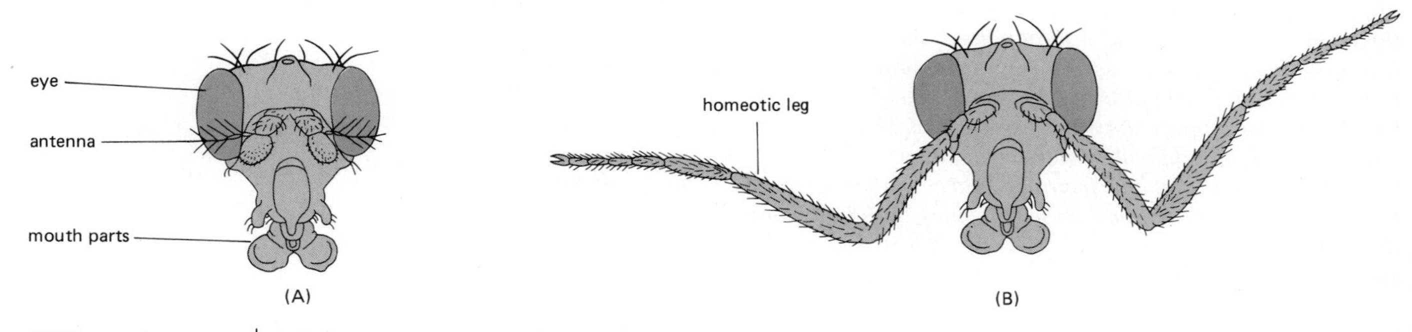

Figure 16–53 The head of a normal adult *Drosophila* (A) compared with that of a fly carrying the homeotic mutation *Antennapedia* (B). The fly shown here displays the mutation in an extreme form; usually only parts of the antennae are converted into leg structures. (*Antennapedia* drawing based on a photograph supplied by Peter Lawrence.)

a whole patch of cells to be misinformed as to their location and consequently to make a structure appropriate to another region.

The homeotic selector genes are only one part of a larger system that creates the normal patchwork pattern of the insect body. In this section we shall describe the system as a whole and some of its molecular mechanisms, using the concepts of pattern formation discussed in the previous section. We shall see that the system consists of three classes of pattern-control genes: (1) The products of the *egg-polarity genes* act first to define the spatial coordinates of the embryo by setting up morphogen gradients in the egg. (2) The *segmentation genes* then serve to interpret the positional information provided by the initial morphogen gradients. These genes mark out the embryo into a series of *segments*—the basic modular units from which all insects are constructed. (3) The products of the segmentation genes influence the expression of the *homeotic selector genes*, which maintain the distinctions between one segment and another; and through the combined activities of the segmentation genes and the homeotic selector genes, the cells in each segment become imprinted with remembered positional values that guide their subsequent behavior. Lastly, within each segmental subdivision of the body, the cells communicate with one another to generate the finest details of the mature structure, apparently governed by the rule of intercalation.

The Insect Body Is Constructed by Modulation of a Fundamental Pattern of Repeating Units[45]

The timetable of *Drosophila* development, from egg to adult, is summarized in Figure 16–54. The period of embryonic development begins at fertilization and takes about a day, at the end of which the embryo hatches out of the egg shell to become a larva. The larva then passes through three stages, or *instars*, separated by molts in which it sheds its old coat of cuticle and lays down a larger one. At the end of the third instar it pupates. Inside the pupa a radical remodeling of the body takes place, and eventually, about nine days after fertilization, an adult fly, or *imago*, emerges.

The fly consists of a head, three thoracic segments (numbered T1 to T3), and nine abdominal segments (numbered A1 to A9). Each segment, although different from the others, is built according to a similar plan. Segment T1, for example, carries a pair of legs; T2 carries a pair of legs plus a pair of wings; and T3 carries a pair of legs plus a pair of halteres—small knob-shaped balancers important in flight, evolved from the second pair of wings that more primitive insects possess. The quasi-repetitive segmentation is more obvious in the larva, where the segments look more similar; and in the embryo it can be seen that the rudiments of the head, or at least the future adult mouth parts, are likewise segmental (Figure 16–55). It is partly a matter of convention where one draws the boundary between one segmental unit and the next; in discussing patterns of gene expression, we shall see that it is convenient to speak in terms of a total of 14 *parasegments* (numbered P1 to P14) that are half a segment out of register with traditionally defined segments (Figure 16–56). Finally, at the two ends of the animal are highly specialized structures that are not segmentally derived.

The groundplan of the whole structure, with its two specialized ends and between them a series of modulated repetitions of a basic segmental unit, is established by processes occurring in the egg and early embryo during the first few hours after fertilization.

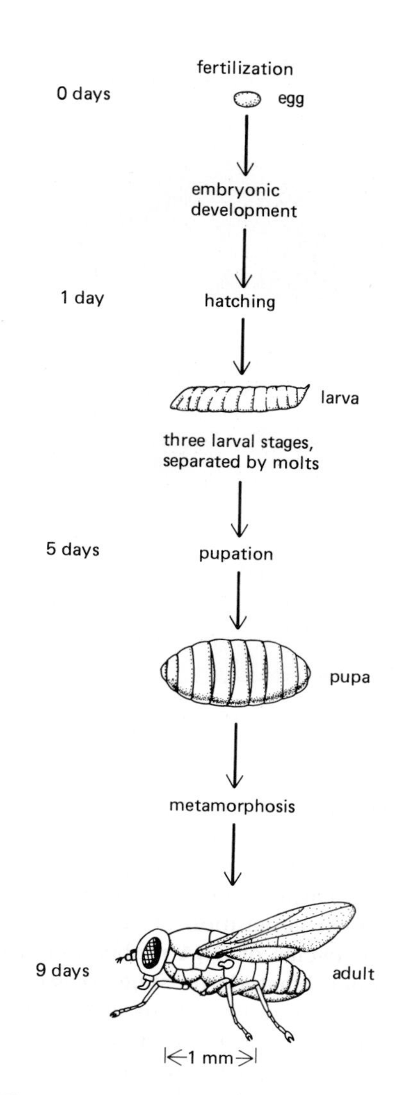

Figure 16–54 Synopsis of *Drosophila* development from egg to adult fly.

Drosophila Begins Its Development as a Syncytium[46]

The egg of *Drosophila* is about 400 μm long and about 160 μm in diameter, with a clearly defined polarity. Like the eggs of other insects, it begins its development in an unusual way: a series of nuclear divisions without cell division creates a syncytium. The early nuclear divisions are synchronous and extremely rapid, occurring about every 8 minutes. The first nine divisions generate a cloud of nuclei, most of which migrate from the middle of the egg toward the surface, where they form a monolayer called the *syncytial blastoderm*. After another four rounds of nuclear division, plasma membranes grow inward from the egg surface to enclose

Signals from the Two Ends of the *Drosophila* Egg Control the Antero-posterior Polarity of the Embryo[48]

The initial polarity of the egg depends on the distribution of materials produced before fertilization, when the egg or oocyte was still in the ovary (see Figure 16–39). If a *Drosophila* egg is carefully punctured at its anterior end, allowing a small amount of the most anterior cytoplasm to leak out, the embryo fails to develop head structures. Moreover, if cytoplasm from the posterior end of another egg is injected into the site from which the anterior cytoplasm has leaked, a second set of abdominal segments will develop, with reversed polarity, in the anterior half of the recipient egg (Figure 16–59).

Mutations have been identified that cause similar disturbances in the pattern in the anterior or the posterior half of the embryo. The **egg-polarity genes** defined by these mutations are the first elements in a hierarchical system that generates the antero-posterior pattern of the body. The egg-polarity genes are among those transcribed from the maternal genome during oogenesis, and the stored products begin to act very soon after fertilization. Thus the phenotype of the embryo is determined by the alleles present in the mother rather than by the combination of maternal and paternal genes possessed by the embryo itself. Genes expressed in this way are called **maternal-effect genes.**

Mothers that are homozygous for the egg-polarity mutation *bicoid* produce embryos that lack head and thoracic structures and have abdominal structures extended over an abnormally large fraction of the body length; conversely, the egg-polarity mutation *oskar* gives rise to embryos that lack all the abdominal segments. (The nonsegmental structures at the two extreme ends of the embryo have a special status: they do not disappear in either of these mutants but are eliminated by the mutation *torso* and certain others.) The *bicoid* and *oskar* mutants are deficient in the corresponding normal gene products, and they can be rescued by injections of normal cytoplasm. A *bicoid* mutant will develop more or less normally if cytoplasm from the anterior end of a normal egg is injected into its anterior end, while an *oskar* mutant will develop more or less normally if cytoplasm from the posterior end of a normal egg is injected into its prospective abdominal region. In each case the normal gene product seems to be localized at one end or the other of the egg and to act as the source of some long-range influence controlling the global pattern of antero-posterior positional values.

Molecular genetic experiments have illuminated the picture. Using *in situ* hybridization with a cloned *bicoid* cDNA probe, it can be shown that *bicoid* mRNA is concentrated at the anterior tip of the egg and that it is originally synthesized in the ovary by the nurse cells connected with the oocyte (see p. 912 and Figure 16–39). As the *bicoid* RNA passes through the cytoplasmic bridges into the oocyte, it becomes anchored to some component of the cytoplasm—perhaps to a part of

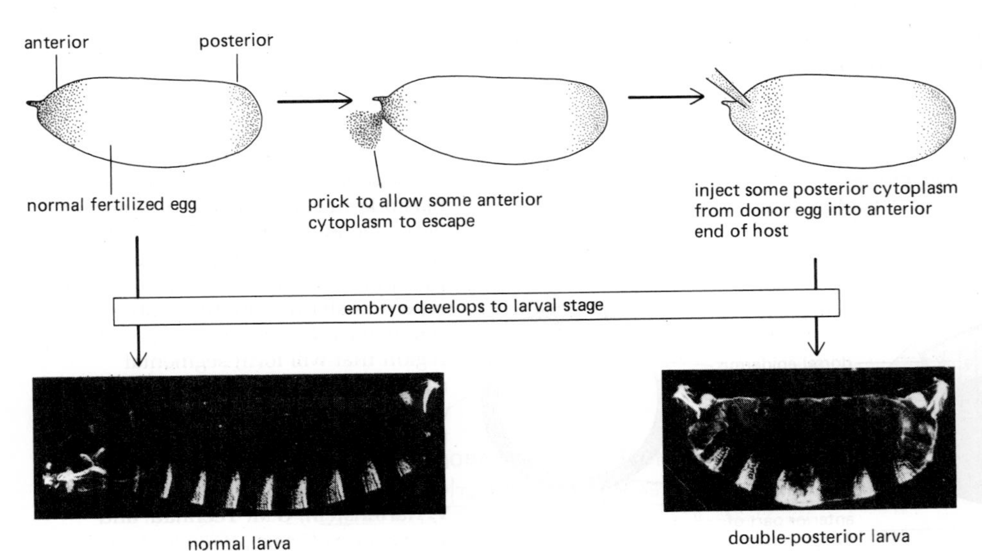

Figure 16–59 Localized determinants at the ends of the *Drosophila* egg control its antero-posterior polarity. A little anterior cytoplasm is allowed to leak out of the anterior end of the egg and is replaced by an injection of posterior cytoplasm. The resulting double-posterior larva (*photograph on right*) is compared with a normal control (*photograph on left*); the substitution of cytoplasm at one end of the egg has had a long-range effect, converting all the more anterior segments into a mirror-image duplicate of the last three abdominal segments. The larvae are shown in dark-field illumination. (From H.G. Frohnhöfer, R. Lehmann, and C. Nüsslein-Volhard, *J. Embryol. Exp. Morphol.* 97[suppl]:169–179, 1986.)

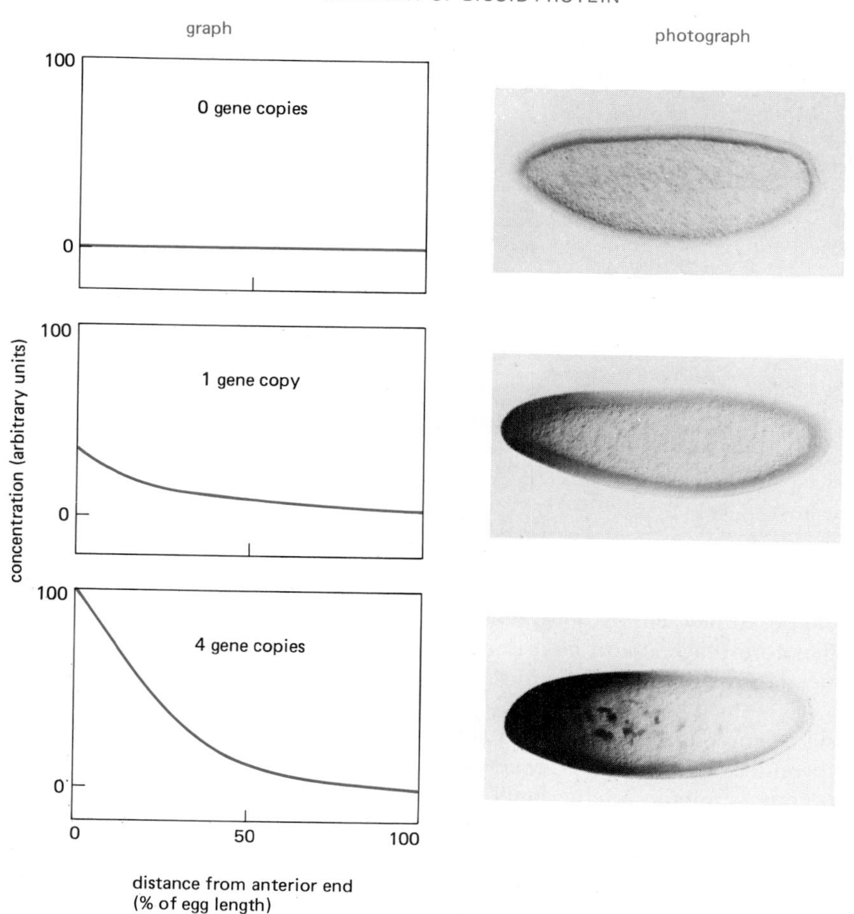

RESULTANT SEGMENTATION PATTERN

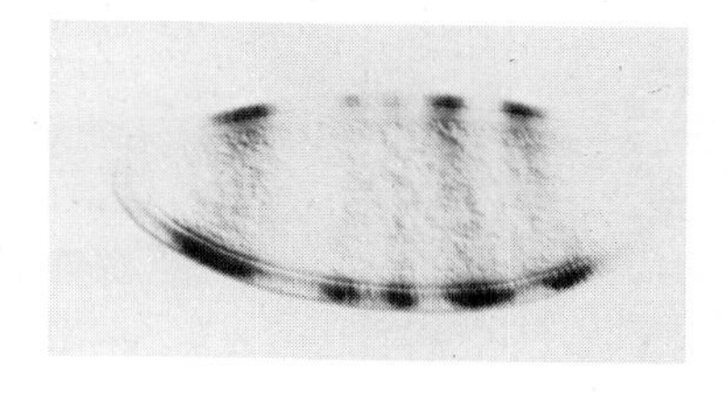

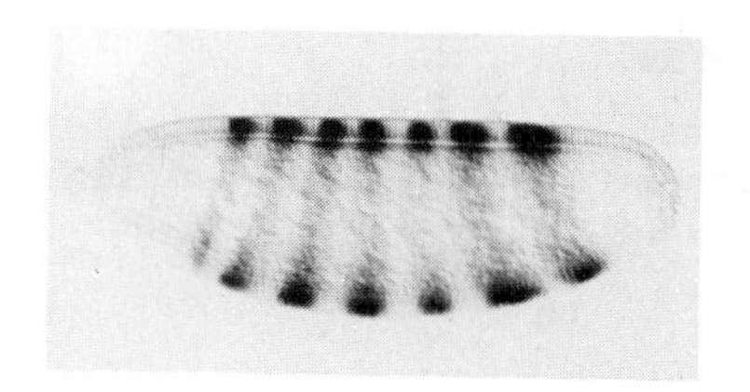

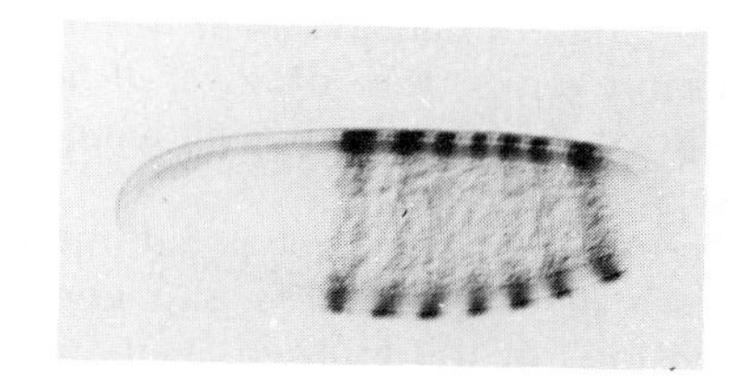

Figure 16–60 The *bicoid* protein gradient in the *Drosophila* egg and its effects on the pattern of segments. The gradient is revealed by staining with an antibody against the *bicoid* protein; the segment pattern is revealed by an antibody against the product of a pair-rule gene, *even-skipped* (see p. 926). Three embryos are compared, containing zero, one, and four copies, respectively, of the normal *bicoid* gene. With zero dosage of *bicoid*, segments with an anterior character do not form; with increasing gene dosage they form progressively farther from the anterior end of the egg, as expected if their position is determined by the local concentration of the *bicoid* protein. Measurements of this concentration, as indicated by the intensity of staining, are shown in the graphs. Despite the considerable differences of position and spacing of the segment rudiments in the embryos with one and four doses of the gene, both embryos will develop into normally proportioned larvae and adults. The mechanism responsible for this regulation is discussed on page 918. (Slightly adapted from W. Driever and C. Nüsslein-Volhard, *Cell* 54:83–104, 1988.)

the cytoskeleton—at the oocyte's anterior end. Translation begins only when the egg is laid, giving rise to a concentration gradient of *bicoid* protein with its high point at the anterior end of the embryo. The concentration gradient can be altered genetically by constructing mutants that contain multiple copies of the normal *bicoid* gene: as the gene dosage increases in the mother, so does the protein concentration increase in the egg. The segments of the resultant embryo are correspondingly shifted toward the posterior pole, as though their locations were determined by positional information derived from the local concentration of *bicoid* protein (Figure 16–60). This protein therefore fits exactly the definition of a morphogen (see p. 913).

Three Classes of Segmentation Genes Subdivide the Embryo[49]

The graded global cues provided by the products of the egg-polarity genes have to guide the creation of a system of discrete segments. This process depends on a collection of about 20 **segmentation genes.** Mutations in these genes alter the number of segments or their basic internal organization without altering the global polarity of the egg. The segmentation genes act at later stages than the egg-polarity genes. Correspondingly, the phenotype of the embryo with regard to them is determined, in whole or in part, by the genotype of the embryo and not purely by the genotype of the mother; that is, they are **zygotic-effect** genes rather than maternal-effect genes.

Most mutations in segmentation genes are lethal: their effects are never seen in the adult fly because the mutant dies before it can develop that far. Lethal mutations of this sort can, however, be propagated if the lethal effects are recessive (as is commonly the case). Heterozygotes, with one mutant copy of the gene and one normal copy, are then viable. When a pair of heterozygous parents breed, one

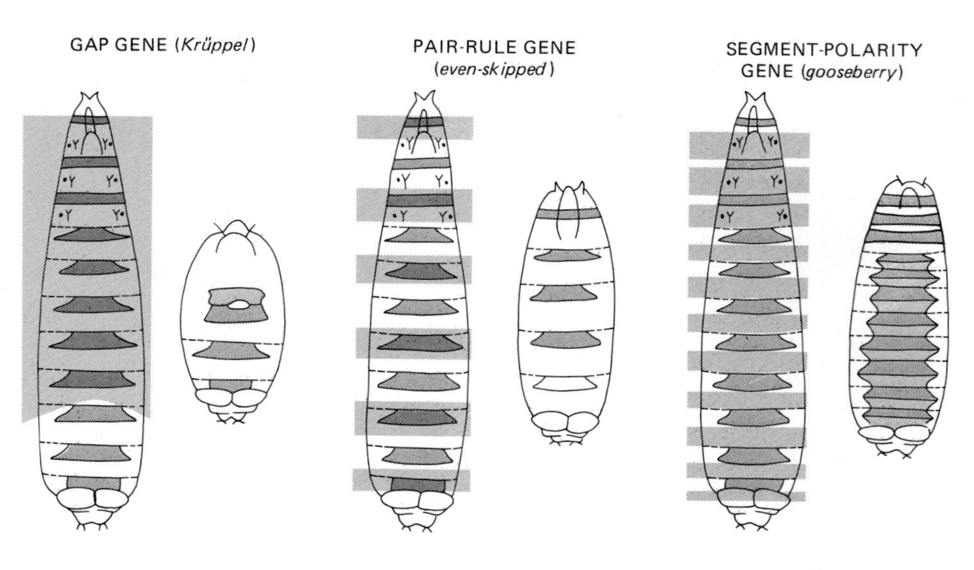

Figure 16–61 Examples of the phenotypes of mutations affecting the three types of segmentation genes. In each case the areas shaded in color on the normal larva (*left*) are deleted in the mutant or are replaced by mirror-image duplicates of the unaffected regions. By convention, dominant mutations are written with an initial capital letter and recessive mutations are written with a lowercase letter. Several of the patterning mutations of *Drosophila* are classed as dominant because they have a perceptible effect on the phenotype of the heterozygote, even though the characteristic major, lethal effects are recessive—that is, visible only in the homozygote. (Modified from C. Nüsslein-Volhard and E. Wieschaus, *Nature* 287:795–801, 1980.)

in four of the progeny are homozygotes, in which both copies of the gene are mutant. These progeny die prematurely, as late embryos or very young larvae, but survive long enough to reveal the mutant phenotype. In fact, almost all the segmentation genes were discovered by exposing flies to mutagens and laboriously screening tens of thousands of dying larvae produced by breeding from the mutant stocks.

The segmentation genes fall into three classes (Figure 16–61). The first to act are a set of at least three **gap genes,** whose products mark out the coarsest subdivisions of the embryo. Mutations in a gap gene eliminate a large block of contiguous segments, and mutations in different gap genes cause different but partially overlapping defects. In the mutant *Krüppel,* for example, the larva lacks eight segments, from T1 to A5 inclusive (approximately parasegments P3 to P10).

The next segmentation genes to act are a set of eight **pair-rule genes.** Mutations in these genes cause a series of deletions affecting alternate segments, leaving the embryo with only half as many segments as usual. While all the pair-rule mutants display this two-segment periodicity, they differ in the precise positioning of the deletions relative to the segmental or parasegmental borders. The pair-rule mutant *even-skipped,* for example, lacks the whole of each even-numbered parasegment, while the pair-rule mutant *fushi tarazu* (*ftz*) lacks the whole of each odd-numbered parasegment, and the pair-rule mutant *hairy* lacks a series of regions that are of similar width but out of register with the parasegmental units.

Finally, there are at least 10 **segment-polarity genes.** Mutations in these genes cause a part of each segment to be lost and replaced by a mirror-image duplicate of all or part of the rest of the segment. In *gooseberry* mutants, for example, the posterior half of each segment (that is, the anterior half of each parasegment) is replaced by an approximate mirror image of the adjacent anterior half-segment (see Figure 16–61).

The phenotypes of the various segmentation mutants suggest that the segmentation genes form a coordinated system that subdivides the embryo progressively into smaller and smaller domains distinguished by different patterns of gene expression. Again, molecular genetics provides the tools to investigate how this system works.

The Localized Expression of Segmentation Genes Is Regulated by a Hierarchy of Positional Signals[44,50]

Several representatives of each of the groups of segmentation genes have been cloned and used as probes to locate the gene transcripts in normal embryos by *in situ* hybridization (see p. 192). We have already seen how this technique has helped to show that the *bicoid* gene transcripts are the source of a positional

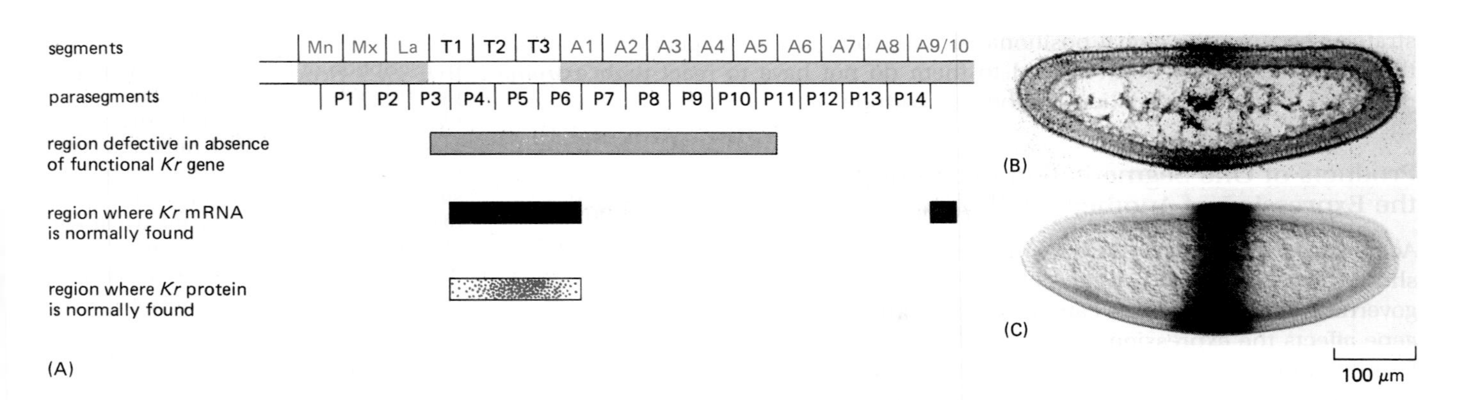

signal: the transcripts are localized at one end of the egg, even though the effects of a mutation in the gene are spread over a large part of the embryo. In a similar way it can be shown that certain segmentation genes—in particular the gap genes—in their turn generate (directly or indirectly) positional signals that help to control the pattern of development in more localized neighborhoods. Mutants that are defective in the gap gene *Krüppel,* for example, show abnormalities extending throughout the region where the gene transcripts are detected in a normal embryo and also for several segments beyond (Figure 16–62). The *Krüppel* gene has been sequenced and is homologous (as is another gap gene, *hunchback*) to a family of genes known to code in vertebrates for DNA-binding regulatory proteins, including the TFIIIA transcription factor of *Xenopus* (see p. 490). It is tempting to suggest, by analogy with *bicoid,* that the *Krüppel* protein spreads out as a diffusible morphogen from the site where *Krüppel* is transcribed, although the observed distribution of the protein seems less extensive than this hypothesis requires.

Some of the pair-rule genes may be involved in spatial signaling on a still finer scale, exerting effects on cells neighboring the regions where they are transcribed; others, by contrast, appear to affect the development only of those regions in which they are transcribed. For example, transcripts of the normal *ftz* gene at the blastoderm stage occur in seven circumferential "zebra stripes" (Figure 16–63), each of the stripes being roughly four cells wide, matching in width and location the rudiments of the even-numbered parasegments that would be missing in a *ftz* mutant.

Taken together, these observations suggest that the products of the egg-polarity genes provide global positional signals that cause particular gap genes to be expressed in particular regions, and the products of the gap genes then provide a second tier of positional signals that act more locally to regulate finer details of patterning by influencing the expression of yet other genes, including the pair-rule genes. In this way the global gradients produced by the egg-polarity genes organize the creation of a fine-grained pattern through a process of sequential subdivision, using a hierarchy of sequential positional controls. This is a reliable

Figure 16–62 The spatial domains of action of the gap gene *Krüppel,* mapped on the *Drosophila* blastoderm. (A) Diagram showing how the defect caused by an absence of functional *Krüppel* product extends far beyond the region where *Krüppel* transcripts are normally found. (B) The normal distribution of *Krüppel* transcripts, as seen by *in situ* hybridization at the blastoderm stage. (C) The normal distribution of *Krüppel* protein, as seen by antibody binding at the same stage. The protein may be distributed more widely at other stages. The phenotype of a mutant that lacks the functional *Krüppel* product is shown in Figure 16–61A. (B, from H. Jäckle, D. Tautz, R. Schuh, E. Siefert, and R. Lehmann, *Nature* 324:668–670, 1986; C, from U. Gaul, E. Seifert, R. Schuh, and H. Jäckle, *Cell* 50:639–647, 1987, copyright Cell Press.)

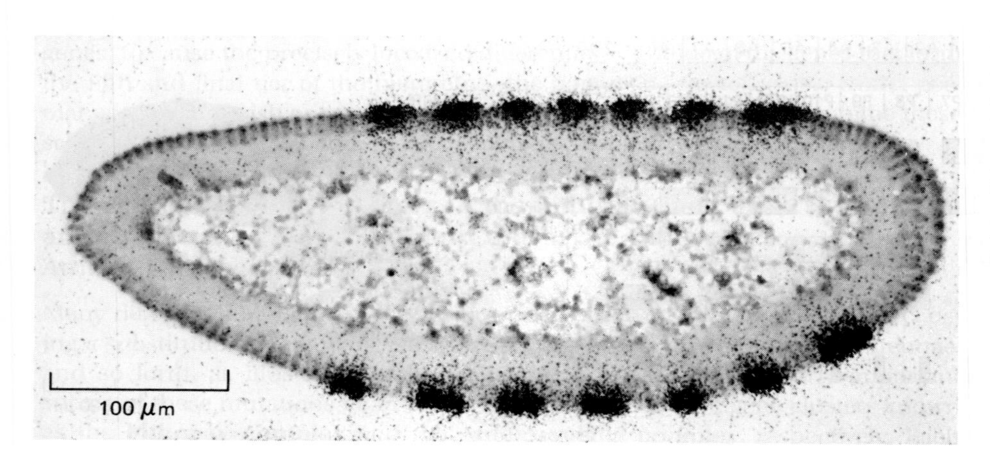

Figure 16–63 *In situ* hybridization of a radioactive *ftz* DNA probe to the *Drosophila* blastoderm, revealing that the gene is transcribed in a pattern of seven stripes corresponding to the pattern of defects in *ftz* mutants. The bands of *ftz* expression appear as black patches of autoradiographic silver grains in this longitudinal section. (Courtesy of Philip Ingham.)

factors controlling its movement are the same as for the migratory cell, involving specific adhesions and so on. But its connection to its cell of origin, its involvement with other nerve fibers, and its capacity to form synapses raise new problems that require special treatment. We therefore leave the construction of the nervous system, the *tour de force* of development, for analysis in Chapter 19.

Summary

The positional values assigned to cells in the course of pattern formation are expressed in the adhesive properties of their surfaces as well as in their internal chemistry. Cells with the same character tend to adhere to one another and remain segregated from differently specified cells, thereby stabilizing the spatial pattern and enabling the cells to sort out spontaneously if they are mixed artificially. Changes in the pattern of adhesive properties underlie morphogenetic movements such as gastrulation, neurulation, and somite formation. Because the pattern of positional values in a given class of cells is expressed on the cell surface, it can guide the migrations of other embryonic cell populations during the assembly of complex tissues and organs. In vertebrates the connective-tissue cells appear to be the primary carriers of positional information. The connective-tissue cells in the dermal layer of the skin, for example, control the regional specialization of the epidermis to form feathers and scales. Similarly, connective-tissue cells in the limb control and coordinate the patterns formed by immigrant cell populations, such as muscle cells (derived from the somites), nerve cell axons (from the central nervous system and peripheral ganglia), and pigment cells (derived from the neural crest). Although many general-purpose cell-adhesion molecules have been identified, and some of them have been shown to be crucial for these phenomena, the molecular mechanisms by which migrant cells are guided along specific paths to precisely defined destinations in the limb are still not known.

References

General

Browder, L. Developmental Biology, 2nd ed. Philadelphia: Saunders, 1984.

Gilbert, S.F. Developmental Biology, 2nd ed. Sunderland, MA: Sinauer, 1988.

Molecular Biology of Development. *Cold Spring Harbor Symp.* 50, 1985.

Slack, J.M.W. From Egg to Embryo: Determinative Events in Early Development. Cambridge, U.K.: Cambridge University Press, 1983.

Spemann, H. Embryonic Development and Induction. New Haven: Yale University Press, 1938. (Reprinted, New York: Garland, 1988.)

Walbot, V.; Holder, N. Developmental Biology. New York: Random House, 1987.

Weiss, P.A. Principles of Development. New York: Holt, 1939.

Cited

1. Browder, L.W., ed. Developmental Biology: A Comprehensive Synthesis. Vol 2: The Cellular Basis of Morphogenesis. New York: Plenum, 1986.

   Gerhart, J.; et al. Amphibian early development. *Bioscience* 36: 541–549, 1986.

   Slack, J.M.W., ed. Early Amphibian Development. *J. Embryol. Exp. Morphol.* Suppl. 89, 1985.

   Trinkaus, J.P. Cells into Organs: The Forces that Shape the Embryo, 2nd ed. Englewood Cliffs, NJ: Prentice Hall, 1984.

2. Gerhart, J.C. Mechanisms regulating pattern formation in the amphibian egg and early embryo. In Biological Regulation and Development (R.F. Goldberger, ed.), Vol. 2, pp. 133–316. New York: Plenum, 1980.

   Gerhart, J.; Ubbels, G.; Black, S.; Hara, K.; Kirschner, M. A reinvestigation of the role of the grey crescent in axis formation in *Xenopus laevis. Nature* 292:511–516, 1981.

   Vincent, J.P.; Oster, G.F.; Gerhart, J.C. Kinematics of gray crescent formation in *Xenopus* eggs: the displacement of subcortical cytoplasm relative to the egg surface. *Dev. Biol.* 113:484–500, 1986.

3. Gilbert, S.F. Developmental Biology, 2nd ed., pp. 73–111. Sunderland, MA: Sinauer, 1988.

   Kirschner, M.; Newport, J.; Gerhart, J. The timing of early developmental events in *Xenopus. Trends Genet.* 1:41–47, 1985.

   Wilson, E.B. The Cell in Development and Heredity, 3rd ed. pp. 980–1034. New York: Macmillan, 1925. (Reprinted, New York: Garland, 1987.)

4. Furshpan, E.J.; Potter, D.D. Low-resistance junctions between cells in embryos and tissue culture. *Curr. Top. Dev. Biol.* 3:95–128, 1968.

   Kalt, M.R. The relationship between cleavage and blastocoel formation in *Xenopus laevis*. II. Electron microscopic observations. *J. Embryol. Exp. Morphol.* 26:51–66, 1971.

   Warner, A. The role of gap junctions in amphibian development. *J.Embryol. Exp. Morphol.*, Suppl. 81:365–380, 1985.

5. Fink, R.D.; McClay, D.R. Three cell recognition changes accompany the ingression of sea urchin primary mesenchyme cells. *Dev. Biol.* 107:66–74, 1985.

Gustafson, T.; Wolpert, L. Cellular movement and contact in sea urchin morphogenesis. *Biol. Rev.* 42:442–498, 1967.

Hardin, J.D.; Cheng, L.Y. The mechanisms and mechanics of archenteron elongation during sea urchin gastrulation. *Dev. Biol.* 115:490–501, 1986.

McClay, D.R.; Wesel, G.M. The surface of the sea urchin embryo at gastrulation: a molecular mosaic. *Trends Genet.* 1:12–00, 1985.

Wilt, F.H. Determination and morphogenesis in the sea urchin embryo. *Development* 100:559–575, 1987.

6. Ettensohn, C.A. Mechanisms of epithelial invagination. *Q. Rev. Biol.* 60:289–307, 1985.

McClay, D.R.; Ettensohn, C.A. Cell adhesion in morphogenesis. *Annu. Rev. Cell Biol.* 3:319–345, 1987.

Odell, G.M.; Oster, G.; Alberch, P.; Burnside B. The mechanical basis of morphogenesis. I. Epithelial folding and invagination. *Dev. Biol.* 85:446–462, 1981.

7. Gerhart, J.; Keller, R. Region-specific cell activities in amphibian gastrulation. *Annu. Rev. Cell Biol.* 2:201–229, 1986.

8. Spemann, H.; Mangold, H. Induction of embryonic primordia by implantation of organizers from a different species. *Roux's Archiv.* 100:599–638, 1924. (English translation in Foundations of Experimental Embryology, 2nd ed. [B.H. Willier, J.M. Oppenheimer, eds.] New York: Hafner, 1974.)

9. Balinsky, B.I. Introduction to Embryology, 5th ed. Philadelphia: Saunders, 1981.

Langman, J. Medical Embryology, 5th ed. Baltimore: Williams & Wilkins, 1985.

Romer, A.S.; Parsons, T. S. The Vertebrate Body, 6th ed. Philadelphia: Saunders, 1986.

10. Kitchin, I.C. The effects of notochordectomy in *Amblystoma mexicanum*. *J. Exp. Zool.* 112:393–411, 1949.

Smith, J.C.; Watt, F.W. Biochemical specificity of *Xenopus* notochord. *Differentiation* 29:109–115, 1985.

11. Le Douarin, N. The Neural Crest. Cambridge, U.K.: Cambridge University Press, 1982.

Newgreen, D.F.; Erickson, C.A. The migration of neural crest cells. *Int. Rev. Cytol.* 103:89–145, 1986.

12. Burnside, B. Microtubules and microfilaments in amphibian neurulation. *Am. Zool.* 13:989–1006, 1973.

Gordon, R. A review of the theories of vertebrate neurulation and their relationship to the mechanics of neural tube birth defects. *J. Embryol. Exp. Morphol.*, Suppl. 89:229–255, 1985.

Karfunkel, P. The mechanisms of neural tube formation. *Int. Rev. Cytol.* 38:245–271, 1974.

13. Blackshaw, S.E.; Warner, A.E. Low resistance junctions between mesoderm cells during development of trunk muscles. *J. Physiol.* 255:209–230, 1976.

Keynes, R.J.; Stern, C.D. Mechanisms of vertebrate segmentation. *Development* 103: 413–429, 1988.

14. Slack, J.M.W. From Egg to Embryo: Determinative Events in Early Development. Cambridge, U.K.: Cambridge University Press, 1983.

15. DiBerardino, M.A.; Orr, N.H.; McKinnell, R.G. Feeding tadpoles cloned from *Rana* erythrocyte nuclei. *Proc. Natl. Acad. Sci. USA* 83:8231–8234, 1986.

Gurdon, J.B. The Control of Gene Expression in Animal Development. Cambridge: Harvard University Press, 1974.

Gurdon, J.B. Transplanted nuclei and cell differentiation. *Sci. Am.* 219(6): 24–35, 1968.

McKinnell, R.G. Cloning—Nuclear Transplantation in Amphibia. Minneapolis: University of Minnesota Press, 1978.

16. Davidson, E.H. Gene Activity in Early Development, 3rd ed., pp. 411–524. Orlando, FL: Academic Press, 1986.

Jeffery, W.R. Spatial distribution of mRNA in the cytoskeletal framework of Ascidian eggs. *Dev. Biol.* 103:482–492, 1984.

Satoh, N. Towards a molecular understanding of differentiation mechanisms in Ascidian embryos. *Bioessays* 7:51–56, 1987.

Wilson, E.B. The Cell in Development and Heredity, 3rd ed., pp. 1035–1121. New York: Macmillan, 1928. (Reprinted, New York: Garland, 1987.)

17. Gurdon, J.B. Embryonic induction—molecular prospects. *Development* 99:285–306, 1987.

Kimelman, D.; Kirschner, M. Synergistic induction of mesoderm by FGF and TGF-beta and the identification of an mRNA coding for FGF in the early *Xenopus* embryo. *Cell* 51:869–877, 1987.

Rosa, F.; et al. Mesoderm induction in amphibians: the role of TGF-β2-like factors. *Science* 239:783–785, 1988.

Slack, J.M.W.; Darlington, B.G.; Heath, J.K.; Godsave, S.F. Mesoderm induction in early *Xenopus* embryos by heparin-binding growth factors. *Nature* 326:197–200, 1987.

Weeks, D.L.; Melton, D.A. A maternal mRNA localized to the vegetal hemisphere in *Xenopus* eggs codes for a growth factor related to TGF-beta. *Cell* 51:861–867, 1987.

18. Austin, C.R.; Short, R.V., eds. Embryonic and Fetal Development, 2nd ed. Reproduction in Mammals, Ser., Book 2. Cambridge, U.K.: Cambridge University Press, 1982.

Hogan, B.; Costantini, F.; Lacy, E. Manipulating the Mouse Embryo: A Laboratory Manual. Cold Spring Harbor, NY: Cold Spring Harbor Laboratory, 1986.

Rugh, R. The Mouse: Its Reproduction and Development. Minneapolis: Burgess, 1968.

19. Gardner, R.L. Clonal analysis of early mammalian development. *Philos. Trans. R. Soc. Lond. (Biol.)* 312:163–178, 1985.

McLaren, A. Mammalian Chimeras. Cambridge, U.K.: Cambridge University Press, 1976.

20. Johnson, M.H.; Chisholm, J.C.; Fleming, T.P.; Houliston, E. A role for cytoplasmic determinants in the development of the mouse early embryo? *J. Embryol. Exp. Morphol.*, Suppl. 97:97–121, 1986.

Kelly, S.J. Studies of the developmental potential of 4- and 8-cell stage mouse blastomeres. *J. Exp. Zool.* 200:365–376, 1977.

Tarkowski, A.K. Experiments on the development of isolated blastomeres of mouse eggs. *Nature* 184:1286–1287, 1959.

21. Illmensee, K.; Stevens, L.C. Teratomas and chimeras. *Sci. Am.* 240(4):120–132, 1979.

Papaioannou, V.E.; Gardner, R.L.; McBurney, M.W.; Babinet, C.; Evans, M.J. Participation of cultured teratocarcinoma cells in mouse embryogenesis. *J. Embryol. Exp. Morphol.* 44:93–104, 1978.

Robertson, E.J. Pluripotential stem cell lines as a route into the mouse germ line. *Trends Genet.* 2:9–13, 1986.

22. Weiss, P.A. Principles of Development, pp. 289–437. New York: Holt, 1939.

23. Spemann, H. Über die Determination der ersten Organanlagen des Amphibienembryo I-VI *Arch. Entw. Mech. Org.* 43:448–555, 1918.

24. Hardeman, E.C.; Chiu, C.-P.; Minty, A; Blau, H.M. The pattern of actin expression in human fibroblast x mouse muscle heterokaryons suggests that human muscle regulatory factors are produced. *Cell* 47:123–130, 1986.

25. Barton, S.C.; Surani, M.A.H.; Norris, M.L. Role of paternal and maternal genomes in mouse development. *Nature* 311: 374–376, 1984.

Monk, M. Memories of mother and father. *Nature* 328:203–204, 1987.

Reik, W.; Collick, A.; Norris, M.L.; Barton, S.C.; Surani, M.A. Genomic imprinting determines methylation of parental alleles in transgenic mice. *Nature* 328:248–251, 1987.

Sapienza, C.; Peterson, A.C.; Rossant, J.; Balling, R. Degree of methylation of transgenes is dependent on gamete of origin. *Nature* 328:251–254, 1987.

Swain, J.L.; Stewart, T.A.; Leder, P. Parental legacy determines methylation and expression of an autosomal transgene: a molecular mechanism for parental imprinting. *Cell* 50:719–727, 1987.

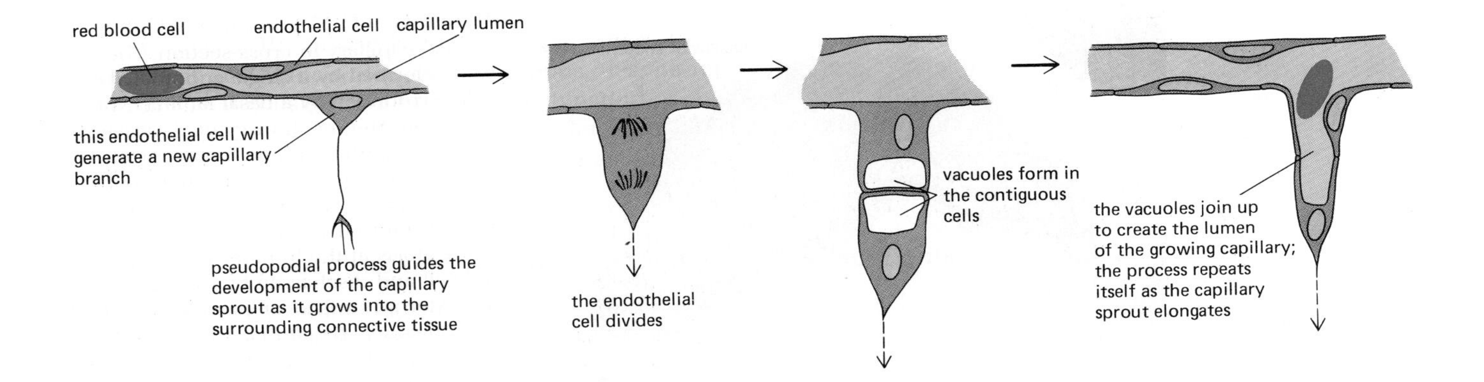

Figure 17–13 A new blood capillary forms by the sprouting of an endothelial cell from the wall of an existing small vessel. This schematic diagram is based on observations of cells in the transparent tail of a living tadpole. (After C.C. Speidel, *Am. J. Anat.* 52:1–79, 1933.)

New Capillaries Form by Sprouting[14,15]

New vessels always originate as capillaries, which sprout from existing small vessels. This process of **angiogenesis** occurs in response to specific signals, and it can be readily observed in rabbits by punching a small hole in the ear and fixing glass cover slips on either side to create a thin transparent viewing chamber into which the cells that surround the wound can grow. Angiogenesis can also be conveniently observed in naturally transparent structures such as the cornea of the eye. Irritants applied to the cornea induce the growth of new blood vessels from the rim of the cornea, which has a rich blood supply, in toward the center, which normally has almost none. Thus the cornea becomes vascularized through an invasion of endothelial cells into the tough collagen-packed corneal tissue.

Observations such as these reveal that endothelial cells that will form a new capillary grow out from the side of a capillary or small venule by extending long processes or pseudopodia (Figure 17–13). The cells at first form a solid sprout, which then hollows out to form a tube. This process continues until the sprout encounters another capillary, with which it connects, allowing blood to circulate. Experiments in culture show that endothelial cells in a medium containing suitable growth factors will spontaneously form capillary tubes even if they are isolated from all other types of cells. The first sign of tube formation in culture is the appearance in a cell of an elongated vacuole that is at first completely encompassed by cytoplasm (Figure 17–14A). Contiguous cells develop similar vacuoles, and eventually the cells arrange their vacuoles end to end so that the vacuoles become continuous from cell to cell, forming a capillary channel (Figure 17–14B). The capillary tubes that develop in a pure culture of endothelial cells do not contain blood, and nothing travels through them, indicating that blood flow and pressure are not required for the formation of a capillary network.

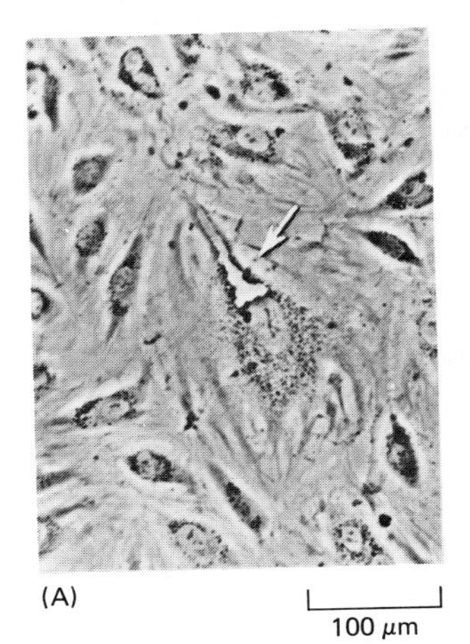

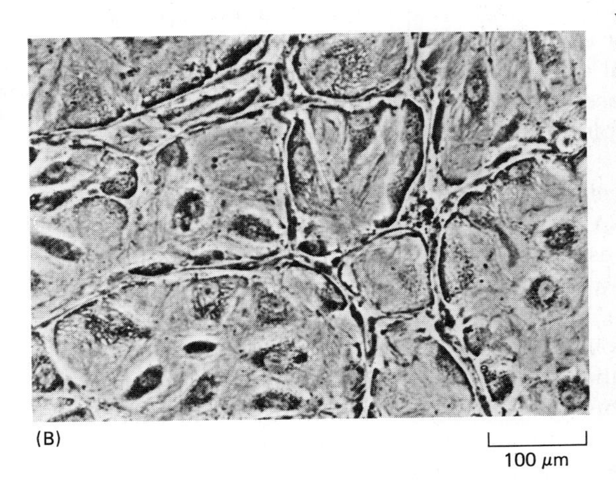

Figure 17–14 Endothelial cells in culture spontaneously develop internal vacuoles that join up, giving rise to a network of capillary tubes. Photographs (A) and (B) show successive stages in the process; the arrow in (A) indicates a vacuole forming initially in a single endothelial cell. The cultures are set up from small patches of two to four endothelial cells taken from short segments of capillary. These cells will settle on the surface of a collagen-coated culture dish and form a small flattened colony that enlarges gradually as the cells proliferate. The colony spreads across the dish, and eventually, after about 20 days, capillary tubes begin to form in the central regions. Once tube formation has started, branches soon appear, and after 5 to 10 more days, an extensive network of tubes is visible, as seen in (B). (From J. Folkman and C. Haudenschild, *Nature* 288:551–556, 1980. © Macmillan Journals Ltd

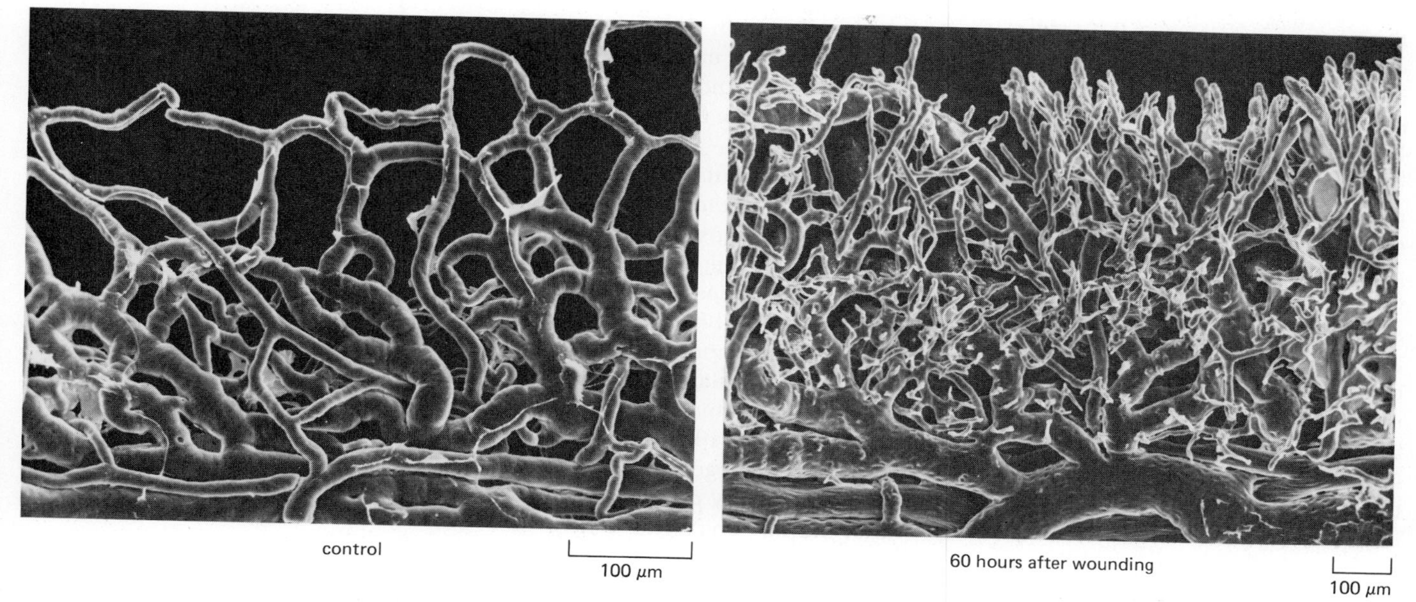

Figure 17–15 Scanning electron micrographs of casts of the system of blood vessels in the margin of the cornea, showing the reaction to wounding. The casts are made by injecting a resin into the vessels and letting the resin set; this reveals the shape of the lumen, as opposed to the shape of the cells. Sixty hours after wounding, many new capillaries have begun to sprout toward the site of injury, which is just above the top of the picture. Their oriented outgrowth reflects a chemotactic response of the endothelial cells to an angiogenic factor released at the wound. (Courtesy of Peter C. Burger.)

Growth of the Capillary Network Is Controlled by Factors Released by the Surrounding Tissues[15]

In living animals, endothelial cells form new capillaries wherever there is a need for them. It is thought that when cells in tissues are deprived of oxygen, they release angiogenic factors that induce new capillary growth. Probably for this reason, nearly all vertebrate cells are located within 50 μm of a capillary. Similarly, after wounding, a burst of capillary growth is stimulated in the neighborhood of the damaged tissue (Figure 17–15). Local irritants or infections also cause a proliferation of new capillaries, most of which regress and disappear when the inflammation subsides.

Angiogenesis is also important in tumor growth. The growth of a solid tumor is limited by its blood supply: if it were not invaded by capillaries, a tumor would be dependent on the diffusion of nutrients from its surroundings and could not enlarge beyond a diameter of a few millimeters. To grow further, a tumor must induce the formation of a capillary network that invades the tumor mass. A small sample of such a tumor implanted in the cornea will cause blood vessels to grow quickly toward the implant from the vascular margin of the cornea (Figure 17–16), and the growth rate of the tumor increases abruptly as soon as the vessels reach it.

In all of these cases the invading endothelial cells must respond to a signal produced by the tissue that requires a blood supply. The response of the endothelial cells includes at least three components. First, the cells must breach the basal lamina that surrounds an existing blood vessel; endothelial cells during angiogenesis have been shown to secrete *proteases*, such as *plasminogen activator*, which enable them to digest their way through the basal lamina of the parent capillary or venule. Second, the endothelial cells must move toward the source of the signal. Third, they must proliferate. In certain circumstances, one or two of

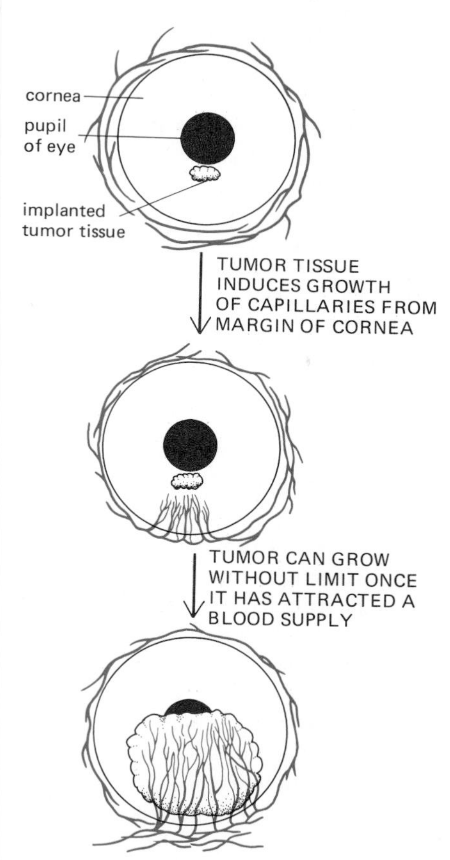

Figure 17–16 Tumor tissue implanted in the cornea releases a factor that causes the ingrowth of capillaries, supplying the tumor with blood-borne nutrients that allow it to grow.

the components of this tripartite response can be elicited in the absence of the others. Some new capillaries form, for example, even when endothelial cell proliferation has been blocked by irradiation, and a factor in fluid from wounds has been shown to attract endothelial cells and to stimulate them to secrete proteases without stimulating their proliferation.

Other factors can elicit on their own all three components of the endothelial cell response. Two examples are *acidic fibroblast growth factor (acidic FGF)* and *basic fibroblast growth factor (basic FGF).* These two proteins, which have been purified independently from several different sources and therefore are known by various other names as well, have similar amino acid sequences (55% identity). In addition to their dramatic effect on endothelial cells, they stimulate the proliferation of fibroblasts and of several other cell types and are important regulators of early embryonic development (see p. 894). Their precise cellular origins are poorly understood. Other substances that can also act as angiogenic factors appear to be released during tissue repair, inflammation, and growth by a variety of cell types, including macrophages, mast cells, and fat cells. Angiogenesis, like the control of cell proliferation in general, seems to be regulated by complex—and perhaps redundant—combinations of signals, rather than by one signal alone.

Summary

Most populations of differentiated cells in vertebrates are subject to turnover through cell death and renewal. In some cases the fully differentiated cells simply divide to produce daughter cells of the same type. The rate of proliferation of hepatocytes, for example, is controlled to maintain appropriate total cell numbers, and if a large part of the liver is destroyed, the remaining hepatocytes increase their division rate to restore the loss. Under normal circumstances cell renewal keeps the numbers of cells of each type in a tissue in appropriate balance. In response to unusual damage, however, repair may be unbalanced, as when the fibroblasts in a repeatedly damaged liver grow too rapidly in relation to the hepatocytes and replace them with connective tissue.

Endothelial cells form a single cell layer that lines all blood vessels and regulates exchanges between the bloodstream and the surrounding tissues. New blood vessels develop from the walls of existing small vessels by the outgrowth of endothelial cells, which have the capacity to form hollow capillary tubes even when isolated in culture. In the living animal, anoxic or damaged tissues stimulate angiogenesis by releasing angiogenic factors, which attract nearby endothelial cells and stimulate them to proliferate and secrete proteases.

Renewal by Stem Cells: Epidermis[8,16]

We turn now from cell populations that are renewed by simple duplication to those that are renewed by means of **stem cells.** These populations vary widely, not only in cell character and rate of turnover, but also in the geometry of cell replacement. In the lining of the small intestine, for example, cells are arranged as a single-layered epithelium. This epithelium covers the surfaces of the *villi* that project into the lumen of the gut, and it lines the deep *crypts* that descend into the underlying connective tissue (Figure 17–17). The stem cells lie in a protected position in the depths of the crypts. The differentiated cells generated from them are carried upward by a sliding movement in the plane of the epithelial sheet until they reach the exposed surfaces of the villi, from whose tips they are finally shed. A contrasting example is found in the epithelium that forms the outer surface of the skin, called the *epidermis.* The epidermis is a many-layered epithelium, and the differentiating cells travel outward from their site of origin in a direction perpendicular to the plane of the cell sheet. In the case of blood cells, the spatial pattern of production is complex and appears chaotic. Before going into such details, however, we must pause to consider what a stem cell is.

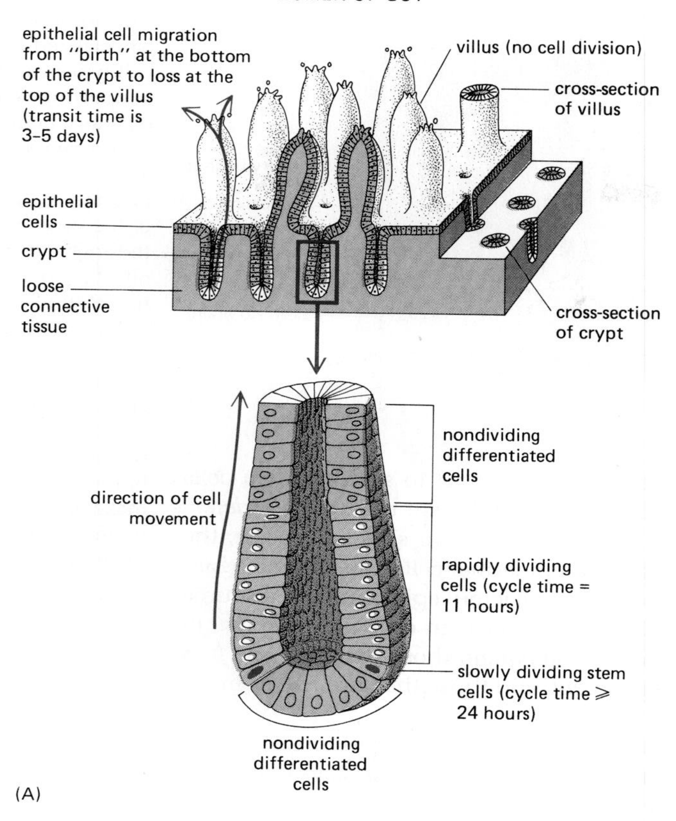

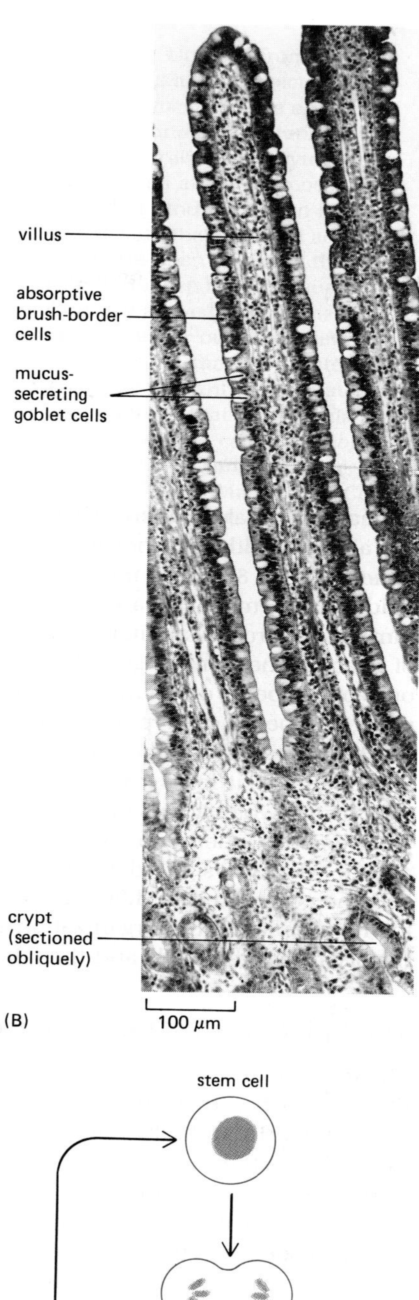

Figure 17–17 (A) The pattern of cell turnover and the proliferation of stem cells in the epithelium that forms the lining of the small intestine. (B) Photograph of a section of part of the lining of the small intestine, showing the villi and crypts. Note how mucus-secreting goblet cells (visible as pale ovals) are interspersed among the absorptive brush-border cells in the epithelium of the villi. See Figure 17–9 for the structure of these cells. (Courtesy of Peter Gould.)

Stem Cells Can Divide Without Limit and Give Rise to Differentiated Progeny[17]

The defining properties of a stem cell are as follows:

1. It is not itself terminally differentiated (that is, it is not at the end of a pathway of differentiation).
2. It can divide without limit (or at least for the lifetime of the animal).
3. When it divides, each daughter has a choice: it can either remain a stem cell, or it can embark on a course leading irreversibly to terminal differentiation (Figure 17–18).

Stem cells are required wherever there is a recurring need to replace differentiated cells that cannot themselves divide. In several tissues the terminal state of cell differentiation is obviously incompatible with cell division. For example, the cell nucleus may be digested, as in the outermost layers of the skin, or be extruded, as in the mammalian red blood cell. Alternatively, the cytoplasm may be heavily encumbered with structures, such as the myofibrils of striated muscle cells, that would hinder mitosis and cytokinesis. In other terminally differentiated cells the chemistry of differentiation may be incompatible with cell division in some more subtle way. In any such case, renewal must depend on stem cells.

The job of the stem cell is not to carry out the differentiated function but rather to produce cells that will. Consequently, stem cells often have a nondescript

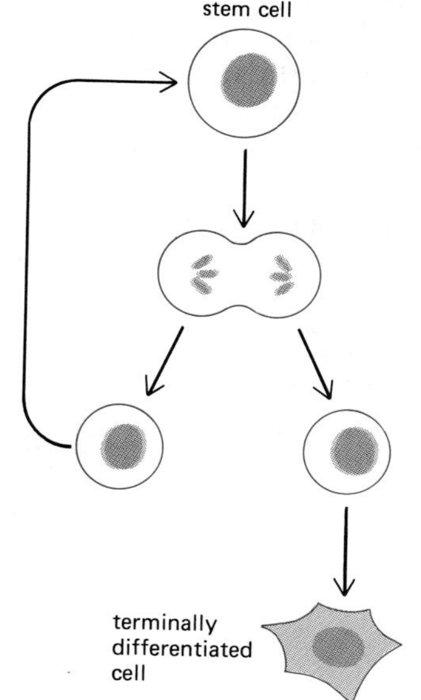

Figure 17–18 The definition of a stem cell. Each daughter produced when a stem cell divides can either remain a stem cell or go on to become terminally differentiated.

Table 17–1 Blood Cells

Type of Cell	Main Functions	Typical Concentration in Human Blood (cells/liter)
Red blood cells (erythrocytes)	transport O_2 and CO_2	5×10^{12}
White blood cells (leucocytes)		
Granulocytes		
Neutrophils (polymorphonuclear leucocytes)	phagocytose and destroy invading bacteria	5×10^{9}
Eosinophils	destroy larger parasites and modulate allergic inflammatory responses	2×10^{8}
Basophils	release histamine and serotonin in certain immune reactions	4×10^{7}
Monocytes	become tissue macrophages, which phagocytose and digest invading microorganisms, foreign bodies, and senescent cells	4×10^{8}
Lymphocytes		
B cells	make antibodies	2×10^{9}
T cells	kill virus-infected cells and regulate activities of other leucocytes	1×10^{9}
Natural killer (NK) cells	kill virus-infected cells and some tumor cells	1×10^{8}
Platelets (cell fragments, arising from megakaryocytes in bone marrow)	initiate blood clotting	3×10^{11}

Humans contain about 5 liters of blood, accounting for 7% of body weight. Red blood cells constitute about 45% of this volume and white cells about 1%, the rest being the liquid *blood plasma.*

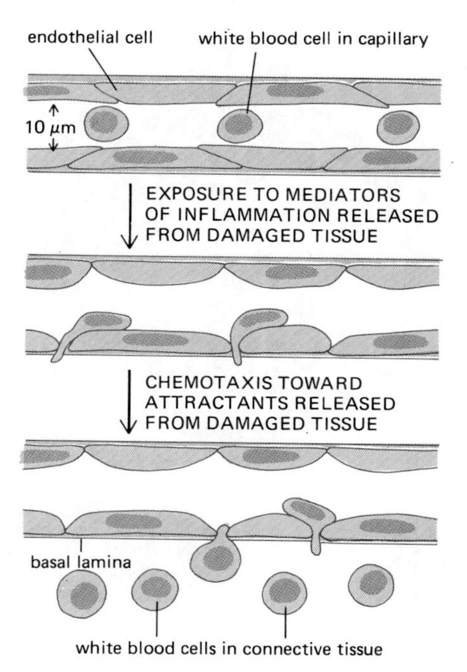

Figure 17–28 White blood cells migrate out of the bloodstream into an injured or infected tissue as part of the inflammatory response. The response is initiated by a variety of signaling molecules produced locally by cells (mainly in the connective tissue) or by complement activation. Some of these mediators act on capillary endothelial cells, causing them to loosen their attachments to their neighbors so that the capillaries become more permeable; the alteration in the endothelial cell surface also causes leucocytes in the blood to adhere to it. Other mediators act as chemoattractants, causing the bound leucocytes to crawl between the capillary endothelial cells into the tissue.

In other circumstances, erythrocyte production is selectively increased—for example, if one goes to live at high altitude, where oxygen is scarce. Thus blood cell formation (**hemopoiesis**) necessarily involves complex controls in which the production of each type of blood cell is regulated individually to meet changing needs. It is a problem of great medical importance to understand how these controls operate.

In intact animals, hemopoiesis is more difficult to analyze than is cell turnover in a tissue such as the epidermal layer of the skin. In epidermis there is a simple, regular spatial organization that makes it easy to follow the process of renewal and to locate the stem cells; this is not true of the hemopoietic tissues. On the other hand, the hemopoietic cells have a nomadic life-style that makes them more accessible to experimental study in other ways. Dispersed hemopoietic cells can be easily transferred, without damage, from one animal to another, and the proliferation and differentiation of individual cells and their progeny can be observed and analyzed in culture. For this reason, more is known about the molecules that control blood cell production than about those that control cell production in other mammalian tissues. Even so, there are still large gaps in our understanding.

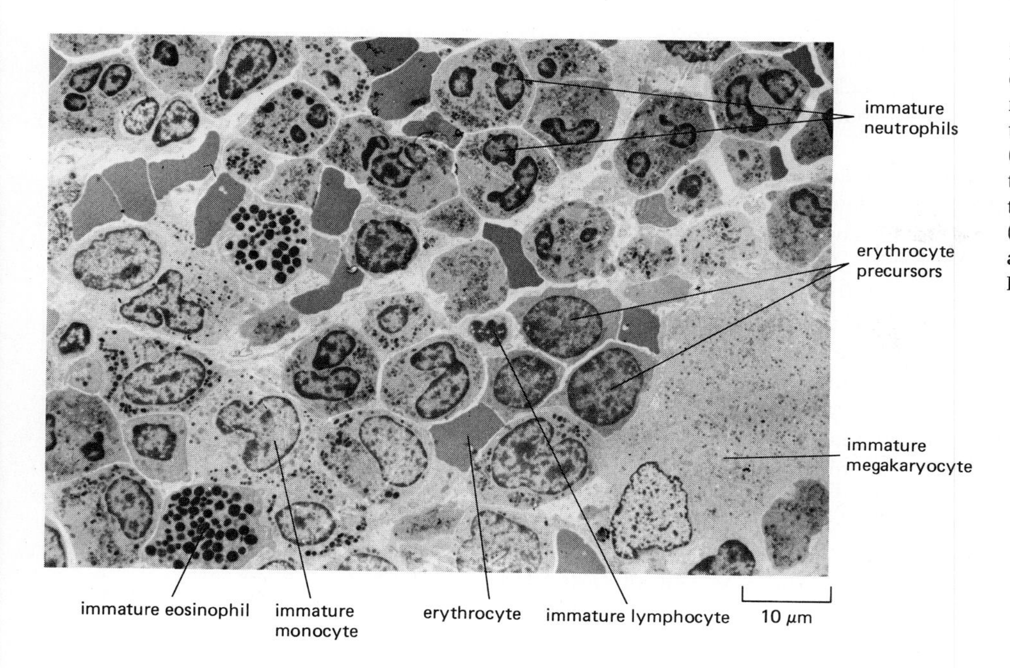

Figure 17–29 Low-magnification electron micrograph of a section of a region of bone marrow. This tissue is the main source of new blood cells (except for T lymphocytes). Note that the immature blood cells of a particular type tend to cluster in "family groups." (From J.A.G. Rhodin, Histology: A Text and Atlas. New York: Oxford University Press, 1974.)

Bone Marrow Contains Hemopoietic Stem Cells[23,26]

The different types of blood cells and their immediate precursors can be recognized in the bone marrow by their distinctive appearances (Figure 17–29). They are intermingled with one another, as well as with fat cells and other *stromal cells* (connective-tissue cells) that form a delicate supporting meshwork of collagen fibers and other extracellular-matrix components. In addition, the whole tissue is richly supplied with thin-walled blood vessels (called *blood sinuses*) into which the new blood cells are discharged. **Megakaryocytes** are also present; these, unlike other blood cells, remain in the bone marrow when mature and are one of its most striking features, being extraordinarily large (diameter up to 60 μm), with a highly polyploid nucleus. They are normally plastered against blood sinuses, and they extend processes through holes in the endothelial lining of these vessels; platelets pinch off from the processes and are swept away into the blood (Figure 17–30).

Because of the complex arrangement of the cells in bone marrow, it is difficult to identify any but the immediate precursors of the mature blood cells. The corresponding cells at still earlier stages of development, before any overt differentiation has begun, are confusingly similar in appearance, and there is no visible feature by which the ultimate stem cells can be recognized. To identify and characterize the stem cells, one needs a functional test, which involves tracing the

Figure 17–30 (A) Schematic drawing of a megakaryocyte among other cells in the bone marrow. The enormous size of the megakarocyte results from its having a highly polyploid nucleus. One megakaryocyte produces about 10,000 platelets, which split off from long processes that extend through holes in the walls of an adjacent blood sinus. (B) Scanning electron micrograph of the interior of a blood sinus in the bone marrow, showing the megakaryocyte processes. (B, from R.G. Kessel and R.H. Kardon, Tissues and Organs: A Text-Atlas of Scanning Electron Microscopy. San Francisco: Freeman, 1979. © 1979, W.H. Freeman and Company.)

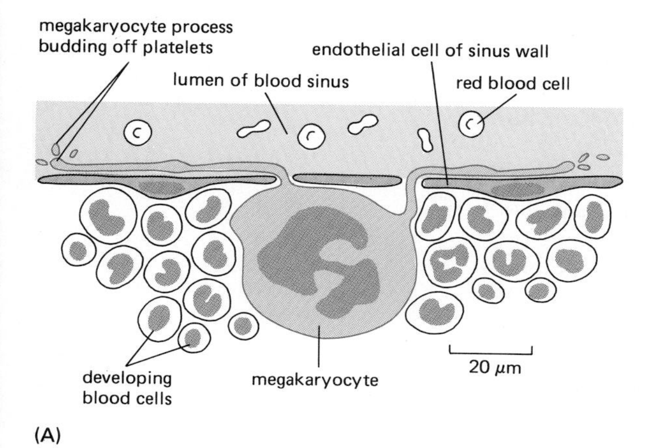

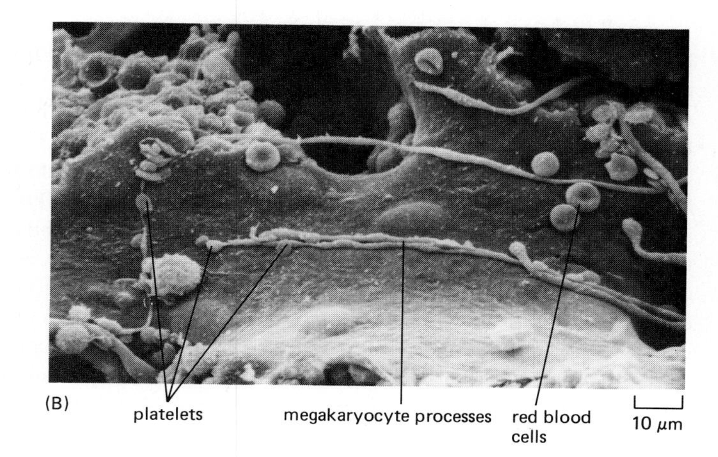

progeny of single cells. As we shall see, this can be done in culture simply by examining the colonies that artificially isolated cells produce. The hemopoietic system, however, can also be manipulated so that such clones of cells can be recognized in the intact animal.

If an animal is exposed to a large dose of x-irradiation, the hemopoietic cells are destroyed and the animal dies within a few days as a result of its inability to manufacture new blood cells. The animal can be saved, however, by a transfusion of cells taken from the bone marrow of a healthy, immunologically compatible donor. Among these cells there are evidently some that can colonize the irradiated host and reequip it with hemopoietic tissue. One of the tissues where colonies develop is the spleen, which in a normal mouse is an important additional site of hemopoiesis. When the spleen of an irradiated mouse is examined a week or two after the transfusion of cells from a healthy donor, a number of distinct nodules are seen in it, each of which is found to contain a colony of myeloid cells (Figure 17–31); after 2 weeks some colonies may contain more than a million cells. The discreteness of the nodules suggests that each might be a clone of cells descended from a single founder cell, like a bacterial colony on a culture plate; and with the help of genetic markers, it can be established that this is indeed the case.

The founder of such a colony is called a **colony-forming cell,** or **CFC** (also known as a colony-forming unit, CFU). The colony-forming cells are heterogeneous. Some give rise to only one type of myeloid cell, while others give rise to mixtures. Some go through many division cycles and form large colonies, while others divide less and form small colonies. Most of the colonies die out after generating a restricted number of terminally differentiated blood cells. A few of the colonies, however, are capable of extensive self-renewal and produce new colony-forming cells in addition to terminally differentiated blood cells. The founders of such self-renewing colonies are assumed to be the hemopoietic stem cells in the transfused bone marrow.

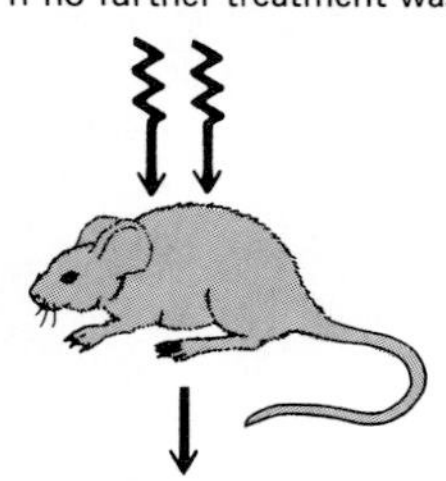

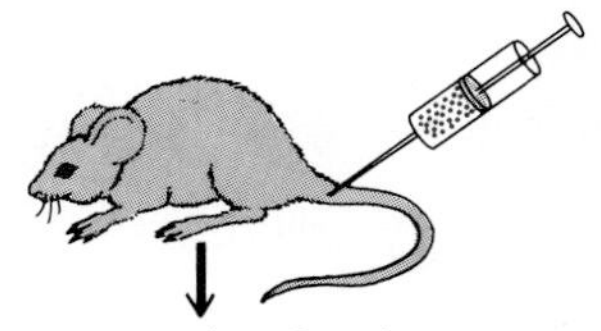

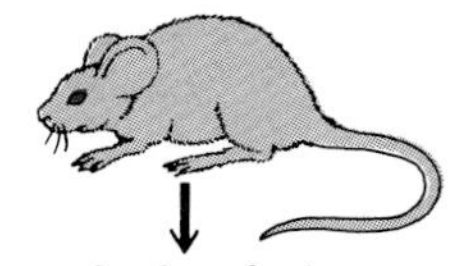

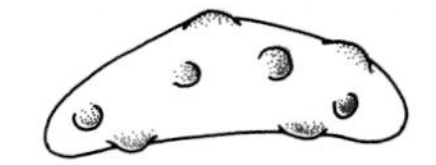

Figure 17–31 The spleen colony assay, in which the spleen of a heavily irradiated animal is seeded with bone marrow cells transfused from a healthy donor. This assay revolutionized the study of hemopoiesis by allowing individual myeloid precursor cells to be analyzed for the first time.

A Pluripotent Stem Cell Gives Rise to All Classes of Blood Cells[27]

All the types of myeloid cells can often be found together in one spleen colony, derived from a single stem cell. The hemopoietic stem cell, therefore, is *pluripotent:* it can give rise to many different cell types. Although the spleen colonies do not seem to contain lymphocytes, another approach shows that these also derive from the same stem cell that gives rise to all of the myeloid cells. The demonstration employs genetic markers that make it possible to identify the members of a clone even after they have been released into the bloodstream. Although several types of clonal markers have been used for this, a specially engineered retrovirus serves the purpose particularly well. The marker virus, like other retroviruses, can insert its own genome into the chromosomes of the cell it infects, but the genes that would enable it to generate infectious virus particles have been removed. The marker, therefore, is confined to the progeny of the cells that were originally infected, and the progeny of one such cell can be distinguished from the progeny of another because the chromosomal sites of insertion of the virus are different. To analyze hemopoietic cell lineages, bone marrow cells are first infected with the retrovirus (see p. 254) *in vitro* and then are transferred into a lethally irradiated recipient; DNA probes (see p. 189) can then be used to trace the progeny of individual infected cells in the various hemopoietic and lymphoid tissues of the host.

These experiments not only confirm that all classes of blood cells—both myeloid and lymphoid—derive from a common stem cell (Figure 17–32), they also make it possible to follow the pedigrees of the blood cells over long periods of time. A month or two after a transfusion of bone marrow cells, most of the blood cells in an irradiated host mouse are found to be descendants of fewer than half a dozen pluripotent stem cells; this is still true several weeks later, but the blood cells are now the progeny of a different handful of stem cells. These observations suggest that an individual stem cell has only a low probability in any given time interval of initiating the formation of a clone of differentiated progeny, but that

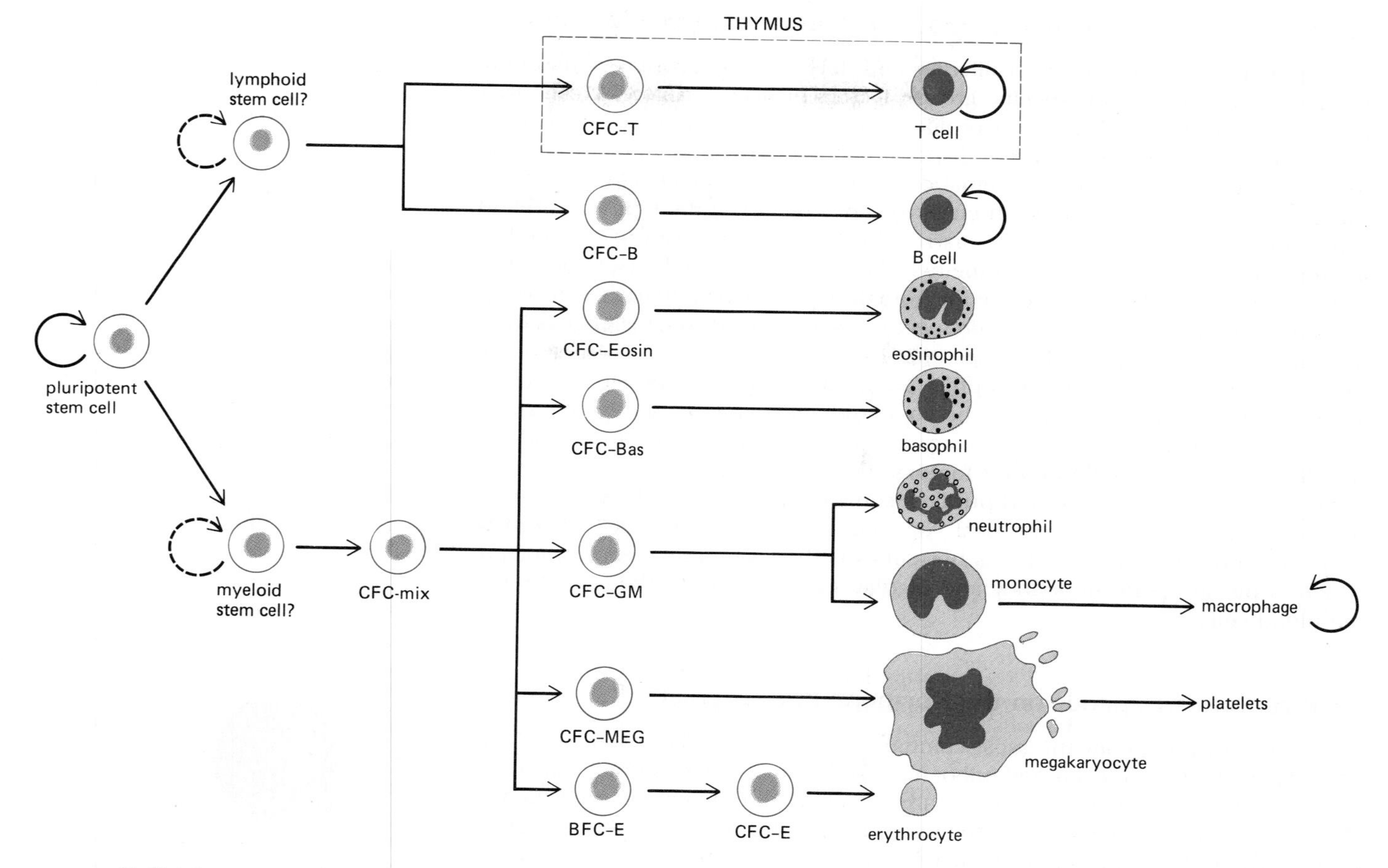

Figure 17–32 A tentative scheme of hemopoiesis. The pluripotent stem cell normally divides infrequently to generate either more pluripotent stem cells (self-renewal) or *committed progenitor cells* (labeled CFC = colony-forming cells), which are irreversibly determined to produce only one or a few types of blood cells. The progenitor cells are stimulated to proliferate by specific growth factors but progressively lose their capacity for division and develop into terminally differentiated blood cells, which usually live for only a few days or weeks.

In adult mammals all of the cells shown develop mainly in the bone marrow—except for T lymphocytes, which develop in the thymus, and macrophages, which develop from monocytes in most tissues. The most controversial part of the scheme is the first branch point: the existence of stem cells committed exclusively to form T and B lymphocytes and of stem cells committed exclusively to form all the other classes of blood cells (myeloid cells), as shown, is uncertain. It is likely that the primary pluripotent stem cells also give rise to various types of tissue cells not shown in this scheme, such as NK cells, mast cells, osteoclasts, and a variety of classes of antigen-presenting cells (see p. 1045); but the pathways by which these cells develop are uncertain.

many cell divisions intervene between this initial step and the final differentiation so that the ultimate clone of progeny is very large—on the order of millions of cells. In spite of the seemingly rare and quantal nature of the initiating events, differentiated blood cells are produced at a smooth and steady rate; this reflects controls that operate at intermediate steps along the pathway to help regulate the final numbers of blood cells of each type.

The Number of Specialized Blood Cells Is Amplified by Divisions of Committed Progenitor Cells[23,28]

Once a cell has differentiated as an erythrocyte or a granulocyte or some other type of blood cell, there is no going back: the state of differentiation is not reversible. Therefore, at some stage in their development, the progeny of the pluripotent stem cell must become irreversibly committed or determined for a particular line of differentiation. It is clear from simple microscopic examination of the bone marrow that this commitment occurs well before the final division in which the mature differentiated cell is formed: one can recognize specialized precursor cells that are still proliferating but already show signs of having begun differentiation. It thus appears that commitment to a particular line of differentiation is followed by a series of cell divisions that amplify the number of cells of a given specialized type.

The hemopoietic system, therefore, can be viewed as a hierarchy of cells. **Pluripotent stem cells** give rise to **committed progenitor cells,** which are irreversibly determined as ancestors of only one or a few blood cell types. The committed progenitors are thought to divide rapidly but a limited number of times. At the end of this series of *amplification divisions,* they develop into **terminally differentiated cells,** which usually divide no further and die after several days or weeks. Studies in culture provide a way to find out how these cellular events are regulated.

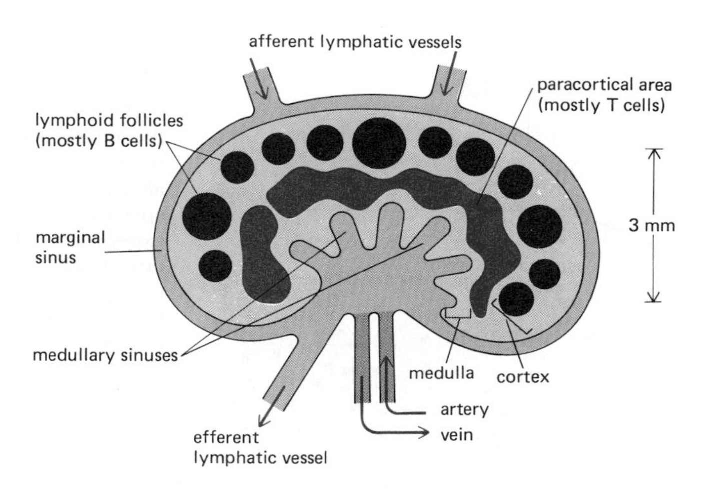

Figure 18–8 A highly simplified drawing of a human lymph node. B lymphocytes are located primarily in the cortex, where they are clustered in structures called *lymphoid follicles.* T lymphocytes are found mainly in the *paracortical area.* Both types of lymphocytes enter the lymph node from the blood via specialized small veins in the paracortical area (not shown). T cells remain in this area, while B cells migrate to the lymphoid follicles. Eventually both T cells and B cells migrate to the medullary sinuses and leave the node via the efferent lymphatic vessel. This vessel ultimately empties into the bloodstream, allowing the lymphocytes to begin another cycle of circulation through a secondary lymphoid organ.

Foreign antigens that enter the lymph node are displayed on the surface of specialized *antigen-presenting cells:* one type presents antigen (in the form of antigen-antibody complexes) to B cells in the lymphoid follicles; another type presents antigen to T cells in the paracortical area (see p. 1045).

Most Lymphocytes Continuously Recirculate[7]

The majority of T and B lymphocytes continuously recirculate between the blood and the secondary lymphoid organs. In a lymph node, for example, lymphocytes leave the bloodstream, squeezing out between specialized endothelial cells; after percolating through the node, they accumulate in small lymphatic vessels that leave the node and connect with other lymphatic vessels, which then pass through other lymph nodes downstream (Figure 18–8). Passing into larger and larger vessels, the lymphocytes eventually enter the main lymphatic vessel (the *thoracic duct*), which carries them back into the blood. This continuous recirculation not only ensures that the appropriate lymphocytes will come into contact with antigen, it also ensures that appropriate lymphocytes encounter each other: we shall see that interactions between specific lymphocytes are a crucial part of most immune responses.

Lymphocyte recirculation depends on specific interactions between the lymphocyte cell surface and the surface of specialized endothelial cells lining small veins (called *postcapillary venules*) in the secondary lymphoid organs: of all the cell types in the blood that come into contact with these endothelial cells, only lymphocytes transiently adhere and then migrate through the postcapillary venules. Monoclonal antibodies (see p. 178) that bind to the surface of lymphocytes and inhibit their ability both to bind to the specialized endothelial cells in tissue sections of secondary lymphoid organs and to recirculate *in vivo* are helping to define the various *"homing receptors"* involved in lymphocyte migration. The majority of T and B cells have one type of glycoprotein on their surface that is required for them to recirculate through lymph nodes and another that is required for them to recirculate through Peyer's patches. Some lymphocytes have only the latter glycoprotein and recirculate selectively through Peyer's patches; they constitute, in effect, a gut-specific subsystem of lymphocytes specialized for responding to antigens that enter the body from the intestine. Other homing receptors on lymphocytes are presumably responsible for the segregation of T and B cells into distinct areas inside a lymphoid organ (see Figure 18–8). When they are activated by antigen, lymphocytes lose the homing receptors that mediate recirculation through lymphoid organs and acquire new ones that guide the activated cells to sites of inflammation.

Immunological Memory Is Due to Clonal Expansion and Lymphocyte Maturation[8]

The immune system, like the nervous system, can remember. This is why we develop lifelong immunity to many common viral diseases after our initial exposure to the virus. The same phenomenon can be demonstrated in experimental animals. If an animal is injected once with antigen A, its immune response (either antibody or cell-mediated) will appear after a lag period of several days, rise rapidly and exponentially, and then, more gradually, fall again. This is the characteristic

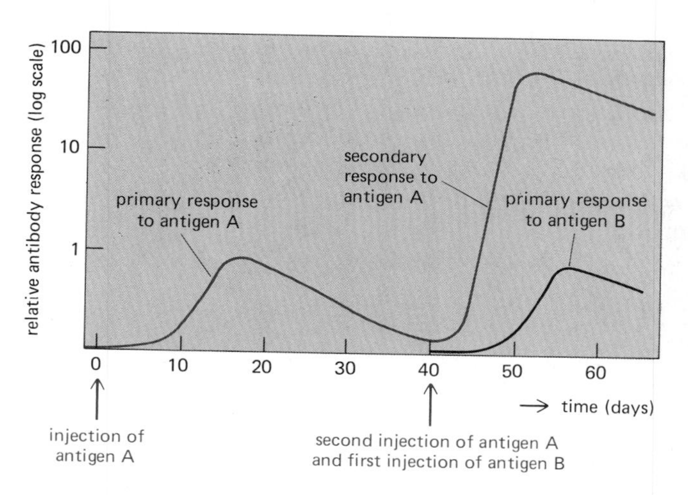

Figure 18–9 Primary and secondary antibody responses induced by a first and second exposure, respectively, to antigen A. Note that the secondary response is faster and greater than the primary response and is specific for A, indicating that the immune system has specifically "remembered" encountering antigen A before. Evidence for the same type of immunological memory is obtained if T-cell-mediated responses rather than B cell antibody responses are measured.

course of a **primary immune response,** occurring on an animal's first exposure to an antigen. If some weeks or months or even years are allowed to pass and the animal is reinjected with antigen A, it will produce a **secondary immune response** that is very different from the primary response: the lag period is shorter, the response is greater, and its duration is longer (Figure 18–9). These differences indicate that the animal has "remembered" its first exposure to antigen A. If the animal is given a different antigen (for example, antigen B) instead of a second injection of antigen A, the response is typical of a primary, and not a secondary, immune response; therefore the secondary response reflects antigen-specific **immunological memory** for antigen A.

The clonal selection theory provides a useful conceptual framework for understanding the cellular basis of immunological memory. In a mature animal the T and B cells in the secondary lymphoid organs are a mixture of cells in at least three stages of maturation, which can be designated *virgin cells, memory cells,* and *active cells.* When **virgin cells** encounter antigen for the first time, some of them are stimulated to multiply and become **active cells,** which we define as cells that are actively engaged in making a response (active T cells carry out cell-mediated responses, while active B cells secrete antibody). Some virgin cells, on the other hand, are stimulated to multiply and mature instead into **memory cells**—cells that do not themselves make a response but are readily induced to become active cells by a later encounter with the same antigen (Figure 18–10). Virgin lymphocytes are thought to survive in secondary lymphoid tissues for only a short time, probably dying within days unless they meet their specific antigen. Memory cells, on the other hand, are thought to live for many months or even years without dividing—continuously recirculating between the blood and secondary lymphoid organs. Moreover, memory cells respond more readily to antigen than do virgin cells. We shall see later (see p. 1026) that one reason for the increased responsiveness of memory B cells is that their receptors have a higher affinity for antigen.

According to this scheme, immunological memory is generated during the primary response because (1) the proliferation of antigen-triggered virgin cells

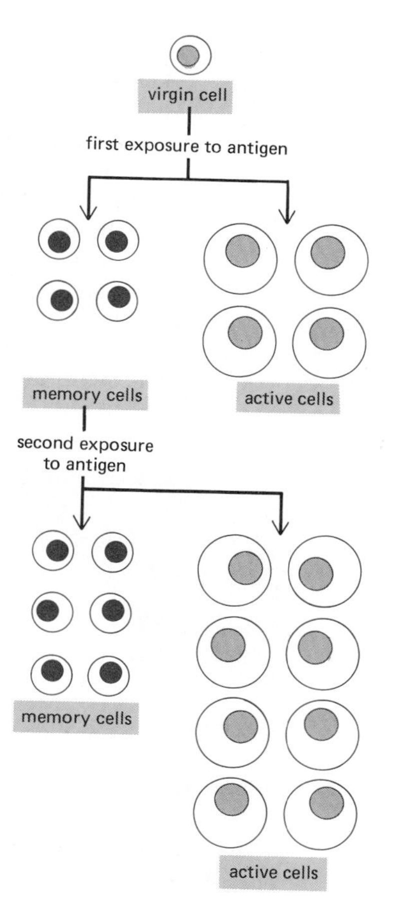

Figure 18–10 When virgin T or B cells are stimulated by their specific antigen, they proliferate and mature; some become activated to make a response, while others become memory cells. During a subsequent exposure to antigen, the memory cells respond more readily than did the virgin cells: they proliferate and give rise to active cells and to more memory cells. In the model shown, an individual virgin cell can give rise to either a memory cell or an activated cell, depending on the conditions. In an alternative model (not shown) the virgin cells that mature into memory cells are different from those that mature into activated cells. It is not known which of these models is correct.

creates many memory cells—a process known as *clonal expansion;* (2) the memory cells have a much longer life-span than virgin cells and recirculate between the blood and secondary lymphoid organs; and (3) each memory cell is able to respond more readily to antigen than does a virgin cell. Because of the changes induced during a primary response, most of the long-lived cells in the recirculating pool of lymphocytes are appropriate to the antigenic environment of the animal and are already primed and ready for action.

The Failure to Respond to Self Antigens Is Due to Acquired Immunological Tolerance[9]

How is the immune system able to distinguish foreign molecules from self molecules? One possibility might be that an animal inherits genes that encode receptors for foreign antigens but not self antigens, so that its immune system is genetically constituted to respond only to foreign antigens. Alternatively, the immune system could be inherently capable of responding to both foreign and self antigens but could "learn" early in development not to respond to self antigens. The latter explanation has been shown to be correct. The first evidence for this was an observation made in 1945. Normally, when tissues are transplanted from one individual to another, they are recognized as foreign by the immune system of the recipient and are destroyed. Dizygotic cattle twins, however, which develop from two fertilized ova and are therefore nonidentical, sometimes exchange blood cells *in utero* as a result of the spontaneous fusion of their placentas; such twins were shown to accept skin grafts from each other. These findings were later reproduced experimentally—in chicks, by allowing the blood vessels of two embryos to fuse, and in mice, by introducing cells from one strain of mouse into a neonatal mouse of another strain, where they survived for most of the recipient animal's life. In both cases, when the animals matured, grafts from the joined or donor animal were accepted (Figure 18–11), while "third-party" grafts from a different animal were rejected. Thus the continuous presence of nonself antigens from before the time the immune system matures leads to a long-lasting unresponsiveness to the specific nonself antigens. The resulting state of antigen-specific immunological unresponsiveness is known as **acquired immunological tolerance.**

There is strong evidence that the unresponsiveness of an animal's immune system to its own macromolecules (*natural immunological tolerance*) is acquired in the same way and is not inborn. Normal mice, for example, cannot make an immune response against their own blood complement protein C5 (see p. 1032), but mutant mice that lack the gene encoding C5 (but are otherwise genetically identical to the normal mice) can make an immune response to this protein. Thus it is clear that the immune system is genetically capable of responding to self but learns not to do so. In some cases at least, the learning process involves eliminating the self-reactive lymphocytes (see p. 1051), although it is not known how this is achieved. Many self-reactive lymphocytes are thought to be eliminated in the pri-

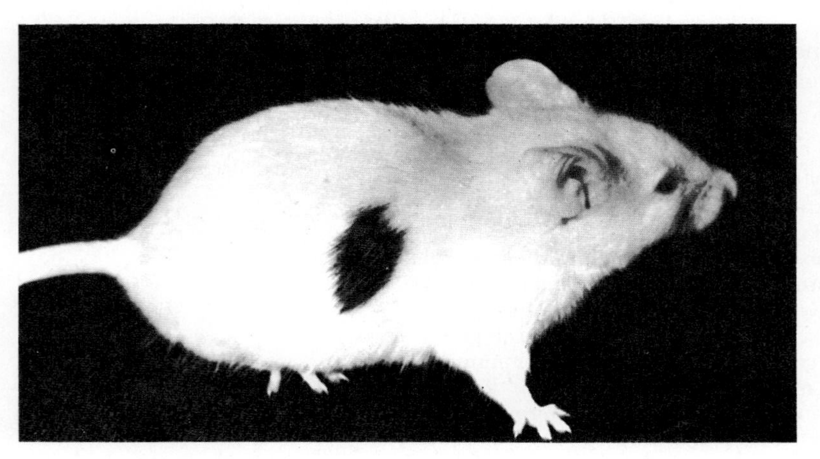

Figure 18–11 The skin graft seen here, transplanted from an adult brown mouse to an adult white mouse, has survived for many weeks only because the latter was made immunologically tolerant by injecting cells from the brown mouse into it at the time of birth. (Courtesy of Leslie Brent, from I. Roitt, Essential Immunology, 6th ed. Oxford, U.K.: Blackwell Scientific, 1988.)

mary lymphoid organs when they encounter their antigen. This negative response to antigen could be due to a special environment in these organs or to a unique responsiveness of a newly formed lymphocyte. Presumably because new self-reactive lymphocytes continue to be produced from stem cells throughout life, maintaining self-tolerance requires the constant presence of the self antigens. If an antigen such as C5 is removed, an animal regains the ability to respond to it within weeks or months.

Tolerance to self antigens sometimes breaks down, causing T or B cells (or both) to react against their own tissue antigens. *Myasthenia gravis* is an example of such an **autoimmune disease.** Affected individuals make antibodies against the acetylcholine receptors on their own skeletal muscle cells (see p. 319); the antibodies interfere with the normal functioning of the receptors so that such patients become weak and can die because they cannot breathe.

Immunological Tolerance to Foreign Antigens Can Be Induced in Adults[10]

It is generally much more difficult to induce immunological tolerance to foreign antigens in adult than in immature animals. But with some antigens it can be done experimentally by injecting the antigen (1) in very high doses, (2) in repeated very low doses, (3) together with an immunosuppressive drug, or (4) intravenously, after the antigen has been chemically coupled to the surface of B lymphocytes or ultracentrifuged to remove all aggregates, so that the normal mechanisms of antigen presentation (see p. 1046) are ineffective. Thus binding an antigen to its complementary receptors on a T or B lymphocyte can stimulate the lymphocyte to divide and mature to become an active cell or a memory cell, or it can eliminate or inactivate the lymphocyte, causing tolerance. The molecular mechanisms that determine the outcome are not well understood, but whether an antigen activates or induces tolerance depends largely on (1) the maturity of the lymphocyte, (2) the nature and concentration of the antigen, and (3) complex interactions between different classes of lymphocytes and between lymphocytes and specialized *antigen-presenting cells,* which will be discussed in a later section.

Summary

The immune system evolved to defend vertebrates against infection. It is composed of millions of lymphocyte clones. The lymphocytes in each clone share a unique cell-surface receptor that enables them to bind a particular "antigenic determinant" consisting of a specific arrangement of atoms on a part of a molecule. There are two classes of lymphocytes: B cells, which make antibodies, and T cells, which make cell-mediated immune responses.

Beginning early in lymphocyte development, many lymphocytes that would react against antigenic determinants on self macromolecules are eliminated or inactivated; as a result, the immune system normally reacts only to foreign antigens. Binding a foreign antigen to a lymphocyte initiates a response by the cell that helps to eliminate the antigen. As part of the response, some lymphocytes proliferate and mature into long-lived memory cells, so that the next time the same antigen is encountered, the immune response to it is faster and stronger.

The Functional Properties of Antibodies[11]

Vertebrates rapidly die of infection if they are unable to make antibodies. Antibodies defend us against infection by inactivating viruses and bacterial toxins and by recruiting the complement system and various types of white blood cells to kill invading microorganisms and larger parasites. Synthesized exclusively by B lymphocytes, antibodies are produced in millions of forms, each with a different amino acid sequence and a different binding site for antigen. Collectively called **immunoglobulins** (abbreviated as **Ig**), they are among the most abundant protein components in the blood, constituting about 20% of the total plasma protein by

PAPAIN CLEAVAGE

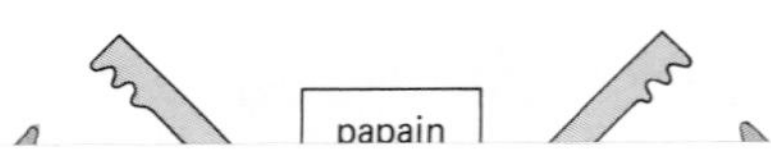

PEPSIN CLEAVAGE

Figure 18–16 The fragments produced when antibody molecules are

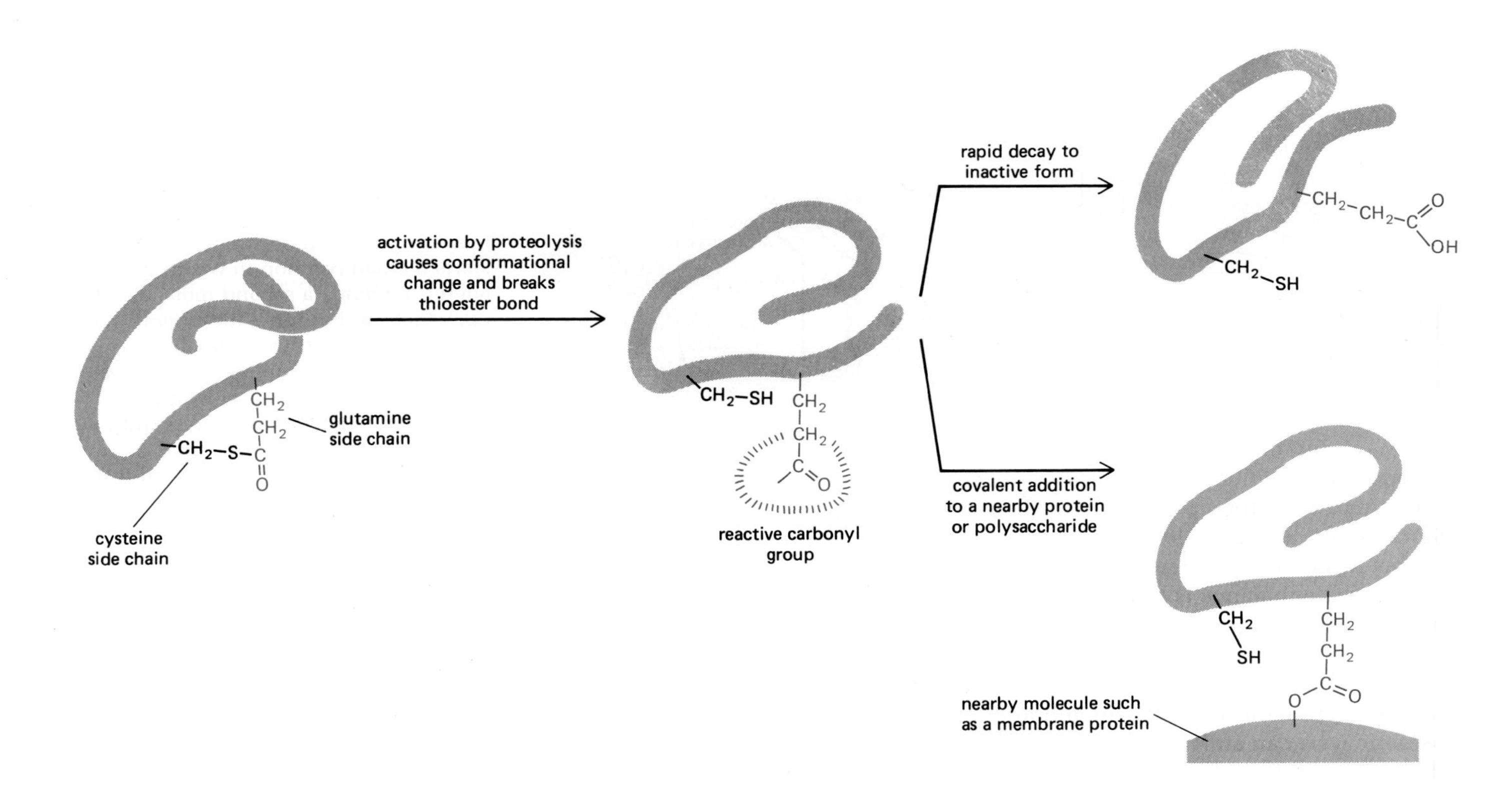

Figure 18–45 The proteolytic activation of either C3 or C4 induces a conformational change in the protein, breaking the unusual intramolecular covalent bond shown. Breaking this thioester bond between protein side chains generates a very reactive carbonyl group that couples covalently to another macromolecule, forming an ester or an amide linkage. However, the ability of the protein to react this way decays with a half-life of 60 microseconds or so, confining the reaction to membranes that are very near the site where complement activation was initiated. Both C3 and C4 are composed of more than one polypeptide chain; the largest chain in each protein undergoes the reactions shown.

of the late components, and therefore are unable to assemble the membrane attack complex, are still protected against infection by all but a few types of bacteria—those that are able to survive inside phagocytic cells, so that complement-mediated lysis is especially important in their control. It is thought that both the alternative and classical pathways evolved from such a primitive complement system. The alternative pathway probably evolved first as a form of nonspecific innate protection against infection, the classical pathway evolving much later to link C3 activation to antibody binding and thereby to specific adaptive immune responses. Consistent with the view that the two pathways are evolutionarily related is the finding that many of their components are homologous, including the serine proteases C1r, C1s, C2, factor B, and factor D.

Summary

The complement system acts on its own and in cooperation with antibodies to defend vertebrates against infection. The early components are blood proenzymes that are sequentially activated in an amplifying series of limited proteolytic reactions, either by the classical pathway—which is triggered by IgG or IgM antibodies binding to antigen—or by the alternative pathway—which can be triggered directly by the cell envelopes of invading microorganisms. The most important complement component is the C3 protein, which is activated by proteolytic cleavage and then binds covalently to nearby membranes. Microorganisms with activated C3 (C3b) on their surface are readily ingested and destroyed by phagocytic cells. In addition, C3b helps to initiate assembly of the late complement components, which form a large membrane attack complex that can cause invading microorganisms to lyse. Complement activation also releases a variety of small, soluble peptide fragments that attract and activate neutrophils and stimulate mast cells to secrete histamine: this results in an inflammatory response at sites of complement activation. The complement proteolytic cascade remains focused on the membranes of invading microorganisms that activated the cascade, mainly because several complement components, including C3b, remain activated for less than 0.1 millisecond and therefore cannot spread the attack to nearby host cells.

T Lymphocytes and Cell-mediated Immunity

The diverse responses of T cells are collectively called *cell-mediated immune reactions.* Like antibody responses, they are important in defending vertebrates against infection, particularly by certain viruses and fungi. Also like antibody responses, they are exquisitely antigen-specific.

T cells differ from B cells, however, in several important ways: (1) While some T cells fight infection directly by killing virus-infected cells, most regulate the activity of other effector cells, such as B cells and macrophages. (2) Both effector and regulatory T cells act mainly at short range, interacting directly with the cells they kill or regulate; B cells, on the other hand, secrete antibodies that can act far away. (3) Presumably for this reason, T cells bind foreign antigen only when it is on the surface of another cell in the body; the antigen is recognized in association with a special class of cell-surface glycoproteins, known as *MHC molecules* because they are encoded by a complex of genes called the *major histocompatibility complex (MHC)* (see p. 1040). This ensures that T cells are activated only when they contact another host cell and is the most important feature distinguishing antigen recognition by T and B cells.

T Cell Receptors Are Antibodylike Heterodimers[32]

Because T cells are activated only through close contact with other cells, their antigen receptors exist only in membrane-bound form. Thus, unlike antibodies, which are secreted (as well as being membrane-bound), T cell receptors for antigen were difficult to isolate, and it took much longer to identify these molecules and the genes that encode them. The receptor proteins were first identified in 1983, after it had become possible to grow pure clones of antigen-specific T cells in culture (see p. 1047), thus making available large numbers of T cells with identical receptors. Monoclonal antibodies could then be raised against the cloned cells, and those that recognized the T cell receptor were identified by their ability to block antigen-induced responses of the original cells but not those of other T cell clones. The antibodies were then used to purify the receptor molecules, which were found to be composed of two disulfide-linked polypeptide chains (called α and β). Each of these chains shares with antibodies the distinctive property of a variable amino-terminal region and a constant carboxyl-terminal region (Figure 18–46).

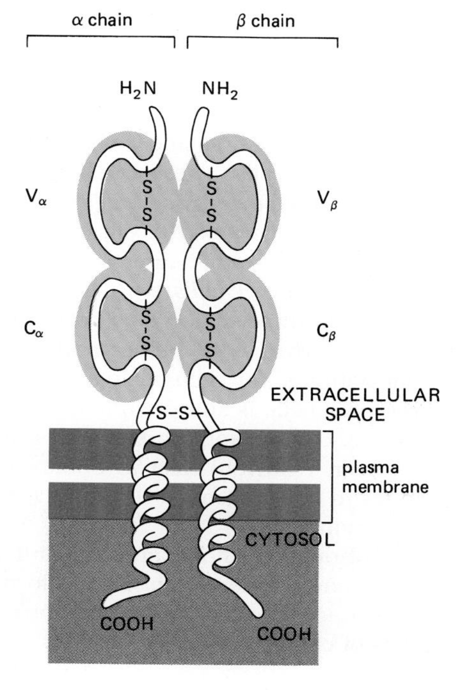

Figure 18–46 A T cell receptor heterodimer composed of an α and a β polypeptide chain, both of which are glycosylated (not shown). Each chain is about 280 amino acid residues long, and its large extracellular part is folded into two immunoglobulinlike domains—one variable (V) and one constant (C). From an analysis of amino acid sequences deduced from cDNA clones, it is thought that an antigen-binding site formed by a V_α and V_β domain is similar in its overall dimensions and geometry to the antigen-binding site of an antibody molecule. Unlike antibodies, however, which have two binding sites for antigen, T cell receptors have only one (probably because they are always bound to the plasma membrane, where they can act cooperatively). The α/β heterodimer shown is noncovalently associated with an invariant set of membrane proteins called the CD3 complex (not shown). A typical T cell has 20,000 to 40,000 α/β receptor proteins on its surface.

This antibodylike feature was an important element in an ingenious strategy used, a year or so later, to isolate the genes that encode T cell receptors. Because T and B lymphocytes are very closely related, most of the genes they transcribe are the same, and so they contain mostly the same mRNAs. Only the T cell, however, contains mRNA for the T cell receptor. By taking the total mRNAs from T cell clones and removing, by subtractive hybridization (see p. 262), the mRNAs also made by B cells, a small population of mRNAs unique to T cells was obtained. A cDNA library was then prepared from the mRNAs (see p. 260), and individual cDNA clones from the library were used to prepare radioactive DNA probes. On the assumption that the variability of the amino-terminal ends of the T cell receptor proteins, like that of antibody V regions, is generated by DNA rearrangements, each DNA probe was then used to examine the corresponding genomic DNA for evidence of rearrangement during T cell development. In this way the gene pools that encode the α and β chains were eventually localized on different chromosomes and shown to contain, like antibody gene pools, separate *V, D, J,* and *C* gene segments, which are brought together by site-specific recombination during T cell development in the thymus.

With one exception, all the mechanisms used by B cells to generate antibody diversity are also used by T cells to generate T cell receptor diversity. It is thought, however, that junctional diversification mechanisms play an especially important part in diversifying T cell receptors. The mechanism that seems not to operate in T cell receptor diversification is antigen-driven somatic hypermutation (see p. 1026). This is presumably because hypermutation would be likely to generate T cells that react against self molecules. This is much less of a problem for B cells, since most self-reactive B cells could not be activated without the aid of self-reactive helper T cells (see p. 1047).

A second type of T cell receptor heterodimer, composed of γ and δ chains, has recently been discovered. These receptors are expressed on subpopulations of cells in the thymus, epidermis, and gut epithelium, whose functions are unknown.

Both α/β and γ/δ T cell receptors are physically associated on the cell surface with an invariant set of polypeptide chains called the *T3 (or CD3) complex*. This complex is present on the surface of all mature T cells and is thought to be involved in passing the signal from an antigen-activated T cell receptor to the cell interior.

Different T Cell Responses Are Mediated by Different Classes of T Cells[33]

T cells kill virus-infected cells, and they help or inhibit the responses of other white blood cells. These three functions are carried out by different classes of T cells—called *cytotoxic T cells, helper (or inducer) T cells,* and *suppressor T cells,* respectively. Cytotoxic T cells, together with B cells, are the main *effector cells* of the immune system; helper T cells and suppressor T cells are collectively referred to as *regulatory T cells.*

Of the three major T cell classes, we know least about suppressor cells. For instance, while both helper and cytotoxic T cells use the same receptors (α/β heterodimers) for recognizing antigen, the nature of the receptors used by suppressor T cells is still uncertain (although at least some seem to use α/β heterodimers). One reason we know so little about suppressor T cells is that, while it has been relatively easy to obtain antigen-specific cytotoxic and helper T cell clones in culture (see p. 1047), it has been exceedingly difficult to obtain suppressor T cell clones.

Although cytotoxic and helper T cells use antigen receptors encoded by the same gene-segment pools, they do not recognize the same MHC molecules on the surface of cells. This difference reflects the different functions of the two types of cells.

Cytotoxic T Cells Kill Virus-infected Cells[34]

Because viruses proliferate inside cells, where they are sheltered from attack by antibodies, the most efficient way to prevent them from spreading to other cells is to kill the infected cell before virus assembly has begun. This is the main function of **cytotoxic T cells.** Given their destructive potential, it is crucial that these lymphocytes confine their attack strictly to infected cells. Microcinematographic analysis has shown that a cytotoxic T cell can focus its attack on one target cell at a time, even when the cytotoxic cell has several target cells bound to it. How does it direct its attack with such accuracy?

The mechanism seems to depend on a cytoskeletal rearrangement in the cytotoxic cell that is triggered by specific contact with the target cell surface. When a cytotoxic T cell, caught in the act of killing its target, is labeled with anti-tubulin antibodies, its centrosome is seen to be oriented toward the point of contact with the target cell (Figure 18–47). Moreover, when labeled with antibodies against *talin,* a protein thought to help link cell-surface receptors to cortical actin filaments (see p. 636), the talin is found concentrated in the cortex of the cytotoxic cell at the contact site. There is evidence that the aggregation of T cell receptors at the contact site leads to a local, talin-dependent accumulation of actin filaments; a microtubule-dependent mechanism then orients the centrosome and associated Golgi apparatus toward the contact site, focusing the killing machinery on the target cell. A similar cytoskeletal polarization is seen when a helper T cell functionally interacts with the cell it helps.

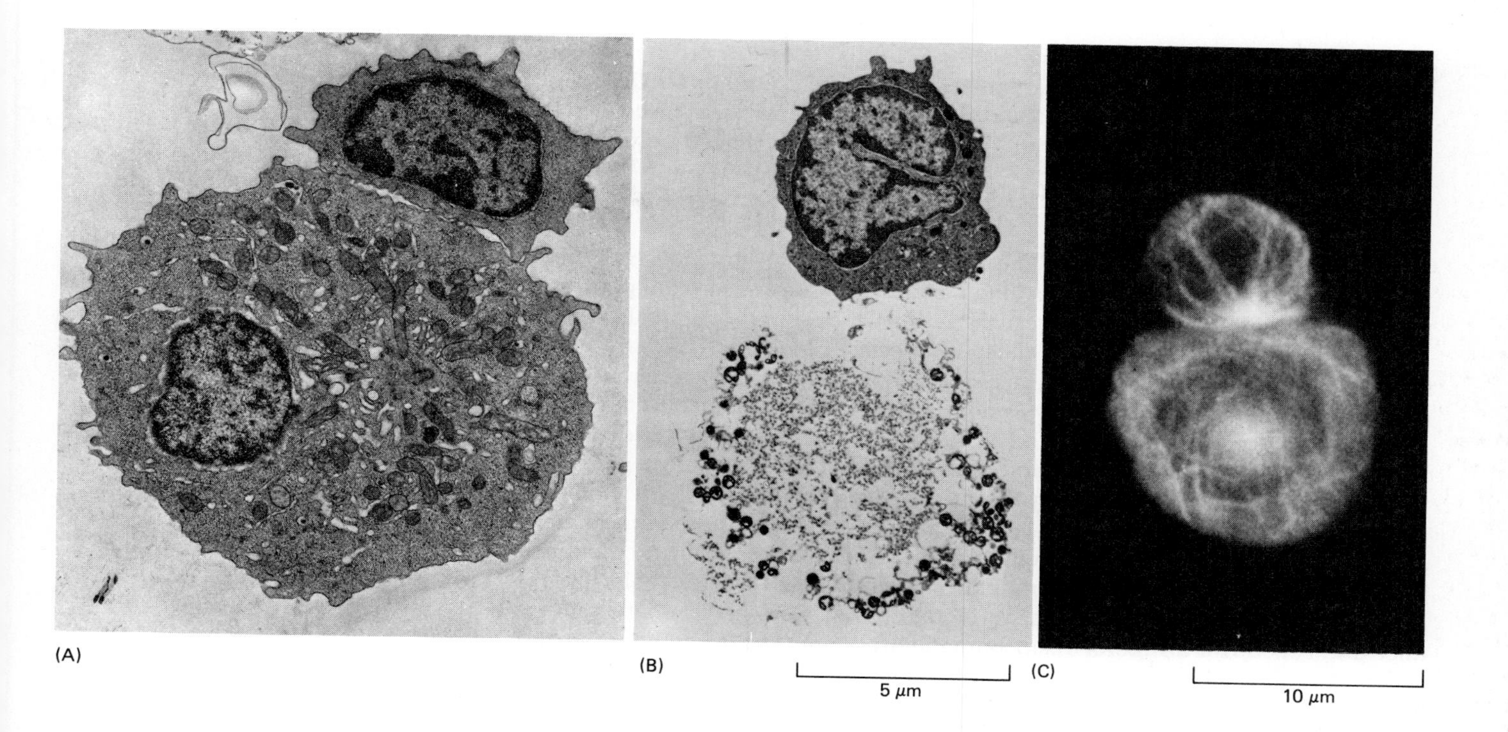

Figure 18–47 Cytotoxic T cells (the small cells) in the process of killing target cells in culture, visualized by electron microscopy in (A) and (B) and by immunofluorescence microscopy after staining with anti-tubulin antibodies in (C). The cytotoxic T cells were obtained from mice immunized with the target cells, which are foreign tumor cells. The T cells are shown binding to the target cell in (A) and (C) and having killed the target cell in (B). Note that the centrosome, and the microtubules radiating from it, are oriented toward the point of cell-cell contact in the T cell but not in the target cell (C). (A and B, from D. Zagury, J. Bernard, N. Thierness, M. Feldman, and G. Berke, *Eur. J. Immunol.* 5:818–822, 1975; C, reproduced from B. Geiger, D. Rosen, and G. Berke, *J. Cell Biol.* 95:137–143, 1982, by copyright permission of the Rockefeller University Press.)

How Do Cytotoxic T Cells Kill Their Targets?[35]

Cytotoxic T cells defend us against virus-induced cancers, just as they do against conventional virus infections. Virus-induced tumors, however, are thought to account for less than 20% of human cancers, and there is little evidence that immune responses protect us against most other forms of cancer: whereas immunosuppressed patients and experimental animals are more susceptible to virus-induced tumors (and tumors suspected of being virus-induced), they are not more susceptible to spontaneous or chemically induced tumors. There is great interest, however, in the possibility that nonimmunological defense mechanisms may be important in defending us against cancer. Two possibilities are the killing of tumor cells by macrophages or by **natural killer (NK) cells.** NK cells are lymphocyte-like cells that are probably the same as *K cells,* which kill antibody-coated eucaryotic cells (see p. 1015). But NK cells can also spontaneously, and relatively nonspecifically, kill a variety of tumor and virus-infected cells in culture in the absence of antibody. How they distinguish abnormal cells from normal ones in these cases is unknown.

It is not known how cytotoxic T cells and NK cells kill their targets. Some NK cell lines and cytotoxic T cell lines, which can be maintained in culture indefinitely, seem to use a mechanism similar to that used by the complement system. Binding to the target stimulates these cytotoxic cells to release pore-forming proteins called **perforins,** which polymerize in the target cell plasma membrane to form transmembrane channels. By causing the membrane to become leaky, the channels are believed to help kill the cell. The perforins, which are homologous to the complement component C9, are stored in secretory vesicles and are released by local exocytosis at the point of contact with the target cell; the secretory vesicles also contain serine esterases, but it is not known if they play a part in target-cell killing. Under the electron microscope the perforin channels in target-cell membranes look very similar to the C9 channels, further supporting the notion that these cytotoxic cell lines and complement kill by related mechanisms. Normal cytotoxic T cells and NK cells, however, can kill target cells by a perforin-independent mechanism, whose molecular basis is not known. One possibility is that these cytotoxic cells activate an internal self-destruct mechanism in the target cell, which consequently commits suicide.

While it is not known what molecules NK cells recognize on the cells they kill, cytotoxic T cells recognize viral molecules that are bound to MHC glycoproteins on the surface of virus-infected cells. But this crucial part played by MHC molecules in "presenting" antigen to T cells has been appreciated only recently.

MHC Molecules Induce Organ Graft Rejection[36]

MHC molecules were recognized long before their normal function was understood. They were initially defined as the main target antigens in **transplantation reactions.** When organ grafts are exchanged between adult individuals of either the same species (*allografts*) or of different species (*xenografts*), they are usually rejected. In the 1950s, experiments involving skin grafting between different strains of mice demonstrated that *graft rejection* is an immune response to the foreign antigens on the surface of the grafted cells. It was later shown that these reactions are mediated mainly by T cells and that they are directed against genetically "foreign" versions of cell-surface glycoproteins called *histocompatibility molecules* (from "histo," meaning tissue). By far the most important of these are the *major histocompatibility molecules,* a family of glycoproteins encoded by a complex of genes called the **major histocompatibility complex (MHC).** MHC molecules are expressed on the cells of all higher vertebrates. They were first demonstrated in mice and called **H-2 antigens** (histocompatibility-2 antigens). In humans they are called **HLA antigens** (human-leucocyte-associated antigens) because they were first demonstrated on leucocytes (white blood cells).

Three remarkable properties of MHC molecules baffled immunologists for a long time. First, MHC molecules are overwhelmingly the preferred target antigens for T-cell-mediated transplantation reactions. Second, an unusually large fraction of T cells is able to recognize foreign MHC molecules: whereas fewer than 0.001% of an individual's T cells respond to a typical viral antigen, more than 0.1% of them respond to a single foreign MHC antigen. Third, many of the loci that code for MHC molecules are the most *polymorphic* known in higher vertebrates; that is, within a species, there is an extraordinarily large number of *alleles* (alternative forms of the same gene) at each locus (often more than 100), each allele being present in a relatively high frequency in the population. For this reason, and because each individual has seven or more loci encoding MHC molecules (see below), it is very rare for two individuals to have an identical set of MHC glycoproteins, making it very difficult to match donor and recipient for organ transplantation in humans (except in the case of genetically identical twins).

But a vertebrate does not need to be protected against invasion by foreign vertebrate cells, so the apparent obsession of its T cells with foreign MHC molecules and the extreme polymorphism of these molecules were not only an obstacle to transplant surgeons, they were also a puzzle to immunologists. The puzzle was solved only after it was discovered that MHC molecules serve to focus T lympho-

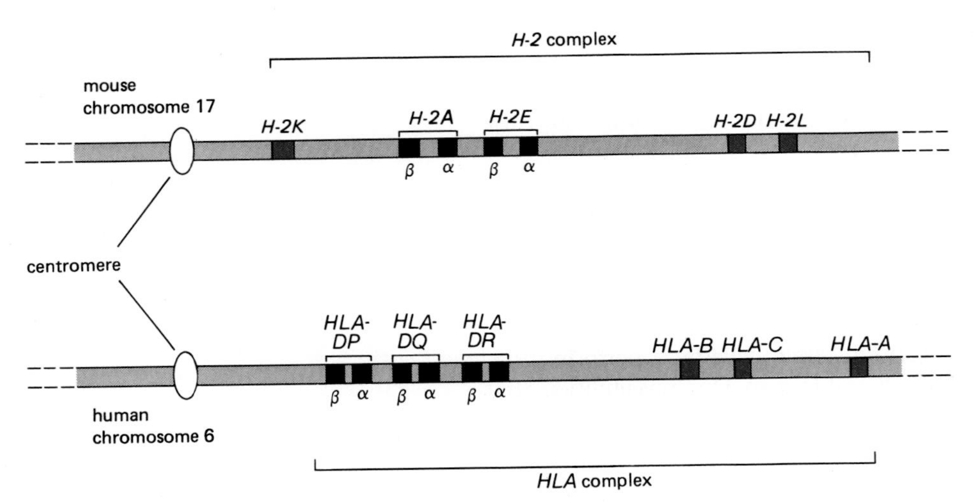

Figure 18–48 Schematic drawing of the *H-2* and *HLA* gene complexes, showing the location of loci that encode class I (*color*) and class II (*black*) MHC glycoproteins. There are three types of class I glycoproteins (H-2K, H-2D, and H-2L in mouse, and HLA-A, HLA-B, and HLA-C in human), each composed of an α chain, encoded by the loci shown, and a β_2-microglobulin chain, encoded on another chromosome. There are two types of class II MHC glycoproteins in the mouse—H-2A and H-2E, each composed of an α and a β chain. Three types of human class II molecules are shown—HLA-DP, HLA-DQ, and HLA-DR (each composed of an α and a β chain), but there are at least one or two others. Human DP and DR molecules are homologous to mouse H-2E, while DQ is homologous to mouse H-2A. All these loci are highly polymorphic except for $H\text{-}2E_\alpha$ and its homologues in humans, DP_α and DR_α, which are much less so. There are many other loci in the complex that encode class-I-MHC-like molecules, but their functions are unknown.

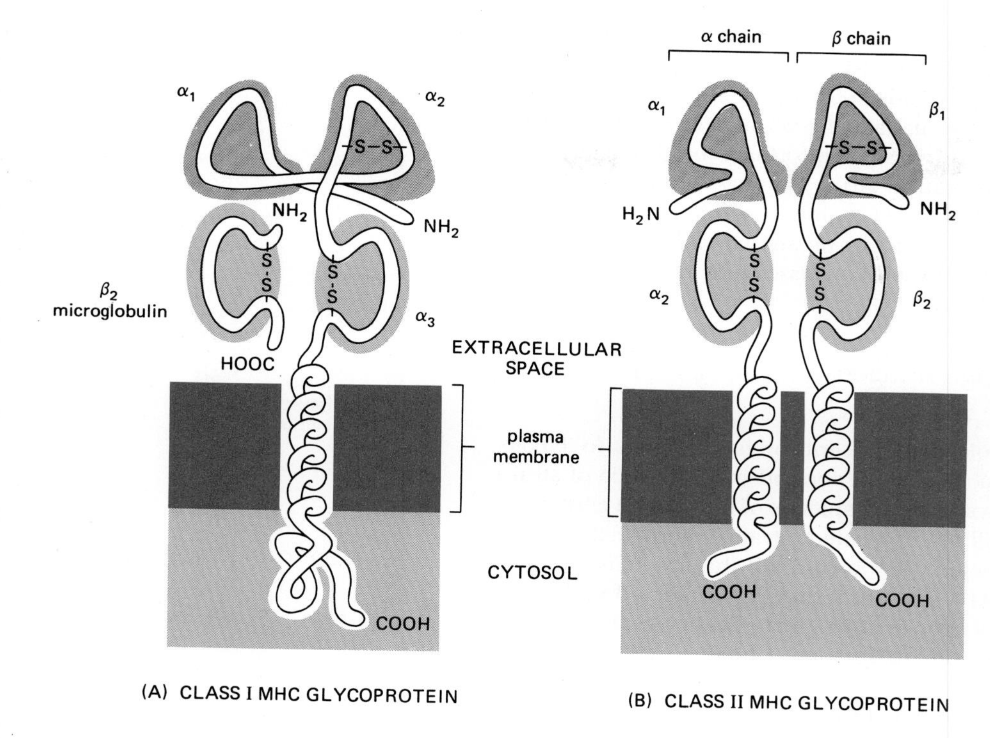

Figure 18–49 (A) Class I and (B) class II MHC glycoprotein molecules. The α chain of the class I molecule, which is about 345 amino acid residues long, has three extracellular domains, α_1, α_2, and α_3, encoded by separate exons. It is noncovalently associated with a smaller polypeptide chain, β_2-microglobulin (96 amino acids), which is not encoded within the MHC. The α_3 domain and β_2-microglobulin are homologous to immunoglobulin domains. While β_2-microglobulin is invariant, the α chain is extremely polymorphic, mainly in the α_1 and α_2 domains. By generating hybrid genes (using genetic engineering techniques) containing a mixture of α_1, α_2, and α_3 exons from different alleles at one locus and transfecting them into cultured fibroblasts, it has been shown that the antigenic determinants recognized by T cells are formed by the interaction of the α_1 and α_2 domains.

In class II MHC molecules, both chains are polymorphic (β more than α), mainly in the α_1 and β_1 domains; the α_2 and β_2 domains are homologous to immunoglobulin domains. Transfection experiments analogous to those described for class I molecules indicate that the antigenic determinants on class II molecules recognized by T cells are formed by the interaction of the α_1 and β_1 domains.

Thus there are striking similarities between class I and class II MHC glycoproteins: both have four extracellular domains, three of which have intrachain disulfide bonds. The two domains closest to the membrane are Ig-like. The other two domains interact to form a complex three-dimensional surface, which, as we shall see later, is thought to bind foreign antigen and present it to T cells. All of the chains are glycosylated except for β_2-microglobulin (not shown).

cytes on those host cells that have foreign antigen on their surface—for example, on virus-infected cells. How this discovery helped resolve much of the MHC puzzle will be discussed later (see p. 1043).

There Are Two Principal Classes of MHC Molecules[37]

There are two main classes of MHC molecules, *class I* and *class II*, each consisting of a set of cell-surface glycoproteins encoded in two linked clusters of genes that together comprise the major histocompatibility complex (Figure 18–48). Both classes of MHC glycoproteins are heterodimers with homologous overall structures; their amino-terminal domains are thought to be specialized for binding antigen for presentation to T cells.

Each **class I MHC gene** encodes a single transmembrane polypeptide chain (called α), most of which is folded into three extracellular globular domains (α_1, α_2, α_3). Each *α chain* is noncovalently associated with an extracellular, nonglycosylated small protein called *β_2-microglobulin,* which does not span the membrane and is separately encoded by a gene on a different chromosome (Figure 18–49A). β_2-microglobulin and the α_3 domain, which are closest to the membrane, are both homologous to an immunoglobulin domain. The two amino-terminal domains of the α chain, which are farthest from the membrane, contain the polymorphic (variable) residues that are recognized by T cells in transplantation reactions.

Like class I MHC molecules, **class II MHC molecules** are heterodimers with two conserved immunoglobulinlike domains close to the membrane and two polymorphic (variable) amino-terminal domains farthest from the membrane. In these molecules, however, both chains are encoded within the MHC, and both span the membrane (Figure 18–49B). The presence of immunoglobulinlike domains in class I and class II glycoproteins suggests that MHC molecules and antibodies have a common evolutionary history (see p. 1053).

There is strong evidence that the polymorphic regions of both classes of MHC molecules interact with foreign antigen and that it is the complex of MHC molecule and foreign antigen that is recognized by the T cell receptor. Before discussing this evidence, however, we shall consider the different parts played by class I and class II molecules in guiding cytotoxic and helper T lymphocytes, respectively, to their appropriate target cells.

The main functional difference between class I and class II MHC molecules is reflected in their tissue distribution. Class I MHC molecules are expressed on virtually all nucleated cells, whereas class II molecules are confined largely to cells involved in immune responses. This presumably is because class I molecules are recognized by cytotoxic T cells, which must be able to focus on any cell in the body that happens to become infected with a virus, whereas class II molecules are recognized by helper T cells, which interact mainly with other cells involved in immune responses, such as B cells and *antigen-presenting cells* (Figure 18–50 and see p. 1045). The principal features of the two classes of MHC glycoproteins are summarized in Table 18–2.

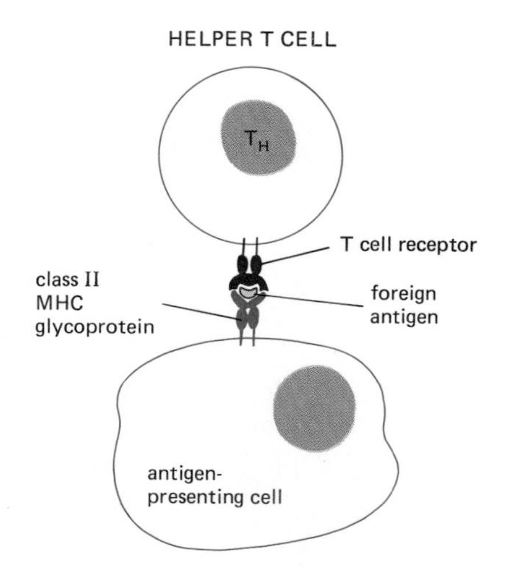

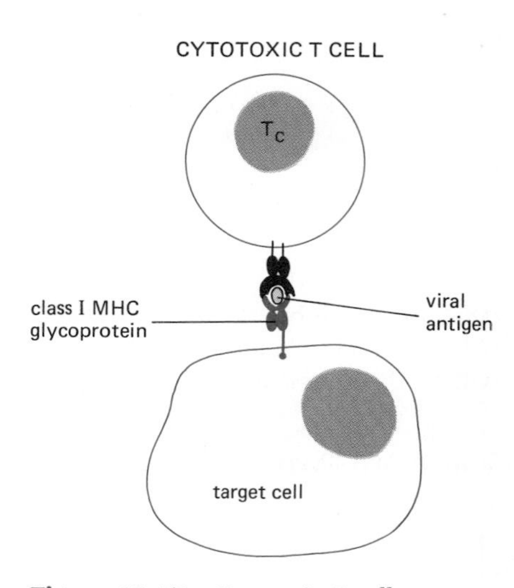

Figure 18–50 Cytotoxic T cells recognize foreign viral antigens in association with class I MHC glycoproteins on the surface of any host cell, whereas helper T cells recognize foreign antigens in association with class II MHC glycoproteins on the surface of an *antigen-presenting cell.* In transplantation reactions, too, helper cells react against foreign class II glycoproteins and cytotoxic cells react against foreign class I glycoproteins.

Cytotoxic T Cells Recognize Foreign Antigens in Association with Class I MHC Molecules[38]

The first clear evidence that MHC molecules present foreign antigens to T cells came from an experiment on cytotoxic T cells performed in 1974. Mice of strain X were infected with virus A. Seven days later, the spleens of these mice contained active cytotoxic T cells that could kill virus-infected, strain-X fibroblasts within several hours in cell culture. As expected, they would kill the fibroblasts only if the fibroblasts were infected with virus A and not if they were infected with virus B; thus the cytotoxic T cells were virus-specific. Unexpectedly, however, the same T cells were unable to kill fibroblasts from strain-Y mice infected with the same virus A (Figure 18–51). The cytotoxic T cells were thus recognizing some difference between the two kinds of fibroblasts and not just the virus. By using special strains of mice (known as *congenic strains*) that were genetically identical except for their class I MHC loci, or genetically different except for their class I MHC loci, it was possible to show that infected target cells would be killed only if they expressed at least one of the same class I MHC molecules expressed by the original infected mouse. This indicated that class I MHC glycoproteins are necessary to present cell-surface-bound viral antigens to cytotoxic T cells. Because the T cells of any individual will recognize antigen only in association with the MHC molecules expressed by that individual, this MHC co-recognition is often known as *MHC restriction.* It was not until 10 years later, however, that a series of experiments on cytotoxic T cells responsive to influenza virus demonstrated the chemical nature of the viral antigens recognized by cytotoxic T cells.

Table 18–2 Properties of Class I and Class II MHC Molecules

	Class I	Class II
Genetic loci	*H-2K, H-2D, H-2L* in mice *HLA-A, HLA-B, HLA-C* in humans	*I-A* and *I-E* clusters in mice; *DP, DQ, DR,* and one or two other clusters in humans
Chain structure	α chain (~45,000 daltons) + β_2-microglobulin (11,500 daltons)	α chain (29,000 to 34,000 daltons) + β chain (25,000 to 28,000 daltons)
Cell distribution	almost all nucleated cells	B cells, antigen-presenting cells, thymus epithelial cells, some others
Involved in presenting antigen mainly to	cytotoxic T cells	helper T cells
Polymorphic domains involved in T cell recognition and antigen binding	$\alpha_1 + \alpha_2$	$\alpha_1 + \beta_1$

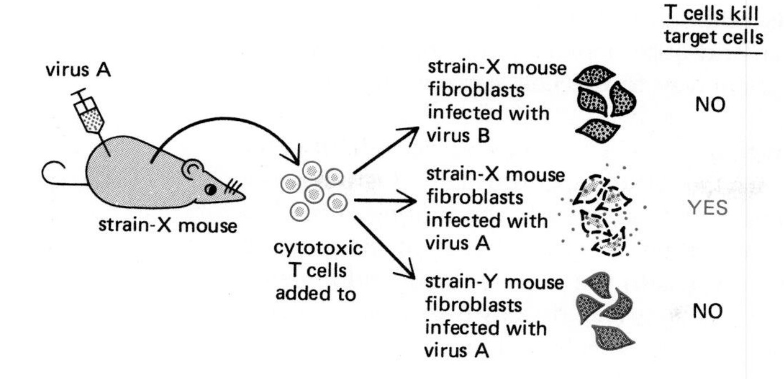

Figure 18–51 The classic experiment showing that a cytotoxic T cell recognizes some aspect of the host target-cell surface in addition to a viral antigen. By repeating this experiment with target cells that differed from the cells of the infected mouse only in limited regions of their genome, the feature of the target-cell surface that the cytotoxic T cell recognizes was shown to be a class I MHC glycoprotein.

The most convenient way of measuring the ability of active cytotoxic T cells to kill uses the radioactive isotope ^{51}Cr, which is taken up by living cells and released only when the cells die. Thus the standard assay for cytotoxic T cells involves incubating them with ^{51}Cr-containing target cells for several hours and measuring the amount of ^{51}Cr radioactivity released from killed target cells.

Cytotoxic T Cells Recognize Fragments of Viral Proteins on the Surface of Virus-infected Cells[39]

It has been known since the 1960s that T cells, unlike B cells (and antibodies), do not usually recognize antigenic determinants on the folded structure of a protein (see Figure 18–23A, p. 1018); instead, they recognize determinants on the unfolded polypeptide chain. The reason for this became clear as evidence accumulated (initially from studies on how helper T cells recognize antigen—see p. 1044, below) that the antigens seen by T cells are normally degraded inside a host cell before they are presented on its surface. The first direct evidence for this mechanism of antigen presentation to cytotoxic T cells came from the observation that some of the cytotoxic T cells activated by influenza virus specifically recognize internal proteins of the virus that would not be accessible in the intact virus particle. Subsequent evidence suggested that the T cells were recognizing degraded fragments of the internal viral proteins. Since viruses are intracellular parasites, whose proteins are synthesized from viral genes inside the infected cell (see p. 250), it is thought that some fragments of viral proteins "leak" onto the surface of an infected cell after they are made, associating with MHC molecules either on the cell surface or somewhere inside the cell (Figure 18–52).

Two types of experiments support this view. First, if normal fibroblasts in culture are briefly exposed to fragments of an internal influenza virus protein (the nucleoprotein, NP—Figure 18–52), the cells are recognized and killed by cytotoxic T cells that originally were activated by influenza-virus-infected fibroblasts, but only if both fibroblasts express the same class I MHC glycoproteins. Second, if a DNA sequence encoding a fragment of the influenza nucleoprotein is introduced

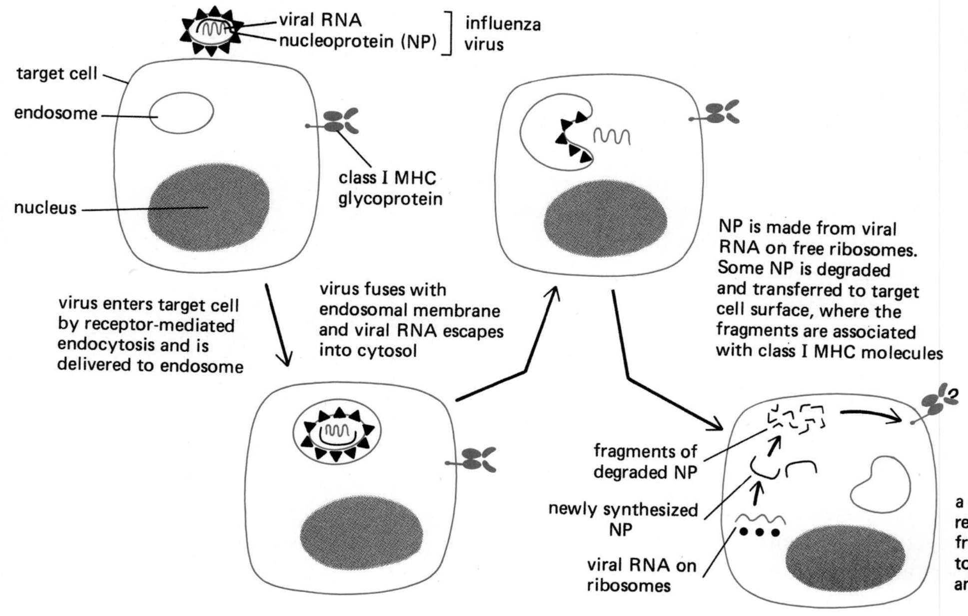

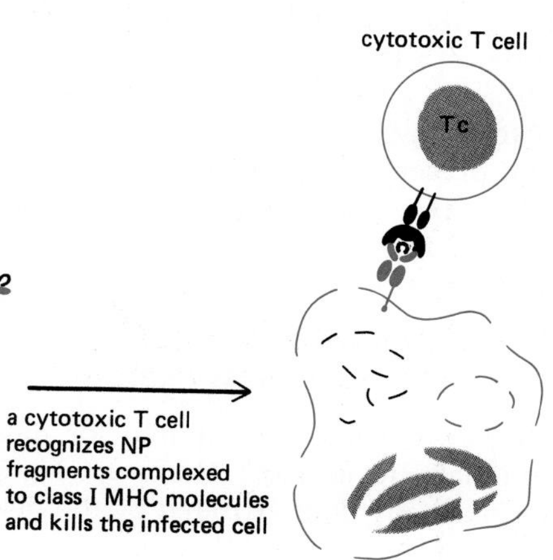

Figure 18–52 A cytotoxic T cell will kill a virus-infected cell when it recognizes fragments of viral protein bound to class I MHC molecules on the surface of the infected cell. In the case shown, the peptide fragments are derived from the nucleoprotein (NP) of the influenza virus; for simplicity, this is the only internal viral protein shown. Only a very small proportion of the viral proteins synthesized in the target cell are degraded. It is not known how they are degraded, how the resulting peptide fragments get to the cell surface, or where the fragments first associate with the MHC glycoproteins.

into normal fibroblasts, the transfected cells are killed by the cytotoxic T cells described in the first experiment. These and other experiments suggest that fragments of viral proteins can both find their way to the cell surface and associate with class I MHC molecules.

It is not difficult to understand how viral proteins can be degraded in infected cells, since almost all cellular proteins are known to be continually degraded (see p. 418). It is more difficult to understand how fragments of the influenza nucleoprotein get to the cell surface, since the protein is synthesized on cytoplasmic ribosomes and would not normally have access to the lumen of the endoplasmic reticulum, where proteins destined for the cell surface usually begin their journey (see p. 412). T cell recognition probably requires very small amounts of antigen, however, so misrouting only a small fraction of the nucleoprotein fragments to the cell surface may create a target cell that an appropriate cytotoxic T lymphocyte can recognize.

X-ray Diffraction Studies Show the Antigen-binding Site of a Class I MHC Glycoprotein[40]

A major advance in our understanding of how MHC molecules present antigen to T cells came in 1987, when the three-dimensional structure of a human class I MHC glycoprotein was obtained by x-ray crystallography. As shown in Figure 18–53A, the protein has a single putative antigen-binding site located at one end of the molecule. The site consists of a deep groove between two long α helices derived from the nearly identical α_1 and α_2 domains; the base of the groove is formed by eight β strands derived from the same two domains. The size of the groove is about 2.5 nm long, 10 nm wide, and 11 nm deep, which is large enough to accommodate a peptide of about 10 to 20 amino acid residues, depending on the extent to which the peptide is compressed by coiling or bending. Remarkably, the groove in the crystallized protein was not empty: it contained a small molecule of unknown origin, suspected to be a peptide, which co-purified and co-crystallized with the MHC glycoprotein (see Figure 18–53B). This finding strongly implicates the groove as the antigen-binding site and suggests that once a peptide binds to this site, it dissociates very slowly; this conclusion is supported by the observation that fibroblasts exposed for a short period to fragments of the influenza virus nucleoprotein remain targets for influenza-specific cytotoxic T cells for at least 3 days.

Most of the polymorphic amino acid residues in the MHC glycoprotein (those that vary between allelic forms of this type of molecule) are located inside the groove, where they would be expected to bind antigen, or on its edges, where they would be accessible for recognition by the T cell receptor. Presumably the variability in class I MHC molecules has been selected to allow them to bind and present many different virus-derived peptides. Nonetheless, it is still surprising that the small number of different antigen-binding sites associated with the class I MHC molecules in an individual (a maximum of six in humans) can bind the large number of virus-derived peptides that T cells can specifically recognize. Even more puzzling in this respect are the class II MHC glycoproteins, which are thought to have a three-dimensional structure very similar to that of class I molecules. Although an individual makes only about 10 to 20 types of class II molecules, each with its own unique antigen-binding site, these molecules seem to be able to bind and present an apparently unlimited variety of foreign peptides to *helper T cells*, which play a crucial part in almost all immune responses.

Helper T Cells Recognize Fragments of Foreign Antigens in Association with Class II MHC Glycoproteins on the Surface of Antigen-presenting Cells[41]

Helper T cells are required for most other types of lymphocytes to respond optimally to antigen. The crucial importance of helper T cells in immunity is dramatically demonstrated by the devastating epidemic of *acquired immunodeficiency syndrome (AIDS)*. The disease is caused by a retrovirus (human immuno-

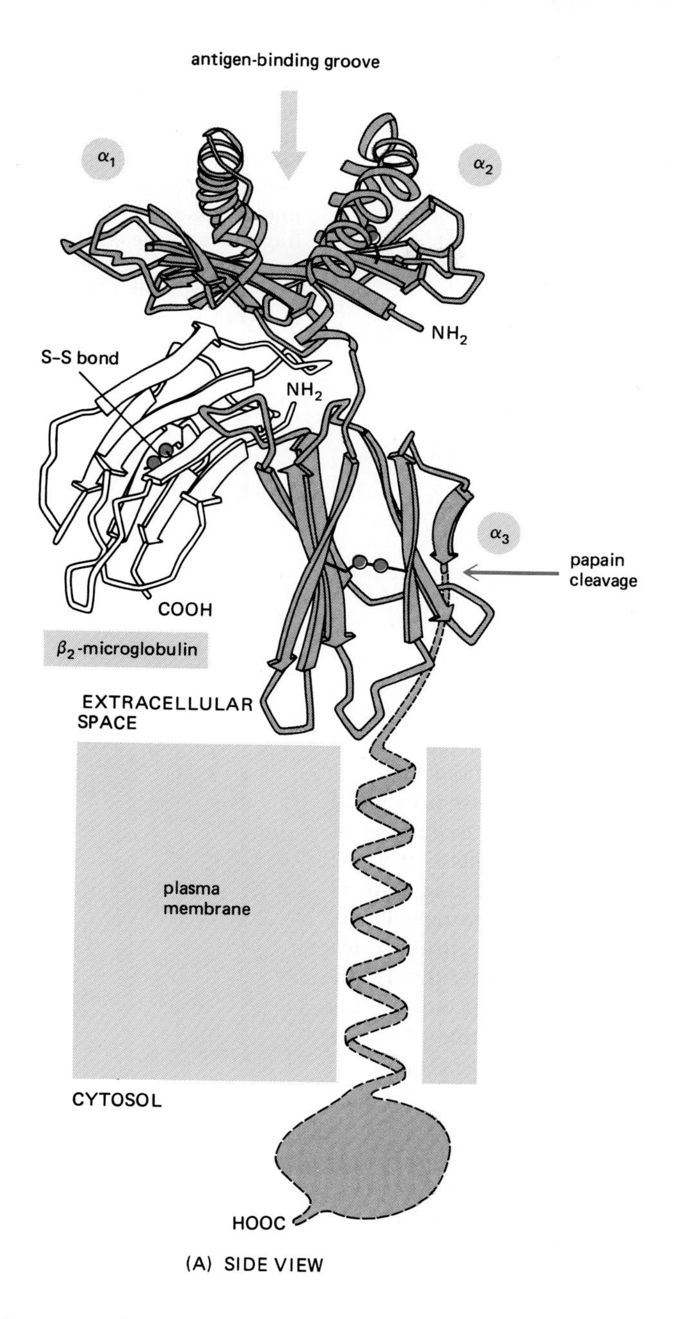

Figure 18–53 (A) The structure of a human class I MHC glycoprotein as determined by x-ray diffraction analysis of crystals of the extracellular part of the molecule. The extracellular part was cleaved from the transmembrane segment by the proteolytic enzyme papain. Each of the two domains closest to the plasma membrane (α_3 and β_2-microglobulin) resembles a typical immunoglobulin domain (see Figure 18–28B), while the two domains farthest from the membrane (α_1 and α_2) are very similar to each other and together form a groove at the top of the molecule that is believed to be the antigen-binding site. Class II MHC molecules are thought to have a very similar structure. (B) The putative antigen-binding groove viewed from above, containing the small molecule (thought to be a peptide) that co-purified with the MHC protein. This is also the part of the molecule that interacts with the T cell receptor. (After P.J. Bjorkman, M.A. Saper, B. Samraoui, W.S. Bennett, J.L. Strominger, and D.C. Wiley, *Nature* 329:506–512, 1987.)

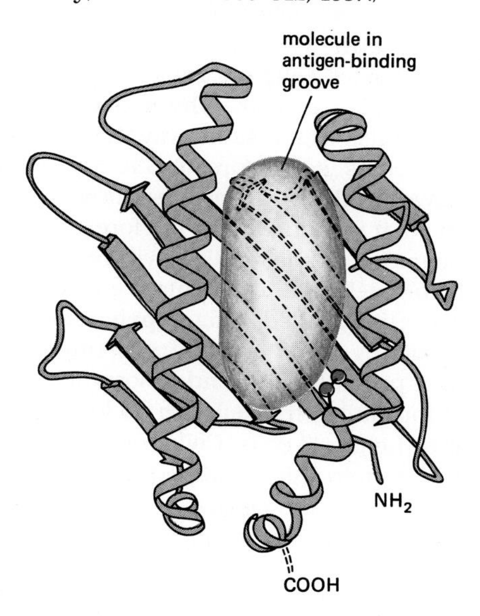

deficiency virus, HIV) that kills helper T cells, thereby crippling the immune system and rendering the patient susceptible to infection by microorganisms that rarely infect normal individuals. As a result, most AIDS patients die of infection within several years of the onset of symptoms.

Before they can help other lymphocytes respond to antigen, helper T cells must first be activated themselves. This activation occurs when a helper T cell recognizes a foreign antigen bound to a class II MHC glycoprotein on the surface of a specialized **antigen-presenting cell.** Antigen-presenting cells are found in most tissues. They are derived from bone marrow and comprise a heterogeneous set of cells, including *dendritic cells* in lymphoid organs, *Langerhans cells* in skin, and certain types of macrophages. Together with B cells, which can also present antigen to helper T cells (see below), and thymus epithelial cells (see p. 1052), these specialized antigen-presenting cells are the main cell types that normally express class II MHC molecules (see Table 18–2).

Many kinds of experiments demonstrate the central importance of class II MHC molecules in presenting foreign antigens to helper T cells. The binding of antibodies to class II molecules on antigen-presenting cells, for example, blocks

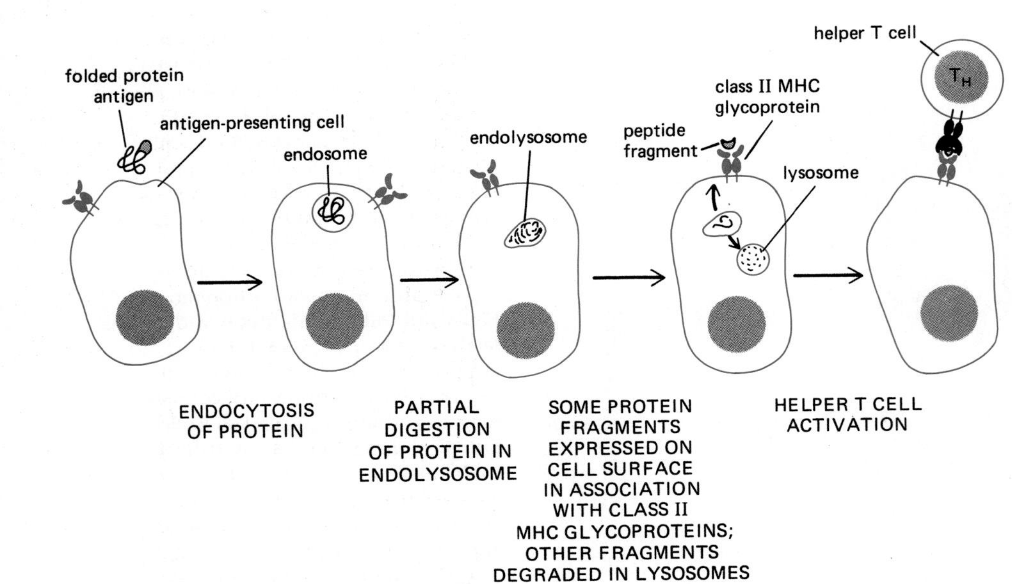

Figure 18–54 How protein antigens are thought to be "processed" and then displayed by an antigen-presenting cell. Since class II MHC glycoproteins have been found to recycle through the endosomal compartment, they may initially associate with peptide fragments in the endolysosomal compartment and then return to the cell surface with a bound peptide (not shown).

the ability of these cells to present foreign antigen to helper T cells. Moreover, fibroblasts, which do not make class II MHC molecules and cannot present foreign antigens to helper T cells, can be converted to effective antigen-presenting cells if they are transfected with a gene that codes for a class II MHC molecule.

Like the viral antigens presented to cytotoxic T cells, the antigens presented to helper T cells on antigen-presenting cells are usually degraded fragments of the foreign protein. These peptides are thought to be bound to class II MHC molecules in the same way that virus-derived peptides are bound to class I MHC molecules (see Figure 18–53). Unlike the virus-infected target of a cytotoxic T cell, however, the antigen-presenting cell does not synthesize the foreign protein. Instead, it is thought that the foreign protein is ingested by endocytosis and partially degraded in the acidic environment of endosomes or endolysosomes (see p. 331) before selected fragments are returned to the cell surface—a sequence of events collectively called **antigen processing** (Figure 18–54). Thus, if endocytosis is blocked by lightly fixing antigen-presenting cells with a chemical such as formaldehyde, or if proteolysis in endolysosomes and lysosomes is inhibited by a drug such as chloroquine, the cells are no longer able to process a foreign protein and present it to helper T cells. Cells treated in these ways, however, are still able to present the protein if it is cleaved into small peptides (10–15 amino acids long) before it is added to the cells.

A remarkable property of an antigen-presenting cell is that it can process and present virtually any antigen to an appropriate helper T cell. This lack of antigen specificity suggests that antigen-presenting cells take up antigen by fluid-phase rather than by receptor-mediated endocytosis (see p. 328). If this is so, then most of the proteins ingested and degraded will be host (self) proteins, whose peptide fragments will occupy the binding site of many of the class II MHC molecules. Presumably the binding of foreign peptides to only a small proportion of MHC molecules is sufficient to activate a helper T cell.

Helper T Cells Stimulate Activated T Lymphocytes to Proliferate by Secreting Interleukin-2[42]

Activation of a helper T cell is a complex process involving various secreted proteins called **interleukins,** which act as local chemical mediators. Activation is thought to begin when the T cell, by unknown means, stimulates the antigen-presenting cell to secrete one or more interleukins. The best characterized of these mediators is **interleukin-1 (IL-1).** The combined action of IL-1 (and probably other interleukins) and antigen binding, however, do not stimulate helper T cell proliferation directly. Instead, they cause the T cell to stimulate its own proliferation by inducing it to secrete a growth factor called **interleukin-2 (IL-2)** as well as to

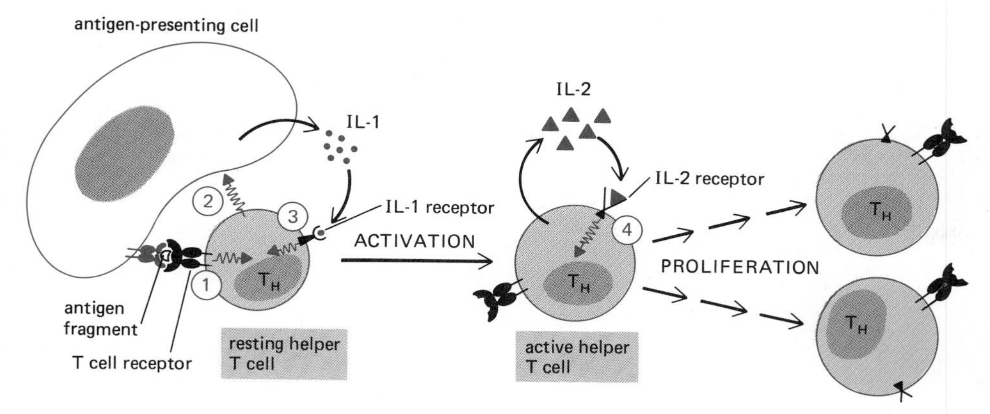

Figure 18–55 The sequence of signaling events believed to occur when antigen stimulates helper T cells to proliferate. Binding of the T cell to the antigen on the surface of an antigen-presenting cell signals the T cell receptor to trigger the inositol phospholipid cell-signaling pathway (see p. 702) (signal 1). This causes the T cell to stimulate the antigen-presenting cell by an unknown mechanism (signal 2). The antigen-presenting cell then secretes interleukins, such as interleukin-1 (IL-1), which help activate the T cell (signal 3). The activated T cell makes interleukin-2 (IL-2) receptors and secretes IL-2; the binding of IL-2 to its receptors (signal 4) stimulates the cell to grow and divide. When the antigen is eliminated, the T cells eventually stop producing IL-2 and IL-2 receptors, so cell proliferation stops.

synthesize cell-surface IL-2 receptors. It is the binding of IL-2 to these receptors that stimulates the T cell to proliferate. In this way the helper T cell can continue to proliferate, through an *autocrine mechanism* (see p. 690), after it has left the surface of the antigen-presenting cell (Figure 18–55). The helper T cell can also help stimulate the proliferation of any other T cells, including cytotoxic T cells, that have first been induced to express IL-2 receptors. Because the expression of IL-2 receptors is strictly dependent on antigen stimulation, however, this does not result in the indiscriminate proliferation of all T cells, but only those that have encountered antigen.

Once the requirements for T cell proliferation were discovered, it was possible to produce indefinitely proliferating, antigen-specific *T cell lines* in culture by continuously administering IL-2 and periodically stimulating the cells with antigen to maintain the expression of IL-2 receptors. Single cells from such lines could then be isolated to generate **T cell clones.** As we have seen, such clones have been critically important in T cell research. They made it possible, for example, to isolate T cell receptors and their genes; they have also been widely used to study the mechanisms of T cell activation and the role of helper T cells in stimulating the responses of other lymphocytes.

Helper T Cells Are Required for Most B Cells to Respond to Antigen[43]

Helper T cells are essential for B cell antibody responses to most antigens. This was first discovered in the mid-1960s through experiments in which either thymus cells or bone marrow cells were injected together with antigen into irradiated mice. Mice that had received only bone marrow or only thymus cells were unable to make antibody; but if a mixture of thymus and bone marrow cells was injected, large amounts of antibody were produced. It was later shown that the thymus provides T cells, while the bone marrow provides B cells (Figure 18–56). The use of a specific chromosome marker to distinguish between the injected T and B cells showed that the antibody-secreting cells are B cells, leading to the conclusion that T cells must help B cells respond to antigen.

There are some antigens, however, including many microbial polysaccharides, that can stimulate B lymphocytes to proliferate and mature without T cell help. Such *T-cell-independent antigens* are usually large polymers with repeating, identical antigenic determinants whose multipoint binding to the membrane-bound antibody molecules that serve as antigen receptors on B cells may generate a strong enough signal to activate B cells directly. There is evidence that the cells that respond to multimeric antigens in this way are mainly a separate subset of B cells that has evolved to react against microbial polysaccharides without T-cell help.

Helper T Cells Help Activate B Cells by Secreting Interleukins[44]

Once activated by foreign antigen on the surface of a specialized antigen-presenting cell, an appropriate helper T cell can help activate a B cell by binding to the same foreign antigen on the B cell surface. The antigen-presenting cell ingests and presents antigens nonspecifically (see p. 1046), but a B cell generally

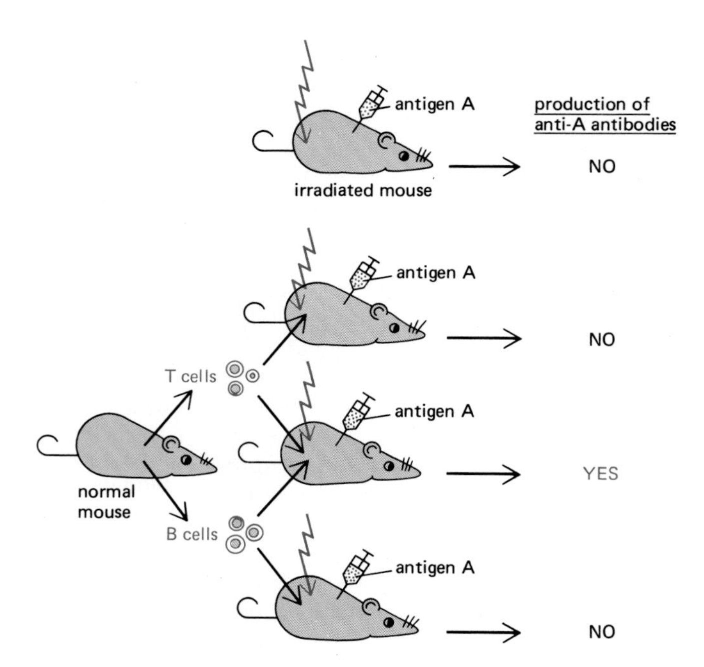

Figure 18–56 The experiment that first suggested that both T cells and B cells are required if an animal is to make antibody responses. The dose of irradiation used kills the T cells and B cells of the irradiated mouse.

presents only an antigen that it specifically recognizes. The antigen is selected by its binding to the specific membrane-bound antibodies (antigen receptors) on the surface of the B cell; it is ingested by receptor-mediated endocytosis (see p. 328) and is then degraded and recycled to the cell surface in the form of peptides bound to class II MHC glycoproteins for recognition by the helper T cell. Thus the helper T cell recognizes the same antigen-MHC complexes on the B cell it helps as on the antigen-presenting cell that initially activated the T cell.

The specific contact between a helper T cell and a B cell initiates an internal rearrangement of the helper cell cytoplasm that orients the centrosome and Golgi apparatus toward the B cell, as described previously for a cytotoxic T cell contacting a target cell (see Figure 18–47). In this case, however, the orientation is thought to enable the helper T cell to direct the secretion of interleukins (and perhaps to focus membrane-bound signaling molecules) onto the B cell surface. These interleukins include *IL-4,* which helps initiate B cell activation, *IL-5,* which stimulates activated B cells to proliferate, and *IL-6,* which induces activated B cells to mature into antibody-secreting cells. Some of these and other interleukins can induce B cells to switch from making one class of antibody to making another (see p. 1029). Some of the signals thought to be involved in the initial activation of a B cell are illustrated in Figure 18–57.

How do signals pass from activated cell-surface receptors to the cell interior when B or T cells are stimulated by antigen and interleukins? The answer is not known for interleukin receptors, but there is strong evidence that receptors for antigen on both B and T cells signal the cell by activating the inositol phospholipid pathway discussed in Chapter 12 (see p. 702).

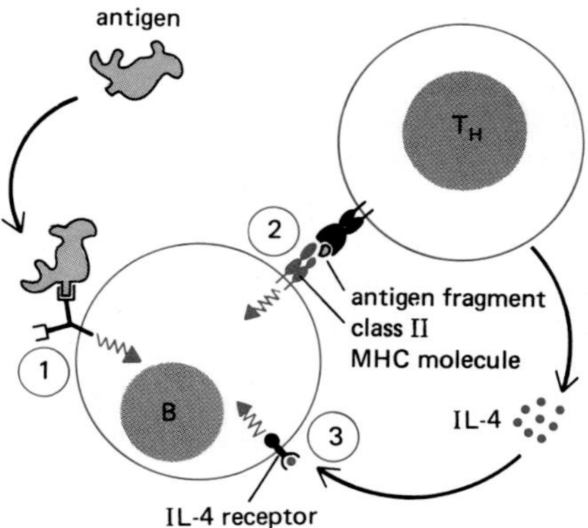

Figure 18–57 At least three types of signals are likely to be involved in the initial stages of B cell activation. The relative importance of these signals is uncertain and may vary depending on the type of B cell and antigen. Signal 1 is caused by antigen binding and is thought to be mediated by the inositol phospholipid cell-signaling pathway (see p. 702); it helps activate the B cell and may induce the expression of receptors for some of the helper-T-cell-derived interleukins. The B cell then ingests and degrades the antigen (not shown) and presents small fragments of the antigen to the helper T cell in association with class II MHC molecules. It is not clear if T cell binding signals the B cell (shown here as signal 2) or only serves to focus the secretion of interleukin-4 (IL-4) and other interleukins (not shown) onto the B cell surface (signal 3). In addition to activating the B cell, signal 3 stimulates the cell to make more class II MHC glycoprotein, thereby increasing the ability of the B cell to receive T cell help. Once the B cell is activated, other helper-T-cell-derived interleukins (such as IL-5, IL-6, and γ-interferon) help induce the cell to proliferate and mature into an antibody-secreting cell (not shown).

Some Helper T Cells Activate Macrophages by Secreting γ-Interferon[45]

Helper T cells do not confine their help to lymphocytes. Those helper T cells that secrete IL-2 when stimulated by antigen also secrete other interleukins, such as *γ-interferon,* that attract macrophages and activate them to become more efficient at phagocytosing and destroying invading microorganisms. The ability of T cells to attract and activate macrophages is especially important in defense against infections by microorganisms that can survive simple phagocytosis by nonactivated macrophages. Tuberculosis is one such infection.

The antigen-triggered secretion of γ-interferon and other macrophage-activating interleukins by helper T cells underlies the familiar tuberculin skin test. If tuberculin (an extract of the bacterium responsible for tuberculosis) is injected into the skin of individuals who have had or have been immunized against tuberculosis, a characteristic immune response occurs in the skin. It is initiated at the site of injection by the secretion of interleukins by memory helper T cells that react to tuberculin. The interleukins attract macrophages and lymphocytes into the site, thereby causing the characteristic swelling of a positive reaction to tuberculin.

Another important effect of γ-interferon is to induce the expression of class II MHC glycoproteins on the surface of some cells (such as endothelial cells) that do not normally express them. This enables these cells to present antigen to helper T cells. In this way helper T cells can recruit extra antigen-presenting cells when the need arises.

There is evidence that there are at least two subclasses of helper T cells. One seems to be concerned mainly with helping B cells and secretes IL-4 and IL-5; the other seems to be concerned mainly with helping other T cells and macrophages and secretes IL-2 and γ-interferon. Some of the interleukins secreted by helper T cells (or antigen-presenting cells) are listed in Table 18–3).

Cell-Cell Adhesion Proteins Stabilize the Interactions Between T Cells and Their Targets[46]

The specific binding of antigen-MHC complexes on the surface of a target cell to α/β antigen receptors on the surface of a T cell is often not strong enough to mediate a functional interaction between the two cells. Various cell-cell adhesion proteins (see Chapter 14, p. 824) on T cells help stabilize such interactions by increasing the overall strength of cell-cell binding. We have already discussed in Chapter 14 (see p. 822) the role of the *lymphocyte-function-associated* protein *LFA-1*

Table 18–3 Properties of Some Interleukins*

Interleukin (IL)	Alternative Name	Approximate Molecular Weight	Source	Target	Action
IL-1	—	15,000	antigen-presenting cells	helper T cells	helps activate
IL-2	T cell growth factor	15,000	some helper T cells	all activated T cells	stimulates proliferation
IL-3	multi-CSF (see p. 980)	25,000	some helper T cells	various hemopoietic cells (see p. 981)	stimulates proliferation
IL-4	B cell stimulating factor-1 (BSF-1)	20,000	some helper T cells	B cells, T cells, mast cells	helps activate and promotes proliferation; increases class II MHC molecules on B cells
IL-5	B cell growth factor-2 (BCGF-2)	50,000 (dimer)	same helper T cells that make IL-4	B cells, eosinophils	promotes proliferation and maturation
IL-6	B cell stimulating factor-2 (BSF-2)	25,000	some helper T cells and macrophages	activated B cells, T cells	promotes B cell maturation to Ig-secreting cells; helps activate T cells
γ-Interferon	—	25,000 (dimer)	same helper T cells that make IL-2	B cells, macrophages, endothelial cells	induces class II MHC molecules and activates macrophages

***Interleukins** are secreted peptides and proteins that mediate local interactions between white blood cells (leucocytes) but do not bind antigen; those secreted by lymphocytes are also called **lymphokines**. The amino acid sequence is known for all the proteins listed. The sources, target cells, and actions listed are those most relevant to the immune system; most of the interleukins have many more sources, targets, and actions than are shown and are therefore more accurately called **cytokines**.

in helping T and B cells (and other white blood cells) to adhere to other cells and to the extracellular matrix. T cells also express a cell-surface protein called *CD2,* which helps them adhere to their target cells by binding to a complementary glycoprotein on the target cell surface called *LFA-3.*

Among the best characterized of the cell-cell adhesion proteins on T cells are the **CD4** and **CD8** glycoproteins, which are expressed on the surface of helper and cytotoxic T cells, respectively. Both glycoproteins have extracellular domains that are homologous to immunoglobulin domains, and they are thought to bind to invariant parts of MHC molecules—CD4 to class II and CD8 to class I MHC glycoproteins (Figure 18–58).

Some of the accessory glycoproteins that are found on the surface of T lymphocytes are summarized in Table 18–4.

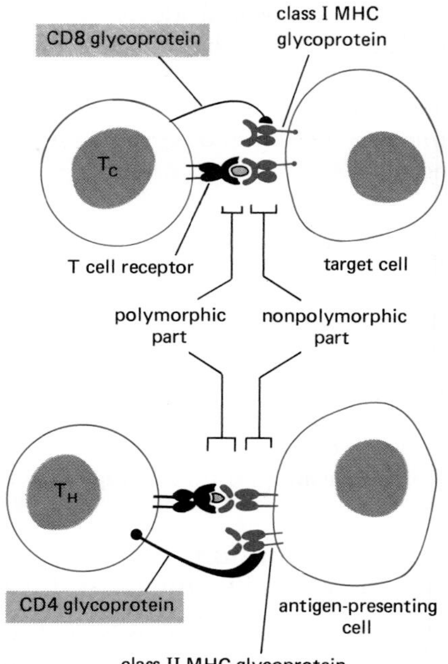

Figure 18–58 The role of two accessory receptor proteins on the surface of T cells. The CD8 glycoprotein on cytotoxic T cells is thought to bind to class I MHC molecules and the CD4 glycoprotein on helper T cells to class II MHC molecules. In both cases the binding is thought to be to nonvariable parts of the MHC molecules. These cell-cell adhesion proteins help stabilize the binding between T cell receptors and antigen-MHC complexes on the target cell, especially when the binding is weak; in these cases antibodies against these accessory receptor proteins inhibit T cell activation.

Antibodies against CD4 and CD8 are widely used to distinguish helper and cytotoxic T cells, respectively. The virus (HIV) that causes AIDS infects helper T cells by initially binding to CD4 molecules on the surface of these cells.

Suppressor T Cells Mainly Suppress Helper T Cells[47]

The discovery that T lymphocytes can *help* B cells make antibody responses was followed several years later by the discovery that they can also *suppress* the response of B cells or other T cells to antigens. Such T cell suppression was first demonstrated in mice that had been made specifically unresponsive (tolerant) to sheep red blood cells (SRBC) by repeated injections of large numbers of SRBCs. When T cells from tolerant mice were injected into normal mice, the latter also became specifically unresponsive to SRBC antigens. This implies that the tolerant state in this case is due to suppression of the response by T cells. Subsequent experiments using surface antigenic markers suggested that the cells responsible are a specialized class of T lymphocytes, called **suppressor T cells.** As we shall discuss below, however, not all forms of immunological tolerance are due to suppressor T cells.

Together, helper T cells and suppressor T cells are thought to control the activity of B cells and cytotoxic T cells, the major effector cells of the immune system. Helper T cells act directly on these effector cells, and suppressor T cells are thought to act indirectly by inhibiting the helper T cells on which the effector cells depend, although the mechanism of inhibition is unknown. How do suppressor T cells recognize the helper T cells they suppress? Because of the way helper cells recognize foreign antigens (see p. 1046), it seems unlikely that they would have sufficient foreign antigen (or fragments of foreign antigen) on their surface for suppressor T cells to recognize. Instead, it seems that suppressor T cells often interact with helper T cells by recognizing antigenic determinants associated with the antigen-binding sites of the helper T cell receptor—so-called *idiotopes* (see p. 1030), as illustrated in Figure 18–59.

The discovery of suppressor T cells raised the question of whether they play a part in self-tolerance by suppressing self-reactive lymphocytes. The available evidence is still controversial, but it suggests that self-tolerance is due mainly to the elimination of self-reactive lymphocytes (so-called *clonal deletion*) and does not depend on suppressor T cells. Since most B cells require helper T cells to respond to antigen, in principle it is necessary only to eliminate self-reactive helper T cells in order to avoid B cell responses to host macromolecules. This is the strategy used for many "self" antigens. Normal mice, for example, will not make antibodies against their own complement component C5. Their B cells, however, can be induced to make such antibodies if they are provided with helper T cells from mutant mice that lack C5 but are otherwise identical. Thus the only reason that normal mice do not make antibodies against this common serum protein is that the helper T cells that recognize C5 are absent or inactivated.

This mechanism, however, will not work for all self antigens. Those self macromolecules that can activate B cells without T cell help, for instance, appear to eliminate the B cells that recognize them, as do those self macromolecules that are present in high concentration. Similarly, cytotoxic T cells that could react against normal host cell-surface molecules must be eliminated, since cytotoxic T cells can be activated by antigen to some extent without helper T cells, although they respond much more vigorously with such help. It is likely that suppressor T cells play primarily a backup role in self-tolerance, being called into play only when the primary mechanism of clonal deletion fails.

Table 18–4 Accessory Glycoproteins on the Surface of T Cells

Protein*	Alternative Name	Approximate Molecular Weight	Expressed on	Putative Function
CD2	T11	50,000	all T cells	promotes adhesion between T cells and their target cells by binding to LFA-3 on target cells
CD3	T3	γ chain = 25,000 δ chain = 20,000 ϵ chain = 20,000 ξ chain = 16,000	all T cells	helps transduce signal when antigen-MHC complex binds to T cell receptors
CD4	T4 in humans L3T4 in mice	50,000	helper T cells	promotes adhesion to antigen-presenting cells and B cells, probably by binding to class II MHC molecules
CD8	T8 in humans Lyt2, Lyt3 in mice	60,000 (homodimer) 70,000 (heterodimer)	cytotoxic T cells	promotes adhesion to virus-infected target cells, probably by binding to class I MHC molecules
LFA-1	—	α chain = 190,000 β chain = 95,000	most white blood cells	promotes cell-cell and cell-matrix adhesion

*CD stands for *cluster of differentiation*, as each of the CD proteins was originally defined as a T cell "differentiation antigen" recognized by multiple monoclonal antibodies. Their identification depended on large-scale collaborative studies in which hundreds of such antibodies, generated in many laboratories, were compared and found to consist of relatively few groups (or "clusters"), each recognizing a single cell-surface protein.

Developing T Cells That React Strongly to Self MHC Molecules Are Eliminated in the Thymus[48]

As discussed previously, the early evidence that MHC glycoproteins are involved in T cell antigen recognition came from experiments showing that T cells can respond to antigen in association with self MHC molecules but not in association with foreign MHC molecules: that is, they showed *MHC restriction* (see p. 1042). Soon after, experiments with thymus transplants suggested that, during their development in the thymus, T cells *learn* to see antigen in association with self rather than foreign MHC molecules. For example, if a Y-strain thymus is transplanted into an X-strain mouse, which has been irradiated to eliminate all its mature T cells and then supplied with fresh bone marrow to provide new ones, new X-strain T cells will develop in the Y-strain thymus. In most cases the mature X-strain T cells produced recognize foreign antigen in association with Y-strain but not X-strain MHC glycoproteins. The simplest interpretation of this finding is that as T cells develop in the thymus, those with receptors that can recognize antigen in association with the types of MHC molecules expressed in the thymus are somehow selected to proliferate. A corollary of this hypothesis is that in the course of this *positive selection,* cytotoxic cells are selected for their recognition of class I MHC molecules while helper cells are selected for their recognition of class II MHC molecules. In support of this view, antibodies against class II MHC molecules specifically block helper T cell development, while antibodies against class I molecules specifically block cytotoxic T cell development.

This interpretation is not entirely satisfying, however, because it fails to explain how the selection is achieved in the absence of the foreign antigens that will later be recognized by the T cells. One possibility is that T cells need to bind weakly to self MHC molecules in order to survive and mature and are thus selected for a low degree of self MHC recognition that is insufficient on its own to activate mature T cells; activation would occur only when the addition of a foreign antigen to a self MHC molecule produces a structure to which the T cell receptor can bind strongly.

The evidence for *positive selection* for weak self-MHC recognition in the thymus is less convincing than the evidence for a *negative selection* process that

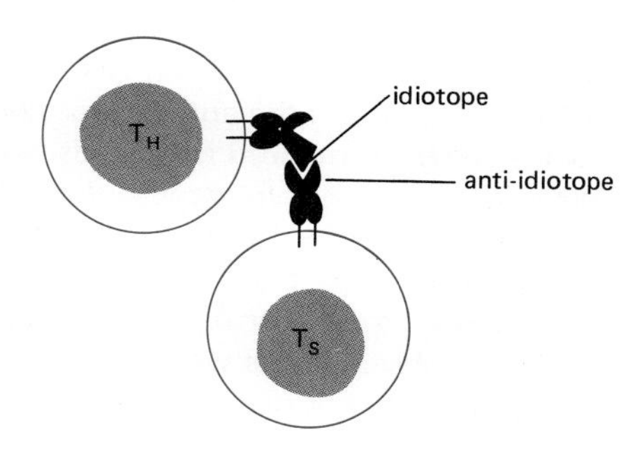

Figure 18–59 An interaction between a helper T cell (T_H) and a suppressor T cell (T_S) in which the receptor on one cell recognizes an idiotope (an antigenic determinant associated with the antigen-binding site) on the receptor of the other cell. An alternative possibility is that the suppressor cell recognizes an idiotope on a fragment of the helper cell receptor that is displayed on the helper cell surface in association with an MHC molecule (not shown). In either case the suppressor cell inhibits the function of the helper cell by an unknown mechanism.

eliminates those cells in the thymus that bind too strongly to self MHC molecules or to self MHC molecules in association with other self molecules. The most compelling evidence that strongly self-MHC-reactive T cells are eliminated in the thymus comes from genetic studies in mice. One such study depends on the fortuitous finding that one of the *V* gene segments that codes for the β chain of the T cell receptor confers on any T cell expressing it strong recognition of a specific class II MHC molecule (called H-2E), regardless of the β chain's associated D and J regions, or of the V region of the α chain. (This finding suggests that T cell receptor *V* gene segments may have been selected in evolution for their ability to encode receptors that bind MHC molecules.) Not all strains of mice, however, express H-2E. In those that do not, T cells expressing the specific β-chain *V* segment are found among both immature and mature thymus lymphocytes. In those that do, such T cells are found only among the immature population of thymus lymphocytes; apparently such T cells are eliminated before they mature in the thymus.

Thymus transplantation experiments suggest that the positive selection proposed to explain MHC restriction and the negative selection conferring self MHC tolerance occur separately. The positive selection seems to occur on the surface of intrinsic thymus epithelial cells, while the negative selection seems to occur on the surface of cells that migrate into the thymus from the bone marrow; both types of cells express both class I and class II MHC molecules on their surface.

The mechanisms responsible for T cell selection in the thymus are unknown. Perhaps positive selection is mediated by growth or survival signals provided by the thymus epithelium to weakly binding T cells. In negative selection the bone-marrow-derived cell may act as a perverse antigen-presenting cell, killing rather than activating any T cell that recognizes it; cells with this property, called *veto cells,* have been demonstrated *in vitro.*

A remarkable feature of T cell development in the thymus is that more than 95% of the cells die without leaving the thymus. This waste is presumably due to the stringent selection that operates on developing T cells.

Some Allelic Forms of MHC Molecules Are Ineffective at Presenting Specific Antigens to T Cells: Immune Response (*Ir*) Genes[49]

Unlike class I MHC genes, which were first recognized by their effects on graft rejection, class II MHC genes were first recognized by their effects on T-cell-dependent immune responses to specific soluble antigens. When animals were immunized with a simple antigen, some made vigorous T-cell-dependent responses while others did not respond at all. Genetic studies indicated that the ability to respond to the antigen was controlled by a single gene, called an **immune response (*Ir*) gene,** and responses to different antigens were often controlled by different *Ir* genes. *Ir* genes that control the response of helper T cells to an antigen were the first to be mapped, and they defined the class II MHC loci; those that controlled the response of cytotoxic T cells to an antigen were later mapped to one or other of the class I MHC loci.

These observations were extremely puzzling until it was recognized that the MHC glycoproteins play a crucial role in presenting antigen to T cells. Now they can be explained by the simple proposal that a genetic nonresponder to a simple antigen (usually one with only a single antigenic determinant) lacks an MHC molecule that can bind and effectively present the antigenic determinant to an appropriate T cell. Strong support for this view has come from *in vitro* studies showing that purified class II MHC molecules from a *responder* animal can bind the relevant antigenic peptide, while those from a genetic *nonresponder* cannot. Further binding experiments have shown that class II molecules have a single antigen-binding site (as do class I MHC molecules—see Figure 18–57) that can bind a wide variety of peptides with an average affinity constant (K_a) of about 10^6 liters/mole (a ΔG of -8.5 kcal/mole, equivalent to the binding energy of about eight hydrogen bonds—see p. 89). Moreover, the binding rate is slow (about 10^5-fold slower than for a typical antibody-antigen reaction), and once bound, the peptide is released with a half-life of more than a day, suggesting that a slow conformational change may have to occur in the MHC molecule for the peptide to be released.

Another mechanism seems to be responsible for some cases of genetic nonresponsiveness to specific antigens. Certain combinations of self MHC molecules and foreign peptides are likely to resemble other self MHC molecules. Because the helper T cells that react to such combinations are eliminated by negative selection during T cell development in the thymus (see p. 1052), an animal may be genetically unable to respond to these foreign peptides.

MHC Co-recognition Provides an Explanation for Transplantation Reactions and MHC Polymorphism

MHC co-recognition provides an explanation for why so many T cells respond to foreign MHC molecules and thereby reject foreign organ grafts. T cells may be obsessed with foreign MHC glycoproteins because these molecules (either alone or complexed with other molecules on the foreign cell surface) resemble various combinations of self MHC molecules complexed with foreign peptides. Thus, for example, some T cell clones that react to a viral antigen in association with a self class I MHC molecule have been shown to react to a foreign class I MHC molecule in the absence of the viral antigen.

MHC co-recognition can also explain the extensive polymorphism of MHC molecules. In the evolutionary war between pathogenic microorganisms and the immune system, microorganisms will tend to change their antigens to avoid associating with MHC molecules. When one succeeds, it will be able to sweep through a population as an epidemic. In such circumstances, the few individuals that produce a new MHC molecule that can associate with an antigen of the altered microorganism will have a large selective advantage. In addition, individuals with two different alleles for each MHC molecule (heterozygotes) will have a better chance of resisting infection than those with identical alleles at any given MHC locus. Thus selection will tend to promote and maintain a large diversity of MHC molecules in the population.

While MHC co-recognition has provided at least tentative answers to many of the questions raised initially by organ transplantation experiments, it has raised another in their place. How do fewer than two dozen different MHC molecules in an animal associate with enough different peptides to ensure that T cells can respond to virtually any protein antigen? The interactions of antigen with antibodies and with class I MHC glycoproteins have been clarified by x-ray diffraction studies of these molecules. Such analyses need to be extended to the interaction between the MHC-antigen complex and the T cell receptor. Recombinant DNA techniques should soon provide abundant amounts of T cell receptors in soluble form, making such projects feasible. Studies using recombinant DNA techniques have already shown that all of these proteins—MHC molecules, T cell receptors, and antibodies—have a common and ancient history.

Immune Recognition Molecules Belong to an Ancient Superfamily[50]

Most of the glycoproteins that mediate cell-cell recognition or antigen recognition in the immune system contain related structural elements, suggesting that the genes that encode them have a common evolutionary history. Included in this **Ig superfamily** are *antibodies, T cell receptors, MHC glycoproteins,* the *CD2, CD4* and *CD8* cell-cell adhesion proteins, some of the polypeptide chains of the *CD3 complex* associated with T cell receptors, and the various *Fc receptors* on lymphocytes and other white blood cells—all of which contain one or more immunoglobulin (Ig)-like domains (*Ig homology units*). Each of these domains is typically about 100 amino acids in length and is thought to be folded into the characteristic sandwichlike structure made of two antiparallel β sheets, usually stabilized by a conserved disulfide bond (see p. 1021). Many of these molecules are dimers or higher oligomers in which Ig homology units of one chain interact with those in another (Figure 18–60).

Each Ig homology unit is usually encoded by a separate exon, and it seems likely that the entire supergene family evolved from a gene coding for a single Ig

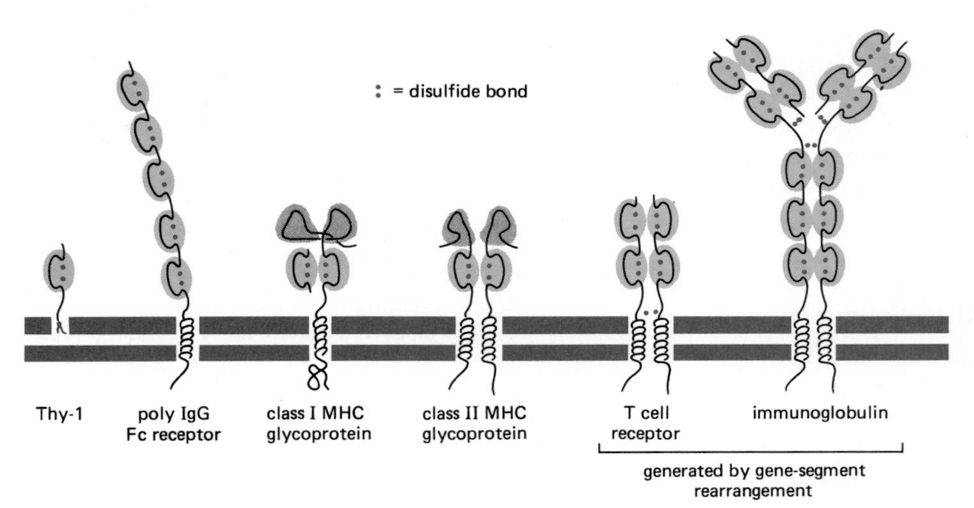

Figure 18–60 Some of the membrane proteins belonging to the *immunoglobulin superfamily.* The homologous immunoglobulin and immunoglobulinlike domains are shown in color; note that disulfide bonds connect the ends of each loop forming such a domain. Most of the domains interact with homologous domains of an associated polypeptide chain. It is also likely that some of the domains on the poly Ig Fc receptor chain interact with one another (not shown); this receptor binds both dimeric IgA (see p. 1016) and polymeric IgM, hence its name. The Thy-1 glycoprotein is linked to the membrane by covalent attachment to a glycosylated phospholipid molecule (see p. 448). Not included in the figure are the CD4 and CD8 accessory proteins on helper and cytotoxic T cells, respectively, the CD3 protein complex associated with T cell receptors, and CD2 (see Table 18–4), all of which also contain immunoglobulinlike domains. The immunoglobulin superfamily also includes cell-surface proteins involved in cell-cell interactions outside the immune system, such as the neural cell adhesion molecule (N-CAM, see p. 830).

homology unit—similar to that encoding Thy-1 or β_2-microglobulin (see Figure 18–54), which may have been involved in mediating cell-cell interactions. Since a Thy-1-like molecule has been isolated from the brain of squids, it is probable that such a primordial gene arose before vertebrates diverged from their invertebrate ancestors some 400 million years ago. New family members presumably arose by exon and gene duplications, and similar duplication events probably gave rise to the multiple gene segments that encode antibodies and T cell receptors.

Its remarkable powers of recognition make the immune system almost unique among cellular systems; only the nervous system is more complex. Both systems are composed of very large numbers of phenotypically distinct cells organized into intricate networks. Within the network, individual cells can interact either positively or negatively, and the response of one cell reverberates through the system by affecting many other cells. While the neural network is relatively fixed in space, the cells constituting the immunological network are constantly changing their locations and interact with one another only transiently. In the next chapter we shall consider the cells of the vertebrate nervous system, which is by far the most complex and sophisticated cellular system known.

Summary

There are at least three functionally distinct subclasses of T cells: (1) cytotoxic T cells, which can kill virus-infected cells directly; (2) helper T cells, which secrete a variety of local chemical mediators (interleukins) that help B cells make antibody responses, stimulate activated T cells to proliferate, and activate macrophages; and (3) suppressor T cells, which are thought mainly to inhibit the responses of helper T cells. Helper and suppressor T cells are the principal regulators of immune responses.

The T cell receptor is an antibodylike heterodimer encoded by genes assembled from multiple gene segments during T cell development in the thymus. T cells are activated when these receptors bind to fragments of foreign antigen that are bound to MHC glycoproteins on the surface of another host cell. This process of MHC co-recognition ensures that T cells recognize a foreign antigen only when it is bound to an appropriate target cell. There are two main classes of MHC molecules: (1) class I molecules are expressed on almost all nucleated somatic cells and present fragments of viral proteins to cytotoxic T cells; (2) class II molecules are expressed on B cells and specialized antigen-presenting cells and present fragments of foreign antigens to helper T cells. The finding that certain allelic forms of class I and class II MHC molecules are ineffective in presenting particular antigenic determinants to T cells probably explains why these molecules are so polymorphic.

References

General

Golub, E.S. Immunology: A Synthesis. Sunderland, MA: Sinauer, 1987.

Hood, L.E.; Weissman, I.L.; Wood, W.B.; Wilson, J.H. Immunology, 2nd ed. Menlo Park, CA: Benjamin-Cummings, 1984.

Male, D.; Champion, B.; Cooke, A. Advanced Immunology. London: Gower, 1987.

Paul, W.E., ed. Fundamental Immunology. New York: Raven, 1984.

Roitt, I.M.; Brostoff, J.; Male, D.K. Immunology. London: Gower, 1985.

Cited

1. Gowans, J.L.; McGregor, D.D. The immunological activities of lymphocytes. *Prog. Allergy* 9:1–78, 1965.
2. Greaves, M.F.; Owen, J.J.T.; Raff, M.C. T and B Lymphocytes: Origins, Properties and Roles in Immune Responses. Amsterdam: Excerpta Medica, 1973.
3. Cooper, M.; Lawton, A. The development of the immune system. *Sci. Am.* 231(5):59–72, 1974.

   Owen, J.J.T. Ontogenesis of lymphocytes. In B and T Cells in Immune Recognition (F. Loor, G.E. Roelants, eds.), pp. 21–34. New York: Wiley, 1977.
4. Möller, G., ed. Functional T Cell Subsets Defined by Monoclonal Antibodies. *Immunol. Rev.*, Vol. 74, 1983.

   Raff, M.C. Cell-surface immunology. *Sci. Am.* 234(5):30–39, 1976.

   Reinherz, E.L.; Schlossman, S.F. The differentiation and function of human T lymphocytes. *Cell* 19:821–827, 1980.
5. Ada, G.L. Antigen binding cells in tolerance and immunity. *Transplant. Rev.* 5:105–129, 1970.

   Ada, G.L.; Nossal, G. The clonal selection theory. *Sci. Am.* 257(2):62–69, 1987.

   Burnet, F.M. The Clonal Selection Theory of Acquired Immunity. Nashville, TN: Vanderbilt University Press, 1959.

   Wigzell, H. Specific fractionation of immunocompetent cells. *Transplant. Rev.* 5:76–104, 1970.
6. Pink, J.R.L.; Askonas, B.A. Diversity of antibodies to crossreacting nitrophenyl haptens in inbred mice. *Eur. J. Immunol.* 4:426–429, 1974.
7. Butcher, E.C.; Weissman, I.L. Lymphoid Tissues and Organs. In Fundamental Immunology (W.E. Paul, ed.), pp. 109–127. New York: Raven, 1984.

   Gallatin, M.; et al. Lymphocyte homing receptors. *Cell* 44:673–680, 1986.

   Gowans, J.L.; Knight E.J. The route of re-circulation of lymphocytes in the rat. *Proc. R. Soc. Lond. (Biol.)* 159:257–282, 1964.

   Sprent, J. Migration and lifespan of lymphocytes. In B and T Cells in Immune Recognition (F. Loor, G.E. Roelants, eds.), pp. 59–82. New York: Wiley, 1977.

   Woodruff, J.J.; Clarke, L.M.; Chin, Y.H. Specific cell-adhesion mechanisms determining migration pathways of recirculating lymphocytes. *Annu. Rev. Immunol.* 5:201–222, 1987.
8. Greaves, M.F.; Owen, J.J.T.; Raff, M.C. T and B Lymphocytes: Origins, Properties and Roles in Immune Responses, pp. 117–186. Amsterdam: Excerpta Medica, 1973.

   Rajewsky, K.; Forster, I.; Cumano, A. Evolutionary and somatic selection of the antibody repertoire in the mouse. *Science* 238:1088–1094, 1987.
9. Billingham, R.E.; Brent, L.; Medawar, P.B. Quantitative studies on tissue transplantation immunity. III. Activity acquired tolerance. *Philos. Trans. R. Soc. Lond. (Biol.)* 239:357–414, 1956.

   Harris, D.E.; Cairns, L.; Rosen, F.S.; Borel, Y. A natural model of immunologic tolerance. Tolerance to murine C5 is mediated by T cells and antigen is required to maintain unresponsiveness. *J. Exp. Med.* 156:567–584, 1982.

   Lindstrom, J. Immunobiology of myasthenia gravis, experimental autoimmune myasthenia gravis and Lambert-Eaton syndrome. *Annu. Rev. Immunol.* 3:109–132, 1985.

   Nossal, G.J.V. Cellular mechanisms of immunologic tolerance. *Annu. Rev. Immunol.* 1:33–62, 1983.

   Owen, R.D. Immunogenetic consequence of vascular anastomoses between bovine twins. *Science* 102:400–401, 1945.
10. Howard, J.G.; Mitchison, N.A. Immunological tolerance. *Prog. Allergy* 18:43–96, 1975.
11. Davies, D.R.; Metzger, H. Structural basis of antibody function. *Annu. Rev. Immunol.* 1:87–118, 1983.

    Kabat, E.A. Structural Concepts in Immunology and Immunochemistry, 2nd ed. New York: Holt, Rinehart & Winston, 1976.

    Nisonoff, A.; Hopper, J.E.; Spring, S.B. The Antibody Molecule. New York: Academic Press, 1975.
12. Möller, G., ed. Lymphocyte Immunoglobulin: Synthesis and Surface Representation. *Transplant. Rev.*, Vol. 14, 1973.
13. Dutton, R.W.; Mishell, R.I. Cellular events in the immune response. The *in vitro* response of normal spleen cells to erythrocyte antigens. *Cold Spring Harbor Symp. Quant. Biol.* 32:407–414, 1967.

    Jerne, N.K.; et al. Plaque forming cells: methodology and theory. *Transplant. Rev.* 18:130–191, 1974.
14. Edelman, G.M. The structure and function of antibodies. *Sci. Am.* 223(2):34–42, 1970.

    Porter, R.R. Structural studies of immunoglobulins. *Science* 180:713–716, 1973.
15. Ishizaka, T.; Ishizaka, K. Biology of immunoglobin E. *Prog. Allergy* 19:60–121, 1975.

    Koshland, M.E. The coming of age of the immunoglobulin J chain. *Annu. Rev. Immunol.* 3:425–454, 1985.

    Morgan, E.L.; Weigle, W.O. Biological activities residing in the Fc region of immunoglobulin. *Adv. Immunol.* 40:61–134, 1987.

    Solari, R.; Kraehenbuhl, J.-P. The biosynthesis of secretory component and its role in the transepithelial transport of IgA dimer. *Immunol. Today* 6:17–20, 1985.

    Underdown, B.J.; Schiff, J.M. Immunoglobulin A: strategic defense initiative at the mucosal surface. *Annu. Rev. Immunol.* 4:389–418, 1986.

    Unkeless, J.C.; Scigliano, E.; Freedman, V. Structure and function of human and murine receptors for IgG. *Annu. Rev. Immunol.* 6:251–282, 1988.
16. Berzofsky, J.A.; Berkover, I.J. Antigen-antibody interactions. In Fundamental Immunology (W.E. Paul, ed.), pp. 595–644. New York: Raven, 1984.
17. Capra, J.D.; Edmundson, A.B. The antibody combining site. *Sci. Am.* 236(1):50–59, 1977.

    Wu, T.T.; Kabat, E.A. An analysis of the sequences of the variable regions of Bence Jones proteins and myeloma light chains and their implications for antibody complementarity. *J. Exp. Med.* 132:211–250, 1970.
18. Sakano, H.; et al. Domains and the hinge region of an immunoglobulin heavy chain are encoded in separate DNA segments. *Nature* 277:627–633, 1979.
19. Alzari, P.M.; Lascombe, M.-B.; Poljak, R.J. Three-dimensional structure of antibodies. *Annu. Rev. Immunol.* 6:555–580, 1988.

    Capra, J.D.; Edmundson, A.B. The antibody combining site. *Sci. Am.* 236(1):50–59, 1977.
20. Dreyer, W.J.; Bennett, J.C. The molecular basis of antibody formation: a paradox. *Proc. Natl. Acad. Sci. USA* 54:864–869, 1965.

Scanning electron micrograph of a sensory hair cell from the inner ear of a frog. (Courtesy of Richard Jacobs; from *Nature* 281(5733), 1979.)

The Nervous System

19

How can we hope to understand the workings of the human brain? This network of some 10^{11} nerve cells, with at least a thousand times that number of interconnections, is more complex, and seems in many ways more powerful, than even the largest of modern computers. Our present understanding of the nervous system is so rudimentary, however, that one can scarcely judge whether the comparison makes sense. We do not know, for example, how many functionally distinct categories of nerve cells the brain contains; nor can we give even an outline of the neural computations involved in hearing a word or reaching for an object, let alone proving a theorem or writing a poem.

And yet, paradoxically, while the brain as a whole remains the most baffling organ in the body, the properties of the individual nerve cells, or *neurons,* are understood better than those of almost any other cell type. At the cellular level at least, simple and general principles can be discerned. With their help, one can begin to see how small parts of the nervous system work. Important progress has been made, for example, in explaining the cellular machinery of simple reflex behavior, and even of visual perception. From a practical point of view, knowledge of the molecular biology of neurons provides a key to the biochemical control of brain function through drugs, and it holds out the promise of more effective treatment for many forms of mental illness.

In this chapter we shall focus on the nerve cell and try to illustrate how its properties give insight into neural organization at higher levels.

The Cells of the Nervous System: An Overview of Their Structure and Function[1]

The nervous system provides for rapid communication between widely separated parts of the body. Through its role as a communications network, it governs reactions to stimuli, processes information, and generates elaborate patterns of signals to control complex behaviors. The nervous system is also capable of learning: as it processes and records sensory information about the external world, it undergoes adjustments that result in altered future patterns of action.

The major neural pathways of communication were mapped out more than a hundred years ago, before the role of individual nerve cells was understood. Figure 19–1 shows the basic plan. Like a big computing facility, the vertebrate

Nerve Cells Convey Electrical Signals[3]

The significance of the signals carried by a neuron depends on the part played by the individual cell in the functioning of the nervous system as a whole. In a *motor neuron* the signals represent commands for the contraction of a particular muscle. In a *sensory neuron* they represent the information that a specific type of stimulus, such as a light, a mechanical force, or a chemical substance, is present at a certain site in the body. In an *interneuron,* forming a connection between one neuron and another, the signals represent parts of elaborate computations that combine information from many different sources and regulate complex behavior.

Despite the varied significance of the signals, their *form* is the same, consisting of changes in the electrical potential across the neuron's plasma membrane. Communication occurs because an electrical disturbance produced in one part of the cell spreads to other parts. Such a disturbance becomes weaker with increasing distance from its source unless energy is expended to amplify it as it travels. Over short distances this attenuation is unimportant, and in fact many small neurons conduct their signals passively, without amplification. For long-distance communication, however, such passive spread is inadequate. Thus the larger neurons employ an active signaling mechanism, which is one of their most striking features: an electrical stimulus that exceeds a certain threshold strength triggers an explosion of electrical activity that is propagated rapidly along the neuron's plasma membrane and is sustained by automatic amplification all along the way. This traveling wave of electrical excitation, known as an *action potential* or *nerve impulse,* can carry a message without attenuation from one end of a neuron to the other at speeds as great as 100 m/sec or more.

Nerve Cells Communicate Chemically at Synapses[4]

Neuronal signals are transmitted from cell to cell at specialized sites of contact known as **synapses**. The usual mechanism of transmission appears surprisingly indirect. The cells are electrically isolated from one another, the *presynaptic cell* being separated from the *postsynaptic cell* by a *synaptic cleft.* A change of electrical potential in the presynaptic cell triggers it to release a chemical known as a *neurotransmitter,* which is stored in membrane-bounded *synaptic vesicles* and released by exocytosis. The neurotransmitter then diffuses across the synaptic cleft and provokes an electrical change in the postsynaptic cell (Figure 19–4). As we shall see, transmission via such *chemical synapses* is far more versatile and adaptable than direct electrical coupling via gap junctions (see p. 799), which is also used, but to a much lesser extent.

The chemical synapse is a site of intense biochemical activity, involving continual degradation, turnover, and secretion of proteins and other molecules. The biosynthetic center of the neuron, however, is in the cell body, where the ultimate instructions for protein synthesis lie. The neuron must therefore have an efficient intracellular transport system to convey molecules from the cell body to the outermost reaches of the axon and dendrites. How is this transport system organized, and what molecules are actually transported?

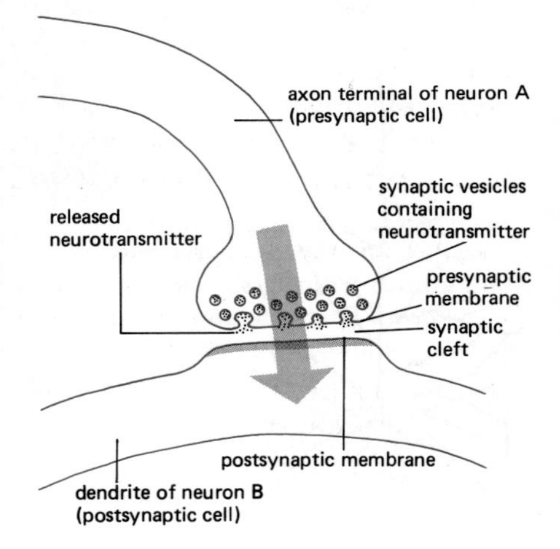

Figure 19–4 Schematic diagram of a typical synapse. An electrical signal arriving at the axon terminal of neuron A triggers the release of a chemical messenger (the neurotransmitter), which crosses the synaptic cleft and causes an electrical change in the membrane of a dendrite of neuron B. A broad arrow indicates the direction of signal transmission.

Slow and Fast Transport Mechanisms Carry Newly Synthesized Materials from the Nerve Cell Body into the Axon and Dendrites[5]

Electron microscopy reveals that the cell body of a typical large neuron contains vast numbers of ribosomes, some crowded together in the cytosol, some attached to rough endoplasmic reticulum (ER) (Figure 19–5A). Although dendrites often contain some ribosomes, there are no ribosomes in the axon, and its proteins must therefore be provided by the many ribosomes in the cell body (Figure 19–5B). The needs of the axon are considerable: a large motor neuron in a human being, for example, may have an axon 15 μm in diameter and a meter long, corresponding to a volume of about 0.2 mm^3, which is about 10,000 times the volume of a liver

Figure 19–5 The organization of the cytoplasm in a typical large nerve cell (a motor neuron in the spinal cord). (A) A sketch of the cell body at low magnification, showing how regions of cytoplasm rich in ribosomes are packed in the gaps between bundles of neurofilaments and other cytoskeletal proteins. (B) An electron micrograph of one such ribosome-rich region; some of the ribosomes are free, others are attached to rough ER. (C) An electron micrograph showing part of a cross-section of the axon; large numbers of neurofilaments and microtubules can be seen, but no ribosomes are present. The membranous vesicles in the axon are probably traveling along the adjacent microtubules by fast axonal transport. (B, courtesy of Jennifer La Vail; C, courtesy of John Hopkins.)

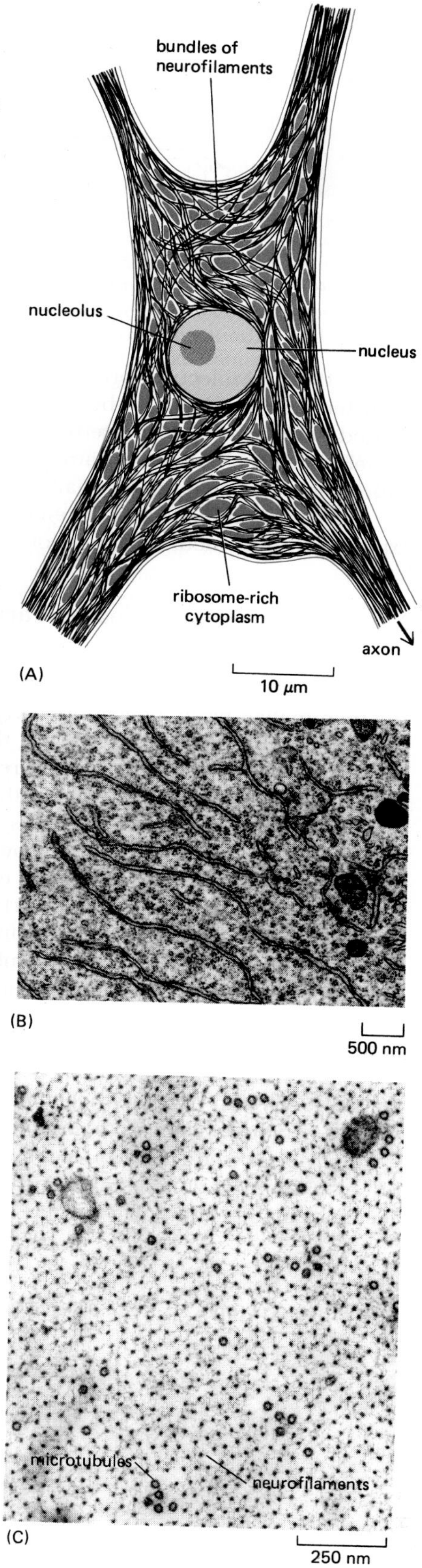

cell. Because such a neuron contains only a single nucleus, its ratio of cytoplasm to DNA is far greater than that of any nonneuronal cell type in the human body.

The most plentiful proteins in the axon are those that form microtubules, neurofilaments (a class of intermediate filaments), and actin filaments (Figure 19–5C). These cytoskeletal proteins are exported from the cell body and move along the axon at speeds of 1 to 5 mm per day by the process of **slow axonal transport.** (A similar transport occurs in the dendrites, which contain a slightly different set of microtubule-associated proteins—see p. 659.) Other cytosolic proteins, including many enzymes, are also carried by slow axonal transport, whose mechanism is not understood.

Noncytosolic materials required at the synapse, such as secreted proteins and membrane-bound molecules, move outward from the cell body by a much faster mode of transport. These proteins and lipids pass from their sites of synthesis in the endoplasmic reticulum to the Golgi apparatus, which lies close to the nucleus, often facing the base of the axon. From here, packaged in membrane vesicles, they are carried by **fast axonal transport,** at speeds of up to 400 mm per day, along tracks formed by microtubules in the axon or the dendrites (see p. 660); mitochondria are conveyed by the same means. Since different populations of proteins are sent out in this way along axons and dendrites, the transported molecules are presumed to be sorted in the cell body into separate and distinctive types of transport vesicles (see p. 468).

Among the proteins rapidly transported along the axon are those to be secreted at the synapse, such as the *neuropeptides* that many neurons release as neurotransmitters, often in conjunction with nonprotein transmitters. From the point of view of their internal organization, neurons can thus be thought of as secretory cells in which the site of secretion has been removed to an enormous distance from the site where proteins and membranes originate (Figure 19–6).

Retrograde Transport Allows the Nerve Terminal to Communicate Chemically with the Cell Body[5,6]

Fast axonal transport is required during development for the growth of axons and dendrites, which elongate by adding new membrane to their tips. Fast axonal transport also occurs in a full-grown neuron, in which there is no net accumulation of membrane at the ends of the axon and dendrites. In this case the fast transport of membrane outward from the cell body, called *fast anterograde transport,* must be exactly balanced by *fast retrograde transport* of membrane back from the ends of the cell processes. The mechanisms of fast transport in the two directions are similar but not identical. The fast retrograde transport has a speed about half that of fast anterograde transport, is driven by a different motor protein (see p. 660), and carries somewhat larger vesicles on average. The structures returning to the cell body consist partly of aging cytoplasmic organelles, such as mitochondria, and partly of vesicles formed by the extensive endocytosis required for membrane retrieval at the axon terminal after neurotransmitter release (see

Figure 19–6 A neuron viewed

brane proteins (see p. 312). When the ion channels open or close, the charge distribution shifts and the membrane potential changes. Neuronal signaling thus depends on channels whose permeability is regulated—so-called **gated ion channels.**

Two classes of gated channels are of crucial importance: (1) *voltage-gated channels*—especially voltage-gated *Na^+ channels*—play the key role in the explosions of electrical activity by which action potentials are propagated along an axon; and (2) *ligand-gated channels,* which convert extracellular chemical signals into electrical signals, play a central part in the operation of synapses. The account in Chapter 6 (pp. 312–319) of ion channels and of their role in electrical signaling forms the basis for the further discussion of neuronal signaling to be given here. Some principles of electrochemistry that are of special relevance to nerve cells are reviewed in Panel 19–1.

Voltage Changes Can Spread Passively Within a Neuron[3,4,8,9]

Action potentials are typically triggered at one end of an axon and propagate along its length. To understand the mechanism it is helpful to consider first how electrical disturbances spread along a nerve cell in the absence of action potentials. As mentioned earlier, such *passive spread* is common, especially in the many neurons that have very short axons or no axon at all; these cells often have few or no voltage-gated Na^+ channels and rely for their signaling entirely on passive spread, manifest as smoothly graded *local potentials.*

In an axon at rest, the membrane potential is uniformly negative, with the interior of the axon everywhere at the same negative potential relative to the external medium. As explained in Chapter 6 (see p. 314), the potential difference depends on the large concentration gradients of Na^+ and K^+, built up by the Na^+-K^+ pump. *K^+ leak channels* make the resting membrane permeable chiefly to K^+, so that the resting potential is close to the K^+ equilibrium potential—typically about -70 mV (see Panel 19–1). An electrical signal may take the form of a *depolarization,* in which the voltage drop across the membrane is reduced, or a *hyperpolarization,* in which it is increased. To illustrate the passive spread of an electrical signal, let us consider what happens when an axon is locally depolarized by injecting current through a microelectrode inserted into it. If the current is small, the depolarization will be *subthreshold:* practically no Na^+ channels open, and no action potentials are triggered. A steady state is quickly reached in which the inflow of current through the microelectrode is exactly balanced by the outflow of current (carried mainly by K^+ ions) across the axonal membrane. Some of this current flows out in the neighborhood of the microelectrode, while some travels down the interior of the axon for some distance in either direction before escaping. The consequence is that the membrane potential is disturbed by an amount that decreases exponentially with the distance from the source of the disturbance (Figure 19–9). This passive spread of an electrical signal along a nerve cell process is analogous to the spread of a signal along an undersea telegraph cable; as the current flows down the central conductor (the cytoplasm), some leaks out through the sheath of insulation (the membrane) into the external medium, so that the signal becomes progressively attenuated. For this reason the electrical characteristics involved in passive spread are often referred to as *cable properties* of the axon.

Axons, though, are much poorer conductors than electric cables, and passive spread is inadequate to transmit a signal over a distance of more than a few millimeters, especially if the signal is brief and transient. This is not only because of current leakage but also because the change in membrane potential that results from current flow is not instantaneous but takes a while to build up. The time required depends on the membrane *capacitance,* that is, on the quantity of charge that has to be accumulated on either side of the membrane to produce a given membrane potential (see Panel 19–1). The membrane capacitance has the effect both of slowing down the passive transmission of signals along the axon and of distorting them, so that a sharp, pulselike stimulus delivered at one point will be detected a few millimeters away as a slow, gradual rise and fall of the potential,

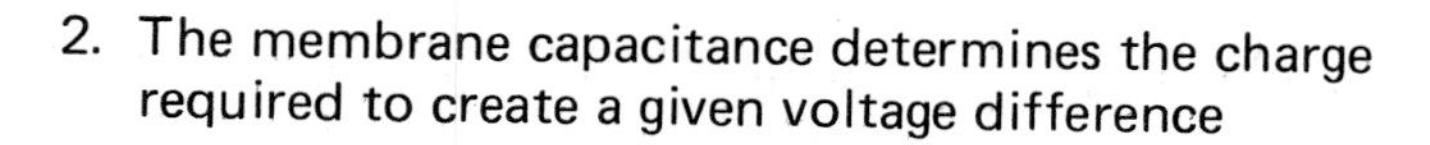

1. Separated layers of charge create a voltage gradient

The voltage gradient across the cell membrane, or *membrane potential*, is created by an excess of positive charge on one side and a matching excess of negative charge on the other. The charge is concentrated in a thin (< 1 nm) layer on each side of the membrane.

2. The membrane capacitance determines the charge required to create a given voltage difference

The amount of charge (in coulombs) required on each side of the membrane to create a voltage difference of 1 V is called the membrane *capacitance* (in farads).

Cell membranes typically have a capacitance of about 1 μF/cm^2, or 0.01 pF/μm^2. Therefore a movement of 0.001 pC of charge across 1 μm^2 of membrane will alter the membrane potential by 100 mV.

UNITS

Charge: couloumb (C) (6.2 $\times$ 10^{18} $\times$ charge on one electron)
Electric potential: volt (V)
Current: ampere (= coulombs per second) (A)
Capacitance: farad (= coulombs per volt) (F)
Conductance: siemens (= amperes per volt) (S)
mV: millivolt (10^{-3} V)
μF: microfarad (10^{-6} F)
nC: nanocoulomb (10^{-9} C)
pS: picosiemens (10^{-12} S)

3. The number of ions that go to form the layer of charge adjacent to the membrane is minute compared with the total number inside the cell

One coulomb is the charge carried by roughly 6 $\times$ 10^{18} univalent ions, so that 0.001 pC is equivalent to 6000 univalent ions. Therefore the movement of 6000 Na^+ ions across 1 μm^2 of membrane will carry sufficient charge to shift the membrane potential by about 100 mV. Because there are about 3 $\times$ 10^7 Na^+ ions in 1 μm^3 of bulk cytoplasm, such a movement of charge will generally have a negligible effect on the ion concentration gradients across the membrane.

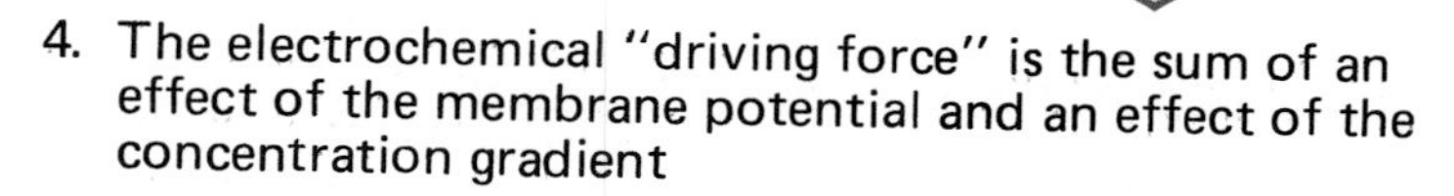

4. The electrochemical "driving force" is the sum of an effect of the membrane potential and an effect of the concentration gradient

For a univalent positive ion, such as Na^+ or K^+, at room temperature, the net "driving force" across the membrane is proportional to

$$V - 58 \log_{10}\left(\frac{C_o}{C_i}\right),$$

where V is the membrane potential in millivolts, and C_o and C_i are, respectively, the extracellular and intracellular concentrations of the ion. The "driving force" for positive ions is zero when

$$V = 58 \log_{10}\left(\frac{C_o}{C_i}\right) \text{ mV}.$$

This is the *Nernst equation* in its simplest form (see p. 315). It defines the *equilibrium potential* for the given positive ion. For the squid axon, the equilibrium potentials V_{Na}, V_K, and V_{Cl} for Na^+, K^+, and Cl^-, are, respectively, about +55 mV, −75 mV, and −65 mV. The net driving forces for each ion are proportional to $V - V_{Na}$, $V - V_K$, and $V - V_{Cl}$.

5. Ion current is proportional to driving force multiplied by membrane conductance

The current of, say, Na^+ ions passing across the membrane (measured in amperes) is

$$i_{Na} = g_{Na} \times (V - V_{Na}),$$

where g_{Na} is the *Na^+ conductance* of the membrane. The Na^+ conductance is proportional to the number of Na^+ channels that are open at any instant. The conductance of a single open Na^+ channel is about 4 pS in the squid axon, and there are about 75 Na^+ channels/μm^2 of membrane.

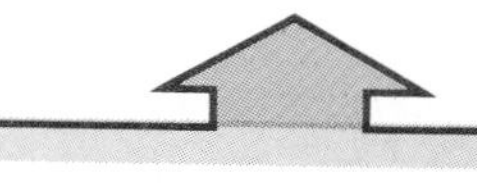

6. A flow of current builds up a charge

For a squid axon membrane at the peak of the action potential,

$$g_{Na} = 300 \text{ pS}/\mu\text{m}^2;$$

the corresponding Na^+ current is roughly

$$i_{Na} = 5 \text{ pA}/\mu\text{m}^2.$$

If no other ions crossed the membrane, the charge transferred by the Na^+ current if it were sustained for 0.2 ms at the peak value seen during the action potential would be 0.001 pC/μm^2. This charge would alter the membrane potential—see (1) and (2), above.

Panel 19–1 Electricity, ions, and membranes: some major principles.

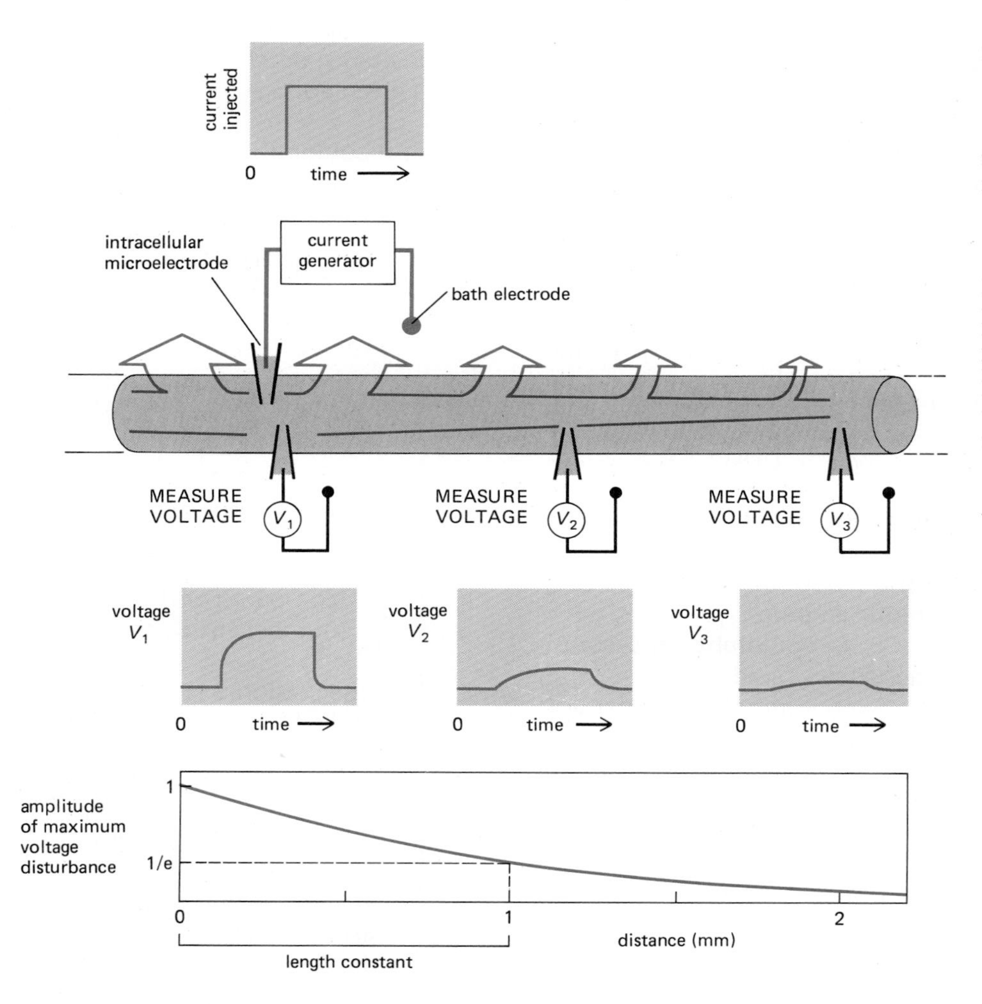

Figure 19–9 Current injected into an axon through a microelectrode flows out again across the plasma membrane; the magnitude of the outflowing current falls off exponentially with distance from the microelectrode. The current flow is assumed to be small, causing a subthreshold depolarization of the membrane. The graphs show how the disturbance of membrane potential produced by injection of a pulse of current falls off with distance from the source of the disturbance. The length constant is the distance over which the amplitude of the disturbance of the membrane potential falls off by a factor of $1/e$. The length constant ranges from about 0.1 mm (for a very small axon with a relatively leaky membrane) to about 5 mm (for a very large axon with a relatively nonleaky membrane). Here it is 1 mm.

with a greatly diminished amplitude (see Figure 19–9). To transmit faithfully over more than a few millimeters, therefore, an axon requires, in addition to its passive cable properties, an active mechanism to maintain the strength and waveform of the signal as it travels. This automatically amplified signal is the *action potential.*

Voltage-gated Na^+ Channels Generate the Action Potential; Voltage-gated K^+ Channels Keep It Brief[3,4,8,10]

The electrochemical mechanism of the **action potential** was first established in the 1940s and 1950s. Techniques for studying electrical events in small single cells had not yet been developed, and the experiments were made possible only by the use of a giant cell, or rather a part of a giant cell: a giant axon from a squid (Figure 19–10). Subsequent work has shown that the neurons of most animals conduct their action potentials in a similar way. Panel 19–2 outlines some of the key original experiments. Despite the many technical advances that have been made since then, the logic of the original analysis continues to serve as a model for present-day work. The crucial insight was that the permeability of the membrane to Na^+ and K^+ is changed by changes in the membrane potential: in other words, the membrane contains channels for Na^+ and K^+ that are voltage-gated. The voltage-

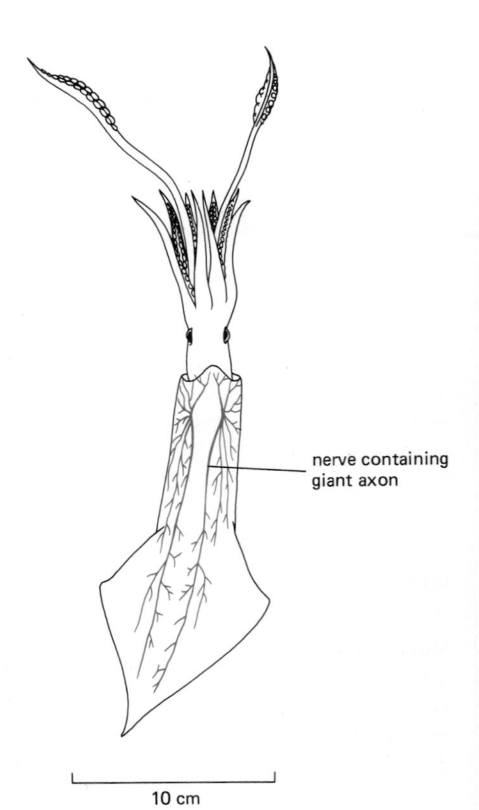

Figure 19–10 A squid, showing the location of the giant axons whose large size made possible the original analysis of the mechanism of the action potential. (From H. Curtis, Biology, 4th ed. New York: Worth, 1983; after Keynes, R.D. The nerve impulse and the squid. *Scientific American,* December 1958. Copyright © 1958 by Scientific American, Inc. All rights reserved.)

1 Action potentials are recorded with an intracellular electrode

The squid giant axon is about 0.5–1 mm in diameter and several centimeters long (Figure 19–10). An electrode in the form of a glass capillary tube containing a conducting solution can be thrust down the axis of the cell so that its tip lies deep in the cytoplasm. With its help, one can measure the voltage difference between the inside and the outside of the cell—that is, the membrane potential—as an action potential sweeps past the electrode. The action potential is triggered by a brief electrical shock to one end of the axon. It does not matter which end, because the excitation can travel in either direction; and it does not matter how big the shock is, as long as it exceeds a certain threshold: the action potential is *all or none.*

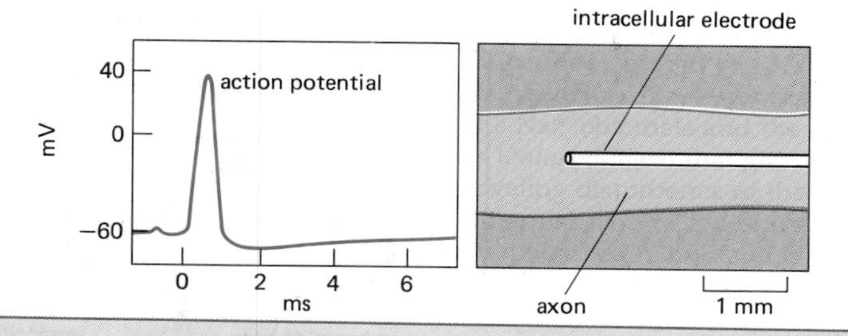

2 Action potentials depend only on the neuronal plasma membrane and on gradients of Na^+ and K^+ across it

The three most plentiful ions, both inside and outside the axon, are Na^+, K^+, and Cl^-. As in other cells, the Na^+-K^+ pump maintains a concentration gradient: the concentration of Na^+ is about 9 times lower inside the axon than outside, while the concentration of K^+ is about 20 times higher inside than outside. Which ions are important for the action potential?

The squid giant axon is so large and robust that it is possible to extrude the cytoplasm from it, like toothpaste from a tube, and then to perfuse it internally with pure artificial solutions of Na^+, K^+, and Cl^- or SO_4^{2-}. Remarkably, if (and only if) the concentrations of Na^+ and K^+ inside and outside approximate those found naturally, the axon will still propagate action potentials of the normal form as shown above. The important part of the cell for electrical signaling, therefore, must be the membrane; the important ions are Na^+ and K^+; and a sufficient source of free energy to power the action potential must be provided by their concentration gradients across the membrane, because all other sources of metabolic energy have presumably been removed by the perfusion.

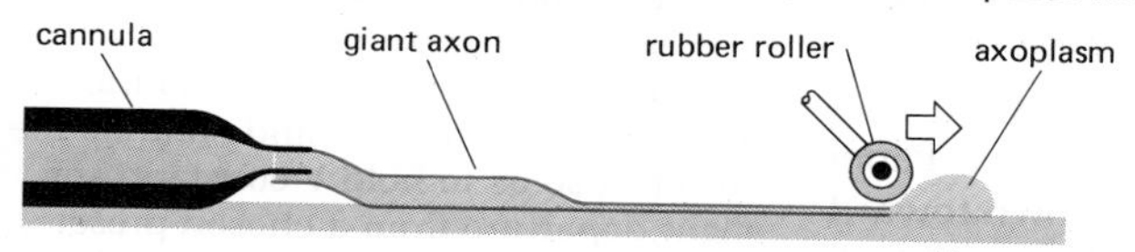

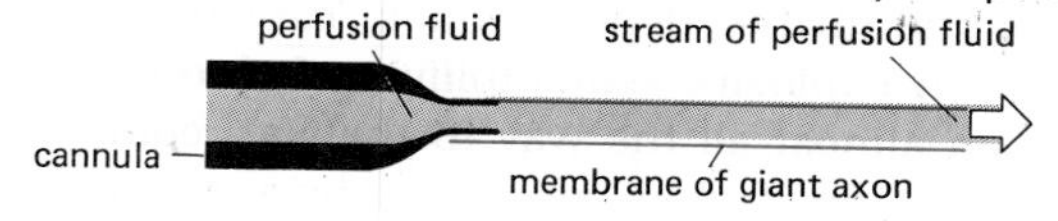

3 At rest, the membrane is chiefly permeable to K^+; during the action potential, it becomes transiently permeable to Na^+

At rest the membrane potential is close to the equilibrium potential for K^+. When the external concentration of K^+ is changed, the resting potential changes roughly in accordance with the Nernst equation for K^+ (see Panel 19–1 and p. 315). At rest, therefore, the membrane is chiefly permeable to K^+: K^+ leak channels provide the main ion pathway through the membrane.

If the external concentration of Na^+ is varied, there is no effect on the resting potential. However, the height of the peak of the action potential varies roughly in accordance with the Nernst equation for Na^+. During the action potential, therefore, the membrane appears to be chiefly permeable to Na^+: Na^+ channels have opened. In the aftermath of the action potential, the membrane potential reverts to a negative value that depends on the external concentration of K^+ and is even closer to the K^+ equilibrium potential than the resting potential is: the membrane has lost its permeability to Na^+ and has become even more permeable to K^+ than before—that is, Na^+ channels have closed, and additional K^+ channels have opened.

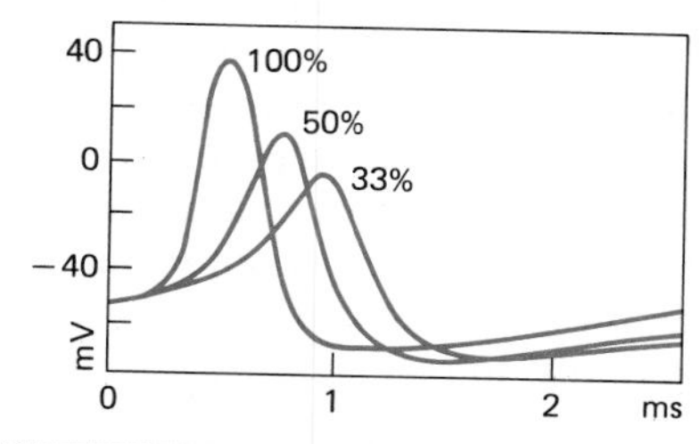

The form of the action potential when the external medium contains 100%, 50%, or 35% of the normal concentration of Na^+

4 Voltage clamping reveals how the membrane potential controls opening and closing of ion channels

The membrane potential can be held constant ("voltage clamped") throughout the axon by passing a suitable current through a bare metal wire inserted along the axis of the axon while monitoring the membrane potential with another intracellular electrode (see Figure 19–11). When the membrane is abruptly shifted from the resting potential and held in a depolarized state (A), Na^+ channels rapidly open until the Na^+ permeability of the membrane is much greater than the K^+ permeability; they then close again spontaneously, even though the membrane potential is clamped and unchanging. K^+ channels also open but with a delay, so that the K^+ permeability becomes large as the Na^+ permeability falls (B). If the experiment is now very promptly repeated, by returning the membrane briefly to the resting potential and then quickly depolarizing it again, the response is different: prolonged depolarization has caused the Na^+ channels to enter an *inactivated* state, so that the second depolarization fails to cause a rise and fall similar to the first. Recovery from this state requires a relatively long time—about 10 milliseconds—spent at the repolarized (resting) membrane potential.

In a normal unclamped axon, an inrush of Na^+ through the opened Na^+ channels produces the spike of the action potential; inactivation of Na^+ channels and opening of K^+ channels bring the membrane back down to the resting potential.

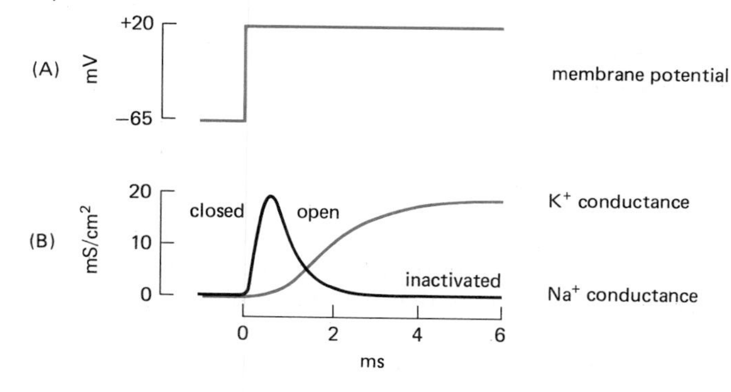

 Panel 19–2 Some classical experiments on the squid giant axon.

large concentration gradients of Na^+ and K^+ across the nerve cell membrane. In the resting neuron the K^+-selective leak channels in the membrane make it more permeable to K^+ than to other ions, and the membrane potential is consequently close to the K^+ equilibrium potential of about −70 mV. An action potential is triggered when a brief depolarizing stimulus causes voltage-gated Na^+ channels to open, making the membrane more permeable to Na^+ and further depolarizing the membrane potential toward the Na^+ equilibrium potential. This positive feedback causes still more Na^+ channels to open, resulting in an all-or-none action potential. In each region of membrane, the action potential is rapidly terminated by the inactivation of the Na^+ channels and, in many neurons, by the opening of voltage-gated K^+ channels.

The propagation of an action potential along a nerve fiber depends on the fiber's passive cable properties: when the membrane is locally depolarized and fires an action potential, the current entering through open Na^+ channels at that site spreads passively to depolarize neighboring regions of the membrane, where action potentials are triggered in turn. In many vertebrate axons the speed and efficiency of propagation of action potentials are increased by insulating sheaths of myelin, which change the cable properties of the axon and leave only small regions of excitable membrane exposed.

Ligand-gated Ion Channels and Fast Synaptic Transmission[13]

The simplest way for one neuron to pass its signal to another is by direct electrical coupling through gap junctions. Such **electrical synapses** have the virtue that transmission occurs without delay. But they are far less rich in possibilities for adjustment and control than are the **chemical synapses** that provide the majority of nerve cell connections. Electrical communication through gap junctions was considered in Chapter 14 (pp. 798–801). Here we shall confine our discussion to chemical synapses.

The principles of chemical communication at a synapse are the same as those of chemical communication by water-soluble hormones, as discussed in Chapter 12. In both cases a cell releases a chemical messenger that acts on another cell, or set of cells, by binding to membrane receptor proteins. Unlike a hormone, however, the chemical messenger at a synapse—the **neurotransmitter**—acts at very close quarters.

Electrical stimulation of the presynaptic cell causes the release of a neurotransmitter by exocytosis (see Figure 19–4); once the neurotransmitter has crossed the gap—typically a small fraction of a micrometer—between the pre- and postsynaptic cells, the chemical signal must be converted back into an electrical one. This conversion is mediated by the receptors in the plasma membrane of the postsynaptic cell, which fall into two distinct categories: *channel-linked receptors* and *non-channel-linked receptors* (Figure 19 -15). Channel-linked receptors can be described equivalently as *ligand-gated channels.* A channel-linked receptor, upon binding neurotransmitter, promptly changes its conformation so as to create an open channel for specific ions to cross the membrane, thereby altering the membrane permeability. This type of receptor underlies the most familiar and the best-understood mode of chemical synaptic signaling, where transmission is very rapid.

Non-channel-linked receptors work by the same mechanisms that mediate responses to water-soluble hormones and local chemical mediators throughout the body (see p. 694). In such receptors the neurotransmitter-binding site is functionally coupled to an enzyme that, in the presence of neurotransmitter, usually catalyzes the production of an intracellular messenger such as cyclic AMP. The intracellular messenger in turn causes changes in the postsynaptic cell, including modifications of the ion channels in its membrane. Compared with channel-linked receptors, these receptors generally provide for neurotransmitter actions that are relatively slow in onset and long in duration. Some of them are believed to mediate the long-lasting neuronal changes that underlie learning and memory (see p. 1094).

(A) CHANNEL-LINKED RECEPTOR

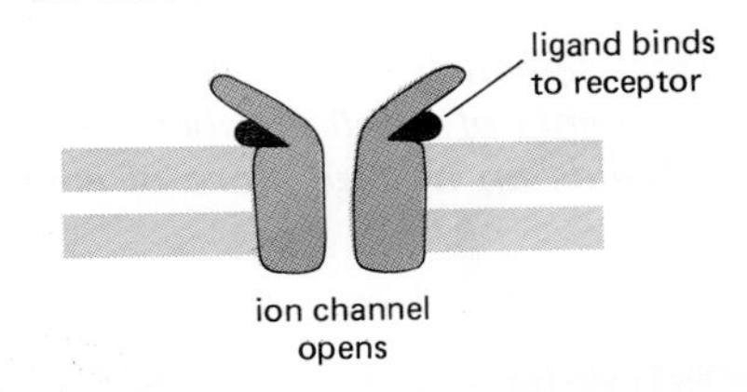

(B) NON-CHANNEL-LINKED RECEPTOR

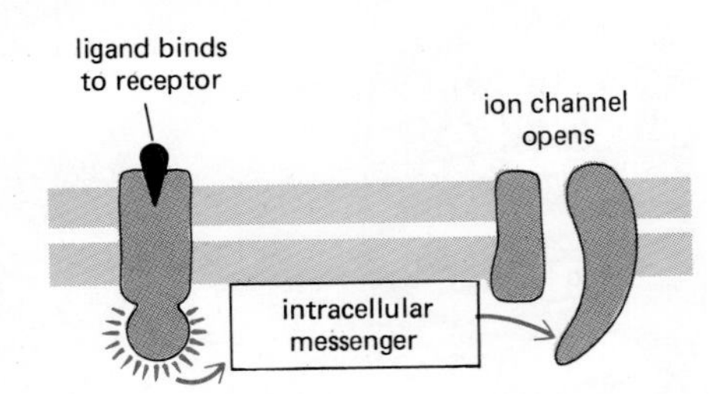

Figure 19–15 A neurotransmitter can exert its effect on a postsynaptic cell by means of two fundamentally different types of receptor proteins: channel-linked receptors and non-channel-linked receptors. Channel-linked receptors are also known as ligand-gated channels.

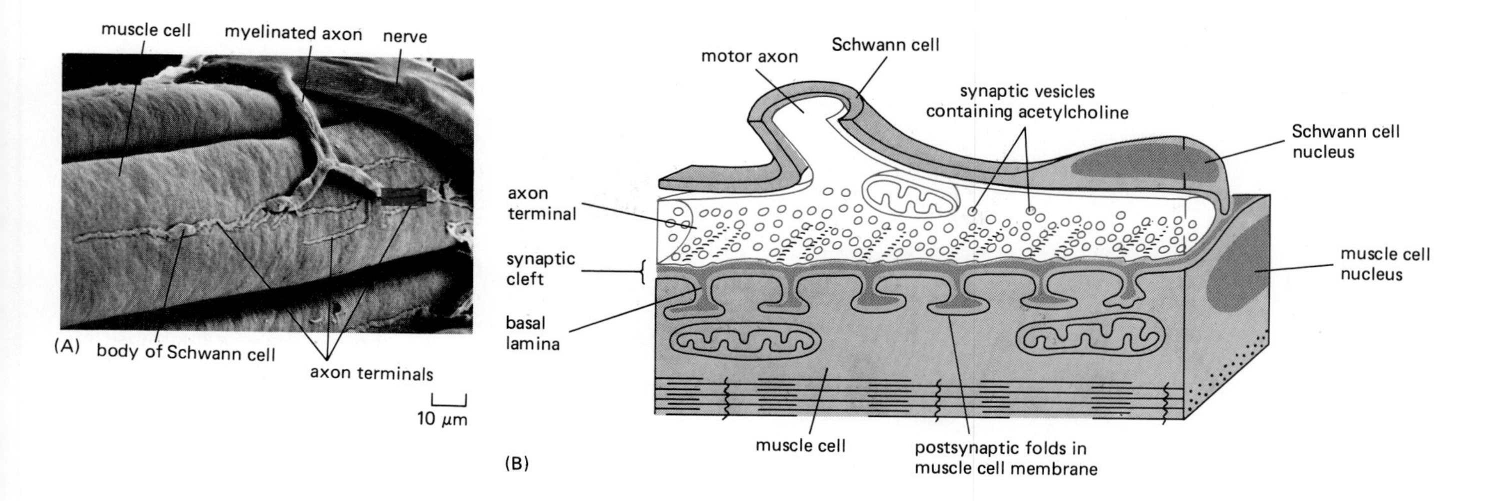

Figure 19–16 A neuromuscular junction in a frog. (A) Low-magnification scanning electron micrograph of the termination of a single axon on a skeletal muscle cell. (B) Schematic drawing of the part of the junction boxed in (A), showing the major features visible by transmission electron microscopy. The pattern of small terminal branches of the axon at the junction varies with the species and with the type of skeletal muscle cell. From its appearance in mammals, the neuromuscular junction is often called the *end plate*. (A, from J. Desaki and Y. Uehara, *J. Neurocytol.* 10:101–110, 1981, by permission of Chapman & Hall.)

In this section we shall discuss rapid synaptic transmission based on ligand-gated ion channels. The special features of synaptic signaling based on non-channel-linked receptors, and their role in long-term synaptic change, will be discussed in a later section (see p. 1091).

The Neuromuscular Junction Is the Best-understood Synapse[14]

The central nervous system is so densely packed with neurons that it is extremely difficult to perform experiments on single synapses within it. Detailed understanding of synaptic function has come instead chiefly from work on the junctions between nerve and skeletal muscle in the frog and, to a lesser extent, on synapses between giant neurons in the squid and other mollusks.

Skeletal muscle cells in vertebrates, like nerve cells, are electrically excitable, and the **neuromuscular junction** (Figure 19–16) has proved to be a valuable model for chemical synapses in general. A motor nerve and its muscle can be dissected free from the surrounding tissue and maintained in a bath of controlled composition. The nerve can be stimulated with extracellular electrodes, and the response of a single muscle cell can be monitored relatively easily with an intracellular microelectrode (Figure 19–17). Figure 19–18 compares the fine structure of a neuromuscular junction with that of a typical synapse between two neurons in the central nervous system.

The neuromuscular junction has been the focus of a long and fruitful series of investigations that began in the 1950s. The background to the early experiments was the discovery, in the early 1920s, that acetylcholine is released upon stimulation of the vagus nerve to the heart and acts on heart muscle to slow its beating. This was the first clear evidence of chemical neurotransmission, and it soon led to the demonstration, in the 1930s, that stimulation of a motor nerve innervating a skeletal muscle also causes the release of acetylcholine and that acetylcholine in turn stimulates skeletal muscle to contract. Acetylcholine was thereby identified as the neurotransmitter at the neuromuscular junction. But how is the release of acetylcholine brought about, and how does it exert its effect on the muscle?

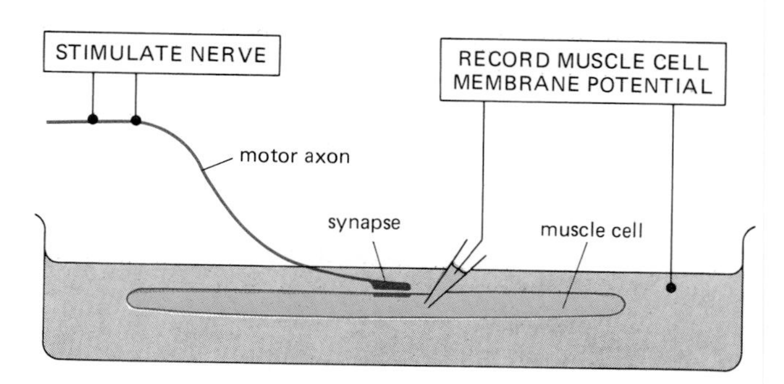

Figure 19–17 An experimental arrangement used to study synaptic transmission at the neuromuscular junction.

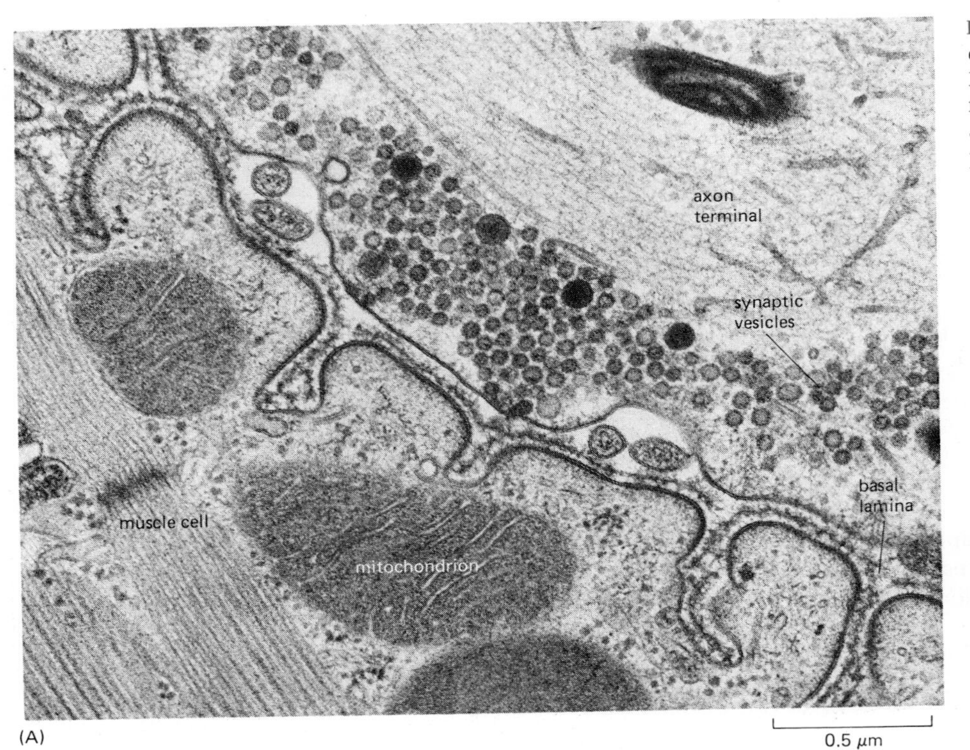

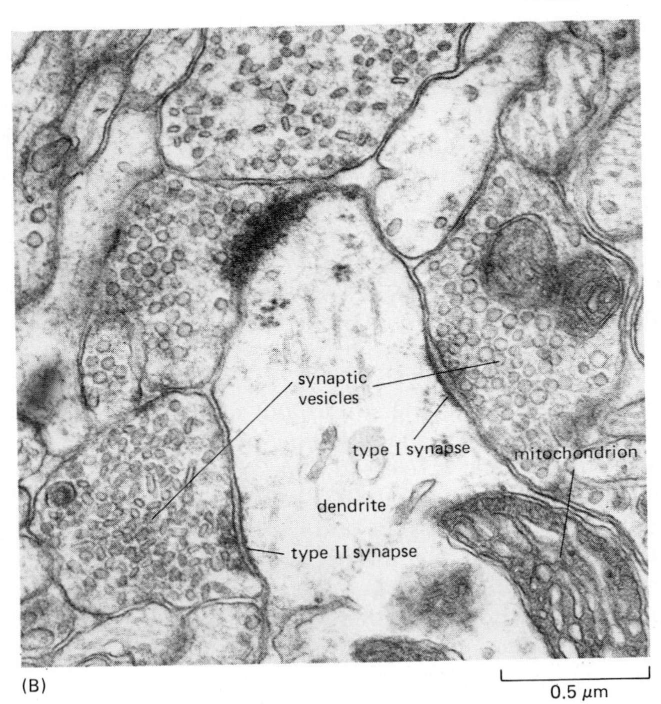

Figure 19–18 (A) Electron micrograph of part of a neuromuscular junction. (B) Electron micrograph of a small region from the brain of a rat. Two synapses are clearly visible in (B), each showing pre- and postsynaptic membranes, a synaptic cleft between them, and synaptic vesicles on the presynaptic side, as in (A). The two synapses labeled in (B) differ from each other in the size and shape of their vesicles: the vesicles at the *type I* synapse are round, whereas those at the *type II* synapse are flattened and are believed to contain a different neurotransmitter. Note the characteristic "thickened" appearance of the postsynaptic membrane and, to a lesser extent, of the presynaptic membrane in both (A) and (B). There is no basal lamina interposed between the pre- and postsynaptic membranes at synapses in the brain, although some extracellular material is faintly apparent in the cleft. The absence of a basal lamina represents the chief structural difference between a synapse in the central nervous system and a neuromuscular junction. (A, courtesy of John Heuser; B, courtesy of G. Campbell and A.R. Lieberman.)

Voltage-gated Ca^{2+} Channels Couple Action Potentials to Neurotransmitter Release[15]

The action potential is propagated along the axon by the opening and closing of Na^+ channels until it reaches the neuromuscular junction. Here the action potential opens *voltage-gated Ca^{2+} channels* in the plasma membrane of the axon terminal, allowing Ca^{2+} to enter and trigger the release of acetylcholine (Figure 19–19).

Three simple observations showed that this influx of Ca^{2+} into the axon terminal is essential for synaptic transmission. First, if there is no Ca^{2+} in the extra-

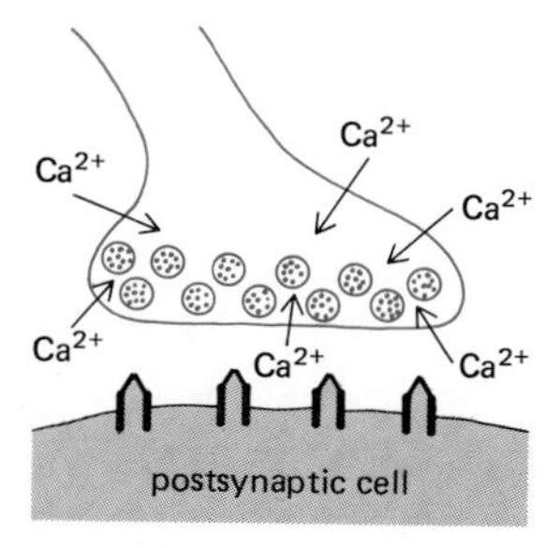

action potential triggers entry of Ca^{2+} into presynaptic terminal

synaptic vesicles fuse with presynaptic membrane releasing transmitter

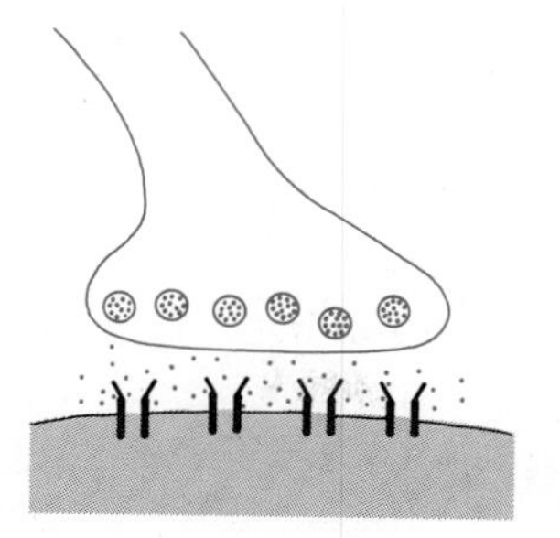

transmitter binds to proteins in the postsynaptic membrane changing their conformation

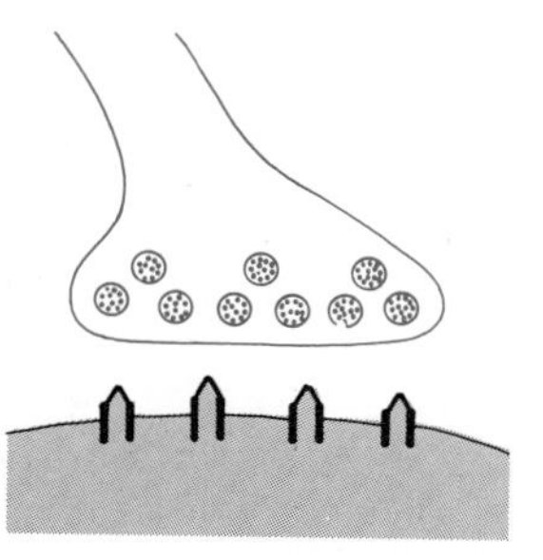

transmitter is removed from cleft and postsynaptic proteins revert to original conformation

Figure 19–19 Summary of the essential events at a chemical synapse following the arrival of an action potential in the axon terminal.

cellular medium bathing the axon terminal when the action potential arrives, no neurotransmitter is released and transmission fails. Second, if Ca^{2+} is injected artificially into the cytoplasm at the axon terminal through a micropipette, transmitter is immediately released even without electrical stimulation of the axon. (This microinjection experiment is difficult to do at the neuromuscular junction because the axon terminal is so small, but it has been done at a synapse between giant neurons in the squid.) Third, artificial depolarization of the axon terminal (again at the squid giant synapse), in the absence of an action potential and with the Na^+ and K^+ channels blocked by specific toxins, leads to Ca^{2+} entry and transmitter release; furthermore, if the depolarization reverses the membrane potential so far as to reduce the electrochemical driving force for Ca^{2+} entry to zero, no transmitter release occurs.

The channel protein that lets the Ca^{2+} into the cell—the **voltage-gated Ca^{2+} channel**—has a uniquely important role. It provides the only known means of converting electrical signals—fleeting depolarizations of the membrane—into chemical changes inside nerve cells. As explained in Panel 19–1, voltage-gated channels for Na^+, K^+, or Cl^- are of no use for this purpose: the ion fluxes driven through them by a single action potential are so small that they do not significantly alter the ion concentrations in the cytosol. The ion flux through voltage-gated Ca^{2+} channels is no larger in absolute terms and generally makes only a small contribution to the electrical current across the membrane; but it is very much larger in relation to the free Ca^{2+} concentration inside the cell, which is normally kept at about 10^{-7} M, corresponding to less than 100 Ca^{2+} ions/μm^3. In 1 millisecond, a single open Ca^{2+} channel would typically pass several hundred Ca^{2+} ions, driven by the membrane potential and the relatively high extracellular concentration of Ca^{2+} (usually ~1–2 mM). Thus a small number of voltage-gated Ca^{2+} channels in the presynaptic terminal, opening in response to an action potential, can easily raise the intracellular concentration of free Ca^{2+} by a factor of 10 to 100. The surge of free Ca^{2+} then acts as an intracellular messenger, triggering the release of neurotransmitter at a rate that increases very steeply with the free Ca^{2+} concentration.

The increase of free Ca^{2+} concentration is short-lived because Ca^{2+}-binding proteins, Ca^{2+}-sequestering vesicles, and mitochondria rapidly take up the Ca^{2+} that has entered the axon terminal, while Ca^{2+} pumps in the plasma membrane, driven either by ATP hydrolysis or by the Na^+ electrochemical gradient, pump it out of the cell (see p. 307 and p. 700). In this way the terminal is ready to transmit another signal as promptly as the axon is ready to deliver one.

Neurotransmitter Is Released Rapidly by Exocytosis[16]

The axon terminal at the neuromuscular junction is crammed with thousands of uniform (~40 nm diameter) secretory vesicles, called *synaptic vesicles*, each containing acetylcholine (see Figure 19–18). The entry of Ca^{2+} induces a synchronized burst of exocytosis in which the vesicles fuse with the presynaptic membrane, discharging their contents into the synaptic cleft to act on the postsynaptic cell.

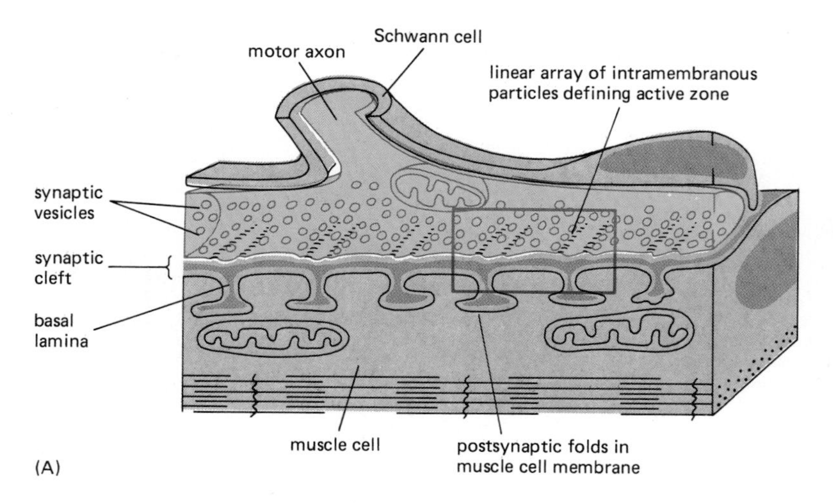

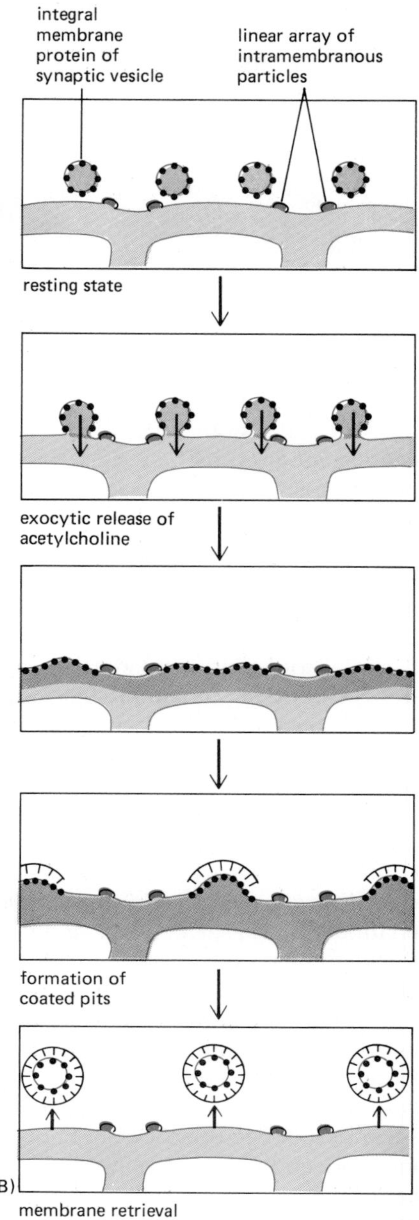

Figure 19–20 The cycle of membrane events in the axon terminal at a neuromuscular synapse following stimulation. To follow the action, samples of tissue are prepared by sudden freezing at measured times after the stimulus. To make the task easier technically, the conditions of excitation are artificially adjusted so as to slow down the normal time course by a factor of 5 or 10 and increase the number of vesicles that undergo exocytosis. (A) Simplified drawing of a neuromuscular junction, showing the active zones where transmitter release occurs. (B) The boxed area in (A) is enlarged and shown schematically in cross-section at a series of different times after stimulation of the nerve.

The exocytosis is restricted to specialized regions known as *active zones,* exactly opposite the receptors on the postsynaptic cell; in this way the delay associated with diffusion of neurotransmitter across the cleft is made negligibly short. The membrane of the discharged synaptic vesicles is subsequently retrieved from the presynaptic plasma membrane by endocytosis.

There is evidence that, as well as triggering exocytosis, the influx of Ca^{2+} into the axon terminal activates a Ca^{2+}/calmodulin-dependent protein kinase (Ca-kinase II—see p. 711), which phosphorylates a number of proteins in the terminal, including *synapsin I,* a protein attached to the surface of synaptic vesicles. Phosphorylation is thought to release the synapsin I and thereby allow the vesicles to dock at the active zone of the presynaptic membrane, where they are needed to replace the vesicles lost from that region by exocytosis. The whole cycle of events that is initiated by a single nerve impulse has been vividly demonstrated by very rapidly freezing the nerve and muscle tissue and then preparing it for electron microscopy. Some of the results are shown in Figure 19–20.

Neurotransmitter Release Is Quantal and Probabilistic[17]

An axon terminal at a neuromuscular junction typically releases a few hundred of its many thousands of synaptic vesicles in response to a single action potential. Each vesicle, by discharging its contents into the synaptic cleft, contributes to the production of a voltage change in the postsynaptic muscle cell, which can be recorded with an intracellular electrode (Figure 19–21). The muscle cell membrane is thus depolarized beyond its threshold and fires an action potential. This excitation sweeps over the cell (Figure 19–22), causing a contraction, as described on page 621.

Even when the axon terminal is electrically quiet, occasional brief depolarizations of the muscle membrane are observed in the neighborhood of the synapse. These **miniature synaptic potentials** typically have an amplitude of only about 1 mV—far below threshold—and they occur at random, with a certain low probability per unit time—typically about once per second (Figure 19–23). Each miniature potential results from a single synaptic vesicle fusing with the presynaptic membrane so as to discharge its contents. The amplitude as recorded in a given muscle cell is roughly uniform because each vesicle contains practically the same number of molecules of acetylcholine, on the order of 5000. This number represents the minimum packet or *quantum* of transmitter release. Larger signals are made up of integral multiples of this basic unit. The Ca^{2+} that enters the axon terminal during an action potential acts for a fraction of a millisecond to increase the rate of occurrence of the exocytic events more than 10,000-fold above the resting spontaneous frequency. Nonetheless, the process remains probabilistic, and identical stimulations of the nerve do not always produce exactly the same postsynaptic effect: if, for example, 300 quanta are released on average, more or less than this number may be released on any particular occasion.

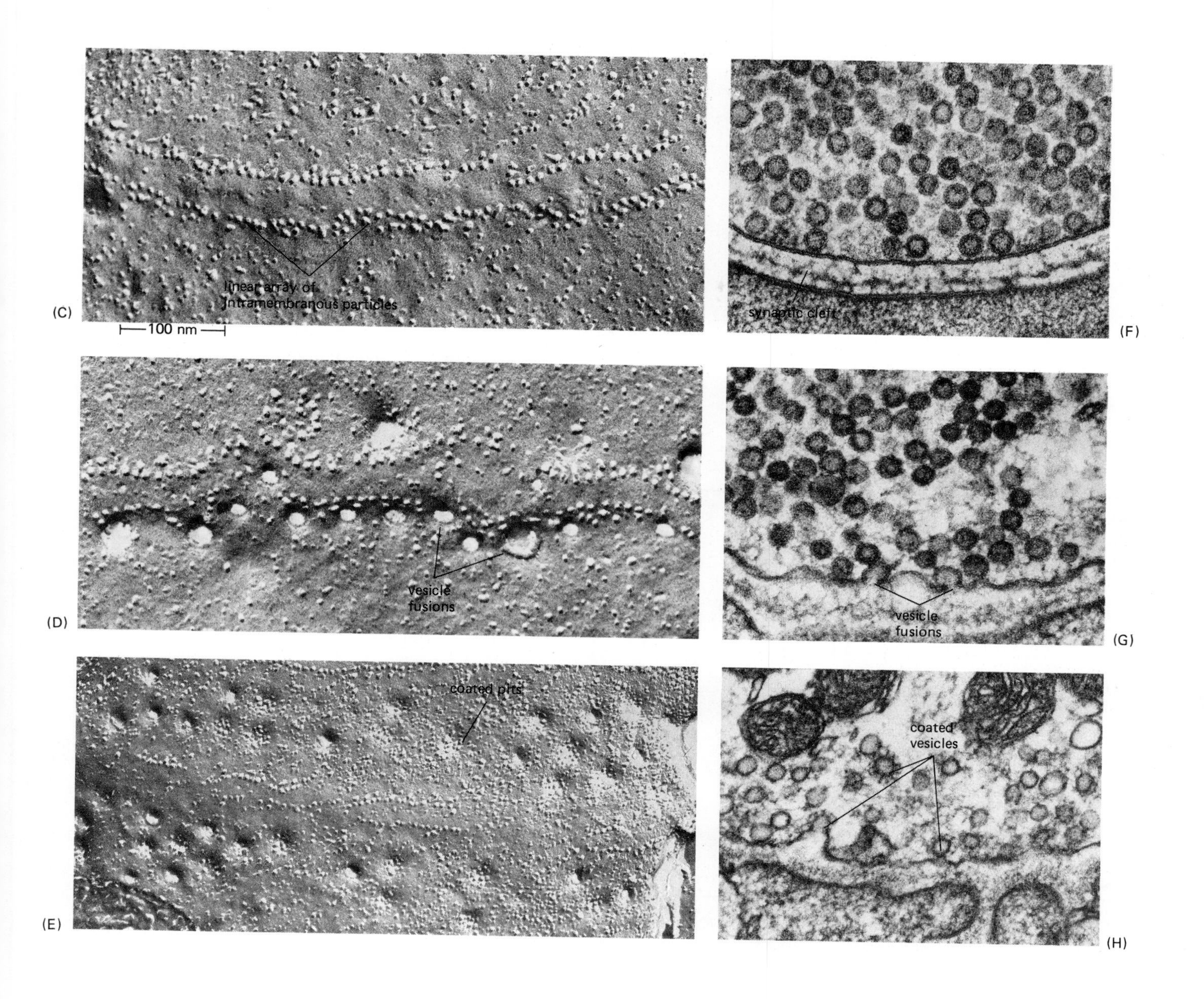

Figure 19–20 *(Cont'd)* (C-H) The actual appearance of the membrane as viewed by electron microscopy. Freeze-fracture electron micrographs of the cytoplasmic half of the presynaptic membrane are shown on the left; thin-section micrographs are shown on the right. (C, F) Resting state. (D, G) Fusion of synaptic vesicles with the plasma membrane at an active zone (marked by the linear arrays of intramembranous particles). (E, H) Retrieval of synaptic vesicle membrane via coated pits and coated vesicles.

Synaptic vesicles can be seen to have begun fusing with the plasma membrane within 5 milliseconds after the stimulus (D, G); each of the openings in the plasma membrane apparent in (D) represents the point of fusion of one synaptic vesicle. Fusion is complete within another 2 milliseconds. The first signs of membrane retrieval become apparent within about 10 seconds as coated pits (see p. 326) form and then, after a further 10 seconds, begin to pinch off by endocytosis to form coated vesicles (E, H). These vesicles include the original membrane proteins of the synaptic vesicle and also contain molecules captured from the external medium. The cycle ends when the coat dissociates from the coated vesicle, which refills with acetylcholine to form a smooth-surfaced, regenerated synaptic vesicle. This scheme probably accounts for the strikingly uniform size of the synaptic vesicles, a size defined by the dimensions of the latticelike coat of clathrin (see p. 327).

Further evidence for this retrieval scheme can be obtained by stimulating the nerve in the presence of electron-dense extracellular markers such as ferritin. These markers quickly appear within coated vesicles and eventually show up in synaptic vesicles.

Note, however, that some experts have been skeptical about these experiments, interpreting some of the phenomena as artefacts. (C–H, courtesy of John Heuser.)

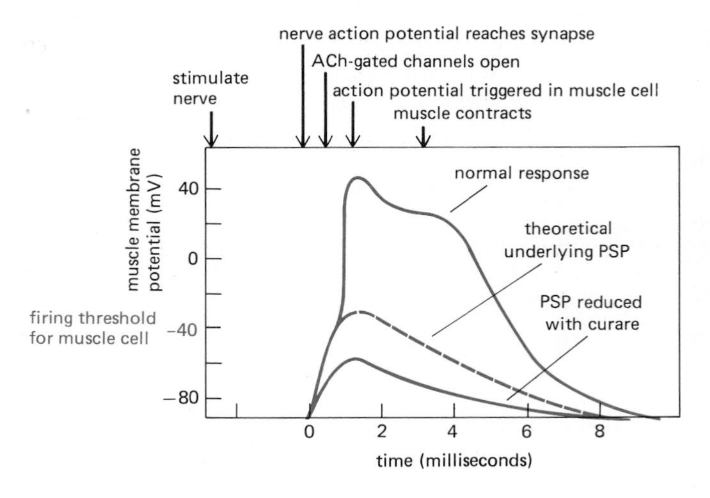

Figure 19–21 The postsynaptic response to a single nerve impulse at the neuromuscular junction: a graph of the voltage change in a frog muscle cell recorded, as in Figure 19–17, with an intracellular electrode close to the synapse. Normally the postsynaptic potential (PSP)—the depolarization directly produced by the neurotransmitter acting on the muscle cell membrane—is large enough to trigger an action potential, which complicates the analysis. A pure PSP, uncomplicated by an action potential, can be obtained by adding a moderate concentration of *curare* to the extracellular medium. This toxin, by binding to some of the receptors and blocking their response to the neurotransmitter, reduces the size of the PSP to the point where no action potential is triggered.

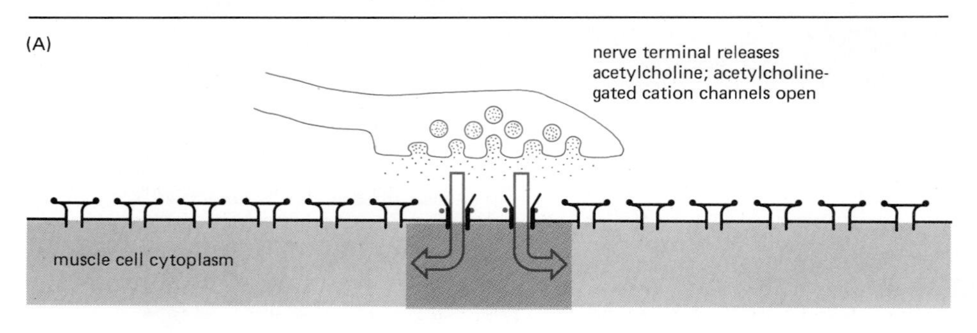

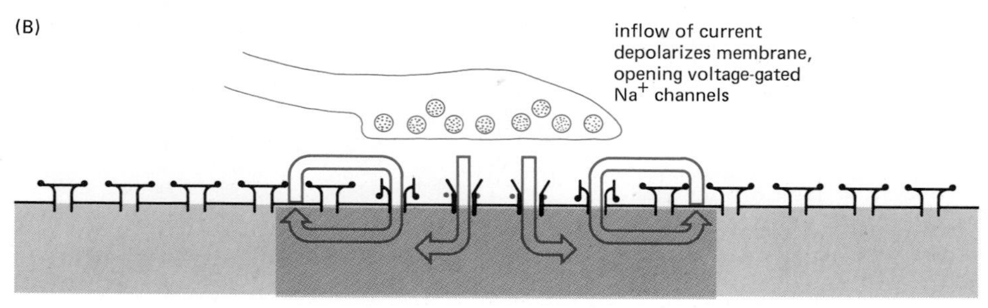

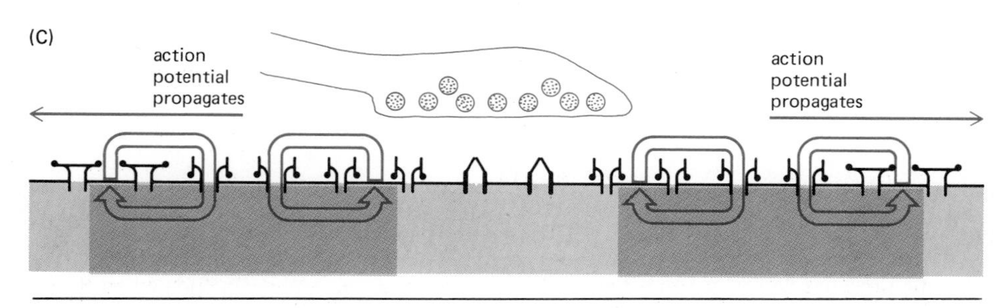

Figure 19–22 Electrical events in the muscle cell membrane at the neuromuscular junction. (A) The opening of ion channels gated by acetylcholine initiates an action potential (B) that propagates along the muscle cell membrane (C), causing contraction of the muscle cell.

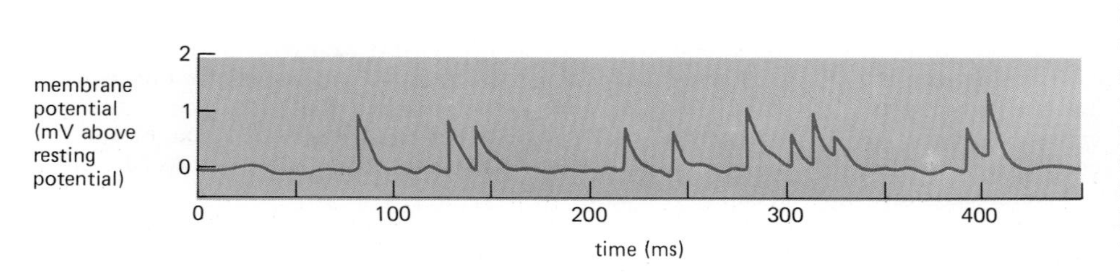

Figure 19–23 Miniature synaptic potentials (often called "miniature end-plate potentials") recorded from a frog muscle cell with an intracellular electrode inserted close to the neuromuscular junction. Each blip in the record is a miniature synaptic potential generated by release of the contents of a single synaptic vesicle from the axon terminal. (Redrawn from P. Fatt and B. Katz, *J. Physiol.* 117:109–128, 1952.)

Ligand-gated Channels Convert the Chemical Signal Back into Electrical Form[18]

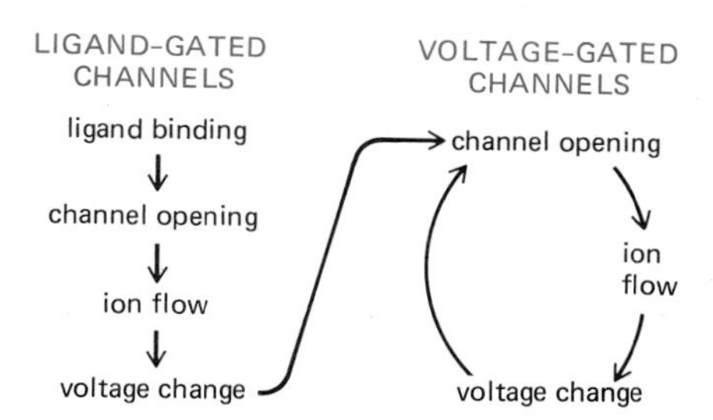

Figure 19–24 Diagram summarizing the function of ligand-gated and voltage-gated channels in the response to a neurotransmitter. The arrows indicate causal connections.

The muscle cell membrane at the synapse behaves as a *transducer* that converts a chemical signal in the form of a neurotransmitter into an electrical signal. The conversion is achieved by **ligand-gated ion channels** (that is, **channel-linked receptors**) in the postsynaptic membrane: when the neurotransmitter binds to these proteins, they change their conformation—opening to let ions cross the membrane—and thereby alter the membrane potential. The shift of membrane potential, if it is large enough, will in turn cause voltage-gated channels to open, thereby triggering an action potential (Figure 19–24). Unlike voltage-gated ion channels, the ligand-gated ion channels are relatively insensitive to the membrane potential. They cannot by themselves, therefore, produce an all-or-none, self-amplifying excitation. Instead they produce an electrical change that is graded according to the intensity and duration of the external chemical signal—that is, according to how much transmitter is released into the synaptic cleft and how long it stays there. This feature of ligand-gated ion channels is important in information processing at synapses, as will be discussed later.

Postsynaptic ligand-gated channels have two other important properties. First, in their role as receptors, they have an enzymelike specificity for particular ligands so that they respond only to one neurotransmitter—the one released from the presynaptic terminal; other transmitters are virtually without effect. Second, in their role as channels, they are characterized by different ion selectivities: some may be selectively permeable to K^+, others to Cl^-, and so on, whereas still others may, for example, be relatively nonselective among the cations but exclude anions. We shall see that the ion selectivity of the ligand-gated channels determines the nature of the postsynaptic response.

The Acetylcholine Receptor Is a Ligand-gated Cation Channel[19]

The channel in the skeletal muscle cell membrane gated by acetylcholine and known as the **acetylcholine receptor** is the best understood of all ligand-gated ion channels, and its molecular properties have already been discussed (see p. 319).

Like the voltage-gated Na^+ channel, the acetylcholine receptor has a number of discrete alternative conformations (Figure 19–25). Upon binding acetylcholine it jumps abruptly from a closed to an open state and then stays open, with the ligand bound, for a randomly variable length of time, averaging about 1 millisecond or even less, depending on the temperature and the species. In the open conformation the channel is indiscriminately permeable to small cations, including Na^+, K^+, and Ca^{2+}, but it is impermeable to anions (Figure 19–26).

Since there is little selectivity among these cations, their relative contributions to the current through the channel depend chiefly on their concentrations and on the electrochemical driving forces. If the muscle cell membrane is at its resting potential, the net driving force for K^+ is near zero because the voltage gradient nearly balances the K^+ concentration gradient across the membrane. For Na^+, on the other hand, the voltage gradient and the concentration gradient both act in the same direction to drive Na^+ into the cell. (The same is true for Ca^{2+}, but the extracellular concentration of Ca^{2+} is so much lower than that of Na^+ that Ca^{2+} makes only a small contribution to the total inward current.) Opening the acetylcholine receptor channel therefore leads chiefly to a large influx of Na^+, causing membrane depolarization.

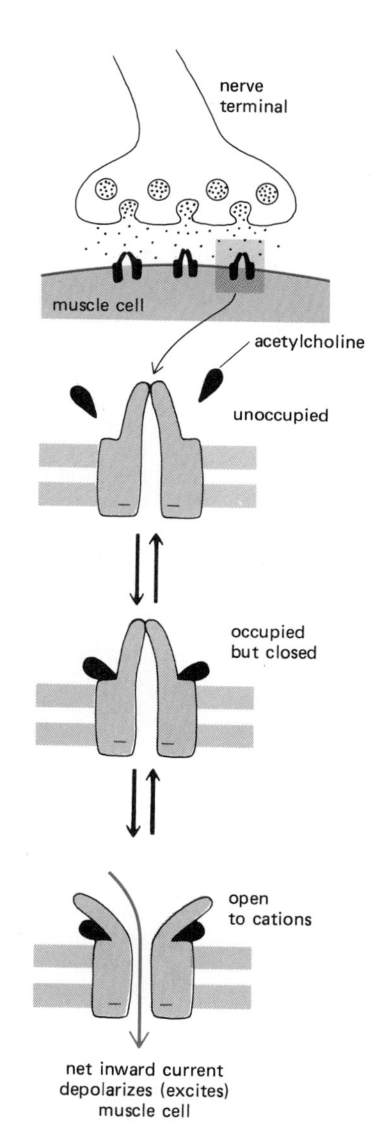

Figure 19–25 The response of the acetylcholine receptor to acetylcholine. Prolonged exposure to high concentrations of acetylcholine causes the receptor to enter yet another state (not shown), in which it is inactivated and will not open even though acetylcholine is present.

Acetylcholine Is Removed from the Synaptic Cleft by Diffusion and by Hydrolysis[20]

If the postsynaptic cell is to be accurately controlled by the pattern of signals sent from the presynaptic cell, the postsynaptic excitation must be switched off promptly when the presynaptic cell falls quiet. At the neuromuscular junction this is achieved by rapidly removing the acetylcholine from the synaptic cleft through two mechanisms. First, the acetylcholine disperses by diffusion, a rapid process because

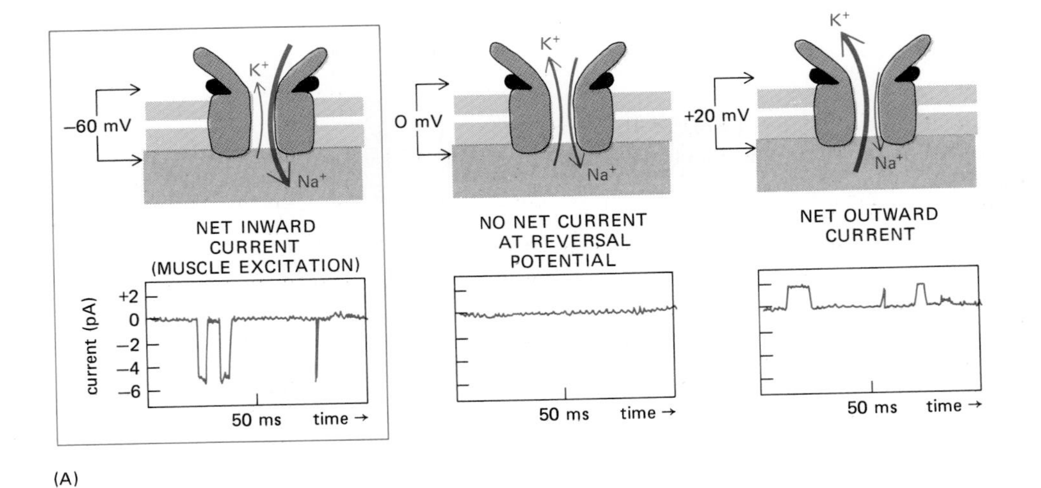

(A)

membrane clamped at 28 mV
membrane clamped at reversal potential
current (pA)
200
0
−200
−400
0
2
4
6
msec
membrane clamped at −120 mV

(B)

Figure 19–26 Measurements of the current through the open acetylcholine-receptor channel at different values of the membrane potential. Such measurements can be used to discover the channel's ion selectivity. A current carried through an open channel by a particular type of ion will vary with the membrane potential in a characteristic way that differs from one type of ion to another according to its concentration gradient across the membrane. Knowing the concentration gradients of the major ions present, one can thus get a useful clue to the channel's ion selectivity simply by measuring its current/voltage relationship; and fuller information can be obtained by repeating the measurements with altered ion concentrations.

(A) Patch-clamp recordings of the current through a single channel kept in a bath containing a fixed concentration of acetylcholine, with the membrane potential clamped at three different voltages. The channel flips randomly between open and closed states in a similar way in each case, but at a certain value of the membrane potential, called the *reversal potential,* the current is zero even when the channel is open. In this particular example the reversal potential happens to be approximately 0 mV. (B) The same phenomenon can be observed by monitoring the total current that flows through the large population of acetylcholine-receptor channels in the postsynaptic membrane at a neuromuscular synapse following a single stimulus to the nerve. The graphs show voltage-clamp recordings of this current made with intracellular electrodes. The channels open during the brief exposure to acetylcholine, but again the current is zero when the membrane potential is clamped at the reversal potential. Because the open channels are permeable to both Na^+ and K^+ and the electrochemical driving forces for these ions are different, zero net current corresponds to balanced nonzero currents of Na^+ and K^+ in opposite directions. (The channels are also permeable to Ca^{2+}, but the Ca^{2+} current is small because Ca^{2+} concentrations are low.) The value of the reversal potential and its sensitivity to ion concentrations in the external medium give a useful indication of the relative permeability of the channel to the different ions. For example, some other ligand-gated channels are selectively permeable to Cl^- (see p. 1083); they can be recognized because they have a reversal potential of about −60 mV, close to the Cl^- equilibrium potential, and the value of this potential depends on the extracellular concentration of Cl^- but not of Na^+ or K^+. (A, data from B. Sakmann, J. Bormann, and O.P. Hamill, *Cold Spring Harbor Symp. Quant. Biol.* 48:247–257, 1983; B, data based on K.L. Magleby and C.F. Stevens, *J. Physiol.* 223:173–197, 1972.)

the dimensions involved are small. Second, the acetylcholine is hydrolyzed to acetate and choline by *acetylcholinesterase.* This enzyme is secreted by the muscle cell and becomes anchored by a short collagenlike "tail" to the basal lamina that lies between the nerve terminal and the muscle cell membrane. Each acetylcholinesterase molecule can hydrolyze up to 10 molecules of acetylcholine per millisecond, so that all of the transmitter is eliminated from the synaptic cleft within a few hundred microseconds after its release from the nerve terminal. Consequently, acetylcholine is available only for a fleeting moment to bind to its receptors and drive them into the open conformation that produces the conductance change in the postsynaptic membrane (Figure 19–27). The sharply defined timing of presynaptic signals is thus preserved in sharply timed postsynaptic responses.

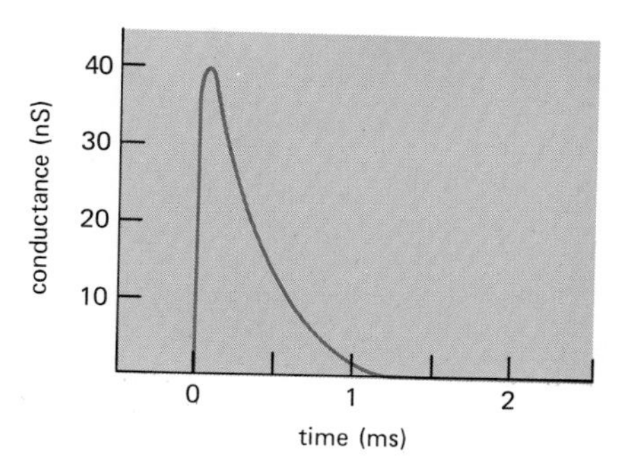

Figure 19–27 The conductance change produced in the postsynaptic membrane by a single quantum (one vesicle) of acetylcholine at the frog neuromuscular junction. About 1600 channels are open at the time of peak conductance, and each channel remains open for an average of 400 microseconds.

Fast Synaptic Transmission Is Mediated by a Small Number of Neurotransmitters[13,21]

Everything about the neuromuscular junction appears to be designed for speed: the large myelinated motor axon; the active zones in the axon terminal, with synaptic vesicles held ready to release their acetylcholine precisely opposite the postsynaptic receptors; the narrow synaptic cleft; the ligand-gated channels in the postsynaptic membrane, ready to open instantly when the transmitter binds; the acetylcholinesterase in the cleft to terminate transmission promptly. The synaptic delay, from the peak of the presynaptic action potential to the peak of the postsynaptic action potential, is on the order of a millisecond or less. Increasing evidence suggests that fast chemical synapses in the central nervous system also employ ligand-gated channels and are constructed on the same principles, with active zones, a narrow cleft, and receptors precisely localized opposite the sites of exocytosis. Moreover, it seems that there are only a handful of neurotransmitters that mediate such rapid signaling. These generalizations must be tentative, however: it is surprisingly difficult to identify conclusively the neurotransmitter acting at a given synapse.

It is probable that the rapid synaptic signaling systems evolved long ago, since the same neurotransmitters are used in the most disparate species of animals, from mollusks to mammals. The rapidly acting neurotransmitters include acetylcholine, γ-aminobutyrate (GABA), glycine, glutamate—and probably aspartate and ATP (Figure 19–28); and in general, a given neuron secretes only one (or occasionally two) of these transmitters—the same at all the synapses it makes. There is direct evidence from patch-clamp studies that the receptors for acetylcholine, GABA, glycine, and glutamate are channel-linked; for the others this is probable but not proved. DNA-sequencing studies indicate that the receptors for acetylcholine, GABA, and glycine are homologous, suggesting that all ligand-gated ion channels share a common evolutionary origin.

Acetylcholine and Glutamate Mediate Fast Excitation; GABA and Glycine Mediate Fast Inhibition[19,22]

Neurotransmitters can be classified according to their actions. We have seen that acetylcholine acting on its receptor in the skeletal muscle cell membrane opens a cation channel and so depolarizes the cell toward the threshold for firing an action potential. This transmitter receptor therefore mediates an *excitatory* effect. **Glutamate** appears to act on a similar type of receptor; it has been shown to be an excitatory transmitter at the neuromuscular junction of an insect and is thought to be the major excitatory transmitter in the central nervous system of vertebrates—a counterpart to acetylcholine, which is the main excitatory transmitter in the vertebrate peripheral nervous system (and also has important central actions). *Aspartate* may act on the same receptors as glutamate, with similar consequences. There is some evidence that *ATP* serves as a fast excitatory transmitter at synapses on certain types of *smooth* muscle.

Counterbalancing these excitatory effects, **GABA** and **glycine** mediate fast *inhibition.* The receptors to which they bind are linked to channels that, when open, admit small negative ions—chiefly Cl^-—but are impermeable to positive ions. The

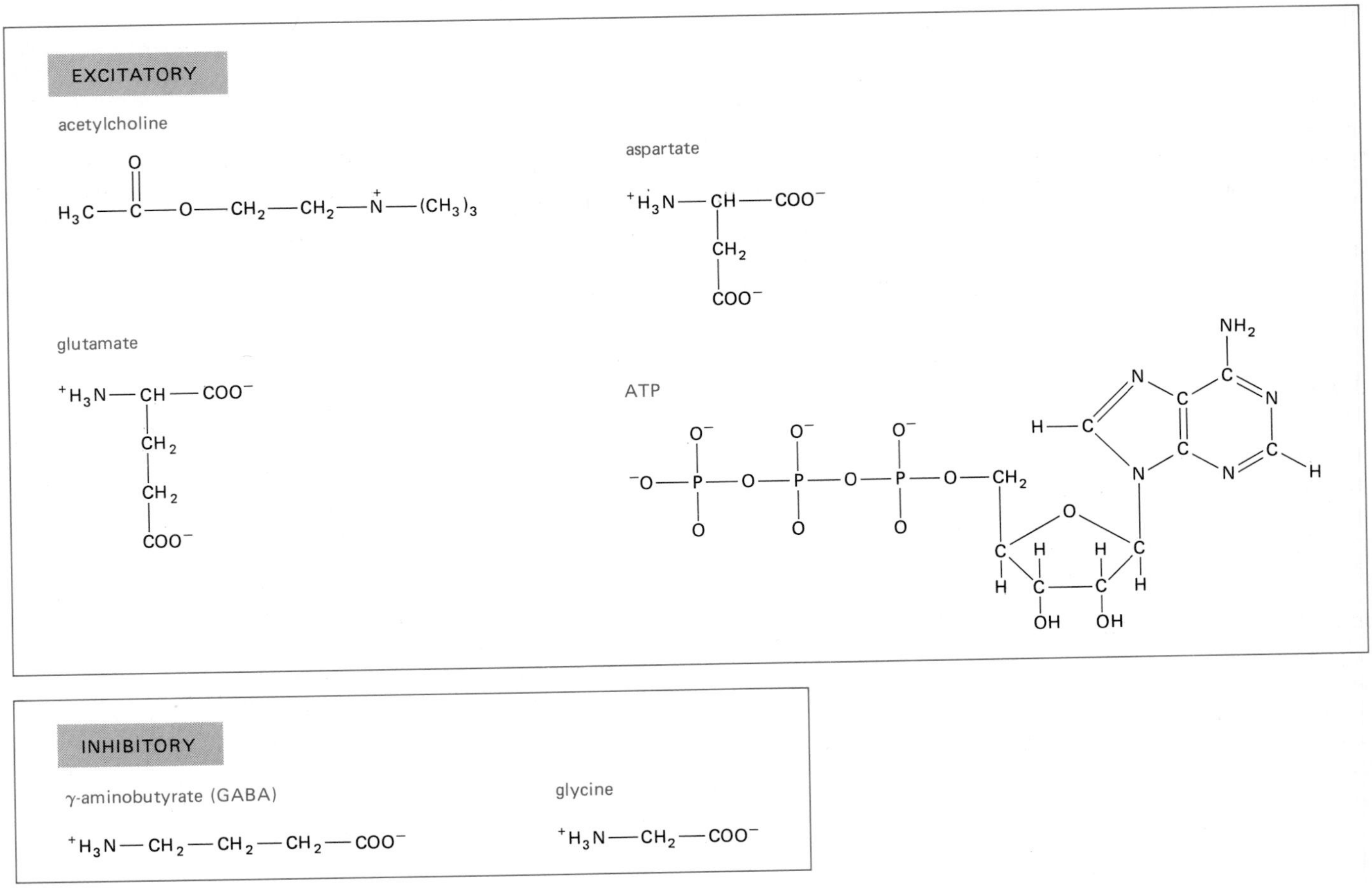

EXCITATORY

acetylcholine

$H_3C-C(=O)-O-CH_2-CH_2-\overset{+}{N}-(CH_3)_3$

aspartate

$^{+}H_3N-CH(CH_2COO^-)-COO^-$

glutamate

$^{+}H_3N-CH(CH_2CH_2COO^-)-COO^-$

ATP

INHIBITORY

γ-aminobutyrate (GABA)

$^{+}H_3N-CH_2-CH_2-CH_2-COO^-$

glycine

$^{+}H_3N-CH_2-COO^-$

Figure 19–28 The chemical structures of the major neurotransmitters believed to act on channel-linked receptors so as to mediate fast synaptic transmission.

concentration of Cl^- is much higher outside the cell than inside, and the equilibrium potential for Cl^- is close to the normal resting potential or even more negative. The opening of these Cl^- channels, therefore, tends to hold the membrane potential at its resting value or even at a hyperpolarized value, making it more difficult to depolarize the membrane and hence to excite the cell (Figure 19–29). GABA and glycine are thought to be the major transmitters that mediate fast inhibition in the vertebrate central nervous system, and GABA is known also to perform the same function at neuromuscular junctions in insects and crustaceans. The importance of the inhibitory transmitters is demonstrated by the effects of toxins that block their action: strychnine, for example, by binding to glycine receptors and blocking the action of glycine, causes muscle spasms, convulsions, and death.

Several Types of Receptors Often Exist for a Single Neurotransmitter[23]

The action of a neurotransmitter is defined not by its own chemistry but by the receptor to which it binds. In fact there are often several types of receptors for the same neurotransmitter. Acetylcholine in vertebrates, for example, acts in opposite ways on skeletal muscle cells and on heart muscle cells, exciting the former and inhibiting the latter. The acetylcholine receptors are different in the two cases; a non-channel-linked receptor is thought to mediate the inhibitory effect, which is much slower than the excitatory effect on skeletal muscle.

The channel-linked receptors that mediate rapid excitatory actions of acetylcholine are called *nicotinic* because they can be activated by nicotine; the non-channel-linked receptors that mediate the slow actions of acetylcholine, which can be either inhibitory or excitatory, are called *muscarinic* because they can be activated by muscarine (a toxin from a fungus). In addition to such receptor-specific activators (so-called *agonists*), there are also potent receptor-specific blockers (so-called *antagonists*) that distinguish the two types of acetylcholine receptors. For example, curare and α-bungarotoxin bind specifically to nicotinic acetylcholine receptors, blocking their activity, whereas atropine acts in a similar way on muscarinic receptors. Other agonists and antagonists distinguish among receptors for other neurotransmitters, and it is common to identify, localize, and assay the different receptors according to the agonists and antagonists that bind to them.

Synapses Are Major Targets for Drug Action[23,24]

The receptors for neurotransmitters are important targets for toxins and drugs. A snake paralyzes its prey by injecting α-bungarotoxin to block the nicotinic acetylcholine receptor. A surgeon can make muscles relax for the duration of an operation by blocking the same receptors with curare. The heart, meanwhile, continues to beat normally because the curare does not bind to muscarinic acetylcholine receptors. Thus the distinct ligand-binding properties of the two acetylcholine receptors allow drug action to be precisely targeted.

Most of the psychoactive drugs exert their effects at synapses, and a large proportion of them act by binding to specific receptors. GABA receptors provide an example. The best-studied type, known as $GABA_A$ receptors, are ligand-gated Cl^- channels, mediating rapid inhibition as described above. They are acted upon both by the benzodiazepine "tranquilizers," such as Valium and Librium, and by the barbiturate drugs used in the treatment of insomnia, anxiety, and epilepsy. GABA, benzodiazepines, and barbiturates bind cooperatively to three different sites on the same receptor protein: the drugs apparently alter behavior by allowing lower concentrations of GABA to open the Cl^- channel, thus *potentiating* the inhibitory action of GABA.

Synaptic transmission can also be disturbed in many other ways—for example, by interfering with the degradation or removal of the transmitter from the synaptic cleft. There are drugs that inhibit acetylcholinesterase activity at neuromuscular junctions so that acetylcholine lingers on the muscle cell for a longer time. This helps to relieve the weakness of patients suffering from myasthenia gravis, who have a shortage of functional acetylcholine receptors (see p. 1011). Other neurotransmitters, such as GABA, are not degraded enzymatically in the synaptic cleft but instead are retrieved by the presynaptic terminals that secreted them or by neighboring glial cells. Typically the nerve terminals and glial cells have specific transport proteins in their plasma membranes for active uptake of the neurotransmitter. Some psychoactive drugs block or potentiate the retrieval mechanism at specific classes of synapse, causing clinically useful effects.

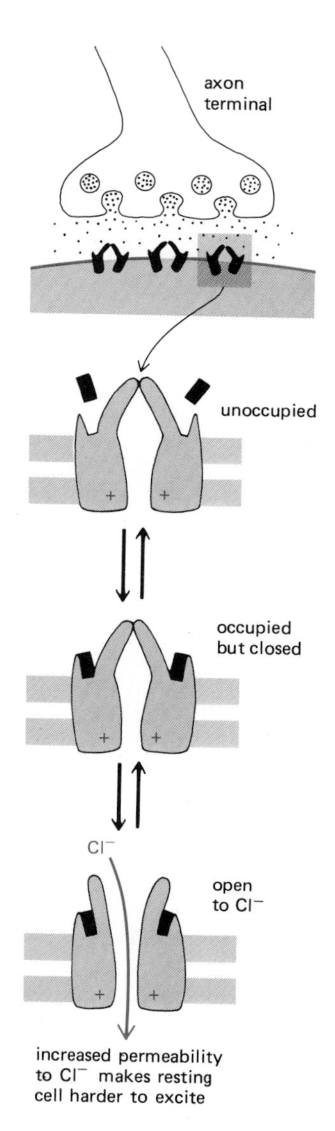

Figure 19–29 The behavior of channel-linked receptors for GABA. In response to binding of GABA, these form an open channel that is selectively permeable to Cl^-. In this way they mediate an inhibitory effect: the open Cl^- channels tend to hold the membrane close to the Cl^- equilibrium potential, which is itself close to the resting potential.

Summary

Neural signals pass from cell to cell at synapses, which can be either electrical (gap junctions) or chemical. At a chemical synapse the depolarization of the presynaptic membrane by an action potential opens voltage-gated Ca^{2+} channels, allowing an influx of Ca^{2+} to trigger exocytic release of neurotransmitter from synaptic vesicles. The neurotransmitter diffuses across the synaptic cleft and binds to receptor proteins in the membrane of the postsynaptic cell; it is rapidly eliminated from the cleft by diffusion, by enzymatic degradation, or by reuptake into nerve terminals or glial cells. The receptors for neurotransmitters can be classified as either channel-linked or non-channel-linked. Channel-linked receptors, also known as ligand-gated ion channels, mediate rapid postsynaptic effects, occurring within a few milliseconds. Only a handful of neurotransmitters are known to act on such receptors. In particular, acetylcholine and glutamate (and probably aspartate and ATP) open ligand-gated channels that are permeable only to cations and thereby produce rapid excitatory postsynaptic potentials, whereas GABA and glycine open homologous channels that are permeable chiefly to Cl^- and thereby produce rapid inhibitory postsynaptic potentials. All these neurotransmitters, as well as many others, may also act on non-channel-linked receptors, with slower and more complex consequences.

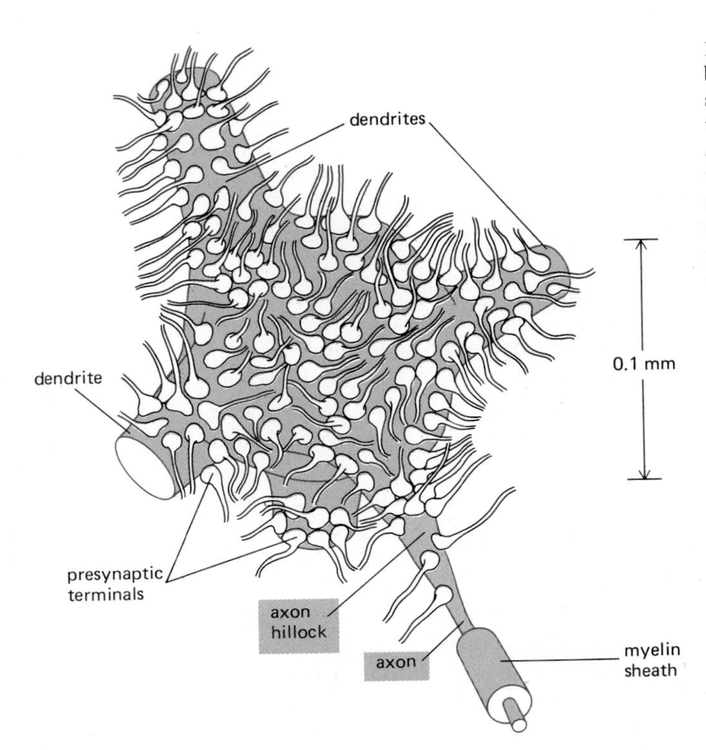

Figure 19–30 A motor neuron cell body in the spinal cord, showing some of the many thousands of nerve terminals that synapse on the cell and deliver signals from other parts of the organism to control its firing. The regions of the motor-neuron plasma membrane that are not covered with synaptic endings are covered by glial cells (not shown).

The Role of Ion Channels in Neuronal Computation[25]

In the central nervous system, neurons typically receive inputs from many presynaptic cells—the number may be anything from one to many thousands. For example, several thousand nerve terminals, from hundreds or perhaps thousands of neurons, make synapses on a typical motor neuron in the spinal cord; its cell body and dendrites are almost completely covered with them (Figure 19–30). Some of these synapses transmit signals from the brain, others bring sensory information from muscles or from the skin, and still others supply the results of computations made by interneurons in the spinal cord. The motor neuron must combine the information received from these many sources and react either by firing signals along its axon or by remaining quiet.

The motor neuron provides a typical example of the way in which neurons play their individual parts in the fundamental task of computing an output from a complex set of inputs. Of the many synapses on the motor neuron, some will tend to excite it, others to inhibit it. Although the motor neuron secretes the same neurotransmitter at all its axon terminals, it makes many different types of receptor proteins, concentrating them at different postsynaptic sites on its surface. At each such site, firing of the presynaptic cell causes a specific set of channels to open or close, leading to a characteristic voltage change or **postsynaptic potential (PSP)** in the motor neuron. An *excitatory PSP* (produced, for example, by the opening of channels permeable to Na^+) is generally a depolarization; an *inhibitory PSP* (produced, for example, by the opening of Cl^- channels) is usually a hyperpolarization. The PSPs generated at the different synapses on a single neuron are highly variable in size and duration. At one synapse on the motor neuron, an incoming nerve impulse might produce a depolarization of less than 0.1 mV, whereas at another there might be a depolarization of 5 mV. However, as we shall now discuss, the properties of the system are such that even small PSPs can combine to produce a large effect.

The Shift of Membrane Potential in the Body of the Postsynaptic Cell Represents a Spatial and Temporal Summation of Many Postsynaptic Potentials[25,26]

The membrane of the dendrites and cell body of most neurons, although rich in receptor proteins, contains few voltage-gated Na^+ channels and so is relatively inexcitable. An individual PSP generally does not trigger the postsynaptic membrane to fire an action potential. Instead, each incoming signal is faithfully reflected in a PSP of graded magnitude, which falls off with distance from the site of the synapse. If signals arrive simultaneously at several synapses in the same region of the dendritic tree, the total PSP in that neighborhood will be roughly the sum of the individual PSPs, with inhibitory PSPs making a negative contribution to the total. At the same time, the net electrical disturbance produced in one postsynaptic region will spread to other regions through the passive cable properties of the dendritic membrane.

The cell body, where the effects of the PSPs converge, is relatively small (generally smaller than 100 μm in diameter) compared with the dendritic tree (whose branches may extend for millimeters). The membrane potential in the cell body and its immediate neighborhood will therefore be roughly uniform and will be a composite of the effects of all the signals impinging on the cell, weighted according to the distances of the synapses from the cell body. The **grand postsynaptic potential** of the cell body is thus said to represent a **spatial summation** of all the stimuli received. If excitatory inputs predominate, it will be a depolarization; if inhibitory inputs predominate, it will usually be a hyperpolarization.

While spatial summation combines the effects of signals received at different sites on the membrane, **temporal summation** combines the effects of signals received at different times. The neurotransmitter released when an action potential arrives at a synapse evokes a PSP in the postsynaptic membrane that rises rapidly to a peak (through the transient opening of ligand-gated ion channels) and then declines to baseline with a roughly exponential time course (which depends on the membrane capacitance). If a second action potential arrives before the first PSP has decayed completely, the second PSP adds to the remaining tail of the first. If, after a period of inactivity, a long train of action potentials is delivered in quick succession, each PSP adds to the tail of the preceding PSP, building up to a large sustained average PSP whose magnitude reflects the rate of firing of the presynaptic neuron (Figure 19–31). This is the essence of temporal summation: it translates the *frequency* of incoming signals into the *magnitude* of a net PSP.

The Grand PSP Is Translated into a Nerve Impulse Frequency for Long-distance Transmission[27]

Temporal and spatial summation together provide the means by which the rates of firing of many presynaptic neurons jointly control the membrane potential in the body of a single postsynaptic cell. The final step in the neuronal computation made by the postsynaptic cell is the generation of an output, usually in the form of action potentials, to relay a signal to other cells that are often far away. The output signal reflects the magnitude of the grand PSP in the cell body. However, while the grand PSP is a continuously graded variable, action potentials are all-or-none and uniform in size. The only variable in signaling by action potentials is the time interval between one action potential and the next. For long-distance transmission, the magnitude of the grand PSP is therefore translated, or *encoded,* into the *frequency* of firing of action potentials (Figure 19–32). This encoding is achieved by a special set of voltage-gated ion channels present at high density at the base of the axon, adjacent to the cell body, in a region known as the **axon hillock** (see Figure 19–30).

Before we discuss how these channels operate, a word of qualification is necessary. The firing of an action potential itself causes drastic changes of the membrane potential of the cell body, which therefore no longer directly reflects the net synaptic stimulation that the cell is receiving. It is therefore a complex

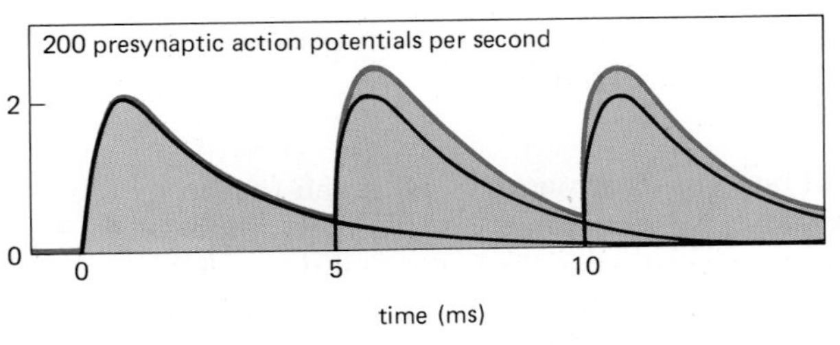

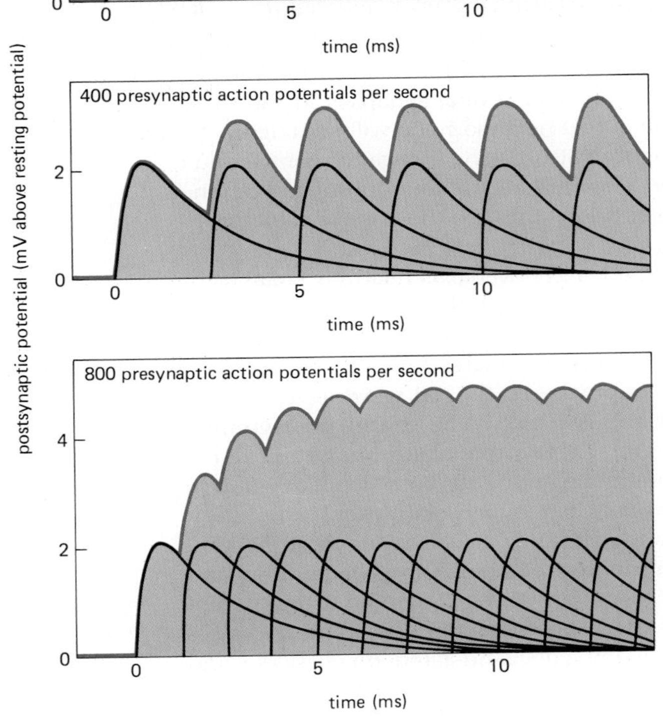

Figure 19–31 Temporal summation. The overlapping curves within the shaded region of each graph represent the individual contributions to the total postsynaptic potential evoked by the arrival of the successive presynaptic action potentials.

problem to give a rigorous analysis of the encoding mechanism. In the nonrigorous, qualitative account that follows, we shall loosely refer to "the strength of synaptic stimulation" or to "the grand PSP," meaning the grand PSP that would be observed if action potentials were somehow prevented from firing; and we shall suppose that this underlying grand PSP is the cause of the firing of action potentials.

Encoding Requires a Combination of Different Ion Channels[28]

The propagation of action potentials depends chiefly, and in many vertebrate axons almost entirely, on voltage-gated Na^+ channels. The membrane of the axon hillock is where action potentials are initiated, and Na^+ channels are plentiful there. But to perform its special function of encoding, the membrane in that neighborhood typically contains in addition at least four other classes of ion channels—three selective for K^+ and one selective for Ca^{2+}. The three varieties of K^+ channels have different properties; we shall refer to them as the *delayed*, the *early*, and the *Ca^{2+}-activated K^+ channels*. The functions of these channels in encoding have been studied most thoroughly in giant neurons of mollusks, but the principles appear to be similar for most other neurons.

To understand the necessity for multiple types of channels, consider first the behavior that would be observed if the only voltage-gated ion channels present in the nerve cell were the Na^+ channels. Below a certain threshold level of synaptic stimulation, the depolarization of the axon hillock membrane would be insufficient to trigger an action potential. With gradually increasing stimulation, the threshold would be crossed: the Na^+ channels would open, and an action potential would fire. The action potential would be terminated in the usual way by inactivation of the Na^+ channels. Before another action potential could fire, these channels would have to recover from their inactivation. But that would require a return of the membrane voltage to a very negative value, which would not occur as long as the

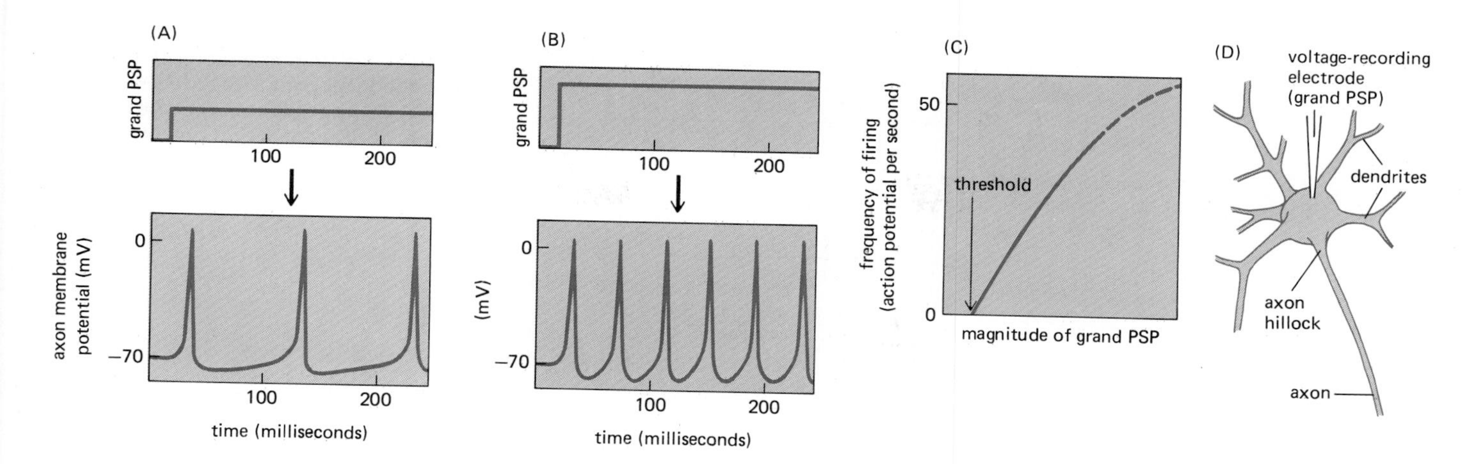

Figure 19–32 The encoding of the grand PSP in the form of the frequency of firing of action potentials by an axon. A comparison of (A) and (B) shows how the firing frequency of an axon increases with an increase in the grand PSP, while (C) summarizes the general relationship. In (D) the experimental setup for measuring the grand PSP is shown. In (A) and (B) the upper graphs (marked "grand PSP") show the net intensity of synaptic stimulation as received by the cell body, while the lower graphs show the resulting trains of action potentials that are transmitted along the axon. The upper graphs can be thought of as representations of the grand PSP that would be observed if the firing of action potentials were somehow blocked.

strong depolarizing stimulus (from PSPs) was maintained. An additional channel type is needed, therefore, to repolarize the membrane after each action potential to prepare the cell to fire another. This task is performed by the **delayed K^+ channels,** which we discussed previously in relation to the propagation of the action potential (see p. 1070). They are voltage-gated and respond to membrane depolarization in much the same way as the Na^+ channels, but with a longer time delay. By opening during the falling phase of the action potential, they permit an efflux of K^+, which short-circuits the effect of even a sustained depolarizing stimulus and drives the membrane back toward the K^+ equilibrium potential. This potential is so far negative that the Na^+ channels recover from their inactivated state. In addition, the K^+ conductance turns itself off: repolarization of the membrane causes the delayed K^+ channels themselves to close again (without ever entering an inactivated state). Once repolarization has occurred, the depolarizing stimulus from synaptic inputs becomes capable of raising the membrane voltage to threshold again so as to cause another action potential to fire. In this way, sustained stimulation of the dendrites and cell body leads to repetitive firing of the axon.

However, repetitive firing in itself is not enough: the frequency of the firing has to reflect the intensity of the stimulation. Detailed calculations show that a simple system of Na^+ channels and delayed K^+ channels is inadequate for this purpose. Below a certain threshold level of steady stimulation, the cell will not fire at all; above that threshold it will abruptly begin to fire at a relatively rapid rate. The **early K^+ channels** (also known as *A channels*) solve the problem. These too are voltage-gated and open when the membrane is depolarized, but their specific voltage sensitivity and kinetics of inactivation are such that they act to reduce the rate of firing at levels of stimulation that are only just above the threshold. Thus they help to remove the discontinuity in the relationship between the firing rate and the intensity of stimulation. The result is a firing rate that is proportional to the strength of the depolarizing stimulus over a very broad range (see Figure 19–32).

Adaptation Lessens the Response to an Unchanging Stimulus[29]

The process of encoding is usually further modulated by the two other types of ion channels in the axon hillock that were mentioned at the outset—*voltage-gated Ca^{2+} channels* and *Ca^{2+}-activated K^+ channels.* The former are similar to the Ca^{2+} channels that mediate release of neurotransmitter at axon terminals: those present in the neighborhood of the axon hillock open when an action potential fires, allowing Ca^{2+} into the axon. The **Ca^{2+}-activated K^+ channel** is different from any of the channel types described earlier. It opens in response to a raised concentration of Ca^{2+} at the cytoplasmic face of the nerve cell membrane.

Suppose that a strong depolarizing stimulus is applied for a long time, triggering a long train of action potentials. Each action potential permits a brief influx of Ca^{2+} through the voltage-gated Ca^{2+} channels, so that the intracellular Ca^{2+}

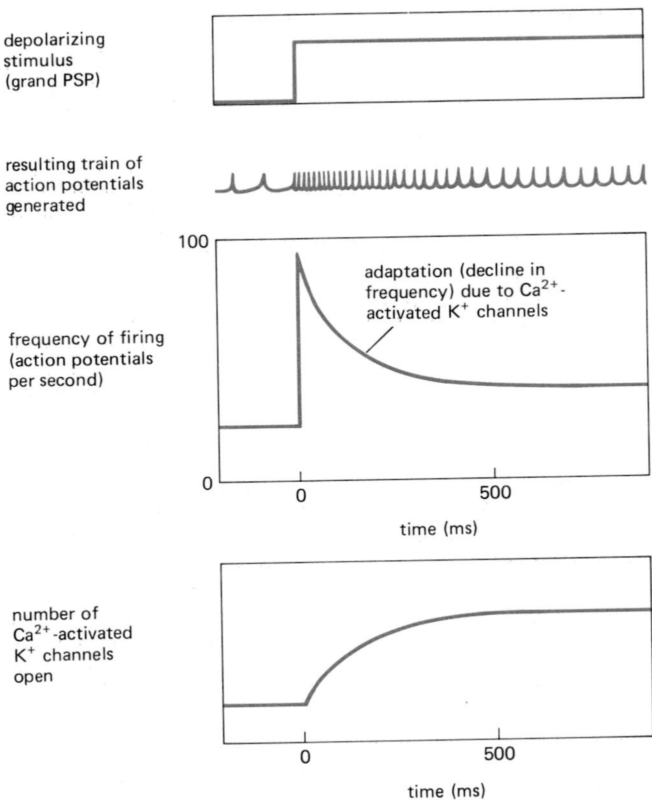

Figure 19–33 Adaptation. When steady stimulation is prolonged, the stimulated cell gradually reduces the strength of its response, as expressed in the rate of firing of action potentials.

concentration gradually builds up to a high level. This opens the Ca^{2+}-activated K^+ channels, and the resulting increased permeability of the membrane to K^+ makes the membrane harder to depolarize and increases the delay between one action potential and the next. In this way a neuron that is stimulated continuously for a prolonged period becomes gradually less responsive to the constant stimulus. The phenomenon, which can also occur by other mechanisms, is known as **adaptation** (Figure 19–33). It allows a neuron, and indeed the nervous system generally, to react sensitively to *change*, even against a high background level of steady stimulation (see p. 1107). It is one of the strategies that help us, for example, to feel a touch on the shoulder and yet ignore the constant pressure of our clothing.

Not All Signals Are Delivered via the Axon[30]

In the typical neuron that we have been describing, there is a clear distinction, in both structure and function, between dendrites and axon. Some neurons, however, do not conform to this model, although the molecular principles of their operation are the same. In most invertebrates, for example, the majority of neurons have a *unipolar* organization: the cell body is connected by a single stalk to a branching system of cell processes, among which it is not always easy to see a structural difference between dendrites and axon (Figure 19–34). The functional distinction can also be blurred, in both vertebrates and invertebrates: processes that are classified structurally as dendrites often form presynaptic as well as postsynaptic specializations and deliver signals to other cells as well as receive them. Conversely, synaptic inputs are sometimes received at strategic sites along the axon—for example, close to the axon terminal, where they can inhibit or facilitate the release of neurotransmitter from that particular terminal without affecting transmission at the terminals of other branches of the same axon (Figure 19–35). We shall discuss later (see p. 1097) an example of this important device of *presynaptic inhibition* or *presynaptic facilitation.*

Synapses at which a dendrite delivers a stimulus to another cell play a large part in communication between neurons that lie close together, within a few millimeters or less. Over such distances, electrical signals can be propagated pas-

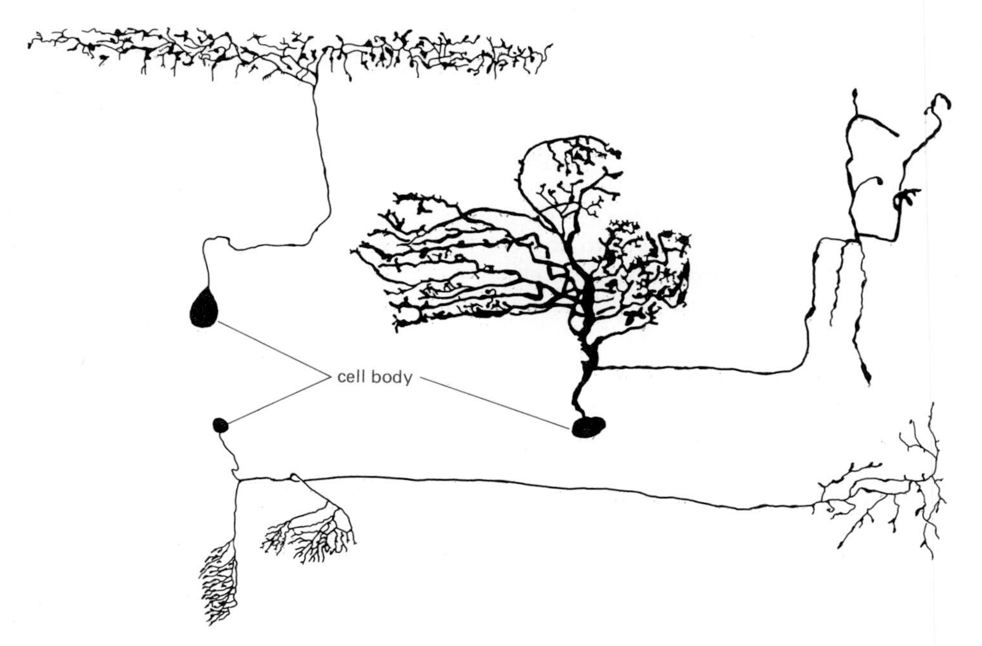

Figure 19–34 Neurons from a fly, showing the structure typical of most neurons in invertebrates, in which the nerve cell body is connected by a stalk to the system of nerve cell processes and does not have dendrites projecting from it directly. The sensory neurons in the spinal ganglia of vertebrates have a similar organization. (From N. Strausfield, Atlas of an Insect Brain. New York: Springer, 1976.)

sively, spreading from postsynaptic sites on the dendritic membrane, where they are received, to presynaptic sites on the same dendritic membrane, where they then control transmitter release. Indeed, there are neurons that possess no axon, do not conduct action potentials, and perform all of their signaling via processes that are conventionally referred to as dendrites. Moreover, if the dendritic tree is large, separate parts of it can behave as more or less independent pathways for communication and for information processing. In some neurons the range of possibilities is still further complicated by the presence of voltage-gated ion channels in the dendritic membrane, which enable the dendrites to conduct action potentials. Thus even a single neuron can behave as a highly complex computational device.

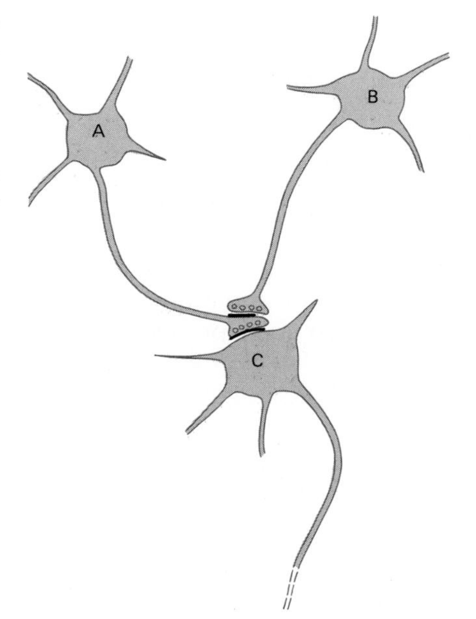

Figure 19–35 An axo-axonic synapse. The neurotransmitter released from the axon terminal of cell B acts on channels in the axon terminal of cell A, thereby altering the number of quanta of neurotransmitter released onto C when A fires. If firing of B causes a reduction in the stimulus delivered by A to C, then B is said to exert a *presynaptic inhibition;* the contrary effect is called *presynaptic facilitation.*

Summary

A typical neuron receives on its dendrites and cell body many different excitatory and inhibitory synaptic inputs, which combine, by spatial and temporal summation, to produce a grand postsynaptic potential in the cell body. The magnitude of the grand postsynaptic potential is translated (encoded) for long-distance transmission into the rate of firing of action potentials, by a system of ion channels in the membrane of the axon hillock. The encoding mechanism often shows adaptation, so that the cell responds weakly to a constant stimulus but strongly to a change of stimulus. There are many variants of this basic scheme: for example, not all neurons produce an output in the form of action potentials, dendrites can be presynaptic as well as postsynaptic, and axons can be postsynaptic as well as presynaptic.

Non-Channel-linked Receptors and Synaptic Modulation[13,31]

At synapses that use channel-linked receptors, the effect of the neurotransmitter is immediate, simple, and brief, and the site of reception of the message is defined with pinpoint accuracy; the transmitter released from one axon terminal acts only on a single postsynaptic cell. By contrast, non-channel-linked receptors allow for effects that are slow, complex, long-lasting, and often spatially diffuse; the transmitter released from one terminal may act on many cells in the neighborhood of that terminal. These slow effects are often described as examples of *neuromodulation* because they modulate the rapid responses mediated by channel-linked receptors on the same cell. The non-channel-linked receptors act by the same molecular mechanisms as the receptors for hormones and local chemical mediators outside the nervous system—indeed, many of them probably are identical.

As discussed in Chapter 12 (see p. 695), non-channel-linked cell-surface receptors for signaling molecules fall into two large families: (1) **catalytic receptors,** most of which are tyrosine-specific protein kinases, which, when activated by ligand binding, directly phosphorylate tyrosine residues on proteins inside the cell; and (2) **G-protein-linked receptors,** which transmit signals into the cell interior by activating a GTP-binding regulatory protein, or *G protein,* which in turn activates or inactivates a membrane-bound enzyme or ion channel. Most of the non-channel-linked receptors for neurotransmitters studied so far seem to be G-protein-linked, employing their G protein in one of at least three ways:

1. The G proteins may activate or inactivate adenylate cyclase, thereby controlling the cyclic AMP level in the postsynaptic cell. The cyclic AMP then regulates the activity of the cyclic-AMP-dependent protein kinase (*A-kinase*—see p. 709), which, among other target proteins, can phosphorylate ion channels in the plasma membrane, altering their properties. Cyclic AMP may also regulate some ion channels by binding to them directly.
2. The G protein may activate the inositol phospholipid pathway (see p. 702), thereby activating protein kinase C (C-kinase) and releasing Ca^{2+} into the cytosol from a Ca^{2+}-sequestering compartment in the postsynaptic cell. The C-kinase may regulate the behavior of ion channels by phosphorylating them. The Ca^{2+} may alter ion channel behavior directly, or indirectly via a Ca^{2+}-dependent protein kinase that phosphorylates the channel (see p. 711).
3. The G protein may interact directly with ion channels, causing them to open or close.

In each case a set of molecules in the postsynaptic cell act as go-betweens or *intracellular messengers,* diffusing within the cell to relay the signal from the receptor to other cell components. The more steps there are in this cascade of intracellular messengers, the more opportunities there are for amplification and regulation of the signal (see p. 713).

More than 50 neurotransmitters have been identified that act on non-channel-linked receptors to produce these varied and complex effects. Some, such as acetylcholine, also bind to channel-linked receptors, whereas others, such as neuropeptides (see below), apparently do not.

Non-Channel-linked Receptors Mediate Slow and Diffuse Responses[32]

Whereas channel-linked receptors take only a few milliseconds or less to produce electrical changes in the postsynaptic cell, non-channel-linked receptors typically take hundreds of milliseconds or longer. This is to be expected, since a series of enzymatic reactions must intervene between the initial signal and the ultimate response. Moreover, the signal itself is often not only temporally but also spatially diffuse.

A clear example is seen in the innervation of smooth muscle by axons releasing *norepinephrine,* which activates adenylate cyclase via a G-protein-linked receptor. Here, the transmitter is released not from nerve terminals but from swellings or *varicosities* along the length of the axon (Figure 19–36). These varicosities contain synaptic vesicles but no active zones to define the exact sites of release. Moreover, the varicosities are not closely apposed to specialized receptive sites on a postsynaptic cell; instead, the transmitter diffuses widely to act on many smooth muscle cells in the neighborhood, in the manner of a local chemical mediator (see p. 682). It is likely that many of the signaling molecules that operate on catalytic and G-protein-linked receptors in the central nervous system also act in this *paracrine* mode. Indeed, many of these neurotransmitters also serve as hormones or as local chemical mediators outside the nervous system: for example, norepinephrine, together with its close relative *epinephrine,* is also released as a hormone from the adrenal gland.

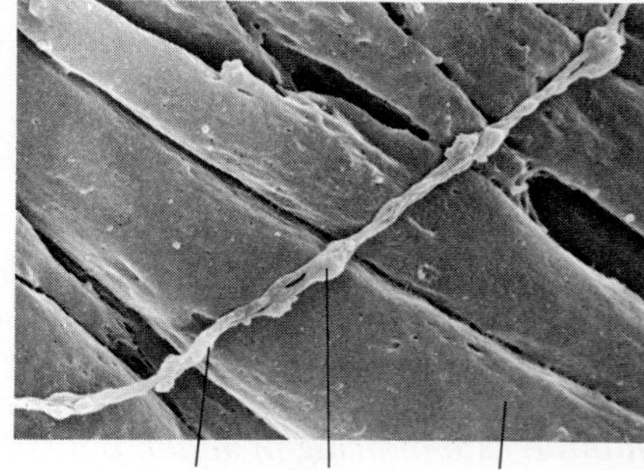

Figure 19–36 Scanning electron micrograph of a small bundle of autonomic motor axons innervating smooth muscle cells in the wall of the ureter. The varicosities (swellings) contain synaptic vesicles loaded with the neurotransmitter norepinephrine. The synapses here are ill-defined structures, with a gap that may be as large as 0.2 μm between the site of release of the neurotransmitter and the nearest muscle cell membrane on which it must act. (From S. Tachibana, M. Takeuchi, and Y. Uehara, *J. Urol.* 134:582–586, 1985. © by Williams & Wilkins, 1985.)

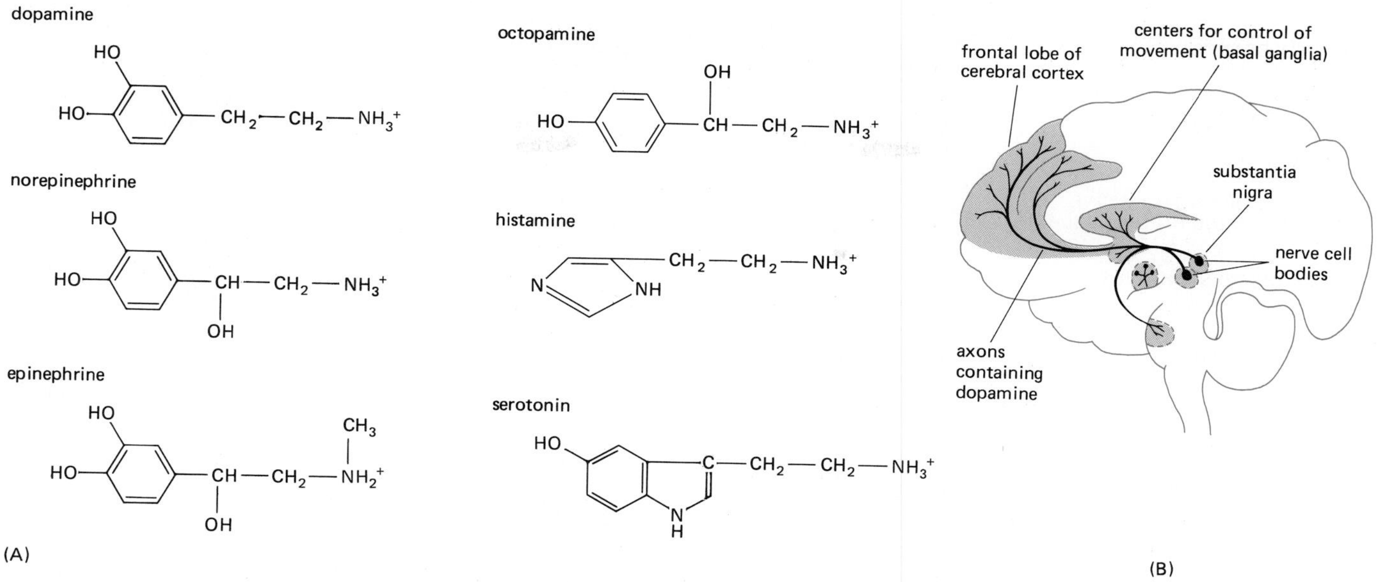

Figure 19–37 (A) The monoamine family of neurotransmitters. (B) Schematized diagram of the distribution of dopamine-containing neurons in the human brain. The movement disorders of Parkinson's disease are due to the death of many of the cells in a particular set of dopamine-containing neurons (those in the substantia nigra); the symptoms can be relieved by drug treatments that boost the synthesis of dopamine and inhibit its breakdown. The distribution of monoamine-containing neurons can be made visible by treating tissue sections with formaldehyde, which reacts with monoamines to give fluorescent products.

Epinephrine and norepinephrine are representatives of the family of **monoamine** neurotransmitters, which have widespread functions both in vertebrates and in invertebrates and are of great medical importance (Figure 19–37A). It is possible to design drugs that interfere with the synthesis, uptake, or breakdown of particular monoamines, or that interact with particular subclasses of monoamine receptors; and some of these drugs have proved to be valuable in the treatment of psychiatric and neurological diseases. Schizophrenia, for instance, can often be treated successfully with drugs that block certain classes of *dopamine* receptors, while drugs that increase the concentrations of dopamine in the brain give dramatic relief from the movement disorders of Parkinson's disease (Figure 19–37B). Drugs that raise synaptic concentrations of noradrenaline and/or serotonin are often effective in the treatment of severe depression.

The Neuropeptides Are by Far the Largest Family of Neurotransmitters[32,33]

Most of the signaling molecules used elsewhere in the body are also employed by neurons. This is true in particular of the array of small protein molecules or peptides that serve as hormones and local chemical mediators to control such bodily functions as the maintenance of blood pressure, the secretion of digestive enzymes, and the proliferation of cells.

Rapid advances in this area over the last ten years or so have largely depended on immunocytochemistry. Once a peptide has been identified in one tissue, it is possible to make antibodies against it and to use these to search elsewhere for that peptide and for others that are structurally related. In this way neurons have been found to contain peptides that were not previously suspected to have a neural function, including many newly discovered varieties. The evidence that these **neuropeptides** (Figure 19–38) serve as neurotransmitters is in most cases persuasive but incomplete. For example, an antipeptide antibody might be shown to label certain neurons and their axon terminals, while the peptide itself, when supplied locally, might mimic the effect of activity of these neurons. Most convincingly, the peptide might be shown to be secreted when the neurons are active, and the effects of activity of the neurons might be shown to be blocked by antibodies against the peptide. Neuropeptides appear to be particularly important in regulating feelings and drives, such as pain, pleasure, hunger, thirst, and sex.

The nonpeptide neurotransmitters are synthesized by enzymes that are usually present both in the cell body and in the axon terminals, so that even if the axon is long, the stores of neurotransmitter at the synapse can be rapidly replen-

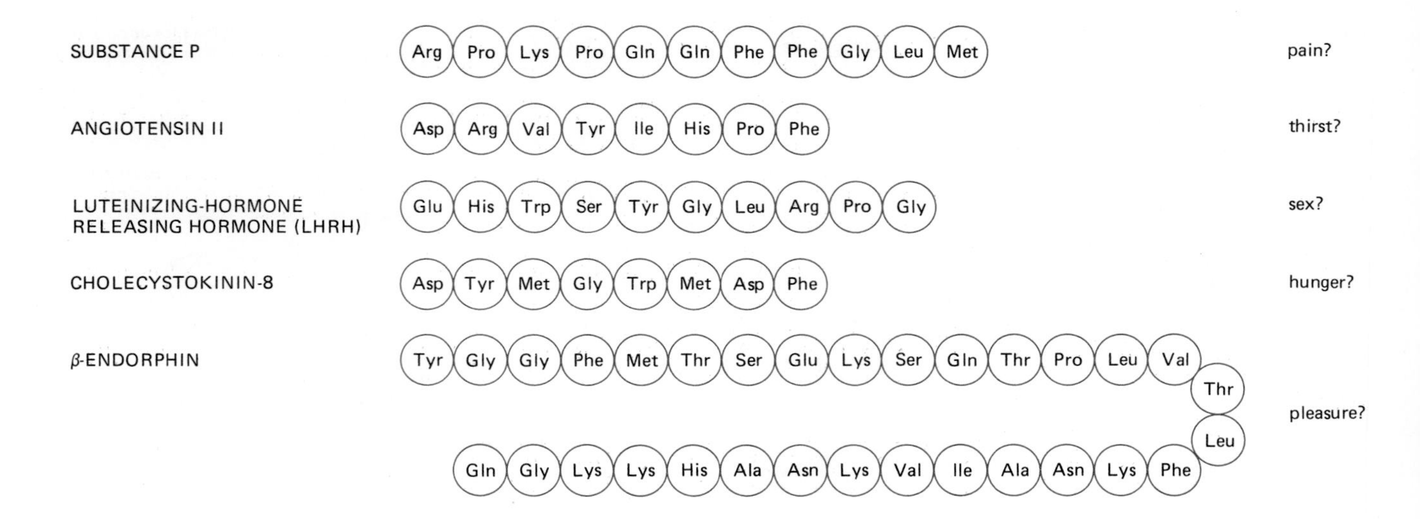

Figure 19–38 A small selection of neuropeptides, with a tentative indication of some of the sensations and drives in which they are thought to be involved.

ished. The neuropeptides, by contrast, are made on ribosomes on rough endoplasmic reticulum in the cell body and must be exported to the axon terminals by fast axonal transport—a journey that may take a day or more for a long axon. Neuropeptides are derived from larger precursor proteins, from which they are cleaved enzymatically; in many cases more than one functional peptide is cleaved from a single precursor molecule, which for this reason is called a *polyprotein.* Synaptic vesicles loaded with neuropeptides can usually be recognized by their large size compared with vesicles containing acetylcholine, amino acid transmitters, or monoamines.

At many of the synapses where neuropeptides are secreted, a nonpeptide neurotransmitter is also released, and the two transmitters act side by side but in different ways. The presynaptic axon terminals in certain autonomic ganglia of a bullfrog, for example, contain both acetylcholine and a peptide that closely resembles the reproductive hormone LHRH (luteinizing-hormone releasing hormone). The postsynaptic cell membrane contains at least three types of receptors: a nicotinic (channel-linked) acetylcholine receptor that mediates a fast response, a muscarinic (G-protein-linked) acetylcholine receptor that mediates a much slower response, and a receptor (probably G-protein-linked) for the LHRH-like peptide that mediates the slowest response of all (Figure 19–39A). The action of the LHRH-like peptide is not only slower than that of acetylcholine but also more diffuse, so that the peptide molecules released at a synapse on one postsynaptic cell also evoke a postsynaptic potential in other cells in the neighborhood (Figure 19–39B).

If, as seems likely, other neuropeptide transmitters have similar properties, one can see why the number of neuropeptides needs to be large. Since the peptides diffuse widely, their site of release does not define their site of action; consequently, if peptides released from different presynaptic terminals in the same neighborhood are to act on different postsynaptic targets, the peptides and their receptors must be chemically different.

Long-lasting Alterations of Behavior Reflect Changes in Specific Synapses[34]

The responses mediated by non-channel-linked receptors are long-lasting as well as slow in onset. Therein lies much of their special importance for the control of behavior: they bring about a persistent change in the rules that govern the immediate reaction of the nervous system to the inputs it receives and thus appear to be the basis for at least some forms of memory. This is most strikingly illustrated by studies on the sea snail *Aplysia,* a type of mollusk (Figure 19–40). In this animal, changes in behavior with experience can be traced to identified neural circuits and their molecular mechanisms can be deciphered.

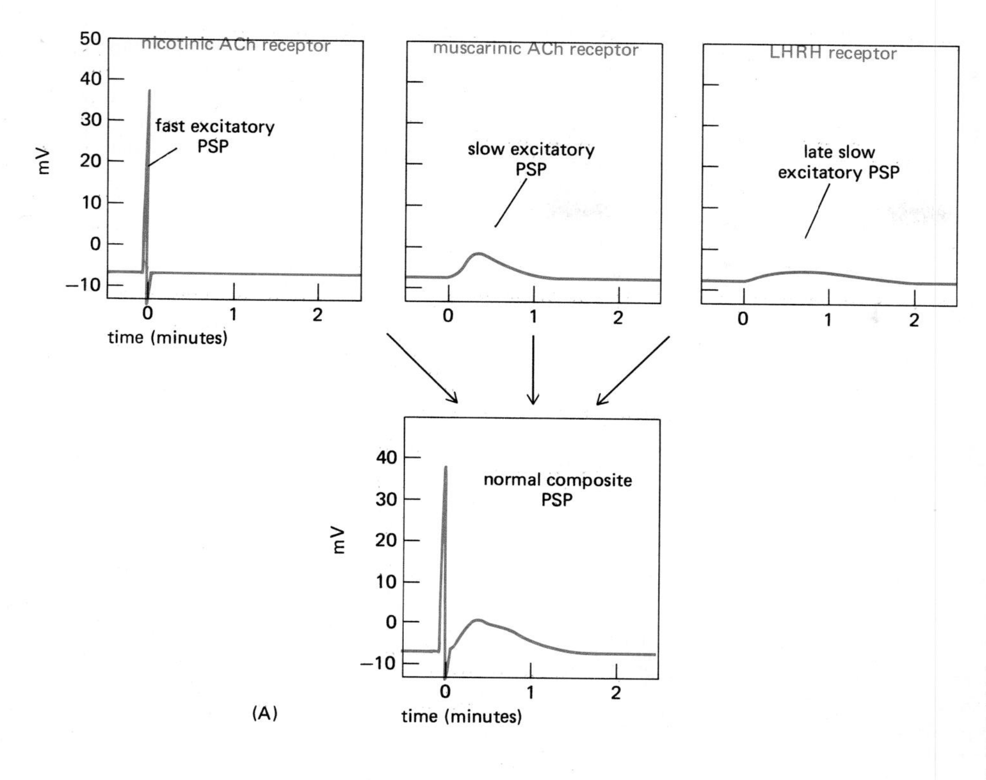

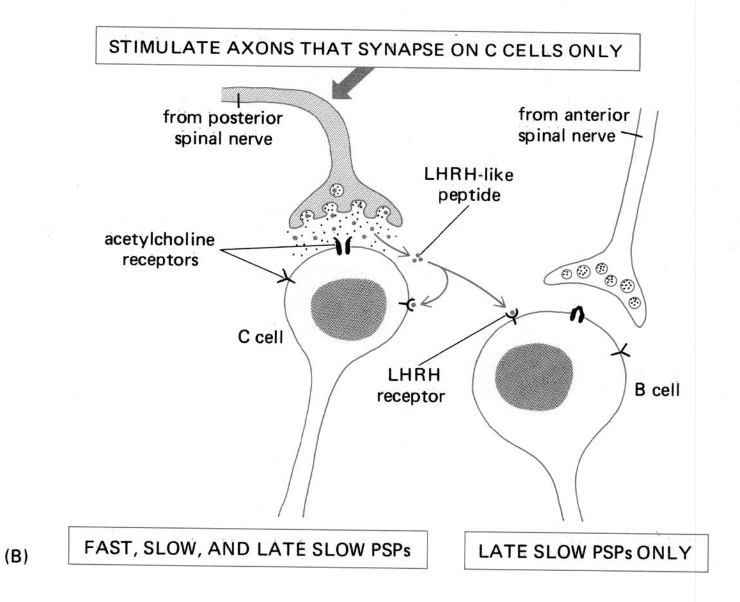

Figure 19–39 The pattern of responses to a peptide neurotransmitter.

(A) The three components of the postsynaptic potential observed in a ganglion cell of a frog following stimulation of the presynaptic nerve. The presynaptic axon terminal releases two neurotransmitters—acetylcholine and a peptide that closely resembles LHRH (luteinizing-hormone releasing hormone). The normal complex PSP is a composite of responses mediated by three kinds of receptors—two for acetylcholine and one for the LHRH-like peptide. Each of the three components can be observed independently by blocking the receptors responsible for the other two components with specific toxins. Only the fast excitatory PSP, mediated by a channel-linked acetylcholine receptor, is large enough to trigger an action potential. The two slow components, mediated presumably by non-channel-linked receptors, serve to modulate the excitability of the cell, making it more responsive to any stimuli that follow soon after the initial stimulus.

(B) Schematic diagram of an experiment on the same ganglion demonstrating the diffuse mode of action of the LHRH-like neuropeptide. This is released together with acetylcholine at synapses made on one population of cells (the C cells), but it diffuses over distances of several tens of micrometers to produce a late slow PSP in other, neighboring cells also (the B cells). (A, after Y.N. Jan *et al., Cold Spring Harbor Symp. Quant. Biol.* 48:363–374, 1983.)

Aplysia withdraws its gill if its siphon is touched (see Figure 19–40). If the siphon is touched repeatedly, the animal becomes **habituated** and ceases to respond. Habituation is similar in function to adaptation, although it operates on a longer time scale and, as we shall see, at a different point in the neural pathway. An unpleasant experience, such as a hard bang or an electrical shock, removes the habituation and leaves the animal **sensitized** so that it responds vigorously again to being touched. The sensitization persists for many minutes or hours, according to the severity of the brief noxious stimulus that caused it, and represents a simple form of *short-term memory*. If the animal is struck or shocked repeatedly on successive days, the sensitization—that is, the memory—becomes

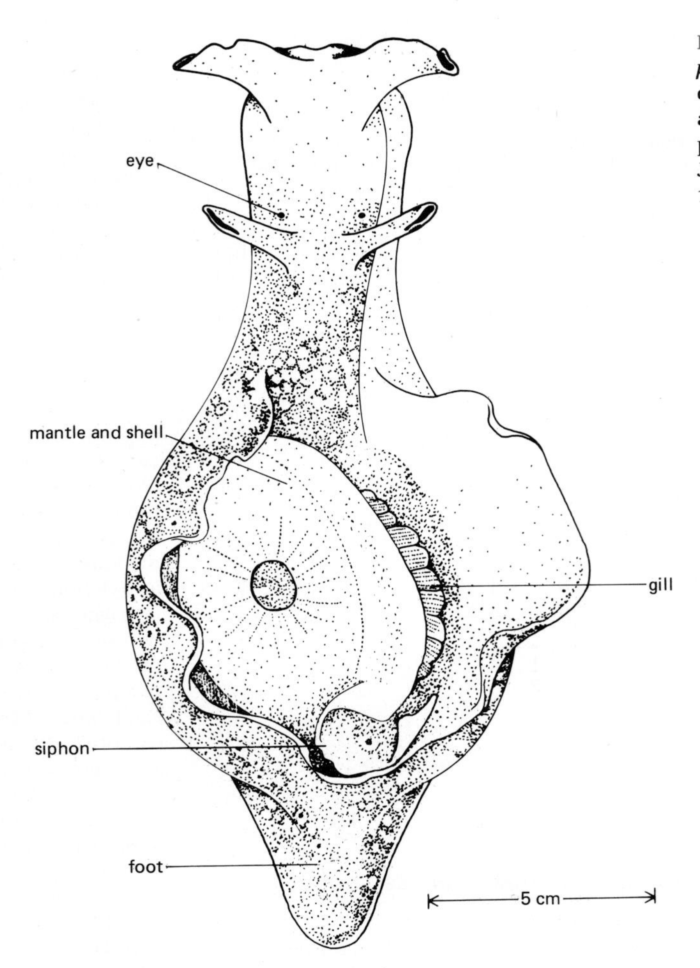

Figure 19–40 The sea snail *Aplysia punctata* viewed from above. An overlying flap of tissue has been drawn aside to reveal the gill under the protective mantle and shell. (After J. Guiart, *Mem. Soc. Zool. France* 14:219, 1901.)

long term, lasting for weeks. These modifications of behavior can be traced to changes occurring in a particular class of synapses in the neural circuit that controls the gill-withdrawal reflex.

The neurons of *Aplysia* are large (~100 μm), relatively few in number (~10^5), and stereotyped in their individual appearance. Touching the siphon stimulates a set of *sensory neurons* to fire. The sensory neurons make excitatory synapses on the *gill-withdrawal motor neurons,* which drive the muscles for gill withdrawal, and changes in these synapses underlie the behavioral phenomena. During habituation, the PSP evoked in the gill-withdrawal neurons is observed to become weaker with repeated firing of the sensory cells. Sensitization has the reverse effect, increasing the PSP. In both cases the changes are due to alterations in the amount of neurotransmitter released from the presynaptic axon terminals of the sensory neurons when they fire. The problem therefore reduces to the question of how transmitter release at these synapses is modulated.

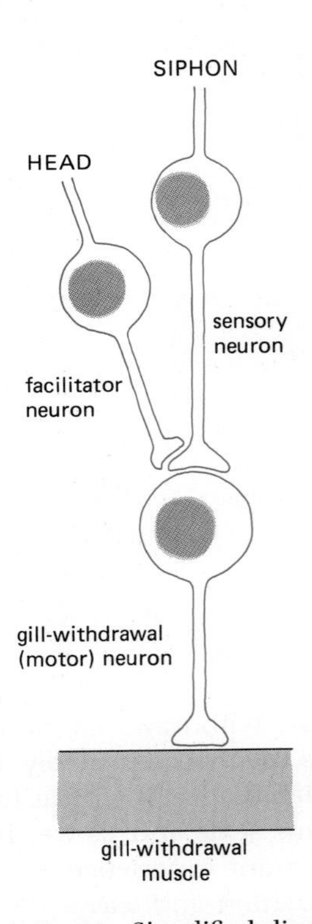

Figure 19–41 Simplified diagram of the neuronal pathways involved in habituation and sensitization of the gill-withdrawal reflex in *Aplysia.* Only one representative neuron of each class is shown.

G-Protein-linked Receptors Mediate Sensitization in *Aplysia*[35]

As mentioned on page 1077, the amount of neurotransmitter released at a synapse is controlled by the amount of Ca^{2+} that enters the terminal during the action potential. In habituation, repeated firing of the sensory cells leads to a modification of channel proteins in the terminals such that Ca^{2+} entry is reduced and the amount of neurotransmitter released decreases; in sensitization, by contrast, Ca^{2+} entry is increased, so that more neurotransmitter is released. The detailed molecular changes underlying these simple forms of memory are best understood in the case of sensitization.

In sensitization—provoked, for example, by shocks to the head—the alteration in transmitter release from the sensory neurons is brought about by the firing of another set of neurons that are responsive to the noxious stimulus. These *facilitator neurons* synapse on the presynaptic terminals of the sensory neurons (Figure 19–41), where they release serotonin (as well as certain neuropeptides). Their actions can be mimicked by applying serotonin directly to the membrane of the *sensory* neurons, whose presynaptic axon terminals contain serotonin receptors. These receptors operate via a G protein: binding of serotonin activates adenylate cyclase, thereby causing a rise in the intracellular concentration of cyclic AMP, which in turn activates A-kinase (see p. 709). It is this protein kinase that alters the electrical properties of the membrane of the sensory neuron by phosphorylating a special class of K^+ channels (Figure 19–42).

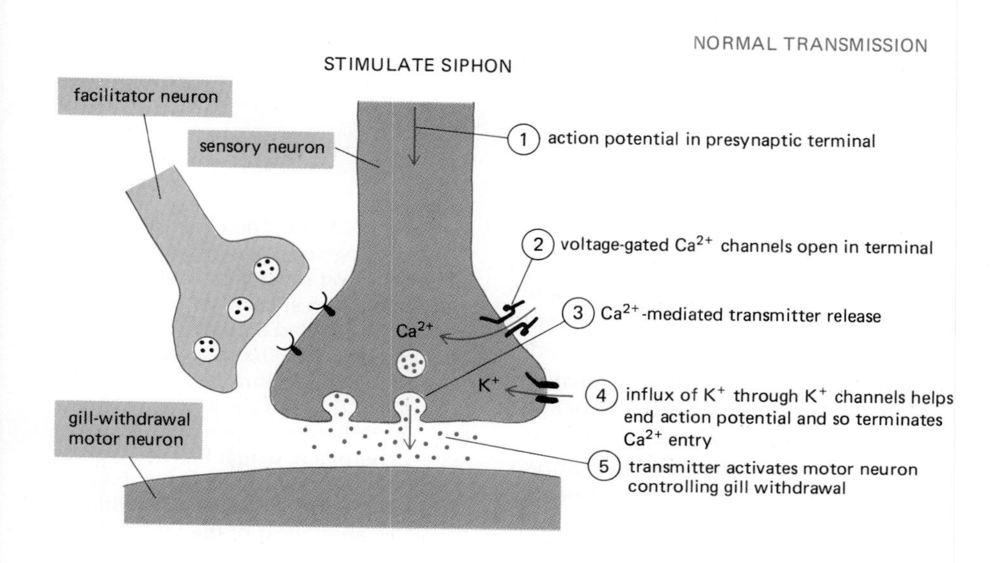

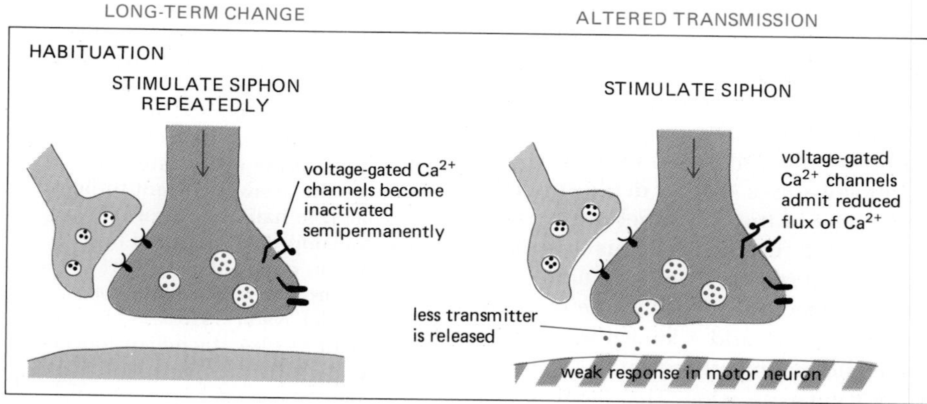

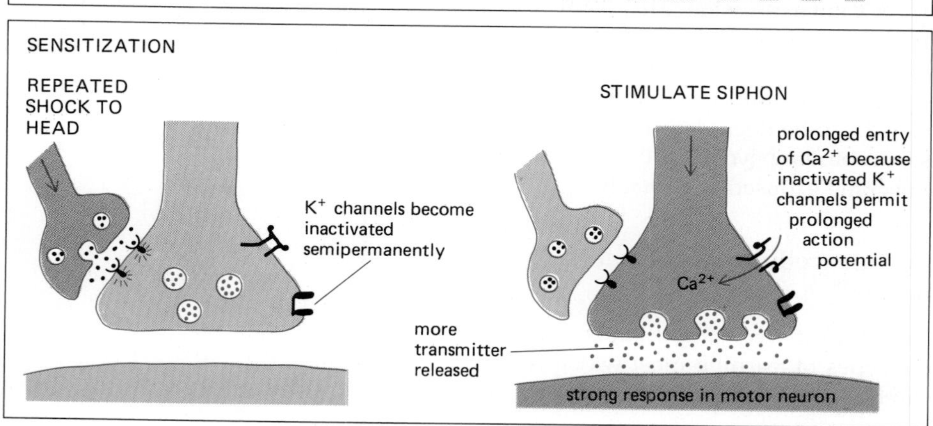

Figure 19–42 The mechanisms that bring about habituation and sensitization of the gill-withdrawal reflex in *Aplysia.* In each diagram the neurons that are electrically active are shown in color. The upper diagram shows the normal mechanism of transmission from the sensory neuron to the gill-withdrawal motor neuron. In each of the lower diagrams, the left-hand drawing shows the nature of the persistent change in the sensory nerve terminal that underlies the memory phenomenon, while the right-hand drawing shows how this change affects synaptic transmission from the sensory neuron to the gill-withdrawal motor neuron. The indicated mechanisms are less certain for habituation than for sensitization.

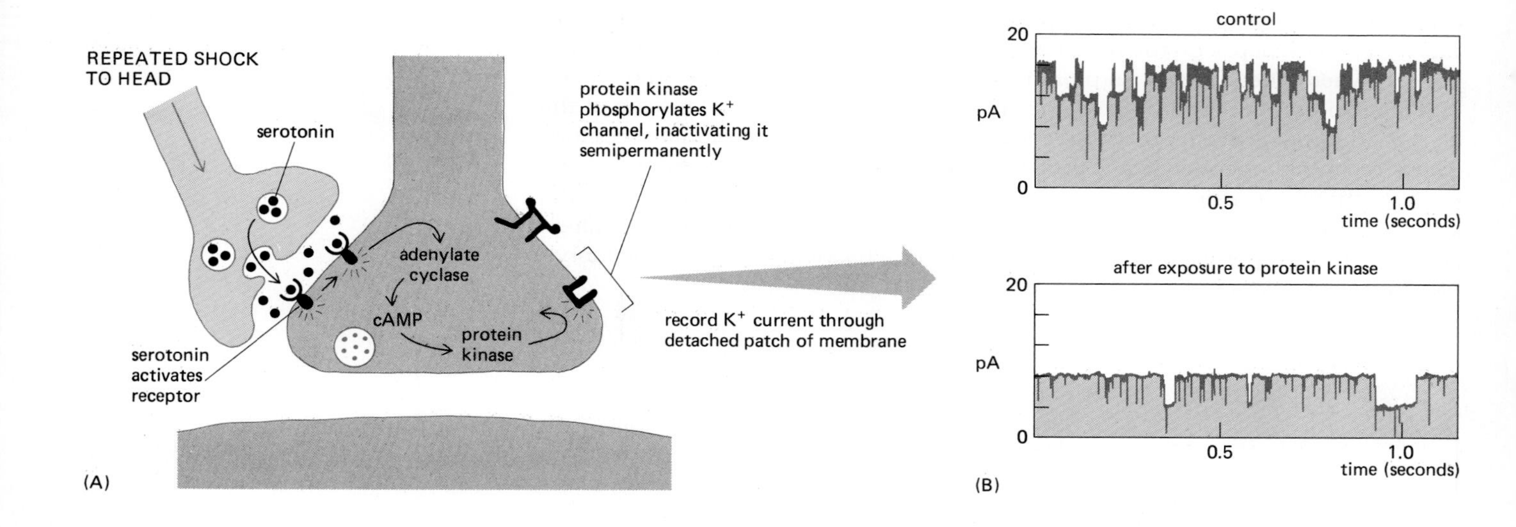

Figure 19–43 (A) The chain of events during sensitization of the gill-withdrawal reflex that leads to inactivation of a special class of K^+ channels (so-called S channels) in the sensory nerve terminal (see Figure 19–42). (B) Patch-clamp recordings of the current through these channels as they flicker between open and closed states. The patch, which has been detached from the cell, contains four channels, and in the control condition these are open most of the time. When the catalytic subunit of the cyclic AMP-dependent protein kinase (A-kinase) is added to the medium bathing the cytosolic side of the patch, two of the four channels become phosphorylated and are thereby locked shut, while the other two continue to spend most of their time in the open state; this reduces the average current through the patch to half its control value. (Patch-clamp data reprinted by permission from M.J. Schuster, J.S. Camardo, S.A. Siegelbaum, and E.R. Kandel, *Nature* 313:392–395, 1985. Copyright © 1985 Macmillan Journals Limited.)

The behavior of these K^+ channels, called *S channels*, can be analyzed by patch-clamp recording (see p. 318). They become locked shut when serotonin is applied to the exterior of the cell (Figure 19–43). Moreover, they close in the same way when the patch of membrane containing them is detached from the cell and transferred to a bath of artificial medium in which the channels are directly subjected to phosphorylation by the catalytic subunit of the A-kinase. This strongly suggests that phosphorylation of the S channels (or of proteins tightly associated with them) is the cause of their long-lasting closure. Because it is the flow of K^+ ions that normally helps to restore the resting potential, closing the S channels prolongs action potentials invading the axon terminal. The prolonged action potentials hold the voltage-gated Ca^{2+} channels open for a longer time, permitting a greater influx of Ca^{2+}, which in turn triggers the release of a larger number of synaptic vesicles; and this produces a larger postsynaptic potential, causing a more vigorous withdrawal of the gill.

These experiments demonstrate how a G-protein-linked receptor can enable a transient signal to cause a persistent change in the electrical properties of a synapse, and hence in the behavior of the animal. The phosphorylation of the S channels represents a form of memory, but it is only a short-term memory, easily erased by the action of phosphoprotein phosphatases (which dephosphorylate the S channels) and limited by the finite lifetime of the S-channel proteins. The mechanism of the *long-term memory* that follows repeated noxious stimulation is not known, but unlike the short-term memory, it requires new RNA and protein synthesis and seems to involve changes in the structure as well as the chemistry of the presynaptic terminals (see p. 1132). Cyclic AMP and A-kinase seem to mediate these changes too, presumably by phosphorylating other proteins in the cell and thereby probably altering the pattern of gene expression. The details are not yet understood, but an intermediate step in the creation of the long-term memory trace appears to be a prolonged activation of the A-kinase itself as a result of a reduction in the concentration of the regulatory subunits that inhibit it (see p. 710). These regulatory subunits are thought to be degraded during the period when cyclic AMP levels are high because, on binding cyclic AMP, they dissociate from the catalytic subunits and thereby become exposed to proteolysis.

Ca^{2+} and Cyclic AMP Are Important Intracellular Messengers in Associative Learning in Invertebrates[36]

Habituation and sensitization as presented above are very simple kinds of learning. An essential feature of the more complex types of learning most widely studied by psychologists is that they are *associative:* thus, in Pavlov's famous experiments, the dog learned to associate the sound of a bell with food. *Aplysia* is also capable of associative learning. For example, if a sensitizing stimulus (a severe electrical shock, as before) is repeatedly paired in time with a particular mild stimulus that normally excites only a weak withdrawal reflex, the animal behaves as though it has learned that the specific mild stimulus is associated with the shock and becomes strongly and specifically sensitized to the mild stimulus. The same classes of neurons are thought to be involved as in the simple sensitization described earlier. The paired stimuli to different parts of the body cause sensory neurons and facilitator neurons to fire at the same time. Thus, while an action potential is invading the sensory axon terminals, causing their voltage-gated Ca^{2+} channels to open, serotonin (or a neuropeptide) is being released onto their exterior from the facilitator neurons, causing an increase in the intra-axonal cyclic AMP concentration. The cyclic AMP by itself would cause simple sensitization; the simultaneous influx of Ca^{2+} is thought to intensify this effect, producing a much stronger sensitization than would result from firing of facilitator neurons while the sensory neuron was quiet.

It is not clear how far one can extrapolate from these findings in *Aplysia*. Whether memories in other animals are generally recorded in presynaptic changes or in postsynaptic changes, in synaptic chemistry or in synaptic structure, or indeed in synapses at all, are open questions. Experiments on mutants of the fruit fly *Drosophila*, however, suggest that molecular mechanisms like those described above in *Aplysia* may operate in many other forms of learning. In particular, normal *Drosophila* can be trained to avoid a specific odor if they repeatedly receive an electrical shock in association with the odor. Flies that rapidly forget or fail to learn the association can easily be picked out because they will stray into regions where the smell is strong. In this way it has been possible to isolate dim-witted and forgetful mutants. Two of them, *dunce* (*dnc*) and *rutabaga* (*rut*), are capable of learning but have a drastically reduced memory span—on the order of tens of seconds in the case of *dunce*. In *dunce* the mutation turns out to be in a phosphodiesterase that breaks down cyclic AMP; in *rutabaga* it is in a Ca^{2+}-dependent adenylate cyclase, which makes cyclic AMP. It seems that either too much or too little cyclic AMP can interfere with memory formation. Another mutant, called *Ddc*, seems to be unable to learn in the first place; here the deficiency is in a gene for the enzyme dopa decarboxylase, which catalyzes an essential step in the production of serotonin and dopamine. All these mutants with defects in associative learning also show defects in their susceptibility to sensitization. Evidently the two processes share some common mechanisms, and it seems that these mechanisms, like sensitization in *Aplysia*, involve a monoamine neurotransmitter at an initial step and protein phosphorylations—controlled by cyclic AMP and by Ca^{2+}—for the production of a lasting effect.

Learning in the Mammalian Hippocampus Depends on Ca^{2+} Entry Through a Doubly Gated Channel[37]

Practically all animals can learn, but mammals seem to learn exceptionally well (or so we like to think). This may reflect the operation of some unique molecular mechanisms. In a mammal's brain the *hippocampus*, a part of the cerebral cortex, seems to play a special role in learning: when it is destroyed on both sides of the brain, the ability to form new memories is largely lost, although previous long-established memories remain. Correspondingly, some synapses in the hippocampus show dramatic functional alterations with repeated use. Whereas occasional single action potentials in the presynaptic cells leave no lasting trace, a short burst of repetitive firing causes **long-term potentiation,** such that subsequent single action potentials in the presynaptic cells evoke a greatly enhanced response in the postsynaptic cells. The effect lasts hours, days, or weeks, according to the number and intensity of the bursts of repetitive firing. Only the synapses that were activated show the potentiation; synapses that have remained quiet on the same postsynaptic cell are not affected. But if, while the cell is receiving a burst of repetitive stimulation via one set of synapses, a single action potential is delivered at *another* synapse on its surface, that latter synapse also will undergo long-term potentiation, even though a single action potential delivered there at another time would leave no such lasting trace. Clearly this provides a basis for associative learning.

The underlying rule in the hippocampus seems to be that *long-term potentiation occurs on any occasion where a presynaptic cell fires (once or more) at a time when the postsynaptic membrane is strongly depolarized* (either through recent repetitive firing of the same presynaptic cell or by other means). There is good evidence that this rule reflects the behavior of a particular class of ion channels in the postsynaptic membrane. Most of the depolarizing current responsible for the excitatory PSP is carried in the ordinary way by ligand-gated ion channels that bind glutamate. But the current has in addition a second and more intriguing component, which is mediated by a distinct subclass of channel-linked glutamate receptors, known as **NMDA receptors** because they are selectively activated by the artificial glutamate analog N-methyl-D-aspartate. The NMDA-receptor channels are doubly gated, opening only when two conditions are satisfied simultaneously: the membrane must be strongly depolarized (the channels are subject to a peculiar form of voltage gating that depends on extracellular Mg^{2+} ions), and the neurotransmitter glutamate must be bound to the receptor. The NMDA receptors are critical for long-term potentiation. When they are selectively blocked with a specific inhibitor, long-term potentiation does not occur, even though ordinary synaptic transmission continues. An animal treated with this inhibitor fails in learning tasks of the type thought to depend on the hippocampus but behaves almost normally otherwise.

How do the NMDA receptors mediate such a remarkable effect? The answer seems to be that these channels, when open, are highly permeable to Ca^{2+}, which acts as an intracellular messenger close to its site of entry into the postsynaptic cell, triggering the local changes responsible for long-term potentiation. Long-term potentiation is prevented when Ca^{2+} levels are held artificially low in the postsynaptic cell by injecting the Ca^{2+} chelator EGTA into it and can be induced by transiently raising extracellular Ca^{2+} levels artificially high. The nature of the long-term changes triggered by Ca^{2+} is uncertain, but they are thought to involve structural alterations in the synapse.

Despite the differences between this example of a memory mechanism in mammals and the previous examples in invertebrates, there is a common theme. Neurotransmitters released at synapses, besides relaying transient electrical signals, can also alter concentrations of intracellular messenger molecules, which activate enzymatic cascades that bring about lasting changes in the efficacy of synaptic transmission. But several major mysteries remain, and we do not yet know how these changes endure for weeks, months, or a lifetime in the face of the normal turnover of cell constituents. We shall see later that the development of the nervous system raises some closely related problems.

Summary

Unlike channel-linked receptors, non-channel-linked neurotransmitter receptors respond to their ligand by initiating a cascade of enzymatic reactions in the postsynaptic cell. In most cases studied so far, the first step in this cascade is the activation of a G protein, which may either interact directly with ion channels or control the production of intracellular messengers such as cyclic AMP or Ca^{2+}. *These in turn regulate ion channels directly or activate kinases that phosphorylate various proteins, including ion channels. At many synapses both channel-linked and non-channel-linked receptors are present, responding either to the same or to different neurotransmitters. Responses mediated by non-channel-linked receptors have a characteristically slow onset and long duration, and they may modulate the efficacy of subsequent synaptic transmission, thus providing the basis for at least some forms of memory. Channel-linked receptors that allow* Ca^{2+} *to enter the cell, such as the NMDA receptor, can also mediate long-term memory effects.*

Sensory Input[38]

We have seen how nerve cells conduct electrical signals, compute with them, record them, and transmit them to muscles to bring about movement. But how do the signals originate? There are two types of source: spontaneous firing and sensory input. Examples of spontaneously active neurons, such as those in the brain that generate the rhythm of breathing, are common: quite complex patterns of spontaneous firing can be produced in a single cell by appropriate combinations of gated ion channels of the types we have already encountered in discussing neuronal computation. Sensory input likewise involves principles that are already familiar, but they are embodied in cells of the most diverse and remarkable types.

Sense organs have evolved to meet exceptionally stringent engineering specifications: they discriminate precisely between stimuli of different types, operate over phenomenally wide ranges of stimulus intensity, and approach the utmost sensitivity that the laws of physics will allow. An olfactory cell in the male gypsy moth can detect a single molecule of a specific sexual attractant (a so-called *pheromone*) released into the air by a female a mile away. The human eye can see both in bright sunlight and on a starlit night, where the illumination is 10^{12} times fainter; and five photons absorbed in the human retina are perceived as a flash.

We shall concentrate on just two examples where the cellular mechanisms of sensory input are beginning to be understood: the ears and the eyes of vertebrates. At each of these gateways into the nervous system, there stands a highly specialized type of **sensory cell,** very different in the two cases but in both cases remarkable for its selectivity, its operating range, and its sensitivity. Before going into details, however, it will be helpful to discuss some general principles.

Stimulus Magnitude Is Reflected in the Receptor Potential[38,39]

Any signal that is to be fed into the nervous system must first be converted to an electrical form. The conversion of one kind of signal into another is known as transduction, and all sensory cells are therefore **transducers.** Indeed, in a general sense almost every neuron is a transducer, receiving chemical signals at synapses and converting these into electrical signals. Thus, although some sensory cells respond to light, some to temperature, some to a particular chemical, some to a mechanical force or displacement, and so on, transduction in all of them involves many of the same basic principles that were discussed earlier for synaptic activation by neurotransmitters. In some sense organs the transducer is part of a *sensory neuron* that propagates action potentials. In others it is part of a *sensory cell* specialized for transduction but not for long-distance communication; such a cell then passes its signal to an adjacent neuron via a synapse (Figure 19–44).

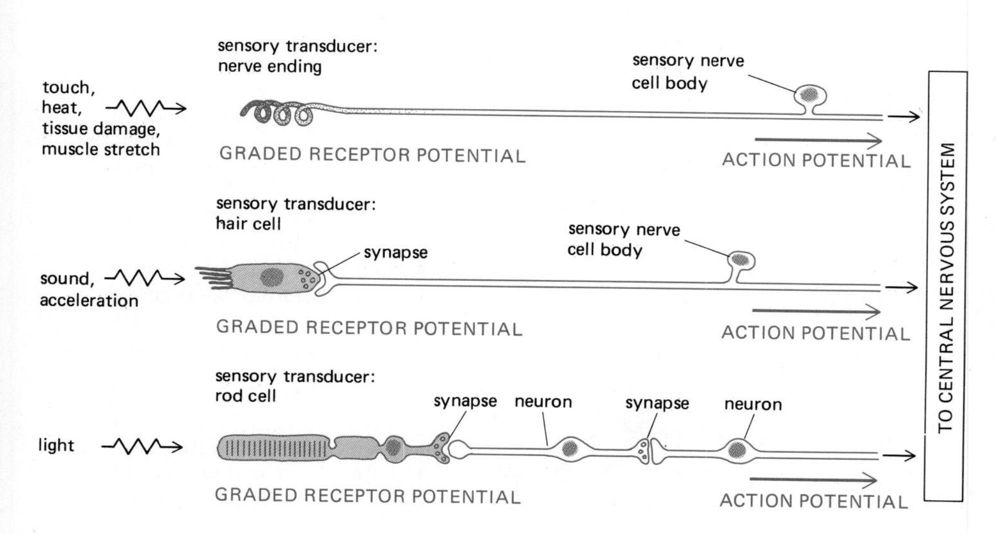

Figure 19–44 Different ways in which sensory stimuli are transmitted to the nervous system. In some cases the sensory transducer is part of a neuron (top drawing); in other cases it is a separate sensory cell (lower two drawings). In all three cases a graded receptor potential (indicated in color) is evoked in the sensory transducer and translated into the frequency of firing of action potentials, which carry the signal rapidly into the CNS.

But in every case the effect of the external stimulus is to cause a voltage change, called the **receptor potential,** in the transducer cell. This is analogous to a postsynaptic potential and likewise serves ultimately to control the release of neurotransmitter from another part of the cell.

Moreover, just as at a synapse, the external stimulus can exert its electrical effect either directly, by acting on an ion channel, or indirectly, by acting on a receptor that generates an intracellular messenger that affects ion channels. Although there are still some uncertainties, it seems that the sensory cells in the ear use the direct mode, based on channel-linked receptors, whereas those in the eye use the indirect mode, based on G-protein-linked receptors.

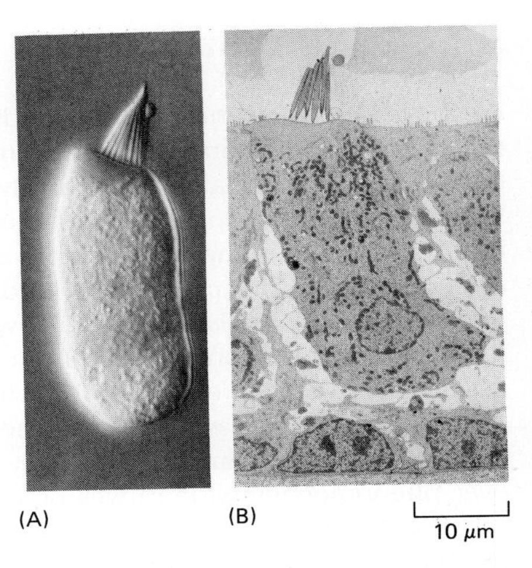

Figure 19–45 (A) Photograph of a sensory hair cell isolated from the inner ear of a bullfrog, showing the bundle of stereocilia on its apical surface. (B) Low-magnification transmission electron micrograph of the hair cell in its normal context, amid supporting cells. (From A.J. Hudspeth, *Science* 230:745–752, 1985. Copyright 1985 by the AAAS.)

Hair Cells in the Ear Respond to Tilting of Their Stereocilia[40]

The ear is not only for hearing. It also provides information on acceleration and on the direction of gravity and so is important for balance and coordination of movements. All these sensory functions of the ear depend on *mechanoreception*—that is, the detection of small movements produced by forces acting in the environment of the ear's sensory cells. The movements are rapid oscillations in the case of sound and slower, more sustained displacements in the case of gravity and acceleration. The cells responsible for the various types of mechanoreception in the ear all have a similar and characteristic form: each of them has a tuft of giant microvilli, confusingly called *stereocilia,* projecting from its upper surface (Figure 19–45 and see p. 675). They are consequently known as **hair cells.**

The hair cells in higher vertebrates all lie in the epithelium of the *membranous labyrinth* of the inner ear, where they are grouped in several separate sensory patches. The hair cells in each group are held in place by a framework of interposed *supporting cells,* while above them lies a sheet of gelatinous extracellular matrix attached to the tips of the tufts of stereocilia (Figure 19–46). The movement of this overlying sheet of matrix tilts the stereocilia and produces a mechanical deformation of the hair cells that gives rise to the receptor potential (Figure 19–47). The specific functions of the different groups of hair cells are determined mainly by the nature of the surrounding structures that transmit forces to them. In the case of those hair cells that respond to linear acceleration and to the force of gravity, the overlying matrix is weighted with dense crystals of calcium carbonate: when the head is accelerated or tilted, the weighted matrix shifts relative to the hair cells and the stereocilia are deflected. In contrast, the hair cells that sense rotational acceleration are arranged so that a sideways force is exerted on their overlying matrix by the swirling of fluid in the semicircular canals of the inner ear when the head is turned.

The most elaborate mechanical setting is provided for the hair cells that detect sound in the ears of mammals (see Figure 19–46). These *auditory hair cells* are arrayed on a thin, resilient sheet of tissue—the *basilar membrane*—that forms a long, narrow dividing partition between two fluid-filled spiral channels running in parallel within the portion of the inner ear known as the *cochlea.* Airborne sounds cause vibrations of the eardrum, which are conveyed via the tiny bones in the middle ear to the fluid-filled channels of the inner ear, where they result

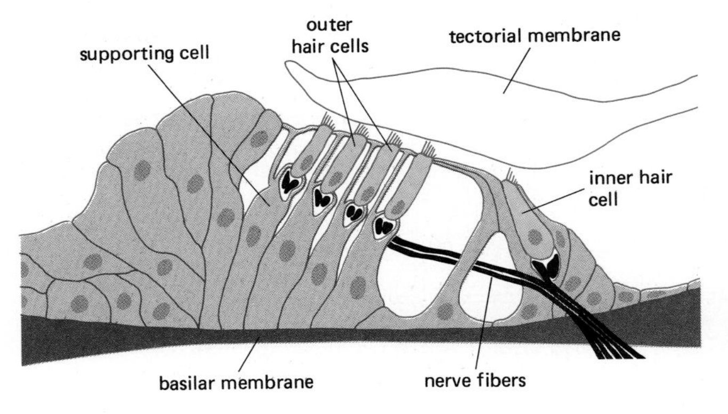

Figure 19–46 Diagrammatic cross-section of the auditory apparatus (the organ of Corti) in the inner ear of a mammal, showing the auditory hair cells held in an elaborate structure of supporting cells and overlaid by the tectorial membrane (a mass of extracellular matrix). The inner hair cells are thought to be the receptors that are primarily responsible for hearing, through the transduction mechanism discussed in the text; they synapse with neurons that convey the auditory signals inward from the ear to the brain. The outer hair cells, by contrast, are richly innervated by an additional set of axons that convey signals outward from the brain, and their function is still a puzzle. There is some evidence to suggest that they are capable of acting (by an unknown mechanism) as transducers in a reverse direction—as loudspeakers rather than microphones—and that they serve as part of a feedback system to modulate the mechanical stimulus delivered to the inner hair cells.

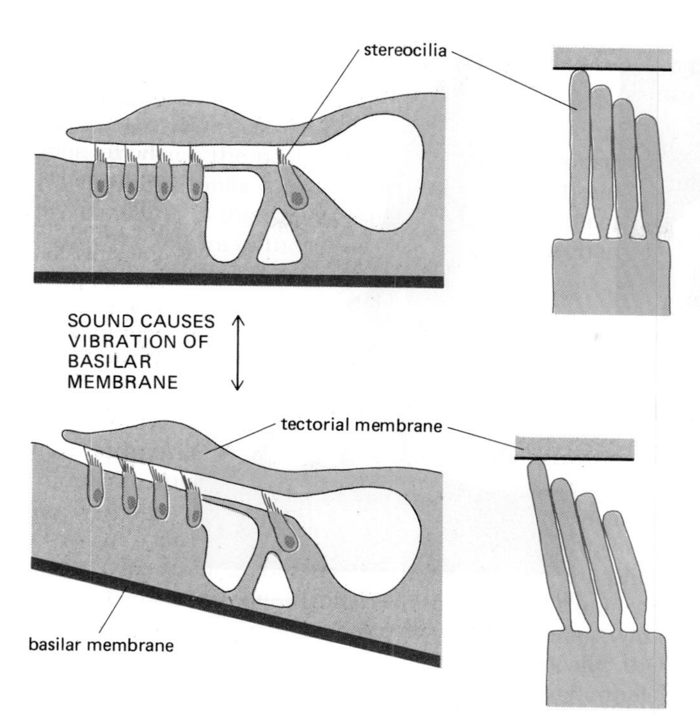

Figure 19–47 How a relative movement of the overlying extracellular matrix (the tectorial membrane) tilts the stereocilia of auditory hair cells in the inner ear of a mammal. The stereocilia behave as rigid rods hinged at the base. The tips of the bundles of stereocilia can be mechanically coupled to the overlying matrix by direct attachment, or indirectly through viscous drag via the intervening fluid.

in vibrations of the basilar membrane and hence of the stereocilia of the auditory hair cells. Hair cells in different positions report on sounds of different pitch because of the mechanics of the cochlea, which resonates most strongly at different positions along its length according to the pitch of the incident sound.

Mechanically Gated Cation Channels at the Tips of Stereocilia Open When the Stereociliary Bundles Tilt[40,41]

When the sheet of matrix overlying a patch of hair cells is abruptly shifted sideways so as to tilt the stereocilia by a few degrees, the hair cells respond by changing their membrane permeability so that a current, called the *receptor current,* flows into them (Figure 19–48). The response reaches a plateau within 100–500 microseconds, which is about the same as the speed of opening of the acetylcholine-activated cation channel at the neuromuscular junction and much faster than the electrical changes produced by any known non-channel-linked receptor. It seems very probable, therefore, that the mechanical stimulus directly opens an ion channel. Studies in which the extracellular ion concentrations are varied have shown that this *mechanically gated ion channel,* like the acetylcholine receptor, is rather

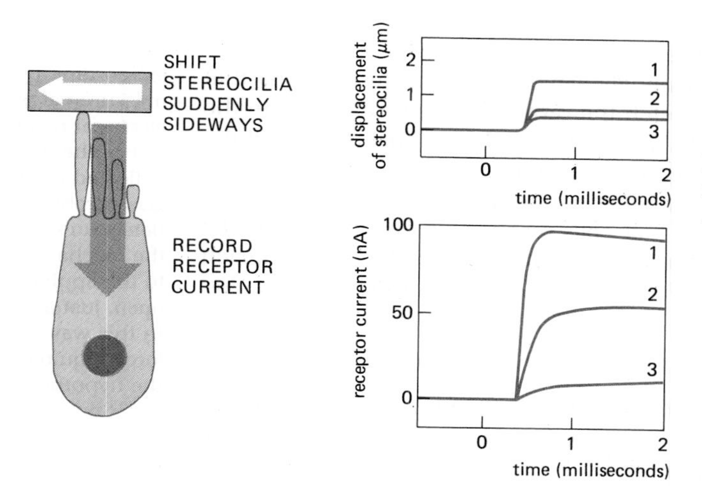

Figure 19–48 Recordings of the receptor current that enters hair cells in the inner ear of a bullfrog in response to a sudden deflection of the bundle of stereocilia. The amount of current is larger for larger deflections. (Data from D.P. Corey and A.J. Hudspeth, *J. Neurosci.* 3:962–976, 1983.)

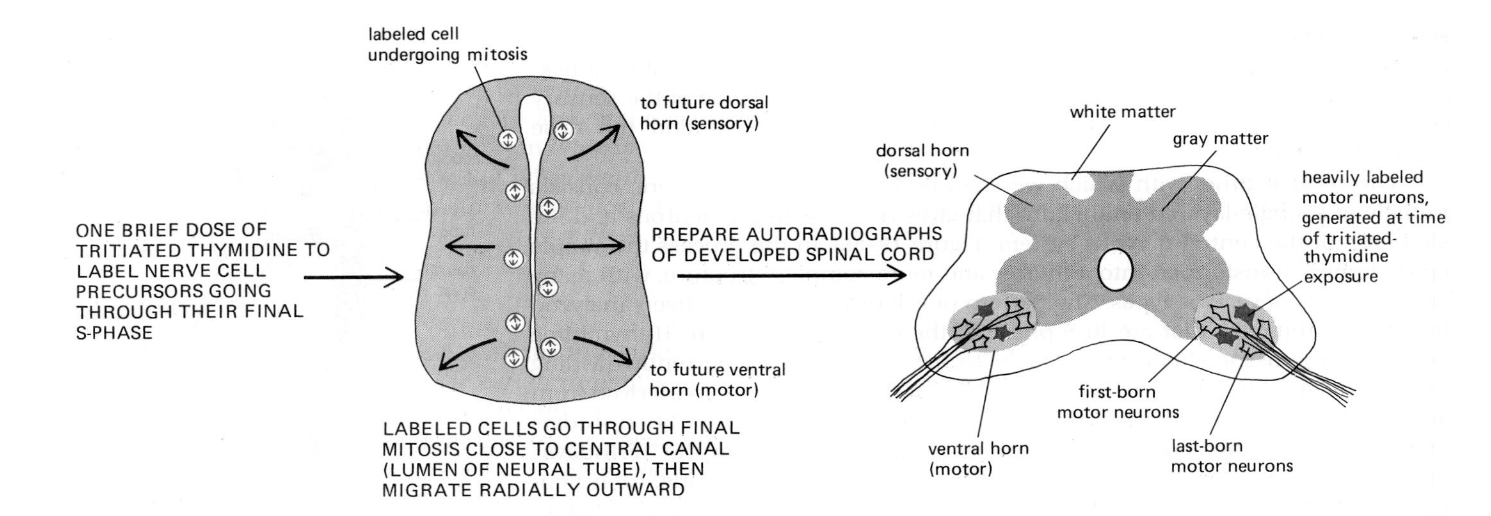

descended from them can later be seen elsewhere. The motor neurons that will innervate the limbs, for example, undergo their final division close to the lumen of the neural tube and then move outward to settle in the *ventral horn* of the future spinal cord (Figure 19–57).

Nerve cell bodies are guided in their migrations by a specialized class of cells in the neural tube—the *radial glial cells* (Figure 19–58A). These can be considered as persisting cells of the original columnar epithelium of the neural tube that become extraordinarily stretched as the wall of the tube thickens: each cell extends from the inner to the outer surface of the tube, a distance that may be as much as 2 cm in the cerebral cortex of the developing brain of a primate. Three-dimensional reconstructions from serial electron microscope sections reveal that the immature migrating neurons cling closely to the radial glial cells and evidently crawl along them (Figure 19–58B and C).

The radial glial cells remain for many days—in some species for months—as a nondividing population, clearly distinct from the neurons and their precursors. Eventually, toward the end of development, they disappear from most regions of the brain and spinal cord; it has been suggested that many of them transform into astrocytes, but this has yet to be directly demonstrated. Thus the radial glial cells can be viewed as a developmental apparatus, necessary—like scaffolding—for the complex process of construction but not retained in most parts of the completed structure.

Figure 19–57 The origins of motor neurons in the spinal cord, as revealed by autoradiography following a brief dose of tritiated thymidine given at an early stage. The diagrams represent cross-sections of the early neural tube (*on the left*) and of the relatively mature spinal cord that develops from it (*on the right*); radioactively labeled cells are shown in color. Cells that are heavily labeled at the late stage are those that were going through their final round of DNA synthesis in the early embryo when tritiated thymidine was given. For simplicity, only the motor neurons are indicated in the mature spinal cord, whose *gray matter* (*shaded*) also contains many other nerve cell bodies. The *white matter* (*unshaded*) consists chiefly of bundles of axons traveling along the length of the spinal cord and connecting one region of gray matter to another. (These regions appear white in the adult because they contain large amounts of myelin.) For an account of the production of glial cells during development, see Chapter 16, p. 909.

The Character and Future Connections of a Neuron Depend on Its Birthday[50,51]

There is a regular relationship between the birthday of a neuron in the vertebrate central nervous system and the site where it comes to rest (an echo, perhaps, of the rigid relationship between cell lineage and cell location that one sees in invertebrates such as nematodes—see p. 902). In the cerebral cortex, for example, the neurons are arranged in layers according to their birthdays through a migration in which the cells that are born later migrate outward past those born earlier. The cells in the successive layers of the cortex, as they mature, will come to differ in their shape, size, and patterns of connections with other cells. Thus small pyramidal cells, born late, lie in an outer layer and send their axons to other regions of the cerebral cortex, whereas large pyramidal and irregularly shaped cells, born earlier, lie in inner layers and send their axons to regions outside the cerebral cortex.

Is it the birthday or the final location that governs these differences? The *reeler* mouse provides an answer. In this mutant, named for its uncoordinated gait, there is a defect in the mechanism of nerve cell migration, so that the cells born late settle in an inner layer and the cells born early settle in an outer one.

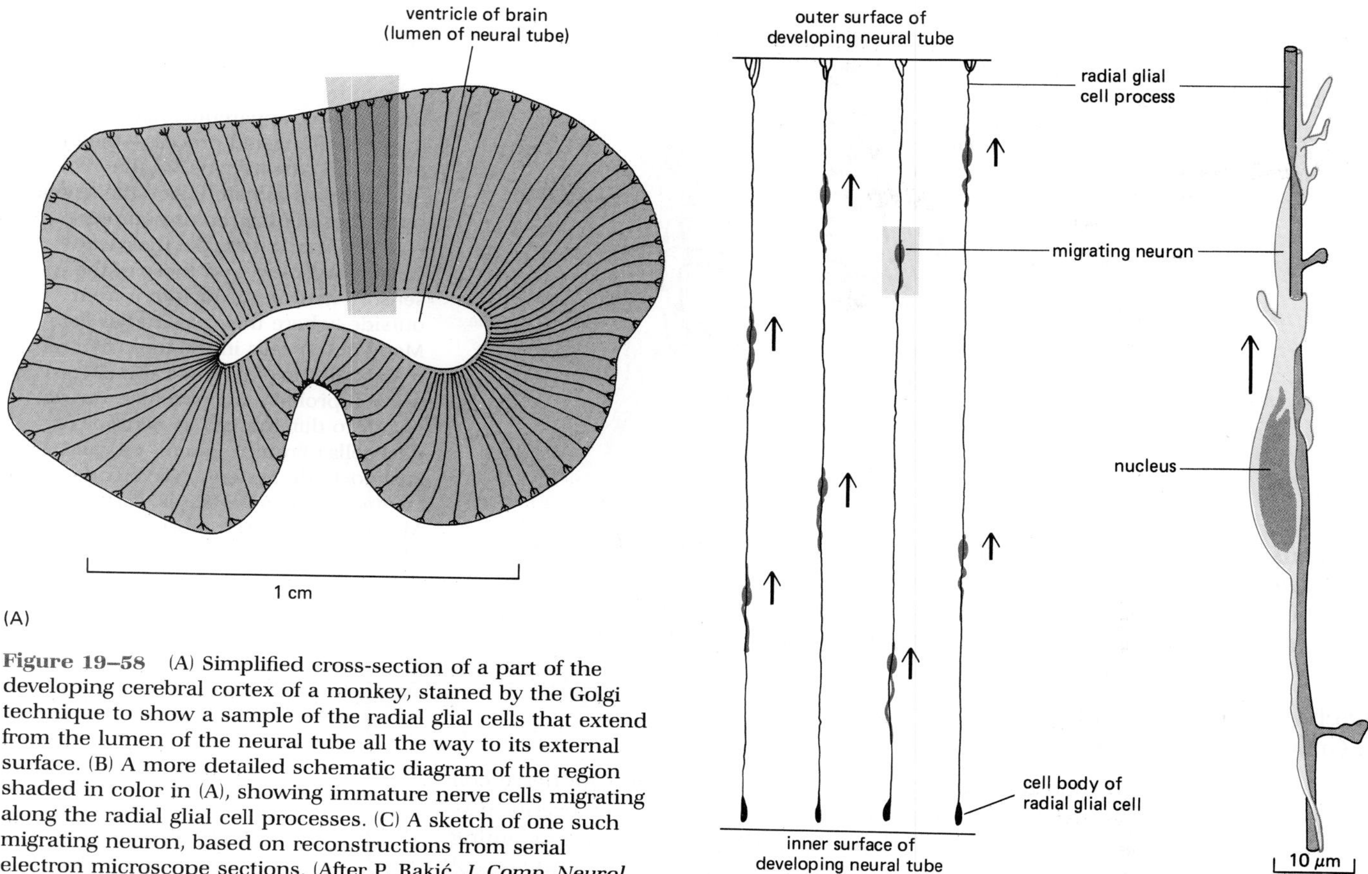

Figure 19–58 (A) Simplified cross-section of a part of the developing cerebral cortex of a monkey, stained by the Golgi technique to show a sample of the radial glial cells that extend from the lumen of the neural tube all the way to its external surface. (B) A more detailed schematic diagram of the region shaded in color in (A), showing immature nerve cells migrating along the radial glial cell processes. (C) A sketch of one such migrating neuron, based on reconstructions from serial electron microscope sections. (After P. Rakić, *J. Comp. Neurol.* 145:61–84, 1972.)

Despite this inversion of their normal positions, the cortical cells differentiate according to their birthdays: late-born cells become small pyramidal neurons, whereas early-born cells become large pyramidal or irregularly shaped neurons. In this system, therefore, it is the birth date rather than the final location that determines cell character (Figure 19–59). Indeed, it seems that in general the character of a neuron is dictated largely by its ancestry and the place and time of its birth.

The intrinsic character of the cell in turn governs the connections it will form—an important general principle about which more will be said later (see p. 1118). Thus the misplaced neurons in the *reeler* mouse, with relatively few errors, make the connections appropriate to their birthdays rather than to their

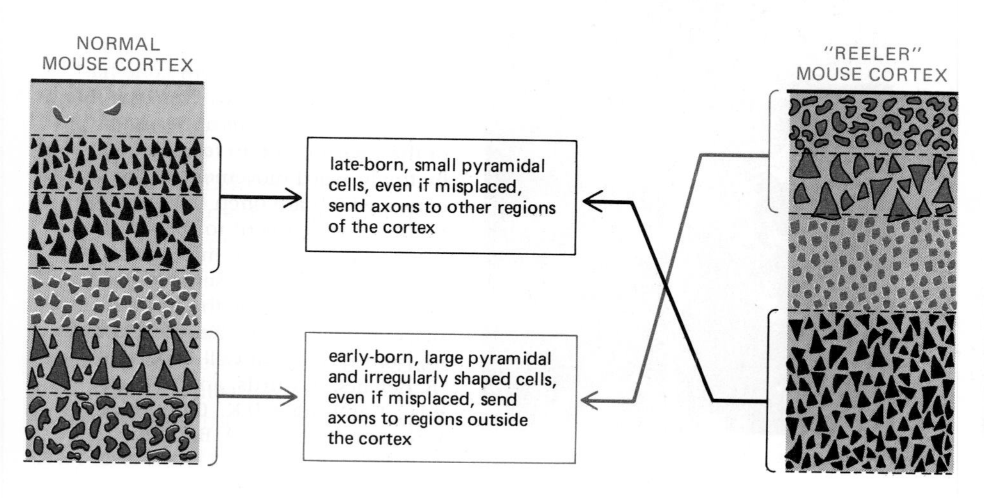

Figure 19–59 Comparison of the layering of neurons in the cortex of normal and *reeler* mice. In the *reeler* mutant an abnormality of cell migration causes an approximate inversion of the normal relationship between neuronal birthday and position. The misplaced neurons nevertheless differentiate according to their birthdays and make the connections appropriate to their birthdays.

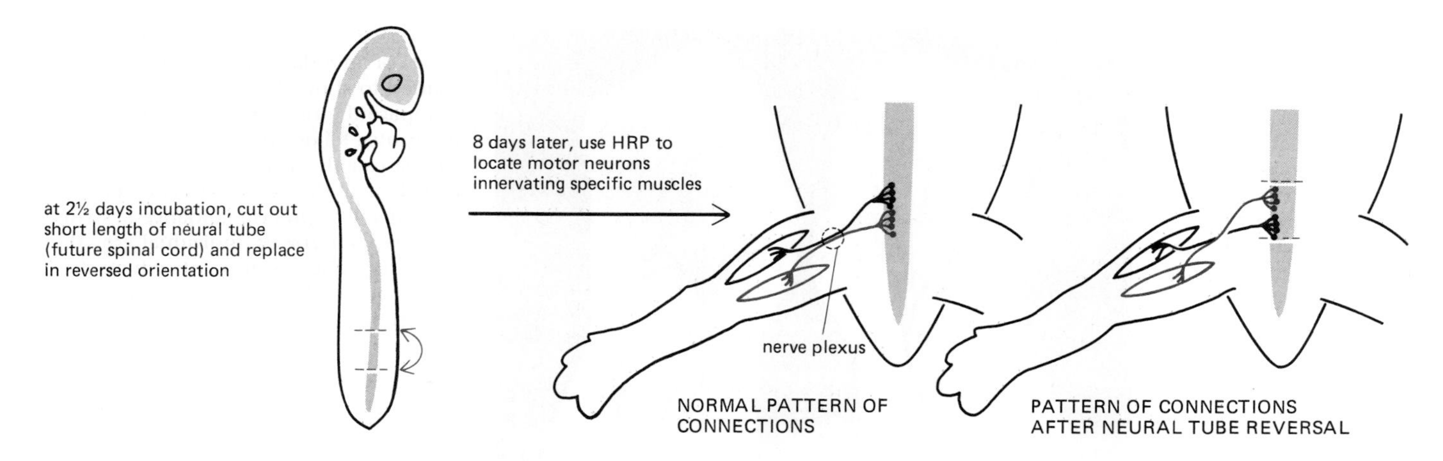

Figure 19–69 An experiment on a chick embryo demonstrating that motor neurons, even when misplaced, nevertheless send their axons to the muscles appropriate to their original positions in the embryonic spinal cord. Note that the axons from motor neurons at different levels along the spinal cord are funneled together into a *plexus* at the base of the limb and then separate again to innervate their separate targets. A growth cone passing through the region of the plexus has a large choice of targets available to it.

of the neurons originally destined to innervate muscle A were put in the place of those originally destined to innervate muscle B, and vice versa. Provided that the shift of position was not too extreme, the growth cones of the misplaced neurons traveled out by altered routes to connect with the muscle appropriate to their *original* position in the neural tube (Figure 19–69). This implies that the motor neurons destined to innervate different muscles are nonequivalent (see p. 915): like the neurons in the *reeler* mouse cerebral cortex, they are distinguished from one another not simply by their positions but by their intrinsic chemical characters. Such nonequivalence among neurons is commonly referred to as **neuronal specificity.** As discussed in Chapter 16, connective tissue cells in different regions of the limb bud are also nonequivalent and may provide the markers that enable a specific growth cone to select a specific branch of the highway system.

In the central nervous system there is also evidence, from both vertebrates and invertebrates, that particular subsets of neurons or glial cells display specific labels that are recognized by other neurons and so help to guide the formation of selective nerve connections. But so far little is known about the molecules involved in either the central or peripheral nervous system.

Target Tissues Release Neurotrophic Factors That Control Nerve Cell Growth and Survival[58]

During the initial part of its journey, the growth cone is generally guided by the tissues through which it is passing; as it nears its destination, it comes under the influence of the target itself, often even before cell-to-cell contact has been made, through the action of *neurotrophic factors* that emanate from the target cells. As we have seen in the example of the trigeminal ganglion innervating the embryonic jaw, such factors may serve as chemotactic attractants for growth cones. More fundamentally, however, they control the *survival* of growth cones, of axon branches, and of entire neurons.

The first neurotrophic factor to be identified, and by far the best characterized, is known simply as **nerve growth factor,** or **NGF.** It was discovered by accident in the course of experiments in which foreign tissues and tumors were transplanted into chick embryos. Transplants of one particular tumor became exceptionally densely innervated and caused a striking enlargement of certain groups of peripheral neurons in the vicinity of the graft. Just two classes of neurons were affected: *sensory neurons* and *sympathetic neurons* (a subclass of the peripheral autonomic neurons that control contractions of smooth muscle and secretion from exocrine glands). Soluble extracts from the tumor also stimulated neurite outgrowth from these neurons in culture. Further work showed that one particular tissue, the salivary gland of the male mouse, produced the same factor in enormous quantities. This quirk of nature is still puzzling, since bulk production of NGF by male mouse salivary gland cells bears no obvious relation to the major functions of the factor, but it made it possible to purify NGF in large enough quantities to discover its chemistry and explore its functions. The activity was found to lie in a protein dimer composed of two identical polypeptide chains 118 amino acids long. Once NGF had been purified, it was possible to raise antibodies that would block its activity. If anti-NGF antibodies are administered to mice while the nervous system is still developing, most sympathetic neurons and some sensory neurons die.

Likewise in culture, sympathetic neurons and some sensory neurons die in the absence of NGF; if NGF is present, they survive and send out neurites (Figure 19–70). Neuronal survival and neurite production represent two distinct effects of NGF. This has been neatly shown by placing the cells in the central compartment of a three-chambered culture dish whose two side compartments are separated from the central one by barriers that prevent mixing of the media in the three compartments but allow neurites to pass (Figure 19–71). If NGF is present in all three compartments, neurites extend into all three. If NGF is absent from one of the side compartments, no neurites will extend into it; and if all NGF is removed from one of the side compartments when neurites are already there, they will wither and retract as far as the barrier. The cells in the central compartment will not survive or send out neurites unless NGF is present there initially; but if NGF is provided initially in all three compartments and then withdrawn from the central one after the neurites have extended into the side compartments, the cells survive and neurite outgrowth continues in the side compartments.

Thus NGF acts both locally at the periphery of the cell, maintaining and stimulating those neurites and growth cones that are exposed to it, and centrally as a survival factor for the cell as a whole. The local effect on growth cones is direct, rapid, and independent of communication with the cell body; when medium devoid of NGF is substituted for medium containing NGF, the deprived growth cones halt their movements within a minute or two. Besides responding directly to NGF, the growth cones of NGF-sensitive cells take up NGF by endocytosis into vesicles, which are carried by retrograde transport back to the cell body, where the NGF (or some intracellular messenger) presumably exerts its effect on cell survival.

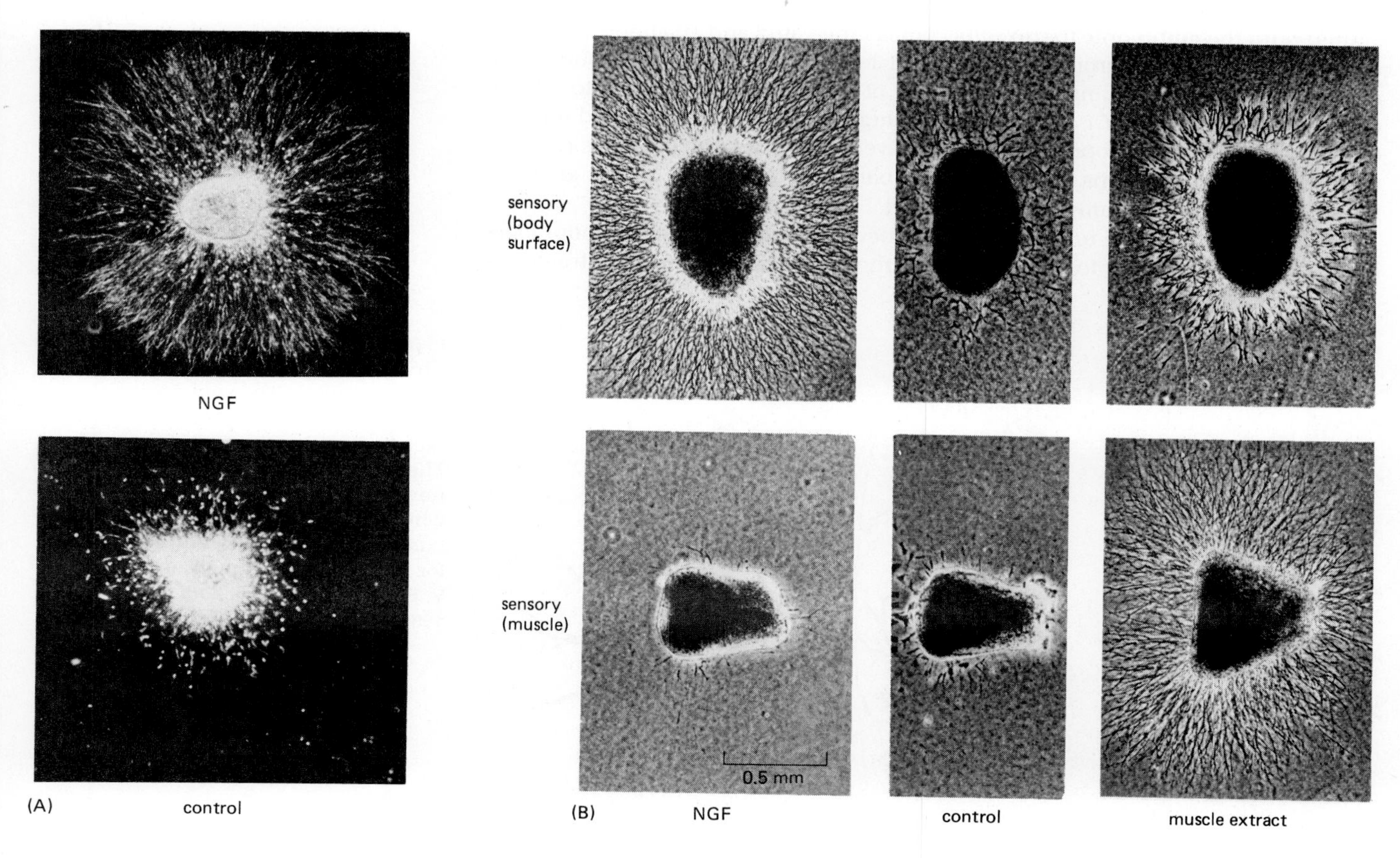

Figure 19–70 (A) Dark-field photomicrographs of a sympathetic ganglion cultured for 48 hours with (*above*) or without (*below*) NGF. Neurites grow out from the sympathetic neurons only if NGF is present in the medium. Each culture also contains Schwann cells that have migrated out of the ganglion; these are not affected by NGF. (B) Phase-contrast photomicrographs showing the behavior of sensory neurons cultured for 24 hours either with NGF (*left*), or in control medium without NGF (*center*), or in medium containing an extract of skeletal muscle (*right*). The sensory neurons in the upper row of pictures are from a group that normally innervates chiefly the body surface (but also sends a few fibers to skeletal muscle and elsewhere); most of these cells respond to NGF like the sympathetic neurons in (A). The sensory neurons in the lower row are from a group that normally innervates skeletal muscle (to provide sensory feedback); these are not responsive to NGF but respond strongly to an extract prepared from skeletal muscle. (A, courtesy of Naomi Kleitman; B, from A.M. Davies, *Dev. Biol.* 115:56–67, 1986.)

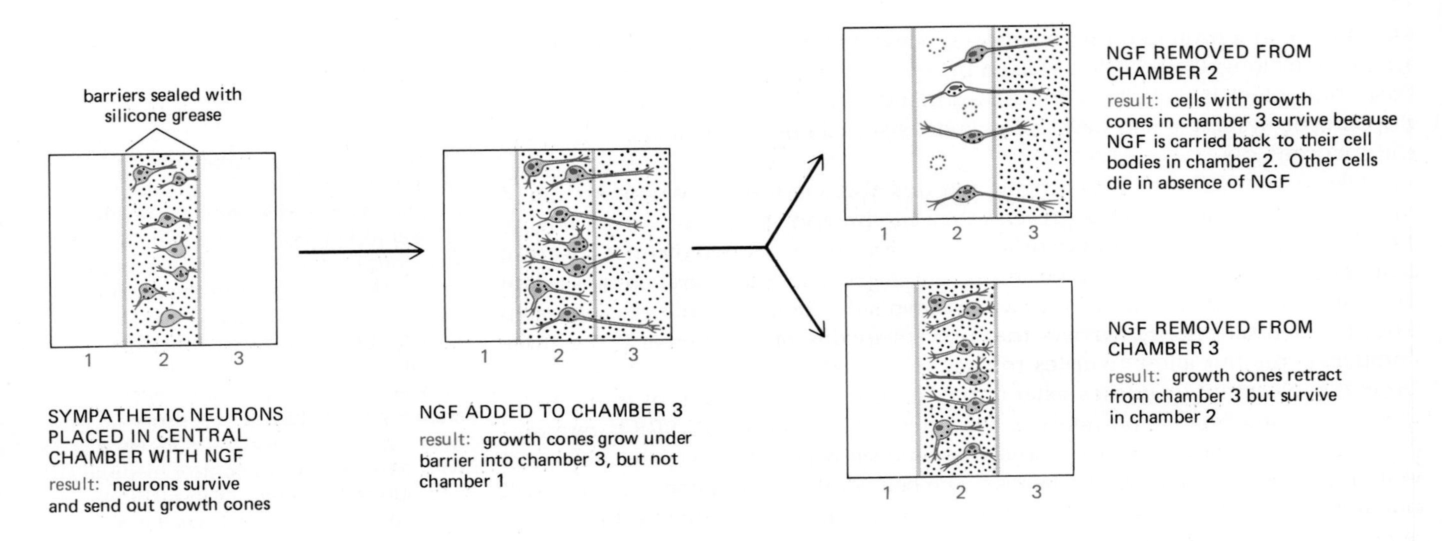

Figure 19–71 Schematic diagram of tissue-culture experiments showing that sympathetic neurons, besides requiring NGF in order to survive, can send out growth cones and maintain neurites only in regions where NGF is present. Note that cell survival does not require NGF in the neighborhood of the cell body as long as the cell has a neurite extending into a region that contains NGF.

Cell Death Adjusts the Number of Surviving Neurons According to the Amount of Target Tissue[59]

In a vertebrate the spinal sensory ganglia are generated in a regular segmental pattern corresponding to the series of vertebrae. Each ganglion consists of a cluster of sensory neurons derived from the neural crest, each of which sends one neurite outward to the periphery of the body and one neurite inward to the spinal cord. The rudiments of the ganglia at first are all similar in size, but in the mature animal the ganglia that innervate the body segments that have limbs attached are much bigger and contain more neurons than the ganglia that innervate the thoracic segments, where there are no limbs (Figure 19–72). This disparity is brought about chiefly by cell death: a larger proportion of the ganglion neurons at the thoracic levels die. If a limb bud is cut off at an early stage, the adjacent ganglia are reduced to the size of thoracic ganglia; conversely, if an extra limb bud is grafted onto the embryonic thorax, it becomes innervated and an abnormally large number of ganglion neurons survive at that level. The control of the survival of ganglion neurons according to the quantity of target tissue is thought to be mediated in large part by NGF secreted by the target. If extra NGF is injected into the embryo during the appropriate period of development, a large proportion of the thoracic ganglion neurons that would ordinarily die are saved, as are ganglion neurons adjacent to an amputated limb bud.

It may seem wasteful to generate excess neurons and then adjust the numbers by cell death according to the amount of the target tissue. Yet this strategy is

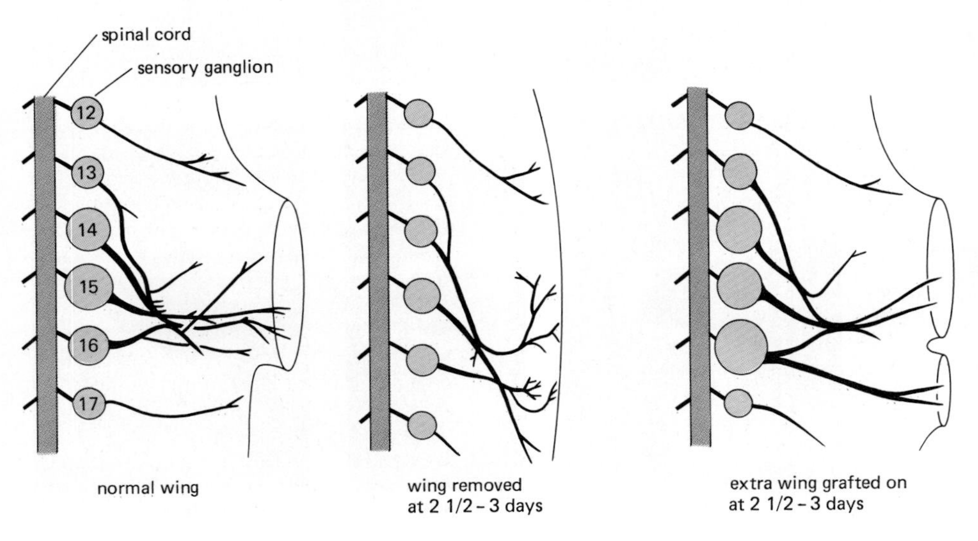

Figure 19–72 The control of nerve cell survival in the spinal sensory ganglia of a chick embryo. The spinal ganglia (*colored*) and nerves (*black*) are sketched from embryos 8–9 days old. The size of each ganglion reflects the number of neurons that have survived, which in turn is governed by the quantity of tissue that is available for the ganglion to innervate. (After V. Hamburger, *J. Exp. Zool.* 68:449–494, 1934 and *J. Exp. Zool.* 80:347–389, 1939.)

commonplace throughout the nervous system in vertebrates—both in sensory and in motor cell groups, and in the central nervous system as well as in the periphery. About 50% of all the motor neurons that send axons to skeletal muscles, for example, die in the course of embryonic development within a few days after making contact with their target muscles. A variety of target-cell-derived trophic factors analogous to NGF appear to regulate neuronal survival in these systems. The strategy has several major advantages. First, it provides an automatic device to correct for variations in the relative sizes of different parts of the body. Second, it facilitates evolution: if mutations alter the size of one part of the body, the numbers of neurons connecting with it will be automatically adjusted without other mutations being needed to alter the programs that generate the neurons. Finally, a small number of neurotrophic factors such as NGF can regulate the individual quantitative matching of a large number of paired targets and sources of innervation, even if the system of connections is very intricate. Through axonal transport, the factor produced by a given target is delivered selectively to the neurons that innervate that target and not to other neurons that may have similarly located cell bodies and similar receptors but send their axons elsewhere. Thus cell death regulated by neurotrophic growth factors can help to set up precise and detailed correspondences between the numbers of cells in different parts of the nervous system.

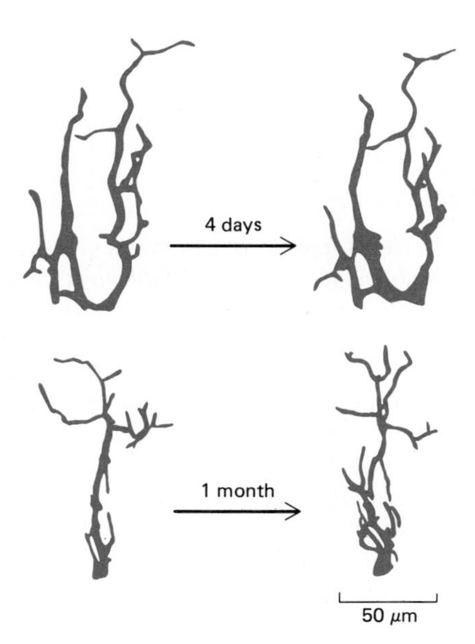

Figure 19–73 Remodeling of the dendrites of neurons in the superior cervical (autonomic) ganglion of a mouse. The ganglion is exposed by careful dissection in the anesthetized animal, and a fluorescent dye is microinjected into one of the nerve cell bodies so as to make its dendrites visible. The wound is then stitched up, and after an interval of some days or weeks, the dissection is repeated and the dye is again injected into the same neuron. The upper and lower pictures show two neurons left for different periods of time. The longer the interval between the first and second injections, the greater is the change seen in the pattern of the cell's dendrites. (Reprinted by permission from D. Purves and R.D. Hadley, *Nature* 315:404–406, 1985. Copyright © 1985 Macmillan Journals Limited.)

Neural Connections Are Made and Broken Throughout Life[60]

Even in normal, undamaged nervous tissue, there is evidence that dendrites and axon terminals continually retract and regrow. In a mature autonomic ganglion of a mouse, for example, identified individual neurons can be seen to withdraw some dendrite branches and to sprout others over the course of a month (Figure 19–73). Such remodeling occurs slowly and on a limited scale in normal circumstances, but it is called into play in a striking way when a proportion of the target cells in a tissue are deprived of innervation. In the case of a skeletal muscle, this can be done by cutting some but not all of the axons that innervate it. The denervated muscle fibers then apparently secrete a diffusible "sprouting factor," which stimulates profuse sprouting of new growth cones from the surviving axon terminals on neighboring innervated muscle fibers (Figure 19–74). The sprouting factors produced by denervated skeletal muscle have not yet been identified, but for smooth muscle, NGF has been shown to play an exactly analogous role. Denervation leads to an increase in the amount of NGF available from the smooth muscle (at least in part because there are fewer nerve terminals transporting the NGF away), and the excess NGF stimulates growth of axons toward the muscle so as to restore its innervation.

Evidently NGF acts in the intact animal just as it does in a culture dish, both as a survival factor, to determine whether cells shall live or die, and as a local stimulus for growth cone activity, to control the sprouting of axon terminals. The first action is prominent during development; the second is important throughout life. But both actions contribute to the same end: they adjust the supply of innervation according to the requirements of the target. Evidence for the existence of other neurotrophic growth factors, performing similar functions in relation to

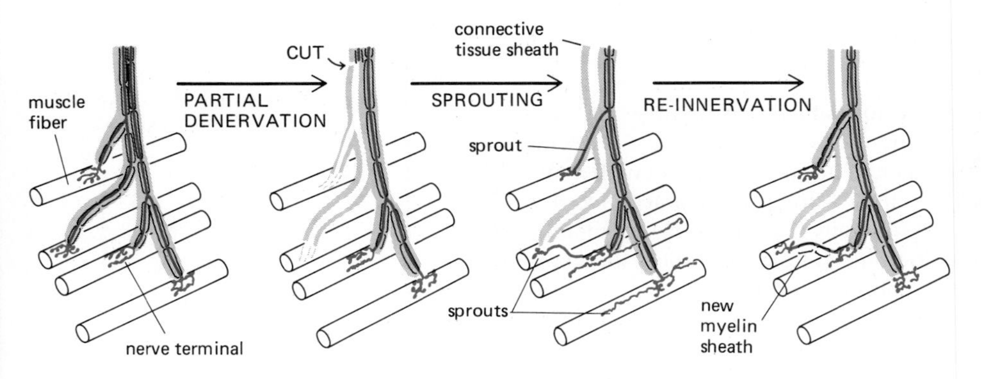

Figure 19–74 By cutting some of the axons that innervate a skeletal muscle, it is possible to deprive some of the muscle cells of innervation while leaving other, adjacent muscle cells with their innervation intact. The cut axons degenerate; the surviving axons, although they have suffered no direct disturbance, are provoked to sprout where they lie close to denervated muscle fibers. Within a month or two, those sprouts that have found their way to vacated sites on the denervated muscle fibers have formed stable synapses on them, restoring their innervation, and the other sprouts have been retracted. Such phenomena suggest that denervated muscle fibers release a diffusible "sprouting factor." (Reproduced with permission from M.C. Brown, R.L. Holland, and W.G. Hopkins, *Ann. Rev. Neurosci.* 4:17–42, 1981. © 1981 by Annual Reviews Inc.)

other classes of nerve cells, is rapidly accumulating (see Figure 19–70). In the next section we shall see that such factors may well mediate some of the important effects of electrical activity on the developing pattern of nerve connections.

Summary

The development of a nervous system can be conveniently divided into three phases, which partly overlap. In the first phase, the neurons are generated by finite programs of cell proliferation and the newborn cells migrate from their birthplaces to settle in an orderly fashion in other locations. In the second phase, axons and dendrites extend from the cell bodies by means of growth cones. The growth cones travel along precisely specified paths, guided for the most part by contact interactions with other cell surfaces or with components of the extracellular matrix. Neurons destined to connect with different targets behave as though they have intrinsically different characters (neuronal specificity), expressed in distinctive surface characteristics that enable their growth cones to select different paths. At the end of its path, a growth cone encounters the target cell with which it is to synapse and falls under the influence of target-derived neurotrophic factors. These govern the sprouting and movement of the growth cone in the neighborhood of the target and also control the survival of the neuron from which the growth cone originates. In these two ways neurotrophic factors, such as nerve growth factor (NGF), regulate the density of innervation of target tissues. In the third phase of neural development, to be discussed in the next section, synapses are formed and the pattern of connections is adjusted by mechanisms that depend on electrical activity.

Synapse Formation and Elimination[61]

The encounter of a growth cone with its target cell is a crucial moment in neuronal development: both the growth cone and the target cell undergo a transformation, and synaptic communication can begin. But the developmental process does not end there: many of the synapses formed initially are later eliminated, and new synapses form elsewhere on the same target cell. This local remodeling of the pattern of synaptic connections provides an opportunity for error correction and fine tuning: first the system is roughed out through pathway guidance as growth cones migrate along specific routes to the vicinity of their target cells; then tentative synaptic connections are made, allowing pre- and postsynaptic cells to communicate; and lastly, the initial connections are revised and adjusted by mechanisms that involve both neurotrophic factors and electrical signals in the form of action potentials and synaptic transmission. Thus external stimuli that excite electrical activity in the nervous system can influence the development of the pattern of nerve connections.

In this section we examine the molecular events of synapse formation, the rules that determine whether synapses are to be formed or eliminated, and the part played by electrical activity in controlling these processes. We begin with synapses between motor neurons and skeletal muscle cells because most is known about them.

Synaptic Contact Induces Specializations for Signaling in Both the Growing Axon and the Target Cell[62]

The early events of neuromuscular synapse formation can be observed best in culture. Here it can be seen that much of the molecular machinery for synaptic transmission is present even before a growth cone has made contact with a muscle cell. As the growth cone crawls forward, it releases tiny pulses of acetylcholine in response to electrical excitation of the nerve cell body (Figure 19–75). Its membrane already contains voltage-gated Ca^{2+} channels to couple excitation to secretion, and these channels also serve to propagate action potentials along the embryonic neurite (which at first lacks Na^{+} channels). Before the muscle cell is innervated,

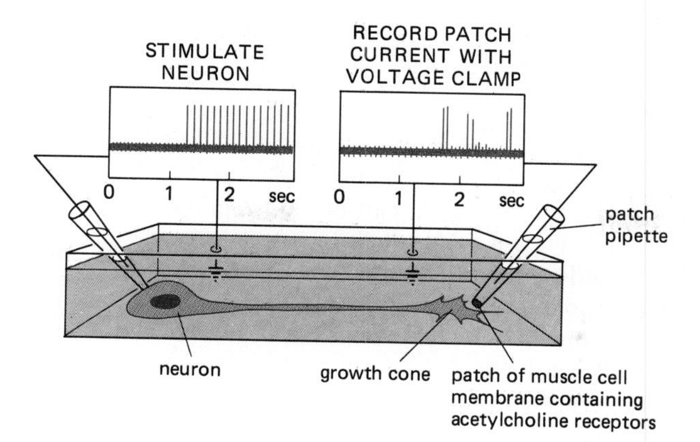

Figure 19–75 An experiment showing that a growing motor neuron in culture discharges pulses of acetylcholine from its growth cone in response to stimulation of the cell body. The minute quantities of acetylcholine released are detected by measuring their effect on the current through a detached patch of muscle cell membrane, rich in acetylcholine receptors, covering the mouth of a patch pipette. The release of acetylcholine from the growth cone is much less plentiful and less reliable than the release from a mature synaptic terminal.

it already has acetylcholine receptors (of an embryonic type) and responds to acetylcholine by depolarization and contraction.

A relatively inefficient form of synaptic transmission can be demonstrated within minutes after the first contact of growth cone with muscle cell. To form a mature synapse, however, both the growth cone and the target cell must develop structural and biochemical specializations—a process that typically takes several days. The growth cone halts its movements, accumulates synaptic vesicles in its interior, and constructs "active zones" for rapid and localized release of acetylcholine (see p. 1078). The muscle cell concentrates its acetylcholine receptors at the synapse and removes them from other regions of its plasma membrane. How is this rearrangement of neurotransmitter receptors achieved? The question is as relevant to neurons as it is to muscle cells, since neurons also, as we have seen, must be able to concentrate particular classes of receptors and ion channels in specific regions of their plasma membranes in order to function in signaling and computation.

Acetylcholine Receptors Diffuse in the Muscle Cell Membrane and Become Tethered at the Forming Synapse[63]

In an adult muscle cell the concentration of acetylcholine receptors at the synapse is more than a thousand times greater than elsewhere in the plasma membrane. Fluorescence bleaching experiments (see p. 295) show that the receptors at the synapse are tethered in place and not free to diffuse in the plane of the membrane. In the uninnervated embryonic muscle cell, by contrast, the receptors are initially spread out over the whole surface and diffuse more freely. When a motor axon makes contact with the muscle cell, these acetylcholine receptors begin to aggregate beneath the axon terminal; moreover, newly synthesized acetylcholine receptors are now preferentially inserted at the developing synapse (Figure 19–76). The receptors become locked into place—perhaps by adhering to one another, perhaps by anchoring to the underlying cytoskeleton or to the overlying extracellular matrix. Some important clues as to how the axon terminal marks out the site of the synapse come from studies of the regeneration of neuromuscular connections.

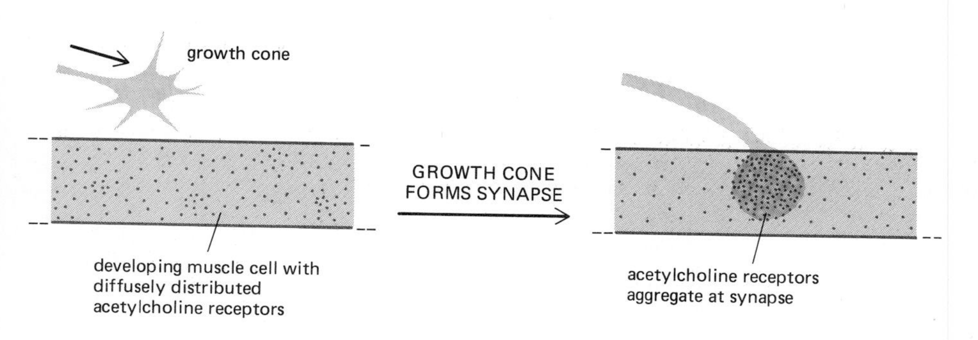

Figure 19–76 The aggregation of acetylcholine receptors in the membrane of a developing muscle cell at the site where a motor axon terminal makes contact to form a synapse. The aggregation depends partly on diffusion of receptors toward that site from neighboring regions of the muscle cell membrane and partly on the insertion of newly synthesized receptors into the membrane there. The aggregation at the synapse seems to be independent of neurotransmitter release from the nerve terminal, for it occurs even in the presence of agents that block action potentials in the nerve cells and even when the extracellular medium contains high concentrations of α-bungarotoxin, a poison from snake venom that binds to the acetylcholine receptors and blocks their interaction with acetylcholine. The receptors aggregated at the synapse are somehow trapped there; they have a much slower rate of turnover than receptors elsewhere in the membrane, surviving for 5 days or more before they are degraded and replaced.

The Site of a Neuromuscular Synapse Is Marked by a Persistent Specialization of the Basal Lamina[64]

Each muscle cell in an adult muscle is enveloped in a basal lamina (see Figures 19–16 and 19–18A). If the muscle is badly damaged, it degenerates and dies, and macrophages move in to clear away the debris. The basal lamina, however, remains and provides a scaffolding within which new muscle fibers can be constructed from surviving stem cells (see p. 986). Moreover, even if a muscle fiber and its axon terminal have both been destroyed, the site of the old neuromuscular junction is still recognizable from the corrugated appearance of the basal lamina there. This *junctional basal lamina* has a specialized chemical character, and it is possible to make antibodies that bind selectively to it. Remarkably, it is the junctional basal lamina that controls the localization of the other components of the synapse.

The importance of the basal lamina at the neuromuscular junction has been demonstrated in a series of experiments on amphibians. By destroying both the nerve and the muscle cells, leaving only empty shells of basal lamina, it is easily shown that the acetylcholinesterase molecules that hydrolyze the acetylcholine released by the axon terminal are tethered in the junctional basal lamina. Moreover, the junctional basal lamina holds the nerve terminal in place: if the muscle cell but not the nerve is destroyed, the nerve terminal remains attached to the basal lamina for many days. On the other hand, removing the basal lamina with collagenase causes the nerve terminal to detach even if the muscle cell is still present.

Indeed, it appears that the basal lamina by itself can guide the regeneration of an axon terminal. This has been demonstrated by destroying both the muscle and the nerve and then allowing the nerve to regenerate while the basal lamina remains empty: a regenerating axon regularly seeks out the original synaptic site and differentiates there into a synaptic ending. The junctional basal lamina also controls the localization of the acetylcholine receptors at the junctional region. If the muscle and the nerve are both destroyed, but now the muscle is allowed to regenerate while the nerve is prevented from doing so, the acetylcholine receptors synthesized by the regenerated muscle localize predominantly in the region of the old junctions, even though the nerve is absent (Figure 19–77). As might be expected, extracts prepared from junctional basal lamina contain a protein, called *agrin*, that promotes receptor clustering in cultured muscle cells.

Evidently, where an axon terminal contacts a muscle cell, it deposits, or causes the muscle cell to deposit, specialized macromolecules, including agrin, that sta-

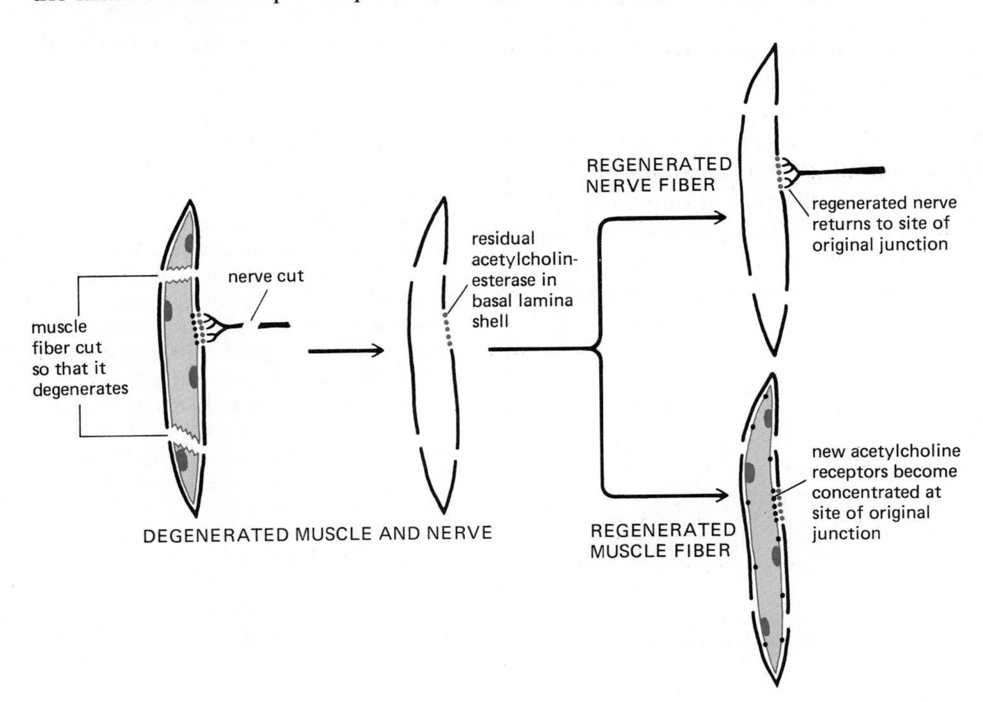

Figure 19–77 Experiment showing that the specialized character of the basal lamina at the neuromuscular junction controls the localization of the other components of the synapse.

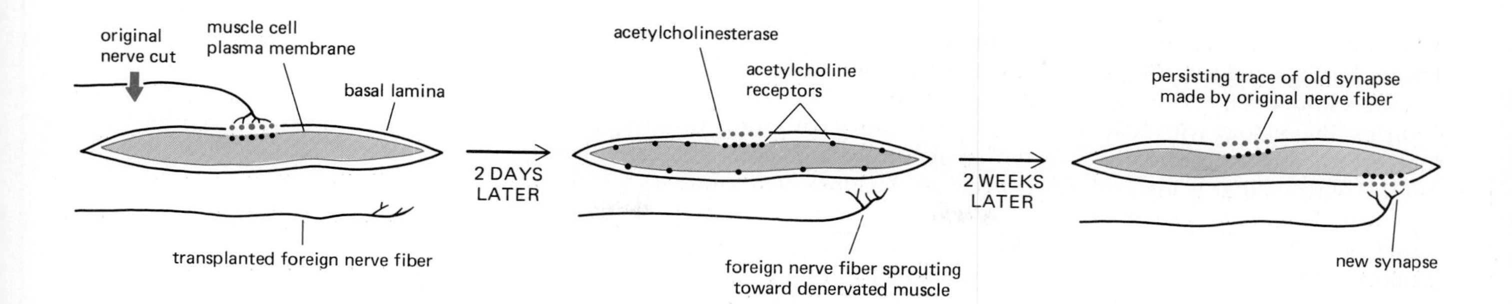

Figure 19–78 An experiment on the soleus muscle of a rat, showing how a muscle cell becomes receptive to synapse formation by a transplanted foreign nerve fiber when (and only when) the original nerve is cut. Note that the distribution of acetylcholine receptors in the muscle cell membrane changes as a result of denervation: new extrajunctional receptors become distributed over the cell's entire surface, although the concentration of receptors remains especially high at the site of the old neuromuscular junction. The electrical excitability of the membrane also changes following denervation, through the appearance in the membrane of a new class of voltage-gated channels that are relatively resistant to tetrodotoxin.

bilize the synaptic connection. The role of the basal lamina, however, is only part of the story of neuromuscular synapse formation. For not all encounters between axon and muscle cell lead to formation of a synapse, nor are all newly formed synapses absolutely stable.

The Receptivity of a Muscle Cell Is Controlled by Its Electrical Activity[63,65]

If a nerve in a rat is cut and the cut end is deflected so that it lies over an adjacent normal healthy muscle, the severed axons will regenerate and grow over the surface of this muscle; but as long as the muscle's own nerve supply is intact, these foreign axons do not make contact with the individual muscle cells or form synapses on them. If now the normal nerve to the muscle is cut, striking changes occur. Within a few days the muscle cells alter their membrane properties and metabolism: in particular, large quantities of new acetylcholine receptors are synthesized and inserted in the membrane of each muscle cell over its whole surface, making it supersensitive to acetylcholine. At the same time the muscle cells become receptive to new synapse formation by the foreign axons that had grown over the surface of the muscle. Although the axons show a preference for sites where synapses existed before, they can also make synapses at new locations on the muscle cells. Once synapses have formed, the diffuse distribution of acetylcholine receptors disappears, as in embryonic development, leaving a high concentration of receptors only at the sites of the synapses (Figure 19–78).

A denervated muscle cell is deprived of stimulation from its nerve, and it is principally the lack of electrical activity in the muscle cell that brings about the changes described above, as well as evoking release of the "sprouting factor" mentioned previously (see p. 1121). All of these effects of denervation, which make the muscle more receptive to synapse formation, can be mimicked by applying a local anesthetic to the intact nerve, thereby blocking the stimulation of the muscle. Conversely, if a denervated muscle is stimulated artifically through implanted electrodes, the extrajunctional sensitivity to acetylcholine is suppressed and new synapses are prevented from forming. Normally the electrical activity triggered by a neuron that has already established a synapse prevents the muscle cell from receiving unwanted additional innervation.

In a related fashion, electrical activity regulates the elimination of synapses during development. In the vertebrate embryo, where many nerve terminals encounter an uninnervated muscle cell more or less simultaneously, many superfluous nerve connections are initially formed. The adult pattern, in which each muscle cell normally receives only one synapse, is achieved in two distinct steps separated in time. The first involves the death of surplus motor neurons (*neuronal death*); the second involves pruning of axon branches (*synapse elimination*).

Electrical Activity in Muscle Influences the Survival of Embryonic Motor Neurons[59,66]

As mentioned earlier, about 50% of embryonic motor neurons die shortly after making synaptic contact with muscle cells. This death of surplus neurons is prevented if neuromuscular transmission is blocked by means of a toxin such as α-bungarotoxin and is increased if the muscle is given direct electrical stimulation. These findings suggest that electrical activity in the muscle controls production of a muscle-derived neurotrophic factor necessary for survival of embryonic motor neurons. By analogy with NGF, this might be identical with the "sprouting factor" thought to cause sprouting of axon terminals toward a muscle cell that is denervated. A muscle that is inactive, either because of a block of synaptic transmission or because it is not innervated, would produce the factor in large quantities as a signal of its need for innervation; electrical activation of the muscle, either by artificial stimulation or by the normal spontaneous firing of motor neurons that innervate it, would depress production of the factor, and in the embryo some of the young motor neurons would die in the competition for what little there was.

Electrical Activity Regulates the Competitive Elimination of Synapses According to a Temporal Firing Rule[61,67]

Even after half the embryonic motor neurons have died, the developing muscles are left with a large excess of synaptic inputs. Each motor neuron branches profusely, making synapses on many muscle cells; and a typical muscle cell becomes innervated by branches from several neurons. To attain the adult configuration, all but one of the synapses on each muscle cell must be eliminated. The process of **synapse elimination** during development has been well studied in the soleus muscle of the rat leg, where about three motor neurons on average innervate each muscle cell at birth. During the next 2 or 3 weeks, each neuron retracts a large proportion of its terminal branches until each muscle cell is innervated by a single branch of one motor axon (Figure 19–79).

If the surplus axon branches were eliminated at random, some muscle cells would be left with no synapse at all while others would retain several. The fact that each muscle cell retains exactly one synapse implies that the process of synapse elimination is competitive. Indeed, competitive synapse elimination throughout the nervous system is one of the most important processes governing the development of neural connections and, as we shall see later, their subsequent modification by environmental input. Although the molecular mechanisms of competitive synapse elimination are not understood, the competition in most cases seems to be governed by a simple and general set of principles, applicable both to neuromuscular synapses and to synapses of neuron on neuron.

First, when competition occurs, it involves an element of chance; but the final outcome is clear-cut, and each synapse either survives or is completely eliminated. Second, competition generally occurs only between synapses that are relatively close together and on the same target cell. Thus, in the normal development of a typical mammalian skeletal muscle cell, the incoming nerve terminals all initially synapse in the same small "end-plate" region, and they then compete until only

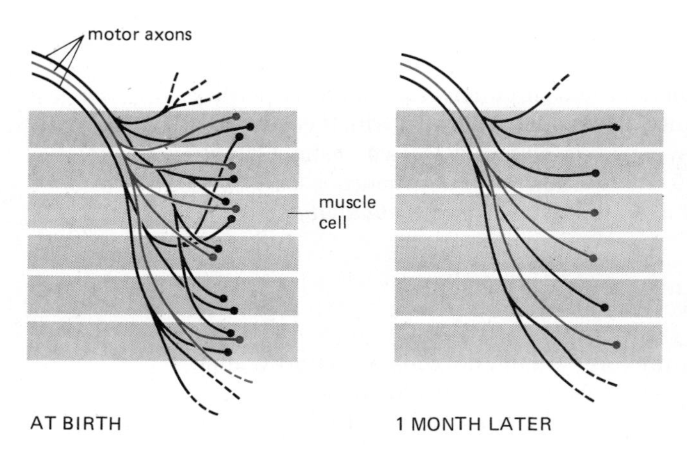

Figure 19–79 Elimination of surplus synapses in a mammalian skeletal muscle in the period after birth. In this schematic diagram the number of terminal branches of each motor axon is underrepresented for the sake of clarity; in reality a single motor axon in a mature muscle typically branches to innervate several hundred muscle cells. All the axon branches innervating one immature muscle cell normally make their synapses in the same small neighborhood on the muscle cell and compete at close range until only one synapse is left.

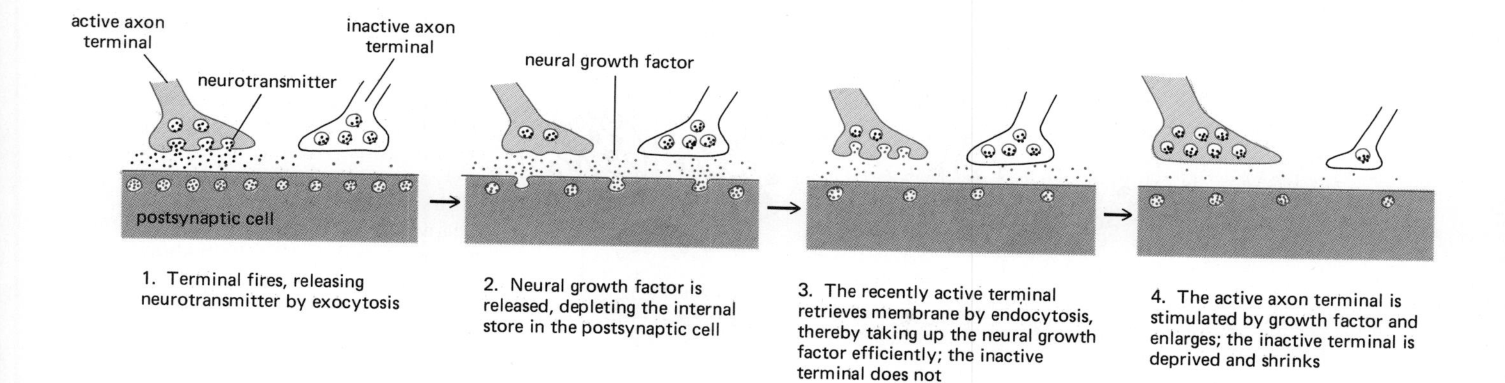

Figure 19–80 One of several molecular mechanisms that have been tentatively suggested to underlie the firing rule. In this view, maintenance of a synapse depends on a neural growth factor released from the postsynaptic cell. Release is triggered locally by the electrical stimulation that follows delivery of an action potential at the synapse; at other times there is some spontaneous release, at a lower rate. The factor is taken up into recently active axon terminals during the endocytic retrieval of membrane that immediately follows the exocytosis of neurotransmitter (see p. 1079). Those axon terminals that endocytose the factor respond by enlarging and depositing materials that reinforce the synaptic bond. The more frequently the muscle is stimulated, the more its internal store of the factor is depleted and the lower is the rate of spontaneous release in the absence of stimulation. Thus inactive axon terminals competing with active ones fail to get adequate growth factor and consequently shrink and are eventually withdrawn. Two axon terminals that are active at different times will compete for the limited amount of growth factor that the postsynaptic cell contains. If large terminals take up more growth factor, which in turn makes them larger still, the outcome of such competition may depend on slight differences in initial size.

An alternative hypothesis proposes that stimulation of the postsynaptic cell causes local release or activation of a protease that tends to destroy inactive synapses but against which recently active synapses are somehow protected.

one is left. When multiple synapses are artificially induced to form on the same cell at sites separated by 1 mm or more, the multiple innervation is retained.

Third, and most important to the theme of this section, the competitive elimination of synapses depends on electrical activity in both the axons and the target cells that they innervate. For example, synapse elimination is delayed if excitation of a developing muscle is blocked by applying local anesthetic to the nerve or α-bungarotoxin to the neuromuscular junction. And in most systems that have been studied, if some of the innervating axons are paralyzed while others remain active, the active axons gain control of more target cells. It seems as though in the neighborhood of an active synapse, the stimulated target cell either produces something that tends to destroy other synapses on that region of its surface or fails to produce something that is required for their maintenance.

But this presents a paradox. If synaptic stimulation of the target cell drives off synaptic contacts, how can the synapse through which the cell gets its stimulation remain and consolidate itself? The answer seems to lie in the following **firing rule:**

> Each excitation of the target cell tends to consolidate any synapse where the presynaptic axon terminal has just been active and to cause rejection of any synapse where the presynaptic axon terminal has just been quiet.

Thus the relative timing of activity is all-important, and where several independently active neurons make neighboring contacts with a single target cell, each of them tends to consolidate its own synapse while promoting elimination of the synapses made by the others.

The molecular mechanisms underlying the firing rule are unknown; Figure 19–80 outlines one speculative suggestion that has been put forward, and the legend mentions another. Nonetheless, there is evidence for the firing rule from many different systems, and we shall now examine some of its applications to synapses other than the neuromuscular junction.

Synchronously Firing Axon Terminals Make Mutually Supportive Synapses[68]

One of the corollaries of the firing rule is illustrated in the submandibular ganglion of the rat, where each neuron is innervated at birth by axons from about five presynaptic neurons located in the brainstem. By the end of the first month of postnatal life, through competitive synapse elimination, each neuron in the ganglion is innervated by only one such axon. But meanwhile that axon has formed many new terminal branches, synapsing on the same cell at many sites, so that the total number of synapses is larger finally than it was initially (Figure 19–81). The branches of a single axon have one obvious property in common that distinguishes them from branches of other axons of the same type: they all fire at the same time. In accordance with the firing rule, neighboring axon terminals that fire synchronously have collaborated in forming synapses, whereas terminals that fire asynchronously have competed.

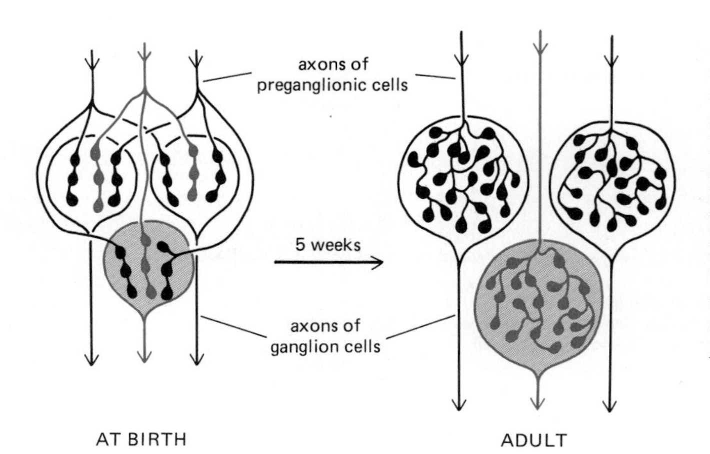

Figure 19–81 The changes that occur soon after birth in the pattern of synapses on neurons in the submandibular ganglion of a rat. Initially each cell is innervated by several axons. These compete until, by synapse elimination, one axon is left in sole command; this one axon, meanwhile, increases the number of its synapses on the cell, which show no sign of competing with one another. (After D. Purves and J.W. Lichtman, *Physiol. Rev.* 58:821–862, 1978.)

The Number of Surviving Inputs Depends on the Number of Dendrites on the Postsynaptic Neuron[69]

Because the competition among synapses for survival depends in part on the distances between them, the final outcome depends on the structure of the postsynaptic cell. The submandibular ganglion neuron is a somewhat atypical neuron in that it has no dendrites and, following a synaptic competition fought out at close quarters on the cell body, retains input from only one axon. Most other neurons have multiple dendrites and continue in adult life to receive inputs from multiple sources, this being essential for their integrative function. The role of dendrites in regulating synapse elimination is illustrated in the ciliary ganglion of a rabbit, where some of the neurons have many dendrites while others have few or none (Figure 19–82). At birth all the neurons are similarly innervated by about four or five presynaptic axons. But in the adult the cells without dendrites receive input from just one axon, while the number of axons providing input to the other cells increases in direct proportion to the number of main dendrites. The synapses on a single dendrite, however, tend to be made by branches of a single axon. Thus

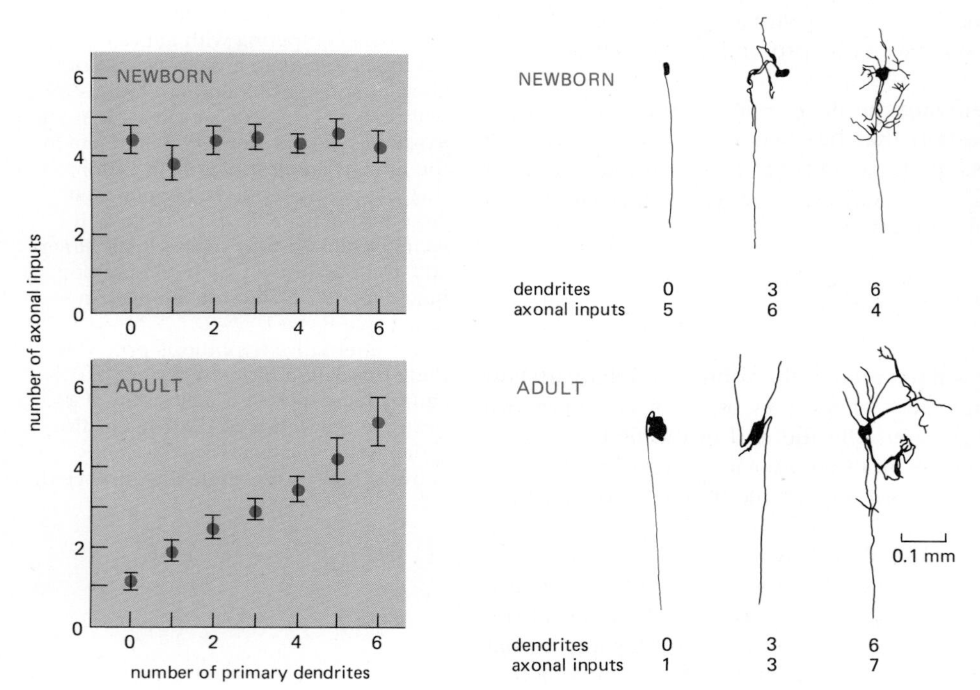

Figure 19–82 The developmental relationship between the number of primary dendrites and the number of axonal inputs to individual cells in the ciliary ganglion of a rabbit. At birth the average number of inputs is independent of the number of dendrites; in the adult the average number of inputs that have survived the period of competitive synapse elimination is proportional to the number of dendrites. At the right the relationship is illustrated by drawings of selected individual ganglion cells. (After D. Purves and R.I. Hume, *J. Neurosci.* 1:441–452, 1981; and R.I. Hume and D. Purves, *Nature* 293:469–471, 1981.)

it seems that each dendrite provides a separate and independent territory, such that synapses on one dendrite do not compete with those on another. As in skeletal muscle, the competition is local, and it obeys the predictions of the firing rule.

Perhaps the most profound implications of the firing rule, however, concern the ways in which stimuli from the external world control adjustments of the anatomical connections between neurons. This is especially clear from studies on the development of the vertebrate visual system. We shall concentrate here on the evidence from mammals.

Visual Connections in Young Mammals Are Adjustable and Sensitive to Visual Experience[70]

The visual system of a mammal is not mature at birth. The first few years of postnatal life (in humans), or the first few months (in cats or monkeys), are a *sensitive* (or *critical*) *period*, during which the pattern of neural connections is still adjustable, and abnormal visual experience can have drastic and irreversible consequences. A common example is the "lazy eye" that can result from a childhood squint. Children with a squint frequently fall into the habit of using one eye only and neglecting the input from the other eye, which is perpetually misdirected and rarely receives a sharply focused image on its retina. If the squint is corrected early and the child is taught to use both eyes, both eyes will continue to function normally. But if the squint goes uncorrected throughout childhood, the unused eye becomes almost completely blind in a permanent way that no lens can correct—a condition known as *amblyopia*. The eye itself remains normal: the defect lies in the brain. Before explaining the nature of the defect, we must outline some of the anatomy of the adult mammalian visual system.

Active Synapses Tend to Displace Inactive Synapses in the Mammalian Visual System[71]

Each eye in a mammal such as a human or a cat sees almost the same visual field, and the two views are combined in the brain to provide binocular stereoscopic vision. This is possible because axons relaying input from equivalent regions in the two retinas make synapses in the same region of the brain (Figure 19–83). Thus, on the *primary visual cortex* of the left side of the adult brain, there are two orderly maps of the right half of the visual field—one from the left eye, the other from the right eye. These two maps, however, are not precisely superimposed on

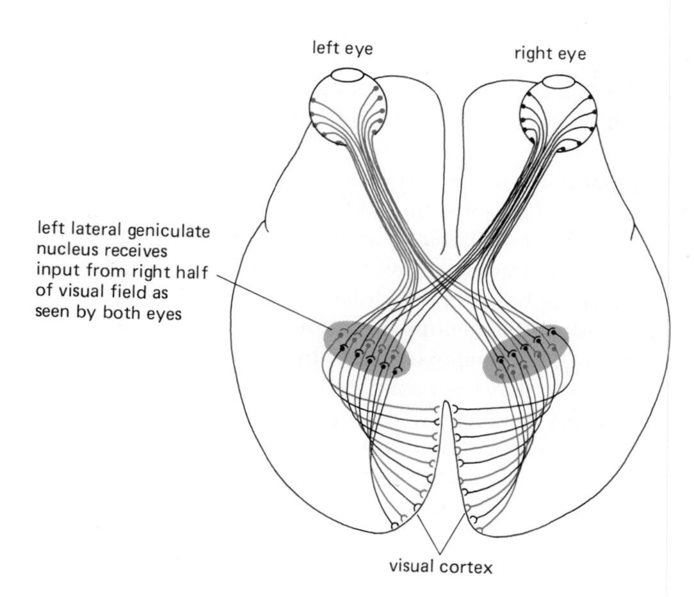

Figure 19–83 The major human visual pathway, showing how the inputs from the right and left eyes are distributed so that related streams of information are brought together in the same region of the brain. Note that all the information obtained by the left side of each eye (relating to the right side of the visual field) is relayed to the left side of the brain, and vice versa.

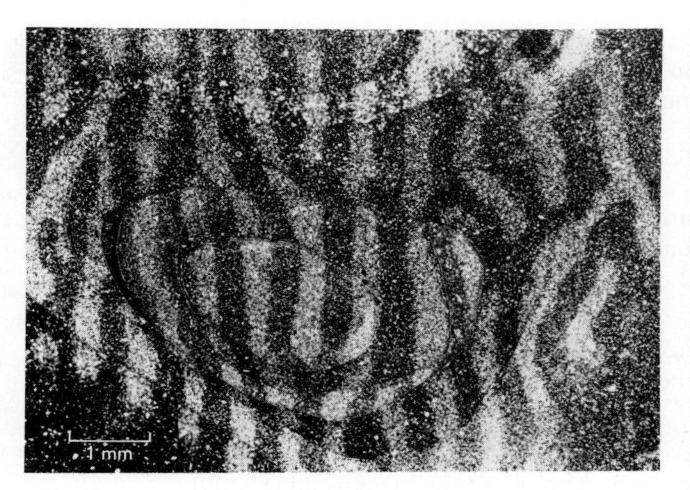

Figure 19–84 Ocular dominance columns in the visual cortex of a normal monkey. Radioactive proline is injected into one eye and the animal is then allowed to survive for 10 days, during which the radioactive label is transported to the parts of the cortex that receive their input from that eye. Sections of the cortex are cut tangentially to its surface, and autoradiographs are prepared. With dark-field illumination the silver grains covering the radioactive regions appear bright against a dark background. The picture is a montage composed of photographs of several successive sections cut at slightly different depths through the thickness of the cortex. The ocular dominance columns connected to the labeled eye (*bright bands*) are of the same width as those connected to the unlabeled eye (*dark bands*). (From D.H. Hubel, T.N. Wiesel, and S. Le Vay, *Philos. Trans. R. Soc.* [*Biol.*] 278:377–409, 1977.)

each other: instead the inputs arriving from the two eyes are segregated in a pattern of narrow, alternating stripes known as **ocular dominance columns.** This arrangement is represented schematically in Figure 19–83 and can be demonstrated directly by a labeling technique that involves injecting radioactive amino acids into one eye. The labeled molecules are taken up by the retinal neurons and incorporated into proteins, which are carried by axonal transport toward the visual cortex; the labeled proteins somehow pass from one neuron to the next at a synaptic relay station (the lateral geniculate nucleus) on the way. Autoradiographs of sections of the visual cortex of the adult monkey brain, for example, show labeled bands about half a millimeter wide, receiving their input from the labeled eye, alternating with unlabeled bands of equal width, receiving their input from the unlabeled eye (Figure 19–84).

During development, however, when the visual connections are first made, no ocular dominance columns can be seen: the projections from the two eyes overlap completely. It is only later (typically during the first few weeks after birth) that the projections segregate into the adult pattern of alternating stripes by means of the competitive elimination of overlapping axon terminals. The pattern apparently develops through the operation of the firing rule: axons relaying excitations from adjacent sites in one eye tend to fire in close synchrony with one another but often out of synchrony with axons relaying excitations from the other eye. The axons that fire in synchrony with one another collaborate in establishing their own set of synapses on a given cortical cell while driving off the synapses from other axons. The segregation into alternating stripes can be halted either by artificially stimulating both optic nerves so as to force the axons from the left eye to fire in strict synchrony with those from the right, or by suppressing electrical activity with injections of tetrodotoxin (which blocks voltage-gated Na^+ channels) into both eyes.

The most striking effects, from a functional point of view, are seen when one eye is simply kept covered and thereby deprived of visual stimulation during the sensitive period. When the cover is removed, the animal behaves as though blind or semiblind in the deprived eye. Autoradiographic tracing shows that the ocular dominance columns connected to the deprived eye have shrunk drastically, while those connected to the experienced eye have widened to occupy the vacated space (Figure 19–85). Again in accordance with the firing rule, synapses made by inactive axons have been eliminated, whereas active axons have consolidated their synapses and made more. In this way cortical territory is allocated to axons that carry information and is not wasted on those that are silent. The effect is irreversible once the sensitive period has ended. Thus the stimulation that a sensory pathway receives in early life determines the amount of cortical machinery—the number of neurons and synapses—that will be available to deal with that input in the adult.

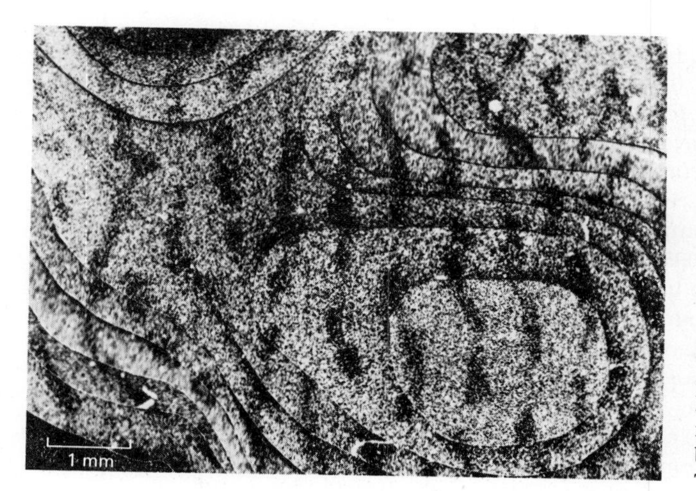

Figure 19–85 Ocular dominance columns in the visual cortex of a monkey that has had one eye covered during the sensitive period of development. The other eye has received an injection of radioactive proline, and autoradiographs have been prepared as described in the caption to Figure 19–84. The ocular dominance columns connected to the eye deprived of visual experience (*dark bands*) are abnormally narrow, whereas those connected to the other eye are abnormally wide. If the deprived eye is labeled, a converse picture is seen, with narrow bright bands alternating with broad dark bands. (From D.H. Hubel, T.N. Wiesel, and S. Le Vay, *Philos. Trans. R. Soc.* [*Biol.*] 278:377–409, 1977.)

Convergent Binocular Visual Connections Depend on Synchronized Binocular Stimulation[70,71,72]

Early visual experience is also important is subtler ways in establishing the nerve connections that enable us to see. For example, some children with an uncorrected squint, instead of neglecting one eye, will use both eyes but in alternation rather than together. Both eyes then remain functional, but the ability to see depth (stereopsis) is permanently lost. Studies of the behavior of single cells in the brain show that this phenomenon too can be explained by the firing rule.

Stereopsis depends on *binocularly driven* neurons—that is, neurons that receive and respond to convergent synaptic inputs from both eyes at once. Such neurons can be identified in experimental animals by inserting a recording microelectrode into the visual cortex of the brain and observing the firing of single cells in response to visual stimuli presented to the two eyes. Such cells are found in particular layers of the visual cortex, above and below the layer containing monocularly driven neurons arranged in clear-cut ocular dominance columns. In a normal animal, binocularly driven neurons are plentiful. But there are scarcely any to be found in an animal that has been specifically deprived of synchronous binocular stimulation during the sensitive period (either by covering different eyes on alternate days or as a result of a severe squint). Evidently, the inputs from each eye to a binocularly driven neuron are maintained only if the two inputs frequently fire in synchrony. When synchronous stimulation is prevented, the axons bringing input from one eye compete with those bringing input to the same neuron from the other eye, in accordance with the firing rule for synapse elimination and maintenance, until one eye gains sole control of that neuron and the possibility of stereopsis is lost.

The Firing Rule Guides the Organization of the Nervous System in the Light of Experience[73]

The development of binocular vision illustrates a general organizing principle: synchronous firing tends to establish convergent connections. This principle, which follows from the firing rule, may help to explain how the brain comes to contain neurons that are specifically responsive to particular complex combinations of sensations that are repeatedly evoked by common objects in the world around us. The brains of primates, for example, contain neurons that appear to fire specifically in response to the sight of a particular face. More generally, one can begin to understand how the brain may be adjusted in the light of experience so as to represent in its own structure and function the existence of relationships between one external phenomenon and another. In this sense the rules for synapse formation and elimination early in life provide a basis for early learning and memory.

It was suggested earlier in the chapter that memory in the adult involves modulation of synaptic transmission through long-lasting chemical changes initiated by neurotransmitters binding to certain types of receptors. Is there any connection between this and the developmental phenomena we have just described, involving relatively gross structural changes in the pattern of synaptic connections? In some cases at least, chemical and structural synaptic changes seem to be intimately related. For example, when *Aplysia* is subjected to long-term habituation or sensitization (see p. 1095) by repeated training on successive days, the chemical modulation of synaptic transmission becomes reinforced by alterations in the size of the presynaptic structures. It is noteworthy also that ocular dominance columns in a frog are altered by exposure to agonists or antagonists of the NMDA receptor, which is thought to play a part in memory in the adult hippocampus (see p. 1100).

Beyond such hints, however, one can only speculate; memory and the mechanisms of synapse formation and elimination are not yet adequately understood. But it is clear that they are among the problems that go to the heart of nerve cell biology and whose solutions promise to illuminate and unify our understanding of the brain at almost every level.

Summary

Synapses first form early in development, but the initial pattern of connections undergoes prolonged remodeling through the elimination of old synapses and the creation of new ones. The creation of a synapse between a motor neuron and a muscle cell involves changes in both cells and the deposition of specialized components in the basal lamina that lies between them. The specialized junctional basal lamina survives the destruction of both motor axons and muscle cells and controls the location of synaptic specializations in both muscle and axon terminal when they regenerate.

In a mammal at the time of birth, each muscle cell generally has several synapses on it, all but one of which are subsequently eliminated by a competitive mechanism. Activation of a muscle cell at a synapse tends to cause disappearance of other synapses in the neighborhood of the one that is active if these synapses are not themselves active at the same time; but where pre- and postsynaptic activity occur synchronously, the synapse tends to be consolidated. This "firing rule," whose molecular mechanism is unknown, appears to explain the phenomena of synapse formation and elimination in many different parts of the developing nervous system. In particular, it helps to explain how synaptic connections in the brain become adjusted according to an animal's experience of the world.

References

General

The Brain. *Sci. Am.* 241(3), 1979. (A whole issue on neurobiology.)

Cooke, I; Lipkin, M., eds. Cellular Neurophysiology: A Source Book. New York: Holt, Rinehart and Winston, 1972. (An anthology.)

Hille, B. Ionic Channels of Excitable Membranes. Sunderland, MA: Sinauer, 1984.

Kandel, E.R.; Schwartz, J.H. Principles of Neural Science, 2nd ed. New York: Elsevier, 1985.

Kuffler, S.W.; Nicholls, J.G.; Martin, A.R. From Neuron to Brain, 2nd ed. Sunderland, MA: Sinauer, 1984.

Molecular Neurobiology. *Cold Spring Harbor Symp. Quant. Biol.* 48, 1983.

Patterson, P.H.; Purves, D. Readings in Developmental Neurobiology. Cold Spring Harbor, NY: Cold Spring Harbor Laboratory, 1982. (An anthology.)

Purves, J.; Lichtman, J.W. Principles of Neural Development. Sunderland, MA: Sinauer, 1985.

Ramón y Cajal, S. Histologie du Systéme Nerveux de l'Homme et des Vértebrés. Paris: Maloine, 1909–1911. (Reprinted, Madrid: Consejo Superior de Investigaciones Científicas, Instituto Ramón y Cajal, 1972.)

Cited

1. Bullock, T.H.; Orkand, R.; Grinnell, A. Introduction to Nervous Systems, pp. 6 and 393–496. San Francisco: Freeman, 1977.

   Nauta, W.J.H.; Feirtag, M. Fundamental Neuroanatomy. New York: Freeman/Scientific American Library, 1986.

2. Ramón y Cajal, S. Recollections of My Life (E.H. Craigie, trans.). In Memoirs of the American Philosophical Society, Vol. 8. Philadelphia, 1937. Reprinted, New York: Garland, 1988. (The most entertaining of all introductions to cellular neurobiology, by the founding father of the subject.)

   Stevens, C.F. The neuron. *Sci. Am.* 241(3):48–59, 1979.

3. Hodgkin, A.L. The Conduction of the Nervous Impulse. Liverpool, U.K.: Liverpool University Press, 1964.
4. Katz, B. Nerve, Muscle, and Synapse. New York: McGraw-Hill, 1966.
5. Bunge, M.B. The axonal cytoskeleton: its role in generating and maintaining cell form. *Trends Neurosci.* 9:477–482, 1986.
   Grafstein, B.; Forman, D.S. Intracellular transport in neurons. *Physiol. Rev.* 60:1167–1283, 1980.
   Peters, A.; Palay, S.L.; Webster, H. deF. The Fine Structure of the Nervous System. Philadelphia: Saunders, 1976.
   Schwartz, J.H. The transport of substances in nerve cells. *Sci. Am.* 242(4):152–171, 1980.
6. Jones, E.G. Pathways to progress—the rise of modern neuroanatomical techniques. *Trends Neurosci.* 9:502–505, 1986.
   LaVail, J.H.; LaVail, M.M. Retrograde axonal transport in the central nervous system. *Science* 176:1416–1417, 1972.
   Schnapp, B.J.; Vale, R.D.; Sheetz, M.P.; Reese, T.S. Single microtubules from squid axoplasm support bidirectional movement of organelles. *Cell* 40:455–462, 1985.
   Vale, R.D. Intracellular transport using microtubule-based motors. *Annu. Rev. Cell Biol.* 3:347–378, 1987.
7. Goldstein, G.W.; Betz, A.L. The blood-brain barrier. *Sci. Am.* 255(3):74–83, 1986.
   Kuffler, S.W.; Nicholls, J.G.; Martin, A.R. From Neuron to Brain, 2nd ed., pp. 323–375. Sunderland, MA: Sinauer, 1984.
   Perry, V.H.; Gordon, S. Macrophages and microglia in the nervous system. *Trends Neurosci.* 11:273–277, 1988.
8. Hille, B. Ionic Channels of Excitable Membranes, pp. 21–75. Sunderland, MA: Sinauer, 1984.
   Kandel, E.R.; Schwartz, J.H. Principles of Neural Science, 2nd ed., pp. 49–86. New York: Elsevier, 1985.
   Kuffler, S.W.; Nicholls, J.G.; Martin, A.R. From Neuron to Brain, 2nd ed., pp. 97–206. Sunderland, MA: Sinauer, 1984.
9. Hodgkin, A.L.; Rushton, W.A.H. The electrical constants of a crustacean nerve fibre. *Proc. R. Soc. Lond. (Biol.)* 133:444–479, 1946.
   Roberts, A.; Bush, B.M.H., eds. Neurones Without Impulses. Cambridge, U.K.: Cambridge University Press, 1981.
10. Baker, P.F.; Hodgkin, A.L.; Shaw, T. The effects of changes in internal ionic concentrations on the electrical properties of perfused giant axons. *J. Physiol.* 164:355–374, 1962.
   Chiu, S.Y.; Ritchie, J.M.; Rogart, R.B.; Stagg, D. A quantitative description of membrane currents in rabbit myelinated nerve. *J. Physiol.* 292:149–166, 1979.
   Hodgkin, A.L. Chance and design in electrophysiology: an informal account of certain experiments on nerve carried out between 1934 and 1952.
   Hodgkin, A.L.; Huxley, A.F. Currents carried by sodium and potassium ions through the membrane of the giant axon of Loligo. *J. Physiol.* 116:449–472, 1952.
   Hodgkin, A.L.; Huxley, A.F.; Katz, B. Measurement of current-voltage relations in the membrane of the giant axon of Loligo. *J. Physiol.* 116:424–448, 1952.
   Hodgkin, A.L.; Katz, B. The effect of sodium ions on the electrical activity of the giant axon of the squid. *J. Physiol.* 108:37–77, 1949.
11. Hodgkin, A.L.; Huxley, A.F. A quantitative description of membrane current and its application to conduction and excitation in nerve. *J. Physiol.* 117:500–544, 1952.
12. Bray, G.M.; Rasminsky, M.; Aguayo, A.J. Interactions between axons and their sheath cells. *Annu. Rev. Neurosci.* 4:127–162, 1981.
   French-Constant, C.; Raff, M.C. The oligodendrocyte-type-2 astrocyte cell lineage is specialized for myelination. *Nature* 323:335–338, 1986.
   Morell, P.; Norton, W.T. Myelin. *Sci. Am.* 242(5):88–118, 1980.
13. Brown, D.A. Synaptic mechanisms. *Trends Neurosci.* 9:468–470, 1986.
   Katz, B. Nerve, Muscle and Synapse, pp. 97–158. New York: McGraw-Hill, 1966.
   Kuffler, S.W.; Nicholls, J.G.; Martin, A.R. From Neuron to Brain, 2nd ed., pp. 207–320. Sunderland, MA: Sinauer, 1984.
14. Dale, H.H.; Feldberg, W.; Vogt, M. Release of acetylcholine at voluntary motor nerve endings. *J. Physiol.* 86:353–380, 1936.
   Fatt, P.; Katz, B. An analysis of the end-plate potential recorded with an intracellular electrode. *J. Physiol.* 115:320–370, 1951.
   Feldberg, W. The early history of synaptic and neuromuscular transmission by acetylcholine: reminiscences of an eyewitness. In The Pursuit of Nature (A.L. Hodgkin, et al.), pp. 65–83. Cambridge, U.K.: Cambridge University Press, 1977.
15. Hille, B. Ionic Channels of Excitable Membranes, pp. 76–98. Sunderland, MA: Sinauer, 1984.
   Katz, B.; Miledi, R. The timing of calcium action during neuromuscular transmission. *J. Physiol.* 189:535–544, 1967.
   Llinas, R. Calcium in synaptic transmission. *Sci. Am.* 247(4): 56–65, 1982.
   Miller, R.J. Calcium signalling in neurons. *Trends Neurosci.* 11:415–419, 1988. (Introducing collection of reviews.)
16. Couteaux, R.; Pécot-Dechavassine, M. Vésicules synaptiques et poches au niveau des zones actives de la jonction neuromusculaire. *Comptes Rendus Acad. Sci. (Paris)* D 271:2346–2349, 1970.
   Heuser, J.E.; et al. Synaptic vesicle exocytosis captured by quick freezing and correlated with quantal transmitter release. *J. Cell Biol.* 81:275–300, 1979.
   Heuser, J.E.; Reese, T.S. Structural changes after transmitter release at the frog neuromuscular junction. *J. Cell Biol.* 88:564–580, 1981.
   Smith, S.J.; Augustine, G.J. Calcium ions, active zones and synaptic transmitter release. *Trends Neurosci.* 11:458–464, 1988.
17. Del Castillo, J.; Katz, B. Quantal components of the end-plate potential. *J. Physiol.* 124:560–573, 1954.
   Fatt, P.; Katz, B. Spontaneous subthreshold activity at motor nerve endings. *J. Physiol.* 117:109–128, 1952.
18. Hille, B. Ionic Channels of Excitable Membranes, pp. 117–147. Sunderland, MA: Sinauer, 1984.
19. Lester, H.A. The response to acetylcholine. *Sci. Am.* 236(2): 106–118, 1977.
   Sakmann, B.; Bormann, J.; Hamill, O.P. Ion transport by single receptor channels. *Cold Spring Harbor Symp. Quant. Biol.* 48:247–257, 1983.
20. Massoulié, J.; Bon, S. The molecular forms of cholinesterase and acetylcholinesterase in vertebrates. *Annu. Rev. Neurosci.* 5:57–106, 1982.
   Taylor, P.; Schumacher, M.; MacPhee-Quigley, K.; Friedmann, T.; Taylor, S. The structure of acetylcholinesterase: relationship to its function and cellular disposition. *Trends Neurosci.* 10:93–95, 1987.
21. Barnard, E.A.; Darlison, M.G.; Seeburg, P. Molecular biology of the GABA-A receptor: the receptor/channel superfamily. *Trends Neurosci.* 10:502–509, 1987.
   Grenningloh, G.; et al. The strychnine-binding subunit of the glycine receptor shows homology with nicotinic acetylcholine receptors. *Nature* 328:215–220, 1987.
   Hille, B. Ionic Channels of Excitable Membranes, pp. 371–383. Sunderland, MA: Sinauer, 1984. (Evolution of channels.)
22. Gottlieb, D.I. GABAergic neurons. *Sci. Am.* 258(2):38–45, 1988.
   Mayer, M.L.; Westbrook, G.L. The physiology of excitatory amino acids in the vertebrate central nervous system. *Prog. Neurobiol.* 28:197–276, 1987.
   Sneddon, P.; Westfall, D.P. Pharmacological evidence that adenosine triphosphate and noradrenaline are co-transmitters in the guinea pig vas deferens. *J. Physiol.* 347:561–580, 1984.

23. Snyder, S.H. Drug and neurotransmitter receptors in the brain. *Science* 224:22–31, 1984.
Nathanson, N.M. Molecular properties of the muscarinic acetylcholine receptor. *Annu. Rev. Neurosci.* 10:195–236, 1987.
24. Bormann, J. Electrophysiology of GABA-A and GABA-B receptor subtypes. *Trends Neurosci.* 11:112–116, 1988.
Snyder, S.H. Drugs and the Brain. New York: W.H. Freeman/ Scientific American Books, 1987.
Tallman, J.F.; Gallager, D.W. The GABAergic system: a locus of benzodiazepine action. *Annu. Rev. Neurosci.* 8:21–44, 1985.
25. Kuffler, S.W.; Nicholls, J.G.; Martin, A.R. From Neuron to Brain, 2nd ed., pp. 407–430. Sunderland, MA: Sinauer, 1984.
26. Barrett, J.N. Motoneuron dendrites: role in synaptic integration. *Fed. Proc.* 34:1398–1407, 1975.
27. Coombs, J.S.; Curtis, D.R.; Eccles, J.C. The generation of impulses in motoneurones. *J. Physiol.* 139:232–249, 1957.
Fuortes, M.G.F.; Frank, K.; Becker, M.C. Steps in the production of motoneuron spikes. *J. Gen. Physiol.* 40:735–752, 1957.
28. Connor, J.A.; Stevens, C.F. Prediction of repetitive firing behaviour from voltage clamp data on an isolated neurone soma. *J. Physiol.* 213:31–53, 1971.
Hille, B. Ionic Channels of Excitable Membranes, pp. 99–116. Sunderland, MA: Sinauer, 1984.
Rogawski, M.A. The A-current: how ubiquitous a feature of excitable cells is it? *Trends Neurosci.* 8:214–219, 1985.
29. Meech, R.W. Calcium-dependent potassium activation in nervous tissues. *Annu. Rev. Biophys. Bioeng.* 7:1–18, 1978.
Tsien, R.W.; Lipscombe, D.; Madison, D.V.; Bley, K.R.; Fox, A.P. Multiple types of neuronal calcium channels and their selective modulation. *Trends Neurosci.* 11:431–438, 1988.
30. Bullock, T.H.; Horridge, G.A. Structure and Function in the Nervous System of Invertebrates, pp. 38–124. San Francisco: Freeman, 1965.
Llinas, R.; Sugimori, M. Electrophysiological properties of *in vitro* Purkinje cell dendrites in mammalian cerebellar slices. *J. Physiol.* 305:197–213, 1980.
Shepherd, G.M. Microcircuits in the nervous system. *Sci. Am.* 238(2):92–103, 1978.
31. Breitwieser, G.E.; Szabo, G. Uncoupling of cardiac muscarinic and beta-adrenergic receptors from ion channels by a guanine nucleotide analogue. *Nature* 317:538–540, 1985.
Levitan, I.B. Modulation of ion channels in neurons and other cells. *Annu. Rev. Neurosci.* 11:119–136, 1988.
Nairn, A.C.; Hemmings, H.C.; Greengard, P. Protein kinases in the brain. *Annu. Rev. Biochem.* 54:931–976, 1985.
Pfaffinger, P.J.; Martin, J.M.; Hunter, D.D.; Nathanson, N.M.; Hille, B. GTP-binding proteins couple cardiac muscarinic receptors to a K channel. *Nature* 317:536–538, 1985.
32. Iversen, L.L. The chemistry of the brain. *Sci. Am.* 241(3):118–129, 1979.
Snyder, S.H. Drugs and the Brain. New York: W.H. Freeman/ Scientific American Library, 1988.
33. Bloom, F.E. Neuropeptides. *Sci. Am.* 245(4):148–168, 1981.
Hokfelt, T.; Johansson, O.; Goldstein, M. Chemical anatomy of the brain. *Science* 225:1326–1334, 1984.
Jan, Y.N.; Bowers, C.W.; Branton, D.; Evans, L.; Jan, L.Y. Peptides in neuronal function: studies using frog autonomic ganglia. *Cold Spring Harbor Symp. Quant. Biol.* 48:363–374, 1983.
Scheller, R.H.; Axel, R. How genes control an innate behavior. *Sci. Am.* 250(3):44–52, 1984. (Neuropeptides in Aplysia).
34. Farley, J.; Alkon, D. Cellular mechanisms of learning, memory, and information storage. *Annu. Rev. Psychol.* 36:419–494, 1985.
Kandel, E.R. Small systems of neurons. *Sci. Am.* 241(3):60–70, 1979.
Morris, R.G.M.; Kandel, E.R.; Squire, L.R., eds. Learning and Memory. *Trends Neurosci.* 11:125–181, 1988.
35. Greenberg, S.M.; Castellucci, V.F.; Bayley, H.; Schwartz, J.H. A molecular mechanism for long-term sensitization in Aplysia. *Nature* 329:62–65, 1987.
Montarolo, P.G.; et al. A critical period for macromolecular synthesis in long-term heterosynaptic facilitation in Aplysia. *Science* 234:1249–1254, 1986.
Schwartz, J.H.; Greenberg, S.M. Molecular mechanisms for memory: second-messenger induced modifications of protein kinases in nerve cells. *Annu. Rev. Neurosci.* 10:459–476, 1987.
Siegelbaum, S.A.; Camardo, J.S.; Kandel, E.R. Serotonin and cyclic AMP close single K^+ channels in Aplysia sensory neurons. *Nature* 299:413–417, 1982.
36. Dudai, Y. Neurogenetic dissection of learning and short-term memory in Drosophila. *Annu. Rev. Neurosci.* 11:537–563, 1988.
Kandel, E.R.; et al. Classical conditioning and sensitization share aspects of the same molecular cascade in Aplysia. *Cold Spring Harbor Symp. Quant. Biol.* 48:821–830, 1983.
37. Bliss, T.V.P.; Lomo, T. Long-lasting potentiation of synaptic transmission in the dentate area of the anaesthetized rabbit following stimulation of the perforant path. *J. Physiol.* 232:331–356, 1973.
Collingridge, G.L.; Bliss, T.V.P. NMDA receptors—their role in long-term potentiation. *Trends Neurosci.* 10:288–293, 1987. (The same issue of the journal contains an excellent collection of reviews of various aspects of NMDA receptors.)
Cotman, C.W.; Monaghan, D.T.; Ganong, A.H. Excitatory amino acid neurotransmission: NMDA receptors and Hebb-type synaptic plasticity. *Annu. Rev. Neurosci.* 11:61–80, 1988.
Lisman, J.E.; Goldring, M.A. Feasibility of long-term storage of graded information by the Ca^{2+}/calmodulin-dependent protein kinase molecules of the postsynaptic density. *Proc. Natl. Acad. Sci. USA* 85:5320–5324, 1988.
Mishkin, M.; Appenzeller, T. The anatomy of memory. *Sci. Am.* 256(6):80–89, 1987.
Morris, R.G.; Anderson, E.; Lynch, G.; Baudry, M. Selective impairment of learning and blockade of long-term potentiation by an N-methyl-D-aspartate receptor antagonist, AP5. *Nature* 319:774–776, 1986.
38. Barlow, H.B.; Mollon, J.D., eds. The Senses. Cambridge, U.K.: Cambridge University Press, 1982.
Schmidt, R.F., ed. Fundamentals of Sensory Physiology. New York: Springer, 1978.
Shepherd, G. Neurobiology, 2nd ed., pp. 205–353. New York: Oxford University Press, 1988.
39. Katz, B. Depolarization of sensory terminals and the initiation of impulses in the muscle spindle. *J. Physiol.* 111:261–282, 1950.
40. Hudspeth, A.J. The cellular basis of hearing: the biophysics of hair cells. *Science* 230:745–752, 1985.
Roberts, W.M.; Howard, J.; Hudspeth, A.J. Hair cells: transduction, tuning, and transmission in the inner ear. *Annu. Rev. Cell Biol.* 4:63–92, 1988.
von Bekesy, G. The ear. *Sci. Am.* 197(2):66 - 78, 1957.
41. Corey, D.P.; Hudspeth, A.J. Kinetics of the receptor current in bullfrog saccular hair cells. *J. Neurosci.* 3:962–976, 1983.
Howard, J.; Hudspeth, A.J. Compliance of the hair bundle associated with mechanoelectrical transduction channels in the bullfrog's saccular hair cell. *Neuron* 1:189–199, 1988.
Pickles, J.O. Recent advances in cochlear physiology. *Prog. Neurobiol.* 24:1–42, 1985.
42. Barlow, H.B.; Mollon, J.D., eds. The Senses, pp. 102–164. Cambridge, U.K.: Cambridge University Press, 1982.
43. Baylor, D.A.; Lamb, T.D.; Yau, K.-W. Responses of retinal rods to single photons. *J. Physiol.* 288:613–634, 1979.
Schnapf, J.L.; Baylor, D.A. How photoreceptor cells respond to light. *Sci. Am.* 256(4):40–47, 1987.

44. Stryer, L. The molecules of visual excitation. *Sci. Am.* 257(1): 32–40, 1987.

45. Fesenko, E.E.; Kolesnikov, S.S.; Lyubarsky, A.L. Induction by cyclic GMP of cationic conductance in plasma membrane of retinal rod outer segment. *Nature* 313:310–313, 1985.

Stryer, L. The cyclic GMP cascade of vision. *Annu. Rev. Neurosci.* 9:87–119, 1986.

46. Koch, K.-W.; Stryer, L. Highly cooperative feedback control of retinal rod guanylate cyclase by calcium ions. *Nature* 334: 64–66, 1988.

Matthews, H.R.; Murphy, R.L.W.; Fain, G.L.; Lamb, T.D. Photoreceptor light adaptation is mediated by cytoplasmic calcium concentration. *Nature* 334:67–69, 1988.

Nakatani, K.; Yau, K.-W. Calcium and light adaptation in retinal rods and cones. *Nature* 334:69–71, 1988.

47. Hubel, D.H. Eye, Brain, and Vision. New York: W.H. Freeman/ Scientific American Library, 1988.

Kuffler, S.W.; Nicholls, J.G.; Martin, A.R. From Neuron to Brain, 2nd ed., pp. 19–96. Sunderland, MA: Sinauer, 1984.

Masland, R.H. The functional architecture of the retina. *Sci. Am.* 255(6):102–111, 1986.

48. Cowan, W.M. The development of the brain. *Sci. Am.* 241(3):106–117, 1979.

Hopkins, W.G.; Brown, M.C. Development of Nerve Cells and their Connections. Cambridge, U.K.: Cambridge University Press, 1984.

Parnavelas, J.G.; Stern, C.D.; Stirling, R.V., eds. The Making of the Nervous System. Oxford, U.K.: Oxford University Press, 1988.

Purves, D.; Lichtman, J.W. Principles of Neural Development. Sunderland, MA: Sinauer, 1985.

49. Alvarez-Buylla, A.; Nottebohm, F. Migration of young neurons in adult avian brain. *Nature* 335:353–354, 1988. (An example of neural stem cells persisting in adult life.)

Le Douarin, N.M.; Smith, J. Development of the peripheral nervous system from the neural crest. *Annu. Rev. Cell Biol.* 4:375–404, 1988.

Stent, G.S.; Weisblat, D.A. Cell lineage in the development of invertebrate nervous systems. *Annu. Rev. Neurosci.* 8:45–70, 1985.

Williams, R.W.; Herrup, K. The control of neuron number. *Annu. Rev. Neurosci.* 11:423–453, 1988.

50. Hollyday, M.; Hamburger, V. An autoradiographic study of the formation of the lateral motor column in the chick embryo. *Brain Res.* 132:197–208, 1977.

Rakic, P. Mode of cell migration to the superficial layers of the fetal monkey neocortex. *J. Comp. Neurol.* 145:61–84, 1972.

51. Caviness, V.S. Neocortical histogenesis in normal and reeler mice: a developmental study based on [3H]thymidine autoradiography. *Dev. Brain Res.* 4:293–302, 1982.

McConnell, S.K. Development and decision-making in the mammalian cerebral cortex. *Brain Res. Rev.* 13:1–23, 1988.

Rakic, P. Specification of cerebral cortical areas. *Science* 241: 170–176, 1988.

52. Bray, D.; Hollenbeck, P.J. Growth cone motility and guidance. *Annu. Rev. Cell Biol.* 4:43–61, 1988.

Harrison, R.G. The outgrowth of the nerve fiber as a mode of protoplasmic movement. *J. Exp. Zool.* 9:787–846, 1910.

Yamada, K.M.; Spooner, B.S.; Wessells, N.K. Ultrastructure and function of growth cones and axons of cultured nerve cells. *J. Cell Biol.* 49:614–635, 1971.

53. Bamburg, J.R. The axonal cytoskeleton: stationary or moving matrix? *Trends Neurosci.* 11:248–249, 1988.

Bamburg, J.R.; Bray, D.; Chapman, K. Assembly of microtubules at the tip of growing axons. *Nature* 321:788–790, 1986.

Bray, D. Surface movements during the growth of single explanted neurons. *Proc. Natl. Acad. Sci. USA* 65:905–910, 1970.

Letourneau, P.C.; Ressler, A.H. Inhibition of neurite initiation and growth by taxol. *J. Cell Biol.* 98:1355–1362, 1984.

54. Davies, A.M. Molecular and cellular aspects of patterning sensory neurone connections in the vertebrate nervous system. *Development* 101:185–208, 1987.

Kater, S.B.; Mattson, M.P.; Cohan, C.; Connor, J. Calcium regulation of the neuronal growth cone. *Trends Neurosci.* 11:315–321, 1988.

Letourneau, P.C. Cell-to-substratum adhesion and guidance of axonal elongation. *Dev. Biol.* 44:92–101, 1975.

Lumsden, A.G.S.; Davies, A.M. Earliest sensory nerve fibres are guided to peripheral targets by attractants other than Nerve Growth Factor. *Nature* 306: 786–788, 1983.

Patel, N.; Poo, M.-M. Orientation of neurite growth by extracellular electric fields. *J. Neurosci.* 2:483–496, 1982.

55. Bentley, D.; Caudy, M. Navigational substrates for peripheral pioneer growth cones: limb-axis polarity cues, limb-segment boundaries, and guidepost neurons. *Cold Spring Harbor Symp. Quant. Biol.* 48:573–585, 1983.

Berlot, J.; Goodman, C.S. Guidance of peripheral pioneer neurons in the grasshopper: adhesive hierarchy of epithelial and neuronal surfaces. *Science* 223:493–496, 1984.

Goodman, C.S.; Bastiani, M.J. How embryonic nerve cells recognize one another. *Sci. Am.* 251(6):58–66, 1984.

56. Bixby, J.L.; Pratt, R.S.; Lilien, J.; Reichardt, L.F. Neurite outgrowth on muscle cell surfaces involves extracellular matrix receptors as well as Ca^{2+}-dependent and -independent cell adhesion molecules. *Proc. Natl. Acad. Sci. USA* 84:2555–2559, 1987.

Chang, S.; Rathjen, F.G.; Raper, J.A. Extension of neurites on axons is impaired by antibodies against specific neural cell adhesion molecules. *J. Cell Biol.* 104:355–362, 1987.

Jessell, T.M. Adhesion molecules and the hierarchy of neural development. *Neuron* 1:3-13, 1988.

Sanes, J.R.; Schachner, M.; Covault, J. Expression of several adhesive macromolecules (N-CAM, L1, J1, NILE, uvomorulin, laminin, fibronectin, and a heparan sulfate proteoglycan) in embryonic, adult, and denervated adult skeletal muscle. *J. Cell Biol.* 102:420–431, 1986.

Tomaselli, K.J.; et al. N-cadherin and integrins: two receptor systems that mediate neuronal process outgrowth on astrocyte surfaces. *Neuron* 1:33–43, 1988.

57. Lance-Jones, C.; Landmesser, L. Motoneurone projection patterns in the chick hindlimb following early partial reversals of the spinal cord. *J. Physiol.* 302:581–602, 1980.

Landmesser, L. The development of specific motor pathways in the chick embryo. *Trends Neurosci.* 7:336–339, 1984.

Sperry, R.W. Chemoaffinity in the orderly growth of nerve fiber patterns and connections. *Proc. Natl. Acad. Sci. USA* 50:703–710, 1963.

Udin, S.B.; Fawcett, J.W. Formation of topographic maps. *Annu. Rev. Neurosci.* 11:289–327, 1988.

58. Campenot, R.B. Local control of neurite development by nerve growth factor. *Proc. Natl. Acad. Sci. USA* 74:4516–4519, 1977.

Davies, A.; Lumsden, A. Relation of target encounter and neuronal death to nerve growth factor responsiveness in the developing mouse trigeminal ganglion. *J. Comp. Neurol.* 223:124–137, 1984.

Greene, L.A. The importance of both early and delayed responses in the biological actions of nerve growth factor. *Trends Neurosci.* 7:91–94, 1984.

Levi-Montalcini, R.; Calissano, P. The nerve growth factor. *Sci. Am.* 240(6):68–77, 1979.

59. Cowan, W.M.; Fawcett, J.W.; O'Leary, D.D.M.; Stanfield, B.B. Regressive events in neurogenesis. *Science* 225:1258–1265, 1984.

Davies, A.M. Role of neurotrophic factors in development. *Trends Genet.* 4:139–144, 1988.

Hamburger, V.; Levi-Montalcini, R. Proliferation, differentiation and degeneration in the spinal ganglia of the chick embryo under normal and experimental conditions. *J. Exp. Zool.* 162:133–160, 1949.

Hamburger, V.; Yip, J.W. Reduction of experimentally induced neuronal death in spinal ganglia of the chick embryo by nerve growth factor. *J. Neurosci.* 4:767–774, 1984.

Williams, R.W.; Herrup, K. The control of neuron number. *Annu. Rev. Neurosci.* 11:423–453, 1988.

60. Brown, M.C.; Holland, R.L.; Hopkins, W.G. Motor nerve sprouting. *Annu. Rev. Neurosci.* 4:17–42, 1981.

Davies, A.M. The survival and growth of embryonic proprioceptive neurons is promoted by a factor present in skeletal muscle. *Dev. Biol.* 115:56–67, 1986.

Ebendal, T.; Olson, L.; Seiger, A.; Hedlund, K.-O. Nerve growth factors in the rat iris. *Nature* 286:25–28, 1980.

Purves, D.; Voyvodic, J.T. Imaging mammalian nerve cells and their connections over time in living animals. *Trends Neurosci.* 10:398–404, 1987.

61. Purves, D.; Lichtman, J.W. Principles of Neural Development. Sunderland, MA: Sinauer, 1985.

62. Hume, R.I.; Role, L.W.; Fischbach, G.D. Acetylcholine release from growth cones detected with patches of acetylcholine receptor-rich membranes. *Nature* 305:632–634, 1983.

Jaramillo, F.; Vicini, S.; Schuetze, S.M. Embryonic acetylcholine receptors guarantee spontaneous contractions in rat developing muscle. *Nature* 335:66–68, 1988.

Spitzer, N.C. Ion channels in development. *Annu. Rev. Neurosci.* 2:363–397, 1979.

Young, S.H.; Poo, M.-M. Spontaneous release of transmitter from growth cones of embryonic neurones. *Nature* 305: 634–637, 1983.

63. Poo, M.-M. Mobility and localization of proteins in excitable membranes. *Annu. Rev. Neurosci.* 8:369–406, 1985.

Schuetze, S.M.; Role, L.W. Developmental regulation of nicotinic acetylcholine receptors. *Annu. Rev. Neurosci.* 10:403–457, 1987.

64. Burden, S.J.; Sargent, P.B.; McMahan, U.J. Acetylcholine receptors in regenerating muscle accumulate at original synaptic sites in the absence of the nerve. *J. Cell Biol.* 82: 412–425, 1979.

Nitkin, R.M.; et al. Identification of agrin, a synaptic organizing protein from Torpedo electric organ. *J. Cell Biol.* 105: 2471–2478, 1987.

Sanes, J.R.; Hall, Z.W. Antibodies that bind specifically to synaptic sites on muscle fiber basal lamina. *J. Cell Biol.* 83: 357–370, 1979.

65. Frank, E.; Jansen, J.K.S.; Lømo, T.; Westgaard, R.H. The interaction between foreign and original motor nerves innervating the soleus muscle of rats. *J. Physiol.* 247:725–743, 1975.

Jones, R.; Vrbová, G. Two factors responsible for denervation hypersensitivity. *J. Physiol.* 236:517–538, 1974.

Lømo, T.; Jansen, J.K.S. Requirements for the formation and maintenance of neuromuscular connections. *Curr. Top. Dev. Biol.* 16:253–281, 1980.

Lømo, T.; Rosenthal, J. Control of ACh sensitivity by muscle activity in the rat. *J. Physiol.* 221:493–513, 1972.

66. Oppenheim, R.W. Cell death during neural development. In Handbook of Physiology, Vol. 1: Neuronal Development, (W.M. Cowan, ed.). Washington, DC: American Physiological Society, 1988.

Pittman, R.; Oppenheim, R.W. Cell death of motoneurons in the chick embryo spinal cord. IV. Evidence that a functional neuromuscular interaction is involved in the regulation of naturally occurring cell death and the stabilization of synapses. *J. Comp. Neurol.* 187:425–446, 1979.

67. Brown, M.C.; Jansen, J.K.S.; Van Essen, D. Polyneuronal innervation of skeletal muscle in new-born rats and its elimination during maturation. *J. Physiol.* 261:387–422, 1976.

Callaway, E.M.; Soha, J.M.; Van Essen, D.C. Competition favouring inactive over active motor neurons during synapse elimination. *Nature* 328:422–426, 1987. (Describes an apparent exception to the general rule.)

O'Brien, R.A.D.; Ostberg, A.J.C.; Vrbová, G. Protease inhibitors reduce the loss of nerve terminals induced by activity and calcium in developing rat soleus muscles in vitro. *Neurosci.* 12:637–646, 1984.

Ribchester, R.R.; Taxt, T. Motor unit size and synaptic competition in rat lumbrical muscles reinnervated by active and inactive motor axons. *J. Physiol.* 344:89–111, 1983.

68. Lichtman, J.W. The reorganization of synaptic connexions in the rat submandibular ganglion during post-natal development. *J. Physiol.* 273:155–177, 1977.

Purves, D.; Lichtman, J.W. Principles of Neural Development, pp. 271–328. Sunderland, MA: Sinauer, 1985.

69. Purves, D. Modulation of neuronal competition by postsynaptic geometry in autonomic ganglia. *Trends Neurosci.* 6:10–16, 1983.

70. Barlow, H. Visual experience and cortical development. *Nature* 258:199–204, 1975.

Wiesel, T.N. Postnatal development of the visual cortex and the influence of environment. *Nature* 299:583–591, 1982.

71. Fawcett, J.W. Retinotopic maps, cell death, and electrical activity in the retinotectal and retinocollicular projections. In The Making of the Nervous System (J.G. Parnavelas; C.D. Stern; R.V. Stirling, eds.), pp. 395–416. Oxford, U.K.: Oxford University Press, 1988.

Hubel, D.H.; Wiesel, T.N. Ferrier lecture: functional architecture of macaque monkey visual cortex. *Proc. R. Soc. Lond. Biol.* 198:1–59, 1977.

Hubel, D.H.; Wiesel, T.N.; Le Vay, S. Plasticity of ocular dominance columns in monkey striate cortex. *Philos. Trans. R. Soc. Lond. Biol.* 278:377–409, 1977.

Rakic, P. Prenatal genesis of connections subserving ocular dominance in the rhesus monkey. *Nature* 261:467–471, 1976.

Stryker, M.P.; Harris, W.A. Binocular impulse blockade prevents the formation of ocular dominance columns in cat visual cortex. *J. Neurosci.* 6:2117–2133, 1986.

72. Hubel, D.H.; Wiesel, T.N. Binocular interaction in striate cortex of kittens reared with artificial squint. *J. Neurophysiol.* 28:1041–1059, 1965.

Le Vay, S.; Wiesel, T.N.; Hubel, D.H. The development of ocular dominance columns in normal and visually deprived monkeys. *J. Comp. Neurol.* 191:1–51, 1980.

73. Baylis, G.C.; Rolls, E.T.; Leonard, C.M. Selectivity between faces in the responses of a population of neurons in the cortex in the superior temporal sulcus of the monkey. *Brain Res.* 342:91–102, 1985.

Cline, H.T.; Debski, E.A.; Constantine-Paton, M. N-methyl-D-aspartate receptor antagonist desegregates eye-specific stripes. *Proc. Natl. Acad. Sci. USA* 84:4342–4345, 1987.

Greenough, W.T.; Bailey, C.H. The anatomy of a memory: convergence of results across a diversity of tests. *Trends Neurosci.* 11:142–147, 1988.

Perrett, D.I.; Mistlin, A.J.; Chitty, A.J. Visual neurones responsive to faces. *Trends Neurosci.* 10:358–364, 1987.

Special Features of Plant Cells

20

Anyone can distinguish a flowering plant from a mammal. Even deciding whether an individual cell is of plant or animal origin is usually a simple matter, although there are problematic cases. But as the level of analysis extends into the cell—to cytoplasm, organelle, and molecule—the similarities between the two kingdoms begin to outweigh the differences. It requires sophisticated procedures to distinguish higher-plant mitochondria, nuclei, ribosomes, or cytoskeletal components from their animal counterparts. Indeed, the disparities between plants and animals tend not to be in fundamental molecular features such as DNA replication, protein synthesis, mitochondrial ATP production, or the basic design of cell membranes. Rather they relate to higher-order functions of cells and tissues.

Comparisons of ribosomal RNA sequences suggest that plants, yeasts, invertebrates, and vertebrates diverged from one another relatively late in the evolution of eucaryotic organisms (see Figure 1–16, p. 13). The differences between plants and animals apparently arose over a period of about 600 million years, and most of them can be traced to two fundamental acquisitions by the progenitors of plants: the ability to fix carbon dioxide by photosynthesis (discussed in Chapter 7), and the production of a rigid *cell wall.* The first acquisition gave plants an internal source of carbon compounds to use in growth and metabolism; the second placed severe constraints on how plants behave. In this chapter we shall consider the special features of plant cells that result from these two acquisitions.

The Importance of the Cell Wall

The plant cell wall is an elaborate extracellular matrix that encloses each cell in a plant. While animal cells also have extracellular matrix components on their surface (see p. 802), the plant cell wall is generally much thicker, stronger, and—most important—more rigid. In fact, most of the differences between plants and animals—in nutrition, digestion, osmoregulation, growth, reproduction, intercellular communication, defense mechanisms, as well as in morphology—can be traced to the plant cell wall. In evolving relatively rigid cell walls, which vary from 0.1 μm to many micrometers in thickness, plants cells forfeited the ability to crawl about; this nonmotile life-style has persisted in multicellular plants. It was the thick cell walls of cork, visible under a microscope, that in 1663 enabled Robert Hooke to distinguish cells clearly and to name them as such.

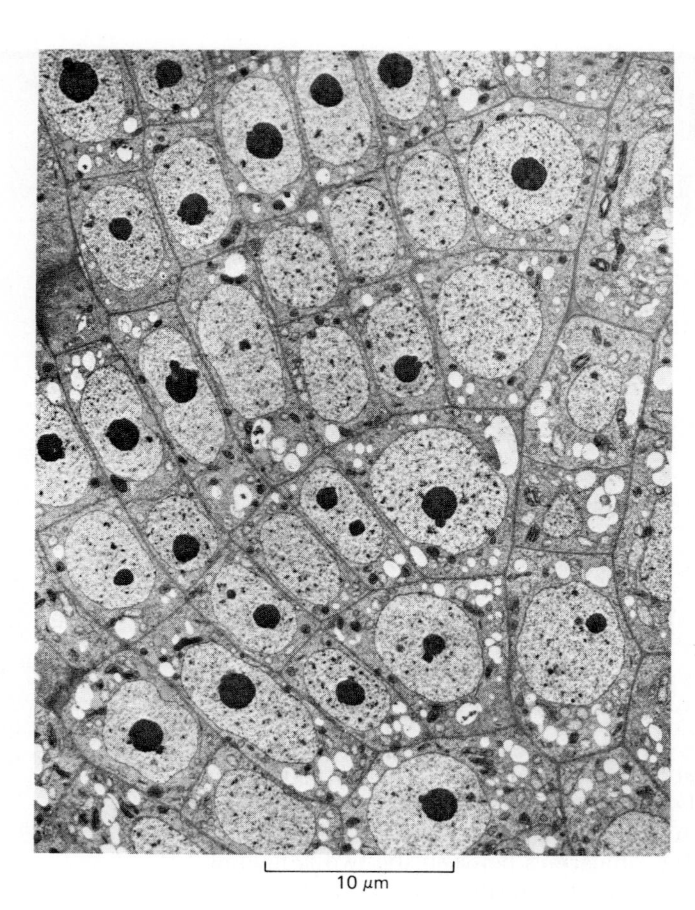

Figure 20–1 Electron micrograph of the root tip of a rush, showing the organized pattern of cells that results from an ordered sequence of cell divisions in cells with rigid cell walls. (Courtesy of Brian Gunning.)

The cell wall provides a protective home for the plant cell within. Each cell wall binds to that of its neighbors, cementing the cells together to form the intact plant (Figure 20–1). Even though each plant cell is encased within its own "wooden box," direct cell-cell communication is still possible through the *plasmodesmata*. Thousands of these plasma-membrane-lined channels of cytoplasm cross the cell wall, connecting adjacent cells and allowing small molecules to move from cell to cell. In addition, fluids percolate along and through cell walls. The plant cell wall thus has transport functions as well as protective and skeletal functions.

When plant cells become specialized, they generally produce walls that are particularly well adapted for one or another of these functions. Thus the different types of cells in a plant can be recognized and classified by the shape and nature of their cell walls. In this section we shall examine the ways in which plant cells have evolved to exploit their walled environment. The starting point is a description of the nature of the wall itself.

The Cell Wall Is Composed of Cellulose Fibers Embedded in a Matrix of Polysaccharide and Protein[1]

Most newly formed cells in a multicellular plant are produced in special regions of the plant called *meristems*, as will be explained later (see p. 1173). These new cells are generally small in comparison to their final size. To accommodate subsequent cell growth, the walls of such cells, called **primary cell walls** (Figure 20–2), are thin and only semirigid. Once growth stops, the wall no longer needs to be able to expand. Although the mature nongrowing cell may simply retain its primary cell wall, far more commonly it augments it by producing a **secondary cell wall,** either by thickening the primary wall or by depositing inside it new, tough wall layers with a different composition (see p. 1146).

Although the primary cell walls of higher plants vary greatly in both composition and detailed organization, like all extracellular matrices they are constructed

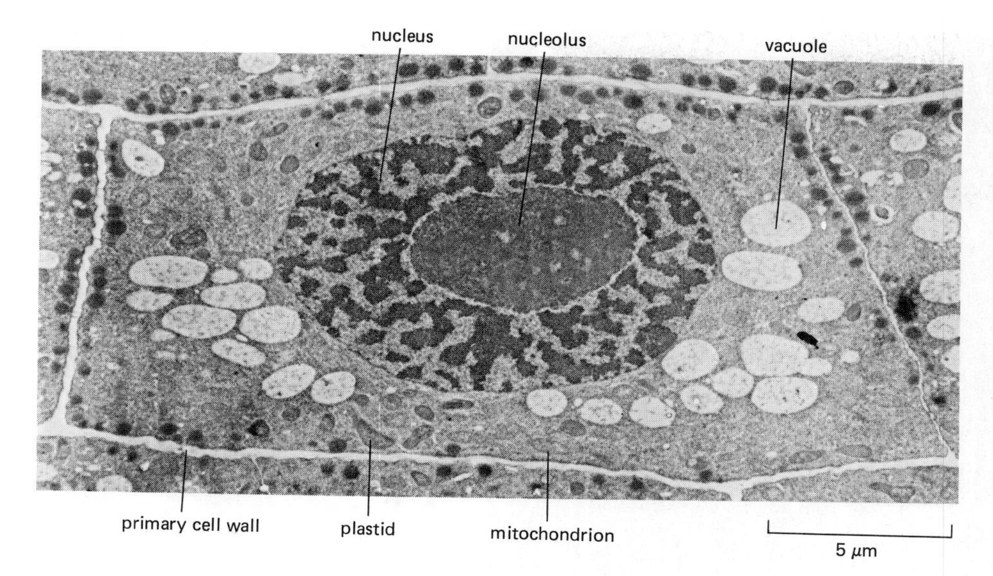

Figure 20–2 Electron micrograph of a young root tip cell of onion, showing the main organelles and the thin primary cell wall. (Courtesy of Brian Wells.)

according to a common principle: they derive their tensile strength from long fibers held together by a matrix of protein and polysaccharide, which is highly resistant to compression. The same architectural principle (strong fibers resistant to tension embedded in an amorphous matrix resistant to compression) is used in the construction of animal bones (see p. 989) and in such common building materials as fiber glass and reinforced concrete. In the cell walls of higher plants, the fibers are generally made from the polysaccharide *cellulose,* the most abundant organic macromolecule on earth. The matrix is composed predominantly of two other sorts of polysaccharide—*hemicellulose* and *pectin*—together with structural glycoproteins (Figure 20–3). The fibers and matrix molecules are cross-linked by a combination of covalent bonds and noncovalent forces into a highly complex structure whose composition is generally cell-specific (Figure 20–4). The major molecules that form the cell wall are known for many cell types, but it is not known what all the component molecules are in any cell type, nor how they are cross-linked and organized in three dimensions.

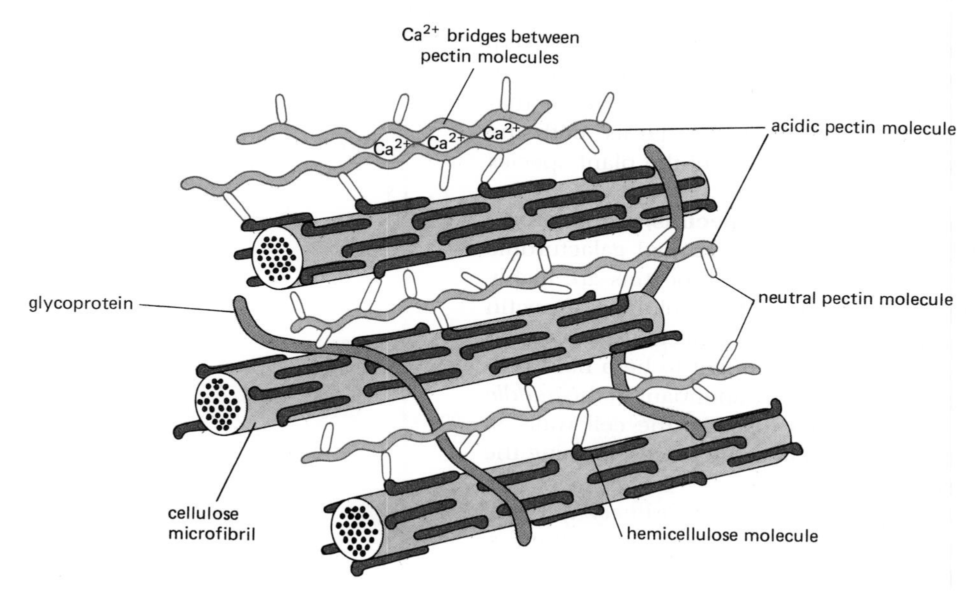

Figure 20–3 Interconnections between the two major components of the primary cell wall, the cellulose microfibrils and the matrix. Hemicellulose molecules (for example, xyloglucans) are linked by hydrogen bonds to the surface of the cellulose microfibrils. Some of these hemicellulose molecules are cross-linked in turn to acidic pectin molecules (for example, rhamnogalacturonans) through short neutral pectin molecules (for example, arabinogalactans). Cell-wall glycoproteins are tightly woven into the texture of the wall to complete the matrix.

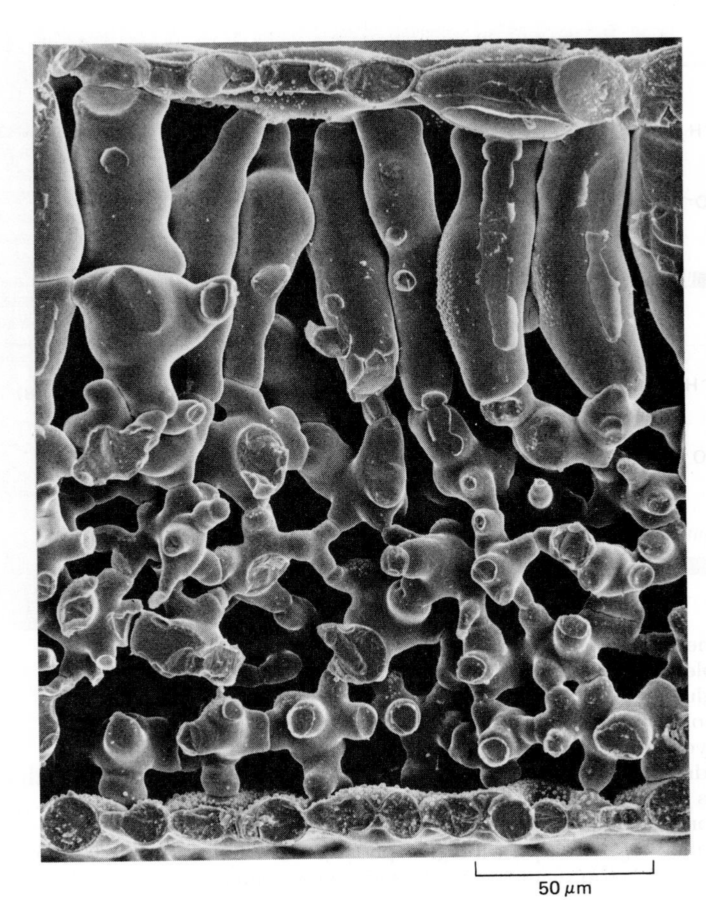

Figure 20–8 Scanning electron micrograph of the inside of a bean leaf that has been rapidly frozen and snapped open to expose the cells within. At the top is the single sheet of cells that forms the upper epidermis, and at the bottom is the lower epidermis. Between them lie large numbers of mesophyll cells. As these photosynthetic cells of the leaf grow and expand, their walls separate at defined regions along the middle lamella, resulting in an open network of cells, all of which have ready access to the carbon dioxide in the large air spaces that surround them. (From C.E. Jeffree, N.D. Read, V.A.C. Smith, and J.E. Dale, *Planta* 172:20–37, 1987.)

Up to 30% of their residues are hydroxyproline, an amino acid formed by the post-translational hydroxylation of proline residues (as in collagen, see p. 811). Many short oligosaccharide side chains are attached to hydroxyproline and serine side chains, so that more than half the weight of each glycoprotein is carbohydrate. It is difficult to extract these glycoproteins without destroying the structure of the cell wall, suggesting that they are tightly woven into the complex three-dimensional polysaccharide network of the wall. They are thought to act like glue to increase the strength of the wall, and, significantly, the levels of some of the mRNAS that encode these glycoproteins rise dramatically as part of the host response to infection or wounding.

In order for a plant cell to grow or change its shape, the cell wall has to stretch or deform. Because of their crystalline structure, individual cellulose microfibrils are unable to stretch, and therefore such changes must involve either the sliding of microfibrils past one another and/or the separation of adjacent microfibrils. As we shall discuss later, the direction in which the growing cell enlarges depends on the way that strain-resisting cellulose microfibrils are deposited in the primary wall (see p. 1168).

The Limited Porosity of Cell Walls Restricts the Exchange of Macromolecules Between Plant Cells and Their Environment[3]

All cells take in nutrients and expel waste products across their plasma membrane. They also respond to chemical signals in their environment. In the case of plant cells, such molecules and signals must penetrate the cell wall. Since the matrix of the wall is a highly hydrated polysaccharide gel (the primary cell wall being 60% water by weight), water, gases, and small water-soluble molecules diffuse rapidly through it. The cross-linked structure of the cell wall only slightly impedes the diffusion of small molecules such as water, sucrose, or K^+. (Even in the case of a cell with a wall 15 μm thick, only 10% of the resistance to water flow between the cytoplasm and the external medium is contributed by the wall; the remaining 90% is contributed by the plasma membrane.) The average diameter of the spaces between the cross-linked macromolecules in most cell walls, however, is about 5 nm; this is small enough to make the movement of any globular macromolecule with a molecular weight much above 20,000 extremely slow. Therefore plants must subsist on molecules of low molecular weight, and any intercellular signaling molecules that have to pass through the cell wall must also be small and water-soluble. In fact, most of the known plant signaling molecules, such as the growth-regulating substances—auxins, cytokinins, and gibberellins—have molecular weights of less than 500 (see p. 1181).

The Tensile Strength of Cell Walls Allows Plant Cells to Generate an Internal Hydrostatic Pressure Called Turgor[4]

The mechanical strength of the cell wall allows plant cells to survive in an extracellular environment that is hypotonic with respect to the cell interior. The extracellular fluid in higher plants consists of the aqueous phase in all of the cell walls plus the fluid in the long tubes formed by the cell walls of dead *xylem vessel elements* (see p. 1153). These tubes carry water (the *transpiration stream*) from the roots to sites of evaporation, mainly in the leaves. Although the extracellular fluid contains more solutes than does the dilute solution in the plant's external milieu (for example, soil), it is still hypotonic in comparison to the intracellular fluid. This can be readily demonstrated by removing the cell wall with cellulases and other wall-degrading enzymes and then observing the behavior of the wall-less cell, called a *protoplast* (see Figure 20–71). If such a spherical protoplast is exposed to the hypotonic fluid that normally bathes the plant cell, it takes up water by osmosis, swells, and eventually bursts (Figure 20–9). A walled cell placed in the same environment, in contrast, takes up water but can swell to only a limited extent. The cell develops an internal hydrostatic pressure that pushes outward on the cell wall, rather like an inner tube pressing against a bicycle tire. This hydrostatic pressure brings the cell to osmotic equilibrium and prevents any further net influx of water. (For a detailed discussion of osmosis refer to Panel 6–1, p. 308.)

The outward **turgor pressure** (or **turgor**) in all plant cells, caused by the osmotic imbalance between its intracellular and extracellular fluids, is vital to plants. It is the main driving force for cell expansion during growth and provides much of the mechanical rigidity of living plant tissues. Compare the wilted leaf of a dehydrated plant with the turgid leaf of a well-watered one.

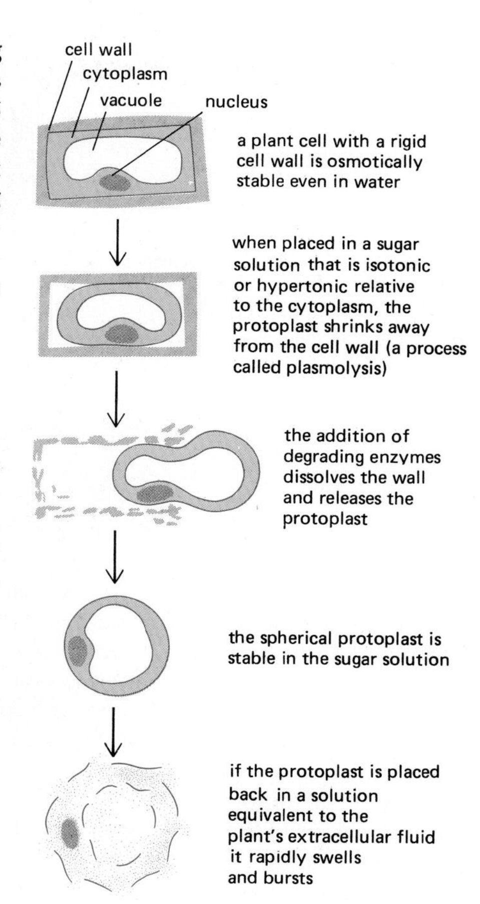

Figure 20–9 A plant cell without its cell wall, called a protoplast, is osmotically unstable and will swell and burst if it is placed in water or the hypotonic extracellular fluid that bathes plant cells. Inside its rigid cell wall, however, it can swell only as far as the boundaries of the wall. The pressure developed by the cell pressing against the wall keeps the cell turgid and in osmotic equilibrium, so that no more water enters the cell (see Panel 6–1, p. 308).

Plant Cell Growth Depends on both Turgor Pressure and a Controlled Yielding of the Wall[5]

Turgor pressure does far more than keep plant tissues distended. Whenever the cell wall yields to the internally generated turgor pressure, an irreversible increase in cell volume results; in other words, the cell grows. Cell growth can occur only if the turgor pressure in the cell exceeds the local tensile strength of the wall. In principle, then, the plant cell could use two strategies to grow: it could increase its turgor, or it could weaken the cell wall in local areas. There is good evidence that plant cells adopt the second strategy and weaken their walls by a variety of mechanisms, including the local secretion of H^+ into the wall by a H^+-pumping ATPase in the plasma membrane. Although the molecular details are not clear, it is thought that the locally lowered pH reduces the number of weak bonds holding the wall together, so that its component macromolecules slip past each other under the influence of turgor pressure. To further facilitate wall growth, other, more complex changes also occur, including the activation of enzymes that hydrolyze glycosidic and other covalent bonds.

In most cases the increase in cell volume is nonuniform with respect to the cell's contents, the major increase being accounted for by enlargement of the *vacuole* rather than of the cytoplasm. The retention of a nearly constant amount of nitrogen-rich cytoplasm reduces the metabolic "cost" of making large cells (Figure 20–10). This is a particular advantage for plants, for which nitrogen is often

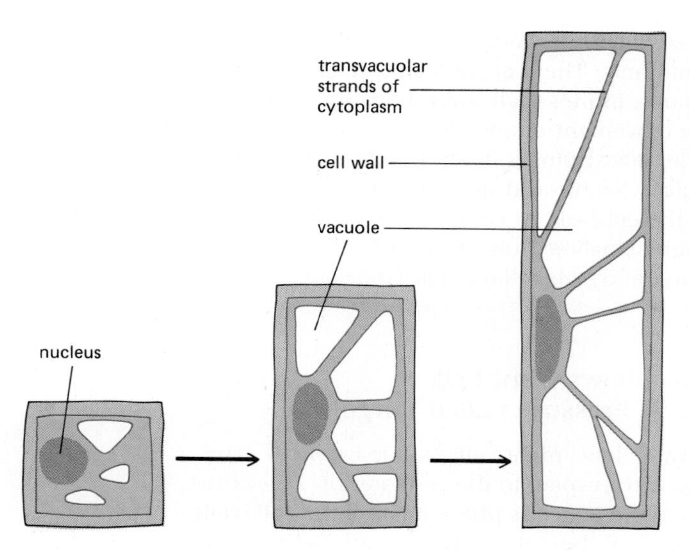

Figure 20–10 A large increase in cell volume can be obtained without increasing the volume of the cytosol. Wall relaxation orients a turgor-driven cell expansion that accompanies the uptake of water into an expanding vacuole. The cytoplasm is eventually confined to a thin peripheral layer, which is connected to the nuclear region by transvacuolar strands of cytoplasm that contain bundles of actin filaments (see Figure 20–53).

a limiting nutrient. In order to maintain the turgor pressure required for continuing cell expansion, solutes must be actively accumulated in the growing vacuole to maintain its osmolarity.

Plant cell growth, however, is more subtle than a simple balloonlike inflation of the cell. Cells also need to be appropriately shaped, and in practice cell growth is a blend of two complex processes. First there is a cyclical process, with appropriate feedback controls, of progressive cell-wall yielding followed by the reestablishment of turgor. Second there is local control of wall yielding, which allows some regions of the wall to remain rigid while others expand, thereby determining the shape of the cell. The basis of this control of cell shape resides in cytoplasmic events to be described later (see p. 1169).

Turgor Is Regulated by Feedback Mechanisms That Control the Concentrations of Intracellular Solutes[6]

In view of the importance of turgor pressure to the plant, it is not surprising that plant cells have evolved sensitive mechanisms for regulating their size. The turgor pressure varies greatly from plant to plant and from cell to cell, from the equivalent of half an atmosphere in some large-celled algae to nearly 50 atmospheres in the stomatal guard cells of some higher plants (Figure 20–11). Cells can increase their turgor pressure by increasing the concentration of osmotically active solutes in the cytosol—either by pumping them in from the extracellular fluid across the plasma membrane or by generating them from osmotically inactive polymeric stores, usually in the vacuole. In both cases, feedback loops monitor the level of turgor and regulate it.

How do such feedback control systems work? Experiments indicate that a "turgor-pressure detector" in the plasma membrane induces inward ion transport,

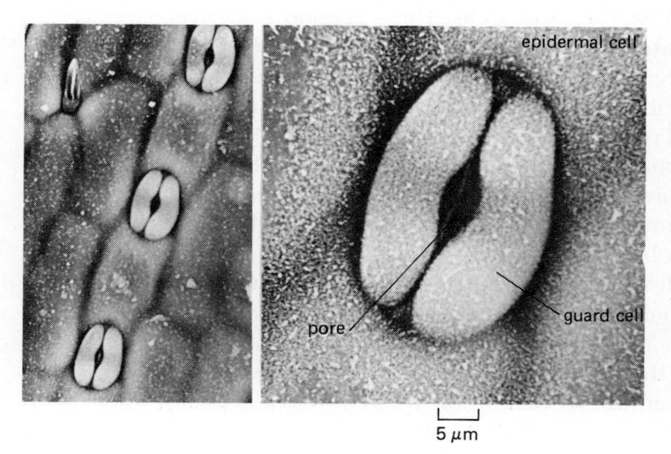

Figure 20–11 Scanning electron micrographs of open stomata in the leaf epidermis of a tropical grass, shown at two magnifications. **Stomata** are pores formed in the leaf surface by two guard cells, whose turgor-regulated movements control the size of the pore and thereby regulate the gas exchange between the leaf and the environment. In most plants the stomata are open during the day to admit carbon dioxide and to allow the products of photorespiration to escape; at night they are usually closed. The epidermal cells are coated by a waterproof cuticle with an outer waxy layer (see also Figure 20–18). (Courtesy of H.W. Woolhouse and G.J. Hills.)

most commonly the active pumping of K^+ into the cell, in response to a sudden fall in turgor pressure, while a sudden increase in turgor brings about a K^+ efflux. These responses are very rapid and reflect changes in the activity of specific transport proteins in the plasma membrane.

Membrane-bound turgor-pressure detectors presumably also induce alterations in the rate at which osmotically active solutes are synthesized in the cytoplasm and vacuole. These changes occur much more slowly but are critical to turgor regulation in plants exposed to environments with extreme or fluctuating osmotic properties. Plants that live in a high-salt habitat, for example, must accumulate very high internal solute concentrations in order to maintain turgor. Since the accumulation of ions such as K^+ to such high levels would probably alter the activities of vital enzymes, plant cells in these environments accumulate specific organic solutes instead: polyhydroxylic compounds such as glycerol or mannitol, amino acids such as proline, or *N*-methylated derivatives of amino acids such as glycinebetaine. These solutes can reach very high concentrations (0.5 M) in the cytosol without affecting the cell's metabolism. The vacuole and its contents are closely involved in the regulation of turgor pressure in response to environmental fluctuations (see p. 1163).

Regulated changes in turgor also generate the limited movements of plants. Stomatal guard cells, for example, control the rate of exchange of gases between leaves and the surrounding air by movements that open or close the pore that they form (see Figure 20–11). During the day, when the stomata are open, light activates K^+-selective pumps in the plasma membrane of the guard cells, causing an influx of K^+ that increases the turgor pressure so that the guard cells swell outward to open the pores. Very rapid changes in turgor in strategically located cells account for more spectacular movements such as trap closure in some carnivorous plants and the rapid movements of flower parts in some pollination mechanisms. The changes in turgor responsible for these phenomena are brought about by drastic increases in the permeability of membranes in critically positioned cells that act as a turgor-regulated "hinge." How a slight touch on a receptive surface induces a turgor collapse in these cells is not understood; it has been suggested that voltage-gated ion channels may be involved.

The Cell Wall Is Modified During the Creation of Specialized Types of Cell[7]

There are large differences between groups of land plants in their structures and reproductive strategies (Figure 20–12). Nevertheless, they are all constructed on similar principles from a very small repertoire of tissues and cell types. Almost all plants are constructed along modular lines, a typical module consisting of a stem,

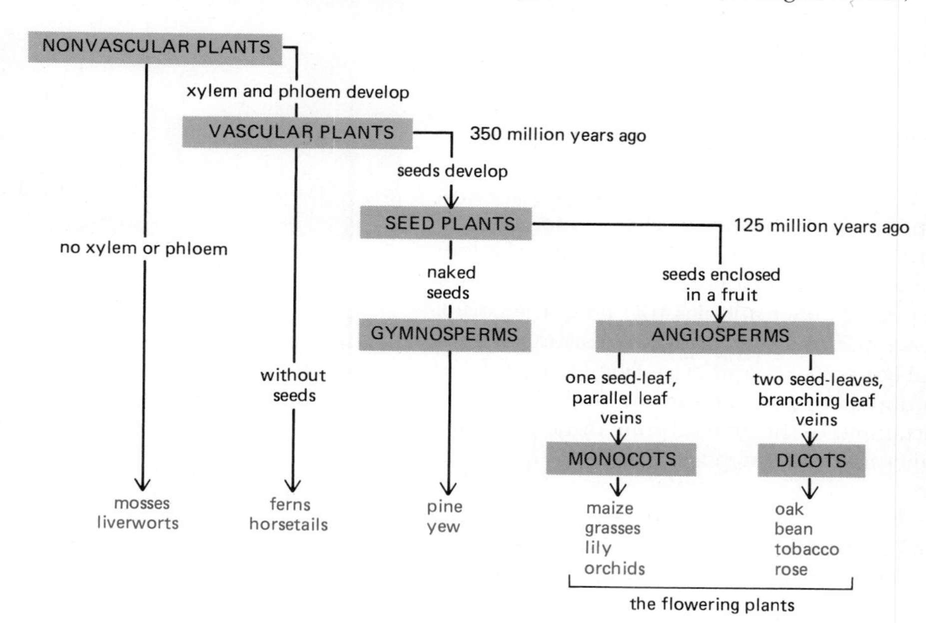

Figure 20–12 The evolution of land plants. A major division in the plant kingdom, excluding the algae, is between the *vascular plants* (which have phloem and xylem tissues for transport) and their progenitors, the *nonvascular plants* (such as mosses), which are small and relatively primitive. As indicated, vascular plants took a further step about 350 million years ago when seeds evolved. The seed provides a protective environment for the developing embryo, which can remain dormant until it encounters the right conditions for further development. Because seed-bearing vascular plants have become the dominant plants on land, they form the main focus of this chapter.

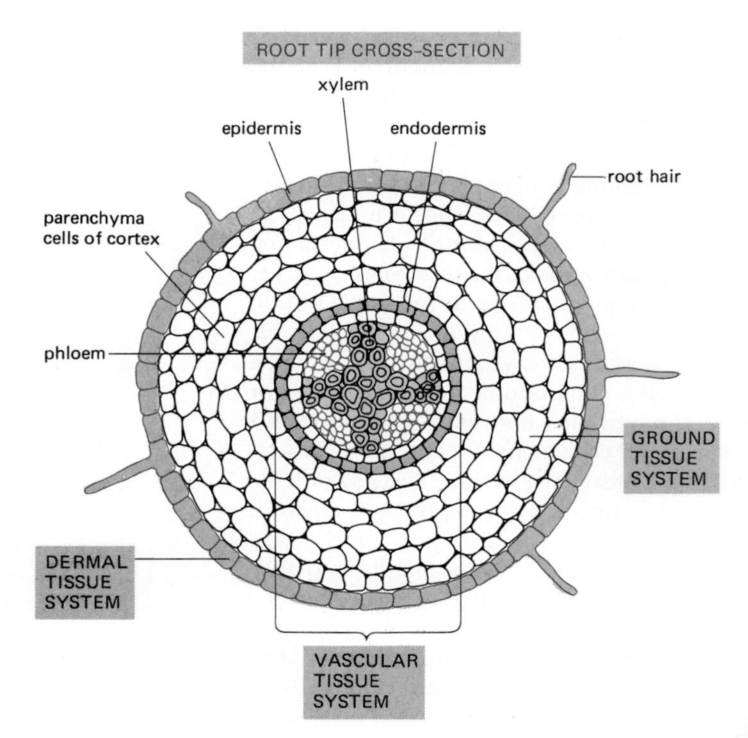

Figure 20–13 The various organs of a higher plant (for example, leaf, stem, and root) are each composed of three easily recognizable *tissue systems*—vascular, ground, and dermal. In this cross-section of a root tip, the *vascular tissue system* is embedded in the *ground tissue system,* which is in turn enclosed by the *dermal tissue system.* The same three tissue systems, in different arrangements, make up all the parts of a higher plant. Each is composed of a relatively small number of common cell types, some of which are illustrated in Panel 20–1.

a leaf, and a bud (see Figure 20–58). Moreover, all contain the same specialized cell types organized into the same three main tissue systems: *dermal* (providing covering), *ground* (providing support and nutrition), and *vascular* (providing fluid transport) (Figure 20–13).

The major cell types are shown in Panel 20–1. They arise from cells with a *primary cell wall* by a process of cell expansion followed by cell differentiation. As cells differentiate, the primary cell wall is elaborated to form the *secondary cell wall.* In some cases this involves simply adding more layers of cellulose, while in other cases new layers of different composition are laid down. The cellulose molecules laid down in secondary walls are in general much longer (~15,000 glucose residues) than those found in primary walls (500 to 5000 glucose residues). Moreover, the highly hydrated pectin components characteristic of the primary cell wall are largely replaced by other polymers, with the result that the secondary wall is considerably more dense and less hydrated than the primary wall.

The secondary cell walls provide most of the plant's mechanical support. They also provide essential components of many animals' diet and are the basis for such products as wood and paper. The form and composition of the final cell wall are closely related to the function of the particular specialized cell type: each cell type is readily distinguishable by its morphology, as exemplified by the highly species-specific secondary cell wall pattern laid down on the surface of mature pollen grains (Figure 20–14).

All major changes in the composition and architecture of both primary and secondary cell walls reflect cytoplasmic events, a principle that is most clearly illustrated by xylem vessel development. During their initial differentiation, xylem vessel elements in young, growing tissues lay down cellulose-rich wall thickenings in patterns defined by bundles of cytoplasmic microtubules that lie in the cortex just under the plasma membrane. These microtubule bundles are arranged generally in helical or hooplike patterns (Figure 20–15) and arise from a regrouping of the more evenly distributed microtubules that constitute the normal cortical array (see p. 1169). This is a specific example of the general rule, discussed later (p. 1169), that extracellular cellulose microfibrils are deposited parallel to cortical cytoplasmic microtubules.

The regions of thickened cellulose wall in developing xylem cells later become strengthened by the deposition of *lignin,* a highly insoluble polymer of aromatic phenolic units that forms an extensive cross-linked network within the cell wall

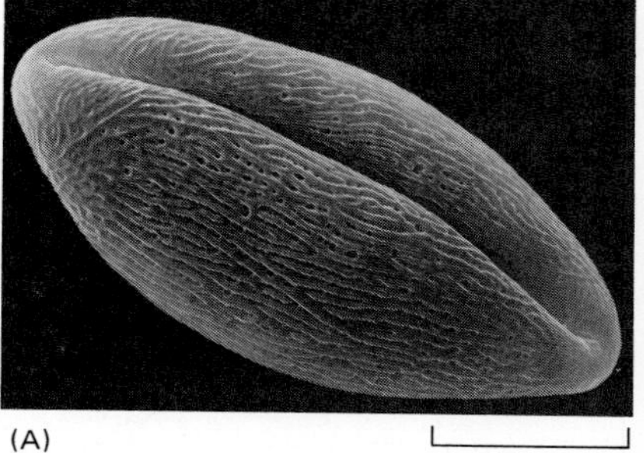

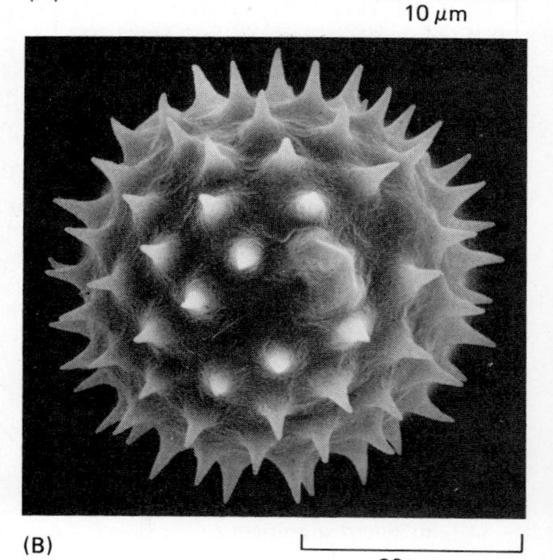

Figure 20–14 Scanning electron micrographs of pollen grains: (A) petunia, (B) sunflower. The sculptured wall is made from sporopollenin, a complex and very tough polymer of hydrocarbons that forms a characteristic and species-specific pattern. Pores exist in the wall, through which the pollen tube will emerge when the grain germinates. (Courtesy of Colin MacFarlane and Chris Jeffree.)

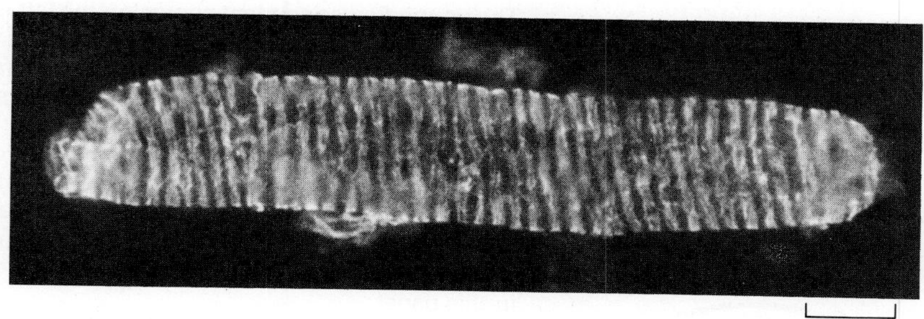

Figure 20–15 Microtubules in the cortex of a cell from the developing xylem of a pea shoot. Immunofluorescence staining shows the grouping of cortical microtubules into a helical array, which defines the area of the cell wall that will be thickened initially with cellulose and then with lignin (see Panel 20–1). The cell is large, and the depth of focus allows only one side of the helical array to be seen. (Courtesy of Ian Roberts.)

and is the main constituent of wood. The localized removal of material from the end walls finally creates the low-resistance, rigid vessels used to transport water in the xylem (Figure 20–16). A similarly dramatic remodeling of the primary cell walls takes place during the development of phloem sieve tubes in the vascular tissues of plants (see Figure 20–16 and Panel 20–1).

The secondary cell wall is usually deposited between the plasma membrane and the primary cell wall, sometimes in successive layers (Figure 20–17). In other cases, however, special macromolecules are also deposited either *within* the primary wall (as is lignin in xylem cells) or on its outer surface. Epidermal cells, for example, which cover the external surface of the plant, usually have thickened primary cell walls whose outer face is coated with a thick, waterproof *cuticle*, which helps to protect the plant from infection, mechanical damage, water loss, and harmful ultraviolet light (see Panel 20–1, pp. 1148–1149). The cuticle is secreted as the epidermal cells differentiate. It is made primarily of *cutin* (or the related compound suberin in bark), which is a polymer of long-chain fatty acids that forms an extensive cross-linked network on the plant surface. The cutin layer is frequently impregnated and overlaid by a complex mixture of *waxes*, which are esters between long-chain alcohols and fatty acids (Figure 20–18). The plant cell cuticle has a very different composition from insect and crustacean cell cuticles, which are composed of proteins and polysaccharides.

Even the Mature Cell Wall Is a Dynamic Structure[8]

The composition and architecture of the cell wall in a mature plant are not fixed: components can be added or removed, and the cross-links between components can change. The localized removal of wall material during the development of the end walls of both xylem vessels and phloem sieve-tube elements provides a particularly striking example of such changes (see Figure 20–16).

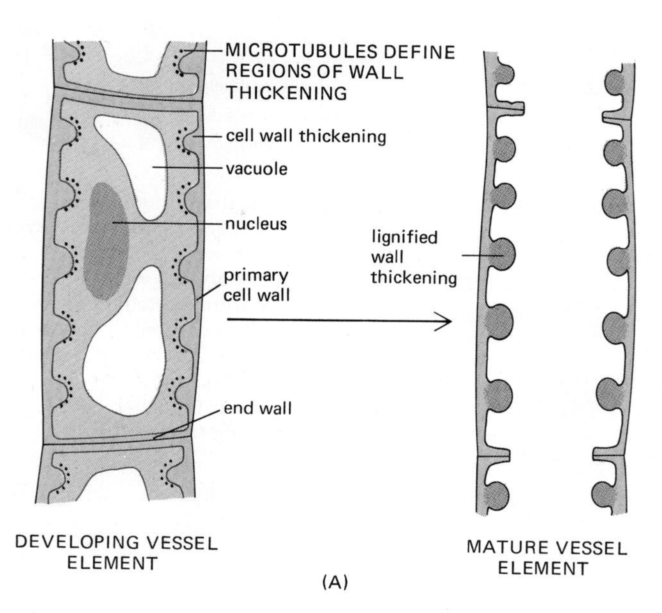

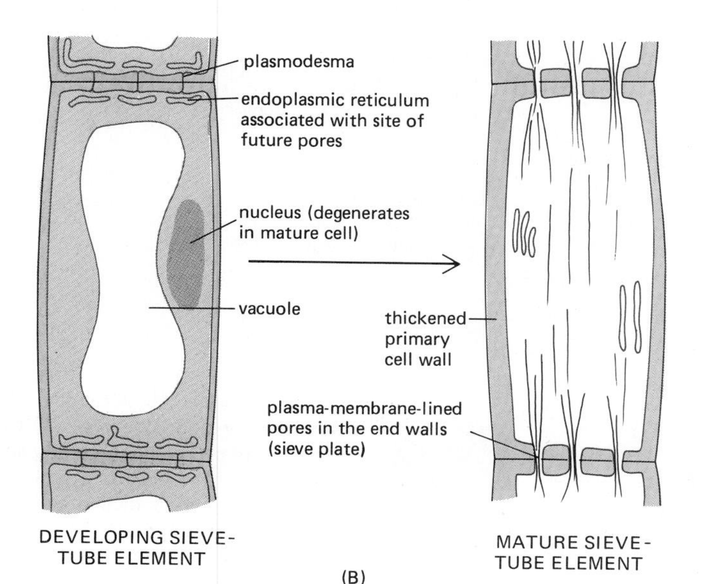

Figure 20–16 Two examples of the ways in which the cell wall is elaborated and modified during the formation of specialized cell types. (A) A schematic longitudinal section of a small developing vessel element of the xylem. This cell forms annular wall thickenings, but many other patterns are also found. Ultimately the protoplast and end walls disappear to create an open-ended tube. The mature element is dead, having lost its protoplast. (B) A schematic longitudinal section of the development of a sieve-tube element in phloem. The primary cell wall becomes thickened, and the end walls become perforated to form the sieve plates that connect adjacent elements of the sieve tube. The mature cell retains its plasma membrane, but the nucleus and much of the cytoplasm are lost.

THE PLANT

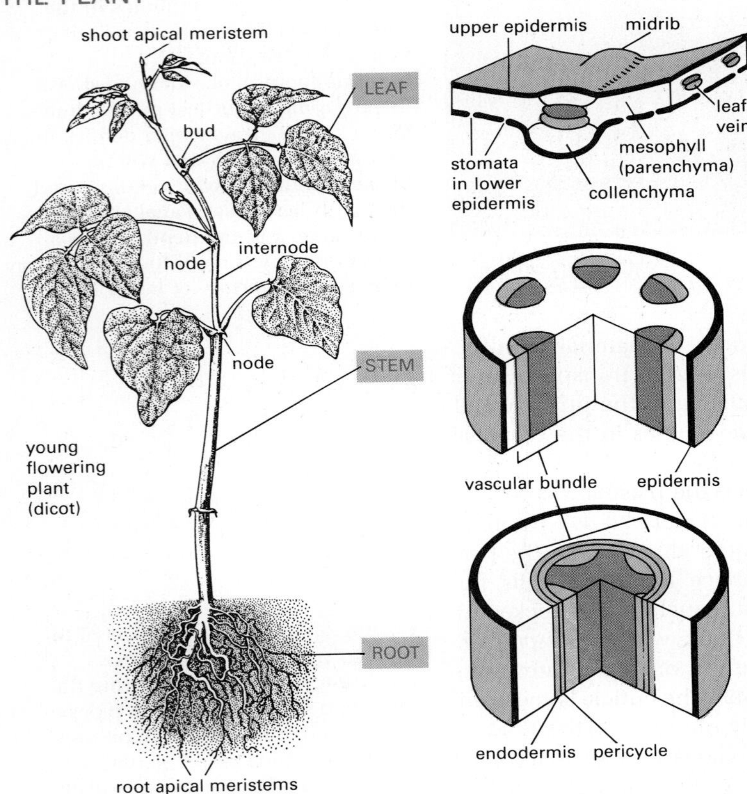

The young flowering plant shown on the left is constructed from three main types of organs: leaves, stems, and roots. Each plant organ is in turn made from three tissue systems: ground, dermal, and vascular (see Figure 20-13).

All three tissue systems derive ultimately from the cell proliferative activity of the shoot or root apical meristems, and each contains a relatively small number of specialized cell types. These three common tissue systems, and the cells that comprise them, are described in this panel.

THE THREE TISSUE SYSTEMS

Cell division, growth, and differentiation give rise to tissue systems with specialized functions.

DERMAL TISSUE ( ) This is the plant's protective outer covering in contact with the environment. It facilitates water and ion uptake in roots and regulates gas exchange in leaves and stems.

VASCULAR TISSUE Together the phloem ( ) and the xylem ( ) form a continuous vascular system throughout the plant. This tissue conducts water and solutes between organs and also provides mechanical support.

GROUND TISSUE ( ) This packing and supportive tissue accounts for much of the bulk of the young plant. It functions also in food manufacture and storage.

GROUND TISSUE

The ground tissue system contains three main cell types called parenchyma, collenchyma, and sclerenchyma.

Parenchyma cells are found in all tissue systems. They are living cells, generally capable of further division, and have a thin primary cell wall. These cells have a variety of functions. The apical and lateral meristematic cells of shoots and roots provide the new cells required for growth. Food production and storage occur in the photosynthetic cells of the leaf and stem (called mesophyll cells); storage parenchyma forms the bulk of most fruit and vegetables. Because of their proliferative capacity, parenchyma cells also serve as stem cells for wound healing and regeneration.

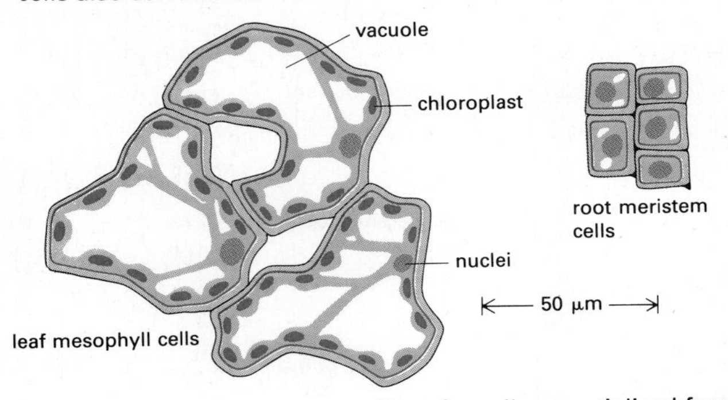

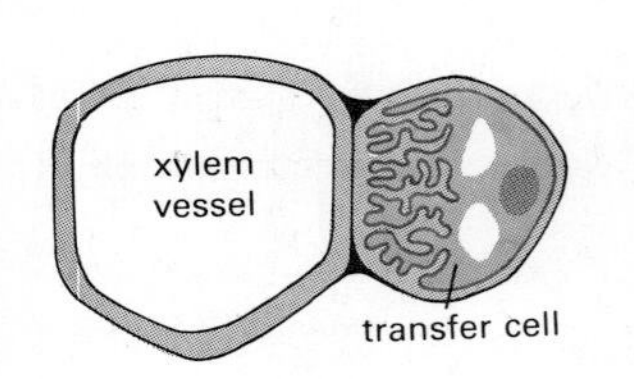

Transfer cell, a specialized form of the parenchyma cell, is readily identified by elaborate ingrowths of the primary cell wall. The increase in the area of the plasma membrane beneath these walls facilitates the rapid transport of solutes to and from cells of the vascular system.

Collenchyma are living cells similar to parenchyma cells except that they have much thicker cell walls and are usually elongated and packed into long ropelike fibers. They are capable of stretching and provide mechanical support in the ground tissue system of the elongating regions of the plant. Collenchyma cells are especially common in subepidermal regions of stems.

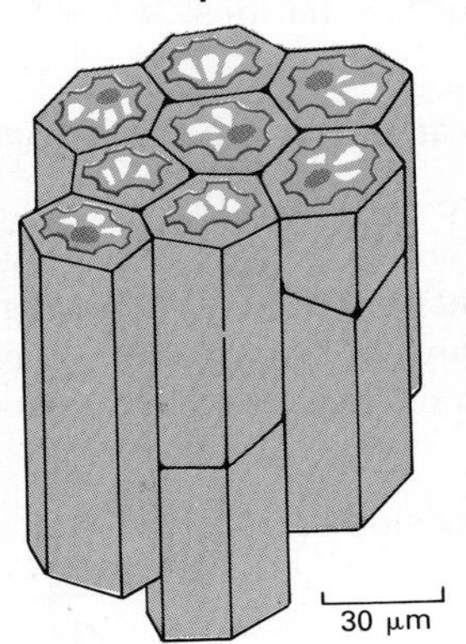

typical locations of supporting groups of cells in a stem

sclerenchyma fibers

vascular bundle

collenchyma

Sclerenchyma, like collenchyma, have strengthening and supporting functions. However, they are usually dead cells with thick, lignified secondary cell walls that prevent them from stretching as the plant grows. Two common types are *fibers* (see Figure 20-17), which often form long bundles, and *sclereids*, which are shorter branched cells found in seed coats and fruit.

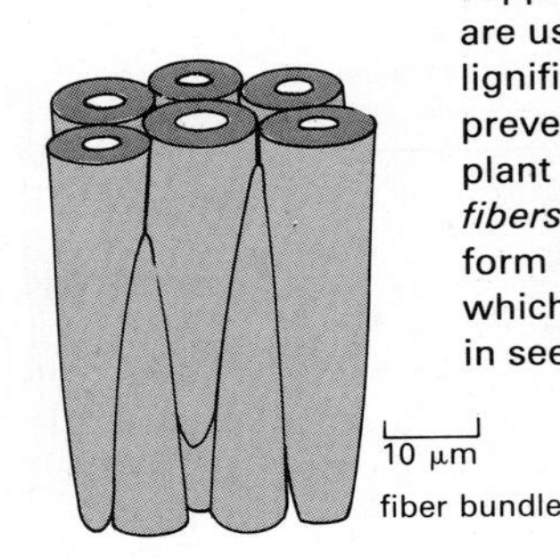

100 µm

sclereid

 Panel 20–1 The cell types and tissues from which higher plants are constructed.

DERMAL TISSUE

The epidermis is the primary outer protective covering of the plant body. Cells of the epidermis are also modified to form stomata and hairs of various kinds.

Epidermis

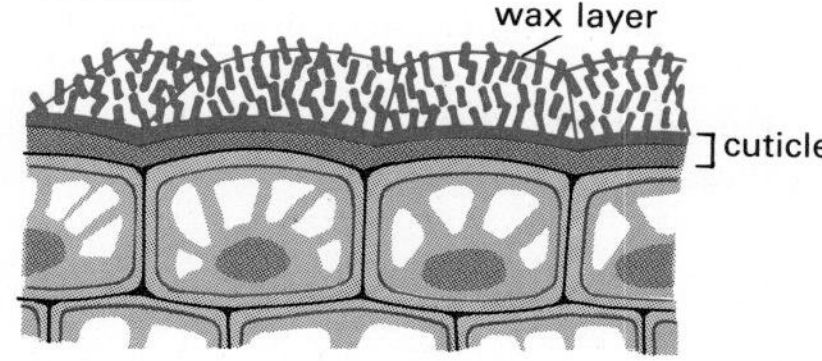

The epidermis (usually one layer of cells deep) covers the entire stem, leaf and root of the young plant. The cells are living, have thick primary cell walls, and are covered on their outer surface by a special cuticle with an outer waxy layer (see Figure 20-18). The cells are tightly interlocked in different patterns.

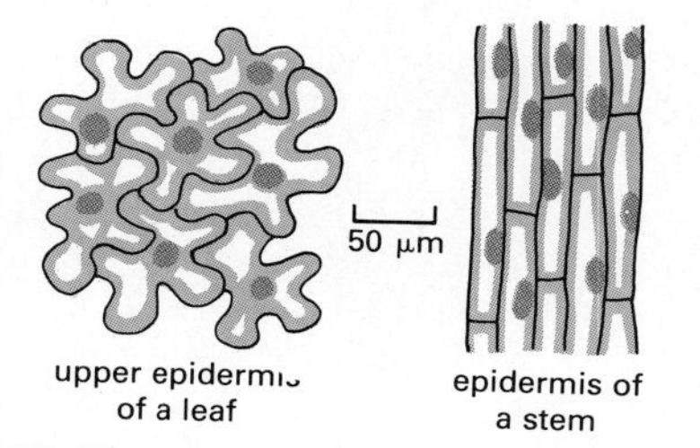

upper epidermis of a leaf

epidermis of a stem

Stomata

guard cells

air space

5 μm

Stomata are openings in the epidermis, mainly on the lower surface of the leaf, that regulate gas exchange in the plant. They are formed by two specialized epidermal cells called guard cells (see Figure 20-11), which regulate the diameter of the pore. Stomata are distributed in a distinct species-specific pattern within each epidermis.

Hairs (or trichomes) are appendages derived from epidermal cells. They exist in a variety of forms and are commonly found in all plant parts. Hairs function in protection, absorption, and secretion.
examples:

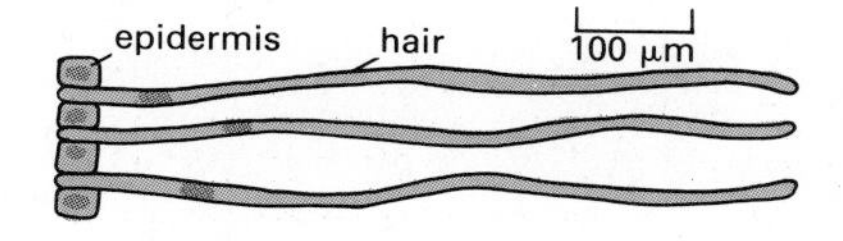

young, single-celled hairs in the epidermis of the cotton seed (see Figure 20-45). When these grow, the walls will be secondarily thickened with cellulose to form cotton fibers.

10 μm

epidermis

root hair

a multicellular secretory hair from a geranium leaf

Single-celled root hairs have an important function in water and ion uptake.

Vascular bundles

In roots there is usually a single vascular bundle, but in stems there are several. These are arranged with strict radial symmetry in dicots, but they are more irregularly dispersed in monocots.

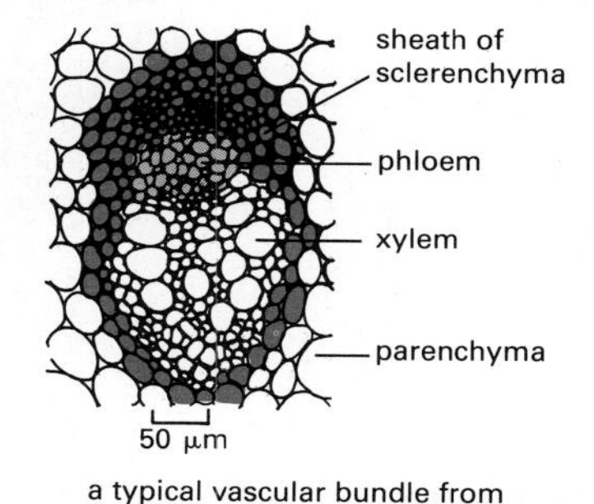

a typical vascular bundle from the young stem of a buttercup

VASCULAR TISSUE

The phloem and the xylem together form a continuous vascular system throughout the plant. In young plants they are usually associated with a variety of other cell types in *vascular bundles.* Both phloem and xylem are complex tissues. Their conducting elements are associated with parenchyma cells that maintain and exchange materials with the elements. In addition, groups of collenchyma and sclerenchyma cells provide mechanical support.

Phloem

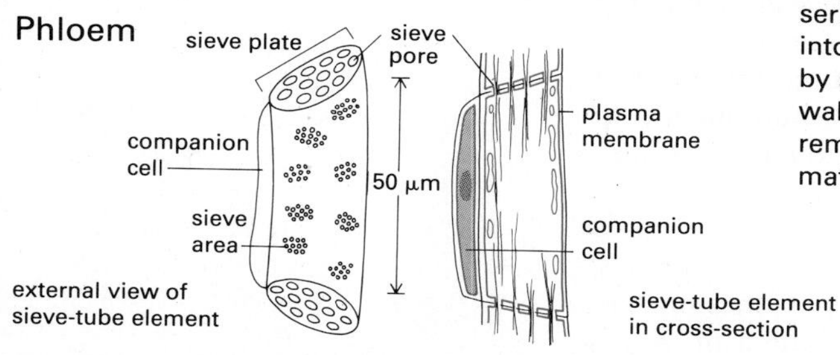

external view of sieve-tube element

sieve-tube element in cross-section

Phloem is involved in the transport of organic solutes in the plant. The main conducting cells (elements) are aligned to form tubes called *sieve tubes.* The sieve-tube elements at maturity are living cells, interconnected by perforations in their end walls formed from enlarged and modified plasmodesmata (sieve plates). These cells retain their plasma membrane, but they have lost their nuclei and much of their cytoplasm; they therefore rely on associated *companion cells* for their maintenance. These companion cells have the additional function of actively transporting soluble food molecules in and out of sieve-tube elements through porous sieve areas in the wall.

Xylem carries water and dissolved ions in the plant. The main conducting cells are the vessel elements shown here, which are dead cells at maturity that lack a plasma membrane. The cell wall has been secondarily thickened and heavily lignified. As shown below, its end wall is largely removed, enabling very long, continuous tubes to be formed.

Xylem

The wall that connects the serially aligned vessel elements into a tube may be perforated by several small pores. This wall, however, is often removed completely in more mature vessels.

small vessel element in root tip

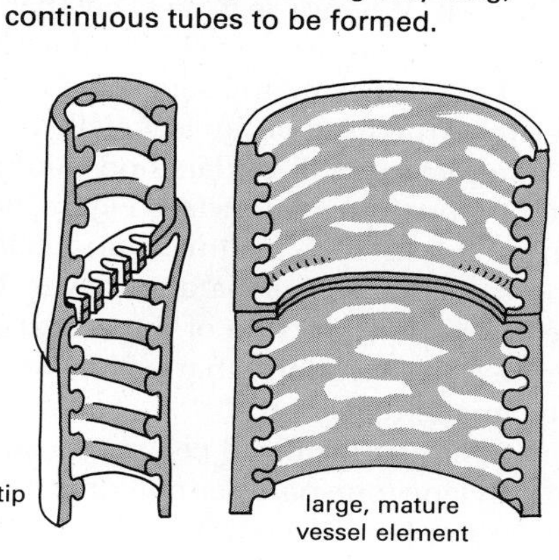

large, mature vessel element

The vessel elements are closely associated with xylem parenchyma cells, which actively transport selected solutes in and out of the elements across their plasma membrane.

xylem parenchyma cells

vessel element

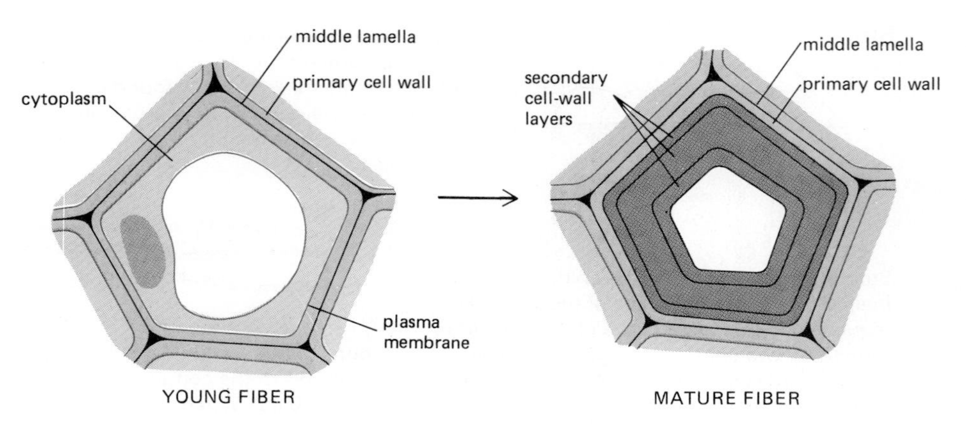

Figure 20–17 Secondary cell-wall deposition shown schematically in a cross-section of a fiber cell. Here, three new cell-wall layers have been laid down underneath the primary cell wall. Because the net orientation of the cellulose microfibrils is different in each layer, a strong plywoodlike effect is created. In many mature fibers, as shown here, the cell inside the wall dies.

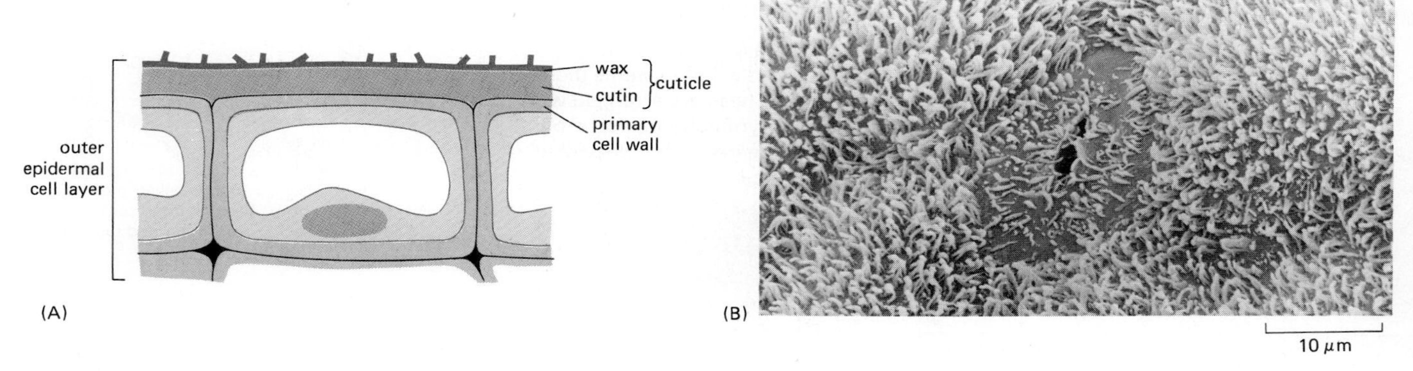

Figure 20–18 (A) Drawing of a section of a typical mature epidermal cell of a leaf. On the outer face of the thick primary cell wall, a waterproof cutin layer has been deposited together with a layer of wax to produce a cuticle. The wax layer often gives the cuticle an elaborate sculptured shape. (B) Scanning electron micrograph of the lower epidermis of a pea leaf, showing the pattern of wax deposition around a stomatal pore. (Courtesy of Paul Linstead.)

The dynamic behavior of mature cell walls is also illustrated by the changes that occur during *abscission*—the controlled loss of a part of the plant, such as a dead leaf. When a leaf dies, cellular polymers are degraded and sugars, amino acids, and ions are recycled to the plant. In addition, *ethylene* gas is produced in small quantities by the senescing leaf. A particular layer of cells in a zone located between the base of the leaf stalk and the stem (the *abscission zone*) responds to a complex and ill-understood combination of ethylene and other endogenous plant growth regulators (see Figure 20–67) by making and secreting wall-degrading enzymes, such as pectinase and cellulase, which act locally to partially dissolve the cell walls in a separation layer (Figure 20–19). At the same time, the layer of cells on the stem side of the abscission zone deposits water-resistant suberin to protect the "wound" that will be left when the leaf finally drops off as a result of enzymatic digestion.

A similar localized cell response occurs during fruit ripening. Here low levels of ethylene (one part per million) stimulate target cells in fruit (oranges or bananas, for example) to secrete pectin-degrading enzymes that weaken the adhesions between adjacent cells. This results in a softening, or "ripening," of the fruit.

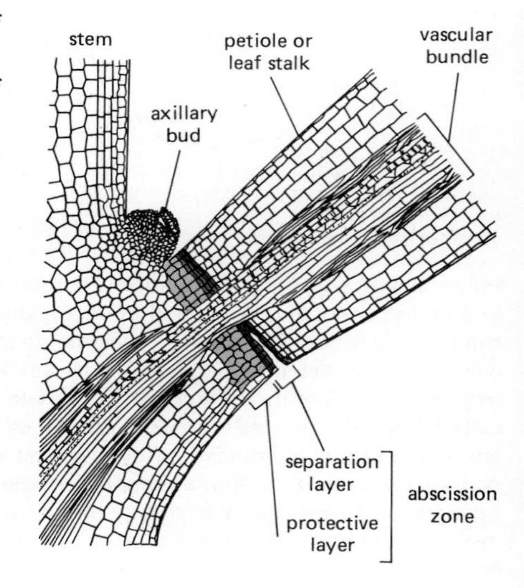

Figure 20–19 Specialized layers of small cells at the base of the leaf stalk are involved in leaf abscission. Two or three layers of small cells secrete wall-degrading enzymes that lead to cell separation. The petiole parts from the stem in this region, where cell walls are thinner, and few lignified cells cross the abscission zone. Cells that remain exposed at abscission deposit suberin and form a protective layer over the wound.

Summary

Higher plants are composed of large numbers of cells cemented together in fixed positions by the rigid cell walls that surround them. Many of the unique features of plants are related directly or indirectly to the presence of this cell wall, whose composition and appearance reflect the different plant cell types and their functions. The underlying structure of all cell walls, however, is remarkably consistent. Tough fibers of cellulose are embedded in a highly cross-linked matrix of polysaccharides, such as pectins and hemicelluloses, and glycoproteins. The result is a primary cell wall that possesses great tensile strength and is permeable only to relatively small molecules. When placed in water, a plant cell without a wall (a protoplast) will take up water by osmosis, swell, and burst. Inside its cell wall, however, it swells and presses against the wall, creating a pressure known as turgor. Turgor is closely regulated and is vital for both cell expansion and the mechanical rigidity of the young plant.

A relatively small number of specialized cell types is used in the construction of the great variety of higher plants. During the formation of these specialized cell types—for example, cells of the two vascular tissues, xylem and phloem—the cell wall is extensively modified. Certain regions of the wall may be strengthened, often by the addition of one or more new layers to form a secondary cell wall. Other regions of the wall may be selectively removed, as are the end walls in the formation of a conducting tube from a long row of cylindrical cells. These changes in the cell wall are controlled by temporal and spatial changes in the cytoplasm of the developing cell. The cell wall is a dynamic structure whose composition and form can change markedly, not only during cell growth and differentiation but also after cells have matured.

Transport Between Cells

In the preceding section we described how the rigid wall of a plant cell limits the exchange of molecules between the cell's cytoplasm and its environment. The cell wall also restricts the ways in which a plant cell can interact and communicate with its neighbors. Plant cells have evolved ingenious ways to overcome these limitations, however. Direct cell-cell communication is just as important in multicellular plants as it is in multicellular animals, and special channels have evolved to connect the cytoplasm of one plant cell to that of its neighbors—thereby allowing the controlled passage of ions and small molecules. In vascular plants, moreover, long columns of cylindrical cells become connected end to end by perforations to generate long tubes through which water and nutrients flow.

Plant Cells Are Connected to Their Neighbors by Special Cytoplasmic Channels Called Plasmodesmata[9]

Except for a very few specialized cell types, every living cell in a higher plant is connected to its living neighbors by fine cytoplasmic channels, each of which is called a **plasmodesma** (plural, **plasmodesmata**), which pass through the intervening cell walls. As shown in Figure 20–20, the plasma membrane of one cell is continuous with that of its neighbor at each plasmodesma. A plasmodesma is a roughly cylindrical, membrane-lined channel with a diameter of 20 to 40 nm. Running from cell to cell through the center of most plasmodesmata is a narrower cylindrical structure, the *desmotubule,* which electron micrographs show to be continuous with elements of the endoplasmic reticulum membrane of each of the connected cells (Figure 20–21). Between the outside of the desmotubule and the inner face of the cylindrical plasma membrane is an annulus of cytosol (see Figure 20–20), which often appears to be constricted at each end of the plasmodesma. These constrictions may be of great significance, for they are located at sites where each cell could, in principle, regulate the flux of molecules through the annulus that joins the two cytosols.

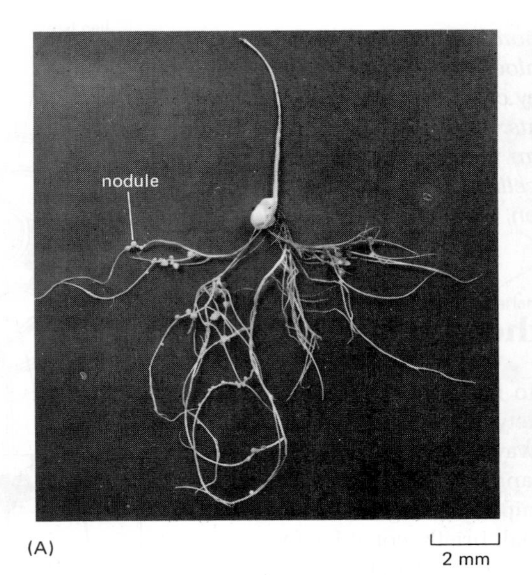

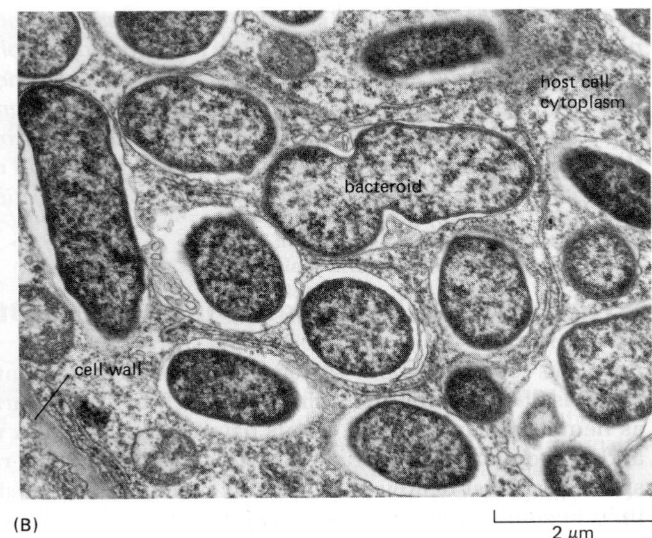

Figure 20–29 (A) A young pea seedling that has already formed a symbiotic association with the nitrogen-fixing bacterium *Rhizobium*. The root nodules containing the bacteria are clearly visible. (B) Electron micrograph of a thin section through a pea root nodule like those shown in (A). The nitrogen-fixing *Rhizobium* bacteroids, surrounded by host-cell-derived membrane, fill the host-cell cytoplasm. (A, courtesy of Andy Johnston; B, courtesy of B. Huang and Q.S. Ma.)

epidermal cells of the host plant. After binding to the root epidermal cells, growing bacteria enter the plant via an *infection thread* and stimulate underlying cortical cells to divide and form a large *root nodule* (Figure 20–29A). The bacteria invade the new cortical cells, colonizing the cytoplasm. Each mature nodule contains about half its weight in intracellular bacteria, which have lost most of their own cell wall. The plasma membrane of each bacterium is surrounded by another membrane derived from the host-cell plasma membrane. It is these altered bacteria, called *bacteroids*, that fix the nitrogen eventually used by the plant (Figure 20–29B).

Nitrogenase, the bacterial enzyme that catalyzes the fixation of nitrogen, is a complex of three polypeptide chains. In symbiotic *Rhizobia* this protein complex catalyzes the conversion of atmospheric nitrogen to ammonia, which is then rapidly released into the cytoplasm of the host cell, where it is converted to glutamine. Eventually the fixed nitrogen is incorporated into all the other amino acids.

Genetic analyses show that establishing and maintaining this symbiotic interaction requires the coordinated expression of many genes in both the bacterium and the host plant cell. The binding and invasion of the host root hair by the bacterium is the first step in a dialogue between host cell and bacterium that eventually activates a set of host genes encoding plant cell proteins called *nodulins*, which are essential for the growth and function of the nodule. The dialogue begins when a flavanoid produced by the host plant cell binds to and activates a protein encoded by a bacterial *nod* gene (Figure 20–30). This activated protein, called *nod*D, then switches on the synthesis of other bacterial *nod* gene products, whose functions include inducing production of host plant nodulins. A large plasmid carried by the bacterium contains most of the bacterial *nod* genes, as well as the *nif* genes that encode the enzymes involved in nitrogen fixation, including the nitrogenase. A *Rhizobium* that normally nodulates only beans can be converted into one that nodulates only peas if the appropriate plasmid genes are replaced with those from the plasmid carried by a pea-specific *Rhizobium*.

The nodulins of the host cell include proteins involved in stimulating root cortical cell division, structural components of the nodule, enzymes that enable the plant to assimilate fixed nitrogen compounds, and specialized proteins that support bacteroid function. One of the most important of these is *leghemoglobin*, a cytoplasmic oxygen-binding protein similar to mammalian myoglobin. The bacterial nitrogenase complex is irreversibly inactivated by free oxygen, and elaborate mechanisms are required to ensure that the bacteroids remain in an environment in the root whose oxygen tension is buffered so as to support the respiration required for the energetically demanding nitrogen-fixing process without inactivating nitrogenase. As part of this mechanism, the *Rhizobium* induce the host cells

Figure 20–30 Various flavones and related compounds released by the host root bind to and activate specific nodulating bacteria. Luteolin, shown here, is a flavone from alfalfa that induces *nod* genes in *Rhizobium meliloti*. This signaling molecule probably binds to the nodD protein, which, in its activated form, then switches on the early bacterial nodulation genes. The flavones are closely related to the anthocyanin pigments of flowers and fruit. Different plants release different combinations of flavanoid molecules, which selectively activate specific *Rhizobium* species.

to produce large amounts of leghemoglobin to act as an oxygen buffer. The globin part of the molecule is encoded by a host gene, while the heme prosthetic group may be supplied by the bacterial partner—a remarkable example of evolutionary co-adaptation.

Nitrogen fixation consumes a great deal of energy, supplied ultimately by the sun through photosynthesis. It is estimated that the fixation of one molecule of nitrogen (N_2) by *Rhizobium* requires between 25 and 35 ATP molecules. These procaryotes contribute vastly more fixed nitrogen for plant growth ($\sim 2 \times 10^8$ tons/year), largely in unmanaged ecosystems, than is provided by nitrate fertilizers.

Figure 20–31 Tumors induced by *Agrobacterium tumefaciens* on a succulent pot plant. (Courtesy of Paul J.J. Hooykaas, from *Genetic Eng.* 1:155, 1979.)

Agrobacterium Is a Plant Pathogen That Transfers Genes to Its Host Plant[16]

Another soil bacterium that is closely related to *Rhizobium* is called *Agrobacterium tumefaciens.* This bacterium causes *crown-gall* disease in plants: when they are exposed to *Agrobacterium,* normal plant cells become transformed into gall-forming tumor cells by a process that involves the transfer of genes from the bacterium to the plant (Figure 20–31).

The tumor cells induced by *Agrobacterium* are remarkable in several ways. First, unlike most normal plant cells, they can be removed from the gall and grown indefinitely in culture in the absence of added growth factors and without the continued presence of *Agrobacterium* cells. Second, they synthesize a variety of unusual chemicals called *opines;* these amino acid derivatives can be catabolized and used preferentially by the strain of bacteria that induced the plant cells to make them—an arrangement with an obvious survival advantage for the *Agrobacterium.*

Explanations for many of the properties of *Agrobacterium*-induced tumor cells have come from molecular genetic studies. The ability of *Agrobacterium* to induce tumors is associated with a large DNA plasmid called Ti (*T*umor-*i*nducing), a small part of which, the **T-DNA** (*transferred DNA*), integrates into the nuclear genome of the plant cell. The bacteria can infect a susceptible plant only at the site of a wound, where plant cells secrete unusual phenolic compounds that include *acetosyringone* (Figure 20–32). This compound, produced in response to wounding, identifies susceptible sites of entry for the pathogen and activates a series of reactions in the bacterium that lead to the excision of T-DNA from the Ti plasmid and its transfer to the host cell genome (Figure 20–33). Once integrated into a host cell chromosome, the T-DNA is transcribed and translated by the host cell to produce three classes of proteins. One is the enzyme that causes the plant to produce a specific opine, while the other two are enzymes that catalyze synthesis of the plant growth regulators *indole-acetic acid* and *cytokinin* (see Figure 20–67). The abnormal balance and increased levels of these two growth regulators, produced as a result of the activity of the integrated T-DNA genes, is responsible for the continued growth and division of the transformed plant cells and explains why these cells can continue to grow in the absence of both added growth regulators and the original bacteria.

The ability of *Agrobacterium* T-DNA to integrate stably into its host's genome has lead to its widespread exploitation as a vector for the genetic transformation of higher plant cells with recombinant DNA molecules (see p. 1183).

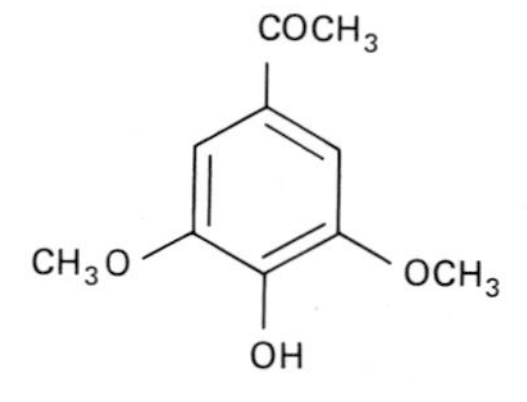

Figure 20–32 Acetosyringone, a signaling molecule found in the exudates of wounded but metabolically active plant cells, specifically activates the virulence genes on the Ti plasmid of *Agrobacterium* that promote the formation of the T-DNA strand that is transferred to the host plant (see Figure 20–33).

Cell-Wall Breakdown Products Often Act as Signals in Plant–Pathogen Interactions[17]

A large number of highly specialized plant metabolites serve as protective agents for plants. Some compounds, such as the glycosides in mustard oil and the wide variety of alkaloids (for example, caffeine, morphine, strychnine, and colchicine), probably act as deterrents to herbivores, although there are always some predators that can tolerate or detoxify such chemical weapons. In addition to these constitutive defense mechanisms, plants have evolved more complex, adaptive defense mechanisms that are activated only when host and pathogen interact.

on the one hand and stem tissues plus leaf primordia on the other; and (2) *lateral meristems*, circumferentially arranged inside the plant, are involved primarily in producing cells that increase the girth of the plant (Figure 20–57). An important lateral meristem in all plants is called the *cambium*—the layers of cells that give rise to the vascular tissues.

The ball of embryonic cells quickly becomes polarized into two groups of meristematic cells—one at the suspensor end of the embryo, which will become a root apical meristem, and one at the opposite pole, which will become a shoot apical meristem. These two meristems define the main root-shoot axis of the future plant. Subsequently, as the embryo elongates, the vascular cambium is formed to provide the future transport tissue between root and shoot. Slightly later in development, the apical meristem of the shoot produces the embryonic seed leaves, or cotyledons—one in the case of monocots and two in the case of dicots (see Figure 20–12). At this stage, further development usually ceases and the embryo becomes packaged in a *seed*, which is specialized for dispersal and for survival in harsh conditions.

The embryo in a seed is stabilized by dehydration, and it can remain dormant for a very long time—even thousands of years, as documented in the case of cereal grains from Egyptian tombs. When rehydrated, the seeds germinate and embryonic development continues. Embryos depend for their development on the large food reserves accumulated in the triploid tissue of the endosperm (see Panel 20–2). During development of the seed, this food material may be transferred to a varying extent to the cotyledons. Thus the seeds of monocots (such as wheat and corn) and some dicots keep large endosperm reserves, whereas the seeds of the majority of dicots (including peas and beans) rely on the nutrients stored in the fleshy cotyledons.

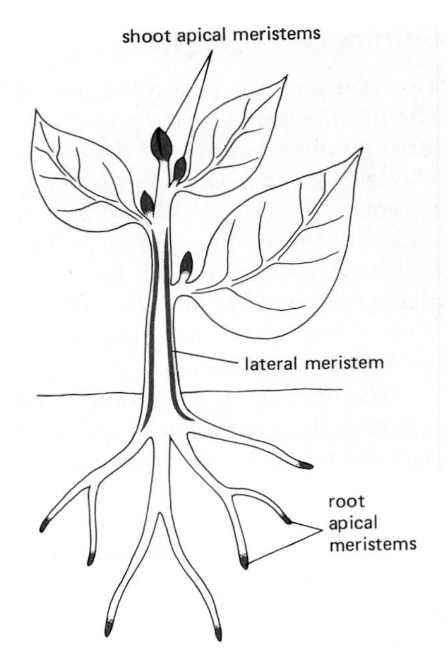

Figure 20–57 Highly schematic diagram of a higher plant, showing the locations of the main meristems, the areas of most rapid and continuous cell division. The activity of the shoot and root apical meristems is responsible for increasing the length, while the activity of a lateral meristem, such as the cambium, is responsible for increasing the girth of the various parts of the plant.

Meristems Continually Produce New Organs and New Meristems, Creating a Repeating Series of Similar Modules[29]

A vertebrate embryo is a small version of the adult, already possessing most of its characteristic organs. The plant embryo inside the seed, in contrast, looks nothing like the adult plant. Each of its two meristems, however, has the capacity to produce a full complement of new organs at the appropriate time. As soon as the seed coat ruptures during germination, rapid cell expansion occurs, with the root emerging first to establish an immediate foothold in the soil. This is followed by rapid and continual cell divisions in the apical meristems: in the apical meristem of a maize root, for example, cells divide every 12 hours, producing 5×10^5 cells per day. The rapidly growing roots and shoots probe the environment—the roots increasing their capacity for taking up water and minerals from the soil, the shoots increasing their capacity for photosynthesis (see Panel 20–2, p. 1154).

Two sorts of plant structures are formed by apical meristems: (1) the main root and shoot system, whose growth perpetuates the meristems themselves, and (2) structures such as leaves and flowers, whose growth is limited. The former are responsible for continuous development and persist for the life of the plant (and in the case of grafts, like the McIntosh or Cox's Orange Pippin apples, for considerably longer). The latter have a short lifetime, invariably undergoing programmed senescence. Contrast the seasonal turnover of leaves with the virtual absence of turnover among the somatic cells of the rest of the plant.

As the shoot elongates, the apical meristem produces an orderly sequence of *nodes* and *internodes*. In this way the continuous activity of the meristem produces an ever-increasing number of similar **modules**, each consisting of a stem, a leaf, and a bud (Figure 20–58). The modules are connected to one another by supportive and transport tissue, and successive modules are precisely located relative to each other, giving rise to the patterned structures that are characteristic of plants (Figure 20–59).

Each individual module has a degree of autonomy, so that in many respects one can regard mature plants as having some of the characteristics of colonial organisms such as sponges and corals. Thus each module begins as an identically sized swelling on the apical meristem, but the local environment it encounters

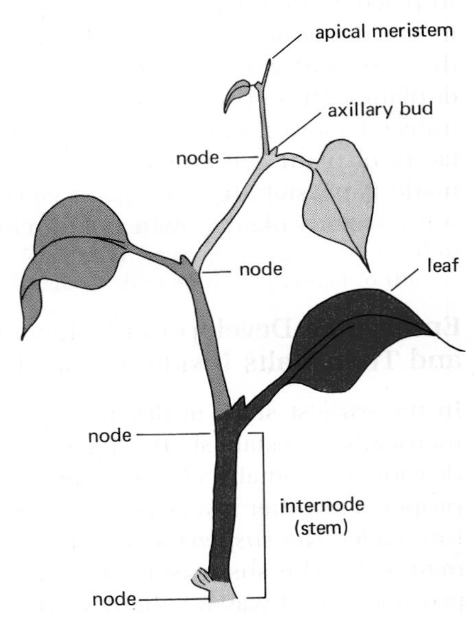

Figure 20–58 A simple example of the modular construction of plants. Each module (shaded in different colors) consists of a stem, a leaf, and a bud containing a potential meristem. Modules arise sequentially from the continuous activity of the apical meristem.

Figure 20–59 Accurate placing of successive modules from a single apical meristem produces these elaborate but regular patterns in leaves (A), flowers (B), and fruits (C). (Courtesy of Andrew Davies.)

during growth can modify its final form. Unusually large, dark green leaves will usually develop in a shady spot, for example, while short, thick leaf stems will develop in a windy one. Experiments show that these local adaptations are properties of each module, not of the entire plant. Moreover, in the acute angle between the stem and leaf branch (the axil), a bud is formed in each module that contains a new apical meristem, which can potentially produce another branch. Many of these new meristems, however, will become active only if the original apical meristems are damaged or removed—a fact well known to gardeners, who pinch off branch tips to stimulate side growth.

The modular nature of plants is emphasized by the experimental practice of meristem culture. Using suitable growth media, one can induce excised apical meristems to grow and regenerate complete normal plants, and this method is now used in horticulture to propagate valuable specimens, roses, orchids, and some crop plants such as strawberries and sugar cane.

The Form of the Plant Is Determined by Pattern-Formation Mechanisms in the Apical Meristems[30]

The repeating pattern of internode shoot growth followed by node differentiation requires that the cells in the apical meristem be able to sense their distance and orientation from the previous node, so as to differentiate appropriately. As in animal development, the pattern-formation mechanisms are required to operate over distances of less than a millimeter and therefore involve only relatively small groups of cells (see p. 914). Thus the precursors of each new node (an incipient branch point) can be detected as successive swellings that form near the very tip of the meristem, each forming from perhaps 100 cells (Figure 20–60). In the sim-

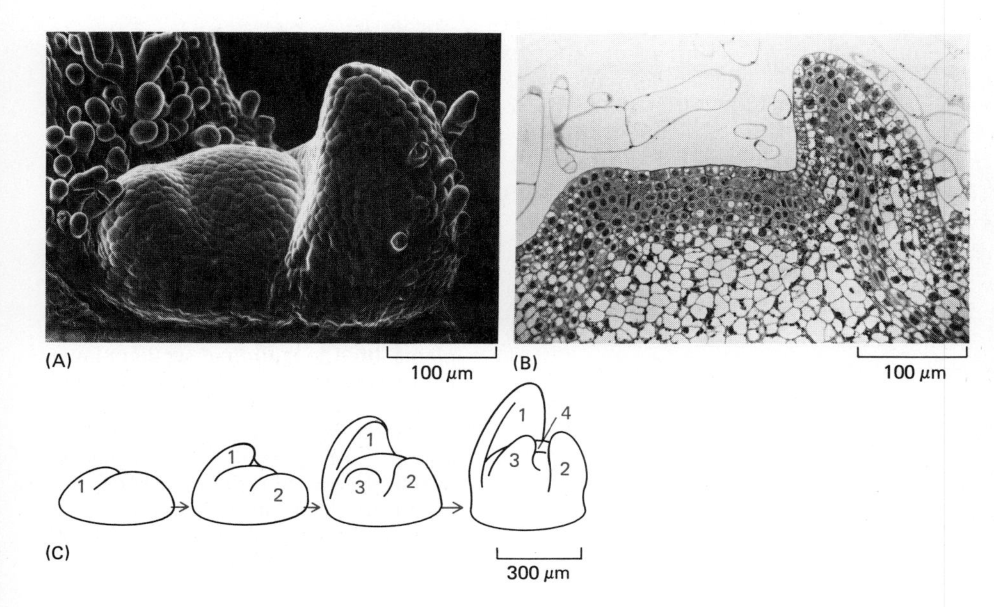

Figure 20–60 A shoot apex from a young tobacco plant. (A) A scanning electron micrograph shows the shoot apex with two sequentially emerging leaf primordia, seen here as lateral swellings on either side of the domed apical meristem. (B) A thin section of a similar apex shows that the youngest leaf primordium arises from a small group of cells (about 100) in the outer four or five layers of cells. (C) A very schematic drawing showing that the sequential appearance of leaf primordia takes place over a small distance and very early in shoot development. Growth of the apex will eventually form internodes that will separate the leaves in order along the stem (see Figure 20–58). (A and B, from R.S. Poethig and I.M. Sussex, *Planta* 165:158–169, 1985.)

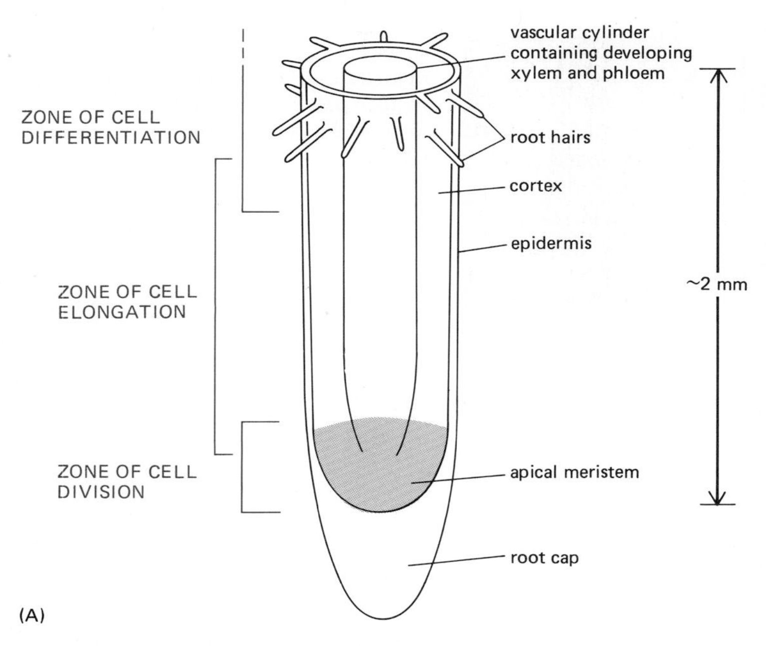

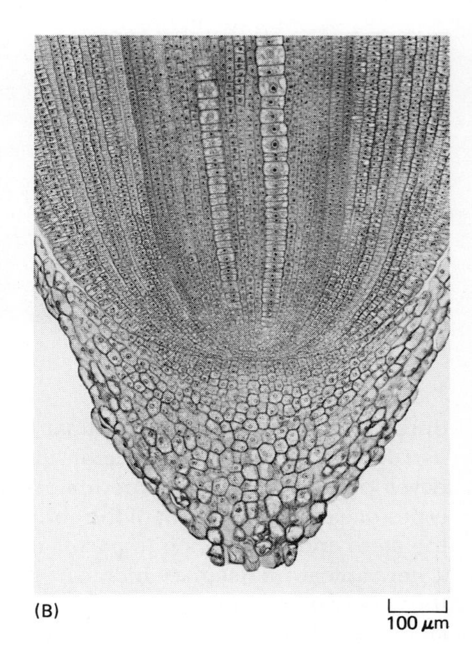

Figure 20–61 (A) The organization of the final 2 mm of a growing root tip. The approximate zones in which cells can be found dividing, elongating, and differentiating are indicated. (B) The apical meristem and root cap of a corn root tip, showing the orderly files of cells produced. (B, from P.H. Raven, R.F. Evert, and S.E. Eichhorn, Biology of Plants, 4th ed. New York: Worth, 1986.)

plest models proposed to explain this process, each new node produces a diffusible *morphogen* that inhibits the formation of the next node until the morphogen is diluted out by a fixed distance of internode growth. Almost nothing is known, however, about the molecular mechanisms that mediate this central patterning mechanism in the plant kingdom.

The Generation of New Structures Depends on the Coordinated Division, Expansion, and Differentiation of Cells[31]

Our discussion of higher plant development has so far been largely at the descriptive level of tissues and organs. What is the cellular basis of these complex events? Since plant cells are immobilized by their cell walls, plant morphogenesis must depend on regulated cell division coupled to strictly oriented cell expansion. Most cells produced in the root tip meristem, for example, go through three distinct phases of development: division, growth (elongation), and differentiation. These three steps, which overlap in both space and time, give rise to the characteristic architecture of a root tip. Although the process of cell differentiation often begins while a cell is still enlarging, it is comparatively easy to distinguish in a root tip a zone of cell division, a zone of cell elongation (which accounts for the growth in length of the root), and a zone of cell differentiation (Figure 20–61). When differentiation is complete, some of the differentiated cell types remain alive (for example, phloem cells) while others die (for example, xylem vessel elements).

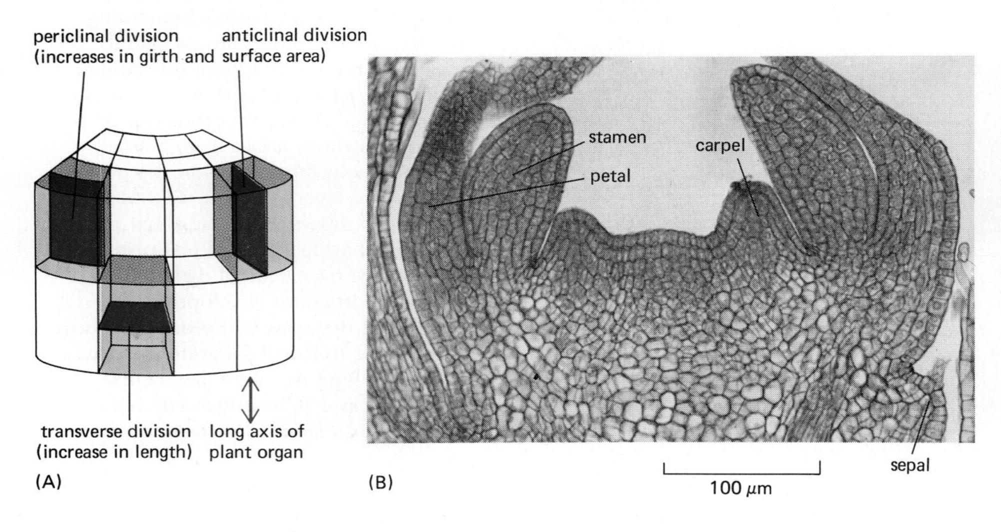

Figure 20–62 The relationship among division plane, cell expansion, and morphogenesis. (A) Three planes of cell division found in a typical plant organ. Variations in the relative proportion of each, combined with oriented cell expansion, can account for the morphogenetic patterns found in plants. (B) A longitudinal section of a young flower bud of a periwinkle. The small domes of cells destined to become the different floral parts have arisen by a combination of new planes of cell division and directional cell expansion determined by the reinforcing hoops of cellulose in the cell wall. (From N.H. Boke, *Am. J. Bot.* 36:535–547, 1949.)

The exact plane in which cells divide is crucial to plant morphogenesis, and many developmental processes, such as the development of stomata or leaf primordia, begin with a change in the plane of cell division. The continual cell divisions in a meristem often give rise to columns of daughter cells oriented parallel to the growth axis (see Figure 20–61), and changes in the plane of division are often associated with morphogenetic events. *Anticlinal* divisions occur frequently in the epidermis, and provide additional cells in the surface layers. These can accommodate the increase in girth that will result from *periclinal* divisions (Figure 20–62A). A combination of local changes in the frequency and orientation of cell divisions is responsible for the formation of organs such as flowers, leaves, and lateral roots (Figure 20–62B).

We have already seen that the turgor-driven expansion of a plant cell, which can often increase its volume fiftyfold or more, is oriented primarily by the orientation of cellulose microfibrils in the cell wall, which in turn is determined by the orientation of the cortical microtubule array. The cytoskeleton also plays a role in defining the plane of cell division.

A Cytoskeletal Framework Determines the Plane of Cell Division[32]

The organized growth of a plant requires that cells in selected sites divide in a particular plane and at a particular time so as to set up correctly oriented cell lineages. Microtubules play an early part in determining the plane of cell division. Thus the first visible sign that a higher plant cell has become committed to divide in a particular plane is seen just after the interphase cortical array of microtubules disappears in preparation for mitosis. At this time a narrow circumferential band of microtubules appears and forms a ring around the entire cell just beneath the plasma membrane (Figure 20–63). Because this array of microtubules appears in G_2 before prophase begins, it is called the **preprophase band.** The band disap-

Figure 20–63 The preprophase band of microtubules in a higher plant cell. (A) Simplified diagram of the changing distribution of microtubules during the cell cycle in higher plant cells, emphasizing four types of arrays. As cells leave interphase and enter prophase, the cortical array of microtubules (1) becomes bunched up into a dense *preprophase band* (2), which is 1–3 μm wide and often contains more than a hundred microtubules. This preprophase band predicts the subsequent plane of cell division. The mitotic spindle (3) then forms and aligns the chromosomes on the metaphase plate (4). Finally, microtubules are organized during cytokinesis into the phragmoplast (5), which lays down the new cell wall (6) (see Figure 13–71, p. 782). The interphase array is reestablished (8) in the daughter cells from microtubules polymerized from amorphous centrosomal material at the surface of the nucleus (7). (B) Immunofluorescence staining of microtubules in the preprophase band of an onion root tip cell. At the left is shown the isolated cell, while on the right two planes of focus clarify the relationship of the band to early spindle formation outside the nucleus. (C) Even though the two cells shown here are of similar shape, the preprophase band predicts that the top cell will divide transversely while the bottom cell will divide longitudinally. (B and C, courtesy of Kim Goodbody and Clive Lloyd.)

1. interphase
2. preprophase
3. prometaphase
4. metaphase
5. telophase
6. cytokinesis
7. early interphase
8. interphase
(A)

(B)
10 μm

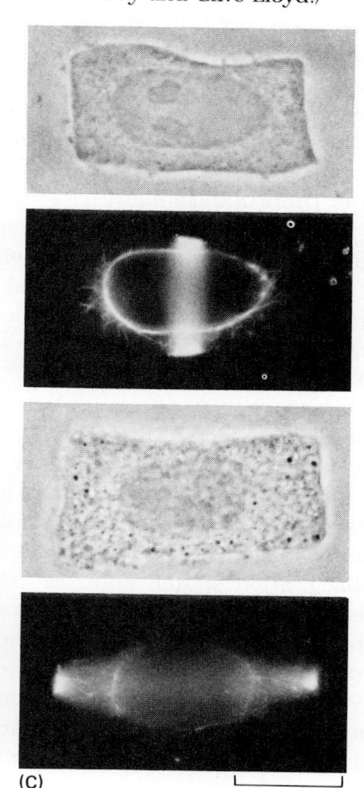

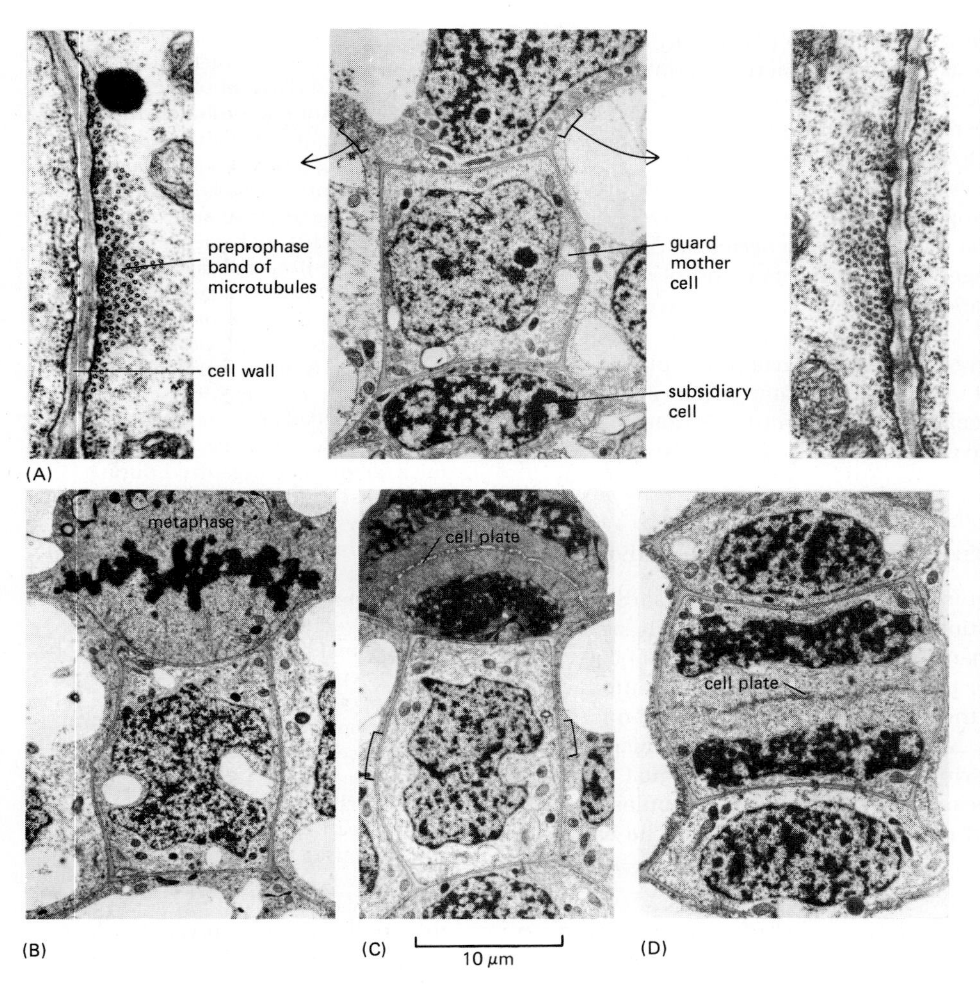

Figure 20–64 Sequence of electron micrographs of cells in the epidermis of a sugar cane leaf, showing how the formation of stomata involves a precise sequence of symmetric and asymmetric cell divisions. The plane of each division is accurately predicted by a preprophase band. (A) The uppermost cell is about to divide asymmetrically to form a subsidiary cell (the one at the bottom has already formed). The position of the preprophase band is indicated by brackets, and the band is shown in greater detail on either side. (B) Somewhat later, the upper cell is in metaphase, and the preprophase band has gone. (C) The cell plate is now forming in the upper cell during cytokinesis and is curving down toward the position of the former preprophase band. Meanwhile, the large central cell, the guard mother cell, is about to divide symmetrically, and the position of its preprophase band is indicated by brackets. (D) Cytokinesis is complete in the upper cell and almost complete in the guard mother cell. The latter division will produce the two guard cells that flank the stomatal pore (see Figure 20–11). (Courtesy of C. Busby.)

pears before metaphase is reached, yet the boundary of the division plane has somehow been imprinted: when the new cell plate forms later during cytokinesis, it grows outward to fuse with the parental wall precisely at the zone that was formerly occupied by the preprophase band (Figures 20–64). Even if the cell contents are displaced by centrifugation after the preprophase band has disappeared, the growing cell plate will tend to find its way back to the plane defined by the former preprophase band.

It is now known that the preprophase band contains numerous actin filaments in addition to microtubules. The actin filaments are not confined to the cell cortex but also form a radial, disclike array of strands, which crosses the vacuole and connects to and supports the central dividing nucleus. After the microtubules in the preprophase band depolymerize, these radial actin strands remain and provide a "memory" of the predetermined division plane. During cytokinesis, as the phragmoplast grows out centrifugally like a circular ripple in a pond, the edges of the growing cell plate are connected to the site of the preprophase band by actin filaments (see Figure 13–73, p. 783).

Regardless of whether the division is symmetric or asymmetric, or whether it is transverse, periclinal, or anticlinal, the preprophase band in a plant cell always specifies where the cell is going to divide *before* it enters mitosis (see Figure 20–63C). Such spatial controls are especially important in the asymmetric cell divisions that create two daughter cells with different developmental fates: stomatal cells, root hair cells, and the generative cells of pollen grains, for example, all develop from the smaller of two daughter cells. In such divisions the nucleus moves to the appropriate position in the cell before mitosis (Figure 20–65). Although the mechanism of nuclear migration is uncertain, there is evidence that both microtubules and actin filaments are involved.

The precise control over cell division planes that we have been describing is possible only if the plant tissue and its constituent cells have a structural polarity that can subsequently be either consolidated or modified. Although the structural basis for the establishment of cell polarity in higher plants is not known, there are well-studied examples among lower plants that may provide relevant models.

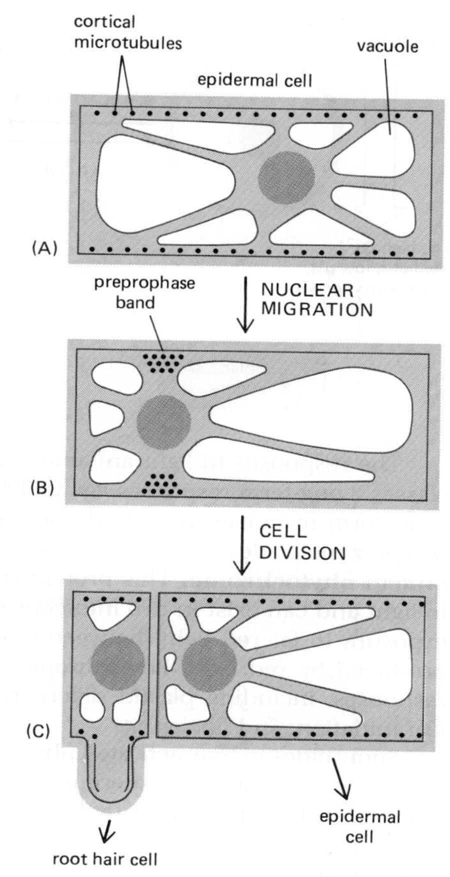

Figure 20–65 Asymmetric division in an elongated epidermal cell from a root. In interphase (A), cortical microtubules are distributed along the length of the cell wall. During preprophase (B), however, they congregate in a discrete band circling the cell. This preprophase band of microtubules accurately predicts where the new cell wall will join the old one when the cell divides (C). This epidermal cell divides asymmetrically to form a large daughter cell that will continue as an epidermal cell and a small daughter cell that will become a root hair cell.

The Polarity of Plant Cells Depends on the Asymmetric Distribution of Membrane-bound Ion Channels and Carrier Proteins[33]

The polarity of a plant and its constituent cells is remarkably stable: small lengths of stem (even upside down) always regenerate roots and shoots at their original basal and apical ends, respectively. It is still unclear how such cell polarity is established, maintained, and transmitted to daughter cells, but there is evidence that asymmetrically distributed ion channels and carrier proteins in the plasma membrane may help to establish cell polarity by generating intracellular ion currents.

A particularly well-studied example is the common brown seaweed *Fucus*. This large alga releases thousands of free-living sperm and eggs into the sea water around it. The large spherical fertilized eggs of *Fucus* are initially homogeneous, with no intrinsic polarity. Within 18 hours, however, the zygotes become polarized and undergo an asymmetric division in which the smaller, basal cell will develop into a structure for anchoring the seaweed to a rock and the larger, apical cell will develop into the thallus (the photosynthetic part). A clue to the mechanism of polarization of the early zygote is a small current that flows across it. This current is generated and maintained in part by a passive Ca^{2+} influx at the basal pole and an active efflux of Ca^{2+} that occurs elsewhere, which presumably reflect an asymmetrical distribution of Ca^{2+} channels and Ca^{2+} pumps, respectively, in the plasma membrane. This current flow through the egg is capable of moving highly charged intracellular components by electrophoresis, and there is evidence that one result of the polarization is an accumulation of secretory vesicles at the basal pole, followed by the localized deposition of specific sulfated polysaccharides in the cell wall there.

The unpolarized *Fucus* zygote becomes polarized in response to external stimuli, usually in the form of a continuous gradient—for example, of light or gravity. The information in the gradient is thought to lead to an initial inhomogeneity in the distribution of Ca^{2+} in the egg (Figure 20–66). Even small inhomogeneities in Ca^{2+} could have large effects, since they could be rapidly amplified and propagated by feedback loops to produce the sustained, self-driven Ca^{2+} flux that is observed. Although it is unclear how the zygote axis becomes stabilized, the process requires both actin filaments and a cell wall. Perhaps the cytoskeletal elements help connect the Ca^{2+} channels to wall fibrils by transmembrane bridges at the future basal pole.

Intracellular Ca^{2+} gradients are thought to exist in many other types of plant cells that exhibit polarized growth. Electrical currents carried in part by Ca^{2+}, for example, have been shown to enter the growing apices of a variety of polarized structures—including root hairs, pollen tubes, and whole roots—and to leave at nongrowing regions. Thus the observations on *Fucus* eggs may have general relevance for the generation of plant cell polarity.

Plant Growth and Development Are Modulated in Response to Environmental Cues[34]

Environmental conditions often have a much more profound influence on the development of plants than they do on the development of animals. Plants have evolved intricate systems for monitoring gravity, nutrient status, temperature, and the intensity and duration of light. The responses to these stimuli are complex and can be either rapid and brief (as in the light-induced chloroplast movements discussed on p. 1171) or slow and long-lasting (as in the long periods of daylight required to induce some plants to flower, or the long periods of cold required before many seeds will germinate).

protoplasts (Figure 20–71). After foreign DNA has been introduced into these naked spherical plant cells, they can often be encouraged to re-form their walls, divide, and regenerate into a new plant.

A common vector used to introduce foreign DNA directly into normal plant cells is the T-DNA of *Agrobacterium* described earlier (see p. 1159). For this purpose the genes in the T-DNA that cause plant tumors are removed by recombinant DNA methods and replaced by a desired new gene. After the modified T-DNA is transferred into an appropriate Ti plasmid in *Agrobacterium*, the bacteria are cocultured with a piece of a leaf. This system allows the integration of the T-DNA into a plant chromosome, thereby adding the new gene permanently to the entire plant (Figure 20–72). At present this method works efficiently only on certain families of dicots.

Recent developments such as these have provided plant cell biologists with the same range of recombinant DNA techniques used by animal cell biologists, greatly facilitating progress in such important research areas as isolating receptors for plant growth regulators, unraveling morphogenetic pathways, and analyzing mechanisms of gene expression. In addition, plant recombinant DNA technology has opened up many new possibilities in agriculture that could benefit both the farmer and the consumer. These possibilities include the ability to modify the lipid, starch, and protein storage reserves in seeds, to impart pest and virus resistance to plants, and to create modified plants that tolerate extreme habitats such as salt marshes or waterlogged soil.

Many of the major advances in understanding animal development have come from studies on invertebrates that are amenable to extensive genetic analysis as well as to experimental manipulation, such as the fruit fly *Drosophila* (see p. 920) and the nematode worm *Caenorhabditis* (see p. 901). Progress in plant developmental biology has been relatively slow in comparison. Many of the organisms that have proved most amenable to genetic analysis, such as maize and tomato, have long life cycles and very large genomes, which have made both classical and molecular genetic analysis time-consuming. Increasing attention is consequently being paid to a small weed, the common wall cress (*Arabidopsis thaliana*), which has several major advantages as a "model plant." It is so small (Figure 20–73) that it can be grown indoors in test tubes in large numbers. With a minimum generation time of only 5 weeks, it can produce thousands of offspring per plant after 8 to 10 weeks. *Arabidopsis* also has the smallest plant genome known (7×10^7

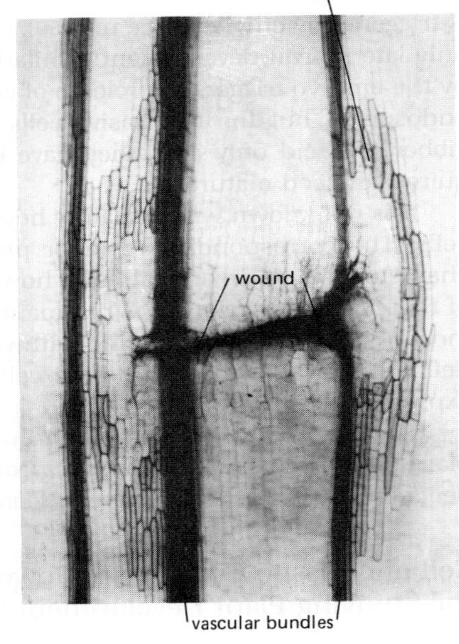

Figure 20–68 Light micrograph showing a wound response by the parenchymal cells that surround the vascular bundle in the common houseplant *Coleus*, seen in a longitudinal section. The vascular bundles have been severed at the wound. Seven days later, after stimulation of division and redifferentiation in the nearby cortical cells, a series of new xylem and phloem cells have been generated that restore vascular continuity around the wound. A supply of auxin is normally carried from the shoot apex to the base of the plant, and it is thought that the wounding provides a newly directed flux of auxin that in turn induces the vascular regeneration shown here. (Courtesy of N.P. Thomson.)

Figure 20–69 The production of self-supporting plants from callus cultures. The *Freesia* callus culture in (A) was induced to produce shoots (B) and then roots (C) by altering the available ratios of an auxin and a cytokinin. A high cytokinin ratio appears to be required for shoot and leaf induction. (Courtesy of Graham Hussey.)

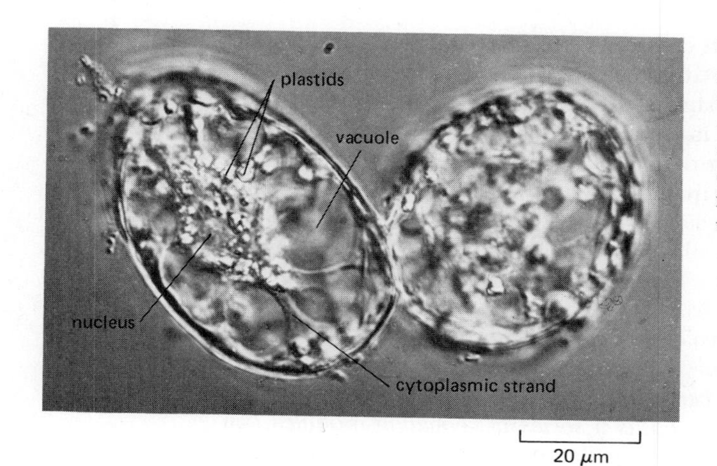

Figure 20–70 Callus cells can be grown as free-living, single cells suspended in liquid media. Two such cells derived from a sycamore callus are shown here. They are highly vacuolated cells with cytoplasmic strands radiating from the region of the nucleus, and they have a primary cell wall.

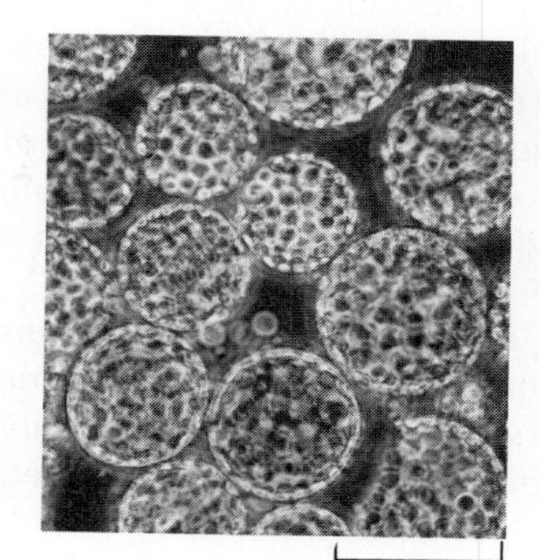

Figure 20–71 Light micrograph of protoplasts prepared from the green leaf cells of a tobacco plant. Without their walls the cells round up and need to be stabilized in a sugar solution that matches the osmotic pressure of their cytoplasm. Numerous chloroplasts can be seen in each protoplast. (Courtesy of J. Burgess.)

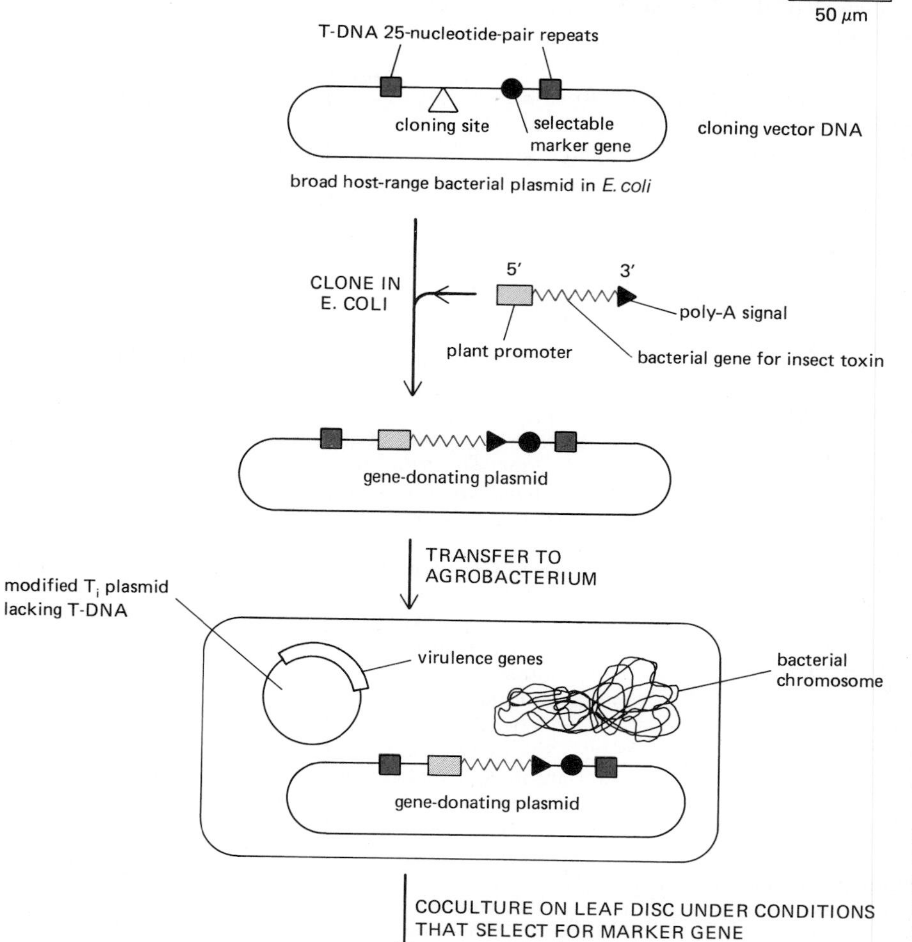

Figure 20–72 A strategy used to obtain transgenic plants. In this example the gene being transferred codes for a bacterial protein that is toxic to insects. To allow the gene to be expressed in a plant cell, the 5′ end of the gene is attached to a plant promoter and the 3′ end to a poly-A addition site. The modified toxin gene is inserted into a plasmid that also contains a marker gene (such as a kanamycin resistance gene) that can be selected for in the plant (see p. 259). The plasmid is designed so that both the marker gene and the toxin gene are flanked by the special 25-nucleotide-pair repeats that normally flank T-DNA. The plasmid is grown in *E. coli* and then transferred to an *Agrobacterium* that contains the virulence genes on a separate plasmid. When the *Agrobacterium* is cocultivated with a leaf disc, the virulence gene products recognize the T-DNA repeats and transfer the DNA that contains the marker and toxin genes to a plant chromosome. All of the cells in the leaf disc are stimulated to proliferate by the growth medium, but only those plant cells that contain the selectable marker gene are able to divide and form a callus. The callus is then used to produce transgenic plants, which express the bacterial gene for the insect toxin and are therefore unusually resistant to insect attack.

nucleotide pairs), comparable to yeast (2×10^7 nucleotide pairs), *C. elegans* (8×10^7 nucleotide pairs), and *Drosophila* (10×10^7 nucleotide pairs). Cell culture and transformation methods have been established, large numbers of interesting mutants have been isolated, and work is underway to create a complete ordered collection of genomic DNA clones (see p. 260). Taken together, the attractive features of *Arabidopsis* suggest that it may soon become the *Drosophila* of plant developmental biology.

Figure 20–73 *Arabidopsis thaliana*, a small member of the mustard—or crucifer—family, is a weed of no economic value but of potentially great value for experiments in the molecular genetics of plants. Its small genome (7×10^7 nucleotide pairs) is comparable in size to the *Drosophila* genome (10×10^7 nucleotide pairs). (Courtesy of Chris Sommerville.)

Summary

In the earliest stages of embryonic development in a higher plant, cell division takes place throughout the body of the embryo. As the embryo grows, however, addition of new cells becomes restricted to certain regions of the plant body known as meristems. The adult plant can be viewed as a series of repeating modules constructed by the pattern-formation mechanisms that occur in these meristems.

Plant morphogenesis depends on the coordinated division, expansion, and differentiation of nonmotile cells. Control over both the plane of cell division and the oriented expansion of plant cells is exerted in part by microtubules associated with the inner face of the plasma membrane. The growth and division of plant cells is influenced by light, gravity, temperature, and other environmental factors, as well as by specific low-molecular-weight growth regulators, such as auxins and cytokinins.

Many aspects of plant cell growth and development can be studied in tissue and cell culture. Plant somatic cells show a plasticity of developmental potential not shown by the somatic cells of animals. A dramatic demonstration of the totipotency of many plant cells is the routine regeneration of a complete plant from a single cultured somatic cell.

Protoplasts are plant cells without their cell walls. They can be manipulated in culture in much the same way that animal cells can, but they have the added property that whole plants can be regenerated from them.

Both protoplasts and intact cells can be manipulated genetically by recombinant DNA methods to introduce any desired new gene into their chromosomes; in many cases these cells can subsequently be used to produce transgenic plants.

References

General

Cutter, E.G. Plant Anatomy, 2nd ed., Part 1, Cells and Tissues; Part 2, Organs. London: Arnold, 1978.

Esau, K. Anatomy of Seed Plants, 2nd ed. New York: Wiley, 1977.

Grierson, D.; Covey, S.N. Plant Molecular Biology, 2nd ed. New York: Methuen, 1988.

Gunning, B.E.S.; Steer, M.W. Ultrastructure and the Biology of Plant Cells. London: Arnold, 1975.

Raven, P.H.; Evert, R.F.; Eichhorn, S.E. Biology of Plants, 4th ed. New York: Worth, 1986.

Roberts, K.; Johnston, A.W.B.; Lloyd, C.W.; Shaw, P.J.; Woolhouse, H.W., eds. The Cell Surface in Plant Growth and Development. *J. Cell. Sci.*, Suppl. 2. Cambridge, U.K.: Company of Biologists Ltd, 1985.

Salisbury, F.B.; Ross, C.W. Plant Physiology, 3rd ed. Belmont, CA: Wadsworth, 1985.

Cited

1. Bacic, A.; Harris, P.J.; Stone, B.A. Structure and Function of Plant Cell Walls. In Biochemistry of Plants: A Comprehensive Treatise (J. Preiss, ed.), Vol. 14: Carbohydrates, pp. 298–371. San Diego, CA: Academic Press, 1988.

   Brett, C.T.; Hillman J.R., eds. Biochemistry of Plant Cell Walls. SEB Seminar Series, 28. New York: Cambridge University Press, 1985.

   McNeil, M.; Darvill, A.G.; Fry, S.C.; Albersheim, P. Structure and function of the primary cell walls of plants. *Annu. Rev. Biochem.* 53:625–663, 1984.

   Rees, D.A. Polysaccharide Shapes. London: Chapman and Hall, 1977.

2. Cassab, G.I.; Varner, J.E. Cell wall proteins. *Annu. Rev. Plant Physiol.* 39:321–353, 1988.

   Fry, S. Cross-linking of matrix polymers in the growing cell walls of angiosperms. *Annu. Rev. Plant Physiol.* 37:165–186, 1986.

   Selvendran, R.R. Developments in the chemistry and biochemistry of pectic and hemicellulosic polymers. *J. Cell Sci.* (Suppl. 2):51–88, 1985.

3. Carpita, N.; Sabularse, D.; Montezinos, D.; Delmer, D.P. Determination of the pore size of cell walls of living plant cells. *Science* 205:1144–1147, 1979.

   Milburn, J.A. Water Flow in Plants. London: Longman, 1979.

4. Street, H.E.; Öpik, H. The Physiology of Flowering Plants, 3rd ed. London, U.K.: Edward Arnold, 1984.

5. Fry, S.C. The Growing Plant Cell Wall: Chemical and Metabolic Analysis. New York: Wiley, 1988.

Masuda, Y.; Yamamoto, R. Cell-wall changes during auxin-induced cell extension. Mechanical properties and constituent polysaccharides of the cell wall. In Biochemistry of Plant Cell Walls; (C.T. Brett, J.R. Hillman, eds.), pp. 269–300. SEB Seminar Series 28. New York: Cambridge University Press, 1985.

6. Baker, D.A.; Hall, J.L. Ion Transport in Plant Cells and Tissues. Amsterdam: North-Holland, 1975.

Cheeseman, J.M. Mechanisms of salinity tolerance in plants. *Plant Physiol.* 87:547–550, 1988.

Morgan, J.M. Osmoregulation and water stress in higher plants. *Annu. Rev. Plant Physiol.* 35:299–319, 1984.

Zimmerman, U. Cell turgor pressure regulation and turgor pressure-mediated transport processes. *Symp. Soc. Exp. Biol.* 31:117–154, 1977.

7. Esau, K. Anatomy of Seed Plants, 2nd ed. New York: Wiley, 1977.

Tanner, W.; Loewus, F.A., eds. Encyclopedia of Plant Physiology, New Series. Vol. 13B: Plant Carbohydrates II, Extracellular Carbohydrates. Heidelberg: Springer-Verlag, 1982.

8. Dugger, W.M.; Bartnicki-Garcia, S., eds. Structure, Function, and Biosynthesis of Plant Cell Walls. *Proc. 7th Annu. Symp. Botany* Rockville, MD: American Society of Plant Physiologists, 1984. (Several relevant papers appear in this collection.)

Fry, S. The Growing Plant Cell Wall: Chemical and Metabolic Analysis. New York: Wiley, 1988.

9. Gunning, B.E.S.; Overall, R.L. Plasmodesmata and cell-to-cell transport in plants. *Bioscience* 33:260–265, 1983.

Gunning, B.E.S.; Robards, A.W., eds. Intercellular Communication in Plants: Studies on Plasmodesmata. New York: Springer-Verlag, 1976.

10. Baron-Epel, O.; Hernandez, D.; Jiang, L-W.; Meiners, S.; Schindler, M. Dynamic continuity of cytoplasmic and membrane compartments between plant cells. *J. Cell Biol.* 106:715–721, 1988.

Gunning, B.E.S.; Hughes, J.E. Quantitative assessment of symplastic transport of pre-nectar into the trichomes of *Abutilon* nectaries. *Aust. J. Plant Physiol.* 3:619–637, 1976.

Terry, B.R.; Robards, A.W. Hydrodynamic radius alone governs the mobility of molecules through plasmodesmata. Planta 171:145–157, 1987.

Zaitlin, M.; Hull, R. Plant virus-host interactions. *Annu. Rev. Plant Physiol.* 38:291–315, 1987.

11. Aloni, R. Differentiation of vascular tissue. *Annu. Rev. Plant Physiol.* 38:179–204, 1987.

Moorby, J. Transport Systems in Plants. New York: Longman, 1981.

12. Baker, D.A. Transport Phenomena in Plants. London, U.K.: Chapman and Hall, 1978.

Clarkson, D.T. Factors affecting mineral nutrient acquisition by plants. *Annu. Rev. Plant Physiol.* 36:77–115, 1985.

Milburn, J.A. Water Flow in Plants. London: Longman, 1979.

Passioura, J.B. Water transport in and to roots. *Annu. Rev. Plant Physiol. Plant Mol. Biol.* 39:245–265, 1988.

13. Cronshaw, J. Phloem structure and function. *Annu. Rev. Plant Physiol.* 32:465–484, 1981.

Cronshaw, J.; Lucas, W.J.; Giaquinta, R.T., eds. Phloem Transport. New York: Liss, 1986.

Gunning, B.E.S. Transfer cells and their roles in transport of solutes in plants. *Sci. Prog. (Oxford)* 64:539–568, 1977.

Ho, L.C. Metabolism and compartmentation of imported sugars in sink organs in relation to sink strength. *Annu. Rev. Plant Physiol. Plant Mol. Biol.* 39:355–378, 1988.

14. Gianinazzi-Pearson, V.; Gianinazzi, S., eds. Physiological and Genetical Aspects of Mycorrhizae. Paris: INRA, 1986.

Smith, S.E.; Gianinazzi-Pearson V. Physiological interactions between symbionts in vesicular-arbuscular mycorrhizal plants. *Annu. Rev. Plant Physiol. Plant Mol. Biol.* 39:221–244, 1988.

15. Downie, J.A.; Johnston, A.W.B. Nodulation of legumes by *Rhizobium*: the recognized root? *Cell* 47:153–154, 1986.

Peters, N.K.; Frost, J.W.; Long, S.R. A plant flavone, luteolin induces expression of *Rhizobium meliloti* nodulation genes. *Science* 233:977–980, 1986.

Rolfe, B.G.; Gresshoff, P.M. Genetic analysis of legume nodule initiation. *Annu. Rev. Plant Physiol. Plant Mol. Biol.* 39:297–319, 1988.

16. Buchanan-Wollaston, V.; Passiatore, J.E.; Cannon, F. The *mob* and *oriT* mobilization functions of a bacterial plasmid promote its transfer to plants. *Nature* 328:172–175, 1987.

Klee, H.; Horsch, R.; Rogers, S. *Agrobacterium*-mediated plant transformation and its further applications to plant biology. *Annu. Rev. Plant Physiol.* 38:467–486, 1987.

Stachel, S.E.; Messens, E.; Van Montagu, M.; Zambryski, P. Identification of the signal molecules produced by wounded plant cells that activate T-DNA transfer in *Agrobacterium tumefaciens*. *Nature* 318:624–629, 1985.

17. Ayers, A.R.; Ebel, J.; Finelli, F.; Berger, N.; Albersheim, P. Host pathogen interactions. *Plant Physiol.* 57:751–759, 1976. (This should be read in conjunction with the three related papers that follow it.)

Darvill, A.G.; Albersheim, P. Phytoalexins and their elicitors—a defence against microbial infection in plants. *Annu. Rev. Plant. Physiol.* 35:243–275, 1984.

Ralton, V.E.; Smart, M.G.; Clarke, A.E. Recognition and infection process in plant pathogen interactions. In Plant-Microbe Interactions. (T. Kosuge, E.W. Nestor, eds.), Vol. 2, pp. 217–252. New York: Macmillan, 1987.

Ryan, C.A. Oligosaccharide signalling in plants. *Annu. Rev. Cell Biol.* 3:295–317, 1987.

18. McDougall, G.J.; Fry, S.C. Inhibition of auxin-stimulated growth of pea stem segments by a specific monosaccharide of xyloglucan. *Planta* 175:412–416, 1988.

Thanh Van, K.T.; et al. Manipulation of the morphogenetic pathways of tobacco explants by oligosaccharins. *Nature* 314:615–617, 1985.

19. Anderson, J.M. Photoregulation of the composition, function, and structure of thylakoid membranes. *Annu. Rev. Plant Physiol.* 37:93–136, 1986.

Mullet, J.E. Chloroplast development and gene expression. *Annu. Rev. Plant Physiol. Plant Mol. Biol.* 39:475–502, 1988.

Thomson, W.W.; Watley, J.M. Development of nongreen plastids. *Annu. Rev. Plant Physiol.* 31:375–394, 1980.

20. Boller, T.; Kende, H. Hydrolytic enzymes in the central vacuole of plant cells. *Plant Physiol.* 63:1123–1132, 1979.

Boller, T.; Wiemken, A. Dynamics of vacuolar compartmentation. *Annu. Rev. Plant Physiol.* 37:137–164, 1986.

Marin, B., ed. Plant Vacuoles: Their Importance in Solute Compartmentation in Cells and Their Applications in Plant Biotechnology. NATO ASI Series, Vol. 134. New York: Plenum, 1987.

Matile, P. Biochemistry and function of vacuoles. *Annu. Rev. Plant Physiol.* 29:193–213, 1978.

21. Mollenhauer, H.H.; Morré, D.J. The Golgi apparatus. In The Biochemistry of Plants—A Comprehensive Treatise. (N.E. Tolbert, ed.), Vol. 1, pp. 437–488. New York: Academic Press, 1980.

Moore, P.J.; Staehelin, L.A. Immunogold localization of the cell-wall-matrix polysaccharides rhamnogalacturonan I and xyloglucan during cell expansion and cytokinesis in *Trifolium pratense* L.; implications for secretory pathways. *Planta* 174:433–445, 1988.

Northcote, D.H. Macromolecular aspects of cell wall differentiation. In Encyclopedia of Plant Physiology, New Series. Vol. 14A: Nucleic Acids and Proteins in Plants I (D. Boulter, B. Parthier, eds.), pp. 637–655. Berlin: Springer-Verlag, 1982.

22. Coleman, J.; Evans, D.; Hawes, C.; Horsley, D.; Cole, L. Structure and molecular organization of higher plant coated vesicles. *J. Cell Sci.* 88:35–45, 1987.
Robinson, D.G.; Depta, H. Coated vesicles. *Annu. Rev. Plant Physol. Plant Mol. Biol.* 39:53–99, 1988.
Tanchak, M.A.; Griffing, L.R.; Mersey, B.G.; Fowke, L.C. Endocytosis of cationized ferritin by coated vesicles of soybean protoplasts. *Planta* 162:481–486, 1984.
23. Brown, R.M. Cellulose microfibril assembly and orientation: recent developments. *J. Cell Sci.*, (Suppl. 2):13–32, 1985.
Delmer, D.P. Cellulose biosynthesis. *Annu. Rev. Plant Physiol.* 38:259–290, 1987.
Schneider, B.; Herth, W. Distribution of plasma membrane rosettes and kinetics of cellulose formation in xylem development of higher plants. *Protoplasma* 131:142–152, 1986.
24. Green, P.B. Organogenesis—a biophysical view. *Annu. Rev. Plant Physiol.* 31:51–82, 1980.
Herth, W. Plant cell wall formation. In Botanical Microscopy (A.W. Robards, ed.), pp. 285–310. New York: Oxford University Press, 1985.
25. Lloyd, C.W., ed. The Cytoskeleton in Plant Growth and Development. New York: Academic Press, 1982. (Chapters 5–8 are particularly relevant.)
26. Kersey, Y.M.; Hepler, P.K.; Palevitz, B.A.; Wessels, N.K. Polarity of actin filaments in characean algae. *Proc. Natl. Acad. Sci. USA* 73:165–167, 1976.
Parthasarathy, M.V. F-actin architecture in coleoptile epidermal cells. *Eur. J. Cell Biol.* 39:1–12, 1985.
Sheetz, M.P.; Spudich, J.A. Movement of myosin-coated fluorescent beads on actin cables *in vitro*. *Nature* 303:31–35, 1983.
Williamson, R.E. Organelle movements along actin filaments and microtubules. *Plant Physiol.* 82:631–634, 1986.
27. Haupt, W. Light-mediated movement of chloroplats. *Annu. Rev. Plant Physiol.* 33:205–233, 1982.
Roberts, I.N.; Lloyd, C.W.; Roberts, K. Ethylene-induced microtubule reorientations: mediation by helical arrays. *Planta* 164:439–447, 1985.
Virgin, H.I. Light and chloroplast movements. *Symp. Soc. Exp. Biol.* 22:329–352, 1968.
Wagner, G.; Klein, K. Mechanism of chloroplast movement in *Mougeotia*. *Protoplasma* 109:169–185, 1981.
28. Cutter, E.G. Plant Anatomy, 2nd ed., Part 2, Organs. London: Arnold, 1978.
Johri, B.M. Embryology of Angiosperms. Berlin: Springer-Verlag, 1984.
Raven, P.H.; Evert, R.F.; Eichhorn, S.E. Biology of Plants, 4th ed. New York: Worth, 1986. (Chapters 19–22 provide a good general outline of plant development.)
29. Harper, J.L., ed. Growth and Form of Modular Organisms. London: Royal Society, 1986. (There are several relevant papers in this collection.)
Walbot, V. On the life strategies of plants and animals. *Trends Genet.* 1:165–169, 1985.
30. Green, P.B. A theory for influorescence development and flower formation based on morphological and biophysical analysis in *Echeveria*. *Planta* 175:153–169, 1988.
McDaniel, C.N.; Poethig, R.S. Cell lineage patterns in the shoot apical meristem of the germinating corn embryo. *Planta* 175:13–22, 1988.
Sachs, T. Controls of cell patterns in plants. In Pattern Formation (G.M. Malacinski, S.V. Bryant, eds.). New York: Macmillan, 1984.
Steeves, T.A.; Sussex, I.M. Patterns in Plant Development, 2nd ed. New York: Cambridge University Press, 1988.
31. Gunning, B.E.S. Microtubules and cytomorphogenesis in a developing organ: the root primordium of *Azolla pinnata*. In Cytomorphogenesis in Plants (O. Kiermayer, ed.), pp. 301–325. New York: Springer, 1981.
Poethig, R.S. Clonal analysis of cell lineage patterns in plant development. *Am. J. Bot.* 74:581–594, 1987.
32. Gunning, B.E.S.; Wick, S.M. Preprophase bands, phragmoplasts and spatial control of cytokinesis. *J. Cell Sci.*, Suppl. 2:157–179, 1985.
Lloyd, C. Actin in plants. *J. Cell Sci.* 90:185–188, 1988.
Lloyd, C.W. The plant cytoskeleton: the impact of fluorescence microscopy. *Annu. Rev. Plant Physiol.* 38:119–139, 1987.
Pickett-Heaps, J.D.; Northcote, D.H. Organization of microtubules and endoplasmic reticulum during mitosis and cytokinesis in wheat meristems. *J. Cell Sci.* 1:109–120, 1966.
Wick, S.M.; Seagull, R.W.; Osborn, M.; Weber, K.; Gunning, B.E.S. Immunofluorescence microscopy of organized microtubule arrays in structurally stabilized meristematic plant cells. *J. Cell Biol.* 89:685–690, 1981.
33. Hepler, P.K.; Wayne, R.O. Calcium and plant development. *Annu. Rev. Plant Physiol.* 36:397–439, 1985.
Kropf, D.L.; Kloareg, B.; Quatrano, R.S. Cell wall is required for fixation of the embryonic axis in *Fucus* zygotes. *Science* 239:187–190, 1988.
Quatrano, R.S.; Griffing, L.R.; Huber-Walchli, V.; Doubet, R.S. Cytological and biochemical requirements for the establishment of a polar cell. *J. Cell Sci.*, Suppl. 2:129–141, 1985.
Schnepf, E. Cellular polarity. *Annu. Rev. Plant Physiol.* 37:23–47, 1986.
34. Jordan, B.R.; Partis, M.D.; Thomas, B. The biology and molecular biology of plytochrome. *Ox. Sur. Plant Mol. Cell Biol.* 3:315–362, 1986.
Kuhlemeier, C.; Green, P.J.; Chua, N. Regulation of gene expression in higher plants. *Annu. Rev. Plant Physiol.* 38:221–257, 1987.
Tobin, E.M.; Silverthorne, J. Light regulation of gene expression in higher plants. *Annu. Rev. Plant Physiol.* 36:569–593, 1985.
Verma, D.P.S.; Goldberg, R.B. Temporal and Spatial Regulation of Plant Genes. New York: Springer-Verlag, 1988.
von Wettstein, D.; Chua, N-H., eds. Plant Molecular Biology. NATO ASI Series A, Vol. 140. New York: Plenum, 1987. (Several relevant papers in this collection.)
35. Hoad, G.V.; Lenton, J.R.; Jackson, M.B.; Atkin, R.K. Hormone Action in Plant Development: A Critical Appraisal. London: Butterworth, 1987.
Salisbury, F.B.; Ross, C.W. Plant Physiology, 3rd ed. Belmont, CA: Wadsworth, 1985. (Chapters 16 and 17 are relevant.)
Theologis, A. Rapid gene regulation by auxin. *Annu. Rev. Plant Physiol.* 37:407–438, 1986.
Trewavas, A.J. Growth substance sensitivity: the limiting factor in plant development. *Physiol. Plant.* 35:60–72, 1982.
36. Applications of Plant Cell and Tissue Culture. CIBA Foundation Symposium 137. New York: Wiley, 1988.
Lee, M.; Phillips, R.L. The chromosomal basis of somaclonal variation. *Annu. Rev. Plant Physol. Plant Mol. Biol.* 39:413–437, 1988.
Vasil, I.K., ed. Perspectives in Plant Cell and Tissue Culture. *Int. Rev. Cytol.*, Suppl. 11A and B, 1980.
37. Bevan, M. Binary *Agrobacterium* vectors for plant transformation. *Nuc. Acids Res.* 12:8711–8721, 1984.
De Block, M.; Herrera-Estrella, L.; Van Montagu, M.; Schell, J.; Zambryski, P. Expression of foreign genes in regenerated plants and in their progeny. *EMBO J.* 3:1681–1689, 1986.
Finkelstein, R.; Estelle, M.; Martinez-Zapater, J.; Somerville, C. *Arabidopsis* as a tool for the identification of genes involved in plant development. In Temporal and Spatial Regulation of Plant Genes (D.P.S. Verma, and R.B. Goldberg, eds.), pp. 1–25. New York: Springer-Verlag, 1988.
Myerowitz, E.M. *Arabidopsis thaliana*. *Annu. Rev. Genet.* 21:93–111, 1987.
Vaeck, M.; et al. Transgenic plants protected from insect attack. *Nature* 328:33–37, 1987.

Cancer

21

Roughly one person in five, in the prosperous countries of the world, will die of cancer; but that is not the reason for devoting a chapter of this book to the subject. Heart disease causes more deaths, and many other illnesses result in just as much distress; in the world as a whole, other health problems, such as malnutrition and parasitic infections, are more serious. In the context of cell biology, however, cancer has a unique importance, for the family of diseases grouped under this heading reflect disturbances of the most fundamental rules of behavior of the cells in a multicellular organism. To understand cancer and to devise rational ways to treat it, we have to understand both the inner workings of cells and their social interactions in the tissues of the body. Thus basic cancer research has been the source of many advances in our knowledge of normal cells. By-products of cancer research have become crucial tools in the current revolution of cell biology—tools such as reverse transcriptase from RNA tumor viruses, used to make cDNA (see p. 183), and myeloma cell lines derived from cancerous B lymphocytes, used to make monoclonal antibodies. One may debate how much the massive resources devoted to laboratory research on cancer have so far contributed directly toward improvements in cancer treatment; but there can be no doubt that, by contributing to progress in cell biology, the cancer research effort has profoundly benefited a much wider area of medical knowledge than that of cancer alone.

We have already discussed how cancer research has begun to reveal the molecular mechanisms underlying the normal controls of cell growth and division (see pp. 752–761). In this concluding chapter we examine the disease itself. In the first section we shall consider the nature of cancer and the natural history of the disease from a cellular standpoint; in the second section we focus on its molecular basis.

Cancer as a Microevolutionary Process[1]

The body of an animal can be viewed as a society or ecosystem whose individual members are cells, reproducing by cell division and organized into collaborative assemblies or tissues. In our earlier discussion of the maintenance of tissues (in Chapter 17), our concerns were similar to those of the ecologist: cell births, deaths, habitats, territorial limitations, the maintenance of population sizes, and the like. The one ecological topic conspicuously absent was that of natural selection: we said nothing of competition or mutation among somatic cells. The reason is that a healthy body is in this respect a very peculiar society, where self-sacrifice, rather from different cell types are, in general, very different diseases. The basal-cell carcinoma, for example, is only locally invasive and rarely forms metastases, whereas the melanoma is much more malignant and rapidly gives rise to many metastases (behavior that recalls the migratory tendencies of the normal pigment-cell precursors during development—see p. 944). The basal-cell carcinoma is usually easy to remove by surgery, leading to complete cure; but the malignant melanoma, once it has metastasized, is often impossible to extirpate and consequently fatal.

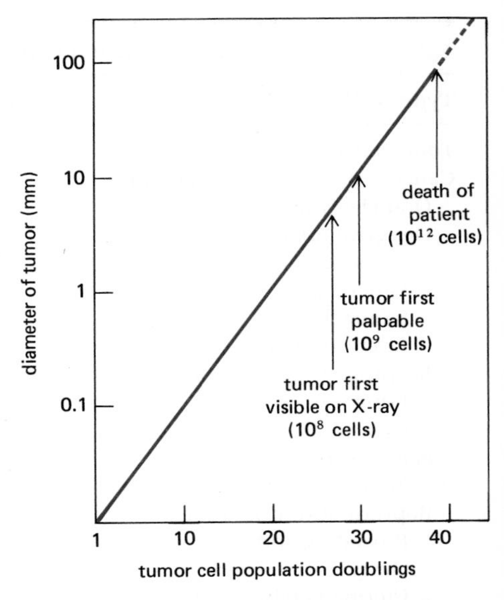

Figure 21–3 The growth of a typical human tumor, with the diameter of the tumor plotted on a logarithmic scale. Years may elapse before the tumor becomes noticeable.

Most Cancers Derive from a Single Abnormal Cell[3]

The origins of most cancers can be traced to a single isolated **primary tumor;** this suggests that they are derived by cell division from a single cell that has undergone some heritable change that enables it to outgrow its neighbors. By the time it is first detected, however, a typical tumor already contains about a billion cells or more (Figure 21–3), often including many normal cells—fibroblasts, for example, in the supporting connective tissue that is associated with a carcinoma. It is not easy to prove that the cancer cells are a clone descended from a single abnormal cell; but where evidence is available, it usually confirms that the cancer has a monoclonal origin. In almost all patients with *chronic myelogenous leukemia,* for example, the leukemic white blood cells are distinguished from the normal cells by a specific chromosomal abnormality (the so-called Philadelphia chromosome, created by a translocation between the long arms of chromosomes 22 and 9, as shown in Figure 21–4). It is unlikely that the genetic accident responsible for this abnormality would have occurred in several cells at once in the same individual; it is much more likely that all the leukemic cells are descendants of the same single mutant cell. Indeed, when the DNA at the site of translocation is cloned and sequenced, it is found that the site of breakage and rejoining of the translocated fragments is identical in all the leukemic cells in any given patient, but differs slightly (by a few hundred or thousand base pairs) from one patient to another, as expected if each case of the leukemia arises from a unique accident occurring in a single cell.

Another way to show that a cancer has a monoclonal origin is by exploiting the phenomenon of X-chromosome inactivation (see p. 577). A normal woman is a random mixture, or mosaic, of two classes of cells—those in which the paternal X chromosome is inactivated and those in which the maternal X chromosome is inactivated. The inactivation of one X chromosome in each cell occurs early in embryonic development, and thereafter the daughters of a dividing somatic cell always have the same X chromosome inactivated as the parent cell. Consequently, the state of X-chromosome inactivation—maternal or paternal—can be used as a heritable marker to trace the lineage of cells in the body. In the great majority of tumors that have been analyzed—both benign and malignant—all the tumor cells have been found to have the same X chromosome inactivated, strongly suggesting that they are derived from a single deranged cell (Figure 21–5).

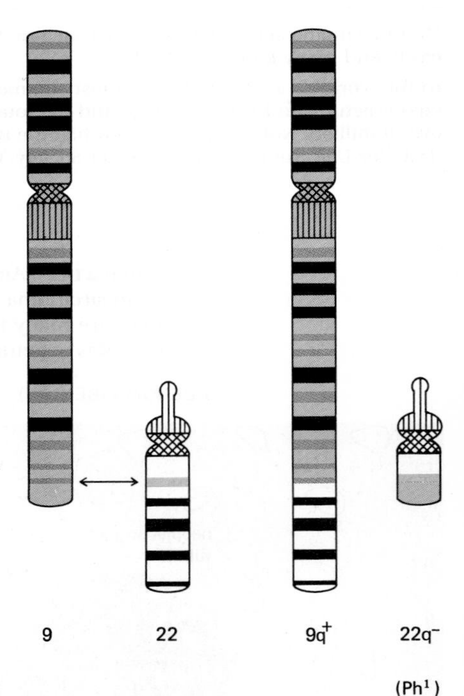

Figure 21–4 The translocation between chromosomes 9 and 22 responsible for chronic myelogenous leukemia. The smaller of the two resulting abnormal chromosomes is called the Philadelphia chromosome, after the city where the abnormality was first recorded.

Most Cancers Are Probably Initiated by a Change in the Cell's DNA Sequence[4]

If a single abnormal cell is to give rise to a tumor, it must pass on its abnormality to its progeny: the aberration has to be heritable. A first problem in understanding a cancer is to discover whether the heritable aberration is due to a genetic change—that is, an alteration in the the cell's DNA sequence—or to an *epigenetic* change—that is, a change in the pattern of gene expression without a change in the DNA sequence. Heritable epigenetic changes, reflecting cell memory (see p. 570 and p. 898), are a familiar feature of normal development, as manifest in the stability of the differentiated state (see p. 952) and in such phenomena as X chromosome inactivation (see p. 577); and there is no obvious a priori reason why they should not be involved in cancer. For one rare and extraordinary type of cancer—the teratocarcinoma (see p. 897)—there is indeed evidence in favor of an epigenetic origin. There are, however, good reasons to think that most cancers are initiated by genetic change (although epigenetic changes may play a part in the subsequent development of the disease). Thus cells of a given cancer can often be shown to have a shared abnormality in their DNA sequence, as we have just seen for chronic myelogenous leukemia; many other examples will be discussed in the second half of this chapter. But this does not prove that genetic change is an essential *first* step in the causation of cancer. A more cogent argument is that most of the agents known to cause cancer cause genetic change; and, conversely, agents that cause genetic change cause cancer. This correlation between **carcinogenesis** (the generation of cancer) and *mutagenesis* is clear for three classes of agents: chemical carcinogens (which typically cause simple local changes in the nucleotide sequence), ionizing radiation such as x-rays (which typically cause chromosome breaks and translocations), and viruses (which introduce foreign DNA into the cell). The role of viruses in cancer will be discussed later; we pause here to discuss **chemical carcinogens.**

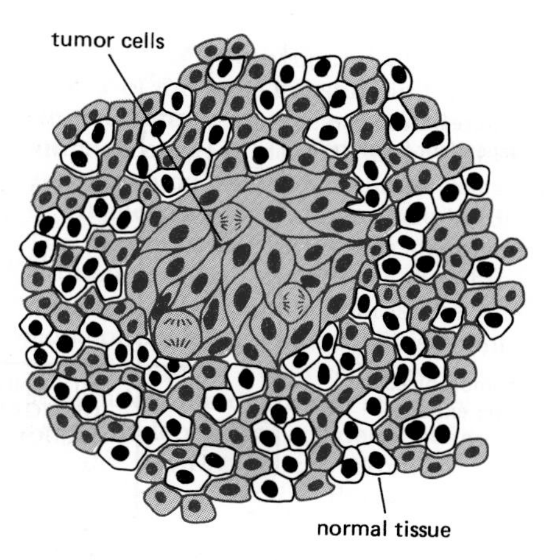

Figure 21–5 Evidence from the analysis of X-inactivation mosaics demonstrating the monoclonal origin of cancers. As a result of a random process that occurs in the early embryo, practically every normal tissue in a woman's body is a mixture of cells with different X chromosomes heritably inactivated (indicated here by the mixture of colored cells and black cells in the normal tissue). When the cells of a cancer are tested for their expression of an X-linked marker gene, however, they are usually all found to have the same X chromosome inactivated. This implies that they are all derived from a single cancerous founder cell.

In general, a given cancer cannot be blamed entirely on a single event or a single cause: as we shall see, cancers as a rule result from the chance occurrence in one cell of several independent accidents, with cumulative effects. The cell's environment influences the frequency of these accidents in a variety of ways, and most cancers should be viewed as the outcome of a random process that is made more probable by a mixture of contributory environmental factors (see p. 1195). There are, however, some unusually carcinogenic agents that increase the likelihood of the critical events to the point where it becomes virtually certain, given a high enough dosage, that at least one cell in the body will turn cancerous. The compound 2-naphthylamine, used in the chemical industry in the early part of this century, is one notorious example: in one British factory, all of the men who had been employed in distilling it (and were thereby subjected to prolonged exposure) eventually developed bladder cancer.

Many quite disparate chemicals have been shown to be likewise carcinogenic when they are fed to experimental animals or painted repeatedly on their skin. Some of these carcinogens act directly on the target cells; many others take effect only after they have been changed to a more reactive form by metabolic processes—notably by a set of intracellular enzymes known as the cytochrome P-450 oxidases, which normally help to convert ingested toxins and foreign lipid-soluble materials into harmless and easily excreted compounds but which fail in this task with certain substances, converting them instead into direct carcinogens (Figure 21–6). Although the known chemical carcinogens are very diverse, most of them have at least one property in common: they cause mutations. The mutagenicity can be demonstrated by various methods, one of the most convenient being the *Ames test,* in which the carcinogen is mixed with an activating extract prepared from rat liver cells and added to a culture of specially designed test bacteria; the resulting mutation rate of the bacteria is then measured (Figure 21–7). Most of the compounds scored as mutagenic by this bacterial assay also cause mutations and/or chromosome aberrations when tested on mammalian cells, and they have chemical structures that can be seen to imply an ability to react with DNA. When mutagenicity data from these various sources are combined and compared with carcinogenicity data from studies of cancer induction *in vivo,* it is found that the majority of known carcinogens are mutagenic and, conversely, that the majority of mutagens are carcinogenic.

Figure 21–6 Many chemical carcinogens have to be activated by a metabolic transformation before they will cause mutations by reacting with DNA. The compound illustrated here is *aflatoxin B1,* a toxin from a mold (*Aspergillus flavus oryzae*) that grows on grain and peanuts when they are stored under humid tropical conditions. It is thought to be a contributory cause of liver cancer in the tropics.

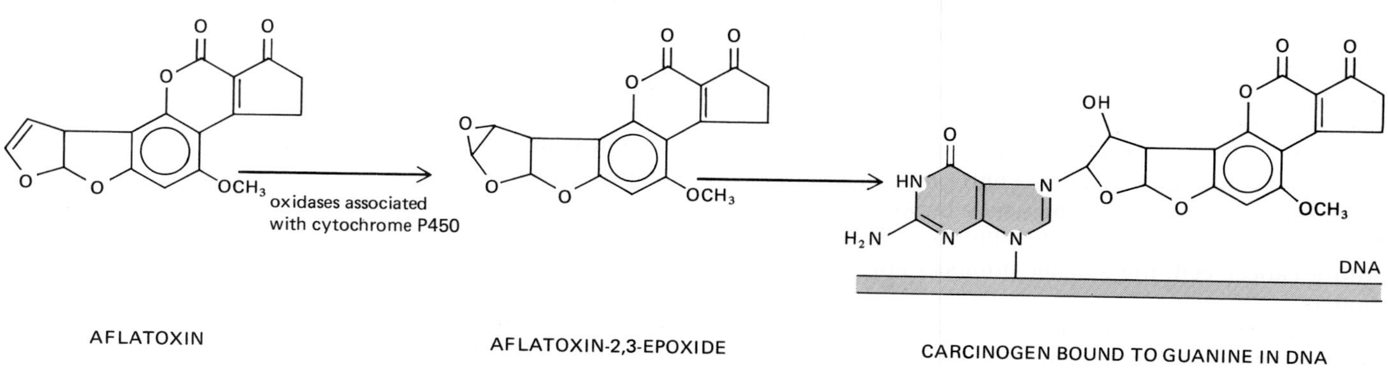

oxidases associated with cytochrome P450

DNA

AFLATOXIN

AFLATOXIN-2,3-EPOXIDE

CARCINOGEN BOUND TO GUANINE IN DNA

continues as such for several years before changing into a much more rapidly progressing illness that usually ends in death within a few months. In the chronic early phase the leukemic cells in the body are distinguished simply by their possession of the chromosomal translocation mentioned previously (see p. 1190). In the subsequent acute phase of the illness, the hemopoietic system is overrun by cells that show not only this chromosomal abnormality but also several others. It appears as though members of the initial mutant clone have undergone further mutations that make them proliferate more rapidly (or divide more times before they terminally differentiate), so that they come to outnumber both the normal hemopoietic cells and their cousins that have only the primary disorder.

Carcinomas and other solid tumors are thought to evolve in a similar way. Although most such cancers in humans are not diagnosed until a relatively late stage, in a few cases it is possible to observe the early steps in the development of the disease. Cancers of the *uterine cervix* (the neck of the womb) provide a typical example. These cancers derive from the multilayered cervical epithelium, which has an organization similar to that of the epidermis of the skin (see p. 968). Normally, proliferation occurs only in the basal layer, generating cells that then move outward toward the surface, differentiating into flattened, keratin-rich, nondividing cells as they go, and finally being sloughed off from the surface (Figure 21–10A). When many specimens of this epithelium from different women are examined, however, it is not unusual to find patches of **dysplasia,** where dividing cells are no longer confined to the basal layer and there is some disorder in the process of differentiation (Figure 21–10B). Cells are sloughed from the surface in abnormally early stages of differentiation, and the presence of the dysplasia can be detected by scraping a sample of cells from the surface and viewing it under the microscope (the "Pap smear" technique—Figure 21–11). Left alone, the dysplastic patches will often remain harmless or even regress spontaneously; more rarely, however, they may progress, over a period of several years, to give rise to patches of so-called **carcinoma *in situ*** (Figure 21–10C). In these more serious lesions (somewhat misleadingly named, since they are not yet fully malignant), the usual pattern of cell division and differentiation is much more severely disrupted, and all the layers of the epithelium consist of undifferentiated proliferating cells, which are often highly variable in size and karyotype; the abnormal cells are still confined, however, to the epithelial side of the basal lamina. At this stage it is still easy to achieve a complete cure by destroying or removing the abnormal tissue surgically. Without such treatment the abnormal patch may still remain harmless or regress; but in an estimated 20–30% of cases it will develop, again over a period of several years, to give rise to a truly malignant cervical carcinoma (Figure 21–10D), whose cells break out of the epithelium by crossing the basal lamina and begin to invade the underlying connective tissue. Surgical cure becomes progressively more difficult as the invasive growth spreads.

Figure 21–10 The stages of progression in the development of cancer of the epithelium of the uterine cervix. In dysplasia, the most superficial cells still show some signs of differentiation; but this is incomplete, and proliferating cells are seen abnormally far above the basal layer. In carcinoma *in situ,* the cells in all the layers are proliferating and apparently undifferentiated. True malignancy begins when the cells cross the basal lamina and begin to invade the underlying connective tissue. Several years may elapse from the first signs of dysplasia to the onset of full-blown malignant cancer.

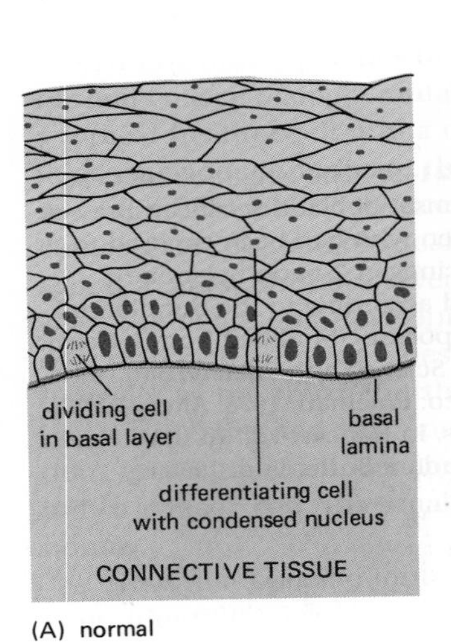

(A) normal

(B) dysplasia

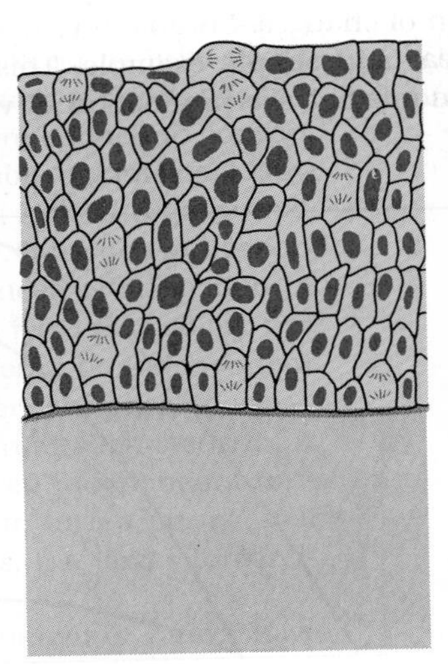

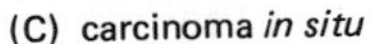

(C) carcinoma *in situ*

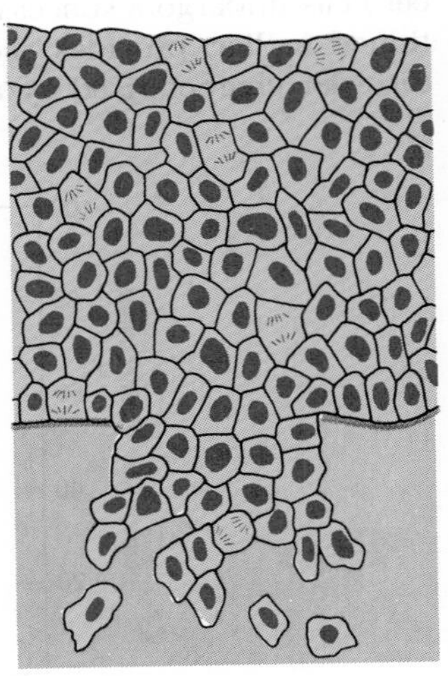

(D) malignant carcinoma

(A)

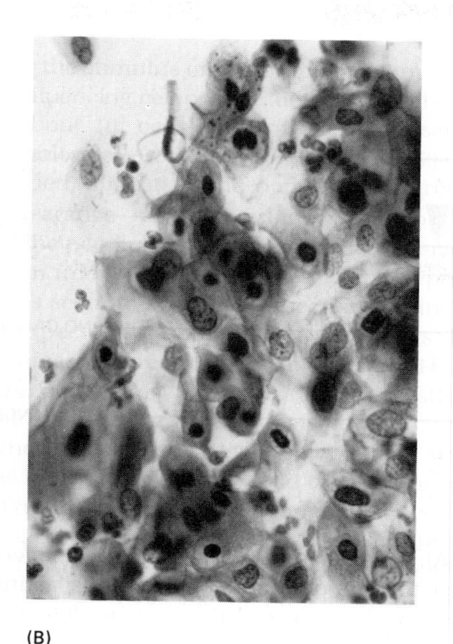

(B)

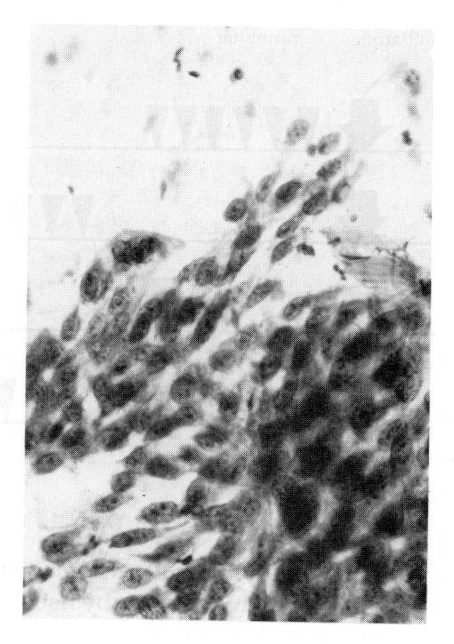

(C)

Figure 21–11 Photographs of cells collected by scraping the surface of the uterine cervix (the Papanicolaou or "Pap smear" technique). (A) Normal; the cells are large and well differentiated, with highly condensed nuclei. (B) Dysplasia; the cells are in a variety of stages of differentiation, some quite immature. (C) Invasive carcinoma; the cells all appear undifferentiated, with scanty cytoplasm and a relatively large nucleus; debris in the background includes blood cells that have leaked out at the site of the ulcer created by the carcinoma. (Courtesy of Edward Miller.)

Tumor Progression Involves Successive Rounds of Mutation and Natural Selection[6,7]

As illustrated by the two very different examples just discussed, cancers in general seem to arise by a process in which an initial population of slightly abnormal cells, descendants of a single mutant ancestor, evolves from bad to worse through successive cycles of mutation and natural selection. This evolution involves a large element of chance and usually takes many years; most of us die of other ailments before cancer has had time to develop. To understand the causation of cancer it is essential to understand the factors that may speed up the process.

In general, the rate of evolution, whether in a population of cells exploiting the opportunities for cancerous behavior in the body or in a population of organisms adapting to a new environment on the surface of the Earth, would be expected to depend on four main parameters: (1) the *mutation rate,* that is, the probability per gene per unit time that any given member of the population will undergo genetic change; (2) the *number of individuals in the population;* (3) the *rate of reproduction,* that is, the average number of generations of progeny produced per unit time; and (4) the *selective advantage* enjoyed by successful mutant individuals, that is, the ratio of the number of surviving fertile progeny they produce per unit time to the number of surviving fertile progeny produced by nonmutant individuals. The selective advantage depends both on the nature of the mutation and on environmental conditions, and further complications arise if heritable epigenetic changes occur, either randomly or in reaction to specific cues.

Experimental studies on the induction of cancer in animals illustrate these evolutionary principles. In the light of such studies, one can begin to make sense of the confusing variety of factors that affect the incidence of human cancers—factors ranging from cigarette smoke (for cancer of the lung) to the age at which a woman has her first baby (for cancer of the breast). The mutation rate per cell is not the only significant variable in the development of cancer.

The Development of a Cancer Can Be Promoted by Factors That Do Not Alter the Cells' DNA Sequence[6,8]

The stages by which an initial mild lesion progresses to become a cancer can be most easily observed in the skin. Skin cancers can be elicited in mice, for example, by repeatedly painting the skin with a mutagenic chemical carcinogen such as benzo[a]pyrene (a constituent of coal tar and tobacco smoke) or the related com-

Table 21–2 Variation Between Countries in the Incidence of Some Common Cancers

Site of Origin of Cancer	High–Incidence Area	Cumulative Incidence (%) in High–Incidence Area	Low–Incidence Area	Ratio of Rates in High– and Low–Incidence Areas
Skin	Australia (Queensland)	20	India (Bombay)	>200
Esophagus	Iran	20	Nigeria	300
Lung	England	11	Nigeria	35
Stomach	Japan	11	Uganda	25
Uterine cervix	Columbia	10	Israel (Jewish)	15
Prostate	United States (blacks)	9	Japan	40
Liver	Mozambique	8	England	100
Breast	Canada	7	Israel (non-Jewish)	7
Colon	United States (Connecticut)	3	Nigeria	10
Uterus	United States (California)	3	Japan	30
Oral cavity	India (Bombay)	2	Denmark	25
Rectum	Denmark	2	Nigeria	20
Bladder	United States (Connecticut)	2	Japan	6
Ovary	Denmark	2	Japan	6
Nasopharynx	Singapore (Chinese)	2	England	40
Pancreas	New Zealand (Maori)	2	India (Bombay)	8
Larynx	Brazil (São Paulo)	2	Japan	10
Pharynx	India (Bombay)	2	Denmark	20
Penis	Parts of Uganda	1	Israel (Jewish)	300

Data for uterine cervix, breast, uterus, and ovary are for women; others are for men. The cumulative incidence is defined as the percentage of the population that would develop the specified cancer by the age of 75, in the absence of other causes of death; the ratio of rates is calculated for the 35- to 64-year age group. (Slightly modified from R. Doll and R. Peto, The Causes of Cancer. New York: Oxford University Press, 1981.)

The Search for Cancer Cures Is Hard but Not Hopeless[10]

The difficulty of curing a cancer is like the difficulty of getting rid of weeds. Cancer cells can be removed surgically or destroyed with toxic chemicals or radiation; but it is hard to eradicate every single one of them. Surgery can rarely ferret out every metastasis, and treatments that kill cancer cells are generally toxic to normal cells as well. If even a few cancerous cells remain, they can proliferate to produce a resurgence of the disease; and unlike the normal cells, they may evolve resistance to the poisons used against them. Yet the outlook is not hopeless. In spite of the difficulties, effective cures using anticancer drugs (alone or in combination with other treatments) have been devised for some formerly highly lethal cancers (notably Hodgkin's lymphoma, testicular cancer, choriocarcinoma, and some leukemias and other cancers of childhood). For several of the more common cancers, moreover, appropriate surgery or local radiotherapy enables a large proportion of patients to recover if the illness is diagnosed at a reasonably early stage; and even where a cure at present seems beyond our reach, there are treatments that will prolong life or at least relieve distress.

A great deal of clinical cancer research centers on the problem of how to kill cancer cells selectively. For the most part, current methods exploit relatively subtle differences between normal and neoplastic cells with respect to proliferation rate, metabolism, and radiosensitivity, and they have unpleasant toxic side effects. A few types of cancer cells are especially vulnerable to selective attack because they depend on specific hormones or because their surfaces have unusual chemical features that can be recognized by antibodies. In general, however, progress with the vexing problem of anticancer selectivity has been slow—a matter of trial and error and guesswork as much as rational calculation.

In the search for better ways of curbing the survival, proliferation, and spread of cancer cells, it is important to examine more closely the strategies by which they thrive and multiply.

Cancerous Growth Often Depends on Derangements of Cell Differentiation[11]

We have so far emphasized that cancer cells defy the normal controls on cell division: this is their central property. But many tissues are organized in such a way that even an uncontrolled increase in the frequency of cell division will not by itself produce a steadily growing tumor. The example of the uterine cervix, discussed above on page 1194, illustrates this point. Like the epidermis of the skin and many other epithelia, the epithelium of the uterine cervix normally renews itself continually by shedding terminally differentiated cells from its outer surface and generating replacements from stem cells in the basal layer (see p. 970). On average, each normal stem cell division generates one daughter stem cell and one cell that is condemned to terminal differentiation and a cessation of cell division. If the stem cell simply divides more rapidly, terminally differentiated cells will be produced and shed more rapidly, and a balance of genesis and destruction will still be maintained. Thus if a transformed stem cell is to generate a steadily growing clone of progeny, the basic rules must be upset: either more than 50% of the daughter cells must remain as stem cells, or the process of differentiation must be deranged so that daughter cells embarked on this route retain an ability to carry on dividing indefinitely and avoid being discarded at the end of the production line (Figure 21–15).

Presumably, the development of such properties underlies the progression from a mild dysplasia of the uterine cervix to carcinoma *in situ* and malignant cancer (see Figure 21–10). Similar considerations apply to the development of cancer in other tissues that rely on stem cells, such as the skin, the lining of the gut, and the hemopoietic system. Several forms of leukemia, for example, seem to arise from a disruption of the normal program of differentiation, such that a committed progenitor of a particular type of blood cell continues to divide indefinitely, instead of differentiating terminally in the normal way after a strictly limited num-

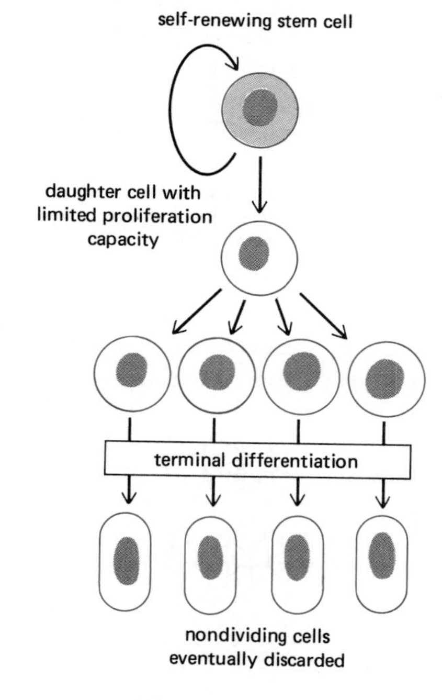

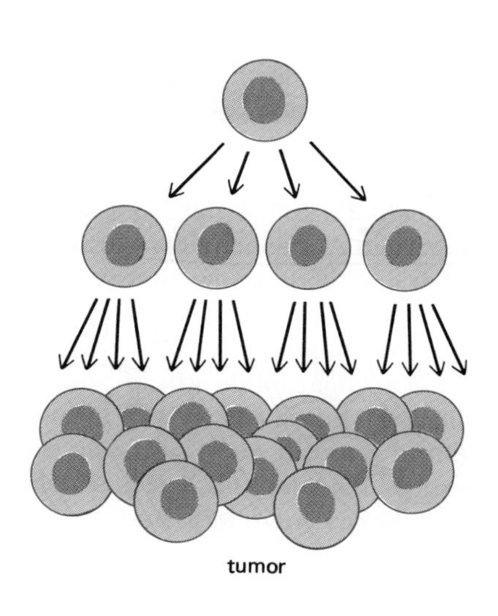

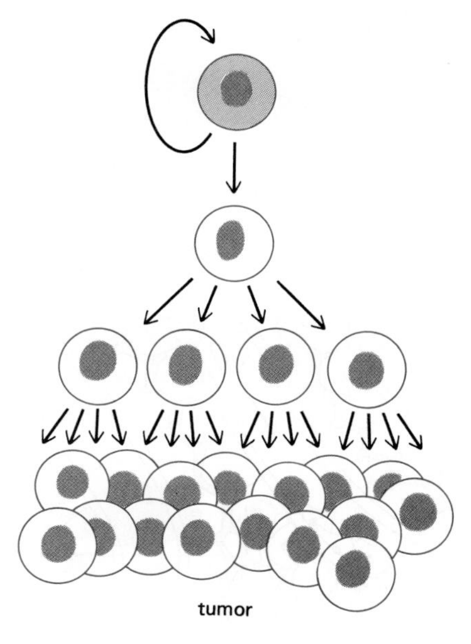

Figure 21–15 The stem-cell strategy for producing new differentiated cells, and two types of derangement that can give rise to the unbridled proliferation characteristic of cancer. Note that an excessive cell-division rate for the stem cells will not by itself have this effect.

ferentiated cell that is poorly adapted for rapid proliferation. Alternatively, it might reflect a heritable deficiency in the machinery or control of DNA replication, repair, or recombination, arising by somatic mutation in any one of the many genes involved in these complex processes. Such a mutation would be liable to increase the likelihood of subsequent mutations in other classes of genes. For this reason one might expect it to be a common feature of cells that undergo the multiple mutations required to become cancerous. Suppose, for example, that three mutations in genes governing the social behavior of cells are required to convert a normal cell into a cancer cell and that each of these mutations normally occurs at a rate of 10^{-4} per cell per human lifetime. Then the probability for a single cell with the normal level of mutability to accumulate these three mutations in the course of a lifetime would be $10^{-4} \times 10^{-4} \times 10^{-4} = 10^{-12}$ per cell. But now let us suppose that each of these mutations occurs at a rate of 10^{-2} per cell per lifetime if there has been a prior mutation in some particular enzyme involved in DNA replication or repair. If this latter mutation itself has a probability of 10^{-4}, cancer cells will arise most frequently by the route that begins with the mutation that increases mutability: this route (given the simplest possible assumptions) involves a combination of four events whose joint probability is on the order of $10^{-4} \times 10^{-2} \times 10^{-2} \times 10^{-2} = 10^{-10}$ per cell per lifetime; it is thus 100 times more likely than the route involving only the minimal three mutations.

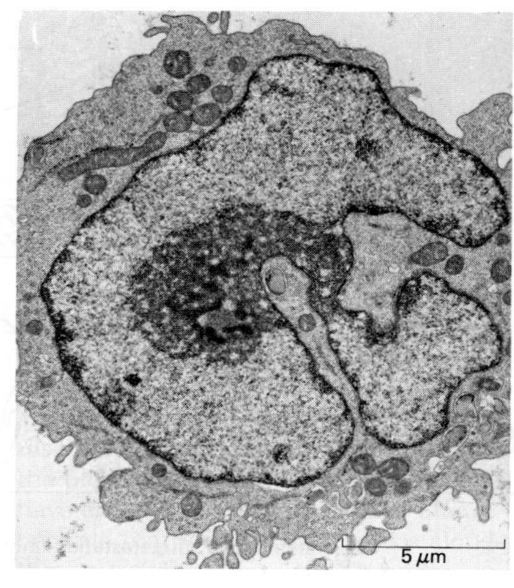

Figure 21–19 Typical abnormalities in the appearance of the nucleus of a cancer cell—in this example, an erythroleukemia cell. The cancer cell nucleus is large in relation to the amount of cytoplasm, with an irregularly indented envelope and a nucleolus that is also abnormally large and complex in its structure. (Courtesy of Daniel Friend.)

The Enhanced Mutability of Cancer Cells Helps Them Evade Destruction by Anticancer Drugs[10,14]

Whatever the origins of the abnormally high mutability of cancer cells, most malignant tumor cell populations are heterogeneous in many respects and capable of evolving at an alarming rate when subjected to new selection pressures. This aggravates the difficulties of cancer therapy. Repeated treatments with drugs that are selectively toxic to dividing cells can be used to kill the majority of neoplastic cells in a cancer patient, but it is rarely possible to kill them all: usually some small proportion are drug-resistant, and the effect of the treatment is to favor the spread and evolution of cells with this trait. To make matters worse, cells that are exposed to one drug often develop a resistance not only to that drug, but also to other drugs to which they have never been exposed.

This phenomenon of **multidrug resistance** is frequently correlated with a curious change in the karyotype: the cell is seen to contain additional pairs of miniature chromosomes—so-called *double minute chromosomes*—or to have a *homogeneously staining region* interpolated in the normal banding pattern of one of its regular chromosomes. Both these aberrations consist of massively amplified numbers of copies of a small segment of the genome (see Figures 21–26 and 21–31, below). Cloning of this amplified DNA has revealed that it often contains a specific gene, known as the *multidrug resistance* (*mdr*1) gene, which codes for a plasma-membrane-bound transport ATPase that is thought to prevent the intracellular accumulation of certain classes of lipophilic drugs by pumping them out of the cell. The amplification of other types of genes can also give the cancer cell a selective advantage: thus the gene for the enzyme dihydrofolate reductase (DHFR) often becomes amplified in response to cancer chemotherapy with the folic-acid antagonist methotrexate; and we shall see that some proto-oncogenes involved in cell-division control are similarly amplified in some cancers (see p. 1214).

While defects in DNA replication, recombination, or repair may help cancer cells to evolve by increasing their mutability, they may also make the cells more vulnerable to certain types of attack. This may explain the observation—exploited in therapy—that the cells of many tumors are killed more easily than normal cells by irradiation or by exposure to specific drugs that interfere with DNA metabolism. As we learn more about the molecular mechanisms of DNA replication, recombination, and repair, it should become possible to devise tests to pinpoint defects in these functions in individual cases of cancer. Using such information we may be better able to kill the delinquent cells by designing drugs that exploit their particular weaknesses.

Summary

Cancer cells, by definition, proliferate in defiance of normal controls (that is, they are neoplastic) and are able to invade and colonize surrounding tissues (that is, they are malignant). By giving rise to secondary tumors, or metastases, they become hard to eradicate surgically. Cancer cells usually retain many features of the specific cell type from which they are derived. Most cancers are thought to originate from a single cell that has undergone a somatic mutation, but the progeny of this cell must undergo further changes, probably requiring several additional mutations, before they become cancerous. This phenomenon of tumor progression, which usually takes many years, reflects the operation of evolution by mutation and natural selection among somatic cells; the rate of the process is accelerated both by mutagenic agents (tumor initiators) and by certain nonmutagenic agents (tumor promoters) that affect gene expression, stimulate cell proliferation, and alter the ecological balance of mutant and nonmutant cells. Thus many factors contribute to the development of a given cancer, and since some of these factors are avoidable features of the environment, a large proportion of cancers are in principle preventable.

Much effort in cancer research has been devoted to the search for ways to cure the disease by exterminating cancer cells while sparing their normal neighbors. A rational approach to this problem requires an understanding of the special properties of cancer cells that enable them to evolve, multiply, and spread. Thus neoplastic cell proliferation often seems to be associated with a block in differentiation, whereby the progeny of a stem cell are enabled to continue dividing instead of entering a terminal nondividing state; in principle, the proliferation could be curbed by promoting cell differentiation. To become malignant, tumor cells must be able to cross basal laminae; antibodies can be designed that interfere with this ability, thereby hindering metastasis. Cancer cells are often found to be abnormally mutable; this hastens evolution of the complex set of properties required for neoplasia and malignancy and helps the cancer cells develop resistance to anticancer drugs. At the same time, however, defects of DNA metabolism underlying such mutability may make the cancer cells uniquely vulnerable to a suitably designed therapeutic attack.

The Molecular Genetics of Cancer[15]

Because cancer is the outcome of a series of random genetic accidents subject to natural selection, no two cases even of the same variety of the disease are likely to be genetically identical. Nevertheless, all cancers can be expected to involve a disruption of the normal restraints on cell proliferation, and for each cell type there is a finite number of ways in which such disruption can occur. Moreover, some parts of the machinery for regulating cell proliferation are likely to be the same in many or all cell types, and similarly vulnerable. In fact, changes in a relatively small set of genes appear to be responsible for much of the deregulation of cell division in cancer. The identification and characterization of many of these genes has been one of the great triumphs of molecular biology in the past decade.

Cell proliferation can be regulated directly, through the mechanism that determines whether a cell passes the restriction point or "Start" of the cell-division cycle (see p. 745), or indirectly—for example, through regulation of the commitment to terminal differentiation (see p. 967). In either case the normal regulatory genes can be loosely classified into those whose products help stimulate cell proliferation and those whose products help inhibit it. Correspondingly, there are two mutational routes toward the uncontrolled cell proliferation that is characteristic of cancer. The first is to make a stimulatory gene hyperactive: this type of mutation has a dominant effect—only one of the cell's two gene copies need undergo the change—and the altered gene is called an **oncogene** (the normal allele being a **proto-oncogene**). The second is to make an inhibitory gene inactive: this type of mutation has a recessive effect—both the cell's gene copies must be

U

V

W

X

Y

Z